The UK Pesticide Guide 2017

Editor: M.A. Lainsbury, BSc(Hons)

BCPC

cabi
www.cabi.org

© 2017 CAB International and BCPC (British Crop Production Council). All rights reserved. No part of this publication may be reproduced in any form or by any means, electronically, mechanically, by photocopying, recording or otherwise, without the prior permission of the copyright owners.

CABI is one of the world's foremost publishers of databases, books and compendia in agriculture and applied life sciences. It has a worldwide reputation for producing high quality, value-added information, drawing on its links with the scientific community. CABI produces CAB Abstracts – the leading agricultural A&I database; together with the Crop Protection Compendium which provides solutions for identifying, preventing and solving crop health problems. CABI is a not-for-profit organisation. For further information, please contact CABI, Nosworthy Way, Wallingford, Oxon OX10 8DE, UK.

Telephone: (01491) 832111
Fax: (01491) 833508
e-mail: enquiries@cabi.org
Web: www.cabi.org

BCPC (British Crop Production Council) promotes the use of good science and technology in understanding and applying effective and sustainable crop production. Its key objectives are to identify developing issues in the science and practice of crop protection and production and provide informed, independent analysis; to publish definitive information for growers, advisors and other stakeholders; to organise conferences and symposia; and to stimulate interest and learning. BCPC is a registered Charity and a Company limited by Guarantee. Further details available from BCPC, 7 Omni Business Centre, Omega Park, Alton, Hampshire GU34 2QD, UK.

Telephone: (01420) 593200
Fax: (01420) 593209
e-mail: md@bcpc.org
Web: www.bcpc.org

ISBN: 978 1 901396 683

Typeset and Printed in the United Kingdom by
Hobbs the Printers Ltd, Totton, Hampshire, SO40 3WX
Website: www.hobbs.uk.com

Contents

Editor's Note	v
Disclaimer	vi
Changes Since 2016 Edition	vii
Introduction	xi
Notes on Contents and Layout	xiii

SECTION 1 CROP/PEST GUIDE — 1

Crop/Pest Guide Index	3
Crop/Pest Guide	5

SECTION 2 PESTICIDE PROFILES — 85

SECTION 3 PRODUCTS ALSO REGISTERED — 585

SECTION 4 ADJUVANTS — 629

SECTION 5 USEFUL INFORMATION — 661

Pesticide Legislation	663
Regulation 1107/2009 – The Replacement for EU 91/414	663
The Water Framework Directive	663
Classification, Labelling and Packaging (CLP) Regulation	664
The Food and Environment Protection Act 1986 (FEPA)	664
Dangerous Preparations Directive (1999/45/EC)	664
The Review Programme	665
Control of Substances Hazardous to Health Regulations 1988 (COSHH)	665
Biocidal Products Regulations	665
Certificates of Competence – the roles of BASIS and NPTC	665
Maximum Residue Levels	667
Approval (On-label and Off-label)	668
Statutory Conditions of Use	668
Types of Approval	668
Withdrawal of Approval	668
Off-label Extension of Use	669
Using Crop Protection Chemicals	670
Use of Herbicides In or Near Water	670
Use of Pesticides in Forestry	671
Pesticides Used as Seed Treatments	672
Aerial Application of Pesticides	675
Resistance Management	676
Preparation in advance	676
Using crop protection products	676
International Action Committees	677
Poisons and Poisoning	678
Chemicals Subject to the Poison Law	678
Part I Poisons	678
Part II Poisons	678
Occupational Exposure Limits	678
First Aid Measures	679
General Measures	679
Reporting of Pesticide Poisoning	679
Additional Information	679
Environmental Protection	681
Environmental land management: 8 steps for arable farmers	681
Protection of Bees	682

Honeybees	682
Wild bees	682
The Campaign Against Accidental or Illegal Poisoning (CAIP) of Wildlife	682
Water Quality	683
Where to get information	683
Protecting surface waters	684
Local Environmental Risk Assessment for Pesticides (LERAPs)	684
Groundwater Regulations	685
Integrated Farm Management (IFM)	685
Campaign for the Farmed Environment	686

SECTION 6 APPENDICES — 687

Appendix 1 Suppliers of Pesticides and Adjuvants	689
Appendix 2 Useful Contacts	698
Appendix 3 Keys to Crop and Weed Growth Stages	701
Appendix 4 Key to Hazard Classifications and Safety Precautions	709
Appendix 5 Key to Abbreviations and Acronyms	717
Appendix 6 Definitions	720
Appendix 7 References	722

INDEX OF PROPRIETARY NAMES OF PRODUCTS — 725

Editor's Note

This is the 30th edition of *The UK Pesticide Guide* and 2016 has been a very busy year as far as changes go. The roll out of the harmonization of hazard precautions with the rest of the world has seen many products re-issued with new MAPP numbers. Where appropriate this has also meant the imposition of buffer zones wider than the standard 5 meters for LERAP and occasionally there has been a refinement of the label claims and application timings. It means that the approval documents of every product in the **Guide** have needed to be checked to identify changes required.

For 2017 the book contains 25 new active ingredient profiles which includes two new actives as full entries (cyantraniliprole for insect control in headed brassicas and halauxifen-methyl formulated with fluroxypyr for weed control in cereals) and five listed as 'Products Also Registered' (PARs) until market availability is confirmed.

The full list of gains and losses is presented here with new actives shown in **bold**:

Herbicide full entry gains are:
- amidosulfuron + iodosulfuron-methyl-sodium + mesosulfuron-methyl
- clodinafop propargyl + cloquintocet-mexyl
- dimethenamid-p + quinmerac
- flufenacet + picolinafen
- fluroxypyr + **halauxifen-methyl**
- **halauxifen-methyl**
- isoxaben
- mesosulfuron-methyl + propoxycarbazone-sodium
- metamitron + quinmerac
- **penthoxamid**
- **tembotrione**

Fungicide full entry gains are:
- azoxystrobin + fluazinam
- bixafen + fluopyram + prothioconazole
- cymoxanil + zoxamide
- dimethomorph + zoxamide
- dithianon + potassium phosphates
- fludioxonil + tebuconazole
- fluopyram + trifloxystrobin
- fluzapyroxad + pyraclostrobin
- **Pythium oligandrum**
- **Pepimo mosaic virus-strain CH2**

Insecticide & PGR full entry gains are:
- **buprofezin**
- coumatetralyl
- **cyantraniliprole**
- prohexadione-calcium + trinexepac-methyl

Herbicides demoted to PAR or lost are:
- bromoxynil + ioxynil
- bromoxynil + terbuthylazine
- carfentrazone-ethyl + flupyrsulfuron
- 2,4-D + dicamba + triclopyr
- fatty acids
- fluroxypyr + metsulfuron-methyl
- imazamox + metazachlor + quinmerac
- ioxynil
- tepraloxydim

Fungicides demoted to PAR or lost are:
- carboxim + thiram
- chlorothalonil + flutriafol
- cymoxanil + famoxadone
- cyproconazole
- difenoconazole + fenpropidin
- fenpyroximate
- fludioxanil + flutriafol
- fluquinconazole
- fluquinconazole + prochloraz
- fosetyl-aluminium
- imazalil
- imazalil + ipconazole
- imazalil + pencycuron
- myclobutanil
- prochloraz
- prochloraz + triticonazole
- spiroxamine + tebuconazole

Insecticides & PGRs demoted to PAR or lost are:
- abermectin
- ethoprophos
- 1-naphthylacetic acid
- spiromesifen
- tetramethrin

The re-issue of products with new MAPP numbers has meant that a large number of new Extension of Authorisation for Minor Use (EAMUs) have also been approved since an EAMU is specific to a MAPP number and not all the old EAMUs have yet been replaced on the latest products. As always, this book is a snapshot of the products with a valid approval on 15 October 2016 when the text went for initial typesetting.

If you cannot see a product in the main section of the book, please check the index because it may be listed in Section 3 – Products also Registered (PAR). For those who must have the most up-to-date information I can recommend the on-line version – *ukpesticideguide.co.uk*. This is a searchable database with much more information than we can squeeze into the book.

We express our gratitude to all those people who have provided the information for the compilation of this guide. Criticisms and suggestions for improvement are always welcome and in particular, if you notice any errors or omissions in the entries, please let me know via the email address below.

Martin Lainsbury, Editor
email: ukpg@bcpc.org

Disclaimer

Every effort has been made to ensure that the information in this book is correct at the time of going to press, but the Editor and the publishers do not accept liability for any error or omission in the content, or for any loss, damage or other accident arising from the use of the products listed herein. Omission of a product does not necessarily mean that it is not approved or that it is unavailable.

It is essential to follow the instructions on the approved label when handling, storing or using any crop protection product. Approved 'off-label' extensions of use now termed Extension of Authorisation for Minor Use (EAMUs) are undertaken entirely at the risk of the user.

The information in this publication has been selected from official sources, from the suppliers product labels and from the product manuals of the pesticides approved for use in the UK under EU Regulation 1107/2009.

Changes Since 2016 Edition

Pesticides and adjuvants added or deleted since the 2016 edition are listed below. Every effort has been made to ensure the accuracy of this list, but late notifications to the Editor are not included.

Products Added

Products new to this edition are shown here. In addition, products that were listed in the previous edition whose PSD/HSE registration number and/or supplier have changed are included.

Product	Reg. No.	Supplier
AdjiFe	A0797	Amega
AdjiMin	A0812	Amega
Antuco	17409	Ventura
Aphox	17401	Syngenta
Ascra Xpro	17623	Bayer CropScience
Azoxystar	17407	Life Scientific
Barclay Gallup Biograde Amenity	17674	Barclay
Beetron 700	17458	AgChem Access
Binder	A0598	Intracrop
Biretta	17408	Ventura
Block	15697	Belchim
Bokken	17557	Syngenta
Bonzi	17576	Syngenta
Brevis	17479	Adama
Bulldog	17538	Sipcam
Carpovirusine EVO 2	17565	Arysta
Centium 360 CS	17327	Headland
Chess WG	17580	Syngenta
Cirran 360	17346	Nufarm UK
Cirrus CS	17328	Headland
Clayton Cast	17727	Clayton
Clayton El Nino	17337	Clayton
Clayton Enigma	17391	Clayton
Clayton Enigma	17435	Clayton
Clayton Gatso	17698	Clayton
Clayton Gatso	17732	Clayton
Clayton Index	17266	Clayton
Clayton Pirimicarb	17704	Clayton
Clayton Sputnik	17344	Clayton
Clayton Tardis	17459	Clayton
Cleancrop Chicane	17330	Headland
Cleancrop Covert	17340	Headland
Cleancrop Gallifrey 3	17399	Dow
Clipless NT	17558	Headland

Product	Reg. No.	Supplier
Clodinastar	17404	Life Scientific
Clopic	17387	UPL Europe
Configure	17523	Fargro
Crescent	17589	Certis
Deacon	17604	Adama
Delan Pro	17374	BASF
Depitone Ultra	17615	Nufarm UK
Difenag Wax Bait	UK14-0851	Killgerm
Difure Pro	17303	Belchim
Dime	17398	Pan Agriculture
Dipel DF	17499	Interfarm
DP911 SX	17493	DuPont
Ecoplug Max	17581	Monsanto
Embutone 24DB	17596	Nufarm UK
Empire	17357	Pan Agriculture
Envidor	17518	Bayer CropScience
Erase	17507	Headland
Eribea	17270	Belchim
Escolta	17397	Bayer CropScience
Esteron T	17432	Everris Ltd
Fandango	17318	Bayer CropScience
Fathom	17452	Headland
Fence	17393	Albaugh UK
Firestorm	17631	Certis
Flurostar 200	17438	Globachem
Fontelis	17465	DuPont
Formula B Rat & Mouse Killer	UK15-0934	Killgerm
Fortify	17718	Rigby Taylor
Freeze NT	17478	Headland
GAL-GONE	17505	Globachem
Gallup XL	17663	Barclay

Product	Reg. No.	Supplier	Product	Reg. No.	Supplier
Garryowen XL	17508	Barclay	Presidium	17372	Gowan
Gavel	17586	Certis	Prevail	17395	Dow
Glam	17707	Pan Amenity	Priaxor EC	17371	BASF
Glotron 700 SC	17308	Belchim	Primo Maaxx II	17509	Syngenta
Gocha	17320	Belchim	Prolonger	17453	UPL Europe
Gorgon	17685	Sipcam	Pronto	17688	Stefes
Gozai	17381	Belchim	Proshield	17525	Everris Ltd
Gyro	17426	UPL Europe	Quidam	17595	UPL Europe
Hamlet	17370	Bayer CropScience	Ratak Cut Wheat	UK12-0313	Killgerm
Hecates	17403	Ventura	Respond	A0486	Amega
Herbi-AKtiv	A0821	Global Adjuvants	Revus	17443	Syngenta
Hurler	17715	Barclay	Rhizopon AA Powder (0.5%)	17570	Fargro
Hyprone P	17418	UPL Europe	Rhizopon AA Powder (1%)	17571	Fargro
Hysward-P	17609	Agrichem			
Intellis	17520	DuPont	Rhizopon AA Powder (2%)	17572	Fargro
Intracrop Agwet GTX	A0646	Intracrop	Rhizopon AA Tablets	17573	Fargro
Intracrop Impetus	A0647	Intracrop	Roundup Biactive GL	17348	Monsanto
Intracrop Zenith AF	A0811	Intracrop			
Kix	17424	Globachem	Roundup Metro XL	17684	Monsanto
Kudos	17578	Fine	Scotia	17415	Adama
Lambdastar	17406	Life Scientific	Sempra XL	17447	UPL Europe
Landscaper Pro Feed Weed Mosskiller	17549	Everris Ltd	Serpent	17427	BASF
			Shiro	17439	UPL Europe
			Solo 500 SC	17622	Sipcam
Laser	17339	BASF	Speedway Total	A0820	Everris Ltd
Magneto	17444	Nufarm UK	Stalwart	17405	UPL Europe
Mamba	UK15-0933	Killgerm	Standon Coolie	17382	Standon
			Standon Mandor	17533	Standon
Master Gly	17594	Generica	Stroby WG	17316	BASF
Master Sil	A0822	Global Adjuvants	SU Wett	A0807	Global Adjuvants
Medax Max	17263	BASF	Surfer	A0800	Dow
Merit Forest	17242	Bayer CropScience	T34 Biocontrol	17290	Fargro
			Tambora	17591	Maxwell
Mesostar	17410	Life Scientific	Tebuzol 250	17417	UPL Europe
Metcostar	17446	Life Scientific	Terrane	17619	Headland
Monitor	17695	Interfarm	Tesoro	17222	Belchim
Multicap	17652	Belcrop	Tharsis	17201	Generica
Naturalis-L	17526	Fargro	Topenco	17539	Globachem
Nelson	A0796	Agrovista	Topkat	17356	BASF
Nevada	17349	Dow	Tricle	17448	UPL Europe
Orius	17414	Adama	Trust	16561	Generica
Ossetia	17741	Rotam	Unicur	17402	Bayer CropScience
Palio	17345	Dow			
Pampa 4 SC	17521	Belchim	Unikat 75WG	14200	Gowan
Pirimate 500	17657	Becesane	Unite	17385	Dow
Pistol	17451	Bayer Environ.	Upbeet	17544	DuPont
Pixxaro EC	17545	Dow			

Product	Reg. No.	Supplier	Product	Reg. No.	Supplier
Vazor Cypermax Plus	H9988	Killgerm	Velvet	A0814	Global Adjuvants
Vazor Cypermethrin 10	H10085	Killgerm	Vendetta	17500	Headland
			Verimark 20 SC	17519	DuPont
Vazor DE	-	Killgerm	Vertisan	17485	DuPont
Vazor Liquid Mosquito Film	-	Killgerm	Xerton	17335	UPL Europe

Products Deleted

The appearance of a product name in the following list may not necessarily mean that it is no longer available. In cases of doubt, refer to Section 3 or the supplier.

Product	Reg. No.	Supplier	Product	Reg. No.	Supplier
A-Cyper 100EC	14175	Goldengrass	Diflanil 500 SC	12489	Q-Chem
Agrovista Paclo	13424	Agrovista	Dursban WG	09153	Dow
Agrovista Penco	13425	Agrovista	Equity	12465	Dow
Antec Durakil 1.5 SC	H7560	Antec Biosentry	Escar-go 3	14044	Chiltern
			Grassland Herbicide	14623	UPL Europe
Antec Durakil 6 SC	H7559	Antec Biosentry	Headland Relay P	08580	Headland
			Headland Relay Turf	08935	Headland
Apollo 50 SC	10590	Adama	Headland Sulphur	12879	Headland
Aramo	10280	BASF	Headland Transfer	11010	Headland
Arenales	15814	Agroquimicos	Headland Trinity	10842	Headland
Artist	11239	Bayer CropScience	Hill-Agro 40	17654	Belchim
			Impact Excel	11547	Headland
Beret Multi	13017	Syngenta	Intracrop Zenith	A0665	Intracrop
Bettix Flo	11959	UPL Europe	Intracrop Zodiac	A0755	Intracrop
Boomerang	14043	BASF	Intruder	15076	UPL Europe
Break-Thru S 240	A0431	PP Products	Kaskara	15593	Arysta
Bronx	15860	AgChem Access	Killgerm Py-Kill W	H4632	Killgerm
			Killgerm ULV 400	H4838	Killgerm
Cavalry	16219	Certis	Killgerm Wasp Nest Destroyer	H9817	Killgerm
Centaur	13852	Bayer CropScience	Landscaper Pro Weed Control + Fertilizer	16609	Everris Ltd
Chlobber	13723	AgChem Access			
			Marshal XL	16334	UPL Europe
Clayton Bestow	14996	Clayton	Masalon	12385	Rigby Taylor
Cleancrop Corsair	14124	Clayton	Mascot Fusion	15967	Rigby Taylor
Confine	15998	Headland	Master Sil	A0790	Generica
Conlara	16032	Agroquimicos	Master Tambora	16230	Generica
Continuum	15600	Clayton	Merit Turf	12415	Bayer Environ.
Cultar	10523	Syngenta	Metric	14214	Belchim
Cuprokylt	16944	Certis	Mitemectin	15804	Becesane
Cyren	11028	Headland	Monceren IM	11426	Bayer CropScience
Cythrin 500 EC	16993	AgChem Access			
			Novall	12031	BASF
Defiant SC	12302	UPL Europe	Nufarm Cropoil	A0447	Nufarm UK
Degesch Plates	07603	Rentokil	Oblix MT	14510	UPL Europe
Detia Gas-Ex-B	06927	Rentokil			

Product	Reg. No.	Supplier
Omarra	13693	AgChem Access
Oryx	13527	BASF
Output	A0644	Nufarm UK
Oxytril CM	14511	Bayer CropScience
Palometa	15833	Agroquimicos
Pentangle	13746	Nufarm UK
Pilot Ultra	1730	Nissan
Pyrethrum Spray	H4636	Killgerm
Pyrifos 480	15991	Becesane
Quartz	A0771	Intracrop
Quorum	A0386	Intracrop
Schooner B	A0270	Intracrop
Segundo	15710	Agroquimicos
Sekator	14746	Interfarm
Sempra	13525	UPL Europe
Sequel	12657	Certis
Spyrale	12566	Syngenta
Standon Mastana	13795	Standon
Standon Tonga	13394	Standon
Taza 500	15989	Becesane
TEB250EW	16811	Goldengrass
Tepra	14160	Goldengrass
Tivmetix OD	16479	GAT Micro
Topenco 100 EC	11972	Globachem
Totril	14633	Bayer CropScience

Introduction

Purpose

The primary aim of this book is to provide a practical handbook of pesticides, plant growth regulators and adjuvants that the farmer or grower can realistically and legally obtain in the UK and to identify the purposes for which they can be used. It is designed to help in the identification of products appropriate to a particular problem. In addition to uses recommended on product labels, details are provided of those uses which do not appear on product labels but which have been granted Extension of Authorisation for Minor Use (EAMUs) – previously known as SOLAs. As well as identifying the products available, the book provides guidance on how to use them safely and effectively but without giving details of doses, volumes, spray schedules or approved tank mixtures. Sections 5 and 6 provide essential background information on a wide range of pesticide-related issues including: legislation, biodiversity, codes of practice, poisons and treatment of poisoning, products for use in special situations and weed and crop growth stage keys.

While we have tried to cover all other important factors, this book does not provide a full statement of product recommendations. Before using any pesticide product it is essential that the user should read the label carefully and comply strictly with the instructions it contains. **This Guide does not substitute for the product label**.

Scope

The *Guide* is confined to pesticides registered in the UK for professional use in arable agriculture, horticulture (including amenity use), forestry and areas in or near water. Within these fields of use there are about 2,013 products in the UK with current approval. The *Guide* is a reference for virtually all of them given in two sections. Section 2 gives full details of the products notified to the editor as available on the market. Products are included in this section only if requested by the supplier and supported by evidence of approval. Section 3 gives brief details of all other registered products with extant approval.

Section 2 lists some 430 individual active ingredients and mixtures. For each entry a list is shown of the available approved products together with the name of the supplier from whom each may be obtained. All types of pesticide covered by Control of Pesticides Regulations or Plant Protection Products Regulations are included. This embraces acaricides, algicides, herbicides, fungicides, insecticides, lumbricides, molluscicides, nematicides and rodenticides together with plant growth regulators used as straw shorteners, sprout inhibitors and for various horticultural purposes. The total number of products included in section 2 (i.e. available in the market in 2017) is about 1,253.

The *Guide* also gives information (in section 4) on around 167 authorised adjuvants which, although not active pesticides themselves, may be added to pesticide products to improve their effectiveness. The *Guide* does not include products solely approved for amateur, home and garden, domestic, food storage, public or animal health uses.

Sources of Information

The information in this edition has been drawn from these authoritative sources:

- approved labels and product manuals received from suppliers of pesticides up to October 2016.
- websites of the Chemicals Regulation Directorate (CRD, www.pesticides.gov.uk) and the Health and Safety Executive (HSE, www.hse.gov.uk).

Criteria for Inclusion

To be included, a pesticide must meet the following conditions:

- it must have extant approval under UK pesticides legislation

Introduction

- information on the approved uses must have been provided by the supplier
- it must be expected to be on the UK market during the currency of this edition

When a company changes its name, whether by merger, takeover or joint venture, it is obliged to re-register its products in the name of the new company, and new MAPP numbers are normally assigned. Where stocks of the previously registered product remain legally available, both old and new identities are included in the *Guide* and remain until approval of the former lapses or stocks are notified as exhausted.

Products that have been withdrawn from the market and whose approval will finally lapse during 2017 are identified in each profile. After the indicated date, sale, storage or use of the product bearing that approval number becomes illegal. Where there is a direct replacement product, this is indicated.

The Voluntary Initiative

The Voluntary Initiative (VI) is a programme of measures, agreed by the crop protection industry and related organisations with Government, to minimise the environmental impacts of crop protection products. The programme has provided a framework of practices and principles to help Government achieve its objective of protection of water quality and enhancement of farmland biodiversity. Many of the environmental protection schemes launched under the VI represent current best practice.

The first five-year phase of the programme formally concluded in March 2006, but it is continuing with a rolling two-year review of proposals and targets; current proposals include a new online crop protection management plan and greater involvement in catchment-sensitive farming (see below).

A key element of the VI has been the provision of environmental information on crop protection products. Members of the Crop Protection Association (CPA) have committed to do this by producing Environmental Information Sheets (EISs) for all their marketed professional products.

EISs reinforce and supplement information on a product label by giving specific environmental impact information in a standardised format. They highlight any situations where risk management is essential to ensure environmental protection. Their purpose is to provide user-friendly information to advisers (including those in the amenity sector), farmers and growers on the environmental impact of crop protection products, to allow planning with a better understanding of the practical implications.

Links are provided to give users rapid access to EISs on the VI website: www.voluntaryinitiative.org.uk.

Catchment-sensitive Farming

Catchment-sensitive Farming (CSF) is the government's response to climate change and the new European Water Directive. Rainfall events are likely to become heavier, with a greater risk of soil, nutrients and pesticides ending up in the waterways. The CSF programme is investigating the effects of specific targeted advice on 20 catchment areas perceived as being at risk, and new entries in this *Guide* can help identify the pesticides that pose the greatest risk of water contamination (see Environmental Safety).

Notes on Contents and Layout

The book consists of six main sections:

1 Crop/Pest Guide
2 Pesticide Profiles
3 Products Also Registered
4 Adjuvants
5 Useful Information
6 Appendices

1 Crop/Pest Guide

This section enables the user to identify which active ingredients are approved for a particular crop/pest combination. The crops are grouped as shown in the Crop Index. For convenience, some crops and pests have been grouped into generic units (for example, 'cereals' are covered as a group, not individually). Indications from the Crop/Pest Guide must always be checked against the specific entry in the pesticide profiles section because a product may not be approved on all crops in the group. Chemicals indicated as having uses in cereals, for example, may be approved for use only on winter wheat and winter barley, not for other cereals. Because of the difference in wording of product labels, it may sometimes be necessary to refer to broader categories of organism in addition to specific organisms (e.g. annual grasses and annual weeds in addition to annual meadow-grass).

2 Pesticide Profiles

Each active ingredient and mixture of ingredients has a separate numbered profile entry. The entries are arranged in alphabetical order using the common names approved by the British Standards Institution and used in *The Pesticide Manual* (17th Edition; BCPC, 2015) now available as an online subscription resource at www.bcpcdata.com/pm. Where an active ingredient is available only in mixtures, this is stated. The ingredients of the mixtures are themselves ordered alphabetically, and the entries appear at the correct point for the first named active ingredient.

Below the profile title, a brief phrase describes the main use of the active ingredient(s). This is followed where appropriate by the mode of action code(s) as published by the Fungicide, Herbicide and Insecticide Resistance Action Committees.

Within each profile entry, a table lists the products approved and available on the market in the following style:

Product name	Main supplier	Active ingredient contents	Formulation type	Registration number
1 Broadsword	United Phosphorus	200:85:65 g/l	EC	09140
2 Greengard	SumiAgro Amenity	200:85:65 g/l	EC	11715
3 Nu-Shot	Nufarm UK	200:85:65 g/l	EC	11148

Many of the **product names** are registered trademarks, but no special indication of this is given. Individual products are indexed under their entry number in the Index of Proprietary Names at the back of the book. The **main supplier** indicates the marketing outlet through which the product may be purchased. Full addresses and contact details of suppliers are listed in Appendix 1. For mixtures, the **active ingredient contents** are given in the same order as they appear in the profile heading. The **formulation types** are listed in full in the Key to Abbreviations and Acronyms (Appendix 5). The **registration number** normally refers to the registration with the Chemicals Regulation Directorate (CRD). In cases where the products registered with the Health and Safety Executive are included, HSE numbers are quoted, e.g. H0462 and approved biocides are numbered UK12-XXXX.

Notes on Contents and Layout

Below the product table, a **Uses** section lists all approved uses notified by suppliers to the Editor by October 2016, giving both the principal target organisms and the recommended crops or situations (identified in ***bold italics***). Where there is an important condition of use, or where the approval is off-label, this is shown in parentheses as (*off-label*). Numbers in square brackets refer to the numbered products in the table above. Thus, in the example shown above, a use approved for Nu-Shot (product 3) but not for Broadsword or Greengard (products 1 and 2) appears as:

Annual dicotyledons in rotational grassland [3]

Any Extensions of Authorisation for Minor Uses (EAMUs), previously known as SOLAs, for products in the profile are detailed below the list of approved uses. Each EAMU has a separate entry and shows the crops to which it applies, the Notice of Approval number, the expiry date (if due to fall during the current edition), and the product reference number in square brackets. EAMUs do not generally appear on the product label and are undertaken entirely at the risk of the user.

Below the EAMU paragraph, **Notes** are listed under the following headings. Unless otherwise stated, any reference to dose is made in terms of product rather than active ingredient. Where a note refers to a particular product rather than to the entry generally, this is indicated by numbers in square brackets as described above.

Approval information	Information of a general nature about the approval status of the active ingredient or products in the profile is given here. Notes on approval for aerial or ULV application are given.
	Inclusion of the active ingredient in Annex I under EC Directive 1107/2009 is shown, as well as any acceptance from the British Beer and Pub Association (BBPA) for its use on barley or hops for brewing or malting.
	Where a product approval will finally expire in 2017, the expiry date is shown here.
Efficacy	This entry identifies factors important in making the most effective use of the treatment. Certain factors, such as the need to apply chemicals uniformly, the need to use appropriate volume rates, and the need to prevent settling out of the active ingredient in the spray tank, are assumed to be important for all treatments and are not emphasised in individual profiles.
Restrictions	Notes are included in this section where products are subject to the Poisons Law, and where the label warns about organophosphorus and/or anticholinesterase compounds. Factors important in optimising performance and minimising the risk of crop damage, including statutory conditions relating to the maximum permitted number of applications, are listed. Any restrictions on crop varieties that may not be sprayed are mentioned here.
Environmental safety	Where the label specifies an environmental hazard classification, it is noted together with the associated risk phrases. Any other special operator precautions and any conditions concerning withholding livestock from treated areas are specified. Where any of the products in the profile are subject to Category A, Category B or broadcast air-assisted buffer zone restrictions under the LERAP scheme, the relevant classification is shown. Where a LERAP requires a buffer zone greater than 5m, detail of the larger buffer zone is given here.
	Other environmental hazards are also noted here, including potential dangers to livestock, game, wildlife, bees and fish. The need to avoid drift onto neighbouring crops and the need to wash out equipment thoroughly after use are important with all pesticide treatments, but may receive special mention here if of particular significance.
Crop-specific information	Instructions about the timing of application or cultivations that are specific to a particular crop, rather than generally applicable, are mentioned here. The latest permitted use and harvest intervals (if any) are shown for individual crops.

Notes on Contents and Layout

Following crops guidance	Any specific label instructions about what may be grown after a treated crop, whether harvested normally or after failure, are shown here.
Hazard classification and safety precautions	The label hazard classifications and precautions are shown using a series of letter and number codes which are explained in Appendix 4. Hazard warnings are now given in full, with the other precautions listed under the following subheadings:

- Risk phrases
- Operator protection
- Environmental protection
- Consumer protection
- Storage and disposal
- Treated seed
- Vertebrate/rodent control products
- Medical advice

This section is given for information only and should not be used for the purpose of making a COSHH assessment without reference to the actual label of the product to be used.

3 Products also Registered

Products with an extant approval for all or part of the year of the edition in which they appear are listed in this section if they have not been requested by their supplier or manufacturer for inclusion in Section 2. Details shown are the same (apart from formulation) as those in the product tables of Section 2 and the approval expiry date is shown. Not all products listed here will be available in the market, but this list, together with the products in Section 2, comprises a comprehensive listing of all approved products in the UK for uses within the scope of the *Guide* at the time of going to print.

4 Adjuvants

Adjuvants are listed in a table ordered alphabetically by product name. For each product, details are shown of the main supplier, the authorisation number and the mode of action (e.g. wetter, sticker, etc.) as shown on the label. A brief statement of the uses of the adjuvant is given. Protective clothing requirements and label precautions are listed using the codes from Appendix 4.

5 Useful Information

This section summarises legislation covering approval, storage, sale and use of pesticides in the UK. There are brief articles on the broader issues concerning the use of crop protection chemicals, including resistance management. Lists are provided of products approved for use in or near water, in forestry, as seed treatments and for aerial application. Chemicals subject to the Poisons Laws are listed, and there is a summary of first aid measures if pesticide poisoning should be suspected. Finally, this section provides guidance on environmental protection issues and covers the protection of bees and the use of pesticides in or near water.

6 Appendices

Appendix 1	Gives contact details of all companies listed as main suppliers in the pesticide profiles.
	Where a supplier is no longer trading under that name (usually following a merger or takeover), but products under that name are still listed in the Guide because they are still available in the supply chain, a cross-reference indicates the new 'parent' company from which technical or commercial information can be obtained.
Appendix 2	Gives contact details of useful contacts, including the National Poisons Information Service.

Notes on Contents and Layout

Appendix 3	Gives details of the keys to crop and weed growth stages, including the publication reference for each. The numeric codes are used in the descriptive sections of the pesticide profiles (Section 2).
Appendix 4	Shows the full text for code letters and numbers used in the pesticide profiles (Section 2) to indicate personal protective equipment requirements and label precautions.
Appendix 5	Shows the full text for the formulation abbreviations used in the pesticide profiles (Section 2). Other abbreviations and acronyms used in the Guide are also explained here.
Appendix 6	Provides full definitions of officially agreed descriptive phrases for crops or situations used in the pesticide profiles (Section 2) where misunderstandings can occur.
Appendix 7	Shows a list of useful reference publications, which amplify the summarised information provided in this section.

SECTION 1
CROP/PEST GUIDE

Crop/Pest Guide Index

Important note: The Crop/Pest Guide Index refers to pages on which the subject can be located.

Arable and vegetable crops
Agricultural herbage
  Grass 5
  Herbage legumes 5
  Herbage seed 5
Brassicas
  Brassica seed crops 6
  Brassicas, general 6
  Fodder brassicas 7
  Leaf and flowerhead brassicas .. 8
  Mustard 9
  Root brassicas 10
  Salad greens 11
Cereals
  Barley 12
  Cereals, general 17
  Maize/sweetcorn 19
  Oats 20
  Rye/triticale 22
  Undersown cereals 25
  Wheat 26
Edible fungi
  Mushrooms 32
Fruiting vegetables
  Aubergines 32
  Cucurbits 32
  Peppers 33
  Tomatoes 33
Herb crops
  Herbs 33
Leafy vegetables
  Endives 35
  Lettuce 35
  Spinach 36
  Watercress 37
Legumes
  Beans (Phaseolus) 37
  Beans (Vicia) 37
  Forage legumes, general 38
  Lupins 38
  Peas 39
Miscellaneous arable
  Industrial crops 40
  Miscellaneous arable crops 40
  Miscellaneous arable situations . 41
Miscellaneous field vegetables
  All vegetables 41
Oilseed crops
  Linseed/flax 41
  Miscellaneous oilseeds 42
  Oilseed rape 43
  Soya 44
  Sunflowers 45

Root and tuber crops
  Beet crops 45
  Carrots/parsnips 47
  Miscellaneous root crops 48
  Potatoes 49
Stem and bulb vegetables
  Asparagus 50
  Celery/chicory 51
  Globe artichokes/cardoons 52
  Onions/leeks/garlic 52
  Vegetables 53

Flowers and ornamentals
Flowers
  Bedding plants, general 54
  Bulbs/corms 54
  Miscellaneous flowers 54
  Pot plants 55
  Protected flowers 55
Miscellaneous flowers and ornamentals
  Miscellaneous uses 55
Ornamentals
  Nursery stock 56
  Trees and shrubs 58
  Woody ornamentals 59

Forestry
Forest nurseries, general
  Forest nurseries 59
Forestry plantations, general
  Forestry plantations 60
Miscellaneous forestry situations
  Cut logs/timber 61
Woodland on farms
  Woodland 61

Fruit and hops
All bush fruit
  All currants 61
  All protected bush fruit 62
  All vines 63
  Bilberries/blueberries/cranberries 63
  Bush fruit, general 64
  Gooseberries 64
  Miscellaneous bush fruit 65
Cane fruit
  All outdoor cane fruit 66
  All protected cane fruit 67
Hops, general
  Hops 67
Miscellaneous fruit situations
  Fruit crops, general 68
  Fruit nursery stock 69

Other fruit
　All outdoor strawberries 69
　Protected miscellaneous fruit . . 70
　Rhubarb 71
Tree fruit
　All nuts 71
　All pome fruit 71
　All stone fruit 73
　Miscellaneous top fruit 73
　Protected fruit 73

Grain/crop store uses
Stored produce
　Food/produce storage 74
Stored seed
　Stored grain/rape/linseed 74

Grass
Turf/amenity grass
　Amenity grassland 74
　Managed amenity turf 75

Non-crop pest control
Farm buildings/yards
　Farm buildings 76
　Farmyards 77
Farmland pest control
　Farmland situations 77
Miscellaneous non-crop pest control
　Manure/rubbish 77
　Miscellaneous pest control
　　situations 77

Protected salad and vegetable crops
Protected brassicas
　Protected brassica vegetables . . 77
　Protected salad brassicas 78
Protected crops, general
　All protected crops 78
　Protected nuts 78
　Protected vegetables, general . . 78
　Soils and compost 78
Protected fruiting vegetables
　Protected aubergines 78
　Protected cucurbits 79
　Protected tomatoes 80
Protected herb crops
　Protected herbs 80
Protected leafy vegetables
　Mustard and cress 81
　Protected leafy vegetables 82
　Protected spinach 82
Protected legumes
　Protected peas and beans 82
Protected root and tuber vegetables
　Protected carrots/parsnips/
　　celeriac 83
　Protected root brassicas 83
Protected stem and bulb vegetables
　Protected celery/chicory 83
　Protected onions/leeks/garlic . . . 83

Total vegetation control
Aquatic situations, general
　Aquatic situations 83
　Miscellaneous non-crop situations 84
　Non-crop farm areas 84
　Paths/roads etc 84

Crop/Pest Guide

Important note: For convenience, some crops and pests or targets have been brought together into genetic groups in this guide, e.g. 'cereals', 'annual grasses'. It is essential to check the profile entry in Section 2 *and* the label to ensure that a product is approved for a specific crop/pest combination, e.g. winter wheat/blackgrass.

Arable and vegetable crops

Agricultural herbage - Grass

Crop control	Desiccation	glyphosate
Diseases	Crown rust	propiconazole
	Drechslera leaf spot	propiconazole
	Powdery mildew	propiconazole
	Rhynchosporium	propiconazole
Pests	Aphids	esfenvalerate
	Birds/mammals	aluminium ammonium sulphate
	Flies	esfenvalerate
	Slugs/snails	ferric phosphate
Plant growth regulation	Growth control	gibberellins
Weeds	Broad-leaved weeds	2,4-D, 2,4-D + dicamba, 2,4-D + MCPA, 2,4-DB, 2,4-DB + MCPA, amidosulfuron, aminopyralid + triclopyr, clopyralid, clopyralid + fluroxypyr + triclopyr, clopyralid + triclopyr, fluroxypyr, fluroxypyr (*off-label*), fluroxypyr + triclopyr, glyphosate, glyphosate (*wiper application*), MCPA, metsulfuron-methyl, thifensulfuron-methyl, tribenuron-methyl, tribenuron-methyl (*off-label*)
	Crops as weeds	fluroxypyr (*off-label*)
	Grass weeds	glyphosate
	Weeds, miscellaneous	aminopyralid + triclopyr, clopyralid + triclopyr, fluroxypyr (*off-label*), glyphosate, MCPA
	Woody weeds/scrub	clopyralid + triclopyr

Agricultural herbage - Herbage legumes

Crop control	Desiccation	diquat
Weeds	Broad-leaved weeds	2,4-DB, diquat (*off-label*), propyzamide
	Grass weeds	diquat (*seed crop*), fluazifop-P-butyl (*off-label*), propyzamide
	Weeds, miscellaneous	diquat (*seed crop*)

Agricultural herbage - Herbage seed

Crop control	Desiccation	diquat
Diseases	Crown rust	boscalid + epoxiconazole (*off-label*), propiconazole
	Damping off	thiram (*seed treatment*)
	Disease control	epoxiconazole (*off-label*), epoxiconazole + fenpropimorph + kresoxim-methyl (*off-label*), epoxiconazole + pyraclostrobin (*off-label*),

Crop/Pest Guide - Arable and vegetable crops

		fenpropimorph + pyraclostrobin (*off-label*), tebuconazole (*off-label*)
	Drechslera leaf spot	fenpropimorph + pyraclostrobin, propiconazole
	Foliar diseases	azoxystrobin (*off-label*), azoxystrobin + cyproconazole (*off-label*), epoxiconazole + fenpropimorph (*off-label*), epoxiconazole + fenpropimorph + kresoxim-methyl (*off-label*), fenpropimorph (*off-label*), pyraclostrobin (*off-label*), trifloxystrobin (*off-label*)
	Leaf spot	prothioconazole + tebuconazole (*off-label*)
	Net blotch	boscalid + epoxiconazole (*off-label*)
	Powdery mildew	azoxystrobin (*off-label*), cyflufenamid (*off-label*), epoxiconazole (*off-label*), propiconazole, prothioconazole + tebuconazole (*off-label*), sulphur (*off-label*), tebuconazole (*off-label*)
	Rhynchosporium	boscalid + epoxiconazole (*off-label*), propiconazole, prothioconazole + tebuconazole (*off-label*)
	Rust	azoxystrobin (*off-label*), boscalid + epoxiconazole (*off-label*), epoxiconazole (*off-label*), fenpropimorph + pyraclostrobin, fenpropimorph + pyraclostrobin (*off-label*), prothioconazole + tebuconazole (*off-label*), tebuconazole (*off-label*)
	Septoria	fenpropimorph + pyraclostrobin, fenpropimorph + pyraclostrobin (*off-label*)
Pests	Aphids	deltamethrin (*off-label*), esfenvalerate (*off-label*), tau-fluvalinate (*off-label*), zeta-cypermethrin (*off-label*)
	Flies	esfenvalerate
	Pests, miscellaneous	deltamethrin (*off-label*), lambda-cyhalothrin (*off-label*), tau-fluvalinate (*off-label*), zeta-cypermethrin (*off-label*)
Plant growth regulation	Growth control	trinexapac-ethyl
Weeds	Broad-leaved weeds	bifenox (*off-label*), bromoxynil + diflufenican (*off-label*), dicamba + MCPA + mecoprop-P, diflufenican (*off-label*), diflufenican + flufenacet (*off-label*), ethofumesate, florasulam (*off-label*), florasulam + fluroxypyr (*off-label*), fluroxypyr, MCPA, mecoprop-P, pendimethalin (*off-label*), propyzamide
	Crops as weeds	bromoxynil + diflufenican (*off-label*), clopyralid (*off-label*), florasulam (*off-label*)
	Grass weeds	clodinafop-propargyl (*off-label*), diflufenican (*off-label*), diflufenican + flufenacet (*off-label*), ethofumesate, fenoxaprop-P-ethyl (*off-label*), pendimethalin (*off-label*), propyzamide
	Weeds, miscellaneous	florasulam (*off-label*), MCPA

Brassicas - Brassica seed crops

Diseases	Alternaria	iprodione, iprodione (*pre-storage only*)
	Botrytis	iprodione
	Damping off	thiram (*off-label*)
Weeds	Broad-leaved weeds	propyzamide
	Grass weeds	propyzamide

Brassicas - Brassicas, general

Diseases	Alternaria	cyprodinil + fludioxonil (*off-label*)

Crop/Pest Guide - Arable and vegetable crops

	Botrytis	Bacillus subtilis (*off-label*), boscalid + pyraclostrobin (*off-label*), cyprodinil + fludioxonil (*off-label*)
	Downy mildew	dimethomorph (*off-label*)
	Pythium	metalaxyl-M (*off-label*)
	Rhizoctonia	boscalid + pyraclostrobin (*off-label*), cyprodinil + fludioxonil (*off-label*)
	Sclerotinia	boscalid + pyraclostrobin (*off-label*)
	Stem canker	cyprodinil + fludioxonil (*off-label*)
Pests	Aphids	pymetrozine (*off-label*), spirotetramat (*off-label*), thiacloprid (*off-label*)
	Caterpillars	Bacillus thuringiensis (*off-label*), deltamethrin (*off-label*), indoxacarb (*off-label*)
	Cutworms	Bacillus thuringiensis (*off-label*)
	Flies	Metarhizium anisopliae (*off-label*)
	Whiteflies	spirotetramat (*off-label*)
Weeds	Broad-leaved weeds	clopyralid (*off-label*), pyridate (*off-label*)
	Crops as weeds	clopyralid (*off-label*)
	Grass weeds	fluazifop-P-butyl (*off-label*)

Brassicas - Fodder brassicas

Diseases	Alternaria	azoxystrobin, cyprodinil + fludioxonil (*off-label*), difenoconazole, iprodione (*off-label*)
	Black leg	fludioxonil (*off-label*)
	Botrytis	cyprodinil + fludioxonil (*off-label*), Gliocladium catenulatum (*off-label*)
	Damping off	Bacillus subtilis (*off-label*), thiram (*off-label*)
	Downy mildew	fluopicolide + propamocarb hydrochloride (*off-label*), metalaxyl-M (*off-label*)
	Fusarium	Gliocladium catenulatum (*off-label*)
	Phytophthora	Gliocladium catenulatum (*off-label*)
	Powdery mildew	azoxystrobin + difenoconazole
	Pythium	Bacillus subtilis (*off-label*)
	Rhizoctonia	cyprodinil + fludioxonil (*off-label*), Gliocladium catenulatum (*off-label*)
	Ring spot	azoxystrobin, difenoconazole
	Stem canker	cyprodinil + fludioxonil (*off-label*)
	White blister	azoxystrobin, azoxystrobin + difenoconazole, boscalid + pyraclostrobin (*off-label*)
Pests	Aphids	esfenvalerate, lambda-cyhalothrin, pymetrozine (*off-label*), thiacloprid (*off-label*), thiamethoxam (*off-label*)
	Beetles	alpha-cypermethrin, thiamethoxam (*off-label*)
	Caterpillars	alpha-cypermethrin, Bacillus thuringiensis, Bacillus thuringiensis (*off-label*), deltamethrin (*off-label*), esfenvalerate, indoxacarb (*off-label*), spinosad (*off-label*)
	Flies	spinosad (*off-label*), thiamethoxam (*off-label*)
	Whiteflies	lambda-cyhalothrin
Weeds	Broad-leaved weeds	clomazone (*off-label*), clopyralid (*off-label*), dimethenamid-p + metazachlor (*off-label*), metazachlor (*off-label*), napropamide, pendimethalin (*off-label*), pyridate (*off-label*)

Crop/Pest Guide - Arable and vegetable crops

	Crops as weeds	clopyralid (*off-label*), fluazifop-P-butyl (*stockfeed only*)
	Grass weeds	dimethenamid-p + metazachlor (*off-label*), fluazifop-P-butyl (*stockfeed only*), metazachlor (*off-label*), napropamide, pendimethalin (*off-label*), S-metolachlor (*off-label*)

Brassicas - Leaf and flowerhead brassicas

Diseases	Alternaria	azoxystrobin, boscalid + pyraclostrobin, chlorothalonil + metalaxyl-M (*moderate control*), cyprodinil + fludioxonil (*off-label*), difenoconazole, iprodione, iprodione (*off-label*), prothioconazole, tebuconazole, tebuconazole + trifloxystrobin
	Black leg	fludioxonil, fludioxonil (*off-label*)
	Botrytis	cyprodinil + fludioxonil (*off-label*), Gliocladium catenulatum (*off-label*), iprodione, iprodione (*off-label*)
	Damping off	Bacillus subtilis (*off-label*), thiram (*seed treatment*), tolclofos-methyl, tolclofos-methyl (*off-label*)
	Downy mildew	chlorothalonil + metalaxyl-M, fluopicolide + propamocarb hydrochloride (*off-label*), metalaxyl-M (*off-label*)
	Fungus diseases	fludioxonil
	Fusarium	Gliocladium catenulatum (*off-label*)
	Light leaf spot	prothioconazole, tebuconazole, tebuconazole + trifloxystrobin
	Phoma leaf spot	tebuconazole + trifloxystrobin
	Phytophthora	Gliocladium catenulatum (*off-label*)
	Powdery mildew	azoxystrobin + difenoconazole, prothioconazole, tebuconazole, tebuconazole + trifloxystrobin
	Pythium	Bacillus subtilis (*off-label*), metalaxyl-M
	Rhizoctonia	cyprodinil + fludioxonil (*off-label*), Gliocladium catenulatum (*off-label*)
	Ring spot	azoxystrobin, boscalid + pyraclostrobin, chlorothalonil + metalaxyl-M (*reduction*), difenoconazole, prothioconazole, tebuconazole, tebuconazole (*off-label*), tebuconazole + trifloxystrobin
	Stem canker	cyprodinil + fludioxonil (*off-label*), prothioconazole
	Storage rots	1-methylcyclopropene (*off-label*), metalaxyl-M (*off-label*)
	White blister	azoxystrobin, azoxystrobin + difenoconazole, boscalid + pyraclostrobin, boscalid + pyraclostrobin (*off-label*), boscalid + pyraclostrobin (*qualified minor use*), chlorothalonil + metalaxyl-M, mancozeb + metalaxyl-M (*off-label*), tebuconazole + trifloxystrobin
Pests	Aphids	acetamiprid (*off-label*), beta-cyfluthrin, cypermethrin, deltamethrin, esfenvalerate, lambda-cyhalothrin, pymetrozine, pymetrozine (*off-label*), pymetrozine (*useful levels of control*), pyrethrins, spirotetramat, thiacloprid, thiacloprid (*off-label*)
	Beetles	alpha-cypermethrin, deltamethrin, pyrethrins
	Birds/mammals	aluminium ammonium sulphate
	Caterpillars	alpha-cypermethrin, Bacillus thuringiensis, Bacillus thuringiensis (*off-label*), beta-cyfluthrin, cypermethrin, deltamethrin, deltamethrin (*off-label*), diflubenzuron, esfenvalerate, indoxacarb, indoxacarb (*off-label*),

	Flies	lambda-cyhalothrin, pyrethrins, spinosad, spinosad (*off-label*) cyantraniliprole, spinosad (*off-label*)
	Slugs/snails	metaldehyde
	Whiteflies	acetamiprid (*off-label*), cypermethrin, deltamethrin, lambda-cyhalothrin, pyrethrins, spirotetramat
Weeds	Broad-leaved weeds	clomazone (*off-label*), clopyralid, clopyralid (*off-label*), dimethenamid-p + metazachlor (*off-label*), dimethenamid-p + pendimethalin (*off-label*), metazachlor, metazachlor (*off-label*), napropamide, napropamide (*off-label*), pendimethalin, pendimethalin (*off-label*), propyzamide (*off-label*), pyridate, pyridate (*off-label*)
	Crops as weeds	clopyralid (*off-label*), cycloxydim, pendimethalin
	Grass weeds	clethodim (*off-label*), cycloxydim, dimethenamid-p + metazachlor (*off-label*), dimethenamid-p + pendimethalin (*off-label*), fluazifop-P-butyl (*off-label*), metazachlor, metazachlor (*off-label*), napropamide, napropamide (*off-label*), pendimethalin, pendimethalin (*off-label*), propyzamide (*off-label*), S-metolachlor (*off-label*)

Brassicas - Mustard

Crop control	Desiccation	diquat (*off-label*), glyphosate
Diseases	Alternaria	azoxystrobin (*off-label*), cyprodinil + fludioxonil (*off-label*), difenoconazole (*off-label*)
	Botrytis	cyprodinil + fludioxonil (*off-label*)
	Disease control	boscalid (*off-label*), mandipropamid, prothioconazole (*off-label*), tebuconazole (*off-label*)
	Downy mildew	fenamidone + fosetyl-aluminium (*off-label*), fluopicolide + propamocarb hydrochloride (*off-label*), mandipropamid
	Powdery mildew	difenoconazole (*off-label*), tebuconazole (*off-label*)
	Rhizoctonia	cyprodinil + fludioxonil (*off-label*)
	Rust	tebuconazole (*off-label*)
	Sclerotinia	azoxystrobin (*off-label*), difenoconazole (*off-label*)
	Stem canker	cyprodinil + fludioxonil (*off-label*)
Pests	Aphids	cypermethrin, spirotetramat (*off-label*), tau-fluvalinate (*off-label*)
	Beetles	deltamethrin, indoxacarb (*off-label*), thiacloprid
	Caterpillars	deltamethrin
	Flies	tefluthrin (*off-label - seed treatment*)
	Pests, miscellaneous	cypermethrin, lambda-cyhalothrin (*off-label*), spinosad (*off-label*), tau-fluvalinate (*off-label*)
	Weevils	deltamethrin
	Whiteflies	spirotetramat (*off-label*)
Weeds	Broad-leaved weeds	bifenox (*off-label*), clopyralid (*off-label*), clopyralid + picloram (*off-label*), glyphosate, metazachlor (*off-label*), napropamide (*off-label*)
	Crops as weeds	glyphosate, napropamide (*off-label*)
	Grass weeds	fluazifop-P-butyl (*off-label*), glyphosate, metazachlor (*off-label*), napropamide (*off-label*)

Crop/Pest Guide - Arable and vegetable crops

	Weeds, miscellaneous	glyphosate, napropamide (*off-label*)

Brassicas - Root brassicas

Diseases	Alternaria	azoxystrobin (*off-label*), azoxystrobin + difenoconazole (*off-label*), fenpropimorph (*off-label*), prothioconazole, tebuconazole
	Black leg	fludioxonil (*off-label*)
	Botrytis	boscalid + pyraclostrobin (*off-label*), Gliocladium catenulatum (*off-label*)
	Crown rot	fenpropimorph (*off-label*)
	Damping off	thiram (*off-label*), thiram (*off-label - seed treatment*), thiram (*seed treatment*), tolclofos-methyl (*off-label*)
	Disease control	azoxystrobin (*off-label*), boscalid + pyraclostrobin (*off-label*)
	Downy mildew	dimethomorph (*off-label*), metalaxyl-M (*off-label*), propamocarb hydrochloride (*off-label*)
	Foliar diseases	azoxystrobin (*off-label*)
	Fusarium	Gliocladium catenulatum (*off-label*)
	Leaf spot	tebuconazole + trifloxystrobin (*off-label*)
	Phytophthora	Gliocladium catenulatum (*off-label*)
	Powdery mildew	azoxystrobin (*off-label*), azoxystrobin + difenoconazole (*off-label*), fenpropimorph (*off-label*), isopyrazam (*off-label*), prothioconazole, sulphur, tebuconazole, tebuconazole (*off-label*), tebuconazole + trifloxystrobin (*off-label*)
	Pythium	metalaxyl-M, metalaxyl-M (*off-label*)
	Rhizoctonia	azoxystrobin (*off-label*), Gliocladium catenulatum (*off-label*), tolclofos-methyl (*off-label*)
	Root rot	tolclofos-methyl (*off-label*)
	Rust	boscalid + pyraclostrobin (*off-label*)
	Sclerotinia	boscalid + pyraclostrobin (*off-label*), cyprodinil + fludioxonil (*off-label*), tebuconazole + trifloxystrobin (*off-label*)
	Seed-borne diseases	thiram (*off-label - seed treatment*)
	White blister	metalaxyl-M (*off-label*)
Pests	Aphids	cypermethrin, esfenvalerate, pirimicarb (*off-label*), pymetrozine (*off-label*), spirotetramat (*off-label*), thiacloprid (*off-label*), thiamethoxam (*off-label*)
	Beetles	deltamethrin, thiamethoxam (*off-label*)
	Caterpillars	Bacillus thuringiensis, Bacillus thuringiensis (*off-label*), chlorantraniliprole (*off-label*), deltamethrin, esfenvalerate, indoxacarb (*off-label*), lambda-cyhalothrin (*off-label*), spinosad (*off-label*)
	Cutworms	Bacillus thuringiensis (*off-label*), cypermethrin, deltamethrin (*off-label*), lambda-cyhalothrin (*off-label*)
	Flies	chlorantraniliprole (*off-label*), cypermethrin, deltamethrin (*off-label*), lambda-cyhalothrin (*off-label*), spinosad (*off-label*), thiamethoxam (*off-label*)
	Pests, miscellaneous	cypermethrin, deltamethrin (*off-label*), lambda-cyhalothrin (*off-label*)
	Slugs/snails	ferric phosphate
	Weevils	lambda-cyhalothrin (*off-label*)

Crop/Pest Guide - Arable and vegetable crops

Plant growth regulation	Quality/yield control	sulphur
Weeds	Broad-leaved weeds	clomazone (*off-label*), clopyralid, dimethenamid-p + metazachlor (*off-label*), glyphosate, metamitron (*off-label*), metazachlor (*off-label*), pendimethalin (*off-label*), prosulfocarb (*off-label*), S-metolachlor (*off-label*)
	Crops as weeds	cycloxydim, fluazifop-P-butyl (*stockfeed only*), glyphosate
	Grass weeds	cycloxydim, dimethenamid p + metazachlor (*off-label*), fluazifop-P-butyl (*off-label*), fluazifop-P-butyl (*stockfeed only*), glyphosate, metamitron (*off-label*), metazachlor (*off-label*), propaquizafop, prosulfocarb (*off-label*), S-metolachlor (*off-label*)
	Weeds, miscellaneous	glyphosate, glyphosate (*off-label*)

Brassicas - Salad greens

Diseases	Alternaria	azoxystrobin (*off-label - for baby leaf production*), cyprodinil + fludioxonil (*off-label*), iprodione (*off-label*)
	Black rot	azoxystrobin (*off-label*)
	Botrytis	cyprodinil + fludioxonil (*off-label*), fenhexamid (*off-label - baby leaf production*), Gliocladium catenulatum (*off-label*), pyrimethanil (*off-label*)
	Damping off	Bacillus subtilis (*off-label*), tolclofos-methyl, tolclofos-methyl (*off-label*)
	Disease control	dimethomorph + mancozeb (*off-label*), mandipropamid
	Downy mildew	azoxystrobin (*off-label - for baby leaf production*), dimethomorph (*off-label*), dimethomorph + mancozeb (*off-label*), fenamidone + fosetyl-aluminium (*off-label*), fluopicolide + propamocarb hydrochloride (*off-label*), mancozeb (*off-label*), mancozeb + metalaxyl-M (*off-label*), mandipropamid, metalaxyl-M (*off-label*)
	Fusarium	Gliocladium catenulatum (*off-label*), Trichoderma asperellum (Strain T34) (*off-label*)
	Phytophthora	Gliocladium catenulatum (*off-label*)
	Powdery mildew	tebuconazole (*off-label*)
	Pythium	metalaxyl-M, Trichoderma asperellum (Strain T34) (*off-label*)
	Rhizoctonia	cyprodinil + fludioxonil (*off-label*), Gliocladium catenulatum (*off-label*)
	Ring spot	tebuconazole (*off-label*)
	Seed-borne diseases	fludioxonil (*off-label*)
	Stem canker	cyprodinil + fludioxonil (*off-label*)
	White blister	azoxystrobin (*off-label*), boscalid + pyraclostrobin (*off-label*)
Pests	Aphids	acetamiprid (*off-label*), esfenvalerate, pymetrozine (*off-label*), pymetrozine (*off-label - for baby leaf production*), spirotetramat (*off-label*), thiacloprid (*off-label*), thiacloprid (*off-label - baby leaf production*)
	Birds/mammals	aluminium ammonium sulphate

Crop/Pest Guide - Arable and vegetable crops

	Caterpillars	Bacillus thuringiensis, Bacillus thuringiensis (*off-label*), diflubenzuron (*off-label - for baby leaf production*), esfenvalerate, indoxacarb (*off-label*)
	Flies	tefluthrin (*off-label - seed treatment*)
	Pests, miscellaneous	lambda-cyhalothrin (*off-label*), spinosad (*off-label*)
	Spider mites	Lecanicillium muscarium (*off-label*)
	Thrips	Lecanicillium muscarium (*off-label*)
	Whiteflies	Lecanicillium muscarium (*off-label*), spirotetramat (*off-label*)
Weeds	Broad-leaved weeds	chlorpropham, chlorpropham (*off-label*), clomazone (*off-label*), clopyralid (*off-label*), dimethenamid-p + metazachlor (*off-label*), dimethenamid-p + pendimethalin (*off-label*), lenacil (*off-label*), metazachlor (*off-label*), napropamide (*off-label*), pendimethalin (*off-label*), propyzamide (*off-label*), S-metolachlor (*off-label*)
	Crops as weeds	clopyralid (*off-label*)
	Grass weeds	chlorpropham, chlorpropham (*off-label*), dimethenamid-p + metazachlor (*off-label*), dimethenamid-p + pendimethalin (*off-label*), fluazifop-P-butyl (*off-label*), lenacil (*off-label*), metazachlor (*off-label*), napropamide (*off-label*), propyzamide (*off-label*), S-metolachlor (*off-label*)

Cereals - Barley

Crop control	Desiccation	diquat (*off-label*), glyphosate
Diseases	Covered smut	clothianidin + prothioconazole (*seed treatment*), fludioxonil (*seed treatment*), fludioxonil + tefluthrin, fluopyram + prothioconazole + tebuconazole, prothioconazole + tebuconazole
	Disease control	chlorothalonil + penthiopyrad, epoxiconazole + folpet, fluxapyroxad + metconazole, isopyrazam
	Ear blight	prothioconazole + tebuconazole
	Eyespot	azoxystrobin + cyproconazole (*reduction*), bixafen + prothioconazole (*reduction of incidence and severity*), boscalid + epoxiconazole (*moderate control*), boscalid + epoxiconazole + pyraclostrobin (*moderate control*), chlorothalonil + cyproconazole, cyprodinil, epoxiconazole (*reduction*), epoxiconazole + fenpropimorph (*reduction*), epoxiconazole + fenpropimorph + kresoxim-methyl (*reduction*), epoxiconazole + metconazole (*reduction*), fluoxastrobin + prothioconazole (*reduction*), fluoxastrobin + prothioconazole + trifloxystrobin (*reduction*), prochloraz + propiconazole, prochloraz + proquinazid + tebuconazole, prochloraz + tebuconazole, prothioconazole, prothioconazole + spiroxamine, prothioconazole + spiroxamine + tebuconazole (*reduction*), prothioconazole + tebuconazole, prothioconazole + tebuconazole (*reduction*), prothioconazole + trifloxystrobin (*reduction*), prothioconazole + trifloxystrobin (*reduction in severity*)
	Foot rot	clothianidin + prothioconazole (*seed treatment*), fludioxonil (*seed treatment*), fludioxonil + tefluthrin

Crop/Pest Guide - Arable and vegetable crops

Fusarium	fluopyram + prothioconazole + tebuconazole, prothioconazole + tebuconazole
Late ear diseases	bixafen + prothioconazole, fluoxastrobin + prothioconazole, fluoxastrobin + prothioconazole + trifloxystrobin, prothioconazole, prothioconazole + spiroxamine + tebuconazole, prothioconazole + tebuconazole
Leaf stripe	clothianidin + prothioconazole (*seed treatment*), fludioxonil (*seed treatment - reduction*), fludioxonil + tefluthrin (*partial control*), fluopyram + prothioconazole + tebuconazole, prothioconazole + tebuconazole
Loose smut	clothianidin + prothioconazole (*seed treatment*), fluopyram + prothioconazole + tebuconazole, prothioconazole + tebuconazole
Net blotch	azoxystrobin, azoxystrobin + chlorothalonil, azoxystrobin + cyproconazole, bixafen + prothioconazole, boscalid + epoxiconazole, boscalid + epoxiconazole + pyraclostrobin, chlorothalonil + cyproconazole, chlorothalonil + cyproconazole + propiconazole (*reduction*), chlorothalonil + penthiopyrad, chlorothalonil + picoxystrobin, cyproconazole + picoxystrobin (*moderate control*), cyproconazole + propiconazole, cyprodinil, cyprodinil + isopyrazam, epoxiconazole, epoxiconazole (*moderate control*), epoxiconazole + fenpropimorph, epoxiconazole + fenpropimorph + kresoxim-methyl, epoxiconazole + fenpropimorph + metrafenone, epoxiconazole + fluxapyroxad, epoxiconazole + fluxapyroxad + pyraclostrobin, epoxiconazole + isopyrazam, epoxiconazole + metconazole, epoxiconazole + metrafenone, epoxiconazole + pyraclostrobin, fenpropimorph + pyraclostrobin, fluopyram + prothioconazole + tebuconazole (*seed borne*), fluoxastrobin + prothioconazole, fluoxastrobin + prothioconazole + trifloxystrobin, fluxapyroxad, fluxapyroxad + metconazole, fluxapyroxad + pyraclostrobin, metconazole (*reduction*), penthiopyrad, penthiopyrad + picoxystrobin, picoxystrobin, prochloraz + proquinazid + tebuconazole (*moderate control*), prochloraz + tebuconazole, prothioconazole, prothioconazole + spiroxamine, prothioconazole + spiroxamine + tebuconazole, prothioconazole + tebuconazole, prothioconazole + trifloxystrobin, pyraclostrobin, tebuconazole, trifloxystrobin
Powdery mildew	azoxystrobin, azoxystrobin (*moderate control*), azoxystrobin + cyproconazole, bixafen + prothioconazole, boscalid + epoxiconazole (*moderate control*), boscalid + epoxiconazole + pyraclostrobin (*moderate control*), chlorothalonil + cyproconazole, chlorothalonil + cyproconazole + propiconazole, chlorothalonil + proquinazid, cyflufenamid, cyproconazole + propiconazole, cyprodinil, cyprodinil + isopyrazam, epoxiconazole, epoxiconazole + fenpropimorph, epoxiconazole + fenpropimorph + kresoxim-methyl, epoxiconazole + fenpropimorph + metrafenone, epoxiconazole + fluxapyroxad (*moderate control*), epoxiconazole + fluxapyroxad + pyraclostrobin (*moderate control*), epoxiconazole +

13

metconazole (*moderate control*), epoxiconazole + metrafenone, epoxiconazole + pyraclostrobin, fenpropimorph, fenpropimorph + pyraclostrobin, fluoxastrobin + prothioconazole, fluoxastrobin + prothioconazole + trifloxystrobin, flutriafol, fluxapyroxad (*reduction*), fluxapyroxad + metconazole (*moderate control*), fluxapyroxad + pyraclostrobin (*moderate control*), metconazole, metconazole (*moderate control*), metrafenone, picoxystrobin, prochloraz + proquinazid + tebuconazole, prochloraz + tebuconazole, propiconazole, proquinazid, prothioconazole, prothioconazole + spiroxamine, prothioconazole + spiroxamine + tebuconazole, prothioconazole + tebuconazole, prothioconazole + trifloxystrobin, quinoxyfen, spiroxamine, sulphur, tebuconazole, tebuconazole (*moderate control*)

Ramularia leaf spots   bixafen + prothioconazole, boscalid + epoxiconazole (*moderate control*), boscalid + epoxiconazole + pyraclostrobin (*moderate control*), chlorothalonil + penthiopyrad, cyprodinil + isopyrazam, epoxiconazole (*reduction*), epoxiconazole + fluxapyroxad, epoxiconazole + fluxapyroxad + pyraclostrobin, epoxiconazole + isopyrazam, epoxiconazole + metconazole, epoxiconazole + metconazole (*moderate control*), epoxiconazole + pyraclostrobin (*moderate control*), epoxiconazole + pyraclostrobin (*reduction*), fluxapyroxad + metconazole, fluxapyroxad + pyraclostrobin, penthiopyrad (*moderate control*)

Rhynchosporium   azoxystrobin, azoxystrobin (*reduction*), azoxystrobin + chlorothalonil, azoxystrobin + chlorothalonil (*moderate control*), azoxystrobin + cyproconazole (*moderate control*), bixafen + prothioconazole, boscalid + epoxiconazole (*moderate control*), boscalid + epoxiconazole + pyraclostrobin, chlorothalonil (*moderate control*), chlorothalonil + cyproconazole, chlorothalonil + cyproconazole + propiconazole (*reduction*), chlorothalonil + penthiopyrad, chlorothalonil + picoxystrobin, chlorothalonil + propiconazole, chlorothalonil + proquinazid (*suppression only*), chlorothalonil + tebuconazole (*useful reduction*), cyproconazole + picoxystrobin (*moderate control*), cyproconazole + propiconazole, cyprodinil, cyprodinil + isopyrazam, epoxiconazole, epoxiconazole + fenpropimorph, epoxiconazole + fenpropimorph + kresoxim-methyl, epoxiconazole + fenpropimorph + metrafenone, epoxiconazole + fluxapyroxad, epoxiconazole + fluxapyroxad + pyraclostrobin, epoxiconazole + folpet, epoxiconazole + isopyrazam, epoxiconazole + metconazole, epoxiconazole + metrafenone, epoxiconazole + pyraclostrobin, fenpropimorph, fenpropimorph + pyraclostrobin, fluoxastrobin + prothioconazole, fluoxastrobin + prothioconazole + trifloxystrobin, flutriafol, fluxapyroxad, fluxapyroxad + metconazole, fluxapyroxad + pyraclostrobin, folpet (*reduction*), metconazole, metconazole (*reduction*), penthiopyrad, penthiopyrad + picoxystrobin, picoxystrobin, prochloraz + propiconazole, prochloraz + proquinazid + tebuconazole (*moderate control*), prochloraz + tebuconazole, propiconazole, prothioconazole,

Crop/Pest Guide - Arable and vegetable crops

		prothioconazole + spiroxamine, prothioconazole + spiroxamine + tebuconazole, prothioconazole + tebuconazole, prothioconazole + trifloxystrobin, pyraclostrobin (*moderate control*), spiroxamine (*reduction*), tebuconazole, tebuconazole (*moderate control*), trifloxystrobin
	Rust	azoxystrobin, azoxystrobin + chlorothalonil, azoxystrobin + cyproconazole, bixafen + prothioconazole, boscalid + epoxiconazole, boscalid + epoxiconazole + pyraclostrobin, chlorothalonil + cyproconazole, chlorothalonil + cyproconazole + propiconazole, chlorothalonil + picoxystrobin, chlorothalonil + propiconazole, cyproconazole + picoxystrobin, cyproconazole + propiconazole, cyprodinil + isopyrazam, epoxiconazole, epoxiconazole + fenpropimorph, epoxiconazole + fenpropimorph + kresoxim-methyl, epoxiconazole + fenpropimorph + metrafenone, epoxiconazole + fluxapyroxad, epoxiconazole + fluxapyroxad + pyraclostrobin, epoxiconazole + isopyrazam, epoxiconazole + metconazole, epoxiconazole + metrafenone, epoxiconazole + pyraclostrobin, fenpropimorph, fenpropimorph + pyraclostrobin, fluoxastrobin + prothioconazole, fluoxastrobin + prothioconazole + trifloxystrobin, flutriafol, fluxapyroxad, fluxapyroxad (*moderate control*), fluxapyroxad + metconazole, fluxapyroxad + pyraclostrobin, metconazole, penthiopyrad, penthiopyrad (*moderate control*), picoxystrobin, prochloraz + proquinazid + tebuconazole, prochloraz + tebuconazole, propiconazole, prothioconazole, prothioconazole + spiroxamine, prothioconazole + spiroxamine + tebuconazole, prothioconazole + tebuconazole, prothioconazole + trifloxystrobin, pyraclostrobin, spiroxamine, tebuconazole, trifloxystrobin
	Seed-borne diseases	fluopyram + prothioconazole + tebuconazole, ipconazole
	Snow mould	fludioxonil (*seed treatment*), fludioxonil + tefluthrin
	Soil-borne diseases	ipconazole
	Sooty moulds	prothioconazole + tebuconazole
	Take-all	azoxystrobin (*reduction*), azoxystrobin + chlorothalonil (*reduction*), azoxystrobin + chlorothalonil (*reduction only*), azoxystrobin + cyproconazole (*reduction*), fluoxastrobin + prothioconazole (*reduction*), silthiofam (*seed treatment*)
Pests	Aphids	alpha-cypermethrin, beta-cyfluthrin, clothianidin (*seed treatment*), clothianidin + prothioconazole (*seed treatment*), cypermethrin, cypermethrin (*autumn sown*), deltamethrin, esfenvalerate, lambda-cyhalothrin, tau-fluvalinate, zeta-cypermethrin
	Beetles	lambda-cyhalothrin
	Birds/mammals	aluminium ammonium sulphate
	Caterpillars	lambda-cyhalothrin
	Flies	alpha-cypermethrin, cypermethrin, cypermethrin (*autumn sown*), fludioxonil + tefluthrin

Crop/Pest Guide - Arable and vegetable crops

	Leafhoppers	clothianidin (*seed treatment*), clothianidin + prothioconazole (*seed treatment*)
	Pests, miscellaneous	lambda-cyhalothrin
	Slugs/snails	clothianidin (*seed treatment*), clothianidin + prothioconazole (*seed treatment*)
	Weevils	lambda-cyhalothrin
	Wireworms	clothianidin (*seed treatment*), clothianidin + prothioconazole (*seed treatment*), fludioxonil + tefluthrin
Plant growth regulation	Growth control	2-chloroethylphosphonic acid, chlormequat, chlormequat + 2-chloroethylphosphonic acid, chlormequat + 2-chloroethylphosphonic acid + mepiquat chloride, ethephon + mepiquat chloride, mepiquat chloride + prohexadione-calcium, prohexadione-calcium + trinexapac-ethyl, trinexapac-ethyl
	Quality/yield control	ethephon + mepiquat chloride (*low lodging situations*), mepiquat chloride + prohexadione-calcium, sulphur
Weeds	Broad-leaved weeds	2,4-D, 2,4-D + MCPA, 2,4-DB, 2,4-DB + MCPA, amidosulfuron, amidosulfuron + iodosulfuron-methyl-sodium, bifenox, bromoxynil, bromoxynil + diflufenican, bromoxynil + diflufenican (*at 1.9 l/ha only*), carfentrazone-ethyl, carfentrazone-ethyl + mecoprop-P, chlorotoluron + diflufenican + pendimethalin, clopyralid, clopyralid + florasulam, clopyralid + florasulam + fluroxypyr, dicamba + MCPA + mecoprop-P, dicamba + mecoprop-P, dichlorprop-P + MCPA + mecoprop-P, diflufenican, diflufenican + florasulam, diflufenican + flufenacet, diflufenican + flufenacet (*off-label*), diflufenican + flufenacet + flurtamone, diflufenican + flupyrsulfuron-methyl, diflufenican + flurtamone, diflufenican + metsulfuron-methyl, diflufenican + pendimethalin, florasulam, florasulam + fluroxypyr, flufenacet, flufenacet + pendimethalin, flufenacet + pendimethalin (*off-label*), flupyrsulfuron-methyl, fluroxypyr, fluroxypyr + halauxifen-methyl, glyphosate, imazosulfuron , isoxaben, MCPA, mecoprop-P, metsulfuron-methyl, metsulfuron-methyl + thifensulfuron-methyl, metsulfuron-methyl + tribenuron-methyl, pendimethalin, pendimethalin + picolinafen, pendimethalin + picolinafen (*off-label*), prosulfocarb, prosulfocarb (*off-label*), thifensulfuron-methyl + tribenuron-methyl, tribenuron-methyl
	Crops as weeds	amidosulfuron + iodosulfuron-methyl-sodium, bromoxynil, bromoxynil + diflufenican, chlorotoluron + diflufenican + pendimethalin, clopyralid + florasulam, diflufenican, diflufenican + florasulam, diflufenican + flufenacet, diflufenican + flupyrsulfuron-methyl, diflufenican + flurtamone, florasulam, florasulam + fluroxypyr, fluroxypyr, glyphosate, imazosulfuron , metsulfuron-methyl, metsulfuron-methyl + tribenuron-methyl, pendimethalin
	Grass weeds	chlorotoluron + diflufenican + pendimethalin, diflufenican, diflufenican + flufenacet, diflufenican + flufenacet (*off-label*), diflufenican + flufenacet +

Crop/Pest Guide - Arable and vegetable crops

		flurtamone, diflufenican + flupyrsulfuron-methyl, diflufenican + flurtamone, diflufenican + pendimethalin, fenoxaprop-P-ethyl, flufenacet, flufenacet + pendimethalin, flufenacet + pendimethalin (*off-label*), flupyrsulfuron-methyl, glyphosate, pendimethalin, pendimethalin + picolinafen, pendimethalin + picolinafen (*off-label*), pinoxaden, prosulfocarb, prosulfocarb (*off-label*), tri-allate
	Weeds, miscellaneous	chlorotoluron + diflufenican + pendimethalin, clopyralid + florasulam, diflufenican, diflufenican + metsulfuron-methyl, florasulam, glyphosate, metsulfuron-methyl, metsulfuron-methyl + thifensulfuron-methyl

Cereals - Cereals, general

Crop control	Desiccation	diquat, glyphosate, glyphosate (*off-label*)
Diseases	Alternaria	bromuconazole + tebuconazole, difenoconazole (*off-label*)
	Blue mould	prothioconazole + tebuconazole
	Bunt	prothioconazole + tebuconazole
	Cladosporium	bromuconazole + tebuconazole, difenoconazole (*off-label*)
	Crown rust	boscalid + epoxiconazole, boscalid + epoxiconazole + pyraclostrobin, epoxiconazole, epoxiconazole + fluxapyroxad, epoxiconazole + fluxapyroxad + pyraclostrobin
	Damping off	thiram (*off-label*)
	Disease control	azoxystrobin + chlorothalonil (*off-label*), chlorothalonil (*off-label*), cyproconazole + propiconazole (*off-label*), fluxapyroxad + metconazole
	Ear blight	metconazole (*reduction*)
	Eyespot	boscalid + epoxiconazole (*moderate control*), boscalid + epoxiconazole + pyraclostrobin (*moderate control*), epoxiconazole + fluxapyroxad (*moderate control*), epoxiconazole + fluxapyroxad + pyraclostrobin (*moderate control*), epoxiconazole + metconazole (*reduction*), fluoxastrobin + prothioconazole + trifloxystrobin (*reduction*), fluxapyroxad + metconazole (*good reduction*), prothioconazole + trifloxystrobin
	Foot rot	fludioxonil (*seed treatment*)
	Fungus diseases	prothioconazole + tebuconazole (*qualified minor use*)
	Fusarium	bromuconazole + tebuconazole, prothioconazole + tebuconazole
	Late ear diseases	fluoxastrobin + prothioconazole + trifloxystrobin, prothioconazole + trifloxystrobin
	Loose smut	prothioconazole + tebuconazole
	Powdery mildew	azoxystrobin, azoxystrobin (*moderate control*), boscalid + epoxiconazole (*moderate control*), boscalid + epoxiconazole + pyraclostrobin (*moderate control*), bromuconazole + tebuconazole (*moderate control*), cyflufenamid (*off-label*), epoxiconazole, epoxiconazole + fluxapyroxad (*moderate control*), epoxiconazole + fluxapyroxad + pyraclostrobin (*moderate control*), epoxiconazole + metconazole

Crop/Pest Guide - Arable and vegetable crops

		(*moderate control*), epoxiconazole + pyraclostrobin, fluoxastrobin + prothioconazole + trifloxystrobin, fluxapyroxad (*reduction*), fluxapyroxad + metconazole (*moderate control*), fluxapyroxad + pyraclostrobin (*moderate control*), metconazole (*moderate control*), prothioconazole + trifloxystrobin, tebuconazole
	Ramularia leaf spots	epoxiconazole + pyraclostrobin (*moderate control*)
	Rhynchosporium	azoxystrobin, azoxystrobin (*reduction*), boscalid + epoxiconazole (*moderate control*), boscalid + epoxiconazole + pyraclostrobin, epoxiconazole, epoxiconazole + fluxapyroxad + pyraclostrobin, epoxiconazole + metconazole (*moderate control*), epoxiconazole + pyraclostrobin, fluxapyroxad, fluxapyroxad + metconazole, fluxapyroxad + pyraclostrobin, metconazole (*moderate control*), penthiopyrad, penthiopyrad (*moderate control*), tebuconazole
	Rust	azoxystrobin, boscalid + epoxiconazole, boscalid + epoxiconazole + pyraclostrobin, bromuconazole + tebuconazole, difenoconazole (*off-label*), epoxiconazole, epoxiconazole + fluxapyroxad, epoxiconazole + fluxapyroxad + pyraclostrobin, epoxiconazole + metconazole, epoxiconazole + pyraclostrobin, fluoxastrobin + prothioconazole + trifloxystrobin, fluxapyroxad, fluxapyroxad (*moderate control*), fluxapyroxad + metconazole, fluxapyroxad + pyraclostrobin, metconazole, penthiopyrad, penthiopyrad (*moderate control*), prothioconazole + trifloxystrobin, tebuconazole
	Seed-borne diseases	fludioxonil (*seed treatment*)
	Septoria	bromuconazole + tebuconazole (*moderate control*), difenoconazole (*off-label*), fluoxastrobin + prothioconazole + trifloxystrobin, metconazole, prothioconazole + trifloxystrobin
	Stripe smut	fludioxonil
	Take-all	azoxystrobin (*reduction*)
	Tan spot	fluoxastrobin + prothioconazole + trifloxystrobin
Pests	Aphids	beta-cyfluthrin, cypermethrin, deltamethrin (*off-label*), dimethoate, lambda-cyhalothrin, tau-fluvalinate (*off-label*), zeta-cypermethrin (*off-label*)
	Beetles	deltamethrin (*off-label*), deltamethrin (*off-label - cereal*), lambda-cyhalothrin
	Caterpillars	lambda-cyhalothrin
	Flies	deltamethrin (*off-label*)
	Midges	beta-cyfluthrin, deltamethrin (*off-label*)
	Pests, miscellaneous	cypermethrin, deltamethrin (*off-label*), lambda-cyhalothrin (*off-label*), zeta-cypermethrin (*off-label*)
	Weevils	lambda-cyhalothrin
Plant growth regulation	Growth control	chlormequat, chlormequat + 2-chloroethylphosphonic acid + mepiquat chloride (*off-label*), chlormequat + imazaquin (*off-label*), mepiquat chloride + prohexadione-calcium, trinexapac-ethyl, trinexapac-ethyl (*off-label*)
	Quality/yield control	mepiquat chloride + prohexadione-calcium
Weeds	Broad-leaved weeds	2,4-DB + MCPA, bromoxynil, bromoxynil (*off-label*), carfentrazone-ethyl (*off-label*), clopyralid + florasulam,

Crop/Pest Guide - Arable and vegetable crops

		dicamba, dicamba + mecoprop-P (*off-label*), dichlorprop-P + MCPA + mecoprop-P, diflufenican, diflufenican + florasulam, diflufenican + flufenacet (*off-label*), florasulam + fluroxypyr, fluroxypyr, fluroxypyr + halauxifen-methyl, mecoprop-P (*off-label*), metsulfuron-methyl + tribenuron-methyl (*off-label*), pendimethalin + picolinafen, prosulfocarb (*off-label*)
	Crops as weeds	carfentrazone-ethyl (*off-label*), clopyralid + florasulam, diflufenican + florasulam, florasulam + fluroxypyr
	Grass weeds	diflufenican + flufenacet (*off-label*), flufenacet (*off-label*), glyphosate, pendimethalin + picolinafen, prosulfocarb (*off-label*)
	Weeds, miscellaneous	clopyralid + florasulam, dicamba, glyphosate (*off-label*)

Cereals - Maize/sweetcorn

Diseases	Damping off	fludioxonil + metalaxyl-M, thiram (*seed treatment*)
	Disease control	azoxystrobin + propiconazole, azoxystrobin + propiconazole (*useful reduction*), epoxiconazole + pyraclostrobin (*off-label*), pyraclostrobin
	Eyespot	azoxystrobin + propiconazole (*useful reduction*), epoxiconazole + pyraclostrobin, epoxiconazole + pyraclostrobin (*off-label*), pyraclostrobin, pyraclostrobin (*off-label*)
	Foliar diseases	epoxiconazole + pyraclostrobin, pyraclostrobin, pyraclostrobin (*moderate control*)
	Fusarium	fludioxonil + metalaxyl-M
	Pythium	fludioxonil + metalaxyl-M
	Rust	epoxiconazole + pyraclostrobin (*off-label*)
Pests	Aphids	cypermethrin, pymetrozine (*off-label*)
	Caterpillars	Bacillus thuringiensis (*off-label*), indoxacarb (*off-label*)
	Cutworms	Bacillus thuringiensis (*off-label*)
	Flies	lambda-cyhalothrin (*off-label*), methiocarb
	Pests, miscellaneous	cypermethrin, lambda-cyhalothrin (*off-label*)
	Wireworms	thiacloprid
Weeds	Bindweeds	dicamba + prosulfuron
	Broad-leaved weeds	bromoxynil, clopyralid, clopyralid (*off-label*), dicamba, dicamba + prosulfuron, dicamba + prosulfuron (*seedlings only*), dimethenamid-p + pendimethalin, dimethenamid-p + pendimethalin (*off-label*), flufenacet + isoxaflutole, fluroxypyr, foramsulfuron + iodosulfuron-methyl-sodium, mesotrione, mesotrione (*off-label*), mesotrione + nicosulfuron, mesotrione + terbuthylazine, mesotrione + terbuthylazine (*off-label*), nicosulfuron, nicosulfuron (*off-label*), pendimethalin, pendimethalin (*off-label*), prosulfuron, rimsulfuron, S-metolachlor, S-metolachlor (*moderately susceptible*)
	Crops as weeds	clopyralid (*off-label*), mesotrione, mesotrione + nicosulfuron, nicosulfuron, nicosulfuron (*off-label*), pendimethalin, rimsulfuron
	Grass weeds	dimethenamid-p + pendimethalin, dimethenamid-p + pendimethalin (*off-label*), flufenacet + isoxaflutole, foramsulfuron + iodosulfuron-methyl-sodium,

Crop/Pest Guide - Arable and vegetable crops

		mesotrione, mesotrione (*off-label*), mesotrione + nicosulfuron, mesotrione + terbuthylazine, mesotrione + terbuthylazine (*off-label*), nicosulfuron, nicosulfuron (*off-label*), pendimethalin, S-metolachlor
	Weed grasses	foramsulfuron + iodosulfuron-methyl-sodium, mesotrione, mesotrione + nicosulfuron, nicosulfuron
	Weeds, miscellaneous	bromoxynil, dicamba, mesotrione + nicosulfuron, nicosulfuron (*off-label*)

Cereals - Oats

Crop control	Desiccation	diquat (*off-label*), diquat (*stockfeed only*), glyphosate
Diseases	Blue mould	prothioconazole + tebuconazole
	Bunt	prothioconazole + tebuconazole
	Covered smut	prothioconazole + tebuconazole
	Crown rust	azoxystrobin, azoxystrobin + cyproconazole, bixafen + prothioconazole, boscalid + epoxiconazole, cyproconazole + picoxystrobin, epoxiconazole + fenpropimorph + metrafenone, epoxiconazole + metrafenone, epoxiconazole + pyraclostrobin, fenpropimorph + pyraclostrobin, fluoxastrobin + prothioconazole, fluxapyroxad, fluxapyroxad + pyraclostrobin, penthiopyrad + picoxystrobin, picoxystrobin, prochloraz + proquinazid + tebuconazole (*qualified minor use recommendation*), propiconazole, prothioconazole, prothioconazole + spiroxamine, prothioconazole + spiroxamine + tebuconazole, prothioconazole + tebuconazole, pyraclostrobin, tebuconazole, tebuconazole (*reduction*)
	Eyespot	bixafen + prothioconazole (*reduction of incidence and severity*), epoxiconazole + fenpropimorph + kresoxim-methyl (*reduction*), fluoxastrobin + prothioconazole, fluoxastrobin + prothioconazole (*reduction of incidence and severity*), prothioconazole, prothioconazole + spiroxamine, prothioconazole + spiroxamine + tebuconazole, prothioconazole + tebuconazole
	Foot rot	clothianidin + prothioconazole (*seed treatment*), difenoconazole + fludioxonil, fludioxonil (*seed treatment*), fludioxonil + tefluthrin
	Fusarium	prothioconazole + tebuconazole
	Leaf spot	difenoconazole + fludioxonil, fludioxonil (*seed treatment*), fludioxonil + tefluthrin
	Loose smut	clothianidin + prothioconazole (*seed treatment*), prothioconazole + tebuconazole
	Powdery mildew	azoxystrobin, azoxystrobin (*moderate control*), azoxystrobin + cyproconazole, bixafen + prothioconazole, cyflufenamid, epoxiconazole, epoxiconazole + fenpropimorph, epoxiconazole + fenpropimorph + kresoxim-methyl, epoxiconazole + fenpropimorph + metrafenone, epoxiconazole + metrafenone, epoxiconazole + pyraclostrobin, fenpropimorph, fenpropimorph + pyraclostrobin, fluoxastrobin + prothioconazole, fluxapyroxad (*reduction*), fluxapyroxad + pyraclostrobin (*moderate control*), metrafenone (*evidence of mildew control on oats is limited*), picoxystrobin, prochloraz +

Crop/Pest Guide - Arable and vegetable crops

		proquinazid + tebuconazole, propiconazole, proquinazid, prothioconazole, prothioconazole + spiroxamine, prothioconazole + spiroxamine + tebuconazole, prothioconazole + tebuconazole, quinoxyfen, sulphur, tebuconazole
	Seed-borne diseases	difenoconazole + fludioxonil
	Snow mould	fludioxonil (*seed treatment*)
	Take-all	azoxystrobin (*reduction*)
Pests	Aphids	clothianidin (*seed treatment*), clothianidin + prothioconazole (*seed treatment*), cypermethrin, deltamethrin, lambda-cyhalothrin, zeta-cypermethrin
	Beetles	lambda-cyhalothrin
	Birds/mammals	aluminium ammonium sulphate
	Caterpillars	lambda-cyhalothrin
	Flies	cypermethrin
	Leafhoppers	clothianidin (*seed treatment*), clothianidin + prothioconazole (*seed treatment*)
	Pests, miscellaneous	lambda-cyhalothrin
	Slugs/snails	clothianidin (*seed treatment*), clothianidin + prothioconazole (*seed treatment*)
	Weevils	lambda-cyhalothrin
	Wireworms	clothianidin (*seed treatment*), clothianidin + prothioconazole (*seed treatment*), fludioxonil + tefluthrin
Plant growth regulation	Growth control	chlormequat, prohexadione-calcium + trinexapac-ethyl, trinexapac-ethyl
	Quality/yield control	sulphur
Weeds	Broad-leaved weeds	2,4-D, 2,4-D + MCPA, 2,4-DB, amidosulfuron, bromoxynil, carfentrazone-ethyl, carfentrazone-ethyl + mecoprop-P, clopyralid, clopyralid + florasulam, clopyralid + florasulam + fluroxypyr, dicamba + MCPA + mecoprop-P, dicamba + mecoprop-P, dichlorprop-P + MCPA + mecoprop-P, diflufenican (*off-label*), diflufenican + flufenacet (*off-label*), diflufenican + flupyrsulfuron-methyl, diflufenican + flupyrsulfuron-methyl (*off-label*), florasulam, florasulam + fluroxypyr, flumioxazin (*off-label*), flupyrsulfuron-methyl, flupyrsulfuron-methyl + thifensulfuron-methyl, fluroxypyr, glyphosate, isoxaben, MCPA, mecoprop-P, metsulfuron-methyl, metsulfuron-methyl + thifensulfuron-methyl, metsulfuron-methyl + tribenuron-methyl, thifensulfuron-methyl + tribenuron-methyl, tribenuron-methyl
	Crops as weeds	bromoxynil, clopyralid + florasulam, diflufenican + flupyrsulfuron-methyl, florasulam, flumioxazin (*off-label*), glyphosate, metsulfuron-methyl, metsulfuron-methyl + tribenuron-methyl
	Grass weeds	diflufenican (*off-label*), diflufenican + flufenacet (*off-label*), diflufenican + flupyrsulfuron-methyl, diflufenican + flupyrsulfuron-methyl (*off-label*), diquat (*animal feed*), flumioxazin (*off-label*), flupyrsulfuron-methyl, flupyrsulfuron-methyl + thifensulfuron-methyl, glyphosate
	Weeds, miscellaneous	clopyralid + florasulam, diquat (*animal feed only*), florasulam, glyphosate, metsulfuron-methyl, metsulfuron-methyl + thifensulfuron-methyl

Crop/Pest Guide - Arable and vegetable crops

Cereals - Rye/triticale

Diseases	Alternaria	bromuconazole + tebuconazole, difenoconazole (*off-label*)
	Blue mould	prothioconazole + tebuconazole
	Bunt	clothianidin + prothioconazole (*seed treatment*), prothioconazole + tebuconazole
	Cladosporium	bromuconazole + tebuconazole, difenoconazole (*off-label*)
	Crown rust	tebuconazole
	Disease control	azoxystrobin + chlorothalonil (*off-label*), chlorothalonil (*off-label*), cyproconazole + propiconazole (*off-label*), epoxiconazole + folpet, fenpropimorph + pyraclostrobin (*off-label*), fluxapyroxad + metconazole, tebuconazole (*off-label*)
	Drechslera leaf spot	fenpropimorph + pyraclostrobin
	Ear blight	epoxiconazole + metconazole (*good reduction*), epoxiconazole + metconazole (*moderate control*), epoxiconazole + pyraclostrobin (*good reduction*), metconazole (*reduction*), tebuconazole, thiophanate-methyl (*reduction*)
	Eyespot	bixafen + prothioconazole (*reduction of incidence and severity*), bixafen + prothioconazole + spiroxamine, bixafen + prothioconazole + tebuconazole (*reduction of incidence and severity*), boscalid + epoxiconazole (*moderate control*), boscalid + epoxiconazole + pyraclostrobin (*moderate control*), epoxiconazole + fenpropimorph + kresoxim-methyl (*reduction*), epoxiconazole + fluxapyroxad (*moderate control*), epoxiconazole + fluxapyroxad + pyraclostrobin (*moderate control*), epoxiconazole + metconazole (*reduction*), fluoxastrobin + prothioconazole (*reduction*), fluoxastrobin + prothioconazole + trifloxystrobin (*reduction*), fluxapyroxad + metconazole (*good reduction*), prochloraz + tebuconazole, prothioconazole, prothioconazole + spiroxamine, prothioconazole + spiroxamine + tebuconazole (*reduction*), prothioconazole + tebuconazole, prothioconazole + tebuconazole (*reduction*), prothioconazole + trifloxystrobin
	Foliar diseases	azoxystrobin + cyproconazole (*off-label*), pyraclostrobin (*off-label*), trifloxystrobin (*off-label*)
	Foot rot	clothianidin + prothioconazole (*seed treatment*), fludioxonil (*seed treatment*)
	Fusarium	bromuconazole + tebuconazole, prothioconazole + tebuconazole, thiophanate-methyl, thiophanate-methyl (*reduction*)
	Late ear diseases	bixafen + fluoxastrobin + prothioconazole, bixafen + prothioconazole, bixafen + prothioconazole + spiroxamine, bixafen + prothioconazole + tebuconazole, fluoxastrobin + prothioconazole + trifloxystrobin, prothioconazole + spiroxamine, prothioconazole + trifloxystrobin
	Loose smut	prothioconazole + tebuconazole
	Powdery mildew	azoxystrobin, azoxystrobin (*moderate control*), azoxystrobin + cyproconazole, bixafen + fluoxastrobin + prothioconazole, bixafen +

Crop/Pest Guide - Arable and vegetable crops

	prothioconazole, bixafen + prothioconazole + spiroxamine, bixafen + prothioconazole + tebuconazole, boscalid + epoxiconazole + pyraclostrobin (*moderate control*), bromuconazole + tebuconazole (*moderate control*), cyflufenamid, epoxiconazole, epoxiconazole + fenpropimorph, epoxiconazole + fenpropimorph + kresoxim-methyl, epoxiconazole + fenpropimorph + metrafenone, epoxiconazole + fluxapyroxad (*moderate control*), epoxiconazole + fluxapyroxad + pyraclostrobin (*moderate control*), epoxiconazole + metconazole (*moderate control*), epoxiconazole + metconazole (*moderate reduction*), epoxiconazole + metrafenone, fenpropimorph, fluoxastrobin + prothioconazole, fluoxastrobin + prothioconazole + trifloxystrobin, fluxapyroxad (*reduction*), fluxapyroxad + metconazole (*moderate control*), fluxapyroxad + pyraclostrobin (*moderate control*), metconazole (*moderate control*), prochloraz + tebuconazole, propiconazole, proquinazid, prothioconazole, prothioconazole + spiroxamine, prothioconazole + spiroxamine + tebuconazole, prothioconazole + tebuconazole, prothioconazole + trifloxystrobin, quinoxyfen, spiroxamine, tebuconazole, tebuconazole (*off-label*)
Ramularia leaf spots	epoxiconazole + pyraclostrobin (*moderate control*)
Rhynchosporium	azoxystrobin, azoxystrobin (*reduction*), azoxystrobin + cyproconazole (*moderate control*), bixafen + fluoxastrobin + prothioconazole, bixafen + prothioconazole, bixafen + prothioconazole + spiroxamine, bixafen + prothioconazole + tebuconazole, epoxiconazole, epoxiconazole + fenpropimorph, epoxiconazole + fenpropimorph + kresoxim-methyl, epoxiconazole + fenpropimorph + metrafenone, epoxiconazole + folpet, epoxiconazole + isopyrazam, epoxiconazole + metrafenone, fluoxastrobin + prothioconazole, penthiopyrad (*moderate control*), prochloraz + tebuconazole, propiconazole, prothioconazole, prothioconazole + spiroxamine, prothioconazole + spiroxamine + tebuconazole, prothioconazole + tebuconazole, spiroxamine (*reduction*), tebuconazole, tebuconazole (*moderate control*)
Rust	azoxystrobin, azoxystrobin + cyproconazole, bixafen + fluoxastrobin + prothioconazole, bixafen + prothioconazole, bixafen + prothioconazole + spiroxamine, bixafen + prothioconazole + tebuconazole, boscalid + epoxiconazole, boscalid + epoxiconazole + pyraclostrobin, bromuconazole + tebuconazole, difenoconazole (*off-label*), epoxiconazole, epoxiconazole + fenpropimorph, epoxiconazole + fenpropimorph + kresoxim-methyl, epoxiconazole + fenpropimorph + metrafenone, epoxiconazole + fluxapyroxad, epoxiconazole + fluxapyroxad + pyraclostrobin, epoxiconazole + folpet, epoxiconazole + isopyrazam, epoxiconazole + metconazole, epoxiconazole + metrafenone, epoxiconazole + pyraclostrobin, fenpropimorph, fenpropimorph + pyraclostrobin, fenpropimorph + pyraclostrobin (*off-label*), fluoxastrobin + prothioconazole, fluoxastrobin + prothioconazole +

Crop/Pest Guide - Arable and vegetable crops

		trifloxystrobin, fluxapyroxad, fluxapyroxad (*moderate control*), fluxapyroxad + metconazole, fluxapyroxad + pyraclostrobin, metconazole, penthiopyrad, penthiopyrad (*moderate control*), prochloraz + tebuconazole, propiconazole, prothioconazole, prothioconazole + spiroxamine, prothioconazole + spiroxamine + tebuconazole, prothioconazole + tebuconazole, prothioconazole + trifloxystrobin, spiroxamine, tebuconazole, tebuconazole (*off-label*)
	Seed-borne diseases	fludioxonil, fludioxonil + tefluthrin (*off-label*)
	Septoria	bixafen + fluoxastrobin + prothioconazole, bixafen + prothioconazole, bixafen + prothioconazole + spiroxamine, bixafen + prothioconazole + tebuconazole, boscalid + epoxiconazole, boscalid + epoxiconazole + pyraclostrobin, bromuconazole + tebuconazole (*moderate control*), difenoconazole (*off-label*), epoxiconazole, epoxiconazole + fenpropimorph, epoxiconazole + fenpropimorph + kresoxim-methyl, epoxiconazole + fenpropimorph + metrafenone, epoxiconazole + fluxapyroxad, epoxiconazole + fluxapyroxad + pyraclostrobin, epoxiconazole + folpet, epoxiconazole + isopyrazam, epoxiconazole + metconazole, epoxiconazole + metrafenone, epoxiconazole + pyraclostrobin, epoxiconazole + pyraclostrobin (*moderate control*), fenpropimorph + pyraclostrobin, fenpropimorph + pyraclostrobin (*off-label*), fluoxastrobin + prothioconazole + trifloxystrobin, fluxapyroxad, fluxapyroxad + metconazole, fluxapyroxad + pyraclostrobin, metconazole, propiconazole, prothioconazole + trifloxystrobin, tebuconazole, tebuconazole (*off-label*)
	Sooty moulds	tebuconazole
	Stripe smut	difenoconazole + fludioxonil
	Take-all	azoxystrobin (*reduction*)
	Tan spot	bixafen + fluoxastrobin + prothioconazole, bixafen + prothioconazole, bixafen + prothioconazole + spiroxamine, bixafen + prothioconazole + tebuconazole, epoxiconazole + pyraclostrobin (*moderate control*), fluoxastrobin + prothioconazole + trifloxystrobin
Pests	Aphids	beta-cyfluthrin, clothianidin (*seed treatment*), clothianidin + prothioconazole (*seed treatment*), cypermethrin, deltamethrin (*off-label*), dimethoate, lambda-cyhalothrin, tau-fluvalinate (*off-label*), zeta-cypermethrin (*off-label*)
	Beetles	deltamethrin (*off-label*), deltamethrin (*off-label - cereal*), lambda-cyhalothrin
	Caterpillars	lambda-cyhalothrin
	Flies	cypermethrin, deltamethrin (*off-label*)
	Leafhoppers	clothianidin (*seed treatment*), clothianidin + prothioconazole (*seed treatment*)
	Midges	beta-cyfluthrin, deltamethrin (*off-label*)
	Pests, miscellaneous	deltamethrin (*off-label*), lambda-cyhalothrin (*off-label*), zeta-cypermethrin (*off-label*)
	Slugs/snails	clothianidin (*seed treatment*), clothianidin + prothioconazole (*seed treatment*)

Crop/Pest Guide - Arable and vegetable crops

	Weevils	lambda-cyhalothrin
	Wireworms	clothianidin (*seed treatment*), clothianidin + prothioconazole (*seed treatment*)
Plant growth regulation	Growth control	2-chloroethylphosphonic acid, chlormequat, chlormequat + 2-chloroethylphosphonic acid + mepiquat chloride (*off-label*), chlormequat + imazaquin (*off-label*), ethephon + mepiquat chloride, mepiquat chloride + prohexadione-calcium, prohexadione-calcium + trinexapac-ethyl, trinexapac-ethyl
	Quality/yield control	mepiquat chloride + prohexadione-calcium
Weeds	Broad-leaved weeds	2,4-D, amidosulfuron, amidosulfuron + iodosulfuron-methyl-sodium, bifenox, bromoxynil, carfentrazone-ethyl, chlorotoluron + diflufenican + pendimethalin, clopyralid (*off-label*), clopyralid + florasulam, dicamba + mecoprop-P (*off-label*), dichlorprop-P + MCPA + mecoprop-P, diflufenican, diflufenican + florasulam, diflufenican + flufenacet, diflufenican + flufenacet (*off-label*), diflufenican + pendimethalin, florasulam, florasulam + fluroxypyr, florasulam + fluroxypyr (*off-label*), florasulam + pyroxsulam, flupyrsulfuron-methyl, fluroxypyr, fluroxypyr + halauxifen-methyl, imazosulfuron , iodosulfuron-methyl-sodium + mesosulfuron-methyl (*off-label*), isoxaben, MCPA, mecoprop-P (*off-label*), metsulfuron-methyl, metsulfuron-methyl + thifensulfuron-methyl, metsulfuron-methyl + tribenuron-methyl, metsulfuron-methyl + tribenuron-methyl (*off-label*), pendimethalin, pendimethalin + picolinafen, prosulfocarb (*off-label*), thifensulfuron-methyl + tribenuron-methyl, tribenuron-methyl
	Crops as weeds	amidosulfuron + iodosulfuron-methyl-sodium, chlorotoluron + diflufenican + pendimethalin, clopyralid + florasulam, diflufenican, diflufenican + florasulam, diflufenican + flufenacet, florasulam, florasulam + fluroxypyr, florasulam + pyroxsulam, imazosulfuron , metsulfuron-methyl, metsulfuron-methyl + tribenuron-methyl, pendimethalin
	Grass weeds	chlorotoluron + diflufenican + pendimethalin, clodinafop-propargyl, diflufenican, diflufenican + flufenacet, diflufenican + flufenacet (*off-label*), diflufenican + pendimethalin, florasulam + pyroxsulam, flufenacet (*off-label*), flupyrsulfuron-methyl, iodosulfuron-methyl-sodium + mesosulfuron-methyl (*off-label*), pendimethalin, pendimethalin + picolinafen, prosulfocarb (*off-label*)
	Weeds, miscellaneous	chlorotoluron + diflufenican + pendimethalin, clopyralid + florasulam, diflufenican, florasulam, metsulfuron-methyl, metsulfuron-methyl + thifensulfuron-methyl

Cereals - Undersown cereals

Diseases	Powdery mildew	fenpropimorph
	Rhynchosporium	fenpropimorph
	Rust	fenpropimorph
Weeds	Broad-leaved weeds	2,4-D, 2,4-DB, 2,4-DB + MCPA, MCPA, MCPA (*red clover or grass*), tribenuron-methyl

Crop/Pest Guide - Arable and vegetable crops

Cereals - Wheat

Crop control	Desiccation	glyphosate
Diseases	Alternaria	bromuconazole + tebuconazole, difenoconazole (*off-label*), tebuconazole
	Blue mould	prothioconazole + tebuconazole
	Bunt	clothianidin + prothioconazole (*seed treatment*), difenoconazole, difenoconazole + fludioxonil, fludioxonil (*seed treatment*), fludioxonil + tefluthrin, prothioconazole + tebuconazole
	Cladosporium	bromuconazole + tebuconazole, difenoconazole (*off-label*), fluxapyroxad + metconazole (*moderate control*), tebuconazole
	Disease control	azoxystrobin + chlorothalonil (*off-label*), chlorothalonil (*off-label*), chlorothalonil + penthiopyrad, cyproconazole + propiconazole (*off-label*), epoxiconazole (*off-label*), fluxapyroxad + metconazole
	Drechslera leaf spot	fenpropimorph + pyraclostrobin
	Ear blight	boscalid + epoxiconazole (*good reduction*), boscalid + epoxiconazole + pyraclostrobin (*good reduction*), chlorothalonil + tebuconazole, dimoxystrobin + epoxiconazole, epoxiconazole, epoxiconazole (*good reduction*), epoxiconazole + fenpropimorph, epoxiconazole + fenpropimorph + kresoxim-methyl (*reduction*), epoxiconazole + fenpropimorph + metrafenone (*reduction*), epoxiconazole + fluxapyroxad (*good reduction*), epoxiconazole + fluxapyroxad + pyraclostrobin (*good reduction*), epoxiconazole + metconazole (*good reduction*), epoxiconazole + metconazole (*moderate control*), epoxiconazole + metrafenone (*reduction*), epoxiconazole + pyraclostrobin (*good reduction*), fluxapyroxad + metconazole (*good reduction*), metconazole (*reduction*), prothioconazole + tebuconazole, tebuconazole, thiophanate-methyl (*reduction*)
	Eyespot	azoxystrobin + cyproconazole (*reduction*), bixafen + fluopyram + prothioconazole (*reduction of incidence and severity*), bixafen + prothioconazole (*reduction of incidence and severity*), bixafen + prothioconazole + spiroxamine, bixafen + prothioconazole + tebuconazole (*reduction of incidence and severity*), boscalid + epoxiconazole (*moderate control*), boscalid + epoxiconazole + pyraclostrobin (*moderate control*), chlorothalonil + cyproconazole, epoxiconazole (*reduction*), epoxiconazole + fenpropimorph (*reduction*), epoxiconazole + fenpropimorph + kresoxim-methyl (*reduction*), epoxiconazole + fluxapyroxad (*moderate control*), epoxiconazole + fluxapyroxad + pyraclostrobin (*moderate control*), epoxiconazole + metconazole (*reduction*), fenpropidin + prochloraz + tebuconazole, fluoxastrobin + prothioconazole (*reduction*), fluoxastrobin + prothioconazole + trifloxystrobin (*reduction*), fluxapyroxad + metconazole (*good reduction*), metrafenone (*reduction*), picoxystrobin (*reduction*), prochloraz + propiconazole, prochloraz + proquinazid + tebuconazole, prochloraz + tebuconazole, prothioconazole, prothioconazole + spiroxamine,

Crop/Pest Guide - Arable and vegetable crops

	prothioconazole + spiroxamine + tebuconazole (*reduction*), prothioconazole + tebuconazole, prothioconazole + tebuconazole (*reduction*), prothioconazole + trifloxystrobin
Foliar diseases	azoxystrobin (*off-label*), azoxystrobin + cyproconazole (*off-label*), epoxiconazole + fenpropimorph (*off-label*), epoxiconazole + fenpropimorph + kresoxim-methyl (*off-label*), fenpropimorph (*off-label*), pyraclostrobin (*off-label*), trifloxystrobin (*off-label*)
Foot rot	clothianidin + prothioconazole (*seed treatment*), difenoconazole + fludioxonil, fludioxonil (*seed treatment*), fludioxonil + tefluthrin, fluoxastrobin + prothioconazole (*reduction*)
Fusarium	bromuconazole + tebuconazole, prothioconazole + tebuconazole, tebuconazole, thiophanate-methyl, thiophanate-methyl (*reduction*)
Late ear diseases	azoxystrobin, bixafen + fluopyram + prothioconazole (*reduction of incidence and severity*), bixafen + fluoxastrobin + prothioconazole, bixafen + prothioconazole, bixafen + prothioconazole + spiroxamine, bixafen + prothioconazole + tebuconazole, cyproconazole + propiconazole, fenpropidin + prochloraz + tebuconazole, fluoxastrobin + prothioconazole, fluoxastrobin + prothioconazole + trifloxystrobin, picoxystrobin, prochloraz + proquinazid + tebuconazole (*reduction only*), prochloraz + tebuconazole, prothioconazole, prothioconazole + spiroxamine, prothioconazole + spiroxamine + tebuconazole, prothioconazole + tebuconazole, prothioconazole + trifloxystrobin, tebuconazole
Loose smut	clothianidin + prothioconazole (*seed treatment*), prothioconazole + tebuconazole
Net blotch	metconazole (*reduction*)
Powdery mildew	azoxystrobin (*off-label*), azoxystrobin + cyproconazole, bixafen + fluopyram + prothioconazole, bixafen + fluoxastrobin + prothioconazole, bixafen + prothioconazole, bixafen + prothioconazole + spiroxamine, bixafen + prothioconazole + tebuconazole, bromuconazole + tebuconazole (*moderate control*), chlorothalonil + cyproconazole, chlorothalonil + cyproconazole + propiconazole (*moderate control*), chlorothalonil + proquinazid, cyflufenamid, epoxiconazole, epoxiconazole + fenpropimorph, epoxiconazole + fenpropimorph + metrafenone, epoxiconazole + fluxapyroxad (*moderate control*), epoxiconazole + fluxapyroxad + pyraclostrobin (*moderate control*), epoxiconazole + metconazole (*moderate control*), epoxiconazole + metrafenone, fenpropidin + prochloraz + tebuconazole, fenpropimorph, fenpropimorph (*off-label*), fluoxastrobin + prothioconazole, fluoxastrobin + prothioconazole + trifloxystrobin, flutriafol, fluxapyroxad (*reduction*), fluxapyroxad + metconazole (*moderate control*), fluxapyroxad + pyraclostrobin (*moderate control*), metconazole (*moderate control*), metrafenone, prochloraz + proquinazid + tebuconazole, prochloraz + tebuconazole, propiconazole, proquinazid,

proquinazid (*off-label*), prothioconazole, prothioconazole + spiroxamine, prothioconazole + spiroxamine + tebuconazole, prothioconazole + tebuconazole, prothioconazole + trifloxystrobin, pyriofenone, quinoxyfen, spiroxamine, sulphur, tebuconazole, tebuconazole (*moderate control*)

Rust — azoxystrobin, azoxystrobin (*off-label*), azoxystrobin + chlorothalonil, azoxystrobin + cyproconazole, bixafen + fluopyram + prothioconazole, bixafen + fluoxastrobin + prothioconazole, bixafen + prothioconazole, bixafen + prothioconazole + spiroxamine, bixafen + prothioconazole + tebuconazole, boscalid + epoxiconazole, boscalid + epoxiconazole + pyraclostrobin, bromuconazole + tebuconazole, chlorothalonil + cyproconazole, chlorothalonil + cyproconazole + propiconazole, chlorothalonil + penthiopyrad, chlorothalonil + picoxystrobin, chlorothalonil + propiconazole, cyproconazole + picoxystrobin, cyproconazole + propiconazole, difenoconazole, difenoconazole (*off-label*), dimoxystrobin + epoxiconazole, epoxiconazole, epoxiconazole + fenpropimorph, epoxiconazole + fenpropimorph + kresoxim-methyl, epoxiconazole + fenpropimorph + metrafenone, epoxiconazole + fluxapyroxad, epoxiconazole + fluxapyroxad + pyraclostrobin, epoxiconazole + folpet, epoxiconazole + isopyrazam, epoxiconazole + metconazole, epoxiconazole + metrafenone, epoxiconazole + pyraclostrobin, fenpropidin + prochloraz + tebuconazole, fenpropimorph, fenpropimorph + pyraclostrobin, fluoxastrobin + prothioconazole, fluoxastrobin + prothioconazole + trifloxystrobin, flutriafol, fluxapyroxad, fluxapyroxad (*moderate control*), fluxapyroxad + metconazole, fluxapyroxad + pyraclostrobin, mancozeb, mancozeb (*useful control*), metconazole, penthiopyrad, penthiopyrad (*moderate control*), penthiopyrad + picoxystrobin, picoxystrobin, prochloraz + proquinazid + tebuconazole, prochloraz + tebuconazole, propiconazole, prothioconazole, prothioconazole + spiroxamine, prothioconazole + spiroxamine + tebuconazole, prothioconazole + tebuconazole, prothioconazole + trifloxystrobin, pyraclostrobin, spiroxamine, tebuconazole, trifloxystrobin

Seed-borne diseases — difenoconazole + fludioxonil, ipconazole

Septoria — azoxystrobin, azoxystrobin + chlorothalonil, azoxystrobin + cyproconazole, bixafen + fluopyram + prothioconazole, bixafen + fluoxastrobin + prothioconazole, bixafen + prothioconazole, bixafen + prothioconazole + spiroxamine, bixafen + prothioconazole + tebuconazole, boscalid + epoxiconazole, boscalid + epoxiconazole + pyraclostrobin, bromuconazole + tebuconazole (*moderate control*), chlorothalonil, chlorothalonil + cyproconazole, chlorothalonil + cyproconazole + propiconazole, chlorothalonil + mancozeb, chlorothalonil + penthiopyrad, chlorothalonil + picoxystrobin, chlorothalonil + propiconazole, chlorothalonil + proquinazid (*suppression only*), chlorothalonil + tebuconazole, cyproconazole +

Crop/Pest Guide - Arable and vegetable crops

	penthiopyrad, cyproconazole + picoxystrobin, cyproconazole + propiconazole, difenoconazole, difenoconazole (*off-label*), difenoconazole + fludioxonil, dimoxystrobin + epoxiconazole, epoxiconazole, epoxiconazole + fenpropimorph, epoxiconazole + fenpropimorph + kresoxim-methyl, epoxiconazole + fenpropimorph + metrafenone, epoxiconazole + fluxapyroxad, epoxiconazole + fluxapyroxad + pyraclostrobin, epoxiconazole + folpet, epoxiconazole + isopyrazam, epoxiconazole + metconazole, epoxiconazole + metrafenone, epoxiconazole + pyraclostrobin, epoxiconazole + pyraclostrobin (*moderate control*), fenpropidin + prochloraz + tebuconazole, fenpropimorph + pyraclostrobin, fludioxonil (*seed treatment*), fludioxonil + tefluthrin, fluoxastrobin + prothioconazole, fluoxastrobin + prothioconazole + trifloxystrobin, flutriafol, fluxapyroxad, fluxapyroxad + metconazole, fluxapyroxad + pyraclostrobin, fluxapyroxad + pyraclostrobin (*moderate control*), folpet (*reduction*), mancozeb, mancozeb (*reduction*), metconazole, penthiopyrad, penthiopyrad + picoxystrobin, picoxystrobin, prochloraz + propiconazole, prochloraz + proquinazid + tebuconazole, prochloraz + proquinazid + tebuconazole (*moderate control*), prochloraz + tebuconazole, propiconazole, prothioconazole, prothioconazole + spiroxamine, prothioconazole + spiroxamine + tebuconazole, prothioconazole + tebuconazole, prothioconazole + trifloxystrobin, pyraclostrobin, tebuconazole, trifloxystrobin
Sharp eyespot	fluoxastrobin + prothioconazole, fluoxastrobin + prothioconazole (*reduction*)
Snow mould	difenoconazole + fludioxonil, fludioxonil (*seed treatment*), fludioxonil + tefluthrin
Soil-borne diseases	ipconazole
Sooty moulds	boscalid + epoxiconazole, boscalid + epoxiconazole + pyraclostrobin (*good reduction*), chlorothalonil + tebuconazole, epoxiconazole (*reduction*), epoxiconazole + fenpropimorph (*reduction*), epoxiconazole + fenpropimorph + kresoxim-methyl (*reduction*), epoxiconazole + fenpropimorph + metrafenone (*reduction*), epoxiconazole + fluxapyroxad (*reduction*), epoxiconazole + fluxapyroxad + pyraclostrobin (*reduction*), epoxiconazole + metconazole (*good reduction*), epoxiconazole + metconazole (*moderate control*), epoxiconazole + metrafenone (*reduction*), epoxiconazole + pyraclostrobin (*reduction*), fluoxastrobin + prothioconazole (*reduction*), mancozeb, propiconazole, prothioconazole + tebuconazole, tebuconazole
Take-all	azoxystrobin (*reduction*), azoxystrobin + chlorothalonil (*reduction*), azoxystrobin + chlorothalonil (*reduction only*), azoxystrobin + cyproconazole (*reduction*), fluoxastrobin + prothioconazole (*reduction*), silthiofam (*seed treatment*)

Crop/Pest Guide - Arable and vegetable crops

	Tan spot	bixafen + fluopyram + prothioconazole, bixafen + fluoxastrobin + prothioconazole, bixafen + prothioconazole, bixafen + prothioconazole + spiroxamine, bixafen + prothioconazole + tebuconazole, boscalid + epoxiconazole (*reduction*), boscalid + epoxiconazole + pyraclostrobin (*moderate control*), chlorothalonil + picoxystrobin, dimoxystrobin + epoxiconazole, epoxiconazole + fluxapyroxad (*moderate control*), epoxiconazole + fluxapyroxad + pyraclostrobin (*moderate control*), epoxiconazole + metconazole (*moderate control*), epoxiconazole + pyraclostrobin, epoxiconazole + pyraclostrobin (*moderate control*), fluoxastrobin + prothioconazole, fluoxastrobin + prothioconazole + trifloxystrobin, fluxapyroxad (*reduction*), fluxapyroxad + metconazole (*good reduction*), fluxapyroxad + pyraclostrobin, penthiopyrad (*reduction*), prothioconazole, prothioconazole + spiroxamine, prothioconazole + tebuconazole
Pests	Aphids	alpha-cypermethrin, beta-cyfluthrin, clothianidin (*seed treatment*), clothianidin + prothioconazole (*seed treatment*), cypermethrin, cypermethrin (*autumn sown*), deltamethrin, deltamethrin (*off-label*), dimethoate, esfenvalerate, flonicamid, lambda-cyhalothrin, tau-fluvalinate, tau-fluvalinate (*off-label*), thiacloprid, zeta-cypermethrin, zeta-cypermethrin (*off-label*)
	Beetles	deltamethrin (*off-label*), deltamethrin (*off-label - cereal*), lambda-cyhalothrin
	Birds/mammals	aluminium ammonium sulphate
	Caterpillars	lambda-cyhalothrin
	Flies	alpha-cypermethrin, cypermethrin, cypermethrin (*autumn sown*), deltamethrin (*off-label*), fludioxonil + tefluthrin, lambda-cyhalothrin
	Leafhoppers	clothianidin (*seed treatment*), clothianidin + prothioconazole (*seed treatment*)
	Midges	beta-cyfluthrin, deltamethrin (*off-label*), lambda-cyhalothrin, thiacloprid
	Pests, miscellaneous	deltamethrin (*off-label*), lambda-cyhalothrin, zeta-cypermethrin (*off-label*)
	Slugs/snails	clothianidin (*seed treatment*), clothianidin + prothioconazole (*seed treatment*)
	Weevils	lambda-cyhalothrin
	Wireworms	clothianidin (*seed treatment*), clothianidin + prothioconazole (*seed treatment*), fludioxonil + tefluthrin
Plant growth regulation	Growth control	2-chloroethylphosphonic acid, 2-chloroethylphosphonic acid (*off-label*), chlormequat, chlormequat + 2-chloroethylphosphonic acid, chlormequat + 2-chloroethylphosphonic acid + mepiquat chloride, chlormequat + 2-chloroethylphosphonic acid + mepiquat chloride (*off-label*), chlormequat + imazaquin, ethephon + mepiquat chloride, mepiquat chloride + prohexadione-calcium, prohexadione-calcium + trinexapac-ethyl, trinexapac-ethyl, trinexapac-ethyl (*off-label*)

Crop/Pest Guide - Arable and vegetable crops

	Quality/yield control	chlormequat + imazaquin, ethephon + mepiquat chloride (*low lodging situations*), mepiquat chloride + prohexadione-calcium, sulphur
Weeds	Broad-leaved weeds	2,4-D, 2,4-D + MCPA, 2,4-DB, 2,4-DB + MCPA, amidosulfuron, amidosulfuron + iodosulfuron-methyl-sodium, amidosulfuron + iodosulfuron-methyl-sodium + mesosulfuron-methyl, bifenox, bifenox (*off-label*), bromoxynil, bromoxynil + diflufenican, carfentrazone-ethyl, carfentrazone-ethyl (*off-label*), carfentrazone-ethyl + mecoprop-P, chlorotoluron + diflufenican + pendimethalin, clodinafop-propargyl + prosulfocarb, clopyralid, clopyralid (*off-label*), clopyralid + florasulam, clopyralid + florasulam + fluroxypyr, dicamba + MCPA + mecoprop-P, dicamba + mecoprop-P, dicamba + mecoprop-P (*off-label*), dichlorprop-P + MCPA + mecoprop-P, diflufenican, diflufenican + florasulam, diflufenican + flufenacet, diflufenican + flufenacet (*off-label*), diflufenican + flufenacet + flurtamone, diflufenican + flupyrsulfuron-methyl, diflufenican + flurtamone, diflufenican + iodosulfuron-methyl-sodium + mesosulfuron-methyl, diflufenican + metsulfuron-methyl, diflufenican + pendimethalin, ethofumesate, florasulam, florasulam + fluroxypyr, florasulam + fluroxypyr (*off-label*), florasulam + pyroxsulam, flufenacet, flufenacet + pendimethalin, flumioxazin, flupyrsulfuron-methyl, flupyrsulfuron-methyl + pyroxsulam, flupyrsulfuron-methyl + thifensulfuron-methyl, fluroxypyr, fluroxypyr + halauxifen-methyl, glyphosate, imazosulfuron, iodosulfuron-methyl-sodium + mesosulfuron-methyl, iodosulfuron-methyl-sodium + mesosulfuron-methyl (*off-label*), isoxaben, MCPA, mecoprop-P, mecoprop-P (*off-label*), metsulfuron-methyl, metsulfuron-methyl + thifensulfuron-methyl, metsulfuron-methyl + thifensulfuron-methyl (*off-label*), metsulfuron-methyl + tribenuron-methyl, metsulfuron-methyl + tribenuron-methyl (*off-label*), pendimethalin, pendimethalin (*off-label*), pendimethalin + picolinafen, pendimethalin + picolinafen (*off-label*), prosulfocarb, prosulfocarb (*off-label*), sulfosulfuron, thifensulfuron-methyl + tribenuron-methyl, tribenuron-methyl
	Crops as weeds	amidosulfuron + iodosulfuron-methyl-sodium, bromoxynil, bromoxynil + diflufenican, carfentrazone-ethyl (*off-label*), chlorotoluron + diflufenican + pendimethalin, clopyralid + florasulam, diflufenican, diflufenican + florasulam, diflufenican + flufenacet, diflufenican + flupyrsulfuron-methyl, diflufenican + flurtamone, diflufenican + iodosulfuron-methyl-sodium + mesosulfuron-methyl, florasulam, florasulam + fluroxypyr, florasulam + pyroxsulam, flumioxazin, flupyrsulfuron-methyl + pyroxsulam (*from seed*), fluroxypyr, glyphosate, imazosulfuron, metsulfuron-methyl, metsulfuron-methyl + tribenuron-methyl, pendimethalin
	Grass weeds	amidosulfuron + iodosulfuron-methyl-sodium + mesosulfuron-methyl, chlorotoluron + diflufenican + pendimethalin, clodinafop-propargyl, clodinafop-propargyl + cloquintocet-mexyl, clodinafop-propargyl + pinoxaden, clodinafop-propargyl + pinoxaden (*from seed*), clodinafop-propargyl + prosulfocarb,

| | | diflufenican, diflufenican + flufenacet, diflufenican + flufenacet (*off-label*), diflufenican + flufenacet + flurtamone, diflufenican + flufenacet + flurtamone (*moderately susceptible*), diflufenican + flupyrsulfuron-methyl, diflufenican + flurtamone, diflufenican + iodosulfuron-methyl-sodium + mesosulfuron-methyl, diflufenican + pendimethalin, ethofumesate, fenoxaprop-P-ethyl, florasulam + pyroxsulam, flufenacet, flufenacet + pendimethalin, flumioxazin, flupyrsulfuron-methyl, flupyrsulfuron-methyl + pyroxsulam, flupyrsulfuron-methyl + pyroxsulam (*moderately susceptible*), flupyrsulfuron-methyl + thifensulfuron-methyl, glyphosate, iodosulfuron-methyl-sodium + mesosulfuron-methyl, iodosulfuron-methyl-sodium + mesosulfuron-methyl (*off-label*), pendimethalin, pendimethalin (*off-label*), pendimethalin + picolinafen, pendimethalin + picolinafen (*off-label*), pinoxaden, propoxycarbazone-sodium, prosulfocarb, prosulfocarb (*off-label*), sulfosulfuron, sulfosulfuron (*moderate control of barren brome*), sulfosulfuron (*moderate control only*), tri-allate |
| | Weeds, miscellaneous | chlorotoluron + diflufenican + pendimethalin, clopyralid + florasulam, diflufenican, diflufenican + metsulfuron-methyl, florasulam, glyphosate, metsulfuron-methyl, metsulfuron-methyl + thifensulfuron-methyl |

Edible fungi - Mushrooms

| Diseases | Bacterial blotch | sodium hypochlorite (commodity substance) |
| Pests | Flies | diflubenzuron (*off-label - other than mushrooms*), Metarhizium anisopliae (*off-label*) |

Fruiting vegetables - Aubergines

Diseases	Botrytis	azoxystrobin (*off-label*)
	Didymella stem rot	azoxystrobin (*off-label*)
	Foot rot	Trichoderma asperellum (Strain T34)
	Phytophthora	azoxystrobin (*off-label*)
	Powdery mildew	azoxystrobin (*off-label*)
	Pythium	Bacillus subtilis (*off-label*)

Fruiting vegetables - Cucurbits

Diseases	Botrytis	mepanipyrim (*off-label*)
	Cladosporium	boscalid + pyraclostrobin (*off-label*)
	Disease control	cyflufenamid, mancozeb
	Leaf spot	boscalid + pyraclostrobin (*off-label*)
	Phytophthora	Bacillus subtilis (*off-label*)
	Powdery mildew	azoxystrobin (*off-label*), boscalid + pyraclostrobin (*off-label*), bupirimate (*off-label*), bupirimate (*outdoor only*), cyflufenamid, penconazole (*off-label*), potassium bicarbonate (commodity substance) (*off-label*), proquinazid (*off-label*)
	Pythium	Bacillus subtilis (*off-label*)
Pests	Aphids	flonicamid (*off-label*), thiacloprid (*off-label*)
	Caterpillars	Bacillus thuringiensis

	Leaf miners	deltamethrin (off-label)
	Pests, miscellaneous	thiacloprid (off-label)
	Spider mites	spirodiclofen (off-label)
	Thrips	deltamethrin (off-label)
	Whiteflies	flonicamid (off-label)
Weeds	Broad-leaved weeds	clomazone (off-label), isoxaben (off-label), propyzamide (off-label)
	Grass weeds	propyzamide (off-label)

Fruiting vegetables - Peppers

Diseases	Foot rot	Trichoderma asperellum (Strain T34)
	Pythium	Bacillus subtilis (off-label)
Pests	Spider mites	spirodiclofen (off-label)

Fruiting vegetables - Tomatoes

Diseases	Botrytis	fenhexamid (off-label)
	Foot rot	Trichoderma asperellum (Strain T34)
	Phytophthora	mancozeb
	Pythium	Bacillus subtilis (off-label)
Pests	Aphids	pyrethrins
	Beetles	pyrethrins
	Caterpillars	pyrethrins
	Spider mites	spirodiclofen (off-label)
	Whiteflies	pyrethrins

Herb crops - Herbs

Crop control	Desiccation	diquat (off-label)
Diseases	Alternaria	boscalid + dimoxystrobin (off-label), cyprodinil + fludioxonil (off-label), fludioxonil (off-label)
	Botrytis	Bacillus subtilis (off-label), cyprodinil + fludioxonil (off-label), fenhexamid (off-label), Gliocladium catenulatum (off-label)
	Damping off	Bacillus subtilis (off-label), fosetyl-aluminium + propamocarb hydrochloride (off-label), thiram (off-label)
	Disease control	boscalid (off-label), dimethomorph + mancozeb (off-label), mandipropamid, tebuconazole (off-label)
	Downy mildew	dimethomorph (off-label), dimethomorph + mancozeb (off-label), fenamidone + fosetyl-aluminium (off-label), fluopicolide + propamocarb hydrochloride (off-label), fosetyl-aluminium + propamocarb hydrochloride (off-label), mancozeb + metalaxyl-M (off-label), mandipropamid, metalaxyl-M (off-label)
	Fusarium	Gliocladium catenulatum (off-label)
	Leaf spot	fludioxonil (off-label)
	Phytophthora	Gliocladium catenulatum (off-label)
	Powdery mildew	azoxystrobin (off-label), prothioconazole (off-label), tebuconazole (off-label)
	Pythium	metalaxyl-M (off-label)
	Rhizoctonia	cyprodinil + fludioxonil (off-label), Gliocladium catenulatum (off-label)

Crop/Pest Guide - Arable and vegetable crops

	Ring spot	azoxystrobin (*off-label*)
	Root rot	thiram (*off-label*)
	Rust	azoxystrobin (*off-label*), azoxystrobin + difenoconazole (*off-label*), tebuconazole (*off-label*)
	Sclerotinia	boscalid + dimoxystrobin (*off-label*), cyprodinil + fludioxonil (*off-label*), prothioconazole (*off-label*)
	Seed-borne diseases	thiram (*seed soak*)
	Septoria	azoxystrobin + difenoconazole (*off-label*)
	Stem canker	cyprodinil + fludioxonil (*off-label*), prothioconazole (*off-label*)
	White blister	boscalid + pyraclostrobin (*off-label*)
Pests	Aphids	acetamiprid (*off-label*), deltamethrin (*off-label*), Metarhizium anisopliae (*off-label*), pirimicarb (*off-label*), spirotetramat (*off-label*)
	Beetles	deltamethrin (*off-label*), lambda-cyhalothrin (*off-label*)
	Caterpillars	Bacillus thuringiensis, Bacillus thuringiensis (*off-label*), deltamethrin (*off-label*), diflubenzuron (*off-label*), spinosad (*off-label*)
	Cutworms	lambda-cyhalothrin (*off-label*)
	Flies	lambda-cyhalothrin (*off-label*), Metarhizium anisopliae (*off-label*), spinosad (*off-label*), tefluthrin (*off-label - seed treatment*)
	Leaf miners	spinosad (*off-label*)
	Midges	Metarhizium anisopliae (*off-label*)
	Pests, miscellaneous	deltamethrin (*off-label*), lambda-cyhalothrin (*off-label*), spinosad, spinosad (*off-label*)
	Thrips	Metarhizium anisopliae (*off-label*), spinosad (*off-label*), thiacloprid (*off-label*)
	Whiteflies	spirotetramat (*off-label*)
Weeds	Broad-leaved weeds	amidosulfuron (*off-label*), bentazone (*off-label*), bromoxynil (*off-label*), carbetamide (*off-label*), chloridazon (*off-label*), chlorpropham (*off-label*), clomazone (*off-label*), clopyralid (*off-label*), clopyralid + picloram (*off-label*), dimethenamid-p + metazachlor (*off-label*), dimethenamid-p + metazachlor + quinmerac (*off-label*), dimethenamid-p + pendimethalin (*off-label*), lenacil (*off-label*), linuron, linuron (*off-label*), metazachlor + quinmerac (*off-label*), napropamide (*off-label*), pendimethalin (*off-label*), phenmedipham (*off-label*), propyzamide (*off-label*), prosulfocarb (*off-label*), pyridate (*off-label*), S-metolachlor (*off-label*)
	Crops as weeds	carbetamide (*off-label*), clethodim (*off-label*), napropamide (*off-label*)
	Grass weeds	carbetamide (*off-label*), chloridazon (*off-label*), chlorpropham (*off-label*), clethodim (*off-label*), dimethenamid-p + metazachlor (*off-label*), dimethenamid-p + metazachlor + quinmerac (*off-label*), dimethenamid-p + pendimethalin (*off-label*), fluazifop-P-butyl (*off-label*), lenacil (*off-label*), linuron, linuron (*off-label*), metazachlor + quinmerac (*off-label*), napropamide (*off-label*), pendimethalin (*off-label*), propyzamide (*off-label*), S-metolachlor (*off-label*)

| | Weeds, miscellaneous | diquat (off-label), napropamide (off-label) |

Leafy vegetables - Endives

Diseases	Alternaria	cyprodinil + fludioxonil (off-label)
	Botrytis	cyprodinil + fludioxonil (off-label), fenhexamid (off-label), Gliocladium catenulatum (off-label)
	Damping off	thiram (off-label)
	Disease control	dimethomorph + mancozeb (off-label), mandipropamid
	Downy mildew	boscalid + pyraclostrobin (off-label), dimethomorph + mancozeb (off-label), fenamidone + fosetyl-aluminium (off-label), mandipropamid
	Fusarium	Gliocladium catenulatum (off-label)
	Phytophthora	Gliocladium catenulatum (off-label)
	Rhizoctonia	cyprodinil + fludioxonil (off-label), Gliocladium catenulatum (off-label)
	Stem canker	cyprodinil + fludioxonil (off-label)
Pests	Aphids	acetamiprid (off-label), deltamethrin (off-label), pymetrozine (off-label), spirotetramat (off-label)
	Beetles	deltamethrin (off-label)
	Caterpillars	Bacillus thuringiensis, Bacillus thuringiensis (off-label), deltamethrin (off-label), diflubenzuron (off-label), indoxacarb (off-label)
	Pests, miscellaneous	deltamethrin (off-label), lambda-cyhalothrin (off-label), spinosad (off-label)
	Whiteflies	spirotetramat (off-label)
Weeds	Broad-leaved weeds	chlorpropham, pendimethalin (off-label), propyzamide, propyzamide (off-label), S-metolachlor (off-label)
	Grass weeds	chlorpropham, diquat, propyzamide, propyzamide (off-label), S-metolachlor (off-label)
	Weeds, miscellaneous	diquat

Leafy vegetables - Lettuce

Diseases	Alternaria	cyprodinil + fludioxonil (off-label), iprodione
	Botrytis	boscalid + pyraclostrobin, boscalid + pyraclostrobin (off-label), cyprodinil + fludioxonil (off-label), fenhexamid (off-label), Gliocladium catenulatum (off-label), iprodione
	Bottom rot	boscalid + pyraclostrobin
	Damping off	Bacillus subtilis (off-label), thiram (off-label)
	Disease control	boscalid + pyraclostrobin (off-label), dimethomorph + mancozeb (off-label), mandipropamid
	Downy mildew	dimethomorph (off-label), dimethomorph + mancozeb (off-label), fenamidone + fosetyl-aluminium (off-label), fluopicolide + propamocarb hydrochloride (off-label), fosetyl-aluminium + propamocarb hydrochloride, mancozeb (off-label), mancozeb + metalaxyl-M (off-label), mandipropamid, potassium bicarbonate (commodity substance) (off-label)
	Fungus diseases	Bacillus subtilis (off-label)

Crop/Pest Guide - Arable and vegetable crops

		pyraclostrobin (*moderate control*), chlorothalonil + cyproconazole, tebuconazole
	Damping off	thiram (*seed treatment*)
	Disease control	azoxystrobin + tebuconazole, azoxystrobin + tebuconazole (*moderate control*), chlorothalonil, metconazole (*off-label*)
	Downy mildew	chlorothalonil + metalaxyl-M, metalaxyl-M (*off-label*)
	Fusarium	Gliocladium catenulatum (*off-label*)
	Phytophthora	Gliocladium catenulatum (*off-label*)
	Powdery mildew	tebuconazole (*off-label*)
	Rhizoctonia	Gliocladium catenulatum (*off-label*)
	Rust	azoxystrobin, boscalid + pyraclostrobin, chlorothalonil + cyproconazole, metconazole, tebuconazole, tebuconazole (*off-label*)
	Sclerotinia	cyprodinil + fludioxonil
Pests	Aphids	esfenvalerate, lambda-cyhalothrin, pirimicarb, zeta-cypermethrin (*off-label*)
	Beetles	lambda-cyhalothrin, lambda-cyhalothrin (*off-label*)
	Birds/mammals	aluminium ammonium sulphate
	Caterpillars	Bacillus thuringiensis (*off-label*), lambda-cyhalothrin
	Cutworms	Bacillus thuringiensis (*off-label*)
	Pests, miscellaneous	lambda-cyhalothrin, lambda-cyhalothrin (*off-label*), zeta-cypermethrin (*off-label*)
	Thrips	lambda-cyhalothrin (*off-label*)
	Weevils	alpha-cypermethrin, cypermethrin, deltamethrin, esfenvalerate, lambda-cyhalothrin, zeta-cypermethrin
Weeds	Broad-leaved weeds	bentazone, carbetamide, clomazone, clomazone (*off-label*), clomazone + linuron, clomazone + pendimethalin, glyphosate, imazamox + pendimethalin, imazamox + pendimethalin (*off-label*), linuron, pendimethalin (*off-label*), propyzamide, prosulfocarb (*off-label*), S-metolachlor (*off-label*)
	Crops as weeds	carbetamide, cycloxydim, fluazifop-P-butyl, glyphosate, pendimethalin (*off-label*), propyzamide, quizalofop-P-ethyl, quizalofop-P-tefuryl
	Grass weeds	carbetamide, clethodim (*off-label*), clomazone + linuron, cycloxydim, diquat, fluazifop-P-butyl, fluazifop-P-butyl (*off-label*), glyphosate, linuron, pendimethalin (*off-label*), propaquizafop, propyzamide, quizalofop-P-ethyl, quizalofop-P-tefuryl, S-metolachlor (*off-label*)
	Weeds, miscellaneous	diquat, glyphosate, pendimethalin (*off-label*)

Legumes - Forage legumes, general

Weeds	Broad-leaved weeds	pendimethalin (*off-label*)
	Grass weeds	pendimethalin (*off-label*)

Legumes - Lupins

Crop control	Desiccation	diquat (*off-label*), glyphosate (*off-label*)
Diseases	Ascochyta	azoxystrobin (*off-label*), metconazole (*qualified minor use*)

Crop/Pest Guide - Arable and vegetable crops

	Botrytis	cyprodinil + fludioxonil (*off-label*), metconazole (*qualified minor use*)
	Damping off	thiram (*off-label - seed treatment*)
	Powdery mildew	azoxystrobin (*off-label*)
	Rust	azoxystrobin (*off-label*), metconazole (*qualified minor use*)
	Seed-borne diseases	thiram (*off-label - seed treatment*)
Pests	Aphids	cypermethrin, lambda-cyhalothrin (*off-label*), zeta-cypermethrin (*off-label*)
	Pests, miscellaneous	cypermethrin, lambda-cyhalothrin (*off-label*), zeta-cypermethrin (*off-label*)
Weeds	Broad-leaved weeds	clomazone (*off-label*), pendimethalin (*off-label*), pyridate (*off-label*)
	Grass weeds	fluazifop-P-butyl (*off-label*), pendimethalin (*off-label*), propaquizafop (*off-label*)
	Weeds, miscellaneous	glyphosate (*off-label*)

Legumes - Peas

Crop control	Desiccation	diquat, glyphosate
Diseases	Alternaria	iprodione (*off-label*)
	Ascochyta	azoxystrobin, boscalid + pyraclostrobin (*moderate control only*), cyprodinil + fludioxonil, metconazole (*reduction*)
	Botrytis	chlorothalonil + cyproconazole, cyprodinil + fludioxonil (*moderate control*), Gliocladium catenulatum (*off-label*), iprodione (*off-label*), metconazole (*reduction*)
	Damping off	thiram (*seed treatment*)
	Downy mildew	cymoxanil + fludioxonil + metalaxyl-M
	Fusarium	Gliocladium catenulatum (*off-label*)
	Mycosphaerella	cyprodinil + fludioxonil, metconazole (*reduction*)
	Phytophthora	Gliocladium catenulatum (*off-label*)
	Powdery mildew	sulphur (*off-label*)
	Rhizoctonia	Gliocladium catenulatum (*off-label*)
	Rust	metconazole
	Sclerotinia	cyprodinil + fludioxonil
Pests	Aphids	alpha-cypermethrin (*reduction*), cypermethrin, esfenvalerate, lambda-cyhalothrin, pirimicarb, thiacloprid, zeta-cypermethrin
	Beetles	lambda-cyhalothrin
	Birds/mammals	aluminium ammonium sulphate
	Caterpillars	alpha-cypermethrin, Bacillus thuringiensis, cypermethrin, deltamethrin, lambda-cyhalothrin, zeta-cypermethrin
	Midges	deltamethrin, lambda-cyhalothrin, thiacloprid
	Pests, miscellaneous	aluminium phosphide, cypermethrin, lambda-cyhalothrin
	Weevils	alpha-cypermethrin, cypermethrin, deltamethrin, esfenvalerate, lambda-cyhalothrin, zeta-cypermethrin
Weeds	Broad-leaved weeds	bentazone, clomazone, clomazone + linuron, clomazone + linuron (*off-label*), clomazone +

Crop/Pest Guide - Arable and vegetable crops

		pendimethalin, flumioxazin (*off-label*), glyphosate, imazamox + pendimethalin, linuron, MCPB, pendimethalin, pendimethalin (*off-label*)
	Crops as weeds	cycloxydim, fluazifop-P-butyl, flumioxazin (*off-label*), glyphosate, pendimethalin, quizalofop-P-ethyl, quizalofop-P-tefuryl
	Grass weeds	clethodim (*off-label*), clomazone + linuron, clomazone + linuron (*off-label*), cycloxydim, diquat (*dry harvested*), fluazifop-P-butyl, flumioxazin (*off-label*), glyphosate, linuron, pendimethalin, pendimethalin (*off-label*), propaquizafop, quizalofop-P-ethyl, quizalofop-P-tefuryl, S-metolachlor (*off-label*)
	Weeds, miscellaneous	clomazone + linuron (*off-label*), diquat (*dry harvested only*), glyphosate, glyphosate (*off-label*)

Miscellaneous arable - Industrial crops

Crop control	Desiccation	glyphosate (*off-label*)
Pests	Aphids	lambda-cyhalothrin (*off-label*)
	Pests, miscellaneous	lambda-cyhalothrin (*off-label*)
Weeds	Broad-leaved weeds	2,4-D (*off-label*), bromoxynil (*off-label*), clopyralid (*off-label*), MCPA (*off-label*), mecoprop-P (*off-label*), metsulfuron-methyl (*off-label*), metsulfuron-methyl + thifensulfuron-methyl (*off-label*), pendimethalin (*off-label*), prosulfocarb (*off-label*), tribenuron-methyl (*off-label*)
	Grass weeds	propoxycarbazone-sodium (*off-label*), prosulfocarb (*off-label*), tri-allate (*off-label*)
	Weeds, miscellaneous	bromoxynil (*off-label*), glyphosate (*off-label*)

Miscellaneous arable - Miscellaneous arable crops

Crop control	Desiccation	glyphosate (*off-label*)
	Miscellaneous non-selective situations	benzoic acid
Diseases	Botrytis	Bacillus subtilis (*off-label*)
	Damping off	thiram (*off-label*)
	Didymella stem rot	Gliocladium catenulatum (*moderate control*)
	Disease control	azoxystrobin (*off-label*), boscalid (*off-label*), boscalid + pyraclostrobin (*off-label*), chlorothalonil (*off-label*), potassium bicarbonate (commodity substance)
	Downy mildew	boscalid + pyraclostrobin (*off-label*)
	Fusarium	Gliocladium catenulatum (*moderate control*)
	Phytophthora	Gliocladium catenulatum (*moderate control*)
	Powdery mildew	azoxystrobin (*off-label*)
	Pythium	Gliocladium catenulatum (*moderate control*)
	Rhizoctonia	Gliocladium catenulatum (*moderate control*)
	Rust	azoxystrobin (*off-label*)
	Sclerotinia	Coniothyrium minitans
Pests	Aphids	lambda-cyhalothrin (*off-label*), maltodextrin, pyrethrins, thiacloprid (*off-label*)
	Beetles	deltamethrin (*off-label*)
	Birds/mammals	aluminium ammonium sulphate

Crop/Pest Guide - Arable and vegetable crops

	Caterpillars	Bacillus thuringiensis, Bacillus thuringiensis (off-label), indoxacarb (off-label), pyrethrins
	Pests, miscellaneous	indoxacarb (off-label), lambda-cyhalothrin (off-label)
	Slugs/snails	metaldehyde
	Spider mites	maltodextrin, pyrethrins
	Thrips	pyrethrins
	Whiteflies	maltodextrin
Plant growth regulation	Growth control	trinexapac-ethyl (off-label)
Weeds	Broad-leaved weeds	2,4-D (off-label), amidosulfuron (off-label), bentazone (off-label), bromoxynil (off-label), carfentrazone-ethyl (before planting), clomazone (off-label), clopyralid (off-label), clopyralid + picloram (off-label), clopyralid + triclopyr (off-label), dicamba + MCPA + mecoprop-P (off-label), diflufenican (off-label), dimethenamid-p + metazachlor (off-label), diquat, diquat (around), florasulam + fluroxypyr (off-label), flufenacet + pendimethalin (off-label), isoxaben (off-label), linuron (off-label), MCPA (off-label), MCPB (off-label), mecoprop-P (off-label), mesotrione (off-label), metsulfuron-methyl (off-label), nicosulfuron (off-label), pendimethalin (off-label), pendimethalin + picolinafen (off-label), propyzamide (off-label), prosulfocarb (off-label), prosulfuron (off-label), pyridate (off-label), thifensulfuron-methyl + tribenuron-methyl (off-label)
	Crops as weeds	carfentrazone-ethyl (before planting), nicosulfuron (off-label), pendimethalin (off-label)
	Grass weeds	diflufenican (off-label), dimethenamid-p + metazachlor (off-label), diquat, diquat (around only), flufenacet + pendimethalin (off-label), linuron (off-label), mesotrione (off-label), nicosulfuron (off-label), pendimethalin (off-label), pendimethalin + picolinafen (off-label), pinoxaden (off-label), propaquizafop (off-label), propyzamide (off-label), prosulfocarb (off-label), tri-allate (off-label)
	Weeds, miscellaneous	2,4-D + glyphosate, bromoxynil (off-label), carfentrazone-ethyl, diquat (around only), glyphosate, glyphosate (before planting), glyphosate (off-label), glyphosate (pre-sowing/planting)

Miscellaneous arable - Miscellaneous arable situations

Weeds	Broad-leaved weeds	citronella oil, glyphosate
	Crops as weeds	fluazifop-P-butyl, glyphosate
	Grass weeds	fluazifop-P-butyl, glyphosate
	Weeds, miscellaneous	cycloxydim, fluazifop-P-butyl, glyphosate, metsulfuron-methyl, thifensulfuron-methyl

Miscellaneous field vegetables - All vegetables

Weeds	Broad-leaved weeds	clomazone (off-label)

Oilseed crops - Linseed/flax

Crop control	Desiccation	diquat, glyphosate
Diseases	Alternaria	difenoconazole (off-label)
	Botrytis	tebuconazole, tebuconazole (reduction)

Crop/Pest Guide - Arable and vegetable crops

	Disease control	boscalid (*off-label*), metconazole (*off-label*), tebuconazole
	Foliar diseases	difenoconazole + paclobutrazol (*off-label*)
	Powdery mildew	difenoconazole (*off-label*), difenoconazole + paclobutrazol (*off-label*), tebuconazole
	Rust	tebuconazole
	Sclerotinia	difenoconazole (*off-label*)
	Septoria	difenoconazole + paclobutrazol (*off-label*)
Pests	Aphids	cypermethrin, tau-fluvalinate (*off-label*)
	Beetles	zeta-cypermethrin
	Pests, miscellaneous	cypermethrin, lambda-cyhalothrin (*off-label*)
Weeds	Broad-leaved weeds	amidosulfuron, bentazone, bifenox (*off-label*), bromoxynil, carbetamide (*off-label*), clopyralid, flupyrsulfuron-methyl (*off-label*), glyphosate, mesotrione (*off-label*), metazachlor (*off-label*), metsulfuron-methyl, napropamide (*off-label*), prosulfocarb (*off-label*)
	Crops as weeds	carbetamide (*off-label*), clethodim (*off-label*), cycloxydim, fluazifop-P-butyl, glyphosate, mesotrione (*off-label*), metsulfuron-methyl, napropamide (*off-label*), quizalofop-P-ethyl, quizalofop-P-tefuryl
	Grass weeds	carbetamide (*off-label*), clethodim (*off-label*), cycloxydim, diquat, fluazifop-P-butyl, fluazifop-P-butyl (*off-label*), flupyrsulfuron-methyl (*off-label*), glyphosate, mesotrione (*off-label*), metazachlor (*off-label*), napropamide (*off-label*), propaquizafop, prosulfocarb (*off-label*), quizalofop-P-ethyl, quizalofop-P-tefuryl, tri-allate (*off-label*)
	Weeds, miscellaneous	bromoxynil, diquat, glyphosate, metsulfuron-methyl, napropamide (*off-label*)

Oilseed crops - Miscellaneous oilseeds

Crop control	Desiccation	diquat (*off-label*), glyphosate (*off-label*)
Diseases	Alternaria	azoxystrobin (*off-label*), boscalid + pyraclostrobin (*off-label*), difenoconazole (*off-label*)
	Botrytis	boscalid (*off-label*), boscalid + pyraclostrobin (*off-label*)
	Damping off	thiram (*off-label - seed treatment*)
	Disease control	azoxystrobin + chlorothalonil (*off-label*), boscalid (*off-label*), chlorothalonil + metalaxyl-M (*off-label*), dimethomorph + mancozeb (*off-label*), propiconazole (*off-label*), tebuconazole (*off-label*)
	Downy mildew	chlorothalonil (*off-label*), dimethomorph + mancozeb (*off-label*), mancozeb + metalaxyl-M (*off-label*)
	Fire	boscalid + pyraclostrobin (*off-label*)
	Foliar diseases	azoxystrobin (*off-label*)
	Powdery mildew	azoxystrobin (*off-label*), difenoconazole (*off-label*), sulphur (*off-label*), tebuconazole (*off-label*)
	Pythium	thiram (*off-label - seed treatment*)
	Rust	tebuconazole (*off-label*)
	Sclerotinia	azoxystrobin (*off-label*), boscalid (*off-label*), difenoconazole (*off-label*)

Crop/Pest Guide - Arable and vegetable crops

Pests	Aphids	deltamethrin (*off-label*), lambda-cyhalothrin (*off-label*), tau-fluvalinate (*off-label*)
	Beetles	lambda-cyhalothrin (*off-label*)
	Caterpillars	deltamethrin (*off-label*)
	Pests, miscellaneous	deltamethrin (*off-label*), lambda-cyhalothrin (*off-label*)
Weeds	Broad-leaved weeds	bifenox (*off-label*), clomazone (*off-label*), clopyralid (*off-label*), clopyralid + picloram (*off-label*), fluroxypyr (*off-label*), mesotrione (*off-label*), metazachlor (*off-label*), napropamide (*off-label*), pendimethalin (*off-label*), prosulfocarb (*off-label*)
	Crops as weeds	clethodim (*off-label*), napropamide (*off-label*), propaquizafop (*off-label*)
	Grass weeds	clethodim (*off-label*), fluazifop-P-butyl (*off-label*), metazachlor (*off-label*), napropamide (*off-label*), propaquizafop (*off-label*), prosulfocarb (*off-label*)
	Weeds, miscellaneous	diquat (*off-label*), glyphosate (*off-label*), napropamide (*off-label*)

Oilseed crops - Oilseed rape

Crop control	Desiccation	diquat, glyphosate
Diseases	Alternaria	azoxystrobin, azoxystrobin + cyproconazole, boscalid, boscalid + dimoxystrobin (*moderate control*), difenoconazole, iprodione, iprodione + thiophanate-methyl, metconazole, tebuconazole
	Phoma and stem canker	bixafen + prothioconazole + tebuconazole, boscalid + dimoxystrobin (*moderate control*), difenoconazole, fluopyram + prothioconazole, iprodione + thiophanate-methyl, prothioconazole, prothioconazole + tebuconazole, tebuconazole
	Botrytis	iprodione, iprodione + thiophanate-methyl
	Damping off	thiram (*seed treatment*)
	Disease control	boscalid + dimoxystrobin , difenoconazole + paclobutrazol, fluopyram + prothioconazole
	Ear blight	thiophanate-methyl (*reduction*)
	Fusarium	thiophanate-methyl
	Light leaf spot	bixafen + prothioconazole + tebuconazole, boscalid + dimoxystrobin (*reduction*), difenoconazole, difenoconazole + paclobutrazol, fluopyram + prothioconazole, iprodione + thiophanate-methyl, metconazole, metconazole (*reduction*), penthiopyrad + picoxystrobin (*moderate control*), prochloraz + propiconazole, propiconazole (*reduction*), prothioconazole, prothioconazole + tebuconazole, prothioconazole + tebuconazole (*moderate control only*), tebuconazole
	Phoma leaf spot	prothioconazole + tebuconazole, tebuconazole
	Powdery mildew	fluopyram + prothioconazole
	Ring spot	tebuconazole, tebuconazole (*reduction*)
	Sclerotinia	azoxystrobin, azoxystrobin + cyproconazole, azoxystrobin + isopyrazam, azoxystrobin + tebuconazole (*moderate control*), bixafen + prothioconazole + tebuconazole, boscalid, boscalid + dimoxystrobin , boscalid + metconazole, cyproconazole + picoxystrobin (*moderate control*), fluopyram + prothioconazole, iprodione (*moderate*

Crop/Pest Guide - Arable and vegetable crops

		control), iprodione + thiophanate-methyl, picoxystrobin, prochloraz + tebuconazole, prothioconazole, prothioconazole + tebuconazole, tebuconazole, tebuconazole (*reduction*), thiophanate-methyl
	Stem canker	bixafen + prothioconazole + tebuconazole, boscalid + dimoxystrobin (*moderate control*), difenoconazole + paclobutrazol, fluopyram + prothioconazole, metconazole (*reduction*), penthiopyrad + picoxystrobin, prochloraz + propiconazole, prochloraz + thiram (*seed treatment*), prothioconazole, prothioconazole + tebuconazole, tebuconazole, tebuconazole (*good reduction*)
Pests	Aphids	acetamiprid, deltamethrin, lambda-cyhalothrin, tau-fluvalinate
	Beetles	alpha-cypermethrin, beta-cyfluthrin, cypermethrin, deltamethrin, etofenprox (*moderate control*), indoxacarb, lambda-cyhalothrin, pymetrozine, tau-fluvalinate, thiacloprid, zeta-cypermethrin
	Birds/mammals	aluminium ammonium sulphate
	Caterpillars	lambda-cyhalothrin
	Midges	alpha-cypermethrin, beta-cyfluthrin, cypermethrin, lambda-cyhalothrin, zeta-cypermethrin
	Pests, miscellaneous	lambda-cyhalothrin
	Weevils	alpha-cypermethrin, beta-cyfluthrin, cypermethrin, deltamethrin, lambda-cyhalothrin, zeta-cypermethrin
Plant growth regulation	Growth control	mepiquat chloride + metconazole, metconazole, tebuconazole
	Quality/yield control	sulphur
Weeds	Broad-leaved weeds	aminopyralid + metazachlor + picloram, aminopyralid + propyzamide, bifenox (*off-label*), carbetamide, clomazone, clomazone + metazachlor, clomazone + napropamide, clopyralid, clopyralid + picloram, clopyralid + picloram (*off-label*), dimethachlor, dimethenamid-p + metazachlor, dimethenamid-p + metazachlor + quinmerac, dimethenamid-p + quinmerac, glyphosate, imazamox + metazachlor, imazamox + quinmerac, imazamox + quinmerac (*cut-leaved*), metazachlor, metazachlor + quinmerac, napropamide, propyzamide, pyridate (*off-label*)
	Crops as weeds	carbetamide, clethodim, cycloxydim, fluazifop-P-butyl, glyphosate, imazamox + quinmerac, propyzamide, quizalofop-P-ethyl, quizalofop-P-tefuryl
	Grass weeds	aminopyralid + propyzamide, carbetamide, clethodim, clomazone + metazachlor, clomazone + napropamide, cycloxydim, dimethachlor, dimethenamid-p + metazachlor + quinmerac, diquat, fluazifop-P-butyl, glyphosate, imazamox + metazachlor, metazachlor, metazachlor + quinmerac, napropamide, propaquizafop, propyzamide, quizalofop-P-ethyl, quizalofop-P-tefuryl
	Weeds, miscellaneous	diquat, glyphosate

Oilseed crops - Soya

Diseases	Botrytis	Gliocladium catenulatum (*off-label*)

	Damping off	thiram (*seed treatment - qualified minor use*)
	Fusarium	Gliocladium catenulatum (*off-label*)
	Phytophthora	Gliocladium catenulatum (*off-label*)
	Rhizoctonia	Gliocladium catenulatum (*off-label*)
Weeds	Broad-leaved weeds	bentazone (*off-label*), clomazone (*off-label*), imazamox + pendimethalin (*off-label*), thifensulfuron-methyl (*off-label*)
	Weeds, miscellaneous	imazamox + pendimethalin (*off-label*)

Oilseed crops - Sunflowers

Crop control	Desiccation	diquat (*off-label*)
Weeds	Broad-leaved weeds	pendimethalin
	Crops as weeds	pendimethalin
	Grass weeds	pendimethalin

Root and tuber crops - Beet crops

Diseases	Alternaria	cyprodinil + fludioxonil (*off-label*), fludioxonil (*off-label*), isopyrazam (*off-label*)
	Anthracnose	fludioxonil (*moderate control*)
	Black leg	hymexazol (*seed treatment*)
	Botrytis	cyprodinil + fludioxonil (*off-label*), Gliocladium catenulatum (*off-label*), iprodione (*off-label*)
	Centrospora	cyproconazole + picoxystrobin (*moderate control*)
	Cercospora leaf spot	azoxystrobin + cyproconazole, cyproconazole + picoxystrobin (*off-label*), cyproconazole + trifloxystrobin, cyproconazole + trifloxystrobin (*off-label*), difenoconazole + propiconazole, epoxiconazole + pyraclostrobin, isopyrazam (*off-label*)
	Damping off	Bacillus subtilis (*off-label*)
	Disease control	boscalid + pyraclostrobin (*off-label*), mandipropamid
	Downy mildew	boscalid + pyraclostrobin (*off-label*), dimethomorph (*off-label*), fenamidone + fosetyl-aluminium (*off-label*), fluopicolide + propamocarb hydrochloride (*off-label*), fosetyl-aluminium + propamocarb hydrochloride (*off-label*), mandipropamid, metalaxyl-M (*off-label*)
	Fusarium	Gliocladium catenulatum (*off-label*)
	Phytophthora	Gliocladium catenulatum (*off-label*)
	Powdery mildew	azoxystrobin + cyproconazole, cyproconazole + picoxystrobin, cyproconazole + picoxystrobin (*off-label*), cyproconazole + trifloxystrobin, cyproconazole + trifloxystrobin (*off-label*), difenoconazole + propiconazole, epoxiconazole, epoxiconazole + pyraclostrobin, epoxiconazole + pyraclostrobin (*off-label*), fenpropimorph (*off-label*), isopyrazam (*off-label*), quinoxyfen, sulphur, sulphur (*off-label*)
	Pythium	metalaxyl-M, metalaxyl-M (*off-label*)
	Ramularia leaf spots	azoxystrobin + cyproconazole, cyproconazole + picoxystrobin (*moderate control*), cyproconazole + picoxystrobin (*off-label*), cyproconazole + trifloxystrobin, cyproconazole + trifloxystrobin (*off-label*), difenoconazole + propiconazole (*moderate control*), epoxiconazole + pyraclostrobin, isopyrazam (*off-label*), propiconazole (*reduction*)

Crop/Pest Guide - Arable and vegetable crops

	Rhizoctonia	azoxystrobin (*off-label*), cyprodinil + fludioxonil (*off-label*), Gliocladium catenulatum (*off-label*)
	Root malformation disorder	azoxystrobin (*off-label*), metalaxyl-M (*off-label*)
	Rust	azoxystrobin + cyproconazole, cyproconazole + picoxystrobin, cyproconazole + picoxystrobin (*off-label*), cyproconazole + trifloxystrobin, cyproconazole + trifloxystrobin (*off-label*), difenoconazole + propiconazole, epoxiconazole, epoxiconazole + pyraclostrobin, epoxiconazole + pyraclostrobin (*off-label*), fenpropimorph (*off-label*), propiconazole
	Sclerotinia	cyprodinil + fludioxonil (*off-label*)
	Seed-borne diseases	thiram (*seed soak*)
	Stem canker	cyprodinil + fludioxonil (*off-label*)
Pests	Aphids	acetamiprid (*off-label*), beta-cyfluthrin, beta-cyfluthrin + clothianidin (*seed treatment*), lambda-cyhalothrin, lambda-cyhalothrin (*off-label*), pirimicarb (*off-label*), pymetrozine (*off-label*), spirotetramat (*off-label*), thiacloprid (*off-label*), thiamethoxam, zeta-cypermethrin (*off-label*)
	Beetles	deltamethrin, lambda-cyhalothrin, lambda-cyhalothrin (*off-label*), tefluthrin (*seed treatment*), thiamethoxam (*seed treatment*)
	Birds/mammals	aluminium ammonium sulphate
	Caterpillars	Bacillus thuringiensis, Bacillus thuringiensis (*off-label*), chlorantraniliprole (*off-label*), lambda-cyhalothrin, lambda-cyhalothrin (*off-label*)
	Cutworms	Bacillus thuringiensis (*off-label*), cypermethrin, lambda-cyhalothrin, lambda-cyhalothrin (*off-label*), zeta-cypermethrin
	Flies	chlorantraniliprole (*off-label*), lambda-cyhalothrin, thiacloprid (*off-label*)
	Free-living nematodes	garlic extract (*off-label*), oxamyl
	Leaf miners	lambda-cyhalothrin, lambda-cyhalothrin (*off-label*), thiamethoxam (*seed treatment*)
	Millipedes	tefluthrin (*seed treatment*), thiamethoxam (*seed treatment*)
	Pests, miscellaneous	lambda-cyhalothrin (*off-label*), spinosad (*off-label*), zeta-cypermethrin (*off-label*)
	Slugs/snails	ferric phosphate
	Springtails	tefluthrin (*seed treatment*), thiamethoxam (*seed treatment*)
	Symphylids	tefluthrin (*seed treatment*), thiamethoxam (*seed treatment*)
	Weevils	lambda-cyhalothrin
	Whiteflies	spirotetramat (*off-label*)
	Wireworms	thiamethoxam (*seed treatment - reduction*)
Plant growth regulation	Quality/yield control	sulphur
Weeds	Broad-leaved weeds	chloridazon, chloridazon (*off-label*), chloridazon + metamitron, chloridazon + quinmerac, chlorpropham (*off-label*), clomazone (*off-label*), clopyralid, clopyralid (*off-label*), desmedipham + ethofumesate + lenacil + phenmedipham, desmedipham + ethofumesate +

Crop/Pest Guide - Arable and vegetable crops

	lenacil + phenmedipham (*off-label*), desmedipham + ethofumesate + phenmedipham, desmedipham + phenmedipham, diquat, ethofumesate, ethofumesate + metamitron, ethofumesate + metamitron + phenmedipham, ethofumesate + phenmedipham, glyphosate, lenacil, lenacil (*off-label*), lenacil + triflusulfuron-methyl, lenacil + triflusulfuron-methyl (*off-label*), metamitron, phenmedipham, phenmedipham (*off-label*), propyzamide, S-metolachlor (*off-label*), triflusulfuron-methyl, triflusulfuron-methyl (*off-label*)
Crops as weeds	clethodim, cycloxydim, fluazifop-P-butyl, glyphosate, glyphosate (*wiper application*), lenacil + triflusulfuron-methyl, propaquizafop (*off-label*), quizalofop-P-ethyl, quizalofop-P-tefuryl
Grass weeds	chloridazon, chloridazon (*off-label*), chloridazon + metamitron, chloridazon + quinmerac, chlorpropham (*off-label*), clethodim, cycloxydim, desmedipham + ethofumesate + lenacil + phenmedipham, desmedipham + ethofumesate + lenacil + phenmedipham (*off-label*), desmedipham + ethofumesate + phenmedipham, diquat, ethofumesate, ethofumesate + metamitron, ethofumesate + metamitron + phenmedipham, ethofumesate + phenmedipham, fluazifop-P-butyl, fluazifop-P-butyl (*off-label*), glyphosate, lenacil, lenacil (*off-label*), metamitron, propaquizafop, propaquizafop (*off-label*), propyzamide, quizalofop-P-ethyl, quizalofop-P-tefuryl, S-metolachlor (*off-label*)
Weeds, miscellaneous	chloridazon + quinmerac, desmedipham + ethofumesate + lenacil + phenmedipham (*off-label*), diquat, glyphosate, glyphosate (*off-label*)

Root and tuber crops - Carrots/parsnips

Diseases		
	Alternaria	azoxystrobin, azoxystrobin (*off-label*), azoxystrobin + difenoconazole, azoxystrobin + difenoconazole (*off-label*), boscalid + pyraclostrobin (*moderate control*), cyprodinil + fludioxonil (*moderate control*), fenpropimorph (*off-label*), fludioxonil, fludioxonil (*off-label*), mancozeb, prothioconazole, tebuconazole, tebuconazole + trifloxystrobin
	Botrytis	cyprodinil + fludioxonil (*moderate control*), Gliocladium catenulatum (*off-label*)
	Cavity spot	Bacillus subtilis (*off-label*), metalaxyl-M, metalaxyl-M (*off-label*), metalaxyl-M (*off-label - reduction*), metalaxyl-M (*reduction only*)
	Crown rot	fenpropimorph (*off-label*)
	Damping off	Bacillus subtilis (*off-label*)
	Disease control	azoxystrobin (*off-label*), mancozeb
	Fusarium	Gliocladium catenulatum (*off-label*)
	Leaf spot	fludioxonil (*off-label*), tebuconazole + trifloxystrobin (*off-label*)
	Phytophthora	Gliocladium catenulatum (*off-label*)
	Powdery mildew	azoxystrobin, azoxystrobin (*off-label*), azoxystrobin + difenoconazole, azoxystrobin + difenoconazole (*off-label*), boscalid + pyraclostrobin, fenpropimorph (*off-label*), prothioconazole, sulphur (*off-label*),

Crop/Pest Guide - Arable and vegetable crops

		tebuconazole, tebuconazole + trifloxystrobin, tebuconazole + trifloxystrobin (*off-label*)
	Pythium	Bacillus subtilis (*off-label*)
	Rhizoctonia	Gliocladium catenulatum (*off-label*)
	Sclerotinia	boscalid + pyraclostrobin (*moderate control*), cyprodinil + fludioxonil (*moderate control*), cyprodinil + fludioxonil (*off-label*), prothioconazole, tebuconazole, tebuconazole + trifloxystrobin, tebuconazole + trifloxystrobin (*off-label*)
	Seed-borne diseases	thiram (*seed soak*)
	Stem canker	isopyrazam, isopyrazam (*off-label*)
Pests	Aphids	cypermethrin, spirotetramat (*off-label*), thiacloprid
	Birds/mammals	aluminium ammonium sulphate
	Cutworms	Bacillus thuringiensis (*off-label*), deltamethrin (*off-label*), lambda-cyhalothrin
	Flies	chlorantraniliprole (*off-label*), cypermethrin, deltamethrin (*off-label*), lambda-cyhalothrin (*off-label*), tefluthrin (*off-label - seed treatment*)
	Free-living nematodes	garlic extract
	Pests, miscellaneous	cypermethrin, deltamethrin (*off-label*), lambda-cyhalothrin (*off-label*)
	Slugs/snails	ferric phosphate
	Stem nematodes	oxamyl
Plant growth regulation	Growth control	maleic hydrazide (*off-label*)
Weeds	Broad-leaved weeds	clomazone, flumioxazin (*off-label*), linuron, metamitron (*off-label*), metribuzin (*off-label*), pendimethalin, pendimethalin (*off-label*), prosulfocarb (*off-label*)
	Crops as weeds	cycloxydim, fluazifop-P-butyl, flumioxazin (*off-label*), metribuzin (*off-label*), pendimethalin
	Grass weeds	clethodim (*off-label*), cycloxydim, fluazifop-P-butyl, fluazifop-P-butyl (*off-label*), flumioxazin (*off-label*), linuron, metamitron (*off-label*), metribuzin (*off-label*), pendimethalin, propaquizafop, prosulfocarb (*off-label*)
	Weeds, miscellaneous	glyphosate (*off-label*)

Root and tuber crops - Miscellaneous root crops

Crop control	Desiccation	diquat (*off-label*)
Diseases	Alternaria	azoxystrobin (*off-label*), azoxystrobin + difenoconazole (*off-label*), fenpropimorph (*off-label*), fludioxonil (*off-label*)
	Botrytis	chlorothalonil (*off-label*), cyprodinil + fludioxonil (*moderate control*), Gliocladium catenulatum (*off-label*)
	Crown rot	fenpropimorph (*off-label*)
	Damping off	Bacillus subtilis (*off-label*)
	Disease control	tebuconazole (*off-label*)
	Fusarium	Gliocladium catenulatum (*off-label*)
	Leaf spot	boscalid + pyraclostrobin (*off-label*), chlorothalonil (*off-label*), fludioxonil (*off-label*), tebuconazole + trifloxystrobin (*off-label*)

Crop/Pest Guide - Arable and vegetable crops

	Phytophthora	azoxystrobin (*off-label*), Gliocladium catenulatum (*off-label*)
	Powdery mildew	azoxystrobin (*off-label*), azoxystrobin + difenoconazole (*off-label*), boscalid + pyraclostrobin (*off-label*), fenpropimorph (*off-label*), tebuconazole (*off-label*), tebuconazole + trifloxystrobin (*off-label*)
	Rhizoctonia	Gliocladium catenulatum (*off-label*)
	Rust	azoxystrobin (*off-label*), boscalid + pyraclostrobin (*off-label*), tebuconazole (*off-label*)
	Sclerotinia	azoxystrobin (*off-label*), boscalid + pyraclostrobin (*off-label*), cyprodinil + fludioxonil (*off-label*), isopyrazam (*off-label*), tebuconazole + trifloxystrobin (*off-label*)
	Septoria	isopyrazam (*off-label*)
Pests	Aphids	cypermethrin, pirimicarb (*off-label*), pymetrozine (*off-label*), thiacloprid (*off-label*)
	Cutworms	Bacillus thuringiensis (*off-label*), lambda-cyhalothrin (*off-label*)
	Flies	chlorantraniliprole (*off-label*), cypermethrin, lambda-cyhalothrin (*off-label*)
	Pests, miscellaneous	cypermethrin, lambda-cyhalothrin (*off-label*)
	Slugs/snails	ferric phosphate
	Weevils	lambda-cyhalothrin (*off-label*)
	Wireworms	tefluthrin (*off-label*)
Weeds	Broad-leaved weeds	clomazone (*off-label*), linuron, linuron (*off-label*), metribuzin (*off-label*), pendimethalin (*off-label*), propyzamide (*off-label*), prosulfocarb (*off-label*), S-metolachlor (*off-label*)
	Crops as weeds	metribuzin (*off-label*)
	Grass weeds	diquat, fluazifop-P-butyl (*off-label*), linuron (*off-label*), metribuzin (*off-label*), propyzamide (*off-label*), prosulfocarb (*off-label*), S-metolachlor (*off-label*)
	Weeds, miscellaneous	diquat, glyphosate (*off-label*), metribuzin (*off-label*)

Root and tuber crops - Potatoes

Crop control	Desiccation	carfentrazone-ethyl, diquat, glufosinate-ammonium, pyraflufen-ethyl
Diseases	Alternaria	azoxystrobin + chlorothalonil (*off-label*), difenoconazole + mandipropamid, fenamidone + propamocarb hydrochloride, mancozeb
	Bacterial blight	Bacillus subtilis (*off-label*)
	Black dot	azoxystrobin, fludioxonil (*some reduction*)
	Black scurf and stem canker	azoxystrobin, fludioxonil, flutolanil (*tuber treatment*), pencycuron (*tuber treatment*)
	Blight	ametoctradin + dimethomorph (*reduction*)
	Disease control	azoxystrobin + fluazinam, difenoconazole
	Dry rot	imazalil, thiabendazole
	Gangrene	imazalil, thiabendazole
	Helminthosporium seedling rot	Bacillus subtilis (*off-label*)
	Phytophthora	ametoctradin + dimethomorph, azoxystrobin + fluazinam, benthiavalicarb-isopropyl + mancozeb, boscalid + pyraclostrobin (*off-label*), chlorothalonil +

Crop/Pest Guide - Arable and vegetable crops

		cymoxanil, cyazofamid, cymoxanil, cymoxanil + famoxadone, cymoxanil + fluazinam, cymoxanil + mancozeb, cymoxanil + mandipropamid, cymoxanil + zoxamide, difenoconazole + mandipropamid, dimethomorph + fluazinam, dimethomorph + mancozeb, dimethomorph + zoxamide, fenamidone + propamocarb hydrochloride, fluazinam, fluopicolide + propamocarb hydrochloride, mancozeb, mancozeb + metalaxyl-M, mancozeb + metalaxyl-M (reduction), mancozeb + zoxamide, mandipropamid
	Rhizoctonia	Bacillus subtilis (off-label)
	Scab	fludioxonil (off-label)
	Silver scurf	fludioxonil (reduction), imazalil, thiabendazole
	Skin spot	imazalil, thiabendazole
Pests	Aphids	acetamiprid, esfenvalerate, flonicamid, lambda-cyhalothrin, pymetrozine, thiacloprid, thiamethoxam
	Beetles	lambda-cyhalothrin
	Caterpillars	lambda-cyhalothrin
	Cutworms	cypermethrin, zeta-cypermethrin
	Cyst nematodes	fosthiazate, oxamyl
	Free-living nematodes	fosthiazate (reduction), oxamyl
	Slugs/snails	metaldehyde
	Weevils	lambda-cyhalothrin
	Wireworms	fosthiazate (reduction)
Plant growth regulation	Growth control	chlorpropham, chlorpropham (thermal fog), ethylene, maleic hydrazide, spearmint oil
Weeds	Broad-leaved weeds	bentazone, carfentrazone-ethyl, clomazone, clomazone + linuron, clomazone + metribuzin, diquat, flufenacet + metribuzin, linuron, metobromuron, metribuzin, pendimethalin, prosulfocarb, pyraflufen-ethyl, rimsulfuron
	Crops as weeds	carfentrazone-ethyl, cycloxydim, metribuzin, pendimethalin, quizalofop-P-ethyl, quizalofop-P-tefuryl, rimsulfuron
	Grass weeds	clomazone + linuron, clomazone + metribuzin, cycloxydim, diquat, diquat (weed control and desiccation), flufenacet + metribuzin, linuron, metobromuron, metribuzin, pendimethalin, propaquizafop, prosulfocarb, quizalofop-P-ethyl, quizalofop-P-tefuryl
	Weeds, miscellaneous	carfentrazone-ethyl, diquat, diquat (weed control or desiccation), glyphosate

Stem and bulb vegetables - Asparagus

Diseases	Botrytis	azoxystrobin + chlorothalonil (qualified minor use recommendation), boscalid + pyraclostrobin (off-label), cyprodinil + fludioxonil (off-label), Gliocladium catenulatum (off-label)
	Downy mildew	metalaxyl-M (off-label)
	Fusarium	Gliocladium catenulatum (off-label)
	Phytophthora	Gliocladium catenulatum (off-label)
	Rhizoctonia	Gliocladium catenulatum (off-label)

Crop/Pest Guide - Arable and vegetable crops

	Rust	azoxystrobin, azoxystrobin + chlorothalonil (*moderate control only*), azoxystrobin + difenoconazole (*off-label*), boscalid + pyraclostrobin (*off-label*)
	Stemphylium	azoxystrobin, azoxystrobin + chlorothalonil (*qualified minor use recommendation*), azoxystrobin + difenoconazole (*off-label*)
Pests	Aphids	cypermethrin
	Beetles	spinosad (*off-label*), thiacloprid (*off-label*)
	Pests, miscellaneous	cypermethrin
	Thrips	spinosad (*off-label*)
Weeds	Broad-leaved weeds	bromoxynil (*off-label*), carfentrazone-ethyl (*off-label*), clomazone (*off-label*), clopyralid (*off-label*), isoxaben (*off-label*), linuron, mesotrione (*off-label*), metribuzin (*off-label*), metribuzin (*off-label - from seed*), pendimethalin (*off-label*), pyridate (*off-label*)
	Crops as weeds	metribuzin (*off-label*)
	Grass weeds	fluazifop-P-butyl (*off-label*), metribuzin (*off-label*)
	Weeds, miscellaneous	clomazone (*off-label - spring-germinating*), glyphosate, glyphosate (*off-label*), metribuzin (*off-label*)

Stem and bulb vegetables - Celery/chicory

Diseases	Alternaria	cyprodinil + fludioxonil (*off-label*)
	Botrytis	azoxystrobin (*off-label*), cyprodinil + fludioxonil (*off-label*), Gliocladium catenulatum (*off-label*), iprodione (*off-label*)
	Damping off	Bacillus subtilis (*off-label*), thiram (*off-label*)
	Disease control	dimethomorph + mancozeb (*off-label*), mandipropamid
	Downy mildew	dimethomorph + mancozeb (*off-label*), fenamidone + fosetyl-aluminium (*off-label*), mandipropamid
	Fusarium	Gliocladium catenulatum (*off-label*)
	Leaf spot	cyprodinil + fludioxonil (*off-label*)
	Phytophthora	fenamidone + fosetyl-aluminium (*off-label*), Gliocladium catenulatum (*off-label*)
	Rhizoctonia	azoxystrobin (*off-label*), Gliocladium catenulatum (*off-label*)
	Sclerotinia	azoxystrobin (*off-label*), cyprodinil + fludioxonil (*off-label*), iprodione (*off-label*)
	Seed-borne diseases	thiram (*seed soak*)
Pests	Aphids	acetamiprid (*off-label*), pirimicarb (*off-label*), pymetrozine (*off-label*)
	Caterpillars	Bacillus thuringiensis, Bacillus thuringiensis (*off-label*), diflubenzuron (*off-label*), lambda-cyhalothrin (*off-label*), spinosad (*off-label*)
	Cutworms	lambda-cyhalothrin (*off-label*)
	Flies	lambda-cyhalothrin (*off-label*), spinosad (*off-label*)
	Leaf miners	spinosad (*off-label*)
	Pests, miscellaneous	lambda-cyhalothrin (*off-label*), spinosad (*off-label*)
	Thrips	spinosad (*off-label*)
	Wireworms	tefluthrin (*off-label*)

Crop/Pest Guide - Arable and vegetable crops

Weeds	Broad-leaved weeds	clomazone (*off-label*), linuron (*off-label*), pendimethalin (*off-label*), propyzamide, propyzamide (*off-label*), prosulfocarb (*off-label*), S-metolachlor (*off-label*), triflusulfuron-methyl (*off-label*)
	Grass weeds	fluazifop-P-butyl (*off-label*), linuron (*off-label*), pendimethalin (*off-label*), propyzamide, propyzamide (*off-label*), prosulfocarb (*off-label*), S-metolachlor (*off-label*)

Stem and bulb vegetables - Globe artichokes/cardoons

Diseases	Downy mildew	azoxystrobin (*off-label*)
	Powdery mildew	azoxystrobin (*off-label*)

Stem and bulb vegetables - Onions/leeks/garlic

Diseases	Alternaria	iprodione, iprodione (*off-label*)
	Blight	fluoxastrobin + prothioconazole (*control*), fluoxastrobin + prothioconazole (*useful reduction*)
	Botrytis	azoxystrobin + chlorothalonil, cyprodinil + fludioxonil (*off-label*), fludioxonil, fludioxonil (*off-label*), fluoxastrobin + prothioconazole (*useful reduction*), Gliocladium catenulatum (*off-label*), iprodione, iprodione (*off-label*)
	Cladosporium	fluoxastrobin + prothioconazole (*useful reduction*), propiconazole (*off-label*)
	Collar rot	iprodione, iprodione (*off-label*)
	Damping off	Bacillus subtilis (*off-label*), thiram (*off-label - seed treatment*), thiram (*seed treatment*)
	Disease control	dimethomorph + mancozeb, dimethomorph + mancozeb (*off-label*), mancozeb, propiconazole (*off-label*)
	Downy mildew	azoxystrobin, azoxystrobin (*off-label*), azoxystrobin + chlorothalonil, benthiavalicarb-isopropyl + mancozeb (*off-label*), dimethomorph + mancozeb, dimethomorph + mancozeb (*off-label*), dimethomorph + pyraclostrobin, fluopicolide + propamocarb hydrochloride (*off-label*), fluoxastrobin + prothioconazole (*control*), mancozeb, mancozeb + metalaxyl-M (*off-label*), mancozeb + metalaxyl-M (*useful control*), metalaxyl-M (*off-label*), propiconazole (*off-label*)
	Fusarium	boscalid + pyraclostrobin (*off-label*), Gliocladium catenulatum (*off-label*)
	Phytophthora	Gliocladium catenulatum (*off-label*)
	Powdery mildew	azoxystrobin (*off-label*), tebuconazole (*off-label*)
	Purple blotch	azoxystrobin, azoxystrobin + difenoconazole (*moderate control*), prothioconazole
	Pythium	metalaxyl-M, metalaxyl-M (*off-label*)
	Rhizoctonia	Gliocladium catenulatum (*off-label*)
	Rhynchosporium	prothioconazole (*useful reduction*)
	Rust	azoxystrobin, azoxystrobin (*off-label*), azoxystrobin + difenoconazole, fenpropimorph, fluoxastrobin + prothioconazole (*useful reduction*), mancozeb, propiconazole (*off-label*), prothioconazole, tebuconazole
	Seed-borne diseases	thiram (*off-label - seed treatment*)

Crop/Pest Guide - Arable and vegetable crops

	Stemphylium	prothioconazole (*useful reduction*)
	White rot	Bacillus subtilis (*off-label*), boscalid + pyraclostrobin (*off-label*), tebuconazole (*off-label*)
	White tip	azoxystrobin + difenoconazole (*qualified minor use*), boscalid + pyraclostrobin (*off-label*), dimethomorph + mancozeb (*reduction*)
Pests	Caterpillars	Bacillus thuringiensis, Bacillus thuringiensis (*off-label*)
	Cutworms	Bacillus thuringiensis (*off-label*), deltamethrin (*off-label*)
	Flies	tefluthrin (*off-label*)
	Free-living nematodes	garlic extract (*off-label*), oxamyl (*off-label*)
	Pests, miscellaneous	deltamethrin (*off-label*), lambda-cyhalothrin (*off-label*)
	Root-knot nematodes	garlic extract (*off-label*)
	Stem nematodes	garlic extract (*off-label*)
	Thrips	lambda-cyhalothrin (*off-label*), spinosad, thiacloprid (*off-label*)
Plant growth regulation	Growth control	maleic hydrazide, maleic hydrazide (*off-label*)
Weeds	Broad-leaved weeds	bentazone (*off-label*), bromoxynil, bromoxynil (*off-label*), chloridazon (*off-label*), chlorpropham, chlorpropham (*off-label*), clopyralid, clopyralid (*off-label*), dimethenamid-p + metazachlor (*off-label*), dimethenamid-p + pendimethalin (*off-label*), flumioxazin (*off-label*), fluroxypyr (*off-label*), glyphosate, linuron (*off-label*), pendimethalin, pendimethalin (*off-label*), prosulfocarb (*off-label*), pyridate, pyridate (*off-label*), S-metolachlor (*off-label*)
	Crops as weeds	bromoxynil, clopyralid (*off-label*), cycloxydim, fluazifop-P-butyl, flumioxazin (*off-label*), fluroxypyr (*off-label*), glyphosate, pendimethalin
	Grass weeds	chloridazon (*off-label*), chlorpropham, clethodim (*off-label*), cycloxydim, dimethenamid-p + metazachlor (*off-label*), dimethenamid-p + pendimethalin (*off-label*), diquat, fluazifop-P-butyl, fluazifop-P-butyl (*off-label*), flumioxazin (*off-label*), glyphosate, linuron (*off-label*), pendimethalin, propaquizafop, propaquizafop (*off-label*), prosulfocarb (*off-label*), S-metolachlor (*off-label*)
	Weeds, miscellaneous	bromoxynil (*off-label*), diquat, fluroxypyr (*off-label*), glyphosate, glyphosate (*off-label*)

Stem and bulb vegetables - Vegetables

Diseases	Botrytis	Bacillus subtilis (*off-label*), Gliocladium catenulatum (*off-label*)
	Downy mildew	azoxystrobin (*off-label*)
	Fusarium	Gliocladium catenulatum (*off-label*)
	Phytophthora	Gliocladium catenulatum (*off-label*)
	Powdery mildew	azoxystrobin (*off-label*), penconazole (*off-label*)
	Rhizoctonia	Gliocladium catenulatum (*off-label*)
Pests	Aphids	deltamethrin (*off-label*), Metarhizium anisopliae (*off-label*)
	Beetles	deltamethrin (*off-label*)
	Caterpillars	Bacillus thuringiensis, deltamethrin (*off-label*)

Crop/Pest Guide - Flowers and ornamentals

	Flies	Metarhizium anisopliae (*off-label*)
	Midges	Metarhizium anisopliae (*off-label*)
	Pests, miscellaneous	deltamethrin (*off-label*)
	Thrips	Metarhizium anisopliae (*off-label*)
Weeds	Grass weeds	fluazifop-P-butyl (*off-label*)

Flowers and ornamentals

Flowers - Bedding plants, general

Diseases	Powdery mildew	bupirimate
	Rust	azoxystrobin (*off-label - in pots*)
Pests	Aphids	thiacloprid
	Beetles	thiacloprid
	Weevils	thiacloprid
	Whiteflies	thiacloprid
Plant growth regulation	Flowering control	paclobutrazol
	Growth control	paclobutrazol

Flowers - Bulbs/corms

Crop control	Desiccation	carfentrazone-ethyl (*off-label*)
Diseases	Basal stem rot	chlorothalonil (*off-label*), cyprodinil + fludioxonil (*off-label*), thiabendazole (*off-label*)
	Botrytis	Bacillus subtilis (*off-label*), boscalid + epoxiconazole (*off-label*), chlorothalonil (*off-label*), Gliocladium catenulatum (*off-label*), mancozeb, tebuconazole (*off-label*), thiabendazole (*off-label*)
	Crown rot	metalaxyl-M (*off-label*)
	Downy mildew	metalaxyl-M (*off-label*)
	Fusarium	cyprodinil + fludioxonil (*off-label*), Gliocladium catenulatum (*off-label*), thiabendazole
	Phytophthora	Gliocladium catenulatum (*off-label*)
	Powdery mildew	bupirimate, tebuconazole (*off-label*)
	Rhizoctonia	Gliocladium catenulatum (*off-label*)
	White mould	boscalid + epoxiconazole (*off-label*)
Weeds	Broad-leaved weeds	bentazone, bromoxynil (*off-label*)
	Grass weeds	S-metolachlor (*off-label*)

Flowers - Miscellaneous flowers

Diseases	Botrytis	Gliocladium catenulatum (*off-label*)
	Crown rot	fenpropimorph (*off-label*)
	Disease control	tebuconazole (*off-label*)
	Downy mildew	fenamidone + fosetyl-aluminium (*off-label*)
	Fusarium	Gliocladium catenulatum (*off-label*)
	Phytophthora	Gliocladium catenulatum (*off-label*)
	Powdery mildew	bupirimate, fenpropimorph (*off-label*), tebuconazole (*off-label*)
	Rhizoctonia	Gliocladium catenulatum (*off-label*)

Crop/Pest Guide - Flowers and ornamentals

	Rust	penconazole, tebuconazole (*off-label*)
Pests	Cutworms	deltamethrin (*off-label*)
	Flies	deltamethrin (*off-label*), lambda-cyhalothrin (*off-label*)
	Pests, miscellaneous	deltamethrin (*off-label*), lambda-cyhalothrin (*off-label*)
Plant growth regulation	Flowering control	paclobutrazol
	Growth control	paclobutrazol
Weeds	Broad-leaved weeds	linuron (*off-label*), metribuzin (*off-label*), pendimethalin (*off-label*)
	Crops as weeds	metribuzin (*off-label*)
	Grass weeds	fluazifop-P-butyl (*off-label*), linuron (*off-label*), metribuzin (*off-label*)

Flowers - Pot plants

Pests	Aphids	thiacloprid
	Beetles	thiacloprid
	Weevils	thiacloprid
	Whiteflies	thiacloprid
Plant growth regulation	Flowering control	paclobutrazol
	Growth control	benzyladenine, paclobutrazol

Flowers - Protected flowers

Diseases	Crown rot	metalaxyl-M (*off-label*)
	Downy mildew	metalaxyl-M (*off-label*)
	Rust	azoxystrobin (*off-label*)
Pests	Spider mites	tebufenpyrad
Plant growth regulation	Flowering control	benzyladenine

Miscellaneous flowers and ornamentals - Miscellaneous uses

Diseases	Botrytis	boscalid + pyraclostrobin (*off-label*)
	Disease control	boscalid + pyraclostrobin (*off-label*), potassium bicarbonate (commodity substance)
	Leaf spot	boscalid + pyraclostrobin (*off-label*)
	Powdery mildew	boscalid + pyraclostrobin (*off-label*), physical pest control
	Rust	boscalid + pyraclostrobin (*off-label*)
	Sclerotinia	Coniothyrium minitans
Pests	Aphids	maltodextrin, physical pest control, pyrethrins
	Birds/mammals	aluminium ammonium sulphate
	Caterpillars	Bacillus thuringiensis (*off-label*), cypermethrin, diflubenzuron, pyrethrins
	Flies	cypermethrin
	Mealybugs	physical pest control
	Scale insects	physical pest control
	Slugs/snails	ferric phosphate, metaldehyde
	Spider mites	maltodextrin, physical pest control, pyrethrins
	Suckers	physical pest control
	Thrips	pyrethrins

Crop/Pest Guide - Flowers and ornamentals

		florasulam (*off-label*), florasulam + fluroxypyr (*off-label*), flumioxazin (*off-label*), imazamox + pendimethalin (*off-label*), isoxaben, linuron, linuron (*off-label*), metamitron (*off-label*), metazachlor, metribuzin (*off-label*), metsulfuron-methyl (*off-label*), napropamide, napropamide (*off-label*), nicosulfuron (*off-label*), pendimethalin (*off-label*), phenmedipham (*off-label*), propyzamide (*off-label*), prosulfocarb (*off-label*), rimsulfuron (*off-label*), S-metolachlor (*off-label*)
	Crops as weeds	florasulam (*off-label*), flumioxazin (*off-label*), metribuzin (*off-label*), nicosulfuron (*off-label*)
	Grass weeds	chlorpropham, cycloxydim, diflufenican (*off-label*), dimethenamid-p + metazachlor (*off-label*), dimethenamid-p + pendimethalin (*off-label*), diquat, fluazifop-P-butyl (*off-label*), flumioxazin (*off-label*), linuron (*off-label*), metamitron (*off-label*), metazachlor, metribuzin (*off-label*), napropamide, napropamide (*off-label*), nicosulfuron (*off-label*), propyzamide (*off-label*), prosulfocarb (*off-label*), S-metolachlor (*off-label*)
	Mosses	maleic hydrazide + pelargonic acid, quinoclamine (*off-label*)
	Weeds, miscellaneous	bromoxynil (*off-label*), diquat, glyphosate, glyphosate (*off-label*), glyphosate + pyraflufen-ethyl, maleic hydrazide + pelargonic acid

Ornamentals - Trees and shrubs

Crop control	Chemical stripping/thinning	glyphosate
Diseases	Botrytis	boscalid + pyraclostrobin (*off-label*)
	Disease control	mancozeb
	Downy mildew	benthiavalicarb-isopropyl + mancozeb (*off-label*)
	Leaf spot	boscalid + pyraclostrobin (*off-label*)
	Phytophthora	Bacillus subtilis (*off-label*), metalaxyl-M (*off-label*)
	Powdery mildew	boscalid + pyraclostrobin (*off-label*), cyflufenamid (*off-label*)
	Rust	boscalid + pyraclostrobin (*off-label*)
Pests	Aphids	deltamethrin, lambda-cyhalothrin (*off-label*)
	Birds/mammals	aluminium ammonium sulphate
	Capsid bugs	deltamethrin
	Caterpillars	Bacillus thuringiensis, deltamethrin, diflubenzuron
	Flies	Metarhizium anisopliae (*off-label*)
	Mealybugs	deltamethrin
	Midges	Metarhizium anisopliae (*off-label*)
	Pests, miscellaneous	lambda-cyhalothrin (*off-label*)
	Scale insects	deltamethrin
	Slugs/snails	ferric phosphate
	Thrips	deltamethrin, Metarhizium anisopliae (*off-label*)
	Weevils	Metarhizium anisopliae (*off-label*)
	Whiteflies	deltamethrin
Plant growth regulation	Growth control	glyphosate, maleic hydrazide (*off-label*)
Weeds	Broad-leaved weeds	clopyralid (*off-label*), isoxaben, propyzamide

Crop/Pest Guide - Forestry

	Grass weeds	propyzamide
	Mosses	maleic hydrazide + pelargonic acid
	Weeds, miscellaneous	2,4-D + glyphosate, flazasulfuron, glyphosate, glyphosate + pyraflufen-ethyl, glyphosate + sulfosulfuron, maleic hydrazide + pelargonic acid, propyzamide

Ornamentals - Woody ornamentals

Weeds	Broad-leaved weeds	napropamide, propyzamide
	Grass weeds	napropamide, propyzamide

Forestry

Forest nurseries, general - Forest nurseries

Crop control	Chemical stripping/ thinning	glyphosate
Diseases	Blight	copper oxychloride (*off-label*)
	Botrytis	boscalid + pyraclostrobin (*off-label*), cyprodinil + fludioxonil, fenhexamid (*off-label*), iprodione (*off-label*), mepanipyrim (*off-label*), pyrimethanil (*off-label*)
	Damping off	fosetyl-aluminium + propamocarb hydrochloride (*off-label*)
	Disease control	azoxystrobin (*off-label*), chlorothalonil (*off-label*), chlorothalonil + metalaxyl-M (*off-label*), cyprodinil (*off-label*), dimethomorph + mancozeb (*off-label*), epoxiconazole (*off-label*), fluoxastrobin + prothioconazole (*off-label*), mancozeb (*off-label*), mancozeb + metalaxyl-M (*off-label*), propiconazole (*off-label*), pyrimethanil (*off-label*)
	Downy mildew	benthiavalicarb-isopropyl + mancozeb (*off-label*), dimethomorph + mancozeb (*off-label*), fosetyl-aluminium + propamocarb hydrochloride (*off-label*)
	Fungus diseases	propamocarb hydrochloride (*off-label*)
	Fusarium	Trichoderma asperellum (Strain T34) (*off-label*)
	Leaf spot	difenoconazole (*off-label*)
	Phytophthora	Bacillus subtilis (*off-label*), difenoconazole (*off-label*), metalaxyl-M (*off-label*), propamocarb hydrochloride (*off-label*)
	Powdery mildew	Ampelomyces quisqualis (Strain AQ10) (*off-label*), azoxystrobin (*off-label*), cyflufenamid (*off-label*), metrafenone (*off-label*), proquinazid (*off-label*), quinoxyfen (*off-label*), spiroxamine (*off-label*)
	Pythium	propamocarb hydrochloride (*off-label*), Trichoderma asperellum (Strain T34) (*off-label*)
	Root rot	Phlebiopsis gigantea
	Rust	azoxystrobin (*off-label*), difenoconazole (*off-label*)
Pests	Aphids	acetamiprid (*off-label*), Beauveria bassiana (*off-label*), deltamethrin (*off-label*), lambda-cyhalothrin (*off-label*), pymetrozine (*off-label*), thiacloprid (*off-label*)
	Beetles	acetamiprid (*off-label*), Beauveria bassiana (*off-label*)
	Birds/mammals	aluminium ammonium sulphate

Crop/Pest Guide - Forestry

	Caterpillars	Bacillus thuringiensis (*off-label*), chlorantraniliprole (*off-label*), indoxacarb (*off-label*)
	Cutworms	Bacillus thuringiensis (*off-label*)
	Flies	acetamiprid (*off-label*)
	Gall, rust and leaf & bud mites	clofentezine (*off-label*)
	Mites	bifenazate (*off-label*), clofentezine (*off-label*)
	Pests, miscellaneous	Beauveria bassiana (*off-label*), deltamethrin (*off-label*), dimethoate (*off-label*), imidacloprid, indoxacarb (*off-label*), lambda-cyhalothrin (*off-label*), spirotetramat (*off-label*), thiacloprid (*off-label*)
	Spider mites	clofentezine (*off-label*)
	Wasps	acetamiprid (*off-label*)
	Weevils	acetamiprid (*off-label*), chlorantraniliprole (*off-label*)
	Whiteflies	Beauveria bassiana (*off-label*), pymetrozine (*off-label*)
Plant growth regulation	Growth control	ethephon + mepiquat chloride (*off-label*), trinexapac-ethyl (*off-label*)
Weeds	Broad-leaved weeds	2,4-D (*off-label*), amidosulfuron (*off-label*), amidosulfuron + iodosulfuron-methyl-sodium (*off-label*), chlorpropham (*off-label*), clomazone (*off-label*), clomazone + metazachlor (*off-label*), clopyralid (*off-label*), clopyralid + picloram (*off-label*), desmedipham + ethofumesate + lenacil + phenmedipham (*off-label*), diflufenican (*off-label*), diflufenican + flupyrsulfuron-methyl (*off-label*), diflufenican + iodosulfuron-methyl-sodium + mesosulfuron-methyl (*off-label*), dimethenamid-p + metazachlor (*off-label*), flumioxazin (*off-label*), isoxaben, linuron, metsulfuron-methyl (*off-label*), napropamide, nicosulfuron (*off-label*), pendimethalin (*off-label*), pendimethalin + picolinafen (*off-label*), propyzamide, propyzamide (*off-label*), prosulfocarb (*off-label*), rimsulfuron (*off-label*), S-metolachlor (*off-label*), tribenuron-methyl (*off-label*)
	Crops as weeds	amidosulfuron + iodosulfuron-methyl-sodium (*off-label*), clopyralid (*off-label*), diflufenican + iodosulfuron-methyl-sodium + mesosulfuron-methyl (*off-label*), flumioxazin (*off-label*), nicosulfuron (*off-label*)
	Grass weeds	chlorpropham (*off-label*), clomazone + metazachlor (*off-label*), cycloxydim, desmedipham + ethofumesate + lenacil + phenmedipham (*off-label*), diflufenican (*off-label*), diflufenican + flupyrsulfuron-methyl (*off-label*), diflufenican + iodosulfuron-methyl-sodium + mesosulfuron-methyl (*off-label*), dimethenamid-p + metazachlor (*off-label*), napropamide, nicosulfuron (*off-label*), pendimethalin + picolinafen (*off-label*), propyzamide, propyzamide (*off-label*), prosulfocarb (*off-label*), S-metolachlor (*off-label*)
	Weeds, miscellaneous	diquat (*off-label*), glyphosate
	Woody weeds/scrub	glyphosate

Forestry plantations, general - Forestry plantations

Crop control	Chemical stripping/thinning	glyphosate, glyphosate (*stump treatment*)

Crop/Pest Guide - Fruit and hops

Diseases	Root rot	Phlebiopsis gigantea
Pests	Birds/mammals	aluminium ammonium sulphate
	Caterpillars	Bacillus thuringiensis (*off-label*), diflubenzuron
	Weevils	acetamiprid (*off-label*), cypermethrin
Plant growth regulation	Growth control	glyphosate
Weeds	Broad-leaved weeds	clopyralid (*off-label*), napropamide (*off-label*), pendimethalin (*off-label*), propyzamide
	Grass weeds	cycloxydim, glyphosate, napropamide (*off-label*), propyzamide
	Weeds, miscellaneous	glyphosate
	Woody weeds/scrub	glyphosate

Miscellaneous forestry situations - Cut logs/timber

Pests	Beetles	cypermethrin

Woodland on farms - Woodland

Crop control	Chemical stripping/thinning	glyphosate
Diseases	Root rot	Phlebiopsis gigantea
Pests	Aphids	lambda-cyhalothrin (*off-label*)
	Beetles	lambda-cyhalothrin (*off-label*)
	Pests, miscellaneous	lambda-cyhalothrin (*off-label*)
	Sawflies	lambda-cyhalothrin (*off-label*)
Plant growth regulation	Growth control	gibberellins (*do not exceed 2.5 g per 5 litres when used as a seed soak*)
Weeds	Broad-leaved weeds	2,4-D (*off-label*), amidosulfuron (*off-label*), clopyralid (*off-label*), flumioxazin (*off-label*), isoxaben (*off-label*), lenacil (*off-label*), MCPA (*off-label*), metsulfuron-methyl (*off-label*), napropamide (*off-label*), pendimethalin (*off-label*), propyzamide, tribenuron-methyl (*off-label*)
	Crops as weeds	fluazifop-P-butyl, flumioxazin (*off-label*)
	Grass weeds	fluazifop-P-butyl, lenacil (*off-label*), napropamide (*off-label*), propyzamide
	Weeds, miscellaneous	glyphosate, glyphosate (*off-label*)
	Woody weeds/scrub	glyphosate

Fruit and hops

All bush fruit - All currants

Diseases	Botrytis	Bacillus subtilis (*off-label*), boscalid + pyraclostrobin, boscalid + pyraclostrobin (*off-label*), cyprodinil + fludioxonil (*qualified minor use recommendation*), fenhexamid, Gliocladium catenulatum (*off-label*), pyrimethanil (*off-label*)
	Disease control	mancozeb + metalaxyl-M (*off-label*), sulphur (*off-label*)
	Fusarium	Gliocladium catenulatum (*off-label*)

Crop/Pest Guide - Fruit and hops

	Leaf spot	boscalid + pyraclostrobin (*moderate control only*), boscalid + pyraclostrobin (*off-label*)
	Phytophthora	Bacillus subtilis (*off-label*), Gliocladium catenulatum (*off-label*)
	Powdery mildew	azoxystrobin (*off-label*), boscalid + pyraclostrobin (*moderate control only*), boscalid + pyraclostrobin (*off-label*), bupirimate, bupirimate (*off-label*), fenpropimorph (*off-label*), kresoxim-methyl, kresoxim-methyl (*off-label*), penconazole, quinoxyfen (*off-label*), sulphur
	Rhizoctonia	Gliocladium catenulatum (*off-label*)
	Rust	azoxystrobin (*off-label*)
Pests	Aphids	pymetrozine (*off-label*), spirotetramat (*off-label*)
	Caterpillars	Bacillus thuringiensis (*off-label*), chlorantraniliprole (*off-label*), diflubenzuron, diflubenzuron (*off-label*), spinosad (*off-label*)
	Cutworms	Bacillus thuringiensis (*off-label*)
	Flies	lambda-cyhalothrin (*off-label*), Metarhizium anisopliae (*off-label*)
	Gall, rust and leaf & bud mites	sulphur, tebufenpyrad (*off-label*)
	Leatherjackets	Metarhizium anisopliae (*off-label*)
	Midges	lambda-cyhalothrin (*off-label*), Metarhizium anisopliae (*off-label*)
	Pests, miscellaneous	Bacillus thuringiensis (*off-label*), lambda-cyhalothrin (*off-label*), spinosad (*off-label*), thiacloprid (*off-label*)
	Sawflies	lambda-cyhalothrin (*off-label*), spinosad (*off-label*)
	Scale insects	thiacloprid (*off-label*)
	Spider mites	spirodiclofen (*off-label*)
	Thrips	lambda-cyhalothrin (*off-label*), Metarhizium anisopliae (*off-label*)
	Weevils	Metarhizium anisopliae
Plant growth regulation	Growth control	maleic hydrazide (*off-label*)
Weeds	Broad-leaved weeds	clopyralid (*off-label*), flufenacet + metribuzin (*off-label*), lenacil (*off-label*), napropamide, napropamide (*off-label*), pendimethalin, pendimethalin (*off-label*), propyzamide
	Crops as weeds	fluazifop-P-butyl, pendimethalin
	Grass weeds	fluazifop-P-butyl, fluazifop-P-butyl (*off-label*), flufenacet + metribuzin (*off-label*), lenacil (*off-label*), napropamide, pendimethalin, propyzamide
	Weeds, miscellaneous	glyphosate (*off-label*)

All bush fruit - All protected bush fruit

Diseases	Anthracnose	boscalid + pyraclostrobin (*off-label*), cyprodinil + fludioxonil (*off-label*)
	Botrytis	boscalid + pyraclostrobin (*off-label*), cyprodinil + fludioxonil (*off-label*), fenhexamid (*off-label*)
	Disease control	mancozeb + metalaxyl-M (*off-label*)
	Fusarium	Trichoderma asperellum (Strain T34) (*off-label*)
	Leaf spot	boscalid + pyraclostrobin (*off-label*)

Crop/Pest Guide - Fruit and hops

Pests	Powdery mildew	Ampelomyces quisqualis (Strain AQ10) (*off-label*)
	Pythium	Trichoderma asperellum (Strain T34) (*off-label*)
	Aphids	pymetrozine (*off-label*), pyrethrins
	Caterpillars	Bacillus thuringiensis (*off-label*), pyrethrins, spinosad (*off-label*), thiacloprid (*off-label*)
	Flies	spinosad (*off-label*)
	Midges	thiacloprid (*off-label*)
	Spider mites	Lecanicillium muscarium (*off-label*)
	Thrips	Lecanicillium muscarium (*off-label*)
	Whiteflies	Lecanicillium muscarium (*off-label*), pymetrozine (*off-label*)

All bush fruit - All vines

Diseases	Black rot	kresoxim-methyl (*off-label*)
	Botrytis	Bacillus subtilis (*off-label*), fenhexamid, fenpyrazamine, Gliocladium catenulatum (*off-label*), potassium bicarbonate (commodity substance) (*off-label*), prohexadione-calcium (*off-label*)
	Disease control	mancozeb, mancozeb + zoxamide
	Downy mildew	ametoctradin + dimethomorph (*off-label*), benthiavalicarb-isopropyl + mancozeb (*off-label*), copper oxychloride, cymoxanil (*off-label*), fenamidone + fosetyl-aluminium (*off-label*), potassium bicarbonate (commodity substance) (*off-label*)
	Fusarium	Gliocladium catenulatum (*off-label*)
	Phytophthora	Gliocladium catenulatum (*off-label*)
	Powdery mildew	Ampelomyces quisqualis (Strain AQ10) (*off-label*), fenbuconazole (*off-label*), kresoxim-methyl (*off-label*), meptyldinocap, penconazole, potassium bicarbonate (commodity substance) (*off-label*), sulphur
	Rhizoctonia	Gliocladium catenulatum (*off-label*)
Pests	Aphids	spirotetramat (*off-label*)
	Capsid bugs	lambda-cyhalothrin (*off-label*)
	Caterpillars	Bacillus thuringiensis, Bacillus thuringiensis (*off-label*), lambda-cyhalothrin (*off-label*), methoxyfenozide (*off-label*)
	Flies	Metarhizium anisopliae (*off-label*)
	Leatherjackets	Metarhizium anisopliae (*off-label*)
	Midges	Metarhizium anisopliae (*off-label*)
	Spider mites	Lecanicillium muscarium (*off-label*)
	Thrips	Lecanicillium muscarium (*off-label*), Metarhizium anisopliae (*off-label*)
	Wasps	lambda-cyhalothrin (*off-label*)
	Whiteflies	Lecanicillium muscarium (*off-label*)
Weeds	Broad-leaved weeds	propyzamide (*off-label*)
	Grass weeds	propyzamide (*off-label*)
	Weeds, miscellaneous	glyphosate

All bush fruit - Bilberries/blueberries/cranberries

Diseases	Anthracnose	boscalid + pyraclostrobin (*off-label*)

Crop/Pest Guide - Fruit and hops

	Botrytis	Bacillus subtilis (*off-label*), boscalid + pyraclostrobin (*off-label*), cyprodinil + fludioxonil (*qualified minor use recommendation*), fenhexamid (*off-label*), Gliocladium catenulatum (*off-label*), pyrimethanil (*off-label*)
	Disease control	mancozeb + metalaxyl-M (*off-label*), sulphur (*off-label*)
	Fusarium	Gliocladium catenulatum (*off-label*)
	Leaf spot	boscalid + pyraclostrobin (*off-label*)
	Phytophthora	Bacillus subtilis (*off-label*), Gliocladium catenulatum (*off-label*)
	Powdery mildew	azoxystrobin (*off-label*), fenpropimorph (*off-label*), kresoxim-methyl (*off-label*), quinoxyfen (*off-label*)
	Rhizoctonia	Gliocladium catenulatum (*off-label*)
	Rust	azoxystrobin (*off-label*)
Pests	Aphids	pymetrozine (*off-label*), spirotetramat (*off-label*)
	Caterpillars	Bacillus thuringiensis (*off-label*), chlorantraniliprole (*off-label*), diflubenzuron (*off-label*), indoxacarb (*off-label*), spinosad (*off-label*)
	Cutworms	Bacillus thuringiensis (*off-label*)
	Flies	lambda-cyhalothrin (*off-label*), Metarhizium anisopliae (*off-label*), spinosad (*off-label*)
	Gall, rust and leaf & bud mites	tebufenpyrad (*off-label*)
	Leatherjackets	Metarhizium anisopliae (*off-label*)
	Midges	Metarhizium anisopliae (*off-label*)
	Pests, miscellaneous	Bacillus thuringiensis (*off-label*), indoxacarb (*off-label*), thiacloprid (*off-label*)
	Scale insects	thiacloprid (*off-label*)
	Thrips	lambda-cyhalothrin (*off-label*), Metarhizium anisopliae (*off-label*)
	Weevils	Metarhizium anisopliae
Plant growth regulation	Growth control	maleic hydrazide (*off-label*)
Weeds	Broad-leaved weeds	clopyralid (*off-label*), flufenacet + metribuzin (*off-label*), lenacil (*off-label*), napropamide (*off-label*), pendimethalin (*off-label*), propyzamide (*off-label*)
	Grass weeds	fluazifop-P-butyl (*off-label*), flufenacet + metribuzin (*off-label*), lenacil (*off-label*), propyzamide (*off-label*)
	Weeds, miscellaneous	glyphosate (*off-label*)

All bush fruit - Bush fruit, general

Pests	Aphids	pyrethrins
	Birds/mammals	aluminium ammonium sulphate
	Caterpillars	pyrethrins
Weeds	Broad-leaved weeds	clopyralid (*off-label*), isoxaben

All bush fruit - Gooseberries

Diseases	Anthracnose	boscalid + pyraclostrobin (*off-label*)
	Botrytis	Bacillus subtilis (*off-label*), boscalid + pyraclostrobin (*off-label*), chlorothalonil + metalaxyl-M (*off-label*), cyprodinil + fludioxonil (*qualified minor use*

Crop/Pest Guide - Fruit and hops

		recommendation), fenhexamid, Gliocladium catenulatum (off-label), pyrimethanil (off-label)
	Disease control	mancozeb + metalaxyl-M (off-label)
	Foliar diseases	chlorothalonil + metalaxyl-M (off-label)
	Fusarium	Gliocladium catenulatum (off-label)
	Leaf spot	boscalid + pyraclostrobin (off-label), chlorothalonil + metalaxyl-M (off-label)
	Phytophthora	Bacillus subtilis (off-label), Gliocladium catenulatum (off-label)
	Powdery mildew	azoxystrobin (off-label), bupirimate, fenpropimorph (off-label), kresoxim-methyl (off-label), quinoxyfen (off-label), sulphur
	Rhizoctonia	Gliocladium catenulatum (off-label)
	Rust	azoxystrobin (off-label)
Pests	Aphids	pymetrozine (off-label), spirotetramat (off-label)
	Caterpillars	Bacillus thuringiensis (off-label), chlorantraniliprole (off-label), diflubenzuron (off-label), spinosad (off-label)
	Cutworms	Bacillus thuringiensis (off-label)
	Flies	spinosad (off-label)
	Gall, rust and leaf & bud mites	tebufenpyrad (off-label)
	Midges	lambda-cyhalothrin (off-label)
	Pests, miscellaneous	Bacillus thuringiensis (off-label), lambda-cyhalothrin (off-label), thiacloprid (off-label)
	Sawflies	lambda-cyhalothrin (off-label)
	Scale insects	thiacloprid (off-label)
	Spider mites	spirodiclofen (off-label)
	Weevils	Metarhizium anisopliae
Plant growth regulation	Growth control	maleic hydrazide (off-label)
Weeds	Broad-leaved weeds	clopyralid (off-label), flufenacet + metribuzin (off-label), lenacil (off-label), napropamide, pendimethalin, propyzamide
	Crops as weeds	fluazifop-P-butyl, pendimethalin
	Grass weeds	fluazifop-P-butyl, flufenacet + metribuzin (off-label), lenacil (off-label), napropamide, pendimethalin, propyzamide
	Weeds, miscellaneous	glyphosate (off-label)

All bush fruit - Miscellaneous bush fruit

Diseases	Alternaria	iprodione (off-label)
	Botrytis	cyprodinil + fludioxonil (off-label), fenhexamid, iprodione (off-label), pyrimethanil (off-label)
	Downy mildew	metalaxyl-M (off-label)
	Powdery mildew	boscalid (off-label), proquinazid (off-label), sulphur
Pests	Capsid bugs	lambda-cyhalothrin (off-label)
	Caterpillars	indoxacarb (off-label)
	Pests, miscellaneous	lambda-cyhalothrin (off-label), spinosad (off-label)
	Wasps	lambda-cyhalothrin (off-label)

Crop/Pest Guide - Fruit and hops

Weeds	Broad-leaved weeds	isoxaben
	Grass weeds	fluazifop-P-butyl (*off-label*)
	Weeds, miscellaneous	glyphosate (*off-label*)

Cane fruit - All outdoor cane fruit

Crop control	Desiccation	carfentrazone-ethyl (*off-label*)
Diseases	Alternaria	iprodione
	Botrytis	Bacillus subtilis (*off-label*), fenhexamid, Gliocladium catenulatum (*off-label*), iprodione, pyrimethanil (*off-label*)
	Botrytis fruit rot	cyprodinil + fludioxonil, tebuconazole (*off-label*)
	Cane blight	boscalid + pyraclostrobin (*off-label*), tebuconazole (*off-label*)
	Disease control	mancozeb + metalaxyl-M (*off-label*)
	Downy mildew	metalaxyl-M (*off-label*)
	Fusarium	Gliocladium catenulatum (*off-label*)
	Phytophthora	Bacillus subtilis (*off-label*), Gliocladium catenulatum (*off-label*), tebuconazole (*off-label*)
	Powdery mildew	azoxystrobin (*off-label*), bupirimate (*outdoor only*), fenpropimorph (*off-label*), tebuconazole (*off-label*)
	Purple blotch	boscalid + pyraclostrobin (*off-label*)
	Rhizoctonia	Gliocladium catenulatum (*off-label*)
	Root rot	dimethomorph, dimethomorph (*moderate control*)
	Rust	azoxystrobin (*off-label*)
Pests	Aphids	pymetrozine (*off-label*), pyrethrins
	Beetles	deltamethrin, deltamethrin (*off-label*)
	Birds/mammals	aluminium ammonium sulphate
	Capsid bugs	lambda-cyhalothrin (*off-label*), thiacloprid (*off-label*)
	Caterpillars	Bacillus thuringiensis, Bacillus thuringiensis (*off-label*), deltamethrin (*off-label*), diflubenzuron (*off-label*), indoxacarb (*off-label*), pyrethrins
	Cutworms	Bacillus thuringiensis (*off-label*)
	Flies	Metarhizium anisopliae (*off-label*), spinosad (*off-label*)
	Gall, rust and leaf & bud mites	tebufenpyrad (*off-label*)
	Leatherjackets	Metarhizium anisopliae (*off-label*)
	Midges	Metarhizium anisopliae (*off-label*)
	Pests, miscellaneous	Bacillus thuringiensis (*off-label*), deltamethrin (*off-label*), indoxacarb (*off-label*), lambda-cyhalothrin (*off-label*), spinosad (*off-label*), thiacloprid (*off-label*)
	Slugs/snails	metaldehyde
	Spider mites	tebufenpyrad (*off-label*)
	Thrips	Metarhizium anisopliae (*off-label*), spinosad (*off-label*)
	Weevils	lambda-cyhalothrin (*off-label*), Metarhizium anisopliae
Plant growth regulation	Growth control	maleic hydrazide (*off-label*)
Weeds	Broad-leaved weeds	isoxaben, lenacil (*off-label*), napropamide, napropamide (*off-label*), pendimethalin, propyzamide, propyzamide (*England only*)
	Crops as weeds	fluazifop-P-butyl, pendimethalin

Cane fruit - All protected cane fruit

Crop control	Desiccation	carfentrazone-ethyl (*off-label*)
Diseases	Botrytis	pyrimethanil (*off-label*)
	Cane blight	boscalid + pyraclostrobin (*off-label*)
	Disease control	mancozeb + metalaxyl-M (*off-label*)
	Fusarium	Trichoderma asperellum (Strain T34) (*off-label*)
	Powdery mildew	azoxystrobin (*off-label*)
	Purple blotch	boscalid + pyraclostrobin (*off-label*)
	Pythium	Trichoderma asperellum (Strain T34) (*off-label*)
	Root rot	dimethomorph (*moderate control*)
	Rust	boscalid + pyraclostrobin (*off-label*)
Pests	Aphids	pymetrozine (*off-label*), pyrethrins
	Capsid bugs	thiacloprid (*off-label*)
	Caterpillars	indoxacarb (*off-label*), pyrethrins
	Gall, rust and leaf & bud mites	tebufenpyrad (*off-label*)
	Mites	abamectin (*off-label*)
	Pests, miscellaneous	indoxacarb (*off-label*), spinosad (*off-label*), thiacloprid (*off-label*)
	Spider mites	Lecanicillium muscarium (*off-label*), tebufenpyrad (*off-label*)
	Thrips	Lecanicillium muscarium (*off-label*), spinosad (*off-label*)
	Whiteflies	Lecanicillium muscarium (*off-label*), pymetrozine (*off-label*)

Grass weeds — fluazifop-P-butyl, fluazifop-P-butyl (*off-label*), napropamide, napropamide (*off-label*), pendimethalin, propyzamide, propyzamide (*England only*)

Hops, general - Hops

Crop control	Chemical stripping/thinning	diquat
	Desiccation	diquat, diquat (*off-label*)
Diseases	Botrytis	Bacillus subtilis (*off-label*), fenhexamid (*off-label*), Gliocladium catenulatum (*off-label*), iprodione (*off-label*)
	Disease control	boscalid + pyraclostrobin (*off-label*), captan (*off-label*), cymoxanil + mancozeb (*off-label*), mancozeb (*off-label*), mancozeb + metalaxyl-M (*off-label*), propiconazole (*off-label*), thiophanate-methyl (*off-label - do not harvest for human or animal consumption (including idling) within 12 months of treatment.*)
	Downy mildew	benthiavalicarb-isopropyl + mancozeb (*off-label*), cymoxanil (*off-label*), mandipropamid (*off-label*), metalaxyl-M (*off-label*)
	Fusarium	Gliocladium catenulatum (*off-label*)
	Phytophthora	Gliocladium catenulatum (*off-label*)
	Powdery mildew	bupirimate, cyflufenamid (*off-label*), fenpropimorph (*off-label*), penconazole (*off-label*), quinoxyfen (*off-label*), sulphur
	Rhizoctonia	Gliocladium catenulatum (*off-label*)

Crop/Pest Guide - Fruit and hops

Pests	Aphids	acetamiprid (*off-label*), flonicamid (*off-label*), lambda-cyhalothrin (*off-label*), pymetrozine (*off-label*), spirotetramat (*off-label*), tebufenpyrad, thiacloprid (*off-label*)
	Beetles	lambda-cyhalothrin (*off-label*)
	Capsid bugs	lambda-cyhalothrin (*off-label*)
	Caterpillars	Bacillus thuringiensis, chlorantraniliprole (*off-label*), indoxacarb (*off-label*), lambda-cyhalothrin (*off-label*), spinosad (*off-label*)
	Free-living nematodes	fosthiazate (*off-label*)
	Pests, miscellaneous	indoxacarb (*off-label*), lambda-cyhalothrin (*off-label*), methoxyfenozide (*off-label*), spinosad (*off-label*), thiacloprid (*off-label*)
	Spider mites	spirodiclofen (*off-label*), tebufenpyrad
Plant growth regulation	Growth control	maleic hydrazide (*off-label*), prohexadione-calcium (*off-label*)
Weeds	Broad-leaved weeds	bentazone (*off-label*), clomazone (*off-label*), clopyralid (*off-label*), flumioxazin (*off-label*), imazamox + pendimethalin (*off-label*), isoxaben, pendimethalin (*off-label*), propyzamide (*off-label*)
	Crops as weeds	fluazifop-P-butyl, flumioxazin (*off-label*)
	Grass weeds	diquat, fluazifop-P-butyl, flumioxazin (*off-label*), propyzamide (*off-label*)
	Weeds, miscellaneous	diquat, glyphosate (*off-label*)

Miscellaneous fruit situations - Fruit crops, general

Diseases	Botrytis	Bacillus subtilis (*off-label*), fenhexamid (*off-label*), iprodione (*off-label*), pyrimethanil (*off-label*)
	Damping off	fosetyl-aluminium + propamocarb hydrochloride (*off-label*)
	Disease control	azoxystrobin (*off-label*), boscalid + pyraclostrobin (*off-label*), mancozeb (*off-label*), propiconazole (*off-label*), pyrimethanil (*off-label*), thiophanate-methyl (*off-label*)
	Downy mildew	benthiavalicarb-isopropyl + mancozeb (*off-label*), fosetyl-aluminium + propamocarb hydrochloride (*off-label*), metalaxyl-M (*off-label*)
	Powdery mildew	cyflufenamid (*off-label*), kresoxim-methyl (*reduction*), proquinazid (*off-label*)
	Scab	kresoxim-methyl
Pests	Aphids	acetamiprid (*off-label*), lambda-cyhalothrin (*off-label*), pymetrozine (*off-label*), thiacloprid (*off-label*)
	Caterpillars	Bacillus thuringiensis (*off-label*), indoxacarb (*off-label*), spinosad (*off-label*)
	Flies	Metarhizium anisopliae (*off-label*)
	Free-living nematodes	fosthiazate (*off-label*)
	Midges	Metarhizium anisopliae (*off-label*)
	Pests, miscellaneous	aluminium phosphide, lambda-cyhalothrin (*off-label*), methoxyfenozide (*off-label*), spinosad (*off-label*), thiacloprid (*off-label*)
	Spider mites	spirodiclofen (*off-label*)
	Thrips	Metarhizium anisopliae (*off-label*)

Crop/Pest Guide - Fruit and hops

Weeds	Broad-leaved weeds	flumioxazin (*off-label*), imazamox + pendimethalin (*off-label*), isoxaben, nicosulfuron (*off-label*), pendimethalin (*off-label*), prosulfocarb (*off-label*)
	Crops as weeds	flumioxazin (*off-label*), nicosulfuron (*off-label*)
	Grass weeds	fluazifop-P-butyl (*off-label*), flumioxazin (*off-label*), nicosulfuron (*off-label*), prosulfocarb (*off-label*)
	Weeds, miscellaneous	glyphosate (*off-label*)

Miscellaneous fruit situations - Fruit nursery stock

Diseases	Disease control	captan (*off-label*)
	Fungus diseases	potassium bicarbonate (commodity substance)
	Powdery mildew	penconazole (*off-label*)
	Sooty moulds	potassium bicarbonate (commodity substance)
Plant growth regulation	Growth control	prohexadione-calcium (*off-label*)
Weeds	Broad-leaved weeds	clopyralid (*off-label*), metazachlor
	Grass weeds	metazachlor

Other fruit - All outdoor strawberries

Diseases	Alternaria	iprodione
	Black spot	azoxystrobin (*off-label*), boscalid + pyraclostrobin, cyprodinil + fludioxonil (*qualified minor use*)
	Botrytis	Bacillus subtilis, boscalid + pyraclostrobin, chlorothalonil (*off-label*), cyprodinil + fludioxonil, fenhexamid, fenpyrazamine, iprodione, mepanipyrim, pyrimethanil
	Botrytis fruit rot	Bacillus subtilis
	Collar rot	fenamidone + fosetyl-aluminium (*moderate control from foliar spray*)
	Crown rot	dimethomorph (*moderate control*), fenamidone + fosetyl-aluminium (*moderate control from foliar spray*)
	Didymella stem rot	Gliocladium catenulatum (*moderate control*)
	Disease control	mancozeb + metalaxyl-M (*off-label*)
	Fusarium	Gliocladium catenulatum (*moderate control*)
	Leaf spot	chlorothalonil (*off-label*)
	Phytophthora	fenamidone + fosetyl-aluminium, Gliocladium catenulatum (*moderate control*)
	Powdery mildew	azoxystrobin (*off-label*), boscalid + pyraclostrobin, bupirimate, cyflufenamid (*off-label*), fenpropimorph (*off-label*), kresoxim-methyl, penconazole, potassium bicarbonate (commodity substance) (*off-label*), quinoxyfen (*off-label*), sulphur
	Pythium	fenamidone + fosetyl-aluminium, Gliocladium catenulatum (*moderate control*)
	Red core	fenamidone + fosetyl-aluminium
	Rhizoctonia	Gliocladium catenulatum (*moderate control*)
	Rust	azoxystrobin (*off-label*)
Pests	Aphids	pymetrozine (*off-label*)
	Birds/mammals	aluminium ammonium sulphate
	Capsid bugs	indoxacarb (*off-label*), thiacloprid (*off-label*)

	Caterpillars	Bacillus thuringiensis, indoxacarb (*off-label*)
	Flies	spinosad (*off-label*)
	Pests, miscellaneous	lambda-cyhalothrin (*off-label*)
	Spider mites	abamectin (*off-label*), spirodiclofen (*off-label*), tebufenpyrad
	Tarsonemid mites	abamectin (*off-label*)
	Thrips	Beauveria bassiana (*off-label*), deltamethrin (*off-label*)
	Weevils	Metarhizium anisopliae
	Whiteflies	Beauveria bassiana (*off-label*)
Plant growth regulation	Growth control	maleic hydrazide (*off-label*)
Weeds	Broad-leaved weeds	clopyralid (*off-label*), dimethenamid-p + pendimethalin (*off-label*), isoxaben, lenacil (*off-label*), metamitron (*off-label*), napropamide, pendimethalin, propyzamide
	Crops as weeds	cycloxydim, fluazifop-P-butyl, pendimethalin
	Grass weeds	cycloxydim, dimethenamid-p + pendimethalin (*off-label*), diquat (*around only*), fluazifop-P-butyl, metamitron (*off-label*), napropamide, pendimethalin, propyzamide, S-metolachlor (*off-label*)
	Weeds, miscellaneous	diquat (*around only*)

Other fruit - Protected miscellaneous fruit

Diseases	Alternaria	iprodione
	Black spot	azoxystrobin (*off-label*), boscalid + pyraclostrobin, cyprodinil + fludioxonil (*qualified minor use*)
	Botrytis	Bacillus subtilis (*reduction of damage to fruit*), boscalid + pyraclostrobin, cyprodinil + fludioxonil, fenpyrazamine, iprodione, iprodione (*off-label*), mepanipyrim, pyrimethanil, pyrimethanil (*moderate control*)
	Botrytis fruit rot	Bacillus subtilis (*reduction of damage to fruit*)
	Crown rot	dimethomorph (*moderate control*)
	Damping off	fosetyl-aluminium + propamocarb hydrochloride (*off-label*)
	Disease control	boscalid + pyraclostrobin (*off-label*), mancozeb + metalaxyl-M (*off-label*)
	Downy mildew	fosetyl-aluminium + propamocarb hydrochloride (*off-label*)
	Fusarium	Trichoderma asperellum (Strain T34) (*off-label*)
	Powdery mildew	Ampelomyces quisqualis (Strain AQ10), boscalid + pyraclostrobin, bupirimate, kresoxim-methyl, meptyldinocap, penconazole, penconazole (*off-label*), quinoxyfen (*off-label*)
	Pythium	Trichoderma asperellum (Strain T34) (*off-label*)
Pests	Aphids	acetamiprid (*off-label*), Beauveria bassiana, pymetrozine (*off-label*)
	Beetles	Beauveria bassiana
	Capsid bugs	thiacloprid (*off-label*)
	Caterpillars	Bacillus thuringiensis (*off-label*), deltamethrin (*off-label*), indoxacarb (*off-label*)
	Mites	bifenazate (*off-label*)

Crop/Pest Guide - Fruit and hops

	Pests, miscellaneous	indoxacarb (off-label), lambda-cyhalothrin (off-label), thiacloprid (off-label)
	Spider mites	abamectin (off-label), bifenazate, etoxazole (off-label), Lecanicillium muscarium (off-label), spirodiclofen (off-label)
	Tarsonemid mites	abamectin (off-label)
	Thrips	Beauveria bassiana (off-label), Lecanicillium muscarium (off-label)
	Whiteflies	Beauveria bassiana, Beauveria bassiana (off-label), Lecanicillium muscarium, Lecanicillium muscarium (off-label), pymetrozine (off-label)

Other fruit - Rhubarb

Diseases	Botrytis	Gliocladium catenulatum (off-label)
	Damping off	thiram (off-label - seed treatment)
	Downy mildew	mancozeb + metalaxyl-M (off-label)
	Fusarium	Gliocladium catenulatum (off-label)
	Phytophthora	Gliocladium catenulatum (off-label)
	Rhizoctonia	Gliocladium catenulatum (off-label)
Pests	Aphids	pirimicarb (off-label)
	Caterpillars	Bacillus thuringiensis (off-label)
	Cutworms	Bacillus thuringiensis (off-label)
Weeds	Broad-leaved weeds	clomazone (off-label), mesotrione (off-label), pendimethalin (off-label), propyzamide
	Grass weeds	pendimethalin (off-label), propyzamide
	Weeds, miscellaneous	glyphosate (off-label), mesotrione (off-label), metribuzin (off-label)

Tree fruit - All nuts

Diseases	Botrytis	Gliocladium catenulatum (off-label)
	Fusarium	Gliocladium catenulatum (off-label)
	Phytophthora	Gliocladium catenulatum (off-label)
	Powdery mildew	tebuconazole (off-label)
	Rhizoctonia	Gliocladium catenulatum (off-label)
	Stem canker	tebuconazole (off-label)
Pests	Aphids	lambda-cyhalothrin (off-label), pymetrozine (off-label)
	Caterpillars	Bacillus thuringiensis (off-label), diflubenzuron (off-label)
	Mites	tebufenpyrad (off-label), thiacloprid (off-label)
	Pests, miscellaneous	aluminium phosphide, lambda-cyhalothrin (off-label)
	Whiteflies	pymetrozine (off-label)
Weeds	Broad-leaved weeds	pendimethalin (off-label), propyzamide (off-label)
	Grass weeds	fluazifop-P-butyl (off-label), propyzamide (off-label)
	Weeds, miscellaneous	glyphosate (off-label)

Tree fruit - All pome fruit

Crop control	Sucker/shoot control	glyphosate
Diseases	Alternaria	cyprodinil + fludioxonil, fludioxonil (reduction)
	Botrytis	fludioxonil (reduction), Gliocladium catenulatum (off-label), iprodione (off-label), pyrimethanil (off-label)

Crop/Pest Guide - Fruit and hops

	Botrytis fruit rot	cyprodinil + fludioxonil
	Disease control	captan (*off-label*), dithianon + potassium phosphates, fludioxonil, mancozeb
	Fungus diseases	fludioxonil (*reduction*), mancozeb + metalaxyl-M (*off-label*), potassium bicarbonate (commodity substance)
	Fusarium	cyprodinil + fludioxonil, Gliocladium catenulatum (*off-label*)
	Penicillium rot	cyprodinil + fludioxonil, fludioxonil (*reduction*)
	Phytophthora	Gliocladium catenulatum (*off-label*)
	Powdery mildew	boscalid + pyraclostrobin, bupirimate, cyflufenamid, dithianon + pyraclostrobin, fenbuconazole (*reduction*), kresoxim-methyl (*reduction*), meptyldinocap (*off-label*), penconazole, penthiopyrad (*suppression*), sulphur, tebuconazole (*off-label*)
	Rhizoctonia	Gliocladium catenulatum (*off-label*)
	Scab	boscalid + pyraclostrobin, captan, difenoconazole, dithianon, dithianon + pyraclostrobin, fenbuconazole, kresoxim-methyl, mancozeb, penthiopyrad, pyrimethanil, pyrimethanil (*off-label*), sulphur
	Sooty moulds	potassium bicarbonate (commodity substance), potassium bicarbonate (commodity substance) (*reduction*)
	Stem canker	tebuconazole (*off-label*)
	Storage rots	captan, cyprodinil + fludioxonil
Pests	Aphids	acetamiprid, deltamethrin, flonicamid, lambda-cyhalothrin, thiacloprid
	Capsid bugs	deltamethrin
	Caterpillars	Bacillus thuringiensis, Bacillus thuringiensis (*off-label*), chlorantraniliprole, Cydia pomonella GV, deltamethrin, diflubenzuron, diflubenzuron (*off-label*), dodeca-8,10-dienyl acetate, fenoxycarb, indoxacarb, lambda-cyhalothrin, methoxyfenozide, spinosad
	Gall, rust and leaf & bud mites	diflubenzuron, spirodiclofen
	Midges	thiacloprid (*off-label*)
	Pests, miscellaneous	potassium bicarbonate (commodity substance), thiacloprid (*off-label*)
	Sawflies	deltamethrin
	Scale insects	spirodiclofen
	Spider mites	clofentezine, spirodiclofen, tebufenpyrad
	Suckers	deltamethrin, diflubenzuron, lambda-cyhalothrin, potassium bicarbonate (commodity substance) (*reduction*), spirodiclofen
	Whiteflies	acetamiprid
Plant growth regulation	Fruiting control	1-methylcyclopropene (*off-label - post harvest only*), 1-methylcyclopropene (*post-harvest use*), benzyladenine, gibberellins, metamitron
	Growth control	2-chloroethylphosphonic acid (*off-label*), benzyladenine, gibberellins, gibberellins (*off-label*), metamitron, prohexadione-calcium, prohexadione-calcium (*off-label*)
	Quality/yield control	gibberellins

Weeds	Broad-leaved weeds	2,4-D, clopyralid (*off-label*), pendimethalin, pendimethalin (*off-label*), propyzamide, propyzamide (*off-label*)
	Crops as weeds	pendimethalin
	Grass weeds	fluazifop-P-butyl (*off-label*), pendimethalin, propyzamide, propyzamide (*off-label*)
	Weeds, miscellaneous	2,4-D + glyphosate, glyphosate, glyphosate (*off-label*)

Tree fruit - All stone fruit

Crop control	Sucker/shoot control	glyphosate
Diseases	Blossom wilt	boscalid + pyraclostrobin (*off-label*), cyprodinil + fludioxonil (*off-label*), fenbuconazole (*off-label*)
	Botrytis	cyprodinil + fludioxonil (*off-label*), fenhexamid (*off-label*), Gliocladium catenulatum (*off-label*)
	Fusarium	Gliocladium catenulatum (*off-label*)
	Phytophthora	Gliocladium catenulatum (*off-label*)
	Rhizoctonia	Gliocladium catenulatum (*off-label*)
	Sclerotinia	boscalid + pyraclostrobin (*off-label*), fenbuconazole (*off-label*)
Pests	Aphids	acetamiprid, pymetrozine (*off-label*), tebufenpyrad (*off-label*), thiacloprid (*off-label*)
	Capsid bugs	indoxacarb (*off-label*)
	Caterpillars	Bacillus thuringiensis (*off-label*), diflubenzuron, indoxacarb (*off-label*), methoxyfenozide (*off-label*)
	Flies	deltamethrin (*off-label*), spinosad (*off-label*)
	Gall, rust and leaf & bud mites	diflubenzuron, spirodiclofen (*off-label*)
	Pests, miscellaneous	lambda-cyhalothrin (*off-label*), thiacloprid (*off-label - under temporary protective rain covers*)
	Spider mites	spirodiclofen (*off-label*)
	Whiteflies	acetamiprid, pymetrozine (*off-label*)
Weeds	Broad-leaved weeds	pendimethalin, propyzamide, propyzamide (*off-label*)
	Crops as weeds	pendimethalin
	Grass weeds	fluazifop-P-butyl (*off-label*), pendimethalin, propyzamide, propyzamide (*off-label*)
	Weeds, miscellaneous	glyphosate, glyphosate (*off-label*)

Tree fruit - Miscellaneous top fruit

Diseases	Botrytis	Gliocladium catenulatum (*off-label*)
	Fusarium	Gliocladium catenulatum (*off-label*)
	Phytophthora	Gliocladium catenulatum (*off-label*)
	Rhizoctonia	Gliocladium catenulatum (*off-label*)
Pests	Birds/mammals	aluminium ammonium sulphate

Tree fruit - Protected fruit

Diseases	Powdery mildew	Ampelomyces quisqualis (Strain AQ10) (*off-label*), penconazole (*off-label*)
Pests	Aphids	Beauveria bassiana, pymetrozine (*off-label*)
	Beetles	Beauveria bassiana

Crop/Pest Guide - Grain/crop store uses

	Caterpillars	Bacillus thuringiensis (*off-label*)
	Spider mites	Lecanicillium muscarium (*off-label*)
	Thrips	Lecanicillium muscarium (*off-label*)
	Whiteflies	Beauveria bassiana, Lecanicillium muscarium (*off-label*), pymetrozine (*off-label*)

Grain/crop store uses

Stored produce - Food/produce storage

Pests	Pests, miscellaneous	aluminium phosphide, magnesium phosphide

Stored seed - Stored grain/rape/linseed

Pests	Birds/mammals	aluminium ammonium sulphate
	Food/grain storage pests	chlorpyrifos-methyl, d-phenothrin + tetramethrin, pirimiphos-methyl
	Mites	chlorpyrifos-methyl, physical pest control, pirimiphos-methyl
	Pests, miscellaneous	aluminium phosphide, chlorpyrifos-methyl, deltamethrin, magnesium phosphide, pirimiphos-methyl

Grass

Turf/amenity grass - Amenity grassland

Diseases	Anthracnose	azoxystrobin, azoxystrobin + propiconazole (*moderate control*), chlorothalonil + fludioxonil + propiconazole (*reduction*), fludioxonil (*reduction*), propiconazole (*qualified minor use*), tebuconazole + trifloxystrobin
	Brown patch	azoxystrobin, chlorothalonil + fludioxonil + propiconazole, iprodione, propiconazole (*qualified minor use*)
	Crown rust	azoxystrobin
	Dollar spot	azoxystrobin + propiconazole, chlorothalonil + fludioxonil + propiconazole, iprodione, propiconazole, tebuconazole + trifloxystrobin
	Drechslera leaf spot	fludioxonil (*useful levels of control*)
	Fairy rings	azoxystrobin
	Fusarium	azoxystrobin, azoxystrobin + propiconazole, chlorothalonil + fludioxonil + propiconazole, fludioxonil, iprodione, propiconazole, tebuconazole + trifloxystrobin, trifloxystrobin
	Melting out	azoxystrobin, iprodione, tebuconazole + trifloxystrobin
	Red thread	iprodione, tebuconazole + trifloxystrobin, trifloxystrobin
	Rust	tebuconazole + trifloxystrobin
	Seed-borne diseases	azoxystrobin + propiconazole, fludioxonil
	Snow mould	iprodione

Crop/Pest Guide - Grass

	Take-all patch	azoxystrobin
Pests	Birds/mammals	aluminium ammonium sulphate, aluminium ammonium sulphate (*anti-fouling*)
	Slugs/snails	metaldehyde
Plant growth regulation	Growth control	maleic hydrazide (*off-label*), trinexapac-ethyl
Weeds	Broad-leaved weeds	2,4-D, 2,4-D + dicamba, 2,4-D + dicamba + fluroxypyr, 2,4-D + florasulam, aminopyralid + fluroxypyr, aminopyralid + triclopyr, citronella oil, clopyralid + florasulam + fluroxypyr, clopyralid + triclopyr, dicamba + MCPA + mecoprop-P, dicamba + MCPA + mecoprop-P (*moderately susceptible*), dicamba + mecoprop-P, ethofumesate, florasulam + fluroxypyr, fluroxypyr + triclopyr, mecoprop-P
	Crops as weeds	2,4-D + dicamba + fluroxypyr, 2,4-D + florasulam, dicamba + MCPA + mecoprop-P (*moderately susceptible*), florasulam + fluroxypyr, pinoxaden (*reduction only*)
	Grass weeds	ethofumesate, pinoxaden (*off-label*)
	Mosses	carfentrazone-ethyl + mecoprop-P, ferrous sulphate
	Weeds, miscellaneous	2,4-D + glyphosate, aminopyralid + triclopyr, carfentrazone-ethyl + mecoprop-P, clopyralid + florasulam + fluroxypyr, dicamba + MCPA + mecoprop-P, florasulam + fluroxypyr, glyphosate
	Woody weeds/scrub	aminopyralid + triclopyr, clopyralid + triclopyr

Turf/amenity grass - Managed amenity turf

Diseases	Anthracnose	azoxystrobin, azoxystrobin + propiconazole (*moderate control*), chlorothalonil + fludioxonil + propiconazole (*reduction*), fludioxonil (*reduction*), iprodione, iprodione + trifloxystrobin (*moderate control*), propiconazole (*qualified minor use*), tebuconazole + trifloxystrobin
	Brown patch	azoxystrobin, chlorothalonil + fludioxonil + propiconazole, iprodione, propiconazole (*qualified minor use*)
	Crown rust	azoxystrobin
	Disease control	carbendazim, iprodione, iprodione + trifloxystrobin
	Dollar spot	azoxystrobin + propiconazole, chlorothalonil + fludioxonil + propiconazole, iprodione, iprodione + trifloxystrobin, propiconazole, pyraclostrobin (*useful reduction*), tebuconazole + trifloxystrobin
	Drechslera leaf spot	fludioxonil (*useful levels of control*)
	Fairy rings	azoxystrobin
	Fusarium	azoxystrobin, azoxystrobin + propiconazole, chlorothalonil + fludioxonil + propiconazole, fludioxonil, iprodione, iprodione + trifloxystrobin, propiconazole, pyraclostrobin (*moderate control*), pyraclostrobin (*moderate control only*), tebuconazole + trifloxystrobin, trifloxystrobin
	Leaf spot	iprodione + trifloxystrobin
	Melting out	azoxystrobin, iprodione, tebuconazole + trifloxystrobin
	Red thread	iprodione, iprodione + trifloxystrobin, pyraclostrobin, tebuconazole + trifloxystrobin, trifloxystrobin

Crop/Pest Guide - Non-crop pest control

	Rust	iprodione, iprodione + trifloxystrobin, tebuconazole + trifloxystrobin
	Seed-borne diseases	azoxystrobin + propiconazole, fludioxonil
	Snow mould	iprodione
	Take-all patch	azoxystrobin
Pests	Birds/mammals	aluminium ammonium sulphate, aluminium ammonium sulphate (*anti-fouling*)
	Earthworms	carbendazim
	Flies	esfenvalerate
	Slugs/snails	ferric phosphate, metaldehyde
Plant growth regulation	Growth control	trinexapac-ethyl
Weeds	Broad-leaved weeds	2,4-D, 2,4-D + dicamba, 2,4-D + dicamba + fluroxypyr, 2,4-D + dicamba + MCPA + mecoprop-P, 2,4-D + florasulam, clopyralid + 2,4-D + MCPA, clopyralid + florasulam + fluroxypyr, clopyralid + fluroxypyr + MCPA, dicamba + MCPA + mecoprop-P, dicamba + MCPA + mecoprop-P (*moderately susceptible*), dicamba + mecoprop-P, dichlorprop-P + ferrous sulphate + MCPA, ferrous sulphate + MCPA + mecoprop-P, florasulam + fluroxypyr, MCPA + mecoprop-P, mecoprop-P
	Crops as weeds	2,4-D + dicamba + fluroxypyr, 2,4-D + dicamba + MCPA + mecoprop-P, 2,4-D + florasulam, clopyralid + 2,4-D + MCPA, dicamba + MCPA + mecoprop-P (*moderately susceptible*), florasulam + fluroxypyr, pinoxaden (*reduction only*)
	Grass weeds	pinoxaden (*off-label*)
	Mosses	carfentrazone-ethyl + mecoprop-P, dichlorprop-P + ferrous sulphate + MCPA, ferrous sulphate, ferrous sulphate + MCPA + mecoprop-P, quinoclamine (*moderate control*)
	Weeds, miscellaneous	carfentrazone-ethyl + mecoprop-P, clopyralid + florasulam + fluroxypyr, dicamba + MCPA + mecoprop-P, florasulam + fluroxypyr, glufosinate-ammonium, glyphosate (*pre-establishment*), glyphosate (*pre-establishment only*)

Non-crop pest control

Farm buildings/yards - Farm buildings

Pests	Ants	cypermethrin + tetramethrin
	Bedbugs	cypermethrin + tetramethrin
	Birds/mammals	aluminium ammonium sulphate, bromadiolone, difenacoum, flocoumafen
	Caterpillars	cypermethrin, cypermethrin + tetramethrin
	Cockroaches	cypermethrin + tetramethrin
	Flies	cypermethrin, cypermethrin + tetramethrin, diflubenzuron, d-phenothrin + tetramethrin
	Pests, miscellaneous	physical pest control, pyrethrins
	Silverfish	cypermethrin + tetramethrin

Crop/Pest Guide - Protected salad and vegetable crops

| | Wasps | cypermethrin, cypermethrin + tetramethrin, d-phenothrin + tetramethrin |

Farm buildings/yards - Farmyards
| Pests | Birds/mammals | bromadiolone, difenacoum |

Farmland pest control - Farmland situations
| Crop control | Desiccation | diquat (off-label) |
| Pests | Birds/mammals | aluminium phosphide |

Miscellaneous non-crop pest control - Manure/rubbish
Pests	Ants	cypermethrin + tetramethrin
	Bedbugs	cypermethrin + tetramethrin
	Caterpillars	cypermethrin + tetramethrin
	Cockroaches	cypermethrin + tetramethrin
	Flies	cypermethrin + tetramethrin, diflubenzuron
	Silverfish	cypermethrin + tetramethrin
	Wasps	cypermethrin + tetramethrin

Miscellaneous non-crop pest control - Miscellaneous pest control situations
| Pests | Birds/mammals | carbon dioxide (commodity substance) |

Protected salad and vegetable crops

Protected brassicas - Protected brassica vegetables
Diseases	Black leg	fludioxonil (off-label)
	Damping off	fosetyl-aluminium + propamocarb hydrochloride, propamocarb hydrochloride
	Downy mildew	fluopicolide + propamocarb hydrochloride (off-label), fosetyl-aluminium + propamocarb hydrochloride, propamocarb hydrochloride
	Fungus diseases	propamocarb hydrochloride
	Fusarium	Trichoderma asperellum (Strain T34) (off-label)
	Phytophthora	propamocarb hydrochloride
	Pythium	propamocarb hydrochloride, Trichoderma asperellum (Strain T34) (off-label)
	Rhizoctonia	azoxystrobin (off-label)
	Root rot	propamocarb hydrochloride
	White blister	boscalid + pyraclostrobin (off-label)
Pests	Aphids	acetamiprid (off-label), pymetrozine (off-label), pyrethrins, spirotetramat (off-label)
	Beetles	pyrethrins
	Caterpillars	Bacillus thuringiensis (off-label), indoxacarb (off-label), pyrethrins
	Flies	chlorpyrifos, chlorpyrifos (off-label)
	Mites	abamectin (off-label)
	Pests, miscellaneous	chlorpyrifos
	Whiteflies	pyrethrins, spirotetramat (off-label)

Protected brassicas - Protected salad brassicas

Diseases	Damping off	fosetyl-aluminium + propamocarb hydrochloride
	Downy mildew	fosetyl-aluminium + propamocarb hydrochloride
	Fusarium	Trichoderma asperellum (Strain T34) (off-label)
	Pythium	Trichoderma asperellum (Strain T34) (off-label)
Pests	Aphids	lambda-cyhalothrin (off-label), pymetrozine (off-label)
	Beetles	lambda-cyhalothrin (off-label)
	Caterpillars	Bacillus thuringiensis (off-label), lambda-cyhalothrin (off-label)
	Flies	chlorpyrifos (off-label)
	Pests, miscellaneous	lambda-cyhalothrin (off-label)

Protected crops, general - All protected crops

Pests	Aphids	Beauveria bassiana, pyrethrins
	Beetles	Beauveria bassiana, pyrethrins
	Caterpillars	Bacillus thuringiensis (off-label), indoxacarb (off-label), pyrethrins
	Mealybugs	pyrethrins
	Pests, miscellaneous	lambda-cyhalothrin (off-label)
	Scale insects	pyrethrins
	Spider mites	pyrethrins
	Thrips	pyrethrins
	Whiteflies	Beauveria bassiana, pyrethrins

Protected crops, general - Protected nuts

Pests	Aphids	pymetrozine (off-label)
	Whiteflies	pymetrozine (off-label)

Protected crops, general - Protected vegetables, general

Diseases	Powdery mildew	Ampelomyces quisqualis (Strain AQ10) (off-label)
Pests	Aphids	pymetrozine (off-label)
	Spider mites	Lecanicillium muscarium (off-label)
	Thrips	Lecanicillium muscarium (off-label)
	Whiteflies	Lecanicillium muscarium (off-label), pymetrozine (off-label)

Protected crops, general - Soils and compost

Diseases	Soil-borne diseases	dazomet
Pests	Soil pests	dazomet

Protected fruiting vegetables - Protected aubergines

Diseases	Botrytis	boscalid + pyraclostrobin (off-label), cyprodinil + fludioxonil (off-label), fenhexamid (off-label), fenpyrazamine, iprodione (off-label), pyrimethanil (off-label)
	Damping off	fosetyl-aluminium + propamocarb hydrochloride (off-label)
	Disease control	isopyrazam
	Powdery mildew	Ampelomyces quisqualis (Strain AQ10), proquinazid (off-label), sulphur (off-label)

Crop/Pest Guide - Protected salad and vegetable crops

Pests	Aphids	acetamiprid, Beauveria bassiana, lambda-cyhalothrin (off-label), pymetrozine (off-label)
	Beetles	Beauveria bassiana
	Caterpillars	Bacillus thuringiensis, indoxacarb
	Leaf miners	deltamethrin (off-label), thiacloprid (off-label)
	Mealybugs	flonicamid (off-label)
	Mites	etoxazole
	Pests, miscellaneous	lambda-cyhalothrin (off-label)
	Thrips	deltamethrin (off-label), spinosad, thiacloprid (off-label)
	Whiteflies	acetamiprid, Beauveria bassiana, flonicamid (off-label), pymetrozine (off-label), thiacloprid (off-label)

Protected fruiting vegetables - Protected cucurbits

Diseases	Botrytis	boscalid + pyraclostrobin (off-label), cyprodinil + fludioxonil (off-label), fenhexamid (off-label), fenpyrazamine
	Cladosporium	boscalid + pyraclostrobin (off-label)
	Damping off	fosetyl-aluminium + propamocarb hydrochloride, fosetyl-aluminium + propamocarb hydrochloride (off-label)
	Disease control	isopyrazam
	Downy mildew	azoxystrobin (off-label), fosetyl-aluminium + propamocarb hydrochloride, metalaxyl-M (off-label)
	Fungus diseases	boscalid + pyraclostrobin (off-label)
	Fusarium	cyprodinil + fludioxonil (off-label), Trichoderma asperellum (Strain T34) (off-label)
	Powdery mildew	Ampelomyces quisqualis (Strain AQ10), Ampelomyces quisqualis (Strain AQ10) (off-label), azoxystrobin (off-label), boscalid + pyraclostrobin (off-label), bupirimate, penconazole, penconazole (off-label), potassium bicarbonate (commodity substance) (off-label), proquinazid (off-label), sulphur (off-label)
	Pythium	fosetyl-aluminium + propamocarb hydrochloride, Trichoderma asperellum (Strain T34) (off-label)
Pests	Aphids	acetamiprid, Beauveria bassiana, deltamethrin, flonicamid (off-label), lambda-cyhalothrin (off-label), pirimicarb (off-label), pymetrozine, pymetrozine (off-label), pyrethrins (off-label)
	Beetles	Beauveria bassiana
	Caterpillars	Bacillus thuringiensis, Bacillus thuringiensis (off-label), deltamethrin, indoxacarb
	Cutworms	Bacillus thuringiensis (off-label)
	Leaf miners	deltamethrin (off-label), thiacloprid (off-label)
	Mealybugs	deltamethrin
	Pests, miscellaneous	lambda-cyhalothrin (off-label)
	Scale insects	deltamethrin
	Spider mites	abamectin, Lecanicillium muscarium (off-label), spirodiclofen (off-label)
	Thrips	abamectin, deltamethrin, deltamethrin (off-label), Lecanicillium muscarium (off-label), spinosad, thiacloprid (off-label)

Crop/Pest Guide - Protected salad and vegetable crops

	Whiteflies	acetamiprid, Beauveria bassiana, deltamethrin, flonicamid (*off-label*), Lecanicillium muscarium, Lecanicillium muscarium (*off-label*), pymetrozine (*off-label*), thiacloprid (*off-label*)

Protected fruiting vegetables - Protected tomatoes

Diseases	Alternaria	iprodione
	Botrytis	azoxystrobin (*off-label*), boscalid + pyraclostrobin (*off-label*), cyprodinil + fludioxonil (*off-label*), fenhexamid (*off-label*), fenpyrazamine, iprodione, pyrimethanil (*off-label*)
	Damping off	propamocarb hydrochloride (*off-label*)
	Didymella stem rot	azoxystrobin (*off-label*)
	Disease control	isopyrazam
	Phytophthora	azoxystrobin (*off-label*), propamocarb hydrochloride (*off-label*)
	Powdery mildew	Ampelomyces quisqualis (Strain AQ10), azoxystrobin (*off-label*), bupirimate (*off-label*), cyflufenamid (*off-label*), penconazole (*off-label*), proquinazid (*off-label*), sulphur (*off-label*)
	Pythium	fosetyl-aluminium + propamocarb hydrochloride
	Root rot	propamocarb hydrochloride (*off-label*)
	Verticillium wilt	thiophanate-methyl (*off-label*)
Pests	Aphids	acetamiprid, Beauveria bassiana, deltamethrin, lambda-cyhalothrin (*off-label*), pymetrozine (*off-label*), pyrethrins
	Beetles	Beauveria bassiana, pyrethrins
	Caterpillars	Bacillus thuringiensis, chlorantraniliprole (*off-label*), deltamethrin, indoxacarb, pyrethrins
	Leaf miners	abamectin, spinosad (*off-label*), thiacloprid (*off-label*)
	Mealybugs	deltamethrin, flonicamid (*off-label*), pyrethrins, pyrethrins (*off-label*)
	Mirid bugs	pyrethrins, pyrethrins (*off-label*)
	Mites	etoxazole
	Pests, miscellaneous	lambda-cyhalothrin (*off-label*), spinosad (*off-label*)
	Scale insects	deltamethrin
	Spider mites	abamectin, spirodiclofen (*off-label*)
	Thrips	deltamethrin, spinosad, spinosad (*off-label*), thiacloprid (*off-label*)
	Whiteflies	acetamiprid, Beauveria bassiana, deltamethrin, flonicamid (*off-label*), Lecanicillium muscarium, pymetrozine (*off-label*), pyrethrins, thiacloprid (*off-label*)
Plant growth regulation	Growth control	2-chloroethylphosphonic acid (*off-label*)

Protected herb crops - Protected herbs

Diseases	Alternaria	cyprodinil + fludioxonil (*off-label*), fludioxonil (*off-label*), iprodione (*off-label*)
	Botrytis	cyprodinil + fludioxonil (*off-label*), fenhexamid (*off-label*), iprodione (*off-label*), pyrimethanil (*off-label*)
	Damping off	fosetyl-aluminium + propamocarb hydrochloride (*off-label*)

Crop/Pest Guide - Protected salad and vegetable crops

	Disease control	isopyrazam, mandipropamid
	Downy mildew	dimethomorph (off-label), mancozeb + metalaxyl-M (off-label), mandipropamid, metalaxyl-M (off-label)
	Fusarium	Trichoderma asperellum (Strain T34) (off-label)
	Leaf spot	fludioxonil (off-label)
	Powdery mildew	Ampelomyces quisqualis (Strain AQ10), Ampelomyces quisqualis (Strain AQ10) (off-label), penconazole (off-label), sulphur (off-label)
	Pythium	Trichoderma asperellum (Strain T34) (off-label)
	Rhizoctonia	azoxystrobin (off-label), cyprodinil + fludioxonil (off-label)
	Stem canker	cyprodinil + fludioxonil (off-label)
	White blister	boscalid + pyraclostrobin (off-label)
Pests	Aphids	acetamiprid (off-label), Beauveria bassiana, pymetrozine (off-label), pyrethrins (off-label), spirotetramat (off-label), thiacloprid (off-label)
	Beetles	Beauveria bassiana
	Caterpillars	Bacillus thuringiensis, Bacillus thuringiensis (off-label), indoxacarb (off-label)
	Cutworms	Bacillus thuringiensis (off-label)
	Mites	abamectin (off-label)
	Pests, miscellaneous	spinosad (off-label)
	Spider mites	Lecanicillium muscarium (off-label), spirodiclofen (off-label)
	Thrips	Lecanicillium muscarium (off-label)
	Whiteflies	Beauveria bassiana, Lecanicillium muscarium (off-label), pymetrozine (off-label), spirotetramat (off-label)
Weeds	Broad-leaved weeds	propyzamide (off-label)
	Grass weeds	propyzamide (off-label)

Protected leafy vegetables - Mustard and cress

Diseases	Alternaria	cyprodinil + fludioxonil (off-label)
	Botrytis	cyprodinil + fludioxonil (off-label), Gliocladium catenulatum (off-label)
	Damping off	thiram (off-label)
	Disease control	dimethomorph + mancozeb (off-label), mandipropamid
	Downy mildew	dimethomorph + mancozeb (off-label), fenamidone + fosetyl-aluminium (off-label), fluopicolide + propamocarb hydrochloride (off-label), mandipropamid
	Fusarium	Gliocladium catenulatum (off-label)
	Phytophthora	Gliocladium catenulatum (off-label)
	Rhizoctonia	cyprodinil + fludioxonil (off-label), Gliocladium catenulatum (off-label)
	Stem canker	cyprodinil + fludioxonil (off-label)
Pests	Aphids	acetamiprid (off-label), deltamethrin (off-label), spirotetramat (off-label)
	Beetles	deltamethrin (off-label)
	Caterpillars	deltamethrin (off-label)
	Pests, miscellaneous	deltamethrin (off-label), spinosad (off-label)

	Whiteflies	spirotetramat (off-label)
Weeds	Broad-leaved weeds	pendimethalin (off-label), propyzamide (off-label)
	Grass weeds	propyzamide (off-label)

Protected leafy vegetables - Protected leafy vegetables

Diseases	Alternaria	cyprodinil + fludioxonil (off-label), iprodione
	Botrytis	boscalid + pyraclostrobin, boscalid + pyraclostrobin (off-label), cyprodinil + fludioxonil (off-label), iprodione, pyrimethanil (off-label)
	Bottom rot	boscalid + pyraclostrobin, tolclofos-methyl
	Disease control	mandipropamid
	Downy mildew	boscalid + pyraclostrobin (off-label), dimethomorph (off-label), fosetyl-aluminium + propamocarb hydrochloride, mandipropamid, potassium bicarbonate (commodity substance) (off-label)
	Pythium	fosetyl-aluminium + propamocarb hydrochloride
	Rhizoctonia	azoxystrobin (off-label), boscalid + pyraclostrobin (off-label), cyprodinil + fludioxonil (off-label)
	Sclerotinia	boscalid + pyraclostrobin (off-label)
	Soft rot	boscalid + pyraclostrobin
	Stem canker	cyprodinil + fludioxonil (off-label)
Pests	Aphids	acetamiprid (off-label), lambda-cyhalothrin (off-label), pymetrozine (off-label), pyrethrins, spirotetramat (off-label), thiacloprid (off-label)
	Beetles	lambda-cyhalothrin (off-label)
	Caterpillars	Bacillus thuringiensis (off-label), lambda-cyhalothrin (off-label), pyrethrins
	Cutworms	Bacillus thuringiensis (off-label)
	Mites	abamectin (off-label)
	Pests, miscellaneous	lambda-cyhalothrin (off-label)
	Whiteflies	spirotetramat (off-label)
Weeds	Broad-leaved weeds	propyzamide (off-label)
	Grass weeds	propyzamide (off-label)

Protected leafy vegetables - Protected spinach

Diseases	Alternaria	cyprodinil + fludioxonil (off-label)
	Botrytis	boscalid + pyraclostrobin (off-label), cyprodinil + fludioxonil (off-label)
	Disease control	mandipropamid
	Downy mildew	mandipropamid, metalaxyl-M (off-label)
	Rhizoctonia	boscalid + pyraclostrobin (off-label), cyprodinil + fludioxonil (off-label)
	Sclerotinia	boscalid + pyraclostrobin (off-label)
	Stem canker	cyprodinil + fludioxonil (off-label)
Pests	Aphids	acetamiprid (off-label), spirotetramat (off-label)
	Caterpillars	Bacillus thuringiensis (off-label)
	Whiteflies	spirotetramat (off-label)

Protected legumes - Protected peas and beans

Diseases	Damping off	fosetyl-aluminium + propamocarb hydrochloride (off-label)

Pests	Caterpillars	Bacillus thuringiensis, Bacillus thuringiensis (*off-label*)
	Cutworms	Bacillus thuringiensis (*off-label*)

Protected root and tuber vegetables - Protected carrots/parsnips/celeriac

Diseases	Alternaria	fludioxonil (*off-label*)
	Leaf spot	fludioxonil (*off-label*)

Protected root and tuber vegetables - Protected root brassicas

Diseases	Alternaria	cyprodinil + fludioxonil (*off-label*)
	Black leg	fludioxonil (*off-label*)
	Damping off	fosetyl-aluminium + propamocarb hydrochloride
	Disease control	boscalid + pyraclostrobin (*off-label*)
	Downy mildew	fosetyl-aluminium + propamocarb hydrochloride
	Leaf spot	cyprodinil + fludioxonil (*off-label*)
	Rhizoctonia	tolclofos-methyl (*off-label*)
	Sclerotinia	cyprodinil + fludioxonil (*off-label*)

Protected stem and bulb vegetables - Protected celery/chicory

Diseases	Alternaria	azoxystrobin (*off-label*)
	Botrytis	azoxystrobin (*off-label*)
	Phytophthora	azoxystrobin (*off-label*), azoxystrobin (*off-label - for forcing*)
	Powdery mildew	azoxystrobin (*off-label*)
	Rhizoctonia	azoxystrobin (*off-label*), tolclofos-methyl (*off-label*)
	Rust	azoxystrobin (*off-label*)
	Sclerotinia	azoxystrobin (*off-label*)
Pests	Aphids	pymetrozine (*off-label*)
	Caterpillars	Bacillus thuringiensis (*off-label*), deltamethrin (*off-label*)
	Cutworms	Bacillus thuringiensis (*off-label*)
	Mites	abamectin (*off-label*)

Protected stem and bulb vegetables - Protected onions/leeks/garlic

Pests	Leaf miners	deltamethrin (*off-label*)
	Pests, miscellaneous	deltamethrin (*off-label*)

Total vegetation control

Aquatic situations, general - Aquatic situations

Pests	Flies	physical pest control
Weeds	Aquatic weeds	glyphosate
	Grass weeds	glyphosate
	Weeds, miscellaneous	glyphosate

Non-crop areas, general - Miscellaneous non-crop situations

Pests	Birds/mammals	difenacoum, flocoumafen
Weeds	Weeds, miscellaneous	2,4-D + glyphosate, flazasulfuron, glyphosate

Non-crop areas, general - Non-crop farm areas

Weeds	Broad-leaved weeds	metsulfuron-methyl, tribenuron-methyl
	Crops as weeds	metsulfuron-methyl
	Weeds, miscellaneous	2,4-D + glyphosate, glyphosate, glyphosate + pyraflufen-ethyl, metsulfuron-methyl

Non-crop areas, general - Paths/roads etc

Pests	Birds/mammals	aluminium ammonium sulphate, aluminium ammonium sulphate (*anti-fouling*)
	Slugs/snails	metaldehyde
Weeds	Grass weeds	glyphosate
	Mosses	maleic hydrazide + pelargonic acid, starch, protein, oil and water
	Weeds, miscellaneous	acetic acid, diflufenican + glyphosate, flazasulfuron, glufosinate-ammonium, glyphosate, glyphosate + pyraflufen-ethyl, glyphosate + sulfosulfuron, maleic hydrazide + pelargonic acid, starch, protein, oil and water

SECTION 2
PESTICIDE PROFILES

1 abamectin

A selective acaricide and insecticide for use in ornamentals and other protected crops
IRAC mode of action code: 6

Products

1	Clayton Abba	Clayton	18 g/l	EC	13808
2	Killermite	Pan Agriculture	18 g/l	EC	16453
3	Smite	Pan Agriculture	18 g/l	EC	16255

Uses
- Leaf miner in **protected tomatoes** [1-3]
- Mites in **protected blackberries** (off-label), **protected cress** (off-label), **protected frise** (off-label), **protected herbs (see appendix 6)** (off-label), **protected lamb's lettuce** (off-label), **protected leaf brassicas** (off-label), **protected lettuce** (off-label), **protected radicchio** (off-label), **protected raspberries** (off-label), **protected scarole** (off-label) [1]
- Red spider mites in **protected strawberries** (off-label), **strawberries** (off-label) [1]
- Tarsonemid mites in **protected strawberries** (off-label), **strawberries** (off-label) [1]
- Two-spotted spider mite in **ornamental plant production**, **protected cucumbers**, **protected tomatoes** [1-3]; **protected ornamentals** [1]
- Western flower thrips in **ornamental plant production**, **protected cucumbers** [1-3]; **protected ornamentals** [1]

Extension of Authorisation for Minor Use (EAMUs)
- *protected blackberries* 20101940 [1]
- *protected cress* 20101939 [1]
- *protected frise* 20101939 [1]
- *protected herbs (see appendix 6)* 20101939 [1]
- *protected lamb's lettuce* 20101939 [1]
- *protected leaf brassicas* 20101939 [1]
- *protected lettuce* 20101939 [1]
- *protected radicchio* 20101939 [1]
- *protected raspberries* 20101940 [1]
- *protected scarole* 20101939 [1]
- *protected strawberries* 20090342 [1]
- *strawberries* 20090342 [1]

Approval information
- Abamectin included in Annex I under EC Regulation 1107/2009

Efficacy guidance
- Abamectin controls adults and immature stages of the two-spotted spider mite, the larval stages of leaf miners and the nymphs of Western Flower Thrips, plus a useful reduction in adults
- Treat at first sign of infestation. Repeat sprays may be required
- For effective control total cover of all plant surfaces is essential, but avoid run-off
- Target pests quickly become immobilised but 3-5 d may be required for maximum mortality
- Indoor applications should be made through a hydraulic nozzle applicator or a knapsack applicator. Outdoors suitable high volume hydraulic nozzle applicators should be used
- Limited data shows abamectin only slightly harmful to Anthocorid bugs and so compatible with biological control systems in which Anthocorid bugs are important

Restrictions
- Number of treatments 6 on protected tomatoes and cucumbers (only 4 of which can be made when flowers or fruit present); not restricted on flowers but rotation with other products advised
- Maximum concentration must not exceed 50 ml per 100 l water
- Do not mix with wetters, stickers or other adjuvants
- Do not use on ferns (*Adiantum* spp) or Shasta daisies
- Do not treat protected crops which are in flower or have set fruit between 1 Nov and 28 Feb
- For use on all varieties of pears
- Do not treat cherry tomatoes (but protected cherry tomatoes may be treated off-label)
- Consult manufacturer for list of plant varieties tested for safety

SEE SECTION 3 FOR PRODUCTS ALSO REGISTERED

Pesticide Profiles

- There is insufficient evidence to support product compatibility with integrated and biological pest control programmes
- Unprotected persons must be kept out of treated areas until the spray has dried

Crop-specific information
- HI 3 d for protected edible crops
- On tomato or cucumber crops that are in flower, or have started to set fruit, treat only between 1 Mar and 31 Oct. Seedling tomatoes or cucumbers that have not started to flower or set fruit may be treated at any time
- Some spotting or staining may occur on carnation, kalanchoe and begonia foliage

Environmental safety
- Dangerous for the environment
- Very toxic to aquatic organisms
- High risk to bees. Do not apply to crops in flower or to those in which bees are actively foraging. Do not apply when flowering weeds are present
- Keep in original container, tightly closed, in a safe place, under lock and key
- Where bumble bees are used in tomatoes as pollinators, keep them out for 24 h after treatment

Hazard classification and safety precautions
Hazard Harmful, Dangerous for the environment
Transport code 9
Packaging group III
UN Number 3082
Risk phrases R22a, R50, R53a
Operator protection A, C, D, H, K, M; U02a, U05a, U20a
Environmental protection E02a (until spray has dried); E12a, E12e, E15b, E34, E38
Storage and disposal D01, D02, D05, D09b, D10c, D12a
Medical advice M04a

2 acetamiprid

A neonicotinoid insecticide for use in top fruit and horticulture
IRAC mode of action code: 4A

Products

1	Gazelle SG	Certis	20% w/w	SG	13725
2	Insyst	Certis	20% w/w	SP	13414
3	Sangue	Agroquimicos	20% w/w	SG	16042
4	Vulcan	Pan Agriculture	20% w/w	SG	16689

Uses
- Aphids in *apples, cherries, ornamental plant production, pears, plums, protected aubergines, protected ornamentals, protected peppers, protected tomatoes* [1, 3, 4]; *brussels sprouts* (off-label), *seed potatoes, spring oilseed rape, ware potatoes, winter oilseed rape* [2]; *chard* (off-label), *cress* (off-label), *forest nurseries* (off-label), *frise* (off-label), *herbs (see appendix 6)* (off-label), *hops* (off-label), *lamb's lettuce* (off-label), *leaf brassicas* (off-label), *lettuce* (off-label), *parsley* (off-label), *protected cress* (off-label), *protected forest nurseries* (off-label), *protected herbs (see appendix 6)* (off-label), *protected hops* (off-label), *protected lamb's lettuce* (off-label), *protected leaf brassicas* (off-label), *protected lettuce* (off-label), *protected parsley* (off-label), *protected rocket* (off-label), *protected soft fruit* (off-label), *protected spinach* (off-label), *radicchio* (off-label), *rocket* (off-label), *scarole* (off-label), *soft fruit* (off-label), *spinach* (off-label), *spinach beet* (off-label), *top fruit* (off-label) [1]
- Common oak thelaxid in *forest nurseries* (off-label), *protected forest nurseries* (off-label) [1]
- Gall wasp in *forest nurseries* (off-label), *protected forest nurseries* (off-label) [1]
- Great spruce bark beetle in *forest nurseries* (off-label), *protected forest nurseries* (off-label) [1]
- Green aphid in *forest nurseries* (off-label), *protected forest nurseries* (off-label) [1]
- Large pine weevil in *forest* (off-label) [1]
- Pine weevil in *forest nurseries* (off-label) [1]

FOR FULL CONDITIONS OF USE ALWAYS READ THE PRODUCT LABEL

acetamiprid

- Whitefly in **apples, cherries, ornamental plant production, pears, plums, protected aubergines, protected ornamentals, protected peppers, protected tomatoes** [1, 3, 4]; **brussels sprouts** *(off-label)* [2]
- Woolly aphid in **forest nurseries** *(off-label)*, **protected forest nurseries** *(off-label)* [1]

Extension of Authorisation for Minor Use (EAMUs)
- *brussels sprouts* 20072866 [2]
- *chard* 20113144 [1]
- *cress* 20101101 [1]
- *forest* 20121068 [1]
- *forest nurseries* 20111313 [1], 20160084 [1], 20162653 [1]
- *frise* 20111994 [1]
- *herbs (see appendix 6)* 20101101 [1]
- *hops* 20082857 [1]
- *lamb's lettuce* 20111994 [1]
- *leaf brassicas* 20111994 [1]
- *lettuce* 20111994 [1]
- *parsley* 20111994 [1]
- *protected cress* 20101101 [1]
- *protected forest nurseries* 20162653 [1]
- *protected herbs (see appendix 6)* 20101101 [1]
- *protected hops* 20082857 [1]
- *protected lamb's lettuce* 20111994 [1]
- *protected leaf brassicas* 20111994 [1]
- *protected lettuce* 20111994 [1]
- *protected parsley* 20111994 [1]
- *protected rocket* 20111994 [1]
- *protected soft fruit* 20082857 [1]
- *protected spinach* 20101101 [1]
- *radicchio* 20111994 [1]
- *rocket* 20111994 [1]
- *scarole* 20111994 [1]
- *soft fruit* 20082857 [1]
- *spinach* 20101101 [1]
- *spinach beet* 20113144 [1]
- *top fruit* 20082857 [1]

Approval information
- Acetamiprid included in Annex I under EC Regulation 1107/2009

Efficacy guidance
- Best results obtained from application at the first sign of pest attack or when appropriate thresholds are reached
- Thorough coverage of foliage is essential to ensure best control. Acetamiprid has contact, systemic and translaminar activity

Restrictions
- Maximum number of treatments 1 per yr for cherries, ware potatoes; 2 per yr or crop for apples, pears, plums, ornamental plant production, protected crops, seed potatoes
- Do not use more than two applications of any neonicotinoid insecticide (e.g. acetamiprid, clothianidin, imidacloprid, thiacloprid) on any crop. Previous soil or seed treatment with a neonicotinoid counts as one such treatment
- 2 treatments are permitted on seed potatoes but must not be used consecutively [2]

Crop-specific information
- HI 3 d for protected crops; 14 d for apples, cherries, pears, plums, potatoes

Environmental safety
- Harmful to aquatic organisms
- Acetamiprid is slightly toxic to predatory mites and generally slightly toxic to other beneficials
- Broadcast air-assisted LERAP [1, 3, 4] (18 m); LERAP Category B [1-4]

SEE SECTION 3 FOR PRODUCTS ALSO REGISTERED

Pesticide Profiles

Hazard classification and safety precautions
  Hazard Dangerous for the environment
  UN Number N/C
  Risk phrases R22a [1, 3, 4]; R52, R53a [1-4]
  Operator protection A [1-4]; H [1, 3, 4]; U02a, U19a [1-4]; U20a [2]; U20b [1, 3, 4]
  Environmental protection E15b, E16a, E16b [1-4]; E17b [1, 3, 4] (18 m); E38 [2]
  Storage and disposal D05 [1, 3, 4]; D09a, D10b [1-4]; D12a [2]

3 acetic acid

A non-selective herbicide for non-crop situations

Products
  New Way Weed Spray   Headland Amenity   240 g/l   SL   15319

Uses
- General weed control in **hard surfaces**, **natural surfaces not intended to bear vegetation**, **permeable surfaces overlying soil**

Approval information
- Acetic acid is included in Annex 1 under EC Regulation 1107/2009

Efficacy guidance
- Best results obtained from treatment of young tender weeds less than 10 cm high
- Treat in spring and repeat as necessary throughout the growing season
- Treat survivors as soon as fresh growth is seen
- Ensure complete coverage of foliage to the point of run-off
- Rainfall after treatment may reduce efficacy

Restrictions
- No restriction on number of treatments

Following crops guidance
- There is no residual activity in the soil and sowing or planting may take place as soon as treated weeds have died

Environmental safety
- Harmful to aquatic organisms
- High risk to bees
- Do not apply to crops in flower or to those in which bees are actively foraging. Do not apply when flowering weeds are present
- Keep people and animals off treated dense weed patches until spray has dried. This is not necessary for areas with occasional low growing prostrate weeds such as on pathways

Hazard classification and safety precautions
  Hazard Irritant
  UN Number N/C
  Risk phrases R36, R37, R38, R52
  Operator protection A, C, D, H; U05a, U09a, U19a, U20b
  Environmental protection E12a, E12e, E15a
  Storage and disposal D01, D02, D09a, D12a

4 alpha-cypermethrin

A contact and ingested pyrethroid insecticide for use in arable crops and agricultural buildings
IRAC mode of action code: 3

Products
  Eribea   Belchim   100 g/l   EC   17270

FOR FULL CONDITIONS OF USE ALWAYS READ THE PRODUCT LABEL

alpha-cypermethrin

Uses
- Brassica pod midge in **winter oilseed rape**
- Cabbage seed weevil in **spring oilseed rape, winter oilseed rape**
- Cabbage stem flea beetle in **winter oilseed rape**
- Caterpillars in **broccoli, brussels sprouts, cabbages, calabrese, cauliflowers, kale**
- Cereal aphid in **spring barley, spring wheat, winter barley, winter wheat**
- Flea beetle in **broccoli, brussels sprouts, cabbages, calabrese, cauliflowers, kale**
- Pea and bean weevil in **broad beans, combining peas, spring field beans, vining peas, winter field beans**
- Pea aphid in **combining peas** (reduction), **vining peas** (reduction)
- Pea moth in **combining peas, vining peas**
- Pollen beetle in **spring oilseed rape, winter oilseed rape**
- Rape winter stem weevil in **winter oilseed rape**
- Yellow cereal fly in **winter barley, winter wheat**

Approval information
- Alpha-cypermethrin included in Annex I under EC Regulation 1107/2009
- Accepted by BBPA for use on malting barley

Efficacy guidance
- For cabbage stem flea beetle control spray oilseed rape when adult or larval damage first seen and about 1 mth later
- For flowering pests on oilseed rape apply at any time during flowering, against pollen beetle best results achieved at green to yellow bud stage (GS 3.3-3.7), against seed weevil between 20 pods set and 80% petal fall (GS 4.7-5.8)
- Spray cereals in autumn for control of cereal aphids, in spring/summer for grain aphids. (See label for details)
- For flea beetle, caterpillar and cabbage aphid control on brassicas apply when the pest or damage first seen or as a preventive spray. Repeat if necessary
- For pea and bean weevil control in peas and beans apply when pest attack first seen and repeat as necessary

Restrictions
- Apply up to 2 sprays on cereals in autumn and spring, 1 in summer between 1 Apr and 31 Aug. See label for details of rates and maximum total dose
- Only 1 aphicide treatment may be applied in cereals between 1 Apr and 31 Aug and spray volume must not be reduced in this period
- Do not apply to a cereal crop if any product containing a pyrethroid or dimethoate has been applied after the start of ear emergence (GS 51)

Crop-specific information
- Latest use: before the end of flowering for oilseed rape; before 31 Aug for cereals
- HI vining peas 1 d; brassicas, combining peas, broad beans, field beans 7 d
- For summer cereal application do not spray within 6 m from edge of crop and do not reduce volume when used after 31 Mar

Environmental safety
- Very toxic to aquatic organisms
- Extremely dangerous to fish or other aquatic life. Do not contaminate surface waters or ditches with chemical or used container
- Dangerous to bees. Where possible spray oilseed rape crops in the late evening or early morning or in dull weather. Give local beekeepers warning when using in flowering crops
- Do not spray within 6 m of the edge of a cereal crop after 31 Mar in yr of harvest
- Buffer zone requirement 12m [1]
- LERAP Category B

Hazard classification and safety precautions
Hazard Harmful, Dangerous for the environment, Flammable liquid and vapour, Toxic if swallowed, Harmful if inhaled, Very toxic to aquatic organisms
Transport code 6.1
Packaging group III
UN Number 2903

Pesticide Profiles

Risk phrases H304, H315, H317, H318, H335, H336, H373, R43
Operator protection A; U05a, U14, U19c
Environmental protection E12c, E16a, E38, H410
Storage and disposal D01, D02, D05, D12a
Medical advice M04a, M05b

5   aluminium ammonium sulphate

An inorganic bird and animal repellent

Products

1	Curb Crop Spray Powder	Sphere	88% w/w	WP	02480
2	Liquid Curb Crop Spray	Sphere	83 g/l	SC	03164
3	Rezist	Sphere	88% w/w	WP	08576
4	Rezist Liquid	Sphere	83 g/l	SC	14643
5	Sphere ASBO	Sphere	83 g/l	SC	15064

Uses

- Animal repellent in *agricultural premises, all top fruit, broad beans, bush fruit, cane fruit, carrots, flowerhead brassicas, forest nursery beds, forestry plantations, grain stores, leaf brassicas, peas, permanent grassland, spring barley, spring field beans, spring oats, spring oilseed rape, spring wheat, strawberries, sugar beet, winter barley, winter field beans, winter oats, winter oilseed rape, winter wheat* [1, 2]; *all edible crops (outdoor), all non-edible crops (outdoor), forest* [5]; *amenity grassland, hard surfaces, managed amenity turf* [4, 5]; *amenity vegetation* [1]
- Bird repellent in *agricultural premises, all top fruit, broad beans, bush fruit, cane fruit, carrots, flowerhead brassicas, forest nursery beds, forestry plantations, grain stores, leaf brassicas, peas, permanent grassland, spring barley, spring field beans, spring oats, spring oilseed rape, spring wheat, strawberries, sugar beet, winter barley, winter field beans, winter oats, winter oilseed rape, winter wheat* [1, 2]; *all edible crops (outdoor), all non-edible crops (outdoor), forest* [5]; *amenity grassland, hard surfaces, managed amenity turf* [4, 5]; *amenity vegetation* [1]
- Dogs in *amenity grassland* (anti-fouling), *hard surfaces* (anti-fouling), *managed amenity turf* (anti-fouling) [3]
- Moles in *amenity grassland, hard surfaces, managed amenity turf* [3]

Approval information

- aluminium ammonium sulphate is included in Annex 1 under EC Regulation 1107/2009

Efficacy guidance

- Apply as overall spray to growing crops before damage starts or mix powder with seed depending on type of protection required
- Spray deposit protects growth present at spraying but gives little protection to new growth
- Product must be sprayed onto dry foliage to be effective and must dry completely before dew or frost forms. In winter this may require some wind

Crop-specific information

- Latest use: no restriction

Hazard classification and safety precautions
   UN Number N/C
   Operator protection U05a [1, 2, 5]; U15 [4]; U20a [1-5]
   Environmental protection E15a [1-5]; E19b [1, 2, 4, 5]
   Storage and disposal D01, D02, D05 [1, 2, 4, 5]; D09a [3, 4]; D10b [4]; D11a [1-5]; D12a [1, 2, 5]
   Medical advice M03 [1, 2, 4, 5]

FOR FULL CONDITIONS OF USE ALWAYS READ THE PRODUCT LABEL

6 aluminium phosphide

A phosphine generating compound used against vertebrates and grain store pests. Under new regulations any person wishing to purchase or use aluminium phosphide must be suitably qualified and licensed by January 2015. A new licensing body has been set up.
IRAC mode of action code: 24A

Products

1	Degesch Fumigation Tablets	Rentokil	56% w/w	GE	17035
2	Detia Gas-Ex-B	Rentokil	57% w/w	GE	17097
3	Phostoxin	Rentokil	56% w/w	GE	17000
4	Talunex	Killgerm	56% w/w	GE	17001

Uses
- Insect pests in **carob, cocoa, coffee, herbal infusions, stored dried spices, stored grain, stored linseed, stored nuts, stored oilseed rape, stored pulses, tea, tobacco** [1]; **coconuts, dried fruit, grain stores, nuts, pulses** [2]; **crop handling & storage structures, processed consumable products** [1, 2].
- Moles in **all situations** [3, 4]; **farmland** [3]
- Rabbits in **all situations** [3, 4]; **farmland** [3]
- Rats in **all situations** [3, 4]; **farmland** [3]

Approval information
- Aluminium phosphide is included in Annex 1 under EC Regulation 1107/2009
- Accepted by BBPA for use in stores for malting barley

Efficacy guidance
- Product releases poisonous hydrogen phosphide gas in contact with moisture
- Place fumigation tablets in grain stores as directed [1, 2]
- Place pellets in burrows or runs and seal hole by heeling in or covering with turf. Do not cover pellets with soil. Inspect daily and treat any new or re-opened holes [3, 4]

Restrictions
- Aluminium phosphide is subject to the Poisons Rules 1982 and the Poisons Act 1972. See Section 5 for more information
- Only to be used by operators instructed or trained in the use of aluminium phosphide and familiar with the precautionary measures to be taken. See label and HSE Guidance Notes for full precautions
- Only open container outdoors [3, 4] and for immediate use. Keep away from liquid or water as this causes immediate release of gas. Do not use in wet weather
- Do not use within 3 m of human or animal habitation. Before application ensure that no humans or domestic animals are in adjacent buildings or structures. Allow a minimum airing-off period of 4 h before re-admission

Environmental safety
- Product liberates very toxic, highly flammable gas
- Dangerous to fish or other aquatic life. Do not contaminate surface waters or ditches with chemical or used container
- Prevent access by livestock, pets and other non-target mammals and birds to buildings under fumigation and ventilation
- Pellets must never be placed or allowed to remain on ground surface
- Do not use adjacent to watercourses
- Take particular care to avoid gassing non-target animals, especially those protected under the Wildlife and Countryside Act (e.g. badgers, polecat, reptiles, natterjack toads, most birds). Do not use in burrows where there is evidence of badger or fox activity, or when burrows might be occupied by birds
- Dust remaining after decomposition is harmless and of no environmental hazard
- Keep in original container, tightly closed, in a safe place, under lock and key
- Dispose of empty containers as directed on label

SEE SECTION 3 FOR PRODUCTS ALSO REGISTERED

Pesticide Profiles

Hazard classification and safety precautions
Hazard Very toxic, In contact with water releases flammable gases which may ignite spontaneously, Fatal if swallowed, Toxic in contact with skin, Fatal if inhaled, Very toxic to aquatic organisms [1-4]; Harmful [2-4]; Highly flammable [2, 3]; Dangerous for the environment [1-3]
Transport code 4.1 6.1 [4]; 4.3 [1-3]
Packaging group I
UN Number 1397
Risk phrases H315, H318 [1, 3, 4]
Operator protection A, D, H; U01, U07, U13, U19a, U20a [1-4]; U05a [2-4]; U05b, U18 [1]
Environmental protection E02a [1] (4 h min); E02a [2-4] (4 h); E02b [1]; E13b, E34 [1-4]
Storage and disposal D01, D02, D07, D09b [1-4]; D11b [2-4]; D11c [1]
Vertebrate/rodent control products V04a [2-4]
Medical advice M04a

7 ametoctradin

A potato blight fungicide available only in mixtures
FRAC mode of action code: 45

See also ametoctradin + dimethomorph
    ametoctradin + mancozeb

8 ametoctradin + dimethomorph

A systemic and protectant fungicide for potato blight control
FRAC mode of action code: 40 + 45

Products
1	Percos	BASF	300:225 g/l	SC	15248
2	Zampro DM	BASF	300:225 g/l	SC	15013

Uses
- Blight in **potatoes** [1, 2]
- Downy mildew in **ornamental plant production** *(off-label)*, **protected ornamentals** *(off-label)*, **table grapes** *(off-label)*, **wine grapes** *(off-label)* [1]
- Tuber blight in **potatoes** *(reduction)* [1, 2]

Extension of Authorisation for Minor Use (EAMUs)
- *ornamental plant production* 20130819 [1]
- *protected ornamentals* 20130819 [1]
- *table grapes* 20150254 [1]
- *wine grapes* 20150254 [1]

Approval information
- Ametoctradin and dimethomorph are included in Annex 1 under EC Regulation 1107/2009.

Efficacy guidance
- Application to dry foliage is rainfast within 1 hour of drying on the leaf.
- Will give effective control of phenylamide-resistant strains of blight

Crop-specific information
- HI 7 days for potatoes

Hazard classification and safety precautions
Hazard Harmful, Dangerous for the environment, Harmful if swallowed
UN Number N/C
Operator protection U05a
Environmental protection E15b, E34, E38, H412
Storage and disposal D01, D02, D09a, D10c
Medical advice M03, M05a

FOR FULL CONDITIONS OF USE ALWAYS READ THE PRODUCT LABEL

9 amidosulfuron

A post-emergence sulfonylurea herbicide for cleavers and other broad-leaved weed control in cereals
HRAC mode of action code: B

Products

1	Eagle	Interfarm	75% w/w	WG	16490
2	Squire Ultra	Interfarm	75% w/w	WG	16491

Uses
- Annual dicotyledons in **durum wheat**, **farm forestry** *(off-label)*, **forest nurseries** *(off-label)*, **game cover** *(off-label)*, **linseed**, **spring barley**, **spring oats**, **spring rye**, **spring wheat**, **triticale**, **winter barley**, **winter oats**, **winter rye**, **winter wheat** [1]; **grassland** [2]; **ornamental plant production** *(off-label)* [1, 2]
- Charlock in **corn gromwell** *(off-label)*, **durum wheat**, **farm forestry** *(off-label)*, **forest nurseries** *(off-label)*, **game cover** *(off-label)*, **linseed**, **spring barley**, **spring oats**, **spring rye**, **spring wheat**, **triticale**, **winter barley**, **winter oats**, **winter rye**, **winter wheat** [1]; **ornamental plant production** *(off-label)* [1, 2]
- Cleavers in **corn gromwell** *(off-label)*, **durum wheat**, **farm forestry** *(off-label)*, **forest nurseries** *(off-label)*, **game cover** *(off-label)*, **linseed**, **spring barley**, **spring oats**, **spring rye**, **spring wheat**, **triticale**, **winter barley**, **winter oats**, **winter rye**, **winter wheat** [1]; **ornamental plant production** *(off-label)* [1, 2]
- Docks in **grassland** [2]
- Forget-me-not in **corn gromwell** *(off-label)*, **durum wheat**, **farm forestry** *(off-label)*, **forest nurseries** *(off-label)*, **game cover** *(off-label)*, **linseed**, **spring barley**, **spring oats**, **spring rye**, **spring wheat**, **triticale**, **winter barley**, **winter oats**, **winter rye**, **winter wheat** [1]; **ornamental plant production** *(off-label)* [1, 2]
- Shepherd's purse in **corn gromwell** *(off-label)*, **durum wheat**, **farm forestry** *(off-label)*, **forest nurseries** *(off-label)*, **game cover** *(off-label)*, **linseed**, **spring barley**, **spring oats**, **spring rye**, **spring wheat**, **triticale**, **winter barley**, **winter oats**, **winter rye**, **winter wheat** [1]; **ornamental plant production** *(off-label)* [1, 2]

Extension of Authorisation for Minor Use (EAMUs)
- *corn gromwell* 20150903 [1]
- *farm forestry* 20142508 [1]
- *forest nurseries* 20142508 [1]
- *game cover* 20142508 [1]
- *ornamental plant production* 20142508 [1], 20142511 [2]

Approval information
- Accepted by BBPA for use on malting barley
- Amidosulfuron included in Annex I under EC Regulation 1107/2009

Efficacy guidance
- For best results apply in spring (from 1 Feb) in warm weather when soil moist and weeds growing actively. When used in grassland following cutting or grazing, docks should be allowed to regrow before treatment
- Weed kill is slow, especially under cool, dry conditions. Weeds may sometimes only be stunted but will have little or no competitive effect on crop
- May be used on all soil types unless certain sequences are used on linseed. See label
- Spray is rainfast after 1 h
- Cleavers controlled from emergence to flower bud stage. If present at application charlock (up to flower bud), shepherds purse (up to flower bud) and field forget-me-not (up to 6 leaves) will also be controlled
- Amidosulfuron is a member of the ALS-inhibitor group of herbicides and products should be used in a planned Resistance Management strategy. See Section 5 for more information

Restrictions
- Maximum number of treatments 1 per crop [1]

SEE SECTION 3 FOR PRODUCTS ALSO REGISTERED

Pesticide Profiles

- Use after 1 Feb and do not apply to rotational grass after 30 Jun, or to permanent grassland after 15 Oct [2]
- Do not apply to crops undersown or due to be undersown with clover or alfalfa [1]
- Do not spray crops under stress, suffering drought, waterlogged, grazed, lacking nutrients or if soil compacted
- Do not spray if frost expected
- Do not roll or harrow within 1 wk of spraying
- Specific restrictions apply to use in sequence or tank mixture with other sulfonylurea or ALS-inhibiting herbicides. See label for details. There are no recommendations for mixtures with metsulfuron-methyl products on linseed
- Certain mixtures with fungicides are expressly forbidden. See label for details

Crop-specific information
- Latest use: before first spikelets just visible (GS 51) for cereals; before flower buds visible for linseed; 15 Oct for grassland
- Broadcast cereal crops should be sprayed post-emergence after plants have a well established root system

Following crops guidance
- If a treated crop fails cereals may be sown after 15 d and thorough cultivation
- After normal harvest of a treated crop only cereals, winter oilseed rape, mustard, turnips, winter field beans or vetches may be sown in the same year as treatment and these must be preceded by ploughing or thorough cultivation
- Only cereals may be sown within 12 mth of application to grassland [2]
- Cereals or potatoes must be sown as the following crop after use of permitted mixtures or sequences with other sulfonylurea herbicides in cereals. Only cereals may be sown after the use of such sequences in linseed

Environmental safety
- Take care to wash out sprayers thoroughly. See label for details
- Avoid drift onto neighbouring broad-leaved plants or onto surface waters or ditches

Hazard classification and safety precautions
Hazard Dangerous for the environment
Transport code 9
Packaging group III
UN Number 3077
Risk phrases R50, R53a
Operator protection U20a [1]; U20b [2]
Environmental protection E07b [2] (1 week); E15a, E38 [1, 2]; E41 [2]
Storage and disposal D10a, D12a

10   amidosulfuron + iodosulfuron-methyl-sodium

A post-emergence sulfonylurea herbicide mixture for cereals
HRAC mode of action code: B + B

See also iodosulfuron-methyl-sodium

Products

1	Chekker	Bayer CropScience	12.5:1.25% w/w	WG	16495
2	Sekator OD	Interfarm	100:25 g/l	OD	16494

Uses
- Annual dicotyledons in **forest nurseries** *(off-label)* [1]; **spring barley**, **spring rye**, **spring wheat**, **triticale**, **winter barley**, **winter rye**, **winter wheat** [1, 2]
- Chickweed in **forest nurseries** *(off-label)* [1]; **spring barley**, **spring rye**, **spring wheat**, **triticale**, **winter barley**, **winter rye**, **winter wheat** [1, 2]
- Cleavers in **forest nurseries** *(off-label)* [1]; **spring barley**, **spring rye**, **spring wheat**, **triticale**, **winter barley**, **winter rye**, **winter wheat** [1, 2]

amidosulfuron + iodosulfuron-methyl-sodium

- Mayweeds in **forest nurseries** *(off-label)* [1]; **spring barley**, **spring rye**, **spring wheat**, **triticale**, **winter barley**, **winter rye**, **winter wheat** [1, 2]
- Volunteer oilseed rape in **forest nurseries** *(off-label)* [1]; **spring barley**, **spring rye**, **spring wheat**, **triticale**, **winter barley**, **winter rye**, **winter wheat** [1, 2]

Extension of Authorisation for Minor Use (EAMUs)
- *forest nurseries* 20142722 [1]

Approval information
- Amidosulfuron and iodosulfuron-methyl-sodium included in Annex I under EC Regulation 1107/2009
- Accepted by BBPA for use on malting barley

Efficacy guidance
- Best results obtained from treatment in warm weather when soil is moist and the weeds are growing actively
- Weeds must be present at application to be controlled
- Dry conditions resulting in moisture stress may reduce effectiveness
- Weed control is slow especially under cool dry conditions
- Occasionally weeds may only be stunted but they will normally have little or no competitive effect on the crop
- Amidosulfuron and iodosulfuron are members of the ALS-inhibitor group of herbicides and products should be used in a planned Resistance Management strategy. See Section 5 for more information

Restrictions
- Maximum number of treatments 1 per crop
- Must only be applied between 1 Feb in yr of harvest and specified latest time of application
- Do not apply to crops undersown or to be undersown with grass, clover or alfalfa
- Do not roll or harrow within 1 wk of spraying
- Do not spray crops under stress from any cause or if the soil is compacted
- Do not spray if rain or frost expected
- Do not apply in mixture or in sequence with any other ALS inhibitor

Crop-specific information
- Latest use: before first spikelet of inflorescence just visible (GS 51)
- Treat drilled crops after the 2-leaf stage; treat broadcast crops after the plants have a well-established root system
- Applications to spring barley may cause transient crop yellowing

Following crops guidance
- Cereals, winter oilseed rape and winter field beans may be sown in the same yr as treatment provided they are preceded by ploughing or thorough cultivation. Any crop may be sown in the spring of the yr following treatment
- A minimum of 3 mth must elapse between treatment and sowing winter oilseed rape

Environmental safety
- Dangerous for the environment
- Toxic to aquatic organisms
- Take extreme care to avoid damage by drift onto broad-leaved plants outside the target area or onto ponds, waterways and ditches
- Observe carefully label instructions for sprayer cleaning
- LERAP Category B

Hazard classification and safety precautions
Hazard Irritant, Dangerous for the environment
Transport code 9
Packaging group III
UN Number 3077 [1]; 3082 [2]
Risk phrases H317, H319 [2]; R36, R50, R53a [1]
Operator protection A, C, H; U05a, U08, U11, U14, U15, U20b

SEE SECTION 3 FOR PRODUCTS ALSO REGISTERED

Environmental protection E15a, E16a, E16b, E38 [1, 2]; H410 [2]
Storage and disposal D01, D02, D10a, D12a

11   amidosulfuron + iodosulfuron-methyl-sodium + mesosulfuron-methyl

A herbicide mixture for weed control in winter wheat
HRAC mode of action code: B + B + B

See also amidosulfuron
amidosulfuron + iodosulfuron-methyl-sodium
iodosulfuron-methyl-sodium
mesosulfuron-methyl

Products
Pacifica Plus   Bayer CropScience   0.5:0.1:0.3% w/w   WG   17272

Uses
- Annual dicotyledons in **winter wheat**
- Annual grasses in **winter wheat**
- Blackgrass in **winter wheat**

Approval information
- Amidosulfuron, iodosulfuron-methyl-sodium and mesosulfuron-methyl included in Annex I under EC Regulation 1107/2009
- Accepted by BBPA for use on malting barley

Efficacy guidance
- Apply as early as possible in spring and before stem extension stage of any grass weeds
- Monitor efficacy and investigate areas of poor control

Restrictions
- Do not apply in tank-mixture or sequence with any product containing any other sulfonylurea or ALS inhibiting herbicide
- Do not use on crops undersown with grasses, clover, legumes or any other broad-leaved crop
- take care to avoid drift on to non-target plants

Following crops guidance
- Winter wheat and winter barley may be sown in the same year as harvest of a crop treated with 0.5 kg/ha. Winter oilseed rape may be sown after ploughing following a crop treated with 0.4 kg/ha. In the event of crop failure, sow only winter or spring wheat.

Environmental safety
- LERAP Category B

Hazard classification and safety precautions
Hazard Very toxic to aquatic organisms
Transport code 9
Packaging group III
UN Number 3077
Risk phrases H317, H318
Operator protection A, C, H; U05a, U11, U20a
Environmental protection E15a, E16a, H410
Storage and disposal D01, D02, D05, D09a, D10b

12   aminopyralid

A pyridine carboxylic acid herbicide available only in mixtures
HRAC mode of action code: O

FOR FULL CONDITIONS OF USE ALWAYS READ THE PRODUCT LABEL

13 aminopyralid + fluroxypyr

A foliar acting herbicide mixture for use in grassland
HRAC mode of action code: O + O

See also fluroxypyr

Products

Synero Nomix Enviro 30:100 g/l EW 14708

Uses
- Buttercups in *amenity grassland*
- Chickweed in *amenity grassland*
- Dandelions in *amenity grassland*
- Docks in *amenity grassland*
- Stinging nettle in *amenity grassland*
- Thistles in *amenity grassland*

Approval information
- Aminopyralid and fluroxypyr are included in Annex I under EC Regulation 1107/2009.

Efficacy guidance
- For best results and to avoid crop check, grass and weeds must be growing actively
- Allow 2-3 wk after cutting for hay or silage for sufficient regrowth to occur before spraying and leave 7 d afterwards to allow maximum translocation
- Where there is a high reservoir of weed seed or a historically high weed population a programmed approach may be needed involving a second treatment in the following yr
- Control may be reduced if rain falls within 1 h of spraying

Restrictions
- Maximum number of treatments 1 per yr.
- Do not apply to leys less than 1 year old.
- Do not apply by hand-held equipment.
- Do not use on grassland that will be used for animal feed, bedding, composting or mulching within 1 calender year of application.
- Do not use on grassland that will be grazed by animals other than cattle or sheep.
- Use of an antifoam is compulsory
- Do not use any treated plant material, or manure from animals fed on treated crops, for composting or mulching
- Do not use on crops grown for seed
- Manure from animals fed on pasture or silage treated with this herbicide should not leave the farm

Crop-specific information
- Treatment will kill clover
- Treatment may occasionally cause transient yellowing of the sward which is quickly outgrown
- Late treatments may lead to a slight transient leaning of grass that does not affect yield

Following crops guidance
- Do not drill clover or other legumes within 4 mth of treatment, or potatoes in the spring following treatment in the previous autumn
- In the event of failure of newly seeded treated grassland, grass may be re-seeded immediately, or wheat may be sown provided 4 mth have elapsed since application
- Residues in plant tissues, including manure, may affect succeeding susceptible crops of peas, beans, other legumes, carrots, other Umbelliferae, potatoes, tomatoes, lettuce and other Compositae. These crops should not be sown within 3 mth of ploughing up treated grassland

Environmental safety
- Dangerous for the environment
- Toxic to aquatic organisms

SEE SECTION 3 FOR PRODUCTS ALSO REGISTERED

Pesticide Profiles

- Keep livestock out of treated areas for at least 7 d following treatment and until poisonous weeds, such as ragwort, have died down and become unpalatable
- Avoid damage by drift onto susceptible crops, non-target plants or waterways

Hazard classification and safety precautions
Hazard Irritant, Dangerous for the environment
Transport code 9
Packaging group III
UN Number 3082
Risk phrases H304, H315, H318, H336
Operator protection A, C; U05a, U08, U11, U14, U23a
Environmental protection E07a, E15b, E38, H411
Consumer protection C01
Storage and disposal D01, D02, D05, D09a, D10b, D12a

14   aminopyralid + metazachlor + picloram

A herbicide mixture for weed control in oilseed rape.

See also metazachlor

Products

Ralos	Dow	5.3:500:13.3 g/l	SC	16737

Uses
- Chickweed in **winter oilseed rape**
- Dead nettle in **winter oilseed rape**
- Fat hen in **winter oilseed rape**
- Penny cress in **winter oilseed rape**
- Poppies in **winter oilseed rape**
- Scentless mayweed in **winter oilseed rape**

Approval information
- Aminopyralid, metazachlor and picloram are included in Annex 1 under 1107/2009.

Restrictions
- Applications shall be limited to a total dose of not more than 1.0 kg metazachlor/ha in a three year period on the same field.
- Do not use on sands and very light soils or on soils with more than 10% organic matter.
- Straw from treated crops must not be baled and must stay on the field.
- Do not use treated plant material for composting or mulching.
- Do not use digestate from anaerobic digesters on crops such as peas, beans and other legumes, carrots and other Umbelliferae, potatoes, lettuce or other Compositae, glasshouse or protected crops.

Following crops guidance
- Only wheat, barley, oats, maize or oilseed rape should be planted within 4 months (120 days) of application. All other crops should wait until 36 months after application.
- In the event of crop failure, only spring wheat, spring barley, spring oats, maize or ryegrass should be planted after ploughing or thorough cultivation of the soil.

Environmental safety
- Metazachlor stewardship guidelines advise a maximum dose of 750 g.a.i/ha/annum. Applications to drained fields should be complete by 15th Oct but, if drains are flowing, complete applications by 1st Oct.
- LERAP Category B

Hazard classification and safety precautions
Hazard Dangerous for the environment
Transport code 9
Packaging group III

FOR FULL CONDITIONS OF USE ALWAYS READ THE PRODUCT LABEL

aminopyralid + propyzamide

UN Number 3082
Risk phrases H351
Operator protection A; U05a, U20a
Environmental protection E06c (1 week); E15b, E16a, E34, H410
Storage and disposal D01, D02, D09a, D10c
Medical advice M03

15   aminopyralid + propyzamide

A herbicide mixture for weed control in winter oilseed rape
HRAC mode of action code: K1 + O

Products

1	AstroKerb	Dow	6.3:500 g/l	SC	16184
2	Clayton Propel Plus	Clayton	6.3:500 g/l	SC	17227

Uses
- Annual dicotyledons in *winter oilseed rape*
- Annual grasses in *winter oilseed rape*
- Mayweeds in *winter oilseed rape*
- Poppies in *winter oilseed rape*

Approval information
- Aminopyralid and propyzamide are included in Annex 1 under EC Regulation 1107/2009.

Efficacy guidance
- Allow soil temperatures to drop before application to winter oilseed rape to optimise activity of propyzamide component

Restrictions
- Users must have received adequate instruction, training and guidance on the safe and efficient use of the product.
- Must not be used on land where vegetation will be cut for animal feed, fodder, bedding nor for composting or mulching within one calendar year of treatment.
- Do not remove oilseed rape straw from the field unless it is for burning for heat or electricity production. Do not use the oilseed rape straw for animal feed, animal bedding, composting or mulching.

Crop-specific information
- Only cereals may follow application of aminopyralid + propyzamide and these should not be planted within 30 weeks of treatment.
- Aminopyralid residues in plant tissue which have not completely decayed may affect susceptible crops such as legumes (e.g. peas and beans), sugar and fodder beet, carrots and other Umbelliferae, potatoes, tomatoes, lettuce and other Compositae.

Following crops guidance
- May only be followed by spring or winter cereals. Treated land must be mouldboard ploughed to a depth of at least 15 cm before a following cereal crop is planted.

Hazard classification and safety precautions
Hazard Harmful, Dangerous for the environment
Transport code 9
Packaging group III
UN Number 3082
Risk phrases H351
Operator protection A, H; U20c
Environmental protection E07b (1); E15b, E38, E39, H411
Storage and disposal D09a, D11a, D12a

SEE SECTION 3 FOR PRODUCTS ALSO REGISTERED

Pesticide Profiles

16 aminopyralid + triclopyr

A foliar acting herbicide mixture for broad-leaved weed control in grassland
HRAC mode of action code: O + O

See also triclopyr

Products

1	Forefront T	Dow	30:240g/l	EW	15568
2	Garlon Ultra	Nomix Enviro	12:120 g/l	SL	16211
3	Icade	Dow	12:120 g/l	SL	16182

Uses

- Annual dicotyledons in **amenity grassland** [2, 3]
- Brambles in **amenity grassland** [3]
- Broom in **amenity grassland** [3]
- Buddleia in **amenity grassland** [3]
- Buttercups in **grassland** [1]
- Common mugwort in **amenity grassland** [3]
- Common nettle in **amenity grassland** [3]; **grassland** [1]
- Creeping thistle in **amenity grassland** [3]
- Dandelions in **grassland** [1]
- Docks in **grassland** [1]
- Gorse in **amenity grassland** [3]
- Hogweed in **amenity grassland** [3]
- Japanese knotweed in **amenity grassland** [3]
- Perennial dicotyledons in **amenity grassland** [3]
- Rosebay willowherb in **amenity grassland** [3]
- Thistles in **grassland** [1]

Approval information

- Aminopyralid and triclopyr are included in Annex I under EC Regulation 1107/2009.

Efficacy guidance

- For best results and to avoid crop check, grass and weeds must be growing actively
- Use adeqate water volume to ensure good weed coverage. Increase water volume if necessary where the weed population is high and where the grass is dense
- Allow 2-3 wk after cutting for hay or silage for sufficient regrowth to occur before spraying and leave 7 d afterwards to allow maximum translocation
- Where there is a high reservoir of weed seed or a historically high weed population a programmed approach may be needed involving a second treatment in the following yr
- Control may be reduced if rain falls within 1 h of spraying

Restrictions

- Maximum number of treatments 1 per yr.
- Do not apply to leys less than 1 year old.
- Do not apply by hand-held equipment.
- Do not use on grassland that will be used for animal feed, bedding, composting or mulching within 1 calender year of application.
- Do not use on grassland that will be grazed by animals other than cattle or sheep.
- Do not use any treated plant material, or manure from animals fed on treated crops, for composting or mulching
- Do not use on crops grown for seed
- Manure from animals fed on pasture or silage from treated crops should not leave the farm

Crop-specific information

- Latest use: 7 d before grazing or harvest for grassland
- Late applications may lead to transient leaning of grass which does not affect final yield

FOR FULL CONDITIONS OF USE ALWAYS READ THE PRODUCT LABEL

Following crops guidance
- Ensure that all plant remains of a treated crop have completely decayed before planting susceptible crops such as peas, beans and other legumes, sugar beet, carrots and other Umbelliferae, potatoes and tomatoes, lettuce and other Compositae
- Do not plant potatoes, sugar beet, vegetables, beans or other leguminous crops in the calendar yr following application

Environmental safety
- Dangerous for the environment
- Toxic to aquatic organisms
- Keep livestock out of treated areas for at least 7 d after treatment or until foliage of any poisonous weeds such as ragwort has died and become unpalatable
- To protect groundwater do not apply to grass leys less than 1 yr old
- Take extreme care to avoid drift onto susceptible crops, non-target plants or waterways. All conifers, especially pine and larch, are very sensitive and may be damaged by vapour drift in hot conditions
- LERAP Category B [1]

Hazard classification and safety precautions
Hazard Harmful [2, 3]; Irritant [1]; Dangerous for the environment [1-3]
Transport code 9 [1]
Packaging group III [1]
UN Number 3082 [1]; N/C [2, 3]
Risk phrases H317 [1]; H319, H335 [2, 3]
Operator protection A, H [1-3]; M [2, 3]; U05a, U08 [1-3]; U14, U23a [1]
Environmental protection E06a [1] (7 days); E07c [2, 3] (7 days); E15b, E38 [1-3]; E15c, E16a, H411 [1]; E39 [2, 3]
Consumer protection C01
Storage and disposal D01, D02, D09a [1-3]; D05, D10c [2, 3]; D10b, D12a [1]
Medical advice M05a [2, 3]

17 Ampelomyces quisqualis (Strain AQ10)

A biological fungicide that combats powdery mildew in fruit and ornamentals

Products
AQ 10   Fargro   58% w/w   WG   17102

Uses
- Powdery mildew in **protected apple** (off-label), **protected aubergines**, **protected blackcurrants** (off-label), **protected blueberry** (off-label), **protected chilli peppers**, **protected chilli peppers** (off-label), **protected courgettes**, **protected courgettes** (off-label), **protected crabapple** (off-label), **protected cucumbers**, **protected forest nurseries** (off-label), **protected gooseberries** (off-label), **protected herbs (see appendix 6)** (off-label), **protected melons**, **protected ornamentals** (off-label), **protected other small fruit and berries** (off-label), **protected pear** (off-label), **protected peppers**, **protected peppers** (off-label), **protected pumpkins**, **protected quince** (off-label), **protected redcurrants** (off-label), **protected squashes**, **protected strawberries**, **protected summer squash** (off-label), **protected table grapes** (off-label), **protected tomatoes**, **protected watermelon** (off-label), **protected wine grapes** (off-label), **protected winter squash**

Extension of Authorisation for Minor Use (EAMUs)
- *protected apple* 20152646
- *protected blackcurrants* 20152646
- *protected blueberry* 20152646
- *protected chilli peppers* 20152646
- *protected courgettes* 20152646
- *protected crabapple* 20152646
- *protected forest nurseries* 20152646
- *protected gooseberries* 20152646

SEE SECTION 3 FOR PRODUCTS ALSO REGISTERED

Pesticide Profiles

- **protected herbs (see appendix 6)** *20152646*
- **protected ornamentals** *20152646*
- **protected other small fruit and berries** *20152646*
- **protected pear** *20152646*
- **protected peppers** *20152646*
- **protected quince** *20152646*
- **protected redcurrants** *20152646*
- **protected summer squash** *20152646*
- **protected table grapes** *20152646*
- **protected watermelon** *20152646*
- **protected wine grapes** *20152646*

Approval information
- Ampelomyces quisqualis is included in Annex 1 under EC Regulation 1107/2009

Hazard classification and safety precautions
UN Number N/C
Operator protection A, D, H; U14, U20b
Environmental protection E15b, E34
Storage and disposal D01, D02, D05, D09a, D10a, D16, D20
Medical advice M03

18   azoxystrobin

A systemic translaminar and protectant strobilurin fungicide for a wide range of crops
FRAC mode of action code: 11

Products

1	Amistar	Syngenta	250 g/l	SC	10443
2	Aubrac	AgChem Access	250 g/l	SC	13483
3	Azaka	Headland	250 g/l	SC	16422
4	Azoxy 250	Becesane	250 g/l	SC	15980
5	Azoxystar	Life Scientific	250 g/l	SC	17407
6	Azzox	Goldengrass	250 g/l	SC	14296
7	Clayton Bastille	Clayton	250 g/l	SC	17012
8	Clayton Belfry	Clayton	250 g/l	SC	12886
9	Clayton Putter	Clayton	50% w/w	WG	15827
10	Conclude	Globachem	250 g/l	SC	16905
11	Globaztar SC	Globachem	250 g/l	SC	15575
12	Heritage	Syngenta	50% w/w	WG	13536
13	Heritage Maxx	Syngenta	95 g/l	DC	14787
14	Reconcile	Becesane	250 g/l	SC	15379
15	Tazer	Nufarm UK	250 g/l	SC	15495
16	Valiant	Headland	250 g/l	SC	16776

Uses
- Alternaria in **broccoli**, **brussels sprouts**, **cabbages**, **calabrese**, **carrots**, **cauliflowers**, **collards**, **kale** [1, 2, 4, 6-8, 14]; **chicory grown outside for forcing** *(off-label)*, **crambe** *(off-label)*, **mustard** *(off-label)*, **protected chicory** *(off-label)* [11]; **chicory root** *(off-label)*, **horseradish** *(off-label)*, **parsnips** *(off-label)*, **salad brassicas** *(off-label - for baby leaf production)* [1]; **spring oilseed rape**, **winter oilseed rape** [1, 2, 4-8, 14]
- Anthracnose in **amenity grassland**, **managed amenity turf** [9, 12, 13]
- Ascochyta in **broad beans** *(off-label)*, **lupins** *(off-label)* [1]; **combining peas**, **vining peas** [1, 2, 4-8, 14]
- Black dot in **potatoes** [1, 2, 4, 6-8, 14]
- Black rot in **chinese cabbage** *(off-label)*, **choi sum** *(off-label)*, **komatsuna** *(off-label)*, **pak choi** *(off-label)*, **tatsoi** *(off-label)* [1]
- Black scurf and stem canker in **potatoes** [1, 2, 4, 6-8, 14]
- Black spot in **protected strawberries** *(off-label)*, **strawberries** *(off-label)* [1]

FOR FULL CONDITIONS OF USE ALWAYS READ THE PRODUCT LABEL

azoxystrobin

- Botrytis in **broad beans** *(off-label)*, **celery (outdoor)** *(off-label)*, **dwarf beans** *(off-label)*, **navy beans** *(off-label)*, **protected celery** *(off-label)*, **runner beans** *(off-label)* [1]
- Brown patch in **amenity grassland**, **managed amenity turf** [9, 12, 13]
- Brown rust in **rye** [3, 10, 11, 15, 16]; **spring barley**, **spring wheat**, **triticale**, **winter barley**, **winter wheat** [1-8, 10, 11, 14-16]; **spring rye**, **winter rye** [1, 2, 4-8, 14]
- Crown rust in **amenity grassland**, **managed amenity turf** [9, 12, 13]; **spring oats**, **winter oats** [1-8, 10, 11, 14, 16]
- Dark leaf spot in **spring oilseed rape**, **winter oilseed rape** [3, 16]
- Didymella in **inert substrate aubergines** *(off-label)*, **inert substrate tomatoes** *(off-label)* [1]
- Didymella stem rot in **aubergines** *(off-label)*, **protected tomatoes** *(off-label)* [1]
- Disease control in **all edible seed crops grown outdoors** *(off-label)*, **all non-edible seed crops grown outdoors** *(off-label)*, **forest nurseries** *(off-label)*, **ornamental plant production** *(off-label)*, **protected forest nurseries** *(off-label)*, **protected ornamentals** *(off-label)*, **protected soft fruit** *(off-label)*, **soft fruit** *(off-label)* [1, 2]; **horseradish** *(off-label)*, **parsnips** *(off-label)* [11]
- Downy mildew in **artichokes** *(off-label)*, **garlic** *(off-label)*, **inert substrate courgettes** *(off-label)*, **inert substrate cucumbers** *(off-label)*, **inert substrate gherkins** *(off-label)*, **protected cucumbers** *(off-label)*, **protected gherkins** *(off-label)*, **protected marrows** *(off-label)*, **salad brassicas** *(off-label - for baby leaf production)*, **salad onions** *(off-label)*, **shallots** *(off-label)* [1]; **bulb onions** [1, 2, 4, 6-8, 14]; **globe artichoke** *(off-label)*, **protected summer squash** *(off-label)* [11]; **protected courgettes** *(off-label)*, **protected melons** *(off-label)*, **protected pumpkins** *(off-label)*, **protected squashes** *(off-label)* [1, 11]
- Ear diseases in **spring wheat**, **winter wheat** [3, 10, 11, 15, 16]
- Fairy rings in **amenity grassland**, **managed amenity turf** [9, 12, 13]
- Foliar disease control in **crambe** *(off-label)*, **durum wheat** *(off-label)*, **grass seed crops** *(off-label)*, **radishes** *(off-label)* [1]
- Fusarium patch in **amenity grassland**, **managed amenity turf** [9, 12, 13]
- Glume blotch in **spring wheat**, **winter wheat** [1-8, 10, 11, 14-16]
- Grey mould in **aubergines** *(off-label)*, **inert substrate aubergines** *(off-label)*, **inert substrate tomatoes** *(off-label)*, **protected tomatoes** *(off-label)* [1]
- Late blight in **aubergines** *(off-label)*, **inert substrate aubergines** *(off-label)*, **inert substrate tomatoes** *(off-label)*, **protected tomatoes** *(off-label)* [1]
- Late ear diseases in **spring wheat**, **winter wheat** [1, 2, 4-8, 14]
- Melting out in **amenity grassland**, **managed amenity turf** [9, 12, 13]
- Net blotch in **spring barley**, **winter barley** [1-8, 10, 11, 14-16]
- Phytophthora in **chicory grown outside for forcing** *(off-label)*, **protected chicory** *(off-label)* [11]; **protected chicory** *(off-label - for forcing)* [1]
- Powdery mildew in **all edible seed crops grown outdoors** *(off-label)*, **all non-edible seed crops grown outdoors** *(off-label)*, **bilberries** *(off-label)*, **blackcurrants** *(off-label)*, **blueberries** *(off-label)*, **chicory grown outside for forcing** *(off-label)*, **cranberries** *(off-label)*, **durum wheat** *(off-label)*, **forest nurseries** *(off-label)*, **garlic** *(off-label)*, **globe artichoke** *(off-label)*, **gooseberries** *(off-label)*, **grass seed crops** *(off-label)*, **loganberries** *(off-label)*, **lupins** *(off-label)*, **ornamental plant production** *(off-label)*, **protected chicory** *(off-label)*, **protected forest nurseries** *(off-label)*, **protected ornamentals** *(off-label)*, **protected summer squash** *(off-label)*, **redcurrants** *(off-label)*, **rubus hybrids** *(off-label)*, **shallots** *(off-label)*, **strawberries** *(off-label)*, **summer squash** *(off-label)* [11]; **artichokes** *(off-label)*, **aubergines** *(off-label)*, **chicory root** *(off-label)*, **chives** *(off-label)*, **herbs (see appendix 6)** *(off-label)*, **inert substrate aubergines** *(off-label)*, **inert substrate courgettes** *(off-label)*, **inert substrate cucumbers** *(off-label)*, **inert substrate gherkins** *(off-label)*, **inert substrate tomatoes** *(off-label)*, **parsley** *(off-label)*, **protected blackberries** *(off-label)*, **protected cayenne peppers** *(off-label)*, **protected cucumbers** *(off-label)*, **protected gherkins** *(off-label)*, **protected marrows** *(off-label)*, **protected peppers** *(off-label)*, **protected raspberries** *(off-label)*, **protected tomatoes** *(off-label)* [1]; **blackberries** *(off-label)*, **courgettes** *(off-label)*, **horseradish** *(off-label)*, **parsnips** *(off-label)*, **poppies for morphine production** *(off-label)*, **protected courgettes** *(off-label)*, **protected melons** *(off-label)*, **protected pumpkins** *(off-label)*, **protected squashes** *(off-label)*, **raspberries** *(off-label)* [1, 11]; **carrots** [1, 2, 4, 6-8, 14]; **rye** [3]; **rye** *(moderate control)*, **triticale** *(moderate control)* [10, 11, 15, 16]; **spring barley**, **spring oats**, **winter barley**, **winter oats**, **winter rye** [1, 2, 4-8, 14]; **spring barley** *(moderate control)*, **winter barley** *(moderate control)* [3, 10, 11, 15, 16]; **spring oats** *(moderate control)*, **winter oats** *(moderate control)* [3, 10, 11, 16]; **triticale** [1-8, 14]

SEE SECTION 3 FOR PRODUCTS ALSO REGISTERED

Pesticide Profiles

- Purple blotch in *leeks* [1, 2, 4, 6-8, 14]
- Rhizoctonia in *celery (outdoor)* (off-label), *protected celery* (off-label), *protected herbs (see appendix 6)* (off-label), *protected leaf brassicas* (off-label), *protected lettuce* (off-label), *protected radicchio* (off-label), *protected scarole* (off-label) [1]; *red beet* (off-label) [2]; *swedes* (off-label), *turnips* (off-label) [1, 2]
- Rhynchosporium in *rye* [3]; *rye* (reduction), *triticale* (reduction) [10, 11, 15, 16]; *spring barley*, *spring rye*, *winter barley*, *winter rye* [1, 2, 4-8, 14]; *spring barley* (reduction), *winter barley* (reduction) [3, 10, 11, 15, 16]; *triticale* [1-8, 14]
- Ring spot in *broccoli*, *brussels sprouts*, *cabbages*, *calabrese*, *cauliflowers*, *collards*, *kale* [1, 2, 4, 6-8, 14]; *chives* (off-label), *herbs (see appendix 6)* (off-label), *parsley* (off-label) [1]
- Root malformation disorder in *red beet* (off-label) [1]
- Rust in *all edible seed crops grown outdoors* (off-label), *all non-edible seed crops grown outdoors* (off-label), *bilberries* (off-label), *blackberries* (off-label), *blackcurrants* (off-label), *blueberries* (off-label), *chicory grown outside for forcing* (off-label), *cranberries* (off-label), *durum wheat* (off-label), *forest nurseries* (off-label), *garlic* (off-label), *gooseberries* (off-label), *grass seed crops* (off-label), *loganberries* (off-label), *ornamental plant production* (off-label), *protected chicory* (off-label), *protected forest nurseries* (off-label), *protected ornamentals* (off-label), *raspberries* (off-label), *redcurrants* (off-label), *rubus hybrids* (off-label), *shallots* (off-label), *strawberries* (off-label) [11]; *asparagus*, *leeks*, *spring field beans*, *winter field beans* [1, 2, 4, 6-8, 14]; *chicory root* (off-label), *chives* (off-label), *herbs (see appendix 6)* (off-label), *parsley* (off-label) [1]; *lupins* (off-label) [1, 11]
- Sclerotinia in *celeriac* (off-label), *celery (outdoor)* (off-label), *lettuce* (off-label), *protected celery* (off-label) [1]; *crambe* (off-label) [11]; *mustard* (off-label) [1, 11]
- Sclerotinia stem rot in *spring oilseed rape*, *winter oilseed rape* [1-8, 14, 16]
- Septoria leaf blotch in *spring wheat*, *winter wheat* [1-8, 10, 11, 14-16]
- Stemphylium in *asparagus* [1, 2, 4, 6-8, 14]
- Take-all in *rye* (reduction), *triticale* (reduction) [3, 10, 11, 15, 16]; *spring barley* (reduction), *spring wheat* (reduction), *winter barley* (reduction), *winter wheat* (reduction) [1-8, 10, 11, 14-16]; *spring oats* (reduction), *winter oats* (reduction) [3]
- Take-all patch in *amenity grassland*, *managed amenity turf* [9, 12, 13]
- White blister in *broccoli*, *brussels sprouts*, *cabbages*, *calabrese*, *cauliflowers*, *collards*, *kale* [1, 2, 4, 6-8, 14]; *chinese cabbage* (off-label), *choi sum* (off-label), *komatsuna* (off-label), *pak choi* (off-label), *tatsoi* (off-label) [1]
- White rust in *chrysanthemums* (off-label - in pots) [1]; *protected chrysanthemums* (off-label) [1, 11]
- Yellow rust in *spring wheat*, *winter wheat* [1-8, 10, 11, 14-16]

Extension of Authorisation for Minor Use (EAMUs)
- *all edible seed crops grown outdoors* 20090443 [1], 20100372 [2], 20151861 [11]
- *all non-edible seed crops grown outdoors* 20090443 [1], 20100372 [2], 20151861 [11]
- *artichokes* 20051814 [1]
- *aubergines* 20021533 [1]
- *bilberries* 20151861 [11]
- *blackberries* 20030365 [1], 20151861 [11]
- *blackcurrants* 20151861 [11]
- *blueberries* 20151861 [11]
- *broad beans* 20032311 [1]
- *celeriac* 20031862 [1]
- *celery (outdoor)* 20011041 [1]
- *chicory grown outside for forcing* 20151864 [11]
- *chicory root* 20051813 [1]
- *chinese cabbage* 20140571 [1]
- *chives* 20021293 [1]
- *choi sum* 20140571 [1]
- *chrysanthemums* (in pots) 20011684 [1]
- *courgettes* 20051985 [1], 20151862 [11]
- *crambe* 20081341 [1], 20151852 [11]
- *cranberries* 20151861 [11]
- *durum wheat* 20061722 [1], 20151860 [11]

FOR FULL CONDITIONS OF USE ALWAYS READ THE PRODUCT LABEL

azoxystrobin

- **dwarf beans** *20032311* [1]
- **forest nurseries** *20090443* [1], *20100372* [2], *20151861* [11]
- **garlic** *20061724* [1], *20151856* [11]
- **globe artichoke** *20151855* [11]
- **gooseberries** *20151861* [11]
- **grass seed crops** *20061722* [1], *20151860* [11]
- **herbs (see appendix 6)** *20021293* [1]
- **horseradish** *20061721* [1], *20151858* [11]
- **inert substrate aubergines** *20011685* [1]
- **inert substrate courgettes** *20011685* [1]
- **inert substrate cucumbers** *20011685* [1]
- **inert substrate gherkins** *20011685* [1]
- **inert substrate tomatoes** *20011685* [1]
- **komatsuna** *20140571* [1]
- **lettuce** *20011465* [1]
- **loganberries** *20151861* [11]
- **lupins** *20061723* [1], *20151857* [11]
- **mustard** *20111165* [1], *20151851* [11]
- **navy beans** *20032311* [1]
- **ornamental plant production** *20090443* [1], *20100372* [2], *20151861* [11]
- **pak choi** *20140571* [1]
- **parsley** *20021293* [1]
- **parsnips** *20061721* [1], *20151858* [11]
- **poppies for morphine production** *20031137* [1], *20151859* [11]
- **protected blackberries** *20051194* [1]
- **protected cayenne peppers** *20021295* [1]
- **protected celery** *20011041* [1]
- **protected chicory** *(for forcing)* *20051813* [1], *20151864* [11]
- **protected chrysanthemums** *20011684* [1], *20151863* [11]
- **protected courgettes** *20021533* [1], *20151850* [11]
- **protected cucumbers** *20021533* [1]
- **protected forest nurseries** *20090443* [1], *20100372* [2], *20151861* [11]
- **protected gherkins** *20021533* [1]
- **protected herbs (see appendix 6)** *20120876* [1]
- **protected leaf brassicas** *20120876* [1]
- **protected lettuce** *20120876* [1]
- **protected marrows** *20070371* [1]
- **protected melons** *20070371* [1], *20151850* [11]
- **protected ornamentals** *20090443* [1], *20100372* [2], *20151861* [11]
- **protected peppers** *20021295* [1]
- **protected pumpkins** *20070371* [1], *20151850* [11]
- **protected radicchio** *20120876* [1]
- **protected raspberries** *20051194* [1]
- **protected scarole** *20120876* [1]
- **protected soft fruit** *20090443* [1], *20100372* [2]
- **protected squashes** *20070371* [1], *20151850* [11]
- **protected strawberries** *20021294* [1]
- **protected summer squash** *20151850* [11]
- **protected tomatoes** *20021533* [1]
- **radishes** *20081448* [1]
- **raspberries** *20030365* [1], *20151861* [11]
- **red beet** *20040614* [1], *20100371* [2]
- **redcurrants** *20151861* [11]
- **rubus hybrids** *20151861* [11]
- **runner beans** *20032311* [1]
- **salad brassicas** *(for baby leaf production)* *20011465* [1]
- **salad onions** *20021687* [1]
- **shallots** *20061724* [1], *20151856* [11]
- **soft fruit** *20090443* [1], *20100372* [2]

SEE SECTION 3 FOR PRODUCTS ALSO REGISTERED

Pesticide Profiles

- **strawberries** 20021294 [1], 20151861 [11]
- **summer squash** 20151862 [11]
- **swedes** 20040614 [1], 20100371 [2]
- **tatsoi** 20140571 [1]
- **turnips** 20040614 [1], 20100371 [2]

Approval information
- Azoxystrobin included in Annex I under EC Regulation 1107/2009
- Accepted by BBPA for use on malting barley

Efficacy guidance
- Best results obtained from use as a protectant or during early stages of disease establishment or when a predictive assessment indicates a risk of disease development
- Azoxystrobin inhibits fungal respiration and should always be used in mixture with fungicides with other modes of action
- Treatment under poor growing conditions may give less reliable results
- For good control of *Fusarium* patch in amenity turf and grass repeat treatment at minimum intervals of 2 wk
- Azoxystrobin is a member of the QoI cross resistance group. Product should be used preventatively and not relied on for its curative potential
- Use product in cereals as part of an Integrated Crop Management strategy incorporating other methods of control, including where appropriate other fungicides with a different mode of action. Do not apply more than two foliar applications of QoI containing products to any cereal crop
- There is a significant risk of widespread resistance occurring in *Septoria tritici* populations in UK. Failure to follow resistance management action may result in reduced levels of disease control
- On cereal crops product must always be used in mixture with another product, recommended for control of the same target disease, that contains a fungicide from a different cross resistance group and is applied at a dose that will give robust control
- Strains of barley powdery mildew resistant to QoI's are common in the UK

Restrictions
- Maximum number of treatments 1 per crop for potatoes; 2 per crop for brassicas, peas, cereals, oilseed rape; 4 per crop or yr for onions, carrots, leeks, amenity turf [1, 8]
- Maximum total dose ranges from 2-4 times the single full dose depending on crop and product. See labels for details
- On turf the maximum number of treatments is 4 per yr but they must not exceed one third of the total number of fungicide treatments applied
- Do not use where there is risk of spray drift onto neighbouring apple crops
- The same spray equipment should not be used to treat apples

Crop-specific information
- Latest use: at planting for potatoes; grain watery ripe (GS 71) for cereals; before senescence for asparagus
- HI: 10 d for carrots;14 d for broccoli, Brussels sprouts, bulb onions, cabbages, calabrese, cauliflowers, collards, kale, vining peas; 21 d for leeks, spring oilseed rape, winter oilseed rape; 36 d for combining peas, 35 d for field beans [1, 8]
- In cereals control of established infections can be improved by appropriate tank mixtures or application as part of a programme. Always use in mixture with another product from a different cross-resistance group
- In turf use product at full dose rate in a disease control programme, alternating with fungicides of different modes of action
- In potatoes when used as incorporated treatment apply overall to the entire area to be planted, incorporate to 15 cm and plant on the same day. In-furrow spray should be directed at the furrow and not the seed tubers
- Applications to brassica crops must only be made to a developed leaf canopy and not before growth stages specified on the label
- Heavy disease pressure in brassicae and oilseed rape may require a second treatment
- All crops should be treated when not under stress. Check leaf wax on peas if necessary
- Consult processor before treating any crops for processing

FOR FULL CONDITIONS OF USE ALWAYS READ THE PRODUCT LABEL

- Treat asparagus after the harvest season. Where a new bed is established do not treat within 3 wk of transplanting out the crowns
- Do not apply to turf when ground is frozen or during drought

Environmental safety
- Dangerous for the environment
- Very toxic to aquatic organisms
- Avoid spray drift onto surrounding areas or crops, especially apples, plums or privet
- Buffer zone requirement 5 m in winter wheat, winter barley and winter oilseed rape
- LERAP Category B [3, 5]; LERAP Category B [1, 2, 4, 6-8, 14] (potatoes only)

Hazard classification and safety precautions
Hazard Harmful [10, 11, 15]; Dangerous for the environment [1-16]; Harmful if inhaled [10]; Very toxic to aquatic organisms [5, 10, 12, 15]
Transport code 9
Packaging group III
UN Number 3077 [9, 12]; 3082 [1-8, 10, 11, 13-16]
Risk phrases R20 [11]; R50, R53a [1, 2, 4, 6-9, 11, 13, 14, 16]
Operator protection A [1-7]; A [8] (during treatment of potatoes); A [10, 11, 13-16]; H [3]; U05a, U20b [1-16]; U09a, U19a [1-8, 10, 11, 14-16]
Environmental protection E15a [1, 2, 4-8, 14]; E15b [3, 9-13, 15, 16]; E16a [1, 2] (potatoes only); E16a [3]; E16a [4] (potatoes only); E16a [5]; E16a [6-8, 14] (potatoes only); E16b [1, 2, 4, 6-8, 14] (potatoes only); E38 [1-16]; H410 [3, 5, 10, 12, 15]
Storage and disposal D01, D02, D09a, D12a [1-16]; D03 [9, 12, 13]; D05 [1, 2, 4-14]; D10b [8, 9, 12, 13]; D10c [1-7, 10, 11, 14-16]
Medical advice M05a [9, 12, 13]

19 azoxystrobin + chlorothalonil

A preventative and systemic fungicide mixture for cereals
FRAC mode of action code: 11 + M5

See also chlorothalonil

Products

1	Amistar Opti	Syngenta	100:500 g/l	SC	14582
2	Aubrac Optimum	AgChem Access	80:400 g/l	SC	15921
3	Curator	Syngenta	80:400 g/l	SC	14955
4	Olympus	Syngenta	80:400 g/l	SC	13797

Uses
- Alternaria in **potatoes** (off-label) [4]
- Botrytis in **asparagus** (qualified minor use recommendation), **bulb onions**, **garlic**, **shallots** [2, 4]; **salad onions** [4]
- Brown rust in **spring barley**, **spring wheat**, **winter barley**, **winter wheat** [1, 3]
- Disease control in **durum wheat** (off-label), **poppies for morphine production** (off-label), **rye** (off-label), **triticale** (off-label) [1]
- Downy mildew in **bulb onions**, **garlic**, **shallots** [2, 4]; **salad onions** [4]
- Glume blotch in **spring wheat**, **winter wheat** [1, 3]
- Net blotch in **spring barley**, **winter barley** [1, 3]
- Rhynchosporium in **spring barley**, **winter barley** [3]; **spring barley** (moderate control), **winter barley** (moderate control) [1]
- Rust in **asparagus** (moderate control only) [2, 4]
- Septoria leaf blotch in **spring wheat**, **winter wheat** [1, 3]
- Stemphylium in **asparagus** (qualified minor use recommendation) [2, 4]
- Take-all in **spring barley** (reduction), **spring wheat** (reduction), **winter barley** (reduction) [1]; **spring barley** (reduction only), **spring wheat** (reduction only), **winter barley** (reduction only) [3]; **winter wheat** (reduction) [1, 3]
- Yellow rust in **spring barley**, **winter barley** [3]; **spring wheat**, **winter wheat** [1, 3]

Pesticide Profiles

Extension of Authorisation for Minor Use (EAMUs)
- *durum wheat* 20120472 [1]
- *poppies for morphine production* 20120473 [1]
- *potatoes* 20090711 [4]
- *rye* 20120472 [1]
- *triticale* 20120472 [1]

Approval information
- Azoxystrobin and chlorothalonil included in Annex I under EC Regulation 1107/2009
- Accepted by BBPA for use on malting barley

Efficacy guidance
- Best results obtained from applications made as a protectant treatment or in earliest stages of disease development. Further applications may be needed if disease attack is prolonged
- Reduction of barley spotting occurs when used as part of a programme with other fungicides
- Control of *Septoria* and rust diseases may be improved by mixture with a triazole fungicide
- Azoxystrobin is a member of the QoI cross resistance group. Product should be used preventatively and not relied on for its curative potential
- Use product in cereals as part of an Integrated Crop Management strategy incorporating other methods of control, including where appropriate other fungicides with a different mode of action. Do not apply more than two foliar applications of QoI containing products to any cereal crop
- There is a significant risk of widespread resistance occurring in *Septoria tritici* populations in UK. Failure to follow resistance management action may result in reduced levels of disease control

Restrictions
- Maximum number of treatments 2 per crop
- Maximum total dose on barley equivalent to one full dose treatment
- Do not use where there is risk of spray drift onto neighbouring apple crops
- The same spray equipment should not be used to treat apples

Crop-specific information
- Latest use: before beginning of heading (GS 51) for barley; before caryopsis watery ripe (GS 71) for wheat

Environmental safety
- Dangerous for the environment
- Very toxic to aquatic organisms
- Buffer zone requirement 18m
- LERAP Category B

Hazard classification and safety precautions
    Hazard Toxic [1]; Harmful [2-4]; Dangerous for the environment [1-4]
    Transport code 9
    Packaging group III
    UN Number 3082
    Risk phrases R20 [2-4]; R22a, R23a [1]; R37, R40, R41, R43, R50, R53a [1-4]
    Operator protection A, C, H; U02a, U05a, U09a, U11, U14, U15, U19a [1-4]; U20a [3]; U20b [1, 2, 4]
    Environmental protection E15a [1, 2, 4]; E15b [3]; E16a, E34, E38 [1-4]
    Storage and disposal D01, D02, D09a, D10c, D12a [1-4]; D05 [1, 2, 4]
    Medical advice M04a [1, 2, 4]

20   azoxystrobin + cyproconazole

A contact and systemic broad spectrum fungicide mixture for cereals, beet crops and oilseed rape
FRAC mode of action code: 11 + 3

See also cyproconazole

FOR FULL CONDITIONS OF USE ALWAYS READ THE PRODUCT LABEL

azoxystrobin + cyproconazole

Products
Priori Xtra   Syngenta   200:80 g/l   SC   11518

Uses
- Alternaria in **spring oilseed rape**, **winter oilseed rape**
- Brown rust in **spring barley**, **spring rye**, **spring wheat**, **winter barley**, **winter rye**, **winter wheat**
- Cercospora leaf spot in **fodder beet**, **sugar beet**
- Crown rust in **spring oats**, **winter oats**
- Eyespot in **spring barley** (reduction), **spring wheat** (reduction), **winter barley** (reduction), **winter wheat** (reduction)
- Foliar disease control in **durum wheat** (off-label), **grass seed crops** (off-label), **triticale** (off-label)
- Glume blotch in **spring wheat**, **winter wheat**
- Net blotch in **spring barley**, **winter barley**
- Powdery mildew in **fodder beet**, **spring barley**, **spring oats**, **spring rye**, **spring wheat**, **sugar beet**, **winter barley**, **winter oats**, **winter rye**, **winter wheat**
- Ramularia leaf spots in **fodder beet**, **sugar beet**
- Rhynchosporium in **spring barley** (moderate control), **spring rye** (moderate control), **winter barley** (moderate control), **winter rye** (moderate control)
- Rust in **fodder beet**, **sugar beet**
- Sclerotinia stem rot in **spring oilseed rape**, **winter oilseed rape**
- Septoria leaf blotch in **spring wheat**, **winter wheat**
- Take-all in **spring barley** (reduction), **spring wheat** (reduction), **winter barley** (reduction), **winter wheat** (reduction)
- Yellow rust in **spring wheat**, **winter wheat**

Extension of Authorisation for Minor Use (EAMUs)
- *durum wheat* 20052891
- *grass seed crops* 20052891
- *triticale* 20052891

Approval information
- Azoxystrobin and cyproconazole included in Annex I under EC Regulation 1107/2009
- Accepted by BBPA for use on malting barley

Efficacy guidance
- Best results obtained from treatment during the early stages of disease development
- A second application may be needed if disease attack is prolonged
- Azoxystrobin is a member of the QoI cross resistance group. Product should be used preventatively and not relied on for its curative potential
- Use product as part of an Integrated Crop Management strategy incorporating other methods of control, including where appropriate other fungicides with a different mode of action. Do not apply more than two foliar applications of QoI containing products to any cereal crop
- There is a significant risk of widespread resistance occurring in *Septoria tritici* populations in UK. Failure to follow resistance management action may result in reduced levels of disease control
- Strains of wheat and barley powdery mildew resistant to QoI's are common in the UK. Control of wheat powdery mildew can only be relied upon from the triazole component
- Where specific control of wheat mildew is required this should be achieved through a programme of measures including products recommended for the control of mildew that contain a fungicide from a different cross-resistance group and applied at a dose that will give robust control
- Cyproconazole is a DMI fungicide. Resistance to some DMI fungicides has been identified in Septoria leaf blotch which may seriously affect performance of some products. For further advice contact a specialist advisor and visit the Fungicide Resistance Action Group (FRAG)-UK website

Restrictions
- Maximum total dose equivalent to two full dose treatments
- Do not use where there is risk of spray drift onto neighbouring apple crops
- The same spray equipment should not be used to treat apples

SEE SECTION 3 FOR PRODUCTS ALSO REGISTERED

Pesticide Profiles

Crop-specific information
- Latest use: up to and including anthesis complete (GS 69) for rye and wheat; up to and including emergence of ear complete (GS 59) for barley and oats; BBCH79 (nearly all pods at final size) or 30 days before harvest, whichever is sooner for oilseed rape

Environmental safety
- Dangerous for the environment
- Very toxic to aquatic organisms

Hazard classification and safety precautions
Hazard Harmful, Dangerous for the environment
Transport code 9
Packaging group III
UN Number 3082
Risk phrases R22a, R50, R53a, R63
Operator protection A; U05a, U09a, U19a, U20b
Environmental protection E15b, E34, E38
Storage and disposal D01, D02, D05, D09a, D10c, D12a
Medical advice M03

21 azoxystrobin + difenoconazole

A broad spectrum fungicide mixture for field crops
FRAC mode of action code: 11 + 3

See also difenoconazole

Products
Amistar Top   Syngenta   200:125 g/l   SC   12761

Uses
- Alternaria blight in **carrots**, **horseradish** *(off-label)*, **parsley root** *(off-label)*, **parsnips** *(off-label)*, **salsify** *(off-label)*
- Powdery mildew in **broccoli**, **brussels sprouts**, **cabbages**, **calabrese**, **carrots**, **cauliflowers**, **collards**, **horseradish** *(off-label)*, **kale**, **parsley root** *(off-label)*, **parsnips** *(off-label)*, **salsify** *(off-label)*
- Purple blotch in **leeks** *(moderate control)*
- Rust in **asparagus** *(off-label)*, **herbs (see appendix 6)** *(off-label)*, **leeks**
- Septoria leaf blotch in **herbs (see appendix 6)** *(off-label)*
- Stemphylium in **asparagus** *(off-label)*
- White blister in **broccoli**, **brussels sprouts**, **cabbages**, **calabrese**, **cauliflowers**, **collards**, **kale**
- White tip in **leeks** *(qualified minor use)*

Extension of Authorisation for Minor Use (EAMUs)
- *asparagus* 20070831
- *herbs (see appendix 6)* 20121838
- *horseradish* 20061476
- *parsley root* 20061476
- *parsnips* 20061476
- *salsify* 20061476

Approval information
- Azoxystrobin and difenoconazole included in Annex I under EC Regulation 1107/2009

Efficacy guidance
- Best results obtained from applications made in the earliest stages of disease development or as a protectant treatment following a disease risk assessment
- Ensure the crop is free from any stress caused by environmental or agronomic effects
- Azoxystrobin is a member of the QoI cross resistance group. Product should be used preventatively and not relied on for its curative potential

FOR FULL CONDITIONS OF USE ALWAYS READ THE PRODUCT LABEL

azoxystrobin + fluazinam

- Use as part of an Integrated Crop Management strategy incorporating other methods of control, including where appropriate other fungicides with a different mode of action. Do not apply more than two foliar applications of QoI containing products

Restrictions
- Maximum number of treatments 2 per crop
- Do not apply where there is a risk of spray drift onto neighbouring apple crops
- Consult processors before treating a crop destined for processing

Crop-specific information
- HI 14 d for carrots; 21 d for brassicas, leeks
- Minimum spray interval of 14 d must be observed on brassicas

Environmental safety
- Dangerous for the environment
- Very toxic to aquatic organisms

Hazard classification and safety precautions
  Hazard Irritant, Dangerous for the environment
  Transport code 9
  Packaging group III
  UN Number 3082
  Risk phrases R20, R43, R50, R53a
  Operator protection A, H; U05a, U09a, U19a, U20b
  Environmental protection E15b, E38
  Storage and disposal D01, D02, D05, D09a, D10c, D12a

22 azoxystrobin + fluazinam

A blight fungicide for use in potatoes
FRAC mode of action code: 11 + 29

Products
  Vendetta   Headland   150:375 g/l   SC   17500

Uses
- Blight in *potatoes*
- Leaf blight in *potatoes*

Approval information
- Azoxystrobin and fluazinam included in Annex I under EC Regulation 1107/2009

Efficacy guidance
- Applications must begin prior to blight development. The first application should be applied at the first blight warning or when local weather conditions are favourable for diseasedevelopment, whichever is the sooner. In the absence of weather conducive to disease development, the first application should be made just before the crop meets in the rows

Restrictions
- Do not use more than 3 consecutive QoI-containing sprays
- Certain apple varieties are highly sensitive. As a precaution, do not apply when there is a risk of spray drift onto neighbouring apple crops. Spray equipment used for application should not be used to treat apples

Environmental safety
- Buffer zone requirement 7m [1]

Hazard classification and safety precautions
  Hazard Very toxic to aquatic organisms
  Transport code 9
  Packaging group III
  UN Number 3082

SEE SECTION 3 FOR PRODUCTS ALSO REGISTERED

Pesticide Profiles

Risk phrases H317, H361
Operator protection A, H
Environmental protection H410

23 azoxystrobin + isopyrazam

A fungicide mixture for use in oilseed rape

See also azoxystrobin

Products
Symetra	Syngenta	200:125 g/l	SC	16701

Uses
- Sclerotinia stem rot in **spring oilseed rape**, **winter oilseed rape**

Approval information
- Azoxystrobin and isopyrazam are included in Appendix 1 under EC Regulation 1107/2009

Efficacy guidance
- Rainfast within 1 hour of application.

Restrictions
- Do not use application equipment used to apply Symetra on apples or damage will occur.
- Contains a member of the QoI cross resistance group and a member of the SDHI cross resistance group. Should be used preventatively and should not be relied on for its curative potential.

Environmental safety
- LERAP Category B

Hazard classification and safety precautions
Hazard Toxic, Dangerous for the environment
Transport code 9
Packaging group III
UN Number 3082
Risk phrases R22a, R23a, R50, R53a, R63
Operator protection A, H; U05a, U19a
Environmental protection E16a, E38
Storage and disposal D01, D02, D12a

24 azoxystrobin + propiconazole

A contact and systemic broad spectrum fungicide mixture for use on maize or grass
FRAC mode of action code: 11 + 3

Products
1	Headway	Syngenta	62.5:104 g/l	EC	14396
2	PureProgress	Pure Amenity	62.5:104 g/l	EC	15471
3	Quilt Xcel	Syngenta	141.4:122.4 g/l	SE	16450

Uses
- Anthracnose in **amenity grassland** *(moderate control)*, **managed amenity turf** *(moderate control)* [1, 2]
- Disease control in **forage maize**, **grain maize** [3]
- Dollar spot in **amenity grassland**, **managed amenity turf** [1, 2]
- Eyespot in **forage maize** *(useful reduction)*, **grain maize** *(useful reduction)* [3]
- Fusarium diseases in **amenity grassland**, **managed amenity turf** [1, 2]
- Leaf blight in **forage maize** *(useful reduction)*, **grain maize** *(useful reduction)* [3]
- Microdochium nivale in **amenity grassland**, **managed amenity turf** [1, 2]

FOR FULL CONDITIONS OF USE ALWAYS READ THE PRODUCT LABEL

azoxystrobin + tebuconazole

Approval information
- Azoxystrobin and propiconazole included in Annex I under EC Regulation 1107/2009

Environmental safety
- LERAP Category B [1, 2]

Hazard classification and safety precautions
    Hazard Dangerous for the environment [1, 2]; Harmful if swallowed, Harmful if inhaled [3]
    Transport code 9
    Packaging group III
    UN Number 3082
    Risk phrases H319 [3]; R50, R53a [1, 2]
    Operator protection A [1-3]; C, H [3]; U05a, U20a [1, 2]; U08, U11, U20b [3]; U09a, U19a [1-3]
    Environmental protection E15b, E34 [1-3]; E16a, E16b, E38 [1, 2]; H410 [3]
    Storage and disposal D01, D05, D09a, D10c [1-3]; D02, D12a [1, 2]; D12b [3]
    Medical advice M03 [3]

25   azoxystrobin + tebuconazole

A strobilurin/triazole fungicide mixture for disease control in oilseed rape
FRAC mode of action code: 11 + 3

Products
    Custodia   Adama   120:200 g/l   SC   16393

Uses
- Chocolate spot in **spring field beans** *(moderate control)*, **winter field beans**
- Disease control in **spring field beans**, **winter field beans** *(moderate control)*
- Sclerotinia in **spring oilseed rape** *(moderate control)*, **winter oilseed rape** *(moderate control)*

Approval information
- Azoxystrobin and tebuconazole included in Annex I under EC Regulation 1107/2009

Efficacy guidance
- Azoxystrobin is a member of the QoI cross resistance group. Product should be used preventatively and not relied on for its curative potential
- Use product as part of an Integrated Crop Management strategy incorporating other methods of control, including where appropriate other fungicides with a different mode of action. Do not apply more than two foliar applications of QoI containing products to any crop

Restrictions
- Avoid drift on to neighbouring crops since damage may occur especially to broad-leaved plants
- Newer authorisations for tebuconazole products require application to cereals only after GS 30 and applications to oilseed rape and linseed after GS20 - check label

Environmental safety
- LERAP Category B

Hazard classification and safety precautions
    Hazard Harmful, Dangerous for the environment, Harmful if swallowed, Very toxic to aquatic organisms
    Transport code 9
    Packaging group III
    UN Number 3082
    Risk phrases H361
    Operator protection A, H; U02a, U04a, U05a, U20b
    Environmental protection E15b, E16a, E34, E38, H410
    Storage and disposal D01, D02, D09a, D10b, D12a
    Medical advice M03, M05a

SEE SECTION 3 FOR PRODUCTS ALSO REGISTERED

Pesticide Profiles

26 Bacillus subtilis

A bacterial fungicide for the control of Botrytis cinerea
FRAC mode of action code: 44

Products

Serenade ASO   Bayer CropScience   13.96 g/l   SC   16139

Uses
- Botrytis in **bilberries** *(off-label)*, **blackcurrants** *(off-label)*, **blueberries** *(off-label)*, **bulb vegetables** *(off-label)*, **canary grass** *(off-label)*, **cane fruit** *(off-label)*, **figs** *(off-label)*, **fruiting vegetables** *(off-label)*, **gooseberries** *(off-label)*, **herbs (see appendix 6)** *(off-label)*, **hops** *(off-label)*, **leafy vegetables** *(off-label)*, **legumes** *(off-label)*, **ornamental plant production** *(off-label)*, **redcurrants** *(off-label)*, **ribes hybrids** *(off-label)*, **root & tuber crops** *(off-label)*, **stem vegetables** *(off-label)*, **table grapes** *(off-label)*, **top fruit** *(off-label)*, **vegetable brassicas** *(off-label)*, **wine grapes** *(off-label)*
- Botrytis fruit rot in **protected strawberries** *(reduction of damage to fruit)*, **strawberries**
- Butt rot in **lettuce** *(off-label)*
- Cavity spot in **carrots** *(off-label)*, **parsnips** *(off-label)*
- Damping off in **baby leaf crops** *(off-label)*, **broccoli** *(off-label)*, **brussels sprouts** *(off-label)*, **bulb onions** *(off-label)*, **cabbages** *(off-label)*, **calabrese** *(off-label)*, **carrots** *(off-label)*, **cauliflowers** *(off-label)*, **celeriac** *(off-label)*, **celery (outdoor)** *(off-label)*, **chard** *(off-label)*, **collards** *(off-label)*, **garlic** *(off-label)*, **herbs (see appendix 6)** *(off-label)*, **kale** *(off-label)*, **leeks** *(off-label)*, **lettuce** *(off-label)*, **parsnips** *(off-label)*, **salad onions** *(off-label)*, **shallots** *(off-label)*, **spinach** *(off-label)*
- Grey mould in **protected strawberries** *(reduction of damage to fruit)*, **strawberries**
- Helminthosporium in **potatoes** *(off-label)*
- Phytophthora in **amenity vegetation** *(off-label)*, **blackberries** *(off-label)*, **blackcurrants** *(off-label)*, **blueberries** *(off-label)*, **courgettes** *(off-label)*, **cranberries** *(off-label)*, **cucumbers** *(off-label)*, **forest nurseries** *(off-label)*, **gooseberries** *(off-label)*, **loganberries** *(off-label)*, **marrows** *(off-label)*, **pumpkins** *(off-label)*, **raspberries** *(off-label)*, **redcurrants** *(off-label)*, **rubus hybrids** *(off-label)*, **squashes** *(off-label)*
- Pythium in **aubergines** *(off-label)*, **broccoli** *(off-label)*, **brussels sprouts** *(off-label)*, **cabbages** *(off-label)*, **calabrese** *(off-label)*, **carrots** *(off-label)*, **cauliflowers** *(off-label)*, **chillies** *(off-label)*, **collards** *(off-label)*, **courgettes** *(off-label)*, **cucumbers** *(off-label)*, **kale** *(off-label)*, **marrows** *(off-label)*, **parsnips** *(off-label)*, **peppers** *(off-label)*, **pumpkins** *(off-label)*, **squashes** *(off-label)*, **tomatoes (outdoor)** *(off-label)*
- Rhizoctonia in **lettuce** *(off-label)*, **potatoes** *(off-label)*
- Streptomyces in **potatoes** *(off-label)*
- White rot in **bulb onions** *(off-label)*, **garlic** *(off-label)*, **leeks** *(off-label)*, **salad onions** *(off-label)*, **shallots** *(off-label)*

Extension of Authorisation for Minor Use (EAMUs)
- *amenity vegetation* 20130704
- *aubergines* 20150306
- *baby leaf crops* 20150306
- *bilberries* 20130706
- *blackberries* 20150306
- *blackcurrants* 20130706, 20150306
- *blueberries* 20130706, 20150306
- *broccoli* 20150306
- *brussels sprouts* 20150306
- *bulb onions* 20150306
- *bulb vegetables* 20130706
- *cabbages* 20150306
- *calabrese* 20150306
- *canary grass* 20130706
- *cane fruit* 20130706
- *carrots* 20150306
- *cauliflowers* 20150306
- *celeriac* 20150306

FOR FULL CONDITIONS OF USE ALWAYS READ THE PRODUCT LABEL

Bacillus subtilis

- *celery (outdoor)* 20150306
- *chard* 20150306
- *chillies* 20150306
- *collards* 20150306
- *courgettes* 20150306
- *cranberries* 20150306
- *cucumbers* 20150306
- *figs* 20130706
- *forest nurseries* 20130704
- *fruiting vegetables* 20130706
- *garlic* 20150306
- *gooseberries* 20130706, 20150306
- *herbs (see appendix 6)* 20130706, 20150306
- *hops* 20130706
- *kale* 20150306
- *leafy vegetables* 20130706
- *leeks* 20150306
- *legumes* 20130706
- *lettuce* 20150306
- *loganberries* 20150306
- *marrows* 20150306
- *ornamental plant production* 20130706
- *parsnips* 20150306
- *peppers* 20150306
- *potatoes* 20150306
- *pumpkins* 20150306
- *raspberries* 20150306
- *redcurrants* 20130706, 20150306
- *ribes hybrids* 20130706
- *root & tuber crops* 20130706
- *rubus hybrids* 20150306
- *salad onions* 20150306
- *shallots* 20150306
- *spinach* 20150306
- *squashes* 20150306
- *stem vegetables* 20130706
- *table grapes* 20130706
- *tomatoes (outdoor)* 20150306
- *top fruit* 20130706
- *vegetable brassicas* 20130706
- *wine grapes* 20130706

Approval information
- Bacillus subtilis included in Annex I under EC Regulation 1107/2009

Efficacy guidance
- Alternating applications with fungicides using a different mode of action is recommended for resistance management.
- Do not apply using irrigation equipment.
- For maximum effectiveness, start applications before disease development.
- Apply in a minimum water volume of 400 l/ha.

Restrictions
- Consult processor before using on crops grown for processing

Hazard classification and safety precautions
Hazard Irritant
UN Number N/C
Risk phrases R43
Operator protection A, D, H; U05a, U11, U14, U20b

SEE SECTION 3 FOR PRODUCTS ALSO REGISTERED

Pesticide Profiles

Environmental protection E15b
Storage and disposal D01, D02, D05, D10a, D16

27 Bacillus thuringiensis

A bacterial insecticide for control of caterpillars
IRAC mode of action code: 11

Products

1	Dipel DF	Interfarm	54% w/w	WG	17499
2	Lepinox Plus	Fargro	37.5% w/w	WP	16269

Uses

- Bramble shoot moth in *bilberries* (off-label), *blackberries* (off-label), *blackcurrants* (off-label), *blueberries* (off-label), *cranberries* (off-label), *gooseberries* (off-label), *loganberries* (off-label), *redcurrants* (off-label), *rubus hybrids* (off-label) [2]
- Cabbage moth in *baby leaf crop* (off-label), *choi sum* (off-label), *collards* (off-label), *kohlrabi* (off-label), *oriental cabbage* (off-label), *protected chicory* (off-label), *protected cress* (off-label), *protected lamb's lettuce* (off-label), *protected rocket* (off-label), *protected watercress* (off-label) [2]
- Cabbage white butterfly in *baby leaf crop* (off-label), *choi sum* (off-label), *collards* (off-label), *kohlrabi* (off-label), *oriental cabbage* (off-label), *protected chicory* (off-label), *protected cress* (off-label), *protected lamb's lettuce* (off-label), *protected rocket* (off-label), *protected watercress* (off-label) [2]
- Caterpillars in *all edible seed crops grown outdoors* (off-label), *all non-edible crops (outdoor)* (off-label), *all protected non-edible crops* (off-label), *amenity vegetation*, *apples* (off-label), *apricots* (off-label), *baby leaf crops* (off-label), *calabrese* (off-label), *cauliflowers*, *celery (outdoor)* (off-label), *cherries* (off-label), *chestnuts* (off-label), *choi sum* (off-label), *cob nuts* (off-label), *collards* (off-label), *edible podded peas* (off-label), *endives* (off-label), *fennel* (off-label), *filberts* (off-label), *forest* (off-label), *hazel nuts* (off-label), *herbs (see appendix 6)* (off-label), *kale* (off-label), *kiwi fruit* (off-label), *kohlrabi* (off-label), *leeks*, *lettuce* (off-label), *nectarines* (off-label), *olives* (off-label), *oriental cabbage* (off-label), *ornamental plant production*, *peaches* (off-label), *pears* (off-label), *plums* (off-label), *protected apricots* (off-label), *protected baby leaf crops* (off-label), *protected broad beans* (off-label), *protected broccoli* (off-label), *protected brussels sprouts* (off-label), *protected cabbages* (off-label), *protected calabrese* (off-label), *protected cauliflowers* (off-label), *protected celery* (off-label), *protected chestnuts* (off-label), *protected chilli peppers* (off-label), *protected choi sum* (off-label), *protected collards* (off-label), *protected cucumbers* (off-label), *protected dwarf french beans* (off-label), *protected edible crops* (off-label), *protected endives* (off-label), *protected hazelnuts* (off-label), *protected herbs (see appendix 6)* (off-label), *protected kale* (off-label), *protected kiwi fruit* (off-label), *protected lettuce* (off-label), *protected nectarines* (off-label), *protected olives* (off-label), *protected oriental cabbage* (off-label), *protected peaches* (off-label), *protected plums* (off-label), *protected quince* (off-label), *protected rhubarb* (off-label), *protected runner beans*, *protected spinach* (off-label), *protected spinach beet* (off-label), *protected tatsoi* (off-label), *protected walnuts* (off-label), *protected watercress* (off-label), *protected wine grapes* (off-label), *quinces* (off-label), *raspberries*, *rhubarb* (off-label), *spinach* (off-label), *spinach beet* (off-label), *spring cabbage* (off-label), *tatsoi* (off-label), *vining peas*, *walnuts* (off-label), *watercress* (off-label), *wine grapes* (off-label) [1]; *apples, brussels sprouts, celery (outdoor), chicory, chinese cabbage, courgettes, dwarf beans, endives, french beans, herbs (see appendix 6), hops, kale, lettuce, pears, protected melons, protected watermelon, pumpkins, radishes, spinach, spinach beet, turnips, wine grapes, winter squash* [2]; *broccoli, cabbages, calabrese, combining peas, globe artichoke, protected aubergines, protected peppers, protected tomatoes, strawberries* [1, 2]
- Cherry fruit moth in *apricots* (off-label), *cherries* (off-label), *nectarines* (off-label), *peaches* (off-label), *plums* (off-label) [2]
- Cutworms in *baby leaf crop* (off-label), *bilberries* (off-label), *blackberries* (off-label), *blackcurrants* (off-label), *blueberries* (off-label), *broad beans* (off-label), *cranberries* (off-label), *forest nurseries* (off-label), *gooseberries* (off-label), *loganberries* (off-label), *protected chicory* (off-label), *protected chilli peppers* (off-label), *protected courgettes* (off-label), *protected*

FOR FULL CONDITIONS OF USE ALWAYS READ THE PRODUCT LABEL

Bacillus thuringiensis

cress *(off-label)*, **protected dwarf french beans** *(off-label)*, **protected edible podded peas** *(off-label)*, **protected lamb's lettuce** *(off-label)*, **protected peppers** *(off-label)*, **protected rocket** *(off-label)*, **protected runner beans** *(off-label)*, **protected summer squash** *(off-label)*, **protected watercress** *(off-label)*, **redcurrants** *(off-label)*, **rhubarb** *(off-label)*, **rubus hybrids** *(off-label)*, **sweetcorn** *(off-label)* [2]; **bulb onions** *(off-label)*, **carrots** *(off-label)*, **celeriac** *(off-label)*, **garlic** *(off-label)*, **horseradish** *(off-label)*, **mooli** *(off-label)*, **parsley root** *(off-label)*, **parsnips** *(off-label)*, **radishes** *(off-label)*, **red beet** *(off-label)*, **salsify** *(off-label)*, **shallots** *(off-label)*, **swedes** *(off-label)*, **turnips** *(off-label)* [1]; **leeks** *(off-label)* [1, 2]
- Diamond-back moth in **baby leaf crop** *(off-label)*, **broccoli**, **brussels sprouts**, **cabbages**, **calabrese**, **chinese cabbage**, **choi sum** *(off-label)*, **collards** *(off-label)*, **kohlrabi** *(off-label)*, **oriental cabbage** *(off-label)*, **protected chicory** *(off-label)*, **protected cress** *(off-label)*, **protected lamb's lettuce** *(off-label)*, **protected rocket** *(off-label)*, **protected watercress** *(off-label)*, **turnips** [2]
- Flax tortrix moth in **broad beans** *(off-label)*, **protected dwarf french beans** *(off-label)*, **protected edible podded peas** *(off-label)*, **protected runner beans** *(off-label)* [2]
- Leek moth in **leeks** *(off-label)*, **rhubarb** *(off-label)* [2]
- Light brown apple moth in **apricots** *(off-label)*, **cherries** *(off-label)*, **nectarines** *(off-label)*, **peaches** *(off-label)*, **plums** *(off-label)* [2]
- Oak Processionary Moth in **forest** *(off-label)* [1]
- Plum fruit moth in **apricots** *(off-label)*, **cherries** *(off-label)*, **nectarines** *(off-label)*, **peaches** *(off-label)*, **plums** *(off-label)* [2]
- Plum tortrix moth in **apricots** *(off-label)*, **cherries** *(off-label)*, **nectarines** *(off-label)*, **peaches** *(off-label)*, **plums** *(off-label)* [2]
- Plutella xylostella in **broccoli**, **brussels sprouts**, **cabbages**, **calabrese**, **chinese cabbage**, **turnips** [2]
- Raspberry moth in **bilberries** *(off-label)*, **blackberries** *(off-label)*, **blackcurrants** *(off-label)*, **blueberries** *(off-label)*, **cranberries** *(off-label)*, **gooseberries** *(off-label)*, **loganberries** *(off-label)*, **redcurrants** *(off-label)*, **rubus hybrids** *(off-label)* [2]
- Silver Y moth in **baby leaf crop** *(off-label)*, **protected chicory** *(off-label)*, **protected chilli peppers** *(off-label)*, **protected courgettes** *(off-label)*, **protected cress** *(off-label)*, **protected dwarf french beans** *(off-label)*, **protected edible podded peas** *(off-label)*, **protected lamb's lettuce** *(off-label)*, **protected peppers** *(off-label)*, **protected rocket** *(off-label)*, **protected runner beans** *(off-label)*, **protected summer squash** *(off-label)*, **protected watercress** *(off-label)* [2]; **broad beans** *(off-label)*, **sweetcorn** *(off-label)* [1, 2]; **dwarf beans** *(off-label)*, **french beans** *(off-label)*, **runner beans** *(off-label)* [1]
- Strawberry tortrix in **bilberries** *(off-label)*, **blackberries** *(off-label)*, **blackcurrants** *(off-label)*, **blueberries** *(off-label)*, **cranberries** *(off-label)*, **gooseberries** *(off-label)*, **loganberries** *(off-label)*, **redcurrants** *(off-label)*, **rubus hybrids** *(off-label)* [2]
- Summer-fruit tortrix moth in **apricots** *(off-label)*, **cherries** *(off-label)*, **nectarines** *(off-label)*, **peaches** *(off-label)*, **plums** *(off-label)* [2]
- Tomato moth in **protected chilli peppers** *(off-label)*, **protected courgettes** *(off-label)*, **protected peppers** *(off-label)*, **protected summer squash** *(off-label)*, **sweetcorn** *(off-label)* [2]
- Tortrix moths in **apples**, **pears**, **strawberries** [2]
- Vapourer moth in **forest nurseries** *(off-label)* [2]
- White ermine moth in **forest nurseries** *(off-label)* [2]
- Winter moth in **apricots** *(off-label)*, **cherries** *(off-label)*, **nectarines** *(off-label)*, **peaches** *(off-label)*, **plums** *(off-label)* [2]; **bilberries** *(off-label)*, **blackcurrants** *(off-label)*, **blueberries** *(off-label)*, **cranberries** *(off-label)*, **prorected vaccinium spp.** *(off-label)*, **protected bilberries** *(off-label)*, **protected blackcurrants** *(off-label)*, **protected blueberry** *(off-label)*, **protected cranberries** *(off-label)*, **protected gooseberries** *(off-label)*, **protected redcurrants** *(off-label)*, **redcurrants** *(off-label)*, **vaccinium spp.** *(off-label)* [1]

Extension of Authorisation for Minor Use (EAMUs)
- **all edible seed crops grown outdoors** *20162634* [1]
- **all non-edible crops (outdoor)** *20162634* [1]
- **all protected non-edible crops** *20162634* [1]
- **apples** *20162635* [1]
- **apricots** *20162632* [1], *20142700* [2]
- **baby leaf crop** *20142704* [2]

SEE SECTION 3 FOR PRODUCTS ALSO REGISTERED

Pesticide Profiles

- **baby leaf crops** *20162627* [1]
- **bilberries** *20162633* [1], *20142706* [2]
- **blackberries** *20142706* [2]
- **blackcurrants** *20162633* [1], *20142706* [2]
- **blueberries** *20162633* [1], *20142706* [2]
- **broad beans** *20162625* [1], *20142702* [2]
- **bulb onions** *20162623* [1]
- **calabrese** *20162626* [1]
- **carrots** *20162631* [1]
- **celeriac** *20162631* [1]
- **celery (outdoor)** *20162626* [1]
- **cherries** *20162635* [1], *20142700* [2]
- **chestnuts** *20162632* [1]
- **choi sum** *20162626* [1], *20142701* [2]
- **cob nuts** *20162632* [1]
- **collards** *20162626* [1], *20142701* [2]
- **cranberries** *20162633* [1], *20142706* [2]
- **dwarf beans** *20162625* [1]
- **endives** *20162627* [1]
- **fennel** *20162626* [1]
- **filberts** *20162632* [1]
- **forest** *20160931* [1]
- **forest nurseries** *20142705* [2]
- **french beans** *20162625* [1]
- **garlic** *20162623* [1]
- **gooseberries** *20142706* [2]
- **hazel nuts** *20162632* [1]
- **herbs (see appendix 6)** *20162627* [1]
- **horseradish** *20162631* [1]
- **kale** *20162626* [1]
- **kiwi fruit** *20162632* [1]
- **kohlrabi** *20162626* [1], *20142701* [2]
- **leeks** *20162629* [1], *20142703* [2]
- **lettuce** *20162627* [1]
- **loganberries** *20142706* [2]
- **mooli** *20162631* [1]
- **nectarines** *20162632* [1], *20142700* [2]
- **olives** *20162632* [1]
- **oriental cabbage** *20162626* [1], *20142701* [2]
- **parsley root** *20162631* [1]
- **parsnips** *20162631* [1]
- **peaches** *20162632* [1], *20142700* [2]
- **pears** *20162635* [1]
- **plums** *20162632* [1], *20142700* [2]
- **prorected vaccinium spp.** *20162633* [1]
- **protected apricots** *20162632* [1]
- **protected baby leaf crops** *20162627* [1]
- **protected bilberries** *20162633* [1]
- **protected blackcurrants** *20162633* [1]
- **protected blueberry** *20162633* [1]
- **protected broccoli** *20162626* [1]
- **protected brussels sprouts** *20162626* [1]
- **protected cabbages** *20162626* [1]
- **protected calabrese** *20162626* [1]
- **protected cauliflowers** *20162626* [1]
- **protected celery** *20162626* [1]
- **protected chestnuts** *20162632* [1]
- **protected chicory** *20142704* [2]
- **protected chilli peppers** *20142707* [2]

FOR FULL CONDITIONS OF USE ALWAYS READ THE PRODUCT LABEL

Bacillus thuringiensis

- *protected choi sum* 20162626 [1]
- *protected collards* 20162626 [1]
- *protected courgettes* 20142707 [2]
- *protected cranberries* 20162633 [1]
- *protected cress* 20142704 [2]
- *protected dwarf french beans* 20142702 [2]
- *protected edible crops* 20162634 [1]
- *protected edible podded peas* 20142702 [2]
- *protected endives* 20162627 [1]
- *protected gooseberries* 20162633 [1]
- *protected hazelnuts* 20162632 [1]
- *protected herbs (see appendix 6)* 20162627 [1]
- *protected kale* 20162626 [1]
- *protected kiwi fruit* 20162632 [1]
- *protected lamb's lettuce* 20142704 [2]
- *protected lettuce* 20162627 [1]
- *protected nectarines* 20162632 [1]
- *protected olives* 20162632 [1]
- *protected oriental cabbage* 20162626 [1]
- *protected peaches* 20162632 [1]
- *protected peppers* 20142707 [2]
- *protected plums* 20162632 [1]
- *protected quince* 20162632 [1]
- *protected redcurrants* 20162633 [1]
- *protected rhubarb* 20162626 [1]
- *protected rocket* 20142704 [2]
- *protected runner beans* 20142702 [2]
- *protected spinach* 20162627 [1]
- *protected spinach beet* 20162627 [1]
- *protected summer squash* 20142707 [2]
- *protected tatsoi* 20162626 [1]
- *protected walnuts* 20162632 [1]
- *protected watercress* 20162628 [1], 20142704 [2]
- *protected wine grapes* 20162632 [1]
- *quinces* 20162632 [1]
- *radishes* 20162631 [1]
- *red beet* 20162624 [1]
- *redcurrants* 20162633 [1], 20142706 [2]
- *rhubarb* 20162626 [1], 20142703 [2]
- *rubus hybrids* 20142706 [2]
- *runner beans* 20162625 [1]
- *salsify* 20162631 [1]
- *shallots* 20162623 [1]
- *spinach* 20162627 [1]
- *spinach beet* 20162627 [1]
- *spring cabbage* 20162626 [1]
- *swedes* 20162631 [1]
- *sweetcorn* 20162630 [1], 20142707 [2]
- *tatsoi* 20162626 [1]
- *turnips* 20162631 [1]
- *vaccinium spp.* 20162633 [1]
- *walnuts* 20162632 [1]
- *watercress* 20162628 [1]
- *wine grapes* 20162632 [1]

Approval information
- Bacillus thuringiensis included in Annex I under EC Regulation 1107/2009

SEE SECTION 3 FOR PRODUCTS ALSO REGISTERED

Pesticide Profiles

Efficacy guidance
- Pest control achieved by ingestion by caterpillars of the treated plant vegetation. Caterpillars cease feeding and die in 1-3 d
- Apply as soon as larvae appear on crop and repeat every 7-10 d until the end of the hatching period
- Good coverage is essential, especially of undersides of leaves. Spray onto dry foliage and do not apply if rain expected within 6 h

Restrictions
- Apply spray mixture as soon as possible after preparation

Crop-specific information
- HI zero

Environmental safety
- Store out of direct sunlight

Hazard classification and safety precautions
UN Number N/C
Operator protection A, C, D, H; U05a, U15, U20c [1]; U11, U12, U14, U16b, U20b [2]; U19a [1, 2]
Environmental protection E15a [1]; E15b, E34 [2]
Storage and disposal D01, D02, D09a [1, 2]; D05, D11a [1]; D10b [2]
Medical advice M03, M05a [2]

28 Bacillus thuringiensis israelensis

A bacterial insecticide for control of larvae of chironomid midges
IRAC mode of action code: 11

Products
Vectobac 12 AS   Resource Chemicals   12 g/l   ZZ   HSE 6205

Uses
- Chironomid midge in **protected watercress** *(off-label)*, **watercress** *(off-label)*

Extension of Authorisation for Minor Use (EAMUs)
- *protected watercress* 20162667
- *watercress* 20162667

Hazard classification and safety precautions
UN Number N/C
Operator protection A, H

29 Beauveria bassiana

It is an entomopathogenic fungus causing white muscardine disease. It can be used as a biological insecticide to control a number of pests such as termites, whitefly and some beetles

Products
1	Botanigard WP	Certis	22% w/w	WP	17054
2	Naturalis-L	Belchim	7.16% w/w	OD	14655
3	Naturalis-L	Fargro	7.16% w/w	OD	17526

Uses
- Aphids in **protected aubergines, protected chilli peppers, protected courgettes, protected cucumbers, protected melons, protected nursery fruit trees, protected peppers, protected strawberries, protected summer squash, protected tomatoes** [1]; **protected edible crops, protected forest nurseries** *(off-label)* [2, 3]; **protected ornamentals** [1-3]
- Beetles in **protected aubergines, protected chilli peppers, protected courgettes, protected cucumbers, protected melons, protected nursery fruit trees, protected peppers, protected**

FOR FULL CONDITIONS OF USE ALWAYS READ THE PRODUCT LABEL

bentazone

  strawberries, protected summer squash, protected tomatoes [1]; *protected edible crops, protected forest nurseries* (off-label) [2, 3]; *protected ornamentals* [1-3]
- Insect pests in *protected forest nurseries* (off-label) [2, 3]
- Thrips in *protected strawberries* (off-label), *strawberries* (off-label) [1]
- Whitefly in *protected aubergines, protected chilli peppers, protected courgettes, protected cucumbers, protected melons, protected nursery fruit trees, protected peppers, protected strawberries, protected strawberries* (off-label), *protected summer squash, protected tomatoes, strawberries* (off-label) [1]; *protected edible crops, protected forest nurseries* (off-label) [2, 3]; *protected ornamentals* [1-3]

Extension of Authorisation for Minor Use (EAMUs)
- *protected forest nurseries* 20162194 [2], 20162195 [3]
- *protected strawberries* 20160110 [1]
- *strawberries* 20160110 [1]

Approval information
- Beauveria bassiana included in Annex 1 under EC Regulation 1107/2009

Hazard classification and safety precautions
UN Number N/C
Risk phrases H317, H334 [1]
Operator protection A, D, H; U05a, U11, U14, U15, U16b, U19a, U20b
Environmental protection E15b, E34
Storage and disposal D01, D02, D05, D09a, D10a
Medical advice M03, M04a

30 bentazone

A post-emergence contact benzothiadiazinone herbicide
HRAC mode of action code: C3

Products

1	Basagran SG	BASF	87% w/w	SG	08360
2	Benta 480 SL	Sharda	480 g/l	SL	14940
3	Clayton Dent 480	Clayton	480 g/l	SL	15546
4	Tanaru	Agroquimicos	480 g/l	SL	15726
5	Troy 480	UPL Europe	480 g/l	SL	16954

Uses
- Annual dicotyledons in *broad beans, linseed, potatoes, runner beans, spring field beans, winter field beans* [1-5]; *bulb onions* (off-label), *chives* (off-label), *garlic* (off-label), *herbs (see appendix 6)* (off-label), *leeks* (off-label), *salad onions* (off-label), *shallots* (off-label), *soya beans* (off-label) [1]; *combining peas, dwarf beans, vining peas* [4, 5]; *french beans, peas* [1-3]; *narcissi* [1-4]; *navy beans* [1, 4]; *ornamental plant production* [5]
- Chickweed in *game cover* (off-label), *hops* (off-label), *ornamental plant production* (off-label) [1]
- Cleavers in *game cover* (off-label), *hops* (off-label), *ornamental plant production* (off-label) [1]
- Common storksbill in *leeks* (off-label) [1]
- Fool's parsley in *leeks* (off-label) [1]
- Groundsel in *game cover* (off-label), *herbs (see appendix 6)* (off-label), *hops* (off-label), *ornamental plant production* (off-label) [1]
- Mayweeds in *bulb onions* (off-label), *game cover* (off-label), *garlic* (off-label), *herbs (see appendix 6)* (off-label), *hops* (off-label), *ornamental plant production* (off-label), *shallots* (off-label) [1]

Extension of Authorisation for Minor Use (EAMUs)
- *bulb onions* 20061631 [1]
- *chives* 20102994 [1]
- *game cover* 20082819 [1]
- *garlic* 20061631 [1]

SEE SECTION 3 FOR PRODUCTS ALSO REGISTERED

Pesticide Profiles

- *herbs (see appendix 6)* 20120744 [1]
- *hops* 20082819 [1]
- *leeks* 20061630 [1]
- *ornamental plant production* 20082819 [1]
- *salad onions* 20102994 [1]
- *shallots* 20061631 [1]
- *soya beans* 20061629 [1]

Approval information
- Bentazone included in Annex I under EC Regulation 1107/2009

Efficacy guidance
- Most effective control obtained when weeds are growing actively and less than 5 cm high or across. Good spray cover is essential
- The addition of specified adjuvant oils is recommended for use on some crops to improve fat hen control. Do not use under hot or humid conditions. See label for details
- Split dose application may be made in all recommended crops except peas and generally gives better weed control. See label for details

Restrictions
- Maximum number of treatments normally 2 per crop but check label
- Crops must be treated at correct stage of growth to avoid danger of scorch. See label for details
- Not all varieties of recommended crops are fully tolerant. Use only on tolerant varieties named in label. Do not use on forage or mange-tout varieties of peas
- Do not use on crops which have been affected by drought, waterlogging, frost or other stress conditions
- Do not apply insecticides within 7 d of treatment
- Leave 14 d after using a post-emergence grass herbicide and carry out a leaf wax test where relevant or 7 d where treatment precedes the grass herbicide
- Do not spray at temperatures above 21°C. Delay spraying until evening if necessary
- Do not apply if rain or frost expected, if unseasonably cold, if foliage wet or in drought
- A minimum of 6 h (preferably 12 h) free from rain is required after application
- May be used on selected varieties of maincrop and second early potatoes (see label for details), not on seed crops or first earlies

Crop-specific information
- Latest use: before shoots exceed 15 cm high for potatoes and spring field beans (or 6-7 leaf pairs); 4 leaf pairs (6 pairs or 15 cm high with split dose) for broad beans; before flower buds visible for French, navy, runner and winter field beans, and linseed; before flower buds can be found enclosed in terminal shoot for peas
- Best results in narcissi obtained by using a suitable pre-emergence herbicide first
- Consult processor before using on crops for processing
- A satisfactory wax test must be carried out before use on peas
- Do not treat narcissi during flower bud initiation

Environmental safety
- Dangerous for the environment
- Harmful to aquatic organisms
- Some pesticides pose a greater threat of contamination of water than others and bentazone is one of these pesticides. Take special care when applying bentazone near water and do not apply if heavy rain is forecast

Hazard classification and safety precautions
 Hazard Harmful [1-3]; Irritant [4, 5]; Harmful if swallowed [1, 3]
 Packaging group III [1]
 UN Number 2588 [1]; N/C [2-5]
 Risk phrases H317 [1, 3]; H318 [1]; R22a [2]; R36 [4, 5]; R43 [2, 4, 5]
 Operator protection A [1-5]; C [1, 4, 5]; H [4, 5]; U05a, U08, U19a [1-5]; U11, U20b [1, 4, 5]; U14 [1]; U20a [2, 3]
 Environmental protection E15a [2, 3]; E15b, E34 [1, 4, 5]; E38 [5]; H412 [3]

FOR FULL CONDITIONS OF USE ALWAYS READ THE PRODUCT LABEL

Storage and disposal D01, D02, D09a [1-5]; D10b [1]; D10c [2-5]
Medical advice M03 [1, 4, 5]; M05a [1]

31 benthiavalicarb-isopropyl

An amino acid amide carbamate fungicide available only in mixtures
FRAC mode of action code: 40

32 benthiavalicarb-isopropyl + mancozeb

A fungicide mixture for potatoes
FRAC mode of action code: 40 + M3

See also mancozeb

Products
Valbon   Certis   1.75:70% w/w   WG   14868

Uses
- Blight in **potatoes**
- Downy mildew in **amenity vegetation** *(off-label)*, **bulb onions** *(off-label)*, **forest nurseries** *(off-label)*, **garlic** *(off-label)*, **hops** *(off-label)*, **ornamental plant production** *(off-label)*, **shallots** *(off-label)*, **soft fruit** *(off-label)*, **table grapes** *(off-label)*, **wine grapes** *(off-label)*

Extension of Authorisation for Minor Use (EAMUs)
- *amenity vegetation* 20142325
- *bulb onions* 20142326
- *forest nurseries* 20142328
- *garlic* 20142326
- *hops* 20142328
- *ornamental plant production* 20142325
- *shallots* 20142326
- *soft fruit* 20142328
- *table grapes* 20142327
- *wine grapes* 20142327

Approval information
- Benthiavalicarb-isopropyl and mancozeb included in Annex I under EC Regulation 1107/2009

Efficacy guidance
- Apply as a protectant spray commencing before blight enters the crop irrespective of growth stage following a blight warning or where there is a local source of infection
- In the absence of a warning commence the spray programme before the foliage meets in the rows
- Repeat treatment every 7-10 d depending on disease pressure
- Increase water volume as necessary to ensure thorough coverage of the plant
- Use of air assisted sprayers or drop legs may help to improve coverage

Restrictions
- Maximum number of treatments 6 per crop
- Do not use more than three consecutive treatments. Include products with a different mode of action in the programme

Crop-specific information
- HI 7 d for potatoes

Environmental safety
- Very toxic to aquatic organisms
- Risk to non-target insects or other arthropods. Avoid spraying within 6 m of field boundary
- LERAP Category B

SEE SECTION 3 FOR PRODUCTS ALSO REGISTERED

Pesticide Profiles

Hazard classification and safety precautions
Hazard Irritant
Transport code 9
Packaging group III
UN Number 3077
Risk phrases R40, R43, R50, R53a
Operator protection A, D, H; U14
Environmental protection E15a, E16a, E22c
Storage and disposal D01, D02, D05, D09a, D11a, D12b

33 benzoic acid

An organic horticultural disinfectant

Products

Menno Florades	Fargro	90 g/kg	SL	13985

Uses
- Bacteria in *all protected non-edible crops*
- Fungal spores in *all protected non-edible crops*
- Viroids in *all protected non-edible crops*
- Virus in *all protected non-edible crops*

Approval information
- Benzoic acid is included in Annex 1 under EC Regulation 1107/2009

Hazard classification and safety precautions
Hazard Irritant, Flammable
Transport code 3
Packaging group III
UN Number 1987
Risk phrases R41, R67
Operator protection A, C, H; U05a, U11, U13, U14, U15
Environmental protection E15a
Storage and disposal D01, D02, D09a, D10a
Medical advice M05a

34 benzyladenine

A cytokinin plant growth regulator for use in apples and pears

Products

1	Configure	Fargro	20 g/l	SC	17523
2	Exilis	Fine	20 g/l	SL	15706
3	Globaryll 100	Globachem	100 g/l	SL	16097
4	MaxCel	Interfarm	20 g/l	SL	15708

Uses
- Fruit thinning in *apples* [2, 4]; *pears* [2]
- Growth regulation in *apples*, *pears* [3]; *kalanchoes*, *ornamental plant production* [1]
- Increasing flowering in *phalaenopsis spp.*, *schlumbergera spp.*, *sempervivium spp.* [1]

Approval information
- Benzyladenine is included in Annex 1 under EC Directive 1107/2009

Efficacy guidance
- A post-bloom thinner for apples and pears with excessive blossom [2, 3]
- Apply when the temperature will exceed 15°C on day of application but note that temperatures above 28°C may result in excessive thinning.

FOR FULL CONDITIONS OF USE ALWAYS READ THE PRODUCT LABEL

beta-cyfluthrin

Restrictions
- Use in pears is a Qualified Minor Use recommendation. Evidence of efficacy and crop safety in pears is limited to the pear variety 'Conference' [2]

Environmental safety
- Broadcast air-assisted LERAP [4] (5 m)

Hazard classification and safety precautions
    Hazard Harmful, Dangerous for the environment [3]
    UN Number N/C
    Risk phrases H317 [1]; H318, H361 [3]; R52, R53a [2, 4]
    Operator protection A [1, 2, 4]; C [3]; H [1]; U02a, U09a, U19a, U20b [2, 4]; U05a [2-4]; U11, U14, U15, U20c [3]
    Environmental protection E15a, H412 [3]; E15b [1, 2, 4]; E17b [4] (5 m); E38 [2-4]
    Storage and disposal D01 [1, 3]; D02, D09a [1-4]; D03, D05, D12a [2, 4]; D10a [1]; D10b [2-4]
    Medical advice M04a [3]; M05a [2, 4]

35   beta-cyfluthrin

A non-systemic pyrethroid insecticide for insect control in cereals, oilseed rape, cabbages, cauliflowers and sugar beet
IRAC mode of action code: 3

Products
    Gandalf   Adama   25 g/l   EC   12865

Uses
- Aphids in *oats, rye, spring barley, spring wheat, sugar beet, triticale, winter barley, winter wheat*
- Brassica pod midge in *oilseed rape*
- Cabbage moth in *cabbages, cauliflowers*
- Cabbage seed weevil in *oilseed rape*
- Cabbage stem flea beetle in *oilseed rape*
- Cabbage stem weevil in *oilseed rape*
- Peach-potato aphid in *cabbages, cauliflowers*
- Wheat-blossom midge in *rye, spring wheat, triticale, winter wheat*

Approval information
- Beta-cyfluthrin included in Annex I under EC Regulation 1107/2009

Hazard classification and safety precautions
    Hazard Harmful, Dangerous for the environment
    Transport code 9
    Packaging group III
    UN Number 3082
    Risk phrases R20, R22a, R22b, R36, R38, R50, R53a, R66, R67
    Operator protection A, C, H; U04a, U05a, U10, U12, U14, U19a, U20a
    Environmental protection E12c, E12e, E16e, E34, E38
    Storage and disposal D01, D02, D09a, D10a, D12a
    Medical advice M03, M05b

36   beta-cyfluthrin + clothianidin

An insecticidal seed treatment mixture for beet
IRAC mode of action code: 3 + 4A

See also clothianidin

Products
    Poncho Beta   Bayer CropScience   53:400 g/l   FS   12076

SEE SECTION 3 FOR PRODUCTS ALSO REGISTERED

Pesticide Profiles

Uses
- Beet virus yellows vectors in **fodder beet** *(seed treatment)*, **sugar beet** *(seed treatment)*

Approval information
- Beta-cyfluthrin and clothianidin included in Annex I under EC Regulation 1107/2009

Efficacy guidance
- In addition to control of aphid virus vectors, product improves crop establishment by reducing damage caused by symphylids, springtails, millipedes, wireworms and leatherjackets
- Additional control measures should be taken where very high populations of soil pests are present
- Product reduces direct feeding damage by foliar pests such as pygmy mangold beetle, beet flea beetle, mangold fly
- Product is not active against nematodes
- Treatment does not alter physical characteristics of pelleted seed and no change to standard drill settings should be necessary

Restrictions
- Maximum number of treatments: 1 per seed batch
- Product must be co-applied with a colouring dye
- Product must only be applied as a coating to pelleted seed using special treatment machinery

Crop-specific information
- Latest use: pre-drilling

Environmental safety
- Extremely dangerous to fish or other aquatic life. Do not contaminate surface waters or ditches with chemical or used container

Hazard classification and safety precautions
Hazard Harmful
Transport code 9
Packaging group III
UN Number 3082
Risk phrases R22a
Operator protection A, H; U04a, U05a, U07, U13, U14, U20b, U24
Environmental protection E13a, E34, E36a
Storage and disposal D01, D02, D05, D09a, D14
Treated seed S02, S03, S04a, S04b, S05, S06a, S06b, S07
Medical advice M03

37 bifenazate

A bifenazate acaricide for use on protected strawberries
IRAC mode of action code: 25

Products
Floramite 240 SC   Certis   240 g/l   SC   13686

Uses
- Mites in **protected forest nurseries** *(off-label)*, **protected hops** *(off-label)*
- Two-spotted spider mite in **protected strawberries**

Extension of Authorisation for Minor Use (EAMUs)
- *protected forest nurseries* 20142220
- *protected hops* 20142220

Approval information
- Bifenazate included in Annex I under EC Regulation 1107/2009

Efficacy guidance
- Bifenazate acts by contact knockdown after approximately 4 d and a subsequent period of residual control

FOR FULL CONDITIONS OF USE ALWAYS READ THE PRODUCT LABEL

bifenox

- All mobile stages of mites are controlled with occasionally some ovicidal activity
- Best results obtained from a programme of two treatments started as soon as first spider mites are seen
- To minimise possible development of resistance use in a planned Resistance Management strategy. Alternate at least two products with different modes of action between treatment programmes of bifenazate. See Section 5 for more information

Restrictions
- Maximum number of treatments 2 per yr for protected strawberries
- After use in protected environments avoid re-entry for at least 2 h while full ventilation is carried out allowing spray to dry on leaves
- Personnel working among treated crops should wear suitable long-sleeved garments and gloves for 14 d after treatment
- Do not apply as a low volume spray

Crop-specific information
- HI: 7 d for protected strawberries
- Carry out small scale tolerance test on the strawberry cultivar before large scale use

Environmental safety
- Dangerous for the environment
- Toxic to aquatic organisms
- Harmful to non-target predatory mites. Avoid spray drift onto field margins, hedges, ditches, surface water and neighbouring crops

Hazard classification and safety precautions
   Hazard Irritant, Dangerous for the environment
   Transport code 9
   Packaging group III
   UN Number 3082
   Risk phrases R43, R51, R53a
   Operator protection A, H; U02a, U04a, U05a, U10, U14, U15, U19a, U20b
   Environmental protection E15b, E38
   Storage and disposal D01, D02, D05, D09a, D10b, D12b
   Medical advice M03

38   bifenox

A diphenyl ether herbicide for use in cereals and (off-label) in oilseed rape.
HRAC mode of action code: E

Products

1	Clayton Belstone	Clayton	480 g/l	SC	14033
2	Cleancrop Diode	Clayton	480 g/l	SC	14620
3	Fox	Adama	480 g/l	SC	11981

Uses
- Annual dicotyledons in *canary flower (echium spp.)* (off-label) [1, 3]; *durum wheat* (off-label), *grass seed crops* (off-label) [1-3]; *evening primrose* (off-label), *honesty* (off-label), *linseed* (off-label), *mustard* (off-label), *oilseed rape* (off-label) [3]; *triticale, winter barley, winter rye, winter wheat* [1]
- Cleavers in *triticale, winter barley, winter rye, winter wheat* [3]
- Field pansy in *triticale, winter barley, winter rye, winter wheat* [3]
- Field speedwell in *triticale, winter barley, winter rye, winter wheat* [3]
- Forget-me-not in *triticale, winter barley, winter rye, winter wheat* [3]
- Geranium species in *canary flower (echium spp.)* (off-label), *durum wheat* (off-label), *evening primrose* (off-label), *grass seed crops* (off-label), *honesty* (off-label), *linseed* (off-label) [2]
- Ivy-leaved speedwell in *triticale, winter barley, winter rye, winter wheat* [3]
- Poppies in *triticale, winter barley, winter rye, winter wheat* [3]
- Red dead-nettle in *triticale, winter barley, winter rye, winter wheat* [3]

SEE SECTION 3 FOR PRODUCTS ALSO REGISTERED

Pesticide Profiles

Extension of Authorisation for Minor Use (EAMUs)
- *canary flower (echium spp.)* 20121422 [1], 20121423 [2], 20142222 [3], 20142318 [3]
- *durum wheat* 20121421 [1], 20121424 [2], 20142221 [3], 20142319 [3]
- *evening primrose* 20121423 [2], 20142222 [3], 20142318 [3]
- *grass seed crops* 20121421 [1], 20121424 [2], 20142221 [3], 20142319 [3]
- *honesty* 20121423 [2], 20142222 [3], 20142318 [3]
- *linseed* 20121423 [2], 20142222 [3], 20142318 [3]
- *mustard* 20142222 [3], 20142318 [3]
- *oilseed rape* 20142222 [3], 20142318 [3]

Approval information
- Accepted by BBPA for use on malting barley
- Bifenox included in Annex I under EC Regulation 1107/2009

Efficacy guidance
- Bifenox is absorbed by foliage and emerging roots of susceptible species
- Best results obtained when weeds are growing actively with adequate soil moisture

Restrictions
- Maximum number of treatments: one per crop
- Do not apply to crops suffering from stress from whatever cause
- Do not apply if the crop is wet or if rain or frost is expected
- Avoid drift onto broad-leaved plants outside the target area

Crop-specific information
- Latest use: before 2nd node detectable (GS 32) for all crops

Environmental safety
- Dangerous for the environment
- Very toxic to aquatic organisms
- Do not empty into drains

Hazard classification and safety precautions
    Hazard Dangerous for the environment [1, 3]
    Transport code 9
    Packaging group III
    UN Number 3082
    Operator protection A [1, 3]; U05a, U08, U20b [1, 3]
    Environmental protection E15a, E19b, E34, E38, H410 [1, 3]
    Storage and disposal D01, D02, D09a, D12a [1, 3]
    Medical advice M03 [1, 3]

39 bixafen

A succinate dehydrogenase inhibitor (SDHI) fungicide available only in mixtures
FRAC mode of action code: 7

40 bixafen + fluopyram + prothioconazole

A fungicide mixture for disease control in wheat
FRAC mode of action code: 7 + 3

See also bixafen + prothioconazole

Products
    Ascra Xpro   Bayer CropScience   65:65:130 g/l   EC   17623

Uses
- Brown rust in **spring wheat**, **winter wheat**
- Ear diseases in **spring wheat** *(reduction of incidence and severity)*, **winter wheat** *(reduction of incidence and severity)*

FOR FULL CONDITIONS OF USE ALWAYS READ THE PRODUCT LABEL

bixafen + fluoxastrobin + prothioconazole

- Eyespot in **spring wheat** (reduction of incidence and severity), **winter wheat** (reduction of incidence and severity)
- Glume blotch in **spring wheat**, **winter wheat**
- Powdery mildew in **spring wheat**, **winter wheat**
- Septoria leaf blotch in **spring wheat**, **winter wheat**
- Tan spot in **spring wheat**, **winter wheat**
- Yellow rust in **spring wheat**, **winter wheat**

Approval information
- Bixafen, fluopyram and prothioconazole are included in Annex 1 under EC Regulation 1107/2009.

Efficacy guidance
- Applications to upper leaves where S. tritici symptoms are present are likely to be less effective.

Restrictions
- Bixafen is an SDH respiration inhibitor; Do not apply more than two foliar applications of products containing an SDH inhibitor to any cereal crop
- Resistance to some DMI fungicides has been identified in Septoria leaf blotch (*Mycosphaerella graminicola*) which may seriously affect the performance of some products. For further advice on resistance management in DMIs contact your agronomist or specialist advisor, and visit the FRAG-UK website.

Environmental safety
- LERAP Category B

Hazard classification and safety precautions
Hazard Harmful if swallowed
Transport code 9
Packaging group III
UN Number 3082
Risk phrases H317, H318
Operator protection A, C, H; U05a, U09b, U11, U20a
Environmental protection E15b, E16a, E34, H410
Storage and disposal D01, D02, D05, D09a, D10b, D12b

41   bixafen + fluoxastrobin + prothioconazole

An SDHI + strobilurin + triazole mixture for disease control in cereals
FRAC mode of action code: 7 + 11 + 3

Products

| Variano Xpro | Bayer CropScience | 40:50:100 g/l | EC | 15218 |

Uses
- Brown rust in **spring wheat**, **triticale**, **winter rye**, **winter wheat**
- Ear diseases in **spring wheat**, **triticale**, **winter rye**, **winter wheat**
- Glume blotch in **spring wheat**, **triticale**, **winter rye**, **winter wheat**
- Leaf blotch in **triticale**, **winter rye**
- Powdery mildew in **spring wheat**, **triticale**, **winter rye**, **winter wheat**
- Rhynchosporium in **winter rye**
- Tan spot in **spring wheat**, **triticale**, **winter rye**, **winter wheat**
- Yellow rust in **spring wheat**, **triticale**, **winter rye**, **winter wheat**

Approval information
- Bixafen, fluroxystrobin and prothioconazole are included in Annex 1 under EC Regulation 1107/2009.

Restrictions
- Bixafen is an SDH respiration inhibitor; Do not apply more than two foliar applications of products containing an SDH inhibitor to any cereal crop

SEE SECTION 3 FOR PRODUCTS ALSO REGISTERED

Pesticide Profiles

Environmental safety
- LERAP Category B

Hazard classification and safety precautions
Hazard Irritant, Dangerous for the environment
Transport code 9
Packaging group III
UN Number 3082
Risk phrases H317, H319
Operator protection A, H; U05b, U08, U14, U20b
Environmental protection E15a, E16a, E16b, E34, E38, H410
Storage and disposal D01, D02, D05, D09a, D10b, D12a
Medical advice M03

42 bixafen + prothioconazole

Succinate dehydrogenase inhibitor (SDHI) + triazole fungicide mixture for disease control in cereals
FRAC mode of action code: 3 + 7

Products
1	Aviator 235 Xpro	Bayer CropScience	75:160 g/l	EC	15026
2	Siltra Xpro	Bayer CropScience	60:200 g/l	EC	15082

Uses
- Brown rust in *spring barley*, *winter barley* [2]; *spring wheat*, *triticale*, *winter rye*, *winter wheat* [1]
- Crown rust in *spring oats*, *winter oats* [2]
- Ear diseases in *spring barley*, *winter barley* [2]; *spring wheat*, *triticale*, *winter wheat* [1]
- Eyespot in *spring barley* (reduction of incidence and severity), *spring oats* (reduction of incidence and severity), *winter barley* (reduction of incidence and severity), *winter oats* (reduction of incidence and severity) [2]; *spring wheat* (reduction of incidence and severity), *triticale* (reduction of incidence and severity), *winter wheat* (reduction of incidence and severity) [1]
- Glume blotch in *spring wheat*, *triticale*, *winter wheat* [1]
- Net blotch in *spring barley*, *winter barley* [2]
- Powdery mildew in *spring barley*, *spring oats*, *winter barley*, *winter oats* [2]; *spring wheat*, *triticale*, *winter rye*, *winter wheat* [1]
- Ramularia leaf spots in *spring barley*, *winter barley* [2]
- Rhynchosporium in *spring barley*, *winter barley* [2]; *winter rye* [1]
- Septoria leaf blotch in *spring wheat*, *triticale*, *winter wheat* [1]
- Tan spot in *spring wheat*, *triticale*, *winter wheat* [1]
- Yellow rust in *spring barley*, *winter barley* [2]; *spring wheat*, *triticale*, *winter wheat* [1]

Approval information
- Bixafen and prothioconazole are included in Annex I under EC Regulation 1107/2009.
- Accepted by BBPA for use on malting barley

Efficacy guidance
- Resistance to some DMI fungicides has been identified in Septoria leaf blotch (*Mycosphaerella graminicola*) which may seriously affect the performance of some products. For further advice on resistance management in DMIs contact your agronomist or specialist advisor, and visit the FRAG-UK website.

Restrictions
- Bixafen is an SDH respiration inhibitor; Do not apply more than two foliar applications of products containing an SDH inhibitor to any cereal crop
- Resistance to some DMI fungicides has been identified in Septoria leaf blotch (*Mycosphaerella graminicola*) which may seriously affect the performance of some products. For further advice on resistance management in DMIs contact your agronomist or specialist advisor, and visit the FRAG-UK website.

FOR FULL CONDITIONS OF USE ALWAYS READ THE PRODUCT LABEL

Environmental safety
- LERAP Category B

Hazard classification and safety precautions
Hazard Harmful, Dangerous for the environment [1, 2]; Harmful if swallowed [2]
Transport code 9
Packaging group III
UN Number 3082
Risk phrases H317 [2]; H319 [1, 2]
Operator protection U05a, U09a, U20b
Environmental protection E15b, E16a, E16b, E34, E38, H410
Storage and disposal D01, D02, D09a, D10b, D12a

43 bixafen + prothioconazole + spiroxamine

A broad-spectrum fungicide mixture for disease control in cereals
FRAC mode of action code: 3 + 5 + 7

Products

Boogie Xpro   Bayer CropScience   50:100:250 g/l   EC   15061

Uses
- Brown rust in *spring wheat*, *triticale*, *winter rye*, *winter wheat*
- Ear diseases in *spring wheat*, *triticale*, *winter rye*, *winter wheat*
- Eyespot in *spring wheat*, *triticale*, *winter rye*, *winter wheat*
- Glume blotch in *spring wheat*, *triticale*, *winter rye*, *winter wheat*
- Powdery mildew in *spring wheat*, *triticale*, *winter rye*, *winter wheat*
- Rhynchosporium in *triticale*, *winter rye*
- Septoria leaf blotch in *spring wheat*, *triticale*, *winter rye*, *winter wheat*
- Tan spot in *spring wheat*, *triticale*, *winter rye*, *winter wheat*
- Yellow rust in *spring wheat*, *triticale*, *winter rye*, *winter wheat*

Approval information
- Bixafen, prothioconazole and spiroxamine included in Annex I under EC Regulation 1107/2009.

Efficacy guidance
- Resistance to some DMI fungicides has been identified in Septoria leaf blotch (*Mycosphaerella graminicola*) which may seriously affect the performance of some products. For further advice on resistance management in DMIs contact your agronomist or specialist advisor, and visit the FRAG-UK website.

Restrictions
- No more than two applications of SDH inhibitors must be applied to the same cereal crop.

Environmental safety
- LERAP Category B

Hazard classification and safety precautions
Hazard Harmful, Dangerous for the environment, Harmful if swallowed, Harmful if inhaled
Risk phrases H318
Operator protection A, C, H; U05a, U09b, U11, U20a, U26
Environmental protection E15b, E16a, E16b, E34, E38, H410
Storage and disposal D01, D02, D05, D09a, D10b, D12a
Medical advice M03

44 bixafen + prothioconazole + tebuconazole

A succinate dehydrogenase inhibitor (SDHI) + triazole fungicide mixture for disease control in cereals
FRAC mode of action code: 3 + 7

SEE SECTION 3 FOR PRODUCTS ALSO REGISTERED

Pesticide Profiles

Products

1	Skyway 285 Xpro	Bayer CropScience	75:110:100 g/l	EC	15028
2	Sparticus Xpro	Bayer CropScience	75:110:90 g/l	EC	15162

Uses
- Brown rust in *spring wheat*, *triticale*, *winter rye*, *winter wheat*
- Ear diseases in *spring wheat*, *triticale*, *winter wheat*
- Eyespot in *spring wheat* (reduction of incidence and severity), *triticale* (reduction of incidence and severity), *winter wheat* (reduction of incidence and severity)
- Glume blotch in *spring wheat*, *triticale*, *winter wheat*
- Light leaf spot in *winter oilseed rape*
- Phoma in *winter oilseed rape*
- Powdery mildew in *spring wheat*, *triticale*, *winter rye*, *winter wheat*
- Rhynchosporium in *winter rye*
- Sclerotinia in *winter oilseed rape*
- Septoria leaf blotch in *spring wheat*, *triticale*, *winter wheat*
- Stem canker in *winter oilseed rape*
- Tan spot in *spring wheat*, *triticale*, *winter wheat*
- Yellow rust in *spring wheat*, *triticale*, *winter wheat*

Approval information
- Bixafen, prothioconazole and tebuconazole included in Annex I under EC Regulation 1107/2009.

Efficacy guidance
- Resistance to some DMI fungicides has been identified in Septoria leaf blotch (*Mycosphaerella graminicola*) which may seriously affect the performance of some products. For further advice on resistance management in DMIs contact your agronomist or specialist advisor, and visit the FRAG-UK website.

Restrictions
- Bixafen is an SDH respiration inhibitor; Do not apply more than two foliar applications of products containing an SDH inhibitor to any cereal crop
- Newer authorisations for tebuconazole products require application to cereals only after GS 30 and applications to oilseed rape and linseed after GS20 - check label

Environmental safety
- LERAP Category B

Hazard classification and safety precautions
Hazard Harmful, Dangerous for the environment, Harmful if swallowed
Transport code 9
Packaging group III
UN Number 3082
Risk phrases H315, H317, H361
Operator protection A, H; U05a, U09a, U20b
Environmental protection E15b, E16a, E16b, E34, E38, H410
Storage and disposal D01, D02, D09a, D10b, D12a

45 boscalid

A translocated and translaminar anilide fungicide
FRAC mode of action code: 7

Products

1	Filan	BASF	50% w/w	WG	11449
2	Fulmar	AgChem Access	50% w/w	WG	15767

Uses
- Alternaria in *spring oilseed rape*, *winter oilseed rape* [1, 2]
- Botrytis in *poppies for morphine production* (off-label) [1]

FOR FULL CONDITIONS OF USE ALWAYS READ THE PRODUCT LABEL

boscalid

- Disease control in **all edible seed crops grown outdoors** *(off-label)*, **all non-edible seed crops grown outdoors** *(off-label)*, **borage for oilseed production** *(off-label)*, **canary flower (echium spp.)** *(off-label)*, **corn gromwell** *(off-label)*, **evening primrose** *(off-label)*, **honesty** *(off-label)*, **linseed** *(off-label)*, **mustard** *(off-label)* [1]
- Powdery mildew in **grapevines** *(off-label)* [1]
- Sclerotinia in **poppies for morphine production** *(off-label)* [1]
- Sclerotinia stem rot in **spring oilseed rape**, **winter oilseed rape** [1, 2]

Extension of Authorisation for Minor Use (EAMUs)
- **all edible seed crops grown outdoors** *20093231* [1]
- **all non-edible seed crops grown outdoors** *20093231* [1]
- **borage for oilseed production** *20093229* [1], *20131293* [1]
- **canary flower (echium spp.)** *20093229* [1], *20131293* [1]
- **corn gromwell** *20131293* [1]
- **evening primrose** *20093229* [1], *20131293* [1]
- **grapevines** *20131947* [1]
- **honesty** *20093229* [1], *20131293* [1]
- **linseed** *20093229* [1], *20131293* [1]
- **mustard** *20093229* [1], *20131293* [1]
- **poppies for morphine production** *20093222* [1]

Approval information
- Boscalid included in Annex I under EC Regulation 1107/2009
- Accepted by BBPA on malting barley before ear emergence only and for use on hops

Efficacy guidance
- For best results apply as a protectant spray before symptoms are visible
- Applications against *Sclerotinia* should be made in high disease risk situations at early to full flower
- *Sclerotinia* control may be reduced when high risk conditions occur after flowering, leading to secondary disease spread
- Applications against *Alternaria* should be made at full flowering. Later applications may result in reduced levels of control
- Ensure adequate spray penetration and good coverage

Restrictions
- Maximum total dose equivalent to two full dose treatments
- Do no treat crops to be used for seed production later than full flowering stage
- Avoid spray drift onto neighbouring crops

Crop-specific information
- Latest use: up to and including when 50% of pods have reached final size (GS 75) for oilseed rape

Environmental safety
- Dangerous for the environment
- Toxic to aquatic organisms
- Product represents minimal hazard to bees when used as directed. However local bee-keepers should be notified if crops are to be sprayed when in flower

Hazard classification and safety precautions
Hazard Dangerous for the environment
Transport code 9
Packaging group III
UN Number 3077
Risk phrases R51, R53a [2]
Operator protection U05a, U20c
Environmental protection E15a, E38 [1, 2]; H411 [1]
Storage and disposal D01, D02, D05, D09a, D10c, D12a

SEE SECTION 3 FOR PRODUCTS ALSO REGISTERED

Pesticide Profiles

46 boscalid + dimoxystrobin

A fungicide mixture for disease control in oilseed rape
FRAC mode of action code: 7 + 11

See also boscalid

Products
    Pictor   BASF   200:200 g/l   SC   16783

Uses
- Alternaria in **corn gromwell** *(off-label)*, **winter oilseed rape** *(moderate control)*
- Dark leaf and pod spot in **corn gromwell** *(off-label)*
- Dark leaf spot in **winter oilseed rape** *(moderate control)*
- Disease control in **winter oilseed rape**
- Light leaf spot in **winter oilseed rape** *(reduction)*
- Phoma in **winter oilseed rape** *(moderate control)*
- Sclerotinia in **corn gromwell** *(off-label)*, **winter oilseed rape**
- Stem canker in **winter oilseed rape** *(moderate control)*

Extension of Authorisation for Minor Use (EAMUs)
- **corn gromwell** 20151106

Approval information
- Boscalid and dimoxystrobin included in Annex I under EC Regulation 1107/2009

Efficacy guidance
- Treat light leaf spot and phoma/stem canker at first sign of infection in the spring. Treat sclerotinia at early - full flower in high disease risk situations. Treat altenaria from full flower to when 50% of pods have reached full size.

Environmental safety
- LERAP Category B

Hazard classification and safety precautions
Hazard Harmful if swallowed, Harmful if inhaled, Very toxic to aquatic organisms
Transport code 9
Packaging group III
UN Number 3082
Risk phrases H317, H351, H361
Operator protection A, H; U05a
Environmental protection E15b, E16a, E34, H410
Storage and disposal D01, D02, D09a, D10c, D12b
Medical advice M05a

47 boscalid + epoxiconazole

A broad spectrum fungicide mixture for cereals
FRAC mode of action code: 7 + 3

See also epoxiconazole

Products

1	Chord	BASF	210:75 g/l	SC	16065
2	Enterprise	BASF	140:50 g/l	OD	15228
3	Kingdom	BASF	140:50 g/l	OD	15240
4	Tracker	BASF	233:67 g/l	SC	16048

Uses
- Botrytis in **narcissi** *(off-label)* [4]
- Brown rust in **durum wheat** [1, 4]; **grass seed crops** *(off-label)* [4]; **rye**, **spring barley**, **spring wheat**, **triticale**, **winter barley**, **winter wheat** [1-4]

FOR FULL CONDITIONS OF USE ALWAYS READ THE PRODUCT LABEL

boscalid + epoxiconazole

- Crown rust in **grass seed crops** *(off-label)* [4]; **oats** [2, 3]; **spring oats, winter oats** [1, 4]
- Eyespot in **durum wheat** *(moderate control)* [1, 4]; **rye** *(moderate control)*, **spring barley** *(moderate control)*, **spring wheat** *(moderate control)*, **winter barley** *(moderate control)*, **winter wheat** *(moderate control)* [1-4]; **triticale** *(moderate control)* [1-3]
- Fusarium ear blight in **spring wheat** *(good reduction)*, **winter wheat** *(good reduction)* [2, 3]
- Glume blotch in **spring wheat**, **triticale**, **winter wheat** [2, 3]
- Net blotch in **grass seed crops** *(off-label)* [4]; **spring barley, winter barley** [1-4]
- Powdery mildew in **oats** *(moderate control)*, **rye** *(moderate control)*, **spring barley** *(moderate control)*, **winter barley** *(moderate control)* [2, 3]
- Ramularia leaf spots in **spring barley** *(moderate control)*, **winter barley** *(moderate control)* [1-4]
- Rhynchosporium in **grass seed crops** *(off-label)* [4]; **rye** *(moderate control)*, **spring barley** *(moderate control)*, **winter barley** *(moderate control)* [1-4]
- Septoria leaf blotch in **durum wheat** [1, 4]; **spring wheat**, **triticale**, **winter wheat** [1-4]
- Smoulder in **narcissi** *(off-label)* [4]
- Sooty moulds in **spring wheat**, **winter wheat** [2, 3]
- Tan spot in **durum wheat** *(reduction)* [1, 4]; **spring wheat** *(reduction)*, **winter wheat** *(reduction)* [1-4]
- White mould in **narcissi** *(off-label)* [4]
- Yellow rust in **rye**, **spring barley**, **spring wheat**, **triticale**, **winter barley**, **winter wheat** [2, 3]

Extension of Authorisation for Minor Use (EAMUs)
- **grass seed crops** *20150655* [4]
- **narcissi** *20132458* [4]

Approval information
- Boscalid and epoxiconazole included in Annex I under EC Regulation 1107/2009
- Accepted by BBPA for use on malting barley (before ear emergence only)

Efficacy guidance
- Apply at the start of foliar or stem based disease attack
- Optimum effect against eyespot achieved by spraying between leaf-sheath erect and second node detectable stages (GS 30-32)
- Epoxiconazole is a DMI fungicide. Resistance to some DMI fungicides has been identified in Septoria leaf blotch which may seriously affect performance of some products. For further advice contact a specialist advisor and visit the Fungicide Resistance Action Group (FRAG)-UK website

Crop-specific information
- Latest use: before cereal ear emergence

Environmental safety
- Dangerous for the environment
- Toxic to aquatic organisms
- Avoid drift onto neighbouring crops. May cause damage to broad-leaved plant species
- LERAP Category B

Hazard classification and safety precautions
    Hazard Harmful, Dangerous for the environment
    Transport code 9
    Packaging group III
    UN Number 3082
    Risk phrases R36, R51 [1, 4]; R38, R41, R43, R50 [2, 3]; R40, R53a, R62, R63 [1-4]
    Operator protection A, C [1-4]; H [1, 4]; U05a [1-4]; U14, U20a [2, 3]; U20b [1, 4]
    Environmental protection E15a [1, 4]; E15b [2, 3]; E16a, E38 [1-4]
    Storage and disposal D01, D02, D10c, D12a [1-4]; D05, D09a [2, 3]; D08 [1, 4]
    Medical advice M03 [1-4]; M05a [2, 3]

SEE SECTION 3 FOR PRODUCTS ALSO REGISTERED

Pesticide Profiles

48   boscalid + epoxiconazole + pyraclostrobin

A systemic and protectant fungicide mixture for disease control in cereals
FRAC mode of action code: 3 + 7 + 11

Products
Nebula XL   BASF   140:50:60 g/l   OD   15555

Uses
- Brown rust in *rye*, *spring wheat*, *triticale*, *winter barley*, *winter wheat*
- Crown rust in *oats*
- Eyespot in *rye* (moderate control), *spring wheat* (moderate control), *triticale* (moderate control), *winter barley* (moderate control), *winter wheat* (moderate control)
- Fusarium ear blight in *spring wheat* (good reduction), *winter wheat* (good reduction)
- Glume blotch in *spring wheat*, *triticale*, *winter wheat*
- Net blotch in *winter barley*
- Powdery mildew in *oats* (moderate control), *rye* (moderate control), *triticale* (moderate control), *winter barley* (moderate control)
- Ramularia leaf spots in *winter barley* (moderate control)
- Rhynchosporium in *rye*, *winter barley*
- Septoria leaf blotch in *spring wheat*, *triticale*, *winter wheat*
- Sooty moulds in *spring wheat* (good reduction), *winter wheat* (good reduction)
- Tan spot in *spring wheat* (moderate control), *winter wheat* (moderate control)
- Yellow rust in *rye*, *spring wheat*, *triticale*, *winter barley*, *winter wheat*

Approval information
- Boscalid, epoxiconazole and pyraclostrobin included in Annex I under EC Regulation 1107/2009

Following crops guidance
- Onions, oats, sugar beet, oilseed rape, cabbages, carrots, sunflowers, barley, lettuce, ryegrass, dwarf french beans, peas, potatoes, clover, wheat, field beans and maize may be sown as the following crop.

Environmental safety
- LERAP Category B

Hazard classification and safety precautions
   Hazard Harmful, Dangerous for the environment
   Transport code 9
   Packaging group III
   UN Number 3082
   Risk phrases R20, R38, R40, R43, R50, R53a, R62, R63
   Operator protection A, H; U05a, U14, U20a
   Environmental protection E15b, E16a, E34, E38
   Storage and disposal D01, D02, D05, D09a, D10c, D12a
   Medical advice M03, M05a

49   boscalid + metconazole

An anilide and triazole fungicide mixture for disease control in oilseed rape
FRAC mode of action code: 3 + 7

Products
1 Highgate   BASF   133:60 g/l   SC   15251
2 Tectura   BASF   133:60 g/l   SC   15232

Uses
- Sclerotinia stem rot in *oilseed rape*

Approval information
- Boscalid and metconazole included in Annex I under EC Regulation 1107/2009

FOR FULL CONDITIONS OF USE ALWAYS READ THE PRODUCT LABEL

Following crops guidance
- Any crop can follow normally-harvested oilseed rape treated with [1] or [2]

Environmental safety
- LERAP Category B

Hazard classification and safety precautions
Hazard Harmful
UN Number N/C
Risk phrases R40, R52, R53a, R63
Operator protection A, H; U05a, U20a
Environmental protection E15b, E16a, E16b, E34, E38
Storage and disposal D01, D02, D05, D09a, D10c, D12a
Medical advice M03, M05a

50   boscalid + pyraclostrobin

A protectant and systemic fungicide mixture
FRAC mode of action code: 7 + 11

See also pyraclostrobin

Products
1	Bellis	BASF	25.2:12.8% w/w	WG	12522
2	Signum	BASF	26.7:6.7% w/w	WG	11450

Uses
- Alternaria in **cabbages**, **calabrese**, **carrots** *(moderate control)*, **cauliflowers**, **poppies for morphine production** *(off-label)* [2]
- American gooseberry mildew in **blackcurrants** *(moderate control only)* [2]
- Anthracnose in **blueberries** *(off-label)*, **gooseberries** *(off-label)*, **protected blueberry** *(off-label)*, **protected gooseberries** *(off-label)* [2]
- Black spot in **protected strawberries**, **strawberries** [2]
- Blight in **potatoes** *(off-label)* [2]
- Blossom wilt in **cherries** *(off-label)*, **mirabelles** *(off-label)*, **plums** *(off-label)*, **protected cherries** *(off-label)*, **protected mirabelles** *(off-label)*, **protected plums** *(off-label)* [2]
- Botrytis in **amenity vegetation** *(off-label)*, **asparagus** *(off-label)*, **beans without pods (fresh)** *(off-label)*, **blueberries** *(off-label)*, **broad beans** *(off-label)*, **forest nurseries** *(off-label)*, **gooseberries** *(off-label)*, **interior landscapes** *(off-label)*, **lamb's lettuce** *(off-label)*, **ornamental plant production** *(off-label)*, **poppies for morphine production** *(off-label)*, **protected aubergines** *(off-label)*, **protected blueberry** *(off-label)*, **protected chard** *(off-label)*, **protected forest nurseries** *(off-label)*, **protected gooseberries** *(off-label)*, **protected lettuce** *(off-label)*, **protected ornamentals** *(off-label)*, **protected peppers** *(off-label)*, **protected spinach beet** *(off-label)*, **protected tomatoes** *(off-label)*, **redcurrants** *(off-label)*, **swedes** *(off-label)*, **turnips** *(off-label)*, **whitecurrants** *(off-label)* [2]
- Bottom rot in **lettuce**, **protected lettuce** [2]
- Brown rot in **apricots** *(off-label)*, **nectarines** *(off-label)*, **peaches** *(off-label)* [2]
- Cane blight in **blackberries** *(off-label)*, **protected blackberries** *(off-label)*, **raspberries** *(off-label)* [2]
- Chocolate spot in **spring field beans** *(moderate control)*, **winter field beans** *(moderate control)* [2]
- Cladosporium in **protected cucumbers** *(off-label)* [2]
- Cladosporium leaf blotch in **courgettes** *(off-label)*, **summer squash** *(off-label)* [2]
- Colletotrichum orbiculare in **courgettes** *(off-label)*, **summer squash** *(off-label)* [2]
- Disease control in **all edible crops (outdoor and protected)** *(off-label)*, **all non-edible crops (outdoor)** *(off-label)*, **all protected non-edible crops** *(off-label)*, **beetroot** *(off-label)*, **kohlrabi** *(off-label)*, **lamb's lettuce** *(off-label)*, **protected hops** *(off-label)*, **protected radishes** *(off-label)*, **protected soft fruit** *(off-label)*, **protected top fruit** *(off-label)*, **radishes** *(off-label)*, **red beet** *(off-label)*, **soft fruit** *(off-label)*, **top fruit** *(off-label)* [2]; **hops** *(off-label)* [1, 2]

SEE SECTION 3 FOR PRODUCTS ALSO REGISTERED

Pesticide Profiles

- Downy mildew in **bulls blood** *(off-label)*, **chard** *(off-label)*, **endives** *(off-label)*, **protected endives** *(off-label)*, **spinach** *(off-label)*, **spinach beet** *(off-label)* [2]
- Fusarium in **bulb onion sets** *(off-label)* [2]
- Grey mould in **blackcurrants, lettuce, protected lettuce, protected strawberries, strawberries** [2]
- Gummy stem blight in **protected cucumbers** *(off-label)* [2]
- Late blight in **potatoes** *(off-label)* [2]
- Leaf and pod spot in **combining peas** *(moderate control only)*, **vining peas** *(moderate control only)* [2]
- Leaf spot in **amenity vegetation** *(off-label)*, **black salsify** *(off-label)*, **blackcurrants** *(moderate control only)*, **blueberries** *(off-label)*, **gooseberries** *(off-label)*, **interior landscapes** *(off-label)*, **parsley root** *(off-label)*, **protected blueberry** *(off-label)*, **protected gooseberries** *(off-label)*, **redcurrants** *(off-label)*, **salsify** *(off-label)*, **whitecurrants** *(off-label)* [2]
- Poppy fire in **poppies for morphine production** *(off-label)* [2]
- Powdery mildew in **amenity vegetation** *(off-label)*, **black salsify** *(off-label)*, **carrots, interior landscapes** *(off-label)*, **parsley root** *(off-label)*, **protected cucumbers** *(off-label)*, **protected strawberries, pumpkins** *(off-label)*, **redcurrants** *(off-label)*, **salsify** *(off-label)*, **strawberries, whitecurrants** *(off-label)* [2]; **apples** [1]
- Purple blotch in **blackberries** *(off-label)*, **protected blackberries** *(off-label)*, **raspberries** *(off-label)* [2]
- Rhizoctonia in **protected chard** *(off-label)*, **protected lettuce** *(off-label)*, **protected spinach beet** *(off-label)* [2]
- Ring spot in **broccoli, brussels sprouts, cabbages, calabrese, cauliflowers** [2]
- Rust in **amenity vegetation** *(off-label)*, **asparagus** *(off-label)*, **black salsify** *(off-label)*, **interior landscapes** *(off-label)*, **kohlrabi** *(off-label)*, **parsley root** *(off-label)*, **protected blackberries** *(off-label)*, **salsify** *(off-label)*, **spring field beans, winter field beans** [2]
- Scab in **apples, pears** [1]
- Sclerotinia in **black salsify** *(off-label)*, **carrots** *(moderate control)*, **celeriac** *(off-label)*, **horseradish** *(off-label)*, **parsley root** *(off-label)*, **protected chard** *(off-label)*, **protected lettuce** *(off-label)*, **protected spinach beet** *(off-label)*, **salsify** *(off-label)* [2]
- Soft rot in **lettuce, protected lettuce** [2]
- White blister in **broccoli, brussels sprouts, cabbages** *(qualified minor use)*, **calabrese** *(qualified minor use)*, **chinese cabbage** *(off-label)*, **choi sum** *(off-label)*, **collards** *(off-label)*, **herbs (see appendix 6)** *(off-label)*, **kale** *(off-label)*, **leaf brassicas** *(off-label)*, **pak choi** *(off-label)*, **protected herbs (see appendix 6)** *(off-label)*, **protected leaf brassicas** *(off-label)*, **tatsoi** *(off-label)* [2]
- White rot in **bulb onions** *(off-label)*, **garlic** *(off-label)*, **salad onions** *(off-label)*, **shallots** *(off-label)* [2]
- White rust in **kohlrabi** *(off-label)* [2]
- White tip in **leeks** *(off-label)* [2]

Extension of Authorisation for Minor Use (EAMUs)
- *all edible crops (outdoor and protected)* 20102111 [2]
- *all non-edible crops (outdoor)* 20102111 [2]
- *all protected non-edible crops* 20102111 [2]
- *amenity vegetation* 20122317 [2]
- *apricots* 20121721 [2]
- *asparagus* 20102105 [2]
- *beans without pods (fresh)* 20121009 [2]
- *beetroot* 20121717 [2]
- *black salsify* 20121720 [2]
- *blackberries* 20102110 [2]
- *blueberries* 20121722 [2]
- *broad beans* 20121009 [2]
- *bulb onion sets* 20103122 [2]
- *bulb onions* 20102108 [2]
- *bulls blood* 20102136 [2]
- *celeriac* 20141059 [2]
- *chard* 20102136 [2]
- *cherries* 20102109 [2]

FOR FULL CONDITIONS OF USE ALWAYS READ THE PRODUCT LABEL

boscalid + pyraclostrobin

- **chinese cabbage** *20130285* [2]
- **choi sum** *20130285* [2]
- **collards** *20130285* [2]
- **courgettes** *20142855* [2]
- **endives** *20102136* [2]
- **forest nurseries** *20102119* [2]
- **garlic** *20102108* [2]
- **gooseberries** *20121722* [2]
- **herbs (see appendix 6)** *20102115* [2]
- **hops** *20112732* [1], *20102111* [2]
- **horseradish** *20093375* [2]
- **interior landscapes** *20122317* [2]
- **kale** *20130285* [2]
- **kohlrabi** *20121719* [2]
- **lamb's lettuce** *20121718* [2]
- **leaf brassicas** *20102115* [2]
- **leeks** *20102134* [2]
- **mirabelles** *20102109* [2]
- **nectarines** *20121721* [2]
- **ornamental plant production** *20122141* [2]
- **pak choi** *20130285* [2]
- **parsley root** *20121720* [2]
- **peaches** *20121721* [2]
- **plums** *20102109* [2]
- **poppies for morphine production** *20101233* [2]
- **potatoes** *20110394* [2]
- **protected aubergines** *20120427* [2]
- **protected blackberries** *20102102* [2]
- **protected blueberry** *20121722* [2]
- **protected chard** *20131807* [2]
- **protected cherries** *20102109* [2]
- **protected cucumbers** *20151178* [2]
- **protected endives** *20102136* [2]
- **protected forest nurseries** *20102119* [2]
- **protected gooseberries** *20121722* [2]
- **protected herbs (see appendix 6)** *20102115* [2]
- **protected hops** *20102111* [2]
- **protected leaf brassicas** *20102115* [2]
- **protected lettuce** *20131807* [2]
- **protected mirabelles** *20102109* [2]
- **protected ornamentals** *20122141* [2]
- **protected peppers** *20120427* [2]
- **protected plums** *20102109* [2]
- **protected radishes** *20121717* [2]
- **protected soft fruit** *20102111* [2]
- **protected spinach beet** *20131807* [2]
- **protected tomatoes** *20120427* [2]
- **protected top fruit** *20102111* [2]
- **pumpkins** *20152651* [2]
- **radishes** *20121717* [2]
- **raspberries** *20102110* [2]
- **red beet** *20121717* [2]
- **redcurrants** *20102114* [2]
- **salad onions** *20102107* [2]
- **salsify** *20121720* [2]
- **shallots** *20102108* [2]
- **soft fruit** *20102111* [2]
- **spinach** *20102136* [2]
- **spinach beet** *20102136* [2]

SECTION 2

SEE SECTION 3 FOR PRODUCTS ALSO REGISTERED

Pesticide Profiles

- ***summer squash*** *20142855* [2]
- ***swedes*** *20151961* [2]
- ***tatsoi*** *20130285* [2]
- ***top fruit*** *20102111* [2]
- ***turnips*** *20151961* [2]
- ***whitecurrants*** *20102114* [2]

Approval information
- Boscalid and pyraclostrobin included in Annex I under EC Regulation 1107/2009
- Accepted by BBPA on malting barley before ear emergence only and for use on hops

Efficacy guidance
- On brassicas apply as a protectant spray or at the first sign of disease and repeat at 3-4 wk intervals depending on disease pressure [2]
- Ensure adequate spray penetration and coverage by increasing water volume in dense crops [2]
- For best results on strawberries apply as a protectant spray at the white bud stage. Applications should be made in sequence with other products as part of a fungicide spray programme during flowering at 7-10 day intervals [2]
- On carrots and field beans apply as a protectant spray or at the first sign of disease with a repeat treatment if needed, as directed on the label [2]
- On lettuce apply as a protectant spray 1-2 wk after planting [2]
- Optimum results on apples and pears obtained from a protectant treatment from bud burst [1]
- Application as the final 2 sprays on apples and pears gives reduction in storage rots [1]
- Pyraclostrobin is a member of the QoI cross-resistance group of fungicides and should be used in programmes with fungicides with a different mode of action

Restrictions
- Maximum total dose equivalent to three full dose treatments on brassicas; two full dose treatments on all other field crops [2]
- Maximum number of treatments (including other QoI treatments) on apples and pears 4 per yr if total number of applications is 12 or more, or 3 per yr if the total number is fewer than 12 [1]
- Do not use more than 2 consecutive treatments on apples and pears, and these must be separated by a minimum of 2 applications of a fungicide with a different mode of action [1]
- Use a maximum of three applications per yr on brassicas and no more than two per yr on other field crops [2]
- Do not use consecutive treatments; apply in alternation with fungicides from a different cross resistance group and effective against the target diseases [2]
- Do not apply more than 6 kg/ha to the same area of land per yr [2]
- Consult processor before use on crops for processing
- Applications to lettuce and protected salad crops may only be made between 1 Apr and 31 Oct [2]

Crop-specific information
- HI 21 d for field beans; 14 d for brassica crops, carrots, lettuce; 7 d for apples, pears; 3 d for strawberries

Environmental safety
- Dangerous for the environment
- Very toxic to aquatic organisms
- Broadcast air-assisted LERAP [1] (40 m); LERAP Category B [2]

Hazard classification and safety precautions
    Hazard Harmful, Dangerous for the environment
    Transport code 9
    Packaging group III
    UN Number 3077
    Risk phrases R22a, R50, R53a
    Operator protection A; U05a, U20c [1, 2]; U13 [1]
    Environmental protection E15a, E38 [1, 2]; E16a [2]; E17b [1] (40 m); E34 [1]
    Storage and disposal D01, D02, D10c, D12a [1, 2]; D08 [1]; D09a [2]
    Medical advice M03 [1]; M05a [1, 2]

FOR FULL CONDITIONS OF USE ALWAYS READ THE PRODUCT LABEL

bromadiolone

51 bromadiolone

An anti-coagulant coumarin-derivative rodenticide

Products

1	Bromag	Killgerm	0.005% w/w	RB	UK15-0932
2	Bromag Fresh Bait	Killgerm	0.005% w/w	RB	UK12-0657
3	Bromag Wax Blocks	Killgerm	0.005% w/w	RB	UK12-0755
4	Formula B Rat & Mouse Killer	Killgerm	0.005% w/w	RB	UK15-0934
5	Mamba	Killgerm	0.005% w/w	RB	UK15-0933
6	Sakarat Bromabait	Killgerm	0.005% w/w	RB	UK15-0931
7	Sakarat Bromakill	Killgerm	0.005% w/w	RB	UK15-0935

Uses
- Mice in **farm buildings**, **farmyards** [3, 6, 7]; **farm buildings/yards** [1, 2, 4, 5]
- Rats in **farm buildings**, **farmyards** [3, 6, 7]; **farm buildings/yards** [1, 2, 4, 5]

Approval information
- Bromadiolone included in Annex I under EC Directive 1107/2009

Efficacy guidance
- Ready-to-use baits are formulated on a mould-resistant, whole-wheat base
- Use in baiting programme. Place baits in protected situations, sufficient for continuous feeding between treatments
- Chemical is effective against warfarin- and coumatetralyl-resistant rats and mice and does not induce bait shyness
- Use bait bags where loose baiting inconvenient (eg behind ricks, silage clamps etc)
- The resistance status of the rodent population should be assessed when considering the choice of product to use. Resistance to bromadiolone in rats is now widespread in the UK.

Restrictions
- For use only by professional operators

Environmental safety
- Access to baits by children, birds and animals, particularly cats, dogs, pigs and poultry, must be prevented
- Baits must not be placed where food, feed or water could become contaminated
- Remains of bait and bait containers must be removed after treatment and burned or buried
- Rodent bodies must be searched for and burned or buried. They must not be placed in refuse bins or on rubbish tips
- Take extreme care to prevent domestic animals having access to the bait

Hazard classification and safety precautions
Operator protection A [1]; U13, U20b
Storage and disposal D05, D07, D09a, D11a
Vertebrate/rodent control products V01b, V02, V03b, V04b
Medical advice M03

52 bromoxynil

A contact acting HBN herbicide
HRAC mode of action code: C3

Products

1	Buctril	Bayer CropScience	225 g/l	EC	16597
2	Butryflow	Nufarm UK	402 g/l	SC	14056
3	Maya	Nufarm UK	402 g/l	SC	16760

SEE SECTION 3 FOR PRODUCTS ALSO REGISTERED

Pesticide Profiles

Uses

- Annual dicotyledons in *asparagus* (off-label), *chives* (off-label) [1]; *bulb onion sets* (off-label), *millet* (off-label), *ornamental bulbs* (off-label), *salad onions* (off-label), *shallots* (off-label) [2]; *bulb onions, forage maize, grain maize, linseed, shallots, spring barley, spring oats, spring wheat, sweetcorn, winter barley, winter oats, winter wheat* [2, 3]; *leeks* (off-label) [1, 2]
- Black bindweed in *forage maize, grain maize, linseed, sweetcorn* [3]; *game cover* (off-label), *millet* (off-label), *miscanthus* (off-label), *ornamental bulbs* (off-label), *ornamental plant production* (off-label) [2]
- Black nightshade in *bulb onions* (off-label), *garlic* (off-label), *rye, salad onions* (off-label), *shallots* (off-label), *spring barley, spring oats, spring wheat, triticale, winter barley, winter oats, winter wheat* [1]; *forage maize, grain maize* [1, 3]; *game cover* (off-label), *millet* (off-label), *miscanthus* (off-label), *ornamental plant production* (off-label) [2]; *linseed, sweetcorn* [3]
- Charlock in *bulb onions* (off-label), *garlic* (off-label), *salad onions* (off-label), *shallots* (off-label) [1]
- Chickweed in *bulb onions* (off-label), *garlic* (off-label), *salad onions* (off-label), *shallots* (off-label) [1]; *game cover* (off-label), *millet* (off-label), *miscanthus* (off-label), *ornamental plant production* (off-label) [2]
- Common orache in *forage maize, grain maize, linseed, sweetcorn* [3]; *game cover* (off-label), *miscanthus* (off-label), *ornamental plant production* (off-label) [2]
- Docks in *bulb onion sets* (off-label), *leeks* (off-label), *salad onions* (off-label), *shallots* (off-label) [2]
- Fat hen in *bulb onions, linseed, shallots, sweetcorn* [3]; *bulb onions* (off-label), *garlic* (off-label), *rye, salad onions* (off-label), *shallots* (off-label), *triticale* [1]; *forage maize, grain maize, spring barley, spring oats, spring wheat, winter barley, winter oats, winter wheat* [1, 3]; *game cover* (off-label), *millet* (off-label), *miscanthus* (off-label), *ornamental bulbs* (off-label), *ornamental plant production* (off-label) [2]
- Field bindweed in *forage maize, grain maize, linseed, sweetcorn* [3]; *game cover* (off-label), *miscanthus* (off-label), *ornamental plant production* (off-label) [2]
- Field speedwell in *bulb onions* (off-label), *forage maize, garlic* (off-label), *grain maize, rye, salad onions* (off-label), *shallots* (off-label), *spring barley, spring oats, spring wheat, triticale, winter barley, winter oats, winter wheat* [1]
- Groundsel in *bulb onions, shallots, spring barley, spring oats, spring wheat, winter barley, winter oats, winter wheat* [3]; *bulb onions* (off-label), *garlic* (off-label), *salad onions* (off-label), *shallots* (off-label) [1]; *ornamental bulbs* (off-label) [2]
- Knotgrass in *forage maize, grain maize, linseed, sweetcorn* [3]; *game cover* (off-label), *miscanthus* (off-label), *ornamental plant production* (off-label) [2]
- Mayweeds in *bulb onions* (off-label), *forage maize, garlic* (off-label), *grain maize, rye, salad onions* (off-label), *shallots* (off-label), *spring barley, spring oats, spring wheat, triticale, winter barley, winter oats, winter wheat* [1]
- Penny cress in *bulb onions* (off-label), *forage maize, garlic* (off-label), *grain maize, rye, salad onions* (off-label), *shallots* (off-label), *spring barley, spring oats, spring wheat, triticale, winter barley, winter oats, winter wheat* [1]
- Poppies in *bulb onions* (off-label), *forage maize, garlic* (off-label), *grain maize, rye, salad onions* (off-label), *shallots* (off-label), *spring barley, spring oats, spring wheat, triticale, winter barley, winter oats, winter wheat* [1]
- Redshank in *game cover* (off-label), *millet* (off-label), *miscanthus* (off-label), *ornamental plant production* (off-label) [2]
- Runch in *bulb onions* (off-label), *garlic* (off-label), *salad onions* (off-label), *shallots* (off-label) [1]
- Shepherd's purse in *bulb onions* (off-label), *forage maize, garlic* (off-label), *grain maize, rye, salad onions* (off-label), *shallots* (off-label), *spring barley, spring oats, spring wheat, triticale, winter barley, winter oats, winter wheat* [1]
- Volunteer oilseed rape in *bulb onions, shallots, spring barley, spring oats, spring wheat, winter barley, winter oats, winter wheat* [3]
- Wild mustard in *forage maize, grain maize, rye, spring barley, spring oats, spring wheat, triticale, winter barley, winter oats, winter wheat* [1]
- Wild radish in *forage maize, grain maize, rye, spring barley, spring oats, spring wheat, triticale, winter barley, winter oats, winter wheat* [1]

FOR FULL CONDITIONS OF USE ALWAYS READ THE PRODUCT LABEL

bromoxynil

Extension of Authorisation for Minor Use (EAMUs)
- *asparagus* 20152332 [1]
- *bulb onion sets* 20162683 [2]
- *bulb onions* 20152400 [1]
- *chives* 20152331 [1]
- *game cover* 20131620 [2]
- *garlic* 20152400 [1]
- *leeks* 20152332 [1], 20162683 [2]
- *millet* 20120593 [2]
- *miscanthus* 20131620 [2]
- *ornamental bulbs* 20130517 [2]
- *ornamental plant production* 20140561 [2]
- *salad onions* 20152400 [1], 20162683 [2]
- *shallots* 20152400 [1], 20162683 [2]

Approval information
- Bromoxynil included in Annex I under EC Regulation 1107/2009
- Accepted by BBPA for use on malting barley

Efficacy guidance
- Spray when main weed flush has germinated and the largest are at the 4 leaf stage
- Weed control can be enhanced by using a split treatment spraying each application when the weeds are seedling to 2 true leaves.

Restrictions
- Maximum number of treatments 1 per crop or yr or maximum total dose equivalent to one full dose treatment
- Do not apply with oils or other adjuvants
- Do not apply to bulb onion or shallot grown from sets [2]
- Do not apply using hand-held equipment or at concentrations higher than those recommended
- Do not apply during frosty weather, drought, when soil is waterlogged, when rain expected within 4 h or to crops under any stress
- Take particular care to avoid drift onto neighbouring susceptible crops or open water surfaces
- Do not apply if air temp is above 25°C in cereals or 20°C in maize as some necrosis may occur [1]

Crop-specific information
- Latest use: before 10 fully expanded leaf stage of crop for maize, sweetcorn; before crop 20 cm tall and before flower buds visible for linseed; before 2nd node detectable (GS 32) for cereals
- Foliar scorch, which rapidly disappears without affecting growth, will occur if treatment made in hot weather or during rapid growth

Following crops guidance
- There are no restrictions on the establishment of succeeding or replacement crops

Environmental safety
- Very toxic to aquatic organisms
- Keep livestock out of treated areas for at least 6 wk after treatment
- High risk to bees. Do not apply to crops in flower or to those in which bees are actively foraging
- Do not contaminate surface waters or ditches with chemical or used container
- LERAP Category B

Hazard classification and safety precautions
Hazard Harmful, Dangerous for the environment, Harmful if swallowed, Harmful if inhaled [1-3]; Flammable, Flammable liquid and vapour, Very toxic to aquatic organisms [1]
Transport code 3 [1]; 9 [2, 3]
Packaging group III
UN Number 1993 [1]; 3082 [2, 3]
Risk phrases H304, H315, H319, H336 [1]; H317, H361 [1-3]
Operator protection A, H [1-3]; C [1]; K [2, 3]; U05a [1-3]; U12, U12b [1]; U14, U15, U23a [2, 3]
Environmental protection E15a, E16a, E34, E38, H410
Storage and disposal D01, D02, D09a, D12a [1-3]; D05, D10c [2, 3]; D10a [1]
Medical advice M03 [1-3]; M05b [1]

SEE SECTION 3 FOR PRODUCTS ALSO REGISTERED

Pesticide Profiles

53 bromoxynil + diflufenican

A contact and residual herbicide mixture for weed control in cereals
HRAC mode of action code: C3 + F1

See also diflufenican

Products
Cyclops   Nufarm UK   160:26.7 g/l   EC   16676

Uses
- Chickweed in **grass seed crops** *(off-label)*, **winter barley**, **winter wheat**
- Fat hen in **grass seed crops** *(off-label)*, **winter barley** *(at 1.9 l/ha only)*
- Field pansy in **grass seed crops** *(off-label)*, **winter barley**, **winter wheat**
- Groundsel in **grass seed crops** *(off-label)*, **winter barley** *(at 1.9 l/ha only)*
- Ivy-leaved speedwell in **grass seed crops** *(off-label)*, **winter barley** *(at 1.9 l/ha only)*
- Volunteer beans in **winter barley**, **winter wheat**
- Volunteer oilseed rape in **grass seed crops** *(off-label)*, **winter barley**, **winter wheat**

Extension of Authorisation for Minor Use (EAMUs)
- *grass seed crops* 20161452

Approval information
- Bromoxynil and diflufenican included in Annex I under EC Regulation 1107/2009

Restrictions
- Do not allow spray to drift on to non-target plants.
- Do not treat stressed crops.
- Do not overlap spray swathes.

Following crops guidance
- In the event of crop failure, winter wheat can be drilled immediately after cultivation but winter barley can only be sown after ploughing. Before spring crops of wheat, barley, oilseed rape, peas, sugar beet, field beans, potatoes, carrots, edible brassicas or onions are sown, 12 weeks should elapse since application and the land must be ploughed. Do not plant any crop other than those listed here before the subsequent autumn.

Environmental safety
- Buffer zone requirement 7 m [1]
- LERAP Category B

Hazard classification and safety precautions
  Hazard Corrosive, Dangerous for the environment, Harmful if swallowed
  Transport code 9
  Packaging group III
  UN Number 3082
  Risk phrases H317, H361, R70
  Operator protection A, C, H; U04b, U05a, U09a, U11, U19a, U20a
  Environmental protection E07a, E15b, E16a, E34, E38, H410
  Storage and disposal D01, D02, D05, D09a, D10b, D12a, D12b
  Medical advice M03, M05a

54 bromuconazole

A systemic triazole fungicide for cereals only available in mixtures
FRAC mode of action code: 3

FOR FULL CONDITIONS OF USE ALWAYS READ THE PRODUCT LABEL

55 bromuconazole + tebuconazole

A triazole mixture for disease control in wheat, rye and triticale
FRAC mode of action code: 3

See also tebuconazole

Products
Soleil	Interfarm	167:107 g/l	EC	16869

Uses
- Alternaria in *rye*, *spring wheat*, *triticale*, *winter wheat*
- Brown rust in *rye*, *spring wheat*, *triticale*, *winter wheat*
- Cladosporium in *rye*, *spring wheat*, *triticale*, *winter wheat*
- Fusarium in *rye*, *spring wheat*, *triticale*, *winter wheat*
- Powdery mildew in *rye* (moderate control), *spring wheat* (moderate control), *triticale* (moderate control), *winter wheat* (moderate control)
- Septoria leaf blotch in *rye* (moderate control), *spring wheat* (moderate control), *triticale* (moderate control), *winter wheat* (moderate control)

Approval information
- Bromuconazole and tebuconazole included in Annex 1 under EC Directive 1107/2009

Restrictions
- Newer authorisations for tebuconazole products require application to cereals only after GS 30 and applications to oilseed rape and linseed after GS20 - check label

Following crops guidance
- Only sugar beet, cereals, oilseed rape, field beans, peas, potatoes, linseed and Italian rye-grass may be sown as the following crop. The effect on other crops has not been assessed.
- Some effects may be seen on sugar beet crops grown following a cereal crop but the sugar beet will usually recover completely.

Environmental safety
- LERAP Category B

Hazard classification and safety precautions
Hazard Harmful, Dangerous for the environment
Transport code 9
Packaging group III
UN Number 3082
Risk phrases R22c, R38, R41, R63, R67
Operator protection A, C, H; U11, U19a, U19f, U20b
Environmental protection E16a, E34, E38, H410
Storage and disposal D01, D02, D05, D09a, D11a, D12a, D12b
Medical advice M03

56 bupirimate

A systemic aminopyrimidinol fungicide active against powdery mildew
FRAC mode of action code: 8

Products
1	Clayton Maritime	Clayton	250 g/l	EC	16593
2	Jatapu	Agroquimicos	250 g/l	EC	15776
3	Nimrod	Adama	250 g/l	EC	13046

SEE SECTION 3 FOR PRODUCTS ALSO REGISTERED

Pesticide Profiles

Uses
- Powdery mildew in **apples**, **blackcurrants**, **courgettes** *(outdoor only)*, **gooseberries**, **hops**, **marrows** *(outdoor only)*, **pears**, **protected cucumbers**, **protected strawberries**, **raspberries** *(outdoor only)*, **strawberries** [1-3]; **begonias**, **chrysanthemums**, **roses** [2, 3]; **ornamental plant production**, **protected ornamentals** [1]; **protected tomatoes** *(off-label)*, **pumpkins** *(off-label)*, **redcurrants** *(off-label)*, **squashes** *(off-label)*, **whitecurrants** *(off-label)* [3]

Extension of Authorisation for Minor Use (EAMUs)
- *protected tomatoes* 20070997 [3], 20141225 [3]
- *pumpkins* 20070994 [3], 20141224 [3]
- *redcurrants* 20082082 [3], 20141223 [3]
- *squashes* 20070994 [3], 20141224 [3]
- *whitecurrants* 20082082 [3], 20141223 [3]

Approval information
- Bupirimate included in Annex 1 under EC Regulation 1107/2009
- Accepted by BBPA for use on hops

Efficacy guidance
- On apples during periods that favour disease development lower doses applied weekly give better results than higher rates fortnightly
- Not effective in protected crops against strains of mildew resistant to bupirimate

Restrictions
- Maximum number of treatments or maximum total dose depends on crop and dose (see label for details)

Crop-specific information
- HI: 1 d for apples, pears, strawberries; 2 d for cucurbits; 7 d for blackcurrants; 8 d for raspberries; 14 d for gooseberries, hops
- Apply before or at first signs of disease and repeat at 7-14 d intervals. Timing and maximum dose vary with crop. See label for details
- With apples, hops and ornamentals cultivars may vary in sensitivity to spray. See label for details
- If necessary to spray cucurbits in winter or early spring spray a few plants 10-14 d before spraying whole crop to test for likelihood of leaf spotting problem
- On roses some leaf puckering may occur on young soft growth in early spring or under low light intensity. Avoid use of high rates or wetter on such growth
- Never spray flowering begonias (or buds showing colour) as this can scorch petals
- Do not mix with other chemicals for application to begonias, cucumbers or gerberas

Environmental safety
- Dangerous for the environment
- Toxic to aquatic organisms
- Flammable
- Product has negligible effect on *Phytoseiulus* and *Encarsia* and may be used in conjunction with biological control of red spider mite

Hazard classification and safety precautions
Hazard Harmful [1]; Irritant [2, 3]; Flammable, Dangerous for the environment [1-3]
Transport code 3
Packaging group III
UN Number 1993
Risk phrases R22b, R38, R51, R53a [1-3]; R37, R67 [1]
Operator protection A, C; U05a, U20b [1-3]; U09a, U14, U19a [1]
Environmental protection E15a [1-3]; E34, E38 [1]
Storage and disposal D01, D02, D09a, D10c [1-3]; D12a [1]
Medical advice M05b

FOR FULL CONDITIONS OF USE ALWAYS READ THE PRODUCT LABEL

57 captan

A protectant phthalimide fungicide with horticultural uses
FRAC mode of action code: M4

Products

1	Captan 80 WDG	Adama	80% w/w	WG	16293
2	Clayton Core	Clayton	80% w/w	WG	16934
3	Multicap	Belcrop	80% w/w	WG	17652
4	PP Captan 80 WG	Arysta	80% w/w	WG	16294

Uses

- Botrytis in **ornamental plant production** *(off-label)* [1]
- Disease control in **hops in propagation** *(off-label)*, **nursery fruit trees** *(off-label)*, **quinces** *(off-label)* [1]
- Downy mildew in **ornamental plant production** *(off-label)* [1]
- Gloeosporium rot in **apples** [1-4]; **pears** [1]
- Scab in **apples** [1-4]; **ornamental plant production** *(off-label)* [1]; **pears** [1, 3, 4]
- Shot-hole in **ornamental plant production** *(off-label)* [1]

Extension of Authorisation for Minor Use (EAMUs)

- **hops in propagation** *20142510* [1]
- **nursery fruit trees** *20142510* [1]
- **ornamental plant production** *20151919* [1], *20151920* [1]
- **quinces** *20141276* [1]

Approval information

- Captan included in Annex I under EC Regulation 1107/2009

Restrictions

- Maximum number of treatments 12 per yr on apples and pears as pre-harvest sprays
- Do not use on apple cultivars Bramley, Monarch, Winston, King Edward, Spartan, Kidd's Orange or Red Delicious or on pear cultivar D'Anjou
- Do not mix with alkaline materials or oils
- Do not use on fruit for processing
- Powered visor respirator with hood and neck cape must be used when handling concentrate

Crop-specific information

- HI apples, pears 14 d; strawberries 7 d
- For control of scab apply at bud burst and repeat at 10-14 d intervals until danger of scab infection ceased
- For suppression of fruit storage rots apply from late Jul and repeat at 2-3 wk intervals
- For black spot control in roses apply after pruning with 3 further applications at 14 d intervals or spray when spots appear and repeat at 7-10 d intervals
- For grey mould in strawberries spray at first open flower and repeat every 7-10 d
- Do not leave diluted material for more than 2 h. Agitate well before and during spraying

Environmental safety

- Dangerous for the environment
- Very toxic to aquatic organisms
- LERAP Category B [3, 4]

Hazard classification and safety precautions

Hazard Harmful, Dangerous for the environment [1-4]; Very toxic to aquatic organisms [1-3]
Transport code 9
Packaging group III
UN Number 3077
Risk phrases H317, H319, H351 [1-3]; R36, R40, R41, R43, R50, R53a [4]
Operator protection A, D, E, H; U05a, U09a, U11, U19a, U20c
Environmental protection E15a [1-4]; E16a [3, 4]; H412 [1-3]
Storage and disposal D01, D02, D09a, D11a, D12a

SEE SECTION 3 FOR PRODUCTS ALSO REGISTERED

Pesticide Profiles

58 carbendazim

A systemic benzimidazole fungicide with curative and protectant activity
FRAC mode of action code: 1

Products

1	Mascot Systemic	Rigby Taylor	500 g/l	SC	14488
2	Ringer	Barclay	500 g/l	SC	13289

Uses
- Disease control in **managed amenity turf** [1]
- Wormcast formation in **managed amenity turf** [2]

Approval information
- Carbendazim included in Annex I under EC Regulation 1107/2009
- Approval expiry 31 Aug 2017 [1, 2]

Efficacy guidance
- Mostly applied as spray or drench. Spray treatments normally applied at first sign of disease and repeated after 1 mth if required
- Apply as drench rather than spray where red spider mite predators are being used
- Do not apply during drought conditions. Rain or irrigation after treatment may improve control
- Leave 4 mth intervals where used only for worm cast control [2]
- To delay appearance of resistant strains alternate treatment with non-MBC fungicide. To avoid the development of resistance, a maximum of 2 applications of any MBC product (thiophanate-methyl or carbendazim) are recommended in any crop. Avoid using MBC fungicides alone.

Restrictions
- Maximum number of treatments (including applications of any product containing benomyl, carbendazim or thiophanate-methyl) varies with crop treated and product used - see labels for details
- Do not treat crops or turf suffering from drought or other physical or chemical stress
- Not compatible with alkaline products such as lime sulphur

Crop-specific information
- On turf apply as preventative treatment in spring or autumn during periods of high disease risk [2]

Environmental safety
- Dangerous for the environment
- Very toxic to aquatic organisms

Hazard classification and safety precautions
   Hazard Toxic, Dangerous for the environment
   Transport code 9
   Packaging group III
   UN Number 3082
   Risk phrases R46, R50, R53a, R60, R61, R68
   Operator protection A, C, H, M [2]; U14, U19a, U20b
   Environmental protection E14a, E34, E38
   Storage and disposal D05, D08, D09a, D10a, D12a
   Medical advice M04a

59 carbetamide

A residual pre- and post-emergence carbamate herbicide for a range of field crops
HRAC mode of action code: K2

Products

	Crawler	Adama	60% w/w	WG	16856

FOR FULL CONDITIONS OF USE ALWAYS READ THE PRODUCT LABEL

Uses
- Annual dicotyledons in **corn gromwell** *(off-label)*, **linseed** *(off-label)*, **winter field beans**, **winter oilseed rape**
- Annual grasses in **winter field beans**, **winter oilseed rape**
- Annual meadow grass in **corn gromwell** *(off-label)*, **linseed** *(off-label)*
- Blackgrass in **corn gromwell** *(off-label)*, **linseed** *(off-label)*
- Chickweed in **corn gromwell** *(off-label)*, **linseed** *(off-label)*
- Cleavers in **corn gromwell** *(off-label)*, **linseed** *(off-label)*
- Couch in **corn gromwell** *(off-label)*, **linseed** *(off-label)*
- Fat hen in **corn gromwell** *(off-label)*, **linseed** *(off-label)*
- Fumitory in **corn gromwell** *(off-label)*, **linseed** *(off-label)*
- Knotgrass in **corn gromwell** *(off-label)*, **linseed** *(off-label)*
- Ryegrass in **corn gromwell** *(off-label)*, **linseed** *(off-label)*
- Small nettle in **corn gromwell** *(off-label)*, **linseed** *(off-label)*
- Speedwells in **corn gromwell** *(off-label)*, **linseed** *(off-label)*
- Volunteer cereals in **corn gromwell** *(off-label)*, **linseed** *(off-label)*, **winter field beans**, **winter oilseed rape**
- Wild oats in **corn gromwell** *(off-label)*, **linseed** *(off-label)*
- Yorkshire fog in **corn gromwell** *(off-label)*, **linseed** *(off-label)*

Extension of Authorisation for Minor Use (EAMUs)
- *corn gromwell* 20151828
- *linseed* 20152544

Approval information
- Carbetamide included in Annex I under EC Regulation 1107/2009

Efficacy guidance
- Best results pre- or early post-emergence of weeds under cool, moist conditions. Adequate soil moisture is essential
- May be applied pre or post emergence up to end of Feb in year of harvest
- Maintain an interval of at least 7 days when using split applications
- Dicotyledons controlled include chickweed, cleavers and speedwell
- Weed growth stops rapidly after treatment but full effects may take 6-8 wk to develop
- Various tank mixes are recommended to broaden the weed spectrum. See label for details
- Always follow WRAG guidelines for preventing and managing herbicide resistant weeds. See Section 5 for more information

Restrictions
- Maximum number of treatments 1 per crop for all crops
- Do not treat any crop on waterlogged soil
- Do not use on soils with more than 10% organic matter as residual activity is impaired
- Do not apply during prolonged periods of cold weather when weeds are dormant
- Do not tank-mix with prothioconazole or liquid manganese

Crop-specific information
- HI 6 wk for all crops
- Apply to brassicas from mid-Oct to end-Feb provided crop has at least 4 true leaves (spring cabbage, spring greens), 3-4 true leaves (seed crops, oilseed rape)
- Apply to established lucerne or sainfoin from Nov to end-Feb
- Apply to established red or white clover from Feb to mid-Mar

Following crops guidance
- Succeeding crops may be sown 2 wk after treatment for brassicas, field beans, 8 wk after treatment for peas, runner beans, 16 wk after treatment for cereals, maize
- Ploughing is not necessary before sowing subsequent crops

Environmental safety
- Do not graze crops for at least 6 wk after treatment
- Some pesticides pose a greater threat of contamination of water than others and carbetamide is one of these pesticides. Take special care when applying carbetamide near water and do not apply if heavy rain is forecast

SEE SECTION 3 FOR PRODUCTS ALSO REGISTERED

Pesticide Profiles

Hazard classification and safety precautions
  Hazard Dangerous for the environment
  Transport code 9
  Packaging group III
  UN Number 3077
  Risk phrases H319, H361
  Operator protection A, C, H; U20c
  Environmental protection E15a, E38, H411
  Storage and disposal D01, D02, D09a, D11a

60 carbon dioxide (commodity substance)

A gas for the control of trapped rodents and other vertebrates. Approval valid until 31/8/2019

Products
  carbon dioxide   various   99.9%   GA   -

Uses
- Birds in *traps*
- Mice in *traps*
- Rats in *traps*

Approval information
- Carbon dioxide included in Annex 1 under EC Regulation 1107/2009

Efficacy guidance
- Use to destroy trapped rodent pests
- Use to control birds covered by general licences issued by the Agriculture and Environment Departments under Section 16(1) of the Wildlife and Countryside Act (1981) for the control of opportunistic bird species, where birds have been trapped or stupefied with alphachloralose/seconal

Restrictions
- Operators must wear self-contained breathing apparatus when carbon dioxide levels are greater than 0.5% v/v
- Operators must be suitably trained and competent

Environmental safety
- Unprotected persons and non-target animals must be excluded from the treatment enclosures and surrounding areas unless the carbon dioxide levels are below 0.5% v/v

Hazard classification and safety precautions
  Operator protection G

61 carfentrazone-ethyl

A triazolinone contact herbicide
HRAC mode of action code: E

Products
1	Aurora 40 WG	Headland	40% w/w	WG	17278
2	Headlite	AgChem Access	60 g/l	ME	13624
3	Shark	Headland	60 g/l	ME	17256
4	Shylock	AgChem Access	60 g/l	ME	15405
5	Spotlight Plus	Headland	60 g/l	ME	17241

Uses
- Annual and perennial weeds in *all edible crops (outdoor)*, *all non-edible crops (outdoor)*, *potatoes* [5]
- Annual dicotyledons in *all edible crops (outdoor)* *(before planting)*, *all non-edible crops (outdoor)* *(before planting)*, *potatoes* [3, 4]; *asparagus* *(off-label)* [3]; *durum wheat* *(off-label)*,

FOR FULL CONDITIONS OF USE ALWAYS READ THE PRODUCT LABEL

carfentrazone-ethyl

- rye *(off-label)*, spring barley, spring oats, spring wheat, triticale, winter barley, winter oats, winter wheat [1]
- Black bindweed in **durum wheat** *(off-label)*, **rye** *(off-label)* [1]; **potatoes** [3, 4]
- Cleavers in **durum wheat** *(off-label)*, **rye** *(off-label)* [1]; **potatoes** [3, 4]
- Desiccation in **blackberries** *(off-label)*, **loganberries** *(off-label)*, **protected blackberries** *(off-label)*, **protected loganberries** *(off-label)*, **protected raspberries** *(off-label)*, **protected rubus hybrids** *(off-label)*, **raspberries** *(off-label)*, **rubus hybrids** *(off-label)* [3]; **narcissi** *(off-label)*, **potatoes** [5]
- Fat hen in **durum wheat** *(off-label)*, **rye** *(off-label)* [1]; **potatoes** [3, 4]
- Field pansy in **durum wheat** *(off-label)*, **rye** *(off-label)* [1]
- Groundsel in **durum wheat** *(off-label)*, **rye** *(off-label)* [1]
- Haulm destruction in **seed potatoes**, **ware potatoes** [2]
- Ivy-leaved speedwell in **potatoes** [3, 4]
- Knotgrass in **durum wheat** *(off-label)*, **rye** *(off-label)* [1]; **potatoes** [3, 4]
- Redshank in **durum wheat** *(off-label)*, **rye** *(off-label)* [1]; **potatoes** [3, 4]
- Small nettle in **durum wheat** *(off-label)*, **rye** *(off-label)* [1]
- Speedwells in **durum wheat** *(off-label)*, **rye** *(off-label)* [1]
- Volunteer oilseed rape in **all edible crops (outdoor)** *(before planting)*, **all non-edible crops (outdoor)** *(before planting)*, **potatoes** [3, 4]; **durum wheat** *(off-label)*, **rye** *(off-label)* [1]

Extension of Authorisation for Minor Use (EAMUs)
- *asparagus* 20160098 [3]
- *blackberries* 20160097 [3]
- *durum wheat* 20152860 [1]
- *loganberries* 20160097 [3]
- *narcissi* 20152644 [5]
- *protected blackberries* 20160097 [3]
- *protected loganberries* 20160097 [3]
- *protected raspberries* 20160097 [3]
- *protected rubus hybrids* 20160097 [3]
- *raspberries* 20160097 [3]
- *rubus hybrids* 20160097 [3]
- *rye* 20152860 [1]

Approval information
- Carfentrazone-ethyl included in Annex I under EC Regulation 1107/2009
- Accepted by BBPA for use on malting barley and hops

Efficacy guidance
- Best weed control results achieved from good spray cover applied to small actively growing weeds
- Carfentrazone-ethyl acts by contact only; see label for optimum timing on specified weeds. Weeds emerging after application will not be controlled
- For weed control in cereals use as two spray programme with one application in autumn and one in spring
- Efficacy of haulm destruction will be reduced where flailed haulm covers the stems at application
- For potato crops with very dense vigorous haulm or where regrowth occurs following a single application a second application may be necessary to achieve satisfactory desiccation. A minimum interval between applications of 7 d should be observed to achieve optimum performance

Restrictions
- Maximum number of treatments 2 per crop for cereals (1 in Autumn and 1 in Spring); 1 per crop for weed control in potatoes; 1 per yr for pre-planting treatments
- Maximum total dose for potato haulm destruction equivalent to 1.6 full dose treatments
- Do not treat cereal crops under stress from drought, waterlogging, cold, pests, diseases, nutrient or lime deficiency or any factors reducing plant growth
- Do not treat cereals undersown with clover or other legumes
- Allow at least 2-3 wk between application and lifting potatoes to allow skins to set if potatoes are to be stored

SEE SECTION 3 FOR PRODUCTS ALSO REGISTERED

Pesticide Profiles

- Follow label instructions for sprayer cleaning
- Do not apply through knapsack sprayers
- Contact processor before using a split dose on potatoes for processing
- Do not apply within 10 days of an application of iodosulfuron-methyl-sodium + mesosulfuron-methyl formulations, e.g. MAPP 12478

Crop-specific information
- Latest use: 1 mth before planting edible and non-edible crops; before 3rd node detectable (GS 33) on cereals
- HI 7 d for potatoes
- For weed control in cereals treat from 2 leaf stage
- When used as a treatment prior to planting a subsequent crop apply before weeds exceed maximum sizes indicated in the label
- If treated potato tubers are to be stored then allow at least 2-3 wk between the final application and lifting to allow skins to set

Following crops guidance
- No restrictions apply on the planting of succeeding crops 1 mth after application to potatoes for haulm destruction or as a pre-planting treatment, or 3 mth after application to cereals for weed control
- In the event of failure of a treated cereal crop, all cereals, ryegrass, maize, oilseed rape, peas, sunflowers, *Phacelia*, vetches, carrots or onions may be planted within 1 mth of treatment

Environmental safety
- Dangerous for the environment
- Very toxic to aquatic organisms
- Some non-target crops are sensitive. Avoid drift onto broad-leaved plants outside the treated area, or onto ponds waterways or ditches

Hazard classification and safety precautions
Hazard Irritant, Dangerous for the environment [1-5]; Very toxic to aquatic organisms [1, 3, 5]
Transport code 9
Packaging group III
UN Number 3077 [1]; 3082 [2-5]
Risk phrases H317 [3, 5]; R38 [2]; R43, R50, R53a [2, 4]
Operator protection A, H; U05a, U14 [1-5]; U08, U13 [1]; U20a [2-5]
Environmental protection E15a [1, 3, 4]; E15b [2, 5]; E34 [1]; E38 [1-5]; H410 [1, 3, 5]
Storage and disposal D01, D02, D09a, D12a [1-5]; D10b [1]; D10c [2-5]
Medical advice M05a [1]

62   carfentrazone-ethyl + mecoprop-P

A foliar applied herbicide for cereals
HRAC mode of action code: E + O

See also mecoprop-P

Products

1	Jewel	Everris Ltd	1.5:60% w/w	WG	16212
2	Pan Glory	Pan Amenity	1.5:60% w/w	WG	14487
3	Platform S	Headland	1.5:60% w/w	WG	17259

Uses
- Annual and perennial weeds in **amenity grassland, managed amenity turf** [1, 2]
- Charlock in **spring barley, spring oats, spring wheat, winter barley, winter oats, winter wheat** [3]
- Chickweed in **spring barley, spring oats, spring wheat, winter barley, winter oats, winter wheat** [3]
- Cleavers in **spring barley, spring oats, spring wheat, winter barley, winter oats, winter wheat** [3]

FOR FULL CONDITIONS OF USE ALWAYS READ THE PRODUCT LABEL

- Field speedwell in **spring barley, spring oats, spring wheat, winter barley, winter oats, winter wheat** [3]
- Ivy-leaved speedwell in **spring barley, spring oats, spring wheat, winter barley, winter oats, winter wheat** [3]
- Moss in **amenity grassland, managed amenity turf** [1, 2]
- Red dead-nettle in **spring barley, spring oats, spring wheat, winter barley, winter oats, winter wheat** [3]

Approval information
- Carfentrazone-ethyl and mecoprop-P included in Annex I under EC Regulation 1107/2009
- Accepted by BBPA for use on malting barley

Efficacy guidance
- Best results obtained when weeds have germinated and growing vigorously in warm moist conditions
- Treatment of large weeds and poor spray coverage may result in reduced weed control

Restrictions
- Maximum number of treatments 2 per crop. The total amount of mecoprop-P applied in a single yr must not exceed the maximum total dose approved for any single product for the crop per situation
- Do not treat crops suffering from stress from any cause
- Do not treat crops undersown or to be undersown
- Do not apply between 1st October and 1st March

Crop-specific information
- Latest use: before 3rd node detectable (GS 33)
- Can be used on all varieties of wheat and barley in autumn or spring from the beginning of tillering

Following crops guidance
- In the event of crop failure, any cereal, maize, oilseed rape, peas, vetches or sunflowers may be sown 1 mth after a spring treatment. Any crop may be planted 3 mth after treatment

Environmental safety
- Dangerous for the environment
- Very toxic to aquatic organisms
- Keep livestock out of treated areas for at least two weeks following treatment and until poisonous weeds, such as ragwort, have died down and become unpalatable
- Some pesticides pose a greater threat of contamination of water than others and mecoprop-P is one of these pesticides. Take special care when applying mecoprop-P near water and do not apply if heavy rain is forecast

Hazard classification and safety precautions
Hazard Harmful, Dangerous for the environment [1-3]; Harmful if swallowed, Very toxic to aquatic organisms [3]
Transport code 9
Packaging group III
UN Number 3077
Risk phrases H317, H318 [3]; R22a, R41, R43, R50, R53a [1, 2]
Operator protection A, C, H [1-3]; M [3]; U05a, U11, U13, U14, U20b [1-3]; U08 [3]; U09a [1, 2]
Environmental protection E07a, E34, E38 [1-3]; E15a, H411 [3]; E15b [1, 2]
Storage and disposal D01, D02, D09a, D10b [1-3]; D12a [3]
Medical advice M05a

SEE SECTION 3 FOR PRODUCTS ALSO REGISTERED

Pesticide Profiles

63 chlorantraniliprole

An ingested and contact insecticide for insect pest control in apples and pears and, in Ireland only, for colorado beetle control in potatoes.
IRAC mode of action code: 28

Products

Coragen   DuPont   200 g/l   SC   14930

Uses
- Carnation tortrix moth in **bilberries** *(off-label)*, **blackcurrants** *(off-label)*, **blueberries** *(off-label)*, **cranberries** *(off-label)*, **gooseberries** *(off-label)*, **redcurrants** *(off-label)*
- Carrot fly in **carrots** *(off-label)*, **celeriac** *(off-label)*, **horseradish** *(off-label)*, **parsley root** *(off-label)*, **parsnips** *(off-label)*, **red beet** *(off-label)*, **swedes** *(off-label)*, **turnips** *(off-label)*
- Caterpillars in **forest nurseries** *(off-label)*, **horseradish** *(off-label)*, **red beet** *(off-label)*, **swedes** *(off-label)*, **turnips** *(off-label)*
- Codling moth in **apples**, **pears**
- Currant pug moth in **hops** *(off-label)*
- European corn borer in **hops** *(off-label)*
- Large pine weevil in **forest nurseries** *(off-label)*
- Light brown apple moth in **bilberries** *(off-label)*, **blackcurrants** *(off-label)*, **blueberries** *(off-label)*, **cranberries** *(off-label)*, **gooseberries** *(off-label)*, **redcurrants** *(off-label)*
- Rosy rustic moth in **hops** *(off-label)*
- Silver Y moth in **hops** *(off-label)*, **horseradish** *(off-label)*, **red beet** *(off-label)*, **swedes** *(off-label)*, **turnips** *(off-label)*
- South American Tomato Moth in **protected tomatoes** *(off-label)*

Extension of Authorisation for Minor Use (EAMUs)
- *bilberries* 20151087
- *blackcurrants* 20151087
- *blueberries* 20151087
- *carrots* 20131156
- *celeriac* 20131156
- *cranberries* 20151087
- *forest nurseries* 20162003
- *gooseberries* 20151087
- *hops* 20161590
- *horseradish* 20140302
- *parsley root* 20131156
- *parsnips* 20131156
- *protected tomatoes* 20162080
- *red beet* 20140302
- *redcurrants* 20151087
- *swedes* 20140302
- *turnips* 20140302

Approval information
- Chlorantraniliprole is included in Annex 1 under EC Regulation 1107/2009

Efficacy guidance
- Can be used as part of an Integrated Pest Management programme.
- For best fruit protection in apples and pears, apply before egg hatch.
- Best applied early morning or late evening to avoid applications when bees may be present.

Restrictions
- Maximum number of applications is two per year

Crop-specific information
- Latest time of application is 14 days before harvest

FOR FULL CONDITIONS OF USE ALWAYS READ THE PRODUCT LABEL

Environmental safety
- To protect bees and pollinating insects, do not apply to crops when in flower. Do not apply when bees are actively foraging or when flowering plants are present.
- Broadcast air-assisted LERAP (10 m); LERAP Category B

Hazard classification and safety precautions
Hazard Dangerous for the environment
Transport code 9
Packaging group III
UN Number 3082
Risk phrases R50, R53a
Environmental protection E12e, E15b, E16a, E22b, E38; E17b (10 m)
Storage and disposal D01, D02, D05, D09a, D12a

64 chloridazon

A residual pyridazinone herbicide for beet crops
HRAC mode of action code: C1

Products
1	Better DF	Sipcam	65% w/w	WG	16101
2	Pyramin DF	BASF	65% w/w	WG	16768
3	Takron	BASF	430 g/l	SC	16769

Uses
- Annual dicotyledons in **bulb onions** (off-label), **garlic** (off-label), **herbs (see appendix 6)** (off-label), **red beet** (off-label), **shallots** (off-label) [2]; **fodder beet**, **mangels**, **sugar beet** [1-3]
- Annual meadow grass in **bulb onions** (off-label), **garlic** (off-label), **shallots** (off-label) [1, 2]; **fodder beet**, **mangels**, **sugar beet** [1-3]; **herbs (see appendix 6)** (off-label), **red beet** (off-label) [2]
- Charlock in **bulb onions** (off-label), **garlic** (off-label), **shallots** (off-label) [1]
- Fat hen in **bulb onions** (off-label), **garlic** (off-label), **shallots** (off-label) [1]
- Knotgrass in **bulb onions** (off-label), **garlic** (off-label), **shallots** (off-label) [1]
- Mayweeds in **bulb onions** (off-label), **garlic** (off-label), **shallots** (off-label) [1]
- Poppies in **bulb onions** (off-label), **garlic** (off-label), **shallots** (off-label) [1]
- Small nettle in **bulb onions** (off-label), **garlic** (off-label), **shallots** (off-label) [1]

Extension of Authorisation for Minor Use (EAMUs)
- **bulb onions** 20140395 [1], 20152349 [2]
- **garlic** 20140395 [1], 20152349 [2]
- **herbs (see appendix 6)** 20152350 [2]
- **red beet** 20151219 [2]
- **shallots** 20140395 [1], 20152349 [2]

Approval information
- Chloridazon included in Annex I under EC Regulation 1107/2009

Efficacy guidance
- Absorbed by roots of germinating weeds and best results achieved pre-emergence of weeds or crop when soil moist and adequate rain falls after application
- Application rate for some products depends on soil type. Check label for details

Restrictions
- Maximum number of treatments generally 1 per crop (pre-emergence) for fodder beet and mangels; 1 (pre-emergence) + 3 (post-emergence) per crop for sugar beet, but labels vary slightly. Check labels for maximum total dose for the crop to be treated
- Do not use on Coarse Sands, Sands or Fine Sands or where organic matter exceeds 5%
- A maximum total dose of 2.6 kg/ha chloridazon may only be applied every third year on the same field

SEE SECTION 3 FOR PRODUCTS ALSO REGISTERED

Pesticide Profiles

Crop-specific information
- Latest use: pre-emergence for fodder beet and mangels; normally before leaves of crop meet in row for sugar beet, but labels vary slightly
- Where used pre-emergence spray as soon as possible after drilling in mid-Mar to mid-Apr on fine, firm, clod-free seedbed
- Where crop drilled after mid-Apr or soil dry apply pre-drilling and incorporate to 2.5 cm immediately afterwards
- Various tank mixes recommended on sugar beet for pre- and post-emergence use and as repeated low dose treatments. See label for details
- Crop vigour may be reduced by treatment of crops growing under unfavourable conditions including poor tilth, drilling at incorrect depth, soil capping, physical damage, pest or disease damage, excess seed dressing, trace-element deficiency or a sudden rise in temperature after a cold spell

Following crops guidance
- Winter cereals or any spring sown crop may follow a treated beet crop harvested at the normal time and after ploughing to at least 150 mm
- In the event of failure of a treated crop only a beet crop or maize may be drilled, after cultivation

Environmental safety
- Dangerous for the environment
- Very toxic to aquatic organisms

Hazard classification and safety precautions
Hazard Harmful [1, 2]; Dangerous for the environment [1-3]; Harmful if swallowed [2, 3]; Harmful if inhaled [2]
Transport code 9
Packaging group III
UN Number 3077 [1, 2]; 3082 [3]
Risk phrases H317 [2, 3]; R20 [1, 2]; R22a [1, 3]; R43, R50, R53a [1]
Operator protection A; U05a, U08, U14, U20b [1-3]; U19a [3]
Environmental protection E15a, E38 [1-3]; H411 [2, 3]
Storage and disposal D01, D02, D09a, D12a [1-3]; D10b [3]; D11a [1, 2]
Medical advice M05a

65   chloridazon + metamitron

A contact and residual herbicide mixture for sugar beet
HRAC mode of action code: C1 + C1

See also metamitron

Products
Volcan Combi FL   Sipcam   300:280 g/l   SC   16681

Uses
- Annual dicotyledons in **fodder beet**, **sugar beet**
- Annual meadow grass in **fodder beet**, **sugar beet**

Approval information
- Chloridazon and metamitron included in Annex I under EC Regulation 1107/2009

Efficacy guidance
- Best results achieved from a sequential programmme of sprays
- Pre-emergence use improves efficacy of post-emergence programme. Best results obtained from application to a moist seed bed
- First post-emergence treatment should be made when weeds at early cotyledon stage and subsequent applications made when new weed flushes reach this stage
- Weeds surviving an earlier treatment should be treated again after 7-10 d even if no new weeds have appeared

FOR FULL CONDITIONS OF USE ALWAYS READ THE PRODUCT LABEL

chloridazon + quinmerac

Restrictions
- Maximum number of treatments 1 pre-emergence followed by 3 post-emergence
- Take advice if light soils have a high proportion of stones

Crop-specific information
- Latest use: when leaves of crop meet in rows
- Tolerance of crops growing under stress from any cause may be reduced
- Crops treated pre-emergence and subsequently subjected to frost may be checked and recovery may not be complete

Following crops guidance
- After the last application only sugar beet or mangels may be sown within 4 mth; cereals may be sown after 16 wk. Land should be mouldboard ploughed to 15 cm and thoroughly cultivated before any succeeding crop

Environmental safety
- Harmful to fish or other aquatic life. Do not contaminate surface waters or ditches with chemical or used container

Hazard classification and safety precautions
Hazard Irritant, Harmful if swallowed
Transport code 9
Packaging group III
UN Number 3082
Risk phrases H317
Operator protection A, C, H; U08, U14, U15, U20a
Environmental protection E13c, E34, H410
Storage and disposal D09a, D10b

66 chloridazon + quinmerac

A herbicide mixture for use in beet crops
HRAC mode of action code: C1 + O

See also quinmerac

Products
Fiesta T   BASF   360:60 g/l   SC   17153

Uses
- Annual dicotyledons in **sugar beet**
- Annual meadow grass in **sugar beet**
- Chickweed in **sugar beet**
- Cleavers in **sugar beet**
- Common field speedwell in **sugar beet**
- Poppies in **sugar beet**
- Small nettle in **sugar beet**

Approval information
- Chloridazon and quinmerac included in Annex I under EC Regulation 1107/2009

Efficacy guidance
- Best results obtained when adequate soil moisture is present at application and afterwards to form an active herbicidal layer in the soil
- A programme of pre-emergence treatment followed by post-emergence application(s) in mixture with a contact herbicide optimises weed control and is essential for some difficult weed species such as cleavers
- Treatment pre-emergence only may not provide sufficient residual activity to give season-long weed control
- Effectiveness may be reduced under conditions of low pH
- Always follow WRAG guidelines for preventing and managing herbicide resistant weeds. See Section 5 for more information

SEE SECTION 3 FOR PRODUCTS ALSO REGISTERED

Pesticide Profiles

Restrictions
- Maximum total dose on sugar beet equivalent to one treatment pre-emergence plus two treatments post-emergence, all at maximum individual dose; maximum total dose on fodder beet and mangels equivalent to one full dose pre-emergence treatment
- Heavy rain shortly after treatment may check crop growth particularly when it leaves water standing in surface depressions
- Treatment of stressed crops or those growing in unfavourable conditions may depress crop vigour and possibly reduce stand
- Do not use on Sands or soils of high organic matter content
- Where rates of nitrogen higher than those generally recommended are considered necessary, apply at least 3 wk before drilling
- A maximum total dose of 2.6 kg/ha chloridazon may only be applied every third year on the same field

Crop-specific information
- Latest use: before plants meet between the rows for sugar beet; pre-emergence for fodder beet, mangels

Following crops guidance
- In the spring following normal harvest of a treated crop sow only cereals, beet or mangel crops, potatoes or field beans, after ploughing
- In the event of failure of a treated crop only a beet crop may be drilled, after cultivation

Environmental safety
- Dangerous for the environment
- Toxic to aquatic organisms
- To reduce movement to groundwater do not apply to dry soil or when heavy rain is forecast

Hazard classification and safety precautions
Hazard Irritant, Dangerous for the environment
UN Number N/C
Risk phrases H317
Operator protection A, H; U05a, U08, U14, U19a, U20b
Environmental protection E15a, E38
Storage and disposal D01, D02, D08, D09a, D10c, D12a
Medical advice M03, M05a

67   chlormequat

A plant-growth regulator for reducing stem growth and lodging

Products

1	3C Chlormequat 750	BASF	750 g/l	SL	16690
2	Adjust	Taminco	620 g/l	SL	17141
3	Agrovista 3 See 750	Agrovista	750 g/l	SL	15975
4	Barleyquat B	Taminco	620 g/l	SL	17204
5	Belcocel	Taminco	720 g/l	SL	16652
6	Bettequat B	Taminco	620 g/l	SL	17208
7	CleanCrop Transist	Agrii	720 g/l	SL	16684
8	Jadex-o-720	SFP Europe	720 g/l	SL	16284
9	K2	Taminco	620 g/l	SL	17206
10	Manipulator	Taminco	620 g/l	SL	17207
11	Selon	Taminco	620 g/l	SL	17209
12	Stabilan 750	Nufarm UK	750 g/l	SL	09303

Uses
- Growth regulation in **ornamental plant production** [3]; **spring barley** [2, 4, 6, 9-11]; **spring oats**, **spring wheat**, **winter barley**, **winter oats** [2, 4-7, 9-11]; **triticale** [12]; **winter wheat** [2, 4-11]

FOR FULL CONDITIONS OF USE ALWAYS READ THE PRODUCT LABEL

chlormequat

- Lodging control in **oats**, **rye** [1]; **spring barley** [1, 2, 4, 6, 9-11]; **spring oats**, **winter oats** [2-7, 9-12]; **spring rye** [3]; **spring wheat**, **winter barley** [1-7, 9-12]; **triticale** [1, 3, 12]; **winter rye** [3, 12]; **winter wheat** [1-12]

Approval information
- Chlormequat included in Annex 1 under EC Regulation 1107/2009
- Accepted by BBPA for use on malting barley

Efficacy guidance
- Most effective results on cereals normally achieved from application from Apr onwards, on wheat and rye from leaf sheath erect to first node detectable (GS 30-31), on oats at second node detectable (GS 32), on winter barley from mid-tillering to leaf sheath erect (GS 25-30). However, recommendations vary with product. See label for details
- Influence on growth varies with crop and growth stage. Risk of lodging reduced by application at early stem extension. Root development and yield can be improved by earlier treatment
- Results on barley can be variable
- In tank mixes with other pesticides on cereals optimum timing for herbicide action may differ from that for growth reduction. See label for details of tank mix recommendations
- Most products recommended for use on oats require addition of approved non-ionic wetter. Check label
- Some products are formulated with trace elements to help compensate for increased demand during rapid growth

Restrictions
- Maximum number of treatments or maximum total dose varies with crop and product and whether split dose treatments are recommended. Check labels
- Do not use on very late sown spring wheat or oats or on crops under stress
- Mixtures with liquid nitrogen fertilizers may cause scorch and are specifically excluded on some labels
- Do not use on soils of low fertility unless such crops regularly receive adequate dressings of nitrogen
- At least 6 h, preferably 24 h, required before rain for maximum effectiveness. Do not apply to wet crops
- Check labels for tank mixtures known to be incompatible
- Must only be applied between growth stages BBCH 31 to BBCH 39

Crop-specific information
- Latest use varies with crop and product. See label for details
- May be used on cereals undersown with grass or clovers
- Ornamentals to be treated must be well established and growing vigorously. Do not treat in strong sunlight or when temperatures are likely to fall below 10°C
- Temporary yellow spotting may occur on poinsettias. It can be minimised by use of a non-ionic wetting agent - see label

Environmental safety
- Wash equipment thoroughly with water and wetting agent immediately after use and spray out. Spray out again before storing or using for another product. Traces can cause harm to susceptible crops sprayed later
- Do not use straw from treated cereals as horticultural growth medium or mulch

Hazard classification and safety precautions
  Hazard Harmful [1, 3, 5, 7, 12]; Harmful if swallowed [1, 2, 4, 6, 8-12]
  Transport code 8
  Packaging group III
  UN Number 1760
  Risk phrases H290 [1, 2, 4, 6, 9-11]; H312 [1, 8, 12]; R21, R22a [3, 5, 7]
  Operator protection A [1-12]; C [8]; H [5, 7, 8]; U05a, U20b [1-12]; U08 [1-7, 9-11]; U09a [12]; U13 [8]; U19a [1-7, 9-12]
  Environmental protection E15a [1-7, 9-12]; E15b, E40e, H411 [8]; E34 [1-12]; H412 [2, 4, 6, 9-12]

SEE SECTION 3 FOR PRODUCTS ALSO REGISTERED

Storage and disposal D01, D02 [1-12]; D05 [1, 3, 12]; D09a [2, 4-12]; D10a [12]; D10b [1-11]
Medical advice M03 [1-12]; M05a [5, 7, 12]

68   chlormequat + 2-chloroethylphosphonic acid

A plant growth regulator for use in cereals

See also 2-chloroethylphosphonic acid

Products

1	Chlormephon UK	SFP Europe	305:155 g/l	SL	16215
2	Spatial Plus	Interfarm	300:150 g/l	SL	16665
3	Vivax	Interfarm	300:150 g/l	SL	16682

Uses
- Lodging control in **spring barley**, **winter barley** [2, 3]; **winter wheat** [1-3]

Approval information
- Chlormequat and 2-chloroethylphosphonic acid (ethephon) included in Annex I under EC Regulation 1107/2009
- Accepted by BBPA for use on malting barley
- All products containing 2-chloroethylphosphonic acid carry the warning: '2-chloroethylphosphonic acid is an anticholinesterase organophosphate. Handle with care'.

Efficacy guidance
- Best results obtained when crops growing vigorously
- Recommended dose varies with growth stage. See labels for details and recommendations for use of sequential treatments

Restrictions
- 2-chloroethylphosphonic acid is an anticholinesterase organophosphorus compound. Do not use if under medical advice not to work with such compounds
- Maximum number of treatments 1 per crop; maximum total dose equivalent to one full dose treatment
- Product must always be used with specified non-ionic wetter - see labels
- Do not use on any crop in sequence with any other product containing 2-chloroethylphosphonic acid
- Do not spray when crop wet or rain imminent
- Do not spray during cold weather or periods of night frost, when soil is very dry, when crop diseased or suffering pest damage, nutrient deficiency or herbicide stress
- If used on seed crops grown for certification inform seed merchant beforehand
- Do not use on wheat variety Moulin or on any winter varieties sown in spring
- Do not use on spring barley variety Triumph
- Do not treat barley on soils with more than 10% organic matter
- Do not use in programme with any other product containing 2-chloroethylphosphonic acid
- Only crops growing under conditions of high fertility should be treated

Crop-specific information
- Latest use: before flag leaf ligule/collar just visible (GS 39) or 1st spikelet visible (GS 51) for wheat or barley at top dose; or before flag leaf sheath opening (GS 47) for winter wheat at reduced dose
- Apply before lodging has started

Environmental safety
- Harmful to fish or other aquatic life. Do not contaminate surface waters or ditches with chemical or used container
- Do not use straw from treated cereals as a horticultural growth medium or as a mulch

Hazard classification and safety precautions
Hazard Harmful
Transport code 8

FOR FULL CONDITIONS OF USE ALWAYS READ THE PRODUCT LABEL

Packaging group III
UN Number 3265
Risk phrases R21, R36 [1]; R22a, R52 [1-3]; R37, R53a [2, 3]
Operator protection A; U05a, U08, U19a, U20b [1-3]; U14 [1]
Environmental protection E15a, E34, E38
Storage and disposal D01, D02, D09a, D10b [1-3]; D12a [1]
Medical advice M01, M03, M05a

69 chlormequat + 2-chloroethylphosphonic acid + mepiquat chloride

A plant growth regulator for reducing lodging in cereals

See also 2-chloroethylphosphonic acid
mepiquat chloride

Products
Cyclade	BASF	230:155:75 g/l	SL	16350

Uses
- Growth regulation in **durum wheat** *(off-label)*, **rye** *(off-label)*, **triticale** *(off-label)*
- Lodging control in **durum wheat** *(off-label)*, **rye** *(off-label)*, **spring barley**, **triticale** *(off-label)*, **winter barley**, **winter wheat**

Extension of Authorisation for Minor Use (EAMUs)
- **durum wheat** 20140216 expires 03 Dec 2017
- **rye** 20140216 expires 03 Dec 2017
- **triticale** 20140216 expires 03 Dec 2017

Approval information
- Chlormequat, 2-chloroethylphosphonic acid (ethephon) and mepiquat chloride included in Annex I under EC Regulation 1107/2009
- Accepted by BBPA for use on malting barley
- All products containing 2-chloroethylphosphonic acid carry the warning: '2-chloroethylphosphonic acid is an anticholinesterase organophosphate. Handle with care'.
- Approval expiry 03 Dec 2017 [1]

Efficacy guidance
- Best results achieved in a vigorous, actively growing crop with adequate fertility and moisture
- Optimum timing on all crops is from second node detectable stage (GS 32)
- Recommended for use as part of an intensive growing system which includes provision for optimum fertilizer treatment and disease control

Restrictions
- 2-chloroethylphosphonic acid is an anticholinesterase organophosphorus compound. Do not use if under medical advice not to work with such compounds
- Maximum number of treatments 1 per crop
- Maximum total dose depends on spraying regime adopted. See label
- Must be used with a non-ionic wetting agent
- Do not apply to stressed crops or those on soils of low fertility unless receiving adequate dressings of fertilizer
- Do not apply in temperatures above 21°C or if crop is wet or if rain expected
- Do not treat variety Moulin nor any winter varieties sown in spring
- Do not use in a programme with any other product containing 2-chloroethylphosphonic acid
- Do not apply to barley on soils with more than 10% organic matter (winter wheat may be treated)
- Notify seed merchant in advance if use on a seed crop is proposed

Crop-specific information
- Latest use: before first spikelet of ear visible (GS 51) using reduced dose on winter barley; before flag leaf sheath opening (GS 47) using reduced dose on winter wheat; before flag leaf just visible on spring barley

---
SEE SECTION 3 FOR PRODUCTS ALSO REGISTERED

Pesticide Profiles

- May be applied to crops undersown with grasses or clovers
- Treatment may cause some delay in ear emergence

Environmental safety
- Harmful to aquatic organisms
- Do not use straw from treated crops as a horticultural growth medium

Hazard classification and safety precautions
Hazard Harmful, Corrosive
Transport code 8
Packaging group III
UN Number 3265
Risk phrases R20, R22a, R37, R52, R67
Operator protection A; U05a, U08, U19a, U20b
Environmental protection E15a, E34
Storage and disposal D01, D02, D09a, D10c, D12b
Medical advice M01, M03, M05a

70   chlormequat + imazaquin

A plant growth regulator mixture for winter wheat

See also imazaquin

Products

1	Meteor	BASF	368:0.8 g/l	SL	16800
2	Upright	BASF	368:0.8 g/l	SL	16884

Uses
- Increasing yield in **winter wheat** [1, 2]
- Lodging control in *rye (off-label)*, *triticale (off-label)* [1]; **winter wheat** [1, 2]

Extension of Authorisation for Minor Use (EAMUs)
- *rye* 20142647 [1]
- *triticale* 20142647 [1]

Approval information
- Chlormequat and imazaquin included in Annex I under EC Regulation 1107/2009

Efficacy guidance
- Apply to crops during good growing conditions or to those at risk from lodging
- On soils of low fertility, best results obtained where adequate nitrogen fertilizer used

Restrictions
- Maximum number of applications 1 per crop (2 per crop at split dose)
- Do not treat durum wheat
- Do not apply to undersown crops
- Do not apply when crop wet or rain imminent

Crop-specific information
- Latest use: before second node detectable (GS 31)
- Apply as single dose from leaf sheath lengthening up to and including 1st node detectable or as split dose, the first from tillers formed to leaf sheath lengthening, the second from leaf sheath erect up to and including 1st node detectable

Environmental safety
- Dangerous for the environment
- Toxic to aquatic organisms
- Do not use straw from treated cereals as horticultural growth medium or mulch

Hazard classification and safety precautions
Hazard Harmful, Dangerous for the environment
Transport code 8

FOR FULL CONDITIONS OF USE ALWAYS READ THE PRODUCT LABEL

2-chloroethylphosphonic acid

Packaging group III
UN Number 1760
Risk phrases R20, R22a, R41, R51, R53a [1, 2]; R36 [2]
Operator protection A, C; U05a, U08, U11, U13, U15, U19a, U20b
Environmental protection E15a, E34, E38
Storage and disposal D01, D02, D05, D06c, D09a, D10b, D12a
Medical advice M03, M05a

71   2-chloroethylphosphonic acid

A plant growth regulator for cereals and various horticultural crops

See also chlormequat + 2-chloroethylphosphonic acid
      chlormequat + 2-chloroethylphosphonic acid + imazaquin
      chlormequat + 2-chloroethylphosphonic acid + mepiquat chloride

Products

1	Cerone	Nufarm UK	480 g/l	SL	15944
2	Ethe 480	Agroquimicos	480 g/l	SL	16887
3	Ethefon 480	Goldengrass	480 g/l	SL	16102
4	Ipanema	SFP Europe	480 g/l	SL	15961

Uses
- Growth regulation in **apples** *(off-label)*, **durum wheat** *(off-label)*, **ornamental plant production** *(off-label)*, **protected tomatoes** *(off-label)* [1]
- Lodging control in **spring barley**, **winter barley**, **winter wheat** [1-4]; **triticale**, **winter rye** [1-3]

Extension of Authorisation for Minor Use (EAMUs)
- **apples** 20122367 [1]
- **durum wheat** 20122368 [1]
- **ornamental plant production** 20122366 [1]
- **protected tomatoes** 20122369 [1]

Approval information
- 2-chloroethylphosphonic acid (ethephon) included in Annex I under EC Regulation 1107/2009
- All products containing 2-chloroethylphosphonic acid carry the warning: '2-chloroethylphosphonic acid is an anticholinesterase organophosphate. Handle with care'
- Accepted by BBPA for use on malting barley

Efficacy guidance
- Best results achieved on crops growing vigorously under conditions of high fertility
- Optimum timing varies between crops and products. See labels for details
- Do not spray crops when wet or if rain imminent

Restrictions
- 2-chloroethylphosphonic acid is an anticholinesterase organophosphorus compound. Do not use if under medical advice not to work with such compounds
- Maximum number of treatments 1 per crop or yr
- Do not spray crops suffering from stress caused by any factor, during cold weather or period of night frost nor when soil very dry
- Do not apply to cereals within 10 d of herbicide or liquid fertilizer application
- Do not spray wheat or triticale where the leaf sheaths have split and the ear is visible

Crop-specific information
- Latest use: before 1st spikelet visible (GS 51) for spring barley, winter barley, winter rye; before flag leaf sheath opening (GS 47) for triticale, winter wheat
- HI cider apples, tomatoes 5 d

Environmental safety
- Harmful to aquatic organisms
- Avoid accidental deposits on painted objects such as cars, trucks, aircraft

SEE SECTION 3 FOR PRODUCTS ALSO REGISTERED

Pesticide Profiles

Hazard classification and safety precautions
Hazard Harmful [1-3]; Irritant [4]
Transport code 8
Packaging group III
UN Number 3265
Risk phrases R20, R41, R52, R53a [1-4]; R38 [4]
Operator protection A, C; U05a, U08, U11, U13, U19c, U20b
Environmental protection E13c, E38
Storage and disposal D01, D02, D09a, D10b, D12a
Medical advice M01

72   chlorothalonil

A protectant chlorophenyl fungicide for use in many crops and turf
FRAC mode of action code: M5

*See also azoxystrobin + chlorothalonil
           carbendazim + chlorothalonil*

Products

1	Abringo	Belcrop	500 g/l	SC	15607
2	Barclay Avoca	Barclay	500 g/l	SC	15549
3	Bravo 500	Syngenta	500 g/l	SC	14548
4	Clayton Turret	Clayton	500 g/l	SC	17220
5	CTL 500	Goldengrass	500 g/l	SC	14812
6	Damocles	AgChem Access	500 g/l	SC	15689
7	Damocles	AgChem Access	500 g/l	SC	15799
8	Joules	Nufarm UK	500 g/l	SC	15730
9	Life Scientific Chlorothalonil	Life Scientific	500 g/l	SC	15369
10	Piper	Nufarm UK	500 g/l	SC	15786
11	Premium 500	Agroquimicos	500 g/l	SC	16137
12	Pronto	Stefes	500 g/l	SC	17688
13	Rover 500	Sipcam	500 g/l	SC	15496
14	Stefonil	Stefes	500 g/l	SC	16862
15	Supreme	Mitsui	500 g/l	SC	15395
16	UPL Chlorothalonil	UPL Europe	500 g/l	SC	15472

Uses

- Basal stem rot in **narcissi** *(off-label)* [3, 9]
- Botrytis in **celeriac** *(off-label)*, **strawberries** *(off-label)* [3]
- Disease control in **all edible seed crops grown outdoors** *(off-label)*, **all non-edible seed crops grown outdoors** *(off-label)*, **durum wheat** *(off-label)*, **forest nurseries** *(off-label)*, **ornamental plant production** *(off-label)*, **rye** *(off-label)*, **triticale** *(off-label)* [3, 9]; **spring field beans**, **winter field beans** [13]
- Downy mildew in **poppies for morphine production** *(off-label)* [3]
- Glume blotch in **spring wheat**, **winter wheat** [1-16]
- Leaf spot in **celeriac** *(off-label)*, **strawberries** *(off-label)* [3]
- Neck rot in **narcissi** *(off-label)* [3, 9]
- Rhynchosporium in **spring barley** *(moderate control)*, **winter barley** *(moderate control)* [1-16]
- Septoria leaf blotch in **spring wheat**, **winter wheat** [1-16]

Extension of Authorisation for Minor Use (EAMUs)

- *all edible seed crops grown outdoors* 20111130 [3], 20120380 [9]
- *all non-edible seed crops grown outdoors* 20111130 [3], 20120380 [9]
- *celeriac* 20110948 [3]
- *durum wheat* 20110944 [3], 20120381 [9]
- *forest nurseries* 20111130 [3], 20120380 [9]
- *narcissi* 20110943 [3], 20120379 [9]
- *ornamental plant production* 20111130 [3], 20120380 [9]

FOR FULL CONDITIONS OF USE ALWAYS READ THE PRODUCT LABEL

chlorothalonil

- *poppies for morphine production* 20110947 [3]
- *rye* 20110944 [3], 20120381 [9]
- *strawberries* 20110949 [3]
- *triticale* 20110944 [3], 20120381 [9]

Approval information
- Chlorothalonil included in Annex I under EC Regulation 1107/2009.
- Accepted by BBPA for use on malting barley and hops

Efficacy guidance
- For some crops products differ in diseases listed as controlled. See label for details and for application rates, timing and number of sprays
- Apply as protective spray or as soon as disease appears and repeat as directed
- On cereals activity against Septoria may be reduced where serious mildew or rust present. In such conditions mix with suitable mildew or rust fungicide
- May be used for preventive and curative treatment of turf but treatment at a late stage of disease development may not be so successful and can leave bare patches of soil requiring renovation.

Restrictions
- Maximum number of treatments and maximum total doses vary with crop and product - see labels for details
- Operators of vehicle mounted equipment must use a vehicle fitted with a cab and a forced air filtration unit with a pesticide filter complying with HSE Guidance Note PM74 or to an equivalent or higher standard when making broadcast or air-assisted applications

Crop-specific information
- Latest time of application to cereals varies with product. Consult label
- Latest use for winter oilseed rape before flowering; end Aug in yr of harvest for blackcurrants, gooseberries, redcurrants,
- HI 8 wk for field beans; 6 wk for combining peas; 28 d or before 31 Aug for post-harvest treatment for blackcurrants, gooseberries, redcurrants; 14 d for onions, strawberries; 10 d for hops; 7-14 d for broccoli, Brussels sprouts, cabbages, cauliflowers, celery, onions, potatoes; 3 d for cane fruit; 12-48 h for protected cucumbers, protected tomatoes
- For Botrytis control in strawberries important to start spraying early in flowering period and repeat at least 3 times at 10 d intervals
- On strawberries some scorching of calyx may occur with protected crops

Environmental safety
- Dangerous for the environment
- Very toxic to aquatic organisms
- Do not spray from the air within 250 m horizontal distance of surface waters or ditches
- Broadcast air-assisted LERAP [1-6, 8-10, 12, 14, 15] (18 m); LERAP Category B [1-16]

Hazard classification and safety precautions
Hazard Harmful, Dangerous for the environment [1-16]; Harmful if swallowed [8]; Fatal if inhaled [10]; Harmful if inhaled [4, 8, 12, 14]; Very toxic to aquatic organisms [12, 14]
Transport code 9
Packaging group III
UN Number 3082
Risk phrases H317, H335 [4, 8, 10, 12, 14]; H318 [8, 10]; H319 [4, 12, 14]; H351 [4, 10, 12, 14]; R20, R37, R40, R43, R50, R53a [1-3, 5-7, 9, 11, 13, 15, 16]; R36 [1-3, 5, 6, 9, 15, 16]; R38, R41 [7, 11, 13]
Operator protection A, H [1-16]; C [1, 3-16]; J, M [1, 3, 4, 6, 9, 14, 16]; U02a, U05a, U09a, U19a [1-16]; U11 [3, 4, 7, 11-14]; U13, U14, U15, U20b [7, 11, 13]; U20a [1-6, 8-10, 12, 14-16]
Environmental protection E15a, E16a, E38 [1-16]; E16b [1-6, 8-10, 12, 14, 15]; E17b [1-6, 8-10, 12, 14, 15] (18 m); E18 [1-6, 8-10, 12, 14-16]; H410 [4, 8, 10, 12, 14]
Storage and disposal D01, D02, D09a, D12a [1-16]; D05, D10c [1-6, 8-10, 12, 14-16]; D10b [7, 11, 13]
Medical advice M03 [2]

SEE SECTION 3 FOR PRODUCTS ALSO REGISTERED

Pesticide Profiles

73   chlorothalonil + cymoxanil

A protectant and systemic fungicide for potato blight
FRAC mode of action code: M5 + 27

See also cymoxanil

Products
   Mixanil   Sipcam   375:50 g/l   SC   15675

Uses
- Blight in **potatoes**

Approval information
- Cymoxanil and chlorothalonil included in Annex I under EC Regulation 1107/2009

Environmental safety
- LERAP Category B

Hazard classification and safety precautions
   Hazard Harmful, Dangerous for the environment
   Transport code 9
   Packaging group III
   UN Number 3082
   Risk phrases R20, R37, R40, R43, R50, R53a
   Operator protection A, H; U04a, U05b, U08, U20b
   Environmental protection E15b, E16a, E16b, E38
   Storage and disposal D01, D02, D09a, D12a, D19
   Medical advice M04a, M05a

74   chlorothalonil + cyproconazole

A systemic protectant and curative fungicide mixture for cereals, combining peas and some products can also be used in field beans
FRAC mode of action code: M5 + 3

See also cyproconazole

Products

1	Alto Elite	Syngenta	375:40 g/l	SC	08467
2	Baritone	AgChem Access	375:40 g/l	SC	15910
3	Octolan	Syngenta	375:40 g/l	SC	11675
4	SAN 703	Syngenta	375:40 g/l	SC	11676

Uses
- Brown rust in **spring barley**, **winter barley**, **winter wheat** [1-4]
- Chocolate spot in **spring field beans**, **winter field beans** [1-4]
- Eyespot in **winter barley**, **winter wheat** [1-4]
- Glume blotch in **winter wheat** [1, 2]
- Grey mould in **combining peas** [1-4]
- Net blotch in **spring barley**, **winter barley** [1-4]
- Powdery mildew in **spring barley**, **winter barley**, **winter wheat** [1-4]
- Rhynchosporium in **spring barley**, **winter barley** [1-4]
- Rust in **spring field beans**, **winter field beans** [1, 2]
- Septoria leaf blotch in **winter wheat** [1-4]
- Yellow rust in **spring barley**, **winter barley**, **winter wheat** [1-4]

Approval information
- Chlorothalonil and cyproconazole included in Annex I under EC Regulation 1107/2009
- Accepted by BBPA for use on malting barley

FOR FULL CONDITIONS OF USE ALWAYS READ THE PRODUCT LABEL

chlorothalonil + cyproconazole + propiconazole

Efficacy guidance
- Apply at first signs of infection or as soon as disease becomes active
- A repeat application may be made if re-infection occurs
- For established mildew tank-mix with an approved mildewicide
- When applied prior to third node detectable (GS 33) a useful reduction of eyespot will be obtained
- Cyproconazole is a DMI fungicide. Resistance to some DMI fungicides has been identified in Septoria leaf blotch which may seriously affect performance of some products. For further advice contact a specialist advisor and visit the Fungicide Resistance Action Group (FRAG)-UK website

Restrictions
- Maximum total dose equivalent to 2 full dose treatments
- Do not apply at concentrations higher than recommended

Crop-specific information
- Latest use: before beginning of anthesis (GS 60) for barley; before caryopsis watery ripe (GS 71) for wheat
- HI peas, field beans 6 wk
- If applied to winter wheat in spring at GS 30-33 straw shortening may occur but yield is not reduced

Environmental safety
- Dangerous for the environment
- Very toxic to aquatic organisms
- LERAP Category B

Hazard classification and safety precautions
   Hazard Harmful, Dangerous for the environment
   Transport code 9
   Packaging group III
   UN Number 3082
   Risk phrases R20, R37, R40, R41, R43, R50, R53a
   Operator protection A, C, H, M; U05a, U19a [1-4]; U11, U20a [1, 2]; U12, U20b [3, 4]
   Environmental protection E15a, E16b [1, 2]; E15b [3, 4]; E16a, E38 [1-4]
   Storage and disposal D01, D02, D09a, D10c, D12a [1-4]; D05 [1, 2]
   Medical advice M04a

75   chlorothalonil + cyproconazole + propiconazole

A broad-spectrum fungicide mixture for cereals
FRAC mode of action code: M5 + 3 + 3

See also cyproconazole
   propiconazole

Products
   Cherokee   Syngenta   375:50:62.5 g/l   SE   13251

Uses
- Brown rust in **spring barley**, **winter barley**, **winter wheat**
- Net blotch in **spring barley** *(reduction)*, **winter barley** *(reduction)*
- Powdery mildew in **spring barley**, **winter barley**, **winter wheat** *(moderate control)*
- Rhynchosporium in **spring barley** *(reduction)*, **winter barley** *(reduction)*
- Septoria leaf blotch in **winter wheat**
- Yellow rust in **winter wheat**

Approval information
- Chlorothalonil, propiconazole and cyproconazole included in Annex I under EC Regulation 1107/2009

SEE SECTION 3 FOR PRODUCTS ALSO REGISTERED

Pesticide Profiles

Efficacy guidance
- Best results obtained from treatment at the early stages of disease development
- Cyproconazole and propiconazole are DMI fungicides. Resistance to some DMI fungicides has been identified in Septoria leaf blotch which may seriously affect performance of some products. For further advice contact a specialist advisor and visit the Fungicide Resistance Action Group (FRAG)-UK website
- Product should be used as part of an Integrated Crop Management strategy incorporating other methods of control and including, where appropriate, other fungicides with a different mode of action
- Product should be used preventatively and not relied on for its curative potential
- Always follow FRAG guidelines for resistance management. See Section 5 for more information

Restrictions
- Maximum total dose equivalent to two full dose treatments on wheat and barley
- Do not treat crops under stress
- Spray solution must be used as soon as possible after mixing

Crop-specific information
- Latest use: before caryopsis watery ripe (GS 71) for winter wheat; before first spikelet just visible (GS 51) for barley
- When applied to winter wheat in the spring straw shortening may occur which causes no loss of yield

Environmental safety
- Dangerous for the environment
- Very toxic to aquatic organisms
- LERAP Category B

Hazard classification and safety precautions
Hazard Harmful, Dangerous for the environment
Transport code 9
Packaging group III
UN Number 3082
Risk phrases R20, R36, R37, R40, R43, R50, R53a
Operator protection A, C, H; U02a, U05a, U09a, U10, U11, U14, U15, U19a, U20b
Environmental protection E15b, E16a, E16b, E34, E38
Storage and disposal D01, D02, D05, D07, D09a, D10c, D12a
Medical advice M04a

76   chlorothalonil + fludioxonil + propiconazole

A fungicide treatment for amenity grassland
FRAC mode of action code: M5 + 12 + 3

See also fludioxonil
propiconazole

Products
Instrata	Syngenta	362:14.5:56.9 g/l	SE	16458

Uses
- Anthracnose in **amenity grassland** *(reduction)*, **managed amenity turf** *(reduction)*
- Brown patch in **amenity grassland**, **managed amenity turf**
- Dollar spot in **amenity grassland**, **managed amenity turf**
- Fusarium patch in **amenity grassland**, **managed amenity turf**

Approval information
- Chlorothalonil, fludioxonil and propiconazole included in Annex I under EC Regulation 1107/2009

Environmental safety
- Buffer zone requirement 12m [1]
- LERAP Category B

FOR FULL CONDITIONS OF USE ALWAYS READ THE PRODUCT LABEL

Hazard classification and safety precautions
  Hazard Toxic, Dangerous for the environment
  Transport code 9
  Packaging group III
  UN Number 3082
  Risk phrases R23a, R36, R37, R40, R43, R50, R53a
  Operator protection A, C, H; U02a, U05a, U09a, U10, U11, U19a, U19c, U20b
  Environmental protection E15b, E16a, E34, E38
  Storage and disposal D01, D02, D09a, D12a
  Medical advice M04a

77 chlorothalonil + mancozeb

A multi-site protectant fungicide mixture for disease control in winter wheat
FRAC mode of action code: M5 + M3

See also mancozeb

Products
  Guru   Interfarm   286:194 g/l   SC   15268

Uses
- Glume blotch in *winter wheat*
- Septoria leaf blotch in *winter wheat*

Approval information
- Chlorothalonil and mancozeb included in Annex I under EC Regulation 1107/2009

Efficacy guidance
- Best results on winter wheat achieved from protective applications to the flag leaf. If disease is already present on the lower leaves, treat as soon as the flag leaf is just visible (GS 37).
- Activity on Septoria may be reduced in the presence of severe mildew infection.

Restrictions
- Maximum total dose on winter wheat equivalent to 1.5 x full dose treatment.
- Do not treat crops under stress from frost, drought, water-logging, trace element deficiency or pest attack.
- Broadcast air assisted applications must only be made by equipment fitted with a cab with a forced air filtration unit plus a pesticide filter complying with HSE Guidance Note PM 74 or an equivalent or higher standard

Crop-specific information
- Latest use: before grain watery ripe stage (GS 71)

Environmental safety
- Dangerous for the environment
- Very toxic to aquatic organisms
- LERAP Category B

Hazard classification and safety precautions
  Hazard Irritant, Dangerous for the environment, Fatal if inhaled, Very toxic to aquatic organisms
  Transport code 9
  Packaging group III
  UN Number 3082
  Risk phrases H317, H318, H335, H351, H361
  Operator protection A, C, H, M; U02a, U05a, U08, U10, U11, U13, U14, U19a, U20b
  Environmental protection E16a, E16b, E38, H410
  Storage and disposal D01, D02, D05, D09a, D10b, D12a

SEE SECTION 3 FOR PRODUCTS ALSO REGISTERED

78 chlorothalonil + metalaxyl-M

A systemic and protectant fungicide for various crops
FRAC mode of action code: M5 + 4

See also metalaxyl-M

Products

1	Broadsheet	AgChem Access	500:37.5 g/l	SC	15906
2	Folio Gold	Syngenta	500:37.5 g/l	SC	14368

Uses
- Alternaria in **brussels sprouts** *(moderate control)*, **cauliflowers** *(moderate control)* [1, 2]
- Botrytis in **gooseberries** *(off-label)* [2]
- Disease control in **forest nurseries** *(off-label)*, **ornamental plant production** *(off-label)*, **poppies for morphine production** *(off-label)*, **protected ornamentals** *(off-label)* [2]
- Downy mildew in **broad beans**, **brussels sprouts**, **cauliflowers** [1, 2]
- Drepanopeziza ribis in **gooseberries** *(off-label)* [2]
- Leaf spot in **gooseberries** *(off-label)* [2]
- Ring spot in **brussels sprouts** *(reduction)*, **cauliflowers** *(reduction)* [1, 2]
- White blister in **brussels sprouts**, **cauliflowers** [1, 2]

Extension of Authorisation for Minor Use (EAMUs)
- *forest nurseries* 20111128 [2]
- *gooseberries* 20161257 [2]
- *ornamental plant production* 20111128 [2]
- *poppies for morphine production* 20090251 [2]
- *protected ornamentals* 20120032 [2]

Approval information
- Chlorothalonil and metalaxyl-M included in Annex I under EC Regulation 1107/2009

Efficacy guidance
- Apply at first signs of disease or when weather conditions favourable for disease pressure
- Best results obtained when used in a full and well-timed programme. Repeat treatment at 14-21 d intervals if necessary
- Treatment of established disease will be less effective
- Evidence of effectiveness in bulb onions, shallots and leeks is limited

Restrictions
- Maximum total dose equivalent to 2 full doses on broad beans, field beans, cauliflowers, calabrese; 3 full doses on Brussels sprouts, leeks, onions and shallots

Crop-specific information
- HI 14 d for all crops

Environmental safety
- Dangerous for the environment
- Very toxic to aquatic organisms
- Buffer zone requirement 20 m in gooseberries [2]
- LERAP Category B

Hazard classification and safety precautions
Hazard Harmful, Dangerous for the environment
Transport code 9
Packaging group III
UN Number 3082
Risk phrases R20, R36, R37, R38, R40, R43, R50, R53a
Operator protection A, C, H; U05a, U09a, U15, U20a
Environmental protection E15a, E16a, E38
Consumer protection C02a (14 d)
Storage and disposal D01, D02, D05, D09a, D10c, D12a

FOR FULL CONDITIONS OF USE ALWAYS READ THE PRODUCT LABEL

79 chlorothalonil + penthiopyrad

A protectant and systemic fungicide mixture for disease control in wheat
FRAC mode of action code: 3 + 7

See also penthiopyrad

Products

1	Aylora	DuPont	250:100 g/l	SL	16187
2	Intellis Plus	DuPont	250:100 g/l	SL	16419
3	Treoris	DuPont	250:100 g/l	SL	16126
4	Trust	Generica	250:100 g/l	SL	16561

Uses
- Brown rust in *spring wheat*, *winter wheat* [1-4]
- Disease control in *spring barley*, *spring wheat*, *winter barley*, *winter wheat* [1, 2]
- Net blotch in *spring barley*, *winter barley* [1-4]
- Ramularia leaf spots in *spring barley*, *winter barley* [1-4]
- Rhynchosporium in *spring barley*, *winter barley* [1-4]
- Septoria leaf blotch in *spring wheat*, *winter wheat* [1-4]

Approval information
- Chlorothalonil and penthiopyrad are included in Annex 1 under EC Regulation 1107/2009

Restrictions
- To reduce the risk of resistance, only two applications of any SDHI fungicide may be made to a cereal crop.
- A maximum total dose of not more than 500 g/ha penthiopyrad may be applied in a two-year period to the same field.

Environmental safety
- Buffer zone requirement 6 m [1-4]
- LERAP Category B

Hazard classification and safety precautions
  Hazard Harmful, Dangerous for the environment [1, 3, 4]
  Transport code 9
  Packaging group III
  UN Number 3082
  Risk phrases R20, R37, R50, R53a, R70 [1-4]; R40b [3, 4]; R45 [1, 2]
  Operator protection A, H; U02a, U05a, U09a, U11, U19a, U20a [1-4]; U09c [2]
  Environmental protection E15b, E16a [1-4]; E34 [1, 3, 4]; E38 [1, 2]
  Storage and disposal D01, D02, D09a, D10c, D12a [1-4]; D06e, D12b [1, 2]
  Medical advice M03a

80 chlorothalonil + picoxystrobin

A broad spectrum fungicide mixture for cereals
FRAC mode of action code: M5 + 11

See also picoxystrobin

Products

1	Credo	DuPont	500:100 g/l	SC	14577
2	Zimbrail	DuPont	500:100 g/l	SC	14735

Uses
- Brown rust in *spring barley*, *spring wheat*, *winter barley*, *winter wheat*
- Glume blotch in *spring wheat*, *winter wheat*
- Net blotch in *spring barley*, *winter barley*
- Rhynchosporium in *spring barley*, *winter barley*

SEE SECTION 3 FOR PRODUCTS ALSO REGISTERED

Pesticide Profiles

- Septoria leaf blotch in **spring wheat**, **winter wheat**
- Tan spot in **spring wheat**, **winter wheat**
- Yellow rust in **spring wheat**, **winter wheat**

Approval information
- Chlorothalonil and picoxystrobin included in Annex I under EC Regulation 1107/2009
- Accepted by BBPA for use on malting barley

Efficacy guidance
- Best results obtained as a protectant treatment, or when disease first seen in crop, applied in good growing conditions with adequate soil moisture
- Disease control for 4-6 wk normally achieved during stem elongation
- Results may be less reliable when used on crops under stress
- Treatments for protection against ear disease should be made at ear emergence
- Picoxystrobin is a member of the QoI cross resistance group. Product should be used preventatively and not relied on for its curative potential
- Use product as part of an Integrated Crop Management strategy incorporating other methods of control, including where appropriate other fungicides with a different mode of action. Do not apply more than two foliar applications of QoI containing products to any cereal crop
- Control of many diseases may be improved by use in mixture with an appropriate triazole
- There is a significant risk of widespread resistance occurring in *Septoria tritici* populations in UK. Failure to follow resistance management action may result in reduced levels of disease control
- Strains of barley powdery mildew resistant to QoI's are common in the UK

Restrictions
- Maximum number of treatments 2 per crop of wheat or barley

Crop-specific information
- Latest use: before first spikelet just visible (GS 51) for barley; before grain watery ripe (GS 71) for wheat

Environmental safety
- Dangerous for the environment
- Very toxic to aquatic organisms
- LERAP Category B

Hazard classification and safety precautions
   Hazard Harmful, Dangerous for the environment, Harmful if inhaled, Very toxic to aquatic organisms
   Transport code 9
   Packaging group III
   UN Number 3082
   Risk phrases H317, H335, H351
   Operator protection A, H; U05a, U09a, U14, U15, U19a, U20a
   Environmental protection E15a, E16a, E16b, E38, H410
   Storage and disposal D01, D02, D05, D09a, D10c, D12a

81   chlorothalonil + propiconazole

A systemic and protectant fungicide for winter wheat and barley
FRAC mode of action code: M5 + 3

See also propiconazole

Products
Prairie	Syngenta	250:62.5 g/l	SC	13994

Uses
- Brown rust in **spring barley**, **spring wheat**, **winter barley**, **winter wheat**
- Glume blotch in **spring wheat**, **winter wheat**
- Rhynchosporium in **spring barley**, **winter barley**

FOR FULL CONDITIONS OF USE ALWAYS READ THE PRODUCT LABEL

chlorothalonil + proquinazid

- Septoria leaf blotch in **spring wheat**, **winter wheat**
- Yellow rust in **spring wheat**, **winter wheat**

Approval information
- Chlorothalonil and propiconazole included in Annex I under EC Regulation 1107/2009
- Accepted by BBPA for use on malting barley

Efficacy guidance
- On wheat apply from start of flag leaf emergence up to and including when ears just fully emerged (GS 37-59), on barley at any time to ears fully emerged (GS 59)
- Best results achieved from early treatment, especially if weather wet, or as soon as disease develops

Restrictions
- Maximum number of treatments 2 per crop or 1 per crop if other propiconazole based fungicide used in programme

Crop-specific information
- Latest use: up to and including emergence of ear just complete (GS 59).
- HI 42 d

Environmental safety
- Irritating to eyes, skin and respiratory system
- Risk of serious damage to eyes
- Dangerous to fish or other aquatic life. Do not contaminate surface waters or ditches with chemical or used container
- LERAP Category B

Hazard classification and safety precautions
  Hazard Harmful, Dangerous for the environment
  Transport code 9
  Packaging group III
  UN Number 3082
  Risk phrases R20, R36, R37, R40, R43, R50, R53a
  Operator protection A, C, H; U02a, U05a, U11, U20b
  Environmental protection E15b, E16a, E16b, E38
  Storage and disposal D01, D02, D09a, D10c, D12a

82 chlorothalonil + proquinazid

A chlorophenyl and quinazolinone fungicide mixture for powdery mildew control in cereals
FRAC mode of action code: M5 + U7

Products
Fielder SE   DuPont   500:25 g/l   SE   15120

Uses
- Powdery mildew in **spring barley**, **spring wheat**, **winter barley**, **winter wheat**
- Rhynchosporium in **spring barley** *(suppression only)*, **winter barley** *(suppression only)*
- Septoria leaf blotch in **spring wheat** *(suppression only)*, **winter wheat** *(suppression only)*

Approval information
- Chlorothalonil and proquinazid included in Annex 1 under EC Regulation 1107/2009

Efficacy guidance
- If disease has actively spread to new growth, application should be in tank-mixture with a fungicide that has an alternative mode of action with curative activity.
- If two applications are used consecutively in a crop, the second application should be in tank-mixture with a fungicide with an alternative mode of action against mildew, i.e. not a quinazolinone or quinoline fungicide.
- Should be considered as a member of the quinoline fungicides for resistance management.

SEE SECTION 3 FOR PRODUCTS ALSO REGISTERED

Pesticide Profiles

Restrictions
- Maximum number of applications: 2 per crop.

Crop-specific information
- Latest use in wheat is before full flowering (GS65)
- Latest use in barley is before first awns visible

Environmental safety
- LERAP Category B

Hazard classification and safety precautions
Hazard Harmful, Dangerous for the environment, Fatal if inhaled, Very toxic to aquatic organisms
Transport code 9
Packaging group III
UN Number 3082
Risk phrases H317, H318, H335, H351
Operator protection A, C, H; U05a, U09a, U12, U14, U15, U20a
Environmental protection E15a, E15b, E16a, E38, E40b, H410
Storage and disposal D01, D02, D05, D09a, D10c, D12a
Medical advice M05a

83   chlorothalonil + tebuconazole

A protectant and systemic fungicide mixture for winter wheat
FRAC mode of action code: M5 + 3

See also tebuconazole

Products

1	Crafter	Nufarm UK	250:90 g/l	SC	15895
2	Fezan Plus	Sipcam	166:60 g/l	SC	15490
3	Pentangle	Nufarm UK	500:180 g/l	SC	16792
4	Timpani	Nufarm UK	250:90 g/l	SC	14651

Uses
- Fusarium ear blight in **winter wheat** [2]
- Rhynchosporium in **winter barley** *(useful reduction)* [1, 4]
- Septoria leaf blotch in **winter wheat** [1-4]
- Sooty moulds in **winter wheat** [2]

Approval information
- Chlorothalonil and tebuconazole included in Annex I under EC Regulation 1107/2009

Efficacy guidance
- Latest time of application before anthesis (GS61)
- Maximum individual dose 1.0 l/ha. Where disease pressure is severe, a second application may be required but applications after flag leaf visible (GS39) may only result in a reduction of disease severity

Restrictions
- Newer authorisations for tebuconazole products require application to cereals only after GS 30 and applications to oilseed rape and linseed after GS20 - check label

Environmental safety
- Buffer zone requirement 8 m [3]
- Buffer zone requirement 6 m [1, 4]
- LERAP Category B

Hazard classification and safety precautions
Hazard Very toxic [1, 3, 4]; Harmful [2]; Dangerous for the environment [1-4]; Fatal if inhaled [1]; Harmful if inhaled [3, 4]; Very toxic to aquatic organisms [1, 3]
Transport code 6.1 [3]; 9 [1, 2, 4]
Packaging group II [3]; III [1, 2, 4]

FOR FULL CONDITIONS OF USE ALWAYS READ THE PRODUCT LABEL

chlorotoluron

UN Number 2902 [3]; 3082 [1, 2, 4]
Risk phrases H317, H335, H351, H361 [1, 3, 4]; H318 [3]; H319 [1, 4]; R20, R40, R50, R53a, R63 [2]
Operator protection A, H [1-4]; C [1, 3, 4]; U02a, U09a, U10, U11, U13, U14, U15, U19b [1, 3, 4]; U05a, U19a [1-4]
Environmental protection E15a, E16b, H410 [1, 3, 4]; E15b [2]; E16a, E34, E38 [1-4]
Storage and disposal D01, D02, D09a, D12a [1-4]; D05, D10c [2]; D10b [1, 3, 4]
Medical advice M03 [2]; M04a [1, 3, 4]

84   chlorotoluron

A contact and residual urea herbicide for cereals only available in mixtures
HRAC mode of action code: C2

85   chlorotoluron + diflufenican + pendimethalin

A herbicide mixture for weed control in cereals
HRAC mode of action code: C2 + F1 + K1

Products
1	Tower	Adama	250:40:300 g/l	SC	16586
2	Tribal	Adama	250:40:300 g/l	SC	17075

Uses
- Annual meadow grass in *spring barley, spring wheat, triticale, winter barley, winter rye, winter wheat*
- Charlock in *spring barley, spring wheat, triticale, winter barley, winter rye, winter wheat*
- Chickweed in *spring barley, spring wheat, triticale, winter barley, winter rye, winter wheat*
- Cleavers in *spring barley, spring wheat, triticale, winter barley, winter rye, winter wheat*
- Crane's-bill in *spring barley, spring wheat, triticale, winter barley, winter rye, winter wheat*
- Dead nettle in *spring barley, spring wheat, triticale, winter barley, winter rye, winter wheat*
- Field pansy in *spring barley, spring wheat, triticale, winter barley, winter rye, winter wheat*
- Forget-me-not in *spring barley, spring wheat, triticale, winter barley, winter rye, winter wheat*
- Fumitory in *spring barley, spring wheat, triticale, winter barley, winter rye, winter wheat*
- Loose silky bent in *spring barley, spring wheat, triticale, winter barley, winter rye, winter wheat*
- Mayweeds in *spring barley, spring wheat, triticale, winter barley, winter rye, winter wheat*
- Penny cress in *spring barley, spring wheat, triticale, winter barley, winter rye, winter wheat*
- Poppies in *spring barley, spring wheat, triticale, winter barley, winter rye, winter wheat*
- Runch in *spring barley, spring wheat, triticale, winter barley, winter rye, winter wheat*
- Shepherd's purse in *spring barley, spring wheat, triticale, winter barley, winter rye, winter wheat*
- Speedwells in *spring barley, spring wheat, triticale, winter barley, winter rye, winter wheat*
- Volunteer oilseed rape in *spring barley, spring wheat, triticale, winter barley, winter rye, winter wheat*

Approval information
- Chlorotoluron, diflufenican and pendimethalin included in Annex I under EC Regulation 1107/2009

Efficacy guidance
- Minimum water volume for full dose (2.0 l/ha) is 200 l/ha while reducing the dose to 1.0 l/ha allows a lowest water volume of 100 l/ha.

Restrictions
- Maximum number of treatments 1 per crop

SEE SECTION 3 FOR PRODUCTS ALSO REGISTERED

Pesticide Profiles

- Use only on listed crop varieties when applying post-emergence to winter wheat. Do not apply to undersown crops
- Substances containing the chlorotoluron active agent may not be applied more than once per year on the same area
- Application must not be made after 31 October in the year of drilling
- Do not use if it is frosty
- DO NOT apply to soils with greater than 10% organic matter

Crop-specific information
- Following normal harvest, there are no restrictions. In the event of crop failure, plough before drilling the following crop in the spring. Only winter wheat can be re-drilled in the same autumn following crop failure

Environmental safety
- Buffer zone requirement 6 m [1, 2]
- LERAP Category B

Hazard classification and safety precautions
Transport code 9
Packaging group III
UN Number 3082
Risk phrases H351, H361
Operator protection A; U02a, U04a, U20b
Environmental protection E15b, E16a, E34, E38, H410
Storage and disposal D01, D09a, D10b, D12a
Medical advice M03

86 chlorpropham

A residual carbamate herbicide and potato sprout suppressant
HRAC mode of action code: K2

See also chloridazon + chlorpropham + metamitron

Products

1	Aceto Chlorpropham 50M	Aceto	500 g/l	HN	14134
2	Aceto Sprout Nip	Aceto	100% w/w	HN	14156
3	Aceto Sprout Nip Pellet	Aceto	100% w/w	HN	15397
4	Aceto Sprout Nip Ultra	Aceto	500 g/l	HN	15456
5	CIPC Fog 300	AgChem Access	300 g/l	HN	15904
6	Cleancrop Amigo 2	UPL Europe	400 g/l	EC	15419
7	Gro-Stop Ready	Certis	120 g/l	EW	15109
8	Gro-Stop Solid	Certis	100% w/w	BR	14103
9	Intruder	UPL Europe	400 g/l	EC	15076
10	Prolonger	UPL Europe	500 g/l	HN	17453

Uses
- Annual dicotyledons in **baby leaf crops** [9]; **bulb onions**, **chard** *(off-label)*, **endives**, **herbs (see appendix 6)** *(off-label)*, **lettuce**, **ornamental plant production**, **salad onions**, **shallots**, **spinach** *(off-label)* [6, 9]; **forest nurseries** *(off-label)*, **leaf brassicas** *(off-label)*, **leeks** *(off-label)* [6]
- Annual grasses in **chard** *(off-label)*, **herbs (see appendix 6)** *(off-label)*, **spinach** *(off-label)* [9]
- Annual meadow grass in **baby leaf crops** [9]; **bulb onions**, **endives**, **lettuce**, **ornamental plant production**, **salad onions**, **shallots** [6, 9]; **chard** *(off-label)*, **forest nurseries** *(off-label)*, **herbs (see appendix 6)** *(off-label)*, **leaf brassicas** *(off-label)*, **spinach** *(off-label)* [6]
- Chickweed in **chard** *(off-label)*, **forest nurseries** *(off-label)*, **herbs (see appendix 6)** *(off-label)*, **leaf brassicas** *(off-label)*, **leeks** *(off-label)* [6]
- Polygonums in **chard** *(off-label)*, **forest nurseries** *(off-label)*, **herbs (see appendix 6)** *(off-label)*, **leaf brassicas** *(off-label)*, **leeks** *(off-label)* [6]

FOR FULL CONDITIONS OF USE ALWAYS READ THE PRODUCT LABEL

chlorpropham

- Small nettle in **chard** *(off-label)*, **forest nurseries** *(off-label)*, **herbs (see appendix 6)** *(off-label)*, **leaf brassicas** *(off-label)*, **leeks** *(off-label)* [6]
- Sprout suppression in **potatoes** [1-4, 7]; **ware potatoes** *(thermal fog)* [5, 8, 10]
- Wild oats in **baby leaf crops** [9]; **bulb onions**, **endives**, **lettuce**, **ornamental plant production**, **salad onions**, **shallots** [6, 9]; **chard** *(off-label)*, **forest nurseries** *(off-label)*, **herbs (see appendix 6)** *(off-label)*, **leaf brassicas** *(off-label)*, **spinach** *(off-label)* [6]

Extension of Authorisation for Minor Use (EAMUs)
- **chard** *20120728* [6], *20110596* [9]
- **forest nurseries** *20120727* [6]
- **herbs (see appendix 6)** *20120728* [6], *20110596* [9]
- **leaf brassicas** *20120728* [6]
- **leeks** *20121841* [6]
- **spinach** *20120728* [6], *20110596* [9]

Approval information
- Chlorpropham included in Annex I under EC Regulation 1107/2009
- Some products are formulated for application by thermal fogging. See labels for details
- Approval expiry 31 Aug 2017 [1-5, 7, 8, 10]

Efficacy guidance
- For sprout suppression apply with suitable fogging or rotary atomiser equipment over dry tubers before sprouting commences. Repeat applications may be needed. See labels for details
- Best results on potatoes obtained in purpose-built box stores with suitable forced draft ventilation. Potatoes in bulk stores should not be stacked more than 3 m high. Positive ventilation systems are a requirement for future use after 2017
- Blockage of air spaces between tubers prevents circulation of vapour and consequent loss of efficacy
- It is important to treat potatoes before the eyes open to obtain best results
- Effectiveness of fogging reduced in non-dedicated stores without proper insulation and temperature controls. Best results obtained at 5-10°C and 75-80% humidity
- When treating potatoes in store with CIPC best practice guidelines are available on-line at www.BeCIPCcompliant.co.uk - some potatoes have recently been found to exceed the MRL and if more are found then the registration of CIPC is in jeopardy.

Restrictions
- Only clean, mature, disease-free potatoes should be treated for sprout suppression. Use of chlorpropham can inhibit tuber wound healing and the severity of skin spot infection in store may be increased if damaged tubers are treated
- Do not fog potatoes with a high level of skin spot
- Do not use on potatoes for seed. Do not handle, dry or store seed potatoes or any other seed or bulbs in boxes or buildings in which potatoes are being or have been treated
- Do not remove treated potatoes from store for sale or processing for at least 21 d after application

Crop-specific information
- Cure potatoes according to label instructions before treatment and allow 2 days - 4 wks between completion of loading into store and first treatment - check labels.

Following crops guidance
- There is a risk of damage to seed potatoes which are handled or stored in boxes or buildings previously treated with chlorpropham

Environmental safety
- Dangerous for the environment
- Toxic to aquatic organisms
- Keep unprotected persons out of treated stores for at least 24 h after application
- Keep in original container, tightly closed, in a safe place, under lock and key

Hazard classification and safety precautions
Hazard Harmful [5-7, 9]; Dangerous for the environment [1-10]; Highly flammable liquid and vapour [1, 10]; Toxic if swallowed, Toxic in contact with skin, Toxic if inhaled [1]

SEE SECTION 3 FOR PRODUCTS ALSO REGISTERED

Pesticide Profiles

Transport code 6.1 [1, 5, 7, 10]; 9 [4, 6, 8, 9]
Packaging group III [1, 4-10]
UN Number 2902 [1, 5, 7, 10]; 3077 [4, 8]; 3082 [6, 9]
Risk phrases H351, H373 [1-4, 8, 10]; H370 [10]; R22a, R22b, R36, R38, R67 [6, 9]; R39 [7]; R40, R51, R53a [5-7, 9]; R42 [5]; R43 [5, 6, 9]; R48 [6, 7, 9]
Operator protection A, H [1-10]; C [1, 4, 6, 9, 10]; D, M [1, 4, 5, 8, 10]; E [4, 5, 8, 10]; J [4, 8, 10]; U01, U19b [8]; U02a [5, 6, 8, 9]; U04a, U15, U20a [5]; U05a [1, 4-6, 8-10]; U08 [1, 4, 5, 8, 10]; U09a, U11, U16b [6, 9]; U14 [5, 7]; U19a [1-6, 8-10]; U20b [1, 4, 6, 8-10]
Environmental protection E02a [1, 4, 10] (24 h); E15a [1, 4, 5, 10]; E15b [6-9]; E22c [6, 9]; E34 [1, 4-10]; E38 [1-4, 6-9]; H410 [10]; H411 [1-4, 8]
Consumer protection C02b, C12 [5]; C02c [1, 4, 10]
Storage and disposal D01, D02 [1, 4-10]; D05 [1, 4, 5, 7, 8, 10]; D07 [5, 6, 8, 9]; D09a, D10c [6, 9]; D09b [1, 4, 5, 10]; D10a, D12b [1, 4, 10]; D11a [5, 7, 8]; D12a [2, 3, 5-9]
Medical advice M03, M05b [6, 9]; M04a [1, 4, 5, 8, 10]

87   chlorpyrifos

A contact and ingested organophosphorus insecticide with all uses except application to Brassica seedlings via gantry sprayers revoked March 2016
IRAC mode of action code: 1B

Products

1	Chlobber	AgChem Access	480 g/l	EC	13723
2	Cyren	Headland	480 g/l	EC	11028
3	Dursban WG	Dow	75% w/w	WG	09153
4	Equity	Dow	480 g/l	EC	12465
5	Pyrifos 480	Becesane	480 g/l	EC	15991

Uses

- Cabbage root fly in **protected broccoli** [2, 5]; **protected brussels sprouts**, **protected cabbages**, **protected calabrese**, **protected cauliflowers** [2-5]; **protected choi sum** *(off-label)*, **protected collards** *(off-label)*, **protected kale** *(off-label)*, **protected oriental cabbage** *(off-label)* [3]
- Insect pests in **protected broccoli** [1, 2, 5]; **protected brussels sprouts**, **protected cabbages**, **protected calabrese**, **protected cauliflowers** [1-5]

Extension of Authorisation for Minor Use (EAMUs)
- *protected choi sum* 20161146 [3]
- *protected collards* 20161146 [3]
- *protected kale* 20161146 [3]
- *protected oriental cabbage* 20161146 [3]

Approval information
- Chlorpyrifos included in Annex I under EC Regulation 1107/2009
- Accepted by BBPA for use on malting barley
- Following the environmental review of chlorpyrifos uses on ware potatoes, carrots, conifers (drench), forestry trees (dip) and cereals (for aphids and yellow cereal fly) were revoked in September 2005. In addition the maximum permitted number of treatments, or the maximum total dose, on several other crops was reduced.
- In 2006 CRD required that all products containing this active ingredient should carry the following warning in the main area of the container label: "Chlorpyrifos is an anticholinesterase organophosphate. Handle with care"
- In 2016 CRD revoked all uses except for treatment of brassica crops in peat blocks via gantry-mounted sprayers only.

Efficacy guidance
- Apply via gantry-mounted sprayer only.

FOR FULL CONDITIONS OF USE ALWAYS READ THE PRODUCT LABEL

Restrictions
- Contains an anticholinesterase organophosphate compound. Do not use if under medical advice not to work with such compounds

Environmental safety
- Dangerous for the environment
- Very toxic to aquatic organisms
- Flammable [4]
- LERAP Category A [1-5]

Hazard classification and safety precautions
Hazard Harmful, Dangerous for the environment [1-5]; Flammable [4, 5]; Flammable liquid and vapour [1, 4]; Toxic if swallowed [2]; Harmful if swallowed [1, 3, 4]; Harmful if inhaled [1, 2, 4]
Transport code 3 [1]; 6.1 [2, 4, 5]; 9 [3]
Packaging group III
UN Number 3017 [1, 4, 5]; 3018 [2]; 3077 [3]
Risk phrases H304 [1, 2]; H315, H335, H336 [1, 4]; H319 [1, 2, 4]; R20, R22a, R22b, R36, R37, R38, R50, R53a [5]
Operator protection A [1-5]; C, H [1, 2]; P [4, 5]; U02a [1]; U05a [1, 2, 4, 5]; U08, U19a, U20b [1-5]; U11 [4, 5]; U14 [1, 2]
Environmental protection E06a [1-5] (14 d); E12a [2, 3]; E12c [1, 4, 5]; E12e, E16c, E16d, E34 [1-5]; E13a [1, 2]; E15a [3-5]; E17a [2] (18 m); E17b [4] (18 m); E38 [1, 3-5]; H410 [1-4]
Storage and disposal D01, D09a, D10b [1-5]; D02, D12a [1, 3-5]; D05 [1, 2]; D12b [2]
Medical advice M01 [3-5]; M03 [2-5]; M05b [1, 2, 4, 5]

88 chlorpyrifos-methyl

An organophosphorus insecticide and acaricide for grain store use
IRAC mode of action code: 1B

Products
Reldan 22	Dow	225 g/l	EC	16710

Uses
- Grain storage pests in **stored grain**
- Insect pests in **grain stores**, **stored grain**
- Mites in **grain stores**, **stored grain**
- Pre-harvest hygiene in **grain stores**

Approval information
- Chlorpyrifos-methyl included in Annex I under EC Regulation 1107/2009
- Accepted by BBPA for use in stores for malting barley
- In 2006 CRD required that all products containing this active ingredient should carry the following warning in the main area of the container label: "Chlorpyrifos-methyl is an anticholinesterase organophosphate. Handle with care"

Efficacy guidance
- Apply to grain after drying to moisture content below 14%, cooling and cleaning
- Insecticide may become depleted at grain surface if grain is being cooled by continuous extraction of air from the base leading to reduced control of grain store pests especially mites
- Resistance to organophosphorus compounds sometimes occurs in insect and mite pests of stored products

Restrictions
- Contains an anticholinesterase organophosphorus compound. Do not use if under medical advice not to work with such compounds
- Maximum number of treatments 1 per batch or 1 per store, prior to storage
- Only treat grain in good condition
- Do not treat grain intended for sowing

SEE SECTION 3 FOR PRODUCTS ALSO REGISTERED

Pesticide Profiles

Crop-specific information
- May be applied pre-harvest to surfaces of empty store and grain handling machinery and as admixture with grain
- May be used in wheat, barley, oats, rye, triticale or oilseed rape stores

Environmental safety
- Dangerous for the environment
- Very toxic to aquatic organisms

Hazard classification and safety precautions
    Hazard Harmful, Dangerous for the environment
    Transport code M6
    Packaging group III
    UN Number 3082
    Risk phrases R22b, R41, R50, R53a
    Operator protection A, C, H, M; U05a, U08, U19a, U20b
    Environmental protection E15a, E34
    Storage and disposal D01, D02, D05, D09a, D10b
    Medical advice M01, M03, M05b

89 citronella oil

A natural plant extract herbicide

Products
    Barrier H   Barrier   22.9% w/w   OD   17145

Uses
- Ragwort in *amenity grassland, land temporarily removed from production*

Approval information
- Plant oils such as citronella oil are included in Annex 1 under EC Regulation 1107/2009

Efficacy guidance
- Best results obtained from spot treatment of ragwort in the rosette stage, during dry still conditions
- Aerial growth of ragwort is rapidly destroyed. Longer term control depends on overall management strategy
- Check for regrowth after 28 d and re-apply as necessary

Crop-specific information
- Contact with grasses will result in transient scorch which is outgrown in good growing conditions

Environmental safety
- Apply away from bees
- Harmful to fish or other aquatic life. Do not contaminate surface waters or ditches with chemical or used container
- Keep livestock out of treated areas for at least 2 wk and until foliage of any poisonous weeds such as ragwort has died and become unpalatable

Hazard classification and safety precautions
    Hazard Irritant
    UN Number N/C
    Risk phrases H317
    Operator protection A, C; U05a, U20c
    Environmental protection E07b (2 wk); E12g, E13c
    Storage and disposal D01, D02, D05, D09a, D11a

FOR FULL CONDITIONS OF USE ALWAYS READ THE PRODUCT LABEL

clethodim

90 clethodim

A post-emergence grass herbicide for use in listed broad-leaved crops
HRAC mode of action code: A

Products

1	Centurion Max	Interfarm	120 g/l	EC	16310
2	Chellist	Chem-Wise	120 g/l	EC	16263
3	Clayton Gatso	Clayton	120 g/l	EC	17698
4	Clayton Gatso	Clayton	120 g/l	EC	17732

Uses
- Annual grasses in **broad beans** *(off-label)*, **brussels sprouts** *(off-label)*, **bulb onions** *(off-label)*, **cabbages** *(off-label)*, **carrots** *(off-label)*, **dwarf beans** *(off-label)*, **french beans** *(off-label)*, **garlic** *(off-label)*, **runner beans** *(off-label)*, **shallots** *(off-label)*, **vining peas** *(off-label)* [1]; **sugar beet**, **winter oilseed rape** [1-4]
- Annual meadow grass in **corn gromwell** *(off-label)*, **crambe** *(off-label)*, **linseed** *(off-label)* [1]; **sugar beet**, **winter oilseed rape** [1-4]
- Blackgrass in **corn gromwell** *(off-label)*, **crambe** *(off-label)*, **linseed** *(off-label)* [1]; **sugar beet**, **winter oilseed rape** [1-4]
- Couch in **broad beans** *(off-label)*, **brussels sprouts** *(off-label)*, **bulb onions** *(off-label)*, **cabbages** *(off-label)*, **carrots** *(off-label)*, **dwarf beans** *(off-label)*, **french beans** *(off-label)*, **garlic** *(off-label)*, **runner beans** *(off-label)*, **shallots** *(off-label)*, **vining peas** *(off-label)* [1]
- Volunteer barley in **corn gromwell** *(off-label)*, **crambe** *(off-label)*, **linseed** *(off-label)* [1]; **sugar beet**, **winter oilseed rape** [1-4]
- Volunteer wheat in **corn gromwell** *(off-label)*, **crambe** *(off-label)*, **linseed** *(off-label)* [1]; **sugar beet**, **winter oilseed rape** [1-4]
- Wild oats in **corn gromwell** *(off-label)*, **crambe** *(off-label)*, **linseed** *(off-label)* [1]

Extension of Authorisation for Minor Use (EAMUs)
- *broad beans* 20162041 [1]
- *brussels sprouts* 20162041 [1]
- *bulb onions* 20162041 [1]
- *cabbages* 20162041 [1]
- *carrots* 20162041 [1]
- *corn gromwell* 20151656 [1]
- *crambe* 20161841 [1]
- *dwarf beans* 20162041 [1]
- *french beans* 20162041 [1]
- *garlic* 20162041 [1]
- *linseed* 20152306 [1]
- *runner beans* 20162041 [1]
- *shallots* 20162041 [1]
- *vining peas* 20162041 [1]

Approval information
- Clethodim included in Annex I under EC Regulation 1107/2009

Efficacy guidance
- Optimum efficacy is achieved from applications made when grasses are actively growing and have 3 lvs up to beginning of tillering

Restrictions
- To avoid the build-up of resistance, do not apply products containing an ACCase inhibitor more than twice to any crop.
- Label requires a 14 day interval before and after application before other products are applied, particularly in oilseed rape.
- Do not apply in tank mix with other products.
- Do not apply to winter oilseed rape after end of October in year of sowing

SEE SECTION 3 FOR PRODUCTS ALSO REGISTERED

Pesticide Profiles

Crop-specific information
- Consult processors before use on crops grown for processing.
- Do not apply to crops suffering from stress for any reason.

Following crops guidance
- Broad-leaved crops can be sown at any time after application but when planting cereals, maize or grasses, it is advisable to wait at least 4 weeks after application and to cultivate to at least 20 cm deep before sowing.

Hazard classification and safety precautions
Hazard Harmful, Dangerous for the environment
Transport code 9
Packaging group III
UN Number 3082
Risk phrases H304, H336 [1, 3, 4]; R22b, R51, R53a, R66, R67 [2]
Operator protection A, H; U02a, U04a, U05a, U08, U19a, U20a
Environmental protection E15b, E38 [1-4]; H410 [1]; H411 [3, 4]
Storage and disposal D01, D02, D05, D09a, D10b, D12a
Medical advice M05b

91 clodinafop-propargyl

A foliar acting herbicide for annual grass weed control in wheat, triticale and rye
HRAC mode of action code: A

Products

1	Clodinastar	Life Scientific	240 g/l	EC	17404
2	Sword	Adama	240 g/l	EC	15099
3	Topik	Syngenta	240 g/l	EC	15123
4	Tuli	AgChem Access	240 g/l	EC	16401

Uses
- Blackgrass in *durum wheat*, *spring rye*, *spring wheat*, *triticale*, *winter rye*, *winter wheat* [1-4]; *grass seed crops* (off-label) [3]
- Rough-stalked meadow grass in *durum wheat*, *spring rye*, *spring wheat*, *triticale*, *winter rye*, *winter wheat* [1-4]; *grass seed crops* (off-label) [3]
- Wild oats in *durum wheat*, *spring rye*, *spring wheat*, *triticale*, *winter rye*, *winter wheat* [1-4]; *grass seed crops* (off-label) [3]

Extension of Authorisation for Minor Use (EAMUs)
- *grass seed crops* 20132310 [3]

Approval information
- Clodinafop-propargyl included in Annex I under EC Regulation 1107/2009

Efficacy guidance
- Spray when majority of weeds have germinated but before competition reduces yield
- Products contain a herbicide safener (cloquintocet-mexyl) that improves crop tolerance to clodinafop-propargyl
- Optimum control achieved when all grass weeds emerged. Wait for delayed germination on dry or cloddy seedbed
- A mineral oil additive is recommended to give more consistent control of very high blackgrass populations or for late season treatments. See label for details
- Weed control not affected by soil type, organic matter or straw residues
- Control may be reduced if rain falls within 1 h of treatment
- Clodinafop-propargyl is an ACCase inhibitor herbicide. To avoid the build up of resistance do not apply products containing an ACCase inhibitor herbicide more than twice to any crop. In addition do not use any product containing clodinafop-propargyl in mixture or sequence with any other product containing the same ingredient

FOR FULL CONDITIONS OF USE ALWAYS READ THE PRODUCT LABEL

- Use these products as part of a resistance management strategy that includes cultural methods of control and does not use ACCase inhibitors as the sole chemical method of grass weed control
- Applying a second product containing an ACCase inhibitor to a crop will increase the risk of resistance development; only use a second ACCase inhibitor to control different weeds at a different timing
- Always follow WRAG guidelines for preventing and managing herbicide resistant weeds. See Section 5 for more information

Restrictions
- Maximum number of treatments 1 per crop
- Do not use on barley or oats
- Do not treat crops under stress or suffering from waterlogging, pest attack, disease or frost
- Do not treat crops undersown with grass mixtures
- Do not mix with products containing MCPA, mecoprop-P, 2,4-D or 2,4-DB
- MCPA, mecoprop, 2,4-D or 2,4-DB should not be applied within 21 d before, or 7 d after, treatment

Crop-specific information
- Latest use: before second node detectable stage (GS 32) for durum wheat, triticale, rye; before flag leaf sheath extending (GS 41) for wheat
- Spray in autumn, winter or spring from 1 true leaf stage (GS 11) to before second node detectable (GS 32) on durum, rye, triticale; before flag leaf sheath extends (GS 41) on wheat

Following crops guidance
- Any broad leaved crop or cereal (except oats) may be sown after failure of a treated crop provided that at least 3 wk have elapsed between application and drilling a cereal
- After normal harvest of a treated crop any broad leaved crop or wheat, durum wheat, rye, triticale or barley should be sown. Oats and grass should not be sown until the following spring

Environmental safety
- Dangerous for the environment
- Very toxic to aquatic organisms

Hazard classification and safety precautions
Hazard Dangerous for the environment [1-4]; Harmful if swallowed, Very toxic to aquatic organisms [2]
Transport code 9
Packaging group III
UN Number 3082
Risk phrases H304, H373 [1, 2]; R50, R53a [3, 4]
Operator protection A, C, H, K; U02a, U05a, U09a, U20a [1-4]; U14 [1, 2]
Environmental protection E15a, E38 [1-4]; H410 [1, 2]
Storage and disposal D01, D02, D05, D09a, D10c, D12a

92 clodinafop-propargyl + cloquintocet-mexyl

A grass herbicide for use in winter wheat and durum wheat
HRAC mode of action code: A

Products

Buguis	UPL Europe	100: 25 g/l	EC	17151

Uses
- Annual grasses in **durum wheat, winter wheat**
- Blackgrass in **durum wheat, winter wheat**

Approval information
- Clodinafop-propargyl included in Annex I under EC Regulation 1107/2009 but cloquintocet-mexyl not yet listed

SEE SECTION 3 FOR PRODUCTS ALSO REGISTERED

Pesticide Profiles

Hazard classification and safety precautions
  Transport code 9
  Packaging group III
  UN Number 3082
  Risk phrases H304, H317, H319, H373
  Operator protection A, H
  Environmental protection H411

93 clodinafop-propargyl + pinoxaden

A foliar acting herbicide mixture for winter wheat
HRAC mode of action code: A + A

See also pinoxaden

Products
  Traxos   Syngenta   100:100 g/l   EC   12742

Uses
- Italian ryegrass in **winter wheat**
- Perennial ryegrass in **winter wheat** *(from seed)*
- Wild oats in **winter wheat**

Approval information
- Clodinafop-propargyl and pinoxaden included in Annex I under EC Regulation 1107/2009
- Approved for use on crops for brewing by BBPA

Efficacy guidance
- Products contain a herbicide safener (cloquintocet-mexyl) that improves crop tolerance to clodinafop-propargyl
- Best results obtained from treatment when all grass weeds have emerged. There is no residual activity
- Broad-leaved weeds are not controlled
- Treat before emerged weed competition reduces yield
- For ryegrass control use as part of a programme including other products with activity against ryegrass
- For blackgrass control must be used as part of an integrated control strategy
- Grass weed control may be reduced if rain falls within 1 hr of application
- Clodinafop-propargyl and pinoxaden are ACCase inhibitor herbicides. To avoid the build up of resistance do not apply products containing an ACCase inhibitor herbicide more than twice to any crop. In addition do not use any product containing clodinafop-propargyl or pinoxaden in mixture or sequence with any other product containing the same ingredient
- Use as part of a resistance management strategy that includes cultural methods of control and does not use ACCase inhibitors as the sole chemical method of grass weed control
- Applying a second product containing an ACCase inhibitor to a crop will increase the risk of resistance development; only use a second ACCase inhibitor to control different weeds at a different timing
- Always follow WRAG guidelines for preventing and managing herbicide resistant weeds. See Section 5 for more information

Restrictions
- Maximum number of treatments 1 per crop
- Product must always be used with specified adjuvant. See label
- Do not use on other cereals
- Do not treat crops under stress from any cause
- Do not treat crops undersown with grass or grass mixtures
- Avoid use of hormone-containing herbicides in mixture or in sequence. See label for restrictions on tank mixes

Crop-specific information
- Latest use: before flag leaf sheath extended (before GS 41) for winter wheat

FOR FULL CONDITIONS OF USE ALWAYS READ THE PRODUCT LABEL

- All varieties of winter wheat may be treated

Following crops guidance
- There are no restrictions on succeeding crops in a normal rotation
- In the event of failure of a treated crop ryegrass, maize, oats or any broad-leaved crop may be planted after a minimum interval of 4 wk from application

Environmental safety
- Dangerous for the environment
- Toxic to aquatic organisms

Hazard classification and safety precautions
    Hazard Irritant, Dangerous for the environment
    Transport code 9
    Packaging group III
    UN Number 3082
    Risk phrases R36, R38, R43, R51, R53a
    Operator protection A, C, H, K; U05a, U09a, U20b
    Environmental protection E15b, E38
    Storage and disposal D01, D02, D05, D09a, D10c, D11a, D12a

94 clodinafop-propargyl + prosulfocarb

A foliar acting herbicide mixture for grass and broad-leaved weed control in winter wheat
HRAC mode of action code: A + N

See also prosulfocarb

Products
    Grapple   Syngenta   10:800 g/l   EC   16523

Uses
- Annual dicotyledons in **winter wheat**
- Annual grasses in **winter wheat**

Approval information
- Clodinafop-propargyl and prosulfocarb included in Annex I under EC Regulation 1107/2009

Efficacy guidance
- To avoid the build up of resistance do not apply products containing an ACCase inhibitor herbicide more than twice to any crop. In addition, do not use this product in mixture or sequence with any other product containing clodinafop-propargyl.

Restrictions
- Do not use on barley or oats
- Do not spray crops undersown with grass mixtures
- When it is applied first, leave 7 days before applying hormone herbicides. If mecoprop-P or 2,4-DB containing products are applied first, leave 14 days before it is applied. If MCPA or 2,4-D containing products are applied first, leave 21 days before it is applied.
- The cereal seed must be covered by 3 cm of soil and for best results apply to a firm, moist seedbed free of clods.

Crop-specific information
- Only one application per crop
- Latest use before 5 tiller stage (GS25)

Environmental safety
- LERAP Category B

Hazard classification and safety precautions
    Hazard Harmful, Dangerous for the environment
    Transport code 9
    Packaging group III

SEE SECTION 3 FOR PRODUCTS ALSO REGISTERED

Pesticide Profiles

UN Number 3082
Risk phrases H304, H319
Operator protection A, H; U02a, U05a, U08, U20c
Environmental protection E15b, E16a, E16b, E38, H410
Storage and disposal D01, D02, D05, D09a, D10c, D12a
Medical advice M05b

95   clofentezine

A selective ovicidal tetrazine acaricide for use in top fruit
IRAC mode of action code: 10A

Products

1	Apollo 50 SC	Adama	500 g/l	SC	17187
2	Clayton Sputnik	Clayton	500 g/l	SC	17344

Uses
- Conifer spinning mite in **forest nurseries** *(off-label)*, **ornamental plant production** *(off-label)* [1]
- Red spider mites in **apples**, **pears** [1, 2]
- Rust mite in **forest nurseries** *(off-label)*, **ornamental plant production** *(off-label)* [1]
- Spider mites in **forest nurseries** *(off-label)*, **ornamental plant production** *(off-label)* [1]

Extension of Authorisation for Minor Use (EAMUs)
- **forest nurseries** 20162082 [1]
- **ornamental plant production** 20162082 [1]

Approval information
- Clofentezine included in Annex I under EC Regulation 1107/2009

Efficacy guidance
- Acts on eggs and early motile stages of mites. For effective control total cover of plants is essential, particular care being needed to cover undersides of leaves

Restrictions
- Maximum number of treatments 1 per yr for apples, pears, cherries, plums

Crop-specific information
- HI apples, pears 28 d; cherries, plums 8 wk
- For red spider mite control spray apples and pears between bud burst and pink bud, plums and cherries between white bud and first flower. Rust mite is also suppressed
- On established infestations apply in conjunction with an adult acaricide

Environmental safety
- Harmful to aquatic organisms
- Product safe on predatory mites, bees and other predatory insects

Hazard classification and safety precautions
  UN Number N/C
  Operator protection U05a, U08, U20b
  Environmental protection E15a, H412
  Storage and disposal D01, D02, D05, D09a, D11a

96   clomazone

An isoxazolidinone residual herbicide for oilseed rape, field beans, combining and vining peas
HRAC mode of action code: F3

Products

1	Centium 360 CS	Headland	360 g/l	CS	17327
2	Cirrus CS	Headland	371 g/l	CS	17328
3	Clayton Chrome CS	Clayton	360 g/l	CS	16383

FOR FULL CONDITIONS OF USE ALWAYS READ THE PRODUCT LABEL

clomazone

Products – continued

4	Cleancrop Chicane	Headland	360 g/l	CS	17330
5	Cleancrop Covert	Headland	360 g/l	CS	17340
6	Gamit 36 CS	Headland	360 g/l	CS	17314
7	Standon Soulmate	Standon	360 g/l	CS	16879
8	Titan	Goldengrass	360 g/l	CS	16471

Uses

- Annual dicotyledons in **carrots, combining peas, potatoes, spring field beans, vining peas, winter field beans, winter oilseed rape** [7, 8]
- Chickweed in **asparagus** (off-label), **cucumbers** (off-label), **gherkins** (off-label), **melons** (off-label), **okra** (off-label), **pumpkins** (off-label), **watermelons** (off-label), **winter squash** (off-label) [6]; **baby leaf crops** (off-label), **borage** (off-label), **broad beans** (off-label), **broccoli** (off-label), **brussels sprouts** (off-label), **cabbages** (off-label), **calabrese** (off-label), **cauliflowers** (off-label), **celeriac** (off-label), **celery (outdoor)** (off-label), **choi sum** (off-label), **collards** (off-label), **courgettes** (off-label), **fennel** (off-label), **french beans** (off-label), **herbs (see appendix 6)** (off-label), **kale** (off-label), **lupins** (off-label), **oriental cabbage** (off-label), **poppies for morphine production** (off-label), **rhubarb** (off-label), **runner beans** (off-label), **soya beans** (off-label), **spinach** (off-label), **spring cabbage** (off-label), **summer squash** (off-label), **swedes** (off-label), **sweet potato** (off-label) [5, 6]; **carrots, potatoes** [3, 5, 6]; **combining peas, spring field beans, vining peas, winter oilseed rape** [1-4]; **forest nurseries** (off-label), **game cover** (off-label), **hops** (off-label), **ornamental plant production** (off-label) [1, 2, 5, 6]; **spring oilseed rape, winter field beans** [1, 2, 4]
- Cleavers in **asparagus** (off-label), **cucumbers** (off-label), **gherkins** (off-label), **melons** (off-label), **okra** (off-label), **pumpkins** (off-label), **watermelons** (off-label), **winter squash** (off-label) [6]; **baby leaf crops** (off-label), **borage** (off-label), **broad beans** (off-label), **broccoli** (off-label), **brussels sprouts** (off-label), **cabbages** (off-label), **calabrese** (off-label), **cauliflowers** (off-label), **celeriac** (off-label), **celery (outdoor)** (off-label), **choi sum** (off-label), **collards** (off-label), **courgettes** (off-label), **fennel** (off-label), **french beans** (off-label), **herbs (see appendix 6)** (off-label), **kale** (off-label), **lupins** (off-label), **oriental cabbage** (off-label), **poppies for morphine production** (off-label), **rhubarb** (off-label), **runner beans** (off-label), **soya beans** (off-label), **spinach** (off-label), **spring cabbage** (off-label), **summer squash** (off-label), **swedes** (off-label), **sweet potato** (off-label) [5, 6]; **carrots, potatoes** [3, 5-7]; **combining peas, spring field beans, vining peas, winter oilseed rape** [1-4, 7]; **forest nurseries** (off-label), **game cover** (off-label), **hops** (off-label), **ornamental plant production** (off-label) [1, 2, 5, 6]; **spring oilseed rape** [1, 2, 4]; **winter field beans** [1, 2, 4, 7]
- Fool's parsley in **asparagus** (off-label), **cucumbers** (off-label), **gherkins** (off-label), **melons** (off-label), **okra** (off-label), **pumpkins** (off-label), **watermelons** (off-label), **winter squash** (off-label) [6]; **baby leaf crops** (off-label), **borage** (off-label), **broad beans** (off-label), **broccoli** (off-label), **brussels sprouts** (off-label), **cabbages** (off-label), **calabrese** (off-label), **cauliflowers** (off-label), **celeriac** (off-label), **celery (outdoor)** (off-label), **choi sum** (off-label), **collards** (off-label), **courgettes** (off-label), **fennel** (off-label), **french beans** (off-label), **herbs (see appendix 6)** (off-label), **kale** (off-label), **lupins** (off-label), **oriental cabbage** (off-label), **poppies for morphine production** (off-label), **rhubarb** (off-label), **runner beans** (off-label), **soya beans** (off-label), **spinach** (off-label), **spring cabbage** (off-label), **summer squash** (off-label), **swedes** (off-label), **sweet potato** (off-label) [5, 6]; **carrots, potatoes** [3, 5, 6]; **combining peas, spring field beans, vining peas** [1-4]; **forest nurseries** (off-label), **game cover** (off-label), **hops** (off-label), **ornamental plant production** (off-label) [1, 2, 5, 6]; **spring oilseed rape** [1, 2, 4]; **winter oilseed rape** [3]
- Ivy-leaved speedwell in **baby leaf crops** (off-label), **borage** (off-label), **broad beans** (off-label), **courgettes** (off-label), **herbs (see appendix 6)** (off-label), **lupins** (off-label), **spinach** (off-label), **summer squash** (off-label) [5, 6]; **broccoli** (off-label), **brussels sprouts** (off-label), **cabbages** (off-label), **calabrese** (off-label), **carrots, cauliflowers** (off-label), **celeriac** (off-label), **celery (outdoor)** (off-label), **choi sum** (off-label), **collards** (off-label), **fennel** (off-label), **forest nurseries** (off-label), **french beans** (off-label), **game cover** (off-label), **hops** (off-label), **kale** (off-label), **oriental cabbage** (off-label), **ornamental plant production** (off-label), **poppies for morphine production** (off-label), **potatoes, rhubarb** (off-label), **runner beans** (off-label), **soya beans** (off-label), **spring cabbage** (off-label), **swedes** (off-label), **sweet potato** (off-label) [5]; **cucumbers**

SEE SECTION 3 FOR PRODUCTS ALSO REGISTERED

Pesticide Profiles

(off-label), **gherkins** (off-label), **melons** (off-label), **okra** (off-label), **pumpkins** (off-label), **watermelons** (off-label), **winter squash** (off-label) [6]
- Perennial weeds in **asparagus** (off-label - spring-germinating) [5, 6]
- Red dead-nettle in **asparagus** (off-label), **cucumbers** (off-label), **gherkins** (off-label), **melons** (off-label), **okra** (off-label), **pumpkins** (off-label), **watermelons** (off-label), **winter squash** (off-label) [6]; **baby leaf crops** (off-label), **borage** (off-label), **broad beans** (off-label), **broccoli** (off-label), **brussels sprouts** (off-label), **cabbages** (off-label), **calabrese** (off-label), **cauliflowers** (off-label), **celeriac** (off-label), **celery (outdoor)** (off-label), **choi sum** (off-label), **collards** (off-label), **courgettes** (off-label), **fennel** (off-label), **french beans** (off-label), **herbs (see appendix 6)** (off-label), **kale** (off-label), **lupins** (off-label), **oriental cabbage** (off-label), **poppies for morphine production** (off-label), **rhubarb** (off-label), **runner beans** (off-label), **soya beans** (off-label), **spinach** (off-label), **spring cabbage** (off-label), **summer squash** (off-label), **swedes** (off-label), **sweet potato** (off-label) [5, 6]; **carrots, potatoes** [3, 5, 6]; **combining peas, spring field beans, vining peas, winter oilseed rape** [1-4]; **forest nurseries** (off-label), **game cover** (off-label), **hops** (off-label), **ornamental plant production** (off-label) [1, 2, 5, 6]; **spring oilseed rape, winter field beans** [1, 2, 4]
- Shepherd's purse in **asparagus** (off-label), **cucumbers** (off-label), **gherkins** (off-label), **melons** (off-label), **okra** (off-label), **pumpkins** (off-label), **watermelons** (off-label), **winter squash** (off-label) [6]; **baby leaf crops** (off-label), **borage** (off-label), **broad beans** (off-label), **broccoli** (off-label), **brussels sprouts** (off-label), **cabbages** (off-label), **calabrese** (off-label), **cauliflowers** (off-label), **celeriac** (off-label), **celery (outdoor)** (off-label), **choi sum** (off-label), **collards** (off-label), **courgettes** (off-label), **fennel** (off-label), **french beans** (off-label), **herbs (see appendix 6)** (off-label), **kale** (off-label), **lupins** (off-label), **oriental cabbage** (off-label), **poppies for morphine production** (off-label), **rhubarb** (off-label), **runner beans** (off-label), **soya beans** (off-label), **spinach** (off-label), **spring cabbage** (off-label), **summer squash** (off-label), **swedes** (off-label), **sweet potato** (off-label) [5, 6]; **carrots, potatoes** [3, 5, 6]; **combining peas, spring field beans, vining peas, winter oilseed rape** [1-4]; **forest nurseries** (off-label), **game cover** (off-label), **hops** (off-label), **ornamental plant production** (off-label) [1, 2, 5, 6]; **spring oilseed rape, winter field beans** [1, 2, 4]
- Speedwells in **carrots, combining peas, potatoes, spring field beans, vining peas, winter field beans, winter oilseed rape** [7]

Extension of Authorisation for Minor Use (EAMUs)
- **asparagus** (spring-germinating) 20152992 [5], (spring-germinating) 20152817 [6], 20152817 [6]
- **baby leaf crops** 20160076 [5], 20152830 [6]
- **borage** 20152994 [5], 20152829 [6]
- **broad beans** 20152999 [5], 20152828 [6]
- **broccoli** 20152997 [5], 20152820 [6]
- **brussels sprouts** 20152997 [5], 20152820 [6]
- **cabbages** 20152997 [5], 20152820 [6]
- **calabrese** 20152997 [5], 20152820 [6]
- **cauliflowers** 20152997 [5], 20152820 [6]
- **celeriac** 20152993 [5], 20152815 [6]
- **celery (outdoor)** 20160074 [5], 20152811 [6]
- **choi sum** 20152997 [5], 20152820 [6]
- **collards** 20152997 [5], 20152820 [6]
- **courgettes** 20152990 [5], 20152831 [6]
- **cucumbers** 20152831 [6]
- **fennel** 20160072 [5], 20152810 [6]
- **forest nurseries** 20152924 [1], 20152921 [2], 20160075 [5], 20152813 [6]
- **french beans** 20152991 [5], 20152819 [6]
- **game cover** 20152924 [1], 20152921 [2], 20160075 [5], 20152813 [6]
- **gherkins** 20152831 [6]
- **herbs (see appendix 6)** 20160073 [5], 20152824 [6]
- **hops** 20152924 [1], 20152921 [2], 20160075 [5], 20152813 [6]
- **kale** 20152997 [5], 20152820 [6]
- **lupins** 20152998 [5], 20152827 [6]
- **melons** 20152831 [6]
- **okra** 20152831 [6]

FOR FULL CONDITIONS OF USE ALWAYS READ THE PRODUCT LABEL

clomazone

- *oriental cabbage* 20152997 [5], 20152820 [6]
- *ornamental plant production* 20152924 [1], 20152921 [2], 20160075 [5], 20152813 [6]
- *poppies for morphine production* 20160070 [5], 20152821 [6]
- *pumpkins* 20152831 [6]
- *rhubarb* 20160077 [5], 20152816 [6]
- *runner beans* 20152985 [5], 20152818 [6]
- *soya beans* 20152982 [5], 20152823 [6]
- *spinach* 20160076 [5], 20152830 [6]
- *spring cabbage* 20152997 [5], 20152820 [6]
- *summer squash* 20152990 [5], 20152831 [6]
- *swedes* 20152925 [5], 20152814 [6]
- *sweet potato* 20160071 [5], 20152812 [6]
- *watermelons* 20152831 [6]
- *winter squash* 20152831 [6]

Approval information
- Clomazone included in Annex I under EC Regulation 1107/2009

Efficacy guidance
- Best results obtained from application as soon as possible after sowing crop and before emergence of crop or weeds
- Uptake is via roots and shoots. Seedbeds should be firm, level and free from clods. Loose puffy seedbeds should be consolidated before spraying
- Efficacy is reduced on organic soils, on dry cloddy seedbeds and if prolonged dry weather follows application
- Clomazone acts by inhibiting synthesis of chlorophyll pigments. Susceptible weeds emerge but are chlorotic and die shortly afterwards
- Season-long control of weeds may not be achieved
- Always follow WRAG guidelines for preventing and managing herbicide resistant weeds. See Section 5 for more information

Restrictions
- Maximum number of treatments one per crop
- Crops must be covered by a minimum of 20 mm settled soil. Do not apply to broadcast crops. Direct-drilled crops should be harrowed across the slits to cover seed before spraying
- Do not use on compacted soils or soils of poor structure that may be liable to waterlogging
- Do not use on Sands or Very Light soils or those with more than 10% organic matter
- Do not treat two consecutive crops of carrots with clomazone in one calendar yr
- Consult manufacturer or your advisor before use on potato seed crops
- Do not overlap spray swaths. Crop plants emerged at time of treatment may be severely damaged
- Application must be made using a coarse spray quality.

Crop-specific information
- Latest use: pre-emergence of crop
- Severe, but normally transient, crop damage may occur in overlaps on field beans
- Some transient crop bleaching may occur under certain climatic conditions and can be severe where heavy rain follows application. This is normally rapidly outgrown and has no effect on final crop yield. Overlapping spray swaths may cause severe damage to field beans

Following crops guidance
- Following normal harvest of a spring or autumn treated crop, cereals, oilseed rape, field beans, combining peas, potatoes, maize, turnips, linseed or sugar beet may be sown
- In the event of failure of an autumn treated crop, winter cereals or winter beans may be sown in the autumn if 6 wk have elapsed since treatment. In the spring following crop failure combining peas, field beans or potatoes may be sown if 6 wk have elapsed since treatment, and spring cereals, maize, turnips, onions, carrots or linseed may be sown if 7 mth have elapsed since treatment
- In the event of a failure of a spring treated crop a wide range of crops may be sown provided intervals of 6-9 wk have elapsed since treatment. See label for details

SEE SECTION 3 FOR PRODUCTS ALSO REGISTERED

Pesticide Profiles

- Prior to resowing any listed replacement crop the soil should be ploughed and cultivated to 15 cm

Environmental safety
- Take extreme care to avoid drift outside the target area, or on to ponds, waterways or ditches as considerable damage may occur. Apply using a coarse quality spray

Hazard classification and safety precautions
Transport code 9 [3, 7, 8]
Packaging group III [3, 7, 8]
UN Number 3082 [3, 7]; N/C [1, 2, 4-6, 8]
Operator protection A [1-6, 8]; U05a, U14, U20b [1-6, 8]
Environmental protection E15b [1-6, 8]; E39 [3]; H413 [1, 2, 4-6]
Storage and disposal D01, D02, D09a [1-6, 8]; D10b [1, 2, 4-6, 8]; D10c [3]

97 clomazone + linuron

An isoxazolidinone and urea residual herbicide for peas, field beans and potatoes
HRAC mode of action code: F3 + C2

See also linuron

Products

1	Lingo	Belchim	45:250 g/l	CC	16564
2	Linzone	Belchim	45:250 g/l	CC	16578

Uses
- Annual dicotyledons in **combining peas**, **potatoes**, **spring field beans**, **vining peas**, **winter field beans** [1, 2]; **edible podded peas** *(off-label)* [1]
- Annual meadow grass in **combining peas**, **potatoes**, **spring field beans**, **vining peas**, **winter field beans** [1, 2]; **edible podded peas** *(off-label)* [1]
- Thorn apple in **edible podded peas** *(off-label)* [1]

Extension of Authorisation for Minor Use (EAMUs)
- *edible podded peas* 20151443 [1]

Approval information
- Clomazone and linuron included in Annex I under EC Regulation 1107/2009

Restrictions
- Do not apply to broadcast crops
- Do not apply on light soils if heavy rain is expected within 24 hours of application
- Transient crop bleaching may occur under some conditions. This is normally outgrown with no effect on final yield
- Application must be made using a coarse spray quality. Drift reduction technology required for application within 30m of water.

Crop-specific information
- Do not apply residual herbicides to fields with a poor soil structure prone to waterlogging
- Do not overlap spray boom
- Ensure seed is well covered (2 cm) at application and do not apply when the crop is close to emergence

Following crops guidance
- Do not sow or plant lettuce in the same season as application.
- Following normal harvest of an autumn treated crop, cereals, oilseed rape, field beans, combining peas, potatoes, maize, turnip, linseed and sugar beet may be sown. Prior to planting, soil should be ploughed and cultivated to a minimum depth of 15 cm.
- Following normal harvest of a spring treated crop, cereals, oilseed rape, field beans, combining peas, potatoes, maize, linseed, sugar beet and turnip may be sown. Prior to planting, soil should be ploughed and cultivated to a minimum depth of 15 cm.

FOR FULL CONDITIONS OF USE ALWAYS READ THE PRODUCT LABEL

- In the event of crop failure of autumn treated crops: Only winter cereals and winter beans may be sown, providing at least six weeks has elapsed since application. In the spring following crop failure, only spring cereals, combining peas, field beans, maize, onion, carrots, potatoes and linseed may be sown. Prior to re-sowing spring cereals, maize, onion, carrots and linseed, a period of at least 7 months should have elapsed since application. Prior to re-sowing combining peas, field beans and potatoes a period of at least six weeks should have elapsed since application. Prior to re-sowing any of the listed replacement crops soil should be ploughed and cultivated to a minimum depth of 25cm.
- In the event of crop failure of spring treated crops: Only combining peas, field beans, potatoes, maize, dwarf French beans, carrots, sugar beet, onions and linseed may be sown. Prior to re-sowing combining peas, field beans, potatoes, carrots, sugar beet, onions and linseed, a period of at least six weeks should have elapsed since application. Prior to re-sowing maize and dwarf French beans, a period of at least nine weeks should have elapsed since application. For all spring replacement crops, soil should be ploughed and cultivated to a minimum depth of 25cm.

Environmental safety
- Buffer zone requirement 6 m [1, 2]
- LERAP Category B

Hazard classification and safety precautions
Hazard Toxic, Dangerous for the environment
Transport code 9
Packaging group III
UN Number 3082
Risk phrases H351, H360, H373 [2]; R22a, R40, R48, R50, R53a, R61, R62 [1]
Operator protection A, H; U02a, U05a, U13, U14, U15, U19a
Environmental protection E15a, E16a, E23, E34, E38 [1, 2]; H410 [2]
Storage and disposal D01, D02, D12a
Medical advice M04a

98 clomazone + metazachlor

A mixture of isoxazolidinone and residual anilide herbicides for weed control in oilseed rape
HRAC mode of action code: F3 + K3

See also metazachlor

Products

1	Centium Plus	Belchim	360:500 g/l	CC	16611
2	Nimbus CS	BASF	33.3:250 g/l	SC	16573

Uses
- Annual dicotyledons in **forest nurseries** *(off-label)* [2]; **spring oilseed rape** [1]; **winter oilseed rape** [1, 2]
- Annual grasses in **forest nurseries** *(off-label)*, **winter oilseed rape** [2]
- Annual meadow grass in **forest nurseries** *(off-label)* [2]
- Cleavers in **forest nurseries** *(off-label)*, **winter oilseed rape** [2]

Extension of Authorisation for Minor Use (EAMUs)
- **forest nurseries** 20142721 [2]

Approval information
- Clomazone and metazachlor included in Annex I under EC Regulation 1107/2009

Restrictions
- Applications shall be limited to a total dose of not more than 1.0 kg metazachlor/ha in a three year period on the same field.
- Do not apply in mixture with phosphate liquid fertilisers
- Metazachlor stewardship requires that all autumn applications should be made before the end of September to reduce the risk to water
- Application must be made using a coarse spray quality.

SEE SECTION 3 FOR PRODUCTS ALSO REGISTERED

Environmental safety
- Metazachlor stewardship guidelines advise a maximum dose of 750 g.a.i/ha/annum. Applications to drained fields should be complete by 15th Oct but, if drains are flowing, complete applications by 1st Oct.
- LERAP Category B

Hazard classification and safety precautions
Hazard Harmful, Dangerous for the environment [1, 2]; Irritant, Very toxic to aquatic organisms [2]
Transport code 9
Packaging group III
UN Number 3082
Risk phrases H317, H351 [2]; R22a, R40, R50, R53a [1]; R43 [1, 2]
Operator protection A, H [1, 2]; C [1]; U05a, U19a [1, 2]; U08, U14, U20b [2]
Environmental protection E07a, E15b, E19b, H410 [2]; E16a, E38 [1, 2]
Storage and disposal D01, D02, D12a [1, 2]; D08, D09a, D10c [2]
Medical advice M04a [1]; M05a [2]

99   clomazone + metribuzin

A mixture of isoxazolidinone and triazinone residual herbicides for use in potatoes
HRAC mode of action code: F3 + C1

See also metribuzin

Products
Metric	Belchim	60:233 g/l	CC	16720

Uses
- Annual dicotyledons in **potatoes**
- Annual meadow grass in **potatoes**

Approval information
- Clomazone and metribuzin included in Annex I under EC Regulation 1107/2009

Restrictions
- For use on specified varieties of potato only
- Safety to daughter tubers has not been tested; consult manufacturer before treating seed crops
- Application must be made using a coarse spray quality.

Following crops guidance
- Before drilling or planting any succeeding crop, soil MUST be mouldboard ploughed to a depth of at least 15 cm (6") taking care to ensure that the furrow slice is inverted. Ploughing should be carried out as soon as possible (preferably within 3-4 weeks) after lifting the potato crop, but certainly no later than the end of December.
- In the same year: Provided at least 16 weeks have elapsed after the application of the recommended rate cereals and winter beans may be grown as following crops.In the following year: Do not grow any vegetable brassica crop (including cauliflower, calabrese, Brussels sprout and cabbage), lettuce or radish on land treated in the previous year.
- Cereals, oilseed rape, field beans, combining peas, potatoes, maize, turnip, linseed and sugar beet may be sown from spring onwards in the year following use.

Environmental safety
- Buffer zone requirement 10 m [1]
- LERAP Category B

Hazard classification and safety precautions
Hazard Dangerous for the environment, Very toxic to aquatic organisms
Transport code 9
Packaging group III
UN Number 3082
Operator protection A, H; U05a, U19a, U20b

FOR FULL CONDITIONS OF USE ALWAYS READ THE PRODUCT LABEL

Environmental protection E15b, E16a, E16b, E34, E38, H410
Storage and disposal D01, D02, D09a, D10b, D12a
Medical advice M05a

100   clomazone + napropamide

A herbicide mixture for use in winter oilseed rape
HRAC mode of action code: F3 + K3

See also clomazone
napropamide

Products

Altiplano DAMtec	Headland	3.5:40% w/w	WG	17189

Uses
- Annual dicotyledons in **winter oilseed rape**
- Annual grasses in **winter oilseed rape**
- Chickweed in **winter oilseed rape**
- Cleavers in **winter oilseed rape**
- Crane's-bill in **winter oilseed rape**
- Dead nettle in **winter oilseed rape**
- Mayweeds in **winter oilseed rape**
- Poppies in **winter oilseed rape**

Approval information
- Clomazone and napropamide included in Annex I under EC Regulation 1107/2009

Restrictions
- Do not use on sands, very light soils or on soils with more than 10% organic matter.
- Do not roll after application.
- Do not irrigate within 3 weeks of application.

Crop-specific information
- Ensure seed is covered by a minimum of 2 cm settled soil before application

Environmental safety
- LERAP Category B

Hazard classification and safety precautions
Hazard Dangerous for the environment, Very toxic to aquatic organisms
Transport code 9
Packaging group III
UN Number 3077
Operator protection A, H
Environmental protection E16a, E38, H410
Storage and disposal D12a

101   clomazone + pendimethalin

A residual herbicide mixture for weed control in combining peas and field beans
HRAC mode of action code: F3 + K1

Products

Stallion Sync TEC	Headland	30:333 g/l	CS	17243

Uses
- Annual dicotyledons in **combining peas, spring field beans**
- Cleavers in **combining peas, spring field beans**

Approval information
- Clomazone and pendimethalin included in Annex I under EC Regulation 1107/2009

Restrictions
- Application must be made using a coarse spray quality.

Environmental safety
- LERAP Category B

Hazard classification and safety precautions
Hazard Dangerous for the environment, Very toxic to aquatic organisms
Transport code 9
Packaging group III
UN Number 3082
Risk phrases H317
Operator protection A; U02a, U05a, U08, U13, U14, U19a, U20b
Environmental protection E15b, E16a, E38, H410
Storage and disposal D01, D02, D05, D09a, D10b, D12a

102 clopyralid

A foliar translocated picolinic herbicide for a wide range of crops
HRAC mode of action code: O

See also 2,4-D + clopyralid + MCPA
bromoxynil + clopyralid

Products
1	Dow Shield 400	Dow	400 g/l	SL	14984
2	Glopyr 400	Globachem	400 g/l	SL	15009
3	Life Scientific Clopyralid	Life Scientific	200 g/l	SL	16259
4	Vivendi 200	UPL Europe	200 g/l	SL	16966

Uses
- Annual dicotyledons in **broccoli, bulb onions, cabbages, calabrese, cauliflowers** [2]; **brussels sprouts, forage maize, linseed** [1, 2]; **fodder beet, mangels, red beet, spring oilseed rape, sugar beet, swedes, turnips, winter oilseed rape** [1-4]; **grassland, ornamental plant production, spring barley, spring oats, spring wheat, winter barley, winter oats, winter wheat** [1, 2, 4]
- Black bindweed in **forest** *(off-label)* [1]
- Clovers in **grass seed crops** *(off-label)* [1]
- Corn marigold in **broccoli, bulb onions, cabbages, calabrese, cauliflowers** [2]; **brussels sprouts, fodder beet, forage maize, grassland, linseed, mangels, ornamental plant production, red beet, spring barley, spring oats, spring oilseed rape, spring wheat, sugar beet, swedes, turnips, winter barley, winter oats, winter oilseed rape, winter wheat** [1, 2]; **forest** *(off-label)* [1]
- Creeping thistle in **broccoli, bulb onions, cabbages, calabrese, cauliflowers** [2]; **brussels sprouts, fodder beet, forage maize, grassland, linseed, mangels, ornamental plant production, red beet, spring barley, spring oats, spring oilseed rape, spring wheat, sugar beet, swedes, turnips, winter barley, winter oats, winter oilseed rape, winter wheat** [1, 2]; **forest** *(off-label)*, **sweetcorn** *(off-label)* [1]
- Dandelions in **forest** *(off-label)* [1]
- Groundsel in **all edible seed crops grown outdoors** *(off-label)*, **all non-edible seed crops grown outdoors** *(off-label)*, **bilberries** *(off-label)*, **blackcurrants** *(off-label)*, **blueberries** *(off-label)*, **chard** *(off-label)*, **corn gromwell** *(off-label)*, **cranberries** *(off-label)*, **elderberries** *(off-label)*, **farm forestry** *(off-label)*, **forest** *(off-label)*, **garlic** *(off-label)*, **gooseberries** *(off-label)*, **hemp grown for fibre production** *(off-label)*, **hops in propagation** *(off-label)*, **miscanthus** *(off-label)*, **mulberries** *(off-label)*, **nursery fruit trees** *(off-label)*, **outdoor leaf herbs** *(off-label)*, **redcurrants** *(off-label)*, **rose hips** *(off-label)*, **shallots** *(off-label)*, **spinach** *(off-label)*, **spinach beet** *(off-label)*, **strawberries** *(off-label)*, **sweetcorn** *(off-label)* [1]; **forest nurseries** *(off-label)*, **game cover** *(off-label)* [1, 4]

FOR FULL CONDITIONS OF USE ALWAYS READ THE PRODUCT LABEL

clopyralid

- Mayweeds in **all edible seed crops grown outdoors** *(off-label)*, **all non-edible seed crops grown outdoors** *(off-label)*, **apples** *(off-label)*, **bilberries** *(off-label)*, **blackcurrants** *(off-label)*, **blueberries** *(off-label)*, **borage for oilseed production** *(off-label)*, **canary flower (echium spp.)** *(off-label)*, **chard** *(off-label)*, **corn gromwell** *(off-label)*, **crab apples** *(off-label)*, **cranberries** *(off-label)*, **durum wheat** *(off-label)*, **elderberries** *(off-label)*, **evening primrose** *(off-label)*, **farm forestry** *(off-label)*, **forest** *(off-label)*, **garlic** *(off-label)*, **gooseberries** *(off-label)*, **hemp grown for fibre production** *(off-label)*, **honesty** *(off-label)*, **hops in propagation** *(off-label)*, **leeks** *(off-label)*, **miscanthus** *(off-label)*, **mulberries** *(off-label)*, **mustard** *(off-label)*, **nursery fruit trees** *(off-label)*, **outdoor leaf herbs** *(off-label)*, **pears** *(off-label)*, **poppies for morphine production** *(off-label)*, **redcurrants** *(off-label)*, **rose hips** *(off-label)*, **salad onions** *(off-label)*, **shallots** *(off-label)*, **spinach** *(off-label)*, **spinach beet** *(off-label)*, **spring rye** *(off-label)*, **strawberries** *(off-label)*, **triticale** *(off-label)*, **winter rye** *(off-label)* [1]; **broccoli**, **bulb onions**, **cabbages**, **calabrese**, **cauliflowers** [2]; **brussels sprouts**, **forage maize**, **linseed** [1, 2]; **fodder beet**, **mangels**, **red beet**, **spring oilseed rape**, **sugar beet**, **swedes**, **turnips**, **winter oilseed rape** [1-4]; **forest nurseries** *(off-label)*, **game cover** *(off-label)* [1, 4]; **grassland**, **ornamental plant production**, **spring barley**, **spring oats**, **spring wheat**, **winter barley**, **winter oats**, **winter wheat** [1, 2, 4]
- Perennial dicotyledons in **leeks** *(off-label)*, **salad onions** *(off-label)* [1]
- Ragwort in **forest** *(off-label)* [1]
- Sowthistle in **bilberries** *(off-label)*, **blackcurrants** *(off-label)*, **blueberries** *(off-label)*, **cranberries** *(off-label)*, **elderberries** *(off-label)*, **forest** *(off-label)*, **garlic** *(off-label)*, **gooseberries** *(off-label)*, **mulberries** *(off-label)*, **poppies for morphine production** *(off-label)*, **redcurrants** *(off-label)*, **rose hips** *(off-label)*, **shallots** *(off-label)* [1]; **forest nurseries** *(off-label)*, **game cover** *(off-label)* [4]
- Thistles in **apples** *(off-label)*, **asparagus** *(off-label)*, **borage for oilseed production** *(off-label)*, **canary flower (echium spp.)** *(off-label)*, **chinese cabbage** *(off-label)*, **choi sum** *(off-label)*, **collards** *(off-label)*, **corn gromwell** *(off-label)*, **crab apples** *(off-label)*, **durum wheat** *(off-label)*, **evening primrose** *(off-label)*, **forest** *(off-label)*, **garlic** *(off-label)*, **honesty** *(off-label)*, **kale** *(off-label)*, **leeks** *(off-label)*, **mustard** *(off-label)*, **pak choi** *(off-label)*, **pears** *(off-label)*, **poppies for morphine production** *(off-label)*, **salad onions** *(off-label)*, **shallots** *(off-label)*, **spring greens** *(off-label)*, **spring rye** *(off-label)*, **strawberries** *(off-label)*, **tatsoi** *(off-label)*, **trees** *(off-label)*, **triticale** *(off-label)*, **winter rye** *(off-label)* [1]; **fodder beet**, **mangels**, **red beet**, **spring oilseed rape**, **sugar beet**, **swedes**, **turnips**, **winter oilseed rape** [3, 4]; **forest nurseries** *(off-label)* [1, 4]; **game cover** *(off-label)*, **grassland**, **ornamental plant production**, **spring barley**, **spring oats**, **spring wheat**, **winter barley**, **winter oats**, **winter wheat** [4]
- Volunteer potatoes in **chinese cabbage** *(off-label)*, **choi sum** *(off-label)*, **collards** *(off-label)*, **forest nurseries** *(off-label)*, **garlic** *(off-label)*, **kale** *(off-label)*, **leeks** *(off-label)*, **pak choi** *(off-label)*, **salad onions** *(off-label)*, **shallots** *(off-label)*, **spring greens** *(off-label)*, **sweetcorn** *(off-label)*, **tatsoi** *(off-label)* [1]

Extension of Authorisation for Minor Use (EAMUs)
- *all edible seed crops grown outdoors* 20150007 [1]
- *all non-edible seed crops grown outdoors* 20150007 [1]
- *apples* 20102080 [1]
- *asparagus* 20102079 [1]
- *bilberries* 20161629 [1]
- *blackcurrants* 20161629 [1]
- *blueberries* 20161629 [1]
- *borage for oilseed production* 20102086 [1]
- *canary flower (echium spp.)* 20102086 [1]
- *chard* 20102081 [1]
- *chinese cabbage* 20131710 [1]
- *choi sum* 20131710 [1]
- *collards* 20131710 [1]
- *corn gromwell* 20121046 [1]
- *crab apples* 20102080 [1]
- *cranberries* 20161629 [1]
- *durum wheat* 20102085 [1]
- *elderberries* 20161629 [1]
- *evening primrose* 20102086 [1]

SEE SECTION 3 FOR PRODUCTS ALSO REGISTERED

Pesticide Profiles

- **farm forestry** *20150007* [1]
- **forest** *20152633* [1]
- **forest nurseries** *20130514* [1], *20150007* [1], *20151529* [4]
- **game cover** *20150007* [1], *20151529* [4]
- **garlic** *20130292* [1]
- **gooseberries** *20161629* [1]
- **grass seed crops** *20102084* [1]
- **hemp grown for fibre production** *20122041* [1], *20150007* [1]
- **honesty** *20102086* [1]
- **hops in propagation** *20150007* [1]
- **kale** *20131710* [1]
- **leeks** *20140400* [1]
- **miscanthus** *20150007* [1]
- **mulberries** *20161629* [1]
- **mustard** *20102086* [1]
- **nursery fruit trees** *20150007* [1]
- **outdoor leaf herbs** *20113236* [1]
- **pak choi** *20131710* [1]
- **pears** *20102080* [1]
- **poppies for morphine production** *20102082* [1]
- **redcurrants** *20161629* [1]
- **rose hips** *20161629* [1]
- **salad onions** *20140400* [1]
- **shallots** *20130292* [1]
- **spinach** *20102081* [1]
- **spinach beet** *20102081* [1]
- **spring greens** *20131710* [1]
- **spring rye** *20102085* [1]
- **strawberries** *20131822* [1]
- **sweetcorn** *20152626* [1]
- **tatsoi** *20131710* [1]
- **trees** *20102083* [1]
- **triticale** *20102085* [1]
- **winter rye** *20102085* [1]

Approval information
- Clopyralid included in Annex I under EC Regulation 1107/2009
- Accepted by BBPA for use on malting barley

Efficacy guidance
- Best results achieved by application to young actively growing weed seedlings. Treat creeping thistle at rosette stage and repeat 3-4 wk later as directed
- High activity on weeds of Compositae family. For most crops recommended for use in tank mixes. See label for details

Restrictions
- Maximum total dose varies between the equivalent of one and two full dose treatments, depending on the crop treated. See labels for details
- Do not apply to cereals later than the second node detectable stage (GS 32)
- Do not apply when crop damp or when rain expected within 6 h
- Do not use straw from treated cereals in compost or any other form for glasshouse crops. Straw may be used for strawing down strawberries
- Straw from treated grass seed crops or linseed should be baled and carted away. If incorporated do not plant winter beans in same year
- Do not use on onions at temperatures above 20°C or when under stress
- Do not treat maiden strawberries or runner beds or apply to early leaf growth during blossom period or within 4 wk of picking. Aug or early Sep sprays may reduce yield
- Applications must be made earlier than 1 March in the year of harvest [3]

FOR FULL CONDITIONS OF USE ALWAYS READ THE PRODUCT LABEL

Crop-specific information
- Latest use: 7 d before cutting grass for hay or silage; before 3rd node detectable (GS 33) for cereals; before flower buds visible from above for oilseed rape, linseed
- HI grassland 7 d; apples, pears, strawberries 4 wk; maize, sweetcorn, onions, Brussels sprouts, broccoli, cabbage, cauliflowers, calabrese, kale, fodder rape, oilseed rape, swedes, turnips, sugar beet, red beet, fodder beet, mangels, sage, honesty 6 wk
- Timing of application varies with weed problem, crop and other ingredients of tank mixes. See labels for details
- Apply as directed spray in woody ornamentals, avoiding leaves, buds and green stems. Do not apply in root zone of families Compositae or Leguminosae

Following crops guidance
- Do not plant susceptible autumn-sown crops in same year as treatment. Do not apply later than Jul where susceptible crops are to be planted in spring. See label for details

Environmental safety
- Harmful to aquatic organisms [1, 2]
- Wash spray equipment thoroughly with water and detergent immediately after use. Traces of product can damage susceptible plants sprayed later
- Keep livestock out of treated areas for at least 7 d and until foliage of any poisonous weeds such as ragwort has died and become unpalatable
- Some pesticides pose a greater threat of contamination of water than others and clopyralid is one of these pesticides. Take special care when applying clopyralid near water and do not apply if heavy rain is forecast

Hazard classification and safety precautions
UN Number N/C
Risk phrases R52, R53a [1, 2]
Operator protection A, C [1-4]; H [3]; U05a [3, 4]; U08, U19a, U20b [1-4]
Environmental protection E07a [1, 2]; E15a [1-4]; E34 [3, 4]
Storage and disposal D01, D05, D09a, D10b [1-4]; D02 [3, 4]; D12a [1-3]

103   clopyralid + 2,4-D + MCPA

A translocated herbicide mixture for managed amenity turf
HRAC mode of action code: O + O + O

See also 2,4-D
     MCPA

Products
1	Esteem	Vitax	35:150:175 g/l	SL	15181
2	Redeem	Headland Amenity	35:150:175 g/l	SL	17096

Uses
- Annual dicotyledons in **managed amenity turf** [1, 2]
- Buttercups in **managed amenity turf** [2]
- Clover in **managed amenity turf** [2]
- Daisies in **managed amenity turf** [2]
- Dandelions in **managed amenity turf** [2]
- Perennial dicotyledons in **managed amenity turf** [1, 2]
- Plantains in **managed amenity turf** [2]

Approval information
- Clopyralid, 2,4-D and MCPA included in Annex I under EC Regulation 1107/2009

Efficacy guidance
- Apply when soil is moist and weeds growing actively, normally between Apr and Oct
- Ensure sufficient leaf area for uptake and select appropriate water volume to achieve good spray coverage
- To allow maximum translocation do not cut grass for 3 d after treatment

---
SEE SECTION 3 FOR PRODUCTS ALSO REGISTERED

Pesticide Profiles

Restrictions
- Maximum number of treatments on managed amenity turf 1 per yr
- Do not spray in periods of drought unless irrigation is applied
- Do not apply if night temrperatures are low, if ground frost is imminent or in periods of prolonged cold weather
- Do not treat turf less than 1 yr old
- Do not use any treated plant materials for composting or mulching
- Do not mow for 3 d before or after treatment

Crop-specific information
- Consult manufacturer or carry out small scale safety test on turf grass cultivars not previously treated

Following crops guidance
- If reseeding of treated turf is required allow at least 6 wk after treatment before drilling grasses into the sward
- Where treated land is subsequently to be sown with broad-leaved plants allow an interval of at least 6 mth

Environmental safety
- Dangerous for the environment
- Harmful to aquatic organisms
- Wash spray equipment thoroughly with water and detergent immediately after use. Traces of product can damage susceptible plants sprayed later
- Do not spray outside the target area and avoid drift onto non-target plants
- Some pesticides pose a greater threat of contamination of water than others and clopyralid is one of these pesticides. Take special care when applying clopyralid near water and do not apply if heavy rain is forecast

Hazard classification and safety precautions
Hazard Harmful, Dangerous for the environment [1, 2]; Harmful if swallowed [2]
Transport code 9 [2]; M6 [1]
Packaging group III
UN Number 3082
Risk phrases H318 [2]; R22a, R37, R41, R52, R53a [1]
Operator protection A, C, H; U05a, U11, U15
Environmental protection E15a, E34 [1, 2]; H411 [2]
Storage and disposal D01, D02, D12a

104 clopyralid + florasulam

A translocated herbicide mixture for weed control in cereals
HRAC mode of action code: O + B

Products
Gartrel   Dow   300:25 g/l   SC   16828

Uses
- Black bindweed in *durum wheat, spelt, spring barley, spring oats, spring wheat, triticale, winter barley, winter rye, winter wheat*
- Charlock in *durum wheat, spelt, spring barley, spring oats, spring wheat, triticale, winter barley, winter rye, winter wheat*
- Chickweed in *durum wheat, spelt, spring barley, spring oats, spring wheat, triticale, winter barley, winter rye, winter wheat*
- Cleavers in *durum wheat, spelt, spring barley, spring oats, spring wheat, triticale, winter barley, winter rye, winter wheat*
- Mayweeds in *durum wheat, spelt, spring barley, spring oats, spring wheat, triticale, winter barley, winter rye, winter wheat*
- Runch in *durum wheat, spelt, spring barley, spring oats, spring wheat, triticale, winter barley, winter rye, winter wheat*

FOR FULL CONDITIONS OF USE ALWAYS READ THE PRODUCT LABEL

- Shepherd's purse in *durum wheat, spelt, spring barley, spring oats, spring wheat, triticale, winter barley, winter rye, winter wheat*
- Volunteer oilseed rape in *durum wheat, spelt, spring barley, spring oats, spring wheat, triticale, winter barley, winter rye, winter wheat*
- Wild radish in *durum wheat, spelt, spring barley, spring oats, spring wheat, triticale, winter barley, winter rye, winter wheat*

Approval information
- Clopyralid and florasulam are all included in Annex I under EC Regulation 1107/2009

Efficacy guidance
- Best results obtained when weeds are small and actively growing
- Effectiveness may be reduced when soil is very dry
- Use adequate water volume to achieve complete spray coverage of the weeds
- Florasulam is a member of the ALS-inhibitor group of herbicides

Restrictions
- Do not roll or harrow for 7 d before or after application
- Do not use any treated plant material for composting or mulching
- Do not use manure from animals fed on treated crops for composting
- Specific restrictions apply to use in sequence or tank mixture with other sulfonylurea or ALS-inhibiting herbicides. See label for details

Following crops guidance
- Where residues of a treated crop have not completely decayed by the time of planting a succeeding crop, avoid planting peas, beans and other legumes, carrots and other Umbelliferae, potatoes, lettuce and other Compositae, glasshouse and protected crops
- Where the product has been used in mixture with certain named products (see label) only cereals or grass may be sown in the autumn following harvest. Otherwise cereals, oilseed rape, grass or vegetable brassicas as transplants may be sown as a following crop in the same calendar yr as treatment. Oilseed rape may show some temporary reduction of vigour after a dry summer, but yields are not affected
- In addition to the above, field beans, linseed, peas, sugar beet, potatoes, maize, clover (for use in grass/clover mixtures) or carrots may be sown in the calendar yr following treatment
- In the event of failure of a treated crop in spring, only spring wheat, spring barley, spring oats, maize or ryegrass may be sown

Environmental safety
- LERAP Category B

Hazard classification and safety precautions
Hazard Very toxic to aquatic organisms
Transport code 9
Packaging group III
UN Number 3082
Operator protection U05a, U19a, U20a
Environmental protection E15a, E16a, E34, H410
Storage and disposal D05, D09a, D10b, D12a
Medical advice M03

105   clopyralid + florasulam + fluroxypyr

A translocated herbicide mixture for cereals
HRAC mode of action code: O + B + O

See also florasulam
fluroxypyr

Products

1	Dakota	Dow	80:2.5:100 g/l	EC	16121
2	Galaxy	Dow	80:2.5:100 g/l	EC	14085

SEE SECTION 3 FOR PRODUCTS ALSO REGISTERED

Products – continued

3	Polax	Pan Amenity	80:2.5;100 g/l	EC	15915
4	Praxys	Dow	80:2.5;100 g/l	EC	13912

Uses
- Annual dicotyledons in **amenity grassland, lawns, managed amenity turf** [3, 4]; **spring barley, spring oats, spring wheat, winter barley, winter oats, winter wheat** [1, 2]
- Black medick in **amenity grassland** [4]
- Bristly oxtongue in **amenity grassland** [3, 4]; **lawns, managed amenity turf** [3]
- Buttercups in **amenity grassland** [3, 4]; **lawns, managed amenity turf** [3]
- Chickweed in **spring barley, spring oats, spring wheat, winter barley, winter oats, winter wheat** [1, 2]
- Cleavers in **spring barley, spring oats, spring wheat, winter barley, winter oats, winter wheat** [1, 2]
- Common mouse-ear in **amenity grassland** [3, 4]; **lawns, managed amenity turf** [3]
- Creeping thistle in **spring barley, spring oats, spring wheat, winter barley, winter oats, winter wheat** [1]
- Daisies in **amenity grassland** [3, 4]; **lawns, managed amenity turf** [3]
- Dandelions in **amenity grassland** [3, 4]; **lawns, managed amenity turf** [3]
- Mayweeds in **spring barley, spring oats, spring wheat, winter barley, winter oats, winter wheat** [1]
- Plantains in **amenity grassland** [3, 4]; **lawns, managed amenity turf** [3]
- Self-heal in **amenity grassland** [4]
- Slender speedwell in **amenity grassland** [4]

Approval information
- Clopyralid, florasulam and fluroxypyr are all included in Annex I under EC Regulation 1107/2009

Efficacy guidance
- Best results obtained when weeds are small and actively growing
- Effectiveness may be reduced when soil is very dry
- Use adequate water volume to achieve complete spray coverage of the weeds
- Florasulam is a member of the ALS-inhibitor group of herbicides

Restrictions
- Maximum number of treatments 1 per yr for all crops
- Do not spray when crops are under stress from any cause
- Do not apply through CDA applicators
- Do not roll or harrow for 7 d before or after application
- Do not use any treated plant material for composting or mulching
- Do not use manure from animals fed on treated crops for composting
- Specific restrictions apply to use in sequence or tank mixture with other sulfonylurea or ALS-inhibiting herbicides. See label for details

Crop-specific information
- Latest use: before second node detectable for oats; before third node detectable for spring barley and spring wheat; before flag leaf detectable in winter barley and winter wheat.

Following crops guidance
- Where residues of a treated crop have not completely decayed by the time of planting a succeeding crop, avoid planting peas, beans and other legumes, carrots and other Umbelliferae, potatoes, lettuce and other Compositae, glasshouse and protected crops
- Where the product has been used in mixture with certain named products (see label) only cereals or grass may be sown in the autumn following harvest. Otherwise cereals, oilseed rape, grass or vegetable brassicas as transplants may be sown as a following crop in the same calendar yr as treatment. Oilseed rape may show some temporary reduction of vigour after a dry summer, but yields are not affected
- In addition to the above, field beans, linseed, peas, sugar beet, potatoes, maize, clover (for use in grass/clover mixtures) or carrots may be sown in the calendar yr following treatment
- In the event of failure of a treated crop in spring, only spring wheat, spring barley, spring oats, maize or ryegrass may be sown

FOR FULL CONDITIONS OF USE ALWAYS READ THE PRODUCT LABEL

Environmental safety
- Dangerous for the environment
- Very toxic to aquatic organisms
- Take extreme care to avoid drift outside the target area
- Some pesticides pose a greater threat of contamination of water than others and clopyralid is one of these pesticides. Take special care when applying clopyralid near water and do not apply if heavy rain is forecast

Hazard classification and safety precautions
Hazard Harmful, Dangerous for the environment
Transport code 9
Packaging group III
UN Number 3082
Risk phrases R20, R36, R38, R50, R53a
Operator protection A, H [1-4]; C [1, 2]; U05a, U11, U19a, U20a
Environmental protection E15b, E34, E38
Storage and disposal D01, D02, D09a, D10b, D12a [1-4]; D05 [3, 4]
Medical advice M03 [1-4]; M05a [1, 2]

106   clopyralid + fluroxypyr + MCPA

A translocated herbicide mixture for use in sports and amenity turf
HRAC mode of action code: O + O + O

See also fluroxypyr
  MCPA

Products
  Greenor   Rigby Taylor   20:40:200 g/l   ME   15204

Uses
- Annual dicotyledons in **managed amenity turf**

Approval information
- Clopyralid, fluroxypyr and MCPA included in Annex I under EC Regulation 1107/2009

Efficacy guidance
- Best results achieved when weeds actively growing and turf grass competitive
- Treatment should normally be between Apr-Sep when the soil is moist
- Do not apply during drought unless irrigation is applied
- Allow 3 d before or after mowing established turf to ensure sufficient weed leaf surface present to allow uptake and movement

Restrictions
- Maximum number of treatments 2 per yr
- Do not treat grass under stress from frost, drought, waterlogging, trace element deficiency, disease or pest attack
- Do not treat if night temperatures are low, when frost is imminent or during prolonged cold weather

Crop-specific information
- Treat young turf only in spring when at least 2 mth have elapsed since sowing
- Allow 5 d after mowing young turf before treatment
- Product selective on a number of turf grass species (see label) but consultation or testing recommended before treatment of any cultivar

Environmental safety
- Dangerous for the environment
- Very toxic to aquatic organisms
- Wash spray equipment thoroughly with water and detergent immediately after use. Traces of product can damage susceptible plants sprayed later

SEE SECTION 3 FOR PRODUCTS ALSO REGISTERED

Pesticide Profiles

- Some pesticides pose a greater threat of contamination of water than others and clopyralid is one of these pesticides. Take special care when applying clopyralid near water and do not apply if heavy rain is forecast
- LERAP Category B

Hazard classification and safety precautions
Hazard Irritant, Dangerous for the environment
Transport code 9
Packaging group III
UN Number 3082
Risk phrases R36, R43, R50, R53a
Operator protection A, C; U05a, U08, U14, U19a, U20b
Environmental protection E15a, E16a, E38
Storage and disposal D01, D02, D05, D09a, D10b, D12a

107   clopyralid + fluroxypyr + triclopyr

A foliar acting herbicide mixture for grassland
HRAC mode of action code: O + O + O

See also fluroxypyr
       triclopyr

Products

1	Governer	Dow	50:75:100 g/l	EC	16849
2	Pastor Pro	Dow	50:75:100 g/l	EC	16465

Uses
- Docks in **grassland**
- Stinging nettle in **grassland**
- Thistles in **grassland**

Approval information
- Clopyralid, fluroxypyr and triclopyr included in Annex I under EC Regulation 1107/2009
- Approval expiry 31 Oct 2017 [1, 2]

Efficacy guidance
- Apply in spring or autumn depending on weeds present or as a split treatment in spring followed by autumn
- Treatment must be made when weeds and grass are actively growing
- It is important to ensure sufficient leaf area is present for uptake especially on established docks and thistles
- On large well established docks and where there is a large soil seed reservoir further treatment in the following year may be needed
- To allow maximum translocation in the weeds do not cut grass for 4 wk after treatment

Restrictions
- Maximum number of treatments 1 per yr at full dose or 2 per yr at half dose
- Do not spray in drought, very hot or very cold weather
- Do not treat sports or amenity turf
- Product kills or severely checks clover and should not be used where clover is an important constituent of the sward
- Do not roll or harrow 10 d before or 7 d after treatment
- Do not allow spray or drift to reach other crops, amenity plantings, gardens, ponds, lakes or water courses
- Do not use on newly-sown leys
- Only apply between 1st March and 31st Oct.

Crop-specific information
- HI 7 d

FOR FULL CONDITIONS OF USE ALWAYS READ THE PRODUCT LABEL

- Application during active growth ensures minimal check to grass. Product may be used in established grassland which is under non-rotational setaside arrangements
- Very occasionally some yellowing of the sward may occur after treatment but is quickly outgrown

Following crops guidance
- Residues in incompletely decayed plant tissue may affect succeeding susceptible crops such as peas, beans and other legumes, carrots and related crops, potatoes, tomatoes, lettuce
- Do not plant susceptible autumn-sown crops in the same yr as treatment with product. Spring sown crops may follow if treatment was before end July in the previous yr

Environmental safety
- Dangerous for the environment
- Toxic to aquatic organisms
- Keep livestock out of treated areas for at least 7 d following treatment and until foliage of poisonous weeds such as ragwort has died and become unpalatable
- Wash spray equipment thoroughly with water and detergent immediately after use. Traces of product can damage susceptible plants sprayed later
- Some pesticides pose a greater threat of contamination of water than others and clopyralid is one of these pesticides. Take special care when applying clopyralid near water and do not apply if heavy rain is forecast
- LERAP Category B

Hazard classification and safety precautions
Hazard Harmful, Flammable, Dangerous for the environment, Flammable liquid and vapour, Very toxic to aquatic organisms
Transport code 3
Packaging group III
UN Number 1993
Risk phrases H304, H317, H319, H335, H336
Operator protection A, C; U02a, U05a, U08, U11, U14, U19a, U20b
Environmental protection E07a, E15a, E16a, E34, E38, H410
Consumer protection C01
Storage and disposal D01, D02, D05, D09a, D10b, D12a
Medical advice M05b

108 clopyralid + picloram

A post-emergence herbicide mixture for oilseed rape
HRAC mode of action code: O + O

See also picloram

Products
1	Clopic	UPL Europe	267:67 g/l	SL	17387
2	Galera	Dow	267:67 g/l	SL	16413
3	Legara	AgChem Access	267:67 g/l	SL	16789
4	Piccant	Goldengrass	267:67 g/l	SL	16794

Uses
- Annual dicotyledons in **corn gromwell** *(off-label)*, **crambe** *(off-label)*, **forest nurseries** *(off-label)*, **game cover** *(off-label)*, **mustard** *(off-label)*, **ornamental plant production** *(off-label)* [2]
- Cleavers in **corn gromwell** *(off-label)*, **crambe** *(off-label)*, **forest nurseries** *(off-label)*, **game cover** *(off-label)*, **mustard** *(off-label)*, **ornamental plant production** *(off-label)*, **spring oilseed rape** *(off-label)* [2]; **winter oilseed rape** [1-4]
- Mayweeds in **corn gromwell** *(off-label)*, **crambe** *(off-label)*, **forest nurseries** *(off-label)*, **game cover** *(off-label)*, **mustard** *(off-label)*, **ornamental plant production** *(off-label)* [2]; **winter oilseed rape** [1-4]

Extension of Authorisation for Minor Use (EAMUs)
- **corn gromwell** 20160559 [2]

- ***crambe*** *20161622* [2]
- ***forest nurseries*** *20142720* [2]
- ***game cover*** *20142720* [2]
- ***mustard*** *20152939* [2]
- ***ornamental plant production*** *20150968* [2]
- ***spring oilseed rape*** *20140827* [2]

Approval information
- Clopyralid and picloram included in Annex I under EC Regulation 1107/2009

Efficacy guidance
- Best results obtained from treatment when weeds are small and actively growing
- Cleavers that germinate after treatment will not be controlled

Restrictions
- Maximum total dose equivalent to one full dose treatment
- Do not treat crops under stress from cold, drought, pest damage, nutrient deficiency or any other cause
- Do not roll or harrow for 7 d before or after spraying
- Do not apply through CDA applicators
- Do not use any treated plant material for composting or mulching
- Do not use manure from animals fed on treated crops for composting
- Chop and incorporate all treated plant remains in early autumn, or as soon as possible after harvest, to release any residues into the soil. Ensure that all treated plant remains have completely decayed before planting susceptible crops

Crop-specific information
- Latest use: before flower buds visible above crop canopy for winter oilseed rape

Following crops guidance
- Wheat, barley, oats, maize, or oilseed rape may be sown 120 days after application, all other crops should only be sown 36 months after application [2]
- Ploughing or thorough cultivation should be carried out before planting leguminous crops
- Do not attempt to plant peas, beans, other legumes, carrots, other umbelliferous crops, potatoes, lettuce, other Compositae, or any glasshouse or protected crops if treated crop remains have not fully decayed by the time of planting
- In the event of failure of an autumn treated crop only oilseed rape, wheat, barley, oats, maize or ryegrass may be sown in the spring and only after ploughing or thorough cultivation

Environmental safety
- Dangerous for the environment
- Toxic to aquatic organisms
- Take extreme care to avoid drift onto crops and non-target plants outside the target area
- Some pesticides pose a greater threat of contamination of water than others and clopyralid is one of these pesticides. Take special care when applying clopyralid near water and do not apply if heavy rain is forecast

Hazard classification and safety precautions
   Hazard Dangerous for the environment
   UN Number N/C
   Risk phrases R51, R53a [2-4]
   Operator protection A, C; U05a
   Environmental protection E15a, E34 [1-4]; E38 [1, 2, 4]; H413 [1]
   Storage and disposal D01, D02, D05, D07, D09a [1-4]; D12a [1, 2, 4]

FOR FULL CONDITIONS OF USE ALWAYS READ THE PRODUCT LABEL

109   clopyralid + triclopyr

A perennial and woody weed herbicide for use in grassland
HRAC mode of action code: O + O

See also triclopyr

Products

1	Blaster Pro	Headland Amenity	60:240 g/l	EC	15752
2	Grazon Pro	Dow	60:240 g/l	EC	15785
3	Prevail	Dow	200:200 g/l	SL	17395
4	Thistlex	Dow	200:200 g/l	SL	16123

Uses
- Annual and perennial weeds in **grassland** [2]
- Annual dicotyledons in **game cover** *(off-label)* [4]
- Brambles in **amenity grassland** [1]; **grassland** [2]
- Broom in **amenity grassland** [1]; **grassland** [2]
- Creeping thistle in **grassland** [2-4]; **rotational grass** [3]
- Docks in **amenity grassland** [1]; **grassland** [2]
- Gorse in **amenity grassland** [1]; **grassland** [2]
- Mayweeds in **game cover** *(off-label)* [4]
- Perennial dicotyledons in **amenity grassland** [1]
- Stinging nettle in **amenity grassland** [1]
- Thistles in **amenity grassland** [1]; **game cover** *(off-label)* [4]

Extension of Authorisation for Minor Use (EAMUs)
- **game cover** 20142512 [4]

Approval information
- Clopyralid and triclopyr included in Annex I under EC Regulation 1107/2009

Efficacy guidance
- Must be applied to actively growing weeds
- Correct timing crucial for good control. Spray stinging nettle before flowering, docks in rosette stage in spring, creeping thistle before flower stems 15-20 cm high, brambles, broom and gorse in Jun-Aug
- Allow 2-3 wk regrowth after grazing or mowing before spraying perennial weeds
- Where there is a large reservoir of weed seed in the soil further treatment in the following yr may be needed

Restrictions
- Maximum number of treatments 1 per yr
- Only use on permanent pasture or rotational grassland established for at least 1 yr
- Do not apply where clover is an important constituent of sward
- Do not roll or harrow within 10 d before or 7 d after spraying
- Do not cut grass for 21 d before or 28 d after spraying
- Do not use any treated plant material for composting or mulching, and do not use manure for composting from animals fed on treated crops
- Do not apply by hand-held rotary atomiser equipment
- Do not allow drift onto other crops, amenity plantings or gardens, ponds, lakes or water courses. All conifers, especially pine and larch, are very sensitive

Crop-specific information
- Latest use: 7 d before grazing or cutting grass
- Some transient yellowing of treated swards may occur but is quickly outgrown

Following crops guidance
- Residues in plant tissues which have not completely decayed may affect succeeding susceptible crops such as peas, beans, other legumes, carrots, parsnips, potatoes, tomatoes, lettuce, glasshouse and protected crops

---

SEE SECTION 3 FOR PRODUCTS ALSO REGISTERED

Pesticide Profiles

- Do not plant susceptible autumn-sown crops (eg winter beans) in same year as treatment and allow at least 9 mth from treatment before planting a susceptible crop in the following yr
- Do not direct drill kale, swedes, turnips, grass or grass mixtures within 6 wk of spraying
- Do not spray after end Jul where susceptible crops are to be planted in the next spring

Environmental safety
- Dangerous for the environment
- Very toxic to aquatic organisms
- Keep livestock out of treated areas for at least 7 d after spraying and until foliage of any poisonous weeds such as ragwort or buttercup has died down and become unpalatable
- Some pesticides pose a greater threat of contamination of water than others and clopyralid is one of these pesticides. Take special care when applying clopyralid near water and do not apply if heavy rain is forecast
- LERAP Category B

Hazard classification and safety precautions
Hazard Harmful [1, 2]; Irritant [1, 3, 4]; Dangerous for the environment [1-4]
Transport code 3 [1, 2]
Packaging group III [1, 2]
UN Number 1993 [1, 2]; N/C [3, 4]
Risk phrases H318, H335, H373 [3]; R22a, R38, R43 [1, 2]; R36, R37, R51, R67 [1]; R41, R53a [1, 2, 4]; R50 [2]; R52 [4]
Operator protection A, C [1-4]; H, M [1, 2]; U02a, U05a, U11, U20b [1-4]; U08, U14, U19a, U23b [1, 2]; U15 [1, 3, 4]; U23a [3, 4]
Environmental protection E07a, E15a, E23, E34, E38 [1, 2]; E07c [3, 4] (7 days); E16a [1-4]; H410 [3]
Consumer protection C01 [1, 2]
Storage and disposal D01, D02, D05, D09a, D10b, D12a
Medical advice M03 [3, 4]; M05b [1]

110 clothianidin

A nitromethylene neonicotinoid insecticide seed treatment for control of virus vectors, aphids and a reduction of damage by slugs and wireworms in winter cereals
IRAC mode of action code: 4A

See also beta-cyfluthrin + clothianidin

Products
Deter	Bayer CropScience	250 g/l	FS	12411

Uses
- Leafhoppers in **durum wheat** *(seed treatment)*, **triticale** *(seed treatment)*, **winter barley** *(seed treatment)*, **winter oats** *(seed treatment)*, **winter rye** *(seed treatment)*, **winter wheat** *(seed treatment)*
- Slugs in **durum wheat** *(seed treatment)*, **triticale** *(seed treatment)*, **winter barley** *(seed treatment)*, **winter oats** *(seed treatment)*, **winter rye** *(seed treatment)*, **winter wheat** *(seed treatment)*
- Virus vectors in **durum wheat** *(seed treatment)*, **triticale** *(seed treatment)*, **winter barley** *(seed treatment)*, **winter oats** *(seed treatment)*, **winter rye** *(seed treatment)*, **winter wheat** *(seed treatment)*
- Wireworm in **durum wheat** *(seed treatment)*, **triticale** *(seed treatment)*, **winter barley** *(seed treatment)*, **winter oats** *(seed treatment)*, **winter rye** *(seed treatment)*, **winter wheat** *(seed treatment)*

Approval information
- Clothianidin included in Annex I under EC Regulation 1107/2009
- Accepted by BBPA for use on malting barley

FOR FULL CONDITIONS OF USE ALWAYS READ THE PRODUCT LABEL

clothianidin + prothioconazole

Efficacy guidance
- May only be used in conjunction with manufacturer's approved seed treatment application equipment or by following procedures given in the operating instructions
- Treated seed should preferably be drilled in the same season
- Evenness of seed cover improved by simultaneous application with a small volume (1.5-3 litres/tonne) of water
- Calibrate drill for treated seed and drill at 2.5-4 cm into firm, well prepared seedbed
- Use minimum 125 kg treated seed per ha
- Seed should be drilled to a depth of 40 mm into a well prepared and firm seedbed. If seed is present on the surface, or if spills have occurred, the field should be harrowed and rolled if conditions are appropriate
- When aphid activity is unusually late, or is heavy and prolonged in areas of high risk, and mild weather predominates, follow-up foliar aphicide may be required
- Incidental suppression of leafhoppers in early spring (Mar/Apr) may be achieved but if a specific attack develops an additional foliar insecticide may be required

Restrictions
- Maximum number of seed treatments one per batch
- Product must be fully re-dispersed and homogeneous before use
- Do not use on seed with more than 16% moisture content, or on sprouted, cracked or skinned seed
- All seed batches should be tested to ensure they are suitable for treatment
- Do not sow crops treated with clothianidin between 1st January and 30th June

Crop-specific information
- Latest use: pre-drilling

Environmental safety
- Harmful to game and wildlife. Treated seed should not be left on the surface. Bury spillages.

Hazard classification and safety precautions
Hazard Irritant, Very toxic to aquatic organisms
UN Number N/C
Risk phrases H317
Operator protection A, H; U04a, U05a, U07, U13, U14, U20b
Environmental protection E03, E15b, E34, E36a, H410
Storage and disposal D01, D02, D05, D09a, D14
Treated seed S01, S02, S03, S04b, S05, S06a, S06b, S07, S08
Medical advice M03

111 clothianidin + prothioconazole

A combined fungicide and insecticide seed dressing for cereals
IRAC mode of action code: 4A

See also prothioconazole

Products
Redigo Deter   Bayer CropScience   250:50 g/l   FS   12423

Uses
- Bunt in **durum wheat** (seed treatment), **triticale** (seed treatment), **winter rye** (seed treatment), **winter wheat** (seed treatment)
- Covered smut in **winter barley** (seed treatment)
- Fusarium foot rot and seedling blight in **durum wheat** (seed treatment), **triticale** (seed treatment), **winter barley** (seed treatment), **winter oats** (seed treatment), **winter rye** (seed treatment), **winter wheat** (seed treatment)
- Leaf stripe in **winter barley** (seed treatment)
- Leafhoppers in **durum wheat** (seed treatment), **triticale** (seed treatment), **winter barley** (seed treatment), **winter oats** (seed treatment), **winter rye** (seed treatment), **winter wheat** (seed treatment)

SEE SECTION 3 FOR PRODUCTS ALSO REGISTERED

- Loose smut in **durum wheat** (seed treatment), **winter barley** (seed treatment), **winter oats** (seed treatment), **winter wheat** (seed treatment)
- Slugs in **durum wheat** (seed treatment), **triticale** (seed treatment), **winter barley** (seed treatment), **winter oats** (seed treatment), **winter rye** (seed treatment), **winter wheat** (seed treatment)
- Virus vectors in **durum wheat** (seed treatment), **triticale** (seed treatment), **winter barley** (seed treatment), **winter oats** (seed treatment), **winter rye** (seed treatment), **winter wheat** (seed treatment)
- Wireworm in **durum wheat** (seed treatment), **triticale** (seed treatment), **winter barley** (seed treatment), **winter oats** (seed treatment), **winter rye** (seed treatment), **winter wheat** (seed treatment)

Approval information
- Clothianidin and prothioconazole included in Annex I under EC Regulation 1107/2009
- Accepted by BBPA for use on malting barley

Efficacy guidance
- May only be used in conjunction with manufacturer's approved seed treatment application equipment or by following procedures given in the operating instructions
- Treated wheat seed should preferably be drilled in the same season. Treated seed of other cereals must be used in the same season
- Evenness of seed cover improved by simultaneous application with a small volume (1.5-3 litres/tonne) of water
- Calibrate drill for treated seed and drill at 2.5-4 cm into firm, well prepared seedbed
- Use minimum 125 kg treated seed per ha
- When aphid activity unusually late, or is heavy and prolonged in areas of high risk, and mild weather predominates, follow-up foliar aphicide may be required
- Incidental suppression of leafhoppers in early spring (Mar/Apr) may be achieved but if a specific attack develops an additional foliar insecticide may be required
- Protection against slugs only applies to germinating seeds and not to aerial parts after emergence
- Where very high populations of aphids, wireworms or slugs occur additional specific control measures may be needed

Restrictions
- Maximum number of seed treatments one per batch for all cereals
- Product must be fully re-dispersed and homogeneous before use
- Do not use on seed with more than 16% moisture content, or on sprouted, cracked or skinned seed
- All seed batches should be tested to ensure they are suitable for treatment
- Do not sow treated crops between 1st Jan and 30th June

Crop-specific information
- Latest use: pre-drilling of all cereals

Environmental safety
- Harmful to aquatic organisms

Hazard classification and safety precautions
    Hazard Irritant, Very toxic to aquatic organisms
    UN Number N/C
    Risk phrases H317
    Operator protection A, H; U14, U19a, U20b
    Environmental protection E15a, E34, E38, H410
    Storage and disposal D01, D02, D09a, D11a, D12a
    Treated seed S01, S02, S03, S04b, S05, S06a, S07, S08

FOR FULL CONDITIONS OF USE ALWAYS READ THE PRODUCT LABEL

112 Coniothyrium minitans

A fungal parasite of sclerotia in soil

Products
    Contans WG   Bayer CropScience   1000 IU/mg   WG   17085

Uses
- Sclerotinia in *all edible crops (outdoor), all non-edible crops (outdoor)*

Approval information
- Coniothyrium minitans included in Annex 1 under EC Regulation 1107/2009

Efficacy guidance
- *C.minitans* is a soil acting biological fungicide with specific action against the resting bodies (sclerotia) of *Sclerotinia sclerotiorum* and *S.minor*
- Treat 3 mth before disease protection is required to allow time for the infective sclerotia in the soil to be reduced
- A post-harvest treatment of soil and debris prevents further contamination of the soil with sclerotia produced by the previous crop
- For best results the soil should be moist and the temperature between 12-20°C
- Application should be followed by soil incorporation into the surface layer to 10 cm with a rotovator or rotary harrow. Thorough spray coverage of the soil is essential to ensure uniform distribution
- If soil temperature drops below 0°C or rises above 27°C fungicidal activity is suspended but restarts when soil temperature returns within this range
- In glasshouses untreated areas at the margins should be covered by a film to avoid spread of spores from untreated sclerotia

Restrictions
- Maximum number of treatments 1 per crop or situation
- Do not plough or cultivate after treatment
- Do not mix with other pesticides, acids, alkalines or any product that attacks organic material
- Store product under cool dry conditions away from direct sunlight and away from heat sources

Crop-specific information
- Latest use: pre-planting
- HI: zero

Following crops guidance
- There are no restrictions on following crops and no waiting interval is specified

Hazard classification and safety precautions
    Hazard Harmful
    UN Number N/C
    Operator protection A, C, D; U05a, U19a
    Environmental protection E15b, E34
    Storage and disposal D01, D02, D12a
    Medical advice M03

113 copper oxychloride

A protectant copper fungicide and bactericide
FRAC mode of action code: M1

Products
    Cuprokylt   Certis   87% w/w (copper)   WP   17079

Uses
- Dothistroma needle blight in *forest nurseries* (off-label)
- Downy mildew in *table grapes*, *wine grapes*

SEE SECTION 3 FOR PRODUCTS ALSO REGISTERED

Extension of Authorisation for Minor Use (EAMUs)
- *forest nurseries* 20152416

Approval information
- Copper oxychloride included in Annex 1 under EC Regulation 1107/2009
- Accepted by BBPA for use on hops

Efficacy guidance
- Spray crops at high volume when foliage dry but avoid run off. Do not spray if rain expected soon
- Spray interval commonly 10-14 d but varies with crop, see label for details

Environmental safety
- Dangerous for the environment
- Very toxic to aquatic organisms
- Keep all livestock out of treated areas for at least 3 wk
- LERAP Category B

Hazard classification and safety precautions
Hazard Dangerous for the environment, Harmful if swallowed, Harmful if inhaled, Very toxic to aquatic organisms
Transport code 9
Packaging group III
UN Number 3077
Operator protection U20a
Environmental protection E06a (3 wk); E13c, E16a, E34, E38, H410, H411
Storage and disposal D01, D09a, D10b

114 cyantraniliprole

An ingested and contact insecticide for insect pest control in brassica crops
IRAC mode of action code: 28

Products
Verimark 20 SC   DuPont   200 g/l   SC   17519

Uses
- Cabbage root fly in **broccoli**, **brussels sprouts**, **cabbages**, **cauliflowers**

Approval information
- Cyantraniliprole is included in Annex 1 under EC Regulation 1107/2009
- Approval expiry 31 Dec 2017 [1]

Efficacy guidance
- May be used as part of an Integrated Pest Management (IPM) program

Restrictions
- To protect bees and pollinating insects avoid application to crops when in flower.
- Consult processor when the product is to be applied to crops grown for processing

Hazard classification and safety precautions
Hazard Very toxic to aquatic organisms
Transport code 9
Packaging group III
UN Number 3082
Operator protection U05a, U09a, U19a, U20a
Environmental protection E12e, E15b, H410
Storage and disposal D01, D02, D09a, D12a, D22
Medical advice M03

cyazofamid

115 cyazofamid

A cyanoimidazole sulfonamide protectant fungicide for potatoes
FRAC mode of action code: 21

Products

1	Linford	AgChem Access	400 g/l	SC	13824
2	Ranman Top	Belchim	160 g/l	SC	14753
3	Swallow	AgChem Access	400 g/l	SC	14078

Uses
- Blight in *potatoes*

Approval information
- Cyazofamid included in Annex I under EC Regulation 1107/2009

Efficacy guidance
- Apply as a protectant treatment before blight enters the crop and repeat every 5 -10 d depending on severity of disease pressure
- Commence spray programme immediately the risk of blight in the locality occurs, usually when the crop meets along the rows
- Product must always be used with organosilicone adjuvant provided in the twin pack
- To minimise the chance of development of resistance no more than three applications should be made consecutively (out of a permissible total of six) in the blight control programme. For more information on Resistance Management see Section 5

Restrictions
- Maximum number of treatments 6 per crop (no more than three of which should be consecutive)
- Mixed product must not be allowed to stand overnight
- Consult processor before using on crops intended for processing

Crop-specific information
- HI 7 d

Environmental safety
- Dangerous for the environment
- Very toxic to aquatic organisms
- Do not empty into drains

Hazard classification and safety precautions
  Hazard Harmful [1, 3] (adjuvant); Irritant [2]; Dangerous for the environment [1-3]
  Transport code 9
  Packaging group III
  UN Number 3082
  Risk phrases H315 [2]; R20, R36, R48 [1, 3] (adjuvant); R41, R50, R53a [1, 3]
  Operator protection A, C; U02a, U05a, U11, U15 [1-3]; U19a [1, 3] (adjuvant); U20a [1, 3]; U20b [2]
  Environmental protection E15a, E34 [1, 3]; E15b, H410 [2]; E19b, E38 [1-3]
  Storage and disposal D01, D02, D09a, D10c, D12a [1-3]; D05, D12b [2]
  Medical advice M03

116 cycloxydim

A translocated post-emergence cyclohexanedione oxime herbicide for grass weed control
HRAC mode of action code: A

Products

Laser	BASF	200 g/l	EC	17339

SEE SECTION 3 FOR PRODUCTS ALSO REGISTERED

Pesticide Profiles

Uses
- Annual grasses in **carrots, cauliflowers, christmas trees, combining peas, dwarf beans, early potatoes, fodder beet, forest, forest nurseries, leeks, linseed, maincrop potatoes, mangels, ornamental plant production, salad onions, spring field beans, spring oilseed rape, strawberries, sugar beet, swedes, vining peas, winter field beans, winter oilseed rape**
- Black bent in **carrots, cauliflowers, combining peas, dwarf beans, early potatoes, fodder beet, leeks, linseed, maincrop potatoes, mangels, salad onions, spring field beans, spring oilseed rape, strawberries, sugar beet, swedes, vining peas, winter field beans, winter oilseed rape**
- Blackgrass in **carrots, cauliflowers, combining peas, dwarf beans, early potatoes, fodder beet, leeks, linseed, maincrop potatoes, mangels, salad onions, spring field beans, spring oilseed rape, strawberries, sugar beet, swedes, vining peas, winter field beans, winter oilseed rape**
- Couch in **carrots, cauliflowers, combining peas, dwarf beans, early potatoes, fodder beet, leeks, linseed, maincrop potatoes, mangels, salad onions, spring field beans, spring oilseed rape, strawberries, sugar beet, swedes, vining peas, winter field beans, winter oilseed rape**
- Creeping bent in **carrots, cauliflowers, combining peas, dwarf beans, early potatoes, fodder beet, leeks, linseed, maincrop potatoes, mangels, salad onions, spring field beans, spring oilseed rape, strawberries, sugar beet, swedes, vining peas, winter field beans, winter oilseed rape**
- Green cover in **land temporarily removed from production**
- Onion couch in **carrots, cauliflowers, combining peas, dwarf beans, early potatoes, fodder beet, leeks, linseed, maincrop potatoes, mangels, salad onions, spring field beans, spring oilseed rape, strawberries, sugar beet, swedes, vining peas, winter field beans, winter oilseed rape**
- Perennial grasses in **christmas trees, forest, forest nurseries, ornamental plant production**
- Volunteer cereals in **carrots, cauliflowers, combining peas, dwarf beans, early potatoes, fodder beet, leeks, linseed, maincrop potatoes, mangels, salad onions, spring field beans, spring oilseed rape, strawberries, sugar beet, swedes, vining peas, winter field beans, winter oilseed rape**
- Wild oats in **carrots, cauliflowers, combining peas, dwarf beans, early potatoes, fodder beet, leeks, linseed, maincrop potatoes, mangels, salad onions, spring field beans, spring oilseed rape, strawberries, sugar beet, swedes, vining peas, winter field beans, winter oilseed rape**

Approval information
- Cycloxydim included in Annex 1 under EC Regulation 1107/2009

Efficacy guidance
- Best results achieved when weeds small and have not begun to compete with crop. Effectiveness reduced by drought, cool conditions or stress. Weeds emerging after application are not controlled
- Foliage death usually complete after 3-4 wk but longer under cool conditions, especially late treatments to winter oilseed rape
- Perennial grasses should have sufficient foliage to absorb spray and should not be cultivated for at least 14 d after treatment
- On established couch pre-planting cultivation recommended to fragment rhizomes and encourage uniform emergence
- Split applications to volunteer wheat and barley at GS 12-14 will often give adequate control in winter oilseed rape. See label for details
- Apply to dry foliage when rain not expected for at least 2 h
- Cycloxydim is an ACCase inhibitor herbicide. To avoid the build up of resistance do not apply products containing an ACCase inhibitor herbicide more than twice to any crop. In addition do not use any product containing cycloxydim in mixture or sequence with any other product containing the same ingredient
- Use these products as part of a resistance management strategy that includes cultural methods of control and does not use ACCase inhibitors as the sole chemical method of grass weed control

FOR FULL CONDITIONS OF USE ALWAYS READ THE PRODUCT LABEL

Cydia pomonella GV

- Applying a second product containing an ACCase inhibitor to a crop will increase the risk of resistance development; only use a second ACCase inhibitor to control different weeds at a different timing
- Always follow WRAG guidelines for preventing and managing herbicide resistant weeds. See Section 5 for more information

Restrictions
- Maximum number of treatments 1 per crop or yr in most situations. See label
- Must be used with authorised adjuvant oil. See label
- Do not apply to crops damaged or stressed by adverse weather, pest or disease attack or other pesticide treatment
- Prevent drift onto other crops, especially cereals and grass

Crop-specific information
- HI cabbage, cauliflower, calabrese, salad onions 4 wk; peas, dwarf beans 5 wk; bulb onions, carrots, parsnips, strawberries 6 wk; sugar and fodder beet, leeks, mangels, potatoes, field beans, swedes, Brussels sprouts, winter field beans 8 wk; oilseed rape, soya beans, linseed 12 wk
- Recommended time of application varies with crop. See label for details
- On peas a crystal violet wax test should be done if leaf wax likely to have been affected by weather conditions or other chemical treatment. The wax test is essential if other products are to be sprayed before or after treatment
- May be used on ornamental bulbs when crop 5-10 cm tall. Product has been used on tulips, narcissi, hyacinths and irises but some subjects may be more sensitive and growers advised to check tolerance on small number of plants before treating the rest of the crop
- May be applied to land temporarily removed from production where the green cover is made up predominantly of tolerant crops listed on label. Use on industrial crops of linseed and oilseed rape on land temporarily removed from production also permitted.

Following crops guidance
- Guideline intervals for sowing succeeding crops after failed treated crop: field beans, peas, sugar beet, rape, kale, swedes, radish, white clover, lucerne 1 wk; dwarf French beans 4 wk; wheat, barley, maize 8 wk
- Oats should not be sown after failure of a treated crop

Environmental safety
- Dangerous for the environment
- Toxic to aquatic organisms
- Harmful to fish or other aquatic life. Do not contaminate surface waters or ditches with chemical or used container

Hazard classification and safety precautions
Hazard Harmful, Dangerous for the environment
Transport code 9
Packaging group III
UN Number 3082
Risk phrases H304, H315, H318, H336, H361
Operator protection A, C; U05a, U08, U20b
Environmental protection E15a, E38, H411
Storage and disposal D01, D02, D09a, D10c, D12a
Medical advice M05b

117 Cydia pomonella GV

An baculovirus insecticide for codling moth control

Products

1	Carpovirusine	Arysta	1.0 x10^{13} GV/l	SC	15243
2	Carpovirusine EVO 2	Arysta	1.0 x10^{13} GV/l	SC	17565
3	Cyd-X Xtra	Certis	1% w/w virus	SC	17020

SEE SECTION 3 FOR PRODUCTS ALSO REGISTERED

Pesticide Profiles

Uses
- Codling moth in **apples, pears**

Approval information
- Cydia pomonella included in Annex I under EC Regulation 1107/2009

Efficacy guidance
- Optimum results require a precise spraying schedule commencing soon after egg laying and before first hatch
- First generation of codling moth normally occurs in first 2 wk of Jun in UK
- Apply in sufficient water to achieve good coverage of whole tree
- Normally 3 applications at intervals of 8 sunny days are needed for each codling moth generation
- Fruit damage is reduced significantly in the first yr and will reduce the population of codling moth in the following yr

Restrictions
- Maximum number of treatments 3 per season
- Do not tank mix with copper or any products with a pH lower than 5 or higher than 8
- Do not use as the exclusive measure for codling moth control. Adopt an anti-resistance strategy. See Section 5 for more information
- Consult processor before use on crops for processing

Crop-specific information
- HI for apples and pears 1 day [1]
- Evidence of efficacy and crop safety on pears is limited [3]

Environmental safety
- Product has no adverse effect on predatory insects and is fully compatible with IPM programmes

Hazard classification and safety precautions
Hazard Harmful [3]
UN Number N/C
Risk phrases H317 [2]; R42, R43 [3]
Operator protection A, D, H; U05a, U16b [1, 2]; U14, U19a [1-3]
Environmental protection E15a [3]; E15b, E34 [1, 2]
Storage and disposal D01, D02, D09a, D10b [1, 2]; D12a [1-3]
Medical advice M05a [1, 2]

118   cyflufenamid

An amidoxime fungicide for cereals
FRAC mode of action code: U6

Products

1	Cosine	Certis	50 g/l	EW	16404
2	Cyflamid	Certis	50 g/l	EW	12403
3	Takumi SC	Certis	100 g/l	SC	16000
4	Vegas	Certis	50 g/l	EW	15238

Uses
- Disease control in **courgettes, cucumbers, melons, summer squash** [3]
- Powdery mildew in **amenity vegetation** *(off-label)*, **forest nurseries** *(off-label)*, **grass seed crops** *(off-label)*, **hops** *(off-label)*, **rye** *(off-label)*, **soft fruit** *(off-label)* [2]; **apples, pears** [1]; **courgettes, cucumbers, melons, protected tomatoes** *(off-label)*, **strawberries** *(off-label)*, **summer squash** [3]; **durum wheat, spring barley, spring oats, spring wheat, triticale, winter barley, winter oats, winter rye, winter wheat** [2, 4]; **ornamental plant production** *(off-label)* [2, 3]

Extension of Authorisation for Minor Use (EAMUs)
- **amenity vegetation** *20070512* [2]
- **forest nurseries** *20082915* [2]
- **grass seed crops** *20060602* [2]

- **hops** *20082915* [2]
- **ornamental plant production** *20070512* [2], *20131294* [3]
- **protected tomatoes** *20140800* [3]
- **rye** *20060602* [2]
- **soft fruit** *20082915* [2]
- **strawberries** *20162055* [3]

Approval information
- Cyflufenamid included in Annex 1 under EC Regulation 1107/2009
- Accepted by BBPA on malting barley

Efficacy guidance
- Best results obtained from treatment at the first visible signs of infection by powdery mildew
- Sustained disease pressure may require a second treatment
- Disease spectrum may be broadened by appropriate tank mixtures. See label
- Product is rainfast within 1 h
- Must be used as part of an Integrated Crop Management programme that includes alternating use, or mixture, with fungicides with a different mode of action effective against powdery mildew

Restrictions
- Maximum number of treatments 2 per crop on all recommended cereals
- Apply only in the spring
- Do not apply to crops under stress from drought, waterlogging, cold, pests or diseases, lime or nutrient deficiency or other factors affecting crop growth

Crop-specific information
- Latest use: before start of flowering of cereal crops

Following crops guidance
- No restrictions

Environmental safety
- Dangerous for the environment
- Very toxic to aquatic organisms
- Avoid drift onto ponds, waterways or ditches

Hazard classification and safety precautions
Hazard Harmful [1]; Dangerous for the environment [1-4]
Transport code 9 [1, 2, 4]
Packaging group III [1, 2, 4]
UN Number 3082 [1, 2, 4]; N/C [3]
Risk phrases R22a [1]
Operator protection A; U05a
Environmental protection E15b, E34, H411 [1-4]; E17a [1] (5 m); E38 [1]
Storage and disposal D01, D02, D05, D10c, D12b
Medical advice M05b [1]

119 cymoxanil

A cyanoacetamide oxime fungicide for potatoes
FRAC mode of action code: 27

See also chlorothalonil + cymoxanil
cyazofamid + cymoxanil

Products
1	Cymbal 45	Belchim	45% w/w	WG	16647
2	Drum	Belchim	45% w/w	WG	16750
3	Option	DuPont	60% w/w	WG	16959
4	Sacron WG	UPL Europe	45% w/w	WG	16433

SEE SECTION 3 FOR PRODUCTS ALSO REGISTERED

Pesticide Profiles

Products – continued

5 Sipcam C50 WG   Sipcam   50% w/w   WG   16743

Uses
- Blight in *early potatoes* [5]; *potatoes* [1-5]
- Downy mildew in *hops* (off-label), *wine grapes* (off-label) [3]

Extension of Authorisation for Minor Use (EAMUs)
- *hops* 20161101 [3]
- *wine grapes* 20160542 [3]

Approval information
- Cymoxanil included in Annex I under EC Regulation 1107/2009
- Accepted by BBPA for use on hops

Efficacy guidance
- Product to be used in mixture with specified mixture partners in order to combine systemic and protective activity. See labels for details
- Commence spray programme as soon as weather conditions favourable for disease development occur and before infection appears. At latest the first treatment should be made as the foliage meets along the rows
- Repeat treatments at 7-14 day intervals according to disease incidence and weather conditions
- Treat irrigated crops as soon as possible after irrigation and repeat at 10 d intervals

Restrictions
- Minimum spray interval 7 d
- Do not allow packs to become wet during storage and do not keep opened packs from one season to another
- Cymoxanil must be used in tank mixture with other fungicides

Crop-specific information
- HI for potatoes 1 d [3], 7 d [1, 2, 5], 14 d [4]
- Consult processor before treating potatoes grown for processing

Environmental safety
- Dangerous for the environment
- Very toxic to aquatic organisms

Hazard classification and safety precautions
Hazard Harmful [1, 3-5]; Irritant [2]; Dangerous for the environment [1-4]; Harmful if swallowed, Very toxic to aquatic organisms [3]
Transport code 9 [1-4]
Packaging group III [1-4]
UN Number 3077 [1-4]; N/C [5]
Risk phrases H317 [2, 3]; H319, H361, H373 [2]; R22a, R43, R51, R53a [1, 4, 5]; R48, R62, R63 [1, 4]
Operator protection A [1-5]; H [1, 2, 4]; U05a [1-5]; U08 [3, 5]; U09a [1, 2, 4]; U14 [1, 2, 4, 5]; U19a [3]; U20a [1]; U20b [2-5]
Environmental protection E13a, E34 [3]; E13c [5]; E15a [1, 2, 4]; E38 [1-4]; H410 [2, 3]
Storage and disposal D01, D02, D09a, D11a [1-5]; D12a [1-4]
Medical advice M03 [3, 5]; M05a [1, 4]

120 cymoxanil + famoxadone

A preventative and curative fungicide mixture for potatoes
FRAC mode of action code: 27 + 11

See also famoxadone

Products
Tanos   DuPont   25:25% w/w   WG   17160

FOR FULL CONDITIONS OF USE ALWAYS READ THE PRODUCT LABEL

cymoxanil + fluazinam

Uses
- Blight in **potatoes**

Approval information
- Cymoxanil and famoxadone included in Annex I under EC Regulation 1107/2009

Efficacy guidance
- Commence spray programme before infection appears as soon as weather conditions favourable for disease development to occur. At latest the first treatment should be made as the foliage meets along the rows
- Repeat treatments at 7-10 day intervals according to disease incidence and weather conditions
- Reduce spray interval if conditions are conducive to the spread of blight
- Spray as soon as possible after irrigation
- Increase water volume in dense crops
- Product combines active substances with different modes of action and is effective against strains of potato blight that are insensitive to phenylamide fungicides
- To minimise the likelihood of development of resistance to QoI fungicides these products should be used in a planned Resistance Management strategy. See Section 5 for more information

Restrictions
- Maximum number of treatments 6 per crop
- Consult processors before using on crops for processing
- Do not apply more than 3 consecutive treatments containing QoI active fungicides

Crop-specific information
- HI 14 d for potatoes

Environmental safety
- Dangerous for the environment
- Very toxic to aquatic organisms
- Dangerous to fish or other aquatic life. Do not contaminate surface waters or ditches with chemical or used container
- Buffer zone requirement 12m [1]
- LERAP Category B

Hazard classification and safety precautions
  Hazard Harmful, Dangerous for the environment, Harmful if swallowed, Very toxic to aquatic organisms
  Transport code 9
  Packaging group III
  UN Number 3077
  Risk phrases H317, H361, H373
  Operator protection A, D; U05a, U20b
  Environmental protection E13b, E16a, E16b, E34, E38, H410
  Storage and disposal D01, D02, D09a, D10b, D12a
  Medical advice M03

121 cymoxanil + fluazinam

A fungicide mixture for blight control in potatoes
FRAC mode of action code: 27 + 29

Products

1	Grecale	Sipcam	200:300 g/l	SC	17002
2	Kunshi	Belchim	25:37.5% w/w	WG	16991
3	Plexus	Headland	200:300 g/l	SC	17099

Uses
- Blight in **potatoes**
- Late blight in **potatoes**
- Phytophthora in **potatoes**

SEE SECTION 3 FOR PRODUCTS ALSO REGISTERED

Approval information
- Cymoxanil and fluazinam included in Annex 1 under EC Regulation 1107/2009

Restrictions
- Use as part of a programme with fungicides acting via a different mode of action to reduce the risk of resistance

Following crops guidance
- No restrictions on following crops

Environmental safety
- Buffer zone requirement 6 m [1, 2], 7 m [3]
- LERAP Category B [3]

Hazard classification and safety precautions
Hazard Dangerous for the environment [1]; Harmful if swallowed [1, 3]; Very toxic to aquatic organisms [1-3]
Transport code 9
Packaging group III
UN Number 3077 [2]; 3082 [1, 3]
Risk phrases H317, H361 [1-3]; H319 [2]; H373 [1, 2]
Operator protection A, C, H [1, 3]; U05a, U20d [3]
Environmental protection E15b, E16a, E38 [3]; H410 [1-3]
Medical advice M03 [3]

122 cymoxanil + fludioxonil + metalaxyl-M

A fungicide seed dressing for peas
FRAC mode of action code: 27 + 12 + 4

See also fludioxonil
metalaxyl-M

Products
Wakil XL   Syngenta   10:5:17.5% w/w   WS   17217

Uses
- Downy mildew in **combining peas**, **vining peas**

Approval information
- Cymoxanil, fludioxonil and metalaxyl-M included in Annex I under EC Regulation 1107/2009

Efficacy guidance
- Apply through continuous flow seed treaters which should be calibrated before use

Restrictions
- Max number of treatments 1 per seed batch
- Ensure moisture content of treated seed satisfactory and store in a dry place
- Check calibration of seed drill with treated seed before drilling and sow as soon as possible after treatment
- Consult before using on crops for processing

Crop-specific information
- Latest use: pre-drilling

Environmental safety
- Harmful to aquatic organisms
- Do not use treated seed as food or feed

Hazard classification and safety precautions
UN Number N/C
Risk phrases H361; H373
Operator protection A D H U02a, U05a, U20b

FOR FULL CONDITIONS OF USE ALWAYS READ THE PRODUCT LABEL

cymoxanil + mancozeb

Environmental protection H410
Storage and disposal D01, D02, D05, D07, D09a, D11a, D12a
Treated seed S02, S04b, S05, S07

123 cymoxanil + mancozeb

A protectant and systemic fungicide for potato blight control
FRAC mode of action code: 27 + M3

See also mancozeb

Products

1	Curzate M WG	DuPont	4.5:68% w/w	WG	11901
2	Nautile DG	UPL Europe	5:68% w/w	WG	16653
3	Profilux WG	Belchim	4.5:68% w/w	WG	16125
4	Rhapsody	DuPont	4.5:68% w/w	WG	11958
5	Video DG	UPL Europe	5:68% w/w	WG	16685
6	Zetanil WG	Sipcam	4.5:65% w/w	WG	15488

Uses
- Blight in **potatoes** [1-6]
- Disease control in **hops** *(off-label)* [1]
- Late blight in **potatoes** [3]

Extension of Authorisation for Minor Use (EAMUs)
- **hops** 20162684 [1]

Approval information
- Cymoxanil and mancozeb included in Annex I under EC Regulation 1107/2009

Efficacy guidance
- Apply immediately after blight warning or as soon as local conditions dictate and repeat at 7-14 d intervals until haulm dies down or is burnt off
- Spray interval should not be more than 7-10 d in irrigated crops (see product label). Apply treatment after irrigation
- To minimise the likelihood of development of resistance these products should be used in a planned Resistance Management strategy. See Section 5 for more information

Restrictions
- Not specified for most products
- Do not apply at less than 7 d intervals
- Do not allow packs to become wet during storage

Crop-specific information
- Destroy and remove any haulm that remains after harvest of early varieties to reduce blight pressure on neighbouring maincrop potatoes

Following crops guidance
- H I 7 days [3]

Environmental safety
- Dangerous for the environment
- Very toxic to aquatic organisms
- Keep product away from fire or sparks
- Buffer zone requirement 6m [2, 5]

Hazard classification and safety precautions
  Hazard Harmful [1, 3, 4, 6]; Irritant [2, 5]; Dangerous for the environment [1-6]; Very toxic to aquatic organisms [2, 3, 5]
  Transport code 9
  Packaging group III
  UN Number 3077

SEE SECTION 3 FOR PRODUCTS ALSO REGISTERED

Pesticide Profiles

Risk phrases H317, H361 [2, 3, 5]; R36, R42 [1, 4]; R37, R43, R50, R53a [1, 4, 6]; R63 [6]
Operator protection A [1-6]; C [1, 3, 4, 6]; H [2, 5, 6]; U05a, U08, U14, U19a, U20b [1-6]; U11 [2, 5, 6]
Environmental protection E13c [1, 3, 4]; E15a, E38 [1-6]; H410 [2, 5]; H411 [3]
Storage and disposal D01, D02, D09a [1-6]; D06b, D12a [2, 5, 6]; D11a [1, 3, 4]
Medical advice M05a [5]

124 cymoxanil + mandipropamid

A fungicide mixture for use in potatoes

See also cymoxanil

Products
Carial Flex   Syngenta   18:25% w/w   WG   16629

Uses
- Blight in **potatoes**
- Late blight in **potatoes**

Approval information
- Cymoxanil and mandipropamid are included in Annex 1 under EC Regulation 1107/2009

Efficacy guidance
- Where possible, use an alternating strategy using fungicides from different mode of action groups.
- Where CAA fungicides are applied as a mixture (co-formulated or as a tank mix) up to six applications (or max. of 50% of the total number of applications) may be made per crop or season.
- No more than 3 applications of any CAA fungicide should be made consecutively

Restrictions
- Application may only be made between 1 May and 31 August [1]

Hazard classification and safety precautions
Transport code 9
Packaging group III
UN Number 3077
Risk phrases R22a, R43, R48, R52, R53a, R62, R63
Operator protection A, H; U05a, U08, U20b
Environmental protection E15b, E34, E38
Storage and disposal D01, D02, D09a, D10c, D12a, D19
Medical advice M03, M05a

125 cymoxanil + zoxamide

A protectant and systemic fungicide for blight control in potatoes
FRAC mode of action code: 27 + 22

See also cymoxanil
dimethomorph + zoxamide

Products
Lieto   Sipcam   33:33% w/w   WG   16703

Uses
- Blight in **potatoes**

Approval information
- Cymoxanil and zoxamide included in Annex I under EC Regulation 1107/2009

FOR FULL CONDITIONS OF USE ALWAYS READ THE PRODUCT LABEL

cypermethrin

Environmental safety
- Buffer zone requirement 6 m [1]

Hazard classification and safety precautions
Hazard Harmful, Dangerous for the environment
Transport code 9
Packaging group III
UN Number 3077
Risk phrases R22a, R43, R50, R53a
Operator protection A, C, H; U05a, U11, U14, U15, U19a
Environmental protection E38
Storage and disposal D01, D02, D12a, D12b
Medical advice M05a

126 cypermethrin

A contact and stomach acting pyrethroid insecticide
IRAC mode of action code: 3

Products

1	Cythrin 500 EC	Arysta	500 g/l	EC	16993
2	Forester	Fargro	100 g/l	EW	13164
3	Vazor Cypermethrin 10	Killgerm	100 g/l	EW	H10085

Uses
- Aphids in *asparagus, broccoli, brussels sprouts, cabbages, calabrese, carrots, cauliflowers, celeriac, dwarf beans, edible podded peas, forage maize, french beans, grain maize, horseradish, linseed, lupins, mustard, ornamental plant production, parsley root, parsnips, radishes, runner beans, salsify, spelt, spring barley* (autumn sown), *spring oats, spring rye, spring wheat* (autumn sown), *triticale, vining peas, winter barley, winter oats, winter rye, winter wheat* [1]
- Barley yellow dwarf virus vectors in *spring barley* (autumn sown), *spring oats, spring rye, spring wheat* (autumn sown), *triticale, winter barley, winter oats, winter rye, winter wheat* [1]
- Beetles in *cut logs* [2]
- Bladder pod midge in *spring oilseed rape, winter oilseed rape* [1]
- Cabbage stem flea beetle in *winter oilseed rape* [1]
- Carrot fly in *carrots, celeriac, horseradish, parsley root, parsnips* [1]
- Caterpillars in *broccoli, brussels sprouts, cabbages, calabrese, cauliflowers* [1]
- Cutworms in *fodder beet, mangels, potatoes, red beet, sugar beet, swedes, turnips* [1]
- Flies in *agricultural premises, farm buildings, outdoor crops* [3]
- Insect pests in *asparagus, carrots, celeriac, dwarf beans, edible podded peas, french beans, grain maize, horseradish, linseed, lupins, mustard, ornamental plant production, parsley root, parsnips, radishes, runner beans, salsify, spelt* [1]
- Large pine weevil in *forest* [2]
- Moths in *agricultural premises, farm buildings, outdoor crops* [3]
- Pea and bean weevil in *combining peas, spring field beans, vining peas, winter field beans* [1]
- Pea moth in *vining peas* [1]
- Pollen beetle in *spring oilseed rape, winter oilseed rape* [1]
- Rape winter stem weevil in *winter oilseed rape* [1]
- Seed weevil in *spring oilseed rape, winter oilseed rape* [1]
- Wasps in *agricultural premises, farm buildings, outdoor crops* [3]
- Whitefly in *broccoli, brussels sprouts, cabbages, calabrese, cauliflowers* [1]
- Yellow cereal fly in *spring barley* (autumn sown), *spring oats, spring rye, spring wheat* (autumn sown), *triticale, winter barley, winter oats, winter rye, winter wheat* [1]

Approval information
- Cypermethrin included in Annex I under EC Regulation 1107/2009
- Accepted by BBPA for use on malting barley and hops

SEE SECTION 3 FOR PRODUCTS ALSO REGISTERED

Efficacy guidance
- Products combine rapid action, good persistence, and high activity on Lepidoptera.
- As effect is mainly via contact, good coverage is essential for effective action. Spray volume should be increased on dense crops
- A repeat spray after 10-14 d is needed for some pests of outdoor crops, several sprays at shorter intervals for whitefly and other glasshouse pests
- Rates and timing of sprays vary with crop and pest. See label for details [2]
- Where aphids in hops, pear suckers or glasshouse whitefly resistant to cypermethrin occur control is unlikely to be satisfactory

Restrictions
- Maximum number of treatments varies with crop and product. See label or approval notice for details [2]
- Apply by knapsack or hand-held sprayer [2]

Crop-specific information
- HI vining peas 7 d; other crops 0 d
- Test spray sample of new or unusual ornamentals or trees before committing whole batches [2]
- Post-planting forestry applications should be made before damage is seen or at the onset of damage during the first 2 yrs after transplanting [2]

Environmental safety
- Dangerous for the environment
- Very toxic to aquatic organisms
- High risk to bees. Do not apply to crops in flower, or to those in which bees are actively foraging, except as directed. Do not apply when flowering weeds are present
- Give local beekeepers warning when using in flowering crops
- Flammable
- Do not spray cereals after 31 Mar within 6 m of the edge of the growing crop [2]
- Buffer zone requirement 18 m, LERAP Category A [1]

Hazard classification and safety precautions
Hazard Harmful, Dangerous for the environment [1, 2]; Flammable [1]; Harmful if swallowed [3]
Transport code 3 [1]; 9 [2, 3]
Packaging group III
UN Number 1993 [1]; 3082 [2, 3]
Risk phrases H317 [3]; R22a, R38, R50, R53a [1, 2]; R22b, R36, R37, R41, R66, R67 [1]; R43 [2]
Operator protection A [1-3]; C, D [3]; H [1, 2]; M [2]; U02a, U04a, U14, U16b [2]; U05a, U19a [1-3]; U08, U20b [1]; U10 [2, 3]
Environmental protection E12c, E12f, E34 [1, 2]; E13a, E16c, E16d, E16h [1]; E15a, E38 [2]; E17b [1] (18 m); H410 [3]
Storage and disposal D01, D02, D05, D10b [1, 2]; D09a, D12a [1-3]
Medical advice M03 [2]; M03a [3]; M05b [1]

127 cypermethrin + tetramethrin

A pyrethroid insecticide mixture for control of crawling and flying insects in a variety of situations
IRAC mode of action code: 3

See also cypermethrin

Products
Vazor Cypermax Plus   Killgerm   10% + 5% w/w   ME   H9988

Uses
- Ants in **agricultural premises, manure heaps, poultry houses, refuse tips**
- Bedbugs in **agricultural premises, manure heaps, poultry houses, refuse tips**
- Cockroaches in **agricultural premises, manure heaps, poultry houses, refuse tips**
- Mosquitoes in **agricultural premises, manure heaps, poultry houses, refuse tips**
- Moths in **agricultural premises, manure heaps, poultry houses, refuse tips**
- Silverfish in **agricultural premises, manure heaps, poultry houses, refuse tips**

FOR FULL CONDITIONS OF USE ALWAYS READ THE PRODUCT LABEL

cyproconazole + penthiopyrad

- Wasps in *agricultural premises*, *manure heaps*, *poultry houses*, *refuse tips*

Approval information
- Cypermethrin is included but tetramethrin not included in Annex 1 under EC Regulation 1107/2009

Hazard classification and safety precautions
Transport code 9
Packaging group III
UN Number 3082
Operator protection A, C, D, H; U20b
Environmental protection E05c, E12b, H410
Storage and disposal D09a, D12a

128   cyproconazole + penthiopyrad

An SDHI and triazole mixture for disease control in wheat
FRAC mode of action code: 3 + 7

See also penthiopyrad

Products
Cielex	DuPont	60:150 g/l	SC	16540

Uses
- Septoria leaf blotch in *spring wheat*, *winter wheat*

Approval information
- Cyproconazole and penthiopyrad are included in Annex 1 under EC Regulation 1107/2009

Efficacy guidance
- Disease control is more effective when used as a protectant treatment or during the earliest stages of disease development

Restrictions
- To reduce the risk of resistance, only two applications of any SDHI fungicide may be made to a cereal crop.

Hazard classification and safety precautions
Transport code 9
Packaging group III
UN Number 3082
Risk phrases R51, R53a, R63
Operator protection A; U05a, U09b, U11, U20a
Environmental protection E15b, E38
Storage and disposal D01, D02, D09a, D09b, D10c, D12a, D12b
Medical advice M03a

129   cyproconazole + picoxystrobin

A broad spectrum triazole + strobilurin fungicide mixture for use in cereals
FRAC mode of action code: 3 + 11

See also cyproconazole

Products
Furlong	DuPont	80:200 g/l	SC	17184

Uses
- Brown rust in *spring barley*, *spring wheat*, *winter barley*, *winter wheat*
- Centrospora in *sugar beet* *(moderate control)*
- Cercospora leaf spot in *fodder beet* *(off-label)*

SEE SECTION 3 FOR PRODUCTS ALSO REGISTERED

Pesticide Profiles

- Crown rust in **spring oats, winter oats**
- Glume blotch in **spring wheat, winter wheat**
- Net blotch in **spring barley** *(moderate control)*, **winter barley** *(moderate control)*
- Powdery mildew in **fodder beet** *(off-label)*, **sugar beet**
- Ramularia leaf spots in **fodder beet** *(off-label)*, **sugar beet** *(moderate control)*
- Rhynchosporium in **spring barley** *(moderate control)*, **winter barley** *(moderate control)*
- Rust in **fodder beet** *(off-label)*, **sugar beet**
- Sclerotinia in **spring oilseed rape** *(moderate control)*, **winter oilseed rape** *(moderate control)*
- Septoria leaf blotch in **spring wheat, winter wheat**
- Yellow rust in **spring wheat, winter wheat**

Extension of Authorisation for Minor Use (EAMUs)
- *fodder beet* 20161187

Approval information
- Picoxystrobin and cyproconazole included in Annex I under EC Regulation 1107/2009
- Accepted by BBPA for use on malting barley

Restrictions
- To reduce the risk of resistance developing in target diseases the total number of applications of products containing QoI fungicides made to any cereal crop must not exceed two.
- Maximum number of applications is 2 per crop

Hazard classification and safety precautions
  Hazard Harmful, Dangerous for the environment, Very toxic to aquatic organisms
  Transport code 9
  Packaging group III
  UN Number 3082
  Risk phrases H317, H361
  Operator protection A, H
  Environmental protection E15a, E34, E38, H410
  Storage and disposal D01, D02, D05, D09a, D12a

130   cyproconazole + propiconazole

A broad-spectrum mixture of conazole fungicides for wheat and barley
FRAC mode of action code: 3 + 3

See also propiconazole

Products
  Menara   Syngenta   160:250 g/l   EC   14398

Uses
- Brown rust in **spring barley, winter barley, winter wheat**
- Disease control in **durum wheat** *(off-label)*, **rye** *(off-label)*, **triticale** *(off-label)*
- Glume blotch in **winter wheat**
- Late ear diseases in **winter wheat**
- Net blotch in **spring barley, winter barley**
- Powdery mildew in **spring barley, winter barley**
- Rhynchosporium in **spring barley, winter barley**
- Septoria leaf blotch in **winter wheat**
- Yellow rust in **winter wheat**

Extension of Authorisation for Minor Use (EAMUs)
- *durum wheat* 20161510 expires 30 Nov 2017
- *rye* 20161510 expires 30 Nov 2017
- *triticale* 20161510 expires 30 Nov 2017

Approval information
- Propiconazole and cyproconazole included in Annex I under EC Regulation 1107/2009

FOR FULL CONDITIONS OF USE ALWAYS READ THE PRODUCT LABEL

- Accepted by BBPA for use on malting barley
- Approval expiry 30 Nov 2017 [1]

Efficacy guidance
- Treatment should be made at first sign of disease
- Useful reduction of eyespot when applied in spring during stem extension (GS 30-32)
- Cyproconazole and propiconazole are DMI fungicides. Resistance to some DMI fungicides has been identified in Septoria leaf blotch which may seriously affect performance of some products. For further advice contact a specialist advisor and visit the Fungicide Resistance Action Group (FRAG)-UK website

Restrictions
- Maximum total dose equivalent to one full dose treatment for barley; two full dose treatments for winter wheat

Crop-specific information
- Latest use: before beginning of anthesis (GS 60) for barley; before grain watery ripe (GS 71) for wheat
- Application to winter wheat in spring may cause straw shortening but does not cause loss of yield

Environmental safety
- Dangerous for the environment
- Very toxic to aquatic organisms

Hazard classification and safety precautions
Hazard Harmful, Dangerous for the environment
Transport code 9
Packaging group III
UN Number 3082
Risk phrases R50, R53a, R63
Operator protection A, C, H; U05a, U08, U20b
Environmental protection E15a, E38
Storage and disposal D01, D02, D05, D07, D09a, D10c, D12a

131 cyproconazole + trifloxystrobin

A conazole and strobilurin fungicide mixture for disease control in sugar beet
FRAC mode of action code: 3 + 11

See also trifloxystrobin

Products

1	Clayton Cast	Clayton	160:375 g/l	SC	17727
2	Escolta	Bayer CropScience	160:375 g/l	SC	17397

Uses
- Cercospora leaf spot in **fodder beet**, **sugar beet** [1, 2]; **red beet** (off-label) [2]
- Powdery mildew in **fodder beet**, **sugar beet** [1, 2]; **red beet** (off-label) [2]
- Ramularia leaf spots in **fodder beet**, **sugar beet** [1, 2]; **red beet** (off-label) [2]
- Rust in **fodder beet**, **sugar beet** [1, 2]; **red beet** (off-label) [2]

Extension of Authorisation for Minor Use (EAMUs)
- **red beet** 20161089 [2]

Approval information
- Cyproconazole and trifloxystrobin included in Annex I under EC Regulation 1107/2009

Efficacy guidance
- Best results obtained from treatment at early stages of disease development. Further treatment may be needed if disease attack is prolonged

---

SEE SECTION 3 FOR PRODUCTS ALSO REGISTERED

Pesticide Profiles

- Trifloxystrobin is a member of the QoI cross resistance group. Product should be used preventatively and not relied on for its curative potential
- Use product as part of an Integrated Crop Management strategy incorporating other methods of control, including where appropriate other fungicides with a different mode of action.
- Strains of barley powdery mildew resistant to QoI's are common in the UK

Restrictions
- Maximum total dose per crop equivalent to two full dose treatments
- First application not before BBCH 31 (10% ground cover); second application not before BBCH 40 (full canopy & start of harvestable root development); maintain at least 21 day interval between sprays.

Crop-specific information
- HI 35 d

Environmental safety
- Dangerous for the environment
- Very toxic to aquatic organisms
- LERAP Category B

Hazard classification and safety precautions
Hazard Harmful, Dangerous for the environment, Very toxic to aquatic organisms
Transport code 9
Packaging group III
UN Number 3082
Risk phrases H361
Operator protection A; U02a, U05a, U09a, U19a, U20a
Environmental protection E15b, E16a, E38, H410
Storage and disposal D01, D02, D05, D09a, D10b, D12a
Medical advice M03

132 cyprodinil

An anilinopyrimidine systemic broad spectrum fungicide for cereals
FRAC mode of action code: 9

See also cyproconazole + cyprodinil

Products

| 1 | Coracle | AgChem Access | 300 g/l | EC | 15866 |
| 2 | Kayak | Syngenta | 300 g/l | EC | 14847 |

Uses
- Disease control in **forest nurseries** *(off-label)* [2]
- Eyespot in **spring barley**, **winter barley** [1, 2]
- Net blotch in **spring barley**, **winter barley** [1, 2]
- Powdery mildew in **spring barley**, **winter barley** [1, 2]
- Rhynchosporium in **spring barley**, **winter barley** [1, 2]

Extension of Authorisation for Minor Use (EAMUs)
- *forest nurseries* 20122056 [2]

Approval information
- Cyprodinil included in Annex I under EC Regulation 1107/2009
- Accepted by BBPA for use on malting barley

Efficacy guidance
- Best results obtained from treatment at early stages of disease development
- For best control of eyespot spray before or during the period of stem extension in spring. Control may be reduced if very dry conditions follow treatment

FOR FULL CONDITIONS OF USE ALWAYS READ THE PRODUCT LABEL

cyprodinil + fludioxonil

Restrictions
- Maximum number of treatments 2 per crop on wheat and barley [2]

Crop-specific information
- Latest use: before first spikelet of inflorescence just visible stage for barley, before early milk stage for wheat; up to and including first awns visible stage (GS49) in barley [2]

Environmental safety
- Dangerous for the environment
- Very toxic to aquatic organisms
- LERAP Category B

Hazard classification and safety precautions
Hazard Irritant, Dangerous for the environment
Transport code 9
Packaging group III
UN Number 3082
Risk phrases R38, R43, R50, R53a
Operator protection A, H; U05a, U09a, U20a
Environmental protection E15b, E16a, E34, E38
Storage and disposal D01, D02, D05, D07, D09a, D10c, D12a

133 cyprodinil + fludioxonil

A broad-spectrum fungicide mixture for fruit crops, legumes, some vegetables, forest nurseries and ornamentals
FRAC mode of action code: 9 + 12

See also fludioxonil

Products

1	Clayton Gear	Clayton	37.5:25% w/w	WG	17169
2	Switch	Syngenta	37.5:25 % w/w	WG	15129

Uses
- Alternaria in **apples**, **crab apples**, **pears**, **quinces** [1, 2]; **brassica leaves and sprouts** (off-label), **broccoli** (off-label), **brussels sprouts** (off-label), **cabbages** (off-label), **calabrese** (off-label), **cauliflowers** (off-label), **celery (outdoor)** (off-label), **collards** (off-label), **cress** (off-label), **endives** (off-label), **herbs (see appendix 6)** (off-label), **kale** (off-label), **lamb's lettuce** (off-label), **lettuce** (off-label), **protected baby leaf crops** (off-label), **protected cress** (off-label), **protected endives** (off-label), **protected herbs (see appendix 6)** (off-label), **protected lamb's lettuce** (off-label), **protected lettuce** (off-label), **protected radishes** (off-label), **protected red mustard** (off-label), **protected rocket** (off-label), **protected sorrel** (off-label), **protected spinach** (off-label), **protected spinach beet** (off-label), **red mustard** (off-label), **rocket** (off-label), **sorrel** (off-label), **spinach** (off-label), **spinach beet** (off-label), **spring cabbage** (off-label) [2]
- Alternaria blight in **carrots** (moderate control) [1, 2]
- Anthracnose in **protected blueberry** (off-label) [2]
- Ascochyta in **mange-tout peas**, **sugar snap peas**, **vining peas** [1, 2]
- Basal stem rot in **ornamental bulbs** (off-label) [2]
- Black spot in **protected strawberries** (qualified minor use), **strawberries** (qualified minor use) [1, 2]
- Blossom wilt in **apricots** (off-label), **cherries** (off-label), **peaches** (off-label), **plums** (off-label) [2]
- Botrytis in **apricots** (off-label), **asparagus** (off-label), **brassica leaves and sprouts** (off-label), **broccoli** (off-label), **brussels sprouts** (off-label), **bulb onions** (off-label), **cabbages** (off-label), **calabrese** (off-label), **cauliflowers** (off-label), **cherries** (off-label), **chicory** (off-label), **collards** (off-label), **cress** (off-label), **endives** (off-label), **garlic** (off-label), **grapevines** (off-label), **herbs (see appendix 6)** (off-label), **kale** (off-label), **lamb's lettuce** (off-label), **lettuce** (off-label), **lupins** (off-label), **peaches** (off-label), **plums** (off-label), **protected aubergines** (off-label), **protected baby leaf crops** (off-label), **protected blueberry** (off-label), **protected cress** (off-label), **protected cucumbers** (off-label), **protected endives** (off-label), **protected herbs (see appendix 6)** (off-label), **protected lamb's lettuce** (off-label), **protected lettuce** (off-label),

SEE SECTION 3 FOR PRODUCTS ALSO REGISTERED

Pesticide Profiles

protected peppers (off-label), *protected red mustard* (off-label), *protected rocket* (off-label), *protected sorrel* (off-label), *protected spinach* (off-label), *protected spinach beet* (off-label), *protected tomatoes* (off-label), *red mustard* (off-label), *rocket* (off-label), *salad onions* (off-label), *shallots* (off-label), *sorrel* (off-label), *spinach* (off-label), *spinach beet* (off-label), *spring cabbage* (off-label) [2]; *broad beans* (moderate control), *broad beans (dry-harvested)* (moderate control), *carrots* (moderate control), *celeriac* (moderate control), *combining peas* (moderate control), *forest nurseries, french beans (dry-harvested)* (moderate control), *green beans* (moderate control), *mange-tout peas* (moderate control), *ornamental plant production, protected forest nurseries, protected ornamentals, protected strawberries, runner beans* (moderate control), *strawberries, sugar snap peas* (moderate control), *vining peas* (moderate control) [1, 2]
- Botrytis fruit rot in *apples, blackberries, crab apples, pears, quinces, raspberries* [1, 2]
- Botrytis root rot in *bilberries* (qualified minor use recommendation), *blackcurrants* (qualified minor use recommendation), *blueberries* (qualified minor use recommendation), *cranberries* (qualified minor use recommendation), *gooseberries* (qualified minor use recommendation), *redcurrants* (qualified minor use recommendation), *whitecurrants* (qualified minor use recommendation) [1, 2]
- Fusarium in *protected cucumbers* (off-label) [2]
- Fusarium bulb rot in *ornamental bulbs* (off-label) [2]
- Fusarium diseases in *apples, crab apples, pears, quinces* [1, 2]
- Gloeosporium rot in *apples, crab apples, pears, quinces* [1, 2]
- Leaf spot in *celery (outdoor)* (off-label), *protected radishes* (off-label) [2]
- Mycosphaerella in *mange-tout peas, sugar snap peas, vining peas* [1, 2]
- Penicillium rot in *apples, crab apples, pears, quinces* [1, 2]
- Phoma in *brassica leaves and sprouts* (off-label), *broccoli* (off-label), *brussels sprouts* (off-label), *cabbages* (off-label), *calabrese* (off-label), *cauliflowers* (off-label), *collards* (off-label), *cress* (off-label), *endives* (off-label), *herbs (see appendix 6)* (off-label), *kale* (off-label), *lamb's lettuce* (off-label), *lettuce* (off-label), *protected baby leaf crops* (off-label), *protected cress* (off-label), *protected endives* (off-label), *protected herbs (see appendix 6)* (off-label), *protected lamb's lettuce* (off-label), *protected lettuce* (off-label), *protected red mustard* (off-label), *protected rocket* (off-label), *protected sorrel* (off-label), *protected spinach* (off-label), *protected spinach beet* (off-label), *red mustard* (off-label), *rocket* (off-label), *sorrel* (off-label), *spinach* (off-label), *spinach beet* (off-label), *spring cabbage* (off-label) [2]
- Rhizoctonia in *brassica leaves and sprouts* (off-label), *broccoli* (off-label), *brussels sprouts* (off-label), *cabbages* (off-label), *calabrese* (off-label), *cauliflowers* (off-label), *collards* (off-label), *cress* (off-label), *endives* (off-label), *herbs (see appendix 6)* (off-label), *kale* (off-label), *lamb's lettuce* (off-label), *lettuce* (off-label), *protected baby leaf crops* (off-label), *protected cress* (off-label), *protected endives* (off-label), *protected herbs (see appendix 6)* (off-label), *protected lamb's lettuce* (off-label), *protected lettuce* (off-label), *protected red mustard* (off-label), *protected rocket* (off-label), *protected sorrel* (off-label), *protected spinach* (off-label), *protected spinach beet* (off-label), *red mustard* (off-label), *rocket* (off-label), *sorrel* (off-label), *spinach* (off-label), *spinach beet* (off-label), *spring cabbage* (off-label) [2]
- Sclerotinia in *beetroot* (off-label), *celery (outdoor)* (off-label), *ginger* (off-label), *herbs for medicinal uses (see appendix 6)* (off-label), *horseradish* (off-label), *parsley root* (off-label), *parsnips* (off-label), *protected radishes* (off-label), *salsify* (off-label), *spice roots* (off-label), *turmeric* (off-label) [2]; *broad beans, broad beans (dry-harvested), carrots* (moderate control), *combining peas, french beans (dry-harvested), green beans, mange-tout peas, runner beans, sugar snap peas, vining peas* [1, 2]

Extension of Authorisation for Minor Use (EAMUs)
- *apricots* 20103092 [2]
- *asparagus* 20103173 [2]
- *beetroot* 20103087 [2]
- *brassica leaves and sprouts* 20150085 [2]
- *broccoli* 20150085 [2]
- *brussels sprouts* 20150085 [2]
- *bulb onions* 20122285 [2]
- *cabbages* 20150085 [2]
- *calabrese* 20150085 [2]

FOR FULL CONDITIONS OF USE ALWAYS READ THE PRODUCT LABEL

cyprodinil + fludioxonil

- *cauliflowers* 20150085 [2]
- *celery (outdoor)* 20150865 [2]
- *cherries* 20103092 [2]
- *chicory* 20122285 [2]
- *collards* 20150085 [2]
- *cress* 20150085 [2]
- *endives* 20150085 [2]
- *garlic* 20122285 [2]
- *ginger* 20103087 [2]
- *grapevines* 20112098 [2]
- *herbs (see appendix 6)* 20150085 [2]
- *herbs for medicinal uses (see appendix 6)* 20103087 [2]
- *horseradish* 20103087 [2]
- *kale* 20150085 [2]
- *lamb's lettuce* 20150085 [2]
- *lettuce* 20150085 [2]
- *lupins* 20103170 [2]
- *ornamental bulbs* 20152274 [2]
- *parsley root* 20103087 [2]
- *parsnips* 20103087 [2]
- *peaches* 20103092 [2]
- *plums* 20103092 [2]
- *protected aubergines* 20103172 [2]
- *protected baby leaf crops* 20150085 [2]
- *protected blueberry* 20160648 [2]
- *protected cress* 20150085 [2]
- *protected cucumbers* 20103171 [2]
- *protected endives* 20150085 [2]
- *protected herbs (see appendix 6)* 20150085 [2]
- *protected lamb's lettuce* 20150085 [2]
- *protected lettuce* 20150085 [2]
- *protected peppers* 20103172 [2]
- *protected radishes* 20150865 [2]
- *protected red mustard* 20150085 [2]
- *protected rocket* 20150085 [2]
- *protected sorrel* 20150085 [2]
- *protected spinach* 20150085 [2]
- *protected spinach beet* 20150085 [2]
- *protected tomatoes* 20110302 [2]
- *red mustard* 20150085 [2]
- *rocket* 20150085 [2]
- *salad onions* 20103174 [2]
- *salsify* 20103087 [2]
- *shallots* 20122285 [2]
- *sorrel* 20150085 [2]
- *spice roots* 20103087 [2]
- *spinach* 20150085 [2]
- *spinach beet* 20150085 [2]
- *spring cabbage* 20150085 [2]
- *turmeric* 20103087 [2]

Approval information
- Cyprodinil and fludioxonil included in Annex I under EC Regulation 1107/2009

Efficacy guidance
- Best results obtained from application at the earliest stage of disease development or as a protective treatment following disease risk assessment
- Subsequent treatments may follow after a minimum of 10 d if disease pressure remains high
- To minimise the likelihood of development of resistance product should be used in a planned Resistance Management strategy. See Section 5 for more information

SEE SECTION 3 FOR PRODUCTS ALSO REGISTERED

Restrictions
- Maximum number of treatments 2 per crop
- Consult processor before treating any crop for processing

Crop-specific information
- HI broad beans, green beans, mange-tout peas, runner beans, sugar snap peas, vining peas 14 d; protected strawberries, strawberries 3 d
- First signs of disease infection likely to be seen from early flowering for all crops
- Ensure peas are free from any stress before treatment. If necessary check wax with a crystal violet test

Environmental safety
- Dangerous for the environment
- Very toxic to aquatic organisms
- LERAP Category B

Hazard classification and safety precautions
Hazard Dangerous for the environment
Transport code 9
Packaging group III
UN Number 3077
Risk phrases R50, R53a
Operator protection A, H; U05a, U20a
Environmental protection E15b, E16a, E16b, E38
Storage and disposal D01, D02, D05, D07, D09a, D10c, D12a

134   cyprodinil + isopyrazam

A fungicide mixture for disease control in barley
FRAC mode of action code: 7 + 9

See also cyprodinil

Products
1	Bontima	Syngenta	187.5:62.5 g/l	EC	17188
2	Cebara	Syngenta	187.5:62.5 g/l	EC	17286
3	Concorde	Syngenta	150:62.5 g/l	EC	17248

Uses
- Brown rust in **spring barley, winter barley**
- Net blotch in **spring barley, winter barley**
- Powdery mildew in **spring barley, winter barley**
- Ramularia leaf spots in **spring barley, winter barley**
- Rhynchosporium in **spring barley, winter barley**

Approval information
- Cyprodinil and isopyrazam are included in Annex 1 under EC Regulation 1107/2009
- Accepted by BBPA for use on malting barley before ear emergence only

Restrictions
- Isopyrazam is an SDH respiration inhibitor; Do not apply more than two foliar applications of products containing an SDH inhibitor to any cereal crop

Crop-specific information
- There are no restrictions on succeeding crops in a normal rotation

Environmental safety
- LERAP Category B

Hazard classification and safety precautions
Hazard Harmful, Dangerous for the environment, Harmful if inhaled [1-3]; Very toxic to aquatic organisms [3]
Transport code 9

FOR FULL CONDITIONS OF USE ALWAYS READ THE PRODUCT LABEL

Packaging group III
UN Number 3082
Risk phrases H304, H361 [1-3]; H351 [3]
Operator protection A, C, H; U05a, U09a, U20b [1-3]; U11 [1, 2]
Environmental protection E15b, E16a, E16b, E38, H410
Storage and disposal D01, D02, D05, D09a, D10c, D11a, D12a
Medical advice M05b [3]

135   2,4-D

A translocated phenoxycarboxylic acid herbicide for cereals, grass and amenity use
HRAC mode of action code: O

See also amitrole + 2,4-D + diuron
clopyralid + 2,4-D + MCPA

Products

1	Depitone Ultra	Nufarm UK	600 g/l	EC	17615
2	Depitox	Nufarm UK	500 g/l	SL	13258
3	Dioweed 50	UPL Europe	500 g/l	SL	13197
4	Headland Staff 500	Headland	500 g/l	SL	13196
5	Herboxone	Headland	500 g/l	SL	13958
6	HY-D Super	Agrichem	500 g/l	SL	13198

Uses
- Annual dicotyledons in **amenity grassland**, **managed amenity turf**, **spring barley**, **spring wheat**, **winter barley**, **winter oats**, **winter rye**, **winter wheat** [1-6]; **apple orchards**, **farm forestry** *(off-label)*, **forest nurseries** *(off-label)*, **game cover** *(off-label)*, **ornamental plant production** *(off-label)*, **pear orchards** [2]; **grassland** [1-3]; **miscanthus** *(off-label)*, **undersown oats** [6]; **permanent grassland** [4-6]; **spring rye** [1, 2]; **undersown barley**, **undersown rye**, **undersown wheat** [2, 6]; **undersown spring cereals**, **undersown winter cereals** [3]
- Perennial dicotyledons in **amenity grassland**, **managed amenity turf**, **spring barley**, **spring wheat**, **winter barley**, **winter oats**, **winter rye**, **winter wheat** [1, 2, 4-6]; **apple orchards**, **farm forestry** *(off-label)*, **forest nurseries** *(off-label)*, **game cover** *(off-label)*, **ornamental plant production** *(off-label)*, **pear orchards** [2]; **grassland**, **spring rye** [1, 2]; **permanent grassland** [4-6]; **undersown barley**, **undersown rye**, **undersown wheat** [2, 6]; **undersown oats** [6]

Extension of Authorisation for Minor Use (EAMUs)
- *farm forestry* 20082843 [2]
- *forest nurseries* 20082843 [2]
- *game cover* 20082843 [2]
- *miscanthus* 20082914 [6]
- *ornamental plant production* 20082843 [2]

Approval information
- 2,4-D included in Annex I under EC Regulation 1107/2009
- Approved for aquatic weed control [2]. See Section 5 for information on use of herbicides in or near water
- Accepted by BBPA for use on malting barley

Efficacy guidance
- Best results achieved by spraying weeds in seedling to young plant stage when growing actively in a strongly competing crop
- Most effective stage for spraying perennials varies with species. See label for details

Restrictions
- Maximum number of treatments normally 1 per crop and in forestry, 1 per yr in grassland and 2 or 3 per yr in amenity turf. Check individual labels
- Do not use on newly sown leys containing clover
- Do not spray grass seed crops after ear emergence
- Do not spray within 6 mth of laying turf or sowing fine grass

SEE SECTION 3 FOR PRODUCTS ALSO REGISTERED

Pesticide Profiles

- Do not dump surplus herbicide in water or ditch bottoms
- Do not plant conifers until at least 1 mth after treatment
- Do not spray crops stressed by cold weather or drought or if frost expected
- Do not roll or harrow within 7 d before or after spraying
- Do not spray if rain falling or imminent
- Do not mow or roll turf or amenity grass 4 d before or after spraying. The first 4 mowings after treatment must be composted for at least 6 mth before use
- Do not mow grassland or graze for at least 10 d after spraying

Crop-specific information
- Latest use: before 1st node detectable in cereals; end Aug for conifer plantations; before established grassland 25 cm high
- Spray winter cereals in spring when leaf-sheath erect but before first node detectable (GS 31), spring cereals from 5-leaf stage to before first node detectable (GS 15-31)
- Cereals undersown with grass and/or clover, but not with lucerne, may be treated
- Selective treatment of resistant conifers can be made in Aug when growth ceased and plants hardened off, spray must be directed if applied earlier. See label for details

Following crops guidance
- Do not use shortly before or after sowing any crop
- Do not plant succeeding crops within 3 mth [4]
- Do not direct drill brassicas or grass/clover mixtures within 3 wk of application

Environmental safety
- Dangerous for the environment
- Very toxic to aquatic organisms [4]
- Dangerous to aquatic higher plants. Do not contaminate surface waters or ditches with chemical or used container
- 2,4-D is active at low concentrations. Take extreme care to avoid drift onto neighbouring crops, especially beet crops, brassicas, most market garden crops including lettuce and tomatoes under glass, pears and vines
- May be used to control aquatic weeds in presence of fish if used in strict accordance with directions for waterweed control and precautions needed for aquatic use [2]
- Keep livestock out of treated areas for at least 2 wk following treatment and until poisonous weeds, such as ragwort, have died down and become unpalatable
- Water containing the herbicide must not be used for irrigation purposes within 3 wk of treatment or until the concentration in water is below 0.05 ppm

Hazard classification and safety precautions
    Hazard Harmful, Dangerous for the environment [3-6]; Irritant [1, 2]; Harmful if swallowed [2, 3]
    Transport code 9 [3-5]
    Packaging group III [3-5]
    UN Number 3082 [3-5]; N/C [1, 2, 6]
    Risk phrases H315, H317 [1]; H318 [2, 3]; R22a, R41, R53a [4-6]; R43, R51 [6]; R50 [4, 5]
    Operator protection A, C, H [1-6]; M [1, 2]; U05a, U11 [1-6]; U08, U20b [1, 2, 4-6]; U09a, U20a [3]; U14, U15 [6]
    Environmental protection E07a, E34 [1-6]; E15a [1-3]; E15b, E38 [3-6]; H411 [1]
    Storage and disposal D01, D02, D05 [1-6]; D09a [1, 2, 4-6]; D10a [4-6]; D10b [1, 2]; D10c [3]; D12a [3-6]
    Medical advice M03 [1-6]; M05a [3, 6]

136 2,4-D + dicamba

A translocated herbicide for use on turf
HRAC mode of action code: O + O

See also dicamba

FOR FULL CONDITIONS OF USE ALWAYS READ THE PRODUCT LABEL

2,4-D + dicamba

Products

1	Landscaper Pro Weed Control + Fertilizer	Everris Ltd	0.8:0.12% w/w	GR	17239
2	Magneto	Nufarm UK	344:120 g/l	SL	17444
3	Thrust	Nufarm UK	344:120 g/l	SL	15408

Uses
- Annual dicotyledons in *amenity grassland* [2, 3]; *grassland* [2]; *managed amenity turf* [1]; *permanent grassland* [3]
- Buttercups in *managed amenity turf* [1]
- Perennial dicotyledons in *amenity grassland* [2, 3]; *grassland* [2]; *permanent grassland* [3]

Approval information
- 2,4-D and dicamba included in Annex I under EC Regulation 1107/2009
- Accepted by BBPA for use on malting barley

Efficacy guidance
- Best results achieved by application when weeds growing actively in spring or early summer (later with irrigation and feeding)
- More resistant weeds may need repeat treatment after 3 wk
- Improved control of some weeds can be obtained by use of specifed oil adjuvant [2, 3]
- Do not use during drought conditions

Restrictions
- Maximum number of treatments 2 per yr on amenity grass and established grassland [2, 3]
- Do not treat newly sown or turfed areas or grass less than 1 yr old
- Do not treat grass crops intended for seed production [2, 3]
- Do not treat grass suffering from drought, disease or other adverse factors [2, 3]
- Do not roll or harrow for 7 d before or after treatment [2, 3]
- Do not apply when grassland is flowering [2, 3]
- Avoid spray drift onto cultivated crops or ornamentals
- Do not graze grass for at least seven days after spraying [2, 3]
- Do not mow or roll four days before or after application. The first four mowings after treatment must be composted for at least six months before use [2, 3]

Crop-specific information
- Latest use: before grass 25 cm high for amenity grass, established grassland [3]
- The first four mowings after treatment must be composted for at least 6 mth before use

Environmental safety
- Dangerous for the environment [3]
- Toxic to aquatic organisms
- Keep livestock out of treated areas for at least two weeks following treatment and until poisonous weeds, such as ragwort, have died down and become unpalatable [2, 3]

Hazard classification and safety precautions
    Hazard Harmful, Dangerous for the environment [1-3]; Harmful if swallowed [2, 3]
    Transport code 9 [2, 3]
    Packaging group III [2, 3]
    UN Number 3082 [2, 3]
    Risk phrases H317, H318 [2, 3]; R22a, R22b, R36, R43, R51, R53a, R66 [1]
    Operator protection A, C, H, M [2, 3]; U05a [2, 3]; U08, U14, U15, U20b [1]; U11 [1-3]
    Environmental protection E07a, E15a, E38 [1-3]; E22c [2]; E34, H411 [2, 3]
    Storage and disposal D01, D02 [2, 3]; D05, D09a, D10a, D12a [1-3]
    Medical advice M03 [2, 3]; M05b [1]

SEE SECTION 3 FOR PRODUCTS ALSO REGISTERED

Pesticide Profiles

137   2,4-D + dicamba + fluroxypyr

A translocated and contact herbicide mixture for amenity turf
HRAC mode of action code: O + O + O

See also dicamba
fluroxypyr

Products

1	Holster XL	Barclay	285:52.5:105 g/l	EC	16407
2	Mascot Crossbar	Rigby Taylor	285:52.5:105 g/l	EC	16454
3	Speedster	Pan Amenity	285:52.5:105 g/l	EC	17025

Uses
- Annual dicotyledons in **amenity grassland, managed amenity turf** [1-3]
- Buttercups in **amenity grassland, managed amenity turf** [1]
- Clover in **amenity grassland, managed amenity turf** [1]
- Daisies in **amenity grassland, managed amenity turf** [1]
- Dandelions in **amenity grassland, managed amenity turf** [1]
- Yarrow in **amenity grassland, managed amenity turf** [1]

Approval information
- 2,4-D, dicamba and fluroxypyr included in Annex I under EC Regulation 1107/2009

Efficacy guidance
- Apply when weeds actively growing (normally between Apr and Sep) and when soil is moist
- Best results obtained from treatment in spring or early summer before weeds begin to flower
- Do not apply if turf is wet or if rainfall expected within 4 h of treatment. Both circumstances will reduce weed control

Restrictions
- Maximum number of treatments 2 per yr
- Do not mow for 3 d before or after treatment
- Avoid overlapping or overdosing, especially on newly sown turf
- Do not spray in drought conditions or if turf under stress from frost, waterlogging, trace element deficiency, pest or disease attack

Crop-specific information
- Latest use: normally Sep for managed amenity turf
- New turf may be treated in spring provided at least 2 mth have elapsed since sowing

Environmental safety
- Dangerous for the environment
- Toxic to aquatic organisms
- Keep livestock out of treated areas for at least two weeks following treatment and until poisonous weeds, such as ragwort, have died down and become unpalatable
- LERAP Category B

Hazard classification and safety precautions
Hazard Harmful, Dangerous for the environment
Transport code 9
Packaging group III
UN Number 3082
Risk phrases R22a, R22b, R36, R38, R51, R53a, R67
Operator protection A, C, H, M; U05a, U08, U11, U14, U15, U19a, U20b
Environmental protection E07a, E15a, E16a, E34, E38
Storage and disposal D01, D02, D09a, D10a
Medical advice M03, M05a, M05b

FOR FULL CONDITIONS OF USE ALWAYS READ THE PRODUCT LABEL

138 2,4-D + dicamba + MCPA + mecoprop-P

A translocated herbicide mixture for grassland
HRAC mode of action code: O + O + O + O

See also dicamba
MCPA
mecoprop-P

Products

1	Enforcer	Everris Ltd	70:20:70:42 g/l	SL	17274
2	Longbow	Bayer Environ.	70:20:70:42 g/l	SL	16528

Uses
- Buttercups in *managed amenity turf*
- Clovers in *managed amenity turf*
- Daisies in *managed amenity turf*
- Dandelions in *managed amenity turf*
- Self-heal in *managed amenity turf*

Approval information
- 2,4-D, dicamba, MCPA and mecoprop-P included in Annex I under EC Regulation 1107/2009

Environmental safety
- Some pesticides pose a greater threat of contamination of water than others and mecoprop-P is one of these pesticides. Take special care when applying mecoprop-P near water and do not apply if heavy rain is forecast

Hazard classification and safety precautions
UN Number N/C
Risk phrases H319 [1]
Operator protection A, C, H, M
Environmental protection E07a, E15a [1, 2]; H410 [1]
Storage and disposal D05, D09a, D10a, D12a

139 2,4-D + florasulam

A post-emergence herbicide mixture for control of broad leaved weeds in managed amenity turf
HRAC mode of action code: O + B

See also florasulam

Products

1	Esteron T	Everris Ltd	300:6.25 g/l	ME	17432
2	Junction	Rigby Taylor	452:6.25 g/l	ME	12493

Uses
- Clover in *amenity grassland* [1]; *managed amenity turf* [1, 2]
- Daisies in *amenity grassland* [1]; *managed amenity turf* [1, 2]
- Dandelions in *amenity grassland* [1]; *managed amenity turf* [1, 2]
- Plantains in *amenity grassland* [1]; *managed amenity turf* [1, 2]
- Sticky mouse-ear in *amenity grassland* [1]; *managed amenity turf* [1, 2]

Approval information
- 2,4-D and florasulam included in Annex I under EC Regulation 1107/2009

Efficacy guidance
- Best results obtained from treatment of actively growing weeds between Mar and Oct when soil is moist
- Do not apply when rain is imminent or during periods of drought unless irrigation is applied

Pesticide Profiles

Restrictions
- Maximum number of treatments on managed amenity turf: 1 per yr

Crop-specific information
- Avoid mowing turf 3 d before and after spraying
- Ensure newly sown turf has established before treating. Turf sown in late summer or autumn should not be sprayed until growth is resumed in the following spring

Following crops guidance
- An interval of 4 wk must elapse between application and re-seeding turf

Environmental safety
- Dangerous for the environment
- Toxic to aquatic organisms

Hazard classification and safety precautions
Hazard Harmful, Dangerous for the environment [1, 2]; Harmful if swallowed [1]
Transport code 9
Packaging group III
UN Number 3082
Risk phrases H317 [1]; R22a, R43, R51, R53a [2]
Operator protection A, H; U02a, U05a, U08, U14, U20a
Environmental protection E15a, E34, E38 [1, 2]; H410 [1]
Storage and disposal D01, D02, D09a, D10b, D12a

140 2,4-D + glyphosate

A total herbicide for weed control in crops and on railway ballast
HRAC mode of action code: O + G

See also glyphosate

Products
Kyleo   Nufarm UK   160:240 g/l   SL   16567

Uses
- Annual and perennial weeds in **all edible crops (outdoor), all non-edible seed crops grown outdoors, amenity grassland, amenity vegetation, apple orchards, green cover on land temporarily removed from production, pear orchards, railway tracks**

Approval information
- 2,4-D and glyphosate included in Annex I under EC Regulation 1107/2009

Restrictions
- Do not use under polythene or glass
- Application of lime, fertiliser, farmyard manure and pesticides should be delayed until 5 days after application of Kyleo. After application large quantities of decaying foliage, stolons, roots or rhizomes should be dispersed or buried by thorough cultivation before crop drilling
- Do not use in or alongside hedges
- Do not apply by hand-held equipment
- Windfall (fruit fall to ground) must not be used as food or feeding stuff

Environmental safety
- LERAP Category B

Hazard classification and safety precautions
Transport code 9
Packaging group III
UN Number 3082
Risk phrases R36, R38, R50, R53a, R70
Operator protection A, C, H; U05a, U19a
Environmental protection E15a, E16a, E34, E38
Storage and disposal D01, D09a, D10a, D12a, D12b

FOR FULL CONDITIONS OF USE ALWAYS READ THE PRODUCT LABEL

141　2,4-D + MCPA

A translocated herbicide mixture for cereals and grass
HRAC mode of action code: O + O

See also MCPA

Products

1	Cirran 360	Nufarm UK	360:315 g/l	SL	17346
2	Headland Polo	Headland	360:315 g/l	SL	14933
3	Lupo	Nufarm UK	360:315 g/l	SL	14931

Uses

- Annual dicotyledons in **grassland**, **spring barley**, **spring wheat**, **winter barley**, **winter oats**, **winter wheat** [1-3]
- Perennial dicotyledons in **grassland** [1, 3]; **spring barley**, **spring wheat**, **winter barley**, **winter oats**, **winter wheat** [1-3]

Approval information

- 2,4-D and MCPA included in Annex I under EC Regulation 1107/2009
- Accepted by BBPA for use on malting barley

Efficacy guidance

- Best results achieved by spraying weeds in seedling to young plant stage when growing actively in a strongly competing crop
- Most effective stage for spraying perennials varies with species. See label for details

Restrictions

- Maximum number of treatments 1 per crop in cereals, 2 per year in grassland and 3 per year in managed amenity turf [2]; 1 per crop in cereals and 1 per year in grassland [3]
- Do not spray if rain falling or imminent
- Do not cut grass or graze for at least 10 d after spraying
- Do not use on newly sown leys containing clover or other legumes
- Do not spray crops stressed by cold weather or drought or if frost expected
- Do not use shortly before or after sowing any crop
- Do not roll or harrow within 7 d before or after spraying

Crop-specific information

- Spray winter cereals in spring when leaf-sheath erect but before first node detectable (GS 31), spring cereals from 5-leaf stage to before first node detectable (GS 15-31)
- Latest use: before first node detectable (GS 31) in cereals;

Environmental safety

- Dangerous for the environment
- Toxic to aquatic organisms
- Keep livestock out of treated areas for at least two weeks following treatment and until poisonous weeds, such as ragwort, have died down and become unpalatable
- 2,4-D and MCPA are active at low concentrations. Take extreme care to avoid drift onto neighbouring crops, especially beet crops, brassicas, most market garden crops including lettuce and tomatoes under glass, pears and vines

Hazard classification and safety precautions

Hazard Harmful [1-3]; Dangerous for the environment [2]
Transport code 9 [2]
Packaging group III [2]
UN Number 3082 [2]; N/C [1, 3]
Risk phrases H318
Operator protection A, C, H, M; U05a, U08, U11, U14, U20b [1-3]; U15 [2]
Environmental protection E07a, E15a, E34 [1-3]; E38 [1, 3]
Storage and disposal D01, D02, D05, D09a, D10a [1-3]; D12a [1, 3]
Medical advice M03 [1-3]; M05a [2]

SEE SECTION 3 FOR PRODUCTS ALSO REGISTERED

Pesticide Profiles

142 daminozide

A hydrazide plant growth regulator for use in certain ornamentals

Products

1	B-Nine SG	Certis	85% w/w	SG	14435
2	Dazide Enhance	Fargro	85% w/w	SG	16092

Uses
- Growth regulation in **protected ornamentals**

Approval information
- Daminozide included in Annex I under EC Regulation 1107/2009

Efficacy guidance
- Best results obtained by application in late afternoon when glasshouse has cooled down
- Apply a fine spray using compressed air or power sprayers to give good coverage of dry foliage without run off
- Response to treatment differs widely depending on variety, stage of growth and physiological condition of the plant. It is recommended that any new variety is first tested on a small scale to observe if adverse effects occur

Restrictions
- Apply only to turgid, well watered plants. Do not water for 24 h after spraying
- Do not mix with other spray chemicals unless specifically recommended
- Do not store product in metal containers

Crop-specific information
- Do not use on chrysanthemum variety Fandango
- Evidence of effectiveness on Azaleas is limited
- See label for guidance on a range of bedding plant species

Hazard classification and safety precautions
  UN Number N/C
  Operator protection A, H; U02a, U08, U19a, U20b [2]; U05a [1, 2]
  Environmental protection E15b
  Storage and disposal D01, D02, D09a [1, 2]; D10b, D12a [2]
  Medical advice M05a [2]

143 dazomet

A methyl isothiocyanate releasing soil fumigant
HRAC mode of action code: Z

Products

Basamid	Certis	97% w/w	GR	11324

Uses
- Soil pests in **soil and compost**
- Soil-borne diseases in **soil and compost**

Approval information
- Dazomet included in Annex I under EC Regulation 1107/2009

Efficacy guidance
- Dazomet acts by releasing methyl isothiocyanate in contact with moist soil
- Soil sterilization is carried out after harvesting one crop and before planting the next
- The soil must be of fine tilth, free of clods and evenly moist to the depth of sterilization
- Soil moisture must not be less than 50% of water-holding capacity or oversaturated. If too dry, water at least 7-14 d before treatment

FOR FULL CONDITIONS OF USE ALWAYS READ THE PRODUCT LABEL

- In order to obtain short treatment times it is recommended to treat soils when soil temperature is above 7°C. Treatment should be used outdoors before winter rains make soil too wet to cultivate - usually early Nov
- For club root control treat only in summer when soil temperature above 10°C
- Where onion white rot is a problem it is unlikely to give effective control where inoculum level high or crop under stress
- Apply granules with suitable applicators, mix into soil immediately to desired depth and seal surface with polythene sheeting, by flooding or heavy rolling. See label for suitable application and incorporation machinery
- With 'planting through' technique polythene seal is left in place to form mulch into which new crop can be planted

Restrictions
- Maximum number of treatments 1 per batch of soil for protected crops or field crop situation
- Use is limited to one application every third year on the same area
- Crops must not be planted until the cress safety test has been carried out and germination found to be satisfactory
- Do not treat ground where water table may rise into treated layer

Crop-specific information
- Latest use: pre-planting

Following crops guidance
- With 'planting through' technique no gas release cultivations are made and safety test with cress is particularly important. Conduct cress test on soil samples from centre as well as edges of bed. Observe minimum of 30 d from application to cress test in soils at 10°C or above, at least 50 d in soils below 10°C. See label for details
- In all other situations 14-28 d after treatment cultivate lightly to allow gas to disperse and conduct cress test after a further 14-28 d (timing depends on soil type and temperature). Do not treat structures containing live plants or any ground within 1 m of live plants

Environmental safety
- Dangerous for the environment
- Very toxic to aquatic organisms

Hazard classification and safety precautions
Hazard Harmful, Dangerous for the environment, Harmful if swallowed
Transport code 9
Packaging group III
UN Number 3077
Risk phrases H315, H317, H319, H335
Operator protection A, M; U04a, U05a, U09a, U19a, U20a
Environmental protection E15a, E34, E38, H410
Storage and disposal D01, D02, D07, D09a, D11a, D12a
Medical advice M03, M05a

144 2,4-DB

A translocated phenoxycarboxylic acid herbicide
HRAC mode of action code: O

Products

1	DB Straight	UPL Europe	400 g/l	SL	13736
2	Embutone 24DB	Nufarm UK	400 g/l	SL	17596
3	Headland Spruce	Headland	400 g/l	SL	15089

Uses
- Annual dicotyledons in *grassland, lucerne, red clover, spring barley, spring oats, spring wheat, white clover, winter barley, winter oats, winter wheat* [1-3]; *undersown barley, undersown oats, undersown wheat* [2]
- Buttercups in *grassland* [1, 3]

SEE SECTION 3 FOR PRODUCTS ALSO REGISTERED

Pesticide Profiles

- Fat hen in *grassland* [1, 3]
- Penny cress in *grassland* [1, 3]
- Perennial dicotyledons in *grassland, lucerne, red clover, spring barley, spring oats, spring wheat, undersown barley, undersown oats, undersown wheat, white clover, winter barley, winter oats, winter wheat* [2]
- Shepherd's purse in *grassland* [1, 3]

Approval information
- 2,4-DB included in Annex I under EC Regulation 1107/2009
- Accepted by BBPA for use on malting barley

Efficacy guidance
- Best results achieved on young seedling weeds under good growing conditions. Treatment less effective in cold weather and dry soil conditions
- Rain within 12 h may reduce effectiveness

Restrictions
- Do not allow spray drift onto neighbouring crops

Crop-specific information
- Latest use: before first node detectable (GS 31) for undersown cereals; fourth trifoliate leaf for lucerne
- In direct sown lucerne spray when seedlings have reached first trifoliate leaf stage. Optimum time 3-4 trifoliate leaves
- Do not treat any lucerne after fourth trifoliate leaf
- In spring barley and spring oats undersown with lucerne spray from when cereal has 1 leaf unfolded and lucerne has first trifoliate leaf
- In spring wheat undersown with lucerne spray from when cereal has 3 leaves unfolded and lucerne has first trifoliate leaf

Following crops guidance
- Do not sow any crop into soil treated with 2,4-DB for at least 3 months after application.

Environmental safety
- Dangerous for the environment
- Toxic to aquatic organisms
- Keep livestock out of treated areas for at least two weeks following treatment and until poisonous weeds, such as ragwort, have died down and become unpalatable

Hazard classification and safety precautions
    Hazard Harmful, Dangerous for the environment [1-3]; Harmful if swallowed [1, 2]
    Transport code 9 [1, 3]
    Packaging group III [1, 3]
    UN Number 3082 [1, 3]; N/C [2]
    Risk phrases H318 [1, 2]; R22a, R41, R43, R51, R53a [3]
    Operator protection A, C, H; U05a, U09a, U11, U20a
    Environmental protection E07a, E15a, E15b, E34, E38
    Storage and disposal D01, D02, D05, D10c, D12a
    Medical advice M03, M05a

145 2,4-DB + MCPA

A translocated herbicide for cereals, clovers and leys
HRAC mode of action code: O + O

See also MCPA

Products

1	Clovermax	Nufarm UK	240:40 g/l	SL	15780
2	Redlegor	UPL Europe	240:40 g/l	SL	15783

FOR FULL CONDITIONS OF USE ALWAYS READ THE PRODUCT LABEL

2,4-DB + MCPA

Uses
- Annual dicotyledons in **grassland**, **spring wheat**, **winter wheat** [1, 2]; **oats**, **spring barley**, **undersown barley**, **undersown wheat**, **winter barley** [1]; **undersown spring cereals**, **undersown winter cereals** [2]
- Buttercups in **grassland**, **oats**, **spring barley**, **spring wheat**, **undersown barley**, **undersown wheat**, **winter barley**, **winter wheat** [1]
- Fat hen in **grassland**, **oats**, **spring barley**, **spring wheat**, **undersown barley**, **undersown wheat**, **winter barley**, **winter wheat** [1]
- Penny cress in **grassland**, **oats**, **spring barley**, **spring wheat**, **undersown barley**, **undersown wheat**, **winter barley**, **winter wheat** [1]
- Perennial dicotyledons in **grassland**, **spring wheat**, **undersown spring cereals**, **undersown winter cereals**, **winter wheat** [2]
- Plantains in **grassland** [1]
- Polygonums in **grassland**, **spring wheat**, **undersown spring cereals**, **undersown winter cereals**, **winter wheat** [2]
- Shepherd's purse in **grassland**, **oats**, **spring barley**, **spring wheat**, **undersown barley**, **undersown wheat**, **winter barley**, **winter wheat** [1]

Approval information
- 2,4-DB and MCPA included in Annex I under EC Regulation 1107/2009
- Accepted by BBPA for use on malting barley

Efficacy guidance
- Best results achieved on young seedling weeds under good growing conditions
- Spray thistles and other perennials when 10-20 cm high provided clover at correct stage
- Effectiveness may be reduced by rain within 12 h, by very cold conditions or drought

Restrictions
- Maximum number of treatments 1 per crop
- Do not spray established clover crops or lucerne
- Do not roll or harrow within 7 d before or after spraying
- Do not spray immediately before or after sowing any crop
- Do not graze crops within a week before or 2 weeks after applying
- Do not apply during drought, rain or if rain is expected
- Do not apply in very cold conditions as effectiveness may be reduced
- Do not use immediately before or after sowing any crop
- Do not use on cereals undersown with lucerne or in seed mixtures containing lucerne
- Do not spray in windy conditions as the spray drift may cause damage to neighbouring crops. The following crops are particularly susceptible: beet, Brassicae (e.g. turnips, swedes, oilseed rape), and most market garden crops including lettuce and tomatoes under glass, pears and vines

Crop-specific information
- Latest use: before first node detectable (GS 31) for cereals; 4th trifoliate leaf stage of clover for grass re-seeds
- Apply in spring to winter cereals from leaf sheath erect stage, to spring barley or oats from 2-leaf stage (GS 12), to spring wheat from 5-leaf stage (GS 15)
- Spray clovers as soon as possible after first trifoliate leaf, grasses after 2-3 leaf stage
- Red clover may suffer temporary distortion after treatment

Environmental safety
- Harmful to aquatic organisms
- Harmful to fish or other aquatic life. Do not contaminate surface waters or ditches with chemical or used container
- Keep livestock out of treated areas for at least two weeks following treatment and until poisonous weeds, such as ragwort, have died down and become unpalatable

Hazard classification and safety precautions
Hazard Harmful, Dangerous for the environment [1, 2]; Harmful if swallowed [1]
UN Number N/C
Risk phrases H318 [1]; R22a, R38, R41, R52, R53a [2]
Operator protection A, C [1]; U05a, U20b [1, 2]; U08, U11, U13, U14, U15 [2]; U12, U19a [1]
Environmental protection E07a [1, 2]; E13c [2]; E15b, E34, E38, H411 [1]

SEE SECTION 3 FOR PRODUCTS ALSO REGISTERED

Pesticide Profiles

Storage and disposal D01, D02, D05, D09a [1, 2]; D10b, D12a [1]; D10c [2]
Medical advice M03 [1, 2]; M05a [2]

146 deltamethrin

A pyrethroid insecticide with contact and residual activity
IRAC mode of action code: 3

Products

1	Bandu	Headland	25 g/l	EC	16153
2	Decis	Bayer CropScience	25 g/l	EC	16124
3	Decis Protech	Bayer CropScience	15 g/l	EW	16160
4	Grainstore	Pan Agriculture	25 g/l	EC	14932
5	K-Obiol EC 25	Bayer Environ.	25 g/l	EC	13573
6	K-Obiol ULV 6	Bayer Environ.	0.69% w/v	UL	13572

Uses

- American serpentine leaf miner in **gherkins** *(off-label)*, **summer squash** *(off-label)* [2]; **protected aubergines** *(off-label)*, **protected courgettes** *(off-label)* [1, 2]; **protected gherkins** *(off-label)*, **protected squashes** *(off-label)* [1]; **protected ornamentals** *(off-label)* [1-3]
- Aphids in **amenity vegetation**, **apples**, **cress** *(off-label)*, **durum wheat** *(off-label)*, **endives** *(off-label)*, **grass seed crops** *(off-label)*, **lamb's lettuce** *(off-label)*, **land cress** *(off-label)*, **leaves and shoots** *(off-label)*, **lettuce**, **protected peppers**, **rye** *(off-label)*, **spring barley**, **spring oats**, **spring wheat**, **triticale** *(off-label)*, **winter barley**, **winter oats**, **winter wheat** [1-3]; **brussels sprouts**, **cabbages**, **cauliflowers** [2]; **evening primrose** *(off-label)* [3]; **forest nurseries** *(off-label)*, **ornamental plant production**, **protected cucumbers**, **protected forest nurseries** *(off-label)*, **protected ornamentals**, **protected tomatoes** [1, 2]
- Apple sucker in **apples** [1-3]
- Barley yellow dwarf virus vectors in **winter barley**, **winter wheat** [1-3]
- Beet virus yellows vectors in **winter oilseed rape** [1-3]
- Bramble shoot moth in **blackberries** *(off-label)* [1-3]
- Cabbage seed weevil in **mustard** [1]; **spring mustard**, **winter mustard** [2, 3]; **spring oilseed rape**, **winter oilseed rape** [1-3]
- Cabbage stem flea beetle in **quinoa** *(off-label)* [2]; **winter oilseed rape** [1-3]
- Cabbage stem weevil in **mustard** [1]; **spring mustard**, **winter mustard** [2, 3]; **spring oilseed rape**, **winter oilseed rape** [1-3]
- Capsids in **amenity vegetation**, **apples** [1-3]; **ornamental plant production** [1, 2]
- Carrot fly in **carrots** *(off-label)*, **horseradish** *(off-label)*, **mallow (althaea spp.)** *(off-label)*, **parsnips** *(off-label)* [1-3]
- Caterpillars in **amenity vegetation**, **apples**, **blackberries** *(off-label)*, **brussels sprouts**, **cabbages**, **cauliflowers**, **cress** *(off-label)*, **endives** *(off-label)*, **lamb's lettuce** *(off-label)*, **land cress** *(off-label)*, **leaves and shoots** *(off-label)*, **protected peppers**, **swedes**, **turnips** [1-3]; **collards** *(off-label)*, **kale** *(off-label)*, **oriental brassicas** *(off-label)*, **spring greens** *(off-label)* [2]; **evening primrose** *(off-label)*, **lettuce**, **protected celery** *(off-label)*, **protected rhubarb** *(off-label)* [3]; **mustard** [1]; **ornamental plant production**, **protected cucumbers**, **protected ornamentals**, **protected tomatoes** [1, 2]
- Codling moth in **apples** [1-3]
- Cutworms in **carrots** *(off-label)*, **horseradish** *(off-label)*, **leeks** *(off-label)*, **mallow (althaea spp.)** *(off-label)*, **parsnips** *(off-label)* [1-3]; **lettuce** [1, 3]
- Flea beetle in **brussels sprouts**, **cabbages**, **cauliflowers**, **durum wheat** *(off-label - cereal)*, **rye** *(off-label - cereal)*, **swedes**, **triticale** *(off-label - cereal)*, **turnips** [2]; **cress** *(off-label)*, **durum wheat** *(off-label)*, **rye** *(off-label)*, **triticale** *(off-label)* [1, 3]; **endives** *(off-label)*, **lamb's lettuce** *(off-label)*, **land cress** *(off-label)*, **leaves and shoots** *(off-label)*, **sugar beet** [1-3]; **lettuce** [3]; **mustard** [1]
- Gout fly in **durum wheat** *(off-label)*, **rye** *(off-label)*, **triticale** *(off-label)* [1-3]
- Insect pests in **blackberries** *(off-label)*, **carrots** *(off-label)*, **cress** *(off-label)*, **durum wheat** *(off-label)*, **endives** *(off-label)*, **grass seed crops** *(off-label)*, **horseradish** *(off-label)*, **lamb's lettuce** *(off-label)*, **land cress** *(off-label)*, **leaves and shoots** *(off-label)*, **leeks** *(off-label)*, **mallow (althaea spp.)** *(off-label)*, **parsnips** *(off-label)*, **protected ornamentals** *(off-label)*, **protected spring**

FOR FULL CONDITIONS OF USE ALWAYS READ THE PRODUCT LABEL

deltamethrin

onions *(off-label)*, **rye** *(off-label)*, **triticale** *(off-label)* [1]; **evening primrose** *(off-label)* [3]; **forest nurseries** *(off-label)*, **protected forest nurseries** *(off-label)* [1, 2]; **grain** *(off-label)* [6]; **grain stores** [4, 5]; **lettuce** [2]; **stored grain**, **stored pulses** [4-6]
- Leaf miner in **protected spring onions** *(off-label)* [1-3]
- Leaf moth in **leeks** *(off-label)* [1-3]
- Leafhoppers in **lettuce** [3]
- Mealybugs in **amenity vegetation**, **protected peppers** [1-3]; **ornamental plant production**, **protected cucumbers**, **protected ornamentals**, **protected tomatoes** [1, 2]
- Pea and bean weevil in **broad beans**, **combining peas**, **spring field beans**, **vining peas**, **winter field beans** [1-3]
- Pea midge in **combining peas**, **vining peas** [2, 3]
- Pea moth in **combining peas**, **vining peas** [1-3]
- Pear sucker in **pears** [1-3]
- Pollen beetle in **lettuce** [3]; **mustard** [1]; **spring mustard**, **winter mustard** [2, 3]; **spring oilseed rape**, **winter oilseed rape** [1-3]
- Raspberry beetle in **blackberries** *(off-label)*, **raspberries** [1-3]
- Saddle gall midge in **durum wheat** *(off-label)*, **rye** *(off-label)*, **triticale** *(off-label)* [1-3]
- Sawflies in **apples** [1-3]
- Scale insects in **amenity vegetation**, **protected peppers** [1-3]; **ornamental plant production**, **protected cucumbers**, **protected ornamentals**, **protected tomatoes** [1, 2]
- Spotted wing drosophila in **mirabelles** *(off-label)*, **plums** *(off-label)* [2]
- Thrips in **amenity vegetation**, **protected peppers** [1-3]; **lettuce** [3]; **ornamental plant production**, **protected cucumbers**, **protected tomatoes** [1, 2]
- Tortrix moths in **apples** [1-3]
- Wessex flea beetle in **quinoa** *(off-label)* [2]
- Western flower thrips in **gherkins** *(off-label)*, **summer squash** *(off-label)* [2]; **protected aubergines** *(off-label)*, **protected courgettes** *(off-label)*, **protected cucumbers**, **protected tomatoes**, **strawberries** *(off-label)* [1, 2]; **protected gherkins** *(off-label)*, **protected squashes** *(off-label)* [1]; **protected ornamentals** *(off-label)*, **protected peppers** [1-3]
- Whitefly in **amenity vegetation**, **protected peppers** [1-3]; **brussels sprouts**, **cabbages**, **cauliflowers** [2]; **ornamental plant production**, **protected cucumbers**, **protected ornamentals**, **protected tomatoes** [1, 2]
- Yellow cereal fly in **durum wheat** *(off-label)*, **rye** *(off-label)*, **triticale** *(off-label)* [1-3]

Extension of Authorisation for Minor Use (EAMUs)
- **blackberries** *20141106* [1], *20140905* [2], *20131673* [3]
- **carrots** *20141105* [1], *20140900* [2], *20131672* [3]
- **collards** *20152429* [2]
- **cress** *20141104* [1], *20140901* [2], *20131670* [3]
- **durum wheat** *20141100* [1], *(cereal) 20140908* [2], *20140908* [2], *20131674* [3]
- **endives** *20141104* [1], *20140901* [2], *20131670* [3]
- **evening primrose** *20142584* [3]
- **forest nurseries** *20131856* [1], *20131857* [2]
- **gherkins** *20131663* [2]
- **grain** *20112491* [6]
- **grass seed crops** *20141107* [1], *20140909* [2], *20131671* [3]
- **horseradish** *20141105* [1], *20140900* [2], *20131672* [3]
- **kale** *20152429* [2]
- **lamb's lettuce** *20141104* [1], *20140901* [2], *20131670* [3]
- **land cress** *20141104* [1], *20140901* [2], *20131670* [3]
- **leaves and shoots** *20141104* [1], *20140901* [2], *20131670* [3]
- **leeks** *20141103* [1], *20140906* [2], *20131665* [3]
- **mallow (althaea spp.)** *20141105* [1], *20140900* [2], *20131672* [3]
- **mirabelles** *20140492* [2]
- **oriental brassicas** *20152429* [2]
- **parsnips** *20141105* [1], *20140900* [2], *20131672* [3]
- **plums** *20140492* [2]
- **protected aubergines** *20132523* [1], *20131661* [2]
- **protected celery** *20142585* [3]

SEE SECTION 3 FOR PRODUCTS ALSO REGISTERED

Pesticide Profiles

- *protected courgettes* 20132524 [1], 20131663 [2]
- *protected forest nurseries* 20131856 [1], 20131857 [2]
- *protected gherkins* 20132524 [1]
- *protected ornamentals* 20141102 [1], 20141011 [2], 20131662 [3]
- *protected rhubarb* 20142585 [3]
- *protected spring onions* 20141101 [1], 20140902 [2], 20131664 [3]
- *protected squashes* 20132524 [1]
- *quinoa* 20151052 [2]
- *rye* 20141100 [1], *(cereal)* 20140908 [2], 20140908 [2], 20131674 [3]
- *spring greens* 20152429 [2]
- *strawberries* 20132527 [1], 20131643 [2]
- *summer squash* 20131663 [2]
- *triticale* 20141100 [1], *(cereal)* 20140908 [2], 20140908 [2], 20131674 [3]

Approval information
- Deltamethrin included in Annex I under EC Regulation 1107/2009
- Accepted by BBPA for use on malting barley and hops

Efficacy guidance
- A contact and stomach poison with 3-4 wk persistence, particularly effective on caterpillars and sucking insects
- Normally applied at first signs of damage with follow-up treatments where necessary at 10-14 d intervals. Rates, timing and recommended combinations with other pesticides vary with crop and pest. See label for details
- Spray is rainfast within 1 h
- May be applied in frosty weather provided foliage not covered in ice
- Temperatures above 35°C may reduce effectiveness or persistence

Restrictions
- Maximum number of treatments varies with crop and pest, 4 per crop for wheat and barley, only 1 application between 1 Apr and 31 Aug. See label for other crops
- Do not apply more than 1 aphicide treatment to cereals in summer
- Do not spray crops suffering from drought or other physical stress
- Consult processer before treating crops for processing
- Do not apply to a cereal crop if any product containing a pyrethroid insecticide or dimethoate has been applied to that crop after the start of ear emergence (GS 51)
- Do not spray cereals after 31 Mar in the year of harvest within 6 m of the outside edge of the crop
- Reduced volume spraying must not be used on cereals after 31 Mar in yr of harvest

Crop-specific information
- Latest use: early dough (GS 83) for barley, oats, wheat; before flowering for mustard, oilseed rape; before 31 Mar for grass seed crops [1-3]

Environmental safety
- Dangerous for the environment
- Toxic to aquatic organisms
- Flammable [1, 2]
- Extremely dangerous to fish or other aquatic life. Do not contaminate surface waters or ditches with chemical or used container
- High risk to bees. Do not apply to crops in flower or to those in which bees are actively foraging. Do not apply when flowering weeds are present
- Do not apply in tank-mixture with a triazole-containing fungicide when bees are likely to be actively foraging in the crop
- High risk to non-target insects or other arthropods. Do not spray within 6 m of the field boundary
- Buffer zone requirement 7 m [1-3]
- Buffer zone requirement 20 m
- Broadcast air-assisted LERAP [3, 4] (18m); Broadcast air-assisted LERAP [2] (30m raspberry, 50m apple & pear0; LERAP category A [1-4]
- Broadcast air-assisted LERAP [3, 4] (18 m); Broadcast air-assisted LERAP [2] (30m (raspberry); 50m (apple & pear)); LERAP Category A [1-4]

FOR FULL CONDITIONS OF USE ALWAYS READ THE PRODUCT LABEL

desmedipham

Hazard classification and safety precautions
  Hazard Harmful, Flammable [1, 2, 4]; Dangerous for the environment [1-4]
  Transport code 3 [1-4]; 9 [5, 6]
  Packaging group III
  UN Number 1993 [1-4]; 3082 [5, 6]
  Risk phrases R20, R22a, R22b [1, 2, 4]; R37 [2]; R38, R41, R51 [1, 4]; R50 [2, 3]; R53a [1-4]
  Operator protection A, C, H [1-6]; D, M [4-6]; U04a, U05a, U19a [1, 2, 4-6]; U08 [1, 2, 4]; U09a, U20c [5, 6]; U10 [1]; U11, U19c [2]; U14 [3]; U20b [1-4]
  Environmental protection E12a [2, 4]; E12c [1]; E12f [1] (cereals, oilseed rape, peas, beans); E12f [2-4] (cereals, oilseed rape, peas, beans - see label for guidance); E15a, E16d [1, 2, 4]; E15b [3, 5, 6]; E16c [1, 3, 4]; E16j [2]; E17b [2] (30m (raspberry); 50m (apple & pear)); E17b [3, 4] (18 m); E22c [3]; E34 [1, 2, 4-6]; E38 [1-4]
  Storage and disposal D01, D02, D09a, D10c [1-6]; D05 [1, 2, 4]; D12a [2-4]
  Medical advice M03 [1, 2, 4-6]; M05b [1, 2, 4]

147 desmedipham

A contact phenyl carbamate herbicide available only in mixtures
HRAC mode of action code: C1

148 desmedipham + ethofumesate + lenacil + phenmedipham

A selective contact and residual herbicide mixture for weed control in beet
HRAC mode of action code: C1 + N + C1 + C1

Products
  Betanal MaxxPro   Bayer CropScience   47:75:27:60 g/l   OD   15086

Uses
- Annual dicotyledons in **fodder beet**, **forest nurseries** (off-label), **mangels**, **sugar beet**
- Annual meadow grass in **beetroot** (off-label), **fodder beet**, **forest nurseries** (off-label), **mangels**, **sugar beet**
- Black bindweed in **beetroot** (off-label)
- Black nightshade in **beetroot** (off-label)
- Bugloss in **beetroot** (off-label)
- Charlock in **beetroot** (off-label)
- Chickweed in **beetroot** (off-label)
- Cleavers in **beetroot** (off-label)
- Field speedwell in **beetroot** (off-label)
- Fumitory in **beetroot** (off-label)

Extension of Authorisation for Minor Use (EAMUs)
- **beetroot** 20152038
- **forest nurseries** 20162138

Approval information
- Desmedipham, ethofumesate, lenacil and phenmedipham included in Annex I under EC Regulation 1107/2009

Efficacy guidance
- On soils with more than 5% organic matter content, residual activity may be reduced.

Restrictions
- The maximum total dose must not exceed 1.0 kg/ha ethofumesate in any 3 year period.

Following crops guidance
- Beet crops may be sown at any time after the use of Betanal MaxxPro. Any other crop may be sown 3 months after using Betanal MaxxPro. Ploughing (mould board) to a minimum depth of 15 cm should precede preparation of a new seed bed.

SEE SECTION 3 FOR PRODUCTS ALSO REGISTERED

- If a crop is suffering from manganese deficiency it may be checked. To avoid crop check, manganese should ideally be applied to the crop first.
- Crops suffering from lime deficiency may also be checked. Growers should ensure that the lime status of the soil is satisfactory before drilling.
- When the temperature is, or is likely to be, above 21°C (70°F) on the day of spraying, application should be made after 5 pm otherwise crop check may occur. If crops are subjected to substantial day to night temperature changes shortly before or after spraying, a check may occur from which the crop may not fully recover.

Hazard classification and safety precautions
Hazard Irritant, Dangerous for the environment
Transport code 9
Packaging group III
UN Number 3082
Risk phrases R41, R43, R50, R53a
Operator protection A, C, H; U05a, U11, U14, U20b
Environmental protection E15b, E38, E39
Storage and disposal D09a, D10b, D12a
Medical advice M03

149 desmedipham + ethofumesate + phenmedipham

A selective contact and residual herbicide for beet
HRAC mode of action code: C1 + N + C1

See also ethofumesate
phenmedipham

Products

1	Betasana Trio SC	UPL Europe	15:115:75 g/l	SC	15551
2	Beta-Team	UPL Europe	25:150:75 g/l	SE	16975
3	D.E.P. 205	Goldengrass	15:115:75 g/l	SC	16431
4	D.E.P. 251	Goldengrass	25:151:75 g/l	EC	14922
5	Mundua	Agroquimicos	25:151:75 g/l	EC	16103
6	Sniper	Adama	50:200:150 g/l	SE	16133
7	Trilogy	UPL Europe	15:115:75 g/l	SC	15644

Uses
- Annual dicotyledons in **fodder beet** [1-3, 6, 7]; **mangels** [1-3, 7]; **sugar beet** [1-7]
- Annual meadow grass in **fodder beet** [2, 6]; **mangels** [2]; **sugar beet** [2, 4-6]

Approval information
- Desmedipham, ethofumesate and phenmedipham included in Annex I under EC Regulation 1107/2009

Efficacy guidance
- Product recommended for low-volume overall application in a planned spray programme
- Product acts mainly by contact action. A full programme also gives some residual control but this may be reduced on soils with more than 5% organic matter
- Best results achieved from treatments applied at fully expanded cotyledon stage of largest weeds present. Occasional larger weeds will usually be controlled from a full programme of sprays
- Where a pre-emergence band spray has been applied treatment must be timed according to size of the untreated weeds between the rows
- Susceptible weeds may not all be killed by the first spray. Repeat applications as each flush of weeds reaches cotyledon size normally necessary for season long control
- Sequential treatments should be applied when the previous one is still showing an effect on the weeds
- Various mixtures with other beet herbicides are recommended. See label for details

FOR FULL CONDITIONS OF USE ALWAYS READ THE PRODUCT LABEL

desmedipham + phenmedipham

Restrictions
- Maximum total dose 3.9 l/ha [6], 4.5 l/ha [5] & 7.0 l/ha [1, 7]
- Do not spray crops stressed by nutrient deficiency, wind damage, pest or disease attack, or previous herbicide treatments. Stressed crops treated under conditions of high light intensity may be checked and not recover fully
- If temperature likely to exceed 21°C spray after 5 pm
- Crystallisation may occur if spray volume exceeds that recommended or spray mixture not used within 2 h, especially if the water temperature is below 5°C
- Before use, wash out sprayer to remove all traces of previous products, especially hormone and sulfonyl urea weedkillers.
- The maximum total dose must not exceed 1.0 kg/ha ethofumesate in any three year period

Crop-specific information
- Latest use: before crop meets between rows
- Apply first treatment when majority of crop plants have reached the fully expanded cotyledon stage
- Frost within 7 d of treatment may cause check from which the crop may not recover

Following crops guidance
- Beet crops may be sown at any time after treatment. Any other crop may be sown 3 mth after treatment following mouldboard ploughing to 15 cm minimum

Environmental safety
- Dangerous for the environment
- Toxic to aquatic organisms
- LERAP Category B [2, 6]

Hazard classification and safety precautions
Hazard Irritant [1-3, 7]; Dangerous for the environment [1-7]
Transport code 9
Packaging group III
UN Number 3082
Risk phrases R36, R50 [1, 3, 7]; R43 [2]; R51 [2, 4, 5]; R53a [1-5, 7]
Operator protection A, H [2, 4-6]; U08 [1-7]; U14 [2]; U20a [2, 4-6]
Environmental protection E15a [2, 4-6]; E15b [1, 3, 7]; E16a [2, 6]; E38 [1-7]; H410 [6]
Storage and disposal D05, D09a, D12a [1-7]; D10b [2, 4-6]; D10c [1, 3, 7]
Treated seed S06a [1, 3, 7]
Medical advice M03

150 desmedipham + phenmedipham

A mixture of contact herbicides for use in sugar beet
HRAC mode of action code: C1 + C1

See also phenmedipham

Products

1	Beetup Compact SC	UPL Europe	80:80 g/l	SC	15566
2	Betanal Turbo	Bayer CropScience	160:160 g/l	EC	15505
3	Rifle	Adama	160:160 g/l	SE	15705

Uses
- Amaranthus in **sugar beet** [1]
- Annual dicotyledons in **fodder beet** [2, 3]; **mangels** [2]; **sugar beet** [1-3]
- Black nightshade in **fodder beet** [3]; **sugar beet** [1, 3]
- Chickweed in **fodder beet** [3]; **sugar beet** [1, 3]
- Fat hen in **sugar beet** [1]
- Field pansy in **fodder beet** [3]; **sugar beet** [1, 3]
- Field speedwell in **fodder beet**, **sugar beet** [3]
- Groundsel in **fodder beet**, **sugar beet** [3]
- Ivy-leaved speedwell in **fodder beet**, **sugar beet** [3]

SEE SECTION 3 FOR PRODUCTS ALSO REGISTERED

- Mayweeds in **fodder beet**, **sugar beet** [3]
- Penny cress in **fodder beet** [3]; **sugar beet** [1, 3]
- Red dead-nettle in **sugar beet** [1]
- Shepherd's purse in **fodder beet** [3]; **sugar beet** [1, 3]

Approval information
- Desmedipham and phenmedipham included in Annex I under EC Regulation 1107/2009

Efficacy guidance
- Product is recommended for low volume overall spraying in a planned programme involving pre- and/or post-emergence treatments at doses recommended for low volume programmes
- Best results obtained from treatment when earliest germinating weeds have reached cotyledon stage
- Further treatments must be applied as each flush of weeds reaches cotyledon stage but allowing a minimum of 7 d between each spray
- Where a pre-emergence band spray has been applied, the first treatment should be timed according to the size of the weeds in the untreated area between the rows
- Product is absorbed by leaves of emerged weeds which are killed by scorching action in 2-10 d. Apply overall as a fine spray to optimise weed cover and spray retention
- Various tank mixtures and sequences recommended to widen weed spectrum and add residual activity - see label for details

Restrictions
- Maximum total dose equivalent to three full dose treatments
- Do not spray crops stressed by nutrient deficiency, frost, wind damage, pest or disease attack, or previous herbicide treatments. Stressed crops may be checked and not recover fully
- If temperature likely to exceed 21°C spray after 5 pm
- Before use, wash out sprayer to remove all traces of previous products, especially hormone and sulfonyl urea weedkillers
- Crystallisation may occur if spray volume exceeds that recommended or spray mixture not used within 2 h, especially if the water temperature is below 5°C
- Product may cause non-reinforced PVC pipes and hoses to soften and swell. Wherever possible, use reinforced PVC or synthetic rubber hoses

Crop-specific information
- Latest use: before crop leaves meet between rows
- Product safe to use on all soil types

Environmental safety
- Dangerous for the environment
- Toxic to aquatic organisms
- Risk to certain non-target insects or other arthropods. Avoid spraying within 6 m of field boundary
- LERAP Category B

Hazard classification and safety precautions
Hazard Irritant [1]; Dangerous for the environment [1-3]
Transport code 9
Packaging group III
UN Number 3082
Risk phrases H315, H317, H319 [1]; R50, R53a [2]
Operator protection A [1, 2]; U05a, U19a [1-3]; U08 [2]; U09a [1, 3]; U11, U14, U20b [1]; U20a [2, 3]
Environmental protection E15b, E16a [1-3]; E16b, E39 [2]; E38 [2, 3]; H410 [1, 3]
Storage and disposal D01, D02, D09a [1-3]; D05 [1, 3]; D10b, D12a [2, 3]; D10c [1]

FOR FULL CONDITIONS OF USE ALWAYS READ THE PRODUCT LABEL

151 dicamba

A translocated benzoic herbicide available in mixtures or formulated alone for weed control in maize
HRAC mode of action code: O

See also 2,4-D + dicamba
2,4-D + dicamba + dichlorprop-P
2,4-D + dicamba + fluroxypyr
2,4-D + dicamba + MCPA + mecoprop-P
2,4-D + dicamba + triclopyr

Products

1	Antuco	Ventura	70% w/w	SG	17409
2	Oceal	Rotam	70% w/w	SG	15618

Uses
- Amaranthus in **forage maize**, **grain maize** [1]; **maize** [2]
- Fat hen in **forage maize**, **grain maize** [1]; **maize** [2]
- Field pansy in **forage maize**, **grain maize** [1]; **maize** [2]
- Maple-leaved goosefoot in **forage maize**, **grain maize** [1]; **maize** [2]

Approval information
- Dicamba included in Annex I under EC Regulation 1107/2009

Restrictions
- Do not treat maize during cold or frosty conditions, during periods of high day/night temperature variations or during periods of very high temperatures.
- Avoid drift into greenhouses or onto other agricultural or horticultural crops, amenity plantings or gardens. No to be used in glasshouses. Do not allow spray applications to come into contact with desired broad-leaved trees. Beets, all brassicae (including oilseed rape), lettuce, peas, tomatoes, potatoes, all crops and ornamentals are particularly susceptible to dicamba and may be damaged by spray drift.
- Application may only be made between 1 May and 30 September

Following crops guidance
- In the event of crop failure, only redrill with maize but any crop may be sown after a normal harvest date

Hazard classification and safety precautions
Hazard Dangerous for the environment
Transport code 9
Packaging group III
UN Number 3077
Risk phrases R51 [2]
Operator protection U20b
Environmental protection E15b, E38, H411
Storage and disposal D01, D02, D10c, D12b

152 dicamba + MCPA + mecoprop-P

A translocated herbicide for cereals, grassland, amenity grass and orchards
HRAC mode of action code: O + O + O

See also MCPA
mecoprop-P

Products

1	Hyprone P	UPL Europe	16:101:92 g/l	SL	17418
2	Hysward-P	Agrichem	16:101:92 g/l	SL	17609
3	Mircam Plus	Nufarm UK	19.5:245:43.3 g/l	SL	15868

SEE SECTION 3 FOR PRODUCTS ALSO REGISTERED

Pesticide Profiles

Products – continued

4	T2 Green Pro	Nufarm UK	19:245:43.3 g/l	SL	16366
5	Turfmaster	Nufarm UK	19.5:245:43.3 g/l	SL	16344

Uses
- Annual dicotyledons in *amenity grassland* [2, 4]; *canary seed* (off-label), *grass seed crops* [3]; *managed amenity turf* [2-5]; *spring barley, spring oats, spring wheat, winter barley, winter oats, winter wheat* [1, 3]
- Black bindweed in *amenity grassland* [4]; *managed amenity turf* [4, 5]
- Charlock in *amenity grassland* [4]; *managed amenity turf* [4, 5]
- Chickweed in *amenity grassland* [4]; *managed amenity turf* [4, 5]
- Clover in *amenity grassland* (moderately susceptible) [4]; *managed amenity turf* (moderately susceptible) [4, 5]
- Common orache in *amenity grassland* [4]; *managed amenity turf* [4, 5]
- Dandelions in *amenity grassland* (moderately susceptible) [4]; *managed amenity turf* (moderately susceptible) [4, 5]
- Docks in *amenity grassland, managed amenity turf* [2]
- Fat hen in *amenity grassland* [4]; *managed amenity turf* [4, 5]
- Forget-me-not in *amenity grassland* [4]; *managed amenity turf* [4, 5]
- Fumitory in *amenity grassland* [4]; *managed amenity turf* [4, 5]
- Groundsel in *amenity grassland* [4]; *managed amenity turf* [4, 5]
- Knotgrass in *amenity grassland* [4]; *managed amenity turf* [4, 5]
- Mayweeds in *amenity grassland* [4]; *managed amenity turf* [4, 5]
- Pale persicaria in *amenity grassland* [4]; *managed amenity turf* [4, 5]
- Penny cress in *amenity grassland* [4]; *managed amenity turf* [4, 5]
- Perennial dicotyledons in *amenity grassland* [2, 4]; *grass seed crops* [3]; *managed amenity turf* [2-5]; *spring barley, spring oats, spring wheat, winter barley, winter oats, winter wheat* [1, 3]
- Plantains in *amenity grassland* (moderately susceptible) [4]; *managed amenity turf* (moderately susceptible) [4, 5]
- Poppies in *amenity grassland* [4]; *managed amenity turf* [4, 5]
- Redshank in *amenity grassland* [4]; *managed amenity turf* [4, 5]
- Shepherd's purse in *amenity grassland* [4]; *managed amenity turf* [4, 5]
- Wild radish in *amenity grassland* [4]; *managed amenity turf* [4, 5]

Extension of Authorisation for Minor Use (EAMUs)
- *canary seed* 20160549 [3]

Approval information
- Dicamba, MCPA and mecoprop-P included in Annex I under EC Regulation 1107/2009
- Accepted by BBPA for use on malting barley

Efficacy guidance
- Treatment should be made when weeds growing actively. Weeds hardened by winter weather may be less susceptible
- For best results apply in fine warm weather, preferably when soil is moist. Do not spray if rain expected within 6 h or in drought
- Application of fertilizer 1-2 wk before spraying aids weed control in turf
- Where a second treatment later in the season is needed in amenity situations and on grass allow 4-6 wk between applications to permit sufficient foliage regrowth for uptake

Restrictions
- Maximum number of treatments (including other mecoprop-P products) or maximum total dose varies with crop and product. See label for details. The total amount of mecoprop-P applied in a single yr must not exceed the maximum total dose approved for any single product for the crop/situation
- Do not apply to cereals after the first node is detectable (GS 31), or to grass under stress from drought or cold weather
- Do not spray cereals undersown with clovers or legumes, to be undersown with grass or legumes or grassland where clovers or other legumes are important
- Do not spray leys established less than 18 mth or orchards established less than 3 yr

FOR FULL CONDITIONS OF USE ALWAYS READ THE PRODUCT LABEL

- Do not roll or harrow within 7 d before or after treatment, or graze for at least 7 d afterwards (longer if poisonous weeds present)
- Do not use on turf or grass in year of establishment. Allow 6-8 wk after treatment before seeding bare patches
- The first 4 mowings after use should not be used for mulching unless composted for 6 mth
- Turf should not be mown for 24 h before or after treatment (3-4 d for closely mown turf)
- Avoid drift onto all broad-leaved plants outside the target area

Crop-specific information
- Latest use: before first node detectable (GS 31) for cereals; 5-6 wk before head emergence for grass seed crops; mid-Oct for established grass
- HI 7-14 d before cutting or grazing for leys, permanent pasture
- Apply to winter cereals from the leaf sheath erect stage (GS 30), and to spring cereals from the 5 expanded leaf stage (GS 15)
- Spray grass seed crops 4-6 wk before flower heads begin to emerge (timothy 6 wk)
- Turf containing bulbs may be treated once the foliage has died down completely [2]

Environmental safety
- Harmful to aquatic organisms
- Keep livestock out of treated areas for at least 2 wk following treatment and until poisonous weeds, such as ragwort, have died down and become unpalatable
- Harmful to fish or other aquatic life. Do not contaminate surface waters or ditches with chemical or used container
- Some pesticides pose a greater threat of contamination of water than others and mecoprop-P is one of these pesticides. Take special care when applying mecoprop-P near water and do not apply if heavy rain is forecast

Hazard classification and safety precautions
Hazard Harmful [4, 5]; Irritant [1-3]; Dangerous for the environment [3-5]
UN Number N/C
Risk phrases H315 [1, 2]; H318 [1-5]
Operator protection A, C, H [1-5]; M [1-3]; U05a, U08, U11, U20b [1-5]; U15 [3]; U19a [2]
Environmental protection E07a, H411 [1-5]; E13c [1, 2]; E15a, E38 [3]; E15b [4, 5]; E34 [3-5]
Storage and disposal D01, D02, D05, D09a [1-5]; D10b [3-5]; D10c [1, 2]; D12a [3]
Medical advice M03 [3-5]; M05a [1, 2]

153 dicamba + mecoprop-P

A translocated post-emergence herbicide for cereals and grassland
HRAC mode of action code: O + O

See also mecoprop-P

Products

1	Foundation	Headland	84:600 g/l	SL	16414
2	Headland Saxon	Headland	84:600 g/l	SL	11947
3	High Load Mircam	Nufarm UK	80:600 g/l	SL	11930
4	Hyban P	Agrichem	18.7:150 g/l	SL	16799
5	Hygrass P	Agrichem	18.7:150 g/l	SL	16802
6	Prompt	Headland	84:600 g/l	SL	11948
7	Quickfire	Headland Amenity	18.7:150 g/l	SL	17245

Uses
- Annual dicotyledons in *amenity grassland* [1-3, 6]; *durum wheat* (off-label), *rye* (off-label), *triticale* (off-label) [1, 6]; *managed amenity turf* [4, 5, 7]; *spring barley*, *spring oats*, *spring wheat*, *winter barley*, *winter oats*, *winter wheat* [1-4, 6]
- Chickweed in *amenity grassland* [3]; *durum wheat* (off-label), *rye* (off-label), *triticale* (off-label) [1]; *managed amenity turf* [4]; *spring barley*, *spring oats*, *spring wheat*, *winter barley*, *winter oats*, *winter wheat* [1-4, 6]

SEE SECTION 3 FOR PRODUCTS ALSO REGISTERED

Pesticide Profiles

- Cleavers in **amenity grassland** [3]; **durum wheat** *(off-label)*, **rye** *(off-label)*, **triticale** *(off-label)* [1]; **managed amenity turf** [4]; **spring barley**, **spring oats**, **spring wheat**, **winter barley**, **winter oats**, **winter wheat** [1-4, 6]
- Docks in **managed amenity turf** [5, 7]
- Mayweeds in **amenity grassland** [3]; **durum wheat** *(off-label)*, **rye** *(off-label)*, **triticale** *(off-label)* [1]; **managed amenity turf** [4]; **spring barley**, **spring oats**, **spring wheat**, **winter barley**, **winter oats**, **winter wheat** [1-4, 6]
- Perennial dicotyledons in **amenity grassland** [1, 2, 6]; **managed amenity turf** [4, 5, 7]; **spring barley**, **spring oats**, **spring wheat**, **winter barley**, **winter oats**, **winter wheat** [4]
- Plantains in **managed amenity turf**, **spring barley**, **spring oats**, **spring wheat**, **winter barley**, **winter oats**, **winter wheat** [4]
- Polygonums in **amenity grassland** [3]; **durum wheat** *(off-label)*, **rye** *(off-label)*, **triticale** *(off-label)* [1]; **managed amenity turf** [4]; **spring barley**, **spring oats**, **spring wheat**, **winter barley**, **winter oats**, **winter wheat** [1-4, 6]
- Thistles in **managed amenity turf** [5, 7]

Extension of Authorisation for Minor Use (EAMUs)
- *durum wheat* *20140666* [1], *20152161* [6]
- *rye* *20140666* [1], *20152161* [6]
- *triticale* *20140666* [1], *20152161* [6]

Approval information
- Dicamba and mecoprop-P included in Annex I under EC Regulation 1107/2009
- Accepted by BBPA for use on malting barley

Efficacy guidance
- Best results by application in warm, moist weather when weeds are actively growing

Restrictions
- Maximum number of treatments 1 per crop for cereals and 1 or 2 per yr on grass depending on label. The total amount of mecoprop-P applied in a single yr must not exceed the maximum total dose approved for any single product for the crop/situation
- Do not spray in cold or frosty conditions
- Do not spray if rain expected within 6 h
- Do not treat undersown grass until tillering begins
- Do not spray cereals undersown with clover or legume mixtures
- Do not roll or harrow within 7 d before or after spraying
- Do not treat crops suffering from stress from any cause
- Use product immediately following dilution; do not allow diluted product to stand before use [2]
- Avoid treatment when drift may damage neighbouring susceptible crops
- Do not use on new grass for at least 6 months after establishment [7]
- Do not cut grass for at least 1 day after treatment

Crop-specific information
- Latest use: before 1st node detectable for cereals; 7 d before cutting or 14 d before grazing grass
- Apply to winter sown crops from 5 expanded leaf stage (GS 15)
- Apply to spring sown cereals from 5 expanded leaf stage but before first node is detectable (GS 15-31)
- Treat grassland just before perennial weeds flower
- Transient crop prostration may occur after spraying but recovery is rapid

Following crops guidance
- The total amount of mecoprop-P applied in a single year must not exceed the maximum total dose approved for that crop for any single product

Environmental safety
- Dangerous for the environment [2, 3, 6]
- Toxic to aquatic organisms
- Harmful to fish or other aquatic life. Do not contaminate surface waters or ditches with chemical or used container
- Keep livestock out of treated areas for at least 2 wk and until foliage of poisonous weeds such as ragwort has died and become unpalatable

FOR FULL CONDITIONS OF USE ALWAYS READ THE PRODUCT LABEL

- Some pesticides pose a greater threat of contamination of water than others and mecoprop-P is one of these pesticides. Take special care when applying mecoprop-P near water and do not apply if heavy rain is forecast

Hazard classification and safety precautions
Hazard Harmful, Dangerous for the environment [1-3, 6]; Irritant [4, 5, 7]; Harmful if swallowed [3]
UN Number N/C
Risk phrases H315, H318 [3, 7]; R22a, R38 [1, 2, 6]; R41, R53a [1, 2, 4-6]; R51 [1]; R52 [2, 4-6]
Operator protection A, C [1-7]; H, M [1, 2, 4-7]; U05a, U11, U20b [1-7]; U08 [1, 2, 4, 6]; U09a [3, 5, 7]; U15 [4, 5, 7]; U19a [5, 7]
Environmental protection E07a, E38 [1, 2, 4-7]; E13c [3-5, 7]; E15a [1, 2, 6]; E34 [1-7]; H411 [3]; H412 [7]
Storage and disposal D01, D02, D05, D09a [1-7]; D10b [1, 2, 6]; D10c [4, 5, 7]; D12a [1, 2, 4-7]
Medical advice M03 [1-7]; M05a [4, 5, 7]

154 dicamba + prosulfuron

A herbicide mixture for weed control in forage and grain maize
HRAC mode of action code: O + B

Products

1	Casper	Syngenta	50:5% w/w	WG	15573
2	Rosan	Headland	50:5% w/w	WG	16591

Uses
- Annual dicotyledons in **forage maize, grain maize**
- Bindweeds in **forage maize, grain maize**
- Docks in **forage maize** (seedlings only), **grain maize** (seedlings only)

Approval information
- Dicamba and prosulfuron included in Annex I under EC Regulation 1107/2009

Efficacy guidance
- Always apply in mixture with a non-ionic adjuvant.
- Do not apply in mixture with organo-phosphate insecticides

Following crops guidance
- Following normal harvest wheat, barley, rye, triticale and perennial ryegrass may be sown in the autumn, wheat, barley, rye, triticale, combining peas, maize, field beans, forage kale, broccoli and cauliflower may be sown the following spring but sugar beet, sunflowers or lucerne are not recommended.

Environmental safety
- LERAP Category B

Hazard classification and safety precautions
Hazard Dangerous for the environment [1, 2]; Very toxic to aquatic organisms [2]
Transport code 9
Packaging group III
UN Number 3077
Risk phrases R50, R53a [1]
Operator protection A, C, H; U02a, U05a, U14, U20b
Environmental protection E15b, E16a, E38 [1, 2]; H410 [2]
Storage and disposal D01, D02, D09a, D10c, D12a

SEE SECTION 3 FOR PRODUCTS ALSO REGISTERED

Pesticide Profiles

155   dichlorprop-P + ferrous sulphate + MCPA

A herbicide/fertilizer combination for moss and weed control in turf
HRAC mode of action code: O + O

See also ferrous sulphate
MCPA

Products

SHL Granular Feed, Weed & Mosskiller	Sinclair	0.2:10.9:0.3% w/w	GR	10972

Uses
- Annual dicotyledons in **managed amenity turf**
- Buttercups in **managed amenity turf**
- Moss in **managed amenity turf**
- Perennial dicotyledons in **managed amenity turf**

Approval information
- Dichlorprop-P, ferrous sulphate and MCPA included in Annex I under EC Regulation 1107/2009

Efficacy guidance
- Apply between Mar and Sep when grass in active growth and soil moist
- A repeat treatment may be needed after 4-6 wk to control perennial weeds or if moss regrows
- Water in if rainfall does not occur within 48 h

Restrictions
- Do not treat newly sown grass for 6 mth after establishment [1]
- Do not apply during drought or freezing conditions or when rain imminent
- Avoid walking on treated areas until it has rained or turf has been watered
- Do not mow for 3-4 d before or after application
- Do not use first 4 mowings after treatment for mulching. Mowings should be composted for 6 mth before use
- Do not treat areas of fine turf such as golf or bowling greens [1]
- Avoid contact with tarmac surfaces as staining may occur

Environmental safety
- Avoid drift onto nearby plants and borders [1]

Hazard classification and safety precautions
UN Number N/C
Operator protection U20c
Storage and disposal D01, D09a, D11a

156   dichlorprop-P + MCPA + mecoprop-P

A translocated herbicide mixture for winter and spring cereals
HRAC mode of action code: O + O + O

See also MCPA
mecoprop-P

Products

1	Isomec Ultra	Nufarm UK	310:160:130 g/l	SL	16033
2	Optica Trio	Headland	310:160:130 g/l	SL	16113

Uses
- Annual dicotyledons in **durum wheat**, **spelt**, **spring barley**, **spring oats**, **spring wheat**, **winter barley**, **winter oats**, **winter wheat** [1, 2]; **rye**, **triticale** [2]
- Chickweed in **durum wheat**, **spelt**, **spring barley**, **spring oats**, **spring wheat**, **winter barley**, **winter oats**, **winter wheat** [1, 2]; **rye**, **triticale** [2]

FOR FULL CONDITIONS OF USE ALWAYS READ THE PRODUCT LABEL

- Cleavers in **durum wheat, spelt, spring barley, spring oats, spring wheat, winter barley, winter oats, winter wheat** [1, 2]; **rye, triticale** [2]
- Field pansy in **durum wheat, spelt, spring barley, spring oats, spring wheat, winter barley, winter oats, winter wheat** [1, 2]; **rye, triticale** [2]
- Mayweeds in **durum wheat, spelt, spring barley, spring oats, spring wheat, winter barley, winter oats, winter wheat** [1, 2]; **rye, triticale** [2]
- Poppies in **durum wheat, spelt, spring barley, spring oats, spring wheat, winter barley, winter oats, winter wheat** [1, 2]; **rye, triticale** [2]

Approval information
- Dichlorprop-P, MCPA and mecoprop-P included in Annex I under EC Regulation 1107/2009
- Accepted by BBPA for use on malting barley

Efficacy guidance
- Best results obtained if application is made while majority of weeds are at seedling stage but not if temperatures are too low
- Optimum results achieved by spraying when temperature is above 10°C. If temperatures are lower delay spraying until growth becomes more active

Restrictions
- Maximum number of treatments 1 per crop
- Do not spray in windy conditions where spray drift may cause damage to neighbouring crops, especially sugar beet, oilseed rape, peas, turnips and most horticultural crops including lettuce and tomatoes under glass

Crop-specific information
- Latest use: before second node detectable (GS 32) for all crops

Environmental safety
- Harmful to aquatic organisms
- Harmful to fish or other aquatic life. Do not contaminate surface waters or ditches with chemical or used container
- Some pesticides pose a greater threat of contamination of water than others and mecoprop-P is one of these pesticides. Take special care when applying mecoprop-P near water and do not apply if heavy rain is forecast

Hazard classification and safety precautions
Hazard Harmful, Dangerous for the environment [1, 2]; Harmful if swallowed [1]
UN Number N/C
Risk phrases H317, H318 [1]; R22a, R41, R43, R53a [2]; R51 [1, 2]
Operator protection A, C, H, M; U05a, U08, U11, U20b [1, 2]; U14, U15 [1]
Environmental protection E13c, E34, E38
Storage and disposal D01, D02, D05, D09a, D10c [1, 2]; D12a [1]
Medical advice M03 [2]; M05a [1]

157 difenacoum

An anticoagulant coumarin rodenticide

Products

1	Difenag	Killgerm	0.005% w/w	RB	UK12-0497
2	Difenag Fresh Bait	Killgerm	0.005% w/w	RB	UK12-0436
3	Difenag Wax Bait	Killgerm	0.005% w/w	RB	UK14-0851
4	Difenag Wax Blocks	Killgerm	0.005% w/w	RB	UK12-0435
5	Neokil	BASF	0.005% w/w	RB	UK12-0298
6	Neosorexa Bait Blocks	BASF	0.005% w/w	RB	UK12-0360
7	Neosorexa Gold	BASF	0.005% w/w	RB	UK12-0304
8	Neosorexa Gold Ratpacks	BASF	0.005% w/w	RB	UK12-0306
9	Neosorexa Pasta Bait	BASF	0.005% w/w	RB	UK12-0365
10	Ratak Cut Wheat	Killgerm	0.005% w/w	RB	UK12-0313

SEE SECTION 3 FOR PRODUCTS ALSO REGISTERED

Pesticide Profiles

Products – continued

11	Sakarat D Pasta Bait	Killgerm	0.005% w/w	RB	UK12-0371
12	Sakarat D Wax Bait	Killgerm	0.005% w/w	RB	UK12-0370
13	Sakarat D Whole Wheat	Killgerm	0.005% w/w	RB	UK12-0301
14	Sorexa D	BASF	0.005% w/w	RB	UK12-0319
15	Sorexa Gel	BASF	0.005% w/w	RB	UK12-0364

Uses
- Mice in **farm buildings** [5-15]; **farm buildings/yards** [1-4]; **farmyards** [5-13, 15]; **sewers** [5]
- Rats in **farm buildings** [5-10, 13, 14]; **farm buildings/yards** [1-4]; **farmyards** [5-10, 13]; **sewers** [5]

Approval information
- Difenacoum included in Annex I under EC Regulation 1107/2009
- Difenacoum approval was to expire 31/3/2015 but the EU has postponed expiry until 30/6/2018 to allow time for a decision to be reached on the applications for renewal.

Efficacy guidance
- Difenacoum is a chronic poison and rodents need to feed several times before accumulating a lethal dose. Effective against rodents resistant to other commonly used anticoagulants
- Best results achieved by placing baits at points between nesting and feeding places, at entry points, in holes and where droppings are seen
- A minimum of five baiting points normally required for a small infestation; more than 40 for a large infestation
- Inspect bait sites frequently and top up as long as there is evidence of feeding
- Product formulated for application through a skeleton or caulking gun [15]
- Maintain a few baiting points to guard against reinfestation after a successful control campaign
- Resistance to difenacoum in rats is now widespread in the UK. Take local professional advice before relying on these products to control infestations.

Restrictions
- Only for use by farmers, horticulturists and other professional users
- When working in rodent infested areas wear synthetic rubber/PVC gloves to protect against rodent-borne diseases

Environmental safety
- Harmful to wildlife
- Cover baits by placing in bait boxes, drain pipes or under boards to prevent access by children, animals or birds
- Products contain human taste deterrent

Hazard classification and safety precautions
  UN Number N/C
  Operator protection A [15]; U13, U20b
  Environmental protection E10b [5-9, 14]; E15a [15]
  Storage and disposal D09a, D11a
  Vertebrate/rodent control products V01a, V03a, V04a [15]; V01b, V04b [5-9, 11, 12, 14]; V02 [5-9, 11, 12, 14, 15]; V03b [11, 12]; V04c [5-9, 14]
  Medical advice M03

158   difenoconazole

A diphenyl-ether triazole protectant and curative fungicide
FRAC mode of action code: 3

See also azoxystrobin + difenoconazole

Products

1	Difcor 250 EC	Belchim	250 g/l	EC	13917
2	Difend	Q-Chem	30 g/l	FS	16316
3	Difference	Belchim	250 g/l	EC	16129

FOR FULL CONDITIONS OF USE ALWAYS READ THE PRODUCT LABEL

Products – continued

4	Kix	Globachem	250 g/l	EC	17424
5	Narita	Belchim	250 g/l	EC	16210
6	Plover	Syngenta	250 g/l	EC	17288

Uses
- Alternaria in **borage for oilseed production** *(off-label)*, **canary flower (echium spp.)** *(off-label)*, **durum wheat** *(off-label)*, **evening primrose** *(off-label)*, **honesty** *(off-label)*, **linseed** *(off-label)*, **mustard** *(off-label)*, **rye** *(off-label)*, **triticale** *(off-label)* [1]; **broccoli, brussels sprouts, cabbages, calabrese, cauliflowers, spring oilseed rape, winter oilseed rape** [1, 6]; **collards, kale** [6]
- Blight in **forest nurseries** *(off-label)* [1]
- Brown rust in **durum wheat** *(off-label)*, **rye** *(off-label)*, **triticale** *(off-label)*, **winter wheat** [1]
- Bunt in **winter wheat** [2]
- Cladosporium in **durum wheat** *(off-label)*, **rye** *(off-label)*, **triticale** *(off-label)* [1]
- Disease control in **potatoes** [4, 5]
- Leaf spot in **forest nurseries** *(off-label)* [1]
- Light leaf spot in **spring oilseed rape, winter oilseed rape** [1, 6]
- Powdery mildew in **borage for oilseed production** *(off-label)*, **canary flower (echium spp.)** *(off-label)*, **evening primrose** *(off-label)*, **honesty** *(off-label)*, **linseed** *(off-label)*, **mustard** *(off-label)* [1]
- Ring spot in **broccoli, brussels sprouts, cabbages, calabrese, cauliflowers** [1, 6]; **collards, kale** [6]
- Rust in **forest nurseries** *(off-label)* [1]
- Scab in **apples, pears** [3]
- Sclerotinia in **borage for oilseed production** *(off-label)*, **canary flower (echium spp.)** *(off-label)*, **evening primrose** *(off-label)*, **honesty** *(off-label)*, **linseed** *(off-label)*, **mustard** *(off-label)* [1]
- Septoria leaf blotch in **durum wheat** *(off-label)*, **rye** *(off-label)*, **triticale** *(off-label)*, **winter wheat** [1]
- Stem canker in **spring oilseed rape, winter oilseed rape** [1, 6]
- Stinking smut in **winter wheat** [2]

Extension of Authorisation for Minor Use (EAMUs)
- *borage for oilseed production* 20140211 [1]
- *canary flower (echium spp.)* 20140211 [1]
- *durum wheat* 20140212 [1]
- *evening primrose* 20140211 [1]
- *forest nurseries* 20140205 [1]
- *honesty* 20140211 [1]
- *linseed* 20140211 [1]
- *mustard* 20140211 [1]
- *rye* 20140212 [1]
- *triticale* 20140212 [1]

Approval information
- Difenoconazole included in Annex I under EC Regulation 1107/2009

Efficacy guidance
- Improved control of established infections on oilseed rape achieved by mixture with carbendazim. See labels
- In brassicas a 3-spray programme should be used starting at the first sign of disease and repeated at 14-21 d intervals
- Product is fully rainfast 2 h after application
- For most effective control of Septoria, apply as part of a programme of sprays which includes a suitable flag leaf treatment
- Difenoconazole is a DMI fungicide. Resistance to some DMI fungicides has been identified in Septoria leaf blotch which may seriously affect performance of some products. For further advice contact a specialist advisor and visit the Fungicide Resistance Action Group (FRAG)-UK website

SEE SECTION 3 FOR PRODUCTS ALSO REGISTERED

Pesticide Profiles

Restrictions
- Maximum number of treatments 3 per crop for brassicas; 2 per crop for oilseed rape; 1 per crop for wheat [1]
- Maximum total dose equivalent to 3 full dose treatments on brassicas; 2 full dose treatments on oilseed rape; 1 full dose treatment on wheat
- Apply to wheat any time from ear fully emerged stage but before early milk-ripe stage (GS 59-73)

Crop-specific information
- Latest use: before grain early milk-ripe stage (GS 73) for cereals; end of flowering for oilseed rape
- HI brassicas 21 d
- Treat oilseed rape in autumn from 4 expanded true leaf stage (GS 1,4). A repeat spray may be made in spring at the beginning of stem extension (GS 2,0) if visible symptoms develop

Environmental safety
- Dangerous for the environment
- Very toxic to aquatic organisms
- Broadcast air-assisted LERAP [3] (20m); LERAP Category B [1, 4-6]

Hazard classification and safety precautions
Hazard Harmful [3]; Irritant [1]; Dangerous for the environment [1-6]; Harmful if swallowed [1, 4, 5]
Transport code 9
Packaging group III
UN Number 3082
Risk phrases H304 [1, 4-6]; H319 [1, 4, 5]; R22a, R22b, R36, R38, R67 [3]; R51, R53a [2, 3]
Operator protection A, H [1-6]; C [1, 3-6]; U05a [1-6]; U09a [1, 4-6]; U11 [1, 3]; U20b [1, 3-6]
Environmental protection E15a, E16a [1, 4-6]; E15b [3]; E17b [3] (20m); E38 [1-6]; H410 [6]; H411 [1, 4, 5]
Storage and disposal D01, D02 [1-6]; D05 [2, 4-6]; D07, D10c [4-6]; D09a [1, 4-6]; D10b [1]; D12a [1, 2, 4-6]
Treated seed S01, S02, S03, S04a, S05, S06a, S07, S08 [2]
Medical advice M05b [3]

159   difenoconazole + fludioxonil

A triazole + phenylpyrrole seed treatment for use in cereals
FRAC mode of action code: 3 + 12

See also fludioxonil

Products
Celest Extra   Syngenta   25:25 g/l   FS   16630

Uses
- Bunt in **winter wheat**
- Fusarium foot rot and seedling blight in **winter oats**, **winter wheat**
- Microdochium nivale in **winter oats**, **winter wheat**
- Pyrenophora leaf spot in **winter oats**
- Seed-borne diseases in **winter wheat**
- Seedling blight and foot rot in **winter wheat**
- Septoria seedling blight in **winter wheat**
- Snow mould in **winter wheat**
- Stripe smut in **winter rye**

Approval information
- Difenoconazole and fludioxonil included in Annex I under EC Regulation 1107/2009

Efficacy guidance
- Effective against benzimidazole-resistant and benzimidazole-sensitive strains of *Microdochium nivale*

FOR FULL CONDITIONS OF USE ALWAYS READ THE PRODUCT LABEL

difenoconazole + mandipropamid

Crop-specific information
- Under adverse environmental or soil conditions, seed rates should be increased to compensate for a slight drop in germination capacity. Flow rates of treated seed should be checked before drilling commences.

Hazard classification and safety precautions
  Hazard Harmful, Dangerous for the environment
  Transport code 9
  Packaging group III
  UN Number 3082
  Risk phrases R20, R51, R53a
  Operator protection A, H; U05a, U20c
  Environmental protection E15a, E38
  Storage and disposal D01, D02, D09a, D11a, D12a
  Treated seed S01, S02, S03, S04a, S04b, S05, S06a, S07, S08, S09
  Medical advice M05a

160   difenoconazole + mandipropamid

A triazole/mandelamide fungicide mixture for blight control in potatoes
FRAC mode of action code: 3 + 40

Products

1	Amphore Plus	Syngenta	250:250 g/l	SL	16327
2	Carial Star	Syngenta	250:250 g/l	SL	16323

Uses
- Alternaria blight in *potatoes*
- Blight in *potatoes*
- Early blight in *potatoes*

Approval information
- Difenoconazole and mandipropamid included in Annex I under EC Regulation 1107/2009

Efficacy guidance
- Rainfast within 15 minutes
- Spray programme must start before blight enters the crop

Crop-specific information
- Can be used on all varieties of potato, including seed potatoes

Environmental safety
- LERAP Category B

Hazard classification and safety precautions
  Transport code 9
  Packaging group III
  UN Number 3082
  Risk phrases R22a, R50, R53a
  Operator protection A, H; U05a, U20b
  Environmental protection E15b, E16a, E22c, E34, E38
  Storage and disposal D01, D09a, D10c
  Medical advice M03

161   difenoconazole + paclobutrazol

A triazole fungicide mixture for growth regulation and disease control in oilseed rape
FRAC mode of action code: 3

Products

Toprex	Syngenta	250:125 g/l	SC	16456

SEE SECTION 3 FOR PRODUCTS ALSO REGISTERED

Pesticide Profiles

Uses
- Disease control in **winter oilseed rape**
- Kabatiella lini in **linseed** *(off-label)*
- Light leaf spot in **winter oilseed rape**
- Mildew in **linseed** *(off-label)*
- Phoma in **winter oilseed rape**
- Septoria leaf blotch in **linseed** *(off-label)*

Extension of Authorisation for Minor Use (EAMUs)
- **linseed** *20151633*

Approval information
- Difenoconazole and paclobutrazol included in Annex I under EC Regulation 1107/2009

Efficacy guidance
- Contains a growth regulator for height reduction & lodging control and a fungicide for disease control. It should only be used when both growth regulation and disease control are required. If this is not the case, use appropriate alternative products at the required timing

Hazard classification and safety precautions
Hazard Dangerous for the environment
Transport code 9
Packaging group III
UN Number 3082
Risk phrases R50, R53a
Operator protection A, H; U05a
Environmental protection E38
Storage and disposal D01, D02, D12a
Medical advice M05a

162 difenoconazole + propiconazole

A triazole fungicide mixture for disease control in fodder beet and sugar beet
FRAC mode of action code: 3

See also difenoconazole
propiconazole

Products

1	Armure	Syngenta	150:150 g/l	EC	16587
2	Difure Pro	Belchim	150:150 g/l	EC	17303

Uses
- Cercospora leaf spot in **fodder beet**, **sugar beet**
- Powdery mildew in **fodder beet**, **sugar beet**
- Ramularia leaf spots in **fodder beet** *(moderate control)*, **sugar beet** *(moderate control)*
- Rust in **fodder beet**, **sugar beet**

Approval information
- Difenoconazole and propiconazole included in Annex I under EC Regulation 1107/2009

Efficacy guidance
- Use preventatively or at the first sign of disease in the crop

Restrictions
- Do not apply before complete crop cover is achieved.
- Do not apply using hand-held equipment.

Environmental safety
- Avoid spraying/application within 5 metres of the field boundary to reduce effects on non-target insects or other arthropods.
- LERAP Category B

FOR FULL CONDITIONS OF USE ALWAYS READ THE PRODUCT LABEL

diflubenzuron

Hazard classification and safety precautions
 Hazard Irritant, Dangerous for the environment [1, 2]; Harmful if swallowed [2]
 Transport code 9
 Packaging group III
 UN Number 3082
 Risk phrases H304 [1, 2]; H315, H318, H336, H373 [2]; H319, H335 [1]
 Operator protection A, C; U04a, U09a, U20a
 Environmental protection E15b, E16a, E34, E38, H410
 Storage and disposal D01, D02, D05, D09a, D10c, D12a
 Medical advice M03

163   diflubenzuron

A selective, persistent, contact and stomach acting insecticide
IRAC mode of action code: 15

Products
Dimilin Flo   Certis   480 g/l   SC   11056

Uses
- Browntail moth in **amenity vegetation**, **hedges**, **nursery stock**, **ornamental plant production**
- Bud moth in **apples**, **pears**
- Carnation tortrix moth in **amenity vegetation**, **hedges**, **nursery stock**, **ornamental plant production**
- Caterpillars in **broccoli**, **brussels sprouts**, **cabbages**, **calabrese**, **cauliflowers**, **chives** (off-label), **endives** (off-label), **frise** (off-label), **herbs (see appendix 6)** (off-label), **leaf spinach** (off-label), **lettuce** (off-label), **parsley** (off-label), **radicchio** (off-label), **salad brassicas** (off-label - for baby leaf production), **scarole** (off-label)
- Clouded drab moth in **apples**, **pears**
- Codling moth in **apples**, **pears**
- Fruit tree tortrix moth in **apples**, **pears**
- Houseflies in **livestock houses**, **manure heaps**, **refuse tips**
- Lackey moth in **amenity vegetation**, **hedges**, **nursery stock**, **ornamental plant production**
- Moths in **almonds** (off-label), **bilberries** (off-label), **blackberries** (off-label), **chestnuts** (off-label), **crab apples** (off-label), **cranberries** (off-label), **gooseberries** (off-label), **hazel nuts** (off-label), **quinces** (off-label), **redcurrants** (off-label), **walnuts** (off-label), **whitecurrants** (off-label)
- Oak leaf roller moth in **forest**
- Pear sucker in **pears**
- Phorid flies in **edible fungi** (off-label - other than mushrooms)
- Pine beauty moth in **forest**
- Pine looper in **forest**
- Plum fruit moth in **plums**
- Rust mite in **apples**, **pears**, **plums**
- Sciarid flies in **edible fungi** (off-label - other than mushrooms)
- Small ermine moth in **amenity vegetation**, **hedges**, **nursery stock**, **ornamental plant production**
- Tortrix moths in **plums**
- Winter moth in **amenity vegetation**, **apples**, **blackcurrants**, **forest**, **hedges**, **nursery stock**, **ornamental plant production**, **pears**, **plums**

Extension of Authorisation for Minor Use (EAMUs)
- *almonds* 20060571
- *bilberries* 20060573
- *blackberries* 20060573
- *chestnuts* 20060571
- *chives* 20051321
- *crab apples* 20060572
- *cranberries* 20060573
- *edible fungi* (other than mushrooms) 20060574

SEE SECTION 3 FOR PRODUCTS ALSO REGISTERED

Pesticide Profiles

- *endives* 20051321
- *frise* 20051321
- *gooseberries* 20060573
- *hazel nuts* 20060571
- *herbs (see appendix 6)* 20051321
- *leaf spinach* 20051321
- *lettuce* 20051321
- *parsley* 20051321
- *quinces* 20060572
- *radicchio* 20051321
- *redcurrants* 20060573
- *salad brassicas* (for baby leaf production) 20051321
- *scarole* 20051321
- *walnuts* 20060571
- *whitecurrants* 20060573

Approval information
- Approved for aerial application in forestry when average wind velocity does not exceed 18 knots and gusts do not exceed 20 knots. See Section 5 for more information
- Diflubenzuron included in Annex I under EC Regulation 1107/2009

Efficacy guidance
- Most active on young caterpillars and most effective control achieved by spraying as eggs start to hatch
- Dose and timing of spray treatments vary with pest and crop. See label for details
- Addition of wetter recommended for use on brassicas and for pear sucker control in pears

Restrictions
- Maximum number of treatments 3 per yr for apples, pears; 2 per yr for plums, blackcurrants; 2 per crop for brassicas; 1 per yr for forest
- Before treating ornamentals check varietal tolerance on a small sample
- Do not use as a compost drench or incorporated treatment on ornamental crops
- Do not spray protected plants in flower or with flower buds showing colour
- For use only on the food crops specified on the label

Crop-specific information
- HI apples, pears, plums, blackcurrants, brassicas 14 d

Environmental safety
- Dangerous for the environment
- Very toxic to aquatic organisms
- Broadcast air-assisted LERAP [1] (20 m when used in orchards; 10 m when used in blackcurrants, forestry or ornamentals); LERAP Category B [1]

Hazard classification and safety precautions
Hazard Dangerous for the environment
Transport code 9
Packaging group III
UN Number 3082
Risk phrases R50
Operator protection U20c
Environmental protection E05a, E15a, E16a, E16b, E18, E38 [1]; E17b [1] (20 m when used in orchards; 10 m when used in blackcurrants, forestry or ornamentals)
Storage and disposal D05, D09a, D11a, D12b

FOR FULL CONDITIONS OF USE ALWAYS READ THE PRODUCT LABEL

164 diflufenican

A shoot absorbed pyridinecarboxamide herbicide for winter cereals
HRAC mode of action code: F1

See also bromoxynil + diflufenican
bromoxynil + diflufenican + ioxynil
chlorotoluron + diflufenican
chlorotoluron + diflufenican + pendimethalin
clodinafop-propargyl + diflufenican
clopyralid + diflufenican + MCPA

Products

1	Clayton El Nino	Clayton	500 g/l	SC	17337
2	Diflanil 500 SC	Q-Chem	500 g/l	SC	17024
3	Goshawk	Goldengrass	500 g/l	SC	16514
4	Hurricane SC	Adama	500 g/l	SC	16027
5	Ossetia	Rotam	50% w/w	WG	17741
6	Sempra	UPL Europe	500 g/l	SC	16967
7	Sempra XL	UPL Europe	500 g/l	SC	17447
8	Solo 500 SC	Sipcam	500 g/l	SC	17622

Uses
- Annual dicotyledons in **forest nurseries** (off-label), **game cover** (off-label), **ornamental plant production** (off-label), **spring oats** (off-label), **winter oats** (off-label) [4]; **grass seed crops** (off-label) [4, 6]; **oat seed crops** (off-label) [6]; **spring barley, triticale, winter barley, winter rye, winter wheat** [2, 6]
- Annual grasses in **forest nurseries** (off-label), **game cover** (off-label), **ornamental plant production** (off-label), **spring oats** (off-label), **winter oats** (off-label) [4]; **grass seed crops** (off-label) [4, 6]; **oat seed crops** (off-label), **spring barley, triticale, winter barley, winter rye, winter wheat** [6]
- Annual meadow grass in **forest nurseries** (off-label), **game cover** (off-label), **grass seed crops** (off-label), **ornamental plant production** (off-label), **spring oats** (off-label), **winter oats** (off-label) [4]
- Brassica spp. in **durum wheat, spring barley, spring wheat, triticale, winter barley, winter rye, winter wheat** [5]
- Chickweed in **spring barley, triticale, winter barley, winter rye, winter wheat** [2]
- Cleavers in **durum wheat, spring wheat** [1, 3-5, 7]; **rye** [8]; **spring barley, triticale, winter barley, winter wheat** [1-5, 7, 8]; **winter rye** [1-5, 7]
- Field pansy in **durum wheat, spring wheat, winter rye** [1, 3-5, 7]; **rye** [8]; **spring barley, triticale, winter barley, winter wheat** [1, 3-5, 7, 8]
- Field speedwell in **durum wheat, spring wheat** [1, 3-5, 7]; **rye** [8]; **spring barley, triticale, winter barley, winter wheat** [1-5, 7, 8]; **winter rye** [1-5, 7]
- Ivy-leaved speedwell in **durum wheat, spring wheat** [1, 3-5, 7]; **rye** [8]; **spring barley, triticale, winter barley, winter wheat** [1-5, 7, 8]; **winter rye** [1-5, 7]
- Mayweeds in **durum wheat, spring wheat, winter rye** [1, 3-5, 7]; **rye** [8]; **spring barley, triticale, winter barley, winter wheat** [1, 3-5, 7, 8]
- Poppies in **durum wheat, spring wheat, winter rye** [1, 3-5, 7]; **rye** [8]; **spring barley, triticale, winter barley, winter wheat** [1, 3-5, 7, 8]
- Red dead-nettle in **durum wheat, spring wheat, winter rye** [1, 3, 4, 7]; **rye** [8]; **spring barley, triticale, winter barley, winter wheat** [1, 3, 4, 7, 8]
- Volunteer oilseed rape in **spring barley, triticale, winter barley, winter rye, winter wheat** [2]

Extension of Authorisation for Minor Use (EAMUs)
- *forest nurseries* 20130142 [4]
- *game cover* 20130142 [4], 20141325 [4]
- *grass seed crops* 20130141 [4], 20141324 [4], 20160023 [6]
- *oat seed crops* 20160022 [6]
- *ornamental plant production* 20130142 [4]

SEE SECTION 3 FOR PRODUCTS ALSO REGISTERED

Pesticide Profiles

- **spring oats** *20130140* [4], *20141269* [4]
- **winter oats** *20130140* [4], *20141269* [4]

Approval information
- Diflufenican included in Annex I under EC Regulation 1107/2009
- Accepted by BBPA for use on malting barley

Efficacy guidance
- Best results achieved from treatment of small actively growing weeds in early autumn or spring
- Good weed control depends on efficient burial of trash or straw before or during seedbed preparation
- Loose or fluffy seedbeds should be rolled before application
- The final seedbed should be moist, fine and firm with clods no bigger than fist size
- Ensure good even spray coverage and increase spray volume for post-emergence treatments where the crop or weed foliage is dense
- Activity may be slow under cool conditions and final level of weed control may take some time to appear
- Where cleavers are a particular problem a separate specific herbicide treatment may be required
- Efficacy may be impaired on soils with a Kd factor greater than 6
- Always follow WRAG guidelines for preventing and managing herbicide resistant weeds. See Section 5 for more information

Restrictions
- Maximum number of treatments 1 per crop
- Do not treat broadcast crops [2, 4]
- Do not roll treated crops or harrow at any time after treatment
- Do not apply to soils with more than 10% organic matter or on Sands, or very stony or gravelly soils
- Do not treat after a period of cold frosty weather

Crop-specific information
- Latest use: before end of tillering (GS 29) for wheat and barley [2, 4], pre crop emergence for triticale and winter rye [2, 4]
- Treat only named varieties of rye or triticale [2, 4]

Following crops guidance
- Labels vary slightly but in general ploughing to 150 mm and thoroughly mixing the soil is recommended before drilling or planting any succeeding crops either after crop failure or after normal harvest
- In the event of crop failure only winter wheat or winter barley may be re-drilled immediately after ploughing. Spring crops of wheat, barley, oilseed rape, peas, field beans, sugar beet [2, 4], potatoes, carrots, edible brassicas or onions may be sown provided an interval of 12 wk has elapsed after ploughing
- After normal harvest of a treated crop winter cereals, oilseed rape, field beans, leaf brassicas, sugar beet seed crops and winter onions may be drilled in the following autumn. Other crops listed above for crop failure may be sown in the spring after normal harvest
- Successive treatments with any products containing diflufenican can lead to soil build-up and inversion ploughing to 150 mm must precede sowing any following non-cereal crop. Even where ploughing occurs some crops may be damaged

Environmental safety
- Dangerous for the environment
- Very toxic to aquatic organisms
- Buffer zone requirement 6m [1, 7, 8]
- Buffer zone requirement 7 m for spring barley, 8m for winter barley, winter rye, triticale and winter wheat [2, 6]
- Buffer zone requirement 8m for rye and triticale, 14m for spring barley, winter barley, Durum wheat, spring wheat and winter wheat [5]
- Buffer zone requirement 12 m [3, 4]
- LERAP Category B

FOR FULL CONDITIONS OF USE ALWAYS READ THE PRODUCT LABEL

diflufenican + florasulam

Hazard classification and safety precautions
   Hazard Dangerous for the environment [1-8]; Very toxic to aquatic organisms [5, 8]
   Transport code 9
   Packaging group III
   UN Number 3077 [5]; 3082 [1-4, 6-8]
   Risk phrases R50, R53a [2-4, 6]
   Operator protection A [1-8]; C [1-4, 7, 8]; U05a [6, 8]; U08, U13, U19a [1, 3-5, 7, 8]; U15 [1, 3-5, 7]; U20a [1, 3-8]
   Environmental protection E13a [6]; E13c [8]; E15a [2]; E16a [1-8]; E16i [1, 3, 4]; E34, E38 [1, 3-5, 7, 8]; H410 [1, 5, 7, 8]
   Storage and disposal D01, D02 [1, 3-8]; D05 [1-5, 7, 8]; D09a, D10b [1-8]; D12a [1-7]
   Medical advice M05a [1, 3-5, 7]

165 diflufenican + florasulam

A herbicide mixture for weed control in cereals
HRAC mode of action code: B + F1

See also diflufenican
florasulam

Products
   Lector Delta   Headland   500:50 g/l   SC   16808

Uses
- Annual dicotyledons in *rye, spring barley, triticale, winter barley, winter wheat*
- Black bindweed in *rye, spring barley, triticale, winter barley, winter wheat*
- Charlock in *rye, spring barley, triticale, winter barley, winter wheat*
- Chickweed in *rye, spring barley, triticale, winter barley, winter wheat*
- Cleavers in *rye, spring barley, triticale, winter barley, winter wheat*
- Field pansy in *rye, spring barley, triticale, winter barley, winter wheat*
- Forget-me-not in *rye, spring barley, triticale, winter barley, winter wheat*
- Hemp-nettle in *rye, spring barley, triticale, winter barley, winter wheat*
- Mayweeds in *rye, spring barley, triticale, winter barley, winter wheat*
- Penny cress in *rye, spring barley, triticale, winter barley, winter wheat*
- Shepherd's purse in *rye, spring barley, triticale, winter barley, winter wheat*
- Volunteer oilseed rape in *rye, spring barley, triticale, winter barley, winter wheat*

Approval information
- Diflufenican and florasulam included in Annex I under EC Regulation 1107/2009

Restrictions
- Following application to grass seed crops, the treated grass must not be fed to livestock.
- The total amount of florasulam applied to a cereal crop must not exceed 7.5 g.a.i/ha

Following crops guidance
- In the event of crop failure in the spring, cultivate to 20 cms and then only plant spring wheat, spring barley or spring oats.
- Crops that can be sown in the same calendar year following treatment are cereals, oilseed rape, field beans, grass, peas, sugar beet, potatoes, maize and vegetable brassicas as transplants.

Environmental safety
- Buffer zone requirement 6 m for spring barley, 5 m for winter barley, winter wheat, winter rye and triticale [1]
- LERAP Category B

Hazard classification and safety precautions
   Hazard Dangerous for the environment
   Transport code 9
   Packaging group III
   UN Number 3082

SEE SECTION 3 FOR PRODUCTS ALSO REGISTERED

Pesticide Profiles

Operator protection A
Environmental protection E15b, E16a, E38, H410
Storage and disposal D01, D02, D03, D05, D06a, D09a, D10c, D12a

166 diflufenican + flufenacet

A contact and residual herbicide mixture for cereals
HRAC mode of action code: F1 + K3

See also flufenacet

Products

1	Clayton Vista	Clayton	200:400 g/l	SC	16705
2	Firestorm	Certis	100:400 g/l	SC	17631
3	Herold	Adama	200:400 g/l	SC	16195
4	Liberator	Bayer CropScience	100:400 g/l	SC	15206
5	Pincer	Agrii	100:400 g/l	SC	17130
6	Regatta	Bayer CropScience	100:400 g/l	SC	15353
7	Terrane	Headland	100:400 g/l	SC	17619

Uses
- Annual dicotyledons in **durum wheat** *(off-label)*, **rye** *(off-label)*, **triticale** *(off-label)* [3, 4]; **grass seed crops** *(off-label)* [4]; **spring barley** *(off-label)* [3, 4, 6]; **spring wheat** *(off-label)* [3]; **winter oats** *(off-label)* [3, 6]
- Annual grasses in **winter oats** *(off-label)* [4]
- Annual meadow grass in **durum wheat** *(off-label)*, **rye** *(off-label)*, **triticale** *(off-label)* [3, 4]; **grass seed crops** *(off-label)* [4]; **spring barley** *(off-label)* [3, 4, 6]; **spring wheat** *(off-label)* [3]; **triticale, winter rye** [7]; **winter barley, winter wheat** [1-7]; **winter oats** *(off-label)* [3, 6]
- Blackgrass in **durum wheat** *(off-label)*, **grass seed crops** *(off-label)*, **rye** *(off-label)*, **triticale** *(off-label)* [4]; **spring barley** *(off-label)*, **winter oats** *(off-label)* [4, 6]; **triticale, winter rye** [7]; **winter barley, winter wheat** [1-7]
- Chickweed in **triticale, winter rye** [7]; **winter barley, winter wheat** [1-7]
- Field pansy in **triticale, winter rye** [7]; **winter barley, winter wheat** [1-7]
- Field speedwell in **triticale, winter rye** [7]; **winter barley, winter wheat** [1-7]
- Ivy-leaved speedwell in **triticale, winter rye** [7]; **winter barley, winter wheat** [1-7]
- Mayweeds in **triticale, winter rye** [7]; **winter barley, winter wheat** [1-7]
- Red dead-nettle in **winter barley, winter wheat** [2, 4-6]
- Shepherd's purse in **triticale, winter rye** [7]; **winter barley, winter wheat** [1, 3, 7]
- Volunteer oilseed rape in **triticale, winter rye** [7]; **winter barley, winter wheat** [1, 3, 7]
- Wild oats in **winter oats** *(off-label)* [4]

Extension of Authorisation for Minor Use (EAMUs)
- *durum wheat* 20131493 [3], 20141699 [3], 20110297 [4]
- *grass seed crops* 20110297 [4]
- *rye* 20131493 [3], 20141699 [3], 20110297 [4]
- *spring barley* 20131492 [3], 20141698 [3], 20121010 [4], 20121905 [6]
- *spring wheat* 20131492 [3], 20141698 [3]
- *triticale* 20131493 [3], 20141699 [3], 20110297 [4]
- *winter oats* 20131491 [3], 20141693 [3], 20111550 [4], 20121904 [6]

Approval information
- Diflufenican and flufenacet included in Annex I under EC Regulation 1107/2009
- Accepted by BBPA for use on malting barley

Efficacy guidance
- Best results obtained when there is moist soil at and after application and rain falls within 7 d
- Residual control may be reduced under prolonged dry conditions
- Activity may be slow under cool conditions and final level of weed control may take some time to appear
- Good weed control depends on burying any trash or straw before or during seedbed preparation

FOR FULL CONDITIONS OF USE ALWAYS READ THE PRODUCT LABEL

- Established perennial grasses and broad-leaved weeds will not be controlled
- Do not use as a stand-alone treatment for blackgrass control. Always follow WRAG guidelines for preventing and managing herbicide resistant weeds. Section 5 for more information

Restrictions
- Maximum number of treatments 1 per crop
- Do not treat undersown cereals or those to be undersown
- Do not use on waterlogged soils or soils prone to waterlogging
- Do not use on Sands or Very Light soils, or very stony or gravelly soils, or on soils containing more than 10% organic matter
- Do not treat broadcast crops and treat shallow-drilled crops post-emergence only
- Do not incorporate into the soil or disturb the soil after application by rolling or harrowing
- Avoid treating crops under stress from whatever cause and avoid treating during periods of prolonged or severe frosts

Crop-specific information
- Latest use: before 31 Dec in yr of sowing and before 3rd tiller stage (GS 23) for wheat or 4th tiller stage (GS 24) for barley
- For pre-emergence treatments the seed should be covered with a minimum of 32 mm settled soil

Following crops guidance
- In the event of crop failure wheat, barley or potatoes may be sown provided the soil is ploughed to 15 cm, and a minimum of 12 weeks elapse between treatment and sowing spring wheat or spring barley
- After normal harvest wheat, barley or potatoes my be sown without special cultivations. Soil must be ploughed or cultivated to 15 cm before sowing oilseed rape, field beans, peas, sugar beet, carrots, onions or edible brassicae
- Successive treatments with any products containing diflufenican can lead to soil build-up and inversion ploughing to 150 mm must precede sowing any following non-cereal crop. Even where ploughing occurs some crops may be damaged

Environmental safety
- Dangerous for the environment
- Very toxic to aquatic organisms
- Risk to non-target insects or other arthropods. Avoid spraying within 6 m of the field boundary to reduce the effects on non-target insects or other arthropods
- Buffer zone requirement 10 m
- Buffer zone requirement 6 m [7]
- LERAP Category B

Hazard classification and safety precautions
    Hazard Harmful, Dangerous for the environment [1-7]; Harmful if swallowed, Very toxic to aquatic organisms [2]
    Transport code 9
    Packaging group III
    UN Number 3082
    Risk phrases H373 [2, 7]; R22a, R48, R50, R53a [1, 3-6]; R43 [1, 3-7]
    Operator protection A, H; U05a, U14
    Environmental protection E15a [1, 5, 6]; E15b [3, 7]; E16a, E34, E38 [1-7]; E22c [1-4, 7]; H410 [2, 7]
    Storage and disposal D01, D02, D09a, D10b, D12a
    Medical advice M03, M05a

SEE SECTION 3 FOR PRODUCTS ALSO REGISTERED

167 diflufenican + flufenacet + flurtamone

A herbicide mixture for use in wheat and barley
HRAC mode of action code: F1 + K3

Products
1	Movon	Bayer CropScience	90:240:120 g/l	SC	14784
2	Vigon	Bayer CropScience	60:240:120 g/l	SC	14785

Uses
- Annual meadow grass in **winter barley, winter wheat** [1, 2]
- Blackgrass in **winter wheat** *(moderately susceptible)* [1, 2]
- Chickweed in **winter barley, winter wheat** [1, 2]
- Field pansy in **winter barley, winter wheat** [1, 2]
- Field speedwell in **winter barley, winter wheat** [1, 2]
- Forget-me-not in **winter wheat** [1, 2]
- Groundsel in **winter wheat** [1, 2]
- Italian ryegrass in **winter wheat** *(moderately susceptible)* [1, 2]
- Ivy-leaved speedwell in **winter wheat** [1, 2]
- Mayweeds in **winter barley** [1]; **winter wheat** [1, 2]

Approval information
- Diflufenican, flufenacet and flurtamone included in Annex 1 under EC Regulation 1107/2009
- Accepted by BBPA for use on malting barley

Environmental safety
- LERAP Category B

Hazard classification and safety precautions
Hazard Harmful, Dangerous for the environment
Transport code 9
Packaging group III
UN Number 3082
Risk phrases R22a, R50, R53a, R69
Operator protection A; U05a
Environmental protection E15b, E16a, E22c, E34, E38, E39, E40b
Storage and disposal D01, D02, D09a, D10b, D12a
Medical advice M03, M05a

168 diflufenican + flupyrsulfuron-methyl

A pyridinecarboxamide and sulfonylurea herbicide mixture for cereals
HRAC mode of action code: F1 + B

See also flupyrsulfuron-methyl

Products
1	Absolute	DuPont	41.7:8.3% w/w	WG	12558
2	Excalibur	DuPont	44.4:5.6 % w/w	WG	16477

Uses
- Annual dicotyledons in **forest nurseries** *(off-label)* [1]; **winter barley, winter wheat** [1, 2]; **winter oats** [2]
- Blackgrass in **forest nurseries** *(off-label)*, **winter oats** *(off-label)* [1]; **winter barley, winter wheat** [1, 2]; **winter oats** [2]
- Charlock in **winter barley, winter oats, winter wheat** [2]
- Chickweed in **winter barley, winter oats, winter wheat** [2]
- Crane's-bill in **forest nurseries** *(off-label)* [1]; **winter barley, winter oats, winter wheat** [2]; **winter oats** *(off-label)* [1, 2]
- Field pansy in **forest nurseries** *(off-label)* [1]; **winter barley, winter oats, winter wheat** [2]; **winter oats** *(off-label)* [1, 2]

FOR FULL CONDITIONS OF USE ALWAYS READ THE PRODUCT LABEL

diflufenican + flupyrsulfuron-methyl

- Forget-me-not in **winter barley, winter oats, winter wheat** [2]
- Fumitory in **forest nurseries** *(off-label)* [1]; **winter oats** *(off-label)* [1, 2]
- Groundsel in **winter barley, winter oats, winter wheat** [2]
- Knotgrass in **winter barley, winter oats, winter wheat** [2]
- Mayweeds in **winter barley, winter oats, winter wheat** [2]
- Poppies in **winter barley, winter oats, winter oats** *(off-label)*, **winter wheat** [2]
- Red dead-nettle in **winter barley, winter oats, winter wheat** [2]
- Shepherd's purse in **winter barley, winter oats, winter wheat** [2]
- Speedwells in **winter barley, winter oats, winter wheat** [2]
- Volunteer oilseed rape in **winter barley, winter oats, winter wheat** [2]

Extension of Authorisation for Minor Use (EAMUs)
- **forest nurseries** 20142723 [1]
- **winter oats** 20142017 [1], 20142224 [2]

Approval information
- Diflufenican and flupyrsulfuron-methyl included in Annex I under EC Regulation 1107/2009

Efficacy guidance
- Best results achieved from applications made in good growing conditions
- Good spray cover of weeds must be obtained
- Best control of blackgrass obtained from application from the 2-leaf stage but tank mixing or sequencing with a partner product is recommended as a resistance management strategy (see below)
- Growth of weeds is inhibited within hours of treatment but visible symptoms may not be apparent for up to 4 wk
- Product has moderate residual life in soil. Under normal moisture conditions susceptible weeds germinating soon after treatment will be controlled
- Product may be used on all soil types except those with more than 10% organic matter, but residual activity and weed control is reduced by high soil temperatures and when soil is very dry and on highly alkaline soils
- Flupyrsulfuron-methyl is a member of the ALS-inhibitor group of herbicides. To avoid the build up of resistance do not use any product containing an ALS-inhibitor herbicide with claims for control of grass weeds more than once on any crop
- Use these products as part of a resistance management strategy that includes cultural methods of control and does not use ALS inhibitors as the sole chemical method of grass weed control

Restrictions
- Maximum number of treatments 1 per crop
- Do not use on wheat undersown with grasses, clover or other legumes, or any other broad-leaved crop
- Do not apply within 7 d of rolling
- Do not treat broadcast crops. Seed should be covered by at least 25 mm of settled soil
- Do not treat any crop suffering from drought, waterlogging, pest or disease attack, nutrient deficiency, or any other stress factors
- Do not use on soils with more than 10% organic matter
- Specific restrictions apply to use in sequence or tank mixture with other sulfonylurea or ALS-inhibiting herbicides. See label for details

Crop-specific information
- Latest use: Pre-emergence in winter barley and winter oats; before end of March in year of harvest and before GS23 in winter wheat
- Under certain conditions slight chlorosis and stunting may occur from which recovery is rapid

Following crops guidance
- Only cereals, oilseed rape or field beans may be sown in the yr of harvest of a treated crop. Ploughing must be carried out before drilling any non-cereal crop
- An interval of at least 20 wk must elapse after treatment before planting wheat, barley, oilseed rape, peas, field beans, sugar beet, potatoes, carrots, brassicas or onions in the following spring, and only after ploughing [1, 2]

SEE SECTION 3 FOR PRODUCTS ALSO REGISTERED

Pesticide Profiles

- In the event of failure of a treated crop only winter wheat or spring wheat may be sown within 3 mth of treatment, and only after ploughing and cultivating to at least 15 cm. After 3 mth the normal guidance for following crops may be observed
- Successive treatments with any products containing diflufenican can lead to soil build-up and inversion ploughing to 150 mm must precede sowing any following non-cereal crop. Even where ploughing occurs some crops may be damaged

Environmental safety
- Dangerous for the environment
- Very toxic to aquatic organisms
- Take extreme care to avoid damage by drift onto broad-leaved plants outside the target area
- Do not drain or flush spraying equipment onto land planted, or intended for planting, with trees or crops other than cereals
- Poor sprayer cleanout practices may result in inadequate removal of chemical deposits which may lead to damage of subsequently treated non-cereal crops. See label for guidance on cleaning
- LERAP Category B

Hazard classification and safety precautions
  Hazard Dangerous for the environment
  Transport code 9
  Packaging group III
  UN Number 3077
  Risk phrases R50, R53a
  Operator protection A [2]; U05a [2]; U08, U19a, U20b [1, 2]
  Environmental protection E15a [1]; E15b [2]; E16a, E38 [1, 2]
  Storage and disposal D01, D12b [2]; D09a, D12a [1, 2]

169   diflufenican + flurtamone

A contact and residual herbicide mixture for cereals
HRAC mode of action code: F1 + F1

See also flurtamone

Products
   Bacara   Bayer CropScience   100:250 g/l   SC   16448

Uses
- Annual dicotyledons in **spring barley**, **winter barley**, **winter wheat**
- Annual meadow grass in **spring barley**, **winter barley**, **winter wheat**
- Blackgrass in **spring barley**, **winter barley**, **winter wheat**
- Loose silky bent in **spring barley**, **winter barley**, **winter wheat**
- Volunteer oilseed rape in **spring barley**, **winter barley**, **winter wheat**

Approval information
- Diflufenican and flurtamone included in Annex I under EC Regulation 1107/2009
- Accepted by BBPA for use on malting barley

Efficacy guidance
- Apply pre-emergence or from when crop has first leaf unfolded before susceptible weeds pass recommended size
- Best results obtained on firm, fine seedbeds with adequate soil moisture present at and after application. Increase water volume where crop or weed foliage is dense
- Good weed control requires ash, trash and burnt straw to be buried during seed bed preparation
- Loose fluffy seedbeds should be rolled before application and the final seed bed should be fine and firm without large clods
- Speed of control depends on weather conditions and activity can be slow under cool conditions
- Always follow WRAG guidelines for preventing and managing herbicide resistant weeds. See Section 5 for more information

FOR FULL CONDITIONS OF USE ALWAYS READ THE PRODUCT LABEL

Restrictions
- Maximum number of treatments 1 per crop
- Crops should be drilled to a normal depth of 25 mm and the seed well covered. Do not treat broadcast crops as uncovered seed may be damaged
- Do not treat spring sown cereals, durum wheat, oats, undersown cereals or those to be undersown
- Do not treat frosted crops or when frost is imminent. Severe frost after application, or any other stress, may lead to transient discoloration or scorch
- Do not use on Sands or Very Light soils or those that are very stony or gravelly
- Do not use on waterlogged soils, or on crops subject to temporary waterlogging by heavy rainfall, as there is risk of persistent crop damage which may result in yield loss
- Do not use on soils with more than 10% organic matter
- Do not harrow at any time after application and do not roll autumn treated crops until spring

Crop-specific information
- Take particular care to match spray swaths otherwise crop discoloration and biomass reduction may occur which may lead to yield reduction

Following crops guidance
- In the event of crop failure winter wheat may be redrilled immediately after normal cultivation, and winter barley may be sown after ploughing. Fields must be ploughed to a depth of 15 cm and 20 wk must elapse before sowing spring crops of wheat, barley, oilseed rape, peas, field beans or potatoes
- After normal harvest autumn cereals can be drilled after ploughing. Thorough mixing of the soil must take place before drilling field beans, leaf brassicae or winter oilseed rape. For sugar beet seed crops and winter onions complete inversion of the furrow slice is essential
- Do not broadcast or direct drill oilseed rape or other brassica crops as a following crop on treated land. See label for detailed advice on preparing land for subsequent autumn cropping in the normal rotation
- Successive treatments with any products containing diflufenican can lead to soil build-up and inversion ploughing to 150 mm must precede sowing any following non-cereal crop. Even where ploughing occurs some crops may be damaged

Environmental safety
- Dangerous for the environment
- Very toxic to aquatic organisms
- Extremely dangerous to fish or other aquatic life. Do not contaminate surface waters or ditches with chemical or used container
- Buffer zone requirement 5 m for winter wheat and winter barley, 6 m for spring barley
- LERAP Category B

Hazard classification and safety precautions
Transport code 9
Packaging group III
UN Number 3082
Risk phrases R50, R53a
Operator protection A; U20c
Environmental protection E15b, E16a
Storage and disposal D09a, D10a

170 diflufenican + glyphosate

A foliar non-selective herbicide mixture
HRAC mode of action code: F1 + G

See also glyphosate

Products
1	Pistol	Bayer Environ.	40:250 g/l	SC	17451
2	Proshield	Everris Ltd	40:250 g/l	SC	17525

SEE SECTION 3 FOR PRODUCTS ALSO REGISTERED

Uses
- Annual and perennial weeds in **natural surfaces not intended to bear vegetation, permeable surfaces overlying soil**

Approval information
- Diflufenican and glyphosate included in Annex I under EC Regulation 1107/2009

Efficacy guidance
- Treat when weeds actively growing from Mar to end Sep and before they begin to senesce
- Performance may be reduced if application is made to plants growing under stress, such as drought or water-logging
- Pre-emergence activity is reduced on soils containing more than 10% organic matter or where organic debris has collected
- Perennial weeds such as docks, perennial sowthistle and willowherb are best treated before flowering or setting seed
- Perennial weeds emerging from established rootstocks after treatment will not be controlled
- A rainfree period of at least 6 h (preferably 24 h) should follow spraying for optimum control
- For optimum control do not cultivate or rake after treatment

Restrictions
- Maximum number of treatments 1 per yr
- May only be used on porous surfaces overlying soil. Must not be used if an impermeable membrane lies between the porous surface and the soil, and must not be used on any non-porous man-made surfaces
- Do not add any wetting agent or adjuvant oil
- Do not spray in windy weather

Following crops guidance
- A period of at least 6 mth must be allowed after treatment of sites that are to be cleared and grubbed before sowing and planting. Soil should be ploughed or dug first to ensure thorough mixing and dilution of any herbicide residues

Environmental safety
- Dangerous for the environment
- Very toxic to aquatic organisms
- Avoid drift onto non-target plants
- Heavy rain after application may wash product onto sensitive areas such as newly sown grass or areas about to be planted
- Buffer zone requirement 2 m [1, 2]
- LERAP Category B

Hazard classification and safety precautions
Hazard Dangerous for the environment, Very toxic to aquatic organisms
Transport code 9
Packaging group III
UN Number 3082
Operator protection A, C [1, 2]; H [2]; U20b
Environmental protection E15a, E16a, E16b, E38, H410
Storage and disposal D01, D02, D09a, D11a, D12a

171 diflufenican + iodosulfuron-methyl-sodium + mesosulfuron-methyl

A contact and residual herbicide mixture containing sulfonyl ureas for winter wheat
HRAC mode of action code: F1 + B + B

See also iodosulfuron-methyl-sodium
        mesosulfuron-methyl

Products
1	Hamlet	Bayer CropScience	50:2.5:7.5 g/l	OD	17370
2	Othello	Bayer CropScience	50:2.5:7.5 g/l	OD	16149

FOR FULL CONDITIONS OF USE ALWAYS READ THE PRODUCT LABEL

diflufenican + iodosulfuron-methyl-sodium + mesosulfuron-methyl

Uses
- Annual dicotyledons in **forest nurseries** *(off-label)* [2]; **winter wheat** [1, 2]
- Annual meadow grass in **forest nurseries** *(off-label)* [2]; **winter wheat** [1, 2]
- Cleavers in **forest nurseries** *(off-label)* [2]; **winter wheat** [1, 2]
- Mayweeds in **forest nurseries** *(off-label)* [2]; **winter wheat** [1, 2]
- Rough-stalked meadow grass in **winter wheat** [1, 2]
- Volunteer oilseed rape in **forest nurseries** *(off-label)* [2]; **winter wheat** [1, 2]

Extension of Authorisation for Minor Use (EAMUs)
- *forest nurseries* 20142724 [2]

Approval information
- Diflufenican, iodosulfuron-methyl-sodium and mesosulfuron-methyl included in Annex I under EC Regulation 1107/2009

Efficacy guidance
- Optimum control obtained when all weeds are emerged at spraying. Activity is primarily via foliar uptake and good spray coverage of the target weeds is essential
- Translocation occurs readily within the target weeds and growth is inhibited within hours of treatment but symptoms may not be apparent for up to 4 wk, depending on weed species, timing of treatment and weather conditions
- Iodosulfuron-methyl and mesosulfuron-methyl are both members of the ALS-inhibitor group of herbicides. To avoid the build up of resistance do not use any product containing an ALS-inhibitor herbicide with claims for control of grass weeds more than once on any crop
- Use this product as part of a Resistance Management Strategy that includes cultural methods of control and does not use ALS inhibitors as the sole chemical method of weed control in successive crops. See Section 5 for more information

Restrictions
- Maximum number of treatments 1 per crop
- Do not use on crops undersown with grasses, clover or other legumes or any other broad-leaved crop
- Do not use where annual grass weeds other than annual meadow grass and rough meadow grass are present
- Do not use as a stand-alone treatment for control of common chickweed or common poppy. Only use mixtures with non ALS-inhibitor herbicides for these weeds
- Do not use as the sole means of weed control in successive crops
- Specific restrictions apply to use in sequence or tank mixture with other sulfonylurea or ALS-inhibiting herbicides. See label for details
- Do not apply to crops under stress from any cause
- Do not apply when rain is imminent or during periods of frosty weather
- Specified adjuvant must be used. See label

Crop-specific information
- Latest use: before 2nd node detectable for winter wheat
- Transitory crop effects may occur, particularly on overlaps and after late season/spring applications. Recovery is normally complete and yield not affected

Following crops guidance
- In the event of crop failure winter or spring wheat may be drilled after normal cultivation and an interval of 6 wk
- Winter wheat, winter barley or winter oilseed rape may be drilled in the autumn following normal harvest of a treated crop. Spring wheat, spring barley, spring oilseed rape or sugar beet may be drilled in the following spring
- Where the product has been applied in sequence with a permitted ALS-inhibitor herbicide (see label) only winter or spring wheat or barley, or sugar beet may be sown as following crops
- Successive treatments with any products containing diflufenican can lead to soil build-up and inversion ploughing to 150 mm must precede sowing any following non-cereal crop. Even where ploughing occurs some crops may be damaged

Environmental safety
- Dangerous for the environment

SEE SECTION 3 FOR PRODUCTS ALSO REGISTERED

Pesticide Profiles

- Very toxic to aquatic organisms
- Take extreme care to avoid drift outside the target area
- LERAP Category B

Hazard classification and safety precautions
 Hazard Irritant, Dangerous for the environment, Very toxic to aquatic organisms
 Transport code 9
 Packaging group III
 UN Number 3082
 Risk phrases H319
 Operator protection A, C; U05a, U11, U20b
 Environmental protection E15b, E16a, E38, H410
 Storage and disposal D01, D02, D05, D09a, D10a, D12a

172 diflufenican + metsulfuron-methyl

A contact and residual herbicide mixture containing sulfonyl ureas for cereals
HRAC mode of action code: F1 + B

Products
 Pelican Delta   Headland   60:6% w/w   WG   14312

Uses
- Annual dicotyledons in **spring barley, winter barley, winter wheat**
- Charlock in **spring barley, winter barley, winter wheat**
- Corn spurrey in **spring barley, winter barley, winter wheat**
- Docks in **spring barley, winter barley, winter wheat**
- Hemp-nettle in **spring barley, winter barley, winter wheat**
- Mayweeds in **spring barley, winter barley, winter wheat**
- Pale persicaria in **spring barley, winter barley, winter wheat**
- Parsley-piert in **spring barley, winter barley, winter wheat**
- Redshank in **spring barley, winter barley, winter wheat**
- Scarlet pimpernel in **spring barley, winter barley, winter wheat**
- Shepherd's purse in **spring barley, winter barley, winter wheat**

Approval information
- Diflufenican and metsulfuron-methyl included in Annex I under EC Regulation 1107/2009

Efficacy guidance
- Do not apply to frosted crops or scorch may occur
- Can only be applied after 1st Feb up to latest growth stage stipulated

Restrictions
- Do not use on crops grown for seed
- Do not apply in mixture or sequence with any other ALS herbicide

Crop-specific information
- Do not use on soils containing more than 10% organic matter

Following crops guidance
- In the event of crop failure, only sow wheat within 3 months of application

Environmental safety
- Buffer zone requirement 7m [1]
- LERAP Category B

Hazard classification and safety precautions
 Hazard Irritant, Dangerous for the environment
 Transport code 9
 Packaging group III
 UN Number 3077
 Risk phrases R36, R50, R53a

FOR FULL CONDITIONS OF USE ALWAYS READ THE PRODUCT LABEL

Operator protection A, C; U05a, U08, U11, U19a, U20b
Environmental protection E15a, E16a, E34, E38
Storage and disposal D01, D02, D05, D09a, D10b, D12a

173   diflufenican + pendimethalin

A mixture of pyridinecarboxamide and dinitroanaline herbicides, see Section 3
HRAC mode of action code: F1 + K1

See also pendimethalin

Products
1 Bulldog   Sipcam   15.6:313 g/l   SC   17538
2 Omaha 2   Adama   40:400 g/l   SC   16846

Uses
- Annual dicotyledons in *spring barley*, *spring wheat* [2]; *triticale*, *winter barley*, *winter rye*, *winter wheat* [1, 2]
- Annual grasses in *spring barley*, *spring wheat*, *triticale*, *winter barley*, *winter rye*, *winter wheat* [2]
- Annual meadow grass in *triticale*, *winter barley*, *winter rye*, *winter wheat* [1]

Approval information
- Diflufenican and pendimethalin included in Annex I under EC Regulation 1107/2009

Efficacy guidance
- Do not apply pre-emergence to winter crops drilled after 20th November
- Do not treat broadcast crops

Following crops guidance
- Plough to 150 mm and thoroughly mix the soil before planting any following crop

Environmental safety
- Buffer zone requirement 6 m [1, 2]

Hazard classification and safety precautions
  Hazard Dangerous for the environment [1, 2]; Very toxic to aquatic organisms [1]
  Transport code 9
  Packaging group III
  UN Number 3082
  Risk phrases H334 [1]; R50, R53a [2]
  Operator protection A; U05a, U08, U20a [1]
  Environmental protection E38 [1, 2]; H410 [1]
  Storage and disposal D01, D02, D09a, D10c [1]; D12a [1, 2]

174   dimethachlor

A residual anilide herbicide for use in oilseed rape
HRAC mode of action code: K3

Products
1 Clayton Dimethachlor   Clayton   500 g/l   EC   16188
2 Teridox   Syngenta   500 g/l   EC   15876

Uses
- Annual dicotyledons in *winter oilseed rape*
- Chickweed in *winter oilseed rape*
- Docks in *winter oilseed rape*
- Groundsel in *winter oilseed rape*
- Loose silky bent in *winter oilseed rape*
- Mayweeds in *winter oilseed rape*

SEE SECTION 3 FOR PRODUCTS ALSO REGISTERED

Pesticide Profiles

- Red dead-nettle in **winter oilseed rape**
- Scarlet pimpernel in **winter oilseed rape**

Approval information
- Dimethachlor included in Annex 1 under EC Regulation 1107/2009

Efficacy guidance
- Optimum weed control requires moist soils.
- Heavy rain within a few days of application may lead to poor weed control

Restrictions
- Do not use on sands or very light soils, stony soils, organic soil, broadcast crops or late-drilled crops.
- Apply after drilling but before crop germination which can occur within 48 hours of drilling.

Crop-specific information
- High transpiration rates or persistent wet weather during the first weeks after treatment may reduce crop vigour and plant stand.
- Crops must be covered by at least 20mm of settled soil before application.
- Do not cultivate after application.

Following crops guidance
- In the event of autumn crop failure, winter oilseed rape may be re-drilled after cultivating the soil to at least 15 cms.
- If re-drilling the following spring, spring oilseed rape, field beans, peas, maize or potatoes may be sown. If planting spring wheat, barley, oats, spring linseed, sunflowers, sugar beet or grass crops the land should be ploughed before sowing.

Environmental safety
- Buffer zone requirement 10m [1, 2]
- LERAP Category B

Hazard classification and safety precautions
  Hazard Irritant, Dangerous for the environment
  Transport code 9
  Packaging group III
  UN Number 3082
  Risk phrases R38, R43, R50, R53a
  Operator protection A, C, H; U05a, U09a, U19a, U20c
  Environmental protection E15b, E16a, E40b
  Storage and disposal D01, D02, D05, D09a, D10c, D12a
  Medical advice M05a

175 dimethenamid-p

A chloroacetamide herbicide available only in mixtures
HRAC mode of action code: K3

176 dimethenamid-p + metazachlor

A soil acting herbicide mixture for oilseed rape
HRAC mode of action code: K3 + K3

See also metazachlor

Products

1	Muntjac	BASF	200:200 g/l	EC	16893
2	Springbok	BASF	200:200 g/l	EC	16786

FOR FULL CONDITIONS OF USE ALWAYS READ THE PRODUCT LABEL

dimethenamid-p + metazachlor

Uses

- Annual dicotyledons in **broccoli** *(off-label)*, **cabbages** *(off-label)*, **calabrese** *(off-label)*, **cauliflowers** *(off-label)*, **choi sum** *(off-label)*, **collards** *(off-label)*, **kale** *(off-label)*, **kohlrabi** *(off-label)*, **winter oilseed rape** [1, 2]; **brussels sprouts** *(off-label)*, **chinese leaf** *(off-label)*, **chives** *(off-label)*, **forest nurseries** *(off-label)*, **game cover** *(off-label)*, **leeks** *(off-label)*, **oriental cabbage** *(off-label)*, **ornamental plant production** *(off-label)*, **salad onions** *(off-label)*, **swedes** *(off-label)*, **tatsoi** *(off-label)*, **turnips** *(off-label)* [2]; **chinese cabbage** *(off-label)*, **pak choi** *(off-label)* [1]
- Annual grasses in **broccoli** *(off-label)*, **brussels sprouts** *(off-label)*, **cabbages** *(off-label)*, **calabrese** *(off-label)*, **cauliflowers** *(off-label)*, **chinese leaf** *(off-label)*, **chives** *(off-label)*, **choi sum** *(off-label)*, **collards** *(off-label)*, **forest nurseries** *(off-label)*, **game cover** *(off-label)*, **kale** *(off-label)*, **kohlrabi** *(off-label)*, **leeks** *(off-label)*, **oriental cabbage** *(off-label)*, **ornamental plant production** *(off-label)*, **salad onions** *(off-label)*, **swedes** *(off-label)*, **tatsoi** *(off-label)*, **turnips** *(off-label)* [2]
- Annual meadow grass in **broccoli** *(off-label)*, **brussels sprouts** *(off-label)*, **cabbages** *(off-label)*, **calabrese** *(off-label)*, **cauliflowers** *(off-label)*, **chinese leaf** *(off-label)*, **chives** *(off-label)*, **choi sum** *(off-label)*, **collards** *(off-label)*, **kale** *(off-label)*, **kohlrabi** *(off-label)*, **leeks** *(off-label)*, **oriental cabbage** *(off-label)*, **ornamental plant production** *(off-label)*, **salad onions** *(off-label)*, **swedes** *(off-label)*, **tatsoi** *(off-label)*, **turnips** *(off-label)* [2]
- Chickweed in **ornamental plant production** *(off-label)* [2]; **winter oilseed rape** [1, 2]
- Cleavers in **ornamental plant production** *(off-label)* [2]; **winter oilseed rape** [1, 2]
- Common storksbill in **winter oilseed rape** [1, 2]
- Crane's-bill in **chives** *(off-label)*, **forest nurseries** *(off-label)*, **game cover** *(off-label)*, **leeks** *(off-label)*, **ornamental plant production** *(off-label)*, **salad onions** *(off-label)*, **swedes** *(off-label)*, **turnips** *(off-label)* [2]
- Fat hen in **broccoli** *(off-label)*, **brussels sprouts** *(off-label)*, **cabbages** *(off-label)*, **calabrese** *(off-label)*, **cauliflowers** *(off-label)*, **chinese leaf** *(off-label)*, **choi sum** *(off-label)*, **collards** *(off-label)*, **kale** *(off-label)*, **kohlrabi** *(off-label)*, **oriental cabbage** *(off-label)*, **tatsoi** *(off-label)* [2]
- Field speedwell in **winter oilseed rape** [1, 2]
- Groundsel in **broccoli** *(off-label)*, **brussels sprouts** *(off-label)*, **cabbages** *(off-label)*, **calabrese** *(off-label)*, **cauliflowers** *(off-label)*, **chinese leaf** *(off-label)*, **choi sum** *(off-label)*, **collards** *(off-label)*, **kale** *(off-label)*, **kohlrabi** *(off-label)*, **oriental cabbage** *(off-label)*, **ornamental plant production** *(off-label)*, **tatsoi** *(off-label)* [2]
- Mayweeds in **broccoli** *(off-label)*, **brussels sprouts** *(off-label)*, **cabbages** *(off-label)*, **calabrese** *(off-label)*, **cauliflowers** *(off-label)*, **chinese leaf** *(off-label)*, **choi sum** *(off-label)*, **collards** *(off-label)*, **kale** *(off-label)*, **kohlrabi** *(off-label)*, **oriental cabbage** *(off-label)*, **ornamental plant production** *(off-label)*, **tatsoi** *(off-label)* [2]
- Poppies in **winter oilseed rape** [1, 2]
- Redshank in **broccoli** *(off-label)*, **brussels sprouts** *(off-label)*, **cabbages** *(off-label)*, **calabrese** *(off-label)*, **cauliflowers** *(off-label)*, **chinese leaf** *(off-label)*, **choi sum** *(off-label)*, **collards** *(off-label)*, **kale** *(off-label)*, **kohlrabi** *(off-label)*, **oriental cabbage** *(off-label)*, **tatsoi** *(off-label)* [2]
- Scented mayweed in **winter oilseed rape** [1, 2]
- Shepherd's purse in **broccoli** *(off-label)*, **brussels sprouts** *(off-label)*, **cabbages** *(off-label)*, **calabrese** *(off-label)*, **cauliflowers** *(off-label)*, **chinese leaf** *(off-label)*, **chives** *(off-label)*, **choi sum** *(off-label)*, **collards** *(off-label)*, **forest nurseries** *(off-label)*, **game cover** *(off-label)*, **kale** *(off-label)*, **kohlrabi** *(off-label)*, **leeks** *(off-label)*, **oriental cabbage** *(off-label)*, **ornamental plant production** *(off-label)*, **salad onions** *(off-label)*, **swedes** *(off-label)*, **tatsoi** *(off-label)*, **turnips** *(off-label)* [2]; **winter oilseed rape** [1, 2]
- Small nettle in **broccoli** *(off-label)*, **brussels sprouts** *(off-label)*, **cabbages** *(off-label)*, **calabrese** *(off-label)*, **cauliflowers** *(off-label)*, **chinese leaf** *(off-label)*, **choi sum** *(off-label)*, **collards** *(off-label)*, **kale** *(off-label)*, **kohlrabi** *(off-label)*, **oriental cabbage** *(off-label)*, **ornamental plant production** *(off-label)*, **tatsoi** *(off-label)* [2]
- Sowthistle in **ornamental plant production** *(off-label)* [2]

Extension of Authorisation for Minor Use (EAMUs)

- *broccoli* 20140426 [1], 20151540 [2]
- *brussels sprouts* 20151540 [2]
- *cabbages* 20140426 [1], 20151540 [2]
- *calabrese* 20140426 [1], 20151540 [2]
- *cauliflowers* 20140426 [1], 20151540 [2]
- *chinese cabbage* 20140426 [1]

SEE SECTION 3 FOR PRODUCTS ALSO REGISTERED

Pesticide Profiles

- ***chinese leaf*** *20151540* [2]
- ***chives*** *20143008* [2]
- ***choi sum*** *20140426* [1], *20151540* [2]
- ***collards*** *20140426* [1], *20151540* [2]
- ***forest nurseries*** *20143006* [2]
- ***game cover*** *20143006* [2]
- ***kale*** *20140426* [1], *20151540* [2]
- ***kohlrabi*** *20140426* [1], *20151540* [2]
- ***leeks*** *20143008* [2]
- ***oriental cabbage*** *20151540* [2]
- ***ornamental plant production*** *20143006* [2], *20152108* [2]
- ***pak choi*** *20140426* [1]
- ***salad onions*** *20143008* [2]
- ***swedes*** *20143007* [2]
- ***tatsoi*** *20151540* [2]
- ***turnips*** *20143007* [2]

Approval information
- Dimethenamid-p and metazachlor included in Annex I under EC Regulation 1107/2009

Efficacy guidance
- Best results obtained from treatments to fine, firm and moist seedbeds
- Apply pre- or post-emergence of the crop and ideally before weed emergence
- Residual weed control may be reduced under prolonged dry conditions
- Weeds germinating from depth may not be controlled
- Dimethenamid-p is more active than metazachlor under dry soil conditions

Restrictions
- Maximum total dose on winter oilseed rape equivalent to one full dose treatment
- Do not disturb soil after application
- Do not treat broadcast crops until they have attained two fully expanded cotyledons
- Do not use on Sands, Very Light soils, or soils containing 10% organic matter
- Do not apply when heavy rain forecast or on soils waterlogged or prone to waterlogging
- Do not treat crops suffering from stress from any cause
- Do not use pre-emergence when crop seed has started to germinate or if not covered with 15 mm of soil
- Applications shall be limited to a total dose of not more than 1.0 kg metazachlor/ha in a three year period on the same field
- Do not apply in mixture with phosphate liquid fertilisers
- Metazachlor stewardship requires that all autumn applications should be made before the end of September to reduce the risk to water

Crop-specific information
- Latest use: before 7th true leaf for winter oilseed rape
- All varieties of winter oilseed rape may be treated

Following crops guidance
- Any crop may follow a normally harvested treated winter oilseed rape crop. Ploughing is not essential before a following cereal crop but is required for all other crops
- In the event of failure of a treated crop winter wheat (excluding durum) or winter barley may be drilled in the same autumn, and any cereal (excluding durum wheat), spring oilseed rape, peas or field beans may be sown in the following spring. Ploughing to at least 150 mm should precede in all cases

Environmental safety
- Dangerous for the environment
- Very toxic to aquatic organisms
- Take extreme care to avoid spray drift onto non-crop plants outside the target area
- Some pesticides pose a greater threat of contamination of water than others and metazachlor is one of these pesticides. Take special care when applying metazachlor near water and do not apply if heavy rain is forecast

FOR FULL CONDITIONS OF USE ALWAYS READ THE PRODUCT LABEL

- Metazachlor stewardship guidelines advise a maximum dose of 750 g.a.i/ha/annum. Applications to drained fields should be complete by 15th Oct but, if drains are flowing, complete applications by 1st Oct.
- LERAP Category B

Hazard classification and safety precautions
Hazard Harmful, Dangerous for the environment, Harmful if swallowed, Harmful if inhaled, Very toxic to aquatic organisms
Transport code 9
Packaging group III
UN Number 3082
Risk phrases H304, H317, H319, H351
Operator protection A, C, H; U05a, U08, U20c
Environmental protection E15a, E16a, E34, E38
Storage and disposal D01, D02, D09a, D10c, D12a
Medical advice M03, M05a

177 dimethenamid-p + metazachlor + quinmerac

A soil-acting herbicide mixture for use in oilseed rape
HRAC mode of action code: K3 + K3 + O

See also metazachlor
quinmerac

Products

1	Banastar	BASF	100:300:100 g/l	SE	16834
2	Elk	BASF	200:200:100 g/l	SE	16920
3	Katamaran Turbo	BASF	200:200:100 g/l	SE	16921
4	Shadow	BASF	200:200:100 g/l	SE	16804

Uses
- Annual dicotyledons in **winter oilseed rape** [1-4]
- Annual grasses in **winter oilseed rape** [2-4]
- Annual meadow grass in **winter oilseed rape** [1]
- Blackgrass in **winter corn gromwell** *(off-label)* [3]
- Chickweed in **winter corn gromwell** *(off-label)* [3]
- Cleavers in **winter corn gromwell** *(off-label)* [3]; **winter oilseed rape** [1-4]
- Crane's-bill in **winter corn gromwell** *(off-label)* [3]; **winter oilseed rape** [1]
- Fat hen in **winter corn gromwell** *(off-label)* [3]
- Field pansy in **winter corn gromwell** *(off-label)* [3]
- Forget-me-not in **winter corn gromwell** *(off-label)* [3]
- Groundsel in **winter corn gromwell** *(off-label)* [3]
- Ivy-leaved speedwell in **winter corn gromwell** *(off-label)* [3]
- Mayweeds in **winter corn gromwell** *(off-label)* [3]
- Poppies in **winter corn gromwell** *(off-label)* [3]; **winter oilseed rape** [1]
- Red dead-nettle in **winter corn gromwell** *(off-label)* [3]
- Shepherd's purse in **winter corn gromwell** *(off-label)* [3]
- Sowthistle in **winter corn gromwell** *(off-label)* [3]
- Speedwells in **winter oilseed rape** [1]

Extension of Authorisation for Minor Use (EAMUs)
- **winter corn gromwell** *20152318* [3]

Approval information
- Dimethenamid-p, metazachlor and quinmerac included in Annex I under EC Regulation 1107/2009

Efficacy guidance
- Dimethenamid-p is more active than metazachlor under dry soil conditions

SEE SECTION 3 FOR PRODUCTS ALSO REGISTERED

Restrictions

- Applications shall be limited to a total dose of not more than 1.0 kg metazachlor/ha in a three year period on the same field
- Do not apply in mixture with phosphate liquid fertilisers
- Metazachlor stewardship requires that all autumn applications should be made before the end of September to reduce the risk to water

Following crops guidance

- For all situations following or rotational crops must not be planted until four months after application

Environmental safety

- Some pesticides pose a greater threat of contamination of water than others and metazachlor is one of these pesticides. Take special care when applying metazachlor near water and do not apply if heavy rain is forecast.
- Metazachlor stewardship guidelines advise a maximum dose of 750 g.a.i/ha/annum. Applications to drained fields should be complete by 15th Oct but, if drains are flowing, complete applications by 1st Oct.
- LERAP Category B

Hazard classification and safety precautions

Hazard Harmful [3, 4]; Irritant [2]; Dangerous for the environment, Very toxic to aquatic organisms [2-4]
Transport code 9
Packaging group III
UN Number 3082
Risk phrases H317, H351
Operator protection A, H; U05a, U08 [1-4]; U14, U20c [2-4]; U20a [1]
Environmental protection E07a, E38 [2-4]; E07d [1]; E15b, E16a, E34, H410 [1-4]
Storage and disposal D01, D02, D09a, D10c [1-4]; D05 [1]; D12a [2-4]
Medical advice M03, M05a [2-4]

178   dimethenamid-p + pendimethalin

A chloracetamide and dinitroaniline herbicide mixture for weed control in maize crops
HRAC mode of action code: K1 + K3

Products

1	Clayton Launch	Clayton	212.5:250 g/l	EC	16277
2	Dime	Pan Agriculture	212.5:250 g/l	EC	17398
3	Wing - P	BASF	212.5:250 g/l	EC	15425

Uses

- Annual dicotyledons in **baby leaf crops** (off-label), **broccoli** (off-label), **brussels sprouts** (off-label), **bulb onion sets** (off-label), **bulb onions** (off-label), **cabbages** (off-label), **calabrese** (off-label), **cauliflowers** (off-label), **chives** (off-label), **forage maize (under plastic mulches)** (off-label), **garlic** (off-label), **grain maize under plastic mulches** (off-label), **herbs (see appendix 6)** (off-label), **leeks** (off-label), **ornamental plant production** (off-label), **salad onions** (off-label), **shallots** (off-label), **strawberries** (off-label) [3]; **forage maize, grain maize** [1-3]
- Annual grasses in **baby leaf crops** (off-label), **broccoli** (off-label), **brussels sprouts** (off-label), **bulb onion sets** (off-label), **bulb onions** (off-label), **cabbages** (off-label), **calabrese** (off-label), **cauliflowers** (off-label), **chives** (off-label), **forage maize (under plastic mulches)** (off-label), **garlic** (off-label), **grain maize under plastic mulches** (off-label), **herbs (see appendix 6)** (off-label), **leeks** (off-label), **ornamental plant production** (off-label), **salad onions** (off-label), **shallots** (off-label), **strawberries** (off-label) [3]
- Annual meadow grass in **broccoli** (off-label), **brussels sprouts** (off-label), **calabrese** (off-label), **cauliflowers** (off-label) [3]; **forage maize, grain maize** [1-3]
- Wild oats in **ornamental plant production** (off-label) [3]

Extension of Authorisation for Minor Use (EAMUs)

- *baby leaf crops* 20123044 [3]

FOR FULL CONDITIONS OF USE ALWAYS READ THE PRODUCT LABEL

- **broccoli** 20131656 [3]
- **brussels sprouts** 20131656 [3]
- **bulb onion sets** 20122250 [3]
- **bulb onions** 20122250 [3]
- **cabbages** 20122251 [3]
- **calabrese** 20131656 [3]
- **cauliflowers** 20131656 [3]
- **chives** 20122249 [3]
- **forage maize (under plastic mulches)** 20150101 [3]
- **garlic** 20122250 [3]
- **grain maize under plastic mulches** 20111841 [3]
- **herbs (see appendix 6)** 20123044 [3]
- **leeks** 20122248 [3]
- **ornamental plant production** 20130253 [3]
- **salad onions** 20122249 [3]
- **shallots** 20122250 [3]
- **strawberries** 20160933 [3]

Approval information
- Dimethenamid-p and pendimethalin are included in Annex 1 under EC Regulation 1107/2009

Efficacy guidance
- Do not use on soil types with more than 10% organic matter
- Best results obtained when rain falls within 7 days of application

Crop-specific information
- Use on maize crops grown under plastic mulches is by EAMU only; do not use in greenhouses, covers or other forms of protection
- Risk of crop damage if heavy rain falls soon after application to stoney or gravelly soils
- Seed should be covered with at least 5 cm of settled soil

Environmental safety
- LERAP Category B

Hazard classification and safety precautions
Hazard Harmful, Dangerous for the environment
Transport code 9
Packaging group III
UN Number 3082
Risk phrases R22a, R22b, R38, R43, R50, R53a
Operator protection A, H; U05a, U14
Environmental protection E15b, E16a, E34, E38
Storage and disposal D01, D02, D09a, D10c, D12a
Medical advice M03, M05a

179 dimethenamid-p + quinmerac

A herbicide mixture for weed control in oilseed rape

See also dimethenamid-p + metazachlor + quinmerac
quinmerac

Products
1	Tanaris	BASF	333 + 167 g/l	SE	17173
2	Topkat	BASF	333 + 167 g/l	SE	17356

Uses
- Annual dicotyledons in **winter oilseed rape**
- Cleavers in **winter oilseed rape**
- Crane's-bill in **winter oilseed rape**
- Dead nettle in **winter oilseed rape**

SEE SECTION 3 FOR PRODUCTS ALSO REGISTERED

Pesticide Profiles

- Mayweeds in **winter oilseed rape**
- Poppies in **winter oilseed rape**
- Shepherd's purse in **winter oilseed rape**
- Sowthistle in **winter oilseed rape**
- Speedwells in **winter oilseed rape**

Approval information
- Dimethenamid-p and quinmerac included in Annex I under EC Regulation 1107/2009

Environmental safety
- LERAP Category B

Hazard classification and safety precautions
Hazard Very toxic to aquatic organisms
Transport code 9
Packaging group III
UN Number 3082
Risk phrases H317, H319
Operator protection A, C, H; U05a, U08, U20a
Environmental protection E15b, E16a, E34, H410
Storage and disposal D01, D02, D09a, D10c

180 dimethoate

A contact and systemic organophosphorus insecticide and acaricide
IRAC mode of action code: 1B

Products
Danadim Progress   Headland   400 g/l   EC   15890

Uses
- Aphids in **durum wheat**, **ornamental plant production**, **rye**, **spring wheat**, **triticale**, **winter wheat**
- Insect pests in **protected forest nurseries** (off-label)
- Leaf miner in **ornamental plant production**
- Red spider mites in **ornamental plant production**

Extension of Authorisation for Minor Use (EAMUs)
- *protected forest nurseries* 20122130

Approval information
- Dimethoate included in Annex I under EC Regulation 1107/2009
- In 2006 CRD required that all products containing this active ingredient should carry the following warning in the main area of the container label: "Dimethoate is an anticholinesterase organophosphate. Handle with care"

Efficacy guidance
- Chemical has quick knock-down effect and systemic activity lasts for up to 14 d
- With some crops, products differ in range of pests listed as controlled. Uses section above provides summary. See labels for details
- For most pests apply when pest first seen and repeat 2-3 wk later or as necessary. Timing and number of sprays varies with crop and pest. See labels for details
- Best results achieved when crop growing vigorously. Systemic activity reduced when crops suffering from drought or other stress
- In hot weather apply in early morning or late evening
- Where aphids or spider mites resistant to organophosphorus compounds occur control is unlikely to be satisfactory and repeat treatments may result in lower levels of control

Restrictions
- Contains an anticholinesterase organophosphorus compound. Do not use if under medical advice not to work with such compounds

FOR FULL CONDITIONS OF USE ALWAYS READ THE PRODUCT LABEL

- Maximum number of treatments 6 per crop for Brussels sprouts, broccoli, cauliflower, calabrese, lettuce; 4 per crop for grass seed crops, cereals; 2 per crop for beet crops, triticale, rye; 1 per crop for mangel and sugar beet seed crops
- On beet crops only one treatment per crop may be made for control of leaf miners (max 84 g a.i./ha) and black bean aphid (max 400 g a.i./ha)
- In beet crops resistant strains of peach-potato aphid (*Myzus persicae*) are common and dimethoate products must not be used to control this pest
- Test for varietal susceptibility on all unusual plants or new cultivars
- Do not tank mix with alkaline materials. See label for recommended tank-mixes
- Consult processor before spraying crops grown for processing
- Must not be applied to cereals if any product containing a pyrethroid insecticide or dimethoate has been sprayed after the start of ear emergence (GS 51)

Crop-specific information
- Latest use: before 30 Jun in yr of harvest for beet crops; flowering just complete stage (GS 30) for wheat, rye, triticale; before 31 Mar in yr of harvest for aerial application to cereals
- HI brassicas 21 d; grass seed crops, lettuce, cereals 14 d; ornamental plant production 7 d

Environmental safety
- Dangerous for the environment
- Harmful to aquatic organisms
- Flammable
- Harmful to game, wild birds and animals
- Harmful to livestock. Keep all livestock out of treated areas for at least 7 d
- Likely to cause adverse effects on beneficial arthropods
- High risk to non-target insects or other arthropods. Do not treat cereals after 1 Apr within 6 m of edge of crop
- High risk to bees. Do not apply to crops in flower or to those in which bees are actively foraging. Do not apply when flowering weeds are present
- Surface residues may also cause bee mortality following spraying
- Dangerous to fish or other aquatic life. Do not contaminate surface waters or ditches with chemical or used container
- Keep in original container, tightly closed, in a safe place, under lock and key
- LERAP Category A

Hazard classification and safety precautions
Hazard Harmful, Flammable, Flammable liquid and vapour, Harmful if swallowed, Harmful if inhaled
Transport code 6.1
Packaging group III
UN Number 3017
Risk phrases H304, H317
Operator protection A, H, M; U02a, U04a, U08, U14, U15, U19a, U20b
Environmental protection E06c (7 d); E10b, E12a, E12e, E13b, E16c, E16d, E22a, E34, H410
Storage and disposal D01, D02, D05, D09b, D12b
Medical advice M01, M03

181 dimethomorph

A cinnamic acid fungicide with translaminar activity
FRAC mode of action code: 40

See also ametoctradin + dimethomorph

Products
Paraat   BASF   50% w/w   WP   15445

Uses
- Crown rot in **protected strawberries** *(moderate control)*, **strawberries** *(moderate control)*

SEE SECTION 3 FOR PRODUCTS ALSO REGISTERED

Pesticide Profiles

- Downy mildew in **brassica leaves and sprouts** *(off-label)*, **chard** *(off-label)*, **herbs (see appendix 6)** *(off-label)*, **lamb's lettuce** *(off-label)*, **lettuce** *(off-label)*, **protected baby leaf crops** *(off-label)*, **protected herbs (see appendix 6)** *(off-label)*, **protected lamb's lettuce** *(off-label)*, **protected lettuce** *(off-label)*, **protected ornamentals** *(off-label)*, **radishes** *(off-label)*, **rocket** *(off-label)*, **seedling brassicas** *(off-label)*, **spinach** *(off-label)*
- Root rot in **blackberries**, **protected blackberries** *(moderate control)*, **protected raspberries** *(moderate control)*, **raspberries** *(moderate control)*

Extension of Authorisation for Minor Use (EAMUs)
- *brassica leaves and sprouts* 20122244
- *chard* 20122244
- *herbs (see appendix 6)* 20122244
- *lamb's lettuce* 20122244
- *lettuce* 20122244
- *protected baby leaf crops* 20112584
- *protected herbs (see appendix 6)* 20112584
- *protected lamb's lettuce* 20112584
- *protected lettuce* 20112584
- *protected ornamentals* 20112585
- *radishes* 20160190
- *rocket* 20122244
- *seedling brassicas* 20130274
- *spinach* 20122244

Approval information
- Dimethomorph included in Annex I under EC Regulation 1107/2009

Restrictions
- Do not sow clover until the spring following application.
- Dimethomorph is a carboxylic acid amide fungicide. No more than three consecutive applications of CAA fungicides should be made.

Crop-specific information
- After application, allow 10 months before sowing clover [1]

Environmental safety
- LERAP Category B

Hazard classification and safety precautions
Transport code 9
Packaging group III
UN Number 3077
Operator protection A, D, H; U05a
Environmental protection E15b, E16a, E38
Storage and disposal D01, D02, D09a, D10a, D12a

182 dimethomorph + fluazinam

A mixture of cinnamic acid and dinitroaniline fungicides for blight control in potatoes
FRAC mode of action code: 29 + 40

See also fluazinam

Products
Hubble	Adama	200:200 g/l	SC	16089

Uses
- Blight in **potatoes**
- Late blight in **potatoes**

Approval information
- Dimethomorph and fluazinam included in Annex I under EC Regulation 1107/2009

FOR FULL CONDITIONS OF USE ALWAYS READ THE PRODUCT LABEL

Environmental safety
- Buffer zone requirement 6m [1]
- LERAP Category B

Hazard classification and safety precautions
Hazard Harmful, Dangerous for the environment
Transport code 9
Packaging group III
UN Number 3082
Risk phrases H361
Operator protection A, H; U02a, U04a, U05a, U08, U14, U15, U20a
Environmental protection E15b, E16a, E34, E38, H410
Storage and disposal D01, D02, D05, D09a, D12a
Medical advice M03

183 dimethomorph + mancozeb

A systemic and protectant fungicide for potato blight control
FRAC mode of action code: 40 + M3

See also mancozeb

Products
Invader   BASF   7.5:66.7% w/w   WG   15223

Uses
- Blight in **potatoes**
- Disease control in **chives** *(off-label)*, **cress** *(off-label)*, **forest nurseries** *(off-label)*, **frise** *(off-label)*, **herbs (see appendix 6)** *(off-label)*, **lamb's lettuce** *(off-label)*, **leaf brassicas** *(off-label)*, **leeks**, **lettuce** *(off-label)*, **ornamental plant production** *(off-label)*, **poppies for morphine production** *(off-label)*, **radicchio** *(off-label)*, **salad onions** *(off-label)*, **scarole** *(off-label)*, **shallots**
- Downy mildew in **bulb onions**, **chives** *(off-label)*, **cress** *(off-label)*, **forest nurseries** *(off-label)*, **frise** *(off-label)*, **garlic**, **herbs (see appendix 6)** *(off-label)*, **lamb's lettuce** *(off-label)*, **leaf brassicas** *(off-label)*, **lettuce** *(off-label)*, **ornamental plant production** *(off-label)*, **poppies for morphine production** *(off-label)*, **radicchio** *(off-label)*, **salad onions** *(off-label)*, **scarole** *(off-label)*
- White tip in **leeks** *(reduction)*

Extension of Authorisation for Minor Use (EAMUs)
- *chives* 20120111
- *cress* 20120110
- *forest nurseries* 20120109
- *frise* 20120110
- *herbs (see appendix 6)* 20120110
- *lamb's lettuce* 20120110
- *leaf brassicas* 20120110
- *lettuce* 20120110
- *ornamental plant production* 20120109
- *poppies for morphine production* 20120112
- *radicchio* 20120110
- *salad onions* 20120111
- *scarole* 20120110

Approval information
- Dimethomorph and mancozeb included in Annex I under EC Regulation 1107/2009

Efficacy guidance
- Commence treatment as soon as there is a risk of blight infection
- In the absence of a warning treatment should start before the crop meets along the rows
- Repeat treatments every 7-14 d depending in the degree of infection risk
- Irrigated crops should be regarded as at high risk and treated every 7 d

SEE SECTION 3 FOR PRODUCTS ALSO REGISTERED

Pesticide Profiles

- For best results good spray coverage of the foliage is essential
- To minimise the likelihood of development of resistance these products should be used in a planned Resistance Management strategy. See Section 5 for more information

Restrictions
- Maximum total dose equivalent to eight full dose treatments on potatoes

Crop-specific information
- HI 7 d for all crops

Environmental safety
- Dangerous for the environment
- Very toxic to aquatic organisms
- LERAP Category B

Hazard classification and safety precautions
Hazard Irritant, Dangerous for the environment, Very toxic to aquatic organisms
Transport code 9
Packaging group III
UN Number 3077
Risk phrases H317, H319, H335, H361
Operator protection A, C, D, H; U02a, U04a, U05a, U08, U13, U19a, U20b
Environmental protection E16a, E16b, E38, H410
Storage and disposal D01, D02, D05, D09a, D12a
Medical advice M05a

184 dimethomorph + pyraclostrobin

A fungicide mixture for disease control in onions, garlic and shallots
FRAC mode of action code: 11 + 40

Products
Cassiopeia   BASF   72:40 g/l   EC   16522

Uses
- Downy mildew in **bulb onions**, **garlic**, **shallots**

Approval information
- Dimethomorph and pyraclostrobin included in Annex I under EC Regulation 1107/2009

Efficacy guidance
- Use before the disease is established in the crop and use in a programme with other effective fungicides that have a different mode of action to reduce the risk of resistance

Hazard classification and safety precautions
Hazard Harmful, Dangerous for the environment
Transport code 9
Packaging group III
UN Number 3082
Risk phrases R20, R22a, R38, R50, R53a
Operator protection A, H; U05a
Environmental protection E38
Storage and disposal D01, D02, D12a, D12b
Medical advice M05b

FOR FULL CONDITIONS OF USE ALWAYS READ THE PRODUCT LABEL

185 dimethomorph + zoxamide

A fungicide mixture for blight control in potatoes
FRAC mode of action code: 40 + 22

See also dimethomorph
zoxamide

Products
Presidium   Gowan   180:180 g/l   SC   17372

Uses
- Blight in **potatoes**
- Late blight in **potatoes**

Approval information
- Dimethomorph and zoxamide included in Annex I under EC Regulation 1107/2009

Environmental safety
- Buffer zone requirement 7 m.

Hazard classification and safety precautions
Hazard Very toxic to aquatic organisms
Transport code 9
Packaging group III
UN Number 3082
Risk phrases H317
Operator protection A, H
Environmental protection H410

186 dimoxystrobin

A protectant strobilurin fungicide for cereals available only in mixtures
FRAC mode of action code: 11

See also boscalid + dimoxystrobin

187 dimoxystrobin + epoxiconazole

A protectant and curative fungicide mixture for cereals
FRAC mode of action code: 11 + 3

See also epoxiconazole

Products
Swing Gold   BASF   133:50 g/l   SC   16081

Uses
- Brown rust in **durum wheat**, **spring wheat**, **winter wheat**
- Fusarium ear blight in **durum wheat**, **spring wheat**, **winter wheat**
- Glume blotch in **durum wheat**, **spring wheat**, **winter wheat**
- Septoria leaf blotch in **durum wheat**, **spring wheat**, **winter wheat**
- Tan spot in **durum wheat**, **spring wheat**, **winter wheat**

Approval information
- Dimoxystrobin and epoxiconazole included in Annex I under EC Regulation 1107/2009

Efficacy guidance
- For best results apply from the start of ear emergence
- Dimoxystrobin is a member of the QoI cross resistance group. Product should be used preventatively and not relied on for its curative potential

SEE SECTION 3 FOR PRODUCTS ALSO REGISTERED

Pesticide Profiles

- Use product as part of an Integrated Crop Management strategy incorporating other methods of control, including where appropriate other fungicides with a different mode of action. Do not apply more than two foliar applications of QoI containing products to any cereal crop
- There is a significant risk of widespread resistance occurring in *Septoria tritici* populations in UK. Failure to follow resistance management action may result in reduced levels of disease control
- Epoxiconazole is a DMI fungicide. Resistance to some DMI fungicides has been identified in Septoria leaf blotch which may seriously affect performance of some products. For further advice contact a specialist advisor and visit the Fungicide Resistance Action Group (FRAG)-UK website

Restrictions
- Maximum number of treatments 1 per crop
- Do not apply before the start of ear emergence

Crop-specific information
- Latest use: up to and including flowering (GS 69)

Environmental safety
- Dangerous for the environment
- Very toxic to aquatic organisms
- Avoid drift onto neighbouring crops
- LERAP Category B

Hazard classification and safety precautions
Hazard Harmful, Dangerous for the environment
Transport code 9
Packaging group III
UN Number 3082
Risk phrases R20, R22a, R40, R50, R53a, R63
Operator protection A; U05a, U20b
Environmental protection E15a, E16a, E34, E38
Storage and disposal D01, D02, D05, D09a, D10c, D12a
Medical advice M05a

188   diquat

A non-residual bipyridyl contact herbicide and crop desiccant
HRAC mode of action code: D

Products

1	Dessicash 200	Sharda	200 g/l	SL	14848
2	Di Novia	Agroquimicos	200 g/l	SL	15840
3	Diqua	Sharda	200 g/l	SL	14032
4	Dragoon	Hermoo	200 g/l	SL	13927
5	Dragoon Gold	Belcrop	200 g/l	SL	14973
6	Life Scientific Diquat	Life Scientific	200 g/l	SL	15398
7	Mission 200 SL	UPL Europe	200 g/l	SL	16977
8	Quad-Glob 200 SL	Q-Chem	200 g/l	SL	14758
9	Quazar	Belchim	200 g/l	SL	15357
10	Quit	Belchim	200 g/l	SL	14874
11	Reglone	Syngenta	200 g/l	SL	10534
12	Retro	Syngenta	200 g/L	SL	13841
13	Shrike 200	Goldengrass	200 g/l	SL	16197
14	Standon Googly	Standon	200 g/l	SL	12995
15	Woomera	Belcrop	200 g/l	SL	14972

Uses
- Annual and perennial weeds in **all edible crops (outdoor)** *(around only)*, **all non-edible crops (outdoor)** *(around only)*, **bulb onions**, **chicory root**, **combining peas** *(dry harvested only)*, **endives**, **linseed**, **ornamental plant production**, **potatoes**, **potatoes** *(weed control or*

FOR FULL CONDITIONS OF USE ALWAYS READ THE PRODUCT LABEL

diquat

desiccation), **red clover** (seed crop), **spring field beans**, **spring oats** (animal feed only), **spring oilseed rape**, **strawberries** (around only), **sugar beet**, **white clover** (seed crop), **winter field beans**, **winter oats** (animal feed only), **winter oilseed rape** [7]; **forest nurseries** (off-label), **poppies for morphine production** (off-label) [8]; **hops** [5, 7]
- Annual dicotyledons in **all edible crops (outdoor)**, **all non-edible crops (outdoor)**, **potatoes** [9, 10]; **all edible crops (outdoor)** (around), **all non-edible crops (outdoor)** (around), **ornamental plant production**, **sugar beet** [1-6, 8, 11, 13-15]; **fodder beet**, **mangels** [5]; **lucerne** (off-label) [11]
- Annual grasses in **all edible crops (outdoor)**, **all non-edible crops (outdoor)**, **potatoes** [9, 10]
- Black bindweed in **all edible crops (outdoor)** [12]
- Chemical stripping in **hops** [8, 14]
- Chickweed in **all edible crops (outdoor)**, **all non-edible crops (outdoor)**, **ornamental plant production**, **potatoes**, **sugar beet** [12]
- Crane's-bill in **all edible crops (outdoor)** [12]
- Desiccation in **barley seed crops** (off-label), **corn gromwell** (off-label), **crambe** (off-label), **hops** (off-label), **oat seed crops** (off-label), **potato dumps** (off-label) [11]; **borage for oilseed production** (off-label), **lupins** (off-label), **sweet potato** (off-label) [8, 11]; **combining peas**, **linseed**, **oats for animal feed**, **oilseed rape**, **red clover**, **spring field beans**, **white clover**, **winter field beans** [9, 10]; **dwarf beans** (off-label), **french beans** (off-label), **hemp for oilseed production** (off-label), **herbs (see appendix 6)** (off-label), **mustard** (off-label), **poppies for morphine production** (off-label), **poppies for oilseed production** (off-label), **sesame** (off-label), **sunflowers** (off-label) [8]; **hops** [5]; **potatoes** [7, 9, 10]
- Fat hen in **all edible crops (outdoor)** [12]
- General weed control in **chives** (off-label), **herbs (see appendix 6)** (off-label), **parsley** (off-label), **poppies for morphine production** (off-label) [11]; **potatoes** [1-6, 8, 11, 13, 14]
- Grass weeds in **all edible crops (outdoor)** (around only), **all non-edible crops (outdoor)** (around only), **bulb onions**, **chicory root**, **combining peas** (dry harvested), **endives**, **hops**, **linseed**, **ornamental plant production**, **potatoes**, **potatoes** (weed control and desiccation), **red clover** (seed crop), **spring field beans**, **spring oats** (animal feed), **spring oilseed rape**, **strawberries** (around only), **sugar beet**, **white clover** (seed crop), **winter field beans**, **winter oats** (animal feed), **winter oilseed rape** [7]
- Haulm destruction in **potatoes** [1-6, 8, 11, 13, 14]
- Perennial dicotyledons in **lucerne** (off-label) [11]
- Poppies in **all edible crops (outdoor)** [12]
- Pre-harvest desiccation in **broad beans** [1-3, 6, 11, 13]; **brown mustard** (off-label), **french beans** (off-label), **hemp for oilseed production** (off-label), **poppies for morphine production** (off-label), **poppies for oilseed production** (off-label), **sesame** (off-label), **sunflowers** (off-label), **white mustard** (off-label) [11]; **clover seed crops**, **combining peas**, **laid oats** (stockfeed only), **linseed**, **spring field beans**, **spring oilseed rape**, **winter field beans**, **winter oilseed rape** [1-6, 8, 11, 13, 14]
- Sowthistle in **all edible crops (outdoor)** [12]

Extension of Authorisation for Minor Use (EAMUs)
- **barley seed crops** 20111978 [11]
- **borage for oilseed production** 20093348 [8], 20140187 [11]
- **brown mustard** 20062387 [11]
- **chives** 20062388 [11]
- **corn gromwell** 20140187 [11]
- **crambe** 20082092 [11]
- **dwarf beans** 20093346 [8]
- **forest nurseries** 20093350 [8]
- **french beans** 20093346 [8], 20062389 [11]
- **hemp for oilseed production** 20093351 [8], 20062386 [11]
- **herbs (see appendix 6)** 20093345 [8], 20062388 [11]
- **hops** 20102019 [11]
- **lucerne** 20082842 [11]
- **lupins** 20093353 [8], 20073232 [11]
- **mustard** 20093347 [8]
- **oat seed crops** 20111978 [11]

SEE SECTION 3 FOR PRODUCTS ALSO REGISTERED

- *parsley* 20062388 [11]
- *poppies for morphine production* 20093349 [8], 20062385 [11]
- *poppies for oilseed production* 20093347 [8], 20062387 [11]
- *potato dumps* 20111882 [11]
- *sesame* 20093347 [8], 20062387 [11]
- *sunflowers* 20093347 [8], 20062387 [11]
- *sweet potato* 20093352 [8], 20070734 [11]
- *white mustard* 20062387 [11]

Approval information
- Diquat included in Annex I under EC Regulation 1107/2009
- All approvals for aquatic weed control expired in 2004
- NOT accepted by BBPA for use on malting barley but accepted on hops on the basis of historical use

Efficacy guidance
- Acts rapidly on green parts of plants and rainfast in 15 min
- Best results for potato desiccation achieved by spraying in bright light and low humidity conditions
- For weed control in row crops apply as overall spray before crop emergence or before transplanting or as an inter-row treatment using a sprayer designed to prevent contamination of crop foliage by spray
- Use of authorised non-ionic wetter for improved weed control essential for most uses except potato desiccation

Restrictions
- Maximum number of treatments 1 per crop in most situations - see labels for details
- Do not apply potato haulm destruction treatment when soil dry. Tubers may be damaged if spray applied during or shortly after dry periods. See label for details of maximum allowable soil-moisture deficit and varietal drought resistance scores
- Do not add wetters to desiccant sprays for potatoes
- Consult processor before adding non-ionic or other wetter on peas. Staining may be increased
- Do not use treated straw or haulm as animal feed or bedding within 4 d of spraying
- Do not apply through ULV or mistblower equipment

Crop-specific information
- Latest use: before crop emergence for sugar beet and before transplanting ornamentals. Not specified for other crops where use varies according to situation
- HI zero when used as a crop desiccant but see label for advisory intervals
- For pre-harvest desiccation, apply to potatoes when tubers the desired size and to other crops when mature or approaching maturity. See label for details of timing and of period to be left before harvesting potatoes
- If potato tubers are to be stored leave 14 d after treatment before lifting

Environmental safety
- Dangerous for the environment
- Very toxic to aquatic organisms
- Keep all livestock out of treated areas and away from treated water for at least 24 h
- Do not feed treated straw or haulm to livestock within 4 days of spraying
- Do not use on crops if the straw is to be used as animal feed/bedding
- Do not dump surplus herbicide in water or ditch bottoms

Hazard classification and safety precautions
Hazard Toxic, Dangerous for the environment
Transport code 8 [1-13, 15]
Packaging group III [1-13, 15]
UN Number 1760 [1-13, 15]
Risk phrases R22a, R23a, R37, R38, R50, R53a [1-15]; R25 [5, 7, 9, 10, 15]; R36, R43 [5, 7-10, 15]; R48 [1-11, 13-15]; R69 [7, 12]

FOR FULL CONDITIONS OF USE ALWAYS READ THE PRODUCT LABEL

Operator protection A, M [1-15]; C [5, 7, 8, 14, 15]; H [1-7, 9-13, 15]; P [12]; U02a, U05a, U08, U19a, U20b [1-15]; U04a [1-8, 11-15]; U04c [7]; U10 [1-6, 8, 11-15]; U12, U12b [9, 10]; U15 [8-10]; U19c [5, 15]
Environmental protection E06c [1-8] (24 h); E06c [9] (24 hrs); E06c [10] (24 hours); E06c [11-15] (24 h); E08a [8, 14] (4 d); E09 [1-7]; E09 [9] (4 d); E09 [10-13, 15]; E15a [7]; E15b [1-6, 8-11, 13-15]; E34 [1-15]; E38 [1-8, 11-15]
Storage and disposal D01, D02, D09a, D12a [1-15]; D05 [1-6, 8-11, 13-15]; D10a [7]; D10c [1-6, 8-15]
Medical advice M04a

189 dithianon

A protectant and eradicant dicarbonitrile fungicide for scab control
FRAC mode of action code: M9

Products
Dithianon WG   BASF   70% w/w   WG   17018

Uses
- Scab in *apples, pears*

Approval information
- Dithianon included in Annex I under EC Regulation 1107/2009

Efficacy guidance
- Apply at bud-burst and repeat every 7-14 d until danger of scab infection ceases
- Application at high rate within 48 h of a Mills period prevents new infection
- Spray programme also reduces summer infection with apple canker

Restrictions
- Maximum number of treatments on apples and pears 6 per crop
- Do not use on Golden Delicious apples after green cluster
- Do not mix with lime sulphur or highly alkaline products

Crop-specific information
- HI 4 wk for apples, pears

Environmental safety
- Dangerous for the environment
- Very toxic to aquatic organisms

Hazard classification and safety precautions
Hazard Dangerous for the environment, Toxic if swallowed, Very toxic to aquatic organisms
Transport code 9
Packaging group III
UN Number 3077
Risk phrases H317, H318
Operator protection A, D, H; U05a, U11
Environmental protection E15a, E34, E38, H410
Storage and disposal D01, D02, D05, D12a

190 dithianon + potassium phosphates

A fungicide mixture for use in apple and pears

See also dithianon
dithianon + pyraclostrobin

Products
Delan Pro   BASF   125:561 g/l   SC   17374

SEE SECTION 3 FOR PRODUCTS ALSO REGISTERED

Pesticide Profiles

Uses
- Disease control in **apples**, **pears**

Approval information
- Dithianon included in Annex I under EC Regulation 1107/2009 but potassium phosphates are not listed on the database

Efficacy guidance
- Apply at bud-burst and repeat every 7-14 d until danger of scab infection ceases
- Application at high rate within 48 h of a Mills period prevents new infection
- Spray programme also reduces summer infection with apple canker

Environmental safety
- Dangerous for the environment
- Very toxic to aquatic organisms
- LERAP Category B

Hazard classification and safety precautions
Hazard Very toxic to aquatic organisms
Transport code 9
Packaging group III
UN Number 3082
Risk phrases H317, H319, H351
Operator protection A, C, H; U19a, U20a
Environmental protection E16a, H410
Storage and disposal D01, D02, D05

191   dithianon + pyraclostrobin

A protectant fungicide mixture for apples and pears
FRAC mode of action code: M9 + 11

See also pyraclostrobin

Products
Maccani   BASF   12:4% w/w   WG   13227

Uses
- Powdery mildew in **apples**
- Scab in **apples**, **pears**

Approval information
- Pyraclostrobin and dithianon included in Annex I under EC Regulation 1107/2009

Efficacy guidance
- Best resulIts obtained from treatments applied from bud burst before disease development and repeated at 10 d intervals
- Water volume should be adjusted according to size of tree and leaf area to ensure good coverage
- Pyraclostrobin is a member of the QoI cross resistance group. Product should be used preventatively at the full recommended dose and not relied on for its curative potential
- Where field performance may be adversely affected apply QoI containing fungicides in mixtures in strict alternation with fungicides from a different cross-resistance group

Restrictions
- Maximum number of treatments 4 per crop. However where the total number of fungicide treatments is less than 12, the maximum number of applications of any QoI containing products should be three
- Consult processors before use on crops destined for fruit preservation

Crop-specific information
- HI 35 d for apples, pears
- Some apple varieties may be considered to be susceptible to russetting caused by the dithianon component during fruit development

FOR FULL CONDITIONS OF USE ALWAYS READ THE PRODUCT LABEL

Environmental safety
- Dangerous for the environment
- Very toxic to aquatic organisms
- Broadcast air-assisted LERAP (40 m)

Hazard classification and safety precautions
Hazard Harmful, Dangerous for the environment
Transport code 9
Packaging group III
UN Number 3077
Risk phrases R20, R22a, R36, R50, R53a
Operator protection A, C, D; U05a, U11, U20b
Environmental protection E15b, E34, E38; E17b (40 m)
Storage and disposal D01, D02, D05, D08, D11a, D12a
Medical advice M03, M05a

192   dodeca-8,10-dienyl acetate

An insect sex pheromone

Products
Exosex CM	Exosect	0.1% w/w	DP	12103

Uses
- Codling moth in **apple orchards**, **pear orchards**

Approval information
- Dodeca-8,10-dienyl acetate is not included in Annex 1 under EC Regulation 1107/2009

Efficacy guidance
- Store unopened sachets below 4°C for up to 24 months or store at ambient temperatures and use within 5 months.
- Allow at least 20 m between dispensers and use a minimum of 25 dispensers per hectare.
- Dispenser activity lasts 70 - 90 days depending upon temperatures.

Hazard classification and safety precautions
UN Number N/C
Operator protection A
Environmental protection E15a
Storage and disposal D01, D02
Medical advice M03

193   epoxiconazole

A systemic, protectant and curative triazole fungicide for use in cereals
FRAC mode of action code: 3

See also boscalid + epoxiconazole
boscalid + epoxiconazole + pyraclostrobin
chlorothalonil + epoxiconazole
dimoxystrobin + epoxiconazole
epoxiconazole + fluxapyroxad

Products
1	Bassoon EC	BASF	83 g/l	EC	15219
2	Cortez	Adama	125 g/l	SC	16280
3	Epic	BASF	125 g/l	SC	16798
4	Epoxi 125 SC	Goldengrass	125 g/l	SC	16356
5	Ignite	BASF	83 g/l	EC	15205
6	Rubric	Headland	125 g/l	SC	14118
7	Ruby	AgChem Access	125 g/l	SC	16809

SEE SECTION 3 FOR PRODUCTS ALSO REGISTERED

Pesticide Profiles

Products – continued

8	Scholar	Becesane	125 g/l	SC	16402
9	Strand	Headland	125 g/l	SC	14261
10	Zenith	Becesane	125 g/l	SC	16403

Uses
- Brown rust in **durum wheat** [1, 5, 8]; **rye** [1, 5]; **spring barley**, **spring wheat**, **triticale**, **winter barley**, **winter wheat** [1-10]; **spring rye**, **winter rye** [2-4, 6-10]
- Crown rust in **oats** [1, 5]
- Disease control in **durum wheat** *(off-label)*, **forest nurseries** *(off-label)* [2]; **grass seed crops** *(off-label)* [2, 6]
- Eyespot in **winter barley** *(reduction)*, **winter wheat** *(reduction)* [2-4, 6-10]
- Fusarium ear blight in **durum wheat** *(good reduction)*, **winter wheat** *(good reduction)* [1, 5]; **spring wheat** [3]; **spring wheat** *(good reduction)* [1, 2, 5]; **winter wheat** [2, 3]
- Glume blotch in **durum wheat** [1, 5, 8]; **spring wheat**, **triticale**, **winter wheat** [1-10]
- Net blotch in **spring barley** [3, 4, 6-10]; **spring barley** *(moderate control)* [1, 2, 5]; **winter barley** [2-4, 6-10]; **winter barley** *(moderate control)* [1, 5]
- Powdery mildew in **durum wheat** [8]; **fodder beet**, **sugar beet** [4, 6, 7, 9, 10]; **grass seed crops** *(off-label)* [6]; **oats**, **rye** [1, 5]; **spring barley**, **winter barley** [1, 3-10]; **spring oats**, **spring rye**, **winter oats**, **winter rye**, **winter wheat** [2-4, 6-10]; **spring wheat**, **triticale** [3, 4, 6-10]
- Ramularia leaf spots in **spring barley** *(reduction)*, **winter barley** *(reduction)* [1, 5]
- Rhynchosporium in **rye** [1, 5]; **spring barley**, **winter barley** [1-10]; **spring rye**, **winter rye** [2-4, 6-10]
- Rust in **fodder beet**, **sugar beet** [4, 6, 7, 9, 10]; **grass seed crops** *(off-label)* [6]
- Septoria leaf blotch in **durum wheat** [1, 5, 8]; **spring wheat**, **triticale**, **winter wheat** [1-10]
- Sooty moulds in **durum wheat** *(reduction)* [5, 8]; **spring wheat** *(reduction)*, **winter wheat** *(reduction)* [2-10]
- Yellow rust in **durum wheat** [1, 5, 8]; **rye** [1, 5]; **spring barley**, **spring wheat**, **triticale**, **winter barley**, **winter wheat** [1-10]; **spring rye**, **winter rye** [2-4, 6-10]

Extension of Authorisation for Minor Use (EAMUs)
- *durum wheat* 20131995 [2], 20141277 [2]
- *forest nurseries* 20131996 [2], 20141278 [2]
- *grass seed crops* 20131995 [2], 20141277 [2], 20122523 [6]

Approval information
- Accepted by BBPA for use on malting barley (before ear emergence only)
- Epoxiconazole included in Annex I under EC Regulation 1107/2009

Efficacy guidance
- Apply at the start of foliar disease attack
- Optimum effect against eyespot achieved by spraying between leaf-sheath erect and second node detectable stages (GS 30-32)
- Best control of ear diseases of wheat obtained by treatment during ear emergence
- Mildew control improved by use of tank mixtures. See label for details
- For Septoria spray after third node detectable stage (GS 33) when weather favouring disease development has occurred
- Epoxiconazole is a DMI fungicide. Resistance to some DMI fungicides has been identified in Septoria leaf blotch which may seriously affect performance of some products. For further advice contact a specialist advisor and visit the Fungicide Resistance Action Group (FRAG)-UK website

Restrictions
- Maximum total dose equivalent to two full dose treatments
- Product may cause damage to broad-leaved plant species
- Avoid spray drift onto neighbouring crops

Crop-specific information
- Latest use: up to and including flowering just complete (GS 69) in wheat, rye, triticale; up to and including emergence of ear just complete (GS 59) in barley, oats

FOR FULL CONDITIONS OF USE ALWAYS READ THE PRODUCT LABEL

epoxiconazole + fenpropimorph

Environmental safety
- Dangerous for the environment
- Very toxic to aquatic organisms
- LERAP Category B

Hazard classification and safety precautions
    Hazard Harmful, Dangerous for the environment
    Transport code 9
    Packaging group III
    UN Number 3082
    Risk phrases R20 [1, 3, 5]; R38 [1, 3-10]; R40, R50, R53a, R62, R63 [1-10]
    Operator protection A [1-10]; H [1, 3, 5]; U05a, U20b
    Environmental protection E15a, E16a, E38
    Storage and disposal D01, D02, D05, D09a, D10c, D12a [1-10]; D19 [3]
    Medical advice M05a

194   epoxiconazole + fenpropimorph

A systemic, protectant and curative fungicide mixture for cereals
FRAC mode of action code: 3 + 5

See also fenpropimorph

Products

1	Opus Team	BASF	84:250 g/l	SE	11759
2	Oracle Duet	AgChem Access	84:250 g/l	SE	16011

Uses
- Brown rust in **spring barley**, **spring rye**, **spring wheat**, **triticale**, **winter barley**, **winter rye**, **winter wheat** [1, 2]
- Eyespot in **winter barley** *(reduction)*, **winter wheat** *(reduction)* [1, 2]
- Foliar disease control in **durum wheat** *(off-label)*, **grass seed crops** *(off-label)* [1]
- Fusarium ear blight in **spring wheat**, **winter wheat** [1, 2]
- Glume blotch in **spring wheat**, **triticale**, **winter wheat** [1, 2]
- Net blotch in **spring barley**, **winter barley** [1, 2]
- Powdery mildew in **spring barley**, **spring oats**, **spring rye**, **spring wheat**, **triticale**, **winter barley**, **winter oats**, **winter rye**, **winter wheat** [1, 2]
- Rhynchosporium in **spring barley**, **spring rye**, **winter barley**, **winter rye** [1, 2]
- Septoria leaf blotch in **spring wheat**, **triticale**, **winter wheat** [1, 2]
- Sooty moulds in **spring wheat** *(reduction)*, **winter wheat** *(reduction)* [1, 2]
- Yellow rust in **spring barley**, **spring rye**, **spring wheat**, **triticale**, **winter barley**, **winter rye**, **winter wheat** [1, 2]

Extension of Authorisation for Minor Use (EAMUs)
- *durum wheat* 20062645 [1]
- *grass seed crops* 20062645 [1]

Approval information
- Epoxiconazole and fenpropimorph included in Annex I under EC Regulation 1107/2009
- Accepted by BBPA for use on malting barley

Efficacy guidance
- Apply at the start of foliar disease attack
- Optimum effect against eyespot achieved by spraying between leaf-sheath erect and second node detectable stages (GS 30-32)
- Best control of ear diseases obtained by treatment during ear emergence
- For Septoria spray after third node detectable stage (GS 33) when weather favouring disease development has occurred

---

SEE SECTION 3 FOR PRODUCTS ALSO REGISTERED

Pesticide Profiles

- Epoxiconazole is a DMI fungicide. Resistance to some DMI fungicides has been identified in Septoria leaf blotch which may seriously affect performance of some products. For further advice contact a specialist advisor and visit the Fungicide Resistance Action Group (FRAG)-UK website

Restrictions
- Maximum total dose equivalent to two full dose treatments
- Product may cause damage to broad-leaved plant species

Crop-specific information
- Latest use: up to and including flowering just complete (GS 69) in wheat, rye, triticale; up to and including emergence of ear just complete (GS 59) in barley, oats

Environmental safety
- Dangerous for the environment
- Toxic to aquatic organisms
- Avoid spray drift onto neighbouring crops
- LERAP Category B

Hazard classification and safety precautions
Hazard Harmful, Dangerous for the environment
Transport code 9
Packaging group III
UN Number 3082
Risk phrases R20, R40, R43, R51, R53a, R62, R63
Operator protection A, H; U05a, U20b
Environmental protection E15a, E16a, E38
Storage and disposal D01, D02, D05, D09a, D10c, D12a
Medical advice M05a

195   epoxiconazole + fenpropimorph + kresoxim-methyl

A protectant, systemic and curative fungicide mixture for cereals
FRAC mode of action code: 3 + 5 + 11

See also fenpropimorph
kresoxim-methyl

Products
1	Asana	BASF	125:150:125 g/l	SE	11934
2	Mantra	BASF	125:150:125 g/l	SE	11728

Uses
- Brown rust in **spring barley**, **spring rye**, **spring wheat**, **triticale**, **winter barley**, **winter rye**, **winter wheat** [1, 2]
- Disease control in **grass seed crops** *(off-label)* [1]
- Eyespot in **spring oats** *(reduction)*, **spring rye** *(reduction)*, **triticale** *(reduction)*, **winter barley** *(reduction)*, **winter oats** *(reduction)*, **winter rye** *(reduction)*, **winter wheat** *(reduction)* [1, 2]
- Foliar disease control in **durum wheat** *(off-label)*, **grass seed crops** *(off-label)* [2]
- Fusarium ear blight in **spring wheat** *(reduction)*, **winter wheat** *(reduction)* [1, 2]
- Glume blotch in **spring wheat**, **triticale**, **winter wheat** [1, 2]
- Net blotch in **spring barley**, **winter barley** [1, 2]
- Powdery mildew in **spring barley**, **spring oats**, **spring rye**, **triticale**, **winter barley**, **winter oats**, **winter rye** [1, 2]
- Rhynchosporium in **spring barley**, **spring rye**, **winter barley**, **winter rye** [1, 2]
- Septoria leaf blotch in **spring wheat**, **triticale**, **winter wheat** [1, 2]
- Sooty moulds in **spring wheat** *(reduction)*, **winter wheat** *(reduction)* [1, 2]
- Yellow rust in **spring barley**, **spring rye**, **spring wheat**, **triticale**, **winter barley**, **winter rye**, **winter wheat** [1, 2]

FOR FULL CONDITIONS OF USE ALWAYS READ THE PRODUCT LABEL

epoxiconazole + fenpropimorph + metrafenone

Extension of Authorisation for Minor Use (EAMUs)
- *durum wheat* 20061101 [2]
- *grass seed crops* 20122861 [1], 20061101 [2]

Approval information
- Kresoxim-methyl, epoxiconazole and fenpropimorph included in Annex I under EC Regulation 1107/2009
- Accepted by BBPA for use on malting barley (before ear emergence only)

Efficacy guidance
- For best results spray at the start of foliar disease attack and repeat if infection conditions persist
- Optimum effect against eyespot obtained by treatment between leaf sheaths erect and first node detectable stages (GS 30-32)
- For protection against ear diseases apply during ear emergence
- Epoxiconazole is a DMI fungicide. Resistance to some DMI fungicides has been identified in Septoria leaf blotch which may seriously affect performance of some products. For further advice contact a specialist advisor and visit the Fungicide Resistance Action Group (FRAG)-UK website
- Kresoxim-methyl is a member of the QoI cross resistance group. Product should be used preventatively and not relied on for its curative potential
- Use product as part of an Integrated Crop Management strategy incorporating other methods of control, including where appropriate other fungicides with a different mode of action. Do not apply more than two foliar applications of QoI containing products to any cereal crop
- There is a significant risk of widespread resistance occurring in *Septoria tritici* populations in UK. Failure to follow resistance management action may result in reduced levels of disease control
- Strains of barley powdery mildew resistant to QoI's are common in the UK

Restrictions
- Maximum total dose equivalent to two full dose treatments

Crop-specific information
- Latest use: mid flowering (GS 65) for wheat, rye, triticale; completion of ear emergence (GS 59) for barley, oats

Environmental safety
- Dangerous for the environment
- Very toxic to aquatic organisms
- Avoid spray drift onto neighbouring crops. Product may damage broad-leaved species
- LERAP Category B

Hazard classification and safety precautions
Hazard Harmful, Dangerous for the environment
Transport code 9
Packaging group III
UN Number 3082
Risk phrases R40, R50, R53a, R62, R63
Operator protection A, H; U05a, U14, U20b
Environmental protection E15a, E16a, E38
Storage and disposal D01, D02, D05, D09a, D10c, D12a
Medical advice M05a

196   epoxiconazole + fenpropimorph + metrafenone

A broad spectrum fungicide mixture for cereals
FRAC mode of action code: 3 + 5 + U8

See also fenpropimorph
      metrafenone

Products
Capalo	BASF	62.5:200:75 g/l	SE	13170

SEE SECTION 3 FOR PRODUCTS ALSO REGISTERED

Pesticide Profiles

Uses
- Brown rust in **spring barley**, **spring rye**, **spring wheat**, **triticale**, **winter barley**, **winter rye**, **winter wheat**
- Crown rust in **spring oats**, **winter oats**
- Fusarium ear blight in **spring wheat** *(reduction)*, **winter wheat** *(reduction)*
- Glume blotch in **spring wheat**, **triticale**, **winter wheat**
- Net blotch in **spring barley**, **winter barley**
- Powdery mildew in **spring barley**, **spring oats**, **spring rye**, **spring wheat**, **triticale**, **winter barley**, **winter oats**, **winter rye**, **winter wheat**
- Rhynchosporium in **spring barley**, **spring rye**, **winter barley**, **winter rye**
- Septoria leaf blotch in **spring wheat**, **triticale**, **winter wheat**
- Sooty moulds in **spring wheat** *(reduction)*, **winter wheat** *(reduction)*
- Yellow rust in **spring barley**, **spring rye**, **spring wheat**, **triticale**, **winter barley**, **winter rye**, **winter wheat**

Approval information
- Epoxiconazole, fenpropimorph and metrafenone included in Annex I under EC Regulation 1107/2009

Efficacy guidance
- For best results spray at the start of foliar disease attack and repeat if infection conditions persist
- Optimum effect against eyespot obtained by treatment between leaf sheaths erect and first node detectable stages (GS 30-32)
- For protection against ear diseases apply during ear emergence
- Treatment for control of Septoria diseases should normally be made after third node detectable stage if the disease is present and favourable weather for disease development has occurred
- Epoxiconazole is a DMI fungicide. Resistance to some DMI fungicides has been identified in Septoria leaf blotch which may seriously affect performance of some products. For further advice contact a specialist advisor and visit the Fungicide Resistance Action Group (FRAG)-UK website
- Should be used as part of a resistance management strategy that includes mixtures or sequences effective against mildew and non-chemical methods

Restrictions
- Maximum number of treatments 2 per crop
- Do not apply more than two applications of any products containing metrafenone to any one crop

Crop-specific information
- Latest use: up to and including beginning of flowering (GS 61) for rye, wheat, triticale; up to and including ear emergence complete (GS 59) for barley, oats

Following crops guidance
- Cereals. oilseed rape, sugar beet, linseed, maize, clover, field beans, peas, turnips, carrots, cauliflowers, onions, lettuce or potatoes may be sown following normal harvest of a treated crop

Environmental safety
- Dangerous for the environment
- Very toxic to aquatic organisms
- Avoid spray drift onto neighbouring crops. Broad-leaved species may be damaged
- LERAP Category B

Hazard classification and safety precautions
Hazard Harmful, Dangerous for the environment
Transport code 9
Packaging group III
UN Number 3082
Risk phrases R38, R40, R43, R50, R53a, R62, R63
Operator protection A, H; U05a, U08, U14, U15, U19a, U20b
Environmental protection E16a, E34, E38
Storage and disposal D01, D02, D05, D08, D09a, D10c, D12a
Medical advice M03

FOR FULL CONDITIONS OF USE ALWAYS READ THE PRODUCT LABEL

197   epoxiconazole + fluxapyroxad

An SDHI and triazole fungicide mixture for disease control in cereals
FRAC mode of action code: 3 + 7

See also epoxiconazole
fluxapyroxad

Products

1	Adexar	BASF	62.5:62.5 g/l	EC	17109
2	Clayton Index	Clayton	62.5:62.5 g/l	EC	17266
3	Pexan	BASF	62.5:59.4 g/l	EC	17110
4	Standon Coolie	Standon	62.5:62.5 g/l	EC	17382

Uses
- Brown rust in **durum wheat**, **rye**, **spring barley**, **spring wheat**, **triticale**, **winter barley**, **winter wheat**
- Crown rust in **oats**
- Eyespot in **durum wheat** *(moderate control)*, **oats** *(moderate control)*, **rye** *(moderate control)*, **triticale** *(moderate control)*, **winter wheat** *(moderate control)*
- Fusarium ear blight in **durum wheat** *(good reduction)*, **spring wheat** *(good reduction)*, **winter wheat** *(good reduction)*
- Glume blotch in **durum wheat**, **spring wheat**, **triticale**, **winter wheat**
- Net blotch in **spring barley**, **winter barley**
- Powdery mildew in **durum wheat** *(moderate control)*, **oats** *(moderate control)*, **rye** *(moderate control)*, **spring barley** *(moderate control)*, **spring wheat** *(moderate control)*, **triticale** *(moderate control)*, **winter barley** *(moderate control)*, **winter wheat** *(moderate control)*
- Ramularia leaf spots in **spring barley**, **winter barley**
- Rhynchosporium in **spring barley**, **winter barley**
- Septoria leaf blotch in **durum wheat**, **spring wheat**, **triticale**, **winter wheat**
- Sooty moulds in **durum wheat** *(reduction)*, **spring wheat** *(reduction)*, **winter wheat** *(reduction)*
- Tan spot in **durum wheat** *(moderate control)*, **spring wheat** *(moderate control)*, **winter wheat** *(moderate control)*
- Yellow rust in **durum wheat**, **rye**, **spring barley**, **spring wheat**, **triticale**, **winter barley**, **winter wheat**

Approval information
- Epoxiconazole and fluxapyroxad included in Annex I under EC Regulation 1107/2009
- Accepted by BBPA for use on malting barley up to GS 45 only (max. dose, 1 litre/ha)

Restrictions
- Avoid drift on to neighbouring crops since it may damage broad-leaved species.
- Do not apply more that two foliar applications of products containing SDHI fungicides to any cereal crop.

Crop-specific information
- After treating a cereal crop with Adexar, cabbage, carrot, clover, dwarf beans, field beans, lettuce, maize, oats, oilseed rape, onions, peas, potatoes, ryegrass, sugar beet, sunflower, winter barley and winter wheat may be sown as the following crop.

Environmental safety
- LERAP Category B

Hazard classification and safety precautions
Hazard Harmful, Dangerous for the environment, Harmful if swallowed, Very toxic to aquatic organisms
Transport code 9
Packaging group III
UN Number 3082
Risk phrases H317, H319, H351, H360 [1-4]; R40 [3]; R43 [1, 2, 4]
Operator protection A [1-4]; H [3]; U05a, U14, U20b
Environmental protection E15b, E16a, E38, H410

SEE SECTION 3 FOR PRODUCTS ALSO REGISTERED

Pesticide Profiles

Storage and disposal D01, D02, D05, D10c, D12a, D19
Medical advice M05a

198   epoxiconazole + fluxapyroxad + pyraclostrobin

A triazole + SDHI + strobilurin fungicide mixture for disease control in cereals
FRAC mode of action code: 3 + 7 + 11

Products

1	Ceriax	BASF	41.6:41.6:66.6 g/l	EC	17111
2	Vortex	BASF	41.6:41.6:61 g/l	EC	17112

Uses
- Brown rust in **durum wheat**, **rye**, **spring barley**, **spring wheat**, **triticale**, **winter barley**, **winter wheat**
- Crown rust in **oats**
- Eyespot in **durum wheat** *(moderate control)*, **oats** *(moderate control)*, **rye** *(moderate control)*, **triticale** *(moderate control)*, **winter wheat** *(moderate control)*
- Fusarium ear blight in **durum wheat** *(good reduction)*, **spring wheat** *(good reduction)*, **winter wheat** *(good reduction)*
- Glume blotch in **durum wheat**, **spring wheat**, **triticale**, **winter wheat**
- Net blotch in **spring barley**, **winter barley**
- Powdery mildew in **durum wheat** *(moderate control)*, **oats** *(moderate control)*, **rye** *(moderate control)*, **spring barley** *(moderate control)*, **spring wheat** *(moderate control)*, **triticale** *(moderate control)*, **winter barley** *(moderate control)*, **winter wheat** *(moderate control)*
- Ramularia leaf spots in **spring barley**, **winter barley**
- Rhynchosporium in **rye**, **spring barley**, **winter barley**
- Septoria leaf blotch in **durum wheat**, **spring wheat**, **triticale**, **winter wheat**
- Sooty moulds in **durum wheat** *(reduction)*, **spring wheat** *(reduction)*, **winter wheat** *(reduction)*
- Tan spot in **durum wheat** *(moderate control)*, **spring wheat** *(moderate control)*, **winter wheat** *(moderate control)*
- Yellow rust in **durum wheat**, **rye**, **spring barley**, **spring wheat**, **triticale**, **winter barley**, **winter wheat**

Approval information
- Epoxiconazole, fluxapyroxad and pyraclostrobin included in Annex I under EC Regulation 1107/2009

Restrictions
- No more than two applications of any QoI fungicide may be applied to wheat, barley, oats, rye or triticale.
- Do not apply more that two foliar applications of products containing SDHI fungicides to any cereal crop.
- May cause damage to broad-leaved plant species.

Following crops guidance
- Cabbage, carrot, clover, dwarf French beans, lettuce, maize, oats, oilseed rape, onions, peas, potatoes, ryegrass, sugarbeet, sunflower, winter barley and winter wheat may be sown as a following crop but the effect on other crops has not been assessed.

Environmental safety
- LERAP Category B

Hazard classification and safety precautions
Hazard Harmful, Dangerous for the environment, Harmful if swallowed, Harmful if inhaled, Very toxic to aquatic organisms
Transport code 9
Packaging group III
UN Number 3082
Risk phrases H315, H317, H318, H351, H360
Operator protection A, H; U05a, U14, U20b

FOR FULL CONDITIONS OF USE ALWAYS READ THE PRODUCT LABEL

Environmental protection E15b, E16a, E38, H410
Storage and disposal D01, D02, D05, D07, D10c, D12a
Medical advice M05a

199 epoxiconazole + folpet

A systemic and protectant fungicide mixture for disease control in cereals
FRAC mode of action code: 3 + M4

See also epoxiconazole
folpet

Products

1	Manitoba	Adama	50:375 g/l	SC	16539
2	Scotia	Adama	50:375 g/l	SC	17415

Uses
- Brown rust in **spring wheat**, **triticale**, **winter wheat**
- Disease control in **spring barley**, **triticale**, **winter barley**
- Glume blotch in **spring wheat**, **triticale**, **winter wheat**
- Rhynchosporium in **spring barley**, **triticale**, **winter barley**
- Septoria leaf blotch in **spring wheat**, **triticale**, **winter wheat**
- Yellow rust in **spring wheat**, **triticale**, **winter wheat**

Approval information
- Epoxiconazole and folpet included in Annex I under EC Regulation 1107/2009

Environmental safety
- Dangerous for the environment
- Very toxic to aquatic organisms
- Buffer zone requirement 6m

Hazard classification and safety precautions
Hazard Harmful, Dangerous for the environment, Harmful if inhaled, Very toxic to aquatic organisms
Transport code 9
Packaging group III
UN Number 3082
Risk phrases H319, H351, H360
Operator protection A, H; U05a, U14
Environmental protection E38, H412
Storage and disposal D01, D12a

200 epoxiconazole + isopyrazam

A protectant and eradicant fungicide mixture for use in cereals
FRAC mode of action code: 3 + 7

Products

1	Keystone	Syngenta	99:125 g/l	SC	15493
2	Micaraz	Syngenta	90:125 g/l	SC	16106
3	Seguris	Syngenta	90:125 g/l	SC	15246

Uses
- Brown rust in **spring barley**, **triticale**, **winter barley**, **winter rye**, **winter wheat**
- Glume blotch in **winter wheat**
- Net blotch in **spring barley**, **winter barley**
- Ramularia leaf spots in **spring barley**, **winter barley**
- Rhynchosporium in **spring barley**, **winter barley**, **winter rye**

- Septoria leaf blotch in *triticale*, *winter wheat*
- Yellow rust in *winter wheat*

Approval information
- Epoxiconazole and isopyrazam are included in Annex I under EC Regulation 1107/2009
- Approved by BBPA for use on malting barley before ear emergence only
- Approval expiry 30 Sep 2017 [1]

Restrictions
- Isopyrazam is an SDH respiration inhibitor; Do not apply more than two foliar applications of products containing an SDH inhibitor to any cereal crop

Environmental safety
- Buffer zone requirement 6 m [2, 3]
- LERAP Category B

Hazard classification and safety precautions
    Hazard Harmful, Dangerous for the environment [1-3]; Harmful if inhaled, Very toxic to aquatic organisms [2, 3]
    Transport code 9
    Packaging group III
    UN Number 3082
    Risk phrases H317, H319, H351, H361 [2, 3]; R20, R40, R43, R50, R53a, R62, R63 [1]
    Operator protection A, H; U09a, U14, U20b
    Environmental protection E15b, E16a, E16b, E38 [1-3]; H410 [2, 3]
    Storage and disposal D01, D02, D05, D09a, D10c, D11a, D12a
    Medical advice M05a

201 epoxiconazole + metconazole

A triazole mixture for use in cereals
FRAC mode of action code: 3 + 3

See also metconazole

Products

1	Brutus	BASF	37.5:27.5 g/l	EC	14353
2	Icarus	BASF	37.5:25.0 g/l	EC	14471
3	Osiris P	BASF	56.25:41.25 g/l	EC	15627

Uses
- Brown rust in *durum wheat, spring barley, spring wheat, triticale, winter barley, winter wheat* [1-3]; *rye* [3]; *spring rye, winter rye* [1, 2]
- Eyespot in *durum wheat (reduction), triticale (reduction), winter wheat (reduction)* [1-3]; *rye (reduction)* [3]; *spring rye (reduction), winter barley (reduction), winter rye (reduction)* [1, 2]
- Fusarium ear blight in *durum wheat (good reduction), spring wheat (good reduction), triticale (good reduction), winter wheat (good reduction)* [1, 2]; *durum wheat (moderate control), spring wheat (moderate control), triticale (moderate control), winter wheat (moderate control)* [3]
- Glume blotch in *durum wheat, spring wheat, triticale, winter wheat* [1-3]
- Net blotch in *spring barley, winter barley* [1-3]
- Powdery mildew in *durum wheat (moderate control), spring barley (moderate control), spring wheat (moderate control), triticale (moderate control), winter barley (moderate control), winter wheat (moderate control)* [1-3]; *rye (moderate control)* [3]; *spring rye (moderate control), winter rye (moderate control)* [1]; *spring rye (moderate reduction), winter rye (moderate reduction)* [2]
- Ramularia leaf spots in *spring barley* [2]; *spring barley (moderate control)* [1, 3]; *winter barley (moderate control)* [1-3]
- Rhynchosporium in *rye (moderate control)* [3]; *spring barley, winter barley* [1-3]
- Septoria leaf blotch in *durum wheat, spring wheat, triticale, winter wheat* [1-3]

- Sooty moulds in **durum wheat** *(good reduction)*, **spring wheat** *(good reduction)*, **winter wheat** *(good reduction)* [1, 2]; **durum wheat** *(moderate control)*, **spring wheat** *(moderate control)*, **winter wheat** *(moderate control)* [3]
- Tan spot in **durum wheat** *(moderate control)*, **spring wheat** *(moderate control)*, **winter wheat** *(moderate control)* [1-3]
- Yellow rust in **durum wheat**, **spring barley**, **spring wheat**, **triticale**, **winter barley**, **winter wheat** [1-3]; **rye** [3]; **spring rye**, **winter rye** [1, 2]

Approval information
- Epoxiconazole and metconazole included in Annex I under EC Regulation 1107/2009

Following crops guidance
- After trreating a cereal crop, oilseed rape, cereals, sugar beet, linseed, maize, clover, beans, peas, carrots, potatoes, lettuce, cabbage, sunflower, ryegrass or onions may be sown as a following crop. The effect on other crops has not been evaluated.

Environmental safety
- LERAP Category B

Hazard classification and safety precautions
Hazard Harmful, Dangerous for the environment
Transport code 9
Packaging group III
UN Number 3082
Risk phrases R40, R43, R53a [1-3]; R50, R62 [3]; R51, R63 [1, 2]
Operator protection A, H; U05a, U14 [1, 2]; U20b [1-3]
Environmental protection E15b, E16a, E34 [1-3]; E38 [3]
Storage and disposal D01, D02, D09a, D10c [1-3]; D05, D12a, D19 [3]
Medical advice M05a

202 epoxiconazole + metrafenone

A broad spectrum fungicide mixture for cereals
FRAC mode of action code: 3 + U8

See also metrafenone

Products
1	Ceando	Dow	83:100 g/l	SC	16012
2	Cloister	Dow	83:100 g/l	SC	15994

Uses
- Brown rust in **durum wheat** [2]; **spring barley**, **spring rye**, **spring wheat**, **triticale**, **winter barley**, **winter rye**, **winter wheat** [1, 2]
- Crown rust in **spring oats**, **winter oats** [1, 2]
- Fusarium ear blight in **durum wheat** *(reduction)* [2]; **spring wheat** *(reduction)*, **winter wheat** *(reduction)* [1, 2]
- Glume blotch in **durum wheat** [2]; **spring wheat**, **triticale**, **winter wheat** [1, 2]
- Net blotch in **spring barley**, **winter barley** [1, 2]
- Powdery mildew in **durum wheat** [2]; **spring barley**, **spring oats**, **spring rye**, **spring wheat**, **triticale**, **winter barley**, **winter oats**, **winter rye**, **winter wheat** [1, 2]
- Rhynchosporium in **spring barley**, **spring rye**, **winter barley**, **winter rye** [1, 2]
- Septoria leaf blotch in **durum wheat** [2]; **spring wheat**, **triticale**, **winter wheat** [1, 2]
- Sooty moulds in **durum wheat** *(reduction)* [2]; **spring wheat** *(reduction)*, **winter wheat** *(reduction)* [1, 2]
- Yellow rust in **durum wheat** [2]; **spring barley**, **spring rye**, **spring wheat**, **triticale**, **winter barley**, **winter rye**, **winter wheat** [1, 2]

Approval information
- Epoxiconazole and metrafenone included in Annex I under EC Regulation 1107/2009

SEE SECTION 3 FOR PRODUCTS ALSO REGISTERED

Pesticide Profiles

Efficacy guidance
- For best results spray at the start of foliar disease attack and repeat if infection conditions persist
- Optimum effect against eyespot obtained by treatment between leaf sheaths erect and first node detectable stages (GS 30-32)
- For protection against ear diseases apply during ear emergence
- Treatment for control of Septoria diseases should normally be made after third node detectable stage if the disease is present and favourable weather for disease development has occurred
- Epoxiconazole is a DMI fungicide. Resistance to some DMI fungicides has been identified in Septoria leaf blotch which may seriously affect performance of some products. For further advice contact a specialist advisor and visit the Fungicide Resistance Action Group (FRAG)-UK website
- Should be used as part of a resistance management strategy that includes mixtures or sequences effective against mildew and non-chemical methods
- Where mildew well established at time of treatment product should be mixed with fenpropimorph. See label

Restrictions
- Maximum number of treatments 2 per crop
- Do not apply more than two applications of any products containing metrafenone to any one crop

Crop-specific information
- Latest use: up to and including beginning of flowering (GS 61) for rye, wheat, triticale; up to and including ear emergence complete (GS 59) for barley, oats

Following crops guidance
- Cereals. oilseed rape, sugar beet, linseed, maize, clover, field beans, peas, turnips, carrots, cauliflowers, onions, lettuce or potatoes may be sown following normal harvest of a treated crop

Environmental safety
- Dangerous for the environment
- Very toxic to aquatic organisms
- Avoid spray drift onto neighbouring crops. Broad-leaved species may be damaged
- LERAP Category B

Hazard classification and safety precautions
  Hazard Harmful, Dangerous for the environment
  UN Number N/C
  Risk phrases R40, R50, R53a, R62, R63
  Operator protection A, H; U05a, U20b
  Environmental protection E15a, E16a, E34, E38
  Storage and disposal D01, D02, D05, D08, D09a, D10c
  Medical advice M03, M05a

203   epoxiconazole + pyraclostrobin

A protectant, systemic and curative fungicide mixture for cereals
FRAC mode of action code: 3 + 11

See also pyraclostrobin

Products
1	Envoy	BASF	62.5:85 g/l	SE	16297
2	Gemstone	BASF	62.5:80 g/l	SE	16298
3	Ibex	BASF	50:133 g/l	SE	16457
4	Opera	BASF	50:133 g/l	SE	16420

Uses
- Brown rust in **durum wheat**, **rye**, **spring barley**, **spring wheat**, **triticale**, **winter barley**, **winter wheat** [1-4]
- Cercospora leaf spot in **sugar beet** [3, 4]
- Crown rust in **spring oats**, **winter oats** [1-4]

FOR FULL CONDITIONS OF USE ALWAYS READ THE PRODUCT LABEL

epoxiconazole + pyraclostrobin

- Disease control in **grass seed crops** *(off-label)* [4]
- Eyespot in **forage maize**, **forage maize** *(off-label)*, **grain maize**, **grain maize** *(off-label)* [4]
- Fusarium ear blight in **durum wheat** *(good reduction)*, **spring wheat** *(good reduction)*, **winter wheat** *(good reduction)* [1-4]; **triticale** *(good reduction)* [3, 4]
- Glume blotch in **durum wheat**, **spring wheat**, **triticale**, **winter wheat** [1-4]
- Leaf blight in **forage maize** *(off-label)*, **grain maize** *(off-label)* [4]
- Net blotch in **spring barley**, **winter barley** [1-4]
- Northern leaf blight in **forage maize**, **grain maize** [4]
- Powdery mildew in **fodder beet** *(off-label)* [4]; **rye**, **spring barley**, **spring oats**, **winter barley**, **winter oats** [1, 2]; **sugar beet** [3, 4]
- Ramularia leaf spots in **rye** *(moderate control)*, **spring barley** *(moderate control)*, **sugar beet**, **triticale** *(moderate control)*, **winter barley** *(moderate control)* [3, 4]; **spring barley** *(reduction)*, **winter barley** *(reduction)* [1, 2]
- Rhynchosporium in **rye**, **spring barley**, **winter barley** [1-4]
- Rust in **fodder beet** *(off-label)*, **forage maize** *(off-label)*, **grain maize** *(off-label)* [4]; **sugar beet** [3, 4]
- Septoria leaf blotch in **durum wheat**, **spring wheat**, **triticale**, **winter wheat** [1, 2]; **durum wheat** *(moderate control)*, **spring wheat** *(moderate control)*, **triticale** *(moderate control)*, **winter wheat** *(moderate control)* [3, 4]
- Sooty moulds in **spring wheat** *(reduction)*, **winter wheat** *(reduction)* [1, 2]
- Tan spot in **durum wheat**, **spring wheat**, **winter wheat** [1, 2]; **durum wheat** *(moderate control)*, **spring wheat** *(moderate control)*, **triticale** *(moderate control)*, **winter wheat** *(moderate control)* [3, 4]
- Yellow rust in **durum wheat**, **rye**, **spring barley**, **spring wheat**, **triticale**, **winter barley**, **winter wheat** [1-4]

Extension of Authorisation for Minor Use (EAMUs)
- *fodder beet* 20140392 [4]
- *forage maize* 20141592 [4]
- *grain maize* 20141592 [4]
- *grass seed crops* 20140522 [4]

Approval information
- Epoxiconazole and pyraclostrobin included in Annex I under EC Regulation 1107/2009
- Accepted by BBPA for use on malting barley (before ear emergence only)

Efficacy guidance
- For best results apply at the start of foliar disease attack
- Good reduction of Fusarium ear blight can be obtained from treatment during flowering
- Yield response may be obtained in the absence of visual disease symptoms
- Epoxiconazole is a DMI fungicide. Resistance to some DMI fungicides has been identified in Septoria leaf blotch which may seriously affect performance of some products. For further advice contact a specialist advisor and visit the Fungicide Resistance Action Group (FRAG)-UK website
- Pyraclostrobin is a member of the QoI cross resistance group. Product should be used preventatively and not relied on for its curative potential
- Use product as part of an Integrated Crop Management strategy incorporating other methods of control, including where appropriate other fungicides with a different mode of action. Do not apply more than two foliar applications of QoI containing products to any cereal crop
- There is a significant risk of widespread resistance occurring in *Septoria tritici* populations in UK. Failure to follow resistance management action may result in reduced levels of disease control

Restrictions
- Maximum number of treatments 2 per crop
- Do not use on sugar beet crops grown for seed [3, 4]

Crop-specific information
- Latest use: before grain watery ripe (GS 71) for wheat; up to and including emergence of ear just complete (GS 59) for barley and oats
- HI 6 wk for sugar beet [3, 4]

SEE SECTION 3 FOR PRODUCTS ALSO REGISTERED

Pesticide Profiles

Environmental safety
- Dangerous for the environment
- Very toxic to aquatic organisms
- LERAP Category B

Hazard classification and safety precautions
Hazard Harmful, Dangerous for the environment [1-4]; Toxic if swallowed, Harmful if inhaled [4]
Transport code 9
Packaging group III
UN Number 3082
Risk phrases H315, H317, H351, H360 [4]; R20, R22a, R38, R40, R43, R50, R53a [1-3]; R62, R63 [1, 2]
Operator protection A; U05a, U14, U20b
Environmental protection E15a, E16a, E34, E38 [1-4]; E16b [3, 4]; H410 [4]
Storage and disposal D01, D02, D09a, D10c, D12a [1-4]; D05 [3, 4]
Medical advice M03, M05a

204   esfenvalerate

A contact and ingested pyrethroid insecticide
IRAC mode of action code: 3

Products

1	Gocha	Belchim	25 g/l	EC	17320
2	Greencrop Cajole Ultra	Greencrop	25 g/l	EC	12967
3	Standon Hounddog	Standon	25 g/l	EC	13201
4	Sumi-Alpha	Interfarm	25 g/l	EC	14023
5	Sven	Interfarm	25 g/l	EC	14859

Uses
- Aphids in **broccoli, brussels sprouts, cabbages, calabrese, cauliflowers, combining peas, edible podded peas, kale, kohlrabi, potatoes, spring field beans, vining peas, winter field beans** [1, 4, 5]; **chinese cabbage** [4, 5]; **grass seed crops** *(off-label)*, **grassland, ornamental plant production, protected ornamentals** [4]; **oriental cabbage** [1]; **spring barley, spring wheat, winter barley, winter wheat** [1-5]
- Bibionids in **grass seed crops** [5]; **grassland, managed amenity turf** [1, 5]
- Caterpillars in **broccoli, brussels sprouts, cabbages, calabrese, cauliflowers, kale, kohlrabi** [1, 5]; **chinese cabbage** [5]; **oriental cabbage** [1]
- Insect pests in **ornamental plant production** [1, 5]; **protected ornamentals** [5]
- Pea and bean weevil in **combining peas, edible podded peas, spring field beans, vining peas, winter field beans** [1, 5]

Extension of Authorisation for Minor Use (EAMUs)
- *grass seed crops* 20081266 [4]

Approval information
- Esfenvalerate included in Annex I under EC Regulation 1107/2009
- Accepted by BBPA for use on malting barley

Efficacy guidance
- For best reduction of spread of barley yellow dwarf virus winter sown crops at high risk (e.g. after grass or in areas with history of BYDV) should be treated when aphids first seen or by mid-Oct. Otherwise treat in late Oct-early Nov
- High risk winter sown crops will need a second treatment
- Spring sown crops should be treated from the 2-3 leaf stage if aphids are found colonising in the crop and a further application may be needed before the first node stage if they reinfest
- Product also recommended between onset of flowering and milky ripe stages (GS 61-73) for control of summer cereal aphids

Restrictions
- Maximum number of treatments 3 per crop of which 2 may be in autumn

FOR FULL CONDITIONS OF USE ALWAYS READ THE PRODUCT LABEL

ethephon + mepiquat chloride

- Do not use if another pyrethroid or dimethoate has been applied to crop after start of ear emergence (GS 51)
- Do not leave spray solution standing in spray tank

Crop-specific information
- Latest use: 31 Mar in yr of harvest (winter use dose); early milk stage for barley (summer use); late milk stage for wheat (summer use)

Environmental safety
- Dangerous for the environment
- Very toxic to aquatic organisms
- High risk to non-target insects or other arthropods. Do not spray within 6 m of the field boundary
- Give local beekeepers warning when using in flowering crops
- Flammable
- Store product in dark away from direct sunlight
- LERAP Category A

Hazard classification and safety precautions
Hazard Harmful, Flammable, Dangerous for the environment [1-5]; Flammable liquid and vapour, Harmful if swallowed, Harmful if inhaled, Very toxic to aquatic organisms [1]
Transport code 3 [1, 4, 5]; 6.1 [2, 3]
Packaging group III
UN Number 1993 [1, 4, 5]; 3351 [2, 3]
Risk phrases H304, H317, H318, H373 [1]; R20, R22a, R41, R43, R50, R53a [2-5]
Operator protection A, C, H; U04a, U05a, U08, U11, U14, U19a, U20b
Environmental protection E15a, E22a [1, 2, 4, 5]; E15b, E22b [3]; E16c, E16d, E34 [1-5]; E38 [1, 4, 5]; H410 [1]
Storage and disposal D01, D02, D09a [1-5]; D05 [2, 3]; D07, D10b [2]; D10c, D12a [1, 3-5]
Medical advice M03 [1-5]; M05b [3]

205 ethephon + mepiquat chloride

A plant growth regulator for use in cereals

See also mepiquat chloride

Products

1	Clayton Proud	Clayton	155:305 g/l	SL	17067
2	Terpal	BASF	155:305 g/l	SL	16463

Uses
- Growth regulation in **forest nurseries** *(off-label)* [2]
- Increasing yield in **winter barley** *(low lodging situations)*, **winter wheat** *(low lodging situations)* [1, 2]
- Lodging control in **spring barley**, **triticale**, **winter barley**, **winter rye**, **winter wheat** [1, 2]

Extension of Authorisation for Minor Use (EAMUs)
- *forest nurseries* 20142725 [2]

Approval information
- ethephon and mepiquat chloride included in Annex I under EC Regulation 1107/2009
- All products containing 2-chloroethylphosphonic acid carry the warning: '2-chloroethylphosphonic acid is an anticholinesterase organophosphate. Handle with care'
- Accepted by BBPA for use on malting barley

Efficacy guidance
- Best results achieved on crops growing vigorously under conditions of high fertility
- Recommended dose and timing vary with crop, cultivar, growing conditions, previous treatment and desired degree of lodging control. See label for details
- May be applied to crops undersown with grass or clovers
- Do not apply to crops if wet or rain expected as efficacy will be impaired

SEE SECTION 3 FOR PRODUCTS ALSO REGISTERED

Pesticide Profiles

Restrictions
- Ethephon is an anticholinesterase organophosphorus compound. Do not use if under medical advice not to work with such compounds
- Maximum number of treatments 2 per crop
- Add an authorised non-ionic wetter to spray solution. See label for recommended product and rate
- Do not treat crops damaged by herbicides or stressed by drought, waterlogging etc
- Do not treat crops on soils of low fertility unless adequately fertilized
- Do not apply to winter cultivars sown in spring or treat winter barley, triticale or winter rye on soils with more than 10% organic matter (winter wheat may be treated)
- Do not apply at temperatures above 21°C

Crop-specific information
- Latest use: before ear visible (GS 49) for winter barley, spring barley, winter wheat and triticale; flag leaf just visible (GS 37) for winter rye
- Late tillering may be increased with crops subject to moisture stress and may reduce quality of malting barley

Environmental safety
- Do not use straw from treated cereals as a mulch or growing medium

Hazard classification and safety precautions
Hazard Harmful
Transport code 8
Packaging group III
UN Number 3265
Risk phrases R22a, R36, R37, R53a
Operator protection A, C; U20b
Environmental protection E15a
Storage and disposal D01, D02, D08, D09a, D10c
Medical advice M01, M05a

206   ethofumesate

A benzofuran herbicide for grass weed control in various crops
HRAC mode of action code: N

See also chloridazon + ethofumesate
desmedipham + ethofumesate + lenacil + phenmedipham
desmedipham + ethofumesate + metamitron + phenmedipham
desmedipham + ethofumesate + phenmedipham

Products

1	Efeckt	UPL Europe	500 g/l	SC	17177
2	Etho 500 SC	Goldengrass	500 g/l	SC	15256
3	Ethofol 500 SC	UPL Europe	500 g/l	SC	15179
4	Fumesate 500 SC	Hermoo	500 g/l	SC	14321
5	Oblix 500	UPL Europe	500 g/l	SC	16671
6	Palamares	Agroquimicos	500 g/l	SC	15653
7	Xerton	UPL Europe	417 g/l	SC	17335

Uses
- Annual dicotyledons in **amenity grassland** [2, 6]; **fodder beet**, **red beet**, **sugar beet** [1, 2, 4-6]; **grass seed crops**, **mangels** [1, 2, 5, 6]; **winter wheat** [7]
- Annual grasses in **amenity grassland** [2]; **fodder beet**, **red beet**, **sugar beet** [1, 2, 4, 5]; **grass seed crops**, **mangels** [1, 2, 5]; **winter wheat** [7]
- Annual meadow grass in **amenity grassland**, **fodder beet**, **grass seed crops**, **mangels**, **red beet**, **sugar beet** [6]
- Blackgrass in **amenity grassland**, **fodder beet**, **grass seed crops**, **mangels**, **red beet**, **sugar beet** [6]; **winter wheat** [7]
- Cleavers in **fodder beet**, **sugar beet** [3]

FOR FULL CONDITIONS OF USE ALWAYS READ THE PRODUCT LABEL

ethofumesate

Approval information
- Ethofumesate included in Annex I under EC Regulation 1107/2009

Efficacy guidance
- Most products may be applied pre- or post-emergence of crop or weeds but some restricted to pre-emergence or post-emergence use only. Check label
- Some products recommended for use only in mixtures. Check label
- Volunteer cereals not well controlled pre-emergence, weed grasses should be sprayed before fully tillered
- Grass crops may be sprayed during rain or when wet. Not recommended in very dry conditions or prolonged frost

Restrictions
- The maximum total dose must not exceed 1.0 kg/ha ethofumesate in any three year period
- Do not use on Sands or Heavy soils, Very Light soils containing a high percentage of stones, or soils with more than 5-10% organic matter (percentage varies according to label)
- Do not use on swards reseeded without ploughing
- Clovers will be killed or severely checked
- Do not graze or cut grass for 14 d after, or roll less than 7 d before or after spraying

Crop-specific information
- Latest use: before crops meet across rows for beet crops and mangels; not specified for other crops
- Apply in beet crops in tank mixes with other pre- or post-emergence herbicides. Recommendations vary for different mixtures. See label for details
- Safe timing on beet crops varies with other ingredient of tank mix. See label for details
- In grass crops apply to moist soil as soon as possible after sowing or post-emergence when crop in active growth, normally mid-Oct to mid-Dec. See label for details
- May be used in Italian, hybrid and perennial ryegrass, timothy, cocksfoot, meadow fescue and tall fescue. Apply pre-emergence to autumn-sown leys, post-emergence after 2-3 leaf stage. See label for details

Following crops guidance
- In the event of failure of a treated crop only sugar beet, red beet, fodder beet and mangels may be redrilled within 3 mth of application
- Any crop may be sown 3 mth after application of mixtures in beet crops following ploughing, 5 mth after application in grass crops

Environmental safety
- Dangerous for the environment
- Toxic to aquatic organisms
- Do not empty into drains
- LERAP Category B [7]

Hazard classification and safety precautions
Hazard Irritant [3]; Dangerous for the environment [2, 6]
Transport code 9 [1-3, 6]
Packaging group III [1-3, 6]
UN Number 3082 [1-3, 6]; N/C [4, 5, 7]
Risk phrases H317 [7]; R43 [3]; R51 [2, 6]; R52 [4]; R53a [2, 4, 6]
Operator protection A [3]; H [1, 3, 5, 7]; U05a [1, 2, 5, 7]; U08 [2, 6]; U09a [3, 4]; U14 [3]; U19a [2-4, 6]; U20a [6]; U20b [2-4]
Environmental protection E15a, E19b [6]; E15b [1-5, 7]; E16a [7]; E23, H411 [1, 5, 7]; E34 [1, 2, 5-7]; E38 [3-5, 7]
Storage and disposal D01, D02 [1, 3, 5, 7]; D05 [3, 4, 6]; D09a [1-7]; D10a [1, 2, 4, 5, 7]; D10b [6]; D10c [3]; D12a [1, 2, 4-7]
Medical advice M05a [3]

SEE SECTION 3 FOR PRODUCTS ALSO REGISTERED

Pesticide Profiles

207 ethofumesate + metamitron

A contact and residual herbicide mixture for beet crops
HRAC mode of action code: N + C1

See also metamitron

Products

1	Oblix MT	UPL Europe	150:350 g/l	SC	16487
2	Torero	Adama	150:350 g/l	SC	16484
3	Volcano	UPL Europe	150:350 g/l	SC	16532

Uses
- Annual dicotyledons in **fodder beet**, **mangels**, **sugar beet**
- Annual meadow grass in **fodder beet**, **mangels**, **sugar beet**

Approval information
- Ethofumesate and metamitron included in Annex I under EC Regulation 1107/2009

Efficacy guidance
- Best results obtained from a series of treatments applied as an overall fine spray commencing when earliest germinating weeds are no larger than fully expanded cotyledon and the majority of the crop at fully expanded cotyledon
- Apply subsequent sprays as each new flush of weeds reaches early cotyledon and continue until weed emergence ceases
- Product may be used on all soil types but residual activity may be reduced on those with more than 5% organic matter

Restrictions
- Maximum total dose equivalent to three full dose treatments
- The maximum total dose must not exceed 1.0 kg/ha ethofumesate in any three year period

Crop-specific information
- Latest use: before crop leaves meet between rows
- Crop tolerance may be reduced by stress caused by growing conditions, effects of pests, disease or other pesticides, nutrient deficiency etc

Following crops guidance
- Beet crops may be sown at any time after treatment. Any other crop may be sown after mouldboard ploughing to 15 cm and a minimum interval of 3 mth after treatment

Environmental safety
- Dangerous for the environment
- Very toxic to aquatic organisms
- Do not empty into drains

Hazard classification and safety precautions
Hazard Harmful [1]; Dangerous for the environment [1-3]
Transport code 9
Packaging group III
UN Number 3082
Risk phrases R22a [1]; R51, R53a [1, 3]
Operator protection A, H [1, 3]; U05a, U19a, U20a [1-3]; U08, U14, U15 [2, 3]
Environmental protection E13c, E34 [2, 3]; E15a [1]; E19b, E38 [1-3]; H411 [2]
Storage and disposal D01, D02, D09a, D10b, D12a [1-3]; D05 [2, 3]
Medical advice M05a [1]

FOR FULL CONDITIONS OF USE ALWAYS READ THE PRODUCT LABEL

208 ethofumesate + metamitron + phenmedipham

A contact and residual herbicide mixture for sugar beet
HRAC mode of action code: N + C1 + C1

*See also metamitron
phenmedipham*

Products
Phemo   UPL Europe   51:153:51 g/l   SE   16984

Uses
- Annual dicotyledons in **fodder beet**, **mangels**, **red beet**, **sugar beet**
- Annual meadow grass in **fodder beet**, **mangels**, **red beet**, **sugar beet**

Approval information
- Ethofumesate, metamitron and phenmedipham included in Annex I under EC Regulation 1107/2009

Efficacy guidance
- Best results obtained from a series of treatments applied as an overall fine spray commencing when earliest germinating weeds are no larger than fully expanded cotyledon
- Apply subsequent sprays as each new flush of weeds reaches early cotyledon and continue until weed emergence ceases (maximum 3 sprays)
- Product must be applied with Actipron
- Product may follow certain pre-emergence treatments and be used in conjunction with other post-emergence sprays - see label for details
- Where a pre-emergence band spray has been applied, the first treatment should be timed according to the size of the weeds in the untreated area between the rows

Restrictions
- Maximum number of treatments 3 per crop; maximum total dose 6 kg product per ha
- Product may be used on all soil types but residual activity may be reduced on those with more than 5% organic matter
- Crop tolerance may be reduced by stress caused by growing conditions, effects of pests, disease or other pesticides, nutrient deficiency etc
- Only beet crops should be sown within 4 mth of last treatment; winter cereals may be sown after this interval
- Any spring crop may be sown in the year following use
- Mould-board ploughing to 150 mm followed by thorough cultivation recommended before planting any crop.
- The maximum total dose must not exceed 1.0 kg/ha ethofumesate in any three year period

Crop-specific information
- Latest use: before crop leaves meet between rows

Environmental safety
- Irritating to eyes
- Harmful to fish or other aquatic life. Do not contaminate surface waters or ditches with chemical or used container

Hazard classification and safety precautions
Hazard Dangerous for the environment
Transport code 9
Packaging group III
UN Number 3082
Risk phrases R50, R53a
Operator protection A; U05a, U20c
Environmental protection E15a, E34, E38
Storage and disposal D01, D02, D09a, D11a, D11b, D12a

SEE SECTION 3 FOR PRODUCTS ALSO REGISTERED

Pesticide Profiles

209 ethofumesate + phenmedipham

A contact and residual herbicide for use in beet crops
HRAC mode of action code: N + C1

See also phenmedipham

Products

1	Brigida	Agroquimicos	200:200 g/l	SC	15658
2	Teamforce	UPL Europe	100:80 g/l	EC	16978
3	Twin	Adama	94:97 g/l	EC	13998

Uses
- Annual dicotyledons in **fodder beet**, **mangels**, **sugar beet**
- Annual meadow grass in **fodder beet**, **mangels**, **sugar beet**
- Blackgrass in **fodder beet**, **mangels**, **sugar beet**

Approval information
- Ethofumesate and phenmedipham included in Annex I under EC Regulation 1107/2009

Efficacy guidance
- Best results achieved by repeat applications to cotyledon stage weeds. Larger susceptible weeds not killed by first treatment usually checked and controlled by second application
- Apply on all soil types at 5-10 d intervals
- On soils with more than 5-10% organic matter residual activity may be reduced

Restrictions
- Maximum number of treatments normally 3 per crop or maximum total dose equivalent to three full dose treatments, or less - see labels for details
- Do not spray wet foliage or if rain imminent
- Spray in evening if daytime temperatures above 21°C expected
- Avoid or delay treatment if frost expected within 7 days
- Avoid or delay treating crops under stress from wind damage, manganese or lime deficiency, pest or disease attack etc.
- The maximum total dose must not exceed 1.0 kg/ha ethofumesate in any three year period

Crop-specific information
- Latest use: before crop foliage meets in the rows
- Check from which recovery may not be complete may occur if treatment made during conditions of sharp diurnal temperature fluctuation

Following crops guidance
- Beet crops may be sown at any time after treatment. Any other crop may be sown after mouldboard ploughing to 15 cm and a minimum interval of 3 mth after treatment

Environmental safety
- Dangerous for the environment [1, 3]
- Toxic to aquatic organisms
- Do not empty into drains
- Extra care necessary to avoid drift because product is recommended for use as a fine spray

Hazard classification and safety precautions
  Hazard Harmful [2, 3]; Irritant [1, 2]; Dangerous for the environment [1-3]
  Transport code 9
  Packaging group III
  UN Number 3082
  Risk phrases R21, R22a [2]; R37, R40 [2, 3]; R43, R51, R53a [1-3]
  Operator protection A [1-3]; H [1, 2]; U05a, U08, U19a, U20a [1-3]; U14 [1]
  Environmental protection E15a, E19b, E34 [1-3]; E38 [2]
  Storage and disposal D01, D02, D09a, D10b, D12a [1-3]; D05 [1]
  Medical advice M05a

FOR FULL CONDITIONS OF USE ALWAYS READ THE PRODUCT LABEL

ethylene

210 ethylene

A gas used for fruit ripening and potato storage

Products
Biofresh Safestore   Biofresh   99.9% w/w   GA   15729

Uses
- Sprout suppression in **potatoes**

Approval information
- Ethylene is included in Annex 1 under EC Regulation 1107/2009

Efficacy guidance
- For use in stored fruit or potatoes after harvest

Restrictions
- Handling and release of ethylene must only be undertaken by operators suitably trained and competent to carry out the work
- Operators must vacate treated areas immediately after ethylene introduction
- Unprotected persons must be excluded from the treated areas until atmospheres have been thoroughly ventilated for 15 minutes minimum before re-entry
- Ambient atmospheric ethylene concentration must not exceed 1000 ppm. Suitable self-contained breathing apparatus must be worn in atmospheres containing ethylene in excess of 1000 ppm
- A minimum 3 d post treatment period is required before removal of treated crop from storage
- Ethylene treatment must only be undertaken in fully enclosed storage areas that are air tight with appropriate air circulation and venting facilities

Hazard classification and safety precautions
    Hazard Extremely flammable
    Transport code 2
    UN Number 1962
    Risk phrases R52, R67
    Operator protection A, D, H; U19a
    Environmental protection E38
    Storage and disposal D06e, D06f, D06g
    Medical advice M04a

211 etofenprox

An insecticide for use in oilseed rape

Products
Trebon 30 EC   Certis   287.5 g/l   EC   16666

Uses
- Pollen beetle in **spring oilseed rape** (moderate control), **winter oilseed rape** (moderate control)

Approval information
- Etofenprox is included in Annex 1 under EC Regulation 1107/2009

Efficacy guidance
- Apply in a water volume of 100 - 300 l/ha when pollen beetle numbers exceed appropriate threshold levels. Follow up with a non-pyrethroid spray if above threshold numbers survive treatment. To control aphids, use an appropriate aphicide spray.

Environmental safety
- Buffer zone requirement 12m
- LERAP Category B

SEE SECTION 3 FOR PRODUCTS ALSO REGISTERED

Pesticide Profiles

Hazard classification and safety precautions
  Hazard Irritant, Dangerous for the environment
  Transport code 9
  Packaging group III
  UN Number 3082
  Risk phrases R22a, R22b, R36, R38, R50, R53a, R64, R67
  Operator protection A, C, H; U05a, U11, U14, U15, U20a
  Environmental protection E12c, E12e, E15b, E16a, E22b, E38
  Storage and disposal D01, D02, D09a, D10b, D12a
  Medical advice M05b

212 etoxazole

A mite growth inhibitor
IRAC mode of action code: 10B

Products

1	Borneo	Interfarm	110 g/l	SC	13919
2	Clayton Java	Clayton	110 g/l	SC	15155

Uses
- Mites in *protected aubergines*, *protected tomatoes*
- Spider mites in *protected ornamentals* (off-label), *protected strawberries* (off-label)
- Two-spotted spider mite in *protected ornamentals* (off-label), *protected strawberries* (off-label)

Extension of Authorisation for Minor Use (EAMUs)
- *protected ornamentals* 20081216 [1], 20111544 [2]
- *protected strawberries* 20092400 [1], 20111545 [2]

Approval information
- Etoxazole included in Annex I under EC Regulation 1107/2009

Hazard classification and safety precautions
  Hazard Dangerous for the environment
  Transport code 9
  Packaging group III
  UN Number 3082
  Risk phrases R50, R53a
  Operator protection A; U05a, U20b
  Environmental protection E15a, E34
  Storage and disposal D01, D02, D09b, D11a, D12b

213 famoxadone

A strobilurin fungicide available only in mixtures
FRAC mode of action code: 11

See also cymoxanil + famoxadone

214 fenamidone

A strobilurin fungicide available only in mixtures
FRAC mode of action code: 11

See also fenamidone + fosetyl-aluminium
  fenamidone + mancozeb
  fenamidone + propamocarb hydrochloride

215 fenamidone + fosetyl-aluminium

A strobilurin fungicide mixture
FRAC mode of action code: 11 + 33

See also fenamidone
fosetyl-aluminium

Products

Fenomenal   Bayer CropScience   60:600 g/l   WG   15494

Uses
- Collar rot in **strawberries** *(moderate control from foliar spray)*
- Crown rot in **strawberries** *(moderate control from foliar spray)*
- Downy mildew in **baby leaf crops** *(off-label)*, **celery leaves** *(off-label)*, **chives** *(off-label)*, **cress** *(off-label)*, **edible flowers** *(off-label)*, **endives** *(off-label)*, **herbs (see appendix 6)** *(off-label)*, **lamb's lettuce** *(off-label)*, **land cress** *(off-label)*, **lettuce** *(off-label)*, **ornamental plant production** *(off-label)*, **parsley** *(off-label)*, **protected ornamentals** *(off-label)*, **red mustard** *(off-label)*, **spinach** *(off-label)*, **spinach beet** *(off-label)*, **sweet ciceley** *(off-label)*, **wine grapes** *(off-label)*
- Phytophthora in **chicory** *(off-label)*, **ornamental plant production**, **strawberries**
- Pythium in **ornamental plant production**, **strawberries**
- Red core in **strawberries**

Extension of Authorisation for Minor Use (EAMUs)
- **baby leaf crops** 20161308
- **celery leaves** 20161308
- **chicory** 20130448
- **chives** 20161308
- **cress** 20161308
- **edible flowers** 20161308
- **endives** 20161308
- **herbs (see appendix 6)** 20161308
- **lamb's lettuce** 20161308
- **land cress** 20161308
- **lettuce** 20161308
- **ornamental plant production** 20131990
- **parsley** 20161308
- **protected ornamentals** 20131990
- **red mustard** 20161308
- **spinach** 20161308
- **spinach beet** 20161308
- **sweet ciceley** 20161308
- **wine grapes** 20130214

Approval information
- Fenamidone and fosetyl-aluminium included in Annex I under EC Regulation 1107/2009

Efficacy guidance
- Best used preventatively. Do not rely on curative activity
- Hose down dirty plants before dipping
- Due to variations in growing conditions and the range of varieties when treating ornamentals, it is advisable to check the crop tolerance on a small sample before treating the whole crop
- Fenamidone is a member of the QoI cross resistance group. To minimise the likelihood of development of resistance these products should be used in a planned Resistance Management strategy. In addition before application consult and adhere to the latest FRAG-UK resistance guidance on application of QoI fungicides to potatoes. See Section 5 for more information

Crop-specific information
- There is only limited crop safety data on tulip bulbs

SEE SECTION 3 FOR PRODUCTS ALSO REGISTERED

- Container-grown Viburnum, Pieris, Gaultheria, Chamaecyparis and Erica have been successfully treated in trials

Following crops guidance
- No limitations on choice of following crops

Environmental safety
- LERAP Category B

Hazard classification and safety precautions
Hazard Irritant, Dangerous for the environment, Very toxic to aquatic organisms
Transport code 9
Packaging group III
UN Number 3077
Risk phrases H319
Operator protection M; U05a, U11, U20b
Environmental protection E15b, E16a, E22c, E34, E38, H410
Storage and disposal D01, D02, D09a, D11a, D12a

216 fenamidone + propamocarb hydrochloride

A systemic fungicide mixture for potatoes
FRAC mode of action code: 11 + 28

See also propamocarb hydrochloride

Products

1	Consento	Bayer CropScience	75:375 g/l	SC	15150
2	Prompto	Bayer CropScience	75:375 g/l	SC	15156

Uses
- Blight in **potatoes**
- Early blight in **potatoes**

Approval information
- Fenamidone and propamocarb hydrochloride included in Annex I under EC Regulation 1107/2009

Efficacy guidance
- Apply as soon as there is risk of blight infection or immediately after a blight warning
- Fenamidone is a member of the QoI cross resistance group. To minimise the likelihood of development of resistance these products should be used in a planned Resistance Management strategy. In addition before application consult and adhere to the latest FRAG-UK resistance guidance on application of QoI fungicides to potatoes. See Section 5 for more information

Restrictions
- Maximum number of treatments (including any other QoI containing fungicide) 6 per yr but no more than three should be applied consecutively. Use in alternation with fungicides from a different cross-resistance group
- Observe a 7 d interval between treatments
- Use only as a protective treatment. Do not use when blight has become readily visible (1% leaf area destroyed)

Crop-specific information
- HI 7 d for potatoes
- All varieties of early, main and seed crop potatoes may be sprayed [2]

Environmental safety
- Dangerous for the environment
- Very toxic to aquatic organisms
- Dangerous to fish or other aquatic life. Do not contaminate surface waters or ditches with chemical or used container
- LERAP Category B

FOR FULL CONDITIONS OF USE ALWAYS READ THE PRODUCT LABEL

fenbuconazole

Hazard classification and safety precautions
  Hazard Irritant, Dangerous for the environment
  Transport code 9
  Packaging group III
  UN Number 3082
  Risk phrases R36, R50, R53a
  Operator protection A, C; U05a, U08, U11 [1, 2]; U19a, U20b, U20c [2]; U20a [1]
  Environmental protection E13b [1]; E15a [2]; E16a, E38 [1, 2]
  Storage and disposal D01, D02, D09a, D10b, D12a [1, 2]; D05 [1]

217 fenbuconazole

A systemic protectant and curative triazole fungicide for top fruit and grapevines
FRAC mode of action code: 3

Products

1	Agrovista Radni	Agrovista	50 g/l	EW	14404
2	Indar 5 EW	Landseer	50 g/l	EW	09518

Uses
- Blossom wilt in **cherries** (off-label), **mirabelles** (off-label), **plums** (off-label) [2]
- Brown rot in **cherries** (off-label), **mirabelles** (off-label), **plums** (off-label) [2]
- Powdery mildew in **apples** (reduction), **pears** (reduction) [1, 2]; **wine grapes** (off-label) [2]
- Scab in **apples**, **pears** [1, 2]

Extension of Authorisation for Minor Use (EAMUs)
- *cherries* 20031372 [2]
- *mirabelles* 20031372 [2]
- *plums* 20031372 [2]
- *wine grapes* 20032081 [2]

Approval information
- Fenbuconazole included in Annex I under EC Regulation 1107/2009
- Approval expiry 30 Sep 2017 [1]
- Approval expiry 31 Aug 2017 [2]

Efficacy guidance
- Most effective when used as part of a routine preventative programme from bud burst to onset of petal fall
- After petal fall, tank mix with other protectant fungicides to enhance scab control
- See label for recommended spray intervals. In periods of rapid growth or high disease pressure, a 7 d interval should be used

Restrictions
- Maximum total dose equivalent to ten full doses per yr on apples, pears
- Consult processors before using on pears for processing
- Do not harvest for human or animal consumption for at least 4 wk after last application

Crop-specific information
- HI 28 d for apples, pears; 21 d for grapevines; 3 d for cherries, mirabelles, plums
- Safe to use on all main commercial varieties of apples and pears in UK

Environmental safety
- Dangerous for the environment
- Toxic to aquatic organisms

Hazard classification and safety precautions
  Hazard Irritant, Dangerous for the environment
  Transport code 3 [2]; 9 [1]
  Packaging group III
  UN Number 1993 [2]; 3082 [1]
  Risk phrases R36, R51, R53a [1, 2]; R41 [2]

SEE SECTION 3 FOR PRODUCTS ALSO REGISTERED

Pesticide Profiles

Operator protection A, C; U05a, U08, U11 [1, 2]; U20a [2]; U20b [1]
Environmental protection E15a, E38
Consumer protection C02a [1] (4 w); C02a [2] (4 wk)
Storage and disposal D01, D02, D11a, D12a [1, 2]; D05 [2]

218 fenhexamid

A protectant hydroxyanilide fungicide for soft fruit and a range of horticultural crops
FRAC mode of action code: 17

Products

1	Agrovista Fenamid	Agrovista	50% w/w	WG	13733
2	Teldor	Bayer CropScience	50% w/w	WG	11229

Uses
- Botrytis in **bilberries** *(off-label)*, **blackberries**, **blackcurrants**, **blueberries** *(off-label)*, **cherries** *(off-label)*, **gooseberries**, **herbs (see appendix 6)** *(off-label)*, **loganberries**, **plums** *(off-label)*, **protected aubergines** *(off-label)*, **protected chilli peppers** *(off-label)*, **protected courgettes** *(off-label)*, **protected cucumbers** *(off-label)*, **protected gherkins** *(off-label)*, **protected peppers** *(off-label)*, **protected squashes** *(off-label)*, **protected tomatoes** *(off-label)*, **raspberries**, **redcurrants**, **rubus hybrids**, **strawberries**, **tomatoes (outdoor)** *(off-label)*, **whitecurrants** [1, 2]; **frise** *(off-label)*, **hops** *(off-label)*, **lamb's lettuce** *(off-label)*, **leaf brassicas** *(off-label - baby leaf production)*, **protected blueberry** *(off-label)*, **protected forest nurseries** *(off-label)*, **rocket** *(off-label)*, **scarole** *(off-label)*, **soft fruit** *(off-label)*, **wine grapes** [2]; **outdoor grapes** [1]
- Grey mould in **protected aubergines** *(off-label)*, **protected chilli peppers** *(off-label)*, **protected courgettes** *(off-label)*, **protected cucumbers** *(off-label)*, **protected gherkins** *(off-label)*, **protected peppers** *(off-label)*, **protected squashes** *(off-label)*, **protected tomatoes** *(off-label)* [2]

Extension of Authorisation for Minor Use (EAMUs)
- *bilberries* 20080485 [1], 20161214 [2]
- *blueberries* 20080485 [1], 20161214 [2]
- *cherries* 20080479 [1], 20031866 [2]
- *frise* 20082062 [2]
- *herbs (see appendix 6)* 20082056 [1], 20082062 [2]
- *hops* 20082926 [2]
- *lamb's lettuce* 20082062 [2]
- *leaf brassicas* (baby leaf production) 20082062 [2]
- *plums* 20080479 [1], 20031866 [2]
- *protected aubergines* 20080482 [1], 20042087 [2]
- *protected blueberry* 20161214 [2]
- *protected chilli peppers* 20080481 [1], 20042086 [2]
- *protected courgettes* 20080480 [1], 20042085 [2]
- *protected cucumbers* 20080480 [1], 20042085 [2]
- *protected forest nurseries* 20082926 [2]
- *protected gherkins* 20080480 [1], 20042085 [2]
- *protected peppers* 20080481 [1], 20042086 [2]
- *protected squashes* 20080480 [1], 20042085 [2]
- *protected tomatoes* 20080482 [1], 20042087 [2]
- *rocket* 20082062 [2]
- *scarole* 20082062 [2]
- *soft fruit* 20082926 [2]
- *tomatoes (outdoor)* 20080483 [1], 20042399 [2]

Approval information
- Fenhexamid included in Annex I under EC Regulation 1107/2009

Efficacy guidance
- Use as part of a programme of sprays throughout the flowering period to achieve effective control of Botrytis

FOR FULL CONDITIONS OF USE ALWAYS READ THE PRODUCT LABEL

fenoxaprop-P-ethyl

- To minimise possibility of development of resistance, no more than two sprays of the product may be applied consecutively. Other fungicides from a different chemical group should then be used for at least two consecutive sprays. If only two applications are made on grapevines, only one may include fenhexamid
- Complete spray cover of all flowers and fruitlets throughout the blossom period is essential for successful control of Botrytis
- Spray programmes should normally start at the start of flowering

Restrictions
- Maximum number of treatments 2 per yr on grapevines; 4 per yr on other listed crops but no more than 2 sprays may be applied consecutively

Crop-specific information
- HI 1 d for strawberries, raspberries, loganberries, blackberries, Rubus hybrids; 3 d for cherries; 7 d for blackcurrants, redcurrants, whitecurrants, gooseberries; 21d for outdoor grapes

Environmental safety
- Dangerous for the environment
- Harmful to fish or other aquatic life. Do not contaminate surface waters or ditches with chemical or used container

Hazard classification and safety precautions
Hazard Dangerous for the environment
Transport code 9
Packaging group III
UN Number 3077
Risk phrases R53a [2]
Operator protection A, H [1]; U08, U19a, U20b
Environmental protection E13c, E38 [2]; E15a [1]
Storage and disposal D05, D09a, D11a [1, 2]; D12a [2]

219 fenoxaprop-P-ethyl

An aryloxyphenoxypropionate herbicide for use in wheat
HRAC mode of action code: A

See also diclofop-methyl + fenoxaprop-P-ethyl

Products

1	Foxtrot EW	Headland	69 g/l	EW	13243
2	Oskar	Headland	69 g/l	EW	13344
3	Polecat	Interfarm	69 g/l	EC	16112

Uses
- Awned canary grass in **spring wheat**, **winter wheat** [3]
- Blackgrass in **grass seed crops** *(off-label)*, **spring barley**, **winter barley** [1, 2]; **spring wheat**, **winter wheat** [1-3]
- Canary grass in **grass seed crops** *(off-label)* [2]; **spring barley**, **spring wheat**, **winter barley**, **winter wheat** [1, 2]
- Rough-stalked meadow grass in **grass seed crops** *(off-label)*, **spring barley**, **winter barley** [1, 2]; **spring wheat**, **winter wheat** [1-3]
- Wild oats in **grass seed crops** *(off-label)*, **spring barley**, **winter barley** [1, 2]; **spring wheat**, **winter wheat** [1-3]

Extension of Authorisation for Minor Use (EAMUs)
- *grass seed crops* 20150680 [1], 20150866 [2]

Approval information
- Fenoxaprop-P-ethyl included in Annex I under EC Regulation 1107/2009

SEE SECTION 3 FOR PRODUCTS ALSO REGISTERED

Efficacy guidance
- Treat weeds from 2 fully expanded leaves up to flag leaf ligule just visible; for awned canary-grass and rough meadow-grass from 2 leaves to the end of tillering
- A second application may be made in spring where susceptible weeds emerge after an autumn application
- Spray is rainfast 1 h after application
- Dry conditions resulting in moisture stress may reduce effectiveness
- Fenoxaprop-P-ethyl is an ACCase inhibitor herbicide. To avoid the build up of resistance do not apply products containing an ACCase inhibitor herbicide more than twice to any crop. In addition do not use any product containing fenoxaprop-P-ethyl in mixture or sequence with any other product containing the same ingredient
- Use these products as part of a resistance management strategy that includes cultural methods of control and does not use ACCase inhibitors as the sole chemical method of grass weed control
- Applying a second product containing an ACCase inhibitor to a crop will increase the risk of resistance development; only use a second ACCase inhibitor to control different weeds at a different timing
- Always follow WRAG guidelines for preventing and managing herbicide resistant weeds. See Section 5 for more information

Restrictions
- Maximum total dose equivalent to one or two full dose treatments depending on product used
- Do not apply to barley, durum wheat, undersown crops or crops to be undersown
- Do not roll or harrow within 1 wk of spraying
- Do not spray crops under stress, suffering from drought, waterlogging or nutrient deficiency or those grazed or if soil compacted
- Avoid spraying immediately before or after a sudden drop in temperature or a period of warm days/cold nights
- Do not mix with hormone weedkillers

Crop-specific information
- Latest use: before flag leaf sheath extending (GS 41)
- Treat from crop emergence to flag leaf fully emerged (GS 41).
- Product may be sprayed in frosty weather provided crop hardened off but do not spray wet foliage or leaves covered with ice
- Broadcast crops should be sprayed post-emergence after plants have developed well-established root system

Environmental safety
- Dangerous for the environment
- Very toxic to aquatic organisms

Hazard classification and safety precautions
Hazard Irritant, Dangerous for the environment
Transport code 9
Packaging group III
UN Number 3082
Risk phrases H317 [3]; R38, R43, R51, R53a [1, 2]
Operator protection A, C, H; U05a, U08, U14, U20b [1-3]; U11 [3]; U19a [1, 2]
Environmental protection E15a, E38 [1-3]; E34 [1, 2]; H411 [3]
Storage and disposal D01, D02, D05, D09a, D10b [1-3]; D12a [3]

220 fenoxycarb

An insect specific growth regulator for top fruit
IRAC mode of action code: 7B

Products

Insegar WG   Syngenta   25% w/w   WG   09789

FOR FULL CONDITIONS OF USE ALWAYS READ THE PRODUCT LABEL

fenpropidin

Uses
- Summer-fruit tortrix moth in ***apples**, **pears***

Approval information
- Fenoxycarb included in Annex I under EC Regulation 1107/2009

Efficacy guidance
- Best results from application at 5th instar stage before pupation. Product prevents transformation from larva to pupa
- Correct timing best identified from pest warnings
- Because of mode of action rapid knock-down of pest is not achieved and larvae continue to feed for a period after treatment
- Adequate water volume necessary to ensure complete coverage of leaves

Restrictions
- Maximum number of treatments 2 per crop
- Consult processors before use

Crop-specific information
- HI 42 d for apples, pears
- Use on all varieties of apples and pears

Environmental safety
- Dangerous for the environment
- Toxic to aquatic organisms
- High risk to bees. Do not apply to crops in flower or to those in which bees are actively foraging. Do not apply when flowering weeds are present
- Risk to certain non-target insects or other arthropods. See directions for use
- Apply to minimise off-target drift to reduce effects on non-target organisms. Some margin of safety to beneficial arthropods is indicated.
- Broadcast air-assisted LERAP (8 m)

Hazard classification and safety precautions
Hazard Dangerous for the environment
Transport code 9
Packaging group III
UN Number 3077
Risk phrases R51, R53a
Operator protection A, C, H; U05a, U20a, U23a
Environmental protection E12a, E12e, E15a, E22b, E38; E17b (8 m)
Storage and disposal D01, D02, D07, D09a, D11a, D12a

221 fenpropidin

A systemic, curative and protective piperidine (morpholine) fungicide only available in mixtures
FRAC mode of action code: 5

See also difenoconazole + fenpropidin
fenpropidin + prochloraz + tebuconazole

SEE SECTION 3 FOR PRODUCTS ALSO REGISTERED

222 fenpropidin + prochloraz + tebuconazole

A contact and systemic fungicide mixture for cereals
FRAC mode of action code: 5 + 3 + 3

See also fenpropidin
fenpropidin + prochloraz
fenpropidin + tebuconazole
prochloraz
prochloraz + tebuconazole
tebuconazole

Products
  Artemis   Adama   150:200:100 g/l   EC   14178

Uses
- Brown rust in **spring wheat**, **winter wheat**
- Ear diseases in **spring wheat**, **winter wheat**
- Eyespot in **spring wheat**, **winter wheat**
- Glume blotch in **spring wheat**, **winter wheat**
- Powdery mildew in **spring wheat**, **winter wheat**
- Septoria leaf blotch in **spring wheat**, **winter wheat**
- Yellow rust in **spring wheat**, **winter wheat**

Approval information
- Fenpropidin, prochloraz and tebuconazole included in Annex I under EC Regulation 1107/2009

Efficacy guidance
- Best disease control and yield benefit obtained when applied at early stage of disease development before infection spreads to new growth
- To protect the flag leaf and ear from Septoria diseases apply from flag leaf emergence to ear fully emerged (GS 37-59)
- Applications once foliar symptoms of *Septoria tritici* are already present on upper leaves will be less effective
- Resistance to some DMI fungicides has been identified in Septoria leaf blotch which may seriously affect the performance of some products.

Restrictions
- Newer authorisations for tebuconazole products require application to cereals only after GS 30 and applications to oilseed rape and linseed after GS20 - check label

Crop-specific information
- Do not apply when temperatures are high.

Hazard classification and safety precautions
  Hazard Harmful, Dangerous for the environment, Harmful if inhaled, Very toxic to aquatic organisms
  Transport code 9
  Packaging group III
  UN Number 3082
  Risk phrases H315, H318, H361
  Operator protection A, C, H; U02a, U04a, U05a, U09a, U10, U11, U15, U19a, U20b
  Environmental protection E15a, E34, E38, H410
  Storage and disposal D01, D02, D05, D12a
  Medical advice M03, M05a

223   fenpropimorph

A contact and systemic morpholine fungicide
FRAC mode of action code: 5

See also azoxystrobin + fenpropimorph
epoxiconazole + fenpropimorph
epoxiconazole + fenpropimorph + kresoxim-methyl
epoxiconazole + fenpropimorph + metrafenone
epoxiconazole + fenpropimorph + pyraclostrobin

Products

1	Clayton Spigot	Clayton	750 g/l	EC	11560
2	Corbel	BASF	750 g/l	EC	00578
3	Crebol	AgChem Access	750 g/l	EC	13922
4	Fenprop 750	Goldengrass	750 g/l	EC	15388
5	Propimorf 750	Goldengrass	750 g/l	EC	14410
6	Standon Fenpropimorph 750	Standon	750 g/l	EC	08965

Uses
- Alternaria in **carrots** *(off-label)*, **horseradish** *(off-label)*, **parsley root** *(off-label)*, **parsnips** *(off-label)*, **salsify** *(off-label)* [2]
- Brown rust in **spring barley**, **spring wheat**, **triticale**, **winter barley**, **winter wheat** [1-6]; **undersown barley**, **undersown wheat** [5]
- Crown rot in **carrots** *(off-label)*, **horseradish** *(off-label)*, **parsley root** *(off-label)*, **parsnips** *(off-label)*, **salsify** *(off-label)* [1, 2]; **mallow (althaea spp.)** *(off-label)* [1]
- Foliar disease control in **durum wheat** *(off-label)* [2]; **grass seed crops** *(off-label)* [1, 6]
- Powdery mildew in **bilberries** *(off-label)*, **blackcurrants** *(off-label)*, **blueberries** *(off-label)*, **carrots** *(off-label)*, **cranberries** *(off-label)*, **dewberries** *(off-label)*, **gooseberries** *(off-label)*, **hops** *(off-label)*, **horseradish** *(off-label)*, **loganberries** *(off-label)*, **parsley root** *(off-label)*, **parsnips** *(off-label)*, **raspberries** *(off-label)*, **redcurrants** *(off-label)*, **salsify** *(off-label)*, **strawberries** *(off-label)*, **tayberries** *(off-label)*, **whitecurrants** *(off-label)* [1, 2]; **durum wheat** *(off-label)*, **fodder beet** *(off-label)*, **mallow (althaea spp.)** *(off-label)*, **red beet** *(off-label)*, **sugar beet seed crops** *(off-label)* [1]; **spring barley**, **spring oats**, **spring wheat**, **winter barley**, **winter oats**, **winter rye**, **winter wheat** [1-6]; **undersown barley**, **undersown wheat** [5]
- Rhynchosporium in **spring barley**, **winter barley** [1-6]; **undersown barley** [5]
- Rust in **fodder beet** *(off-label)*, **sugar beet seed crops** *(off-label)* [2]; **leeks** [1]; **red beet** *(off-label)* [1, 2]
- Yellow rust in **spring barley**, **spring wheat**, **triticale**, **winter barley**, **winter wheat** [1-6]; **undersown barley**, **undersown wheat** [5]

Extension of Authorisation for Minor Use (EAMUs)
- **bilberries** *20111798* [1], *20040804* [2]
- **blackcurrants** *20111798* [1], *20040804* [2]
- **blueberries** *20111798* [1], *20040804* [2]
- **carrots** *20111796* [1], *20023753* [2]
- **cranberries** *20111798* [1], *20040804* [2]
- **dewberries** *20111798* [1], *20040804* [2]
- **durum wheat** *20111799* [1], *20061460* [2]
- **fodder beet** *20111797* [1], *20023757* [2]
- **gooseberries** *20111798* [1], *20040804* [2]
- **grass seed crops** *20061060* [1], *20061057* [6]
- **hops** *20111794* [1], *20023759* [2]
- **horseradish** *20111796* [1], *20023753* [2]
- **loganberries** *20111798* [1], *20040804* [2]
- **mallow (althaea spp.)** *20111796* [1]
- **parsley root** *20111796* [1], *20023753* [2]
- **parsnips** *20111796* [1], *20023753* [2]
- **raspberries** *20111798* [1], *20040804* [2]

SEE SECTION 3 FOR PRODUCTS ALSO REGISTERED

- **red beet** *20111795* [1], *20023751* [2]
- **redcurrants** *20111798* [1], *20040804* [2]
- **salsify** *20111796* [1], *20023753* [2]
- **strawberries** *20111798* [1], *20040804* [2]
- **sugar beet seed crops** *20111797* [1], *20023757* [2]
- **tayberries** *20111798* [1], *20040804* [2]
- **whitecurrants** *20111798* [1], *20040804* [2]

Approval information
- Accepted by BBPA for use on malting barley and hops
- Fenpropimorph included in Annex I under EC Regulation 1107/2009

Efficacy guidance
- On all crops spray at start of disease attack. See labels for recommended tank mixes. Follow-up treatments may be needed if disease pressure remains high
- Product rainfast after 2 h

Restrictions
- Maximum number of treatments 2 per crop for spring cereals; 3 or 4 per crop for winter cereals; 6 per crop for leeks
- Consult processors before using on crops for processing

Crop-specific information
- HI cereals 5 wk
- Scorch may occur on cereals if applied during frosty weather or in high temperatures
- Some leaf spotting may occur on undersown clovers
- Leeks should be treated every 2-3 wk as required, up to 6 applications

Environmental safety
- Dangerous for the environment
- Very toxic to aquatic organisms

Hazard classification and safety precautions
Hazard Harmful [1-5]; Irritant [6]; Dangerous for the environment [1-6]
Transport code 9
Packaging group III
UN Number 3082
Risk phrases R22a [1]; R38, R53a [1-6]; R50 [2-5]; R51 [1, 6]; R63 [1-5]
Operator protection A [1-6]; C [1, 6]; U05a, U08, U14, U15, U19a, U20b
Environmental protection E15a [1-6]; E38 [2-5]
Storage and disposal D01, D02, D09a [1-6]; D05 [1, 5, 6]; D08 [2-5]; D10b [6]; D10c, D12a [1-5]
Medical advice M05a

224 fenpropimorph + pyraclostrobin

A protectant and curative fungicide mixture for cereals
FRAC mode of action code: 5 +11

See also pyraclostrobin

Products
Jenton BASF 375:100 g/l EC 11898

Uses
- Brown rust in **grass seed crops**, **spring barley**, **spring wheat**, **triticale**, **winter barley**, **winter wheat**
- Crown rust in **spring oats**, **winter oats**
- Disease control in **grass seed crops** *(off-label)*, **triticale** *(off-label)*
- Drechslera leaf spot in **grass seed crops**, **spring wheat**, **triticale**, **winter wheat**
- Glume blotch in **grass seed crops**, **grass seed crops** *(off-label)*, **spring wheat**, **triticale**, **triticale** *(off-label)*, **winter wheat**

FOR FULL CONDITIONS OF USE ALWAYS READ THE PRODUCT LABEL

- Net blotch in *spring barley, winter barley*
- Powdery mildew in *spring barley, spring oats, winter barley, winter oats*
- Rhynchosporium in *spring barley, winter barley*
- Rust in *grass seed crops* (off-label), *triticale* (off-label)
- Septoria leaf blotch in *grass seed crops, grass seed crops* (off-label), *spring wheat, triticale, triticale* (off-label), *winter wheat*
- Yellow rust in *grass seed crops, spring barley, spring wheat, triticale, winter barley, winter wheat*

Extension of Authorisation for Minor Use (EAMUs)
- *grass seed crops* 20122661
- *triticale* 20122661

Approval information
- Fenpropimorph and pyraclostrobin included in Annex I under EC Regulation 1107/2009
- Accepted by BBPA for use on malting barley

Efficacy guidance
- Best results obtained from treatment at the start of foliar disease attack
- Yield response may be obtained in the absence of visual disease
- Pyraclostrobin is a member of the QoI cross resistance group. Product should be used preventatively and not relied on for its curative potential
- Use product as part of an Integrated Crop Management strategy incorporating other methods of control, including where appropriate other fungicides with a different mode of action. Do not apply more than two foliar applications of QoI containing products to any cereal crop
- There is a significant risk of widespread resistance occurring in *Septoria tritici* populations in UK. Strains of barley powdery mildew resistant to QoI's are common in UK. Failure to follow resistance management action may result in reduced levels of disease control

Restrictions
- Maximum number of treatments 2 per crop

Crop-specific information
- Latest use: up to and including emergence of ear just complete (GS 59) for barley

Environmental safety
- Dangerous for the environment
- Very toxic to aquatic organisms
- LERAP Category B

Hazard classification and safety precautions
Hazard Harmful, Dangerous for the environment
Transport code 9
Packaging group III
UN Number 3082
Risk phrases R20, R22a, R36, R38, R50, R53a, R63
Operator protection A, C; U05a, U14, U20b
Environmental protection E15a, E16a, E34, E38
Storage and disposal D01, D02, D05, D09a, D10c, D12a
Medical advice M03, M05a

225 fenpyrazamine

A botriticide for protected crops
FRAC mode of action code: 17

Products

1	Clayton Vitis	Clayton	50% w/w	WG	16974
2	Empire	Pan Agriculture	50% w/w	WG	17357
3	Prolectus	Interfarm	50% w/w	WG	16607

SEE SECTION 3 FOR PRODUCTS ALSO REGISTERED

Pesticide Profiles

Uses
- Botrytis in *protected aubergines, protected courgettes, protected cucumbers, protected peppers, protected tomatoes, strawberries, wine grapes* [1-3]; *protected strawberries* [2, 3]

Approval information
- Fenpyrazamine is included in Annex I under EC Regulation 1107/2009

Restrictions
- Consult processors before use on crops grown for processing
- No more than a third of the intended botryticide applications made per crop, per year, should contain 3-keto reductase (FRAC code 17) fungicides

Hazard classification and safety precautions
Hazard Dangerous for the environment
Transport code 9
Packaging group III
UN Number 3077
Operator protection U05a, U11, U13, U14, U15, U20b [2, 3]
Environmental protection E15b [2, 3]; H410 [1-3]
Storage and disposal D01, D02, D09a, D11a, D12a, D12b [2, 3]
Medical advice M03 [2, 3]

226 ferric phosphate

A molluscicide bait for controlling slugs and snails

Products

1	Derrex	Certis	2.97% w/w	GB	15351
2	Ironmax Pro	De Sangosse	3% w/w	RB	17122
3	Iroxx	Certis	2.97% w/w	GB	16640
4	Slugger	Goldengrass	2.97% w/w	GB	16771
5	Sluxx HP	Certis	2.97% w/w	GB	16571

Uses
- Slugs and snails in *all edible crops (outdoor and protected), all non-edible crops (outdoor)* [1-5]; *amenity vegetation* [1, 3-5]; *carrots, celeriac, fodder beet, grassland, managed amenity turf, radishes, red beet, sugar beet, turnips* [2]

Approval information
- Ferric phosphate included in Annex 1 under EC Regulation 1107/2009
- Accepted by BBPA for use on malting barley

Efficacy guidance
- Treat as soon as damage first seen preferably in early evening. Repeat as necessary to maintain control
- Best results obtained from moist soaked granules. This will occur naturally on moist soils or in humid conditions
- Ferric phosphate does not cause excessive slime secretion and has no requirement to collect moribund slugs from the soil surface
- Active ingredient is degraded by micro-organisms to beneficial plant nutrients

Restrictions
- Maximum total dose equivalent to four full dose treatments

Crop-specific information
- Latest use not specified for any crop

Hazard classification and safety precautions
UN Number N/C
Operator protection A, H [2]; U05a, U20a
Environmental protection E15a [1, 3]; E34 [1-5]
Storage and disposal D01, D09a

FOR FULL CONDITIONS OF USE ALWAYS READ THE PRODUCT LABEL

ferrous sulphate

227　ferrous sulphate

A herbicide/fertilizer combination for moss control in turf

See also dicamba + dichlorprop-P + ferrous sulphate + MCPA
*　　　　 dichlorprop-P + ferrous sulphate + MCPA*

Products

1	Greenmaster Autumn	Everris Ltd	16.3% w/w	GR	16762
2	Greenmaster Mosskiller	Everris Ltd	27% w/w	GR	15796
3	Greentec Mosskiller Pro	Headland Amenity	16.78% w/w	GR	16728
4	Landscaper Pro Moss Control + Fertiliser	Everris Ltd	16.3% w/w	GR	16723

Uses
- Moss in **amenity grassland** [2, 3]; **managed amenity turf** [1-4]

Approval information
- Ferrous sulphate included in Annex 1 under EC Regulation 1107/2009
- Approval expiry 31 Aug 2017 [2]

Efficacy guidance
- For best results apply when turf is actively growing and the soil is moist
- Fertilizer component of most products encourages strong root growth and tillering
- Mow 3 d before treatment and do not mow for 3-4 d afterwards
- Water after 2 d if no rain
- Rake out dead moss thoroughly 7-14 d after treatment. Re-treatment may be necessary for heavy infestations

Restrictions
- Maximum number of treatments - see labels
- Do not apply during drought or when heavy rain expected
- Do not apply in frosty weather or when the ground is frozen
- Do not walk on treated areas until well watered

Crop-specific information
- If spilt on paving, concrete, clothes etc brush off immediately to avoid discolouration
- Observe label restrictions for interval before cutting after treatment

Environmental safety
- Harmful to fish or other aquatic life. Do not contaminate surface waters or ditches with chemical or used container

Hazard classification and safety precautions
Hazard Harmful [3]
UN Number N/C
Risk phrases H315, H318 [3]; H319 [1, 4]
Operator protection U20b [2, 3]; U20c [1, 4]
Environmental protection E13c [1, 4]; E15a [2, 3]
Storage and disposal D01 [2, 3]; D09a, D11a [1-4]

228　ferrous sulphate + MCPA + mecoprop-P

A translocated herbicide and moss killer mixture
HRAC mode of action code: O + O

See also MCPA
*　　　　 mecoprop-P*

Products

1	Landscaper Pro Feed Weed Mosskiller	Everris Ltd	16.3:0.49:0.29% w/w	GR	17549

SEE SECTION 3 FOR PRODUCTS ALSO REGISTERED

Pesticide Profiles

Products – continued

2	Renovator Pro	Everris Ltd	18.2:0.49:0.29% w/w	GR	15815
3	Renovator Pro	Everris Ltd	16.3:0.49:0.29 % w/w	GR	16803

Uses
- Annual dicotyledons in **managed amenity turf**
- Moss in **managed amenity turf**
- Perennial dicotyledons in **managed amenity turf**

Approval information
- Ferrous sulphate, MCPA and mecoprop-P included in Annex I under EC Regulation 1107/2009
- Approval expiry 28 Feb 2017 [2]

Efficacy guidance
- Apply from Apr to Sep when weeds are growing
- For best results apply when light showers or heavy dews are expected
- Apply with a suitable calibrated fertilizer distributor
- Retreatment may be necessary after 6 wk if weeds or moss persist
- For best control of moss scarify vigorously after 2 wk to remove dead moss
- Where regrowth of moss or weeds occurs a repeat treatment may be made after 6 wk
- Avoid treatment of wet grass or during drought. If no rain falls within 48 h water in thoroughly

Restrictions
- Maximum number of treatments 3 per yr
- Do not treat new turf until established for 6 mth
- The first 4 mowings after treatment should not be used to mulch cultivated plants unless composted at least 6 mth
- Avoid walking on treated areas until it has rained or they have been watered
- Do not re-seed or turf within 8 wk of last treatment
- Do not cut grass for at least 3 d before and at least 4 d after treatment
- Do not apply during freezing conditions or when rain imminent

Environmental safety
- Keep livestock out of treated areas for up to two weeks following treatment and until poisonous weeds, such as ragwort, have died down and become unpalatable
- Harmful to fish or other aquatic life. Do not contaminate surface waters or ditches with chemical or used container
- Do not empty into drains
- Some pesticides pose a greater threat of contamination of water than others and mecoprop-P is one of these pesticides. Take special care when applying mecoprop-P near water and do not apply if heavy rain is forecast

Hazard classification and safety precautions
UN Number N/C
Risk phrases H319 [1, 3]
Operator protection A, H [1]; U20b
Environmental protection E07a, E13c, E19b
Storage and disposal D01, D09a, D12a
Medical advice M05a

229 flazasulfuron

A sulfonylurea herbicide for non-crop use
HRAC mode of action code: B

Products

1	Chikara Weed Control	Belchim	25% w/w	WG	14189
2	Katana	Belchim	25% w/w	WG	16162
3	Pacaya	ProKlass	25% w/w	WG	16797

FOR FULL CONDITIONS OF USE ALWAYS READ THE PRODUCT LABEL

flocoumafen

Products – continued

4	Paradise	Pan Agriculture	25% w/w	WG	16829

Uses
- Annual and perennial weeds in **amenity vegetation**, **hard surfaces** [1, 2]; **natural surfaces not intended to bear vegetation**, **permeable surfaces overlying soil** [1-4]; **railway tracks** [3, 4]

Approval information
- Flazasulfuron included in Annex 1 under EC Regulation 1107/2009

Environmental safety
- LERAP Category B

Hazard classification and safety precautions
Hazard Dangerous for the environment [1-4]; Very toxic to aquatic organisms [1, 2]
Transport code 9
Packaging group III
UN Number 3077
Risk phrases R50, R53a [3, 4]
Operator protection A, H, M; U02a, U05a, U20b [1-4]; U08 [1-3]; U09a [4]
Environmental protection E15a, E16a, E16b, E38 [1-4]; H410 [1, 2]
Storage and disposal D01, D02, D09a, D11a, D12a

230 flocoumafen

A second generation anti-coagulant rodenticide

Products

1	Storm Mini Bits	BASF	0.005% w/w	RB	UK15-0915
2	Storm Pasta	BASF	0.005% w/w	RB	UK12-0852
3	Storm Secure	BASF	0.005% w/w	RB	UK15-0850

Uses
- Mice in **farm buildings**, **sewers**
- Rodents in **farm buildings**, **sewers**

Approval information
- Flocoumafen is not included in Annex 1 under EC Regulation 1107/2009

Restrictions
- Not approved for outdoor use except in sewers and covered drains

Hazard classification and safety precautions
Hazard Harmful
UN Number N/C
Operator protection A; U05a, U20a
Environmental protection E38
Storage and disposal D01, D02, D09b, D12b
Medical advice M03

231 flonicamid

A selective feeding blocker aphicide
IRAC mode of action code: 9C

Products

1	Mainman	Belchim	50% w/w	WG	13123
2	Teppeki	Belchim	50% w/w	WG	12402

Uses
- Aphids in **apples**, **pears** [1]; **potatoes**, **winter wheat** [2]
- Damson-hop aphid in **hops** (off-label) [1]

SEE SECTION 3 FOR PRODUCTS ALSO REGISTERED

- Glasshouse whitefly in *courgettes* (off-label), *cucumbers* (off-label), *protected courgettes* (off-label), *protected cucumbers* (off-label), *protected summer squash* (off-label), *summer squash* (off-label) [1]
- Mealybugs in *protected aubergines* (off-label), *protected tomatoes* (off-label) [1]
- Peach-potato aphid in *courgettes* (off-label), *cucumbers* (off-label), *protected courgettes* (off-label), *protected cucumbers* (off-label), *protected summer squash* (off-label), *summer squash* (off-label) [1]
- Tobacco whitefly in *ornamental plant production* (off-label) [1]
- Whitefly in *ornamental plant production* (off-label), *protected aubergines* (off-label), *protected tomatoes* (off-label) [1]

Extension of Authorisation for Minor Use (EAMUs)
- *courgettes* 20142923 [1]
- *cucumbers* 20142923 [1]
- *hops* 20142225 [1]
- *ornamental plant production* 20130045 [1]
- *protected aubergines* 20160191 [1]
- *protected courgettes* 20142923 [1]
- *protected cucumbers* 20142923 [1]
- *protected summer squash* 20142923 [1]
- *protected tomatoes* 20160191 [1]
- *summer squash* 20142923 [1]

Approval information
- Flonicamid included in Annex 1 under EC Regulation 1107/2009
- Accepted by BBPA for use on hops

Efficacy guidance
- Apply when warning systems forecast significant aphid infestations
- Persistence of action is 21 d

Restrictions
- Maximum number of treatments 2 per crop for potatoes, winter wheat
- Must not be applied to winter wheat before 50% ear emerged stage (GS 53)
- Use a maximum of two consecutive applications of [1] in apples and pears. If further treatments are required, use an insecticide with a different mode of action before applying the final application of [1]

Crop-specific information
- HI: potatoes 14 d; winter wheat 28 d;
- HI: apples and pears 21 days [1]

Environmental safety
- Dangerous for the environment
- Harmful to aquatic organisms

Hazard classification and safety precautions
Hazard Dangerous for the environment [2]
UN Number N/C
Risk phrases R52, R53a
Operator protection A, H, M [1, 2]; C [2]; E [1]; U05a [1, 2]; U13, U14, U15 [1]
Environmental protection E15a
Storage and disposal D01, D02, D12b [1, 2]; D03, D08 [1]; D07 [2]

232 florasulam

A triazolopyrimidine herbicide for cereals
HRAC mode of action code: B

See also 2,4-D + florasulam
  clopyralid + florasulam
  clopyralid + florasulam + fluroxypyr
  diflufenican + florasulam

Products

1	Lector	Headland	50 g/l	SC	16565
2	Paramount	Headland	50 g/l	SC	16452

Uses
- Annual dicotyledons in **spring barley**, **spring oats**, **spring wheat**, **triticale**, **winter barley**, **winter oats**, **winter rye**, **winter wheat** [2]
- Charlock in **grass seed crops** (off-label), **ornamental plant production** (off-label) [1, 2]
- Chickweed in **grass seed crops** (off-label), **ornamental plant production** (off-label), **spring barley**, **spring oats**, **spring wheat**, **triticale**, **winter barley**, **winter oats**, **winter rye**, **winter wheat** [1, 2]
- Cleavers in **grass seed crops** (off-label), **ornamental plant production** (off-label), **spring barley**, **spring oats**, **spring wheat**, **triticale**, **winter barley**, **winter oats**, **winter rye**, **winter wheat** [1, 2]
- Groundsel in **ornamental plant production** (off-label) [1, 2]
- Hedge mustard in **grass seed crops** (off-label) [1, 2]; **spring barley**, **spring oats**, **spring wheat**, **triticale**, **winter barley**, **winter oats**, **winter rye**, **winter wheat** [1]
- Mayweeds in **grass seed crops** (off-label), **ornamental plant production** (off-label) [1, 2]; **spring barley**, **spring oats**, **spring wheat**, **triticale**, **winter barley**, **winter oats**, **winter rye**, **winter wheat** [2]
- Runch in **grass seed crops** (off-label) [2]; **spring barley**, **spring oats**, **spring wheat**, **triticale**, **winter barley**, **winter oats**, **winter rye**, **winter wheat** [1]
- Scented mayweed in **spring barley**, **spring oats**, **spring wheat**, **triticale**, **winter barley**, **winter oats**, **winter rye**, **winter wheat** [1]
- Scentless mayweed in **spring barley**, **spring oats**, **spring wheat**, **triticale**, **winter barley**, **winter oats**, **winter rye**, **winter wheat** [1]
- Shepherd's purse in **grass seed crops** (off-label) [1, 2]; **spring barley**, **spring oats**, **spring wheat**, **triticale**, **winter barley**, **winter oats**, **winter rye**, **winter wheat** [1]
- Sowthistle in **ornamental plant production** (off-label) [1, 2]
- Volunteer oilseed rape in **grass seed crops** (off-label), **ornamental plant production** (off-label), **spring barley**, **spring oats**, **spring wheat**, **triticale**, **winter barley**, **winter oats**, **winter rye**, **winter wheat** [1, 2]
- Wild radish in **grass seed crops** (off-label) [1, 2]; **spring barley**, **spring oats**, **spring wheat**, **triticale**, **winter barley**, **winter oats**, **winter rye**, **winter wheat** [1]

Extension of Authorisation for Minor Use (EAMUs)
- *grass seed crops* 20150315 [1], 20150559 [2]
- *ornamental plant production* 20141582 [1], 20141531 [2]

Approval information
- Florasulam included in Annex I under EC Regulation 1107/2009
- Accepted by BBPA for use on malting barley

Efficacy guidance
- Best results obtained from treatment of small actively growing weeds in good conditions
- Apply in autumn or spring once crop has 3 leaves
- Product is mainly absorbed by leaves of weeds and is effective on all soil types
- Florasulam is a member of the ALS-inhibitor group of herbicides

SEE SECTION 3 FOR PRODUCTS ALSO REGISTERED

Restrictions
- Maximum total dose on any crop equivalent to one full dose treatment. The total amount of florasulam applied to a cereal crop must not exceed 7.5 g.a.i/ha [1]
- Do not roll or harrow within 7 d before or after application
- Do not spray when crops under stress from cold, drought, pest damage, nutrient deficiency or any other cause
- Specific restrictions apply to use in sequence or tank mixture with other sulfonylurea or ALS-inhibiting herbicides. See label for details

Crop-specific information
- Latest use: up to and including flag leaf just visible stage for all crops

Following crops guidance
- Where the product has been used in mixture with certain named products (see label) only cereals or grass may be sown in the autumn following harvest
- Unless otherwise restricted cereals, oilseed rape, field beans, grass or vegetable brassicas as transplants may be sown as a following crop in the same calendar yr as treatment. Oilseed rape may show some temporary reduction of vigour after a dry summer, but yields are not affected
- In addition to the above, linseed, peas, sugar beet, potatoes, maize, clover (for use in grass/clover mixtures) or carrots may be sown in the calendar yr following treatment
- In the event of failure of a treated crop in spring only spring wheat, spring barley, spring oats, maize or ryegrass may be sown

Environmental safety
- Dangerous for the environment
- Very toxic to aquatic organisms
- See label for detailed instructions on tank cleaning
- LERAP Category B

Hazard classification and safety precautions
Hazard Dangerous for the environment
Transport code 9
Packaging group III
UN Number 3082
Operator protection U05a
Environmental protection E15b, E16a, E38, H410
Storage and disposal D01, D02, D05, D06a, D09a, D10c, D12b

233   florasulam + fluroxypyr

A post-emergence herbicide mixture for cereals or grass
HRAC mode of action code: B + O

See also fluroxypyr

Products

1	Cabadex	Headland Amenity	2.5:100 g/l	SE	13948
2	Cleave	Adama	2.5:100 g/l	SE	16774
3	GF 184	Dow	2.5:100 g/l	SE	10878
4	Hunter	Dow	2.5:100 g/l	SE	12836
5	Nevada	Dow	5:100 g/l	SE	17349
6	Slalom	Dow	2.5:100 g/l	SE	13772
7	Spitfire	Dow	5:100 g/l	SE	15101
8	Starane XL	Dow	2.5:100 g/l	SE	10921
9	Trafalgar	Pan Amenity	2.5:100 g/l	SE	14888

Uses
- Annual and perennial weeds in **amenity grassland**, **lawns**, **managed amenity turf** [1]
- Annual dicotyledons in **amenity grassland**, **lawns**, **managed amenity turf** [9]; **durum wheat** *(off-label)*, **ornamental plant production** *(off-label)*, **spring rye** *(off-label)*, **triticale** *(off-label)*, **winter rye** *(off-label)* [8]; **game cover** *(off-label)* [3, 8]; **grass seed crops** *(off-label)* [3, 7, 8]; **oats**,

florasulam + fluroxypyr

rye [5, 7]; **spring barley, spring wheat, winter barley, winter wheat** [2-8]; **spring oats, winter oats** [2-4, 6, 8]; **triticale** [2, 5, 7]; **winter rye** [2]
- Black bindweed in **oats, rye, spring barley, spring wheat, triticale, winter barley, winter wheat** [5, 7]
- Black nightshade in **oats, rye, spring barley, spring wheat, triticale, winter barley, winter wheat** [5, 7]
- Buttercups in **amenity grassland, lawns, managed amenity turf** [1]
- Charlock in **oats, rye, spring barley, spring wheat, triticale, winter barley, winter wheat** [5, 7]
- Chickweed in **amenity grassland, lawns, managed amenity turf** [9]; **oats, rye** [5, 7]; **spring barley, spring wheat, winter barley, winter wheat** [2-8]; **spring oats, winter oats** [2-4, 6, 8]; **triticale** [2, 5, 7]; **winter rye** [2]
- Cleavers in **grass seed crops** *(off-label)* [7]; **oats, rye** [5, 7]; **spring barley, spring wheat, winter barley, winter wheat** [2-8]; **spring oats, winter oats** [2-4, 6, 8]; **triticale** [2, 5, 7]; **winter rye** [2]
- Clover in **amenity grassland, lawns, managed amenity turf** [1]
- Daisies in **amenity grassland, lawns, managed amenity turf** [1]
- Dandelions in **amenity grassland, lawns, managed amenity turf** [1]
- Forget-me-not in **oats, rye, spring barley, spring wheat, triticale, winter barley, winter wheat** [5, 7]
- Groundsel in **grass seed crops** *(off-label)* [7]
- Knotgrass in **oats, rye, spring barley, spring wheat, triticale, winter barley, winter wheat** [5, 7]
- Mayweeds in **grass seed crops** *(off-label)* [7]; **oats, rye** [5, 7]; **spring barley, spring wheat, winter barley, winter wheat** [2-8]; **spring oats, winter oats** [2-4, 6, 8]; **triticale** [2, 5, 7]; **winter rye** [2]
- Plantains in **amenity grassland, lawns, managed amenity turf** [1]
- Poppies in **oats, rye, spring barley, spring wheat, triticale, winter barley, winter wheat** [5, 7]
- Redshank in **oats, rye, spring barley, spring wheat, triticale, winter barley, winter wheat** [5, 7]
- Sowthistle in **grass seed crops** *(off-label)* [7]
- Volunteer beans in **oats, rye, spring barley, spring wheat, triticale, winter barley, winter wheat** [5, 7]
- Volunteer oilseed rape in **oats, rye, spring barley, spring wheat, triticale, winter barley, winter wheat** [5, 7]

Extension of Authorisation for Minor Use (EAMUs)
- *durum wheat* 20072815 [8]
- *game cover* 20082858 [3], 20082904 [8]
- *grass seed crops* 20072809 [3], 20132321 [7], 20072815 [8]
- *ornamental plant production* 20082904 [8]
- *spring rye* 20072815 [8]
- *triticale* 20072815 [8]
- *winter rye* 20072815 [8]

Approval information
- Florasulam and fluroxypyr included in Annex I under EC Regulation 1107/2009
- Accepted by BBPA for use on malting barley

Efficacy guidance
- Best results obtained when weeds are small and growing actively
- Products are mainly absorbed through weed foliage. Cleavers emerging after application will not be controlled
- Florasulam is a member of the ALS-inhibitor group of herbicides

Restrictions
- Maximum total dose equivalent to one full dose treatment
- Do not roll or harrow 7 d before or after application
- Do not spray when crops are under stress from cold, drought, pest damage or nutrient deficiency
- Do not apply through CDA applicators

SEE SECTION 3 FOR PRODUCTS ALSO REGISTERED

- Specific restrictions apply to use in sequence or tank mixture with other sulfonylurea or ALS-inhibiting herbicides. See label for details
- The total amount of florasulam applied to a cereal crop must not exceed 7.5 g.a.i/ha

Crop-specific information
- Latest use: before flag leaf sheath extended (before GS 41) for spring barley and spring wheat; before flag leaf sheath opening (before GS 47) for winter barley and winter wheat; before second node detectable (before GS 32) for winter oats

Following crops guidance
- Cereals, oilseed rape, field beans or grass may follow treated crops in the same yr. Oilseed rape may suffer temporary vigour reduction after a dry summer
- In addition to the above, linseed, peas, sugar beet, potatoes, maize or clover may be sown in the calendar yr following treatment
- In the event of failure of a treated crop in the spring, only spring cereals, maize or ryegrass may be planted
- May also be used on crops undersown with grass [5]

Environmental safety
- Dangerous for the environment
- Toxic to aquatic organisms
- Take extreme care to avoid drift onto non-target crops or plants
- LERAP Category B [2, 5, 7]

Hazard classification and safety precautions
Hazard Irritant, Dangerous for the environment [1-9]; Flammable, Flammable liquid and vapour [5, 7]
Transport code 3 [5, 7]; 9 [1-4, 6, 8, 9]
Packaging group III
UN Number 1993 [5, 7]; 3082 [1-4, 6, 8, 9]
Risk phrases H304, H335 [5, 7]; H315, H317, H319, H336 [2, 5, 7]; R36, R38, R53a, R67 [1, 3, 4, 6, 8, 9]; R50 [1]; R51 [3, 4, 6, 8, 9]
Operator protection A [1-9]; C [1-8]; H [5, 7]; U05a, U11, U19a [1-9]; U08 [1]; U09a [5, 7]; U14, U20b [1, 5, 7]
Environmental protection E15a [2-9]; E15b [1]; E16a, H410 [2, 5, 7]; E34, E38 [1-9]
Storage and disposal D01, D02, D09a, D10c, D12a [1-9]; D05 [9]
Medical advice M05a [5, 7, 9]

234 florasulam + pyroxsulam

A mixture of two triazolopirimidine sulfonamides for winter wheat
HRAC mode of action code: B

See also pyroxsulam

Products
1 Broadway Star   Dow   1.4:7.1% w/w   WG   14319
2 Palio   Dow   1.4:7.1% w/w   WG   17345

Uses
- Charlock in *triticale, winter rye, winter wheat*
- Chickweed in *triticale, winter rye, winter wheat*
- Cleavers in *triticale, winter rye, winter wheat*
- Field pansy in *triticale, winter rye, winter wheat*
- Field speedwell in *triticale, winter rye, winter wheat*
- Geranium species in *triticale, winter rye, winter wheat*
- Ivy-leaved speedwell in *triticale, winter rye, winter wheat*
- Mayweeds in *triticale, winter rye, winter wheat*
- Poppies in *triticale, winter rye, winter wheat*
- Ryegrass in *triticale, winter rye, winter wheat*
- Sterile brome in *triticale, winter rye, winter wheat*

FOR FULL CONDITIONS OF USE ALWAYS READ THE PRODUCT LABEL

- Volunteer oilseed rape in *triticale, winter rye, winter wheat*
- Wild oats in *triticale, winter rye, winter wheat*

Approval information
- Florasulam and pyroxsulam included in Annex I under EC Regulation 1107/2009.

Efficacy guidance
- Rainfast within 1 hour of application
- Requires an authorised adjuvant at application and recommended for use in a programme with herbicides employing a different mode of action.

Restrictions
- For autumn planted crops a maximum total dose of 3.75 g.a.i/ha of florasulam must be observed for applications made between crop emergence in the year of planting and February 1st in the year of harvest.
- The total amount of florasulam applied to a cereal crop must not exceed 7.5 g.a.i/ha

Crop-specific information
- Crop injury may occur if applied in tank mixture with plant growth regulators - allow a minimum interval of 7 days.
- Crop injury may occur if applied in tank mixture with OP insecticides, MCPB or dicamba - allow a minimum interval of 14 days

Following crops guidance
- Crop failure before 1st Feb - plough and allow 6 weeks to elapse and then drill spring wheat, spring barley, grass or maize.
- Crop failure after 1st Feb - plough and allow 6 weeks to elapse before drilling grass or maize.

Environmental safety
- Take extreme care to avoid drift on to susceptible crops, non-target plants or waterways
- LERAP Category B

Hazard classification and safety precautions
    Hazard Dangerous for the environment
    Transport code 9
    Packaging group III
    UN Number 3077
    Operator protection U05a, U14, U15, U20a
    Environmental protection E15b, E16a, E34, E38, E39, H410
    Storage and disposal D01, D02, D09a, D12a

235 fluazifop-P-butyl

A phenoxypropionic acid grass herbicide for broadleaved crops
HRAC mode of action code: A

Products

1	Clayton Maximus	Clayton	125 g/l	EC	12543
2	Fusilade Max	Syngenta	125 g/l	EC	11519
3	Greencrop Bantry	Greencrop	125 g/l	EC	12737
4	Salvo	Generica	125 g/l	EC	16228

Uses
- Annual grasses in **almonds** *(off-label)*, **apples** *(off-label)*, **apricots** *(off-label)*, **asparagus** *(off-label)*, **bilberries** *(off-label)*, **blackberries** *(off-label)*, **blueberries** *(off-label)*, **borage** *(off-label)*, **broad beans (dry-harvested)** *(off-label)*, **cabbages** *(off-label)*, **canary flower (echium spp.)** *(off-label)*, **cherries** *(off-label)*, **chicory** *(off-label)*, **choi sum** *(off-label)*, **cob nuts** *(off-label)*, **collards** *(off-label)*, **cranberries** *(off-label)*, **evening primrose** *(off-label)*, **figs** *(off-label)*, **filberts** *(off-label)*, **fodder beet** *(off-label)*, **garlic** *(off-label)*, **globe artichoke** *(off-label)*, **grapevines** *(off-label)*, **haricot beans** *(off-label)*, **hazel nuts** *(off-label)*, **herbs (see appendix 6)** *(off-label)*, **honesty** *(off-label)*, **hybridberry** *(off-label)*, **linseed** *(off-label)*, **lucerne** *(off-label)*, **lupins** *(off-*

Pesticide Profiles

label), **mallow (althaea spp.)** (off-label), **mustard** (off-label), **navy beans** (off-label), **non-edible flowers** (off-label), **ornamental plant production** (off-label), **pak choi** (off-label), **parsley root** (off-label), **parsnips** (off-label), **peaches** (off-label), **pears** (off-label), **plums** (off-label), **protected non-edible flowers** (off-label), **protected ornamentals** (off-label), **quinces** (off-label), **red beet** (off-label), **redcurrants** (off-label), **shallots** (off-label), **spring field beans** (off-label), **spring greens** (off-label), **swedes** (off-label), **tic beans** (off-label), **turnips** (off-label), **walnuts** (off-label), **whitecurrants** (off-label), **winter field beans** (off-label) [2]; **blackcurrants, bulb onions, carrots, combining peas, farm forestry, fodder beet, gooseberries, hops, linseed, raspberries, spring field beans, spring oilseed rape, spring oilseed rape for industrial use, strawberries, sugar beet, swedes** (stockfeed only), **turnips** (stockfeed only), **vining peas, winter field beans, winter oilseed rape, winter oilseed rape for industrial use** [1-4]; **field margins, flax, flax for industrial use, linseed for industrial use** [2, 4]; **kale** (stockfeed only) [1, 3]
- Barren brome in **field margins** [2, 4]
- Blackgrass in **hops, spring field beans, winter field beans** [1-4]
- Green cover in **land temporarily removed from production** [1-4]
- Perennial grasses in **almonds** (off-label), **apples** (off-label), **apricots** (off-label), **asparagus** (off-label), **bilberries** (off-label), **blackberries** (off-label), **blueberries** (off-label), **borage** (off-label), **broad beans (dry-harvested)** (off-label), **cabbages** (off-label), **canary flower (echium spp.)** (off-label), **cherries** (off-label), **chicory** (off-label), **choi sum** (off-label), **cob nuts** (off-label), **collards** (off-label), **cranberries** (off-label), **evening primrose** (off-label), **figs** (off-label), **filberts** (off-label), **fodder beet** (off-label), **garlic** (off-label), **globe artichoke** (off-label), **grapevines** (off-label), **haricot beans** (off-label), **hazel nuts** (off-label), **herbs (see appendix 6)** (off-label), **honesty** (off-label), **hybridberry** (off-label), **linseed** (off-label), **lucerne** (off-label), **lupins** (off-label), **mallow (althaea spp.)** (off-label), **mustard** (off-label), **navy beans** (off-label), **non-edible flowers** (off-label), **ornamental plant production** (off-label), **pak choi** (off-label), **parsley root** (off-label), **parsnips** (off-label), **peaches** (off-label), **pears** (off-label), **plums** (off-label), **protected non-edible flowers** (off-label), **protected ornamentals** (off-label), **quinces** (off-label), **red beet** (off-label), **redcurrants** (off-label), **shallots** (off-label), **spring field beans** (off-label), **spring greens** (off-label), **swedes** (off-label), **tic beans** (off-label), **turnips** (off-label), **walnuts** (off-label), **whitecurrants** (off-label), **winter field beans** (off-label) [2]; **blackcurrants, bulb onions, carrots, combining peas, farm forestry, fodder beet, gooseberries, hops, linseed, raspberries, spring field beans, spring oilseed rape, spring oilseed rape for industrial use, strawberries, sugar beet, swedes** (stockfeed only), **turnips** (stockfeed only), **vining peas, winter field beans, winter oilseed rape, winter oilseed rape for industrial use** [1-4]; **flax, flax for industrial use, linseed for industrial use** [2, 4]; **kale** (stockfeed only) [1, 3]
- Volunteer cereals in **blackcurrants, bulb onions, carrots, combining peas, farm forestry, fodder beet, gooseberries, hops, linseed, raspberries, spring field beans, spring oilseed rape, spring oilseed rape for industrial use, strawberries, sugar beet, swedes** (stockfeed only), **turnips** (stockfeed only), **vining peas, winter field beans, winter oilseed rape, winter oilseed rape for industrial use** [1-4]; **field margins, flax, flax for industrial use, linseed for industrial use** [2, 4]; **kale** (stockfeed only) [1, 3]
- Wild oats in **blackcurrants, bulb onions, carrots, combining peas, farm forestry, fodder beet, gooseberries, hops, linseed, raspberries, spring field beans, spring oilseed rape, spring oilseed rape for industrial use, strawberries, sugar beet, swedes** (stockfeed only), **turnips** (stockfeed only), **vining peas, winter field beans, winter oilseed rape, winter oilseed rape for industrial use** [1-4]; **field margins, flax, flax for industrial use, linseed for industrial use** [2, 4]; **kale** (stockfeed only) [1, 3]

Extension of Authorisation for Minor Use (EAMUs)
- *almonds* 20121261 [2]
- *apples* 20121261 [2]
- *apricots* 20121261 [2]
- *asparagus* 20121323 [2]
- *bilberries* 20121258 [2]
- *blackberries* 20121315 [2]
- *blueberries* 20121258 [2]
- *borage* 20121320 [2]
- *broad beans (dry-harvested)* 20121321 [2]

FOR FULL CONDITIONS OF USE ALWAYS READ THE PRODUCT LABEL

fluazifop-P-butyl

- *cabbages* 20121317 [2]
- *canary flower (echium spp.)* 20121320 [2]
- *cherries* 20121261 [2]
- *chicory* 20121322 [2]
- *choi sum* 20121321 [2]
- *cob nuts* 20121261 [2]
- *collards* 20121321 [2]
- *cranberries* 20121258 [2]
- *evening primrose* 20121320 [2]
- *figs* 20121261 [2]
- *filberts* 20121261 [2]
- *fodder beet* 20121256 [2]
- *garlic* 20121321 [2]
- *globe artichoke* 20121260 [2]
- *grapevines* 20121316 [2]
- *haricot beans* 20121321 [2]
- *hazel nuts* 20121261 [2]
- *herbs (see appendix 6)* 20121319 [2]
- *honesty* 20121320 [2]
- *hybridberry* 20121315 [2]
- *linseed* 20121320 [2]
- *lucerne* 20121318 [2]
- *lupins* 20121255 [2]
- *mallow (althaea spp.)* 20121259 [2]
- *mustard* 20121320 [2]
- *navy beans* 20121321 [2]
- *non-edible flowers* 20121321 [2]
- *ornamental plant production* 20121321 [2]
- *pak choi* 20121321 [2]
- *parsley root* 20121259 [2]
- *parsnips* 20121321 [2]
- *peaches* 20121261 [2]
- *pears* 20121261 [2]
- *plums* 20121261 [2]
- *protected non-edible flowers* 20121321 [2]
- *protected ornamentals* 20121321 [2]
- *quinces* 20121261 [2]
- *red beet* 20121321 [2]
- *redcurrants* 20121258 [2]
- *shallots* 20121257 [2]
- *spring field beans* 20121321 [2]
- *spring greens* 20121321 [2]
- *swedes* 20121321 [2]
- *tic beans* 20121321 [2]
- *turnips* 20121321 [2]
- *walnuts* 20121261 [2]
- *whitecurrants* 20121258 [2]
- *winter field beans* 20121321 [2]

Approval information
- Fluazifop-P-butyl included in Annex I under EC Regulation 1107/2009
- Accepted by BBPA for use on hops

Efficacy guidance
- Best results achieved by application when weed growth active under warm conditions with adequate soil moisture.
- Spray weeds from 2-expanded leaf stage to fully tillered, couch from 4 leaves when majority of shoots have emerged, with a second application if necessary
- Control may be reduced under dry conditions. Do not cultivate for 2 wk after spraying couch
- Annual meadow grass is not controlled

SEE SECTION 3 FOR PRODUCTS ALSO REGISTERED

Pesticide Profiles

- May also be used to remove grass cover crops
- Fluazifop-P-butyl is an ACCase inhibitor herbicide. To avoid the build up of resistance do not apply products containing an ACCase inhibitor herbicide more than twice to any crop. In addition do not use any product containing fluazifop-P-butyl in mixture or sequence with any other product containing the same ingredient
- Use these products as part of a resistance management strategy that includes cultural methods of control and does not use ACCase inhibitors as the sole chemical method of grass weed control

Restrictions
- Maximum number of treatments 1 per crop or yr for all crops
- Do not sow cereals or grass crops for at least 8 wk after application of high rate or 2 wk after low rate
- Do not apply through CDA sprayer, with hand-held equipment or from air
- Avoid treatment before spring growth has hardened or when buds opening
- Do not treat bush and cane fruit or hops between flowering and harvest
- Consult processors before treating crops intended for processing
- Oilseed rape, linseed and flax for industrial use must not be harvested for human or animal consumption nor grazed
- Do not use for forestry establishment on land not previously under arable cultivation or improved grassland
- Treated vegetation in field margins, land temporarily removed from production etc, must not be grazed or harvested for human or animal consumption and unprotected persons must be kept out of treated areas for at least 24 h

Crop-specific information
- Latest use: before 50% ground cover for swedes, turnips; before 5 leaf stage for spring oilseed rape; before flowering for blackcurrants, gooseberries, hops, raspberries, strawberries; before flower buds visible for field beans, peas, linseed, flax, winter oilseed rape; 2 wk before sowing cereals or grass for field margins, land temporarily removed from production
- HI beet crops, kale, carrots 8 wk; onions 4 wk; oilseed rape for industrial use 2 wk
- Apply to sugar and fodder beet from 1-true leaf to 50% ground cover
- Apply to winter oilseed rape from 1-true leaf to established plant stage
- Apply to spring oilseed rape from 1-true leaf but before 5-true leaves
- Apply in fruit crops after harvest. See label for timing details on other crops
- Before using on onions or peas use crystal violet test to check that leaf wax is sufficient

Environmental safety
- Dangerous for the environment
- Very toxic to aquatic organisms

Hazard classification and safety precautions
    Hazard Harmful, Dangerous for the environment
    Transport code 9
    Packaging group III
    UN Number 3082
    Risk phrases R38, R50, R53a, R63
    Operator protection A, C, H, M; U05a, U08, U20b
    Environmental protection E15a [1-4]; E38 [2, 4]
    Storage and disposal D01, D02, D05, D09a, D12a [1-4]; D10b [1]; D10c [2-4]
    Medical advice M05b [1]

FOR FULL CONDITIONS OF USE ALWAYS READ THE PRODUCT LABEL

236 fluazinam

A dinitroaniline fungicide for use in potatoes
FRAC mode of action code: 29

See also azoxystrobin + fluazinam
cymoxanil + fluazinam
dimethomorph + fluazinam

Products

1	Nando 500SC	Nufarm UK	500 g/l	SC	16388
2	Shirlan	Syngenta	500 g/l	SC	16624
3	Tizca	Headland	500 g/l	SC	16289
4	Volley	Adama	500 g/l	SC	16451

Uses
- Blight in **potatoes**

Approval information
- Fluazinam included in Annex I under EC Regulation 1107/2009

Efficacy guidance
- Commence treatment at the first blight risk warning (before blight enters the crop). Products are rainfast within 1 h
- In the absence of a warning, treatment should start before foliage of adjacent plants meets in the rows
- Spray at 5-14 d intervals depending on severity of risk (see label)
- Ensure complete coverage of the foliage and stems, increasing volume as haulm growth progresses, in dense crops and if blight risk increases

Restrictions
- Do not use with hand-held sprayers

Crop-specific information
- HI 0 - 10 d for potatoes. Check label
- Ensure complete kill of potato haulm before lifting and do not lift crops for storage while there is any green tissue left on the leaves or stem bases

Environmental safety
- Dangerous for the environment
- Very toxic to aquatic organisms
- Buffer zone requirement 6m [4]
- Buffer zone requirement 7m [2, 3]
- Buffer zone requirement 8m [1]
- LERAP Category B

Hazard classification and safety precautions

Hazard Harmful [1]; Irritant [2, 3]; Dangerous for the environment [1-4]; Very toxic to aquatic organisms [4]
Transport code 9
Packaging group III
UN Number 3082
Risk phrases H315, H317 [1]; H361 [1, 4]; R36 [2]; R41 [3]; R43, R50, R53a [2, 3]
Operator protection A, C, H [2-4]; U02a, U04a, U08 [1, 2, 4]; U05a, U14 [1-3]; U11 [3]; U15, U20a [1, 2]; U20b [3, 4]
Environmental protection E15a, E16a, E38 [1-4]; E16b, E34 [1, 2, 4]; H410 [1, 4]
Storage and disposal D01, D02, D05, D10c [1-3]; D09a, D12a [1-4]; D10a [4]
Medical advice M03 [1, 2, 4]; M05a [1, 2]

SEE SECTION 3 FOR PRODUCTS ALSO REGISTERED

Pesticide Profiles

237 fludioxonil

A phenylpyrrole fungicide seed treatment for wheat and barley
FRAC mode of action code: 12

See also chlorothalonil + fludioxonil + propiconazole
cymoxanil + fludioxonil + metalaxyl-M
cyprodinil + fludioxonil
difenoconazole + fludioxonil
difenoconazole + fludioxonil + tebuconazole

Products

1	Beret Gold	Syngenta	25 g/l	FS	16430
2	Geoxe	Syngenta	50% w/w	WG	16596
3	Maxim 100FS	Syngenta	100 g/l	FS	15683
4	Maxim 480FS	Syngenta	480 g/l	FS	16725
5	Medallion TL	Syngenta	125 g/l	SC	15287

Uses

- Alternaria in **apples** *(reduction)*, **crab apples** *(reduction)*, **pears** *(reduction)*, **quinces** *(reduction)* [2]; **carrots**, **celeriac** *(off-label)*, **fennel** *(off-label)*, **parsley root** *(off-label)*, **parsnips** *(off-label)*, **protected celeriac** *(off-label)*, **protected fennel** *(off-label)*, **protected parsley root** *(off-label)*, **protected parsnips** *(off-label)*, **red beet** *(off-label)* [4]
- Anthracnose in **amenity grassland** *(reduction)*, **managed amenity turf** *(reduction)* [5]; **spinach** *(moderate control)* [4]
- Black dot in **potatoes** *(some reduction)* [3]
- Black leg in **broccoli** *(off-label)*, **brussels sprouts** *(off-label)*, **cabbages**, **cauliflowers** *(off-label)*, **collards** *(off-label)*, **kale** *(off-label)*, **kohlrabi** *(off-label)*, **oriental cabbage** *(off-label)*, **protected broccoli** *(off-label)*, **protected brussels sprouts** *(off-label)*, **protected cauliflowers** *(off-label)*, **protected collards** *(off-label)*, **protected kale** *(off-label)*, **protected kohlrabi** *(off-label)*, **protected oriental cabbage** *(off-label)*, **protected radishes** *(off-label)*, **radishes** *(off-label)* [4]
- Black scurf in **potatoes** [3]
- Botrytis in **apples** *(reduction)*, **crab apples** *(reduction)*, **pears** *(reduction)*, **quinces** *(reduction)* [2]; **bulb onions** [4]
- Bunt in **spring wheat** *(seed treatment)*, **winter wheat** *(seed treatment)* [1]
- Covered smut in **spring barley** *(seed treatment)*, **winter barley** *(seed treatment)* [1]
- Disease control in **apples**, **crab apples**, **pears**, **quinces** [2]
- Drechslera leaf spot in **amenity grassland** *(useful levels of control)*, **managed amenity turf** *(useful levels of control)* [5]
- Fusarium foot rot and seedling blight in **rye** *(seed treatment)*, **spring barley** *(seed treatment)*, **spring oats** *(seed treatment)*, **spring wheat** *(seed treatment)*, **triticale** *(seed treatment)*, **winter barley** *(seed treatment)*, **winter oats** *(seed treatment)*, **winter wheat** *(seed treatment)* [1]
- Fusarium patch in **amenity grassland**, **managed amenity turf** [5]
- Leaf spot in **celeriac** *(off-label)*, **fennel** *(off-label)*, **parsley root** *(off-label)*, **parsnips** *(off-label)*, **protected celeriac** *(off-label)*, **protected fennel** *(off-label)*, **protected parsley root** *(off-label)*, **protected parsnips** *(off-label)* [4]
- Leaf stripe in **spring barley** *(seed treatment - reduction)*, **winter barley** *(seed treatment - reduction)* [1]
- Leptosphariae maculans in **cabbages** [4]
- Microdochium nivale in **amenity grassland**, **managed amenity turf** [5]; **rye** *(seed treatment)*, **triticale** [1]
- Monilinia spp. in **apples** *(reduction)*, **crab apples** *(reduction)*, **pears** *(reduction)*, **quinces** *(reduction)* [2]
- Neck rot in **salad onions** *(off-label)*, **shallots** *(off-label)* [4]
- Nectria spp in **apples** *(reduction)*, **crab apples** *(reduction)*, **pears** *(reduction)*, **quinces** *(reduction)* [2]
- Penicillium rot in **apples** *(reduction)*, **crab apples** *(reduction)*, **pears** *(reduction)*, **quinces** *(reduction)* [2]

FOR FULL CONDITIONS OF USE ALWAYS READ THE PRODUCT LABEL

fludioxonil

- Phlyctema vagabunda in **apples** *(reduction)*, **crab apples** *(reduction)*, **pears** *(reduction)*, **quinces** *(reduction)* [2]
- Pyrenophora leaf spot in **spring oats** *(seed treatment)*, **winter oats** *(seed treatment)* [1]
- Scab in **seed potatoes** *(off-label)* [3]
- Seed-borne diseases in **baby leaf crops** *(off-label)* [4]
- Septoria seedling blight in **spring wheat** *(seed treatment)*, **winter wheat** *(seed treatment)* [1]
- Silver scurf in **potatoes** *(reduction)* [3]
- Snow mould in **spring barley** *(seed treatment)*, **spring oats** *(seed treatment)*, **spring wheat** *(seed treatment)*, **winter barley** *(seed treatment)*, **winter oats** *(seed treatment)*, **winter wheat** *(seed treatment)* [1]
- Stripe smut in **rye** [1]

Extension of Authorisation for Minor Use (EAMUs)
- *baby leaf crops* 20142846 [4]
- *broccoli* 20161150 [4]
- *brussels sprouts* 20161150 [4]
- *cauliflowers* 20161150 [4]
- *celeriac* 20161149 [4]
- *collards* 20161150 [4]
- *fennel* 20161148 [4]
- *kale* 20161150 [4]
- *kohlrabi* 20161150 [4]
- *oriental cabbage* 20161150 [4]
- *parsley root* 20161149 [4]
- *parsnips* 20152903 [4], 20161148 [4]
- *protected broccoli* 20161150 [4]
- *protected brussels sprouts* 20161150 [4]
- *protected cauliflowers* 20161150 [4]
- *protected celeriac* 20161149 [4]
- *protected collards* 20161150 [4]
- *protected fennel* 20161148 [4]
- *protected kale* 20161150 [4]
- *protected kohlrabi* 20161150 [4]
- *protected oriental cabbage* 20161150 [4]
- *protected parsley root* 20161149 [4]
- *protected parsnips* 20161148 [4]
- *protected radishes* 20161149 [4]
- *radishes* 20161149 [4]
- *red beet* 20152903 [4]
- *salad onions* 20160461 [4]
- *seed potatoes* 20160736 [3]
- *shallots* 20160461 [4]

Approval information
- Fludioxonil included in Annex I under EC Regulation 1107/2009
- Accepted by BBPA for use on malting barley

Efficacy guidance
- Apply direct to seed using conventional seed treatment equipment. Continuous flow treaters should be calibrated using product before use
- Effective against benzimidazole-resistant strains of *Microdochium nivale*

Restrictions
- Maximum number of treatments 1 per seed batch
- Do not apply to cracked, split or sprouted seed
- Sow treated seed within 6 mth

Crop-specific information
- Latest use: before drilling
- Product may reduce flow rate of seed through drill. Recalibrate with treated seed before drilling

SEE SECTION 3 FOR PRODUCTS ALSO REGISTERED

Pesticide Profiles

Environmental safety
- Dangerous for the environment
- Toxic to aquatic organisms
- Do not use treated seed as food or feed
- Treated seed harmful to game and wildlife
- LERAP Category B [2, 5]

Hazard classification and safety precautions
Hazard Irritant [2]; Dangerous for the environment [1-5]
Transport code 9
Packaging group III
UN Number 3077 [2]; 3082 [1, 3-5]
Risk phrases R43 [2]; R50 [2, 4]; R51 [1, 3, 5]; R53a [1-5]
Operator protection A [1-4]; H [1, 2, 4]; U05a [1-5]; U14, U19a, U20a [2]; U20b [1, 3, 5]
Environmental protection E03, E15a [1]; E15b [2-5]; E16a [2, 5]; E16b [5]; E34 [3-5]; E38 [1-5]
Storage and disposal D01, D02, D09a, D12a [1-5]; D05, D11a [1, 3, 4]; D10c [2-5]
Treated seed S01, S03 [3]; S02, S05 [1, 3]; S04d [4]; S07 [1]

238   fludioxonil + metalaxyl-M

A seed dressing for use in forage maize
FRAC mode of action code: 12 + 4

Products
Maxim XL   Syngenta   25:9.69 g/l   FS   16599

Uses
- Damping off in **forage maize**
- Fusarium in **forage maize**
- Pythium in **forage maize**

Approval information
- Fludioxonil and metalaxyl-M included in Annex I under EC Regulation 1107/2009

Restrictions
- For advice on resistance management, refer to the latest Fungicide Resistance Action Group (FRAG) guidelines

Hazard classification and safety precautions
UN Number N/C
Risk phrases R52, R53a
Operator protection A, H; U05a
Environmental protection E34, E38
Storage and disposal D01, D02, D05, D09a, D10c, D11a, D12a
Treated seed S01, S02, S03, S04a, S05, S07

239   fludioxonil + tefluthrin

A fungicide and insecticide seed treatment mixture for cereals

See also tefluthrin

Products
Austral Plus   Syngenta   10:40 g/l   FS   13314

Uses
- Bunt in **spring wheat**, **winter wheat**
- Covered smut in **spring barley**, **winter barley**
- Fusarium foot rot and seedling blight in **spring barley**, **spring oats**, **spring wheat**, **winter barley**, **winter oats**, **winter wheat**
- Leaf stripe in **spring barley** (partial control), **winter barley** (partial control)

FOR FULL CONDITIONS OF USE ALWAYS READ THE PRODUCT LABEL

fludioxonil + tefluthrin

- Pyrenophora leaf spot in **spring oats**, **winter oats**
- Seed-borne diseases in **triticale seed crop** *(off-label)*
- Septoria seedling blight in **spring wheat**, **winter wheat**
- Snow mould in **spring barley**, **spring wheat**, **winter barley**, **winter wheat**
- Wheat bulb fly in **spring barley**, **spring wheat**, **winter barley**, **winter wheat**
- Wireworm in **spring barley**, **spring oats**, **spring wheat**, **winter barley**, **winter oats**, **winter wheat**

Extension of Authorisation for Minor Use (EAMUs)
- **triticale seed crop** *20080982*

Approval information
- Fludioxonil and tefluthrin included in Annex 1 under EC Regulation 1107/2009
- Accepted by BBPA for use on malting barley

Efficacy guidance
- Apply direct to seed using conventional seed treatment equipment. Continuous flow treaters should be calibrated using product before use
- Best results obtained from seed drilled into a firm even seedbed
- Tefluthrin is released into soil after drilling and repels or kills larvae of wheat bulb fly and wireworm attacking below ground. Pest control may be reduced by deep or shallow drilling
- Where egg counts of wheat bulb fly or population counts of wireworm indicate a high risk of severe attack follow up spray treatments may be needed
- Control of leaf stripe in barley may not be sufficient in crops grown for seed certification
- Effective against benzimidazole-resistant strains of *Microdochium nivale*
- Under adverse soil or environmental conditions seed rates should be increased to compensate for possible reduced germination capacity

Restrictions
- Maximum number of treatments 1 per batch
- Store treated seed in cool dry conditions and drill within 3 mth
- Do not use on seed above 16% moisture content or on sprouted, cracked or damaged seed

Crop-specific information
- Latest use: before drilling

Environmental safety
- Dangerous for the environment
- Very toxic to aquatic organisms

Hazard classification and safety precautions
Hazard Dangerous for the environment
Transport code 9
Packaging group III
UN Number 3082
Risk phrases R50, R53a
Operator protection A, C, D, H; U02a, U04a, U05a, U07, U08, U14, U20b
Environmental protection E03, E15b, E34, E38
Storage and disposal D01, D02, D05, D09b, D10c, D11a, D12a
Treated seed S02, S04b, S05, S07, S08

SEE SECTION 3 FOR PRODUCTS ALSO REGISTERED

Pesticide Profiles

240 flufenacet

A broad spectrum oxyacetamide herbicide for weed control in winter cereals
HRAC mode of action code: K3

See also diflufenican + flufenacet
diflufenican + flufenacet + flurtamone
flufenacet + isoxaflutole
flufenacet + metribuzin
flufenacet + pendimethalin

Products

1	Fence	Albaugh UK	480 g/l	SL	17393
2	Gorgon	Sipcam	480 g/l	SL	17685
3	Sunfire	Certis	500 g/l	SL	16745
4	System 50	Certis	500 g/l	SL	16612

Uses
- Annual dicotyledons in **winter barley**, **winter wheat** [1-4]
- Annual meadow grass in **rye** *(off-label)*, **triticale** *(off-label)* [3, 4]
- Blackgrass in **rye** *(off-label)*, **triticale** *(off-label)* [3, 4]; **winter barley**, **winter wheat** [1-4]

Extension of Authorisation for Minor Use (EAMUs)
- *rye* 20162526 [3], 20162019 [4]
- *triticale* 20162526 [3], 20162019 [4]

Approval information
- Flufenacet included in Annex I under EC Regulation 1107/2009

Environmental safety
- LERAP Category B

Hazard classification and safety precautions
Hazard Harmful, Dangerous for the environment, Harmful if swallowed [1-4]; Very toxic to aquatic organisms [3, 4]
Transport code 9
Packaging group III
UN Number 3082
Risk phrases H317 [3, 4]; H373 [1-4]
Operator protection A, H
Environmental protection E16a, H410
Storage and disposal D01, D02 [1, 2]

241 flufenacet + isoxaflutole

A residual herbicide mixture for maize
HRAC mode of action code: K3 + F2

See also isoxaflutole

Products
Amethyst	AgChem Access	48:10% w/w	WG	14079

Uses
- Annual dicotyledons in **forage maize**, **grain maize**
- Grass weeds in **forage maize**, **grain maize**

Approval information
- Flufenacet and isoxaflutole included in Annex I under EC Regulation 1107/2009

FOR FULL CONDITIONS OF USE ALWAYS READ THE PRODUCT LABEL

flufenacet + metribuzin

Efficacy guidance
- Best results obtained from applications to a fine, firm seedbed in the presence of some soil moisture
- Efficacy may be reduced on cloddy seedbeds or under prolonged dry conditions
- Established perennial grasses and broad-leaved weeds growing from rootstocks will not be controlled
- Always follow WRAG guidelines for preventing and managing herbicide resistant weeds. See Section 5 for more information

Restrictions
- Maximum number of treatments 1 per crop
- Do not use on soils both above 70% sand and less than 2% organic matter
- Do not use on maize crops intended for seed production

Crop-specific information
- Latest use: before crop emergence
- Ideally treatment should be made within 4 d of sowing, before the maize seeds have germinated

Following crops guidance
- In the event of failure of a treated crop maize, sweet corn or potatoes may be grown after surface cultivation or ploughing
- After normal harvest of a treated crop oats or barley may be sown after ploughing and wheat, rye, triticale, sugar beet, field beans, peas, soya, sorghum or sunflowers may be sown after surface cultivation or ploughing.

Environmental safety
- Dangerous for the environment
- Very toxic to aquatic organisms
- Take care to avoid drift over other crops. Beet crops and sunflowers are particularly sensitive
- LERAP Category B

Hazard classification and safety precautions
   Hazard Harmful, Dangerous for the environment
   Transport code 9
   Packaging group III
   UN Number 3077
   Risk phrases R22a, R36, R43, R48, R50, R53a, R63
   Operator protection A, C, H; U05a, U11, U20b
   Environmental protection E15b, E16a, E34, E38
   Storage and disposal D01, D02, D05, D09a, D10a, D12a
   Medical advice M03

242   flufenacet + metribuzin

A herbicide mixture for potatoes
HRAC mode of action code: K3 + C1

See also metribuzin

Products
   Artist   Bayer CropScience   24:17.5% w/w   WP   17049

Uses
- Annual dicotyledons in **bilberries** *(off-label)*, **blackcurrants** *(off-label)*, **cranberries** *(off-label)*, **early potatoes**, **gooseberries** *(off-label)*, **maincrop potatoes**, **redcurrants** *(off-label)*, **ribes species** *(off-label)*
- Annual grasses in **bilberries** *(off-label)*, **blackcurrants** *(off-label)*, **cranberries** *(off-label)*, **gooseberries** *(off-label)*, **redcurrants** *(off-label)*, **ribes species** *(off-label)*
- Annual meadow grass in **early potatoes**, **maincrop potatoes**

Extension of Authorisation for Minor Use (EAMUs)
- *bilberries* 20152968

SEE SECTION 3 FOR PRODUCTS ALSO REGISTERED

Pesticide Profiles

- *blackcurrants* 20152968
- *cranberries* 20152968
- *gooseberries* 20152968
- *redcurrants* 20152968
- *ribes species* 20152968

Approval information
- Flufenacet and metribuzin included in Annex I under EC Regulation 1107/2009

Efficacy guidance
- Product acts through root uptake and needs sufficient soil moisture at and shortly after application
- Effectiveness is reduced under dry soil conditions
- Residual activity is reduced on mineral soils with a high organic matter content and on peaty or organic soils
- Ensure application is made evenly to both sides of potato ridges
- Perennial weeds are not controlled

Restrictions
- Maximum total dose equivalent to one full dose treatment
- Potatoes must be sprayed before emergence of crop and weeds
- See label for list of tolerant varieties. Do not treat Maris Piper grown on Sands or Very Light soils
- Do not use on Sands
- On stony or gravelly soils there is risk of crop damage especially if heavy rain falls soon after application

Crop-specific information
- Latest use: before potato crop emergence
- Consult processor before use on crops for processing

Following crops guidance
- Before drilling or planting any succeeding crop soil must be mouldboard ploughed to at least 15 cm as soon as possible after lifting and no later than end Dec
- In W Cornwall on soils with more than 5% organic matter treated early potatoes may be followed by summer planted brassica crops 14 wk after treatment and after mouldboard ploughing. Elsewhere cereals or winter beans may be grown in the same year if at least 16 wk have elapsed since treatment
- In the yr following treatment any crop may be grown except lettuce or radish, or vegetable brassica crops on silt soils in Lincs

Environmental safety
- Dangerous for the environment
- Very toxic to aquatic organisms
- Take care to avoid spray drift onto neighbouring crops, especially lettuce or brassicas
- LERAP Category B

Hazard classification and safety precautions
    Hazard Harmful, Dangerous for the environment, Harmful if swallowed
    Transport code 9
    Packaging group III
    UN Number 3077
    Risk phrases H317, H373
    Operator protection A, C, D, H; U05a, U08, U13, U14, U19a, U20b
    Environmental protection E15a, E16a, E38, H410
    Storage and disposal D01, D02, D09a, D11a, D12a
    Medical advice M03

FOR FULL CONDITIONS OF USE ALWAYS READ THE PRODUCT LABEL

243 flufenacet + pendimethalin

A broad spectrum residual and contact herbicide mixture for winter cereals
HRAC mode of action code: K3 + K1

See also pendimethalin

Products

1	Clayton Glacier	Clayton	60:300 g/l	EC	15920
2	Confluence	Pan Agriculture	60:300 g/l	EC	17199
3	Crystal	BASF	60:300 g/l	EC	13914
4	Rock	Goldengrass	60:300 g/l	EC	14928
5	Shooter	BASF	60:300 g/l	EC	14106
6	Standon Diwana	Standon	60:300 g/l	EC	17221
7	Trooper	BASF	60:300 g/l	EC	13924

Uses
- Annual dicotyledons in **game cover** *(off-label)*, **spring barley** *(off-label)* [3]; **winter barley**, **winter wheat** [1-7]
- Annual grasses in **winter barley**, **winter wheat** [1-7]
- Annual meadow grass in **game cover** *(off-label)*, **spring barley** *(off-label)* [3]; **winter barley**, **winter wheat** [1-7]
- Blackgrass in **game cover** *(off-label)*, **spring barley** *(off-label)* [3]; **winter barley**, **winter wheat** [1-7]
- Chickweed in **winter barley**, **winter wheat** [1-7]
- Corn marigold in **winter barley**, **winter wheat** [1-7]
- Field speedwell in **winter barley**, **winter wheat** [1-7]
- Ivy-leaved speedwell in **winter barley**, **winter wheat** [1-7]
- Wild oats in **spring barley** *(off-label)* [3]

Extension of Authorisation for Minor Use (EAMUs)
- **game cover** 20090450 [3]
- **spring barley** 20130951 [3]

Approval information
- Flufenacet and pendimethalin included in Annex I under EC Regulation 1107/2009
- Accepted by BBPA for use on malting barley

Efficacy guidance
- Best results achieved when applied from pre-emergence of weeds up to 2 leaf stage but post emergence treatment is not recommended on clay soils
- Product requires some soil moisture to be activated ideally from rain within 7 d of application. Prolonged dry conditions may reduce residual control
- Product is slow acting and final level of weed control may take some time to appear
- For effective weed control seed bed preparations should ensure even incorporation of any trash, straw and ash to 15 cm
- Efficacy may be reduced on soils with more than 6% organic matter
- Always follow WRAG guidelines for preventing and managing herbicide resistant weeds. See Section 5 for more information

Restrictions
- Maximum total dose equivalent to one full dose treatment
- For pre-emergence treatments seed should be covered with at least 32 mm settled soil. Shallow drilled crops should be treated post-emergence only
- Do not treat undersown crops
- Avoid spraying during periods of prolonged or severe frosts
- Do not use on stony or gravelly soils or those with more than 10% organic matter
- Pre-emergence treatment may only be used on crops drilled before 30 Nov. All crops must be treated before 31 Dec in yr of planting.
- Concentrated or diluted product may stain clothing or skin

SEE SECTION 3 FOR PRODUCTS ALSO REGISTERED

Pesticide Profiles

Crop-specific information
- Latest use: before third tiller stage (GS 23) and before 31 Dec in yr of planting
- Very wet weather before and after treatment may result in loss of crop vigour and reduced yield, particularly where soils become waterlogged

Following crops guidance
- Any crop may follow a failed or normally harvested treated crop provided ploughing to at least 15 cm is carried out beforehand

Environmental safety
- Dangerous for the environment
- Very toxic to aquatic organisms
- Risk to certain non-target insects or other arthropods
- Some products supplied in small volume returnable packs. Follow instructions for use
- LERAP Category B

Hazard classification and safety precautions
Hazard Harmful, Dangerous for the environment
Transport code 3
Packaging group III
UN Number 1993 [3-7]; 3082 [1, 2]
Risk phrases R22a, R38, R50, R53a [1-7]; R22b, R40 [1-4, 6, 7]
Operator protection A, C, H; U02a, U05a [1-7]; U14, U20b [5]; U20c [1-4, 6, 7]
Environmental protection E15a [1-4, 6, 7]; E15b [5]; E16a, E16b, E22b, E34, E38 [1-7]
Storage and disposal D01, D02, D09a, D12a [1-7]; D05 [4]; D08, D10c [5]; D10a [1-4, 6, 7]
Medical advice M03 [1-7]; M05b [1-4, 6, 7]

244 flumioxazin

A phenyphthalimide herbicide for winter wheat
HRAC mode of action code: E

Products

1	Digital	Interfarm	300 g/l	SC	13561
2	Sumimax	Interfarm	300 g/l	SC	13548

Uses
- Annual dicotyledons in **bulb onions** *(off-label)*, **carrots** *(off-label)*, **ornamental plant production** *(off-label)*, **parsnips** *(off-label)*, **vining peas** *(off-label)*, **winter oats** *(off-label)*, **winter wheat** [1, 2]; **farm forestry** *(off-label)*, **forest nurseries** *(off-label)* [1]; **hops** *(off-label)*, **soft fruit** *(off-label)*, **top fruit** *(off-label)* [2]
- Annual grasses in **bulb onions** *(off-label)*, **winter oats** *(off-label)* [1, 2]; **carrots** *(off-label)*, **hops** *(off-label)*, **ornamental plant production** *(off-label)*, **parsnips** *(off-label)*, **soft fruit** *(off-label)*, **top fruit** *(off-label)*, **vining peas** *(off-label)* [2]
- Annual meadow grass in **winter wheat** [1, 2]
- Blackgrass in **winter oats** *(off-label)* [1, 2]
- Chickweed in **winter wheat** [1, 2]
- Cleavers in **winter wheat** [1, 2]
- Groundsel in **bulb onions** *(off-label)*, **carrots** *(off-label)*, **ornamental plant production** *(off-label)*, **parsnips** *(off-label)*, **vining peas** *(off-label)* [1, 2]; **farm forestry** *(off-label)*, **forest nurseries** *(off-label)* [1]; **hops** *(off-label)*, **soft fruit** *(off-label)*, **top fruit** *(off-label)*, **winter oats** *(off-label)* [2]
- Loose silky bent in **winter wheat** [1, 2]
- Mayweeds in **winter wheat** [1, 2]
- Ryegrass in **winter oats** *(off-label)* [1, 2]
- Speedwells in **winter wheat** [1, 2]
- Volunteer oilseed rape in **bulb onions** *(off-label)*, **carrots** *(off-label)*, **ornamental plant production** *(off-label)*, **parsnips** *(off-label)*, **vining peas** *(off-label)*, **winter wheat** [1, 2]; **farm forestry** *(off-label)*, **forest nurseries** *(off-label)* [1]; **hops** *(off-label)*, **soft fruit** *(off-label)*, **top fruit** *(off-label)*, **winter oats** *(off-label)* [2]

FOR FULL CONDITIONS OF USE ALWAYS READ THE PRODUCT LABEL

flumioxazin

- Volunteer potatoes in **bulb onions** *(off-label)*, **carrots** *(off-label)*, **ornamental plant production** *(off-label)*, **parsnips** *(off-label)*, **vining peas** *(off-label)* [1, 2]; **farm forestry** *(off-label)*, **forest nurseries** *(off-label)* [1]; **hops** *(off-label)*, **soft fruit** *(off-label)*, **top fruit** *(off-label)*, **winter oats** *(off-label)* [2]

Extension of Authorisation for Minor Use (EAMUs)
- **bulb onions** *20091114* [1], *20091108* [2]
- **carrots** *20091112* [1], *20091111* [2]
- **farm forestry** *20082844* [1]
- **forest nurseries** *20082844* [1]
- **hops** *20082881* [2]
- **ornamental plant production** *20082844* [1], *20082881* [2]
- **parsnips** *20091112* [1], *20091111* [2]
- **soft fruit** *20082881* [2]
- **top fruit** *20082881* [2]
- **vining peas** *20091113* [1], *20091109* [2]
- **winter oats** *20093121* [1], *20093122* [2]

Approval information
- Flumioxazine is included in Annex 1 under EC Regulation 1107/2009

Efficacy guidance
- Best results obtained from applications to moist soil early post-emergence of the crop when weeds are germinating up to 1 leaf stage
- Flumioxazin is contact acting and relies on weeds germinating in moist soil and coming into sufficient contact with the herbicide
- Flumioxazin remains in the surface layer of the soil and does not migrate to lower layers
- Contact activity is long lasting during the autumn
- Seed beds should be firm, fine and free from clods. Crops should be drilled to 25 mm and well covered by soil
- Efficacy is reduced on soils with more than 10% organic matter
- Flumioxazin is a member of the protoporphyrin oxidase (PPO) inhibitor herbicides. To avoid the build up of resistance do not use PPO-inhibitor herbicides more than once on any crop
- Use as part of a resistance management strategy that includes cultural methods of control and does not use PPO-inhibitors as the sole chemical method of grass weed control

Restrictions
- Maximum number of treatments 1 per crop for winter wheat
- Only apply to crops that have been hardened by cool weather. Do not treat crops with lush or soft growth
- Do not apply to waterlogged soils or to soils with more than 10% organic matter
- Do not follow an application with another pesticide for 14 d
- Do not treat undersown crops
- Do not treat broadcast crops until they are past the 3 leaf stage
- Do not treat crops under stress for any reason, or during prolonged frosty weather
- Do not roll or harrow for 2 wk before treatment or at any time afterwards

Crop-specific information
- Latest use: before 5th true leaf stage (GS 15) for winter wheat
- Light discolouration of leaf margins can occur after treatment and treatment of soft crops will cause transient leaf bleaching

Following crops guidance
- Any crop may be sown following normal harvest of a treated crop
- In the event of failure of a treated crop spring cereals, spring oilseed rape, sugar beet, maize or potatoes may be redrilled after ploughing

Environmental safety
- Dangerous for the environment
- Very toxic to aquatic organisms
- Take extreme care to avoid drift onto plants outside the target area
- LERAP Category B

SEE SECTION 3 FOR PRODUCTS ALSO REGISTERED

Pesticide Profiles

Hazard classification and safety precautions
Hazard Toxic, Dangerous for the environment
Transport code 9
Packaging group III
UN Number 3082
Risk phrases R50, R53a, R61
Operator protection A, H; U05a, U19a
Environmental protection E15b, E16a, E34, E38
Storage and disposal D01, D02, D05, D07, D09a, D12b
Medical advice M04a

245 fluopicolide

An benzamide fungicide available only in mixtures
FRAC mode of action code: 43

246 fluopicolide + propamocarb hydrochloride

A protectant and systemic fungicide mixture for potato blight
FRAC mode of action code: 43 + 28

See also propamocarb hydrochloride

Products
Infinito   Bayer CropScience   62.5:625 g/l   SC   16335

Uses
- Blight in **potatoes**
- Downy mildew in **baby leaf crops** *(off-label)*, **broccoli** *(off-label)*, **brussels sprouts** *(off-label)*, **bulb onions** *(off-label)*, **cabbages** *(off-label)*, **calabrese** *(off-label)*, **cauliflowers** *(off-label)*, **collards** *(off-label)*, **cress** *(off-label)*, **garlic** *(off-label)*, **herbs (see appendix 6)** *(off-label)*, **kale** *(off-label)*, **lamb's lettuce** *(off-label)*, **land cress** *(off-label)*, **leeks** *(off-label)*, **lettuce** *(off-label)*, **ornamental plant production** *(off-label)*, **protected broccoli** *(off-label)*, **protected brussels sprouts** *(off-label)*, **protected cabbages** *(off-label)*, **protected calabrese** *(off-label)*, **protected cauliflowers** *(off-label)*, **protected collards** *(off-label)*, **protected kale** *(off-label)*, **red mustard** *(off-label)*, **rocket** *(off-label)*, **salad onions** *(off-label)*, **shallots** *(off-label)*, **spinach** *(off-label)*

Extension of Authorisation for Minor Use (EAMUs)
- *baby leaf crops* 20140556
- *broccoli* 20152557
- *brussels sprouts* 20152557
- *bulb onions* 20161552
- *cabbages* 20152557
- *calabrese* 20152557
- *cauliflowers* 20152557
- *collards* 20152557
- *cress* 20140556
- *garlic* 20161552
- *herbs (see appendix 6)* 20140556
- *kale* 20152557
- *lamb's lettuce* 20140556
- *land cress* 20140556
- *leeks* 20161552
- *lettuce* 20140556
- *ornamental plant production* 20142251
- *protected broccoli* 20152557
- *protected brussels sprouts* 20152557
- *protected cabbages* 20152557
- *protected calabrese* 20152557

FOR FULL CONDITIONS OF USE ALWAYS READ THE PRODUCT LABEL

fluopyram

- **protected cauliflowers** *20152557*
- **protected collards** *20152557*
- **protected kale** *20152557*
- **red mustard** *20140556*
- **rocket** *20140556*
- **salad onions** *20161552*
- **shallots** *20161552*
- **spinach** *20140556*

Approval information
- Fluopicolide and propamocarb hydrochloride included in Annex I under EC Regulation 1107/2009

Efficacy guidance
- Commence spray programme before infection appears as soon as weather conditions favourable for disease development occur. At latest the first treatment should be made as the foliage meets along the rows
- Repeat treatments at 7-10 day intervals according to disease incidence and weather conditions
- Reduce spray interval if conditions are conducive to the spread of blight
- Spray as soon as possible after irrigation
- Increase water volume in dense crops
- When used from full canopy development to haulm desiccation as part of a full blight protection programme tubers will be protected from late blight after harvest and tuber blight incidence will be reduced
- To reduce the development of resistance product should be used in single or block applications with fungicides from a different cross-resistance group

Restrictions
- Maximum total dose on potatoes equivalent to four full dose treatments
- Do not apply if rainfall or irrigation is imminent. Product is rainfast in 1 h provided spray has dried on leaf
- Do not apply as a curative treatment when blight is present in the crop
- Do not apply more than 3 consecutive treatments of the product

Crop-specific information
- HI 7 d for potatoes
- All varieties of potatoes, including seed crops, may be treated

Environmental safety
- Dangerous for the environment
- Toxic to aquatic organisms

Hazard classification and safety precautions
Hazard Irritant, Dangerous for the environment, Very toxic to aquatic organisms
Transport code 9
Packaging group III
UN Number 3082
Risk phrases H317
Operator protection A, H; U05a, U08, U11, U19a, U20a
Environmental protection E15a, E38, H410
Storage and disposal D01, D02, D09a, D10b, D12a

247 fluopyram

An SDHI fungicide only available in mixtures
FRAC mode of action code: 7

Pesticide Profiles

248   fluopyram + prothioconazole

A foliar fungicide for use in oilseed rape
FRAC mode of action code: 3 + 7

Products
1	Propulse	Bayer CropScience	125:125 g/l	SE	15735
2	Recital	Bayer CropScience	125:125 g/l	SE	16086

Uses
- Disease control in *spring oilseed rape*, *winter oilseed rape* [1, 2]
- Light leaf spot in *winter oilseed rape* [1]
- Phoma in *winter oilseed rape* [1]
- Powdery mildew in *winter oilseed rape* [1]
- Sclerotinia in *winter oilseed rape* [1]
- Stem canker in *winter oilseed rape* [1]

Approval information
- Fluopyram and prothioconazole included in Annex 1 under EC Regulation 1107/2009.
- Approval expiry 31 Dec 2017 [1, 2]

Environmental safety
- LERAP Category B

Hazard classification and safety precautions
   Hazard Harmful [2]; Dangerous for the environment, Very toxic to aquatic organisms [1, 2]
   Transport code 9
   Packaging group III
   UN Number 3082
   Operator protection A, H; U05a, U09b, U19a, U20c
   Environmental protection E15b, E16a, E34, E38, H410
   Storage and disposal D01, D02, D05, D09a, D10b, D12a
   Medical advice M03

249   fluopyram + prothioconazole + tebuconazole

An SDHI and triazole mixture for seed treatment in winter barley
FRAC mode of action code: 3 + 7

Products
	Raxil Star	Bayer CropScience	20.8:103.1:63.2 g/l	FS	15197

Uses
- Covered smut in *spring barley*, *winter barley*
- Fusarium in *spring barley*, *winter barley*
- Leaf stripe in *spring barley*, *winter barley*
- Loose smut in *spring barley*, *winter barley*
- Microdochium nivale in *spring barley*, *winter barley*
- Net blotch in *spring barley* (seed borne), *winter barley* (seed borne)

Approval information
- Fluopyram, prothioconazole and tebuconazole included in Annex 1 under EC Regulation 1107/2009
- Accepted by BBPA for use on malting barley
- Approval expiry 31 Dec 2017 [1]

Efficacy guidance
- Seed should be drilled to a depth of 40 mm into a well prepared and firm seedbed. If seed is present on the soil surface, or spills have occurred, then, if conditions are appropriate, the field should be harrowed then rolled to ensure good incorporation

FOR FULL CONDITIONS OF USE ALWAYS READ THE PRODUCT LABEL

- Do not use on seed with more than 16% moisture content or on sprouted, cracked, skinned or otherwise damaged seed

Restrictions
- Newer authorisations for tebuconazole products require application to cereals only after GS 30 and applications to oilseed rape and linseed after GS20 - check label

Hazard classification and safety precautions
Hazard Harmful, Dangerous for the environment
Transport code 9
Packaging group III
UN Number 3082
Risk phrases H361
Operator protection A, C, D, H; U05a, U12b, U20b, U20e
Environmental protection E15b, E34, E38, H410
Storage and disposal D01, D02, D05, D09a, D10d, D12a, D14
Treated seed S01, S02, S04b, S05, S07, S08

250 fluoxastrobin

A protectant stobilurin fungicide available in mixtures
FRAC mode of action code: 11

See also bixafen + fluoxastrobin + prothioconazole

251 fluoxastrobin + prothioconazole

A strobilurin and triazole fungicide mixture for cereals
FRAC mode of action code: 11 + 3

See also prothioconazole

Products

1	Fandango	Bayer CropScience	100:100 g/l	EC	12276
2	Fandango	Bayer CropScience	100:100 g/l	EC	17318
3	Firefly 155	Bayer CropScience	45:110 g/l	EC	14818
4	Kurdi	AgChem Access	100:100 g/l	EC	15917
5	Unicur	Bayer CropScience	100:100 g/l	EC	14776
6	Unicur	Bayer CropScience	100:100 g/l	EC	17402

Uses
- Botrytis in **bulb onions** *(useful reduction)*, **shallots** *(useful reduction)* [5, 6]
- Botrytis squamosa in **bulb onions** *(control)*, **shallots** *(control)* [5]; **bulb onions** *(useful reduction)*, **shallots** *(useful reduction)* [6]
- Brown rust in **spring barley**, **winter barley** [1, 2, 4]; **spring wheat**, **winter rye**, **winter wheat** [1-4]
- Cladosporium in **bulb onions** *(useful reduction)*, **shallots** *(useful reduction)* [5]
- Crown rust in **spring oats**, **winter oats** [1, 2, 4]
- Disease control in **forest nurseries** *(off-label)* [1]
- Downy mildew in **bulb onions** *(control)*, **shallots** *(control)* [5, 6]
- Eyespot in **spring barley** *(reduction)*, **spring oats**, **winter barley** *(reduction)*, **winter oats** [1, 2, 4]; **spring oats** *(reduction of incidence and severity)*, **winter oats** *(reduction of incidence and severity)* [3]; **spring wheat** *(reduction)*, **winter rye** *(reduction)*, **winter wheat** *(reduction)* [1-4]
- Fusarium root rot in **spring wheat** *(reduction)*, **winter wheat** *(reduction)* [1, 2, 4]
- Glume blotch in **spring wheat**, **winter wheat** [1-4]
- Late ear diseases in **spring barley**, **spring wheat**, **winter barley**, **winter wheat** [1, 2, 4]
- Net blotch in **spring barley**, **winter barley** [1, 2, 4]
- Powdery mildew in **spring barley**, **spring oats**, **winter barley**, **winter oats** [1, 2, 4]; **spring wheat**, **winter rye**, **winter wheat** [1-4]

SEE SECTION 3 FOR PRODUCTS ALSO REGISTERED

Pesticide Profiles

- Rhynchosporium in **spring barley**, **winter barley** [1, 2, 4]; **winter rye** [1-4]
- Rust in **bulb onions** *(useful reduction)*, **shallots** *(useful reduction)* [5]
- Septoria leaf blotch in **spring wheat**, **winter wheat** [1-4]
- Sharp eyespot in **spring wheat** *(reduction)*, **winter wheat** [1, 2, 4]
- Sooty moulds in **spring wheat** *(reduction)*, **winter wheat** *(reduction)* [1, 2, 4]
- Take-all in **spring wheat** *(reduction)*, **winter barley** *(reduction)*, **winter wheat** *(reduction)* [1, 2, 4]
- Tan spot in **spring wheat**, **winter wheat** [1, 2, 4]
- Yellow rust in **spring wheat**, **winter wheat** [1-4]

Extension of Authorisation for Minor Use (EAMUs)
- *forest nurseries* 20160155 [1]

Approval information
- Fluoxastrobin and prothioconazole included in Annex I under EC Regulation 1107/2009
- Accepted by BBPA for use on malting barley

Efficacy guidance
- Seed treatment products must be applied by manufacturer's recommended treatment application equipment
- Treated cereal seed should preferably be drilled in the same season
- Follow-up treatments will be needed later in the season to give protection against air-borne and splash-borne diseases
- Best results on foliar diseases obtained from treatment at early stages of disease development. Further treatment may be needed if disease attack is prolonged
- Foliar applications to established infections of any disease are likely to be less effective
- Best control of cereal ear diseases obtained by treatment during ear emergence
- Fluoxastrobin is a member of the QoI cross resistance group. Foliar product should be used preventatively and not relied on for its curative potential
- Use product as part of an Integrated Crop Management strategy incorporating other methods of control, including where appropriate other fungicides with a different mode of action. Do not apply more than two foliar applications of QoI containing products to any cereal crop
- There is a significant risk of widespread resistance occurring in *Septoria tritici* populations in UK. Failure to follow resistance management action may result in reduced levels of disease control
- Strains of wheat and barley powdery mildew resistant to QoI's are common in the UK. Control of wheat mildew can only be relied on from the triazole component
- Prothioconazole is a DMI fungicide. Resistance to some DMI fungicides has been identified in Septoria leaf blotch which may seriously affect performance of some products. For further advice contact a specialist advisor and visit the Fungicide Resistance Action Group (FRAG)-UK website
- Where specific control of wheat mildew is required this should be achieved through a programme of measures including products recommended for the control of mildew that contain a fungicide from a different cross-resistance group and applied at a dose that will give robust control

Restrictions
- Maximum total dose of foliar sprays equivalent to two full dose treatments for the crop

Crop-specific information
- Latest use: before grain milky ripe for spray treatments on wheat and rye; beginning of flowering for barley, oats
- Some transient leaf chlorosis may occur after treatment of wheat or barley but this has not been found to affect yield

Environmental safety
- Dangerous for the environment
- Toxic to aquatic organisms
- Risk to non-target insects or other arthropods. Avoid spraying within 6 m of the field boundary to reduce the effects on non-target insects or other arthropods
- LERAP Category B

FOR FULL CONDITIONS OF USE ALWAYS READ THE PRODUCT LABEL

fluoxastrobin + prothioconazole + trifloxystrobin

Hazard classification and safety precautions
Hazard Dangerous for the environment [1-6]; Very toxic to aquatic organisms [2, 6]
Transport code 9
Packaging group III
UN Number 3082
Risk phrases H318 [2]; R51, R53a [1, 3-5]
Operator protection A [1-6]; H [2, 3, 5, 6]; U05a [1-6]; U09a, U20b [3]; U09b, U20a [1, 2, 4-6]
Environmental protection E15a, E16a, E34, E38 [1-6]; E22c [1, 2, 4-6]; H410 [2, 6]
Storage and disposal D01, D05, D09a, D10b, D12a
Medical advice M03

252 fluoxastrobin + prothioconazole + trifloxystrobin

A triazole and strobilurin fungicide mixture for cereals
FRAC mode of action code: 11 + 3 + 11

See also prothioconazole
trifloxystrobin

Products
    Jaunt   Bayer CropScience   75:150:75 g/l   EC   12350

Uses
- Brown rust in **durum wheat**, **rye**, **spring barley**, **spring wheat**, **triticale**, **winter barley**, **winter wheat**
- Eyespot in **durum wheat** (reduction), **rye** (reduction), **spring barley** (reduction), **spring wheat** (reduction), **triticale** (reduction), **winter barley** (reduction), **winter wheat** (reduction)
- Glume blotch in **durum wheat**, **rye**, **spring wheat**, **triticale**, **winter wheat**
- Late ear diseases in **durum wheat**, **rye**, **spring barley**, **spring wheat**, **triticale**, **winter barley**, **winter wheat**
- Net blotch in **spring barley**, **winter barley**
- Powdery mildew in **durum wheat**, **rye**, **spring barley**, **spring wheat**, **triticale**, **winter barley**, **winter wheat**
- Rhynchosporium in **spring barley**, **winter barley**
- Septoria leaf blotch in **durum wheat**, **rye**, **spring wheat**, **triticale**, **winter wheat**
- Tan spot in **durum wheat**, **rye**, **spring wheat**, **triticale**, **winter wheat**
- Yellow rust in **durum wheat**, **rye**, **spring wheat**, **triticale**, **winter wheat**

Approval information
- Fluoxastrobin, prothioconazole and trifloxystrobin included in Annex I under EC Regulation 1107/2009
- Accepted by BBPA for use on malting barley

Efficacy guidance
- Best results obtained from treatment at early stages of disease development. Further treatment may be needed if disease attack is prolonged
- Applications to established infections of any disease are likely to be less effective
- Best control of cereal ear diseases obtained by treatment during ear emergence
- Fluoxastrobin and trifloxystrobin are members of the QoI cross resistance group. Product should be used preventatively and not relied on for its curative potential
- Use product as part of an Integrated Crop Management strategy incorporating other methods of control, including where appropriate other fungicides with a different mode of action. Do not apply more than two foliar applications of QoI containing products to any cereal crop
- There is a significant risk of widespread resistance occurring in *Septoria tritici* populations in UK. Failure to follow resistance management action may result in reduced levels of disease control
- Strains of wheat and barley powdery mildew resistant to QoIs are common in the UK. Control of wheat mildew can only be relied on from the triazole component
- Where specific control of wheat mildew is required this should be achieved through a programme of measures including products recommended for the control of mildew that contain

SEE SECTION 3 FOR PRODUCTS ALSO REGISTERED

Pesticide Profiles

a fungicide from a different cross-resistance group and applied at a dose that will give robust control
- Prothioconazole is a DMI fungicide. Resistance to some DMI fungicides has been identified in Septoria leaf blotch which may seriously affect performance of some products. For further advice contact a specialist advisor and visit the Fungicide Resistance Action Group (FRAG)-UK website

Restrictions
- Maximum total dose equivalent to two full dose treatments

Crop-specific information
- Latest use: before grain milky ripe for winter wheat; up to beginning of anthesis (GS 61) for barley

Environmental safety
- Dangerous for the environment
- Very toxic to aquatic organisms
- Risk to non-target insects or other arthropods. Avoid spraying within 6 m of the field boundary to reduce the effects on non-target insects or other arthropods
- LERAP Category B

Hazard classification and safety precautions
Hazard Irritant, Dangerous for the environment, Very toxic to aquatic organisms
Transport code 9
Packaging group III
UN Number 3082
Operator protection A, C, H; U05a, U09b, U19a, U20b
Environmental protection E15a, E16a, E22c, E34, E38, H410
Storage and disposal D01, D02, D05, D09a, D10b, D12a
Medical advice M03

253 flupyrsulfuron-methyl

A sulfonylurea herbicide for winter wheat and winter barley
HRAC mode of action code: B

See also carfentrazone-ethyl + flupyrsulfuron-methyl
diflufenican + flupyrsulfuron-methyl

Products

1	Bullion	DuPont	50% w/w	WG	14058
2	Ductis SX	DuPont	50% w/w	WG	15426
3	Exceed SX	DuPont	50% w/w	WG	15427
4	Lexus SX	DuPont	50% w/w	WG	12979
5	Staka SX	DuPont	50% w/w	WG	15428

Uses
- Annual dicotyledons in **triticale, winter barley, winter oats, winter rye, winter wheat** [1-5]; **winter linseed** *(off-label)* [1]
- Blackgrass in **triticale, winter barley, winter oats, winter rye, winter wheat** [1-5]; **winter linseed** *(off-label)* [1]

Extension of Authorisation for Minor Use (EAMUs)
- **winter linseed** 20122641 [1]

Approval information
- Flupyrsulfuron-methyl included in Annex I under EC Regulation 1107/2009
- Accepted by BBPA for use on malting barley

Efficacy guidance
- Best results achieved from applications made in good growing conditions
- Good spray cover of weeds must be obtained

FOR FULL CONDITIONS OF USE ALWAYS READ THE PRODUCT LABEL

flupyrsulfuron-methyl

- For pre-emergence control of blackgrass application must be made in tank mixture with specified products (see label). Best control of blackgrass post-emergence in wheat obtained from application from the 1-leaf stage
- Winter barley must be treated pre-emergence and used in mixture. See label
- Growth of weeds is inhibited within hours of treatment but visible symptoms may not be apparent for up to 4 wk
- Flupyrsulfuron-methyl has moderate residual life in soil. Under normal moisture conditions susceptible weeds germinating soon after treatment will be controlled
- May be used on all soil types but residual activity and weed control is reduced on highly alkaline soils
- Flupyrsulfuron-methyl is a member of the ALS-inhibitor group of herbicides. To avoid the build up of resistance do not use any product containing an ALS-inhibitor herbicide with claims for control of grass weeds more than once on any crop. However, approval has now been granted for application of flupyrsulfuron-methyl in mixture with a non-ALS herbicide pre-emergence of blackgrass in sequence with post-emergence applications of iodosulfuron + mesosulfuron products.
- Use these products as part of a resistance management strategy that includes cultural methods of control and does not use ALS inhibitors as the sole chemical method of grass weed control

Restrictions
- Maximum number of treatments 1 per crop
- Do not use on wheat undersown with grasses, clover or other legumes, or any other broad-leaved crop
- Do not apply within 7 d of rolling
- Do not treat any crop suffering from drought, waterlogging, pest or disease attack, nutrient deficiency, or any other stress factors
- Specific restrictions apply to use in sequence or tank mixture with other sulfonylurea or ALS-inhibiting herbicides. See label for details

Crop-specific information
- Latest use: pre-emergence or before 1st node detectable (GS 31) for wheat; pre-emergence for barley
- Latest use for winter oats, rye and triticale is Dec 31st in year before harvest
- Under certain conditions chlorosis and stunting may occur, from which recovery is rapid

Following crops guidance
- Only cereals, oilseed rape, field beans, clover or grass may be sown in the yr of harvest of a treated crop
- In the event of crop failure only winter or spring wheat may be sown within 3 mth of treatment. Land should be ploughed and cultivated to 15 cm minimum before resowing. After 3 mth crops may be sown as shown above

Environmental safety
- Dangerous for the environment
- Very toxic to aquatic organisms
- Take extreme care to avoid drift onto broad-leaved plants outside the target area or onto ponds, waterways or ditches, or onto land intended for cropping
- Spraying equipment should not be drained or flushed onto land planted, or to be planted, with trees or crops other than cereals and should be thoroughly cleansed after use - see label for instructions

Hazard classification and safety precautions
    Hazard Dangerous for the environment
    Transport code 9
    Packaging group III
    UN Number 3077
    Risk phrases R50, R53a
    Operator protection U08, U19a, U20b
    Environmental protection E15a, E38
    Storage and disposal D09a, D11a, D12a

SEE SECTION 3 FOR PRODUCTS ALSO REGISTERED

Pesticide Profiles

254 flupyrsulfuron-methyl + pyroxsulam

A sulfonyl urea and triazolinone herbicide mixture for weed control in winter wheat
HRAC mode of action code: B

Products
    Unite   Dow   3.71:6.94% w/w   WG   17385

Uses
- Annual dicotyledons in *winter wheat*
- Blackgrass in *winter wheat* *(moderately susceptible)*
- Brome grasses in *winter wheat*
- Ryegrass in *winter wheat* *(from seed)*
- Wild oats in *winter wheat*

Approval information
- Flupyrsulfuron-methyl and pyroxsulam included in Annex I under EC Regulation 1107/2009.

Efficacy guidance
- Residual activity in soil will be reduced by high temperatures due to increased degradation.
- Apply in mixture with an authorised adjuvant.
- Do not apply more than one ALS product per crop for grass weed control.
- In some situations slight chlorosis may be observed after treatment. Crops quickly grow out of these symptoms and there is no effect on final grain yield.

Restrictions
- Do not apply in mixture with plant growth regulators - maintain an interval of at least 7 days.
- Do not apply in mixture with an organophosphate insecticide - maintain an interval of at least 14 days.

Following crops guidance
- Only cereals, oilseed rape, field beans, red and white clover or grass may be sown in the year of harvest.
- In the event of crop failure, only sow winter or spring wheat within 3 months of application and, before sowing, plough or thoroughly cultivate to a minimum depth of 15 cm.
- No cultivation restrictions prior to planting following crops.

Environmental safety
- LERAP Category B

Hazard classification and safety precautions
    Hazard Dangerous for the environment
    Transport code 9
    Packaging group III
    UN Number 3077
    Operator protection A; U05a, U14, U15, U20a
    Environmental protection E15b, E16a, E34, E38, H410
    Storage and disposal D01, D02, D09a, D12a

255 flupyrsulfuron-methyl + thifensulfuron-methyl

A sulfonylurea herbicide mixture for winter wheat and winter oats
HRAC mode of action code: B + B

See also thifensulfuron-methyl

Products
    Lexus Millenium   DuPont   10:40% w/w   WG   09206

Uses
- Annual dicotyledons in *winter oats*, *winter wheat*
- Blackgrass in *winter oats*, *winter wheat*

FOR FULL CONDITIONS OF USE ALWAYS READ THE PRODUCT LABEL

flupyrsulfuron-methyl + thifensulfuron-methyl

Approval information
- Flupyrsulfuron-methyl and thifensulfuron-methyl included in Annex I under EC Regulation 1107/2009

Efficacy guidance
- Best results obtained when applied to small actively growing weeds
- Good spray cover of weeds must be obtained
- Increased degradation of active ingredient in high soil temperatures reduces residual activity
- Product has moderate residual life in soil. Under normal moisture conditions susceptible weeds germinating soon after treatment will be controlled
- Product may be used on all soil types but residual activity and weed control are reduced on highly alkaline soils
- Blackgrass should be treated from 1 leaf up to first node stage
- Flupyrsulfuron-methyl and thifensulfuron-methyl are members of the ALS-inhibitor group of herbicides. To avoid the build up of resistance do not use any product containing an ALS-inhibitor herbicide with claims for control of grass weeds more than once on any crop
- Use these products as part of a resistance management strategy that includes cultural methods of control and does not use ALS inhibitors as the sole chemical method of grass weed control

Restrictions
- Maximum number of treatments 1 per crop
- Do not use on wheat undersown with grasses or legumes, or any other broad-leaved crop
- Do not apply within 7 d of rolling
- Do not treat any crop suffering from drought, waterlogging, pest or disease attack, nutrient deficiency, or any other stress factors
- Contact contract agents before use on crops grown for seed
- Specific restrictions apply to use in sequence or tank mixture with other sulfonylurea or ALS-inhibiting herbicides. See label for details

Crop-specific information
- Latest use: before 1st node detectable (GS 31) on wheat; before 31 Dec in yr of sowing on oats [1]
- Slight chlorosis, speckling and stunting may occur in certain conditions. Recovery is rapid and yield not affected

Following crops guidance
- Only cereals, oilseed rape, field beans or grass may be sown in the yr of harvest of a treated crop. Any crop may be sown the following spring
- In the event of crop failure only winter wheat may be sown before normal harvest date. Land should be ploughed and cultivated to 15 cm minimum before resowing

Environmental safety
- Dangerous for the environment
- Very toxic to aquatic organisms
- LERAP Category B

Hazard classification and safety precautions
    Hazard Irritant, Dangerous for the environment, Very toxic to aquatic organisms
    Transport code 9
    Packaging group III
    UN Number 3077
    Risk phrases H317
    Operator protection A, H; U05a, U08, U14, U19a, U20b
    Environmental protection E15a, E16a, E38, H410
    Storage and disposal D01, D02, D09a, D11a, D12a

SEE SECTION 3 FOR PRODUCTS ALSO REGISTERED

Pesticide Profiles

256 fluroxypyr

A post-emergence pyridinecarboxylic acid herbicide
HRAC mode of action code: O

See also 2,4-D + dicamba + fluroxypyr
aminopyralid + fluroxypyr
bromoxynil + fluroxypyr
bromoxynil + fluroxypyr + ioxynil
clopyralid + florasulam + fluroxypyr
clopyralid + fluroxypyr + MCPA
clopyralid + fluroxypyr + triclopyr
florasulam + fluroxypyr

Products

1	Cleancrop Gallifrey 3	Dow	333 g/l	EC	17399
2	Crescent	Certis	200 g/l	EC	17589
3	Flurostar 200	Globachem	200 g/l	EC	17438
4	Flurox 200	Sipcam	200 g/l	EC	13938
5	GAL-GONE	Globachem	200 g/l	EC	17505
6	Hatchet Xtra	Certis	200 g/l	EC	17593
7	Hurler	Barclay	200 g/l	EC	17715
8	Minstrel	UPL Europe	200 g/l	EC	13745
9	Starane Hi-Load HL	Dow	333 g/l	EC	16557

Uses

- Annual dicotyledons in **durum wheat**, **forage maize**, **grassland**, **spring oats**, **spring rye**, **spring wheat**, **triticale**, **winter barley**, **winter oats**, **winter rye**, **winter wheat** [1-9]; **grass seed crops**, **spelt** [1, 9]; **newly sown grass leys** [1, 4]; **rotational grass** [4]; **spring barley** [2-8]
- Black bindweed in **bulb onions** (off-label), **garlic** (off-label), **leeks** (off-label), **newly sown grass leys** (off-label), **salad onions** (off-label), **shallots** (off-label) [9]; **durum wheat**, **forage maize**, **grassland**, **spring barley**, **spring oats**, **spring rye**, **spring wheat**, **triticale**, **winter barley**, **winter oats**, **winter rye**, **winter wheat** [1-9]; **grass seed crops**, **spelt** [1, 9]; **newly sown grass leys** [1, 4]; **rotational grass** [4]
- Black nightshade in **forage maize** [1, 9]
- Chickweed in **bulb onions** (off-label), **garlic** (off-label), **leeks** (off-label), **newly sown grass leys** (off-label), **poppies for morphine production** (off-label), **salad onions** (off-label), **shallots** (off-label) [9]; **durum wheat**, **forage maize**, **grassland**, **spring barley**, **spring oats**, **spring rye**, **spring wheat**, **triticale**, **winter barley**, **winter oats**, **winter rye**, **winter wheat** [1-9]; **grass seed crops**, **spelt** [1, 9]; **newly sown grass leys** [1, 4]; **rotational grass** [4]
- Cleavers in **bulb onions** (off-label), **garlic** (off-label), **leeks** (off-label), **newly sown grass leys** (off-label), **poppies for morphine production** (off-label), **salad onions** (off-label), **shallots** (off-label) [9]; **durum wheat**, **forage maize**, **grassland**, **spring barley**, **spring oats**, **spring rye**, **spring wheat**, **triticale**, **winter barley**, **winter oats**, **winter rye**, **winter wheat** [1-9]; **grass seed crops**, **spelt** [1, 9]; **newly sown grass leys** [1, 4]; **rotational grass** [4]
- Corn spurrey in **bulb onions** (off-label), **garlic** (off-label), **leeks** (off-label), **newly sown grass leys** (off-label), **salad onions** (off-label), **shallots** (off-label) [9]
- Docks in **durum wheat**, **forage maize**, **spring barley**, **spring oats**, **spring rye**, **spring wheat**, **triticale**, **winter barley**, **winter oats**, **winter rye**, **winter wheat** [2-8]; **grassland** [2-7]; **newly sown grass leys**, **rotational grass** [4]
- Forget-me-not in **bulb onions** (off-label), **garlic** (off-label), **leeks** (off-label), **newly sown grass leys** (off-label), **salad onions** (off-label), **shallots** (off-label) [9]; **durum wheat**, **forage maize**, **grassland**, **spring barley**, **spring oats**, **spring rye**, **spring wheat**, **triticale**, **winter barley**, **winter oats**, **winter rye**, **winter wheat** [1-9]; **grass seed crops**, **spelt** [1, 9]; **newly sown grass leys** [1, 4]; **rotational grass** [4]
- Fumitory in **bulb onions** (off-label), **garlic** (off-label), **leeks** (off-label), **newly sown grass leys** (off-label), **salad onions** (off-label), **shallots** (off-label) [9]
- Groundsel in **bulb onions** (off-label), **garlic** (off-label), **leeks** (off-label), **newly sown grass leys** (off-label), **salad onions** (off-label), **shallots** (off-label) [9]

FOR FULL CONDITIONS OF USE ALWAYS READ THE PRODUCT LABEL

fluroxypyr

- Hemp-nettle in **bulb onions** *(off-label)*, **garlic** *(off-label)*, **leeks** *(off-label)*, **newly sown grass leys** *(off-label)*, **salad onions** *(off-label)*, **shallots** *(off-label)* [9]; **durum wheat**, **forage maize**, **grassland**, **spring barley**, **spring oats**, **spring rye**, **spring wheat**, **triticale**, **winter barley**, **winter oats**, **winter rye**, **winter wheat** [1-9]; **grass seed crops**, **spelt** [1, 9]; **newly sown grass leys** [1, 4]; **rotational grass** [4]
- Knotgrass in **bulb onions** *(off-label)*, **garlic** *(off-label)*, **leeks** *(off-label)*, **newly sown grass leys** *(off-label)*, **salad onions** *(off-label)*, **shallots** *(off-label)* [9]
- Pale persicaria in **bulb onions** *(off-label)*, **garlic** *(off-label)*, **leeks** *(off-label)*, **newly sown grass leys** *(off-label)*, **salad onions** *(off-label)*, **shallots** *(off-label)* [9]
- Poppies in **poppies for morphine production** *(off-label)* [9]
- Red dead-nettle in **bulb onions** *(off-label)*, **garlic** *(off-label)*, **leeks** *(off-label)*, **newly sown grass leys** *(off-label)*, **salad onions** *(off-label)*, **shallots** *(off-label)* [9]
- Redshank in **bulb onions** *(off-label)*, **garlic** *(off-label)*, **leeks** *(off-label)*, **newly sown grass leys** *(off-label)*, **salad onions** *(off-label)*, **shallots** *(off-label)* [9]
- Speedwells in **bulb onions** *(off-label)*, **garlic** *(off-label)*, **leeks** *(off-label)*, **newly sown grass leys** *(off-label)*, **salad onions** *(off-label)*, **shallots** *(off-label)* [9]
- Volunteer potatoes in **bulb onions** *(off-label)*, **garlic** *(off-label)*, **leeks** *(off-label)*, **newly sown grass leys** *(off-label)*, **salad onions** *(off-label)*, **shallots** *(off-label)* [9]; **winter barley**, **winter wheat** [2-8]

Extension of Authorisation for Minor Use (EAMUs)
- **bulb onions** *20142851* [9]
- **garlic** *20142851* [9]
- **leeks** *20142851* [9]
- **newly sown grass leys** *20142851* [9]
- **poppies for morphine production** *20161040* [9]
- **salad onions** *20142851* [9]
- **shallots** *20142851* [9]

Approval information
- Fluroxypyr included in Annex I under EC Regulation 1107/2009
- Accepted by BBPA for use on malting barley

Efficacy guidance
- Best results achieved under good growing conditions in a strongly competing crop
- A number of tank mixtures with other herbicides are recommended for use in autumn and spring to extend range of species controlled. See label for details
- Spray is rainfast in 1 h

Restrictions
- Maximum number of treatments 1 per crop or yr or maximum total dose equivalent to one full dose treatment
- Do not apply in any tank-mix on triticale or forage maize
- Do not use on crops undersown with clovers or other legumes
- Do not treat crops suffering stress caused by any factor
- Do not roll or harrow for 7 d before or after treatment
- Do not spray if frost imminent
- Straw from treated crops must not be returned directly to the soil but must be removed and used only for livestock bedding

Crop-specific information
- Latest use: before flag leaf sheath opening (GS 47) for winter wheat and barley; before flag leaf sheath extending (GS 41) for spring wheat and barley; before second node detectable (GS 32) for oats, rye, triticale and durum wheat; before 7 leaves unfolded and before buttress roots appear for maize
- Apply to new leys from 3 expanded leaf stage
- Timing varies in tank mixtures. See label for details
- Crops undersown with grass may be sprayed provided grasses are tillering

SEE SECTION 3 FOR PRODUCTS ALSO REGISTERED

Pesticide Profiles

Following crops guidance
- Clovers, peas, beans and other legumes must not be sown for 12 mth following treatment at the highest dose
- In the event of crop failure, spring cereals, spring oilseed rape, maize, onions, poppies and new leys may be sown 5 weeks after application with no requirements for soil cultivation [9]

Environmental safety
- Dangerous for the environment
- Very toxic to aquatic organisms
- Flammable
- Keep livestock out of treated areas for at least 3 d following treatment and until poisonous weeds, such as ragwort, have died down and become unpalatable
- Wash spray equipment thoroughly with water and detergent immediately after use. Traces of product can damage susceptible plants sprayed later
- LERAP Category B [1-3, 5, 6, 8, 9]

Hazard classification and safety precautions
Hazard Harmful, Flammable [2-8]; Dangerous for the environment [1-9]; Flammable liquid and vapour [2, 4, 6, 7]; Very toxic to aquatic organisms [3-5, 8]
Transport code 3 [1-4, 6-9]; 9 [5]
Packaging group III
UN Number 1993
Risk phrases H304, H336 [2-8]; H315 [2, 3, 5-7]; H317 [1, 2, 4, 6, 7, 9]; H318 [2, 6]; H319 [1, 3-5, 7-9]; H335 [7, 8]
Operator protection A, H [1-9]; C [1-6, 8, 9]; U04c, U12 [1, 9]; U05a [1-7, 9]; U08, U19a, U20b [1-9]; U11 [1-6, 9]; U14 [2-7]
Environmental protection E06a [2] (3 d); E07a [3, 5, 6, 8]; E07c [7] (14 d); E07f [4]; E15a [8]; E15b [1-7, 9]; E16a [1-3, 5, 6, 8, 9]; E34 [1-3, 5-7, 9]; E38, H410 [2-8]; H411 [1, 9]
Storage and disposal D01, D02 [1-7, 9]; D05 [1-3, 5-7, 9]; D09a [1-9]; D10a [7]; D10b [1-6, 8, 9]; D12a [2-8]; D12b [1, 9]
Medical advice M03 [2-7]; M05b [2-8]

257   fluroxypyr + halauxifen-methyl

A herbicide mixture for use in cereals
HRAC mode of action code: O

Products
Pixxaro EC   Dow   280:12 g/l   EC   17545

Uses
- Chickweed in *durum wheat, rye, spelt, spring barley, spring wheat, triticale, winter barley, winter wheat*
- Cleavers in *durum wheat, rye, spelt, spring barley, spring wheat, triticale, winter barley, winter wheat*
- Crane's-bill in *durum wheat, rye, spelt, spring barley, spring wheat, triticale, winter barley, winter wheat*
- Fat hen in *durum wheat, rye, spelt, spring barley, spring wheat, triticale, winter barley, winter wheat*
- Fumitory in *durum wheat, rye, spelt, spring barley, spring wheat, triticale, winter barley, winter wheat*
- Poppies in *durum wheat, rye, spelt, spring barley, spring wheat, triticale, winter barley, winter wheat*

Approval information
- Fluroxypyr and halauxifen-methyl included in Annex I under EC Regulation 1107/2009

Efficacy guidance
- Rainfast one hour after application

FOR FULL CONDITIONS OF USE ALWAYS READ THE PRODUCT LABEL

Following crops guidance
- After application of 0.5 l/ha up to BBCH 45, white mustard, pea, winter oilseed rape, phacelia, ryegrass, winter wheat and winter barley may be sown in the autumn of the same year after the cereal harvest; spring wheat, spring barley, spring oat, ryegrass, maize, spring oilseed rape, sugar beet, potato, field bean, tomato, sunflower, onion, clover, and pea may be sown in the spring of the year after the cereal harvest. Ploughing is recommended prior to drilling maize, alfalfa, broad bean, field bean, soybean and clover in the same year as application.
- In the event of crop failure spring wheat, spring barley, spring oats and ryegrass may be sown one month after application with no cultivation. Maize, spring oilseed rape, peas, broad beans and field beans may be sown two months after application and ploughing. Three months after application sorghum may be sown.

Environmental safety
- LERAP Category B

Hazard classification and safety precautions
Transport code 9
Packaging group III
UN Number 3082
Risk phrases H317, H319, H335
Operator protection A, C, H; U05a, U08, U19a, U20c
Environmental protection E15b, E16a, E34, H410
Storage and disposal D09a, D10c

258   fluroxypyr + triclopyr

A foliar acting herbicide for docks in grassland
HRAC mode of action code: O + O

See also triclopyr

Products

1	Doxstar Pro	Dow	150:150 g/l	EC	15664
2	Pivotal	Dow	150:150 g/l	EC	16943

Uses
- Docks in **amenity grassland**, **grassland** [2]; **established grassland** [1]

Approval information
- Fluroxypyr and triclopyr included in Annex I under EC Regulation 1107/2009

Efficacy guidance
- Seedling docks in established grass only are controlled up to 50 mm diameter. Apply in spring or autumn or, at lower dose, in spring and autumn on docks up to 200 mm. A second application in the subsequent yr may be needed
- Allow 2-3 wk after cutting or grazing to allow sufficient regrowth of docks to occur before spraying
- Control may be reduced if rain falls within 2 h of application
- To allow maximum translocation to the roots of docks do not cut grass for 28 d after spraying

Restrictions
- Maximum total dose equivalent to one full dose treatment
- Do not roll or harrow for 10 d before or 7 d after spraying
- Do not spray in drought, very hot or very cold weather

Crop-specific information
- Latest use: 7 d before grazing or harvest of grass
- Grass less than one yr old may be treated at half dose from the third leaf visible stage
- Clover will be killed or severely checked by treatment

SEE SECTION 3 FOR PRODUCTS ALSO REGISTERED

Pesticide Profiles

Following crops guidance
- Do not sow kale, turnips, swedes or grass mixtures containing clover by direct drilling or minimum cultivation techniques within 6 wk of application

Environmental safety
- Dangerous for the environment
- Toxic to aquatic organisms
- Keep livestock out of treated areas for at least 7 d following treatment and until poisonous weeds, such as ragwort, have died down and become unpalatable
- Do not allow drift to come into contact with crops, amenity plantings, gardens, ponds, lakes or watercourses
- Wash spray equipment thoroughly with water and detergent immediately after use. Traces of product can damage susceptible plants sprayed later

Hazard classification and safety precautions
Hazard Irritant, Dangerous for the environment
Transport code 3
Packaging group III
UN Number 1993
Risk phrases H317
Operator protection A; U14
Environmental protection E38, H410
Storage and disposal D01, D02, D05, D12a
Medical advice M05a

259 flurtamone

A carotenoid synthesis inhibitor available only in mixtures
HRAC mode of action code: F1

See also diflufenican + flufenacet + flurtamone
diflufenican + flurtamone

260 flutolanil

An carboxamide fungicide for treatment of potato seed tubers
FRAC mode of action code: 7

Products
1	Rhino	Certis	460 g/l	SC	14311
2	Rhino DS	Certis	6% w/w	DS	12763

Uses
- Black scurf in **potatoes** *(tuber treatment)*
- Stem canker in **potatoes** *(tuber treatment)*

Approval information
- Flutolanil included in Annex I under EC Regulation 1107/2009

Efficacy guidance
- Apply to clean tubers before chitting, prior to planting, or at planting
- Apply flowable concentrate through canopied, hydraulic or spinning disc equipment (with or without electrostatics) mounted on a rolling conveyor or table [1]
- Flowable concentrate may be diluted with water up to 2.0 l per tonne to improve tuber coverage. Disease in areas not covered by spray will not be controlled [1]
- Dry powder may be applied via an on-planter applicator [2]

Restrictions
- Maximum number of treatments 1 per batch of seed tubers
- Check with processor before use on crops for processing

FOR FULL CONDITIONS OF USE ALWAYS READ THE PRODUCT LABEL

flutriafol

Crop-specific information
- Latest use: at planting
- Seed tubers should be of good quality and free from bacterial rots, physical damage or virus infection, and should not be sprouted to such an extent that mechanical damage to the shoots will occur during treatment or planting

Environmental safety
- Harmful to aquatic organisms

Hazard classification and safety precautions
Hazard Irritant
Transport code 9 [1]
Packaging group III [1]
UN Number 3082 [1]; N/C [2]
Risk phrases H319, R53a [2]; R43, R51 [1]
Operator protection A, H [1, 2]; C [2]; U04a, U05a, U19a, U20a [1, 2]; U14 [1]
Environmental protection E15a, E34, E38 [1, 2]; H412 [2]
Storage and disposal D01, D02, D09a, D12a [1, 2]; D05, D10a [1]; D11a [2]
Treated seed S01, S03, S04a, S05 [1, 2]; S02 [2]
Medical advice M03

261 flutriafol

A broad-spectrum triazole fungicide for cereals
FRAC mode of action code: 3

See also chlorothalonil + flutriafol
fludioxonil + flutriafol

Products
1	Consul	Headland	125 g/l	SC	12976
2	Pointer	Headland	125 g/l	SC	12975

Uses
- Brown rust in **spring barley, winter barley, winter wheat**
- Powdery mildew in **spring barley, winter barley, winter wheat**
- Rhynchosporium in **spring barley, winter barley**
- Septoria leaf blotch in **winter wheat**
- Yellow rust in **spring barley, winter barley, winter wheat**

Approval information
- Flutriafol included in Annex I under EC Regulation 1107/2009
- Accepted by BBPA for use on malting barley

Efficacy guidance
- Best results obtained from treatment in early stages of disease development. See label for detailed guidance on spray timing for specific diseases and the need for repeat treatments
- Tank mix options available on the label to broaden activity spectrum
- Good spray coverage essential for optimum performance
- Flutriafol is a DMI fungicide. Resistance to some DMI fungicides has been identified in Septoria leaf blotch which may seriously affect performance of some products. For further advice contact a specialist advisor and visit the Fungicide Resistance Action Group (FRAG)-UK website

Restrictions
- Maximum number of treatments 2 per crop (including other products containing flutriafol)

Crop-specific information
- Latest use: before early grain milky ripe stage (GS 73)
- Flag leaf tip scorch on wheat caused by stress may be increased by fungicide treatment

Environmental safety
- Harmful to aquatic organisms

SEE SECTION 3 FOR PRODUCTS ALSO REGISTERED

Pesticide Profiles

- Harmful to fish or other aquatic life. Do not contaminate surface waters or ditches with chemical or used container

Hazard classification and safety precautions
  Hazard Harmful
  UN Number N/C
  Risk phrases R43 [1]; R48, R52 [1, 2]
  Operator protection A, H; U05a, U09a, U19a, U20b
  Environmental protection E13c, E38
  Storage and disposal D01, D02, D09a, D10c

262 fluxapyroxad

Also known as Xemium, it is an SDHI fungicide for use in cereals.
FRAC mode of action code: 7

See also epoxiconazole + fluxapyroxad
  epoxiconazole + fluxapyroxad + pyraclostrobin
  fluxapyroxad + metconazole

Products
  Imtrex   BASF   62.5 g/l   EC   17108

Uses
- Brown rust in **durum wheat**, **rye**, **spring barley**, **spring wheat**, **triticale**, **winter barley**, **winter wheat**
- Crown rust in **winter oats**
- Net blotch in **spring barley**, **winter barley**
- Powdery mildew in **durum wheat** (reduction), **rye** (reduction), **spring wheat** (reduction), **triticale** (reduction), **winter barley** (reduction), **winter oats** (reduction), **winter wheat** (reduction)
- Rhynchosporium in **rye**, **spring barley**, **winter barley**
- Septoria leaf blotch in **durum wheat**, **spring wheat**, **triticale**, **winter wheat**
- Tan spot in **durum wheat** (reduction), **spring wheat** (reduction), **winter wheat** (reduction)
- Yellow rust in **durum wheat** (moderate control), **rye** (moderate control), **spring barley** (moderate control), **spring wheat** (moderate control), **triticale** (moderate control), **winter barley** (moderate control), **winter wheat** (moderate control)

Approval information
- Fluxapyroxad included in Annex I under EC Regulation 1107/2009
- Accepted by BBPA for use on malting barley up to GS 45 only (max. dose, 1 litre/ha)

Restrictions
- Do not apply more that two foliar applications of products containing SDHI fungicides to any cereal crop.

Following crops guidance
- Only cereals, cabbages, carrots, chicory, clover, dwarf french beans, field beans, leeks, lettuce, linseed, maize, oats, oilseed rape, onions, peas, potatoes, radishes, ryegrass, soyabean, spinach, sunflowers or sugar beet may be sown as following crops after treatment

Hazard classification and safety precautions
  Hazard Harmful, Dangerous for the environment, Harmful if inhaled
  Transport code 9
  Packaging group III
  UN Number 3082
  Risk phrases H319, H351
  Operator protection A, H; U05a, U20b
  Environmental protection E15b, E34, E38, H411
  Storage and disposal D01, D02, D05, D09a, D10c, D12a
  Medical advice M05a

FOR FULL CONDITIONS OF USE ALWAYS READ THE PRODUCT LABEL

263 fluxapyroxad + metconazole

An SDHI and triazole mixture for disease control in cereals
FRAC mode of action code: 3 + 7

Products

1	Clayton Tardis	Clayton	62.5:45 g/l	EC	17459
2	Librax	BASF	62.5:45 g/l	EC	17107

Uses
- Brown rust in *durum wheat, rye, spring barley, spring wheat, triticale, winter barley, winter wheat*
- Cladosporium in *durum wheat* (moderate control), *spring wheat* (moderate control), *winter wheat* (moderate control)
- Disease control in *durum wheat, rye, spring barley, spring wheat, triticale, winter barley, winter wheat*
- Eyespot in *durum wheat* (good reduction), *rye* (good reduction), *spring wheat* (good reduction), *triticale* (good reduction), *winter wheat* (good reduction)
- Fusarium ear blight in *durum wheat* (good reduction), *spring wheat* (good reduction), *winter wheat* (good reduction)
- Net blotch in *spring barley, winter barley*
- Powdery mildew in *durum wheat* (moderate control), *rye* (moderate control), *spring barley* (moderate control), *spring wheat* (moderate control), *triticale* (moderate control), *winter barley* (moderate control), *winter wheat* (moderate control)
- Ramularia leaf spots in *spring barley, winter barley*
- Rhynchosporium in *rye, spring barley, winter barley*
- Septoria leaf blotch in *durum wheat, spring wheat, triticale, winter wheat*
- Tan spot in *durum wheat* (good reduction), *spring wheat* (good reduction), *winter wheat* (good reduction)
- Yellow rust in *durum wheat, rye, spring barley, spring wheat, triticale, winter barley, winter wheat*

Approval information
- Fluxapyroxad and metconazole included in Annex I under EC Regulation 1107/2009

Efficacy guidance
- Metconazole is a DMI fungicide. Resistance to some DMI fungicides has been identified in Septoria leaf blotch which may seriously affect performance of some products. For further advice contact a specialist advisor and visit the Fungicide Resistance Action Group (FRAG)-UK website

Restrictions
- Do not apply more that two foliar applications of products containing SDHI fungicides to any cereal crop.

Following crops guidance
- Only cereals, cabbages, carrots, clover, dwarf french beans, field beans, lettuce, maize, oats, oilseed rape, onions, peas, potatoes, ryegrass, sunflowers or sugar beet may be sown as following crops after treatment

Environmental safety
- LERAP Category B

Hazard classification and safety precautions
  Hazard Harmful, Dangerous for the environment, Harmful if inhaled, Very toxic to aquatic organisms
  Transport code 9
  Packaging group III
  UN Number 3082
  Risk phrases H317, H319, H351, H361
  Operator protection A, C, H; U05a, U20b

SEE SECTION 3 FOR PRODUCTS ALSO REGISTERED

Pesticide Profiles

Environmental protection E15b, E16a, E34, E38, H410
Storage and disposal D01, D02, D09a, D10c, D12a
Medical advice M03

264 fluxapyroxad + pyraclostrobin

A fungicide mixture for disease control in cereals
FRAC mode of action code: 7 + 11

Products

1	Priaxor EC	BASF	75:150 g/l	EC	17371
2	Serpent	BASF	75:150 g/l	EC	17427

Uses
- Brown rust in *durum wheat*, *rye*, *spring wheat*, *triticale*, *winter wheat* [1]; *spring barley*, *winter barley* [1, 2]
- Crown rust in *spring oats*, *winter oats* [1]
- Glume blotch in *durum wheat* (moderate control), *spring wheat* (moderate control), *winter wheat* (moderate control) [1]
- Net blotch in *spring barley*, *winter barley* [1, 2]
- Powdery mildew in *durum wheat* (moderate control), *rye* (moderate control), *spring oats* (moderate control), *spring wheat* (moderate control), *triticale* (moderate control), *winter oats* (moderate control), *winter wheat* (moderate control) [1]; *spring barley* (moderate control), *winter barley* (moderate control) [1, 2]
- Ramularia leaf spots in *spring barley*, *winter barley* [1, 2]
- Rhynchosporium in *rye* [1]; *spring barley*, *winter barley* [1, 2]
- Septoria leaf blotch in *durum wheat*, *spring wheat*, *triticale*, *winter wheat* [1]
- Tan spot in *durum wheat*, *spring wheat*, *winter wheat* [1]
- Yellow rust in *durum wheat*, *rye*, *spring wheat*, *triticale*, *winter wheat* [1]; *spring barley*, *winter barley* [1, 2]

Approval information
- Fluxapyroxad and pyraclostrobin included in Annex I under EC Regulation 1107/2009

Environmental safety
- LERAP Category B

Hazard classification and safety precautions
Hazard Harmful if swallowed, Harmful if inhaled, Very toxic to aquatic organisms
Transport code 9
Packaging group III
UN Number 3082
Risk phrases H351
Operator protection A, H; U05a, U20a
Environmental protection E15b, E16a, H410
Storage and disposal D01, D02, D10c

265 folpet

A multi-site protectant fungicide for disease control in wheat and barley
FRAC mode of action code: M4

See also epoxiconazole + folpet

Products

1	Arizona	Adama	500 g/l	SC	15318
2	Phoenix	Adama	500 g/l	SC	15259

Uses
- Rhynchosporium in *spring barley* (reduction), *winter barley* (reduction)

FOR FULL CONDITIONS OF USE ALWAYS READ THE PRODUCT LABEL

- Septoria leaf blotch in **spring wheat** *(reduction)*, **winter wheat** *(reduction)*

Approval information
- Folpet included in Annex I under EC Regulation 1107/2009
- Accepted by BBPA for use on cereals

Efficacy guidance
- Folpet is a protectant fungicide and so the first application must be made before the disease becomes established in the crop. A second application should be timed to protect new growth.
- Folpet only has contact activity and so good coverage of the target foliage is essential for good activity.

Environmental safety
- LERAP Category B

Hazard classification and safety precautions
  Hazard Harmful, Dangerous for the environment
  Transport code 9
  Packaging group III
  UN Number 3082
  Risk phrases R40, R50, R53a
  Operator protection A, H; U05a, U08, U11, U14, U15, U20a
  Environmental protection E15b, E16a, E34, E38
  Storage and disposal D01, D02, D09a, D12a
  Medical advice M03

266   foramsulfuron + iodosulfuron-methyl-sodium

A sulfonylurea herbicide mixture for weed control in forage and grain maize.
HRAC mode of action code: B

See also iodosulfuron-methyl-sodium

Products
    Maister WG   Bayer CropScience   0.3:0.01% w/w   SG   16116

Uses
- Annual dicotyledons in **forage maize, grain maize**
- Annual meadow grass in **forage maize, grain maize**
- Black nightshade in **forage maize, grain maize**
- Chickweed in **forage maize, grain maize**
- Cockspur grass in **forage maize, grain maize**
- Couch in **forage maize, grain maize**
- Fat hen in **forage maize, grain maize**
- Knotgrass in **forage maize, grain maize**
- Mayweeds in **forage maize, grain maize**
- Shepherd's purse in **forage maize, grain maize**

Approval information
- Foramsulfuron and iodosulfuron-methyl-sodium included in Annex 1 under EC Regulation 1107/2009

Efficacy guidance
- Do not spray if rain is imminent or if temperature is above 25°C.
- Do not use in a non-CRD approved tank-mixture or sequence with another ALS or sulfonyl urea herbicide.
- Do not use on crops suffering from stress.
- Do not use on crops undersown with grass or a broad-leaved crop.

Environmental safety
- Buffer zone requirement 11m [1]
- LERAP Category B

SEE SECTION 3 FOR PRODUCTS ALSO REGISTERED

Pesticide Profiles

Hazard classification and safety precautions
Hazard Irritant, Dangerous for the environment, Very toxic to aquatic organisms
Transport code 9
Packaging group III
UN Number 3077
Risk phrases H317
Operator protection A, D, H; U05a, U14
Environmental protection E15b, E16a, E38, H410
Storage and disposal D01, D02, D10a, D12a

267   fosetyl-aluminium + propamocarb hydrochloride

A systemic and protectant fungicide mixture for use in horticulture
FRAC mode of action code: 33 + 28

See also propamocarb hydrochloride

Products

1	Avatar	Everris Ltd	310:530 g/l	SL	16608
2	Pan Cradle	Pan Agriculture	310:530 g/l	SL	15923
3	Previcur Energy	Bayer CropScience	310:530 g/l	SL	15367

Uses
- Damping off in **herbs (see appendix 6)** *(off-label)*, **protected aubergines** *(off-label)*, **protected chilli peppers** *(off-label)*, **protected courgettes** *(off-label)*, **protected forest nurseries** *(off-label)*, **protected gherkins** *(off-label)*, **protected hops** *(off-label)*, **protected marrows** *(off-label)*, **protected pumpkins** *(off-label)*, **protected soft fruit** *(off-label)*, **protected squashes** *(off-label)*, **protected sweet peppers** *(off-label)*, **protected top fruit** *(off-label)*, **protected watermelon** *(off-label)* [3]; **protected broccoli, protected brussels sprouts, protected cabbages, protected calabrese, protected cauliflowers, protected chinese cabbage, protected collards, protected kale, protected radishes** [2, 3]; **protected melons** [1-3]
- Downy mildew in **herbs (see appendix 6)** *(off-label)*, **protected forest nurseries** *(off-label)*, **protected hops** *(off-label)*, **protected soft fruit** *(off-label)*, **protected top fruit** *(off-label)*, **spinach** *(off-label)* [3]; **lettuce, protected broccoli, protected brussels sprouts, protected cabbages, protected calabrese, protected cauliflowers, protected chinese cabbage, protected collards, protected kale, protected radishes** [2, 3]; **ornamental plant production** *(off-label)*, **protected ornamentals** *(off-label)* [1, 3]; **protected lettuce, protected melons** [1-3]
- Pythium in **lettuce** [2, 3]; **protected cucumbers, protected lettuce, protected tomatoes** [1-3]; **protected ornamentals** *(off-label)* [3]

Extension of Authorisation for Minor Use (EAMUs)
- *herbs (see appendix 6) 20130983* [3]
- *ornamental plant production 20150100* [1], *20131845* [3]
- *protected aubergines 20111553* [3]
- *protected chilli peppers 20111553* [3]
- *protected courgettes 20111556* [3]
- *protected forest nurseries 20122045* [3]
- *protected gherkins 20111556* [3]
- *protected hops 20122045* [3]
- *protected marrows 20111556* [3]
- *protected ornamentals 20150100* [1], *20111557* [3], *20131845* [3]
- *protected pumpkins 20111556* [3]
- *protected soft fruit 20122045* [3]
- *protected squashes 20111556* [3]
- *protected sweet peppers 20111553* [3]
- *protected top fruit 20122045* [3]
- *protected watermelon 20111556* [3]
- *spinach 20112452* [3]

FOR FULL CONDITIONS OF USE ALWAYS READ THE PRODUCT LABEL

Approval information
- Fosetyl-aluminium and propamocarb hydrochloride included in Annex I under EC Regulation 1107/2009

Hazard classification and safety precautions
Hazard Irritant
UN Number N/C
Risk phrases H317 [3]; R43 [1, 2]
Operator protection A, H, M; U04a, U05a, U08, U14, U19a, U20b
Environmental protection E15a
Storage and disposal D09a, D11a
Medical advice M03

268 fosthiazate

An organophosphorus contact nematicide for potatoes
IRAC mode of action code: 1B

Products
Nemathorin 10G   Syngenta   10% w/w   FG   11003

Uses
- Nematodes in **hops** (off-label), **ornamental plant production** (off-label), **soft fruit** (off-label)
- Potato cyst nematode in **potatoes**
- Spraing vectors in **potatoes** (reduction)
- Wireworm in **potatoes** (reduction)

Extension of Authorisation for Minor Use (EAMUs)
- **hops** 20082912
- **ornamental plant production** 20082912
- **soft fruit** 20082912

Approval information
- Fosthiazate included in Annex I under EC Regulation 1107/2009
- In 2006 CRD required that all products containing this active ingredient should carry the following warning in the main area of the container label: "Fosthiazate is an anticholinesterase organophosphate. Handle with care"

Efficacy guidance
- Application best achieved using equipment such as Horstine Farmery Microband Applicator, Matco or Stocks Micrometer applicators together with a rear mounted powered rotary cultivator
- Granules must not become wet or damp before use.
- For optimum efficacy incorporate evenly in to top 20 cm of soil.
- apply as close as possible to time of planting.

Restrictions
- Contains an anticholinesterase organophosphorus compound. Do not use if under medical advice not to work with such compounds
- Maximum number of treatments 1 per crop
- Product must only be applied using tractor-mounted/drawn direct placement machinery. Do not use air assisted broadcast machinery other than that referenced on the product label
- Do not allow granules to stand overnight in the application hopper
- Do not apply more than once every four years on the same area of land
- Do not use on crops to be harvested less than 17 wk after treatment
- Consult before using on crops intended for processing

Crop-specific information
- Latest use: at planting
- HI 17 wk for potatoes

Environmental safety
- Dangerous for the environment

SEE SECTION 3 FOR PRODUCTS ALSO REGISTERED

Pesticide Profiles

- Toxic to aquatic organisms
- Dangerous to game, wild birds and animals
- Dangerous to livestock. Keep all livestock out of treated areas for at least 13 wk
- Incorporation to 10-15 cm and ridging up of treated soil must be carried out immediately after application. Powered rotary cultivators are preferred implements for incorporation but discs, power, spring tine or Dutch harrows may be used provided two passes are made at right angles
- To protect birds and wild mammals remove spillages
- Failure completely to bury granules immediately after application is hazardous to wildlife
- To protect groundwater do not apply any product containing fosthiazate more than once every four yr

Hazard classification and safety precautions
Hazard Harmful, Dangerous for the environment
Transport code 9
Packaging group III
UN Number 3077
Risk phrases R22a, R43, R51, R53a
Operator protection A, E, G, H, K, M; U02a, U04a, U05a, U09a, U13, U14, U19a, U20a
Environmental protection E06b (13 wk); E15b, E34, E38
Storage and disposal D01, D02, D05, D09a, D11a, D14
Medical advice M01, M03, M05a

269 garlic extract

A naturally occuring nematicide and animal repellent

Products
NEMguard DE   Certis   45% w/w   GR   16749

Uses
- Free-living nematodes in **fodder beet** *(off-label)*, **red beet** *(off-label)*
- Nematodes in **carrots**, **parsnips**
- Root nematodes in **bulb onions** *(off-label)*, **garlic** *(off-label)*, **leeks** *(off-label)*, **shallots** *(off-label)*
- Root-knot nematodes in **bulb onions** *(off-label)*, **garlic** *(off-label)*, **leeks** *(off-label)*, **shallots** *(off-label)*
- Stem and bulb nematodes in **bulb onions** *(off-label)*, **garlic** *(off-label)*, **leeks** *(off-label)*, **shallots** *(off-label)*

Extension of Authorisation for Minor Use (EAMUs)
- *bulb onions* 20151838
- *fodder beet* 20151841
- *garlic* 20151838
- *leeks* 20151838
- *red beet* 20151841
- *shallots* 20151838

Approval information
- Garlic extract included in Annex 1 under EC Regulation 1107/2009

Hazard classification and safety precautions
Hazard Irritant, Dangerous for the environment
UN Number N/C
Risk phrases R36, R38, R43, R51, R53a
Operator protection A, H; U11, U14
Environmental protection E38

FOR FULL CONDITIONS OF USE ALWAYS READ THE PRODUCT LABEL

270 gibberellins

A plant growth regulator for use in top fruit and grassland

Products

1	Gibb 3	Globachem	10% w/w	TB	17013
2	Gibb Plus	Globachem	10 g/l	SL	17251
3	Novagib	Fine	10 g/l	SL	08954
4	Regulex 10 SG	Interfarm	10% w/w	SG	17158
5	Smartgrass	Interfarm	40% w/w	SG	15628

Uses
- Growth regulation in **grassland** [5]; **pears** *(off-label)* [3]
- Improved germination in **nothofagus** *(do not exceed 2.5 g per 5 litres when used as a seed soak)* [4]
- Increasing fruit set in **pears** [1]
- Reducing fruit russeting in **apples** [2, 3]
- Russet reduction in **apples**, **pears** [4]

Extension of Authorisation for Minor Use (EAMUs)
- **pears** 20012756 [3]

Approval information
- Gibberellins included in Annex 1 under EC Regulation 1107/2009

Efficacy guidance
- For optimum results on apples spray under humid, slow drying conditions and ensure good spray cover [3]
- Treat apples and pears immediately after completion of petal fall and repeat as directed on the label [3]
- Nothofagus seeds should be stirred into the solution, soaked for 24 hr and then sown immediately
- Fruit set in pears can be improved when blossom is spare, setting is poor or where frost has killed many flowers [1]

Restrictions
- Maximum total dose varies with crop and product. Check label
- Prepared spray solutions are unstable. Do not leave in the sprayer during meal breaks or overnight
- Return bloom may be reduced in yr following treatment
- Consult processor before treating crops grown for processing [1, 3]
- Avoid storage at temperatures above 32°C [1]

Crop-specific information
- HI zero for apples [3]
- Apply to apples at completion of petal fall and repeat 3 or 4 times at 7-10 d intervals. Number of sprays and spray interval depend on weather conditions and dose (see labels) [3]
- Good results achieved on apple varieties Cox's Orange Pippin, Discovery, Golden Delicious and Karmijn. For other cultivars test on a small number of trees [3]
- Split treatment on pears allows second spray to be omitted if pollinating conditions become very favourable [1]
- Pear variety Conference usually responds well to treatment; Doyenne du Comice can be variable [1]

Hazard classification and safety precautions
UN Number N/C
Operator protection U05a [4]; U08 [1]; U20b [3]; U20c [1, 2, 4]
Environmental protection E15a [1-4]
Storage and disposal D01 [3, 4]; D02 [4]; D03, D05, D07 [3]; D09a [1-4]; D10b [2-4]

SEE SECTION 3 FOR PRODUCTS ALSO REGISTERED

Pesticide Profiles

271 Gliocladium catenulatum

A fungus that acts as an antagonist against other fungi

Products

Prestop   Everris Ltd   32% w/w   WP   17223

Uses
- Botrytis in **almonds** *(off-label)*, **angelica** *(off-label)*, **apples** *(off-label)*, **apricots** *(off-label)*, **asparagus** *(off-label)*, **baby leaf crops** *(off-label)*, **beans without pods (fresh)** *(off-label)*, **bilberries** *(off-label)*, **blackcurrants** *(off-label)*, **blueberries** *(off-label)*, **broad beans** *(off-label)*, **broccoli** *(off-label)*, **brussels sprouts** *(off-label)*, **bulb vegetables** *(off-label)*, **cabbages** *(off-label)*, **calabrese** *(off-label)*, **cane fruit** *(off-label)*, **carrots** *(off-label)*, **cauliflowers** *(off-label)*, **celeriac** *(off-label)*, **celery (outdoor)** *(off-label)*, **celery leaves** *(off-label)*, **cherries** *(off-label)*, **chervil** *(off-label)*, **chestnuts** *(off-label)*, **chicory root** *(off-label)*, **chives** *(off-label)*, **choi sum** *(off-label)*, **collards** *(off-label)*, **cranberries** *(off-label)*, **cress** *(off-label)*, **dwarf beans** *(off-label)*, **edible flowers** *(off-label)*, **edible podded peas** *(off-label)*, **endives** *(off-label)*, **fennel leaves** *(off-label)*, **florence fennel** *(off-label)*, **french beans** *(off-label)*, **fruiting vegetables** *(off-label)*, **globe artichoke** *(off-label)*, **gooseberries** *(off-label)*, **hazel nuts** *(off-label)*, **herbs (see appendix 6)** *(off-label)*, **hops** *(off-label)*, **horseradish** *(off-label)*, **hyssop** *(off-label)*, **jerusalem artichokes** *(off-label)*, **kale** *(off-label)*, **kohlrabi** *(off-label)*, **lamb's lettuce** *(off-label)*, **land cress** *(off-label)*, **leeks** *(off-label)*, **lentils** *(off-label)*, **lettuce** *(off-label)*, **lovage** *(off-label)*, **marjoram** *(off-label)*, **medlar** *(off-label)*, **mint** *(off-label)*, **nectarines** *(off-label)*, **oregano** *(off-label)*, **oriental cabbage** *(off-label)*, **ornamental plant production** *(off-label)*, **parsley** *(off-label)*, **parsley root** *(off-label)*, **parsnips** *(off-label)*, **peaches** *(off-label)*, **pears** *(off-label)*, **plums** *(off-label)*, **quinces** *(off-label)*, **radishes** *(off-label)*, **red beet** *(off-label)*, **redcurrants** *(off-label)*, **rhubarb** *(off-label)*, **rocket** *(off-label)*, **rosemary** *(off-label)*, **runner beans** *(off-label)*, **sage** *(off-label)*, **salad burnet** *(off-label)*, **salsify** *(off-label)*, **savory** *(off-label)*, **seakale** *(off-label)*, **soya beans** *(off-label)*, **spinach** *(off-label)*, **spinach beet** *(off-label)*, **swedes** *(off-label)*, **sweet ciceley** *(off-label)*, **table grapes** *(off-label)*, **tarragon** *(off-label)*, **thyme** *(off-label)*, **turnips** *(off-label)*, **vining peas** *(off-label)*, **walnuts** *(off-label)*, **wine grapes** *(off-label)*
- Didymella in **all protected edible crops** *(moderate control)*, **all protected non-edible crops** *(moderate control)*, **strawberries** *(moderate control)*
- Fusarium in **all protected edible crops** *(moderate control)*, **all protected non-edible crops** *(moderate control)*, **almonds** *(off-label)*, **angelica** *(off-label)*, **apples** *(off-label)*, **apricots** *(off-label)*, **asparagus** *(off-label)*, **baby leaf crops** *(off-label)*, **beans without pods (fresh)** *(off-label)*, **bilberries** *(off-label)*, **blackcurrants** *(off-label)*, **blueberries** *(off-label)*, **broad beans** *(off-label)*, **broccoli** *(off-label)*, **brussels sprouts** *(off-label)*, **bulb vegetables** *(off-label)*, **cabbages** *(off-label)*, **calabrese** *(off-label)*, **cane fruit** *(off-label)*, **carrots** *(off-label)*, **cauliflowers** *(off-label)*, **celeriac** *(off-label)*, **celery (outdoor)** *(off-label)*, **celery leaves** *(off-label)*, **cherries** *(off-label)*, **chervil** *(off-label)*, **chestnuts** *(off-label)*, **chicory root** *(off-label)*, **chives** *(off-label)*, **choi sum** *(off-label)*, **collards** *(off-label)*, **cranberries** *(off-label)*, **cress** *(off-label)*, **dwarf beans** *(off-label)*, **edible flowers** *(off-label)*, **edible podded peas** *(off-label)*, **endives** *(off-label)*, **fennel leaves** *(off-label)*, **florence fennel** *(off-label)*, **french beans** *(off-label)*, **fruiting vegetables** *(off-label)*, **globe artichoke** *(off-label)*, **gooseberries** *(off-label)*, **hazel nuts** *(off-label)*, **herbs (see appendix 6)** *(off-label)*, **hops** *(off-label)*, **horseradish** *(off-label)*, **hyssop** *(off-label)*, **jerusalem artichokes** *(off-label)*, **kale** *(off-label)*, **kohlrabi** *(off-label)*, **lamb's lettuce** *(off-label)*, **land cress** *(off-label)*, **leeks** *(off-label)*, **lentils** *(off-label)*, **lettuce** *(off-label)*, **lovage** *(off-label)*, **marjoram** *(off-label)*, **medlar** *(off-label)*, **mint** *(off-label)*, **nectarines** *(off-label)*, **oregano** *(off-label)*, **oriental cabbage** *(off-label)*, **ornamental plant production** *(off-label)*, **parsley** *(off-label)*, **parsley root** *(off-label)*, **parsnips** *(off-label)*, **peaches** *(off-label)*, **pears** *(off-label)*, **plums** *(off-label)*, **quinces** *(off-label)*, **radishes** *(off-label)*, **red beet** *(off-label)*, **redcurrants** *(off-label)*, **rhubarb** *(off-label)*, **rocket** *(off-label)*, **rosemary** *(off-label)*, **runner beans** *(off-label)*, **sage** *(off-label)*, **salad burnet** *(off-label)*, **salsify** *(off-label)*, **savory** *(off-label)*, **seakale** *(off-label)*, **soya beans** *(off-label)*, **spinach** *(off-label)*, **spinach beet** *(off-label)*, **strawberries** *(moderate control)*, **swedes** *(off-label)*, **sweet ciceley** *(off-label)*, **table grapes** *(off-label)*, **tarragon** *(off-label)*, **thyme** *(off-label)*, **turnips** *(off-label)*, **vining peas** *(off-label)*, **walnuts** *(off-label)*, **wine grapes** *(off-label)*
- Phytophthora in **all protected edible crops** *(moderate control)*, **all protected non-edible crops** *(moderate control)*, **almonds** *(off-label)*, **angelica** *(off-label)*, **apples** *(off-label)*, **apricots** *(off-*

FOR FULL CONDITIONS OF USE ALWAYS READ THE PRODUCT LABEL

Gliocladium catenulatum

label), **asparagus** (off-label), **baby leaf crops** (off-label), **beans without pods (fresh)** (off-label), **bilberries** (off-label), **blackcurrants** (off-label), **blueberries** (off-label), **broad beans** (off-label), **broccoli** (off-label), **brussels sprouts** (off-label), **bulb vegetables** (off-label), **cabbages** (off-label), **calabrese** (off-label), **cane fruit** (off-label), **carrots** (off-label), **cauliflowers** (off-label), **celeriac** (off-label), **celery (outdoor)** (off-label), **celery leaves** (off-label), **cherries** (off-label), **chervil** (off-label), **chestnuts** (off-label), **chicory root** (off-label), **chives** (off-label), **choi sum** (off-label), **collards** (off-label), **cranberries** (off-label), **cress** (off-label), **dwarf beans** (off-label), **edible flowers** (off-label), **edible podded peas** (off-label), **endives** (off-label), **fennel leaves** (off-label), **florence fennel** (off-label), **french beans** (off-label), **fruiting vegetables** (off-label), **globe artichoke** (off-label), **gooseberries** (off-label), **hazel nuts** (off-label), **herbs (see appendix 6)** (off-label), **hops** (off-label), **horseradish** (off-label), **hyssop** (off-label), **jerusalem artichokes** (off-label), **kale** (off-label), **kohlrabi** (off-label), **lamb's lettuce** (off-label), **land cress** (off-label), **leeks** (off-label), **lentils** (off-label), **lettuce** (off-label), **lovage** (off-label), **marjoram** (off-label), **medlar** (off-label), **mint** (off-label), **nectarines** (off-label), **oregano** (off-label), **oriental cabbage** (off-label), **ornamental plant production** (off-label), **parsley** (off-label), **parsley root** (off-label), **parsnips** (off-label), **peaches** (off-label), **pears** (off-label), **plums** (off-label), **quinces** (off-label), **radishes** (off-label), **red beet** (off-label), **redcurrants** (off-label), **rhubarb** (off-label), **rocket** (off-label), **rosemary** (off-label), **runner beans** (off-label), **sage** (off-label), **salad burnet** (off-label), **salsify** (off-label), **savory** (off-label), **seakale** (off-label), **soya beans** (off-label), **spinach** (off-label), **spinach beet** (off-label), **strawberries** (moderate control), **swedes** (off-label), **sweet ciceley** (off-label), **table grapes** (off-label), **tarragon** (off-label), **thyme** (off-label), **turnips** (off-label), **vining peas** (off-label), **walnuts** (off-label), **wine grapes** (off-label)

- Pythium in **all protected edible crops** (moderate control), **all protected non-edible crops** (moderate control), **strawberries** (moderate control)
- Rhizoctonia in **all protected edible crops** (moderate control), **all protected non-edible crops** (moderate control), **almonds** (off-label), **angelica** (off-label), **apples** (off-label), **apricots** (off-label), **asparagus** (off-label), **baby leaf crops** (off-label), **beans without pods (fresh)** (off-label), **bilberries** (off-label), **blackcurrants** (off-label), **blueberries** (off-label), **broad beans** (off-label), **broccoli** (off-label), **brussels sprouts** (off-label), **bulb vegetables** (off-label), **cabbages** (off-label), **calabrese** (off-label), **cane fruit** (off-label), **carrots** (off-label), **cauliflowers** (off-label), **celeriac** (off-label), **celery (outdoor)** (off-label), **celery leaves** (off-label), **cherries** (off-label), **chervil** (off-label), **chestnuts** (off-label), **chicory root** (off-label), **chives** (off-label), **choi sum** (off-label), **collards** (off-label), **cranberries** (off-label), **cress** (off-label), **dwarf beans** (off-label), **edible flowers** (off-label), **edible podded peas** (off-label), **endives** (off-label), **fennel leaves** (off-label), **florence fennel** (off-label), **french beans** (off-label), **fruiting vegetables** (off-label), **globe artichoke** (off-label), **gooseberries** (off-label), **hazel nuts** (off-label), **herbs (see appendix 6)** (off-label), **hops** (off-label), **horseradish** (off-label), **hyssop** (off-label), **jerusalem artichokes** (off-label), **kale** (off-label), **kohlrabi** (off-label), **lamb's lettuce** (off-label), **land cress** (off-label), **leeks** (off-label), **lentils** (off-label), **lettuce** (off-label), **lovage** (off-label), **marjoram** (off-label), **medlar** (off-label), **mint** (off-label), **nectarines** (off-label), **oregano** (off-label), **oriental cabbage** (off-label), **ornamental plant production** (off-label), **parsley** (off-label), **parsley root** (off-label), **parsnips** (off-label), **peaches** (off-label), **pears** (off-label), **plums** (off-label), **quinces** (off-label), **radishes** (off-label), **red beet** (off-label), **redcurrants** (off-label), **rhubarb** (off-label), **rocket** (off-label), **rosemary** (off-label), **runner beans** (off-label), **sage** (off-label), **salad burnet** (off-label), **salsify** (off-label), **savory** (off-label), **seakale** (off-label), **soya beans** (off-label), **spinach** (off-label), **spinach beet** (off-label), **strawberries** (moderate control), **swedes** (off-label), **sweet ciceley** (off-label), **table grapes** (off-label), **tarragon** (off-label), **thyme** (off-label), **turnips** (off-label), **vining peas** (off-label), **walnuts** (off-label), **wine grapes** (off-label)

Extension of Authorisation for Minor Use (EAMUs)

- *almonds* 20152773
- *angelica* 20152773
- *apples* 20152773
- *apricots* 20152773
- *asparagus* 20152773
- *baby leaf crops* 20152773
- *beans without pods (fresh)* 20152773
- *bilberries* 20152773
- *blackcurrants* 20152773

SEE SECTION 3 FOR PRODUCTS ALSO REGISTERED

- **blueberries** 20152773
- **broad beans** 20152773
- **broccoli** 20152773
- **brussels sprouts** 20152773
- **bulb vegetables** 20152773
- **cabbages** 20152773
- **calabrese** 20152773
- **cane fruit** 20152773
- **carrots** 20152773
- **cauliflowers** 20152773
- **celeriac** 20152773
- **celery (outdoor)** 20152773
- **celery leaves** 20152773
- **cherries** 20152773
- **chervil** 20152773
- **chestnuts** 20152773
- **chicory root** 20152773
- **chives** 20152773
- **choi sum** 20152773
- **collards** 20152773
- **cranberries** 20152773
- **cress** 20152773
- **dwarf beans** 20152773
- **edible flowers** 20152773
- **edible podded peas** 20152773
- **endives** 20152773
- **fennel leaves** 20152773
- **florence fennel** 20152773
- **french beans** 20152773
- **fruiting vegetables** 20152773
- **globe artichoke** 20152773
- **gooseberries** 20152773
- **hazel nuts** 20152773
- **herbs (see appendix 6)** 20152773
- **hops** 20152773
- **horseradish** 20152773
- **hyssop** 20152773
- **jerusalem artichokes** 20152773
- **kale** 20152773
- **kohlrabi** 20152773
- **lamb's lettuce** 20152773
- **land cress** 20152773
- **leeks** 20152773
- **lentils** 20152773
- **lettuce** 20152773
- **lovage** 20152773
- **marjoram** 20152773
- **medlar** 20152773
- **mint** 20152773
- **nectarines** 20152773
- **oregano** 20152773
- **oriental cabbage** 20152773
- **ornamental plant production** 20152773
- **parsley** 20152773
- **parsley root** 20152773
- **parsnips** 20152773
- **peaches** 20152773
- **pears** 20152773
- **plums** 20152773

FOR FULL CONDITIONS OF USE ALWAYS READ THE PRODUCT LABEL

- *quinces* 20152773
- *radishes* 20152773
- *red beet* 20152773
- *redcurrants* 20152773
- *rhubarb* 20152773
- *rocket* 20152773
- *rosemary* 20152773
- *runner beans* 20152773
- *sage* 20152773
- *salad burnet* 20152773
- *salsify* 20152773
- *savory* 20152773
- *seakale* 20152773
- *soya beans* 20152773
- *spinach* 20152773
- *spinach beet* 20152773
- *swedes* 20152773
- *sweet ciceley* 20152773
- *table grapes* 20152773
- *tarragon* 20152773
- *thyme* 20152773
- *turnips* 20152773
- *vining peas* 20152773
- *walnuts* 20152773
- *wine grapes* 20152773

Approval information
- Gliocladium catenulatum included in Annex 1 under EC Regulation 1107/2009

Hazard classification and safety precautions
UN Number N/C
Operator protection A, D, H; U05a, U10, U14, U15, U20a
Environmental protection E15b, E34
Storage and disposal D01, D02, D09a, D10a
Medical advice M03

272 glufosinate-ammonium

A non-selective, non-residual phosphinic acid contact herbicide
HRAC mode of action code: H

Products

1	Glam	Pan Amenity	150 g/l	SL	17707
2	Harvest	Bayer CropScience	150 g/l	SL	17236

Uses
- Annual and perennial weeds in **hard surfaces, managed amenity turf, natural surfaces not intended to bear vegetation, permeable surfaces overlying soil** [1]
- Desiccation in **potatoes** [2]

Approval information
- Glufosinate-ammonium included in Annex I under EC Regulation 1107/2009
- Accepted by BBPA for use on malting barley and hops

Efficacy guidance
- Activity quickest under warm, moist conditions. Light rainfall 3-4 h after application will not affect activity. Do not spray wet foliage or if rain likely within 6 h
- For weed control uses treat when weeds growing actively. Deep rooted weeds may require second treatment
- On uncropped headlands apply in May/Jun to prevent weeds invading field

SEE SECTION 3 FOR PRODUCTS ALSO REGISTERED

Pesticide Profiles

- Glufosinate kills all green tissue but does not harm mature bark

Restrictions
- Do not desiccate potatoes in exceptionally wet weather or in saturated soil. See label for details
- Do not spray hedge bottoms or on broadcast crops

Crop-specific information
- Crops can normally be sown/planted immediately after spraying or sprayed post-drilling. On sand, very light or immature peat soils allow at least 3 d before sowing/planting or expected emergence
- For grass destruction apply before winter dormancy occurs. Heavily grazed fields should show active regrowth. Plough from the day after spraying
- For potato haulm desiccation apply to listed varieties (not seed crops) at onset of senescence, 14-21 d before harvest

Environmental safety
- Harmful to fish or other aquatic life. Do not contaminate surface waters or ditches with chemical or used container
- Keep livestock out of treated areas until foliage of any poisonous weeds such as ragwort has died and become unpalatable

Hazard classification and safety precautions
Hazard Harmful [2]; Harmful if swallowed, Toxic in contact with skin [1, 2]
Transport code 6.1
Packaging group III
UN Number 2902
Risk phrases H318, H360, H373
Operator protection A, C, H; U04a, U11, U20a [2]; U04b [1]; U05a, U08, U13, U14, U15, U19a [1, 2]
Environmental protection E07a, E22c [2]; E15a, E34 [1, 2]; E22b [1]
Storage and disposal D01, D02, D09a, D10b [1, 2]; D05 [1]; D12a [2]
Medical advice M03

273 glyphosate

A translocated non-residual glycine derivative herbicide
HRAC mode of action code: G

See also 2,4-D + glyphosate
  diflufenican + glyphosate
  diuron + glyphosate
  flufenacet + glyphosate + metosulam
  glyphosate + pyraflufen-ethyl
  glyphosate + sulfosulfuron

Products

1	Amega Duo	Nufarm UK	540 g/l	SL	13358
2	Amega Duo	Nufarm UK	540 g/l	SL	14131
3	Asteroid	Headland	360 g/l	SL	11118
4	Asteroid Pro	Headland	360 g/l	SL	15373
5	Asteroid Pro 450	Headland	450 g/l	SL	16384
6	Azural	Monsanto	360 g/l	SL	16239
7	Barclay Gallup 360	Barclay	360 g/l	SL	14988
8	Barclay Gallup Amenity	Barclay	360 g/l	SL	13250
9	Barclay Gallup Biograde 360	Barclay	360 g/l	SL	15188
10	Barclay Gallup Biograde Amenity	Barclay	360 g/l	SL	17674
11	Barclay Gallup Hi-Aktiv	Barclay	490 g/l	SL	14987
12	Barclay Glyde 144	Barclay	144 g/l	SL	15362

FOR FULL CONDITIONS OF USE ALWAYS READ THE PRODUCT LABEL

glyphosate

Products – continued

13	Buggy XTG	Sipcam	36% w/w	SG	12951
14	CDA Vanquish Biactive	Bayer Environ.	120 g/l	UL	12586
15	Charger C	AgChem Access	360 g/l	SL	14216
16	Clinic Ace	Nufarm UK	360 g/l	SL	12980
17	Clinic Ace	Nufarm UK	360 g/l	SL	14040
18	Clinic TF	Nufarm UK	360 g/l	SL	16716
19	Credit	Nufarm UK	540 g/l	SL	16775
20	Crestler	Nufarm UK	540 g/l	SL	16555
21	Crestler	Nufarm UK	540 g/l	SL	16663
22	Discman Biograde	Barclay	216 g/l	SL	12856
23	Ecoplug Max	Monsanto	68 % w/w	GR	17581
24	Envision	Headland	450 g/l	SL	10569
25	Frontsweep	Nomix Enviro	360 g/l	SL	14700
26	Gallup XL	Barclay	360 g/l	SL	17663
27	Garryowen XL	Barclay	360 g/l	SL	17508
28	Glyfos	Headland	360 g/l	SL	10995
29	Glyfos Dakar	Headland	68% w/w	SG	13054
30	Glyfos Dakar Pro	Headland Amenity	68% w/w	SG	13147
31	Glyfos Proactive	Nomix Enviro	360 g/l	SL	11976
32	Glyfos Supreme	Headland	450 g/l	SL	12371
33	Glyfosat 36	Goldengrass	360 g/l	SL	14297
34	Hilite	Nomix Enviro	144 g/l	RC	13871
35	Kernel	Headland	480 g/l	SL	10993
36	Liaison	Monsanto	360 g/l	SL	17089
37	Mascot Hi-Aktiv Amenity	Rigby Taylor	490 g/l	SL	15178
38	Master Gly	Generica	360 g/l	SL	17594
39	Master Gly 36T	Generica	360 g/l	SL	15731
40	Mentor	Monsanto	360 g/l	SL	16508
41	Monsanto Amenity Glyphosate	Monsanto	360 g/l	SL	16382
42	Motif	Monsanto	360 g/l	SL	16509
43	Nomix Conqueror Amenity	Nomix Enviro	144 g/l	RC	16039
44	Palomo Green	ProKlass	360 g/l	SL	17081
45	Pitch	Nufarm UK	540 g/l	SL	15485
46	Pitch	Nufarm UK	540 g/l	SL	15693
47	Proliance Quattro	Syngenta	360 g/l	SL	13670
48	Pure Glyphosate 360	Pure Amenity	360 g/l	SL	14834
49	Rattler	Nufarm UK	540 g/l	SL	15522
50	Rattler	Nufarm UK	540 g/l	SL	15692
51	Reaper	AgChem Access	360 g/l	SL	14924
52	Reaper Green	AgChem Access	360 g/l	SL	15935
53	Rodeo	Monsanto	360 g/l	SL	16242
54	Roundup Biactive GL	Monsanto	360 g/l	SL	17348
55	Roundup Energy	Monsanto	450 g/l	SL	12945
56	Roundup Flex	Monsanto	480 g/l	SL	15541
57	Roundup Metro XL	Monsanto	360 g/l	SL	17684
58	Roundup POWERMAX	Monsanto	72% w/w	SG	16373
59	Roundup ProVantage	Monsanto	480 g/l	SL	15534
60	Roundup Ultimate	Monsanto	540 g/l	SL	13081
61	Samurai	Monsanto	360 g/l	SL	16238
62	Sherlock	Synergy	360 g/l	SL	16185
63	Shyfo	Sharda	360 g/l	SL	15040
64	Snapper	Nufarm UK	550 g/l	SL	15489
65	Snapper	Nufarm UK	550 g/l	SL	15695
66	Stirrup	Nomix Enviro	144 g/l	RC	13875
67	Surrender	Agrovista	360 g/l	SL	16169
68	Tambora	Maxwell	360 g/l	SL	17591

SECTION 2

SEE SECTION 3 FOR PRODUCTS ALSO REGISTERED

Pesticide Profiles

Products – continued

69 Tanker	Nufarm UK	540 g/l	SL	15016
70 Tanker	Nufarm UK	540 g/l	SL	15694
71 Touchdown Quattro	Syngenta	360 g/l	SL	10608
72 Trustee Amenity	Barclay	450 g/l	SL	12897
73 Vesuvius	Generica	360 g/l	SL	16232

Uses

- Annual and perennial weeds in **all edible crops (outdoor and protected)** [1, 2, 16-21, 38, 39, 68-70, 73]; **all edible crops (outdoor)** [16-18, 38, 39, 45, 46, 49, 50, 54, 56, 59, 64, 65]; **all edible crops (outdoor)** *(before planting)*, **all non-edible crops (outdoor)** *(before planting)* [7, 9, 26, 27, 29, 32, 33, 44, 48, 51, 52, 57, 62, 63, 67, 71]; **all edible crops (stubble), all non-edible crops (stubble), rotational grass** [54]; **all edible crops (stubble)** *(before planting)*, **all non-edible crops (stubble)** *(before planting)* [26, 27]; **all non-edible crops (outdoor)** [1, 2, 16-21, 38, 39, 45, 46, 49, 50, 54, 56, 59, 64, 65, 68-70, 73]; **all non-edible seed crops grown outdoors** [16-18, 38, 39]; **almonds** *(off-label)*, **apples** *(off-label)*, **apricots** *(off-label)*, **asparagus** *(off-label)*, **cherries** *(off-label)*, **crab apples** *(off-label)*, **cultivated land/soil, fruit trees** *(off-label)*, **grapevines** *(off-label)*, **peaches** *(off-label)*, **pears** *(off-label)*, **plums** *(off-label)*, **quinces** *(off-label)* [3]; **amenity vegetation** [1, 2, 4-6, 12, 14, 16-21, 25, 33, 34, 36, 39-43, 45, 46, 49-51, 53, 56-59, 61, 64-70, 72]; **apple orchards** [1-5, 7, 9, 11, 13, 15-21, 24, 26-29, 32, 33, 35, 38, 39, 44-46, 48-52, 55-57, 59, 60, 62-65, 67, 69, 70]; **apples** [43, 68, 73]; **asparagus** [7, 9, 16-19, 25-27, 33, 38, 39, 44, 48, 51, 56-59, 63, 67, 68, 73]; **beetroot** *(off-label)*, **carrots** *(off-label)*, **garlic** *(off-label)*, **horseradish** *(off-label)*, **leeks** *(off-label)*, **lupins** *(off-label)*, **onions** *(off-label)*, **parsley root** *(off-label)*, **parsnips** *(off-label)*, **rye** *(off-label)*, **salad onions** *(off-label)*, **salsify** *(off-label)*, **shallots** *(off-label)*, **swedes** *(off-label)*, **turnips** *(off-label)* [55, 56, 58]; **bilberries** *(off-label)*, **blackcurrants** *(off-label)*, **blueberries** *(off-label)*, **cranberries** *(off-label)*, **gooseberries** *(off-label)*, **redcurrants** *(off-label)*, **rhubarb** *(off-label)*, **vaccinium spp.** *(off-label)*, **whitecurrants** *(off-label)* [3, 58]; **buckwheat** *(off-label)*, **canary seed** *(off-label)*, **millet** *(off-label)*, **quinoa** *(off-label)*, **sorghum** *(off-label)* [55, 58]; **bulb onions** [1, 2, 16-21, 38, 39, 45, 46, 49, 50, 56, 59, 63-65, 68-70, 73]; **cherries, plums** [1-7, 9, 11, 13, 15-21, 24-29, 32, 33, 35, 36, 38-46, 48-53, 55-65, 67-70, 73]; **chestnuts** *(off-label)*, **hazel nuts** *(off-label)*, **kentish cobnuts** *(off-label)*, **salad onions, table grapes, walnuts** *(off-label)*, **wine grapes** [58]; **combining peas, spring barley, spring field beans, spring wheat, winter barley, winter field beans, winter wheat** [1-3, 16-21, 38, 39, 45, 46, 49, 50, 54, 56, 59, 64, 65, 68-71, 73]; **damsons** [2, 3, 7, 11, 13, 15, 20, 21, 24, 28, 29, 32, 33, 35, 48, 51, 55, 57, 60, 69, 70]; **durum wheat** [1-3, 16-21, 38, 39, 45, 46, 49, 50, 56, 59, 64, 65, 68-70, 73]; **edible podded peas** *(off-label)*, **vining peas** *(off-label)* [55, 56]; **enclosed waters** [4-6, 8, 10, 36, 37, 40, 42, 43, 53, 58, 61, 72]; **farm forestry** [6, 36, 40, 42, 53, 61, 63]; **farm forestry** *(off-label)*, **game cover** *(off-label)*, **hemp** *(off-label)*, **hops** *(off-label)*, **soft fruit** *(off-label)*, **top fruit** *(off-label)*, **woad** *(off-label)* [55, 58, 60]; **forest** [1, 2, 4-13, 15-22, 24-28, 31-36, 38-53, 56-59, 61-70, 72]; **forest nurseries** [1, 16-18, 25, 39, 41, 47, 63, 67, 68]; **forestry plantations** [1, 2, 19-21, 45, 46, 49, 50, 64, 65, 69, 70]; **grassland** [1, 2, 6, 16-21, 25, 36, 38-42, 45, 46, 49, 50, 53, 61, 64, 65, 68-70, 73]; **green cover on land temporarily removed from production** [1, 2, 6, 9, 16-21, 25, 34, 36, 38-42, 44-46, 49, 50, 52-54, 56, 58, 59, 61, 62, 64, 65, 68-70, 73]; **hard surfaces** [1, 4-12, 16-19, 22, 25-27, 32-34, 36-54, 56-59, 61, 62, 64-68, 72, 73]; **industrial sites, road verges** [31]; **land immediately adjacent to aquatic areas** [4-6, 8, 10-12, 36-38, 40, 42, 43, 53, 54, 57, 58, 61, 72]; **land not intended to bear vegetation** [11, 13-15, 24, 28, 31, 35]; **land prior to cultivation** [32]; **leeks, sugar beet, vining peas** [1, 2, 6, 16-21, 25, 36, 38-42, 45, 46, 49, 50, 53, 56, 58, 59, 61, 63-65, 68-70, 73]; **linseed** [1-3, 6, 16-21, 25, 36, 38-42, 45, 46, 49, 50, 53, 56, 58, 59, 61, 64, 65, 68-71, 73]; **managed amenity turf** *(pre-establishment)* [34, 66]; **managed amenity turf** *(pre-establishment only)*, **pears** [43]; **miscanthus** *(off-label)* [55, 56, 58, 60]; **mustard** [1, 2, 16-21, 38, 39, 45, 46, 49, 50, 56, 59, 64, 65, 68-71, 73]; **natural surfaces not intended to bear vegetation, permeable surfaces overlying soil** [1, 2, 4-12, 16-22, 25-27, 32-34, 36-54, 56-59, 61, 62, 64-70, 72, 73]; **oilseed rape** [6, 19, 25, 36, 40-42, 53, 58, 61]; **open waters** [6, 8, 10, 36, 37, 40, 42, 43, 53, 61, 72]; **ornamental plant production** [43, 66]; **ornamental plant production** *(off-label)* [16]; **pear orchards** [1-5, 7, 9, 11, 13, 15-21, 24, 26-29, 32, 33, 35, 38, 39, 44-46, 48-52, 55-57, 59, 60, 62-65, 67-70, 73]; **permanent grassland** [13, 54, 55]; **poppies for morphine production** *(off-label)* [16, 56]; **potatoes** [54, 56, 58, 59]; **spring oats, winter oats** [1-3, 16-21, 38, 39, 45, 46, 49, 50, 54, 56, 59, 64, 65, 68-70, 73]; **spring oilseed rape, winter oilseed rape** [1-3, 16-21, 38, 39, 45, 46, 49, 50, 56, 59, 64, 65, 68-71, 73];

FOR FULL CONDITIONS OF USE ALWAYS READ THE PRODUCT LABEL

glyphosate

- **stubbles** [3, 6, 7, 9, 11, 13, 25, 33, 36, 40-42, 44, 48, 51-53, 55, 57, 58, 60-63, 67, 71]; **swedes**, **turnips** [1, 2, 6, 16-21, 25, 36, 38-42, 45, 46, 49, 50, 53, 54, 56, 58, 59, 61, 63-65, 68-70, 73]; **water or waterside areas** [22]
- Annual dicotyledons in **bulb onions** [6, 7, 9, 25-27, 29, 36, 40-42, 44, 48, 51-53, 55, 57, 58, 60-62, 67]; **combining peas** [7, 9, 26, 27, 29, 44, 48, 51, 52, 57, 62, 67]; **cultivated land/soil** [11]; **durum wheat** [7, 9, 25-27, 29, 41, 44, 48, 51, 52, 55, 57, 60, 62, 67]; **leeks, linseed, mustard, spring field beans, spring oilseed rape, sugar beet, swedes, winter field beans** [7, 9, 26, 27, 29, 44, 48, 51, 52, 55, 57, 60, 62, 67]; **permanent grassland** [55, 60]; **spring barley, spring oats, spring wheat, winter barley, winter oats, winter wheat** [7, 9, 15, 24, 26-29, 32, 35, 44, 48, 51, 52, 55, 57, 60, 62, 67]; **turnips** [9, 26, 27, 29, 44, 52, 55, 60, 62, 67]; **vining peas** [7, 9, 26, 27, 44, 48, 51, 52, 55, 57, 60, 62, 67]; **winter oilseed rape** [7, 9, 26, 27, 29, 33, 44, 48, 51, 52, 55, 57, 60, 62, 63, 67]
- Annual grasses in **bulb onions** [6, 25, 29, 36, 40-42, 53, 55, 58, 60, 61]; **combining peas** [29]; **cultivated land/soil** [11, 15, 24, 28, 35]; **durum wheat** [25, 29, 41, 55, 60]; **leeks, linseed, mustard, spring barley, spring field beans, spring oats, spring oilseed rape, spring wheat, sugar beet, swedes, turnips, winter barley, winter field beans, winter oats, winter oilseed rape, winter wheat** [29, 55, 60]; **permanent grassland, vining peas** [55, 60]
- Aquatic weeds in **enclosed waters** [6, 36, 40, 42, 53, 54, 56, 58, 59, 61, 72]; **land immediately adjacent to aquatic areas** [6, 36, 40, 42, 53, 54, 56-59, 61, 72]; **open waters** [6, 11, 36, 40, 42, 53, 54, 56, 59, 61, 72]
- Biennial weeds in **beetroot** (off-label), **carrots** (off-label), **garlic** (off-label), **horseradish** (off-label), **leeks** (off-label), **onions** (off-label), **parsley root** (off-label), **parsnips** (off-label), **salad onions** (off-label), **salsify** (off-label), **shallots** (off-label), **swedes** (off-label), **turnips** (off-label) [56]
- Black bent in **combining peas, spring field beans, spring oilseed rape, stubbles, winter field beans, winter oilseed rape** [11, 15, 24, 28, 32, 35]; **durum wheat, spring barley, spring oats, spring wheat, winter barley, winter oats, winter wheat** [11]; **linseed** [15, 24, 28, 32, 35]
- Bolters in **sugar beet** (wiper application) [11, 29, 55, 60]
- Bracken in **farm forestry** [63]; **forest** [7, 9, 11, 13, 15, 24, 26-28, 30-33, 35, 44, 48, 51, 52, 56, 57, 59, 62, 63, 67]; **forest nurseries** [63, 67]; **tolerant conifers** [31]
- Brambles in **tolerant conifers** [31]
- Canary grass in **aquatic areas** [29]; **enclosed waters, land immediately adjacent to aquatic areas** [30]
- Chemical thinning in **farm forestry** [6, 36, 40, 42, 53, 61, 63]; **forest** [4-7, 9, 11, 15, 24-28, 30, 32, 33, 35, 36, 40-42, 44, 47, 48, 51-53, 56, 57, 59, 61-63, 67]; **forest** (stump treatment) [13]; **forest nurseries** [25, 41, 47, 63, 67]; **tree stumps** [68]
- Couch in **combining peas, spring field beans, winter field beans** [3, 6, 11, 13, 15, 24, 25, 28, 29, 32, 35, 36, 40-42, 53, 55, 58, 60, 61, 71]; **durum wheat** [3, 6, 11, 25, 29, 36, 40-42, 53, 55, 58, 60, 61]; **forest** [11, 30]; **linseed** [3, 13, 15, 24, 28, 29, 32, 35, 55, 60, 71]; **mustard** [6, 13, 25, 36, 40-42, 53, 55, 58, 60, 61, 71]; **oats** [6, 25, 36, 40-42, 53, 58, 61]; **permanent grassland** [13]; **spring barley, spring wheat, winter barley, winter wheat** [3, 6, 11, 13, 25, 29, 36, 40-42, 53, 55, 58, 60, 61, 71]; **spring oats, winter oats** [3, 11, 13, 29, 55, 60]; **spring oilseed rape, stubbles, winter oilseed rape** [3, 11, 13, 15, 24, 28, 29, 32, 35, 55, 60, 71]
- Creeping bent in **aquatic areas** [29]; **combining peas, spring field beans, spring oilseed rape, stubbles, winter field beans, winter oilseed rape** [11, 15, 24, 28, 32, 35]; **durum wheat, spring barley, spring oats, spring wheat, winter barley, winter oats, winter wheat** [11]; **enclosed waters, land immediately adjacent to aquatic areas** [30]; **linseed** [15, 24, 28, 32, 35]
- Desiccation in **buckwheat** (off-label), **canary seed** (off-label), **hemp** (off-label), **millet** (off-label), **quinoa** (off-label), **sorghum** (off-label) [55, 58]; **combining peas, spring field beans, winter field beans** [54, 56, 59]; **durum wheat, linseed, mustard, spring barley, spring oats, spring oilseed rape, spring wheat, winter barley, winter oats, winter oilseed rape, winter wheat** [56, 59]; **lupins** (off-label), **poppies for morphine production** (off-label), **rye** (off-label) [55, 56]
- Destruction of crops in **all edible crops (outdoor and protected)** [1, 2, 19, 69, 70]; **all edible crops (outdoor)** [6, 16-18, 25, 36, 38-42, 45, 46, 49, 50, 53, 54, 58, 61, 64, 65]; **all edible crops (stubble), all non-edible crops (stubble)** [54, 68]; **all non-edible crops (outdoor)** [6, 19, 25, 36, 40-42, 45, 46, 49, 50, 53, 54, 58, 61, 64, 65]; **all non-edible seed crops grown outdoors** [16-18, 38, 39]; **grassland** [1, 2, 19, 45, 46, 49, 50, 56, 59, 64, 65, 69, 70]; **green cover on land**

SEE SECTION 3 FOR PRODUCTS ALSO REGISTERED

Pesticide Profiles

- *temporarily removed from production* [6, 25, 36, 40-42, 53, 54, 58, 61]; *permanent grassland, rotational grass* [54]
- Destruction of short term leys in *grassland* [34]
- Grass weeds in *forest, hard surfaces, natural surfaces not intended to bear vegetation, permeable surfaces overlying soil, water or waterside areas* [22]
- Green cover in *land immediately adjacent to aquatic areas* [26, 27]; *land temporarily removed from production* [3, 7, 9, 11, 13, 15, 24, 26-29, 32, 33, 35, 44, 48, 51, 52, 55, 57, 60, 62, 63, 67, 71]
- Growth suppression in *forestry plantations* [16-18, 39]; *tree stumps* [2, 16-18, 39, 45, 46, 49, 50, 64, 65]
- Harvest management/desiccation in *combining peas* [1, 2, 6, 7, 9, 13, 16-19, 25-27, 33, 36, 38-42, 44-46, 48-53, 55, 57, 58, 60-65, 67-70, 73]; *durum wheat* [1, 2, 6, 7, 9, 16-19, 25-27, 29, 33, 36, 38-42, 44-46, 48-53, 55, 57, 58, 60-65, 67-70, 73]; *grassland* [6, 25, 36, 40-42, 53, 61, 63]; *linseed* [1, 2, 6, 7, 9, 11, 13, 16-19, 25-27, 33, 36, 38-42, 44-46, 48-53, 55, 57, 58, 60-65, 67-70, 73]; *mustard* [1, 2, 6, 7, 9, 13, 16-19, 25-27, 33, 36, 38-42, 44-46, 48-53, 55, 57, 58, 60-65, 67-71, 73]; *oats, oilseed rape* [6, 25, 36, 40-42, 53, 58, 61]; *poppies for morphine production* (off-label) [16]; *spring barley, spring wheat, winter barley, winter wheat* [1, 2, 6, 7, 9, 13, 15-19, 24-29, 32, 33, 35, 36, 38-42, 44-46, 48-53, 55, 57, 58, 60-65, 67-70, 73]; *spring field beans, winter field beans* [6, 7, 9, 13, 25-27, 33, 36, 40-42, 44-46, 48-53, 55, 57, 58, 60-65, 67]; *spring oats, winter oats* [1, 2, 7, 9, 13, 15-19, 24, 26-29, 32, 33, 35, 38, 39, 44-46, 48-52, 55, 57, 60, 62-65, 67-70, 73]; *spring oilseed rape, winter oilseed rape* [1-3, 7, 9, 11, 13, 15-19, 24, 26-29, 32, 33, 35, 38, 39, 44-46, 48-52, 55, 57, 60, 62-65, 67-71, 73]
- Heather in *farm forestry* [63]; *forest* [7, 9, 11, 13, 15, 24, 26-28, 30-33, 35, 44, 48, 51, 52, 57, 62, 63, 67]; *forest nurseries* [63, 67]
- Perennial dicotyledons in *combining peas, linseed, spring field beans, spring oilseed rape, winter field beans, winter oilseed rape* [15, 24, 28, 29, 32, 35]; *permanent grassland* (wiper application) [55, 60]
- Perennial grasses in *aquatic areas* [3-5, 11, 13, 24, 28, 31, 35]; *combining peas, linseed, spring field beans, stubbles, winter field beans* [29]; *tolerant conifers* [31]
- Pre-harvest desiccation in *combining peas, linseed, spring barley, spring field beans, spring wheat, winter barley, winter field beans, winter wheat* [71]
- Reeds in *aquatic areas* [3-5, 11, 13, 24, 28, 29, 31, 35]; *enclosed waters, land immediately adjacent to aquatic areas* [11, 30, 56, 59]; *open waters* [56, 59]
- Re-growth suppression in *tree stumps* [68]
- Rhododendrons in *farm forestry* [63]; *forest* [7, 9, 11, 13, 15, 24, 26-28, 30-33, 35, 44, 48, 51, 52, 57, 62, 63, 67]; *forest nurseries* [63, 67]
- Rushes in *aquatic areas* [3-5, 11, 13, 24, 28, 29, 31, 35]; *enclosed waters, land immediately adjacent to aquatic areas* [11, 30, 56, 59]; *open waters* [56, 59]
- Sedges in *aquatic areas* [3-5, 11, 13, 24, 28, 29, 31, 35]; *enclosed waters, land immediately adjacent to aquatic areas* [11, 30, 56, 59]; *open waters* [56, 59]
- Sprout suppression in *tree stumps* [56, 59]
- Sucker control in *apple orchards, cherries, pear orchards, plums* [6, 25, 36, 40-42, 53, 55, 58, 61]; *damsons* [55]
- Sucker inhibition in *forest* [4, 5]; *tree stumps* [1, 6, 19, 23, 25, 36, 40-42, 53, 61, 69, 70]
- Sward destruction in *amenity vegetation* [72]; *grassland* [9, 26, 27, 44, 52, 57, 62, 67]; *permanent grassland* [3, 7, 11, 13, 15, 24, 28, 29, 32, 33, 35, 48, 51, 55, 60, 71]; *rotational grass* [7, 33, 48, 51, 71]
- Total vegetation control in *all edible crops (outdoor)* (pre-sowing/planting), *all non-edible crops (outdoor)* (pre-sowing/planting) [55]; *amenity vegetation* [14, 31]; *fencelines, industrial sites, road verges* [31]; *hard surfaces* [30, 31]; *land not intended to bear vegetation* [14, 31, 71]; *natural surfaces not intended to bear vegetation, permeable surfaces overlying soil* [30]
- Volunteer cereals in *bulb onions* [6, 25, 29, 36, 40-42, 53, 55, 58, 60, 61]; *combining peas* [29]; *cultivated land/soil* [11, 15, 24, 28, 35]; *durum wheat* [25, 41, 55, 60]; *leeks, linseed, mustard, spring field beans, sugar beet, swedes, turnips, winter field beans* [29, 55, 60]; *spring barley, spring oats, spring oilseed rape, spring wheat, vining peas, winter barley, winter oats, winter oilseed rape, winter wheat* [55, 60]; *stubbles* [3, 11, 13, 15, 24, 28, 29, 32, 35, 55, 60, 71]
- Volunteer potatoes in *stubbles* [3, 11, 13, 29, 55, 60, 71]

FOR FULL CONDITIONS OF USE ALWAYS READ THE PRODUCT LABEL

glyphosate

- Waterlilies in **aquatic areas** [3-5, 11, 13, 24, 28, 29, 31, 35]; **enclosed waters** [11, 30, 56, 59]; **land immediately adjacent to aquatic areas**, **open waters** [56, 59]
- Weed beet in **sugar beet** *(wiper application)* [29]
- Woody weeds in **farm forestry** [63]; **forest** [7, 9, 11, 13, 15, 24, 26-28, 30-33, 35, 44, 48, 51, 52, 57, 62, 63, 67]; **forest nurseries** [63, 67]; **tolerant conifers** [31]

Extension of Authorisation for Minor Use (EAMUs)
- **almonds** *20102476* [3]
- **apples** *20102476* [3]
- **apricots** *20102476* [3]
- **asparagus** *20071476* [3]
- **beetroot** *20130354* [55], *20132528* [56], *20141305 expires 25 Nov 2017* [58]
- **bilberries** *20071475* [3], *20141316 expires 25 Nov 2017* [58]
- **blackcurrants** *20071475* [3], *20141316 expires 25 Nov 2017* [58]
- **blueberries** *20071475* [3], *20141316 expires 25 Nov 2017* [58]
- **buckwheat** *20071839* [55], *20141313 expires 25 Nov 2017* [58]
- **canary seed** *20071839* [55], *20141313 expires 25 Nov 2017* [58]
- **carrots** *20130354* [55], *20132528* [56], *20141305 expires 25 Nov 2017* [58]
- **cherries** *20102476* [3]
- **chestnuts** *20141317 expires 25 Nov 2017* [58]
- **crab apples** *20102476* [3]
- **cranberries** *20071475* [3], *20141316 expires 25 Nov 2017* [58]
- **edible podded peas** *20141672* [55], *20141671* [56]
- **farm forestry** *20082888* [55], *20141300 expires 25 Nov 2017* [58], *20082892* [60]
- **fruit trees** *20102476* [3]
- **game cover** *20082888* [55], *20141300 expires 25 Nov 2017* [58], *20082892* [60]
- **garlic** *20130354* [55], *20132528* [56], *20141305 expires 25 Nov 2017* [58]
- **gooseberries** *20071475* [3], *20141316 expires 25 Nov 2017* [58]
- **grapevines** *20071479* [3]
- **hazel nuts** *20141317 expires 25 Nov 2017* [58]
- **hemp** *20071840* [55], *20141300 expires 25 Nov 2017* [58], *20141321 expires 25 Nov 2017* [58], *20082892* [60]
- **hops** *20082888* [55], *20141300 expires 25 Nov 2017* [58], *20082892* [60]
- **horseradish** *20130354* [55], *20132528* [56], *20141305 expires 25 Nov 2017* [58]
- **kentish cobnuts** *20141317 expires 25 Nov 2017* [58]
- **leeks** *20130354* [55], *20132528* [56], *20141305 expires 25 Nov 2017* [58]
- **lupins** *20071279* [55], *20131667* [56], *20141306 expires 25 Nov 2017* [58]
- **millet** *20071839* [55], *20141313 expires 25 Nov 2017* [58]
- **miscanthus** *20082888* [55], *20132131* [56], *20141300 expires 25 Nov 2017* [58], *20082892* [60]
- **onions** *20130354* [55], *20132528* [56], *20141305 expires 25 Nov 2017* [58]
- **ornamental plant production** *20122277* [16]
- **parsley root** *20130354* [55], *20132528* [56], *20141305 expires 25 Nov 2017* [58]
- **parsnips** *20130354* [55], *20132528* [56], *20141305 expires 25 Nov 2017* [58]
- **peaches** *20102476* [3]
- **pears** *20102476* [3]
- **plums** *20102476* [3]
- **poppies for morphine production** *20122278* [16], *20081058* [55], *20131668* [56]
- **quinces** *20102476* [3]
- **quinoa** *20071840* [55], *20141321 expires 25 Nov 2017* [58]
- **redcurrants** *20071475* [3], *20141316 expires 25 Nov 2017* [58]
- **rhubarb** *20071478* [3], *20141320 expires 25 Nov 2017* [58]
- **rye** *20111701* [55], *20131666* [56], *20141307 expires 25 Nov 2017* [58]
- **salad onions** *20130354* [55], *20132528* [56], *20141305 expires 25 Nov 2017* [58]
- **salsify** *20130354* [55], *20132528* [56], *20141305 expires 25 Nov 2017* [58]
- **shallots** *20130354* [55], *20132528* [56], *20141305 expires 25 Nov 2017* [58]
- **soft fruit** *20082888* [55], *20141300 expires 25 Nov 2017* [58], *20082892* [60]
- **sorghum** *20071839* [55], *20141313 expires 25 Nov 2017* [58]
- **swedes** *20130354* [55], *20132528* [56], *20141305 expires 25 Nov 2017* [58]
- **top fruit** *20082888* [55], *20141300 expires 25 Nov 2017* [58], *20082892* [60]

SEE SECTION 3 FOR PRODUCTS ALSO REGISTERED

Pesticide Profiles

- ***turnips*** *20130354* [55], *20132528* [56], *20141305 expires 25 Nov 2017* [58]
- ***vaccinium spp.*** *20071475* [3], *20141316 expires 25 Nov 2017* [58]
- ***vining peas*** *20141672* [55], *20141671* [56]
- ***walnuts*** *20141317 expires 25 Nov 2017* [58]
- ***whitecurrants*** *20071475* [3], *20141316 expires 25 Nov 2017* [58]
- ***woad*** *20082888* [55], *20141300 expires 25 Nov 2017* [58], *20082892* [60]

Approval information
- Glyphosate included in Annex I under EC Regulation 1107/2009
- Accepted by BBPA for use on malting barley and hops
- Approval expiry 25 Nov 2017 [23, 58]

Efficacy guidance
- For best results apply to actively growing weeds with enough leaf to absorb chemical
- For most products a rainfree period of at least 6 h (preferably 24 h) should follow spraying
- Adjuvants are obligatory for some products and recommended for some uses with others. See labels
- Mixtures with other pesticides or fertilizers may lead to reduced control.
- Products are formulated as isopropylamine, ammonium, potassium, or trimesium salts of glyphosate and may vary in the details of efficacy claims. See individual product labels
- Some products in ready-to-use formulations for use through hand-held applicators. See label for instructions [14, 34, 66].
- If using to treat hard surfaces, ensure spraying takes place only when weeds are actively growing (normally March to October) and is confined only to visible weeds including those in the 30cm swath covering the kerb edge and road gulley – do not overspray drains.
- When applying products through rotary atomisers the spray droplet spectra must have a minimum Volume Median Diameter (VMD) of 200 microns
- With wiper application weeds should be at least 10 cm taller than crop
- Annual weed grasses should have at least 5 cm of leaf and annual broad-leaved weeds at least 2 expanded true leaves
- Perennial grass weeds should have 4-5 new leaves and be at least 10 cm long when treated. Perennial broad-leaved weeds should be treated at or near flowering but before onset of senescence
- Volunteer potatoes and polygonums are not controlled by harvest-aid rates
- Bracken must be treated at full frond expansion
- Fruit tree suckers best treated in late spring
- Chemical thinning treatment can be applied as stump spray or stem injection
- In order to allow translocation, do not cultivate before spraying and do not apply other pesticides, lime, fertilizer or farmyard manure within 5 d of treatment
- Recommended intervals after treatment and before cultivation vary. See labels

Restrictions
- Maximum total dose per crop or season normally equivalent to one full dose treatment on field and edible crops and no restriction for non-crop uses. However some older labels indicate a maximum number of treatments. Check for details
- Do not treat cereals grown for seed or undersown crops
- Consult grain merchant before treating crops grown on contract or intended for malting
- Do not use treated straw as a mulch or growing medium for horticultural crops
- For use in nursery stock, shrubberies, orchards, grapevines and tree nuts care must be taken to avoid contact with the trees. Do not use in orchards established less than 2 yr and keep off low-lying branches
- Certain conifers may be sprayed overall in dormant season. See label for details
- Use a tree guard when spraying in established forestry plantations
- Do not spray root suckers in orchards in late summer or autumn
- Do not use under glass or polythene as damage to crops may result
- Do not mix, store or apply in galvanised or unlined mild steel containers or spray tanks
- Do not leave diluted chemical in spray tanks for long periods and make sure that tanks are well vented

FOR FULL CONDITIONS OF USE ALWAYS READ THE PRODUCT LABEL

glyphosate

Crop-specific information
- Harvest intervals: 4 wk for blackcurrants, blueberries; 14 d for linseed, oilseed rape; 8 d for mustard; 5-7 d for all other edible crops. Check label for exact details
- Latest use: for most products 2-14 d before cultivating, drilling or planting a crop in treated land; after harvest (post-leaf fall) but before bud formation in the following season for nuts and most fruit and vegetable crops; before fruit set for grapevines. See labels for details

Following crops guidance
- Decaying remains of plants killed by spraying must be dispersed before direct drilling
- Crops may be drilled 48 h after application. Trees and shrubs may be planted 7 d after application.

Environmental safety
- Products differ in their hazard and environmental safety classification. See labels
- Do not dump surplus herbicide in water or ditch bottoms or empty into drains
- Check label for maximum permitted concentration in treated water
- The Environment Agency or Local River Purification Authority must be consulted before use in or near water
- Take extreme care to avoid drift and possible damage to neighbouring crops or plants
- Treated poisonous plants must be removed before grazing or conserving
- Do not use in covered areas such as greenhouses or under polythene
- For field edge treatment direct spray away from hedge bottoms
- Some products require livestock to be excluded from treated areas and do not permit treated forage to be used for hay, silage or bedding. Check label for details
- LERAP Category B [60]

Hazard classification and safety precautions
Hazard Harmful [33, 38, 39, 63, 67, 68, 73]; Irritant [7, 8, 13, 15-18, 23, 25-27, 41, 45, 46, 48-51, 57, 60, 64-66]; Dangerous for the environment [6-8, 15-18, 23, 25-27, 33-36, 38-43, 45, 46, 48-51, 53, 56, 57, 59-61, 63-68, 71, 73]
Transport code 9 [6-8, 16-18, 20, 21, 23, 25-28, 31-36, 38-42, 45, 46, 48-50, 53, 57-61, 64-68, 73]
Packaging group III [6-8, 16-18, 20, 21, 23, 26-28, 31-36, 38-42, 45, 46, 48-50, 53, 57-61, 64-68, 73]
UN Number 3077 [23, 58]; 3082 [6-8, 16-18, 20, 21, 25-28, 31-36, 38-42, 45, 46, 48-50, 53, 57, 59-61, 64-68, 73]; N/C [1-5, 9-15, 19, 22, 24, 29, 30, 37, 43, 44, 47, 51, 52, 54-56, 62, 63, 69-72]
Risk phrases H318 [16, 17, 20, 21, 45, 46, 49, 50, 64, 65, 68]; H319 [27, 36, 53]; R20 [7, 39, 48, 51, 57, 67, 73]; R36 [6, 40, 42, 61, 66]; R38 [60]; R41 [7, 13, 15, 25, 33, 39, 41, 48, 51, 57, 63, 67, 73]; R50 [43, 60]; R51 [6, 7, 15, 25, 33, 35-37, 39-42, 48, 51, 57, 61, 63, 66, 67, 73]; R52 [29, 30, 32, 55, 71]; R53a [6, 7, 15, 25, 29, 30, 32, 33, 35-37, 39-43, 48, 51, 55-57, 59-61, 63, 66, 67, 71, 73]
Operator protection A [1-22, 24-73]; C [3-11, 13, 15-18, 23-28, 31-33, 35-42, 44-46, 48-53, 55, 57, 61-65, 67, 68, 73]; D [3-5, 24, 28, 31, 35]; F [25]; H [1-5, 7-15, 19-22, 24-32, 34, 35, 37, 43-46, 48-52, 54, 56-60, 62, 64-72]; M [1-15, 19-22, 24-32, 34-36, 40-46, 48-52, 54-62, 64-70, 72]; N [15, 25]; U02a [1, 2, 8-12, 16-22, 25, 28-30, 32-35, 38, 39, 41, 43-46, 49, 50, 52, 58, 62-66, 68-70, 72, 73]; U05a [1, 2, 7, 15, 19-21, 23, 26, 27, 34, 37, 43, 45, 46, 48-51, 57, 58, 60, 64-67, 69-71]; U08 [8-11, 23, 28-30, 32, 35, 44, 52, 58, 62, 72]; U09a [12, 15, 22, 34, 43, 66]; U11 [6, 7, 13, 15-18, 23, 25-27, 29, 30, 33, 36-42, 48, 51, 53, 57, 61, 63, 66-68, 73]; U12, U15 [45, 46, 49, 50, 64, 65]; U16b [66]; U19a [8-12, 14, 15, 22, 28-30, 32, 34, 35, 43, 44, 52, 58, 62, 66, 72]; U20a [7, 22, 23, 26, 27, 37, 48, 51, 54, 57, 60, 67, 71]; U20b [1-6, 8-12, 14-21, 24, 25, 28-36, 38-47, 49, 50, 52, 53, 55, 56, 58, 59, 61-66, 68-70, 72, 73]
Environmental protection E06a [47] (2 weeks); E06d, E08b [71]; E07d [69, 70]; E13c [11, 28, 34, 43, 72]; E15a [3-5, 14, 24, 31, 32, 35, 54, 60]; E15b [1, 2, 6-10, 12, 13, 16-23, 25-27, 29, 30, 33, 34, 36, 38-53, 55-59, 61-71, 73]; E16a, E16b [60]; E19a [3-5, 11, 24, 28, 35, 72]; E19b [15]; E34 [14, 19-21, 32, 69, 70]; E36a [19-21, 69, 70]; E38 [6, 7, 13, 16-18, 23, 25-27, 29, 30, 32, 33, 36, 38-42, 45, 46, 48-51, 53, 56-61, 63-68, 71, 73]; H410 [16, 17, 34, 49, 50]; H411 [20, 21, 27, 45, 46, 64, 65]; H412 [18, 19, 23]; H413 [1, 2, 69, 70]
Storage and disposal D01 [1-10, 12-15, 19-23, 26-32, 34-37, 40, 42-46, 48-62, 64-67, 69-71]; D02 [1, 2, 6-10, 12-15, 19-23, 26, 27, 29-31, 34, 36, 37, 40, 42-46, 48-62, 64-67, 69-71]; D05 [1, 2, 6-12, 14, 16-23, 25-27, 31-34, 36, 38-46, 48-57, 59-73]; D09a [3-18, 22-36, 38-68, 71-73]; D10a [8-10, 14, 44, 52, 62]; D10b [3-5, 11, 15, 24, 35, 72]; D10c [1, 2, 6, 7, 16-21, 25-28, 31-33, 36, 38-42,

---

SEE SECTION 3 FOR PRODUCTS ALSO REGISTERED

45-51, 53-57, 59-61, 63-65, 67-71, 73]; D11a [12, 13, 22, 29, 30, 34, 43, 58, 66]; D12a [6, 7, 15-18, 23, 25-27, 29, 30, 32-34, 36, 38-43, 48, 51, 53, 56, 57, 59-61, 63, 66-68, 71, 73]; D12b [35]; D14 [19-21, 69, 70]
Medical advice M03 [71]; M05a [6, 15-18, 25, 33, 36-42, 53, 61, 63, 68, 73]

274 glyphosate + pyraflufen-ethyl

A herbicide mixture for non-selective weed control around amenity plants and on areas not intended to bear vegetation
HRAC mode of action code: G + E

See also glyphosate
pyraflufen-ethyl

Products
Hammer   Everris Ltd   240:0.67 g/l   SE   16498

Uses
- Annual and perennial weeds in *amenity vegetation, hard surfaces, land not intended to bear vegetation, natural surfaces not intended to bear vegetation, ornamental plant production, permeable surfaces overlying soil*

Approval information
- Glyphosate and pyraflufen-ethyl included in Annex I under EC Regulation 1107/2009

Hazard classification and safety precautions
Hazard Dangerous for the environment
Transport code 9
Packaging group III
UN Number 3082
Risk phrases R51, R53a
Operator protection A, H, M; U02a, U08, U20b
Environmental protection E15b, E38
Storage and disposal D01, D02, D05, D09a, D10c, D12a

275 glyphosate + sulfosulfuron

A translocated non-residual glycine derivative herbicide + a sulfonyl urea herbicide
HRAC mode of action code: G + B

See also sulfosulfuron

Products
Nomix Dual   Nomix Enviro   120:2.22 g/l   RC   13420

Uses
- Annual and perennial weeds in *amenity vegetation, hard surfaces, natural surfaces not intended to bear vegetation, permeable surfaces overlying soil*

Approval information
- Glyphosate and sulfosulfuron included in Annex I under EC Regulation 1107/2009

Hazard classification and safety precautions
Hazard Dangerous for the environment
Transport code 9
Packaging group III
UN Number 3082
Risk phrases R50, R53a
Operator protection A, H, M; U02a, U09a, U19a, U20b
Environmental protection E15b
Storage and disposal D01, D05, D09a, D11a

FOR FULL CONDITIONS OF USE ALWAYS READ THE PRODUCT LABEL

276 halauxifen-methyl

A herbicide for use in cereal crops only available in mixtures

See also fluroxypyr + halauxifen-methyl

277 hymexazol

A systemic heteroaromatic fungicide for pelleting sugar beet seed
FRAC mode of action code: 32

Products
Tachigaren 70 WP   Sumi Agro   70% w/w   WP   12568

Uses
- Black leg in **sugar beet** (seed treatment)

Approval information
- Hymexazol included in Annex 1 under EC Regulation 1107/2009

Efficacy guidance
- Incorporate into pelleted seed using suitable seed pelleting machinery

Restrictions
- Maximum number of treatments 1 per batch of seed
- Do not use treated seed as food or feed

Crop-specific information
- Latest use: before planting sugar beet seed

Environmental safety
- Harmful to aquatic organisms
- Harmful to fish or other aquatic life. Do not contaminate surface waters or ditches with chemical or used container
- Treated seed harmful to game and wildlife

Hazard classification and safety precautions
Hazard Flammable solid
Transport code 3
Packaging group III
UN Number 1325
Risk phrases H318
Operator protection A, C, F; U05a, U11, U20b
Environmental protection E03, E13c, H411
Storage and disposal D01, D02, D09a, D11a
Treated seed S01, S02, S03, S04a, S05

278 imazalil

A systemic and protectant imidazole fungicide
FRAC mode of action code: 3

See also guazatine + imazalil

Products
Gavel   Certis   100 g/l   SL   17586

Uses
- Dry rot in **seed potatoes**
- Gangrene in **seed potatoes**
- Silver scurf in **seed potatoes**
- Skin spot in **seed potatoes**

SEE SECTION 3 FOR PRODUCTS ALSO REGISTERED

Pesticide Profiles

Approval information
- Imazalil included in Annex I under EC Regulation 1107/2009

Efficacy guidance
- For best control of skin and wound diseases of ware potatoes treat as soon as possible after harvest, preferably within 7-10 d, before any wounds have healed

Restrictions
- Maximum number of treatments 1 per batch of ware tubers; 2 per batch of seed tubers;
- Consult processor before treating potatoes for processing

Crop-specific information
- Latest use: during storage and before chitting for seed potatoes
- Apply to clean soil-free potatoes post-harvest before putting into store, or at first grading. A further treatment may be applied in early spring before planting
- Apply through canopied hydraulic or spinning disc equipment preferably diluted with up to two litres water per tonne of potatoes to obtain maximum skin cover and penetration
- Use on ware potatoes subject to discharges of imazalil from potato washing plants being within emission limits set by the UK monitoring authority

Environmental safety
- Dangerous for the environment
- Toxic to aquatic organisms
- Do not empty into drains
- Personal protective equipment requirements may vary for each pack size. Check label

Hazard classification and safety precautions
Hazard Irritant
UN Number N/C
Risk phrases H318, H351
Operator protection A, C, H; U04a, U05a, U11, U14, U19a, U20a
Environmental protection E15a, E19b, E34, H410
Storage and disposal D01, D02, D05, D09a, D10c, D12a
Treated seed S01, S02, S03, S04a, S05, S06a

279 imazamox

An imidazolinone contact and residual herbicide available only in mixtures
HRAC mode of action code: B

See also imazamox + metazachlor
imazamox + pendimethalin

280 imazamox + metazachlor

A contact and residual herbicide mixture for weed control in oilseed rape
HRAC mode of action code: B + K3

Products
  Cleranda   BASF   17.5:375 g/l   SC   15036

Uses
- Annual dicotyledons in **winter oilseed rape**
- Annual grasses in **winter oilseed rape**

Approval information
- Imazamox and metazachlor are included in Annex I under EC Regulation 1107/2009

FOR FULL CONDITIONS OF USE ALWAYS READ THE PRODUCT LABEL

Efficacy guidance
- Must only be used for weed control in CLEARFIELD oilseed rape hybrids. Treatment of an oilseed rape variety that is not a 'CLEARFIELD Hybrid' will result in complete crop loss.

Restrictions
- Applications shall be limited to a total dose of not more than 1.0 kg metazachlor/ha in a three year period on the same field
- To avoid the buildup of resistance, do not apply this or any other product containing an ALS inhibitor herbicide with claims for grass weed control more than once to any crop.
- Do not apply in mixture with phosphate liquid fertilisers
- Metazachlor stewardship requires that all autumn applications should be made before the end of September to reduce the risk to water

Following crops guidance
- Wheat, barley, oats, oilseed rape, field beans, combining peas and sugar beet can follow normally harvested oilseed rape treated with [1].
- In the event of crop failure CLEARFIELD oilseed rape may be drilled 4 weeks after application after thorough mixing of the soil to distribute residues. If winter field beans are to be planted, allow 10 weeks after application and plough before planting. Wheat and barley also require ploughing before planting but may be drilled 8 weeks after application.

Environmental safety
- Metazachlor stewardship guidelines advise a maximum dose of 750 g.a.i/ha/annum. Applications to drained fields should be complete by 15th Oct but, if drains are flowing, complete applications by 1st Oct.
- LERAP Category B

Hazard classification and safety precautions
Hazard Irritant, Dangerous for the environment
Transport code 9
Packaging group III
UN Number 3082
Operator protection A, H; U05a
Environmental protection E07d, E15b, E16a, E40b
Storage and disposal D01, D02, D09a, D10c

281 imazamox + pendimethalin

A pre-emergence broad-spectrum herbicide mixture for legumes
HRAC mode of action code: B + K1

See also pendimethalin

Products
Nirvana	BASF	16.7:250 g/l	EC	14256

Uses
- Annual dicotyledons in **broad beans** (off-label), **combining peas**, **hops** (off-label), **ornamental plant production** (off-label), **soft fruit** (off-label), **spring field beans**, **top fruit** (off-label), **vining peas**, **winter field beans**
- Black bindweed in **soya beans** (off-label)
- Charlock in **soya beans** (off-label)
- Chickweed in **soya beans** (off-label)
- Common orache in **soya beans** (off-label)
- Fat hen in **soya beans** (off-label)
- Field speedwell in **soya beans** (off-label)
- Fumitory in **soya beans** (off-label)
- Poppies in **soya beans** (off-label)

Extension of Authorisation for Minor Use (EAMUs)
- **broad beans** 20092891

Pesticide Profiles

- **hops** 20092894
- **ornamental plant production** 20092894
- **soft fruit** 20092894
- **soya beans** 20152425
- **top fruit** 20092894

Approval information
- Imazamox and pendimethalin included in Annex I under EC Regulation 1107/2009

Efficacy guidance
- Best results obtained from applications to fine firm seedbeds in the presence of adequate moisture
- Weed control may be reduced on cloddy seedbeds and on soils with over 6% organic matter
- Residual control may be reduced under prolonged dry conditions

Restrictions
- Maximum number of treatments 1 per crop
- Seed must be drilled to at least 2.5 cm of settled soil
- Do not use on soils containing more than 10% organic matter
- Do not apply to soils that are waterlogged or are prone to waterlogging
- Do not apply if heavy rain is forecast
- Do not soil incorporate the product or disturb the soil after application
- Consult processors before use on crops destined for processing

Crop-specific information
- Latest use: pre-crop emergence for all crops
- Inadequately covered seed may result in cupping of the leaves after application from which recovery is normally complete
- Crop damage may occur on stony or gravelly soils especially if heavy rain follows treatment
- Winter oilseed rape and other brassica crops should not be drilled as the following crop.

Following crops guidance
- Winter wheat or winter barley may be drilled as a following crop provided 3 mth have elapsed since treatment and the land has been at least cultivated by a non-inversion technique such as discing
- Winter oilseed rape or other brassica crops should not be drilled as following crops
- A minimum of 12 mth must elapse between treatment and sowing red beet, sugar beet or spinach

Environmental safety
- Dangerous for the environment
- Very toxic to aquatic organisms
- LERAP Category B

Hazard classification and safety precautions
Hazard Irritant, Dangerous for the environment
Transport code 9
Packaging group III
UN Number 3082
Risk phrases R38, R43, R50, R53a
Operator protection A, H; U05a, U14, U20b
Environmental protection E15b, E16a, E38
Storage and disposal D01, D02, D09a, D10c, D12a
Medical advice M03

282 imazamox + quinmerac

A herbicide mixture for weed control in CLEARFIELD oilseed rape only

Products
Cleravo	BASF	35:250 g/l	SC	16877

FOR FULL CONDITIONS OF USE ALWAYS READ THE PRODUCT LABEL

imazaquin

Uses
- Annual dicotyledons in **spring oilseed rape**, **winter oilseed rape**
- Charlock in **spring oilseed rape**, **winter oilseed rape**
- Cleavers in **spring oilseed rape**, **winter oilseed rape**
- Crane's-bill in **spring oilseed rape** *(cut-leaved)*, **winter oilseed rape** *(cut-leaved)*
- Dead nettle in **spring oilseed rape**, **winter oilseed rape**
- Flixweed in **spring oilseed rape**, **winter oilseed rape**
- Volunteer oilseed rape in **spring oilseed rape**, **winter oilseed rape**

Approval information
- Imazamox and quinmerac included in Annex I under EC Regulation 1107/2009

Efficacy guidance
- Transient crop scorch may occur if applied under frosty conditions.
- Soil moisture is required for efficacy via root uptake.
- Maximum activity is from application before or shortly after weed emergence. Uptake is via cotyledons, roots and shoots.

Restrictions
- Must only be applied to CLEARFIELD oilseed rape hybrids.
- Do not apply if heavy rain is forecast.
- Do not use on sands, very light soils or soils with more than 10% organic matter.
- Use on stony soils may result in a reduction in plant vigour or crop stand.

Crop-specific information
- In the event of crop failure, do not re-drill with non CLEARFIELD oilseed rape or with sugar beet. After ploughing Clearfield rape, maize or peas may be sown 4 weeks after application, wheat, barley or oats 8 weeks after application, field beans 10 weeks after application or sunflower 15 weeks after application.

Hazard classification and safety precautions
Hazard Very toxic to aquatic organisms
Transport code 9
Packaging group III
UN Number 3082
Operator protection U05a, U20a
Environmental protection E15b, E38, H410
Storage and disposal D01, D02, D09a, D10c, D12b

283 imazaquin

An imidazolinone herbicide and plant growth regulator available only in mixtures
HRAC mode of action code: B

See also chlormequat + 2-chloroethylphosphonic acid + imazaquin
chlormequat + imazaquin

284 imazosulfuron

Herbicide for use in winter cereals
HRAC mode of action code: B

Products
Brazzos	Interfarm	50% w/w	WG	16932

Uses
- Charlock in **triticale, winter barley, winter rye, winter wheat**
- Mayweeds in **triticale, winter barley, winter rye, winter wheat**
- Shepherd's purse in **triticale, winter barley, winter rye, winter wheat**
- Volunteer oilseed rape in **triticale, winter barley, winter rye, winter wheat**

SEE SECTION 3 FOR PRODUCTS ALSO REGISTERED

Pesticide Profiles

Approval information
- Imazosulfuron included in Annex I under EC Regulation 1107/2009

Efficacy guidance
- Seedbeds should be moist, firm, fine and free of clods.
- Can be used on all soil types.

Restrictions
- Do not treat broadcast crops until they are past the 3 leaf stage.
- Do not roll or harrow crops 2 weeks before or after treatment.
- Do not disturb the soil after application.
- Do not apply to crops with lush or soft growth.
- Crops should be drilled to 25mm and well covered by soil.
- Extreme care should be taken to avoid damage by drift onto plants outside the target area.
- Do not apply to undersown crops.

Crop-specific information
- Brazzos can be used on all varieties of winter wheat, winter barley, winter rye and winter triticale.

Following crops guidance
- In the year of harvest, winter wheat, winter barley, winter rye, winter triticale and winter beans can be sown following a cereal crop. In the following year after harvest, spring cereals, winter oilseed rape, spring oilseed rape, peas, sugar beet, onions, carrots, lettuce and maize can be sown. Crops other than winter wheat, winter barley, winter rye and winter triticale may only be sown after the field has been ploughed.
- In the event of crop failure for any reason, only winter wheat, winter barley, winter rye and winter triticale may be planted.

Environmental safety
- LERAP Category B

Hazard classification and safety precautions
Hazard Very toxic to aquatic organisms
Transport code 9
Packaging group III
UN Number 3077
Environmental protection E15b, E16a, E34, E38, H410
Storage and disposal D01, D02, D09a, D10c, D12a, D12b

285 imidacloprid

A neonicotinoid insecticide for seed, soil, peat or foliar treatment with a 2 year ban from Dec 2013 on use in many crops due to concerns over effects on bees.
IRAC mode of action code: 4A

See also beta-cyfluthrin + imidacloprid
bitertanol + fuberidazole + imidacloprid
fuberidazole + imidacloprid + triadimenol

Products
1	Imidasect 5GR	Fargro	5 % w/w	GR	16746
2	Merit Forest	Bayer CropScience	70% w/w	WG	17242

Uses
- Aphids in **protected ornamentals** [1]
- Glasshouse whitefly in **protected ornamentals** [1]
- Insect control in **protected forest nurseries** [2]
- Sciarid flies in **protected ornamentals** [1]
- Tobacco whitefly in **protected ornamentals** [1]
- Vine weevil in **protected ornamentals** [1]

FOR FULL CONDITIONS OF USE ALWAYS READ THE PRODUCT LABEL

4-indol-3-ylbutyric acid

Approval information
- Imidacloprid is included in Annex I under EC Regulation 1107/20
- Accepted by BBPA for use on hops

Efficacy guidance
- Treated seed should be drilled within 18 months of purchase
- Product must only be used incorporated into compost using suitable automated equipment [1]

Restrictions
- Maximum number of treatments 1 per batch of seed
- Do not use treated seed as food or feed
- Avoid puddling or run-off of irrigation water after treatment
- Use only inside greenhouses
- Plants grown in treated media may not be put outside until after they have finished flowering

Crop-specific information
- Latest use: before sowing or planting bedding plants, hardy ornamental nursery stock, pot plants

Environmental safety
- High risk to bees. Do not apply to crops in flower or to those in which bees are actively foraging. Do not apply when flowering weeds are present
- Treated seed harmful to game and wildlife

Hazard classification and safety precautions
Hazard Irritant, Harmful if swallowed [2]; Very toxic to aquatic organisms [1]
Transport code 9 [1]
Packaging group III [1]
UN Number 3077 [1]; N/C [2]
Operator protection A [1, 2]; H [2]; U05a [1, 2]; U20b [1]; U20c [2]
Environmental protection E15a [2]; E15b [1]; H410 [1, 2]
Storage and disposal D01, D02, D09a, D11a

286   4-indol-3-ylbutyric acid

A plant growth regulator promoting the rooting of cuttings

Products

1	Rhizopon AA Powder (0.5%)	Fargro	0.5% w/w	DP	17570
2	Rhizopon AA Powder (1%)	Fargro	1% w/w	DP	17571
3	Rhizopon AA Powder (2%)	Fargro	2% w/w	DP	17572
4	Rhizopon AA Tablets	Fargro	50 mg a.i.	WT	17573

Uses
- Rooting of cuttings in *ornamental plant production*

Approval information
- 4-indol-3-ylbutyric acid not included in Annex 1 under EC Regulation 1107/2009

Efficacy guidance
- Dip base of cuttings into powder immediately before planting
- Powders or solutions of different concentration are required for different types of cutting. Lowest concentration for softwood, intermediate for semi-ripe, highest for hardwood
- See label for details of concentration and timing recommended for different species
- Use of planting holes recommended for powder formulations to ensure product is not removed on insertion of cutting. Cuttings should be watered in if necessary

Restrictions
- Maximum number of treatments 1 per situation
- Use of too strong a powder or solution may cause injury to cuttings

SEE SECTION 3 FOR PRODUCTS ALSO REGISTERED

Pesticide Profiles

- No unused moistened powder should be returned to container

Crop-specific information
- Latest use: before cutting insertion for ornamental specimens

Hazard classification and safety precautions
UN Number N/C
Operator protection U19a [1-4]; U20a [2-4]; U20b [1]
Environmental protection E15a
Storage and disposal D09a, D11a

287 indoxacarb

An oxadiazine insecticide for caterpillar control in a range of crops
IRAC mode of action code: 22A

Products

1	Explicit	DuPont	30% w/w	WG	15359
2	Rumo	DuPont	30% w/w	WG	14883
3	Steward	DuPont	30% w/w	WG	13149

Uses
- Cabbage leafroller in **chinese cabbage** *(off-label)*, **choi sum** *(off-label)*, **collards** *(off-label)*, **kale** *(off-label)*, **kohlrabi** *(off-label)*, **pak choi** *(off-label)*, **protected baby leaf crops** *(off-label)*, **protected herbs (see appendix 6)** *(off-label)*, **protected kohlrabi** *(off-label)*, **spring greens** *(off-label)*, **tatsoi** *(off-label)* [1-3]
- Cabbage moth in **chinese cabbage** *(off-label)*, **choi sum** *(off-label)*, **collards** *(off-label)*, **kale** *(off-label)*, **kohlrabi** *(off-label)*, **pak choi** *(off-label)*, **spring greens** *(off-label)*, **tatsoi** *(off-label)* [1-3]; **protected baby leaf crops** *(off-label)*, **protected herbs (see appendix 6)** *(off-label)*, **protected kohlrabi** *(off-label)* [1, 2]
- Cabbage white butterfly in **chinese cabbage** *(off-label)*, **choi sum** *(off-label)*, **collards** *(off-label)*, **kale** *(off-label)*, **kohlrabi** *(off-label)*, **pak choi** *(off-label)*, **protected baby leaf crops** *(off-label)*, **protected herbs (see appendix 6)** *(off-label)*, **protected kohlrabi** *(off-label)*, **spring greens** *(off-label)*, **tatsoi** *(off-label)* [1-3]
- Capsids in **apricots** *(off-label)*, **nectarines** *(off-label)*, **peaches** *(off-label)* [1]; **strawberries** *(off-label)* [1, 3]
- Caterpillars in **all edible seed crops grown outdoors** *(off-label)*, **all non-edible seed crops grown outdoors** *(off-label)*, **blackberries** *(off-label)*, **blueberries** *(off-label)*, **endives** *(off-label)*, **lettuce** *(off-label)*, **protected blackberries** *(off-label)*, **protected raspberries** *(off-label)*, **raspberries** *(off-label)* [3]; **apples**, **cherries** *(off-label)*, **forest nurseries** *(off-label)*, **grapevines** *(off-label)*, **hops** *(off-label)*, **ornamental plant production** *(off-label)*, **pears**, **protected forest nurseries** *(off-label)*, **protected hops** *(off-label)*, **protected soft fruit** *(off-label)*, **protected top fruit** *(off-label)*, **soft fruit** *(off-label)*, **top fruit** *(off-label)* [1, 3]; **broccoli**, **brussels sprouts** *(off-label)*, **cabbages**, **cauliflowers**, **protected aubergines**, **protected courgettes**, **protected cucumbers**, **protected marrows**, **protected melons**, **protected ornamentals**, **protected peppers**, **protected pumpkins**, **protected squashes**, **protected tomatoes**, **sweetcorn** *(off-label)* [1-3]; **chinese cabbage** *(off-label)*, **choi sum** *(off-label)*, **collards** *(off-label)*, **kale** *(off-label)*, **kohlrabi** *(off-label)* [1, 2]; **pak choi** *(off-label)*, **protected all edible seed crops** *(off-label)*, **protected all non-edible seed crops** *(off-label)*, **protected baby leaf crops** *(off-label)*, **protected herbs (see appendix 6)** *(off-label)*, **protected kohlrabi** *(off-label)*, **spring greens** *(off-label)*, **tatsoi** *(off-label)* [1]
- Diamond-back moth in **chinese cabbage** *(off-label)*, **choi sum** *(off-label)*, **collards** *(off-label)*, **kale** *(off-label)*, **kohlrabi** *(off-label)*, **pak choi** *(off-label)*, **protected baby leaf crops** *(off-label)*, **protected herbs (see appendix 6)** *(off-label)*, **protected kohlrabi** *(off-label)*, **spring greens** *(off-label)*, **tatsoi** *(off-label)* [1-3]; **endives** *(off-label)*, **lettuce** *(off-label)* [1, 2]; **sweetcorn** *(off-label)* [3]
- Flax tortrix moth in **endives** *(off-label)*, **lettuce** *(off-label)* [1, 2]
- Garden pebble moth in **chinese cabbage** *(off-label)*, **choi sum** *(off-label)*, **collards** *(off-label)*, **kale** *(off-label)*, **kohlrabi** *(off-label)*, **pak choi** *(off-label)*, **protected baby leaf crops** *(off-label)*,

FOR FULL CONDITIONS OF USE ALWAYS READ THE PRODUCT LABEL

indoxacarb

- **protected herbs (see appendix 6)** *(off-label),* **protected kohlrabi** *(off-label),* **spring greens** *(off-label),* **tatsoi** *(off-label)* [1-3]
- Ghost moth in **endives** *(off-label),* **lettuce** *(off-label)* [1, 2]
- Insect pests in **all edible seed crops grown outdoors** *(off-label),* **all non-edible seed crops grown outdoors** *(off-label),* **forest nurseries** *(off-label),* **hops** *(off-label),* **ornamental plant production** *(off-label),* **protected forest nurseries** *(off-label),* **protected hops** *(off-label)* [2]; **blackberries** *(off-label),* **blueberries** *(off-label),* **protected blackberries** *(off-label),* **protected raspberries** *(off-label),* **raspberries** *(off-label)* [1, 3]
- Peach moth in **apricots** *(off-label),* **nectarines** *(off-label),* **peaches** *(off-label)* [1, 3]; **strawberries** *(off-label)* [1]
- Pollen beetle in **mustard** *(off-label)* [1, 3]; **spring oilseed rape, winter oilseed rape** [1-3]
- Silver Y moth in **endives** *(off-label),* **lettuce** *(off-label),* **protected baby leaf crops** *(off-label),* **protected herbs (see appendix 6)** *(off-label),* **protected kohlrabi** *(off-label)* [1-3]
- Swift moth in **endives** *(off-label),* **lettuce** *(off-label)* [1, 2]
- Winter moth in **apricots** *(off-label),* **nectarines** *(off-label),* **peaches** *(off-label)* [1, 3]; **strawberries** *(off-label)* [1]

Extension of Authorisation for Minor Use (EAMUs)

- **all edible seed crops grown outdoors** *20101537* [2], *20082905* [3]
- **all non-edible seed crops grown outdoors** *20101537* [2], *20082905* [3]
- **apricots** *20141468* [1], *20141031* [3]
- **blackberries** *20131369* [1], *20130988* [3]
- **blueberries** *20131369* [1], *20130988* [3]
- **brussels sprouts** *20112093* [1], *20101537* [2], *20081485* [3]
- **cherries** *20112092* [1], *20092300* [3]
- **chinese cabbage** *20141466* [1], *20141858* [1], *20141465* [2], *20141859* [2], *20141690* [3]
- **choi sum** *20141858* [1], *20141859* [2], *20141690* [3]
- **collards** *20141858* [1], *20141859* [2], *20141690* [3]
- **endives** *20161959* [1], *20161960* [2], *20161566* [3]
- **forest nurseries** *20112090* [1], *20101538* [2], *20082905* [3]
- **grapevines** *20112091* [1], *20072289* [3]
- **hops** *20112090* [1], *20101537* [2], *20082905* [3]
- **kale** *20141466* [1], *20141858* [1], *20141465* [2], *20141859* [2], *20141690* [3]
- **kohlrabi** *20141466* [1], *20141858* [1], *20141465* [2], *20141859* [2], *20141690* [3]
- **lettuce** *20161959* [1], *20161960* [2], *20161566* [3]
- **mustard** *20141836* [1], *20151918* [3]
- **nectarines** *20141468* [1], *20141031* [3]
- **ornamental plant production** *20112090* [1], *20101537* [2], *20082905* [3]
- **pak choi** *20141466* [1], *20141858* [1], *20141465* [2], *20141690* [3]
- **peaches** *20141468* [1], *20141031* [3]
- **protected all edible seed crops** *20112090* [1]
- **protected all non-edible seed crops** *20112090* [1]
- **protected baby leaf crops** *20141467* [1], *20141464* [2], *20141030* [3]
- **protected blackberries** *20131369* [1], *20130988* [3]
- **protected forest nurseries** *20112090* [1], *20101537* [2], *20082905* [3]
- **protected herbs (see appendix 6)** *20141467* [1], *20141464* [2], *20141030* [3]
- **protected hops** *20112090* [1], *20101537* [2], *20082905* [3]
- **protected kohlrabi** *20141467* [1], *20141464* [2], *20141030* [3]
- **protected raspberries** *20131369* [1], *20130988* [3]
- **protected soft fruit** *20112090* [1], *20082905* [3]
- **protected top fruit** *20112090* [1], *20082905* [3]
- **raspberries** *20131369* [1], *20130988* [3]
- **soft fruit** *20112090* [1], *20082905* [3]
- **spring greens** *20141466* [1], *20141858* [1], *20141465* [2], *20141690* [3]
- **strawberries** *20141468* [1], *20141031* [3]
- **sweetcorn** *20112089* [1], *20101536* [2], *20072288* [3]
- **tatsoi** *20141466* [1], *20141858* [1], *20141465* [2], *20141690* [3]
- **top fruit** *20112090* [1], *20082905* [3]

SEE SECTION 3 FOR PRODUCTS ALSO REGISTERED

Pesticide Profiles

Approval information
- Indoxacarb included in Annex I under EC Regulation 1107/2009

Efficacy guidance
- Best results in brassica and protected crops obtained from treatment when first caterpillars are detected, or when damage first seen, or 7-10 d after trapping first adults in pheromone traps
- In apples and pears apply at egg-hatch
- Subsequent treatments in all crops may be applied at 8-14 d intervals
- Indoxacarb acts by ingestion and contact. Only larval stages are controlled but there is some ovicidal action against some species

Restrictions
- Maximum number of treatments 3 per crop or yr for apples, pears, brassica crops; 6 per yr for protected crops and 1 per crop for oilseed rape
- Do not apply to any crop suffering from stress from any cause

Crop-specific information
- HI: brassica crops, protected crops 1 d; apples, pears 7 d

Following crops guidance
- When treating protected crops, observe the maximum concentration permitted.

Environmental safety
- Dangerous for the environment
- Toxic to aquatic organisms
- In accordance with good agricultural practice apply in early morning or late evening when bees are less active
- Broadcast air-assisted LERAP [1-3] (15m); LERAP Category B [1, 2]

Hazard classification and safety precautions
Hazard Harmful, Dangerous for the environment, Harmful if swallowed
Transport code 9
Packaging group III
UN Number 3077
Risk phrases H371
Operator protection A, D; U04a, U05a
Environmental protection E12f, E15b, E34, E38, H410 [1-3]; E16a [1, 2]; E17b [1-3] (15m)
Storage and disposal D01, D02, D09a, D12a
Medical advice M03

288 iodosulfuron-methyl-sodium + mesosulfuron-methyl

A sulfonyl urea herbicide mixture for winter wheat
HRAC mode of action code: B + B

See also mesosulfuron-methyl

Products

1	Atlantis WG	Bayer CropScience	0.6:3.0% w/w	WG	12478
2	Greencrop Doonbeg 2	Clayton	0.6:3.0 % w/w	WG	13025
3	Hatra	Bayer CropScience	2:10 g/l	OD	16190
4	Horus	Bayer CropScience	2:10 g/l	OD	16216
5	Nemo	AgChem Access	0.6:3.0% w/w	WG	14140
6	Ocean	Goldengrass	0.6:3.0 % w/w	WG	14422
7	Pacifica	Bayer CropScience	1.0:3.0% w/w	WG	12049
8	Standon Mimas WG	Standon	0.6:3.0% w/w	WG	17170

Uses
- Annual dicotyledons in **durum wheat** *(off-label)*, **spring rye** *(off-label)*, **triticale** *(off-label)*, **winter rye** *(off-label)* [1, 7]; **spring wheat** *(off-label)* [1]
- Annual grasses in **durum wheat** *(off-label)*, **spring rye** *(off-label)*, **triticale** *(off-label)*, **winter rye** *(off-label)* [1, 7]; **spring wheat** *(off-label)* [1]

FOR FULL CONDITIONS OF USE ALWAYS READ THE PRODUCT LABEL

iodosulfuron-methyl-sodium + mesosulfuron-methyl

- Annual meadow grass in **winter wheat** [1-8]
- Blackgrass in **winter wheat** [1-8]
- Chickweed in **winter wheat** [1-8]
- Italian ryegrass in **winter wheat** [1-8]
- Mayweeds in **winter wheat** [1-8]
- Perennial ryegrass in **winter wheat** [1-8]
- Rough-stalked meadow grass in **winter wheat** [1-8]
- Wild oats in **winter wheat** [1-8]

Extension of Authorisation for Minor Use (EAMUs)
- **durum wheat** *20063529* [1], *20081264* [7]
- **spring rye** *20063529* [1], *20081264* [7]
- **spring wheat** *20082946* [1]
- **triticale** *20063529* [1], *20081264* [7]
- **winter rye** *20063529* [1], *20081264* [7]

Approval information
- Iodosulfuron-methyl-sodium and mesosulfuron-methyl included in Annex I under EC Regulation 1107/2009

Efficacy guidance
- Optimum grass weed control obtained when all grass weeds are emerged at spraying. Activity is primarily via foliar uptake and good spray coverage of the target weeds is essential
- Translocation occurs readily within the target weeds and growth is inhibited within hours of treatment but symptoms may not be apparent for up to 4 wk, depending on weed species, timing of treatment and weather conditions
- Residual activity is important for best results and is optimised by treatment on fine moist seedbeds. Avoid application under very dry conditions
- Residual efficacy may be reduced by high soil temperatures and cloddy seedbeds
- Iodosulfuron-methyl and mesosulfuron-methyl are both members of the ALS-inhibitor group of herbicides. To avoid the build up of resistance do not use any product containing an ALS-inhibitor herbicide with claims for control of grass weeds more than once on any crop
- Use these products as part of a Resistance Management Strategy that includes cultural methods of control and does not use ALS inhibitors as the sole chemical method of grass weed control. See Section 5 for more information

Restrictions
- Maximum number of treatments 1 per crop with a maximum total dose equivalent to one full dose treatment
- Do not use on crops undersown with grasses, clover or other legumes or any other broad-leaved crop
- Do not use as a stand-alone treatment for blackgrass, ryegrass or chickweed control
- Do not use as the sole means of weed control in successive crops
- Do not use in mixture or in sequence with any other ALS-inhibitor herbicide except those (if any) specified on the label
- Do not apply earlier than 1 Feb in the yr of harvest [7]
- Do not apply to crops under stress from any cause
- Do not apply when rain is imminent or during periods of frosty weather
- Specified adjuvant must be used. See label

Crop-specific information
- Latest use: flag leaf just visible (GS 39)
- Winter wheat may be treated from the two-leaf stage of the crop
- Safety to crops grown for seed not established
- Avoid application before a severe frost or transient yellowing of the crop may occur.

Following crops guidance
- In the event of crop failure sow only winter wheat in the same cropping season
- Only winter wheat or winter barley (or winter oilseed rape [1]) may be sown in the year of harvest of a treated crop but ploughing before drilling oilseed rape is advisable, especially in a season where applications are made later than usual.

SEE SECTION 3 FOR PRODUCTS ALSO REGISTERED

Pesticide Profiles

- Spring wheat, spring barley, sugar beet or spring oilseed rape may be drilled in the following spring. Plough before drilling oilseed rape

Environmental safety
- Dangerous for the environment
- Very toxic to aquatic organisms
- Dangerous to fish or other aquatic life. Do not contaminate surface waters or ditches with chemical or used container
- Take extreme care to avoid drift onto plants outside the target area or on to ponds, waterways or ditches
- LERAP Category B

Hazard classification and safety precautions
Hazard Irritant, Dangerous for the environment
Transport code 9
Packaging group III
UN Number 3077 [1, 2, 5-8]; 3082 [3, 4]
Risk phrases R36, R38 [3, 4]; R40 [7]; R41 [1, 2, 5-8]; R50, R53a [1-8]
Operator protection A, C, H; U05a, U11, U20b
Environmental protection E15a [1, 3-8]; E15b [2]; E16a, E38 [1-8]
Storage and disposal D01, D02, D05, D09a, D10a, D12a

289 ipconazole

A triazole fungicide for disease control in cereals
FRAC mode of action code: 3

See also imazalil + ipconazole

Products
Rancona 15 ME   Certis   15.1 g/l   MS   16500

Uses
- Seed-borne diseases in **spring barley**, **spring wheat**, **winter barley**, **winter wheat**
- Soil-borne diseases in **spring barley**, **spring wheat**, **winter barley**, **winter wheat**

Approval information
- Ipconazole included in Annex I under EC Regulation 1107/2009
- Accepted by BBPA for use on malting barley

Hazard classification and safety precautions
UN Number N/C
Operator protection A, H; U05a, U09a, U20b
Environmental protection E03, E15b, E34, H410
Storage and disposal D01, D02, D09a, D10a
Treated seed S01, S02, S03, S04d, S05, S07
Medical advice M03

290 iprodione

A protectant dicarboximide fungicide with some eradicant activity
FRAC mode of action code: 2

See also carbendazim + iprodione

Products

1	Cavron Green	Belcrop	250 g/l	SC	15340
2	Chipco Green	Bayer Environ.	255 g/l	SC	13843
3	Green Turf King	Standon	255 g/l	SC	14348
4	Grisu	Sipcam	500 g/l	SC	15375
5	Mascot Rayzor	Rigby Taylor	255 g/l	SC	14903

FOR FULL CONDITIONS OF USE ALWAYS READ THE PRODUCT LABEL

iprodione

Products – continued

6	Panto	Pan Amenity	255 g/l	SC	14307
7	Pure Turf	Pure Amenity	255 g/l	SC	14851
8	Rovral AquaFlo	BASF	500 g/l	SC	14206
9	Rovral WG	BASF	75% w/w	WG	13811
10	Surpass Pro	Headland Amenity	250 g/l	SC	15368

Uses

- Alternaria in **brassica seed crops, lettuce, oilseed rape, ornamental plant production, protected lettuce, protected strawberries, protected tomatoes, raspberries, salad onions, strawberries** [4]; **brassica seed crops** (pre-storage only), **dwarf beans** (off-label), **french beans** (off-label), **garlic** (off-label), **grapevines** (off-label), **mange-tout peas** (off-label), **protected herbs (see appendix 6)** (off-label), **runner beans** (off-label), **shallots** (off-label), **spring oilseed rape, winter oilseed rape** [9]; **broccoli** (off-label), **brussels sprouts** (off-label), **cabbages** (off-label), **calabrese** (off-label), **cauliflowers** (off-label), **chinese cabbage** (off-label), **collards** (off-label), **kale** (off-label) [8]; **brussels sprouts, cauliflowers** [4, 9]
- Anthracnose in **managed amenity turf** [2]
- Botrytis in **brassica seed crops, brussels sprouts, cauliflowers, oilseed rape** [4]; **bulb onions, cabbages** (off-label), **dwarf beans** (off-label), **forest nurseries** (off-label), **french beans** (off-label), **garlic** (off-label), **grapevines** (off-label), **hops** (off-label), **mange-tout peas** (off-label), **ornamental plant production** (off-label), **pears** (off-label), **protected aubergines** (off-label), **protected forest nurseries** (off-label), **protected herbs (see appendix 6)** (off-label), **protected hops** (off-label), **protected ornamentals** (off-label), **protected rhubarb** (off-label), **protected soft fruit** (off-label), **protected top fruit** (off-label), **red beet** (off-label), **runner beans** (off-label), **shallots** (off-label), **soft fruit** (off-label), **spring oilseed rape, top fruit** (off-label), **winter oilseed rape** [9]; **chicory** (off-label) [8]; **lettuce, ornamental plant production, protected lettuce, protected strawberries, protected tomatoes, raspberries, salad onions, strawberries** [4, 9]
- Brown patch in **amenity grassland** [7]; **managed amenity turf** [1, 7, 10]
- Collar rot in **bulb onions, garlic** (off-label), **salad onions, shallots** (off-label) [9]
- Disease control in **managed amenity turf** [5, 6]; **ornamental plant production** (off-label) [9]
- Dollar spot in **amenity grassland** [2, 7]; **managed amenity turf** [1, 2, 7, 10]
- Fusarium patch in **amenity grassland** [2, 7]; **managed amenity turf** [1-3, 7, 10]
- Melting out in **amenity grassland** [2, 7]; **managed amenity turf** [1, 2, 7, 10]
- Red thread in **amenity grassland** [2, 7]; **managed amenity turf** [1-3, 7, 10]
- Rust in **managed amenity turf** [2]
- Sclerotinia in **chicory** (off-label) [8]; **oilseed rape** (moderate control) [4]
- Seed-borne diseases in **ornamental crop production (seed)** [8]
- Snow mould in **amenity grassland** [2, 7]; **managed amenity turf** [1, 2, 7, 10]

Extension of Authorisation for Minor Use (EAMUs)

- **broccoli** 20090708 [8]
- **brussels sprouts** 20090708 [8]
- **cabbages** 20090708 [8], 20091845 [9]
- **calabrese** 20090708 [8]
- **cauliflowers** 20090708 [8]
- **chicory** 20142764 [8]
- **chinese cabbage** 20090708 [8]
- **collards** 20090708 [8]
- **dwarf beans** 20081760 [9]
- **forest nurseries** 20082902 [9]
- **french beans** 20081760 [9]
- **garlic** 20081768 [9]
- **grapevines** 20081758 [9]
- **hops** 20082902 [9]
- **kale** 20090708 [8]
- **mange-tout peas** 20081760 [9]
- **ornamental plant production** 20113200 [9]
- **pears** 20081761 [9]
- **protected aubergines** 20081767 [9]
- **protected forest nurseries** 20082902 [9]

SEE SECTION 3 FOR PRODUCTS ALSO REGISTERED

Pesticide Profiles

- *protected herbs (see appendix 6)* 20081757 [9]
- *protected hops* 20082902 [9]
- *protected ornamentals* 20113200 [9]
- *protected rhubarb* 20081763 [9]
- *protected soft fruit* 20082902 [9]
- *protected top fruit* 20082902 [9]
- *red beet* 20081762 [9]
- *runner beans* 20081760 [9]
- *shallots* 20081768 [9]
- *soft fruit* 20082902 [9]
- *top fruit* 20082902 [9]

Approval information
- Iprodione included in Annex I under EC Regulation 1107/2009

Efficacy guidance
- Many diseases require a programme of 2 or more sprays at intervals of 2-4 wk. Recommendations vary with disease and crop - see label for details
- Use as a drench to control cabbage storage diseases. Spray ornamental pot plants and cucumbers to run-off
- Apply turf and amenity grass treatments after mowing to dry grass, free of dew. Do not mow again for at least 24 h [2]

Restrictions
- Maximum number of treatments or maximum total dose equivalent to 1 per batch for seed treatments; 1 per crop on cereals, cabbage (as drench), 2 per crop on field beans and stubble turnips; 3 per crop on brassicas (including seed crops), oilseed rape, protected winter lettuce (Oct-Feb); 4 per crop on strawberries, grapevines, salad onions, cucumbers; 5 per crop on raspberries; 6 per crop on bulb onions, tomatoes, turf; 7 per crop on lettuce (Mar-Sep)
- A minimum of 3 wk must elapse between treatments on leaf brassicas
- Treated brassica seed crops not to be used for human or animal consumption
- See label for pot plants showing good tolerance. Check other species before applying on a large scale
- Do not treat oilseed rape seed that is cracked or broken or of low viability
- Do not treat oats

Crop-specific information
- Latest use: pre-planting for seed treatments; at planting for potatoes; before grain watery ripe (GS 69) for wheat and barley; 21 d before consumption by livestock for stubble turnips
- HI strawberries, protected tomatoes 1 d; outdoor tomatoes 2 d; bulb onions, salad onions, raspberries, lettuce 7 d; brassicas, brassica seed crops, oilseed rape, mustard, field beans, stubble turnips 21 d
- Turf and amenity grass may become temporarily yellowed if frost or hot weather follows treatment
- Personal protective equipment requirements may vary for each pack size. Check label

Environmental safety
- Dangerous for the environment
- Very toxic to aquatic organisms
- See label for guidance on disposal of spent drench liquor
- Treatment harmless to *Encarsia* or *Phytoseiulus* being used for integrated pest control
- Broadcast air-assisted LERAP [9] (18 m); LERAP Category B [3, 9]

Hazard classification and safety precautions
Hazard Harmful, Dangerous for the environment
Transport code 9
Packaging group III
UN Number 3077 [9]; 3082 [1-8, 10]
Risk phrases R36 [1, 2, 5-7, 9, 10]; R38 [1, 2, 5-7, 10]; R40 [1, 2, 4-10]; R50, R53a [1-10]
Operator protection A [1-10]; C [1, 2, 5-7, 9, 10]; H [4, 8, 9]; P [3]; U04a, U08 [1, 2, 5-7, 10]; U05a [1-10]; U09a [3]; U11 [4, 8, 9]; U16b, U19a, U20b [9]; U20c [1-8, 10]

FOR FULL CONDITIONS OF USE ALWAYS READ THE PRODUCT LABEL

iprodione + thiophanate-methyl

Environmental protection E15a [1, 2, 4-8, 10]; E15b, E16a [3, 9]; E16b [3]; E17b [9] (18 m); E34 [1-10]; E38 [1, 2, 4-10]
Storage and disposal D01, D02, D09a [1-10]; D05, D10b [1, 2, 5-7, 10]; D10a [3]; D11a [9]; D12a [1, 2, 4-10]
Treated seed S01, S02, S03, S04a, S05 [4, 8]
Medical advice M03 [1, 2, 4-8, 10]; M04a [3]; M05a [3, 4, 8, 9]

291   iprodione + thiophanate-methyl

A protectant and systemic fungicide for oilseed rape
FRAC mode of action code: 2 + 1

See also thiophanate-methyl

Products

| Compass | BASF | 167:167 g/l | SC | 16801 |

Uses
- Alternaria in **spring oilseed rape, winter oilseed rape**
- Grey mould in **spring oilseed rape, winter oilseed rape**
- Light leaf spot in **spring oilseed rape, winter oilseed rape**
- Sclerotinia stem rot in **spring oilseed rape, winter oilseed rape**
- Stem canker in **spring oilseed rape, winter oilseed rape**

Approval information
- Iprodione and thiophanate-methyl included in Annex I under EC Regulation 1107/2009

Efficacy guidance
- Timing of sprays on oilseed rape varies with disease, see label for details
- For season-long control product should be applied as part of a disease control programme

Restrictions
- Maximum number of treatments (refers to total sprays containing benomyl, carbendazim or thiophanate methyl) 2 per crop for oilseed rape, field beans, meadowfoam; 3 per crop for carrots, horseradish, parsnips

Crop-specific information
- Latest use: before end of flowering for oilseed rape, field beans
- HI meadowfoam, field beans, oilseed rape 3 wk; carrots, horseradish, parsnips 28 d
- Treatment may extend duration of green leaf in winter oilseed rape

Environmental safety
- Dangerous for the environment
- Very toxic to aquatic organisms

Hazard classification and safety precautions
Hazard Harmful, Dangerous for the environment
Transport code 9
Packaging group III
UN Number 3082
Risk phrases R40, R43, R50, R53a, R68
Operator protection A, C, H, M; U05a, U20c
Environmental protection E15a, E38
Storage and disposal D01, D02, D08, D09a, D10b, D12a
Medical advice M05a

SEE SECTION 3 FOR PRODUCTS ALSO REGISTERED

292   iprodione + trifloxystrobin

A fungicide for use in managed amenity turf
FRAC mode of action code: 2 + 11

Products
    Interface   Bayer Environ.   256:16 g/l   SC   16060

Uses
- Anthracnose in **managed amenity turf** (moderate control)
- Disease control in **managed amenity turf**
- Dollar spot in **managed amenity turf**
- Fusarium patch in **managed amenity turf**
- Leaf spot in **managed amenity turf**
- Red thread in **managed amenity turf**
- Rust in **managed amenity turf**

Approval information
- Iprodione and trifloxystrobin included in Annex I under EC Regulation 1107/2009

Efficacy guidance
- Do not apply during drought or when turf is frozen
- Do not mow for at least 48 hours after application
- Allow at least 4 weeks between applications

Restrictions
- Do not use on commercial, estate or residential lawns

Environmental safety
- LERAP Category B

Hazard classification and safety precautions
    Hazard Very toxic to aquatic organisms
    Transport code 9
    Packaging group III
    UN Number 3082
    Risk phrases H351
    Operator protection A
    Environmental protection E15a, E16a, E16b, E34, H410
    Storage and disposal D01, D09a, D10b

293   isopyrazam

An SDHI fungicide available for disease control in barley.
FRAC mode of action code: 7

See also azoxystrobin + isopyrazam
         cyprodinil + isopyrazam
         epoxiconazole + isopyrazam

Products
1	Reflect	Syngenta	125 g/l	EC	17228
2	Zulu	Syngenta	125 g/l	EC	16913

Uses
- Alternaria in **red beet** (off-label) [1]
- Canker in **carrots**, **parsnips** (off-label) [1]
- Cercospora leaf spot in **red beet** (off-label) [1]
- Disease control in **protected aubergines**, **protected chilli peppers**, **protected courgettes**, **protected cucumbers**, **protected melons**, **protected peppers**, **protected summer squash**, **protected tomatoes** [1]; **spring barley**, **winter barley** [2]

isoxaben

- Mildew in **red beet** *(off-label)* [1]
- Powdery mildew in **ornamental plant production** *(off-label)*, **protected ornamentals** *(off-label)*, **swedes** *(off-label)*, **turnips** *(off-label)* [1]
- Ramularia leaf spots in **red beet** *(off-label)* [1]
- Sclerotinia in **celeriac** *(off-label)* [1]
- Septoria seedling blight in **celeriac** *(off-label)* [1]

Extension of Authorisation for Minor Use (EAMUs)
- **celeriac** 20160798 [1]
- **ornamental plant production** 20161602 [1]
- **parsnips** 20160798 [1]
- **protected ornamentals** 20161602 [1]
- **red beet** 20160798 [1]
- **swedes** 20160798 [1]
- **turnips** 20160798 [1]

Approval information
- Isopyrazam is included in Appendix 1 under EC Regulation 1107/2009

Efficacy guidance
- Use only in mixture with fungicides with a different mode of action that are effective against the target diseases.

Restrictions
- Isopyrazam is an SDH respiration inhibitor; Do not apply more than two foliar applications of products containing an SDH inhibitor to any cereal crop

Crop-specific information
- There are no restrictions on succeeding crops in a normal rotation

Environmental safety
- LERAP Category B

Hazard classification and safety precautions
Hazard Harmful, Irritant, Dangerous for the environment [2]; Harmful if swallowed [1]
Transport code 9
Packaging group III
UN Number 3082
Risk phrases H319, H351, H361 [1]; R20, R22a, R38, R40, R43, R50, R53a, R63 [2]
Operator protection A, H [1, 2]; C [1]
Environmental protection E16a, H410 [1]; E38 [2]
Storage and disposal D01, D02, D12a, D12b [2]

294 isoxaben

A soil-acting benzamide herbicide for use in cereals, grass and fruit
HRAC mode of action code: L

Products
Flexidor	Landseer	500 g/l	SC	05121

Uses
- Annual dicotyledons in **amenity vegetation**, **bush fruit**, **cane fruit**, **durum wheat**, **farm forestry** *(off-label)*, **forest nurseries**, **grapevines**, **hops**, **ornamental plant production**, **spring barley**, **spring oats**, **spring wheat**, **strawberries**, **top fruit**, **triticale**, **winter barley**, **winter oats**, **winter rye**, **winter wheat**
- Cleavers in **asparagus** *(off-label)*, **courgettes** *(off-label)*, **pumpkins** *(off-label)*, **summer squash** *(off-label)*, **winter squash** *(off-label)*
- Fat hen in **asparagus** *(off-label)*, **courgettes** *(off-label)*, **pumpkins** *(off-label)*, **summer squash** *(off-label)*, **winter squash** *(off-label)*

SEE SECTION 3 FOR PRODUCTS ALSO REGISTERED

Pesticide Profiles

- Groundsel in **asparagus** *(off-label)*, **courgettes** *(off-label)*, **pumpkins** *(off-label)*, **summer squash** *(off-label)*, **winter squash** *(off-label)*
- Knotgrass in **asparagus** *(off-label)*, **courgettes** *(off-label)*, **pumpkins** *(off-label)*, **summer squash** *(off-label)*, **winter squash** *(off-label)*

Extension of Authorisation for Minor Use (EAMUs)
- **asparagus** *20160284*
- **courgettes** *20161368*
- **farm forestry** *910735*
- **pumpkins** *20161368*
- **summer squash** *20161368*
- **winter squash** *20161368*

Approval information
- Isoxaben included in Annex 1 under EC Regulation 1107/2009
- Accepted by BBPA for use on hops

Efficacy guidance
- When used alone apply pre-weed emergence
- Effectiveness is reduced in dry conditions. Weed seeds germinating at depth are not controlled
- Activity reduced on soils with more than 10% organic matter. Do not use on peaty soils
- Various tank mixtures are recommended for early post-weed emergence treatment (especially for grass weeds). See label for details

Restrictions
- Maximum number of treatments 1 per crop for all edible crops; 2 per yr on amenity vegetation and non-edible crops

Crop-specific information
- Latest use: before 1 Apr in yr of harvest for edible crops

Following crops guidance
- See label for details of crops which may be sown in the event of failure of a treated crop

Environmental safety
- Keep all livestock out of treated areas for at least 50 d
- LERAP Category B

Hazard classification and safety precautions
Transport code 9
Packaging group III
UN Number 3082
Operator protection A, C, H, M; U05a, U20a
Environmental protection E06a (50 days); E15b, E16a
Storage and disposal D01, D02, D09a, D11a

295   isoxaflutole

An isoxazole herbicide available only in mixtures
HRAC mode of action code: F2

See also flufenacet + isoxaflutole

FOR FULL CONDITIONS OF USE ALWAYS READ THE PRODUCT LABEL

296 kresoxim-methyl

A protectant strobilurin fungicide for apples
FRAC mode of action code: 11

See also epoxiconazole + fenpropimorph + kresoxim-methyl
   epoxiconazole + kresoxim-methyl
   epoxiconazole + kresoxim-methyl + pyraclostrobin
   fenpropimorph + kresoxim-methyl

Products

   Stroby WG   BASF   50% w/w   WG   17316

Uses

- American gooseberry mildew in **blueberries** *(off-label)*, **cranberries** *(off-label)*, **gooseberries** *(off-label)*, **whitecurrants** *(off-label)*
- Black rot in **wine grapes** *(off-label)*
- Powdery mildew in **apple for cider making** *(reduction)*, **apples** *(reduction)*, **blackcurrants**, **blueberries** *(off-label)*, **cranberries** *(off-label)*, **gooseberries** *(off-label)*, **ornamental plant production**, **protected strawberries**, **redcurrants**, **strawberries**, **whitecurrants** *(off-label)*, **wine grapes** *(off-label)*
- Scab in **apple for cider making**, **apples**

Extension of Authorisation for Minor Use (EAMUs)

- **blueberries** 20160089
- **cranberries** 20160089
- **gooseberries** 20160089
- **whitecurrants** 20160089
- **wine grapes** 20160960

Approval information

- Kresoxim-methyl included in Annex I under EC Regulation 1107/2009
- Approved for use in ULV systems

Efficacy guidance

- Activity is protectant. Best results achieved from treatments prior to disease development. See label for timing details on each crop. Treatments should be repeated at 10-14 d intervals but note limitations below
- To minimise the likelihood of development of resistance to strobilurin fungicides these products should be used in a planned Resistance Management strategy. See Section 5 for more information
- Product may be applied in ultra low volumes (ULV) but disease control may be reduced

Restrictions

- Maximum number of treatments 4 per yr on apples; 3 per yr on other crops. See notes in Efficacy about limitations on consecutive treatments
- Consult before using on crops intended for processing

Crop-specific information

- HI 14 d for blackcurrants, protected strawberries, strawberries; 35 d for apples
- On apples do not spray product more than twice consecutively and separate each block of two consecutive treatments with at least two applications from a different cross-resistance group. For all other crops do not apply consecutively and use a maximum of once in every three fungicide sprays
- Product should not be used as final spray of the season on apples

Environmental safety

- Dangerous for the environment
- Very toxic to aquatic organisms
- Harmless to ladybirds and predatory mites
- Harmless to honey bees and may be applied during flowering. Nevertheless local beekeepers should be notified when treatment of orchards in flower is to occur

---

SEE SECTION 3 FOR PRODUCTS ALSO REGISTERED

Pesticide Profiles

- Broadcast air-assisted LERAP (5 m)

Hazard classification and safety precautions
Hazard Harmful, Dangerous for the environment, Very toxic to aquatic organisms
Transport code 9
Packaging group III
UN Number 3077
Risk phrases H351
Operator protection U05a, U20b
Environmental protection E15a, E38, H410; E17b (5 m)
Storage and disposal D01, D02, D08, D09a, D10c, D12a
Medical advice M05a

297 lambda-cyhalothrin

A quick-acting contact and ingested pyrethroid insecticide
IRAC mode of action code: 3

Products

1	Block	Belchim	50 g/l	EC	15697
2	Clayton Lanark	Clayton	100 g/l	CS	12942
3	Dalda 5	Globachem	50 g/l	EC	13688
4	Hallmark with Zeon Technology	Syngenta	100 g/l	CS	12629
5	Karate 2.5WG	Syngenta	2.5% w/w	GR	14060
6	Karis 10 CS	Headland	100 g/l	CS	16747
7	Lambda 100 CS	Goldengrass	100 g/l	CS	16874
8	Lambda-C 100	Becesane	100 g/l	CS	15987
9	Lambda-C 50	Becesane	50 g/l	EC	15986
10	Lambdastar	Life Scientific	100 g/l	CS	17406
11	Markate 50	Agrovista	50 g/l	EC	13529
12	Seal Z	AgChem Access	100 g/l	CS	14201
13	Sparviero	Sipcam	100 g/l	CS	15687

Uses
- Aphids in **all edible seed crops grown outdoors** *(off-label)*, **all non-edible seed crops grown outdoors** *(off-label)*, **crambe** *(off-label)*, **farm forestry** *(off-label)*, **forest nurseries** *(off-label)*, **hops** *(off-label)*, **miscanthus** *(off-label)*, **ornamental plant production** *(off-label)*, **protected aubergines** *(off-label)*, **protected forest nurseries** *(off-label)*, **protected ornamentals** *(off-label)*, **protected pak choi** *(off-label)*, **protected peppers** *(off-label)*, **protected tatsoi** *(off-label)*, **protected tomatoes** *(off-label)*, **soft fruit** *(off-label)*, **top fruit** *(off-label)*, **willow (short rotation coppice)** *(off-label)* [4]; **broccoli, brussels sprouts, calabrese, kale** [6]; **chestnuts** *(off-label)*, **hazel nuts** *(off-label)*, **walnuts** *(off-label)* [4, 11]; **cob nuts** *(off-label)*, **lupins** *(off-label)* [11]; **combining peas, edible podded peas, spring field beans, spring oilseed rape, vining peas, winter field beans, winter oilseed rape** [1, 3, 9, 11]; **durum wheat** [1-6, 8, 9, 11, 12]; **fodder beet** *(off-label)* [2]; **lettuce, pears, rye, triticale** [3]; **potatoes** [2-4, 6-13]; **spring barley, spring oats** [1-4, 6-8, 10, 12, 13]; **spring wheat** [1-4, 6-13]; **sugar beet** [3, 6, 9, 11]; **winter barley, winter oats** [1-8, 10, 12, 13]; **winter wheat** [1-13]
- Barley yellow dwarf virus vectors in **durum wheat** [2, 4-6, 8, 12]; **winter barley, winter oats, winter wheat** [2, 4-8, 10, 12, 13]
- Beet leaf miner in **sugar beet** [2, 4, 6-8, 10, 12, 13]
- Beet virus yellows vectors in **spring oilseed rape** [2, 4, 6-8, 10, 12, 13]; **winter oilseed rape** [2, 4-8, 10, 12, 13]
- Beetles in **combining peas, durum wheat, edible podded peas, potatoes, spring field beans, spring oilseed rape, spring wheat, sugar beet, vining peas, winter field beans, winter oilseed rape, winter wheat** [3, 9, 11]; **rye, spring barley, spring oats, triticale, winter barley, winter oats** [3]
- Cabbage seed weevil in **spring oilseed rape, winter oilseed rape** [2, 4, 6-8, 10, 12, 13]

FOR FULL CONDITIONS OF USE ALWAYS READ THE PRODUCT LABEL

lambda-cyhalothrin

- Cabbage stem flea beetle in **spring oilseed rape** [2, 4, 6-8, 10, 12, 13]; **winter oilseed rape** [2, 4-8, 10, 12, 13]
- Cabbage stem weevil in **celeriac** *(off-label)*, **radishes** *(off-label)* [2, 11]
- Capsids in **blackberries** *(off-label)*, **dewberries** *(off-label)*, **raspberries** *(off-label)*, **rubus hybrids** *(off-label)* [2, 4]; **grapevines** *(off-label)* [2, 4, 11]; **table grapes** *(off-label)*, **wine grapes** *(off-label)* [2]
- Carrot fly in **carrots** *(off-label)*, **parsnips** *(off-label)* [11]; **celeriac** *(off-label)*, **radishes** *(off-label)* [2, 11]; **celery (outdoor)** *(off-label)*, **horseradish** *(off-label)*, **mallow (althaea spp.)** *(off-label)*, **parsley root** *(off-label)* [2, 4, 11]; **fennel** *(off-label)* [4]
- Caterpillars in **beetroot** *(off-label)* [2, 4]; **broccoli, brussels sprouts, cabbages, calabrese, cauliflowers** [2, 4, 7, 8, 10, 12]; **celery (outdoor)** *(off-label)*, **french beans** *(off-label)*, **navy beans** *(off-label)* [2, 4, 11]; **combining peas, durum wheat, edible podded peas, potatoes, spring field beans, spring oilseed rape, spring wheat, sugar beet, vining peas, winter field beans, winter oilseed rape, winter wheat** [3, 9, 11]; **dwarf beans** *(off-label)*, **protected pak choi** *(off-label)*, **protected tatsoi** *(off-label)*, **red beet** *(off-label)* [11]; **pears, rye, spring barley, spring oats, triticale, winter barley, winter oats** [3]; **runner beans** *(off-label)* [2, 11]; **swedes** *(off-label)*, **turnips** *(off-label)* [4, 11]; **table grapes** *(off-label)*, **wine grapes** *(off-label)* [2]
- Clay-coloured weevil in **blackberries** *(off-label)*, **dewberries** *(off-label)*, **raspberries** *(off-label)*, **rubus hybrids** *(off-label)* [2, 4]
- Common green capsid in **hops** *(off-label)* [4]
- Cutworms in **beetroot** *(off-label)* [2, 4]; **carrots, lettuce, sugar beet** [2, 4, 6-8, 10, 12, 13]; **celeriac** *(off-label)*, **radishes** *(off-label)* [2, 11]; **chicory** *(off-label)*, **fennel** *(off-label)* [4]; **fodder beet** *(off-label)*, **red beet** *(off-label)* [11]; **parsnips** [4, 6-8, 10, 12, 13]
- Flea beetle in **corn gromwell** *(off-label)*, **hops** *(off-label)* [4]; **crambe** *(off-label)*, **fodder beet** *(off-label)*, **protected pak choi** *(off-label)*, **protected tatsoi** *(off-label)* [11]; **poppies for morphine production** *(off-label)* [2, 4, 11]; **spring oilseed rape, sugar beet** [2, 4, 6-8, 10, 12, 13]; **winter oilseed rape** [2, 4-8, 10, 12, 13]
- Frit fly in **sweetcorn** *(off-label)* [2, 4, 11]
- Insect pests in **all edible seed crops grown outdoors** *(off-label)*, **all non-edible seed crops grown outdoors** *(off-label)*, **celeriac** *(off-label)*, **celery (outdoor)** *(off-label)*, **chestnuts** *(off-label)*, **farm forestry** *(off-label)*, **forest nurseries** *(off-label)*, **hazel nuts** *(off-label)*, **hops** *(off-label)*, **lupins** *(off-label)*, **miscanthus** *(off-label)*, **ornamental plant production** *(off-label)*, **protected forest nurseries** *(off-label)*, **protected ornamentals** *(off-label)*, **radishes** *(off-label)*, **soft fruit** *(off-label)*, **spring rye** *(off-label)*, **top fruit** *(off-label)*, **walnuts** *(off-label)*, **winter rye** *(off-label)* [4, 11]; **beetroot** *(off-label)*, **cherries** *(off-label)*, **crambe** *(off-label)*, **mirabelles** *(off-label)*, **plums** *(off-label)*, **poppies for morphine production** *(off-label)*, **protected aubergines** *(off-label)*, **protected pak choi** *(off-label)*, **protected peppers** *(off-label)*, **protected strawberries** *(off-label)*, **protected tatsoi** *(off-label)*, **protected tomatoes** *(off-label)*, **strawberries** *(off-label)*, **willow (short rotation coppice)** *(off-label)* [4]; **blackberries** *(off-label)*, **dewberries** *(off-label)*, **frise** *(off-label)*, **lamb's lettuce** *(off-label)*, **leaf brassicas** *(off-label)*, **raspberries** *(off-label)*, **rubus hybrids** *(off-label)*, **scarole** *(off-label)* [2, 4]; **blackcurrants** *(off-label)*, **borage for oilseed production** *(off-label)*, **bulb onions** *(off-label)*, **canary flower (echium spp.)** *(off-label)*, **evening primrose** *(off-label)*, **fodder beet** *(off-label)*, **garlic** *(off-label)*, **gooseberries** *(off-label)*, **grass seed crops** *(off-label)*, **herbs (see appendix 6)** *(off-label)*, **honesty** *(off-label)*, **horseradish** *(off-label)*, **leeks** *(off-label)*, **mallow (althaea spp.)** *(off-label)*, **mustard** *(off-label)*, **navy beans** *(off-label)*, **parsley root** *(off-label)*, **redcurrants** *(off-label)*, **runner beans** *(off-label)*, **salad onions** *(off-label)*, **shallots** *(off-label)*, **spring linseed** *(off-label)*, **triticale** *(off-label)*, **whitecurrants** *(off-label)*, **winter linseed** *(off-label)* [2, 4, 11]; **broad beans** *(off-label)*, **carrots** *(off-label)*, **cob nuts** *(off-label)*, **dwarf beans** *(off-label)*, **land cress** *(off-label)*, **parsnips** *(off-label)*, **protected all edible seed crops** *(off-label)*, **protected all non-edible seed crops** *(off-label)*, **red beet** *(off-label)*, **short rotation coppice willow** *(off-label)* [11]; **combining peas, durum wheat, edible podded peas, spring barley, spring field beans, spring oats, spring oilseed rape, spring wheat, vining peas, winter barley, winter field beans, winter oats, winter oilseed rape, winter wheat** [1]; **french beans** *(off-label)*, **sweetcorn** *(off-label)* [2, 11]; **grapevines** *(off-label)*, **rye** *(off-label)* [2]
- Leaf midge in **blackcurrants** *(off-label)*, **gooseberries** *(off-label)*, **redcurrants** *(off-label)*, **whitecurrants** *(off-label)* [2, 4, 11]
- Leaf miner in **fodder beet** *(off-label)* [11]
- Mangold fly in **sugar beet** [6]

SEE SECTION 3 FOR PRODUCTS ALSO REGISTERED

Pesticide Profiles

- Pea and bean weevil in **combining peas, edible podded peas, spring field beans, vining peas, winter field beans** [2, 4, 6-8, 10, 12, 13]
- Pea aphid in **combining peas, edible podded peas, vining peas** [2, 4, 6-8, 10, 12, 13]
- Pea midge in **combining peas, edible podded peas, vining peas** [2, 4, 7, 8, 10, 12, 13]
- Pea moth in **combining peas, edible podded peas, vining peas** [2, 4, 6-8, 10, 12, 13]
- Pear midge in **combining peas, edible podded peas, vining peas** [6]
- Pear sucker in **pears** [2, 4, 6-8, 10, 12, 13]
- Pod midge in **spring oilseed rape, winter oilseed rape** [2, 4, 6-8, 10, 12, 13]
- Pollen beetle in **corn gromwell** *(off-label)* [4]; **crambe** *(off-label)* [11]; **poppies for morphine production** *(off-label)* [2, 4, 11]; **spring oilseed rape, winter oilseed rape** [2, 4, 6-8, 10, 12, 13]
- Rosy rustic moth in **hops** *(off-label)* [4]
- Sawflies in **blackcurrants** *(off-label)*, **gooseberries** *(off-label)*, **redcurrants** *(off-label)*, **whitecurrants** *(off-label)* [2, 4, 11]; **short rotation coppice willow** *(off-label)* [11]
- Seed beetle in **broad beans** *(off-label)* [2, 4, 11]
- Silver Y moth in **celery (outdoor)** *(off-label)*, **dwarf beans** *(off-label)*, **navy beans** *(off-label)*, **red beet** *(off-label)* [11]; **french beans** *(off-label)*, **runner beans** *(off-label)* [4, 11]
- Spotted wing drosophila in **bilberries** *(off-label)*, **blueberries** *(off-label)*, **cranberries** *(off-label)*, **ribes hybrids** *(off-label)* [4]
- Thrips in **bilberries** *(off-label)*, **blueberries** *(off-label)*, **cranberries** *(off-label)*, **ribes hybrids** *(off-label)* [4]; **broad beans** *(off-label)*, **bulb onions** *(off-label)*, **garlic** *(off-label)*, **leeks** *(off-label)*, **salad onions** *(off-label)*, **shallots** *(off-label)* [2, 4]
- Wasps in **grapevines** *(off-label)* [2, 4, 11]; **table grapes** *(off-label)*, **wine grapes** *(off-label)* [2]
- Weevils in **combining peas, durum wheat, edible podded peas, potatoes, spring field beans, spring oilseed rape, spring wheat, sugar beet, vining peas, winter field beans, winter oilseed rape, winter wheat** [3, 9, 11]; **rye, spring barley, spring oats, triticale, winter barley, winter oats** [3]
- Wheat-blossom midge in **winter wheat** [2, 4, 6-8, 10, 12, 13]
- Whitefly in **broccoli, brussels sprouts, calabrese** [2, 4, 6-8, 10, 12]; **cabbages, cauliflowers** [2, 4, 7, 8, 10, 12]; **kale** [6]
- Willow aphid in **short rotation coppice willow** *(off-label)* [11]
- Willow beetle in **short rotation coppice willow** *(off-label)* [11]
- Willow sawfly in **short rotation coppice willow** *(off-label)* [11]
- Yellow cereal fly in **winter wheat** [2, 4, 6-8, 10, 12, 13]

Extension of Authorisation for Minor Use (EAMUs)
- *all edible seed crops grown outdoors* 20082944 [4], 20102880 [11]
- *all non-edible seed crops grown outdoors* 20082944 [4], 20102880 [11]
- *beetroot* 20063761 [2], 20060743 [4]
- *bilberries* 20140521 [4]
- *blackberries* 20063755 [2], 20060728 [4]
- *blackcurrants* 20063752 [2], 20060727 [4], 20073269 [11]
- *blueberries* 20140521 [4]
- *borage for oilseed production* 20063748 [2], 20060634 [4], 20073258 [11]
- *broad beans* 20063764 [2], 20060753 [4], 20073234 [11]
- *bulb onions* 20063756 [2], 20060730 [4], 20073256 [11]
- *canary flower (echium spp.)* 20063748 [2], 20060634 [4], 20073258 [11]
- *carrots* 20080201 [11]
- *celeriac* 20063757 [2], 20060731 [4], 20080204 [11]
- *celery (outdoor)* 20063762 [2], 20060744 [4], 20073257 [11]
- *cherries* 20131273 [4]
- *chestnuts* 20060742 [4], 20080206 [11]
- *chicory* 20060740 [4]
- *cob nuts* 20080206 [11]
- *corn gromwell* 20121047 [4]
- *crambe* 20081046 [4], 20102876 [11]
- *cranberries* 20140521 [4]
- *dewberries* 20063755 [2], 20060728 [4]
- *dwarf beans* 20080202 [11]
- *evening primrose* 20063664 [2], 20060634 [4], 20073258 [11]

FOR FULL CONDITIONS OF USE ALWAYS READ THE PRODUCT LABEL

lambda-cyhalothrin

- **farm forestry** *20082944* [4], *20102880* [11]
- **fennel** *20060733* [4]
- **fodder beet** *20063665* [2], *20060637* [4], *20073233* [11]
- **forest nurseries** *20082944* [4], *20102880* [11]
- **french beans** *20063758* [2], *20060739* [4], *20080202* [11]
- **frise** *20063751* [2], *20060636* [4]
- **garlic** *20063756* [2], *20060730* [4], *20073256* [11]
- **gooseberries** *20063752* [2], *20060727* [4], *20073269* [11]
- **grapevines** *20063747* [2], *20060266* [4], *20080205* [11]
- **grass seed crops** *20063749* [2], *20060624* [4], *20073260* [11]
- **hazel nuts** *20060742* [4], *20080206* [11]
- **herbs (see appendix 6)** *20063751* [2], *20060636* [4], *20073259* [11]
- **honesty** *20063748* [2], *20060634* [4], *20073258* [11]
- **hops** *20082944* [4], *20131719* [4], *20102880* [11]
- **horseradish** *20071307* [2], *20071301* [4], *20080201* [11]
- **lamb's lettuce** *20063751* [2], *20060636* [4]
- **land cress** *20102878* [11]
- **leaf brassicas** *20063751* [2], *20060636* [4]
- **leeks** *20063756* [2], *20060730* [4], *20073256* [11]
- **lupins** *20060635* [4], *20102877* [11]
- **mallow (althaea spp.)** *20071307* [2], *20071301* [4], *20080201* [11]
- **mirabelles** *20131273* [4]
- **miscanthus** *20082944* [4], *20102880* [11]
- **mustard** *20063664* [2], *20060634* [4], *20073258* [11]
- **navy beans** *20063758* [2], *20060739* [4], *20080202* [11]
- **ornamental plant production** *20082944* [4], *20102880* [11]
- **parsley root** *20071307* [2], *20071301* [4], *20080201* [11]
- **parsnips** *20080201* [11]
- **plums** *20131273* [4]
- **poppies for morphine production** *20063763* [2], *20060749* [4], *20102875* [11]
- **protected all edible seed crops** *20102880* [11]
- **protected all non-edible seed crops** *20102880* [11]
- **protected aubergines** *20121994* [4]
- **protected forest nurseries** *20082944* [4], *20102880* [11]
- **protected ornamentals** *20082944* [4], *20102880* [11]
- **protected pak choi** *20081263* [4], *20102879* [11]
- **protected peppers** *20121994* [4]
- **protected strawberries** *20111705* [4]
- **protected tatsoi** *20081263* [4], *20102879* [11]
- **protected tomatoes** *20121994* [4]
- **radishes** *20063757* [2], *20060731* [4], *20080204* [11]
- **raspberries** *20063755* [2], *20060728* [4]
- **red beet** *20073254* [11]
- **redcurrants** *20063752* [2], *20060727* [4], *20073269* [11]
- **ribes hybrids** *20140521* [4]
- **rubus hybrids** *20063755* [2], *20060728* [4]
- **runner beans** *20063758* [2], *20060739* [4], *20080202* [11]
- **rye** *20063749* [2]
- **salad onions** *20063756* [2], *20060730* [4], *20073256* [11]
- **scarole** *20063751* [2], *20060636* [4]
- **shallots** *20063756* [2], *20060730* [4], *20073256* [11]
- **short rotation coppice willow** *20073268* [11]
- **soft fruit** *20082944* [4], *20102880* [11]
- **spring linseed** *20063748* [2], *20060634* [4], *20073258* [11]
- **spring rye** *20060624* [4], *20073260* [11]
- **strawberries** *20111705* [4]
- **swedes** *20101856* [4], *20102911* [11]
- **sweetcorn** *20063760* [2], *20060732* [4], *20080203* [11]
- **table grapes** *20063747* [2]

SEE SECTION 3 FOR PRODUCTS ALSO REGISTERED

Pesticide Profiles

- **top fruit** *20082944* [4], *20102880* [11]
- **triticale** *20063749* [2], *20060624* [4], *20073260* [11]
- **turnips** *20101856* [4], *20102911* [11]
- **walnuts** *20060742* [4], *20080206* [11]
- **whitecurrants** *20063752* [2], *20060727* [4], *20073269* [11]
- **willow (short rotation coppice)** *20060748* [4]
- **wine grapes** *20063747* [2]
- **winter linseed** *20063748* [2], *20060634* [4], *20073258* [11]
- **winter rye** *20060624* [4], *20073260* [11]

Approval information
- Lambda-cyhalothrin included in Annex I under EC Regulation 1107/2009
- Accepted by BBPA for use on malting barley and hops

Efficacy guidance
- Best results normally obtained from treatment when pest attack first seen. See label for detailed recommendations on each crop
- Timing for control of barley yellow dwarf virus vectors depends on specialist assessment of the level of risk in the area
- Repeat applications recommended in some crops where prolonged attack occurs, up to maximum total dose. See label for details
- Where strains of aphids resistant to lambda-cyhalothrin occur control is unlikely to be satisfactory
- Addition of wetter recommended for control of certain pests in brassicas and oilseed rape
- Use of sufficient water volume to ensure thorough crop penetration recommended for optimum results
- Use of drop-legged sprayer gives improved results in crops such as Brussels sprouts

Restrictions
- Maximum number of applications or maximum total dose per crop varies - see labels
- Do not apply to a cereal crop if any product containing a pyrethroid insecticide or dimethoate has been applied to the crop after the start of ear emergence (GS 51)
- Do not spray cereals in the spring/summer (i.e. after 1 Apr) within 6 m of edge of crop

Crop-specific information
- Latest use before late milk stage on cereals; before end of flowering for winter oilseed rape
- HI 3 d for radishes, red beet; 7 d for lettuce [2, 4]; 14 d for carrots and parsnips; 25 d for peas, field beans; 6 wk for spring oilseed rape; 8 wk for sugar beet

Environmental safety
- Dangerous for the environment
- Very toxic to aquatic organisms
- Flammable
- Risk to certain non-target insects or other arthropods [4]
- To protect non-target arthropods respect an untreated buffer zone of 6 m to non-crop land
- Broadcast air-assisted LERAP [2, 4, 6-8, 10, 12] (25 m); Broadcast air-assisted LERAP [13] (38 m); LERAP Category A [1, 3, 9, 11]; LERAP Category B [2-8, 10, 12, 13]

Hazard classification and safety precautions
    Hazard Harmful, Dangerous for the environment [1-13]; Corrosive [1, 3, 9, 11]; Toxic if swallowed [3]; Harmful if swallowed, Very toxic to aquatic organisms [1, 6, 10]; Harmful if inhaled [1, 3, 6, 10]
    Transport code 8 [3, 9, 11]; 9 [1, 2, 4-8, 10, 12, 13]
    Packaging group II [3]; III [1, 2, 4-13]
    UN Number 1760 [3, 9, 11]; 3077 [5]; 3082 [1, 2, 4, 6-8, 10, 12, 13]
    Risk phrases H304 [1, 3]; H314, H336 [3]; H318 [1]; R20, R22a, R53a [2, 4, 5, 7-9, 11-13]; R22b, R34, R51 [9, 11]; R36, R38 [5]; R43, R50 [2, 4, 5, 7, 8, 12, 13]
    Operator protection A, H [1-13]; C [3, 5, 9, 11]; J, K, M [5]; U02a, U05a [1-13]; U04a, U19a, U20a [1, 3, 9, 11]; U08 [1-4, 6-13]; U09a [5]; U11 [1, 3, 5, 9, 11]; U14, U20b [2, 4-8, 10, 12, 13]
    Environmental protection E12a, E16d [1, 3, 9, 11]; E12c [2, 4, 5, 7, 8, 10, 12, 13]; E15b, E22a [1, 3, 5, 9, 11]; E16a [2-8, 10, 12, 13]; E16b [2, 4, 6-8, 10, 12, 13]; E16c [1, 9, 11]; E17b [2, 4, 6-8, 10, 12] (25 m); E17b [13] (38 m); E22b [1, 3, 4, 6-12]; E34 [1-4, 6-13]; E38 [1-13]; H410 [1, 3, 6, 10]

FOR FULL CONDITIONS OF USE ALWAYS READ THE PRODUCT LABEL

Storage and disposal D01, D02 [1-13]; D09a [1-4, 6-13]; D10b [1, 3, 9, 11]; D10c [2, 4, 6-8, 10, 12, 13]; D12a [2, 4-8, 10, 12, 13]
Medical advice M03 [2, 4-8, 10, 12, 13]; M04a, M05b [1, 3, 9, 11]

298   Lecanicillium muscarium

A fungal pathogen of whitefly

Products

Mycotal   Koppert   4.8% w/w   WP   16644

Uses

- Spider mites in *leaf brassicas* (off-label), *protected bilberries* (off-label), *protected blackberries* (off-label), *protected blackcurrants* (off-label), *protected blueberry* (off-label), *protected cayenne peppers* (off-label), *protected cranberries* (off-label), *protected gooseberries* (off-label), *protected herbs (see appendix 6)* (off-label), *protected hops* (off-label), *protected loganberries* (off-label), *protected nursery fruit trees* (off-label), *protected raspberries* (off-label), *protected redcurrants* (off-label), *protected ribes hybrids* (off-label), *protected rubus hybrids* (off-label), *protected table grapes* (off-label), *protected wine grapes* (off-label)
- Thrips in *leaf brassicas* (off-label), *protected bilberries* (off-label), *protected blackberries* (off-label), *protected blackcurrants* (off-label), *protected blueberry* (off-label), *protected cayenne peppers* (off-label), *protected cranberries* (off-label), *protected gooseberries* (off-label), *protected herbs (see appendix 6)* (off-label), *protected hops* (off-label), *protected loganberries* (off-label), *protected nursery fruit trees* (off-label), *protected raspberries* (off-label), *protected redcurrants* (off-label), *protected ribes hybrids* (off-label), *protected rubus hybrids* (off-label), *protected table grapes* (off-label), *protected wine grapes* (off-label)
- Whitefly in *leaf brassicas* (off-label), *protected bilberries* (off-label), *protected blackberries* (off-label), *protected blackcurrants* (off-label), *protected blueberry* (off-label), *protected cayenne peppers* (off-label), *protected cranberries* (off-label), *protected cucumbers*, *protected gooseberries* (off-label), *protected herbs (see appendix 6)* (off-label), *protected hops* (off-label), *protected loganberries* (off-label), *protected nursery fruit trees* (off-label), *protected ornamentals*, *protected peppers*, *protected raspberries* (off-label), *protected redcurrants* (off-label), *protected ribes hybrids* (off-label), *protected rubus hybrids* (off-label), *protected strawberries*, *protected table grapes* (off-label), *protected tomatoes*, *protected wine grapes* (off-label)

Extension of Authorisation for Minor Use (EAMUs)

- *leaf brassicas* 20142679
- *protected bilberries* 20142685
- *protected blackberries* 20142684
- *protected blackcurrants* 20142685
- *protected blueberry* 20142685
- *protected cayenne peppers* 20142680
- *protected cranberries* 20142685
- *protected gooseberries* 20142685
- *protected herbs (see appendix 6)* 20142679
- *protected hops* 20142681
- *protected loganberries* 20142684
- *protected nursery fruit trees* 20142681
- *protected raspberries* 20142684
- *protected redcurrants* 20142685
- *protected ribes hybrids* 20142685
- *protected rubus hybrids* 20142684
- *protected table grapes* 20142685
- *protected wine grapes* 20142685

Approval information

- *Lecanicillium* included in Annex I under EC Regulation 1107/2009

SEE SECTION 3 FOR PRODUCTS ALSO REGISTERED

Pesticide Profiles

Efficacy guidance
- *Lecanicillium muscarium* is a pathogenic fungus that infects the target pests and destroys them
- Apply spore powder as spray as part of biological control programme keeping the spray liquid well agitated
- Treat before infestations build to high levels and repeat as directed on the label
- Spray during late afternoon and early evening directing spray onto underside of leaves and to growing points
- Best results require minimum 80% relative humidity (70% if applied with the adjuvant Addit) and 18°C within the crop canopy

Restrictions
- Never use in tank mixture
- Do not use a fungicide within 3 d of treatment. Pesticides containing captan, chlorothalonil, fenarimol, dichlofluanid, imazalil, maneb, prochloraz, quinomethionate, thiram or tolylfluanid may not be used on the same crop
- Keep in a refrigerated store at 2-6°C but do not freeze

Environmental safety
- Products have negligible effects on commercially available natural predators or parasites but consult manufacturer before using with a particular biological control agent for the first time

Hazard classification and safety precautions
UN Number N/C
Operator protection A, D, H; U19a, U20b
Environmental protection E15a
Storage and disposal D09a, D11a

299 lenacil

A residual, soil-acting uracil herbicide for beet crops
HRAC mode of action code: C1

See also desmedipham + ethofumesate + lenacil + phenmedipham

Products

1	Lenazar Flo	Hermoo	440 g/l	SC	14791
2	Pirapama	Agroquimicos	440 g/l	SC	15777
3	Venzar Flowable	DuPont	440 g/l	SC	06907

Uses
- Annual dicotyledons in **baby leaf crops** *(off-label)*, **blackberries** *(off-label)*, **blackcurrants** *(off-label)*, **blueberries** *(off-label)*, **cranberries** *(off-label)*, **farm woodland** *(off-label)*, **gooseberries** *(off-label)*, **herbs (see appendix 6)** *(off-label)*, **raspberries** *(off-label)*, **redcurrants** *(off-label)*, **ribes hybrids** *(off-label)*, **rubus hybrids** *(off-label)*, **spinach** *(off-label)*, **spinach beet** *(off-label)*, **strawberries** *(off-label)*, **whitecurrants** *(off-label)* [3]; **fodder beet**, **mangels**, **red beet**, **sugar beet** [1-3]
- Annual grasses in **baby leaf crops** *(off-label)*, **blackcurrants** *(off-label)*, **blueberries** *(off-label)*, **cranberries** *(off-label)*, **gooseberries** *(off-label)*, **herbs (see appendix 6)** *(off-label)*, **redcurrants** *(off-label)*, **ribes hybrids** *(off-label)*, **spinach** *(off-label)*, **spinach beet** *(off-label)*, **whitecurrants** *(off-label)* [3]
- Annual meadow grass in **farm woodland** *(off-label)* [3]; **fodder beet**, **mangels**, **red beet**, **sugar beet** [1-3]
- Black bindweed in **fodder beet** *(off-label)* [3]

Extension of Authorisation for Minor Use (EAMUs)
- **baby leaf crops** *970703* [3]
- **blackberries** *20093246* [3]
- **blackcurrants** *970704* [3]
- **blueberries** *970704* [3]
- **cranberries** *970704* [3]
- **farm woodland** *971282* [3]

FOR FULL CONDITIONS OF USE ALWAYS READ THE PRODUCT LABEL

lenacil

- **fodder beet** *20111765* [3]
- **gooseberries** *970704* [3]
- **herbs (see appendix 6)** *970703* [3]
- **raspberries** *20093246* [3]
- **redcurrants** *970704* [3]
- **ribes hybrids** *970704* [3]
- **rubus hybrids** *20093246* [3]
- **spinach** *970703* [3]
- **spinach beet** *970703* [3]
- **strawberries** *20093246* [3]
- **whitecurrants** *970704* [3]

Approval information
- Lenacil included in Annex I under EC Regulation 1107/2009

Efficacy guidance
- Best results, especially from pre-emergence treatments, achieved on fine, even, firm and moist soils free from clods. Continuing presence of moisture from rain or irrigation gives improved residual control of later germinating weeds. Effectiveness may be reduced by dry conditions
- On beet crops may be used pre- or post-emergence, alone or in mixture to broaden weed spectrum
- Apply overall or as band spray to beet crops pre-drilling incorporated, pre- or post-emergence
- All labels have limitations on soil types that may be treated. Residual activity reduced on soils with high OM content

Restrictions
- Maximum number of treatments in beet crops 1 pre-emergence + 3 post-emergence per crop; 2 per yr on established woody ornamentals and roses
- See label for soil type restrictions
- Do not use any other residual herbicide within 3 mth of the initial application to fruit or ornamental crops
- Do not treat crops under stress from drought, low temperatures, nutrient deficiency, pest or disease attack, or waterlogging

Crop-specific information
- Latest use: pre-emergence for red beet, fodder beet, spinach, spinach beet, mangels; before leaves meet over rows when used on these crops post-emergence; 24 h after planting new strawberry runners or before flowering for established strawberry crops, blackcurrants, gooseberries, raspberries
- Heavy rain after application to beet crops may cause damage especially if followed by very hot weather
- Reduction in beet stand may occur where crop emergence or vigour is impaired by soil capping or pest attack
- Strawberry runner beds to be treated should be level without depressions around the roots
- New soft fruit cuttings should be planted at least 15 cm deep and firmed before treatment
- Check varietal tolerance of ornamentals before large scale treatment

Following crops guidance
- Succeeding crops should not be planted or sown for at least 4 mth (6 mth on organic soils) after treatment following ploughing to at least 150 mm.
- Only beet crops, mangels or strawberries may be sown within 4 months of treatment and no further applications of lenacil should be made for at least 4 months [3]

Environmental safety
- Dangerous for the environment
- Very toxic to aquatic organisms

Hazard classification and safety precautions
    Hazard Dangerous for the environment
    Transport code 9
    Packaging group III
    UN Number 3082

SEE SECTION 3 FOR PRODUCTS ALSO REGISTERED

Pesticide Profiles

Risk phrases R50, R53a
Operator protection A [1-3]; C [1]; U05a, U08, U19a, U20a
Environmental protection E13c, E38
Storage and disposal D01, D02, D05, D09a, D10a, D12a

300   lenacil + triflusulfuron-methyl

A foliar and residual herbicide mixture for sugar beet
HRAC mode of action code: C1 + B

See also triflusulfuron-methyl

Products
   Safari Lite WSB   DuPont   71.4:5.4% w/w   WG   12169

Uses
- Annual dicotyledons in **fodder beet** *(off-label)*, **sugar beet**
- Volunteer oilseed rape in **sugar beet**

Extension of Authorisation for Minor Use (EAMUs)
- **fodder beet** 20140256

Approval information
- Lenacil and triflusulfuron-methyl included in Annex I under EC Regulation 1107/2009

Efficacy guidance
- Best results obtained when weeds are small and growing actively
- Product recommended for use in a programme of treatments in tank mixture with a suitable herbicide partner to broaden the weed spectrum
- Ensure good spray cover of weeds. Apply when first weeds have emerged provided crop has reached cotyledon stage
- Susceptible plants cease to grow almost immediately after treatment and symptoms can be seen 5-10 d later
- Weed control may be reduced in very dry soil conditions
- Triflusulfuron-methyl is a member of the ALS-inhibitor group of herbicides

Restrictions
- Maximum number of treatments on sugar beet 3 per crop and do not apply more than 4 applications of any product containing triflusulfuron-methyl
- Do not apply to crops suffering from stress caused by drought, water-logging, low temperatures, pest or disease attack, nutrient deficiency or any other factors affecting crop growth
- Do not use on Sands, stony or gravelly soils or on soils with more than 10% organic matter
- Do not apply when temperature above or likely to exceed 21°C on day of spraying or under conditions of high light intensity

Crop-specific information
- Latest use: before crop leaves meet between rows for sugar beet

Following crops guidance
- Only cereals may be sown in the same calendar yr as a treated sugar beet crop. Any crop may be sown in the following spring
- In the event of crop failure sow only sugar beet within 4 mth of treatment

Environmental safety
- Dangerous for the environment
- Very toxic to aquatic organisms
- Take extreme care to avoid drift onto broad-leaved plants outside the target area or onto surface waters or ditches, or land intended for cropping
- Spraying equipment should not be drained or flushed onto land planted, or to be planted, with trees or crops other than cereals and should be thoroughly cleansed after use - see label for instructions
- LERAP Category B

FOR FULL CONDITIONS OF USE ALWAYS READ THE PRODUCT LABEL

linuron

Hazard classification and safety precautions
Hazard Irritant, Dangerous for the environment
Transport code 9
Packaging group III
UN Number 3077
Risk phrases R36, R37, R38, R43, R50, R53a
Operator protection A, C; U05a, U08, U11, U19a, U20b, U22a
Environmental protection E15a, E16a, E38
Storage and disposal D01, D02, D09a, D12a

301 linuron

A contact and residual urea herbicide for various field crops
HRAC mode of action code: C2

See also 2,4-DB + linuron + MCPA
  clomazone + linuron

Products

1	Afalon	Adama	450 g/l	SC	14187
2	Cleancrop Affable	Agrii	450 g/l	SC	16321
3	Datura	UPL Europe	500 g/l	SC	16955
4	Nightjar	AgChem Access	450 g/l	SC	14656

Uses
- Annual dicotyledons in **asparagus, celeriac, chervil, dill, forest nurseries, lovage, ornamental plant production** [3]; **bulb onions** *(off-label)*, **garlic** *(off-label)*, **leeks** *(off-label)*, **shallots** *(off-label)* [1, 4]; **carrots, parsley, parsnips, potatoes** [1-4]; **celeriac** *(off-label)*, **celery (outdoor)** *(off-label)*, **dandelions** *(off-label)*, **dwarf beans** *(off-label)*, **fennel** *(off-label)*, **french beans** *(off-label)*, **game cover** *(off-label)*, **ginseng root** *(off-label)*, **herbs (see appendix 6)** *(off-label)*, **herbs for medicinal uses (see appendix 6)** *(off-label)*, **liquorice** *(off-label)*, **mallow (althaea spp.)** *(off-label)*, **nettle** *(off-label)*, **ornamental plant production** *(off-label)*, **parsley root** *(off-label)*, **runner beans** *(off-label)*, **valerian root** *(off-label)* [1]; **combining peas, spring field beans** [1, 2, 4]
- Annual grasses in **bulb onions** *(off-label)*, **celeriac** *(off-label)*, **celery (outdoor)** *(off-label)*, **dandelions** *(off-label)*, **dwarf beans** *(off-label)*, **fennel** *(off-label)*, **french beans** *(off-label)*, **game cover** *(off-label)*, **garlic** *(off-label)*, **ginseng root** *(off-label)*, **herbs (see appendix 6)** *(off-label)*, **herbs for medicinal uses (see appendix 6)** *(off-label)*, **leeks** *(off-label)*, **liquorice** *(off-label)*, **mallow (althaea spp.)** *(off-label)*, **nettle** *(off-label)*, **ornamental plant production** *(off-label)*, **parsley root** *(off-label)*, **runner beans** *(off-label)*, **shallots** *(off-label)*, **valerian root** *(off-label)* [1]
- Annual meadow grass in **carrots, combining peas, parsley, parsnips, potatoes, spring field beans** [1, 2]; **celery (outdoor)** *(off-label)*, **fennel** *(off-label)* [1]

Extension of Authorisation for Minor Use (EAMUs)
- **bulb onions** *20090875* [1], *20141217* [1], *20100649* [4]
- **celeriac** *20090885* [1], *20141220* [1]
- **celery (outdoor)** *20130273* [1], *20141218* [1]
- **dandelions** *20090861* [1]
- **dwarf beans** *20090859* [1], *20141214* [1]
- **fennel** *20130273* [1], *20141218* [1]
- **french beans** *20090859* [1], *20141214* [1]
- **game cover** *20082802* [1], *20141209* [1]
- **garlic** *20090875* [1], *20141217* [1], *20100649* [4]
- **ginseng root** *20090861* [1]
- **herbs (see appendix 6)** *20090860* [1], *20141222* [1]
- **herbs for medicinal uses (see appendix 6)** *20141215* [1]
- **leeks** *20090864* [1], *20141216* [1], *20101135* [4]
- **liquorice** *20090861* [1]
- **mallow (althaea spp.)** *20090876* [1], *20141219* [1]
- **nettle** *20090861* [1]

SEE SECTION 3 FOR PRODUCTS ALSO REGISTERED

Pesticide Profiles

- **ornamental plant production** *20090877* [1], *20141221* [1]
- **parsley root** *20090876* [1], *20141219* [1]
- **runner beans** *20090859* [1], *20141214* [1]
- **shallots** *20090875* [1], *20141217* [1], *20100649* [4]
- **valerian root** *20090861* [1]

Approval information
- Linuron included in Annex I under EC Regulation 1107/2009
- Accepted by BBPA for use on malting barley

Efficacy guidance
- Many weeds controlled pre-emergence or post-emergence to 2-3 leaf stage, some (annual meadow grass, mayweed) only susceptible pre-emergence. See label for details
- Best results achieved by application to firm, moist soil of fine tilth
- Little residual effect on soil with more than 10% organic matter

Restrictions
- Maximum total dose equivalent to one full dose treatment
- Do not use on undersown cereals or crops grown on Sands or Very Light soils or soils heavier than Sandy Clay Loam or with more than 10% organic matter
- Do not apply to emerged crops of carrots, parsnips or parsley under stress
- Do not apply by hand-held sprayers

Crop-specific information
- Latest use: pre-emergence for most crops (products differ, see label for details).
- HI onions, garlic 8 wk; celeriac 12 wk; leeks 16 wk
- Apply to potatoes well earthed up to a rounded ridge pre-crop emergence and do not cultivate after spraying
- Apply to carrots at any time after drilling on organic soils, and within 4 d of drilling on other soils. Apply post-emergence as soon as weeds appear but after first rough leaf stage.
- Recommendations for parsnips, parsley and celery vary. See label for details

Following crops guidance
- Potatoes, carrots and parsnips may be planted at any time after application. Lettuce should not be grown within 12 mth of treatment. Transplanted brassicas may be grown from 3 mth after treatment

Environmental safety
- Dangerous for the environment
- Very toxic to aquatic organisms
- LERAP Category B

Hazard classification and safety precautions
Hazard Toxic, Dangerous for the environment
Transport code 9
Packaging group III
UN Number 3082
Risk phrases R22a, R40, R48, R50, R53a, R61, R62 [1-4]; R25b, R63 [3]
Operator protection A, H [1-4]; C [4]; U05a, U13, U14, U15, U19a [1-4]; U08, U19c, U20b [3]
Environmental protection E15b, E16a, E38
Storage and disposal D02, D05, D09a, D11a, D12a [1-4]; D08 [1, 2, 4]
Medical advice M04a

302 magnesium phosphide

A phosphine generating compound used to control insect pests in stored commodities

Products

Degesch Plates	Rentokil	56% w/w	GE	17105

FOR FULL CONDITIONS OF USE ALWAYS READ THE PRODUCT LABEL

maleic hydrazide

Uses
- Insect pests in **carob, cocoa, coffee, crop handling & storage structures, herbal infusions, processed consumable products, stored dried spices, stored grain, stored linseed, stored nuts, stored oilseed rape, stored pulses, tea, tobacco**

Approval information
- Magnesium phosphide included in Annex I under EC Regulation 1107/2009
- Accepted by BBPA for use in stores for malting barley

Efficacy guidance
- Product acts as fumigant by releasing poisonous hydrogen phosphide gas on contact with moisture in the air
- Place plates on the floor or wall of the building or on the surface of the commodity. Exposure time varies depending on temperature and pest. See label

Restrictions
- Magnesium phosphide is subject to the Poisons Rules 1982 and the Poisons Act 1972. See Section 5 for more information
- Only to be used by professional operators trained in the use of magnesium phosphide and familiar with the precautionary measures to be observed. See label for full precautions

Environmental safety
- Highly flammable
- Prevent access to buildings under fumigation by livestock, pets and other non-target mammals and birds
- Dangerous to fish or other aquatic life. Do not contaminate surface waters or ditches with chemical or used container
- Keep in original container, tightly closed, in a safe place, under lock and key
- Do not allow plates or their spent residues to come into contact with food other than raw cereal grains
- Remove used plates after treatment. Do not bulk spent plates and residues: spontaneous ignition could result
- Keep livestock out of treated areas

Hazard classification and safety precautions
Hazard Very toxic, Highly flammable, Dangerous for the environment, In contact with water releases flammable gases which may ignite spontaneously, Fatal if swallowed, Fatal if inhaled, Very toxic to aquatic organisms
Transport code 4.3
Packaging group I
UN Number 2011
Risk phrases H312
Operator protection A, D, H; U01, U05b, U07, U13, U19a, U20a
Environmental protection E02a (4 h min); E02b, E13b, E34
Storage and disposal D01, D02, D05, D07, D09b, D11b
Medical advice M04a

303 maleic hydrazide

A pyridazine plant growth regulator suppressing sprout and bud growth

See also fatty acids + maleic hydrazide

Products
1	Clayton Stun	Clayton	60% w/w	SG	17194
2	Fazor	Arysta	60% w/w	SG	13679

Uses
- Growth regulation in **amenity grassland** (off-label), **amenity vegetation** (off-label), **bilberries** (off-label), **blackberries** (off-label), **blackcurrants** (off-label), **blueberries** (off-label), **carrots** (off-label), **cranberries** (off-label), **garlic** (off-label), **gooseberries** (off-label), **hops** (off-label),

SEE SECTION 3 FOR PRODUCTS ALSO REGISTERED

Pesticide Profiles

loganberries *(off-label)*, **ornamental plant production** *(off-label)*, **parsnips** *(off-label)*, **raspberries** *(off-label)*, **redcurrants** *(off-label)*, **rubus hybrids** *(off-label)*, **shallots** *(off-label)*, **strawberries** *(off-label)* [2]; **bulb onions**, **potatoes** [1, 2]

Extension of Authorisation for Minor Use (EAMUs)
- *amenity grassland* 20160184 [2]
- *amenity vegetation* 20160185 [2]
- *bilberries* 20160183 [2]
- *blackberries* 20160183 [2]
- *blackcurrants* 20160183 [2]
- *blueberries* 20160183 [2]
- *carrots* 20160186 [2]
- *cranberries* 20160183 [2]
- *garlic* 20160187 [2]
- *gooseberries* 20160183 [2]
- *hops* 20160183 [2]
- *loganberries* 20160183 [2]
- *ornamental plant production* 20160183 [2]
- *parsnips* 20160186 [2]
- *raspberries* 20160183 [2]
- *redcurrants* 20160183 [2]
- *rubus hybrids* 20160183 [2]
- *shallots* 20160187 [2]
- *strawberries* 20160183 [2]

Approval information
- Maleic hydrazide included in Annex I under EC Regulation 1107/2009

Efficacy guidance
- Apply to grass at any time of yr when growth active, best when growth starting in Apr-May and repeated when growth recommences
- Uniform coverage and dry weather necessary for effective results
- Accurate timing essential for good results on potatoes but rain or irrigation within 24 h may reduce effectiveness on onions and potatoes
- Mow 2-3 d before and 5-10 d after spraying for best results. Need for mowing reduced for up to 6 wk
- When used for suppression of volunteer potatoes treatment may also give some suppression of sprouting in store but separate treatment will be necessary if sprouting occurs

Restrictions
- Maximum number of treatments 2 per yr on amenity grass, land not intended for cropping and land adjacent to aquatic areas; 1 per crop on onions, potatoes and on or around tree trunks
- Do not apply in drought or when crops are suffering from pest, disease or herbicide damage. Do not treat fine turf or grass seeded less than 8 mth previously
- Do not treat potatoes within 3 wk of applying a haulm desiccant or if temperatures above 26°C
- Consult processor before use on potato crops for processing

Crop-specific information
- Latest use: 3 wk before haulm destruction for potatoes; before 50% necking for onions
- HI onions 1 wk; potatoes 3 wk
- Apply to onions at 10% necking and not later than 50% necking stage when the tops are still green
- Only treat onions in good condition and properly cured, and do not treat more than 2 wk before maturing. Treated onions may be stored until Mar but must then be removed to avoid browning
- Apply to second early or maincrop potatoes at least 3 wk before haulm destruction
- Only treat potatoes of good keeping quality; not on seed, first earlies or crops grown under polythene

Environmental safety
- Only apply to grass not to be used for grazing

FOR FULL CONDITIONS OF USE ALWAYS READ THE PRODUCT LABEL

- Do not use treated water for irrigation purposes within 3 wk of treatment or until concentration in water falls below 0.02 ppm
- Maximum permitted concentration in water 2 ppm
- Do not dump surplus product in water or ditch bottoms
- Avoid drift onto nearby vegetables, flowers or other garden plants

Hazard classification and safety precautions
Transport code 9
Packaging group III
UN Number 3077
Operator protection A, H; U08, U20b
Environmental protection E15a [1, 2]; H411 [2]
Storage and disposal D09a, D11a

304 maleic hydrazide + pelargonic acid

A pyridazine plant growth regulator suppressing sprout and bud growth + a naturally occuring 9 carbon chain acid
HRAC mode of action code: Not classified

Products
Finalsan Plus   Certis   30:186.7 g/l   SL   15147

Uses
- Algae in *amenity vegetation, natural surfaces not intended to bear vegetation, ornamental plant production, permeable surfaces overlying soil*
- Annual and perennial weeds in *amenity vegetation, natural surfaces not intended to bear vegetation, ornamental plant production, permeable surfaces overlying soil*
- Moss in *amenity vegetation, natural surfaces not intended to bear vegetation, ornamental plant production, permeable surfaces overlying soil*

Approval information
- Maleic hydrazide and pelargonic acid included in Annex I under EC Regulation 1107/2009

Hazard classification and safety precautions
UN Number N/C
Operator protection A, C, H; U05a, U12, U15
Environmental protection E15a
Storage and disposal D01, D02
Medical advice M05a

305 maltodextrin

A polysaccharide used as a food additive and with activity against red spider mites

Products
1 Eradicoat   Certis   598 g/l   SL   17121
2 Majestik   Certis   598 g/l   SL   17240

Uses
- Aphids in *all edible crops (outdoor), all non-edible crops (outdoor), all protected edible crops, all protected non-edible crops*
- Spider mites in *all edible crops (outdoor), all non-edible crops (outdoor), all protected edible crops, all protected non-edible crops*
- Whitefly in *all edible crops (outdoor), all non-edible crops (outdoor), all protected edible crops, all protected non-edible crops*

Approval information
- Maltodextrin is included in Annex 1 under EC Regulation 1107/2009

SEE SECTION 3 FOR PRODUCTS ALSO REGISTERED

Pesticide Profiles

Hazard classification and safety precautions
  Hazard Irritant
  UN Number N/C
  Risk phrases H319
  Operator protection A, H; U05a, U09a, U14, U19a, U20b
  Environmental protection E12a [2]; E12e [1]; H412 [1, 2]
  Storage and disposal D01, D02, D09a, D10b, D12a

306   mancozeb

A protective dithiocarbamate fungicide for potatoes and other crops
FRAC mode of action code: M3

See also ametoctradin + mancozeb
benalaxyl + mancozeb
benthiavalicarb-isopropyl + mancozeb
chlorothalonil + mancozeb
cymoxanil + mancozeb
dimethomorph + mancozeb
fenamidone + mancozeb

Products

1	AUK 75 WG	Goldengrass	75% w/w	WG	16398
2	Dithane 945	Interfarm	80% w/w	WP	14621
3	Dithane NT Dry Flowable	Interfarm	75% w/w	WG	14704
4	Karamate Dry Flo Neotec	Landseer	75% w/w	WG	14632
5	Laminator Flo	Interfarm	455 g/l	SC	15345
6	Malvi	AgChem Access	75% w/w	WG	14981
7	Manzate 75 WG	UPL Europe	75% w/w	WG	15052
8	Mewati	AgChem Access	80% w/w	WP	15763
9	Penncozeb 80 WP	UPL Europe	80% w/w	WP	16953
10	Penncozeb WDG	UPL Europe	75% w/w	WG	16885
11	Quell Flo	Interfarm	455 g/l	SC	15237

Uses
- Alternaria blight in **carrots**, **parsnips** [3]
- Blight in **potatoes** [1-3, 5-11]; **tomatoes (outdoor)** [6]
- Botrytis in **flower bulbs** [7]
- Brown rust in **durum wheat** *(useful control)*, **spring wheat** *(useful control)*, **winter wheat** *(useful control)* [2, 3, 8]; **spring wheat**, **winter wheat** [5, 11]
- Disease control in **amenity vegetation**, **bulb onions**, **carrots**, **courgettes**, **hops** *(off-label)*, **parsnips**, **shallots**, **soft fruit** *(off-label)*, **top fruit** *(off-label)* [4]; **apples**, **table grapes** [6]; **forest nurseries** *(off-label)* [2, 4]; **ornamental plant production** [4, 10]; **ornamental plant production** *(off-label)* [2]; **wine grapes** [4, 6]
- Downy mildew in **baby leaf crops** *(off-label)*, **lettuce** *(off-label)* [4]; **bulb onions** [1, 7, 9, 10]; **ornamental plant production** [10]
- Early blight in **potatoes** [2, 3, 8]
- Rust in **bulb onions**, **shallots** [3]
- Scab in **apples** [1, 4, 7, 9, 10]; **pears** [4]
- Septoria leaf blotch in **durum wheat** *(reduction)* [2, 3, 8]; **spring wheat**, **winter wheat** [5, 9-11]; **spring wheat** *(reduction)*, **winter wheat** *(reduction)* [1-3, 7, 8]
- Sooty moulds in **spring wheat**, **winter wheat** [5, 11]
- Yellow rust in **winter wheat** [5, 11]

Extension of Authorisation for Minor Use (EAMUs)
- *baby leaf crops* 20101082 [4]
- *forest nurseries* 20152652 [2], 20122047 [4]
- *hops* 20122047 [4]
- *lettuce* 20101082 [4]
- *ornamental plant production* 20152652 [2]

FOR FULL CONDITIONS OF USE ALWAYS READ THE PRODUCT LABEL

mancozeb

- ***soft fruit*** *20122047* [4]
- ***top fruit*** *20122047* [4]

Approval information
- Mancozeb included in Annex I under EC Regulation 1107/2009
- Accepted by BBPA for use on malting barley

Efficacy guidance
- Mancozeb is a protectant fungicide and will give moderate control, suppression or reduction of the cereal diseases listed if treated before they are established but in many cases mixture with carbendazim is essential to achieve satisfactory results. See labels for details
- May be recommended for suppression or control of mildew in cereals depending on product and tank mix. See label for details

Restrictions
- Maximum number of treatments varies with crop and product used - check labels for details
- Check labels for minimum interval that must elapse between treatments
- Avoid treating wet cereal crops or those suffering from drought or other stress
- Keep dry formulations away from fire and sparks
- Use dry formulations immediately. Do not store

Crop-specific information
- Latest use: before early milk stage (GS 73) for cereals;
- HI potatoes 7 d; apples, blackcurrants 28 d
- Apply to potatoes before haulm meets across rows (usually mid-Jun) or at earlier blight warning, and repeat every 7-14 d depending on conditions and product used (see label)
- May be used on potatoes up to desiccation of haulm
- Apply to cereals from 4-leaf stage to before early milk stage (GS 71). Recommendations vary, see labels for details

Environmental safety
- Dangerous for the environment
- Very toxic to aquatic organisms
- Harmful to fish or other aquatic life. Do not contaminate surface waters or ditches with chemical or used container
- Do not empty into drains
- LERAP Category B

Hazard classification and safety precautions
Hazard Harmful [9, 10]; Irritant [1-8, 11]; Dangerous for the environment [1-4, 6-11]; Very toxic to aquatic organisms [2-4]
Transport code 9
Packaging group III
UN Number 3077 [1-4, 6-10]; 3082 [5, 11]
Risk phrases H317, H361 [2-4]; H319 [3, 4]; R36 [6]; R37 [1, 7, 8, 11]; R43 [1, 5, 7-11]; R50, R53a [1, 6-11]; R63 [9, 10]
Operator protection A [1-8, 10, 11]; C [6]; D [1-4, 7, 8, 10]; H [2, 8, 10]; U05a [1-8, 10, 11]; U08 [5, 6, 11]; U11 [6]; U14 [1-4, 7-11]; U19a [6, 9, 10]; U20b [1-8, 11]
Environmental protection E13c [5]; E15a [1-4, 6-8, 11]; E15b [9, 10]; E16a [1-11]; E19b [10]; E34 [5, 6, 11]; E38 [1-4, 6-11]; H410 [2, 3]; H411 [6]
Storage and disposal D01, D12a [1-4, 6-11]; D02 [1-11]; D05 [1-8, 11]; D07 [5, 6]; D09a [1-8, 10, 11]; D10a [6]; D11a [1-5, 7, 8, 11]
Medical advice M05a [9, 10]

Pesticide Profiles

307 mancozeb + metalaxyl-M

A systemic and protectant fungicide mixture
FRAC mode of action code: M3 + 4

See also metalaxyl-M

Products

1	Ensis	AgChem Access	64:4% w/w	WG	15919
2	Fubol Gold WG	Syngenta	64:4% w/w	WG	14605

Uses
- Blight in **potatoes** [1, 2]
- Disease control in **bilberries** *(off-label)*, **blackberries** *(off-label)*, **blackcurrants** *(off-label)*, **blueberries** *(off-label)*, **cranberries** *(off-label)*, **forest nurseries** *(off-label)*, **gooseberries** *(off-label)*, **hops** *(off-label)*, **loganberries** *(off-label)*, **ornamental plant production** *(off-label)*, **protected bilberries** *(off-label)*, **protected blackberries** *(off-label)*, **protected blackcurrants** *(off-label)*, **protected blueberry** *(off-label)*, **protected cranberries** *(off-label)*, **protected forest nurseries** *(off-label)*, **protected gooseberries** *(off-label)*, **protected hops** *(off-label)*, **protected loganberries** *(off-label)*, **protected ornamentals** *(off-label)*, **protected raspberries** *(off-label)*, **protected redcurrants** *(off-label)*, **protected rubus hybrids** *(off-label)*, **protected strawberries** *(off-label)*, **raspberries** *(off-label)*, **redcurrants** *(off-label)*, **rubus hybrids** *(off-label)*, **strawberries** *(off-label)* [2]
- Downy mildew in **baby leaf crops** *(off-label)*, **herbs (see appendix 6)** *(off-label)*, **lettuce** *(off-label)*, **poppies for morphine production** *(off-label)*, **protected herbs (see appendix 6)** *(off-label)*, **rhubarb** *(off-label)*, **salad onions** *(off-label)* [2]; **bulb onions** *(useful control)*, **shallots** *(useful control)* [1, 2]
- Fungus diseases in **apple orchards** *(off-label)* [2]
- Pink rot in **potatoes** *(reduction)* [2]
- White blister in **cabbages** *(off-label)* [2]

Extension of Authorisation for Minor Use (EAMUs)
- **apple orchards** *20132282* [2]
- **baby leaf crops** *20132285* [2]
- **bilberries** *20132288* [2]
- **blackberries** *20132288* [2]
- **blackcurrants** *20132288* [2]
- **blueberries** *20132288* [2]
- **cabbages** *20132281* [2]
- **cranberries** *20132288* [2]
- **forest nurseries** *20132288* [2]
- **gooseberries** *20132288* [2]
- **herbs (see appendix 6)** *20132285* [2]
- **hops** *20132288* [2]
- **lettuce** *20132285* [2]
- **loganberries** *20132288* [2]
- **ornamental plant production** *20132288* [2]
- **poppies for morphine production** *20132286* [2]
- **protected bilberries** *20132288* [2]
- **protected blackberries** *20132288* [2]
- **protected blackcurrants** *20132288* [2]
- **protected blueberry** *20132288* [2]
- **protected cranberries** *20132288* [2]
- **protected forest nurseries** *20132288* [2]
- **protected gooseberries** *20132288* [2]
- **protected herbs (see appendix 6)** *20132287* [2]
- **protected hops** *20132288* [2]
- **protected loganberries** *20132288* [2]
- **protected ornamentals** *20132288* [2]
- **protected raspberries** *20132288* [2]

FOR FULL CONDITIONS OF USE ALWAYS READ THE PRODUCT LABEL

mancozeb + zoxamide

- *protected redcurrants* 20132288 [2]
- *protected rubus hybrids* 20132288 [2]
- *protected strawberries* 20132288 [2]
- *raspberries* 20132288 [2]
- *redcurrants* 20132288 [2]
- *rhubarb* 20132283 [2]
- *rubus hybrids* 20132288 [2]
- *salad onions* 20132284 [2]
- *strawberries* 20132288 [2]

Approval information
- Mancozeb and metalaxyl-M included in Annex I under EC Regulation 1107/2009

Efficacy guidance
- Commence potato blight programme before risk of infection occurs as crops begin to meet along the rows and repeat every 7-14 d according to blight risk. Do not exceed a 14 d interval between sprays
- If infection risk conditions occur earlier than the above growth stage commence spraying potatoes immediately
- Complete the potato blight programme using a protectant fungicide starting no later than 10 d after the last phenylamide spray. At least 2 such sprays should be applied
- To minimise the likelihood of development of resistance these products should be used in a planned Resistance Management strategy. See Section 5 for more information

Crop-specific information
- Latest use: before end of active potato haulm growth or before end Aug, whichever is earlier
- HI 7 d for potatoes
- After treating early potatoes destroy and remove any remaining haulm after harvest to minimise blight pressure on neighbouring maincrop potatoes

Environmental safety
- Dangerous for the environment
- Very toxic to aquatic organisms
- Do not harvest crops for human consumption for at least 7 d after final application
- LERAP Category B

Hazard classification and safety precautions
Hazard Harmful, Dangerous for the environment
Transport code 9
Packaging group III
UN Number 3077
Risk phrases R37, R43, R50, R53a, R63
Operator protection A; U05a, U08, U20b
Environmental protection E15a, E16a, E34, E38
Consumer protection C02a (7 d)
Storage and disposal D01, D02, D05, D09a, D11a, D12a

308 mancozeb + zoxamide

A protectant fungicide mixture for potatoes
FRAC mode of action code: M3 + 22

See also zoxamide

Products

1	Electis 75WG	Gowan	66.7:8.3% w/w	WG	14195
2	Roxam 75WG	Gowan	66.7:8.3% w/w	WG	14191
3	Unikat 75WG	Gowan	66.7:8.3% w/w	WG	14200

Uses
- Blight in *potatoes* [1-3]

SEE SECTION 3 FOR PRODUCTS ALSO REGISTERED

Pesticide Profiles

- Disease control in **ornamental plant production** *(off-label)* [1]; **wine grapes** [1-3]

Extension of Authorisation for Minor Use (EAMUs)
- **ornamental plant production** 20140491 expires 29 Mar 2017 [1]

Approval information
- Mancozeb and zoxamide included in Annex I under EC Regulation 1107/2009

Efficacy guidance
- Apply as protectant spray on potatoes immediately risk of blight in district or as crops begin to meet along the rows and repeat every 7-10 d according to blight risk
- Do not use if potato blight present in crop. Products are not curative
- Spray irrigated potato crops as soon as possible after irrigation once the crop leaves are dry

Restrictions
- Maximum number of treatments 8 per crop for potatoes

Crop-specific information
- HI 7 d for potatoes

Environmental safety
- Dangerous for the environment
- Very toxic to aquatic organisms
- Keep away from fire and sparks
- LERAP Category B

Hazard classification and safety precautions
  Hazard Irritant, Dangerous for the environment, Very toxic to aquatic organisms
  Transport code 9
  Packaging group III
  UN Number 3077
  Risk phrases H317, H361
  Operator protection A, D, H; U05a, U14, U20b [1-3]; U19a [1, 2]
  Environmental protection E15a, E16a, E16b, E38, H410
  Storage and disposal D01, D02, D05, D09a, D11a, D12a

309 mandipropamid

A mandelamide fungicide for the control of potato blight
FRAC mode of action code: 40

See also cymoxanil + mandipropamid
difenoconazole + mandipropamid

Products

1	Revus	Syngenta	250 g/l	SC	17443
2	Standon Mandor	Standon	250 g/l	SC	17533

Uses
- Blight in **potatoes** [2]
- Disease control in **baby leaf crops, celery leaves, chives, protected baby leaf crops, protected chives, protected cress, protected endives, protected lamb's lettuce, protected land cress, protected lettuce, protected red mustard, protected spinach, protected spinach beet** [1]; **chicory, purslane, rocket, salad greens** [2]; **cress, endives, herbs (see appendix 6), lamb's lettuce, land cress, lettuce, protected herbs (see appendix 6), red mustard, spinach, spinach beet** [1, 2]
- Downy mildew in **baby leaf crops, celery leaves, chives, hops** *(off-label)*, **protected baby leaf crops, protected chives, protected cress, protected endives, protected lamb's lettuce, protected land cress, protected lettuce, protected red mustard, protected spinach, protected spinach beet** [1]; **chicory, purslane, rocket, salad greens** [2]; **cress, endives, herbs (see appendix 6), lamb's lettuce, land cress, lettuce, protected herbs (see appendix 6), red mustard, spinach, spinach beet** [1, 2]

FOR FULL CONDITIONS OF USE ALWAYS READ THE PRODUCT LABEL

MCPA

Extension of Authorisation for Minor Use (EAMUs)
- *hops* 20162024 [1]

Approval information
- Mandipropamid is included in Annex 1 under EC Regulation 1107/2009

Efficacy guidance
- Mandipropamid acts preventatively by preventing spore germination and inhibiting mycelial growth during incubation. Apply immediately after blight warning or as soon as local conditions favour disease development but before blight enters the crop
- Spray at 7-10 d intervals reducing the interval as blight risk increases
- Spray programme should include a complete haulm desiccant to prevent tuber infection at harvest
- Eliminate other potential infection sources
- To minimise the likelihood of development of resistance this product should be used in a planned Resistance Management strategy. See Section 5 for more information
- See label for details of tank mixtures that may be used as part of a resistance management strategy
- Rainfast in potatoes within 15 minutes of application

Restrictions
- Maximum number of treatments 4 per crop. Do not apply more than 3 treatments of this, or any other fungicide in the same resistance category, consecutively

Crop-specific information
- HI 3 d for potatoes

Environmental safety
- Buffer zone requirement 10 m in hops [1]

Hazard classification and safety precautions
UN Number N/C
Operator protection A, H; U05a, U20b
Environmental protection E15b, E34, E38, H411
Storage and disposal D01, D02, D05, D09a, D10c, D12a
Medical advice M03, M05a

310 MCPA

A translocated phenoxycarboxylic acid herbicide for cereals and grassland
HRAC mode of action code: O

See also 2,4-D + clopyralid + MCPA
2,4-D + dicamba + MCPA + mecoprop-P
2,4-D + dichlorprop-P + MCPA + mecoprop-P
2,4-D + MCPA
2,4-DB + linuron + MCPA
2,4-DB + MCPA
bentazone + MCPA + MCPB
bifenox + MCPA + mecoprop-P
clopyralid + 2,4-D + MCPA
clopyralid + diflufenican + MCPA
clopyralid + fluroxypyr + MCPA
dicamba + dichlorprop-P + ferrous sulphate + MCPA
dicamba + dichlorprop-P + MCPA
dicamba + MCPA + mecoprop-P
dichlorprop-P + ferrous sulphate + MCPA
dichlorprop-P + MCPA
dichlorprop-P + MCPA + mecoprop-P
ferrous sulphate + MCPA + mecoprop-P

SEE SECTION 3 FOR PRODUCTS ALSO REGISTERED

Pesticide Profiles

Products

1	Agritox	Nufarm UK	500 g/l	SL	14894
2	Agroxone	Headland	500 g/l	SL	14909
3	Easel	Nufarm UK	750 g/l	SL	15548
4	Headland Spear	Headland	500 g/l	SL	14910
5	HY-MCPA	Agrichem	500 g/l	SL	14927
6	Larke	Nufarm UK	750 g/l	SL	14914
7	MCPA 25%	Nufarm UK	250 g/l	SL	14893

Uses
- Annual and perennial weeds in **grass seed crops**, **grassland** [1, 3, 6, 7]
- Annual dicotyledons in **farm forestry** *(off-label)*, **game cover** *(off-label)* [2]; **grass seed crops**, **grassland** [2, 4, 5]; **miscanthus** *(off-label)* [4]; **spring barley**, **spring oats**, **spring wheat**, **winter barley**, **winter oats**, **winter wheat** [1-7]; **spring rye**, **winter rye** [1, 3, 6, 7]; **undersown barley**, **undersown oats**, **undersown rye**, **undersown wheat** [1, 3, 5-7]; **undersown barley** *(red clover or grass)*, **undersown wheat** *(red clover or grass)* [2, 4]
- Charlock in **spring barley**, **spring oats**, **spring wheat**, **winter barley**, **winter oats**, **winter wheat** [1-7]; **spring rye**, **winter rye** [1, 3, 6, 7]; **undersown barley** *(red clover or grass)*, **undersown wheat** *(red clover or grass)* [2, 4]
- Fat hen in **spring barley**, **spring oats**, **spring wheat**, **winter barley**, **winter oats**, **winter wheat** [1-7]; **spring rye**, **winter rye** [1, 3, 6, 7]; **undersown barley** *(red clover or grass)*, **undersown wheat** *(red clover or grass)* [2, 4]
- Hemp-nettle in **spring barley**, **spring oats**, **spring wheat**, **winter barley**, **winter oats**, **winter wheat** [1-7]; **spring rye**, **winter rye** [1, 3, 6, 7]; **undersown barley** *(red clover or grass)*, **undersown wheat** *(red clover or grass)* [2, 4]
- Perennial dicotyledons in **farm forestry** *(off-label)*, **game cover** *(off-label)* [2]; **grass seed crops**, **grassland**, **undersown barley** *(red clover or grass)*, **undersown wheat** *(red clover or grass)* [2, 4]; **miscanthus** *(off-label)* [4]; **spring barley**, **spring oats**, **spring wheat**, **winter barley**, **winter oats**, **winter wheat** [1-7]; **spring rye**, **winter rye** [1, 3, 6, 7]
- Wild radish in **spring barley**, **spring oats**, **spring wheat**, **winter barley**, **winter oats**, **winter wheat** [1-7]; **spring rye**, **winter rye** [1, 3, 6, 7]; **undersown barley** *(red clover or grass)*, **undersown wheat** *(red clover or grass)* [2, 4]

Extension of Authorisation for Minor Use (EAMUs)
- *farm forestry 20122061* [2]
- *game cover 20122061* [2]
- *miscanthus 20122050* [4]

Approval information
- MCPA included in Annex I under EC Regulation 1107/2009
- Accepted by BBPA for use on malting barley

Efficacy guidance
- Best results achieved by application to weeds in seedling to young plant stage under good growing conditions when crop growing actively
- Spray perennial weeds in grassland before flowering. Most susceptible growth stage varies between species. See label for details
- Do not spray during cold weather, drought, if rain or frost expected or if crop wet

Restrictions
- Maximum number of treatments normally 1 per crop or yr except grass (2 per yr) for some products. See label
- Do not treat grass within 3 mth of germination and preferably not in the first yr of a direct sown ley or after reseeding
- Do not use on cereals before undersowing
- Do not roll, harrow or graze for a few days before or after spraying; see label
- Do not use on grassland where clovers are an important part of the sward
- Do not use on any crop suffering from stress or herbicide damage
- Avoid spray drift onto nearby susceptible crops

FOR FULL CONDITIONS OF USE ALWAYS READ THE PRODUCT LABEL

MCPA + mecoprop-P

Crop-specific information
- Latest use: before 1st node detectable (GS 31) for cereals; 4-6 wk before heading for grass seed crops; before crop 15-25 cm high for linseed
- Apply to winter cereals in spring from fully tillered, leaf sheath erect stage to before first node detectable (GS 31)
- Apply to spring barley and wheat from 5-leaves unfolded (GS 15), to oats from 1-leaf unfolded (GS 11) to before first node detectable (GS 31)
- Apply to cereals undersown with grass after grass has 2-3 leaves unfolded
- Recommendations for crops undersown with legumes vary. Red clover may withstand low doses after 2-trifoliate leaf stage, especially if shielded by taller weeds but white clover is more sensitive. See label for details
- Apply to grass seed crops from 2-3 leaf stage to 5 wk before head emergence
- Temporary wilting may occur on linseed but without long term effects

Following crops guidance
- Do not direct drill brassicas or legumes within 6 wk of spraying grassland

Environmental safety
- Harmful to aquatic organisms
- MCPA is active at low concentrations. Take extreme care to avoid drift onto neighbouring crops, especially beet crops, brassicas, most market garden crops including lettuce and tomatoes under glass, pears and vines
- Keep livestock out of treated areas for at least 2 wk and until foliage of poisonous weeds such as ragwort has died and become unpalatable
- LERAP Category B

Hazard classification and safety precautions
Hazard Harmful [1-7]; Dangerous for the environment [2-5]; Harmful if swallowed [1, 3, 6, 7]
UN Number N/C
Risk phrases H318 [1, 3, 6, 7]; R20, R21 [2]; R22a [2, 4-6]; R37, R41 [4, 5]; R50, R53a [2, 4, 5]
Operator protection A, C; U05a, U08, U11, U20b [1-7]; U09a [2]; U12 [3]; U15 [4]
Environmental protection E06a [5] (2 wk); E07a [1-4, 6, 7]; E15a [1, 2, 5-7]; E15b, E38 [3, 4]; E16a, E34 [1-7]; E40b [4]
Storage and disposal D01, D02, D05, D09a [1-7]; D10a [1-3, 5-7]; D10b [4]; D12a [3]
Medical advice M03 [1-7]; M05a [5]

311 MCPA + mecoprop-P

A translocated selective herbicide for amenity grass
HRAC mode of action code: O + O

See also mecoprop-P

Products
Cleanrun Pro   Everris Ltd   0.49:0.29% w/w   GR   15828

Uses
- Annual dicotyledons in **managed amenity turf**
- Perennial dicotyledons in **managed amenity turf**

Approval information
- MCPA and mecoprop-P included in Annex I under EC Regulation 1107/2009

Efficacy guidance
- Apply from Apr to Sep, when weeds growing actively and have large leaf area available for chemical absorption

Restrictions
- The total amount of mecoprop-P applied in a single year must not exceed the maximum total dose approved for any single product for use on turf
- Avoid contact with cultivated plants
- Do not use first 4 mowings as compost or mulch unless composted for 6 mth

SEE SECTION 3 FOR PRODUCTS ALSO REGISTERED

Pesticide Profiles

- Do not treat newly sown or turfed areas for at least 6 mth
- Do not reseed bare patches for 8 wk after treatment
- Do not apply when heavy rain expected or during prolonged drought. Irrigate after 1-2 d unless rain has fallen
- Do not mow within 2-3 d of treatment
- Treat areas planted with bulbs only after the foliage has died down
- Avoid walking on treated areas until it has rained or irrigation has been applied

Crop-specific information
- Granules contain NPK fertilizer to encourage grass growth

Environmental safety
- Take extreme care to avoid drift onto neighbouring crops, especially beet crops, brassicas, most market garden crops including lettuce and tomatoes under glass, pears and vines
- Harmful to fish or other aquatic life. Do not contaminate surface waters or ditches with chemical or used container
- Keep livestock out of treated areas for at least 2 wk and until foliage of any poisonous weeds such as ragwort has died and become unpalatable
- Some pesticides pose a greater threat of contamination of water than others and mecoprop-P is one of these pesticides. Take special care when applying mecoprop-P near water and do not apply if heavy rain is forecast

Hazard classification and safety precautions
UN Number N/C
Operator protection A, C, H, M; U20b
Environmental protection E07a, E13c, E19b
Storage and disposal D01, D09a, D12a
Medical advice M05a

312 MCPB

A translocated phenoxycarboxylic acid herbicide
HRAC mode of action code: O

See also bentazone + MCPA + MCPB
    bentazone + MCPB
    MCPA + MCPB

Products
1	Bellmac Straight	UPL Europe	400 g/l	SL	14448
2	Butoxone	Headland	400 g/l	SL	14406
3	Tropotox	Nufarm UK	400 g/l	SL	14450

Uses
- Annual dicotyledons in **combining peas**, **vining peas** [1-3]; **game cover** *(off-label)* [1, 2]
- Docks in **combining peas**, **vining peas** [3]
- Perennial dicotyledons in **combining peas**, **game cover** *(off-label)*, **vining peas** [1, 2]
- Thistles in **combining peas**, **vining peas** [3]

Extension of Authorisation for Minor Use (EAMUs)
- **game cover** *20122052* [1], *20122053* [2]

Approval information
- MCPB included in Annex I under EC Regulation 1107/2009

Efficacy guidance
- Best results achieved by spraying young seedling weeds in good growing conditions
- Best results on perennials by spraying before flowering
- Effectiveness may be reduced by rain within 12 h, by very cold or dry conditions

Restrictions
- Maximum number of treatments 1 per crop or yr.

FOR FULL CONDITIONS OF USE ALWAYS READ THE PRODUCT LABEL

- Do not roll or harrow for 7-10 d before or after treatment (check label)

Crop-specific information
- Latest use: first node detectable stage (GS 31) for cereals; before flower buds appear in terminal leaf (GS 201) for peas; before flower buds form for clover
- Apply to undersown cereals from 2-leaves unfolded to first node detectable (GS 12-31), and after first trifoliate leaf stage of clover
- Red clover seedlings may be temporarily damaged but later growth is normal
- Apply to white clover seed crops in Mar to early Apr, not after mid-May, and allow 3 wk before cutting and closing up for seed
- Apply to peas from 3-6 leaf stage but before flower bud detectable (GS 103-201). Consult PGRO (see Appendix 2) or label for information on susceptibility of cultivars.
- Do not use on leguminous crops not mentioned on the label
- Apply to cane and bush fruit after harvest and after shoot growth ceased but before weeds are damaged by frost, usually in late Aug or Sep; direct spray onto weeds as far as possible

Environmental safety
- Harmful to aquatic organisms
- Harmful to fish or other aquatic life. Do not contaminate surface waters or ditches with chemical or used container
- Keep livestock out of treated areas until foliage of any poisonous weeds such as ragwort has died and become unpalatable
- Take extreme care to avoid drift onto neighbouring sensitive crops

Hazard classification and safety precautions
Hazard Harmful [1-3]; Dangerous for the environment [2, 3]; Harmful if swallowed [1, 3]
Transport code 9 [3]
Packaging group III [3]
UN Number 3082 [3]; N/C [1, 2]
Risk phrases H315, H318 [1, 3]; R22a, R38, R41 [2]
Operator protection A, C; U05a, U08, U20b [1-3]; U11, U14, U15 [1, 2]; U19a [3]
Environmental protection E07a, E34 [1-3]; E13c [1, 2]; E15a, E38, H411 [3]
Storage and disposal D01, D02, D09a [1-3]; D05 [1, 2]; D10b [2, 3]; D10c [1]
Medical advice M03, M05a

313   mecoprop-P

A translocated phenoxycarboxylic acid herbicide for cereals and grassland
HRAC mode of action code: O

See also 2,4-D + dicamba + MCPA + mecoprop-P
2,4-D + dichlorprop-P + MCPA + mecoprop-P
2,4-D + mecoprop-P
bifenox + MCPA + mecoprop-P
bromoxynil + ioxynil + mecoprop-P
carfentrazone-ethyl + mecoprop-P
dicamba + MCPA + mecoprop-P
dicamba + mecoprop-P
dichlorprop-P + MCPA + mecoprop-P
diflufenican + mecoprop-P
ferrous sulphate + MCPA + mecoprop-P
fluroxypyr + mecoprop-P
MCPA + mecoprop-P

Products

1	Compitox Plus	Nufarm UK	600 g/l	SL	14390
2	Duplosan KV	Nufarm UK	600 g/l	SL	13971
3	Headland Charge	Headland	600 g/l	SL	14394
4	Optica	Headland	600 g/l	SL	14373

SEE SECTION 3 FOR PRODUCTS ALSO REGISTERED

Uses

- Annual dicotyledons in **amenity grassland**, **durum wheat** *(off-label)*, **spring durum wheat** *(off-label)*, **spring rye** *(off-label)*, **spring triticale** *(off-label)*, **triticale** *(off-label)*, **winter rye** *(off-label)* [1, 2]; **game cover** *(off-label)*, **miscanthus** *(off-label)* [2, 4]; **grass seed crops**, **managed amenity turf**, **spring barley**, **spring oats**, **spring wheat**, **winter barley**, **winter oats**, **winter wheat** [1-4]
- Chickweed in **amenity grassland** [1, 2]; **grass seed crops**, **managed amenity turf**, **spring barley**, **spring oats**, **spring wheat**, **winter barley**, **winter oats**, **winter wheat** [1-4]
- Cleavers in **amenity grassland** [1, 2]; **grass seed crops**, **managed amenity turf**, **spring barley**, **spring oats**, **spring wheat**, **winter barley**, **winter oats**, **winter wheat** [1-4]
- Perennial dicotyledons in **amenity grassland**, **grass seed crops**, **managed amenity turf**, **spring barley**, **spring oats**, **spring wheat**, **winter barley**, **winter oats**, **winter wheat** [1, 2]; **game cover** *(off-label)*, **miscanthus** *(off-label)* [2]

Extension of Authorisation for Minor Use (EAMUs)

- **durum wheat** *20093214* [1], *20093211* [2]
- **game cover** *20111129* [2], *20111125* [4]
- **miscanthus** *20111129* [2], *20111125* [4]
- **spring durum wheat** *20093214* [1], *20093211* [2]
- **spring rye** *20093214* [1], *20093211* [2]
- **spring triticale** *20093214* [1], *20093211* [2]
- **triticale** *20093214* [1], *20093211* [2]
- **winter rye** *20093214* [1], *20093211* [2]

Approval information

- Mecoprop-P included in Annex I under EC Regulation 1107/2009
- Accepted by BBPA for use on malting barley

Efficacy guidance

- Best results achieved by application to seedling weeds which have not been frost hardened, when soil warm and moist and expected to remain so for several days

Restrictions

- Maximum number of treatments normally 1 per crop for spring cereals and 1 per yr for newly sown grass; 2 per crop or yr for winter cereals and grass crops. Check labels for details
- The total amount of mecoprop-P applied in a single yr must not exceed the maximum total dose approved for any single product for the crop/situation
- Do not spray cereals undersown with clovers or legumes or to be undersown with legumes or grasses
- Do not spray grass seed crops within 5 wk of seed head emergence
- Do not spray crops suffering from herbicide damage or physical stress
- Do not spray during cold weather, periods of drought, if rain or frost expected or if crop wet
- Do not roll or harrow for 7 d before or after treatment

Crop-specific information

- Latest use: generally before 1st node detectable (GS 31) for spring cereals and before 3rd node detectable (GS 33) for winter cereals, but individual labels vary; 5 wk before emergence of seed head for grass seed crops
- Spray winter cereals from 1 leaf stage in autumn up to and including first node detectable in spring (GS 10-31) or up to second node detectable (GS 32) if necessary. Apply to spring cereals from first fully expanded leaf stage (GS 11) but before first node detectable (GS 31)
- Spray cereals undersown with grass after grass starts to tiller
- Spray newly sown grass leys when grasses have at least 3 fully expanded leaves and have begun to tiller. Any clovers will be damaged

Environmental safety

- Harmful to aquatic organisms
- Harmful to fish or other aquatic life. Do not contaminate surface waters or ditches with chemical or used container
- Keep livestock out of treated areas for at least 2 wk and until foliage of any poisonous weeds, such as ragwort, has died and become unpalatable

FOR FULL CONDITIONS OF USE ALWAYS READ THE PRODUCT LABEL

- Take extreme care to avoid drift onto neighbouring crops, especially beet crops, brassicas, most market garden crops including lettuce and tomatoes under glass, pears and vines
- Some pesticides pose a greater threat of contamination of water than others and mecoprop-P is one of these pesticides. Take special care when applying mecoprop-P near water and do not apply if heavy rain is forecast

Hazard classification and safety precautions
Hazard Harmful, Dangerous for the environment [1-4]; Harmful if swallowed [1, 2]
Transport code 9 [3, 4]
Packaging group III [3, 4]
UN Number 3082 [3, 4]; N/C [1, 2]
Risk phrases H315, H318 [1, 2]; R22a, R38, R41, R51, R52, R53a [3, 4]
Operator protection A, C [1-4]; H [1, 2]; P [3, 4]; U05a, U08, U11, U20b [1-4]; U15 [3, 4]
Environmental protection E07a, E13c [3, 4]; E15b [1, 2]; E34, E38 [1-4]
Storage and disposal D01, D02 [1-4]; D05, D10b [3, 4]; D09a, D10c [1, 2]
Medical advice M03

314   mepanipyrim

An anilinopyrimidine fungicide for use in horticulture
FRAC mode of action code: 9

Products
Frupica SC	Certis	450 g/l	SC	12067

Uses
- Botrytis in **courgettes** (off-label), **forest nurseries** (off-label), **protected strawberries**, **strawberries**

Extension of Authorisation for Minor Use (EAMUs)
- *courgettes* 20093235
- *forest nurseries* 20082853

Approval information
- Mepanipyrim included in Annex I under EC Regulation 1107/2009

Efficacy guidance
- Product is protectant and should be applied as a preventative spray when conditions favourable for Botrytis development occur
- To maintain Botrytis control use as part of a programme with other fungicides that control the disease
- To minimise the possibility of development of resistance adopt resistance management procedures by using products from different chemical groups as part of a mixed spray programme

Restrictions
- Maximum number of treatments 2 per crop (including other anilinopyrimidine products)
- Consult processor before use on crops for processing
- Use spray mixture immediately after preparation

Crop-specific information
- HI 3 d

Environmental safety
- Dangerous for the environment
- Very toxic to aquatic organisms
- LERAP Category B

Hazard classification and safety precautions
Hazard Dangerous for the environment
Transport code 9
Packaging group III

SEE SECTION 3 FOR PRODUCTS ALSO REGISTERED

Pesticide Profiles

UN Number 3082
Risk phrases R50, R53a
Operator protection A, H; U20c
Environmental protection E16a, E16b, E34
Storage and disposal D01, D02, D05, D10c, D11a, D12b

315 mepiquat chloride

A quaternary ammonium plant growth regulator available only in mixtures

See also 2-chloroethylphosphonic acid + mepiquat chloride
chlormequat + 2-chloroethylphosphonic acid + mepiquat chloride
chlormequat + mepiquat chloride
ethephon + mepiquat chloride

316 mepiquat chloride + metconazole

A PGR mixture for growth control in winter oilseed rape

Products
Caryx   BASF   210:30 g/l   SL   16100

Uses
- Growth regulation in **winter oilseed rape**
- Lodging control in **winter oilseed rape**

Approval information
- Metconazole and mepiquat chloride included in Annex I under EC Regulation 1107/2009

Efficacy guidance
- Apply in 200 - 400 l/ha water. Use at lower water volumes has not been evaluated.
- Apply from the beginning of stem extension once the crop is actively growing.

Following crops guidance
- Beans, cabbage, carrots, cereals, clover, lettuce, linseed, maize, oilseed rape, onions, peas, potatoes, ryegrass, sugar beet or sunflowers may be sown as a following crop but the effect on other crops has not been tested.

Environmental safety
- LERAP Category B

Hazard classification and safety precautions
Hazard Harmful, Dangerous for the environment, Harmful if swallowed, Harmful if inhaled
Transport code 9
Packaging group III
UN Number 3082
Risk phrases H317, H318
Operator protection A, C, H; U02a, U05a, U11, U14, U15
Environmental protection E15b, E16a, E38, H411
Storage and disposal D01, D02, D05, D09a, D10c, D19
Medical advice M03, M05a

317 mepiquat chloride + prohexadione-calcium

A growth regulator mixture for cereals

See also prohexadione-calcium

Products
Canopy   BASF   300:50 g/l   SC   16314

FOR FULL CONDITIONS OF USE ALWAYS READ THE PRODUCT LABEL

meptyldinocap

Uses
- Increasing yield in **oats, spring barley, triticale, winter barley, winter rye, winter wheat**
- Lodging control in **oats, spring barley, triticale, winter barley, winter rye, winter wheat**

Approval information
- Mepiquat chloride and prohexadione-calcium included in Annex I under EC Regulation 1107/2009
- Accepted by BBPA for use on malting barley

Efficacy guidance
- Best results obtained from treatments applied to healthy crops from the beginning of stem extension

Restrictions
- Maximum total dose equivalent to one full dose treatment on all crops
- Do not apply to any crop suffering from physical stress caused by waterlogging, drought or other conditions
- Do not treat on soils with a substantial moisture deficit
- Consult grain merchant or processor before use on crops for bread making or brewing. Effects on these processes have not been tested

Crop-specific information
- Latest use: before flag leaf fully emerged on wheat and barley

Following crops guidance
- Any crop may follow a normally harvested treated crop. Ploughing is not essential.

Environmental safety
- Harmful to aquatic organisms
- Avoid spray drift onto neighbouring crops

Hazard classification and safety precautions
Hazard Harmful, Harmful if swallowed
UN Number N/C
Operator protection A; U05a, U20b
Environmental protection E15a, E34, E38, H412
Storage and disposal D01, D02, D05, D09a, D10c
Medical advice M05a

318 meptyldinocap

A protectant dinitrophenyl fungicide for powdery mildew control
FRAC mode of action code: 29

Products
Kindred	Landseer	350 g/l	EC	13891

Uses
- Powdery mildew in **apples** *(off-label)*, **crab apples** *(off-label)*, **pears** *(off-label)*, **protected strawberries**, **quinces** *(off-label)*, **table grapes**, **wine grapes**

Extension of Authorisation for Minor Use (EAMUs)
- *apples* 20092664
- *crab apples* 20092664
- *pears* 20092664
- *quinces* 20092664

Approval information
- Meptyldinocap is included in Annex 1 under EC Regulation 1107/2009

Environmental safety
- LERAP Category B

SEE SECTION 3 FOR PRODUCTS ALSO REGISTERED

Pesticide Profiles

Hazard classification and safety precautions
Hazard Harmful, Flammable, Dangerous for the environment
Transport code 3
Packaging group III
UN Number 1993
Risk phrases R22a, R36, R38, R43, R50, R53a, R67
Operator protection A, C, H; U04a, U05a, U11, U14, U19a, U20a
Environmental protection E15b, E16a, E34, E38
Storage and disposal D01, D02, D09a, D10a, D12a
Medical advice M05a

319　mesosulfuron-methyl

A sulfonyl urea herbicide for cereals available only in mixtures
HRAC mode of action code: B

See also amidosulfuron + iodosulfuron-methyl-sodium + mesosulfuron-methyl
diflufenican + iodosulfuron-methyl-sodium + mesosulfuron-methyl
iodosulfuron-methyl-sodium + mesosulfuron-methyl

320　mesotrione

A foliar applied triketone herbicide for maize
HRAC mode of action code: F2

Products

1	Callisto	Syngenta	100 g/l	SC	12323
2	Clayton Goldcob	Clayton	100 g/l	SC	16299
3	Greencrop Goldcob	Greencrop	100 g/l	SC	12863
4	Mesostar	Life Scientific	100 g/l	SC	17410
5	Standon Cobmajor	Standon	100 g/l	SC	16721
6	Temsa SC	Belchim	100 g/l	SC	17234

Uses
- Annual dicotyledons in **asparagus** *(off-label)*, **game cover** *(off-label)*, **linseed** *(off-label)*, **poppies for morphine production** *(off-label)*, **sweetcorn** *(off-label)* [1]; **forage maize** [1-6]
- Annual grasses in **game cover** *(off-label)*, **sweetcorn** *(off-label)* [1]
- Annual meadow grass in **grain maize** [1, 2, 4-6]; **linseed** *(off-label)* [1]
- Black bindweed in **rhubarb** *(off-label)* [1]
- Charlock in **rhubarb** *(off-label)* [1]
- Cleavers in **rhubarb** *(off-label)* [1]
- Cockspur grass in **forage maize**, **grain maize** [2]
- Field pansy in **rhubarb** *(off-label)* [1]
- Groundsel in **rhubarb** *(off-label)* [1]
- Himalayan balsam in **rhubarb** *(off-label)* [1]
- Mayweeds in **rhubarb** *(off-label)* [1]
- Volunteer oilseed rape in **forage maize** [1-6]; **grain maize** [1, 2, 4-6]; **linseed** *(off-label)* [1]

Extension of Authorisation for Minor Use (EAMUs)
- *asparagus* 20111113 [1]
- *game cover* 20082830 [1]
- *linseed* 20071706 [1]
- *poppies for morphine production* 20113148 [1]
- *rhubarb* 20161761 [1]
- *sweetcorn* 20051893 [1]

Approval information
- Mesotrione included in Annex I under EC Regulation 1107/2009

FOR FULL CONDITIONS OF USE ALWAYS READ THE PRODUCT LABEL

mesotrione + nicosulfuron

Efficacy guidance
- Best results obtained from treatment of young actively growing weed seedlings in the presence of adequate soil moisture
- Treatment in poor growing conditions or in dry soil may give less reliable control
- Activity is mostly by foliar uptake with some soil uptake
- To minimise the possible development of resistance where continuous maize is grown the product should not be used for more than two consecutive seasons

Restrictions
- Maximum number of treatments 1 per crop of forage maize
- Do not use on seed crops or on sweetcorn varieties
- Do not spray when crop foliage wet or when excessive rainfall is expected to follow application
- Do not treat crops suffering from stress from cold or drought conditions, or when wide temperature fluctuations are anticipated

Crop-specific information
- Latest use: 8 leaves unfolded stage (GS 18) for forage maize
- Treatment under adverse conditions may cause mild to moderate chlorosis. The effect is transient and does not affect yield

Following crops guidance
- Winter wheat, durum wheat, winter barley or ryegrass may follow a normally harvested treated crop of maize. Oilseed rape may be sown provided it is preceded by deep ploughing to more than 15 cm
- In the spring following application only forage maize, ryegrass, spring wheat or spring barley may be sown
- In the event of crop failure maize may be re-seeded immediately. Some slight crop effects may be seen soon after emergence but these are normally transient

Environmental safety
- Dangerous for the environment
- Very toxic to aquatic organisms
- Take extreme care to avoid drift onto all plants outside the target area
- LERAP Category B

Hazard classification and safety precautions
    Hazard Irritant, Dangerous for the environment [1-6]; Very toxic to aquatic organisms [4]
    UN Number N/C
    Risk phrases H317, H318 [6]; H319 [4]; R36, R50, R53a [1-3, 5]
    Operator protection A, C; U05a, U20b [1-6]; U08, U19a [3]; U09a [1, 2, 4-6]
    Environmental protection E15b, E16a, E16b [1-6]; E38 [1, 2, 4-6]; H410 [4, 6]
    Storage and disposal D01, D02, D05, D09a, D12a [1-6]; D10b [3]; D10c, D11a [1, 2, 4-6]
    Medical advice M05a [3]

321   mesotrione + nicosulfuron

A herbicide for use in forage and grain maize
HRAC mode of action code: B + F2

Products
    Elumis   Syngenta   75:30 g/l   OD   15800

Uses
- Amaranthus in *forage maize, grain maize*
- Annual dicotyledons in *forage maize, grain maize*
- Annual meadow grass in *forage maize, grain maize*
- Black bindweed in *forage maize, grain maize*
- Black nightshade in *forage maize, grain maize*
- Chickweed in *forage maize, grain maize*
- Cleavers in *forage maize, grain maize*
- Cockspur grass in *forage maize, grain maize*

SEE SECTION 3 FOR PRODUCTS ALSO REGISTERED

Pesticide Profiles

- Fat hen in *forage maize, grain maize*
- Field pansy in *forage maize, grain maize*
- Field speedwell in *forage maize, grain maize*
- Fumitory in *forage maize, grain maize*
- Hemp-nettle in *forage maize, grain maize*
- Mayweeds in *forage maize, grain maize*
- Pale persicaria in *forage maize, grain maize*
- Red dead-nettle in *forage maize, grain maize*
- Redshank in *forage maize, grain maize*
- Ryegrass in *forage maize, grain maize*
- Shepherd's purse in *forage maize, grain maize*
- Volunteer oilseed rape in *forage maize, grain maize*

Approval information
- Mesotrione and nicosulfuron included in Annex 1 under EC Regulation 1107/2009

Efficacy guidance
- Optimum efficacy requires soils to be moist at application.
- To minimise the possible development of resistance where continuous maize is grown the product should not be used for more than two consecutive seasons.

Following crops guidance
- In the event of crop failure, only forage or grain maize may be sown as a replacement after ploughing.
- After normal harvest wheat or barley may be sown the same autumn after ploughing to 15 cm and after 4 months has elapsed since application. The following spring maize, wheat, barley or ryegrass may be planted after ploughing but other crops are not recommended.

Environmental safety
- LERAP Category B

Hazard classification and safety precautions
Hazard Irritant, Dangerous for the environment
Transport code 9
Packaging group III
UN Number 3082
Risk phrases R38, R50, R53a
Operator protection A, H; U02a, U05a, U14, U19a, U20a
Environmental protection E15b, E16a, E23
Storage and disposal D01, D02, D05, D09a, D10c, D12a
Medical advice M05a, M05d

322 mesotrione + terbuthylazine

A foliar and soil acting herbicide mixture for maize
HRAC mode of action code: F2 + C1

See also terbuthylazine

Products

1	Calaris	Syngenta	70:330 g/l	SC	12405
2	Clayton Faize	Clayton	70:330 g/l	SC	13810
3	Destiny	AgChem Access	70:330 g/l	SC	14159
4	Pan Theta	Pan Agriculture	70:330 g/l	SC	14501
5	Standon Cobmaster	Standon	70:330 g/l	SC	16772

Uses
- Annual dicotyledons in *forage maize, grain maize* [1-5]; *sweetcorn* (off-label) [1]
- Annual meadow grass in *forage maize, grain maize* [1-5]; *sweetcorn* (off-label) [1]

FOR FULL CONDITIONS OF USE ALWAYS READ THE PRODUCT LABEL

mesotrione + terbuthylazine

Extension of Authorisation for Minor Use (EAMUs)
- *sweetcorn* 20051892 [1]

Approval information
- Mesotrione and terbuthylazine included in Annex 1 under EC Regulation 1107/2009

Efficacy guidance
- Best results obtained from treatment of young actively growing weed seedlings in the presence of adequate soil moisture
- Treatment in poor growing conditions or in dry soil may give less reliable control
- Residual weed control is reduced on soils with more than 10% organic matter
- To minimise the possible development of resistance where continuous maize is grown the product should not be used for more than two consecutive seasons

Restrictions
- Maximum number of treatments 1 per crop of forage maize
- Do not use on seed crops or on sweetcorn varieties
- Do not spray when crop foliage wet or when excessive rainfall is expected to follow application
- Do not treat crops suffering from stress from cold or drought conditions, or when wide temperature fluctuations are anticipated
- Do not apply on Sands or Very Light soils

Crop-specific information
- Latest use: 8 leaves unfolded stage (GS 18) for forage maize
- Treatment under adverse conditions may cause mild to moderate chlorosis. The effect is transient and does not affect yield

Following crops guidance
- Winter wheat, durum wheat, winter barley or ryegrass may follow a normally harvested treated crop of maize. Oilseed rape may be sown provided it is preceded by deep ploughing to more than 15 cm
- In the spring following application forage maize, ryegrass, spring wheat or spring barley may be sown
- Spinach, beet crops, peas, beans, lettuce and cabbages must not be sown in the yr following application
- In the event of crop failure maize may be re-seeded immediately. Some slight crop effects may be seen soon after emergence but these are normally transient

Environmental safety
- Dangerous for the environment
- Very toxic to aquatic organisms
- Take extreme care to avoid drift onto all plants outside the target area
- LERAP Category B

Hazard classification and safety precautions
Hazard Harmful, Dangerous for the environment
Transport code 9
Packaging group III
UN Number 3082
Risk phrases R22a, R50, R53a
Operator protection A; U05a
Environmental protection E15a, E16a, E16b, E34, E38
Storage and disposal D01, D02, D05, D09a, D10c, D12a

323 metalaxyl-M

A phenylamide systemic fungicide
FRAC mode of action code: 4

See also chlorothalonil + metalaxyl-M
cymoxanil + fludioxonil + metalaxyl-M
fluazinam + metalaxyl-M
fludioxonil + metalaxyl-M
fludioxonil + metalaxyl-M + thiamethoxam
mancozeb + metalaxyl-M

Products

1	Apron XL	Syngenta	339.2 g/l	ES	14654
2	Caveo	Pan Agriculture	465.2 g/l	SL	16361
3	Clayton Tine	Clayton	465 g/l	SL	14072
4	Floreo	Pan Agriculture	465.2 g/l	SL	16358
5	Hobson	AgChem Access	465.2 g/l	SL	16529
6	SL 567A	Syngenta	465.2 g/l	SL	12380
7	Subdue	Fargro	465 g/l	SL	12503

Uses

- Cavity spot in **carrots** [2, 5, 6]; **carrots** *(reduction only)*, **parsnips** *(off-label - reduction)* [3]; **parsnips** *(off-label)* [6]
- Crown rot in **protected water lilies** *(off-label)*, **water lilies** *(off-label)* [6]
- Damping off in **watercress** *(off-label)* [6]
- Downy mildew in **asparagus** *(off-label)*, **baby leaf crops** *(off-label)*, **blackberries** *(off-label)*, **broccoli** *(off-label)*, **calabrese** *(off-label)*, **collards** *(off-label)*, **grapevines** *(off-label)*, **herbs (see appendix 6)** *(off-label)*, **hops** *(off-label)*, **horseradish** *(off-label)*, **kale** *(off-label)*, **protected baby leaf crops** *(off-label)*, **protected cucumbers** *(off-label)*, **protected herbs (see appendix 6)** *(off-label)*, **protected soft fruit** *(off-label)*, **protected spinach** *(off-label)*, **protected spinach beet** *(off-label)*, **protected water lilies** *(off-label)*, **raspberries** *(off-label)*, **rubus hybrids** *(off-label)*, **salad onions** *(off-label)*, **soft fruit** *(off-label)*, **spinach** *(off-label)*, **spinach beet** *(off-label)*, **spring field beans** *(off-label)*, **water lilies** *(off-label)*, **watercress** *(off-label)*, **winter field beans** *(off-label)* [6]
- Phytophthora in **amenity vegetation** *(off-label)*, **forest nurseries** *(off-label)* [7]; **ornamental plant production**, **protected ornamentals** [4]
- Phytophthora root rot in **ornamental plant production**, **protected ornamentals** [7]
- Pythium in **beetroot** *(off-label)*, **brassica leaves and sprouts** *(off-label)*, **broccoli**, **brussels sprouts**, **bulb onions**, **cabbages**, **calabrese**, **cauliflowers**, **chard** *(off-label)*, **chinese cabbage**, **herbs (see appendix 6)** *(off-label)*, **kohlrabi**, **ornamental plant production** *(off-label)*, **radishes** *(off-label)*, **shallots** *(off-label)*, **spinach** [1]; **ornamental plant production**, **protected ornamentals** [4, 7]
- Root malformation disorder in **red beet** *(off-label)* [6]
- Storage rots in **cabbages** *(off-label)* [6]
- White blister in **horseradish** *(off-label)* [6]

Extension of Authorisation for Minor Use (EAMUs)

- *amenity vegetation* 20120383 [7]
- *asparagus* 20051502 [6]
- *baby leaf crops* 20051507 [6]
- *beetroot* 20102539 [1]
- *blackberries* 20072195 [6]
- *brassica leaves and sprouts* 20120526 [1]
- *broccoli* 20112502 [6]
- *cabbages* 20062117 [6]
- *calabrese* 20112502 [6]
- *chard* 20120526 [1]
- *collards* 20112048 [6]
- *forest nurseries* 20120383 [7]

FOR FULL CONDITIONS OF USE ALWAYS READ THE PRODUCT LABEL

metalaxyl-M

- *grapevines* 20051504 [6]
- *herbs (see appendix 6)* 20102539 [1], 20051507 [6]
- *hops* 20051500 [6]
- *horseradish* 20051499 [6], 20061040 [6]
- *kale* 20112048 [6]
- *ornamental plant production* 20102539 [1]
- *parsnips* (reduction) 20111033 [3], 20051508 [6]
- *protected baby leaf crops* 20051507 [6]
- *protected cucumbers* 20051503 [6]
- *protected herbs (see appendix 6)* 20051507 [6]
- *protected soft fruit* 20082937 [6]
- *protected spinach* 20051507 [6]
- *protected spinach beet* 20051507 [6]
- *protected water lilies* 20051501 [6]
- *radishes* 20102539 [1]
- *raspberries* 20072195 [6]
- *red beet* 20051307 [6]
- *rubus hybrids* 20072195 [6]
- *salad onions* 20072194 [6]
- *shallots* 20121635 [1]
- *soft fruit* 20082937 [6]
- *spinach* 20051507 [6]
- *spinach beet* 20051507 [6]
- *spring field beans* 20130917 [6]
- *water lilies* 20051501 [6]
- *watercress* 20072193 [6]
- *winter field beans* 20130917 [6]

Approval information
- Metalaxyl-M included in Annex I under EC Regulation 1107/2009
- Accepted by BBPA for use on hops

Efficacy guidance
- Best results achieved when applied to damp soil or potting media
- Treatments to ornamentals should be followed immediately by irrigation to wash any residues from the leaves and allow penetration to the rooting zone [7]
- Efficacy may be reduced in prolonged dry weather [6]
- Results may not be satisfactory on soils with high organic matter content [6]
- Control of cavity spot on carrots overwintered in the ground or lifted in winter may be lower than expected [6]
- Use in an integrated pest management strategy and, where appropriate, alternate with products from different chemical groups
- Product should ideally be used preventatively and the number of phenylamide applications should be limited to 1-2 consecutive treatments
- Always follow FRAG guidelines for preventing and managing fungicide resistance. See Section 5 for more information

Restrictions
- Maximum number of treatments on ornamentals 1 per situation for media treatment. See label for details of drench treatment of protected ornamentals [7]
- Maximum total dose on carrots equivalent to one full dose treatment [6]
- Do not use where carrots have been grown on the same site within the previous eight yrs [6]
- Consult before use on crops intended for processing [6]
- Do not re-use potting media from treated plants for subsequent crops [7]
- Disinfect pots thoroughly prior to re-use [7]

Crop-specific information
- Latest use: 6 wk after drilling for carrots [6]
- Because of the large number of species and ornamental cultivars susceptibility should be checked before large scale treatment [7]
- Treatment of *Viburnum* and *Prunus* species not recommended [7]

SEE SECTION 3 FOR PRODUCTS ALSO REGISTERED

Pesticide Profiles

Environmental safety
- Harmful to aquatic organisms
- Limited evidence suggests that metalaxyl-M is not harmful to soil dwelling predatory mites

Hazard classification and safety precautions
Hazard Harmful
UN Number N/C
Risk phrases R22a, R52 [1-7]; R37 [2-7]; R53a [1, 3, 5-7]; R53b [2, 4]
Operator protection A [1-7]; C [7]; D [1]; H [1, 7]; U02a, U05a, U20b [1-7]; U04a, U10, U19a [2-7]
Environmental protection E15b, E34, E38
Storage and disposal D01, D02, D05, D09a, D12a [1-7]; D07 [3, 5-7]; D10c [2-7]; D11a [1]
Treated seed S02, S04d, S05, S07, S08, S09 [1]
Medical advice M03 [2-7]; M05a [1]

324 metaldehyde

A molluscicide bait for controlling slugs and snails

Products

1	Allure	Chiltern	1.5% w/w	PT	12651
2	Appeal	Chiltern	1.5% w/w	PT	12713
3	Attract	Chiltern	1.5% w/w	PT	12712
4	Carakol 3	Adama	3% w/w	PT	14309
5	Certis Metaldehyde 3	Certis	3% w/w	PT	14337
6	Condor 3	Certis	3% w/w	PT	14324
7	Desire	Chiltern	1.5% w/w	PT	14048
8	Enzo	Adama	3% w/w	PT	14306
9	ESP3	De Sangosse	3% w/w	RB	16088
10	Gusto 3	Adama	3% w/w	PT	14308
11	Lynx H	De Sangosse	3% w/w	RB	14426
12	Osarex W	De Sangosse	3% w/w	RB	14428
13	Prowler	De Sangosse	3% w/w	RB	16087
14	Super 3	Certis	3% w/w	PT	14370
15	TDS Major	De Sangosse	4% w/w	CB	13462
16	Tempt	Chiltern	1.5% w/w	PT	14227
17	Trigger 3	Certis	3% w/w	GB	14304
18	Trounce	Chiltern	1.5% w/w	PT	14222

Uses
- Slugs in *all edible crops except potatoes and cauliflowers, all non-edible crops (outdoor), cauliflowers, potatoes* [2, 3, 7, 15]; *natural surfaces not intended to bear vegetation* [15]
- Slugs and snails in *all edible crops (outdoor and protected)* [1]; *all edible crops (outdoor)* [6]; *all edible crops except potatoes and cauliflowers* [4, 5, 8-14, 16-18]; *all non-edible crops (outdoor)* [1, 4, 6, 8-14, 16, 18]; *amenity grassland, managed amenity turf* [5, 17]; *cauliflowers* [1, 4, 8-13, 16, 18]; *natural surfaces not intended to bear vegetation* [5, 6, 9, 11-14, 17]; *potatoes* [1, 4, 5, 8-14, 16-18]
- Snails in *all edible crops except potatoes and cauliflowers, all non-edible crops (outdoor), cauliflowers, potatoes* [2, 3, 7, 15]; *natural surfaces not intended to bear vegetation* [15]

Approval information
- Metaldehyde included in Annex I under EC Regulation 1107/2009
- Accepted by BBPA for use on malting barley and hops

Efficacy guidance
- Apply pellets by hand, fiddle drill, fertilizer distributor, by air (check label) or in admixture with seed. See labels for rates and timing.
- Best results achieved from an even spread of granules applied during mild, damp weather when slugs and snails most active. May be applied in standing crops

FOR FULL CONDITIONS OF USE ALWAYS READ THE PRODUCT LABEL

metamitron

- To establish the need for pellet application on winter wheat or winter oilseed rape, monitor for slug activity. Where bait traps are used, use a foodstuff attractive to slugs e.g. chicken layer's mash
- Varieties of oilseed rape low in glucosinolates can be more acceptable to slugs than "single low" varieties and control may not be as good
- To prevent slug build up apply at end of season to brassicas and other leafy crops
- To reduce tuber damage in potatoes apply twice in Jul and Aug
- For information on slug trapping and damage risk assessment refer to AHDB Information Sheet 02 - Integrated slug control, available from the AHDB website (www.ahdb.org.uk/slugcontrol)

Restrictions
- Do not apply when rain imminent or water glasshouse crops within 4 d of application
- Take care to avoid lodging of pellets in the foliage when making late applications to edible crops.
- The maximum total dose of metaldehyde must not exceed 700 g active substance/ha/year.
- The risk to ground water on drained fields or sloping sites can be avoided by substituting metaldehyde products for ferric phosphate products

Crop-specific information
- Put slug traps out before cultivation, when the soil surface is visibly moist and the weather mild (5-25°C) (see label for guidance)
- For winter wheat, a catch of 4 or more slugs/trap indicates a possible risk, where soil and weather conditions favour slug activity
- For winter oilseed rape a catch of 4 or more slugs in standing cereals, or 1 or more in cereal stubble, if other conditions were met, would indicate possible risk of damage

Environmental safety
- Dangerous to game, wild birds and animals
- Some products contain proprietary cat and dog deterrent
- Keep poultry out of treated areas for at least 7 d
- Do not use slug pellets in traps in winter wheat or winter oilseed rape since they are a potential hazard to wildlife and pets
- Some pesticides pose a greater threat of contamination of water than others and metaldehyde is one of these pesticides. Take special care when applying metaldehyde near water and do not apply if heavy rain is forecast

Hazard classification and safety precautions
UN Number N/C
Operator protection A, H; U05a [1-13, 15-18]; U20a [4, 8, 10]; U20b [5, 6, 9, 11-13, 17]; U20c [1-3, 7, 14-16, 18]
Environmental protection E05b [1-5] (7 d); E05b [6, 7] (7 days); E05b [8] (7 d); E05b [9] (7 days); E05b [10] (7 d); E05b [11-13] (7 days); E05b [14-18] (7 d); E06b [7] (7 d); E07a, E34 [6, 9, 11-13]; E10a [1-4, 6-16, 18]; E10c [5, 17]; E15a [1-18]
Storage and disposal D01, D09a [1-18]; D02 [1-13, 15-18]; D05 [5, 17]; D07 [1-3, 5-7, 9, 11-18]; D11a [1-14, 16-18]; D11b [15]
Treated seed S04a [6, 9, 11-13, 15]
Medical advice M04a [5, 17]; M05a [6, 9, 11-13, 15]

325 metamitron

A contact and residual triazinone herbicide for use in beet crops and a fruit thinner for apples and pears.
HRAC mode of action code: C1

See also chloridazon + chlorpropham + metamitron
chloridazon + metamitron
chlorpropham + metamitron
desmedipham + ethofumesate + metamitron + phenmedipham
ethofumesate + metamitron
ethofumesate + metamitron + phenmedipham

SEE SECTION 3 FOR PRODUCTS ALSO REGISTERED

Pesticide Profiles

Products

1	Beetron 700	AgChem Access	700 g/l	SC	17458
2	Bettix Flo	UPL Europe	700 g/l	SC	16559
3	Brevis	Adama	15% w/w	SG	17479
4	Defiant SC	UPL Europe	700 g/l	SC	16531
5	Glotron 700 SC	Belchim	700 g/l	SC	17308
6	Goltix 70 SC	Adama	700 g/l	SC	16638
7	Mitron 700 SC	Belcrop	700 g/l	SC	16908
8	Siskin	Goldengrass	700 g/l	SC	16796
9	Target SC	UPL Europe	700 g/l	SC	16969

Uses

- Annual dicotyledons in **fodder beet**, **mangels**, **red beet**, **sugar beet** [1, 2, 4-9]; **horseradish** (off-label), **parsnips** (off-label), **strawberries** (off-label) [6]
- Annual grasses in **fodder beet**, **mangels**, **red beet**, **sugar beet** [1, 2, 4, 7, 8]
- Annual meadow grass in **fodder beet**, **mangels**, **red beet**, **sugar beet** [1, 2, 4-9]; **horseradish** (off-label), **ornamental plant production** (off-label), **parsnips** (off-label), **protected ornamentals** (off-label), **strawberries** (off-label) [6]
- Fat hen in **fodder beet**, **mangels**, **red beet**, **sugar beet** [1, 2, 4, 7, 8]; **ornamental plant production** (off-label), **protected ornamentals** (off-label) [6]
- Fruit thinning in **apples**, **pears** [3]
- Groundsel in **ornamental plant production** (off-label), **protected ornamentals** (off-label) [6]
- Growth regulation in **apples**, **pears** [3]
- Mayweeds in **ornamental plant production** (off-label), **protected ornamentals** (off-label) [6]
- Red dead-nettle in **ornamental plant production** (off-label), **protected ornamentals** (off-label) [6]
- Scarlet pimpernel in **ornamental plant production** (off-label), **protected ornamentals** (off-label) [6]
- Small nettle in **ornamental plant production** (off-label), **protected ornamentals** (off-label) [6]

Extension of Authorisation for Minor Use (EAMUs)

- **horseradish** 20142918 [6]
- **ornamental plant production** 20151175 [6]
- **parsnips** 20142918 [6]
- **protected ornamentals** 20151175 [6]
- **strawberries** 20142919 [6]

Approval information

- Metamitron included in Annex I under EC Regulation 1107/2009

Efficacy guidance

- May be used pre-emergence alone or post-emergence in tank mixture or with an authorised adjuvant oil
- Low dose programme (LDP). Apply a series of low-dose post-weed emergence sprays, including adjuvant oil, timing each treatment according to weed emergence and size. See label for details and for recommended tank mixes and sequential treatments. On mineral soils the LDP should be preceded by pre-drilling or pre-emergence treatment
- Traditional application. Apply either pre-drilling before final cultivation with incorporation to 8-10 cm, or pre-crop emergence at or soon after drilling into firm, moist seedbed to emerged weeds from cotyledon to first true leaf stage
- On emerged weeds at or beyond 2-leaf stage addition of adjuvant oil advised
- For control of wild oats and certain other weeds, tank mixes with other herbicides or sequential treatments are recommended. See label for details
- When used for fruit thinning in apples and pears, apply post blossom if fruit set is excessive. Apply in temperatures between 10 and 25°C [3]

Restrictions

- Maximum total dose equivalent to three full dose treatments for most products. Check label
- Using traditional method post-crop emergence on mineral soils do not apply before first true leaves have reached 1 cm long

Crop-specific information
- Latest use: before crop foliage meets across rows for beet crops
- HI herbs 6 wk
- HI for apples and pears 60 days
- Crop tolerance may be reduced by stress caused by growing conditions, effects of pests, disease or other pesticides, nutrient deficiency etc

Following crops guidance
- Only sugar beet, fodder beet or mangels may be drilled within 4 mth after treatment. Winter cereals may be sown in same season after ploughing, provided 16 wk passed since last treatment

Environmental safety
- Dangerous for the environment
- Very toxic to aquatic organisms
- Dangerous to fish or other aquatic life. Do not contaminate surface waters or ditches with chemical or used container
- Do not empty into drains

Hazard classification and safety precautions
Hazard Harmful [1, 4-7, 9]; Dangerous for the environment [1-9]; Harmful if swallowed [3, 5]
Transport code 9
Packaging group III
UN Number 3077 [3]; 3082 [1, 2, 4-9]
Risk phrases H318 [3]; R22a, R53a [1, 2, 4-9]; R43 [2, 4, 8]; R50 [1]; R51 [2, 4-9]
Operator protection A [1-8]; C [3]; H [5, 6]; U02a, U04a, U11 [3]; U05a [3, 7, 9]; U08, U19a [7, 9]; U09a, U20c [2, 4, 8]; U14 [2, 4, 7-9]; U20a [7]; U20b [1, 3, 9]
Environmental protection E13b [2, 4, 8]; E15a [1, 7, 9]; E15b, H411 [3]; E19b [1]; E34 [1, 3, 9]; E38 [2, 4-9]; H410 [5]
Storage and disposal D01, D02, D12a [1-9]; D05 [2-4, 8]; D09a [1-4, 7-9]; D10c [2, 4, 8]; D11a [1, 3, 9]
Medical advice M03 [1, 2, 4, 8, 9]; M05a [1, 2, 4-8]

326 Metarhizium anisopliae

A naturally occurring insect parasitic fungus, Metarhizium anisopliae var. anisopliae strain F52

Products
Met52 granular bioinsecticide   Fargro   2% w/w   GR   15168

Uses
- Cabbage root fly in **vegetable brassicas** (off-label)
- Leatherjackets in **bilberries** (off-label), **cranberries** (off-label), **loganberries** (off-label), **ribes hybrids** (off-label), **rubus hybrids** (off-label), **table grapes** (off-label), **wine grapes** (off-label)
- Lettuce root aphid in **herbs (see appendix 6)** (off-label), **leafy vegetables** (off-label)
- Midges in **amenity vegetation** (off-label), **bilberries** (off-label), **blueberries** (off-label), **container-grown ornamentals** (off-label), **cranberries** (off-label), **herbs (see appendix 6)** (off-label), **leafy vegetables** (off-label), **loganberries** (off-label), **ornamental plant production** (off-label), **ribes hybrids** (off-label), **rubus hybrids** (off-label), **table grapes** (off-label), **top fruit** (off-label), **wine grapes** (off-label)
- Sciarid flies in **amenity vegetation** (off-label), **bilberries** (off-label), **blueberries** (off-label), **container-grown ornamentals** (off-label), **cranberries** (off-label), **edible fungi** (off-label), **herbs (see appendix 6)** (off-label), **leafy vegetables** (off-label), **loganberries** (off-label), **ornamental plant production** (off-label), **ribes hybrids** (off-label), **rubus hybrids** (off-label), **table grapes** (off-label), **top fruit** (off-label), **wine grapes** (off-label)
- Thrips in **amenity vegetation** (off-label), **bilberries** (off-label), **blueberries** (off-label), **container-grown ornamentals** (off-label), **cranberries** (off-label), **herbs (see appendix 6)** (off-label), **leafy vegetables** (off-label), **loganberries** (off-label), **ornamental plant production** (off-label), **ribes**

Pesticide Profiles

hybrids *(off-label)*, **rubus hybrids** *(off-label)*, **table grapes** *(off-label)*, **top fruit** *(off-label)*, **wine grapes** *(off-label)*
- Vine weevil in **amenity vegetation** *(off-label)*, **blackberries**, **blackcurrants**, **blueberries**, **gooseberries**, **ornamental plant production**, **raspberries**, **redcurrants**, **strawberries**

Extension of Authorisation for Minor Use (EAMUs)
- *amenity vegetation* 20111568, 20111997
- *bilberries* 20111568, 20111997
- *blueberries* 20111997
- *container-grown ornamentals* 20111997
- *cranberries* 20111568, 20111997
- *edible fungi* 20111568
- *herbs (see appendix 6)* 20111568, 20111997
- *leafy vegetables* 20111568, 20111997
- *loganberries* 20111568, 20111997
- *ornamental plant production* 20111997
- *ribes hybrids* 20111568, 20111997
- *rubus hybrids* 20111568, 20111997
- *table grapes* 20111568, 20111997
- *top fruit* 20111997
- *vegetable brassicas* 20111568
- *wine grapes* 20111568, 20111997

Approval information
- Metarhizium anisopliae included in Annex 1 under EC Regulation 1107/2009

Efficacy guidance
- Incorporate into growing media at any growth stage

Hazard classification and safety precautions
UN Number N/C
Risk phrases R70
Operator protection A, D, H; U14, U20c
Environmental protection E15b, E34, E38
Storage and disposal D01, D09a, D10a, D12a, D20
Medical advice M03, M05a

327 metazachlor

A residual anilide herbicide for use in brassicas, nurseries and forestry
HRAC mode of action code. K3

See also aminopyralid + metazachlor + picloram
clomazone + metazachlor
clomazone + metazachlor + napropamide
dimethenamid-p + metazachlor
dimethenamid-p + metazachlor + quinmerac
imazamox + metazachlor
imazamox + metazachlor + quinmerac

Products

1	Butisan S	BASF	500 g/l	SC	16569
2	Clayton Buzz	Clayton	500 g/l	SC	16928
3	Rapsan 500 SC	Belchim	500 g/l	SC	16592
4	Stalwart	UPL Europe	500 g/l	SC	17405
5	Sultan 50 SC	Adama	500 g/l	SC	16680

Uses
- Annual dicotyledons in **borage for oilseed production** *(off-label)*, **canary flower (echium spp.)** *(off-label)*, **evening primrose** *(off-label)*, **honesty** *(off-label)*, **mustard** *(off-label)* [1, 5]; **broccoli**, **cabbages**, **calabrese**, **cauliflowers**, **nursery fruit trees** [4, 5]; **brussels sprouts**, **ornamental**

FOR FULL CONDITIONS OF USE ALWAYS READ THE PRODUCT LABEL

metazachlor

plant production [1, 4, 5]; *chinese cabbage* (off-label), *choi sum* (off-label), *collards* (off-label), *kale* (off-label), *kohlrabi* (off-label), *linseed* (off-label), *pak choi* (off-label), *tatsoi* (off-label) [1]; *spring linseed* (off-label), *winter linseed* (off-label) [5]; *spring oilseed rape, winter oilseed rape* [1-5]
- Annual meadow grass in *borage for oilseed production* (off-label), *canary flower (echium spp.)* (off-label), *evening primrose* (off-label), *honesty* (off-label), *mustard* (off-label) [1, 5]; *broccoli, cabbages, calabrese, cauliflowers, nursery fruit trees* [4, 5]; *brussels sprouts, ornamental plant production* [1, 4, 5]; *chinese cabbage* (off-label), *choi sum* (off-label), *collards* (off-label), *kale* (off-label), *kohlrabi* (off-label), *linseed* (off-label), *pak choi* (off-label), *tatsoi* (off-label) [1]; *spring linseed* (off-label), *winter linseed* (off-label) [5]; *spring oilseed rape, winter oilseed rape* [1-5]
- Blackgrass in *broccoli, cabbages, calabrese, cauliflowers, nursery fruit trees* [4, 5]; *brussels sprouts, ornamental plant production* [1, 4, 5]; *spring oilseed rape, winter oilseed rape* [1, 3-5]
- Chickweed in *choi sum* (off-label), *collards* (off-label), *kale* (off-label), *kohlrabi* (off-label), *oriental cabbage* (off-label), *tatsoi* (off-label) [5]; *spring oilseed rape, winter oilseed rape* [2]
- Groundsel in *borage for oilseed production* (off-label), *canary flower (echium spp.)* (off-label), *chinese cabbage* (off-label), *evening primrose* (off-label), *honesty* (off-label), *linseed* (off-label), *mustard* (off-label), *pak choi* (off-label) [1]; *choi sum* (off-label), *collards* (off-label), *kale* (off-label), *kohlrabi* (off-label), *tatsoi* (off-label) [1, 5]; *oriental cabbage* (off-label) [5]
- Mayweeds in *borage for oilseed production* (off-label), *canary flower (echium spp.)* (off-label), *chinese cabbage* (off-label), *evening primrose* (off-label), *honesty* (off-label), *linseed* (off-label), *mustard* (off-label), *pak choi* (off-label) [1]; *choi sum* (off-label), *collards* (off-label), *kale* (off-label), *kohlrabi* (off-label), *tatsoi* (off-label) [1, 5]; *oriental cabbage* (off-label) [5]; *spring oilseed rape, winter oilseed rape* [2]
- Shepherd's purse in *borage for oilseed production* (off-label), *canary flower (echium spp.)* (off-label), *chinese cabbage* (off-label), *evening primrose* (off-label), *honesty* (off-label), *linseed* (off-label), *mustard* (off-label), *pak choi* (off-label) [1]; *choi sum* (off-label), *collards* (off-label), *kale* (off-label), *kohlrabi* (off-label), *tatsoi* (off-label) [1, 5]; *oriental cabbage* (off-label) [5]

Extension of Authorisation for Minor Use (EAMUs)
- *borage for oilseed production* 20141053 [1], 20142381 [5]
- *canary flower (echium spp.)* 20141053 [1], 20142381 [5]
- *chinese cabbage* 20141052 [1]
- *choi sum* 20141052 [1], 20161213 [5]
- *collards* 20141052 [1], 20161213 [5]
- *evening primrose* 20141053 [1], 20142381 [5]
- *honesty* 20141053 [1], 20142381 [5]
- *kale* 20141052 [1], 20161213 [5]
- *kohlrabi* 20141052 [1], 20161213 [5]
- *linseed* 20141053 [1]
- *mustard* 20141053 [1], 20142381 [5]
- *oriental cabbage* 20161213 [5]
- *pak choi* 20141052 [1]
- *spring linseed* 20142381 [5]
- *tatsoi* 20141052 [1], 20161213 [5]
- *winter linseed* 20142381 [5]

Approval information
- Metazachlor is included in Annex I under EC Regulation 1107/2009

Efficacy guidance
- Activity is dependent on root uptake. For pre-emergence use apply to firm, moist, clod-free seedbed
- Some weeds (chickweed, mayweed, blackgrass etc) susceptible up to 2- or 4-leaf stage. Moderate control of cleavers achieved provided weeds not emerged and adequate soil moisture present
- Split pre- and post-emergence treatments recommended for certain weeds in winter oilseed rape on light and/or stony soils
- Effectiveness is reduced on soils with more than 10% organic matter

SEE SECTION 3 FOR PRODUCTS ALSO REGISTERED

Pesticide Profiles

- Always follow WRAG guidelines for preventing and managing herbicide resistant weeds. See Section 5 for more information

Restrictions
- Maximum number of treatments 1 per crop for spring oilseed rape, swedes, turnips and brassicas; 2 per crop for winter oilseed rape (split dose treatment); 3 per yr for ornamentals, nursery stock, nursery fruit trees, forestry and farm forestry
- Do not use on sand, very light or poorly drained soils
- Do not treat protected crops or spray overall on ornamentals with soft foliage
- Do not spray crops suffering from wilting, pest or disease
- Do not spray broadcast crops or if a period of heavy rain forecast
- When used on nursery fruit trees any fruit harvested within 1 yr of treatment must be destroyed
- Applications shall be limited to a total dose of not more than 1.0 kg metazachlor/ha in a three year period on the same field
- Do not apply in mixture with phosphate liquid fertilisers
- Metazachlor stewardship requires that all autumn applications should be made before the end of September to reduce the risk to water

Crop-specific information
- Latest use: pre-emergence for swedes and turnips; before 10 leaf stage for spring oilseed rape; before end of Jan for winter oilseed rape
- HI brassicas 6 wk
- On winter oilseed rape may be applied pre-emergence from drilling until seed chits, post-emergence after fully expanded cotyledon stage (GS 1,0) or by split dose technique depending on soil and weeds. See label for details
- On spring oilseed rape may also be used pre-weed-emergence from cotyledon to 10-leaf stage of crop (GS 1,0-1,10)
- With pre-emergence treatment ensure seed covered by 15 mm of well consolidated soil. Harrow across slits of direct-drilled crops
- Ensure brassica transplants have roots well covered and are well established. Direct drilled brassicas should not be treated before 3 leaf stage
- In ornamentals and hardy nursery stock apply after plants established and hardened off as a directed spray or, on some subjects, as an overall spray. See label for list of tolerant subjects. Do not treat plants in containers

Following crops guidance
- Any crop can follow normally harvested treated winter oilseed rape. See label for details of crops which may be planted after spring treatment and in event of crop failure

Environmental safety
- Dangerous for the environment
- Very toxic to aquatic organisms
- Keep livestock out of treated areas until foliage of any poisonous weeds such as ragwort has died and become unpalatable
- Keep livestock out of treated areas of swede and turnip for at least 5 wk following treatment
- Some pesticides pose a greater threat of contamination of water than others and metazachlor is one of these pesticides. Take special care when applying metazachlor near water and do not apply if heavy rain is forecast.
- Metazachlor stewardship guidelines advise a maximum dose of 750 g.a.i/ha/annum. Applications to drained fields should be complete by 15th Oct but, if drains are flowing, complete applications by 1st Oct.
- LERAP Category B [1, 2, 4, 5]

Hazard classification and safety precautions
Hazard Harmful, Dangerous for the environment [1-5]; Harmful if swallowed [1, 2, 4, 5]; Very toxic to aquatic organisms [1, 4]
Transport code 9
Packaging group III
UN Number 3082
Risk phrases H317 [2-5]; H351 [1-5]

FOR FULL CONDITIONS OF USE ALWAYS READ THE PRODUCT LABEL

metazachlor + quinmerac

Operator protection A, H [1-5]; C, M [1, 3-5]; U05a, U19a [1-5]; U08, U20b [1, 2, 4, 5]; U09a, U20a [3]; U14 [1, 4, 5]; U15 [4, 5]
Environmental protection E06a [2-4] (5 wk for swedes, turnips); E07a [1-4]; E15a, E34, H410 [1-5]; E16a [1, 2, 4, 5]; E38 [1, 2]
Consumer protection C02a [3] (5 weeks for swedes and turnips)
Storage and disposal D01, D02, D09a, D10c [1-5]; D05 [2-5]; D12a [1, 2]; D12b [4, 5]; D19 [1]
Medical advice M03 [1-5]; M05a [1, 2]

328   metazachlor + quinmerac

A residual herbicide mixture for oilseed rape
HRAC mode of action code: K3 + O

See also quinmerac

Products

1	Katamaran	BASF	375:125 g/l	SC	16766
2	Legion	Adama	375:125 g/l	SC	16249
3	Naspar Extra	Globachem	375:125 g/l	SC	17038
4	Parsan Extra	Globachem	375:125 g/l	SC	17050
5	Rocket	Adama	375:125 g/l	SC	17093

Uses
- Annual dicotyledons in **spring oilseed rape** [1, 3, 4]; **winter oilseed rape** [1-5]
- Annual meadow grass in **spring oilseed rape** [1, 3, 4]; **winter oilseed rape** [1-5]
- Blackgrass in **corn gromwell** *(off-label)* [2]; **spring oilseed rape** [1, 3, 4]; **winter oilseed rape** [1-5]
- Chickweed in **corn gromwell** *(off-label)* [2]
- Cleavers in **corn gromwell** *(off-label)* [2]; **spring oilseed rape** [1, 3, 4]; **winter oilseed rape** [1-5]
- Mayweeds in **corn gromwell** *(off-label)* [2]; **spring oilseed rape** [3, 4]; **winter oilseed rape** [2-5]
- Poppies in **corn gromwell** *(off-label)* [2]; **spring oilseed rape** [1, 3, 4]; **winter oilseed rape** [1-5]
- Speedwells in **corn gromwell** *(off-label)* [2]

Extension of Authorisation for Minor Use (EAMUs)
- **corn gromwell** 20162047 [2]

Approval information
- Metazachlor and quinmerac included in Annex I under EC Regulation 1107/2009

Efficacy guidance
- Activity is dependent on root uptake. Pre-emergence treatments should be applied to firm moist seedbeds. Applications to dry soil do not become effective until after rain has fallen
- Maximum activity achieved from treatment before weed emergence for some species
- Weed control may be reduced if excessive rain falls shortly after application especially on light soils
- May be used on all soil types except Sands, Very Light Soils, and soils containing more than 10% organic matter. Crop vigour and/or plant stand may be reduced on brashy and stony soils

Restrictions
- Maximum total dose equivalent to one full dose treatment
- Damage may occur in waterlogged conditions. Do not use on poorly drained soils
- Do not treat stressed crops. In frosty conditions transient scorch may occur
- Applications shall be limited to a total dose of not more than 1.0 kg metazachlor/ha in a three year period on the same field
- Do not apply in mixture with phosphate liquid fertilisers
- Metazachlor stewardship requires that all autumn applications should be made before the end of September to reduce the risk to water

Crop-specific information
- Latest use: end Jan in yr of harvest

SEE SECTION 3 FOR PRODUCTS ALSO REGISTERED

Pesticide Profiles

- To ensure crop safety it is essential that crop seed is well covered with soil to 15 mm. Loose or puffy seedbeds must be consolidated before treatment. Do not use on broadcast crops
- Crop vigour and possibly plant stand may be reduced if excessive rain falls shortly after treatment especially on light soils

Following crops guidance
- In the event of crop failure after use, wheat or barley may be sown in the autumn after ploughing to 15 cm. Spring cereals or brassicas may be planted after ploughing in the spring
- For all situations following or rotational crops must not be planted until four months after application [1]

Environmental safety
- Dangerous for the environment
- Very toxic to aquatic organisms
- Keep livestock out of treated areas until foliage of any poisonous weeds such as ragwort has died and become unpalatable
- To reduce risk of movement to water do not apply to dry soil or if heavy rain is forecast. On clay soils create a fine consolidated seedbed
- Some pesticides pose a greater threat of contamination of water than others and metazachlor is one of these pesticides. Take special care when applying metazachlor near water and do not apply if heavy rain is forecast.
- Metazachlor stewardship guidelines advise a maximum dose of 750 g.a.i/ha/annum. Applications to drained fields should be complete by 15th Oct but, if drains are flowing, complete applications by 1st Oct.
- LERAP Category B

Hazard classification and safety precautions
Hazard Harmful [5]; Irritant [1-4]; Dangerous for the environment [1-5]; Very toxic to aquatic organisms [1, 3-5]
Transport code 9
Packaging group III
UN Number 3082
Risk phrases H317 [1, 3, 4]; H351 [1-5]; R43 [3, 4]
Operator protection A [1-5]; H [5]; U05a, U08, U14, U19a, U20b
Environmental protection E07a, E15a, E16a, E38, H410 [1-5]; E19b [5]
Storage and disposal D01, D02, D08, D09a, D10c, D12a
Medical advice M04a [5]; M05a [1-5]

329 metconazole

A triazole fungicide for cereals and oilseed rape
FRAC mode of action code: 3

See also boscalid + metconazole
  epoxiconazole + metconazole
  fluxapyroxad + metconazole
  mepiquat chloride + metconazole

Products

1	Caramba 90	BASF	90 g/l	EC	15524
2	Gringo	AgChem Access	60 g/l	EC	15774
3	Juventus	BASF	90 g/l	EC	15528
4	Metcostar	Life Scientific	60 g/l	EC	17446
5	Redstar	AgChem Access	90 g/l	SL	15938
6	Sirena	Belchim	60 g/l	EC	17103
7	Sunorg Pro	BASF	90 g/l	EC	15433

Uses
- Alternaria in **spring oilseed rape**, **winter oilseed rape** [1-7]

FOR FULL CONDITIONS OF USE ALWAYS READ THE PRODUCT LABEL

metconazole

- Ascochyta in **combining peas** *(reduction)*, **lupins** *(qualified minor use)*, **vining peas** *(reduction)* [1-7]
- Botrytis in **combining peas** *(reduction)*, **lupins** *(qualified minor use)*, **vining peas** *(reduction)* [1-7]
- Brown rust in **durum wheat**, **rye**, **spring barley**, **triticale**, **winter barley**, **winter wheat** [1-7]; **spring wheat** [1, 3, 5, 7]
- Disease control in **broad beans** *(off-label)* [7]; **spring linseed** *(off-label)*, **winter linseed** *(off-label)* [3, 7]
- Fusarium ear blight in **durum wheat** *(reduction)* [2-4, 6, 7]; **rye** *(reduction)* [1, 2, 4-7]; **spring wheat** *(reduction)* [1, 3, 5, 7]; **triticale** *(reduction)*, **winter wheat** *(reduction)* [1-7]
- Growth regulation in **winter oilseed rape** [7]
- Light leaf spot in **spring oilseed rape**, **winter oilseed rape** [1, 2, 4-7]; **spring oilseed rape** *(reduction)*, **winter oilseed rape** *(reduction)* [3]
- Mycosphaerella in **combining peas** *(reduction)*, **vining peas** *(reduction)* [1-7]
- Net blotch in **durum wheat** *(reduction)* [1]; **spring barley** *(reduction)*, **winter barley** *(reduction)* [1-7]
- Phoma in **spring oilseed rape** *(reduction)*, **winter oilseed rape** *(reduction)* [1-7]
- Powdery mildew in **durum wheat** *(moderate control)*, **rye** *(moderate control)*, **spring barley** *(moderate control)*, **triticale** *(moderate control)*, **winter wheat** *(moderate control)* [1-7]; **spring wheat** *(moderate control)* [1, 3, 5, 7]; **winter barley** [1, 2, 4-7]; **winter barley** *(moderate control)* [3]
- Rhynchosporium in **rye** *(moderate control)*, **spring barley** *(reduction)*, **winter barley** *(reduction)* [3]; **spring barley**, **winter barley** [1, 2, 4-7]
- Rust in **combining peas**, **lupins** *(qualified minor use)*, **spring field beans**, **vining peas**, **winter field beans** [1-7]
- Septoria leaf blotch in **durum wheat** [2-4, 6, 7]; **rye** [1, 2, 4-7]; **spring wheat** [1, 3, 5, 7]; **triticale**, **winter wheat** [1-7]
- Yellow rust in **durum wheat** [2-4, 6, 7]; **rye**, **triticale**, **winter wheat** [1-7]; **spring barley**, **winter barley** [3]; **spring wheat** [1, 3, 5, 7]

Extension of Authorisation for Minor Use (EAMUs)
- **broad beans** *20111928* [7]
- **spring linseed** *20130387* [3], *20111929* [7]
- **winter linseed** *20130387* [3], *20111929* [7]

Approval information
- Metconazole included in Annex I under EC Regulation 1107/2009
- Accepted by BBPA for use on malting barley

Efficacy guidance
- Best results from application to healthy, vigorous crops when disease starts to develop
- Good spray cover of the target is essential for best results. Spray volume should be increased to improve spray penetration into dense crops
- Metconazole is a DMI fungicide. Resistance to some DMI fungicides has been identified in Septoria leaf blotch which may seriously affect performance of some products. For further advice contact a specialist advisor and visit the Fungicide Resistance Action Group (FRAG)-UK website

Restrictions
- Maximum total dose equivalent to two full dose treatments on all crops
- Do not apply to oilseed rape crops that are damaged or stressed from previous treatments, adverse weather, nutrient deficiency or pest attack. Spring application may lead to reduction of crop height
- The addition of adjuvants is neither advised nor necessary and can lead to enhanced growth regulatory effects on stressed crops of oilseed rape
- Ensure sprayer is free from residues of previous treatments that may harm the crop, especially oilseed rape. Use of a detergent cleaner is advised before and after use
- Do not apply with pyrethroid insecticides on oilseed rape at flowering
- Maintain an interval of at least 14 days between applications to oilseed rape, peas, beans and lupins, 21 days between applications to cereals [3]

SEE SECTION 3 FOR PRODUCTS ALSO REGISTERED

Pesticide Profiles

- To protect birds, only one application is allowed on cereals before end of tillering (GS29) [3]

Crop-specific information
- Latest use: up to and including milky ripe stage (GS 71) for cereals; 10% pods at final size for oilseed rape
- HI: 14 d for peas, field beans, lupins
- Spring treatments on oilseed rape can reduce the height of the crop
- Treatment for Septoria leaf spot in wheat should be made before second node detectable stage and when weather favouring development of the disease has occurred. If conditions continue to favour disease development a follow-up treatment may be needed
- Treat mildew infections in cereals before 3% infection on any green leaf. A specific mildewicide will improve control of established infections
- Treat yellow rust on wheat before 1% infection on any leaf or as preventive treatment after GS 39
- For brown rust in cereals spray susceptible varieties before any of top 3 leaves has more than 2% infection
- Peas should be treated at the start of flowering and repeat 3-4 wk later if required
- Field beans and lupins should be treated at petal fall and repeat 3-4 wk later if required

Following crops guidance
- Only cereals, oilseed rape, sugar beet, linseed, maize, clover, beans, peas, carrots, potatoes or onions may be sown as following crops after treatment

Environmental safety
- Dangerous for the environment
- Very toxic to aquatic organisms
- Avoid treatment close to field boundary, even if permitted by LERAP assessment, to reduce effects on non-target insects or other arthropods
- LERAP Category B

Hazard classification and safety precautions
Hazard Harmful [2-4, 6]; Irritant [1, 5, 7]; Flammable [2, 4, 6]; Dangerous for the environment [1-7]; Flammable liquid and vapour [4, 6]; Very toxic to aquatic organisms [4]
Transport code 3 [2, 4, 6]; 9 [1, 3, 5, 7]
Packaging group III
UN Number 1993 [2, 4, 6]; 3082 [1, 3, 5, 7]
Risk phrases H304, H315, H317, H318, H335, H361 [4, 6]; R22b, R38, R41, R43, R50 [2]; R36, R51 [1, 3, 5, 7]; R53a, R63 [1-3, 5, 7]
Operator protection A, C [1-7]; H [2-4, 6]; U02a, U05a, U20c [1-7]; U11, U14 [2, 4, 6]
Environmental protection E15a, E16a, E34, E38 [1-7]; E16b [2-4, 6]; H410 [4, 6]
Storage and disposal D01, D02, D05, D09a, D12a [1-7]; D06c [1, 3, 5, 7]; D10a [1, 2, 4-7]; D10c [3]
Medical advice M03, M05a [3]; M05b [2, 4, 6]

330 methiocarb

A stomach acting carbamate molluscicide and insecticide
IRAC mode of action code: 1A

Products
Mesurol	Bayer CropScience	500 g/l	FS	15311

Uses
- Frit fly in **fodder maize**, **grain maize**, **sweetcorn**

Approval information
- Methiocarb included in Annex I under EC Regulation 1107/2009
- In 2006 CRD required that all products containing this active ingredient should carry the following warning in the main area of the container label: "Methiocarb is an anticholinesterase carbamate. Handle with care"
- Accepted by BBPA for use on malting barley

FOR FULL CONDITIONS OF USE ALWAYS READ THE PRODUCT LABEL

methoxyfenozide

Efficacy guidance
- See label for details of suitable application equipment

Restrictions
- This product contains an anticholinesterase carbamate compound. Do not use if under medical advice not to work with such compounds

Environmental safety
- Dangerous for the environment
- Very toxic to aquatic organisms
- Harmful to aquatic organisms
- Dangerous to game, wild birds and animals
- Risk to certain non-target insects or other arthropods
- Avoid surface broadcasting application within 6 m of field boundary to reduce effects on non-target species
- Admixed seed should be drilled and not broadcast, and may not be applied from the air

Hazard classification and safety precautions
Hazard Toxic, Dangerous for the environment
Transport code 6.1
Packaging group III
UN Number 2992
Risk phrases R25, R50, R53a
Operator protection A, D, H; U05a, U14, U19a
Environmental protection E10a, E15a, E22b
Storage and disposal D01, D02, D09a, D10d, D12b
Treated seed S01, S02, S03, S04a, S04b, S05, S06a, S07, S08
Medical advice M04a

331   methoxyfenozide

A moulting accelerating diacylhydrazine insecticide
IRAC mode of action code: 18A

Products

1	Agrovista Trotter	Agrovista	240 g/l	SC	14236
2	Runner	Landseer	240 g/l	SC	15629

Uses
- Codling moth in *apples*, *pears* [1, 2]
- Insect pests in *hops* (off-label), *ornamental plant production* (off-label), *soft fruit* (off-label) [2]
- Plum fruit moth in *plums* (off-label) [2]
- Tortrix moths in *apples*, *pears* [1, 2]; *table grapes* (off-label), *wine grapes* (off-label) [2]
- Winter moth in *apples*, *pears* [1, 2]

Extension of Authorisation for Minor Use (EAMUs)
- *hops* 20120448 [2]
- *ornamental plant production* 20120448 [2]
- *plums* 20161694 [2]
- *soft fruit* 20120448 [2]
- *table grapes* 20161694 [2]
- *wine grapes* 20161694 [2]

Approval information
- Methoxyfenozide included in Annex 1 under EC Regulation 1107/2009

Efficacy guidance
- To achieve best results uniform coverage of the foliage and full spray penetration of the leaf canopy is important, particularly when spraying post-blossom

SEE SECTION 3 FOR PRODUCTS ALSO REGISTERED

Pesticide Profiles

- For maximum effectiveness on winter moth and tortrix spray pre-blossom when first signs of active larvae are seen, followed by a further spray in June if larvae of the summer generation are present
- For codling moth spray post-blossom to coincide with early to peak egg deposition. Follow-up treatments will normally be needed
- Methoxyfenozide is a moulting accelerating compound (MAC) and may be used in an anti-resistance strategy with other top fruit insecticides (including chitin biosynthesis inhibitors and juvenile hormones) which have a different mode of action
- To reduce further the likelihood of resistance development use at full recommended dose in sufficient water volume to achieve required spray penetration

Restrictions
- Maximum number of treatments 3 per yr but no more than two should be sprayed consecutively

Crop-specific information
- HI 14 d for apples, pears

Environmental safety
- Risk to non-target insects or other arthropods
- Broadcast air-assisted LERAP (5 m)

Hazard classification and safety precautions
UN Number N/C
Operator protection U20b
Environmental protection E15b, E16b, E22c; E17b (5 m)
Storage and disposal D05, D09a, D11a

332 1-methylcyclopropene

An inhibitor of ethylene production for use in stored apples

Products
SmartFresh   Landseer   3.3% w/w   SP   11799

Uses
- Ethylene inhibition in **apples** *(post-harvest use)*, **pears** *(off-label - post harvest only)*
- Scald in **apples** *(post-harvest use)*, **pears** *(off-label - post harvest only)*
- Storage rots in **broccoli** *(off-label)*, **brussels sprouts** *(off-label)*, **cabbages** *(off-label)*, **calabrese** *(off-label)*, **cauliflowers** *(off-label)*

Extension of Authorisation for Minor Use (EAMUs)
- *broccoli 20111502*
- *brussels sprouts 20111503*
- *cabbages 20111503*
- *calabrese 20111502*
- *cauliflowers 20111502*
- *pears (post harvest only) 20102424*

Approval information
- Methylcyclopropene included in Annex 1 under EC Regulation 1107/2009

Efficacy guidance
- Best results obtained from treatment of fruit in good condition and of proper quality for long-term storage
- Effects may be reduced in fruit that is in poor condition or ripe prior to storage or harvested late
- Product acts by releasing vapour into store when mixed with water
- Apply as soon as possible after harvest
- Treatment controls superficial scald and maintains fruit firmness and acid content for 3-6 mth in normal air, and 6-9 mth in controlled atmosphere storage
- Ethylene production recommences after removal from storage

metobromuron

Restrictions
- Maximum number of treatments 1 per batch of apples
- Must only be used by suitably trained and competent persons in fumigation operations
- Consult processors before treatment of fruit destined for processing or cider making
- Do not apply in mixture with other products
- Ventilate all areas thoroughly with all refrigeration fans operating at maximum power for at least 15 min before re-entry

Crop-specific information
- Latest use: 7 d after harvest of apples
- Product tested on Granny Smith, Gala, Jonagold, Bramley and Cox. Consult distributor or supplier before treating other varieties

Environmental safety
- Unprotected persons must be kept out of treated stores during the 24 h treatment period
- Prior to application ensure that the store can be properly and promptly sealed

Hazard classification and safety precautions
UN Number N/C
Operator protection U05a
Environmental protection E02a (24 h); E15a, E34
Consumer protection C12
Storage and disposal D01, D02, D07, D14

333 metobromuron

A pre-emergence herbicide for weed control in potatoes

Products
1	Inigo	Belchim	500 g/l	SC	16933
2	Praxim	Belchim	500 g/l	SC	16871
3	Soleto	Belchim	500 g/l	SC	16935

Uses
- Annual dicotyledons in **potatoes**
- Annual meadow grass in **potatoes**
- Charlock in **potatoes**
- Chickweed in **potatoes**
- Fat hen in **potatoes**
- Groundsel in **potatoes**
- Mayweeds in **potatoes**
- Shepherd's purse in **potatoes**

Approval information
- Metobromuron included in Annex I under EC Regulation 1107/2009

Restrictions
- Heavy rain after application may cause transient discolouration of the crop.

Crop-specific information
- May be used on all varieties of potato

Following crops guidance
- Following normal harvest of a treated crop Brassicas (oilseed rape, turnip, brassica vegetables) and sugar beet may be sown, providing the soil has been ploughed. No restrictions apply when other crops are sown.
- In the event of crop failure maize, beans, peas and carrots may be sown after ploughing. It is not recommended to sow Brassicas (oilseed rape, turnip, brassica vegetables) or sugar beet as a replacement crop after crop failure. Potatoes can be re-planted after minimal cultivation.

Hazard classification and safety precautions
Transport code 9

SEE SECTION 3 FOR PRODUCTS ALSO REGISTERED

Packaging group III
UN Number 3082
Risk phrases H351, H373
Operator protection A, H, M; U05a, U09a, U19a, U20a
Environmental protection E15b, E34, E38, H410
Storage and disposal D01, D02, D05, D09a, D12b
Medical advice M05a

334   metrafenone

A benzophenone protectant and curative fungicide for cereals
FRAC mode of action code: U8

*See also epoxiconazole + fenpropimorph + metrafenone
epoxiconazole + metrafenone
fenpropimorph + metrafenone*

Products

1	Attenzo	BASF	300 g/l	SC	11917
2	Flexity	BASF	300 g/l	SC	11775
3	Lexi	AgChem Access	300 g/l	SC	14890

Uses
- Disease control/foliar feed in **ornamental plant production** *(off-label)* [2]
- Eyespot in **spring wheat** *(reduction)*, **winter wheat** *(reduction)* [1-3]
- Powdery mildew in **forest nurseries** *(off-label)* [1]; **ornamental plant production** [2]; **spring barley**, **spring oats** *(evidence of mildew control on oats is limited)*, **spring wheat**, **winter barley**, **winter oats** *(evidence of mildew control on oats is limited)*, **winter wheat** [1-3]

Extension of Authorisation for Minor Use (EAMUs)
- *forest nurseries* 20082814 [1]
- *ornamental plant production* 20082850 [2]

Approval information
- Metrafenone included in Annex I under EC Regulation 1107/2009
- Accepted by BBPA for use on malting barley

Efficacy guidance
- Best results obtained from treatment at the start of foliar disease attack
- Activity against mildew in wheat is mainly protectant with moderate curative control in the latent phase; activity is entirely protectant in barley
- Useful reduction of eyespot in wheat is obtained if treatment applied at GS 30-32
- Should be used as part of a resistance management strategy that includes mixtures or sequences effective against mildew and non-chemical methods

Restrictions
- Maximum number of treatments 2 per crop
- Avoid the use of sequential applications of Flexity unless it is used in tank mixture with other products active against powdery mildew employing a different mode of action

Crop-specific information
- Latest use: beginning of flowering (GS 31) for wheat, barley, oats

Following crops guidance
- Cereals, oilseed rape, sugar beet, linseed, maize, clover, field beans, peas, turnips, carrots, cauliflowers, onions, lettuce or potatoes may follow a treated cereal crop

Environmental safety
- Dangerous for the environment
- Toxic to aquatic organisms

FOR FULL CONDITIONS OF USE ALWAYS READ THE PRODUCT LABEL

metribuzin

Hazard classification and safety precautions
Hazard Irritant, Dangerous for the environment
UN Number N/C
Risk phrases R43, R51, R53a
Operator protection A, H; U05a, U20b
Environmental protection E15a, E34
Storage and disposal D01, D02, D05, D09a, D10c
Medical advice M03

335   metribuzin

A contact and residual triazinone herbicide for use in potatoes
HRAC mode of action code: C1

See also clomazone + metribuzin
flufenacet + metribuzin
mecoprop-P + metribuzin

Products

1	Cleancrop Frizbee	Agrii	70% w/w	WG	16642
2	Sencorex Flow	Nufarm UK	600 g/l	SL	16167
3	Shotput	Adama	70% w/w	WG	15968

Uses
- Annual dicotyledons in **asparagus** *(off-label)*, **carrots** *(off-label)*, **early potatoes**, **maincrop potatoes**, **mallow (althaea spp.)** *(off-label)*, **parsnips** *(off-label)*, **sweet potato** *(off-label)* [1-3]; **celeriac** *(off-label)*, **ornamental plant production** *(off-label)* [2]
- Annual grasses in **asparagus** *(off-label)*, **carrots** *(off-label)*, **mallow (althaea spp.)** *(off-label)*, **parsnips** *(off-label)*, **sweet potato** *(off-label)* [2, 3]; **celeriac** *(off-label)*, **ornamental plant production** *(off-label)* [2]; **early potatoes**, **maincrop potatoes** [1-3]
- Annual meadow grass in **carrots** *(off-label)*, **mallow (althaea spp.)** *(off-label)*, **parsnips** *(off-label)*, **sweet potato** *(off-label)* [3]; **maincrop potatoes** [1, 3]
- Fool's parsley in **asparagus** *(off-label)*, **sweet potato** *(off-label)* [1, 2]; **carrots** *(off-label)*, **mallow (althaea spp.)** *(off-label)*, **parsnips** *(off-label)* [1-3]; **celeriac** *(off-label)* [2]
- Himalayan balsam in **rhubarb** *(off-label)* [2]
- Perennial dicotyledons in **asparagus** *(off-label)*, **asparagus** *(off-label - from seed)* [3]
- Perennial weeds in **asparagus** *(off-label)*, **sweet potato** *(off-label)* [1]
- Volunteer oilseed rape in **asparagus** *(off-label)*, **carrots** *(off-label)*, **celeriac** *(off-label)*, **mallow (althaea spp.)** *(off-label)*, **ornamental plant production** *(off-label)*, **parsnips** *(off-label)* [2]; **early potatoes**, **maincrop potatoes** [1-3]; **sweet potato** *(off-label)* [2, 3]
- Wild mignonette in **asparagus** *(off-label)*, **sweet potato** *(off-label)* [1, 2]; **carrots** *(off-label)*, **mallow (althaea spp.)** *(off-label)*, **parsnips** *(off-label)* [1-3]; **celeriac** *(off-label)* [2]

Extension of Authorisation for Minor Use (EAMUs)
- *asparagus* 20151018 [1], 20150917 [2], *(from seed)* 20142321 [3], 20142230 [3], 20142321 [3]
- *carrots* 20151017 [1], 20150916 [2], 20142229 [3], 20142320 [3]
- *celeriac* 20150916 [2]
- *mallow (althaea spp.)* 20151017 [1], 20150916 [2], 20142229 [3], 20142320 [3]
- *ornamental plant production* 20131867 [2]
- *parsnips* 20151017 [1], 20150916 [2], 20142229 [3], 20142320 [3]
- *rhubarb* 20152218 [2]
- *sweet potato* 20151019 [1], 20150915 [2], 20142228 [3], 20142252 [3]

Approval information
- Metribuzin included in Annex I under EC Regulation 1107/2009

Efficacy guidance
- Best results achieved on weeds at cotyledon to 1-leaf stage
- May be applied pre- or post-emergence of crop on named maincrop varieties and cv. Marfona; pre-emergence only on named early varieties
- Apply to moist soil with well-rounded ridges and few clods

SEE SECTION 3 FOR PRODUCTS ALSO REGISTERED

Pesticide Profiles

- Activity reduced by dry conditions and on soils with high organic matter content
- On fen and moss soils pre-planting incorporation to 10-15 cm gives increased activity. Incorporate thoroughly and evenly
- With named maincrop and second early potato varieties on soils with more than 10% organic matter shallow pre- or post-planting incorporation may be used. See label for details
- Effective control using a programme of reduced doses is made possible by using a spray of smaller droplets, thus improving retention

Restrictions
- Maximum total dose equivalent to one full dose treatment on early potato varieties and one and a third full doses on maincrop varieties
- Only certain varieties may be treated. Apply pre-emergence only on named first earlies, pre- or post-emergence on named second earlies. On named maincrop varieties apply pre-emergence (except for certain varieties on Sands or Very Light soils) or post-emergence. See label
- All post-emergence treatments must be carried out before longest shoots reach 15 cm
- Do not cultivate after treatment
- Some recommended varieties may be sensitive to post-emergence treatment if crop under stress

Crop-specific information
- Latest use: pre-crop emergence for named early potato varieties; before most advanced shoots have reached 15 cm for post-emergence treatment of potatoes
- On stony or gravelly soils there is risk of crop damage, especially if heavy rain falls soon after application
- When days are hot and sunny delay spraying until evening

Following crops guidance
- Ryegrass, cereals or winter beans may be sown in same season provided at least 16 wk elapsed after treatment and ground ploughed to 15 cm and thoroughly cultivated as soon as possible after harvest and no later than end Dec
- In W Cornwall on soil with more than 5% organic matter early potatoes treated as recommended may be followed by summer planted brassica crops provided the soil has been ploughed, spring rainfall has been normal and at least 14 wk have elapsed since treatment
- Do not grow any vegetable brassicas, lettuces or radishes on land treated the previous yr. Other crops may be sown normally in spring of next yr

Environmental safety
- Dangerous for the environment
- Very toxic to aquatic organisms
- Do not empty into drains
- LERAP Category B

Hazard classification and safety precautions
Hazard Harmful, Dangerous for the environment [1-3]; Very toxic to aquatic organisms [1, 3]
Transport code 9
Packaging group III
UN Number 3077 [1, 3]; 3082 [2]
Risk phrases R48, R50, R53a [2]
Operator protection A, H; U05a, U08, U13, U19a [1-3]; U14, U20b [1, 3]; U20a [2]
Environmental protection E15a, E16b, E19b, H410 [1, 3]; E16a [1-3]; E38 [2]
Storage and disposal D01, D02, D09a, D11a, D12a
Medical advice M03 [2]; M05a [1-3]

FOR FULL CONDITIONS OF USE ALWAYS READ THE PRODUCT LABEL

336 metsulfuron-methyl

A contact and residual sulfonylurea herbicide used in cereals, linseed and set-aside
HRAC mode of action code: B

See also carfentrazone-ethyl + metsulfuron-methyl
diflufenican + metsulfuron-methyl
flupyrsulfuron-methyl + metsulfuron-methyl
fluroxypyr + metsulfuron-methyl
fluroxypyr + metsulfuron-methyl + thifensulfuron-methyl
mecoprop-P + metsulfuron-methyl

Products

1	Alias SX	DuPont	20% w/w	SG	13398
2	Answer SX	DuPont	20% w/w	SG	15632
3	Cimarron	Headland	50% w/w	TB	13408
4	Deft Premium	Rotam	20% w/w	SG	16272
5	Finy	UPL Europe	20% w/w	WG	16960
6	Gropper SX	DuPont	20% w/w	SG	16290
7	Jubilee SX	DuPont	20% w/w	SG	12203
8	Lorate	DuPont	20% w/w	SG	12743
9	Metmeth 200	Becesane	20% w/w	SG	15993
10	Savvy	Rotam	20% w/w	WG	14266
11	Savvy Premium	Rotam	20% w/w	SG	14782
12	Simba SX	DuPont	20% w/w	SG	15503
13	Snicket	AgChem Access	20% w/w	SG	15681
14	Standon Metso XT	Standon	20% w/w	SG	13880
15	Standon Mexxon Xtra	Standon	20% w/w	SG	13931
16	Tattler	Goldengrass	20% w/w	WG	16360

Uses

- Amsinkia in **green cover on land temporarily removed from production, linseed, spring barley, spring oats, spring wheat, triticale, winter barley, winter oats, winter wheat** [7]
- Annual dicotyledons in **farm forestry** (off-label), **forest nurseries** (off-label), **ornamental plant production** (off-label) [7]; **game cover** (off-label), **miscanthus** (off-label) [7, 8]; **green cover on land temporarily removed from production** [5, 7, 14-16]; **linseed, spring barley, spring oats, spring wheat, triticale, winter barley, winter oats, winter wheat** [1, 2, 4-16]
- Charlock in **green cover on land temporarily removed from production** [7]; **linseed, spring barley, spring oats, spring wheat, triticale, winter barley, winter oats, winter wheat** [4, 7]
- Chickweed in **green cover on land temporarily removed from production** [7]; **linseed, spring barley, spring oats, spring wheat, triticale, winter barley, winter oats, winter wheat** [1, 2, 4, 6-13]
- Docks in **grassland** [3]; **green cover on land temporarily removed from production, linseed, spring barley, spring oats, spring wheat, triticale, winter barley, winter oats, winter wheat** [7]
- Green cover in **land temporarily removed from production** [1, 2, 4, 6, 8-13]
- Groundsel in **green cover on land temporarily removed from production, linseed, spring barley, spring oats, spring wheat, triticale, winter barley, winter oats, winter wheat** [7]
- Hemp-nettle in **green cover on land temporarily removed from production, linseed, spring barley, spring oats, spring wheat, triticale, winter barley, winter oats, winter wheat** [7]
- Mayweeds in **green cover on land temporarily removed from production** [7]; **linseed, spring barley, spring oats, spring wheat, triticale, winter barley, winter oats, winter wheat** [1, 2, 4, 6-13]
- Mouse-ear chickweed in **green cover on land temporarily removed from production, linseed, spring barley, spring oats, spring wheat, triticale, winter barley, winter oats, winter wheat** [7]
- Scarlet pimpernel in **green cover on land temporarily removed from production, linseed, spring barley, spring oats, spring wheat, triticale, winter barley, winter oats, winter wheat** [7]

SEE SECTION 3 FOR PRODUCTS ALSO REGISTERED

Pesticide Profiles

- Venus looking-glass in **green cover on land temporarily removed from production, linseed, spring barley, spring oats, spring wheat, triticale, winter barley, winter oats, winter wheat** [7]
- Volunteer oilseed rape in **green cover on land temporarily removed from production, linseed, spring barley, spring oats, spring wheat, triticale, winter barley, winter oats, winter wheat** [7]
- Volunteer sugar beet in **green cover on land temporarily removed from production, linseed, spring barley, spring oats, spring wheat, triticale, winter barley, winter oats, winter wheat** [7]

Extension of Authorisation for Minor Use (EAMUs)
- *farm forestry* 20082859 [7]
- *forest nurseries* 20082859 [7]
- *game cover* 20082859 [7], 20082860 [8]
- *miscanthus* 20082859 [7], 20082860 [8]
- *ornamental plant production* 20082859 [7]

Approval information
- Metsulfuron-methyl included in Annex I under EC Regulation 1107/2009
- Accepted by BBPA for use on malting barley

Efficacy guidance
- Best results achieved on small, actively growing weeds up to 6-true leaf stage. Good spray cover is important
- Weed control may be reduced in dry soil conditions
- Commonly used in tank-mixture on wheat and barley with other cereal herbicides to improve control of resistant dicotyledons (cleavers, fumitory, ivy-leaved speedwell), larger weeds and grasses. See label for recommended mixtures
- Metsulfuron-methyl is a member of the ALS-inhibitor group of herbicides and products should be used in a planned Resistance Management strategy. See Section 5 for more information

Restrictions
- Maximum number of treatments 1 per crop or per yr (for set-aside)
- Product must only be used after 1 Feb
- Do not apply within 7 d of rolling
- Do not use on cereal crops undersown with grass or legumes
- Do not use in tank mixture on oats, triticale or linseed. On linseed allow at least 7 d before or after other treatments
- Do not tank mix with chlorpyrifos. Allow at least 14 d before or after chlorpyrifos treatments
- Consult contract agents before use on a cereal crop for seed
- Do not use on any crop suffering stress from drought, waterlogging, frost, deficiency, pest or disease attack
- Specific restrictions apply to use in sequence or tank mixture with other sulfonylurea or ALS-inhibiting herbicides. See label for details
- Spraying equipment should not be drained or flushed onto land planted, or to be planted, with trees or crops other than cereals and should be thoroughly cleansed after use - see label for instructions
- Must only be applied from 1 February in the year of harvest until the specified latest time of application and must only be applied to green cover on land not being used for crop production where a full green cover is established [10]

Crop-specific information
- Latest use: before flag leaf sheath extending stage for cereals (GS 41); before flower buds visible or crop 30 cm tall for linseed [1, 2, 6-8, 12]; before 1 Aug in yr of treatment for land not being used for crop production [1, 2, 6-8, 12]
- Apply after 1 Feb to wheat, oats and triticale from 2-leaf (GS 12), and to barley from 3-leaf (GS 13) until flag-leaf sheath extending (GS 41)
- On linseed allow at least 7 d (10 d if crop is growing poorly or under stress) before or after other treatments
- Use in set-aside when a full green cover is established and made up predominantly of grassland, wheat, barley, oats or triticale. Do not use on seedling grasses

FOR FULL CONDITIONS OF USE ALWAYS READ THE PRODUCT LABEL

metsulfuron-methyl + thifensulfuron-methyl

Following crops guidance
- Only cereals, oilseed rape, field beans or grass may be sown in same calendar year after treating cereals with the product alone. Other restrictions apply to tank mixtures. See label for details
- Only cereals should be planted within 16 mth of applying to a linseed crop or set-aside
- In the event of failure of a treated crop sow only wheat within 3 mth after treatment

Environmental safety
- Dangerous for the environment
- Very toxic to aquatic organisms
- Take extreme care to avoid damage by drift onto broad-leaved plants outside the target area, onto surface waters or ditches or onto land intended for cropping
- A range of broad leaved species will be fully or partially controlled when used in land temporarily removed from production, hence product may not be suitable where wild flower borders or other forms of conservation headland are being developed
- Before use on land temporarily removed from production as part of grant-aided scheme, ensure compliance with the management rules
- Green cover on land temporarily removed from production must not be grazed by livestock or harvested for human or animal consumption or used for animal bedding [7]
- LERAP Category B [2, 4-7, 9-16]

Hazard classification and safety precautions
Hazard Dangerous for the environment [1-16]; Very toxic to aquatic organisms [1, 2, 6-8, 12-15]
Transport code 9
Packaging group III
UN Number 3077
Risk phrases H315, H318 [4, 11]; R50, R53a [5, 9, 16]; R51, R53b [3]
Operator protection A, C, H [4]; U04b, U11 [4]; U05a [5, 16]; U09a [14, 15]; U19a [10, 14, 15]; U20b [3, 4, 10, 14, 15]
Environmental protection E07a [3]; E15a [10, 11]; E15b [1-9, 12-16]; E16a [2, 4-7, 9-16]; E16b [2]; E23 [5, 16]; E34 [14, 15]; E38 [1-15]; H410 [1, 2, 4, 6-8, 10-15]
Storage and disposal D01 [5, 14-16]; D02 [5, 16]; D09a [2-5, 10, 11, 14-16]; D10b [4, 10, 11]; D11a [2, 5, 16]; D12a [1-4, 6-15]; D12b [4, 5, 16]; D17, D18 [3]
Medical advice M05a [14, 15]

337 metsulfuron-methyl + thifensulfuron-methyl

A contact residual and translocated sulfonylurea herbicide mixture for use in cereals
HRAC mode of action code: B + B

See also thifensulfuron-methyl

Products

1	Assuage	Becesane	6.8:68.2% w/w	SG	15801
2	Avro SX	DuPont	2.9:42.9 % w/w	SG	14313
3	Chimera SX	DuPont	2.9:42.9% w/w	SG	13171
4	Concert SX	DuPont	4:40% w/w	SG	12288
5	Ergon	Rotam	6.8:68.2% w/w	WG	14382
6	Finish SX	DuPont	6.7:33.3% w/w	SG	12259
7	Harmony M SX	DuPont	4:40% w/w	SG	12258
8	Mozaic SX	DuPont	4:40% w/w	SG	16108
9	Pennant	Headland	4:40% w/w	SG	13350
10	Presite SX	DuPont	6.7:33.3% w/w	SG	12291
11	Refine Max SX	DuPont	6.7:33.3% w/w	SG	15622

Uses
- Annual dicotyledons in **durum wheat** *(off-label)* [10]; **miscanthus** *(off-label)* [3, 7]; **spring barley**, **spring wheat**, **winter wheat** [1-11]; **spring oats**, **triticale**, **winter rye** [5]; **winter barley** [2, 3, 6, 10, 11]; **winter oats** [6, 10, 11]
- Black bindweed in **spring barley**, **spring wheat**, **winter barley**, **winter wheat** [2]

SEE SECTION 3 FOR PRODUCTS ALSO REGISTERED

Pesticide Profiles

- Charlock in *spring barley*, *spring wheat*, *winter wheat* [1, 2, 5]; *spring oats*, *triticale*, *winter rye* [5]; *winter barley* [2]
- Chickweed in *durum wheat* (off-label) [10]; *spring barley*, *spring wheat*, *winter wheat* [1-11]; *spring oats*, *triticale*, *winter rye* [5]; *winter barley* [2, 3, 6, 10, 11]; *winter oats* [6, 10, 11]
- Creeping thistle in *spring barley*, *spring wheat*, *winter barley*, *winter wheat* [2, 3]
- Fat hen in *spring barley*, *spring wheat*, *winter barley*, *winter wheat* [2]
- Field pansy in *durum wheat* (off-label) [10]; *spring barley*, *spring wheat*, *winter barley*, *winter wheat* [2, 6, 10, 11]; *winter oats* [6, 10, 11]
- Field speedwell in *durum wheat* (off-label) [10]; *spring barley*, *spring wheat*, *winter barley*, *winter wheat* [2, 6, 10, 11]; *winter oats* [6, 10, 11]
- Forget-me-not in *spring barley*, *spring wheat*, *winter barley*, *winter wheat* [2]
- Hemp-nettle in *spring barley*, *spring wheat*, *winter wheat* [1, 2, 5]; *spring oats*, *triticale*, *winter rye* [5]; *winter barley* [2]
- Ivy-leaved speedwell in *durum wheat* (off-label) [10]; *spring barley*, *spring wheat*, *winter barley*, *winter oats*, *winter wheat* [6, 10, 11]
- Knotgrass in *durum wheat* (off-label) [10]; *spring barley*, *spring wheat*, *winter barley*, *winter wheat* [2, 6, 10, 11]; *winter oats* [6, 10, 11]
- Mayweeds in *durum wheat* (off-label) [10]; *spring barley*, *spring wheat*, *winter wheat* [1-11]; *spring oats*, *triticale*, *winter rye* [5]; *winter barley* [2, 3, 6, 10, 11]; *winter oats* [6, 10, 11]
- Parsley-piert in *spring barley*, *spring wheat*, *winter wheat* [1, 5]; *spring oats*, *triticale*, *winter rye* [5]
- Polygonums in *spring barley*, *spring wheat*, *winter wheat* [3, 4, 7-9]; *winter barley* [3]
- Poppies in *spring barley*, *spring wheat*, *winter barley*, *winter wheat* [2]
- Red dead-nettle in *spring barley*, *spring wheat*, *winter wheat* [1-5, 7-9]; *spring oats*, *triticale*, *winter rye* [5]; *winter barley* [2, 3]
- Redshank in *spring barley*, *spring wheat*, *winter wheat* [1, 2, 5]; *spring oats*, *triticale*, *winter rye* [5]; *winter barley* [2]
- Shepherd's purse in *spring barley*, *spring wheat*, *winter wheat* [1-5, 7-9]; *spring oats*, *triticale*, *winter rye* [5]; *winter barley* [2, 3]

Extension of Authorisation for Minor Use (EAMUs)
- *durum wheat* 20070452 [10]
- *miscanthus* 20082835 [3], 20082916 [7]

Approval information
- Metsulfuron-methyl and thifensulfuron-methyl included in Annex I under EC Regulation 1107/2009
- Accepted by BBPA for use on malting barley

Efficacy guidance
- Best results by application to small, actively growing weeds up to 6-true leaf stage
- Ensure good spray cover
- Susceptible weeds stop growing almost immediately but symptoms may not be visible for about 2 wk
- Effectiveness may be reduced by heavy rain or if soil conditions very dry
- Metsulfuron-methyl and thifensulfuron-methyl are members of the ALS-inhibitor group of herbicides and products should be used in a planned Resistance Management strategy. See Section 5 for more information

Restrictions
- Maximum number of treatments 1 per crop
- Products may only be used after 1 Feb
- Do not use on any crop suffering stress from drought, waterlogging, frost, deficiency, pest or disease attack or any other cause
- Do not use on crops undersown with grasses, clover or legumes, or any other broad leaved crop
- Specific restrictions apply to use in sequence or tank mixture with other sulfonylurea or ALS-inhibiting herbicides. See label for details
- Do not apply within 7 d of rolling
- Consult contract agents before use on a cereal crop grown for seed

FOR FULL CONDITIONS OF USE ALWAYS READ THE PRODUCT LABEL

metsulfuron-methyl + tribenuron-methyl

Crop-specific information
- Latest use: before flag leaf sheath extending (GS 39) for barley and wheat, before 2nd node (GS32) for winter oats

Following crops guidance
- Only cereals, oilseed rape, field beans or grass may be sown in same calendar year after treatment
- Additional constraints apply after use of certain tank mixtures. See label
- In the event of crop failure sow only winter wheat within 3 mth after treatment and after ploughing and cultivating to a depth of at least 15 cm

Environmental safety
- Dangerous for the environment
- Very toxic to aquatic organisms
- Take extreme care to avoid damage by drift onto broad-leaved plants outside the target area, or onto ponds, waterways or ditches
- Spraying equipment should not be drained or flushed onto land planted, or to be planted, with trees or crops other than cereals and should be thoroughly cleansed after use - see label for instructions
- LERAP Category B

Hazard classification and safety precautions
Hazard Dangerous for the environment [1-11]; Very toxic to aquatic organisms [2-4, 6-11]
Transport code 9
Packaging group III
UN Number 3077
Risk phrases R50, R53a [1]
Operator protection U05a [6, 10, 11]; U08, U20b [2-4, 6-11]; U19a [1-11]; U20c [1, 5]
Environmental protection E15a [6, 10, 11]; E15b [1, 4, 5, 7-9]; E16a [1-11]; E38 [2-4, 6-11]; H410 [2-11]
Storage and disposal D01, D02 [6, 10, 11]; D09a [1-11]; D10b [1, 5]; D11a [4, 6-11]; D12a [2-4, 6-11]

338   metsulfuron-methyl + tribenuron-methyl

A sulfonylurea herbicide mixture for cereals
HRAC mode of action code: B + B

See also tribenuron-methyl

Products

1	Ally Max SX	DuPont	14.3:14.3% w/w	SG	14835
2	BiPlay SX	DuPont	11.1:22.2% w/w	SG	14836
3	Boudha	Rotam	25:25% w/w	WG	15158
4	DP911 SX	DuPont	11.1:22.2% w/w	SG	17493
5	Traton SX	DuPont	11.1:22.2% w/w	SG	14837

Uses
- Annual dicotyledons in **durum wheat** *(off-label)*, **rye** *(off-label)*, **spring rye** *(off-label)* [1]; **spring barley, spring wheat, triticale, winter barley, winter oats, winter wheat** [1-5]; **spring oats** [1, 3]; **winter rye** [3]
- Chickweed in **spring barley, spring wheat, triticale, winter barley, winter oats, winter wheat** [1-3, 5]; **spring oats** [1, 3]; **winter rye** [3]
- Field pansy in **spring barley, spring wheat, triticale, winter barley, winter oats, winter wheat** [1, 2, 5]
- Field speedwell in **spring oats** [1]
- Hemp-nettle in **spring barley, spring wheat, triticale, winter barley, winter oats, winter wheat** [1, 2, 5]; **spring oats** [1]
- Mayweeds in **spring barley, spring wheat, triticale, winter barley, winter oats, winter wheat** [1-3, 5]; **spring oats** [1, 3]; **winter rye** [3]

SEE SECTION 3 FOR PRODUCTS ALSO REGISTERED

Pesticide Profiles

- Poppies in **spring barley, spring oats, spring wheat, triticale, winter barley, winter oats, winter rye, winter wheat** [3]
- Red dead-nettle in **spring barley, spring wheat, triticale, winter barley, winter oats, winter wheat** [1-3, 5]; **spring oats** [1, 3]; **winter rye** [3]
- Shepherd's purse in **spring barley, spring oats, spring wheat, triticale, winter barley, winter oats, winter rye, winter wheat** [3]
- Volunteer oilseed rape in **spring barley, spring oats, spring wheat, triticale, winter barley, winter oats, winter wheat** [1, 3]; **winter rye** [3]
- Volunteer sugar beet in **spring barley, spring oats, spring wheat, triticale, winter barley, winter oats, winter wheat** [1]

Extension of Authorisation for Minor Use (EAMUs)
- *durum wheat* 20101080 [1]
- *rye* 20101080 [1]
- *spring rye* 20101080 [1]

Approval information
- Metsulfuron-methyl and tribenuron-methyl included in Annex I under EC Regulation 1107/2009
- Accepted by BBPA for use on malting barley

Efficacy guidance
- Best results obtained when applied to small actively growing weeds
- Product acts by foliar and root uptake. Good spray cover essential but performance may be reduced when soil conditions are very dry and residual effects may be reduced by heavy rain
- Weed growth inhibited within hours of treatment and many show marked colour changes as they die back. Full effects may not be apparent for up to 4 wk
- Metsulfuron-methyl and tribenuron-methyl are members of the ALS-inhibitor group of herbicides and products should be used in a planned Resistance Management strategy. See Section 5 for more information

Restrictions
- Maximum number of treatments 1 per crop
- Product must only be used after 1 Feb and after crop has three leaves
- Do not apply to a crop suffering from drought, waterlogging, low temperatures, pest or disease attack, nutrient deficiency, soil compaction or any other stress
- Do not use on crops undersown with grasses, clover or other legumes
- Do not apply within 7 d of rolling
- Specific restrictions apply to use in sequence or tank mixture with other sulfonylurea or ALS-inhibiting herbicides. See label for details

Crop-specific information
- Latest use: before flag leaf sheath extending

Following crops guidance
- Only cereals, field beans, grass or oilseed rape may be sown in the same calendar yr as harvest of a treated crop [1, 2, 5]
- In the event of failure of a treated crop only winter wheat may be sown within 3 mth of treatment, and only after ploughing and cultivation to 15 cm min

Environmental safety
- Dangerous for the environment
- Very toxic to aquatic organisms
- Some non-target crops are highly sensitive. Take extreme care to avoid drift outside the target area, or onto ponds, waterways or ditches
- Spraying equipment should be thoroughly cleaned in accordance with manufacturer's instructions
- LERAP Category B [2-5]

Hazard classification and safety precautions
Hazard Irritant [1, 2, 5]; Dangerous for the environment [1-3, 5]; Very toxic to aquatic organisms [4]
Transport code 9

FOR FULL CONDITIONS OF USE ALWAYS READ THE PRODUCT LABEL

napropamide

Packaging group III
UN Number 3077
Risk phrases R43, R50, R53a [1, 2, 5]
Operator protection A, H [1, 2, 5]; U08, U19a [1-3, 5]; U20b [1, 2, 5]
Environmental protection E15a [1, 2, 5]; E15b, E39 [3]; E16a [2-5]; E38 [1-3, 5]; H410 [3, 4]
Storage and disposal D01, D02, D09a, D11a, D12a [1-3, 5]

339   napropamide

A soil applied alkanamide herbicide for oilseed rape, fruit and woody ornamentals
HRAC mode of action code: K3

See also clomazone + metazachlor + napropamide
clomazone + napropamide

Products

1	AC 650	UPL Europe	450 g/l	SC	11102
2	Colzamid	UPL Europe	450 g/l	SC	16324
3	Devrinol	UPL Europe	450 g/l	SC	09374

Uses
- Annual dicotyledons in **baby leaf crops** *(off-label)*, **bilberries** *(off-label)*, **blackberries** *(off-label)*, **blackcurrants** *(off-label)*, **blueberries** *(off-label)*, **choi sum** *(off-label)*, **collards** *(off-label)*, **farm forestry** *(off-label)*, **forest** *(off-label)*, **lamb's lettuce** *(off-label)*, **oriental cabbage** *(off-label)*, **ornamental plant production** *(off-label)*, **redcurrants** *(off-label)*, **rocket** *(off-label)*, **rubus hybrids** *(off-label)*, **whitecurrants** *(off-label)* [3]; **blackcurrants, broccoli, brussels sprouts, cabbages, calabrese, cauliflowers, forest nurseries, gooseberries, kale, ornamental plant production, raspberries, strawberries, woody ornamentals** [2, 3]; **linseed** *(off-label)* [1]; **winter oilseed rape** [1, 3]
- Annual grasses in **baby leaf crops** *(off-label)*, **blackberries** *(off-label)*, **choi sum** *(off-label)*, **collards** *(off-label)*, **farm forestry** *(off-label)*, **forest** *(off-label)*, **lamb's lettuce** *(off-label)*, **oriental cabbage** *(off-label)*, **ornamental plant production** *(off-label)*, **rocket** *(off-label)*, **rubus hybrids** *(off-label)* [3]; **blackcurrants, broccoli, brussels sprouts, cabbages, calabrese, cauliflowers, forest nurseries, gooseberries, kale, ornamental plant production, raspberries, strawberries, woody ornamentals** [2, 3]; **linseed** *(off-label)* [1]; **winter oilseed rape** [1, 3]
- Annual meadow grass in **baby leaf crops** *(off-label)*, **borage** *(off-label)*, **evening primrose** *(off-label)*, **linseed** *(off-label)*, **mustard** *(off-label)*, **winter oilseed rape** [3]
- Blackgrass in **borage** *(off-label)*, **evening primrose** *(off-label)*, **linseed** *(off-label)*, **mustard** *(off-label)*, **winter oilseed rape** [3]
- Chickweed in **borage** *(off-label)*, **evening primrose** *(off-label)*, **linseed** *(off-label)*, **mustard** *(off-label)* [3]
- Cleavers in **blackcurrants, broccoli, brussels sprouts, cabbages, calabrese, cauliflowers, forest nurseries, gooseberries, kale, ornamental plant production, raspberries, strawberries, woody ornamentals** [2, 3]; **borage** *(off-label)*, **evening primrose** *(off-label)*, **linseed** *(off-label)*, **mustard** *(off-label)* [3]
- Corn spurrey in **borage** *(off-label)*, **evening primrose** *(off-label)*, **linseed** *(off-label)*, **mustard** *(off-label)* [3]
- Fat hen in **borage** *(off-label)*, **evening primrose** *(off-label)*, **linseed** *(off-label)*, **mustard** *(off-label)* [3]
- Field pansy in **borage** *(off-label)*, **evening primrose** *(off-label)*, **linseed** *(off-label)*, **mustard** *(off-label)* [3]
- Field speedwell in **borage** *(off-label)*, **evening primrose** *(off-label)*, **linseed** *(off-label)*, **mustard** *(off-label)* [3]
- Forget-me-not in **borage** *(off-label)*, **evening primrose** *(off-label)*, **linseed** *(off-label)*, **mustard** *(off-label)* [3]
- Groundsel in **blackcurrants, broccoli, brussels sprouts, cabbages, calabrese, cauliflowers, forest nurseries, gooseberries, kale, ornamental plant production, raspberries, strawberries, woody ornamentals** [2, 3]; **borage** *(off-label)*, **evening primrose** *(off-label)*, **linseed** *(off-label)*, **mustard** *(off-label)* [3]

SEE SECTION 3 FOR PRODUCTS ALSO REGISTERED

Pesticide Profiles

- Knotgrass in **baby leaf crops** *(off-label)*, **bilberries** *(off-label)*, **blackberries** *(off-label)*, **blackcurrants** *(off-label)*, **blueberries** *(off-label)*, **borage** *(off-label)*, **evening primrose** *(off-label)*, **lamb's lettuce** *(off-label)*, **linseed** *(off-label)*, **mustard** *(off-label)*, **redcurrants** *(off-label)*, **rocket** *(off-label)*, **rubus hybrids** *(off-label)*, **whitecurrants** *(off-label)* [3]
- Mayweeds in **baby leaf crops** *(off-label)*, **bilberries** *(off-label)*, **blackberries** *(off-label)*, **blackcurrants** *(off-label)*, **blueberries** *(off-label)*, **borage** *(off-label)*, **evening primrose** *(off-label)*, **lamb's lettuce** *(off-label)*, **linseed** *(off-label)*, **mustard** *(off-label)*, **redcurrants** *(off-label)*, **rocket** *(off-label)*, **rubus hybrids** *(off-label)*, **whitecurrants** *(off-label)* [3]
- Pale persicaria in **borage** *(off-label)*, **evening primrose** *(off-label)*, **linseed** *(off-label)*, **mustard** *(off-label)* [3]
- Poppies in **borage** *(off-label)*, **evening primrose** *(off-label)*, **linseed** *(off-label)*, **mustard** *(off-label)* [3]
- Redshank in **borage** *(off-label)*, **evening primrose** *(off-label)*, **linseed** *(off-label)*, **mustard** *(off-label)* [3]
- Shepherd's purse in **borage** *(off-label)*, **evening primrose** *(off-label)*, **linseed** *(off-label)*, **mustard** *(off-label)* [3]
- Small nettle in **borage** *(off-label)*, **evening primrose** *(off-label)*, **linseed** *(off-label)*, **mustard** *(off-label)* [3]
- Sowthistle in **borage** *(off-label)*, **evening primrose** *(off-label)*, **linseed** *(off-label)*, **mustard** *(off-label)* [3]
- Volunteer cereals in **borage** *(off-label)*, **evening primrose** *(off-label)*, **linseed** *(off-label)*, **mustard** *(off-label)* [3]
- Willowherb in **bilberries** *(off-label)*, **blackberries** *(off-label)*, **blackcurrants** *(off-label)*, **blueberries** *(off-label)*, **borage** *(off-label)*, **evening primrose** *(off-label)*, **farm forestry** *(off-label)*, **forest** *(off-label)*, **linseed** *(off-label)*, **mustard** *(off-label)*, **redcurrants** *(off-label)*, **rubus hybrids** *(off-label)*, **whitecurrants** *(off-label)* [3]

Extension of Authorisation for Minor Use (EAMUs)
- **baby leaf crops** *20160104* [3]
- **bilberries** *20160105* [3]
- **blackberries** *20160106* [3]
- **blackcurrants** *20160105* [3]
- **blueberries** *20160105* [3]
- **borage** *20160085* [3]
- **choi sum** *20160107* [3]
- **collards** *20160107* [3]
- **evening primrose** *20160085* [3]
- **farm forestry** *20160108* [3]
- **forest** *20160108* [3]
- **lamb's lettuce** *20160103* [3]
- **linseed** *20151186 expires 31 Dec 2017* [1], *20151431* [1], *20160085* [3]
- **mustard** *20160085* [3]
- **oriental cabbage** *20160107* [3]
- **ornamental plant production** *20162044* [3]
- **redcurrants** *20160105* [3]
- **rocket** *20160103* [3]
- **rubus hybrids** *20160106* [3]
- **whitecurrants** *20160105* [3]

Approval information
- Napropamide included in Annex I under EC Regulation 1107/2009

Efficacy guidance
- Best results obtained from treatment pre-emergence of weeds but product may be used in conjunction with contact herbicides for control of emerged weeds. Otherwise remove existing weeds before application
- Weed control may be reduced where spray is mixed too deeply in the soil
- Manure, crop debris or other organic matter may reduce weed control
- Napropamide broken down by sunlight, so application during such conditions not recommended. Most crops recommended for treatment between Nov and end-Feb

FOR FULL CONDITIONS OF USE ALWAYS READ THE PRODUCT LABEL

napropamide

- Apply to winter oilseed rape pre-emergence or as pre-drilling treatment in tank-mixture with trifluralin and incorporate within 30 min [1]
- Post-emergence use of specific grass weedkilller recommended where volunteer cereals are a serious problem
- Increase water volume to ensure adequate dampening of compost when treating containerised nursery stock with dense leaf canopy [3]

Restrictions
- Maximum number of treatments 1 per crop or yr
- Do not use on Sands
- Do not use on soils with more than 10% organic matter
- Strawberries, blackcurrants, gooseberries and raspberries must be treated between 1 Nov and end Feb [3]
- Applications to strawberries must only be made where the mature foliage has been previously removed [3]
- Applications to ornamentals and forest nurseries must be made after 1 Nov and before end Apr
- Do not treat ornamentals in containers of less than 1 litre [3]
- Some phytotoxicity seen on yellow and golden varieties of conifers and container grown alpines. On any ornamental variety treat only a small number of plants in first season [3]
- Consult processors before use on any crop for processing

Crop-specific information
- Latest use: before transplanting for brassicas; pre-emergence for winter oilseed rape; before end Feb for strawberries, bush and cane fruit; before end of Apr for field and container grown ornamental trees and shrubs
- Where minimal cultivation used to establish oilseed rape, tank-mixture may be applied directly to stubble and mixed into top 25 mm as part of surface cultivations [1]
- Apply up to 14 d prior to drilling winter oilseed rape [1]
- Apply to strawberries established for at least one season or to maiden crops as long as planted carefully and no roots exposed, between 1 Nov and end Feb. Do not treat runners of poor vigour or with shallow roots, or runner beds [3]
- Newly planted ornamentals should have no roots exposed. Do not treat stock of poor vigour or with shallow roots. Treatments made in Mar and Apr must be followed by 25 mm irrigation within 24 h [3]
- Bush and cane fruit must be established for at least 10 mth before spraying and treated between 1 Nov and end Feb [3]

Following crops guidance
- After use in fruit or ornamentals no crop can be drilled within 7 mth of treatment. Leaf, flowerhead, root and fodder brassica crops may be drilled after 7 mth; potatoes, maize, peas or dwarf beans after 9 mth; autumn sown wheat or grass after 18 mth; any crop after 2 yr
- After use in oilseed rape only oilseed rape, swedes, fodder turnips, brassicas or potatoes should be sown within 12 mth of application
- Soil should be mould-board ploughed to a depth of at least 200 mm before drilling or planting any following crop

Environmental safety
- Dangerous for the environment
- Toxic to aquatic organisms

Hazard classification and safety precautions
    Hazard Irritant, Dangerous for the environment
    Transport code 9
    Packaging group III
    UN Number 3082
    Risk phrases R36, R38, R51, R53a
    Operator protection A [1-3]; C [2, 3]; U05a, U09a, U11, U20b
    Environmental protection E15a [2, 3]; E34, E38 [1-3]
    Storage and disposal D01, D02, D05, D09a, D10c [1-3]; D12a [1]; D12b [2, 3]

SEE SECTION 3 FOR PRODUCTS ALSO REGISTERED

340 nicosulfuron

A sulfonylurea herbicide for maize
HRAC mode of action code: B

See also mesotrione + nicosulfuron

Products

1	Accent	DuPont	75% w/w	WG	15392
2	Biretta	Ventura	40.8 g/l	OD	17408
3	Crew	Nufarm UK	40 g/l	OD	15770
4	Entail	Headland	240 g/l	OD	15671
5	Fornet 6 OD	Belchim	60 g/l	OD	15891
6	Milagro 240 OD	Syngenta	240 g/l	OD	15957
7	Nico Pro 4SC	Belchim	40 g/l	OD	16424
8	Pampa 4 SC	Belchim	40 g/l	OD	17521
9	Primero	Rotam	40.8 g/l	OD	16444
10	Samson Extra 6%	Belchim	60 g/l	OD	16418
11	Standon Frontrunner Super 6	Standon	60 g/l	OD	16890
12	Templier	Rotam	75% w/w	WG	16866

Uses

- Amaranthus in **forage maize** [3, 7, 8]; **grain maize** [3]
- Annual dicotyledons in **forage maize** [1, 2, 4-6, 9-12]; **forest nurseries** *(off-label)*, **game cover** *(off-label)*, **ornamental plant production** *(off-label)*, **top fruit** *(off-label)* [5, 7, 10]; **grain maize** [2, 4-6, 9-12]; **sweetcorn** *(off-label)*, **sweetcorn under plastic mulches** *(off-label)* [5, 10]
- Annual grasses in **forest nurseries** *(off-label)*, **game cover** *(off-label)*, **ornamental plant production** *(off-label)*, **top fruit** *(off-label)* [7]
- Annual meadow grass in **forage maize** [1-3, 5, 7-9, 12]; **forest nurseries** *(off-label)*, **game cover** *(off-label)*, **ornamental plant production** *(off-label)*, **sweetcorn** *(off-label)*, **sweetcorn under plastic mulches** *(off-label)*, **top fruit** *(off-label)* [5, 7, 10]; **grain maize** [2, 3, 5, 9, 12]
- Black nightshade in **sweetcorn** *(off-label)*, **sweetcorn under plastic mulches** *(off-label)* [5]
- Blackgrass in **forage maize** [1, 12]; **grain maize** [12]
- Cockspur grass in **forage maize** [1, 12]; **grain maize** [12]
- Common amaranth in **sweetcorn** *(off-label)*, **sweetcorn under plastic mulches** *(off-label)* [7]
- Couch in **forage maize** [1, 12]; **grain maize** [12]
- Grass weeds in **forage maize**, **grain maize** [4, 6, 10, 11]; **forest nurseries** *(off-label)*, **game cover** *(off-label)*, **ornamental plant production** *(off-label)*, **top fruit** *(off-label)* [10]
- Groundsel in **forage maize** [3, 7, 8]; **grain maize** [3]; **sweetcorn** *(off-label)*, **sweetcorn under plastic mulches** *(off-label)* [7]
- Mayweeds in **forage maize** [3, 7, 8]; **grain maize** [3]
- Ryegrass in **forage maize** [1, 3, 7, 8, 12]; **forest nurseries** *(off-label)*, **game cover** *(off-label)*, **ornamental plant production** *(off-label)*, **top fruit** *(off-label)* [5, 10]; **grain maize** [3, 12]; **sweetcorn** *(off-label)*, **sweetcorn under plastic mulches** *(off-label)* [5, 7, 10]
- Shepherd's purse in **forage maize** [3, 7, 8]; **grain maize** [3]; **sweetcorn** *(off-label)*, **sweetcorn under plastic mulches** *(off-label)* [7]
- Volunteer cereals in **forage maize** [1, 12]; **grain maize** [12]
- Wild oats in **forage maize** [1, 12]; **grain maize** [12]

Extension of Authorisation for Minor Use (EAMUs)

- **forest nurseries** *20140942* [5], *20140752* [7], *20140490* [10], *20141054* [10]
- **game cover** *20140942* [5], *20140752* [7], *20140490* [10], *20141054* [10]
- **ornamental plant production** *20140942* [5], *20140752* [7], *20140490* [10], *20141054* [10]
- **sweetcorn** *20140895* [5], *20140941* [7], *20140940* [10]
- **sweetcorn under plastic mulches** *20140895* [5], *20140941* [7], *20140940* [10]
- **top fruit** *20140942* [5], *20140752* [7], *20140490* [10], *20141054* [10]

Approval information

- Nicosulfuron included in Annex I under EC Regulation 1107/2009

FOR FULL CONDITIONS OF USE ALWAYS READ THE PRODUCT LABEL

oxamyl

Efficacy guidance
- Product should be applied post-emergence between 2 and 8 crop leaf stage and to emerged weeds from the 2-leaf stage
- Product acts mainly by foliar activity. Ensure good spray cover of the weeds
- Nicosulfuron is a member of the ALS-inhibitor group of herbicides. To avoid the build up of resistance do not use any product containing an ALS-inhibitor herbicide with claims for control of grass weeds more than once on any crop
- Use these products as part of a resistance management strategy that includes cultural methods of control and does not use ALS inhibitors as the sole chemical method of grass weed control

Restrictions
- Maximum number of treatments 1 per crop
- Do not use if an organophosphorus soil insecticide has been used on the same crop
- Do not mix with foliar or liquid fertilisers or specified herbicides. See label for details
- Do not apply in mixture, or sequence, with any other sulfonyl-urea containing product
- Do not treat crops under stress
- Do not apply if rainfall is forecast to occur within 6 h of application

Crop-specific information
- Latest use: up to and including 8 true leaves for maize
- Some transient yellowing may be seen from 1-2 wk after treatment
- Only healthy maize crops growing in good field conditions should be treated

Following crops guidance
- Winter wheat (not undersown) may be sown 4 mth after treatment; maize may be sown in the spring following treatment; all other crops may be sown from the next autumn
- In normal crop rotation, after ploughing winter wheat, winter barley, winter rye and triticale can be sown and all other crops can be sown the following spring. In the event of crop failure maize or soybeans can be sown after ploughing [5, 7, 10]

Environmental safety
- Dangerous for the environment
- Very toxic to aquatic organisms
- LERAP Category B

Hazard classification and safety precautions
Hazard Harmful [5, 10, 11]; Irritant [3, 4, 6-8]; Dangerous for the environment [1-12]; Harmful if inhaled [3]; Very toxic to aquatic organisms [5, 9, 10]
Transport code 9
Packaging group III
UN Number 3077 [1, 12]; 3082 [2-11]
Risk phrases H315 [3, 4, 7, 8]; H317 [3-5, 10]; H319 [5, 10]; R20, R43 [6, 11]; R36 [11]; R38 [6]; R50 [1, 6, 11, 12]; R53a [1, 2, 6, 11, 12]
Operator protection A [3-8, 10, 11]; C [4, 10, 11]; H [3, 4, 6-8, 10, 11]; U02a, U20b [4, 6, 10, 11]; U05a [1, 3-8, 10, 11]; U10, U11 [4-6, 10, 11]; U14, U19a [3-8, 10, 11]; U15 [3, 7, 8]
Environmental protection E15b [1, 3, 4, 6-8, 10, 11]; E16a [1-12]; E16b [4, 6, 10, 11]; E34 [1, 3, 7, 8]; E38 [1, 3-8, 10-12]; E40b [1]; H410 [2-5, 7-10]; H411 [2]
Storage and disposal D01, D02 [1-11]; D06a [5]; D09a [1, 3, 7, 8]; D09b [2, 9]; D10c [3, 4, 6-8, 10, 11]; D11a [1]; D12a [1, 4-6, 10-12]; D12b [3, 5, 7, 8]
Medical advice M03 [3, 7, 8]; M05a [4-6, 10, 11]

341   oxamyl

A soil-applied, systemic carbamate nematicide and insecticide
IRAC mode of action code: 1A

Products
Vydate 10G	DuPont	10% w/w	GR	16595

Uses
- American serpentine leaf miner in **ornamental plant production** (off-label)

SEE SECTION 3 FOR PRODUCTS ALSO REGISTERED

Pesticide Profiles

- Docking disorder vectors in **fodder beet**, **sugar beet**
- Eelworm in **bulb onions** *(off-label)*, **garlic** *(off-label)*, **shallots** *(off-label)*
- Free-living nematodes in **fodder beet**, **potatoes**, **sugar beet**
- Leaf miner in **ornamental plant production** *(off-label)*
- Potato cyst nematode in **potatoes**
- South American leaf miner in **ornamental plant production** *(off-label)*
- Stem nematodes in **carrots**, **parsnips**

Extension of Authorisation for Minor Use (EAMUs)
- **bulb onions** 20141637
- **garlic** 20141637
- **ornamental plant production** 20141636
- **shallots** 20141637

Approval information
- Oxamyl included in Annex I under EC Regulation 1107/2009
- In 2006 CRD required that all products containing this active ingredient should carry the following warning in the main area of the container label: "Oxamyl is an anticholinesterase carbamate. Handle with care"

Efficacy guidance
- Apply granules with suitable applicator before drilling or planting. See label for details of recommended machines
- In potatoes incorporate thoroughly to 10 cm and plant within 1-2 d
- In sugar beet apply in seed furrow at drilling

Restrictions
- Oxamyl is subject to the Poisons Rules 1982 and the Poisons Act 1972. See Section 5 for more information
- Contains an anticholinesterase carbamate compound. Do not use if under medical advice not to work with such compounds
- Maximum number of treatments 1 per crop or yr
- Keep in original container, tightly closed, in a safe place, under lock and key
- Wear protective gloves if handling treated compost or soil within 2 wk after treatment
- Allow at least 12 h, followed by at least 1 h ventilation, before entry of unprotected persons into treated glasshouses

Crop-specific information
- Latest use: at drilling/planting for vegetables; before drilling/planting for potatoes.
- HI: A minimum of 80 days must elapse between application and initial haulm destruction in potatoes

Environmental safety
- Dangerous for the environment
- Toxic to aquatic organisms
- Dangerous to fish or other aquatic life. Do not contaminate surface waters or ditches with chemical or used container
- Dangerous to game, wild birds and animals. Bury spillages

Hazard classification and safety precautions
Hazard Toxic, Fatal if swallowed, Toxic if inhaled
Transport code 6.1
Packaging group II
UN Number 2757
Operator protection A, B, H, K, M; C (or D); U02a, U04a, U05a, U08, U13, U19a, U20a, U23a
Environmental protection E10a, E13b, E15b, H412
Storage and disposal D01, D02, D06e, D09a, D11b
Medical advice M04a, M04c, M05a

FOR FULL CONDITIONS OF USE ALWAYS READ THE PRODUCT LABEL

paclobutrazol

342 paclobutrazol

A triazole plant growth regulator for ornamentals and fruit
FRAC mode of action code: 3

See also difenoconazole + paclobutrazol

Products

1	Bonzi	Syngenta	4 g/l	SC	17576
2	Pirouette	Fargro	4 g/l	SC	17203

Uses
- Growth regulation in **azaleas**, **bedding plants**, **kalanchoes**, **mini roses**, **pelargoniums**, **poinsettias** [2]; **protected ornamentals** [1]
- Increasing flowering in **azaleas**, **bedding plants**, **kalanchoes**, **mini roses**, **ornamental plant production**, **pelargoniums**, **poinsettias** [2]
- Stem shortening in **ornamental plant production** [2]

Approval information
- Paclobutrazol included in Annex I under EC Regulation 1107/2009

Efficacy guidance
- Chemical is active via both foliage and root uptake. For best results apply in dull weather when relative humidity not high

Restrictions
- Maximum number of treatments 1 per specimen for some species [2]

Following crops guidance
- Chemical has residual soil activity which can affect growth of following crops. Withhold treatment to allow the following intervals between the last treatment and planting the next crop: apples, pears 1 yr; beans, peas, onions 2 yr; stone fruit 3 yr; cereals, grass, oilseed rape, carrots 4 yr; market brassicas 6 yr; potatoes and other crops 7 yr

Environmental safety
- Harmful to aquatic organisms
- Do not use on food crops [2]
- Keep livestock out of treated areas for at least 2 years after treatment

Hazard classification and safety precautions
    Transport code 9 [2]
    Packaging group III [2]
    UN Number 3082 [2]; N/C [1]
    Operator protection A [2]; U05a, U08, U20a
    Environmental protection E13c [2]; E15a, H411 [1, 2]
    Consumer protection C01
    Storage and disposal D01, D02, D05, D09a [1, 2]; D10b [2]; D10c [1]

343 pelargonic acid

A naturally occuring 9 carbon acid terminating in a carboxylic acid only available in mixtures
HRAC mode of action code: Not classified

See also maleic hydrazide + pelargonic acid

SEE SECTION 3 FOR PRODUCTS ALSO REGISTERED

Pesticide Profiles

344   penconazole

A protectant triazole fungicide with antisporulant activity
FRAC mode of action code: 3

See also captan + penconazole

Products

1	Topas	Syngenta	100 g/l	EC	16765
2	Topenco	Globachem	100 g/l	EC	17539
3	Topenco 100 EC	Globachem	100 g/l	EC	17161

Uses
- Powdery mildew in **apples, blackcurrants, crab apples, pears, protected strawberries, redcurrants, strawberries, table grapes, wine grapes** [1, 2]; **courgettes** *(off-label)*, **cucumbers** *(off-label)*, **gherkins** *(off-label)*, **globe artichoke** *(off-label)*, **hops in propagation** *(off-label)*, **nursery fruit trees** *(off-label)*, **protected chilli peppers** *(off-label)*, **protected courgettes** *(off-label)*, **protected cucumbers** *(off-label)*, **protected gherkins** *(off-label)*, **protected hops** *(off-label)*, **protected melons** *(off-label)*, **protected nursery fruit trees** *(off-label)*, **protected peppers** *(off-label)*, **protected pumpkins** *(off-label)*, **protected summer squash** *(off-label)*, **protected tomatoes** *(off-label)*, **protected watermelon** *(off-label)*, **protected winter squash** *(off-label)*, **summer squash** *(off-label)* [1]; **protected courgettes, protected cucumbers, protected summer squash** [3]
- Rust in **roses** [1, 2]

Extension of Authorisation for Minor Use (EAMUs)
- *courgettes* 20152336 [1]
- *cucumbers* 20152336 [1]
- *gherkins* 20152336 [1]
- *globe artichoke* 20152901 [1]
- *hops in propagation* 20152338 [1]
- *nursery fruit trees* 20152338 [1]
- *protected chilli peppers* 20152335 [1]
- *protected courgettes* 20152336 [1]
- *protected cucumbers* 20152336 [1]
- *protected gherkins* 20152336 [1]
- *protected hops* 20152338 [1]
- *protected melons* 20152337 [1]
- *protected nursery fruit trees* 20152338 [1]
- *protected peppers* 20152335 [1]
- *protected pumpkins* 20152337 [1]
- *protected summer squash* 20152336 [1]
- *protected tomatoes* 20152335 [1]
- *protected watermelon* 20152337 [1]
- *protected winter squash* 20152337 [1]
- *summer squash* 20152336 [1]

Approval information
- Accepted by BBPA for use on hops
- Penconazole included in Annex I under EC Regulation 1107/2009

Efficacy guidance
- Use as a protectant fungicide by treating at the earliest signs of disease
- Treat crops every 10-14 d (every 7-10 d in warm, humid weather) at first sign of infection or as a protective spray ensuring complete coverage. See label for details of timing
- Increase dose and volume with growth of hops but do not exceed 2000 l/ha. Little or no activity will be seen on established powdery mildew
- Antisporulant activity reduces development of secondary mildew in apples

Restrictions
- Maximum number of treatments 10 per yr for apples, 6 per yr for hops, 4 per yr for blackcurrants

FOR FULL CONDITIONS OF USE ALWAYS READ THE PRODUCT LABEL

- Check for varietal susceptibility in roses. Some defoliation may occur after repeat applications on Dearest

Crop-specific information
- HI apples, hops 14 d; currants, bilberries, blueberries, cranberries, gooseberries 4 wk

Environmental safety
- Dangerous for the environment
- Toxic to aquatic organisms

Hazard classification and safety precautions
Hazard Irritant, Dangerous for the environment
Transport code 3
Packaging group III
UN Number 1915
Risk phrases H319, H361 [3]; R36, R53a [1, 2]; R51 [1-3]
Operator protection A, C; U02a, U05a, U20b [1-3]; U08 [1, 2]; U11 [3]
Environmental protection E15a [1-3]; E22c, H410 [3]; E38 [1, 2]
Storage and disposal D01, D02, D05, D09a [1-3]; D07, D10b [3]; D10c, D12a [1, 2]
Medical advice M05b

345   pencycuron

A non-systemic urea fungicide for use on seed potatoes
FRAC mode of action code: 20

See also imazalil + pencycuron

Products

1	Monceren DS	Bayer CropScience	12.5% w/w	DS	11292
2	Solaren	AgChem Access	12.5% w/w	DS	15912

Uses
- Black scurf in **potatoes** *(tuber treatment)*
- Stem canker in **potatoes** *(tuber treatment)*

Approval information
- Pencycuron included in Annex I under EC Regulation 1107/2009

Efficacy guidance
- Provides control of tuber-borne disease and gives some reduction of stem canker
- Seed tubers should be of good quality and vigour, and free from soil deposits when treated
- Dust formulations should be applied immediately before, or during, the planting process by treating seed tubers in chitting trays, in bulk bins immediately before planting or in hopper at planting. Liquid formulations may be applied by misting equipment at any time, into or out of store but treatment over a roller table at the end of grading out, or at planting, is usually most convenient
- Whichever method is chosen, complete and even distribution on the tubers is essential for optimum efficacy
- Apply in accordance with detailed guidelines in manufacturer's literature
- If rain interrupts planting, cover dust treated tubers in hopper

Restrictions
- Maximum number of treatments 1 per batch of seed potatoes
- Treated tubers must be used only as seed and not for human or animal consumption
- Do not use on tubers previously treated with a dry powder seed treatment or hot water
- To prevent rapid absorption through damaged 'eyes' seed tubers should only be treated before the 'eyes' begin to open or immediately before or during planting
- Tubers removed from cold storage must be allowed to attain a temperature of at least 8°C before treatment
- Use of suitable dust mask is mandatory when applying dust, filling the hopper or riding on planter

SEE SECTION 3 FOR PRODUCTS ALSO REGISTERED

Pesticide Profiles

- Treated tubers in bulk bins must be allowed to dry before being stored, especially when intended for cold storage
- Some internal staining of boxes used to store treated tubers may occur. Ware tubers should not be stored or supplied to packers in such stained boxes

Crop-specific information
- Latest use: at planting

Environmental safety
- Do not use treated seed as food or feed
- Treated seed harmful to game and wildlife

Hazard classification and safety precautions
UN Number N/C
Operator protection A, F; U20b
Environmental protection E03, E15a
Storage and disposal D09a, D11a
Treated seed S01, S02, S03, S04a, S05, S06a

346 pendimethalin

A residual dinitroaniline herbicide for cereals and other crops
HRAC mode of action code: K1

See also bentazone + pendimethalin
chlorotoluron + diflufenican + pendimethalin
chlorotoluron + pendimethalin
clomazone + pendimethalin
diflufenican + pendimethalin
dimethenamid-p + pendimethalin
flufenacet + pendimethalin
imazamox + pendimethalin
isoproturon + pendimethalin

Products

1	Anthem	Adama	400 g/l	SC	15761
2	Claymore	BASF	400 g/l	SC	13441
3	Cleancrop National	Agrii	400 g/l	SC	17094
4	Most Micro	Sipcam	365 g/l	CS	16063
5	Nighthawk 330 EC	AgChem Access	330 g/l	EC	14083
6	Stomp Aqua	BASF	455 g/l	CS	14664
7	Trample	AgChem Access	330 g/l	EC	14173
8	Yomp	Becesane	330 g/l	EC	15609

Uses
- Annual dicotyledons in **almonds** (off-label), **asparagus** (off-label), **bilberries** (off-label), **blueberries** (off-label), **bog myrtle** (off-label), **broccoli** (off-label), **brussels sprouts** (off-label), **bulb onions** (off-label), **cabbages** (off-label), **calabrese** (off-label), **carrots** (off-label), **cauliflowers** (off-label), **celery (outdoor)** (off-label), **chervil** (off-label), **chestnuts** (off-label), **collards** (off-label), **coriander** (off-label), **cranberries** (off-label), **cress** (off-label), **dill** (off-label), **evening primrose** (off-label), **farm forestry** (off-label), **fennel** (off-label), **forest** (off-label), **forest nurseries** (off-label), **french beans**, **frise** (off-label), **garlic** (off-label), **grass seed crops** (off-label), **hazel nuts** (off-label), **herbs (see appendix 6)** (off-label), **hops** (off-label), **horseradish** (off-label), **kale** (off-label), **lamb's lettuce** (off-label), **leaf brassicas** (off-label), **leeks** (off-label), **lettuce** (off-label), **mallow (althaea spp.)** (off-label), **miscanthus** (off-label), **navy beans**, **ornamental plant production** (off-label), **parsley** (off-label), **parsley root** (off-label), **parsnips** (off-label), **protected forest nurseries** (off-label), **quinces** (off-label), **radicchio** (off-label), **redcurrants** (off-label), **rhubarb** (off-label), **salad onions** (off-label), **scarole** (off-label), **shallots** (off-label), **sweetcorn** (off-label), **sweetcorn under plastic mulches** (off-label), **top fruit** (off-label), **walnuts** (off-label), **whitecurrants** (off-label) [6]; **apple orchards**, **pear orchards** [2, 5];

FOR FULL CONDITIONS OF USE ALWAYS READ THE PRODUCT LABEL

pendimethalin

apples, forage maize (under plastic mulches), grain maize, grain maize under plastic mulches, pears [1, 3, 6, 8]; blackberries, blackcurrants, broccoli, brussels sprouts, cabbages, calabrese, carrots, cauliflowers, cherries, gooseberries, leeks, loganberries, parsnips, plums, raspberries, rubus hybrids, strawberries [1-3, 5, 6, 8]; **broad beans** (off-label), **game cover** (off-label) [1, 3, 6]; **bulb onions** [1-3, 6, 8]; **combining peas, forage maize** [1-3, 5-8]; **durum wheat, potatoes, spring barley, sunflowers, triticale, winter barley, winter rye, winter wheat** [1-8]; **dwarf beans** (off-label), **french beans** (off-label) [4]; **edible podded peas** (off-label), **vining peas** (off-label) [2, 4-6]; **field beans** (off-label), **redcurrants** [3]; **forage soya bean** (off-label) [4, 6]; **lupins** (off-label), **runner beans** (off-label) [2, 4, 6]; **spring field beans** (off-label), **winter field beans** (off-label) [1, 2, 4, 6]; **spring wheat** (off-label) [1, 6]

- Annual grasses in **apples, blackberries, blackcurrants, broccoli, brussels sprouts, bulb onions, cabbages, calabrese, carrots, cauliflowers, cherries, forage maize (under plastic mulches), gooseberries, grain maize, grain maize under plastic mulches, leeks, loganberries, parsnips, pears, plums, raspberries, strawberries** [1, 3, 6, 8]; **bog myrtle** (off-label), **broccoli** (off-label), **brussels sprouts** (off-label), **calabrese** (off-label), **cauliflowers** (off-label), **celery (outdoor)** (off-label), **collards** (off-label), **fennel** (off-label), **grass seed crops** (off-label), **kale** (off-label), **rhubarb** (off-label) [6]; **broad beans** (off-label), **game cover** (off-label) [1, 3]; **combining peas, forage maize** [1, 3, 6-8]; **durum wheat, potatoes, spring barley, sunflowers, triticale, winter barley, winter rye, winter wheat** [1, 3, 4, 6-8]; **dwarf beans** (off-label), **forage soya bean** (off-label), **french beans** (off-label), **lupins** (off-label), **runner beans** (off-label) [4]; **edible podded peas** (off-label), **vining peas** (off-label) [4, 5]; **field beans** (off-label), **redcurrants** [3]; **rubus hybrids** [1, 3, 5, 6, 8]; **spring field beans** (off-label), **winter field beans** (off-label) [1, 4, 5]; **spring wheat** (off-label) [1, 6]
- Annual meadow grass in **apple orchards, blackberries, blackcurrants, broccoli, brussels sprouts, cabbages, calabrese, carrots, cauliflowers, cherries, combining peas, forage maize, gooseberries, leeks, loganberries, parsnips, pear orchards, plums, raspberries, strawberries** [2, 5]; **broad beans** (off-label), **dwarf beans** (off-label), **french beans** (off-label) [4]; **bulb onions, rubus hybrids** [2]; **cabbages** (off-label) [6]; **durum wheat, potatoes, spring barley, sunflowers, triticale, winter barley, winter rye, winter wheat** [2, 4, 5]; **edible podded peas** (off-label), **lupins** (off-label), **runner beans** (off-label), **spring field beans** (off-label), **vining peas** (off-label), **winter field beans** (off-label) [2, 4]; **forage soya bean** (off-label) [4, 6]
- Blackgrass in **apple orchards, blackberries, blackcurrants, broccoli, brussels sprouts, cabbages, calabrese, carrots, cauliflowers, cherries, combining peas, durum wheat, gooseberries, leeks, loganberries, parsnips, pear orchards, plums, raspberries, spring barley, strawberries, sunflowers, triticale, winter barley, winter rye, winter wheat** [2, 5]; **broad beans** (off-label), **game cover** (off-label), **spring field beans** (off-label), **spring wheat** (off-label), **winter field beans** (off-label) [1]; **bulb onions, rubus hybrids** [2]; **forage soya bean** (off-label) [6]
- Chickweed in **broad beans** (off-label) [4]; **celeriac** (off-label) [6]
- Cleavers in **broad beans** (off-label), **spring field beans** (off-label) [1]; **winter field beans** (off-label) [1, 5]
- Common orache in **broad beans** (off-label) [4]
- Corn marigold in **broad beans** (off-label) [4]
- Dead nettle in **broad beans** (off-label) [4]
- Fat hen in **broad beans** (off-label) [4]; **celeriac** (off-label) [6]
- Field pansy in **broad beans** (off-label) [4]
- Forget-me-not in **broad beans** (off-label) [4]
- Fumitory in **broad beans** (off-label) [4]
- Groundsel in **celeriac** (off-label) [6]
- Hemp-nettle in **broad beans** (off-label) [4]
- Knotgrass in **edible podded peas** (off-label), **vining peas** (off-label) [5]
- Mayweeds in **celeriac** (off-label) [6]
- Parsley-piert in **broad beans** (off-label) [4]
- Polygonums in **celeriac** (off-label) [6]
- Poppies in **broad beans** (off-label) [4]
- Rough-stalked meadow grass in **apple orchards, blackberries, blackcurrants, broccoli, brussels sprouts, cabbages, calabrese, carrots, cauliflowers, cherries, combining peas, durum wheat, forage maize, gooseberries, leeks, loganberries, parsnips, pear orchards,**

SEE SECTION 3 FOR PRODUCTS ALSO REGISTERED

Pesticide Profiles

  plums, potatoes, raspberries, spring barley, strawberries, sunflowers, triticale, winter barley, winter rye, winter wheat [2, 5]; *broad beans* (off-label) [4]; *bulb onions, rubus hybrids* [2]
- Scarlet pimpernel in *broad beans* (off-label) [4]
- Shepherd's purse in *broad beans* (off-label) [4]
- Small nettle in *broad beans* (off-label) [4]; *celeriac* (off-label) [6]
- Sowthistle in *broad beans* (off-label) [4]; *celeriac* (off-label) [6]
- Speedwells in *broad beans* (off-label) [4]
- Volunteer oilseed rape in *apple orchards, pear orchards, spring barley, sunflowers* [2, 5]; *apples, forage maize (under plastic mulches), grain maize, grain maize under plastic mulches, pears* [1, 3, 8]; *blackberries, blackcurrants, broccoli, brussels sprouts, cabbages, calabrese, carrots, cauliflowers, cherries, gooseberries, leeks, loganberries, parsnips, plums, raspberries, rubus hybrids* [1-3, 5, 8]; *broad beans* (off-label), *field beans* (off-label), *game cover* (off-label), *redcurrants* [3]; *bulb onions* [1-3, 8]; *combining peas, durum wheat, forage maize, potatoes, triticale, winter barley, winter rye, winter wheat* [1-3, 5, 7, 8]; *strawberries* [2, 5, 8]
- Wild oats in *durum wheat, triticale, winter barley, winter rye, winter wheat* [2, 5]

Extension of Authorisation for Minor Use (EAMUs)
- *almonds* 20092916 [6]
- *asparagus* 20092920 [6]
- *bilberries* 20092907 [6]
- *blueberries* 20092907 [6]
- *bog myrtle* 20120323 [6]
- *broad beans* 20141259 [1], 20152446 [3], 20140230 [4], 20092922 [6]
- *broccoli* 20112451 [6]
- *brussels sprouts* 20112451 [6]
- *bulb onions* 20092914 [6]
- *cabbages* 20100650 [6]
- *calabrese* 20112451 [6]
- *carrots* 20093526 [6]
- *cauliflowers* 20112451 [6]
- *celeriac* 20131625 [6]
- *celery (outdoor)* 20092924 [6], 20111287 [6]
- *chervil* 20092913 [6]
- *chestnuts* 20092916 [6]
- *collards* 20112451 [6]
- *coriander* 20092913 [6]
- *cranberries* 20092907 [6]
- *cress* 20092921 [6]
- *dill* 20092913 [6]
- *dwarf beans* 20140232 [4]
- *edible podded peas* 20152445 [2], 20140233 [4], 20111007 [5], 20092909 [6]
- *evening primrose* 20092915 [6]
- *farm forestry* 20092918 [6]
- *fennel* 20092924 [6], 20111287 [6]
- *field beans* 20152446 [3]
- *forage soya bean* 20140231 [4], 20121634 [6]
- *forest* 20092918 [6]
- *forest nurseries* 20092919 [6]
- *french beans* 20140232 [4], 20092908 [6]
- *frise* 20092921 [6]
- *game cover* 20141260 [1], 20152496 [3], 20092919 [6]
- *garlic* 20092914 [6]
- *grass seed crops* 20092923 [6]
- *hazel nuts* 20092916 [6]
- *herbs (see appendix 6)* 20092921 [6]
- *hops* 20092919 [6]
- *horseradish* 20093526 [6]

FOR FULL CONDITIONS OF USE ALWAYS READ THE PRODUCT LABEL

pendimethalin

- **kale** *20112451* [6]
- **lamb's lettuce** *20092921* [6]
- **leaf brassicas** *20092921* [6]
- **leeks** *20092914* [6]
- **lettuce** *20092921* [6]
- **lupins** *20152445* [2], *20140233* [4], *20092909* [6]
- **mallow (althaea spp.)** *20092921* [6]
- **miscanthus** *20092919* [6]
- **navy beans** *20092908* [6]
- **ornamental plant production** *20092919* [6]
- **parsley** *20092913* [6]
- **parsley root** *20092911* [6], *20093526* [6]
- **parsnips** *20093526* [6]
- **protected forest nurseries** *20092919* [6]
- **quinces** *20092917* [6]
- **radicchio** *20092921* [6]
- **redcurrants** *20092907* [6]
- **rhubarb** *20092924* [6], *20111287* [6]
- **runner beans** *20152445* [2], *20140233* [4], *20092909* [6]
- **salad onions** *20092914* [6]
- **scarole** *20092921* [6]
- **shallots** *20092914* [6]
- **spring field beans** *20141259* [1], *20152445* [2], *20140233* [4], *20092551* [5], *20092909* [6]
- **spring wheat** *20141261* [1], *20092910* [6]
- **sweetcorn** *20092912* [6]
- **sweetcorn under plastic mulches** *20092912* [6]
- **top fruit** *20092919* [6]
- **vining peas** *20152445* [2], *20140233* [4], *20111007* [5], *20092909* [6]
- **walnuts** *20092916* [6]
- **whitecurrants** *20092907* [6]
- **winter field beans** *20141259* [1], *20152445* [2], *20140233* [4], *20092552* [5], *20092909* [6]

Approval information
- Pendimethalin included in Annex I under EC Regulation 1107/2009
- Accepted by BBPA for use on malting barley and hops

Efficacy guidance
- Apply as soon as possible after drilling. Weeds are controlled as they germinate and emerged weeds will not be controlled by use of the product alone
- For effective blackgrass control apply not more than 2 d after final cultivation and before weed seeds germinate
- Best results by application to fine firm, moist, clod-free seedbeds when rain follows treatment. Effectiveness reduced by prolonged dry weather after treatment
- Effectiveness reduced on soils with more than 6% organic matter. Do not use where organic matter exceeds 10%
- Any trash, ash or straw should be incorporated evenly during seedbed preparation
- Do not disturb soil after treatment
- Apply to potatoes as soon as possible after planting and ridging in tank-mix with metribuzin but note that this will restrict the varieties that may be treated
- Always follow WRAG guidelines for preventing and managing herbicide resistant weeds. See Section 5 for more information

Restrictions
- Maximum number of treatments 1 per crop or yr
- Maximum total dose equivalent to one full dose treatment on most crops; 2 full dose treatments on leeks
- May be applied pre-emergence of cereal crops sown before 30 Nov provided seed covered by at least 32 mm soil, or post-emergence to early tillering stage (GS 23)
- Do not undersow treated crops

SEE SECTION 3 FOR PRODUCTS ALSO REGISTERED

Pesticide Profiles

- Do not use on crops suffering stress due to disease, drought, waterlogging, poor seedbed conditions or chemical treatment or on soils where water may accumulate

Crop-specific information
- Latest use: pre-emergence for spring barley, carrots, lettuce, fodder maize, parsnips, parsley, combining peas, potatoes, onions and leeks; before transplanting for brassicas; before leaf sheaths erect for winter cereals; before bud burst for blackcurrants, gooseberries, cane fruit, hops; before flower trusses emerge for strawberries; 14 d after transplanting for leaf herbs
- Do not use on spring barley after end Mar (mid-Apr in Scotland on some labels) because dry conditions likely. Do not apply to dry seedbeds in spring unless rain imminent
- Apply to combining peas as soon as possible after sowing. Do not spray if plumule less than 13 mm below soil surface
- Apply to potatoes up to 7 d before first shoot emerges
- Apply to drilled crops as soon as possible after drilling but before crop and weed emergence
- Apply in top fruit, bush fruit and hops from autumn to early spring when crop dormant
- In cane fruit apply to weed free soil from autumn to early spring, immediately after planting new crops and after cutting out canes in established crops
- Apply in strawberries from autumn to early spring (not before Oct on newly planted bed). Do not apply pre-planting or during flower initiation period (post-harvest to mid-Sep)
- Apply pre-emergence in drilled onions or leeks, not on Sands, Very Light, organic or peaty soils or when heavy rain forecast
- Apply to brassicas after final plant-bed cultivation but before transplanting. Avoid unnecessary soil disturbance after application and take care not to introduce treated soil into the root zone when transplanting. Follow transplanting with specified post-planting treatments - see label
- Do not use on protected crops or in greenhouses

Following crops guidance
- Before ryegrass is drilled after a very dry season plough or cultivate to at least 15 cm. If treated spring crops are to be followed by crops other than cereals, plough or cultivate to at least 15 cm
- In the event of crop failure land must be ploughed or thoroughly cultivated to at least 15 cm. See label for minimum intervals that should elapse between treatment and sowing a range of replacement crops

Environmental safety
- Dangerous for the environment
- Very toxic to aquatic organisms
- Dangerous to fish or other aquatic life. Do not contaminate surface waters or ditches with chemical or used container
- LERAP Category B

Hazard classification and safety precautions
Hazard Harmful [7, 8]; Irritant [4, 5]; Dangerous for the environment [1-8]
Transport code 9 [1, 3-8]
Packaging group III [1, 3-8]
UN Number 3082 [1, 3-8]
Risk phrases H317 [4]; R22a, R36, R38 [7, 8]; R22b, R51 [5]; R50 [2, 7, 8]; R53a [2, 5, 7, 8]
Operator protection A [1-8]; C [5, 7, 8]; H [1, 3, 4, 6]; U02a [1-3, 6]; U05a, U08 [1-8]; U11 [7, 8]; U13 [1-3, 6-8]; U14 [2, 4, 6]; U15 [4]; U19a [1-3, 5-8]; U20b [1-4, 7, 8]
Environmental protection E15a [5]; E15b [1-4, 6-8]; E16a, E38 [1-8]; E16b [1, 3]; E34 [5, 7, 8]; E39 [6]; H410 [1, 3, 6]; H412 [4]
Storage and disposal D01, D02, D09a [1-8]; D05 [1, 3, 5, 6]; D08 [2, 7, 8]; D10b, D12a [1-3, 5-8]; D10c [4]
Medical advice M03, M05b [5, 7, 8]; M05a [1-3]

FOR FULL CONDITIONS OF USE ALWAYS READ THE PRODUCT LABEL

347   pendimethalin + picolinafen

A post-emergence broad-spectrum herbicide mixture for winter cereals
HRAC mode of action code: K1 + F1

See also picolinafen

Products

1	Chronicle	BASF	320:16 g/l	SC	15394
2	Galivor	BASF	320:14.5 g/l	SC	14986
3	Orient	BASF	330:7.5 g/l	SC	12541
4	Parade	BASF	320:16 g/l	SC	17246
5	PicoMax	BASF	320:16 g/l	SC	13456
6	Picona	BASF	320:16 g/l	SC	13428
7	PicoPro	BASF	320:16 g/l	SC	13454

Uses
- Annual dicotyledons in **forest nurseries** *(off-label)* [5]; **game cover** *(off-label)* [6]; **rye** [1]; **spring barley, spring wheat, triticale** [1, 4-7]; **spring barley** *(off-label)*, **spring wheat** *(off-label)* [1-3, 7]; **winter barley, winter wheat** [1-7]; **winter rye** [4-7]
- Annual grasses in **spring barley** *(off-label)*, **spring wheat** *(off-label)* [1]
- Annual meadow grass in **forest nurseries** *(off-label)* [5]; **game cover** *(off-label)* [6]; **rye** [1]; **spring barley, spring wheat, triticale** [1, 4-7]; **spring barley** *(off-label)*, **spring wheat** *(off-label)* [1-3, 7]; **winter barley, winter wheat** [1-7]; **winter rye** [4-7]
- Chickweed in **rye** [1]; **spring barley, spring wheat, triticale** [1, 4-7]; **winter barley, winter wheat** [1-7]; **winter rye** [4-7]
- Cleavers in **rye** [1]; **spring barley, spring wheat, triticale** [1, 4-7]; **winter barley, winter wheat** [1-7]; **winter rye** [4-7]
- Field speedwell in **rye** [1]; **spring barley, spring wheat, triticale** [1, 4-7]; **winter barley, winter wheat** [1-7]; **winter rye** [4-7]
- Ivy-leaved speedwell in **rye** [1]; **spring barley, spring wheat, triticale** [1, 4-7]; **winter barley, winter wheat** [1-7]; **winter rye** [4-7]
- Loose silky bent in **spring barley** [6]
- Rough-stalked meadow grass in **forest nurseries** *(off-label)* [5]; **game cover** *(off-label)* [6]; **rye** [1]; **spring barley, spring wheat, triticale** [1, 4-7]; **spring barley** *(off-label)*, **spring wheat** *(off-label)* [2]; **winter barley, winter wheat** [1-7]; **winter rye** [4-7]

Extension of Authorisation for Minor Use (EAMUs)
- *forest nurseries 20082863* [5]
- *game cover 20082864* [6]
- *spring barley 20130724* [1], *20113143* [2], *20130386* [3], *20130725* [7]
- *spring wheat 20130724* [1], *20113143* [2], *20130386* [3], *20130725* [7]

Approval information
- Pendimethalin and picolinafen included in Annex I under EC Regulation 1107/2009

Efficacy guidance
- Best results obtained on crops growing in a fine, firm tilth and when rain falls within 7 d of application
- Loose or cloddy seed beds must be consolidated prior to application otherwise reduced weed control may occur
- Residual weed control may be reduced on soils with more than 6% organic matter or under prolonged dry conditions
- Always follow WRAG guidelines for preventing and managing herbicide resistant weeds. See Section 5 for more information

Restrictions
- Maximum number of treatments 1 per crop
- Do not treat undersown cereals or those to be undersown
- Do not roll emerged crops before treatment nor autumn treated crops until the following spring

---

SEE SECTION 3 FOR PRODUCTS ALSO REGISTERED

Pesticide Profiles

- Do not use on stony or gravelly soils, on soils that are waterlogged or prone to waterlogging, or on soils with more than 10% organic matter
- Do not treat crops under stress from any cause
- Do not apply pre-emergence to crops drilled after 30 Nov [3]
- Consult processor before treating crops for processing [3]

Crop-specific information
- Latest use: before pseudo-stem erect stage (GS 30)
- Transient bleaching may occur after treatment but it does not lead to yield loss
- Do not use pre-emergence on crops drilled after 30th Nov [5]
- For pre-emergence applications, seed should be covered with a minimum of 3.2 cm settled soil [6, 7]

Following crops guidance
- After normal harvest any crop except ryegrass can be sown. Cultivation or ploughing to at least 15 cm is required before sowing ryegrass.
- In the event of crop failure following autumn application spring wheat, spring barley or winter field beans can be sown after the land is ploughed and at least 8 weeks have elapsed since treatment. The following spring oilseed rape, peas or spring field beans can be sown after non-inversion tillage to at least 5 cm but maize should not be planted until at least 5 months have elapsed since application.
- In the event of crop failure following spring application spring wheat, spring barley, peas, spring field beans, maize or spring oilseed rape can be sown after the land is ploughed to at least 15 cm and at least 8 weeks have elapsed since application.

Environmental safety
- Dangerous for the environment
- Very toxic to aquatic organisms
- Product binds strongly to soil minimising likelihood of movement into groundwater
- LERAP Category B

Hazard classification and safety precautions
    Hazard Dangerous for the environment [1-7]; Very toxic to aquatic organisms [1, 4-6]
    Transport code 9
    Packaging group III
    UN Number 3082
    Risk phrases R50 [2, 3, 5, 7]; R53a [2, 3, 7]
    Operator protection A, H [1, 3-5]; U02a [1-4, 6, 7]; U05a, U20c [1-7]
    Environmental protection E15a [3]; E15b [1, 2, 4-7]; E16a, E34, E38 [1-7]; H410 [1, 4-6]
    Storage and disposal D01, D02, D05, D09a, D10c, D12a
    Medical advice M03

348   penthiopyrad

An SDHI fungicide for disease control in cereals
FRAC mode of action code: 7

See also chlorothalonil + penthiopyrad
         cyproconazole + penthiopyrad

Products

1	Fontelis	DuPont	200 g/l	SC	17465
2	Intellis	DuPont	200 g/l	EC	17520
3	Vertisan	DuPont	200 g/l	EC	17485

Uses
- Brown rust in **rye**, **spring barley**, **spring wheat**, **triticale**, **winter barley**, **winter wheat** [2, 3]
- Leaf blotch in **rye** *(moderate control)*, **triticale** *(moderate control)* [2, 3]
- Net blotch in **spring barley**, **winter barley** [2, 3]
- Powdery mildew in **apples** *(suppression)*, **pears** *(suppression)* [1]
- Ramularia leaf spots in **winter barley** *(moderate control)* [2, 3]

FOR FULL CONDITIONS OF USE ALWAYS READ THE PRODUCT LABEL

penthiopyrad + picoxystrobin

- Rhynchosporium in *rye*, *spring barley*, *winter barley* [2, 3]
- Scab in *apples*, *pears* [1]
- Septoria leaf blotch in *spring wheat*, *winter wheat* [2, 3]
- Tan spot in *spring wheat* (reduction), *winter wheat* (reduction) [2, 3]
- Yellow rust in *rye* (moderate control), *spring barley* (moderate control), *spring wheat* (moderate control), *triticale* (moderate control), *winter barley* (moderate control), *winter wheat* (moderate control) [2, 3]

Approval information
- Penthiopyrad is included in Annex 1 under EC Regulation 1107/2009
- Accepted by BBPA for use on malting barley

Efficacy guidance
- Always use in mixture with another product from a different fungicide resistance group that is used for control of the same target disease.
- Best results achieved from applications made before the disease is established in the crop

Restrictions
- Do not use more than two foliar applications of SDHI fungicides on any cereal crop.
- Must be applied in at least 100 l/ha since it is a skin sensitiser.
- Avoid application in either frosty or hot, sunny weather conditions [1]
- Do not apply to any crop suffering from stress as a result of drought, waterlogging, low temperatures, nutrient or lime deficiency or other factors reducing crop growth.
- In order to prevent contamination of groundwater, the second application to all crops must be made after GS39 [2, 3]

Environmental safety
- LERAP Category B

Hazard classification and safety precautions
Hazard Irritant, Dangerous for the environment [2, 3]; Very toxic to aquatic organisms [1]
Transport code 9
Packaging group III
UN Number 3082
Risk phrases H317, H319 [2, 3]
Operator protection A [1-3]; C, H [2, 3]; U05a, U09a, U11, U14, U15, U20a [2, 3]
Environmental protection E15b, E16a [1-3]; E17a [1] (40 m); E34, H410 [1]; E38, E40b, E40d, H411 [2, 3]
Storage and disposal D01, D02, D10c [2, 3]; D05 [1]; D09a, D12a [1-3]

349 penthiopyrad + picoxystrobin

An SDHI/strobilurin fungicide mixture for disease control in cereals
FRAC mode of action code: 7 + 11

See also penthiopyrad
         picoxystrobin

Products

1	Cypher	DuPont	160:80 g/l	SC	16639
2	Frelizon	DuPont	160:80 g/l	SC	16218
3	Refinzar	DuPont	160:80 g/l	SC	16590
4	Sebrant	DuPont	160:80 g/l	SC	17171

Uses
- Brown rust in *spring wheat*, *winter wheat* [1, 2, 4]
- Crown rust in *spring oats*, *winter oats* [1, 2, 4]
- Light leaf spot in *spring oilseed rape* (moderate control), *winter oilseed rape* (moderate control) [3]
- Net blotch in *spring barley*, *winter barley* [1, 2, 4]
- Phoma in *spring oilseed rape*, *winter oilseed rape* [1-4]

SEE SECTION 3 FOR PRODUCTS ALSO REGISTERED

Pesticide Profiles

- Rhynchosporium in **spring barley, winter barley** [1, 2, 4]
- Septoria leaf blotch in **spring wheat, winter wheat** [1, 2, 4]

Approval information
- Picoxystrobin and penthiopyrad are included in Annex 1 under EC Regulation 1107/2009

Efficacy guidance
- Apply as a protective treatment starting from stem elongation (GS 30)

Restrictions
- Do not use more than two foliar applications of SDHI fungicides on any cereal crop.
- Use as part of an integrated crop management strategy in conjunction with other methods of control or fungicides with different modes of action to reduce the risk of the development of resistance
- Avoid application in either frosty or hot, sunny weather conditions
- Avoid application to any crop suffering from stress as a result of drought, waterlogging, low temperatures, nutrient or lime deficiency or other factors reducing crop growth

Environmental safety
- Buffer zone requirement 10 m in cereals, 8 m in oilseed rape [1, 2]
- Buffer zone requirement 8 m [3]
- LERAP Category B

Hazard classification and safety precautions
Hazard Dangerous for the environment, Very toxic to aquatic organisms
Transport code 9
Packaging group III
UN Number 3082
Operator protection A; U05a [1, 2, 4]
Environmental protection E15b, E16a, E34, E38, H410
Storage and disposal D01, D02 [1, 4]; D09a, D12a [1-4]; D12b [2, 3]
Medical advice M05a [1, 2, 4]

350 phenmedipham

A contact phenyl carbamate herbicide for beet crops and strawberries
HRAC mode of action code: C1

See also desmedipham + ethofumesate + lenacil + phenmedipham
desmedipham + ethofumesate + metamitron + phenmedipham
desmedipham + ethofumesate + phenmedipham
desmedipham + phenmedipham
ethofumesate + metamitron + phenmedipham
ethofumesate + phenmedipham

Products

1	Beetup Flo	UPL Europe	160 g/l	SC	14328
2	Betasana SC	UPL Europe	160 g/l	SC	14209
3	Corzal	UPL Europe	157 g/l	SE	16952

Uses
- Annual dicotyledons in **chard** *(off-label)*, **herbs (see appendix 6)** *(off-label)*, **seakale** *(off-label)*, **spinach** *(off-label)*, **spinach beet** *(off-label)* [3]; **fodder beet, mangels, red beet, sugar beet** [1-3]
- Black bindweed in **ornamental plant production** *(off-label)* [2]
- Chickweed in **fodder beet, mangels, red beet, sugar beet** [1, 2]; **ornamental plant production** *(off-label)* [2]
- Fat hen in **ornamental plant production** *(off-label)* [2]
- Field pansy in **ornamental plant production** *(off-label)* [2]
- Knotgrass in **ornamental plant production** *(off-label)* [2]
- Penny cress in **ornamental plant production** *(off-label)* [2]

FOR FULL CONDITIONS OF USE ALWAYS READ THE PRODUCT LABEL

- Scentless mayweed in *ornamental plant production* (off-label) [2]

Extension of Authorisation for Minor Use (EAMUs)
- *chard* 20151519 [3]
- *herbs (see appendix 6)* 20151519 [3]
- *ornamental plant production* 20152050 [2]
- *seakale* 20151519 [3]
- *spinach* 20151519 [3]
- *spinach beet* 20151519 [3]

Approval information
- Phenmedipham included in Annex I under EC Regulation 1107/2009

Efficacy guidance
- Best results achieved by application to young seedling weeds, preferably cotyledon stage, under good growing conditions when low doses are effective
- If using a low-dose programme 2-3 repeat applications at 7-10 d intervals are recommended on mineral soils, 3-5 applications may be needed on organic soils
- Addition of adjuvant oil may improve effectiveness on some weeds
- Various tank-mixtures with other beet herbicides recommended. See label for details
- Use of certain pre-emergence herbicides is recommended in combination with post-emergence treatment. See label for details

Restrictions
- Maximum number of treatments and maximum total dose varies with crop and product used. See label for details
- At high temperatures (above 21°C) reduce rate and spray after 5 pm
- Do not apply immediately after frost or if frost expected
- Do not spray wet foliage or if rain imminent
- Do not spray crops stressed by wind damage, nutrient deficiency, pest or disease attack etc. Do not roll or harrow for 7 d before or after treatment
- Do not use on strawberries under cloches or polythene tunnels
- Consult processor before use on crops for processing

Crop-specific information
- Latest use: before crop leaves meet between rows for beet crops; before flowering for strawberries
- Apply to beet crops at any stage as low dose/low volume spray or from fully developed cotyledon stage with full rate. Apply to red beet after fully developed cotyledon stage
- Apply to strawberries at any time when weeds in susceptible stage, except in period from start of flowering to picking

Following crops guidance
- Beet crops may follow at any time after a treated crop. 3 mth must elapse from treatment before any other crop is sown and must be preceded by mould-board ploughing to 15 cm

Environmental safety
- Dangerous for the environment
- Very toxic to aquatic organisms
- Harmful to fish or other aquatic life. Do not contaminate surface waters or ditches with chemical or used container
- LERAP Category B

Hazard classification and safety precautions
    Hazard Irritant [1, 2]; Dangerous for the environment [1-3]
    Transport code 9
    Packaging group III
    UN Number 3082
    Risk phrases H317, H319 [1, 2]; R50, R53a [3]
    Operator protection A [1-3]; C, H [1, 2]; U05a [3]; U08 [1-3]; U11, U14, U19a, U20c [1, 2]
    Environmental protection E15b, E16a [1-3]; E16b, E34, E38 [3]; H411 [1, 2]
    Storage and disposal D01, D02, D09a, D12a [1-3]; D05, D10c [1, 2]; D10b [3]

SEE SECTION 3 FOR PRODUCTS ALSO REGISTERED

351   d-phenothrin

A non-systemic pyrethroid insecticide available only in mixtures
IRAC mode of action code: 3A

352   d-phenothrin + tetramethrin

A pyrethroid insecticide mixture for control of flying insects
IRAC mode of action code: 3 + 3

See also tetramethrin

Products

Killgerm ULV 500	Killgerm	36.8:18.4 g/l	UL	H4647

Uses
- Flies in *agricultural premises*
- Grain storage mite in *grain stores*
- Mosquitoes in *agricultural premises*
- Wasps in *agricultural premises*

Approval information
- Product approved for ULV application. See label for details
- D-phenothrin and tetramethrin are not included in Annex 1 under EC Regulation 1107/2009

Efficacy guidance
- Close doors and windows and spray in all directions for 3-5 sec. Keep room closed for at least 10 min

Restrictions
- For use only by professional operators
- Do not use space sprays containing pyrethrins or pyrethroid more than once per week in intensive or controlled environment animal houses in order to avoid development of resistance. If necessary, use a different control method or product

Crop-specific information
- May be used in the presence of poultry and livestock

Environmental safety
- Dangerous for the environment
- Toxic to aquatic organisms
- Do not apply directly to livestock/poultry
- Remove exposed milk and collect eggs before application. Protect milk machinery and containers from contamination

Hazard classification and safety precautions
   Hazard Harmful, Dangerous for the environment
   Transport code 9
   Packaging group III
   UN Number 3082
   Risk phrases R22b, R51, R53a
   Operator protection A, C, D, E, H; U02b, U09b, U14, U19a, U20a, U20b
   Environmental protection E05a, E15a, E38
   Consumer protection C06, C07, C08, C09, C11, C12
   Storage and disposal D01, D05, D06a, D09a, D10a, D12a
   Medical advice M05b

FOR FULL CONDITIONS OF USE ALWAYS READ THE PRODUCT LABEL

353 Phlebiopsis gigantea

A fungal protectant against root and butt rot in conifers

Products

PG Suspension   Forest Research   0.5% w/w   SC   17009

Uses
- Fomes root and butt rot in **farm forestry**, **forest**, **forest nurseries**

Approval information
- Phlebiopsis gigantea included in Annex 1 under EC Regulation 1107/2009

Efficacy guidance
- Store sachets in the original box in a cool dry place between 2°C and 15°C out of the sun.
- Use the diluted solution within 24 hours of mixing. Ensure complete coverage of stumps for optimum efficacy.

Hazard classification and safety precautions
UN Number N/C
Operator protection A, D, H; U20c
Environmental protection E15a
Storage and disposal D11a

354 physical pest control

Products that work by physical action only

Products

1	SB Plant Invigorator	Fargro	-	SL	-
2	Silico-Sec	Interfarm	>90% w/w	DS	-
3	Vazor DE	Killgerm	n/a	PO	-
4	Vazor Liquid Mosquito Film	Killgerm	n/a	RH	-

Uses
- Aphids in **all edible crops (outdoor and protected)**, **ornamental plant production**, **protected ornamentals** [1]
- Insect control in **agricultural premises** [3]
- Mealybugs in **all edible crops (outdoor and protected)**, **ornamental plant production**, **protected ornamentals** [1]
- Mites in **stored grain** [2]
- Mosquitoes in **enclosed waters** [4]
- Powdery mildew in **all edible crops (outdoor and protected)**, **ornamental plant production**, **protected ornamentals** [1]
- Scale insects in **all edible crops (outdoor and protected)**, **ornamental plant production**, **protected ornamentals** [1]
- Spider mites in **all edible crops (outdoor and protected)**, **ornamental plant production**, **protected ornamentals** [1]
- Suckers in **all edible crops (outdoor and protected)**, **ornamental plant production**, **protected ornamentals** [1]
- Whitefly in **all edible crops (outdoor and protected)**, **ornamental plant production**, **protected ornamentals** [1]

Approval information
- Products included in this profile are not subject to the Control of Pesticides Regulations/Plant Protection Products Regulations because they act by physical means only

Efficacy guidance
- Products act by physical means following direct contact with spray. They may therefore be used at any time of year on all pest growth stages

SEE SECTION 3 FOR PRODUCTS ALSO REGISTERED

Pesticide Profiles

- Ensure thorough spray coverage of plant, paying special attention to growing points and the underside of leaves
- Treat as soon as target pests are seen and repeat as often as necessary

Restrictions
- No limit on the number of treatments and no minimum interval between applications
- Before large scale use on a new crop, treat a few plants to check for crop safety
- Do not treat ornamental crops when in flower

Crop-specific information
- HI zero

Environmental safety
- May be used in conjunction with biological control agents. Spray 24 h before they are introduced

Hazard classification and safety precautions
Hazard Highly flammable [3]
UN Number N/C
Risk phrases H319, H336, R36, R67 [3]
Operator protection A, H [3]; F [2]; U10, U14, U15 [1]
Storage and disposal D01, D02, D05, D06h [1]

355 picoxystrobin

A broad spectrum strobilurin fungicide for cereals
FRAC mode of action code: 11

See also chlorothalonil + picoxystrobin
cyproconazole + picoxystrobin
cyprodinil + picoxystrobin
penthiopyrad + picoxystrobin

Products

1	Flanker	DuPont	250 g/l	SC	14760
2	Galileo	DuPont	250 g/l	SC	13252
3	Oranis	DuPont	250 g/l	SC	15185

Uses
- Brown rust in **spring barley, spring wheat, winter barley, winter wheat** [1-3]
- Crown rust in **spring oats, winter oats** [1-3]
- Eyespot in **spring wheat** *(reduction)*, **winter wheat** *(reduction)* [1-3]
- Foliar disease control in **ornamental plant production** *(off-label)* [2]
- Glume blotch in **spring wheat, winter wheat** [1-3]
- Late ear diseases in **spring wheat, winter wheat** [1-3]
- Net blotch in **spring barley, winter barley** [1-3]
- Powdery mildew in **spring barley, spring oats, winter barley, winter oats** [1-3]
- Rhynchosporium in **spring barley, winter barley** [1-3]
- Sclerotinia in **spring oilseed rape, winter oilseed rape** [1-3]
- Septoria leaf blotch in **spring wheat, winter wheat** [1-3]
- Yellow rust in **spring wheat, winter wheat** [1-3]

Extension of Authorisation for Minor Use (EAMUs)
- *ornamental plant production* 20082855 [2]

Approval information
- Picoxystrobin included in Annex I under EC Regulation 1107/2009
- Accepted by BBPA for use on malting barley

Efficacy guidance
- Best results achieved from protectant treatments made before disease establishes in the crop
- Use of tank mixtures is recommended where disease has become established at the time of treatment

FOR FULL CONDITIONS OF USE ALWAYS READ THE PRODUCT LABEL

- Persistence of yellow rust control is less than for other foliar diseases but can be improved by use of an appropriate tank mixture
- Eyespot control is not reliable and an appropriate tank mix should be used where crops are at risk
- Picoxystrobin is a member of the QoI cross resistance group. Product should be used preventatively and not relied on for its curative potential
- Use product as part of an Integrated Crop Management strategy incorporating other methods of control, including where appropriate other fungicides with a different mode of action. Do not apply more than two foliar applications of QoI containing products to any cereal crop
- There is a significant risk of widespread resistance occurring in *Septoria tritici* populations in UK. Failure to follow resistance management action may result in reduced levels of disease control
- On cereal crops product must always be used in mixture with another product, recommended for control of the same target disease, that contains a fungicide from a different cross resistance group and is applied at a dose that will give robust control
- Strains of barley powdery mildew resistant to QoIs are common in the UK

Restrictions
- Maximum number of treatments 2 per crop per year for ornamental plant production, 2 per year for cereals and 1 per year for oilseed rape

Crop-specific information
- Latest use: grain watery ripe (GS71) for wheat, barley and oats, flowers declining (majority of petals fallen) for oilseed rape.

Environmental safety
- Dangerous for the environment
- Very toxic to aquatic organisms

Hazard classification and safety precautions
Hazard Harmful, Dangerous for the environment
Transport code 9
Packaging group III
UN Number 3082
Risk phrases R50, R53a
Operator protection A; U05a, U09a, U14, U15, U20a
Environmental protection E15a, E38
Storage and disposal D01, D02, D05, D09a, D10c, D12a

356 pinoxaden

A phenylpyrazoline grass weed herbicide for cereals
HRAC mode of action code: A

See also clodinafop-propargyl + pinoxaden
     florasulam + pinoxaden

Products

1	Axial	Syngenta	100 g/l	EC	12521
2	Clayton Rugby	Clayton	100 g/l	EC	16941
3	Greencrop Helvick	Greencrop	100 g/l	EC	13052
4	Pinball	Pan Agriculture	100 g/l	EC	16070
5	Pivot	Goldengrass	100 g/l	EC	16560
6	PureRecovery	Pure Amenity	45 g/l	EC	15137
7	Rescue	Syngenta	45 g/l	EC	14518
8	Standon Pinoxy	Standon	100 g/l	EC	16355

Uses
- Annual grasses in **game cover** *(off-label)* [1]
- Blackgrass in **spring barley, winter barley** [1, 2, 4, 5, 8]
- Italian ryegrass in **spring barley, winter barley, winter wheat** [1-5, 8]; **spring wheat** [1, 2, 4, 5, 8]
- Perennial grasses in **game cover** *(off-label)* [1]

SEE SECTION 3 FOR PRODUCTS ALSO REGISTERED

Pesticide Profiles

- Perennial ryegrass in **amenity grassland** *(off-label)*, **managed amenity turf** *(off-label)* [7]; **spring barley**, **winter barley**, **winter wheat** [1-5, 8]; **spring wheat** [1, 2, 4, 5, 8]
- Ryegrass in **amenity grassland** *(reduction only)*, **managed amenity turf** *(reduction only)* [6, 7]
- Wild oats in **spring barley**, **winter barley**, **winter wheat** [1-5, 8]; **spring wheat** [1, 2, 4, 5, 8]

Extension of Authorisation for Minor Use (EAMUs)
- **amenity grassland** *20111290* [7]
- **game cover** *20082815* [1]
- **managed amenity turf** *20111290* [7]

Approval information
- Pinoxaden included in Annex 1 under EC Regulation 1107/2009

Efficacy guidance
- Best results obtained from treatment when all grass weeds have emerged. There is no residual activity
- Broad-leaved weeds are not controlled
- Treat before emerged weed competition reduces yield
- Blackgrass in winter and spring barley is also controlled when used as part of an integrated control strategy. Product is not recommended for blackgrass control in winter wheat
- Grass weed control may be reduced if rain falls within 1 hr of application
- Pinoxaden is an ACCase inhibitor herbicide. To avoid the build up of resistance do not apply products containing an ACCase inhibitor herbicide more than twice to any crop. In addition do not use any product containing pinoxaden in mixture or sequence with any other product containing the same ingredient
- Use these products as part of a resistance management strategy that includes cultural methods of control and does not use ACCase inhibitors as the sole chemical method of grass weed control
- Applying a second product containing an ACCase inhibitor to a crop will increase the risk of resistance development; only use a second ACCase inhibitor to control different weeds at a different timing
- Always follow WRAG guidelines for preventing and managing herbicide resistant weeds. See Section 5 for more information

Restrictions
- Maximum number of treatments 1 per crop
- Do not spray crops under stress or suffering from waterlogging, pest attack, disease or frost damage
- Do not spray crops undersown with grass mixtures
- Avoid the use of hormone-containing herbicides in mixture or in sequence. Allow 21 d after, or 7 d before, hormone application
- Product must be used with prescribed adjuvant. See label

Crop-specific information
- Latest use: before flag leaf extending stage (GS 41)
- Spray cereals in autumn, winter or spring from the 2-leaf stage (GS 12)

Following crops guidance
- There are no restrictions on succeeding crops in a normal rotation
- In the event of failure of a treated crop ryegrass, maize, oats or any broad-leaved crop may be planted after a minimum interval of 4 wk from application

Environmental safety
- Dangerous for the environment
- Toxic to aquatic organisms
- Do not allow spray to drift onto neighbouring crops of oats, ryegrass or maize

Hazard classification and safety precautions
  Hazard Harmful [6, 7]; Irritant [1-5, 8]; Dangerous for the environment [1-8]
  Transport code 9
  Packaging group III
  UN Number 3082

FOR FULL CONDITIONS OF USE ALWAYS READ THE PRODUCT LABEL

pirimicarb

Risk phrases H304, H317 [1, 2, 6-8]; R22b [7]; R36, R38, R51, R53a [3-5]
Operator protection A, H [1-8]; C [1-5, 8]; U05a, U20b [1-8]; U09a [1-5, 8]
Environmental protection E15b, E38 [1-8]; H410 [1, 2, 6-8]
Storage and disposal D01, D02, D05, D09a, D10c, D12a [1-8]; D11a [1-5, 8]
Medical advice M05b [6, 7]

357   pirimicarb

A carbamate insecticide for aphid control
IRAC mode of action code: 1A

See also lambda-cyhalothrin + pirimicarb

Products

1	Aphox	Syngenta	50% w/w	WG	17401
2	Clayton Pirimicarb	Clayton	50% w/w	WG	17704
3	Pirimate 500	Becesane	50% w/w	WG	17657

Uses
- Aphids in **broad beans, combining peas, spring field beans, vining peas, winter field beans** [1-3]; **celeriac** *(off-label)*, **celery (outdoor)** *(off-label)*, **florence fennel** *(off-label)*, **horseradish** *(off-label)*, **parsley root** *(off-label)*, **protected courgettes** *(off-label)*, **protected gherkins** *(off-label)*, **protected summer squash** *(off-label)*, **radishes** *(off-label)*, **red beet** *(off-label)*, **rhubarb** *(off-label)* [1]

Extension of Authorisation for Minor Use (EAMUs)
- *celeriac* 20161609 [1]
- *celery (outdoor)* 20161610 [1]
- *florence fennel* 20161606 [1]
- *horseradish* 20161605 [1]
- *parsley root* 20161605 [1]
- *protected courgettes* 20161611 [1]
- *protected gherkins* 20161611 [1]
- *protected summer squash* 20161611 [1]
- *radishes* 20161605 [1]
- *red beet* 20161605 [1]
- *rhubarb* 20161610 [1]

Approval information
- Pirimicarb included in Annex I under EC Regulation 1107/2009
- Approved for aerial application on cereals. See Section 5 for more information
- Accepted by BBPA for use on malting barley
- In 2006 CRD required that all products containing this active ingredient should carry the following warning in the main area of the container label: "Pirimicarb is an anticholinesterase carbamate. Handle with care"

Efficacy guidance
- Chemical has contact, fumigant and translaminar activity
- Best results achieved under warm, calm conditions when plants not wilting and spray does not dry too rapidly. Little vapour activity at temperatures below 15°C
- Apply as soon as aphids seen or warning issued and repeat as necessary
- Addition of non-ionic wetter recommended for use on brassicas
- On cucumbers and tomatoes a root drench is preferable to spraying when using predators in an integrated control programme
- Where aphids resistant to pirimicarb occur control is unlikely to be satisfactory

Restrictions
- Contains an anticholinesterase carbamate compound. Do not use if under medical advice not to work with such compounds
- Maximum number of treatments not specified in some cases but normally 2-6 depending on crop

SEE SECTION 3 FOR PRODUCTS ALSO REGISTERED

Pesticide Profiles

- Spray equipment must only be used where the operator's normal working position is within a closed cab on a tractor or self-propelled sprayer when making air-assisted applications to apples or other top fruit

Crop-specific information
- Latest use in accordance with harvest intervals below
- HI protected courgettes and gherkins 24 h; other edible crops 3 d;

Environmental safety
- Dangerous for the environment
- Very toxic to aquatic organisms
- Dangerous to fish or other aquatic life. Do not contaminate surface waters or ditches with chemical or used container
- Chemical has little effect on bees, ladybirds and other insects and is suitable for use in integrated control programmes on apples and pears
- Keep all livestock out of treated areas for at least 7 d. Bury or remove spillages
- LERAP Category B

Hazard classification and safety precautions
Hazard Toxic, Dangerous for the environment, Toxic if swallowed, Harmful if inhaled
Transport code 6.1
Packaging group III
UN Number 2757
Risk phrases H319
Operator protection A, C, D, H, J, M; U05a, U08, U19a, U20b
Environmental protection E06c (7 d); E15b, E16a, E34, E38, H410
Storage and disposal D01, D02, D09a, D11a, D12a
Medical advice M02, M03, M04a

358   pirimiphos-methyl

A contact, fumigant and translaminar organophosphorus insecticide
IRAC mode of action code: 1B

Products

1	Actellic 50 EC	Syngenta	500 g/l	EC	12726
2	Actellic Smoke Generator No. 20	Syngenta	22.5% w/w	FU	15739
3	Clayton Galic	Clayton	500 g/l	EC	15374
4	Flycatcher	AgChem Access	500 g/l	EC	14135
5	Pan PMT	Pan Agriculture	500 g/l	EC	13996

Uses
- Flour beetle in **stored grain** [1, 3-5]
- Flour moth in **stored grain** [1, 3-5]
- Grain beetle in **stored grain** [1, 3-5]
- Grain storage mite in **stored grain** [1, 3-5]
- Grain storage pests in **grain stores** [2]
- Grain weevil in **stored grain** [1, 3-5]
- Insect pests in **grain stores** [2]
- Mites in **grain stores** [2]
- Warehouse moth in **stored grain** [1, 3-5]

Approval information
- Pirimiphos-methyl included in Annex I under EC Regulation 1107/2009
- In 2006 PSD required that all products containing this active ingredient should carry the following warning in the main area of the container label: "Pirimiphos-methyl is an anticholinesterase organophosphate. Handle with care"
- Accepted by BBPA for use in stores for malting barley

FOR FULL CONDITIONS OF USE ALWAYS READ THE PRODUCT LABEL

pirimiphos-methyl

Efficacy guidance
- Chemical acts rapidly and has short persistence in plants, but persists for long periods on inert surfaces
- Best results for protection of stored grain achieved by cleaning store thoroughly before use and employing a combination of pre-harvest and grain/seed treatments [1, 2]
- Best results for admixture treatment obtained when grain stored at 15% moisture or less. Dry and cool moist grain coming into store but then treat as soon as possible, ideally as it is loaded [1]
- Surface admixture can be highly effective on localised surface infestations but should not be relied on for long-term control unless application can be made to the full depth of the infestation [1]
- Where insect pests resistant to pirimiphos-methyl occur control is unlikely to be satisfactory

Restrictions
- Contains an organophosphorus anticholinesterase compound. Do not use if under medical advice not to work with such compounds
- Maximum number of treatments 2 per grain store [1, 2]; 1 per batch of stored grain [1]
- Do not apply surface admixture or complete admixture using hand held equipment [1]
- Surface admixture recommended only where the conditions for store preparation, treatment and storage detailed in the label can be met [1]

Crop-specific information
- Latest use: before storing grain [1, 2]
- Disinfect empty grain stores by spraying surfaces and/or fumigation and treat grain by full or surface admixture. Treat well before harvest in late spring or early summer and repeat 6 wk later or just before harvest if heavily infested. See label for details of treatment and suitable application machinery
- Treatment volumes on structural surfaces should be adjusted according to surface porosity. See label for guidelines
- Treat inaccessible areas with smoke generating product used in conjunction with spray treatment of remainder of store
- Predatory mites (*Cheyletus*) found in stored grain feeding on infestations of grain mites may survive treatment and can lead to rejection by buyers. They do not remain for long but grain should be inspected carefully before selling [1, 2]

Environmental safety
- Dangerous for the environment
- Very toxic to aquatic organisms [1]
- Toxic to aquatic organisms [2]
- Highly flammable [2]
- Flammable [1]
- Ventilate fumigated or fogged spaces thoroughly before re-entry
- Unprotected persons must be kept out of fumigated areas within 3 h of ignition and for 4 h after treatment
- Wildlife must be excluded from buildings during treatment [1]
- Keep away from combustible materials [2]

Hazard classification and safety precautions
Hazard Harmful, Dangerous for the environment [1-5]; Highly flammable [2]; Flammable [1, 3-5]
Transport code 3 [1, 3-5]; 9 [2]
Packaging group III
UN Number 1993 [1, 3-5]; 3077 [2]
Risk phrases R20, R50, R53a [1-5]; R22a, R22b, R36, R37 [1, 3-5]
Operator protection A, C, J, M [1, 3-5]; D, H [1-5]; U04a, U09a, U16a, U24 [1, 3-5]; U05a, U19a, U20b [1-5]; U14 [2]
Environmental protection E02a [2] (4 h); E15a, E34, E38 [1-5]
Consumer protection C09 [1, 3-5]; C12 [1-5]
Storage and disposal D01, D02, D09a, D12a [1-5]; D05, D10c [1, 3-5]; D07, D11a [2]
Treated seed S06a [1, 3-5]
Medical advice M01 [1-5]; M03 [2]; M05b [1, 3-5]

SEE SECTION 3 FOR PRODUCTS ALSO REGISTERED

359   potassium bicarbonate (commodity substance)

An inorganic fungicide for use in horticulture.

Products

1	Atilla	Q-Chem	85% w/w	SP	16026
2	Karma	Certis	85.42% w/w	WP	16363
3	potassium bicarbonate	various	100% w/w	AP	00000

Uses
- Botrytis in **table grapes** *(off-label)*, **wine grapes** *(off-label)* [2]
- Disease control in **all edible crops (outdoor and protected)**, **all non-edible crops (outdoor)**, **all protected non-edible crops** [3]
- Downy mildew in **lettuce** *(off-label)*, **protected lettuce** *(off-label)*, **table grapes** *(off-label)*, **wine grapes** *(off-label)* [2]
- Fungus diseases in **apples**, **nursery fruit trees and bushes** [2]
- Insect pests in **pears** [1]
- Pear sucker in **pears** *(reduction)* [1]
- Powdery mildew in **courgettes** *(off-label)*, **cucumbers** *(off-label)*, **melons** *(off-label)*, **protected courgettes** *(off-label)*, **protected cucumbers** *(off-label)*, **protected melons** *(off-label)*, **strawberries** *(off-label)*, **table grapes** *(off-label)*, **wine grapes** *(off-label)* [2]
- Sooty moulds in **apples**, **nursery fruit trees and bushes** [2]; **pears** *(reduction)* [1]

Extension of Authorisation for Minor Use (EAMUs)
- *courgettes* 20151941 [2]
- *cucumbers* 20151941 [2]
- *lettuce* 20151940 [2]
- *melons* 20151941 [2]
- *protected courgettes* 20151941 [2]
- *protected cucumbers* 20151941 [2]
- *protected lettuce* 20151940 [2]
- *protected melons* 20151941 [2]
- *strawberries* 20150901 [2]
- *table grapes* 20161327 [2]
- *wine grapes* 20161327 [2]

Approval information
- Approval for the use of potassium bicarbonate as a commodity substance was granted on 26 July 2005 by Ministers under regulation 5 of the Control of Pesticides Regulations 1986
- Potassium bicarbonate is being supported for review in the fourth stage of the EC Review Programme under EC Regulation 1107/2009. It was included in Annex 1 under the alternative name of potassium hydrogen carbonate and now products containing it will need to gain approval in the normal way if they carry label claims for pesticidal activity
- Accepted by BBPA for use on hops

Efficacy guidance
- Adjuvants authorised by CRD may be used in conjunction with potassium bicarbonate
- Crop phytotoxicity can occur

Restrictions
- Food grade potassium bicarbonate must only be used

Hazard classification and safety precautions
UN Number N/C
Operator protection C [2]; U05a [1, 2]; U19a [2]
Environmental protection E15a [1, 2]
Storage and disposal D09b [1, 2]

360 prochloraz + propiconazole

A broad spectrum fungicide mixture for wheat, barley and oilseed rape
FRAC mode of action code: 3 + 3

See also propiconazole

Products

1	Cirkon	Adama	400:90 g/l	EC	16643
2	Greencrop Twinstar	Greencrop	400:90 g/l	EC	09516

Uses
- Eyespot in **spring barley, spring wheat** [1]; **winter barley, winter wheat** [1, 2]
- Light leaf spot in **spring oilseed rape, winter oilseed rape** [1]
- Phoma in **spring oilseed rape, winter oilseed rape** [1]
- Rhynchosporium in **spring barley** [1]; **winter barley** [1, 2]
- Septoria leaf blotch in **spring wheat** [1]; **winter wheat** [1, 2]

Approval information
- Propiconazole and prochloraz included under EC Regulation 1107/2009
- Accepted by BBPA for use on malting barley

Efficacy guidance
- Best results obtained from treatment when disease is active but not well established
- Treat wheat normally from flag leaf ligule just visible stage (GS 39) but earlier if there is a high risk of Septoria
- Treat barley from the first node detectable stage (GS 31)
- If disease pressure persists a second application may be necessary
- Prochloraz and propiconazole are DMI fungicides. Resistance to some DMI fungicides has been identified in Septoria leaf blotch which may seriously affect performance of some products. For further advice contact a specialist advisor and visit the Fungicide Resistance Action Group (FRAG)-UK website

Restrictions
- Maximum number of treatments 2 per crop

Crop-specific information
- Latest use: before grain watery ripe (GS 71) for cereals; before most seeds green for winter oilseed rape

Environmental safety
- Dangerous for the environment
- Very toxic to aquatic organisms

Hazard classification and safety precautions
Hazard Irritant, Dangerous for the environment
Transport code 9
Packaging group III
UN Number 3082
Risk phrases R36, R53a [1, 2]; R50 [2]; R51 [1]
Operator protection A, C; U05a, U08, U20b [1, 2]; U11, U15 [1]
Environmental protection E15a
Storage and disposal D01, D02, D05, D09a, D10b
Medical advice M05a [1]

SEE SECTION 3 FOR PRODUCTS ALSO REGISTERED

361 prochloraz + proquinazid + tebuconazole

A broad spectrum systemic fungicide mixture for cereals
FRAC mode of action code: 3 + U7 + 3

See also proquinazid
tebuconazole

Products
Vareon	DuPont	320:40:160 g/l	EC	14376

Uses
- Brown rust in **spring barley**, **spring wheat**, **winter barley**, **winter wheat**
- Crown rust in **spring oats** (qualified minor use recommendation), **winter oats** (qualified minor use recommendation)
- Ear diseases in **spring wheat** (reduction only), **winter wheat** (reduction only)
- Eyespot in **spring barley**, **spring wheat**, **winter barley**, **winter wheat**
- Glume blotch in **spring wheat**, **winter wheat**
- Net blotch in **spring barley** (moderate control), **winter barley** (moderate control)
- Powdery mildew in **spring barley**, **spring oats**, **spring wheat**, **winter barley**, **winter oats**, **winter wheat**
- Rhynchosporium in **spring barley** (moderate control), **winter barley** (moderate control)
- Septoria leaf blotch in **spring wheat** (moderate control), **winter wheat** (moderate control)
- Yellow rust in **spring barley**, **spring wheat**, **winter barley**, **winter wheat**

Approval information
- Prochloraz, proquinazid and tebuconazole included in Annex 1 under EC Regulation 1107/2009
- Accepted by BBPA for use on malting barley

Efficacy guidance
- Resistance to DMI fungicides has been identified in Septoria leaf blotch and may seriously affect the performance of some products. Advice on resistance management is available from the Fungicide Action Group (FRAG) web site

Restrictions
- Maximum number of applications is 2 per crop
- Newer authorisations for tebuconazole products require application to cereals only after GS 30 and applications to oilseed rape and linseed after GS20 - check label

Crop-specific information
- Latest application in wheat is before full flowering (GS65)
- Latest application in barley is before beginning of heading (GS49)

Environmental safety
- Avoid spraying within 5 m of the field boundary to reduce the effect on non-target insects or other arthropods
- LERAP Category B

Hazard classification and safety precautions
Hazard Harmful, Dangerous for the environment
Transport code 9
Packaging group III
UN Number 3082
Risk phrases H317, H351, H361, R63
Operator protection A, C, H
Environmental protection E15a, E16a, E22c, E34, E38, H410
Storage and disposal D01, D02, D09a, D12a

362 prochloraz + tebuconazole

A broad spectrum systemic fungicide mixture for cereals, oilseed rape and amenity use
FRAC mode of action code: 3 + 3

See also tebuconazole

Products

1	Agate EW	Adama	267:133 g/l	EW	16515
2	Monkey	Adama	267:133 g/l	EW	12906
3	Orius P	Adama	267:133 g/l	EW	12880

Uses
- Brown rust in **spring barley, spring rye, spring wheat, winter barley, winter rye, winter wheat**
- Eyespot in **spring barley, spring rye, spring wheat, winter barley, winter rye, winter wheat**
- Glume blotch in **spring wheat, winter wheat**
- Late ear diseases in **spring wheat, winter wheat**
- Net blotch in **spring barley, winter barley**
- Powdery mildew in **spring barley, spring rye, spring wheat, winter barley, winter rye, winter wheat**
- Rhynchosporium in **spring barley, spring rye, winter barley, winter rye**
- Sclerotinia stem rot in **spring oilseed rape, winter oilseed rape**
- Septoria leaf blotch in **spring wheat, winter wheat**
- Yellow rust in **spring barley, spring rye, spring wheat, winter barley, winter rye, winter wheat**

Approval information
- Prochloraz and tebuconazole included in Annex I under EC Regulation 1107/2009
- Accepted by BBPA for use on malting barley

Efficacy guidance
- Complete protection of winter cereals will usually require a programme of at least two treatments [2]
- Protection of flag leaf and ear from Septoria diseases requires treatment at flag leaf emergence and repeated when ear fully emerged [2]
- Two treatments separated by a 2 wk gap may be needed for difficult Fusarium patch attacks
- To avoid resistance, do not apply repeated applications of the single product on the same crop against the same disease
- Optimum application timing is normally when disease first seen but varies with main target disease - see labels
- Prochloraz and tebuconazole are DMI fungicides. Resistance to some DMI fungicides has been identified in Septoria leaf blotch which may seriously affect performance of some products. For further advice contact a specialist advisor and visit the Fungicide Resistance Action Group (FRAG)-UK website

Restrictions
- Maximum total dose on cereals equivalent to two full dose treatments
- Maximum number of treatments on managed amenity turf 2 per yr
- Newer authorisations for tebuconazole products require application to cereals only after GS 30 and applications to oilseed rape and linseed after GS 20 - check label

Crop-specific information
- Latest use: before grain milky ripe (GS 73) for cereals [2]; most seeds green for oilseed rape [2]
- HI 6 wk for cereals [2]
- Occasionally transient leaf speckling may occur after treating wheat. Yield responses should not be affected

Environmental safety
- Dangerous for the environment
- Very toxic to aquatic organisms

SEE SECTION 3 FOR PRODUCTS ALSO REGISTERED

Pesticide Profiles

Hazard classification and safety precautions
Hazard Harmful, Dangerous for the environment, Harmful if swallowed, Very toxic to aquatic organisms
Transport code 9
Packaging group III
UN Number 3082
Risk phrases H319, H361
Operator protection A, C, H; U05a, U09b, U14, U20a [1-3]; U11 [1]
Environmental protection E13b, E34, E38, H410
Storage and disposal D01, D02, D05, D09a, D10c, D12a
Medical advice M03

363 prochloraz + thiram

A seed treatment fungicide mixture for oilseed rape
FRAC mode of action code: 3 + M3

See also thiram

Products
Agrichem Hy-Pro Duet   Agrichem   150:333 g/l   FS   14018

Uses
- Phoma in **winter oilseed rape** *(seed treatment)*

Approval information
- Prochloraz and thiram included in Annex I under EC Regulation 1107/2009

Efficacy guidance
- Can be applied through most types of seed treatment machinery
- Good coverage of the seed surface essential for best results
- Product is active against seed-borne disease. It will not control leaf spotting from leaf spores on trash on the seed bed

Restrictions
- Maximum number of treatments 1 per crop
- Seed must be of satisfactory quality and moisture content
- Sow seed as soon as possible after treatment
- Do not store treated seed from one season to the next
- Do not co-apply product with water or mix with any other seed treatment product

Crop-specific information
- Latest use: pre-sowing
- Product will also control soil-borne fungi that cause damping-off of rape seedlings

Following crops guidance
- Any crop may be grown following a treated winter oilseed rape crop

Hazard classification and safety precautions
Hazard Harmful, Dangerous for the environment
Transport code 9
Packaging group III
UN Number 3082
Risk phrases R20, R22a, R36, R38, R43, R50, R53a
Operator protection A, C, H; U02a, U08, U14, U15, U20a
Environmental protection E15a, E36b
Storage and disposal D01, D02, D05, D08, D09a, D12b, D13, D15
Treated seed S01, S02, S03, S04d, S05, S08
Medical advice M04a

FOR FULL CONDITIONS OF USE ALWAYS READ THE PRODUCT LABEL

364 prohexadione-calcium

A cyclohexanecarboxylate growth regulator for use in apples

See also mepiquat chloride + prohexadione-calcium

Products

1	Kudos	Fine	10% w/w	WG	17578
2	Regalis Plus	BASF	10% w/w	WG	16485

Uses
- Botrytis in **table grapes** *(off-label)*, **wine grapes** *(off-label)* [2]
- Control of shoot growth in **apples** [1, 2]; **hops in propagation** *(off-label)*, **nursery fruit trees** *(off-label)*, **ornamental plant production** *(off-label)*, **pears** *(off-label)*, **protected ornamentals** *(off-label)* [2]

Extension of Authorisation for Minor Use (EAMUs)
- *hops in propagation* 20150175 [2]
- *nursery fruit trees* 20150175 [2]
- *ornamental plant production* 20150175 [2]
- *pears* 20150182 [2]
- *protected ornamentals* 20150181 [2]
- *table grapes* 20150180 [2]
- *wine grapes* 20150180 [2]

Approval information
- Prohexadione-calcium included in Annex I under EC Regulation 1107/2009

Efficacy guidance
- For a standard orchard treat at the start of active growth and again after 3-5 wk depending on growth conditions
- Alternatively follow the first application with four reduced rate applications at two wk intervals

Restrictions
- Maximum total dose equivalent to two full dose treatments
- Do not apply in conjunction with calcium based foliar fertilisers

Crop-specific information
- HI 55 d for apples

Environmental safety
- Harmful to aquatic organisms
- Avoid spray drift onto neighbouring crops and other non-target plants
- Product does not harm natural insect predators when used as directed

Hazard classification and safety precautions
 UN Number N/C
 Risk phrases H317
 Operator protection U02a, U08, U19a, U20b [1]; U05a [1, 2]
 Environmental protection E15b, H411 [1]; E38 [1, 2]
 Storage and disposal D01, D02, D09a, D12a [1, 2]; D10b [1]; D10c [2]
 Medical advice M05a [1]

365 prohexadione-calcium + trinexapac-ethyl

A plant growth regulator for use in cereal crops

See also trinexapac-ethyl

Products

	Medax Max	BASF	0.5:0.75% w/w	SG	17263

SEE SECTION 3 FOR PRODUCTS ALSO REGISTERED

Pesticide Profiles

Uses
- Growth regulation in *durum wheat, spring barley, spring oats, spring wheat, triticale, winter barley, winter oats, winter rye, winter wheat*

Hazard classification and safety precautions
Transport code 9
Packaging group III
UN Number 3077
Operator protection A; U08, U19a, U20a
Environmental protection E15b, E34, H412
Storage and disposal D01, D02, D05, D09a, D10c
Medical advice M03

366 propamocarb hydrochloride

A translocated protectant carbamate fungicide
FRAC mode of action code: 28

See also chlorothalonil + propamocarb hydrochloride
 fenamidone + propamocarb hydrochloride
 fluopicolide + propamocarb hydrochloride
 fosetyl-aluminium + propamocarb hydrochloride
 mancozeb + propamocarb hydrochloride

Products
Proplant   Fargro   722 g/l   SL   15422

Uses
- Damping off in **protected broccoli, protected brussels sprouts, protected calabrese, protected cauliflowers, protected tomatoes** *(off-label)*
- Downy mildew in **protected broccoli, protected brussels sprouts, protected calabrese, protected cauliflowers, protected watercress** *(off-label)*, **radishes** *(off-label)*
- Peronospora spp in **forest nurseries** *(off-label)*, **ornamental plant production** *(off-label)*, **protected broccoli, protected brussels sprouts, protected calabrese, protected cauliflowers, protected ornamentals** *(off-label)*
- Phytophthora in **forest nurseries** *(off-label)*, **ornamental plant production** *(off-label)*, **protected broccoli, protected brussels sprouts, protected calabrese, protected cauliflowers, protected ornamentals** *(off-label)*, **protected tomatoes** *(off-label)*, **protected watercress** *(off-label)*
- Pythium in **forest nurseries** *(off-label)*, **ornamental plant production** *(off-label)*, **protected broccoli, protected brussels sprouts, protected calabrese, protected cauliflowers, protected ornamentals** *(off-label)*, **protected watercress** *(off-label)*
- Root rot in **protected broccoli, protected brussels sprouts, protected calabrese, protected cauliflowers, protected tomatoes** *(off-label)*

Extension of Authorisation for Minor Use (EAMUs)
- *forest nurseries* 20160785
- *ornamental plant production* 20160785
- *protected ornamentals* 20160785
- *protected tomatoes* 20160784
- *protected watercress* 20160783
- *radishes* 20160782

Approval information
- Propamocarb hydrochloride included in Annex I under EC Regulation 1107/2009

Efficacy guidance
- Chemical is absorbed through roots and translocated throughout plant
- Incorporate in compost before use or drench moist compost or soil before sowing, pricking out, striking cuttings or potting up

FOR FULL CONDITIONS OF USE ALWAYS READ THE PRODUCT LABEL

- Concentrated solution is corrosive to all metals other than stainless steel
- May also be applied in trickle irrigation systems

Restrictions
- Maximum number of treatments 1 per crop for listed brassicas;
- When applied over established seedlings rinse off foliage with water and do not apply under hot, dry conditions
- Do not apply in a recirculating irrigation/drip system
- Store away from seeds and fertilizers
- Consult processor before use on crops grown for processing

Crop-specific information
- Latest use: before transplanting for brassicas

Hazard classification and safety precautions
Hazard Irritant
UN Number N/C
Risk phrases H317
Operator protection A, H, M; U02a, U05a, U08, U19a, U20b
Environmental protection E15a, E34
Storage and disposal D01, D02, D05, D09a, D10c

367 propaquizafop

A phenoxy alkanoic acid foliar acting herbicide for grass weeds in a range of crops
HRAC mode of action code: A

Products

1	Clayton Enigma	Clayton	100 g/l	EC	17391
2	Clayton Enigma	Clayton	100 g/l	EC	17435
3	Clayton Satchmo	Clayton	100 g/l	EC	17061
4	Falcon	Adama	100 g/l	EC	16459
5	Longhorn	AgChem Access	100 g/l	EC	16951
6	Shogun	Adama	100 g/l	EC	16527
7	Snipe	Goldengrass	100 g/l	EC	16601

Uses
- Annual grasses in *all edible seed crops grown outdoors* (off-label), *all non-edible seed crops grown outdoors* (off-label), *game cover* (off-label), *poppies for morphine production* (off-label) [4, 6]; *broad beans, bulb onions, carrots, combining peas, dwarf beans, early potatoes, fodder beet, french beans, linseed, maincrop potatoes, spring field beans, spring oilseed rape, sugar beet, swedes, turnips, winter field beans, winter oilseed rape* [1-7]; *garlic* (off-label), *lupins* (off-label), *red beet* (off-label), *shallots* (off-label) [6]
- Annual meadow grass in *all edible seed crops grown outdoors* (off-label), *all non-edible seed crops grown outdoors* (off-label), *game cover* (off-label), *poppies for morphine production* (off-label), *red beet* (off-label) [6]; *garlic* (off-label), *lupins* (off-label), *shallots* (off-label) [4, 6]
- Grass weeds in *garlic* (off-label), *lupins* (off-label), *shallots* (off-label) [4]
- Perennial grasses in *all edible seed crops grown outdoors* (off-label), *all non-edible seed crops grown outdoors* (off-label), *game cover* (off-label), *poppies for morphine production* (off-label) [4, 6]; *broad beans, bulb onions, carrots, combining peas, dwarf beans, early potatoes, fodder beet, french beans, linseed, maincrop potatoes, spring field beans, spring oilseed rape, sugar beet, swedes, turnips, winter field beans, winter oilseed rape* [1-7]; *garlic* (off-label), *lupins* (off-label), *red beet* (off-label), *shallots* (off-label) [6]
- Volunteer cereals in *poppies for morphine production* (off-label), *red beet* (off-label) [4, 6]
- Volunteer rye in *red beet* (off-label) [4]

Extension of Authorisation for Minor Use (EAMUs)
- *all edible seed crops grown outdoors* 20140751 [4], 20141283 [4], 20141927 [6], 20142509 [6]
- *all non-edible seed crops grown outdoors* 20140751 [4], 20141283 [4], 20141927 [6], 20142509 [6]

Pesticide Profiles

- **game cover** *20140751* [4], *20141283* [4], *20141927* [6], *20142509* [6]
- **garlic** *20140701* [4], *20141281* [4], *20141926* [6]
- **lupins** *20140702* [4], *20141282* [4], *20141928* [6]
- **poppies for morphine production** *20152056* [4], *20160285* [6]
- **red beet** *20151827* [4], *20160286* [6]
- **shallots** *20140701* [4], *20141281* [4], *20141926* [6]

Approval information
- Propaquizafop included in Annex 1 under EC Regulation 1107/2009

Efficacy guidance
- Apply to emerged weeds when they are growing actively with adequate soil moisture
- Activity is slower under cool conditions
- Broad-leaved weeds and any weeds germinating after treatment are not controlled
- Annual meadow grass up to 3 leaves checked at low doses and severely checked at highest dose
- Spray barley cover crops when risk of wind blow has passed and before there is serious competition with the crop
- Various tank mixtures and sequences recommended for broader spectrum weed control in oilseed rape, peas and sugar beet. See label for details
- Severe couch infestations may require a second application at reduced dose when regrowth has 3-4 leaves unfolded
- Products contain surfactants. Tank mixing with adjuvants not required or recommended
- Propaquizafop is an ACCase inhibitor herbicide. To avoid the build up of resistance do not apply products containing an ACCase inhibitor herbicide more than twice to any crop. In addition do not use any product containing propaquizafop in mixture or sequence with any other product containing the same ingredient
- Use these products as part of a resistance management strategy that includes cultural methods of control and does not use ACCase inhibitors as the sole chemical method of grass weed control
- Applying a second product containing an ACCase inhibitor to a crop will increase the risk of resistance development; only use a second ACCase inhibitor to control different weeds at a different timing
- Always follow WRAG guidelines for preventing and managing herbicide resistant weeds. See Section 5 for more information

Restrictions
- Maximum number of treatments 1 or 2 per crop or yr. See label for details
- See label for list of tolerant tree species
- Do not treat seed potatoes

Crop-specific information
- Latest use: before crop flower buds visible for winter oilseed rape, linseed, field beans; before 8 fully expanded leaf stage for spring oilseed rape; before weeds are covered by the crop for potatoes, sugar beet, fodder beet, swedes, turnips; when flower buds visible for peas
- HI onions, carrots, early potatoes, parsnips 4 wk; combining peas, early potatoes 7 wk; sugar beet, fodder beet, maincrop potatoes, swedes, turnips 8 wk; field beans 14 wk
- Application in high temperatures and/or low soil moisture content may cause chlorotic spotting especially on combining peas and field beans
- Overlaps at the highest dose can cause damage from early applications to carrots and parsnips

Following crops guidance
- In the event of a failed treated crop an interval of 2 wk must elapse between the last application and redrilling with winter wheat or winter barley. 4 wk must elapse before sowing oilseed rape, peas or field beans, and 16 wk before sowing ryegrass or oats

Environmental safety
- Dangerous for the environment
- Toxic to aquatic organisms
- Risk to certain non-target insects or other arthropods. See directions for use
- LERAP Category B

FOR FULL CONDITIONS OF USE ALWAYS READ THE PRODUCT LABEL

propiconazole

Hazard classification and safety precautions
Hazard Harmful, Dangerous for the environment
Transport code 9
Packaging group III
UN Number 3082
Risk phrases R20, R36, R38, R40, R51, R53a
Operator protection A, C [1-7]; H [5, 6]; U02a, U05a, U08, U11, U13, U14, U15, U19a, U20b
Environmental protection E15a, E16a, E22b, E34, E38
Storage and disposal D01, D02, D05, D09a, D12a, D14 [1-7]; D10b [1-4, 7]; D12b [5, 6]
Medical advice M05b

368 propiconazole

A systemic, curative and protectant triazole fungicide
FRAC mode of action code: 3

See also azoxystrobin + propiconazole
chlorothalonil + cyproconazole + propiconazole
chlorothalonil + fludioxonil + propiconazole
chlorothalonil + propiconazole
cyproconazole + propiconazole
difenoconazole + propiconazole
fenbuconazole + propiconazole
fenpropidin + propiconazole
fenpropidin + propiconazole + tebuconazole
prochloraz + propiconazole
propiconazole + tebuconazole
propiconazole + trifloxystrobin

Products

1	Banner Maxx	Syngenta	156 g/l	EC	13167
2	Bounty	AgChem Access	250 g/l	EC	14616
3	Bumper 250 EC	Adama	250 g/l	EC	14399
4	GF Propiconazole 250	AgChem Access	250 g/l	EC	17154
5	Mascot Exocet	Rigby Taylor	156 g/l	EC	15598
6	Propi 25 EC	Sharda	250 g/l	EC	14939
7	Propicon 250	Goldengrass	250 g/l	EC	15255
8	Span	ProKlass	156 g/l	EC	17165
9	Spaniel	Pan Amenity	156 g/l	EC	15530

Uses
- Anthracnose in **amenity grassland** (qualified minor use), **managed amenity turf** (qualified minor use) [1, 5, 8, 9]
- Brown patch in **amenity grassland** (qualified minor use), **managed amenity turf** (qualified minor use) [1, 5, 8, 9]
- Brown rust in **spring barley**, **spring wheat**, **winter barley**, **winter rye**, **winter wheat** [2-4, 6, 7]
- Cladosporium leaf blotch in **garlic** (off-label), **onions** (off-label), **shallots** (off-label) [3]
- Crown rust in **grass seed crops**, **permanent grassland**, **spring oats**, **winter oats** [2-4, 6, 7]
- Disease control in **forest nurseries** (off-label), **garlic** (off-label), **honesty** (off-label), **hops** (off-label), **onions** (off-label), **ornamental plant production** (off-label), **protected forest nurseries** (off-label), **protected ornamentals** (off-label), **shallots** (off-label), **soft fruit** (off-label) [3]
- Dollar spot in **amenity grassland**, **managed amenity turf** [1, 5, 8, 9]
- Downy mildew in **garlic** (off-label), **onions** (off-label), **shallots** (off-label) [3]
- Drechslera leaf spot in **grass seed crops**, **permanent grassland** [2-4, 6, 7]
- Fusarium patch in **amenity grassland**, **managed amenity turf** [1, 5, 8, 9]
- Glume blotch in **spring wheat**, **winter rye**, **winter wheat** [2-4, 6, 7]
- Light leaf spot in **spring oilseed rape** (reduction), **winter oilseed rape** (reduction) [2-4, 6, 7]
- Mildew in **grass seed crops**, **permanent grassland** [2-4, 6, 7]

SEE SECTION 3 FOR PRODUCTS ALSO REGISTERED

Pesticide Profiles

- Powdery mildew in **spring barley**, **spring oats**, **spring wheat**, **winter barley**, **winter oats**, **winter rye**, **winter wheat** [2-4, 6, 7]
- Ramularia leaf spots in **sugar beet** *(reduction)* [2-4, 6, 7]
- Rhynchosporium in **grass seed crops**, **permanent grassland**, **spring barley**, **winter barley**, **winter rye** [2-4, 6, 7]
- Rust in **garlic** *(off-label)*, **onions** *(off-label)*, **ornamental plant production** *(off-label)*, **protected ornamentals** *(off-label)*, **shallots** *(off-label)* [3]; **sugar beet** [2-4, 6, 7]
- Septoria leaf blotch in **spring wheat**, **winter rye**, **winter wheat** [2-4, 6, 7]
- Sooty moulds in **spring wheat**, **winter wheat** [2-4, 6, 7]
- White rust in **ornamental plant production** *(off-label)*, **protected ornamentals** *(off-label)* [3]
- Yellow rust in **spring barley**, **spring wheat**, **winter barley**, **winter wheat** [2-4, 6, 7]

Extension of Authorisation for Minor Use (EAMUs)
- **forest nurseries** *20111133* [3], *20141275* [3]
- **garlic** *20090705* [3], *20141272* [3]
- **honesty** *20090706* [3], *20141273* [3]
- **hops** *20111133* [3], *20141275* [3]
- **onions** *20090705* [3], *20141272* [3]
- **ornamental plant production** *20090707* [3], *20141274* [3]
- **protected forest nurseries** *20111133* [3], *20141275* [3]
- **protected ornamentals** *20090707* [3], *20141274* [3]
- **shallots** *20090705* [3], *20141272* [3]
- **soft fruit** *20111133* [3], *20141275* [3]

Approval information
- Propiconazole included in Annex I under EC Regulation 1107/2009
- Accepted by BBPA for use on malting barley

Efficacy guidance
- Best results achieved by applying at early stage of disease development. Recommended spray programmes vary with crop, disease, season, soil type and product.
- For optimum turf quality and control use in conjunction with turf management practices that promote good plant health [1]
- Propiconazole is a DMI fungicide. Resistance to some DMI fungicides has been identified in Septoria leaf blotch which may seriously affect performance of some products. For further advice contact a specialist advisor and visit the Fungicide Resistance Action Group (FRAG)-UK website

Restrictions
- On oilseed rape do not apply during flowering.
- Grass seed crops must be treated in yr of harvest
- A minimum interval of 14 d must elapse between treatments on leeks and 21 d on sugar beet
- Avoid spraying crops or turf under stress, eg during cold weather or periods of frost

Crop-specific information
- Latest use: before 4 pairs of true leaves for honesty
- HI cereals, grass seed crops 35 d; oilseed rape, sugar beet, garlic, onions, shallots, grass for ensiling 28 d
- May be used on all common turf grass species [1]

Environmental safety
- Dangerous for the environment
- Toxic to aquatic organisms
- Risk to non-target insects or other arthropods [1]
- When treating turf with vehicle mounted or drawn hydraulic boom sprayers avoid spraying within 5 m of unmanaged land to reduce effects on non-target arthropods [1]

Hazard classification and safety precautions
  Hazard Harmful [1, 5, 8, 9]; Dangerous for the environment [1-9]; Very toxic to aquatic organisms [4]
  Transport code 9
  Packaging group III

FOR FULL CONDITIONS OF USE ALWAYS READ THE PRODUCT LABEL

propoxycarbazone-sodium

UN Number 3082
Risk phrases R20, R43, R52 [1, 5, 8, 9]; R22a [7]; R51 [2, 3, 6, 7]; R53a [1-3, 5-9]
Operator protection A [1-5, 7-9]; C [2-4, 7]; H [1, 5, 7-9]; U02a, U05a, U19a, U20b [1-9]; U08 [2-4, 6, 7]; U09a, U14 [1, 5, 8, 9]
Environmental protection E15a, E34 [2-4, 6, 7]; E15b, E22c, E38 [1, 5, 8, 9]; H410 [4]
Storage and disposal D01, D02, D05, D09a [1-9]; D10b [2-4, 6, 7]; D10c, D12a [1, 5, 8, 9]
Treated seed S06a [2-4, 6, 7]
Medical advice M05a [2-4, 6, 7]

369   propoxycarbazone-sodium

A sulfonylaminocarbonyl triazolinone residual grass weed herbicide for winter wheat
HRAC mode of action code: B

See also iodosulfuron-methyl-sodium + propoxycarbazone-sodium
       mesosulfuron-methyl + propoxycarbazone-sodium

Products
  Attribut   Interfarm   70% w/w   SG   14749

Uses
- Annual grasses in **miscanthus** *(off-label)*
- Blackgrass in **winter wheat**
- Couch in **miscanthus** *(off-label)*, **winter wheat**

Extension of Authorisation for Minor Use (EAMUs)
- **miscanthus** 20093179

Approval information
- Propoxycarbazone-sodium included in Annex I under EC Regulation 1107/2009

Efficacy guidance
- Activity by root and foliar absorption but depends on presence of sufficient soil moisture to ensure root uptake by weeds. Control is enhanced by use of an adjuvant to encourage foliar uptake
- Best results obtained from treatments applied when weed grasses are growing actively. Symptoms may not become apparent for 3-4 wk after application
- Ensure good even spray coverage
- To achieve best control of couch and reduction of infestation in subsequent crop treat between 2 true leaves and first node stage of the weed
- Blackgrass should be treated after using specialist blackgrass herbicides with a different mode of action
- Propoxycarbazone-sodium is a member of the ALS-inhibitor group of herbicides. To avoid the build up of resistance do not use any product containing an ALS-inhibitor herbicide with claims for control of grass weeds more than once on any crop
- Use these products as part of a resistance management strategy that includes cultural methods of control and does not use ALS inhibitors as the sole chemical method of grass weed control

Restrictions
- Maximum total dose equivalent to one full dose treatment
- Do not use in a programme with other aceto-lactase synthesis (ALS) inhibitors
- Do not apply when temperature near or below freezing
- Avoid treatment under dry soil conditions

Crop-specific information
- Latest use: third node detectable in wheat (GS 33)

Following crops guidance
- Only winter wheat, field beans or winter barley may be sown in the autumn following spring treatment
- Any crop may be grown in the spring on land treated during the previous calendar yr

SEE SECTION 3 FOR PRODUCTS ALSO REGISTERED

Pesticide Profiles

Environmental safety
- Dangerous for the environment
- Very toxic to aquatic organisms
- Take extreme care to avoid drift onto adjacent plants or land as this could result in severe damage
- LERAP Category B

Hazard classification and safety precautions
Hazard Dangerous for the environment
Transport code 9
Packaging group III
UN Number 3077
Risk phrases R50, R53a
Operator protection A; U05a, U20b
Environmental protection E15b, E16a, E16b, E38
Storage and disposal D01, D02, D09a, D11a, D12a

370 propyzamide

A residual benzamide herbicide for use in a wide range of crops
HRAC mode of action code: K1

See also aminopyralid + propyzamide
  clopyralid + propyzamide

Products

1	Artax Flo	Stefes	400 g/l	SC	16317
2	Barclay Propyz	Barclay	400 g/l	SC	15083
3	Careca	UPL Europe	500 g/l	SC	16976
4	Cohort	Adama	400 g/l	SC	15035
5	Conform	AgChem Access	400 g/l	SC	14602
6	Judo	Headland	400 g/l	SC	15346
7	Kerb Flo	Dow	400 g/l	SC	13716
8	Kerb Flo 500	Dow	500 g/l	SC	15586
9	Kerb Granules	Barclay	4% w/w	GR	14213
10	Levada	Certis	400 g/l	SC	15743
11	Master Prop 40	Generica	400 g/l	SC	15829
12	Pizza 400 SC	Goldengrass	400 g/l	SC	14430
13	Pizza Flo	Goldengrass	400 g/l	SC	16600
14	Proper Flo	Globachem	400 g/l	SC	15102
15	Relva	Belcrop	400 g/l	SC	14873
16	Setanta 50 WP	Mitsui	50% w/w	WP	16181
17	Setanta Flo	Certis	400 g/l	SC	15791
18	Solitaire	Certis	400 g/l	SC	15792
19	Standon Santa Fe 50 WP	Standon	50% w/w	WP	14966
20	Stroller	AgChem Access	400 g/l	SC	14603
21	Stymie	Becesane	400 g/l	SC	16328
22	Zamide 400	Becesane	400 g/l	SC	16002

Uses
- Annual and perennial weeds in **amenity vegetation** [10, 13, 17, 18]
- Annual dicotyledons in **all edible seed crops grown outdoors** *(off-label)*, **all non-edible seed crops grown outdoors** *(off-label)*, **almonds** *(off-label)*, **bilberries** *(off-label)*, **blueberries** *(off-label)*, **bog myrtle** *(off-label)*, **broccoli** *(off-label)*, **calabrese** *(off-label)*, **cauliflowers** *(off-label)*, **cherries** *(off-label)*, **chestnuts** *(off-label)*, **chicory root** *(off-label)*, **cob nuts** *(off-label)*, **corn gromwell** *(off-label)*, **courgettes** *(off-label)*, **cranberries** *(off-label)*, **cress** *(off-label)*, **endives** *(off-label)*, **frise** *(off-label)*, **game cover** *(off-label)*, **hazel nuts** *(off-label)*, **herbs (see appendix 6)** *(off-label)*, **hops** *(off-label)*, **lamb's lettuce** *(off-label)*, **leaf brassicas** *(off-label)*, **marrows** *(off-label)*, **mirabelles** *(off-label)*, **protected endives** *(off-label)*, **protected forest nurseries** *(off-label)*, **protected herbs (see appendix 6)** *(off-label)*, **protected lettuce** *(off-label)*, **pumpkins** *(off-label)*, **quinces** *(off-label)*, **radicchio** *(off-label)*, **salad brassicas** *(off-label)*, **scarole** *(off-label)*,

FOR FULL CONDITIONS OF USE ALWAYS READ THE PRODUCT LABEL

propyzamide

squashes *(off-label)*, **table grapes** *(off-label)*, **walnuts** *(off-label)*, **wine grapes** *(off-label)* [7]; **amenity vegetation** [2-5, 7, 12, 19-22]; **apple orchards, pear orchards, plums, rhubarb** [2-5, 7, 10, 12, 13, 17-22]; **blackberries, blackcurrants, gooseberries, loganberries, raspberries** (England only), **redcurrants, strawberries** [2, 4, 5, 7, 10, 12, 13, 17-22]; **chicory, endives, witloof** [3]; **clover seed crops** [2-5, 7, 10-13, 17-22]; **farm forestry, forest nurseries** [1-5, 7, 10-13, 16-22]; **fodder rape seed crops, kale seed crops, sugar beet seed crops, turnip seed crops** [2-5, 7, 10-14, 17-22]; **forest** [1-5, 7, 9-14, 16-22]; **hedges** [2-5, 7, 10-13, 16-22]; **lettuce** [1, 3-5, 7, 10-14, 16-22]; **lucerne** [4, 5, 7, 10-13, 17-22]; **red clover, white clover** [14]; **trees and shrubs, woody ornamentals** [9]; **winter field beans** [1, 4, 5, 7, 8, 10-14, 16-22]; **winter oilseed rape** [1-8, 10-22]
- Annual grasses in **all edible seed crops grown outdoors** *(off-label)*, **all non-edible seed crops grown outdoors** *(off-label)*, **almonds** *(off-label)*, **bilberries** *(off-label)*, **blueberries** *(off-label)*, **bog myrtle** *(off-label)*, **broccoli** *(off-label)*, **calabrese** *(off-label)*, **cauliflowers** *(off-label)*, **cherries** *(off-label)*, **chestnuts** *(off-label)*, **chicory root** *(off-label)*, **cob nuts** *(off-label)*, **corn gromwell** *(off-label)*, **courgettes** *(off-label)*, **cranberries** *(off-label)*, **cress** *(off-label)*, **endives** *(off-label)*, **frise** *(off-label)*, **game cover** *(off-label)*, **hazel nuts** *(off-label)*, **herbs (see appendix 6)** *(off-label)*, **hops** *(off-label)*, **lamb's lettuce** *(off-label)*, **leaf brassicas** *(off-label)*, **marrows** *(off-label)*, **mirabelles** *(off-label)*, **ornamental plant production** *(off-label)*, **protected endives** *(off-label)*, **protected forest nurseries** *(off-label)*, **protected herbs (see appendix 6)** *(off-label)*, **protected lettuce** *(off-label)*, **pumpkins** *(off-label)*, **quinces** *(off-label)*, **radicchio** *(off-label)*, **salad brassicas** *(off-label)*, **scarole** *(off-label)*, **squashes** *(off-label)*, **table grapes** *(off-label)*, **walnuts** *(off-label)*, **wine grapes** *(off-label)* [7]; **amenity vegetation** [2-5, 7, 12, 19-22]; **apple orchards, pear orchards, plums, rhubarb** [2-5, 7, 10, 12, 13, 17-22]; **blackberries, blackcurrants, gooseberries, loganberries, raspberries** (England only), **redcurrants, strawberries** [2, 4, 5, 7, 10, 12, 13, 17-22]; **chicory, endives, witloof** [3]; **clover seed crops** [2-5, 7, 10-13, 17-22]; **farm forestry, forest nurseries** [1-5, 7, 10-13, 16-22]; **fodder rape seed crops, kale seed crops, sugar beet seed crops, turnip seed crops** [2-5, 7, 10-14, 17-22]; **forest** [1-5, 7, 9-14, 16-22]; **hedges** [2-5, 7, 10-13, 16-22]; **lettuce** [1, 3-5, 7, 10-14, 16-22]; **lucerne** [4, 5, 7, 10-13, 17-22]; **red clover, white clover** [14]; **trees and shrubs, woody ornamentals** [9]; **winter field beans** [1, 4, 5, 7, 10-14, 16-22]; **winter oilseed rape** [1-7, 10-22]
- Annual meadow grass in **winter field beans, winter oilseed rape** [8]
- Blackgrass in **winter field beans, winter oilseed rape** [8]
- Fat hen in **ornamental plant production** *(off-label)* [7]
- Perennial grasses in **all edible seed crops grown outdoors** *(off-label)*, **all non-edible seed crops grown outdoors** *(off-label)*, **almonds** *(off-label)*, **bilberries** *(off-label)*, **blueberries** *(off-label)*, **bog myrtle** *(off-label)*, **broccoli** *(off-label)*, **calabrese** *(off-label)*, **cauliflowers** *(off-label)*, **cherries** *(off-label)*, **chestnuts** *(off-label)*, **chicory root** *(off-label)*, **cob nuts** *(off-label)*, **corn gromwell** *(off-label)*, **courgettes** *(off-label)*, **cranberries** *(off-label)*, **cress** *(off-label)*, **endives** *(off-label)*, **frise** *(off-label)*, **game cover** *(off-label)*, **hazel nuts** *(off-label)*, **herbs (see appendix 6)** *(off-label)*, **hops** *(off-label)*, **lamb's lettuce** *(off-label)*, **leaf brassicas** *(off-label)*, **marrows** *(off-label)*, **mirabelles** *(off-label)*, **protected endives** *(off-label)*, **protected forest nurseries** *(off-label)*, **protected herbs (see appendix 6)** *(off-label)*, **protected lettuce** *(off-label)*, **pumpkins** *(off-label)*, **quinces** *(off-label)*, **radicchio** *(off-label)*, **salad brassicas** *(off-label)*, **scarole** *(off-label)*, **squashes** *(off-label)*, **table grapes** *(off-label)*, **walnuts** *(off-label)*, **wine grapes** *(off-label)* [7]; **amenity vegetation** [2-5, 7, 12, 19-22]; **apple orchards, pear orchards, plums, rhubarb** [2-5, 7, 10, 12, 13, 17-22]; **blackberries, blackcurrants, gooseberries, loganberries, raspberries** (England only), **redcurrants, strawberries** [2, 4, 5, 7, 10, 12, 13, 17-22]; **chicory, endives, witloof** [3]; **clover seed crops, farm forestry, forest nurseries, hedges** [2-5, 7, 10-13, 17-22]; **fodder rape seed crops, kale seed crops, sugar beet seed crops, turnip seed crops** [2-5, 7, 10-14, 17-22]; **forest** [2-5, 7, 9-14, 17-22]; **lettuce** [3-5, 7, 10-14, 17-22]; **lucerne** [4, 5, 7, 10-13, 17-22]; **red clover, white clover** [14]; **trees and shrubs, woody ornamentals** [9]; **winter field beans** [4, 5, 7, 10-14, 17-22]; **winter oilseed rape** [2-7, 10-15, 17-22]
- Volunteer cereals in **winter field beans, winter oilseed rape** [8]
- Wild oats in **winter field beans, winter oilseed rape** [8]

Extension of Authorisation for Minor Use (EAMUs)
- **all edible seed crops grown outdoors** 20082942 [7]
- **all non-edible seed crops grown outdoors** 20082942 [7]
- **almonds** 20082420 [7]

SEE SECTION 3 FOR PRODUCTS ALSO REGISTERED

Pesticide Profiles

- ***bilberries*** *20082419* [7]
- ***blueberries*** *20082419* [7]
- ***bog myrtle*** *20120509* [7]
- ***broccoli*** *20091902* [7]
- ***calabrese*** *20091902* [7]
- ***cauliflowers*** *20091902* [7]
- ***cherries*** *20082418* [7]
- ***chestnuts*** *20082420* [7]
- ***chicory root*** *20091530* [7]
- ***cob nuts*** *20082420* [7]
- ***corn gromwell*** *20122395* [7]
- ***courgettes*** *20082416* [7]
- ***cranberries*** *20082419* [7]
- ***cress*** *20082410* [7]
- ***endives*** *20082410* [7]
- ***frise*** *20082411* [7]
- ***game cover*** *20082942* [7]
- ***hazel nuts*** *20082420* [7]
- ***herbs (see appendix 6)*** *20082412* [7]
- ***hops*** *20082414* [7]
- ***lamb's lettuce*** *20082410* [7]
- ***leaf brassicas*** *20082410* [7]
- ***marrows*** *20082416* [7]
- ***mirabelles*** *20082418* [7]
- ***ornamental plant production*** *20130207* [7]
- ***protected endives*** *20082415* [7]
- ***protected forest nurseries*** *20082942* [7]
- ***protected herbs (see appendix 6)*** *20082412* [7]
- ***protected lettuce*** *20082415* [7]
- ***pumpkins*** *20082416* [7]
- ***quinces*** *20082413* [7]
- ***radicchio*** *20082411* [7]
- ***salad brassicas*** *20082410* [7]
- ***scarole*** *20082411* [7]
- ***squashes*** *20082416* [7]
- ***table grapes*** *20082417* [7]
- ***walnuts*** *20082420* [7]
- ***wine grapes*** *20082417* [7]

Approval information
- Propyzamide included in Annex I under EC Regulation 1107/2009
- Some products may be applied through CDA equipment. See labels for details
- Accepted by BBPA for use on hops

Efficacy guidance
- Active via root uptake. Weeds controlled from germination to young seedling stage, some species (including many grasses) also when established
- Best results achieved by winter application to fine, firm, moist soil. Rain is required after application if soil dry
- Uptake is slow and and may take up to 12 wk
- Excessive organic debris or ploughed-up turf may reduce efficacy
- For heavy couch infestations a repeat application may be needed in following winter
- Always follow WRAG guidelines for preventing and managing herbicide resistant weeds. See Section 5 for more information

Restrictions
- Maximum number of treatments 1 per crop or yr
- Maximum total dose equivalent to one full dose treatment for all crops
- Do not treat protected crops

FOR FULL CONDITIONS OF USE ALWAYS READ THE PRODUCT LABEL

propyzamide

- Apply to listed edible crops only between 1 Oct and the date specified as the latest time of application (except lettuce)
- Do not apply in windy weather and avoid drift onto non-target crops
- Do not use on soils with more than 10% organic matter except in forestry

Crop-specific information
- Latest use: labels vary but normally before 31 Dec in year before harvest for rhubarb, lucerne, strawberries and winter field beans; before 31 Jan for other crops
- HI: 6 wk for edible crops
- Apply as soon as possible after 3-true leaf stage of oilseed rape (GS 1,3) and seed brassicas, after 4-leaf stage of sugar beet for seed, within 7 d after sowing but before emergence for field beans, after perennial crops established for at least 1 season, strawberries after 1 yr
- Only apply to strawberries on heavy soils. Do not use on matted row crops
- Only apply to field beans on medium and heavy soils
- Only apply to established lucerne not less than 7 d after last cut
- In lettuce lightly incorporate in top 25 mm pre-drilling or irrigate on dry soil
- See label for lists of ornamental and forest species which may be treated

Following crops guidance
- Following an application between 1 Apr and 31 Jul at any dose the following minimum intervals must be observed before sowing the next crop: lettuce 0 wk; broad beans, chicory, clover, field beans, lucerne, radishes, peas 5 wk; brassicas, celery, leeks, oilseed rape, onions, parsley, parsnips 10 wk
- Following an application between 1 Aug and 31 Mar at any dose the following minimum intervals must be observed before sowing the next crop: lettuce 0 wk; broad beans, chicory, clover, field beans, lucerne, radishes, peas 10 wk; brassicas, celery, leeks, oilseed rape, onions, parsley, parsnips 25 wk or after 15 Jun, whichever occurs sooner
- Cereals or grasses or other crops not listed may be sown 30 wk after treatment up to 840 g.a.i/ha between 1 Aug and 31 Mar or 40 wk after treatment at higher doses at any time and after mouldboard ploughing to at least 15 cm
- A period of at least 9 mth must elapse between applications of propyzamide to the same land

Environmental safety
- Dangerous for the environment
- Very toxic to aquatic organisms
- Some pesticides pose a greater threat of contamination of water than others and propyzamide is one of these pesticides. Take special care when applying propyzamide near water and do not apply if heavy rain is forecast

Hazard classification and safety precautions
Hazard Harmful, Dangerous for the environment [1-22]; Very toxic to aquatic organisms [10]
Transport code 9
Packaging group III
UN Number 3077 [9, 16, 19]; 3082 [1-8, 10-15, 17, 18, 20-22]
Risk phrases H351 [6-8, 10, 18]; R40, R53a [1-5, 9, 11-17, 19-22]; R50 [1, 4, 5, 11-17, 20-22]; R51 [2, 3, 9, 19]
Operator protection A, H [1-22]; D [2, 4-7, 11, 12, 16, 19-22]; M [2, 4-7, 10-13, 15-22]; U05a [1, 3, 9, 11]; U20c [1-8, 10-22]
Environmental protection E15a [1-9, 11, 12, 14, 20-22]; E15b [10, 13, 15-19]; E34 [3, 14, 16, 19]; E38 [1-15, 17, 18, 20-22]; E39 [10, 13, 15, 17, 18]; H410 [10, 18]; H411 [6-8, 18]
Consumer protection C02a [1-7, 11, 12, 14, 16, 19-22] (6 wk)
Storage and disposal D01, D02 [1, 3, 9, 11]; D05, D11a [1-8, 11, 12, 14, 16, 19-22]; D07 [3, 14]; D09a [1-9, 11, 12, 14, 16, 19-22]; D12a [1-9, 11, 12, 14, 20-22]
Medical advice M03 [9]; M05a [11]

SEE SECTION 3 FOR PRODUCTS ALSO REGISTERED

371 proquinazid

A quinazolinone fungicide for powdery mildew control in cereals
FRAC mode of action code: U7

See also chlorothalonil + proquinazid
prochloraz + proquinazid + tebuconazole

Products

1	Justice	DuPont	200 g/l	EC	12835
2	Talius	DuPont	200 g/l	EC	12752

Uses
- Powdery mildew in **courgettes** *(off-label)*, **cucumbers** *(off-label)*, **forest nurseries** *(off-label)*, **marrows** *(off-label)*, **protected aubergines** *(off-label)*, **protected courgettes** *(off-label)*, **protected cucumbers** *(off-label)*, **protected marrows** *(off-label)*, **protected pumpkins** *(off-label)*, **protected squashes** *(off-label)*, **protected tomatoes** *(off-label)*, **pumpkins** *(off-label)*, **soft fruit** *(off-label)*, **squashes** *(off-label)*, **top fruit** *(off-label)* [2]; **durum wheat** *(off-label)*, **spring barley, spring oats, spring rye, spring wheat, triticale, winter barley, winter oats, winter rye, winter wheat** [1, 2]; **grapevines** *(off-label)* [1]

Extension of Authorisation for Minor Use (EAMUs)
- **courgettes** *20152627* [2]
- **cucumbers** *20152627* [2]
- **durum wheat** *20090418* [1], *20090419* [2]
- **forest nurseries** *20090420* [2]
- **grapevines** *20111763* [1]
- **marrows** *20152627* [2]
- **protected aubergines** *20152627* [2]
- **protected courgettes** *20152627* [2]
- **protected cucumbers** *20152627* [2]
- **protected marrows** *20152627* [2]
- **protected pumpkins** *20152627* [2]
- **protected squashes** *20152627* [2]
- **protected tomatoes** *20152627* [2]
- **pumpkins** *20152627* [2]
- **soft fruit** *20090420* [2]
- **squashes** *20152627* [2]
- **top fruit** *20090420* [2]

Approval information
- Proquinazid included in Annex 1 under EC Regulation 1107/2009
- Accepted by BBPA on malting barley

Efficacy guidance
- Best results obtained from preventive treatment before disease is established in the crop
- Where mildew has already spread to new growth a tank mix with a curative fungicide with an alternative mode of action should be used
- Use as part of an integrated crop management (ICM) strategy incorporating other methods of control or fungicides with different modes of action

Restrictions
- Maximum number of treatments 2 per crop
- Do not apply to any crop suffering from stress from any cause
- Avoid application in either frosty or hot, sunny conditions

Crop-specific information
- Latest use: before beginning of heading (GS 49) for barley, oats, rye, triticale; before full flowering (GS 65) for wheat

Environmental safety
- Dangerous for the environment

FOR FULL CONDITIONS OF USE ALWAYS READ THE PRODUCT LABEL

prosulfocarb

- Toxic to aquatic organisms
- Dangerous to fish or other aquatic life. Do not contaminate surface waters or ditches with chemical or used container
- LERAP Category B

Hazard classification and safety precautions
Hazard Harmful, Dangerous for the environment
Transport code 9
Packaging group III
UN Number 3082
Risk phrases H315, H318, H351
Operator protection A, C, H; U05a, U11
Environmental protection E13b, E16a, E16b, E34, E38, H410
Storage and disposal D01, D02, D05, D09a, D10b, D11a, D12a
Medical advice M03, M05a

372   prosulfocarb

A thiocarbamate herbicide for grass and broad-leaved weed control in cereals and potatoes
HRAC mode of action code: N

See also clodinafop-propargyl + prosulfocarb

Products

1	Bokken	Syngenta	800 g/l	EC	17557
2	Clayton Obey	Clayton	800 g/l	EC	16405
3	Clayton Obey	Clayton	800 g/l	EC	16626
4	Defy	Syngenta	800 g/l	EC	16202
5	Quidam	UPL Europe	800 g/l	EC	17595

Uses
- Annual dicotyledons in **carrots** *(off-label)*, **celeriac** *(off-label)*, **celery (outdoor)** *(off-label)*, **durum wheat** *(off-label)*, **forest nurseries** *(off-label)*, **game cover** *(off-label)*, **garlic** *(off-label)*, **horseradish** *(off-label)*, **leeks** *(off-label)*, **miscanthus** *(off-label)*, **onion sets** *(off-label)*, **ornamental plant production** *(off-label)*, **parsley root** *(off-label)*, **parsnips** *(off-label)*, **poppies for morphine production** *(off-label)*, **rye** *(off-label)*, **salsify** *(off-label)*, **shallots** *(off-label)*, **spring barley** *(off-label)*, **spring field beans** *(off-label)*, **spring onions** *(off-label)*, **top fruit** *(off-label)*, **triticale** *(off-label)*, **winter field beans** *(off-label)*, **winter linseed** *(off-label)* [4]; **potatoes**, **winter barley**, **winter wheat** [1-5]
- Annual grasses in **carrots** *(off-label)*, **celeriac** *(off-label)*, **celery (outdoor)** *(off-label)*, **durum wheat** *(off-label)*, **forest nurseries** *(off-label)*, **game cover** *(off-label)*, **garlic** *(off-label)*, **horseradish** *(off-label)*, **leeks** *(off-label)*, **miscanthus** *(off-label)*, **onion sets** *(off-label)*, **ornamental plant production** *(off-label)*, **parsley root** *(off-label)*, **parsnips** *(off-label)*, **rye** *(off-label)*, **salsify** *(off-label)*, **shallots** *(off-label)*, **spring onions** *(off-label)*, **top fruit** *(off-label)*, **triticale** *(off-label)*, **winter linseed** *(off-label)* [4]
- Annual meadow grass in **carrots** *(off-label)*, **celeriac** *(off-label)*, **celery (outdoor)** *(off-label)*, **durum wheat** *(off-label)*, **forest nurseries** *(off-label)*, **game cover** *(off-label)*, **horseradish** *(off-label)*, **miscanthus** *(off-label)*, **ornamental plant production** *(off-label)*, **parsley root** *(off-label)*, **parsnips** *(off-label)*, **poppies for morphine production** *(off-label)*, **rye** *(off-label)*, **salsify** *(off-label)*, **spring barley** *(off-label)*, **top fruit** *(off-label)*, **triticale** *(off-label)*, **winter linseed** *(off-label)* [4]; **potatoes**, **winter barley**, **winter wheat** [1-5]
- Black nightshade in **herbs (see appendix 6)** *(off-label)* [4]
- Chickweed in **celeriac** *(off-label)* [4]; **potatoes**, **winter barley**, **winter wheat** [1-5]
- Cleavers in **carrots** *(off-label)*, **herbs (see appendix 6)** *(off-label)*, **horseradish** *(off-label)*, **parsley root** *(off-label)*, **parsnips** *(off-label)*, **salsify** *(off-label)*, **spring field beans** *(off-label)*, **winter field beans** *(off-label)* [4]; **potatoes**, **winter barley**, **winter wheat** [1-5]
- Fat hen in **carrots** *(off-label)*, **celeriac** *(off-label)*, **herbs (see appendix 6)** *(off-label)*, **horseradish** *(off-label)*, **parsley root** *(off-label)*, **parsnips** *(off-label)*, **salsify** *(off-label)* [4]
- Field speedwell in **potatoes**, **winter barley**, **winter wheat** [1-5]

SEE SECTION 3 FOR PRODUCTS ALSO REGISTERED

Pesticide Profiles

- Fumitory in **herbs (see appendix 6)** *(off-label)* [4]
- Ivy-leaved speedwell in **potatoes**, **winter barley**, **winter wheat** [1-5]
- Knotgrass in **carrots** *(off-label)*, **herbs (see appendix 6)** *(off-label)*, **horseradish** *(off-label)*, **parsley root** *(off-label)*, **parsnips** *(off-label)*, **salsify** *(off-label)* [4]
- Loose silky bent in **potatoes**, **winter barley**, **winter wheat** [1-5]
- Mayweeds in **carrots** *(off-label)*, **celeriac** *(off-label)*, **herbs (see appendix 6)** *(off-label)*, **horseradish** *(off-label)*, **parsley root** *(off-label)*, **parsnips** *(off-label)*, **salsify** *(off-label)* [4]
- Polygonums in **celeriac** *(off-label)* [4]
- Rough-stalked meadow grass in **potatoes**, **winter barley**, **winter wheat** [1-5]
- Thistles in **celeriac** *(off-label)* [4]

Extension of Authorisation for Minor Use (EAMUs)
- **carrots** *20131354* [4]
- **celeriac** *20131355* [4]
- **celery (outdoor)** *20131353* [4]
- **durum wheat** *20131427* [4]
- **forest nurseries** *20131432* [4]
- **game cover** *20131430* [4]
- **garlic** *20131357* [4]
- **herbs (see appendix 6)** *20152809* [4]
- **horseradish** *20131354* [4]
- **leeks** *20131357* [4]
- **miscanthus** *20131430* [4]
- **onion sets** *20131357* [4]
- **ornamental plant production** *20131431* [4]
- **parsley root** *20131354* [4]
- **parsnips** *20131354* [4]
- **poppies for morphine production** *20131385* [4]
- **rye** *20131429* [4]
- **salsify** *20131354* [4]
- **shallots** *20131357* [4]
- **spring barley** *20131373* [4]
- **spring field beans** *20131356* [4]
- **spring onions** *20131357* [4]
- **top fruit** *20131433* [4]
- **triticale** *20131429* [4]
- **winter field beans** *20131356* [4]
- **winter linseed** *20131428* [4]

Approval information
- Prosulfocarb included in Annex I under EC Regulation 1107/2009
- Accepted by BBPA for use on malting barley

Efficacy guidance
- Best results obtained in cereals from treatment of crops in a firm moist seedbed, free from clods
- Pre-emergence use will reduce blackgrass populations but the product should only be used against this weed as part of a management strategy involving sequences with products of alternative modes of action
- Always follow WRAG guidelines for preventing and managing herbicide resistant weeds. See Section 5 for more information
- Can be used up to 10% emergence in potatoes but must be mixed with diquat [4]

Restrictions
- Maximum number of treatments 1 per crop for winter barley, winter wheat and potatoes
- Do not apply to crops under stress from any cause. Transient yellowing can occur from which recovery is complete
- Winter cereals must be covered by 3 cm of settled soil

Crop-specific information
- Latest use: at emergence (soil rising over emerging potato shoots) for potatoes, up to and including early tillering (GS 21) for winter barley, winter wheat

FOR FULL CONDITIONS OF USE ALWAYS READ THE PRODUCT LABEL

- When applied pre-emergence to cereals crop emergence may occasionally be slowed down but yield is not affected
- For potatoes, complete ridge formation before application and do not disturb treated soil afterwards
- Peas are severely damaged or killed.

Following crops guidance
- Do not sow field or broad beans within 12 mth of treatment
- In the event of failure of a treated cereal crop, winter wheat, winter barley or transplanted brassicas may be re-sown immediately. Seed brassicas can be sown 14 weeks after application provided that the land is ploughed first. In the following spring sunflowers, maize, flax, spring cereals, peas, oilseed rape or soya beans may be sown without ploughing, and carrots, lettuce, onions, sugar beet or potatoes may be sown or planted after ploughing

Environmental safety
- Dangerous for the environment
- Very toxic to aquatic organisms
- LERAP Category B

Hazard classification and safety precautions
Hazard Irritant, Dangerous for the environment
Transport code 9
Packaging group III
UN Number 3082
Risk phrases H304, H315, H317, H319 [1, 4, 5]; R38, R43, R50, R53a [2, 3]
Operator protection A, C, H; U02a, U05a, U08, U14, U15, U20b
Environmental protection E15b, E16a, E16b, E38 [1-5]; H410 [1, 4, 5]
Storage and disposal D01, D02, D05, D09a, D10c, D12a
Medical advice M05a

373 prosulfuron

A contact and residual sulfonyl urea herbicide for use in maize crops
HRAC mode of action code: B

See also bromoxynil + prosulfuron
dicamba + prosulfuron

Products
1	Clayton Kibo	Clayton	75% w/w	WG	15822
2	Peak	Syngenta	75% w/w	WG	15521

Uses
- Annual dicotyledons in **forage maize**, **grain maize** [1, 2]; **game cover** *(off-label)* [2]
- Black bindweed in **forage maize**, **grain maize** [1, 2]
- Chickweed in **forage maize**, **grain maize** [1, 2]
- Fumitory in **forage maize**, **grain maize** [1, 2]
- Groundsel in **forage maize**, **grain maize** [1, 2]
- Knotgrass in **forage maize**, **grain maize** [1, 2]
- Mayweeds in **forage maize**, **grain maize** [1, 2]
- Redshank in **forage maize**, **grain maize** [1, 2]
- Scarlet pimpernel in **forage maize**, **grain maize** [1, 2]
- Shepherd's purse in **forage maize**, **grain maize** [1, 2]
- Sowthistle in **forage maize**, **grain maize** [1, 2]

Extension of Authorisation for Minor Use (EAMUs)
- *game cover* 20120906 [2]

Approval information
- Prosulfuron included in Annex I under EC Regulation 1107/2009

SEE SECTION 3 FOR PRODUCTS ALSO REGISTERED

Pesticide Profiles

Efficacy guidance
- For optimum efficacy, apply when weeds are at the 2 - 4 leaf stage

Restrictions
- Do not apply to forage maize or grain maize grown for seed production
- Do not apply with organo-phosphate insecticides

Following crops guidance
- In the event of crop failure, wait 4 weeks after treatment and then re-sow
- After normal harvest wheat, barley and winter beans may be sown as a following crop in the autumn once the soil has been ploughed to 15 cms. In spring, wheat, barley, peas or beans may be sown but do not sow any other crop at this time.

Environmental safety
- LERAP Category B

Hazard classification and safety precautions
    Hazard Harmful, Dangerous for the environment
    Transport code 9
    Packaging group III
    UN Number 3077
    Risk phrases R22a, R50, R53a
    Operator protection A, H; U05a, U14, U20c
    Environmental protection E15b, E16a, E38
    Storage and disposal D01, D02, D05, D09b, D10c, D12a

374   prothioconazole

A systemic, protectant and curative triazole fungicide
FRAC mode of action code: 3

See also bixafen + fluopyram + prothioconazole
        bixafen + fluoxastrobin + prothioconazole
        bixafen + prothioconazole
        bixafen + prothioconazole + spiroxamine
        bixafen + prothioconazole + tebuconazole
        clothianidin + prothioconazole
        clothianidin + prothioconazole + tebuconazole + triazoxide
        fluopyram + prothioconazole
        fluopyram + prothioconazole + tebuconazole
        fluoxastrobin + prothioconazole
        fluoxastrobin + prothioconazole + tebuconazole
        fluoxastrobin + prothioconazole + trifloxystrobin

Products

1	Proline 275	Bayer CropScience	275 g/l	EC	14790
2	Rudis	Bayer CropScience	480 g/l	SC	14122

Uses
- Alternaria in **broccoli, brussels sprouts, cabbages, calabrese, cauliflowers** [2]
- Alternaria blight in **carrots, parsnips, swedes, turnips** [2]
- Brown rust in **spring barley, spring wheat, winter barley, winter rye, winter wheat** [1]
- Crown rust in **spring oats, winter oats** [1]
- Disease control in **mustard** *(off-label)* [1]
- Eyespot in **spring barley, spring oats, spring wheat, winter barley, winter oats, winter rye, winter wheat** [1]
- Glume blotch in **spring wheat, winter wheat** [1]
- Late ear diseases in **spring barley, spring wheat, winter barley, winter wheat** [1]
- Leaf blotch in **leeks** *(useful reduction)* [2]
- Light leaf spot in **broccoli, brussels sprouts, cabbages, calabrese, cauliflowers** [2]; **winter oilseed rape** [1]

FOR FULL CONDITIONS OF USE ALWAYS READ THE PRODUCT LABEL

- Net blotch in *spring barley, winter barley* [1]
- Phoma in *broccoli, brussels sprouts, cabbages, calabrese, cauliflowers* [2]; *corn gromwell* (off-label), *winter oilseed rape* [1]
- Powdery mildew in *broccoli, brussels sprouts, cabbages, calabrese, carrots, cauliflowers, parsnips, swedes, turnips* [2]; *corn gromwell* (off-label), *spring barley, spring oats, spring wheat, winter barley, winter oats, winter rye, winter wheat* [1]
- Purple blotch in *leeks* [2]
- Rhynchosporium in *spring barley, winter barley, winter rye* [1]
- Ring spot in *broccoli, brussels sprouts, cabbages, calabrese, cauliflowers* [2]
- Rust in *leeks* [2]
- Sclerotinia in *corn gromwell* (off-label) [1]
- Sclerotinia rot in *carrots, parsnips* [2]
- Sclerotinia stem rot in *winter oilseed rape* [1]
- Septoria leaf blotch in *spring wheat, winter wheat* [1]
- Stem canker in *winter oilseed rape* [1]
- Stemphylium in *leeks* (useful reduction) [2]
- Tan spot in *spring wheat, winter wheat* [1]
- Yellow rust in *spring wheat, winter barley, winter wheat* [1]

Extension of Authorisation for Minor Use (EAMUs)
- *corn gromwell* 20152427 [1]
- *mustard* 20111003 [1]

Approval information
- Accepted by BBPA for use on malting barley
- Prothioconazole included in Annex I under EC Regulation 1107/2009

Efficacy guidance
- Seed treatments must be applied by manufacturer's recommended treatment application equipment
- Treated cereal seed should preferably be drilled in the same season
- Follow-up treatments will be needed later in the season to give protection against air-borne and splash-borne diseases
- Best results on cereal foliar diseases obtained from treatment at early stages of disease development. Further treatment may be needed if disease attack is prolonged [1]
- Foliar applications to established infections of any disease are likely to be less effective [1]
- Best control of cereal ear diseases obtained by treatment during ear emergence [1]
- Treat oilseed rape at early to full flower [1]
- Prothioconazole is a DMI fungicide. Resistance to some DMI fungicides has been identified in Septoria leaf blotch which may seriously affect performance of some products. For further advice contact a specialist advisor and visit the Fungicide Resistance Action Group (FRAG)-UK website

Restrictions
- Maximum number of seed treatments one per batch
- Maximum total dose equivalent to two full dose treatments on barley, oats and oilseed rape, three full dose treatments on wheat and rye [1]
- Seed treatment must be fully re-dispersed and homogeneous before use
- Do not use on seed with more than 16% moisture content, or on sprouted, cracked or skinned seed
- All seed batches should be tested to ensure they are suitable for treatment
- Treated winter barley seed must be used within the season of treatment; treated winter wheat seed should preferably be drilled in the same season
- Do not make repeated treatments of the product alone to the same crop against pathogens such as powdery mildew. Use tank mixtures or alternate with fungicides having a different mode of action

Crop-specific information
- Latest use: pre-drilling for seed treatments; before grain milky ripe for foliar sprays on winter rye, winter wheat; beginning of flowering for barley, oats [1]
- HI 56 d for winter oilseed rape [1]

SEE SECTION 3 FOR PRODUCTS ALSO REGISTERED

Pesticide Profiles

Environmental safety
- Dangerous for the environment [1]
- Toxic to aquatic organisms [1]
- Harmful to aquatic organisms
- LERAP Category B

Hazard classification and safety precautions
Hazard Irritant [1]; Dangerous for the environment [1, 2]
Transport code 9
Packaging group III
UN Number 3082
Risk phrases R36 [1]; R51, R53a [1, 2]
Operator protection A, H [1, 2]; C [1]; U05a, U20b [1, 2]; U09b [1]
Environmental protection E15a [1]; E15b [2]; E16a, E34, E38 [1, 2]
Storage and disposal D01, D02, D09a, D10b, D12a [1, 2]; D05 [1]
Medical advice M03

375 prothioconazole + spiroxamine

A broad spectrum fungicide mixture for cereals
FRAC mode of action code: 3 + 5

See also spiroxamine

Products

1	Hale	Pan Agriculture	160:300 g/l	EC	15924
2	Helix	Bayer CropScience	160:300 g/l	EC	12264
3	Spiral	AgChem Access	160:300 g/l	EC	15437

Uses
- Brown rust in **spring barley, winter barley, winter rye, winter wheat**
- Crown rust in **spring oats, winter oats**
- Eyespot in **spring barley, spring oats, winter barley, winter oats, winter rye, winter wheat**
- Glume blotch in **winter wheat**
- Late ear diseases in **winter rye, winter wheat**
- Net blotch in **spring barley, winter barley**
- Powdery mildew in **spring barley, spring oats, winter barley, winter oats, winter rye, winter wheat**
- Rhynchosporium in **spring barley, winter barley, winter rye**
- Septoria leaf blotch in **winter wheat**
- Tan spot in **winter wheat**
- Yellow rust in **spring barley, winter barley, winter wheat**

Approval information
- Prothioconazole and spiroxamine included in Annex I under EC Regulation 1107/2009
- Accepted by BBPA for use on malting barley

Efficacy guidance
- Best results obtained from treatment at early stages of disease development. Further treatment may be needed if disease attack is prolonged
- Applications to established infections of any disease are likely to be less effective
- Best control of cereal ear diseases obtained by treatment during ear emergence
- Prothioconazole is a DMI fungicide. Resistance to some DMI fungicides has been identified in Septoria leaf blotch which may seriously affect performance of some products. For further advice contact a specialist advisor and visit the Fungicide Resistance Action Group (FRAG)-UK website

Restrictions
- Maximum total dose equivalent to two full dose treatments on barley and oats and three full dose treatments on wheat and rye

FOR FULL CONDITIONS OF USE ALWAYS READ THE PRODUCT LABEL

Crop-specific information
- Latest use: before grain watery ripe for rye, oats, winter wheat; up to beginning of anthesis for barley

Environmental safety
- Dangerous for the environment
- Very toxic to aquatic organisms
- LERAP Category B

Hazard classification and safety precautions
Hazard Harmful, Dangerous for the environment
Transport code 9
Packaging group III
UN Number 3082
Risk phrases R20, R22a, R36, R38, R50, R53a
Operator protection A, C, H; U05a, U09b, U11, U19a, U20b
Environmental protection E15a, E16a, E34, E38
Storage and disposal D01, D02, D05, D09a, D10b, D12a
Medical advice M03

376 prothioconazole + spiroxamine + tebuconazole

A broad spectrum fungicide mixture for cereals
FRAC mode of action code: 3 + 5 + 3

See also spiroxamine
    tebuconazole

Products
Cello	Bayer CropScience	100:250:100 g/l	EC	13178

Uses
- Brown rust in **spring barley, winter barley, winter rye, winter wheat**
- Crown rust in **spring oats, winter oats**
- Eyespot in **spring barley** (reduction), **spring oats, winter barley** (reduction), **winter oats, winter rye** (reduction), **winter wheat** (reduction)
- Glume blotch in **winter wheat**
- Late ear diseases in **spring barley, winter barley, winter wheat**
- Net blotch in **spring barley, winter barley**
- Powdery mildew in **spring barley, spring oats, winter barley, winter oats, winter rye, winter wheat**
- Rhynchosporium in **spring barley, winter barley, winter rye**
- Septoria leaf blotch in **winter wheat**
- Yellow rust in **spring barley, winter barley, winter wheat**

Approval information
- Prothioconazole, spiroxamine and tebuconazole included in Annex I under EC Regulation 1107/2009

Efficacy guidance
- Best results obtained from treatment at early stages of disease development. Further treatment may be needed if disease attack is prolonged
- Applications to established infections of any disease are likely to be less effective
- Best control of cereal ear diseases obtained by treatment during ear emergence
- Prothioconazole and tebuconazole are DMI fungicides. Resistance to some DMI fungicides has been identified in Septoria leaf blotch which may seriously affect performance of some products. For further advice contact a specialist advisor and visit the Fungicide Resistance Action Group (FRAG)-UK website

Restrictions
- Maximum total dose equivalent to two full dose treatments

Pesticide Profiles

- Newer authorisations for tebuconazole products require application to cereals only after GS 30 and applications to oilseed rape and linseed after GS 20 - check label

Crop-specific information
- Latest use: before grain milky ripe stage for rye, wheat; up to beginning of anthesis for barley and oats

Environmental safety
- Dangerous for the environment
- Very toxic to aquatic organisms
- LERAP Category B

Hazard classification and safety precautions
Hazard Harmful, Dangerous for the environment
Transport code 9
Packaging group III
UN Number 3082
Risk phrases R20, R36, R37, R38, R43, R50, R53a, R63
Operator protection A, C, H; U05a, U09b, U11, U14, U19a, U20b
Environmental protection E15a, E16a, E34, E38
Storage and disposal D01, D02, D05, D09a, D10b, D12a
Medical advice M03

377   prothioconazole + tebuconazole

A triazole fungicide mixture for cereals
FRAC mode of action code: 3 + 3

See also tebuconazole

Products

1	Clayton Zorro Pro	Clayton	125:125 g/l	EC	17073
2	Corinth	Bayer CropScience	80:160 g/l	EC	16742
3	Kestrel	Bayer CropScience	160:80 g/l	EC	16751
4	Prosaro	Bayer CropScience	125:125 g/l	EC	16732
5	Redigo Pro	Bayer CropScience	150:20 g/l	FS	15145
6	Standon Mastana	Standon	125:125 g/l	EC	17039

Uses
- Black stem rust in **grass seed crops** *(off-label)* [4]
- Blue mould in **durum wheat**, **rye**, **spring oats**, **spring wheat**, **triticale**, **winter oats**, **winter wheat** [5]
- Brown rust in **spring barley**, **spring wheat**, **winter barley**, **winter rye**, **winter wheat** [1, 3, 4, 6]
- Bunt in **durum wheat**, **rye**, **spring oats**, **spring wheat**, **triticale**, **winter oats**, **winter wheat** [5]
- Covered smut in **spring barley**, **spring oats**, **winter barley**, **winter oats** [5]
- Crown rust in **spring oats**, **winter oats** [1, 3, 4, 6]
- Ergot in **rye** *(qualified minor use)* [5]
- Eyespot in **spring barley**, **winter barley**, **winter rye**, **winter wheat** [3]; **spring barley** *(reduction)*, **winter barley** *(reduction)*, **winter rye** *(reduction)*, **winter wheat** *(reduction)* [1, 4, 6]; **spring oats**, **spring wheat**, **winter oats** [1, 3, 4, 6]
- Fusarium ear blight in **spring barley**, **spring wheat**, **winter barley**, **winter wheat** [3]
- Fusarium seedling blight in **durum wheat**, **rye**, **spring barley**, **spring oats**, **spring wheat**, **triticale**, **winter barley**, **winter oats**, **winter wheat** [5]
- Glume blotch in **spring wheat**, **winter wheat** [1, 3, 4, 6]
- Late ear diseases in **spring barley**, **spring wheat**, **winter barley**, **winter wheat** [1, 4, 6]
- Leaf spot in **grass seed crops** *(off-label)* [4]
- Leaf stripe in **spring barley**, **winter barley** [5]

FOR FULL CONDITIONS OF USE ALWAYS READ THE PRODUCT LABEL

prothioconazole + tebuconazole

- Light leaf spot in **oilseed rape** [3]; **oilseed rape** *(moderate control only)* [2]; **spring oilseed rape, winter oilseed rape** [1, 4, 6]
- Loose smut in **durum wheat, rye, spring barley, spring oats, spring wheat, triticale, winter barley, winter oats, winter wheat** [5]
- Net blotch in **spring barley, winter barley** [1, 3, 4, 6]
- Phoma in **spring oilseed rape, winter oilseed rape** [1, 4, 6]
- Phoma leaf spot in **oilseed rape** [2, 3]
- Powdery mildew in **grass seed crops** *(off-label)* [4]; **spring barley, spring oats, spring wheat, winter barley, winter oats, winter rye, winter wheat** [1, 3, 4, 6]
- Rhynchosporium in **grass seed crops** *(off-label)* [4]; **spring barley, winter barley, winter rye** [1, 3, 4, 6]
- Sclerotinia stem rot in **oilseed rape** [2, 3]; **spring oilseed rape, winter oilseed rape** [1, 4, 6]
- Septoria leaf blotch in **spring wheat, winter wheat** [1, 3, 4, 6]
- Sooty moulds in **spring barley, spring wheat, winter barley, winter wheat** [3]
- Stem canker in **oilseed rape** [2]
- Tan spot in **spring wheat, winter wheat** [1, 3, 4, 6]
- Yellow rust in **spring barley, spring wheat, winter barley, winter wheat** [1, 3, 4, 6]

Extension of Authorisation for Minor Use (EAMUs)
- *grass seed crops* 20160083 [4]

Approval information
- Prothioconazole and tebuconazole included in Annex I under EC Regulation 1107/2009
- Accepted by BBPA for use on malting barley

Efficacy guidance
- Best results on cereal foliar diseases obtained from treatment at early stages of disease development. Further treatment may be needed if disease attack is prolonged
- On oilseed rape apply a protective treatment in autumn/winter for Phoma followed by a further spray in early spring from the onset of stem elongation, if necessary. For control of Sclerotinia apply at early to full flower
- Applications to established infections of any disease are likely to be less effective
- Best control of cereal ear diseases obtained by treatment during ear emergence
- Prothioconazole and tebuconazole are DMI fungicides. Resistance to some DMI fungicides has been identified in Septoria leaf blotch which may seriously affect performance of some products. For further advice contact a specialist advisor and visit the Fungicide Resistance Action Group (FRAG)-UK website
- Treated seed should be drilled to a depth of 40mm. Ensure no seed is left on the soil surface. After drilling if conditions allow, field should be harrowed and then rolled to ensure good incorporation [5]

Restrictions
- Maximum total dose equivalent to two full dose treatments on barley, oats, oilseed rape; three full dose treatments on wheat and rye
- Newer authorisations for tebuconazole products require application to cereals only after GS 30 and applications to oilseed rape and linseed after GS20 - check label

Crop-specific information
- Latest use: before grain milky ripe for rye, wheat; beginning of flowering for barley, oats
- HI 56 d for oilseed rape

Environmental safety
- Dangerous for the environment
- Toxic to aquatic organisms
- LERAP Category B [1-4, 6]

Hazard classification and safety precautions
Hazard Harmful [2, 3]; Irritant [1, 4, 6]; Dangerous for the environment [1-4, 6]; Harmful if inhaled [2]
Transport code 9
Packaging group III
UN Number 3082

SEE SECTION 3 FOR PRODUCTS ALSO REGISTERED

Pesticide Profiles

Risk phrases H315, H317, H361 [1-4, 6]; H319 [2]; R43 [4]
Operator protection A, H [1-6]; C [2, 3]; D [5]; U05a [1-6]; U09b, U20b [1, 3, 4, 6]; U11, U20c [2]; U19a [1, 2, 4, 6]; U20a, U20e [5]
Environmental protection E15a, E16a, H411 [1-4, 6]; E15b, H410 [5]; E22b [2]; E22c [1, 3, 4, 6]; E34, E38 [1-6]
Storage and disposal D01, D02, D10b [1-4, 6]; D05, D09a, D12a [1-6]; D10d, D21 [5]
Treated seed S01, S02, S03, S04a, S04b, S05, S07, S08 [5]
Medical advice M03 [1-4, 6]

378   prothioconazole + trifloxystrobin

A triazole and strobilurin fungicide mixture for cereals
FRAC mode of action code: 3 + 11

See also trifloxystrobin

Products

1	Mobius	Bayer CropScience	175:150 g/l	SC	13395
2	Zephyr	Bayer CropScience	175:88 g/l	SC	13174

Uses
- Brown rust in *durum wheat*, *rye*, *spring barley*, *triticale*, *winter barley*, *winter wheat* [1, 2]; *spring wheat* [2]
- Ear diseases in *durum wheat*, *rye*, *triticale*, *winter wheat* [1]
- Eyespot in *durum wheat*, *rye*, *triticale*, *winter wheat* [1, 2]; *spring barley* (reduction), *spring wheat*, *winter barley* (reduction) [2]; *spring barley* (reduction in severity), *winter barley* (reduction in severity) [1]
- Glume blotch in *durum wheat*, *rye*, *triticale*, *winter wheat* [1, 2]; *spring wheat* [2]
- Net blotch in *spring barley*, *winter barley* [1, 2]
- Powdery mildew in *durum wheat*, *rye*, *spring barley*, *triticale*, *winter barley*, *winter wheat* [1, 2]; *spring wheat* [2]
- Rhynchosporium in *spring barley*, *winter barley* [1, 2]
- Septoria leaf blotch in *durum wheat*, *rye*, *triticale*, *winter wheat* [1, 2]; *spring wheat* [2]
- Yellow rust in *durum wheat*, *rye*, *spring barley*, *triticale*, *winter barley*, *winter wheat* [1, 2]; *spring wheat* [2]

Approval information
- Prothioconazole and trifloxystrobin included in Annex I under EC Regulation 1107/2009
- Accepted by BBPA for use on malting barley

Efficacy guidance
- Best results obtained from treatment at early stages of disease development. Further treatment may be needed if disease attack is prolonged
- Applications to established infections of any disease are likely to be less effective
- Best control of cereal ear diseases obtained by treatment during ear emergence
- Prothioconazole is a DMI fungicide. Resistance to some DMI fungicides has been identified in Septoria leaf blotch which may seriously affect performance of some products. For further advice contact a specialist advisor and visit the Fungicide Resistance Action Group (FRAG)-UK website
- Trifloxystrobin is a member of the QoI cross resistance group. Product should be used preventatively and not relied on for its curative potential
- Use product as part of an Integrated Crop Management strategy incorporating other methods of control, including where appropriate other fungicides with a different mode of action. Do not apply more than two foliar applications of QoI containing products to any cereal crop
- There is a significant risk of widespread resistance occurring in *Septoria tritici* populations in UK. Failure to follow resistance management action may result in reduced levels of disease control
- Strains of wheat and barley powdery mildew resistant to QoIs are common in the UK. Control of wheat mildew can only be relied on from the triazole component

FOR FULL CONDITIONS OF USE ALWAYS READ THE PRODUCT LABEL

- Where specific control of wheat mildew is required this should be achieved through a programme of measures including products recommended for the control of mildew that contain a fungicide from a different cross-resistance group and applied at a dose that will give robust control

Restrictions
- Maximum total dose equivalent to two full dose treatments

Crop-specific information
- Latest use: before grain milky ripe for wheat; beginning of flowering for barley

Environmental safety
- Dangerous for the environment
- Very toxic to aquatic organisms
- LERAP Category B

Hazard classification and safety precautions
Hazard Irritant, Dangerous for the environment, Very toxic to aquatic organisms
Transport code 9
Packaging group III
UN Number 3082
Risk phrases H317
Operator protection A, C, H; U05a, U09b, U19a [1, 2]; U20b [2]; U20c [1]
Environmental protection E15a, E16a, E38, H410 [1, 2]; E34 [2]
Storage and disposal D01, D02, D09a, D10c, D12a
Medical advice M03

379   pymetrozine

A novel azomethine insecticide
IRAC mode of action code: 9B

Products

1	Chess WG	Syngenta	50% w/w	WG	17580
2	Plenum WG	Syngenta	50% w/w	WG	10652

Uses
- Aphids in **beetroot** (off-label), **bilberries** (off-label), **blackberries** (off-label), **blackcurrants** (off-label), **blueberries** (off-label), **celeriac** (off-label), **celery (outdoor)** (off-label), **chard** (off-label), **chinese cabbage** (off-label), **choi sum** (off-label), **collards** (off-label), **cranberries** (off-label), **forest nurseries** (off-label), **frise** (off-label), **gooseberries** (off-label), **kale** (off-label), **lamb's lettuce** (off-label), **lettuce** (off-label), **loganberries** (off-label), **pak choi** (off-label), **radicchio** (off-label), **raspberries** (off-label), **redcurrants** (off-label), **rubus hybrids** (off-label), **salad brassicas** (off-label - for baby leaf production), **seed potatoes**, **soft fruit** (off-label), **spinach** (off-label), **spinach beet** (off-label), **strawberries** (off-label), **swedes** (off-label), **sweetcorn** (off-label), **turnips** (off-label), **vaccinium spp.** (off-label), **ware potatoes**, **whitecurrants** (off-label) [2]; **brassicas** (off-label), **ornamental plant production**, **protected almonds** (off-label), **protected apple** (off-label), **protected apricots** (off-label), **protected aubergines** (off-label), **protected bilberries** (off-label), **protected blackberries** (off-label), **protected blackcurrants** (off-label), **protected blueberry** (off-label), **protected celery** (off-label), **protected cherries** (off-label), **protected chestnuts** (off-label), **protected chilli peppers** (off-label), **protected choi sum** (off-label), **protected courgettes** (off-label), **protected cranberries** (off-label), **protected cucumbers**, **protected endives** (off-label), **protected forest nurseries** (off-label), **protected gherkins** (off-label), **protected gooseberries** (off-label), **protected hazelnuts** (off-label), **protected hops** (off-label), **protected loganberries** (off-label), **protected medlars** (off-label), **protected melons** (off-label), **protected nectarines** (off-label), **protected oriental cabbage** (off-label), **protected ornamentals** (off-label), **protected peaches** (off-label), **protected pear** (off-label), **protected peppers** (off-label), **protected plums** (off-label), **protected pumpkins** (off-label), **protected quince** (off-label), **protected raspberries** (off-label), **protected redcurrants** (off-label), **protected rubus hybrids** (off-label), **protected strawberries** (off-label), **protected**

SEE SECTION 3 FOR PRODUCTS ALSO REGISTERED

Pesticide Profiles

summer squash *(off-label)*, **protected tomatoes** *(off-label)*, **protected walnuts** *(off-label)*, **protected winter squash** *(off-label)* [1]
- Cabbage aphid in **chinese cabbage** *(off-label)*, **choi sum** *(off-label)*, **collards** *(off-label)*, **kale** *(off-label)*, **pak choi** *(off-label)* [2]
- Cotton aphids in **protected aubergines** *(off-label)*, **protected chilli peppers** *(off-label)*, **protected courgettes** *(off-label)*, **protected cucumbers** *(off-label)*, **protected gherkins** *(off-label)*, **protected melons** *(off-label)*, **protected okra** *(off-label)*, **protected peppers** *(off-label)*, **protected pumpkins** *(off-label)*, **protected summer squash** *(off-label)*, **protected tomatoes** *(off-label)*, **protected winter squash** *(off-label)* [1]
- Damson-hop aphid in **hops** *(off-label)* [2]
- Glasshouse whitefly in **protected almonds** *(off-label)*, **protected apple** *(off-label)*, **protected apricots** *(off-label)*, **protected aubergines** *(off-label)*, **protected bilberries** *(off-label)*, **protected blackberries** *(off-label)*, **protected blackcurrants** *(off-label)*, **protected blueberry** *(off-label)*, **protected cherries** *(off-label)*, **protected chestnuts** *(off-label)*, **protected chilli peppers** *(off-label)*, **protected courgettes** *(off-label)*, **protected cranberries** *(off-label)*, **protected cucumbers** *(off-label)*, **protected forest nurseries** *(off-label)*, **protected gherkins** *(off-label)*, **protected gooseberries** *(off-label)*, **protected hazelnuts** *(off-label)*, **protected hops** *(off-label)*, **protected loganberries** *(off-label)*, **protected medlars** *(off-label)*, **protected melons** *(off-label)*, **protected nectarines** *(off-label)*, **protected okra** *(off-label)*, **protected ornamentals** *(off-label)*, **protected peaches** *(off-label)*, **protected pear** *(off-label)*, **protected peppers** *(off-label)*, **protected plums** *(off-label)*, **protected pumpkins** *(off-label)*, **protected quince** *(off-label)*, **protected raspberries** *(off-label)*, **protected redcurrants** *(off-label)*, **protected rubus hybrids** *(off-label)*, **protected strawberries** *(off-label)*, **protected summer squash** *(off-label)*, **protected tomatoes** *(off-label)*, **protected walnuts** *(off-label)*, **protected winter squash** *(off-label)* [1]
- Green peach aphid in **protected aubergines** *(off-label)*, **protected chilli peppers** *(off-label)*, **protected courgettes** *(off-label)*, **protected cucumbers** *(off-label)*, **protected gherkins** *(off-label)*, **protected melons** *(off-label)*, **protected okra** *(off-label)*, **protected peppers** *(off-label)*, **protected pumpkins** *(off-label)*, **protected summer squash** *(off-label)*, **protected tomatoes** *(off-label)*, **protected winter squash** *(off-label)* [1]
- Green potato aphids in **protected aubergines** *(off-label)*, **protected chilli peppers** *(off-label)*, **protected courgettes** *(off-label)*, **protected cucumbers** *(off-label)*, **protected gherkins** *(off-label)*, **protected melons** *(off-label)*, **protected okra** *(off-label)*, **protected peppers** *(off-label)*, **protected pumpkins** *(off-label)*, **protected summer squash** *(off-label)*, **protected tomatoes** *(off-label)*, **protected winter squash** *(off-label)* [1]
- Mealy aphid in **broccoli** *(useful levels of control)*, **brussels sprouts** *(useful levels of control)*, **cabbages** *(useful levels of control)*, **calabrese** *(useful levels of control)*, **cauliflowers** *(useful levels of control)*, **chinese cabbage** *(off-label)*, **choi sum** *(off-label)*, **collards** *(off-label)*, **kale** *(off-label)*, **pak choi** *(off-label)* [2]
- Peach-potato aphid in **broccoli**, **brussels sprouts**, **cabbages**, **calabrese**, **cauliflowers**, **chinese cabbage** *(off-label)*, **choi sum** *(off-label)*, **collards** *(off-label)*, **kale** *(off-label)*, **pak choi** *(off-label)* [2]; **protected almonds** *(off-label)*, **protected apple** *(off-label)*, **protected apricots** *(off-label)*, **protected aubergines** *(off-label)*, **protected bilberries** *(off-label)*, **protected blackberries** *(off-label)*, **protected blackcurrants** *(off-label)*, **protected blueberry** *(off-label)*, **protected cherries** *(off-label)*, **protected chestnuts** *(off-label)*, **protected chilli peppers** *(off-label)*, **protected cranberries** *(off-label)*, **protected forest nurseries** *(off-label)*, **protected gooseberries** *(off-label)*, **protected hazelnuts** *(off-label)*, **protected herbs (see appendix 6)** *(off-label)*, **protected hops** *(off-label)*, **protected lamb's lettuce** *(off-label)*, **protected leaf brassicas** *(off-label)*, **protected lettuce** *(off-label)*, **protected loganberries** *(off-label)*, **protected medlars** *(off-label)*, **protected nectarines** *(off-label)*, **protected ornamentals** *(off-label)*, **protected peaches** *(off-label)*, **protected pear** *(off-label)*, **protected peppers** *(off-label)*, **protected plums** *(off-label)*, **protected quince** *(off-label)*, **protected raspberries** *(off-label)*, **protected redcurrants** *(off-label)*, **protected rubus hybrids** *(off-label)*, **protected strawberries** *(off-label)*, **protected tomatoes** *(off-label)*, **protected walnuts** *(off-label)* [1]
- Pollen beetle in **oilseed rape** [2]
- Red aphids in **protected aubergines** *(off-label)*, **protected chilli peppers** *(off-label)*, **protected courgettes** *(off-label)*, **protected cucumbers** *(off-label)*, **protected gherkins** *(off-label)*, **protected melons** *(off-label)*, **protected okra** *(off-label)*, **protected peppers** *(off-label)*, **protected pumpkins** *(off-label)*, **protected summer squash** *(off-label)*, **protected tomatoes** *(off-label)*, **protected winter squash** *(off-label)* [1]

FOR FULL CONDITIONS OF USE ALWAYS READ THE PRODUCT LABEL

pymetrozine

- Tobacco whitefly in **protected almonds** *(off-label)*, **protected apple** *(off-label)*, **protected apricots** *(off-label)*, **protected aubergines** *(off-label)*, **protected bilberries** *(off-label)*, **protected blackberries** *(off-label)*, **protected blackcurrants** *(off-label)*, **protected blueberry** *(off-label)*, **protected cherries** *(off-label)*, **protected chestnuts** *(off-label)*, **protected chilli peppers** *(off-label)*, **protected cranberries** *(off-label)*, **protected forest nurseries** *(off-label)*, **protected gooseberries** *(off-label)*, **protected hazelnuts** *(off-label)*, **protected hops** *(off-label)*, **protected loganberries** *(off-label)*, **protected medlars** *(off-label)*, **protected nectarines** *(off-label)*, **protected ornamentals** *(off-label)*, **protected peaches** *(off-label)*, **protected pear** *(off-label)*, **protected peppers** *(off-label)*, **protected plums** *(off-label)*, **protected quince** *(off-label)*, **protected raspberries** *(off-label)*, **protected redcurrants** *(off-label)*, **protected rubus hybrids** *(off-label)*, **protected strawberries** *(off-label)*, **protected tomatoes** *(off-label)*, **protected walnuts** *(off-label)* [1]

Extension of Authorisation for Minor Use (EAMUs)
- *beetroot 20131567* [2]
- *bilberries 20061702* [2]
- *blackberries 20061633* [2]
- *blackcurrants 20060946* [2]
- *blueberries 20061702* [2]
- *brassicas 20161259* [1]
- *celeriac 20051062* [2]
- *celery (outdoor) 20051062* [2]
- *chard 20111664* [2]
- *chinese cabbage 20082246* [2]
- *choi sum 20082246* [2]
- *collards 20082081* [2]
- *cranberries 20061702* [2]
- *forest nurseries 20082920* [2]
- *frise 20070060* [2]
- *gooseberries 20061702* [2]
- *hops 20031423* [2]
- *kale 20082081* [2]
- *lamb's lettuce 20070060* [2]
- *lettuce 20070060* [2]
- *loganberries 20061702* [2]
- *pak choi 20082246* [2]
- *protected almonds 20161261* [1], *20161273* [1]
- *protected apple 20161261* [1], *20161273* [1]
- *protected apricots 20161261* [1], *20161273* [1]
- *protected aubergines 20161271* [1], *20161272* [1]
- *protected bilberries 20161261* [1], *20161273* [1]
- *protected blackberries 20161258* [1], *20161261* [1], *20161273* [1]
- *protected blackcurrants 20161248* [1], *20161261* [1], *20161273* [1]
- *protected blueberry 20161261* [1], *20161273* [1]
- *protected celery 20161267* [1]
- *protected cherries 20161261* [1], *20161273* [1]
- *protected chestnuts 20161261* [1], *20161273* [1]
- *protected chilli peppers 20161271* [1], *20161272* [1]
- *protected choi sum 20161262* [1]
- *protected courgettes 20161263* [1], *20161272* [1]
- *protected cranberries 20161261* [1], *20161273* [1]
- *protected cucumbers 20161272* [1]
- *protected endives 20161264* [1]
- *protected forest nurseries 20161261* [1], *20161273* [1]
- *protected gherkins 20161263* [1], *20161272* [1]
- *protected gooseberries 20161248* [1], *20161261* [1], *20161273* [1]
- *protected hazelnuts 20161261* [1], *20161273* [1]
- *protected herbs (see appendix 6) 20161268* [1]
- *protected hops 20161261* [1], *20161273* [1]

SEE SECTION 3 FOR PRODUCTS ALSO REGISTERED

Pesticide Profiles

- *protected lamb's lettuce* 20161268 [1]
- *protected leaf brassicas* 20161268 [1]
- *protected lettuce* 20161268 [1]
- *protected loganberries* 20161248 [1], 20161261 [1], 20161273 [1]
- *protected medlars* 20161261 [1], 20161273 [1]
- *protected melons* 20161263 [1], 20161272 [1]
- *protected nectarines* 20161261 [1], 20161273 [1]
- *protected okra* 20161272 [1]
- *protected oriental cabbage* 20161262 [1]
- *protected ornamentals* 20161261 [1], 20161273 [1]
- *protected peaches* 20161261 [1], 20161273 [1]
- *protected pear* 20161261 [1], 20161273 [1]
- *protected peppers* 20161271 [1], 20161272 [1]
- *protected plums* 20161261 [1], 20161273 [1]
- *protected pumpkins* 20161263 [1], 20161272 [1]
- *protected quince* 20161261 [1], 20161273 [1]
- *protected raspberries* 20161258 [1], 20161261 [1], 20161273 [1]
- *protected redcurrants* 20161248 [1], 20161261 [1], 20161273 [1]
- *protected rubus hybrids* 20161248 [1], 20161261 [1], 20161273 [1]
- *protected strawberries* 20161249 [1], 20161261 [1], 20161273 [1]
- *protected summer squash* 20161263 [1], 20161272 [1]
- *protected tomatoes* 20161271 [1], 20161272 [1]
- *protected walnuts* 20161261 [1], 20161273 [1]
- *protected winter squash* 20161263 [1], 20161272 [1]
- *radicchio* 20070060 [2]
- *raspberries* 20061633 [2]
- *redcurrants* 20061702 [2]
- *rubus hybrids* 20061702 [2]
- *salad brassicas* (for baby leaf production) 20070060 [2]
- *soft fruit* 20082920 [2]
- *spinach* 20111664 [2]
- *spinach beet* 20111664 [2]
- *strawberries* 20060461 [2]
- *swedes* 20131567 [2]
- *sweetcorn* 20041318 [2]
- *turnips* 20131567 [2]
- *vaccinium spp.* 20061702 [2]
- *whitecurrants* 20061702 [2]

Approval information
- Pymetrozine included in Annex I under EC Regulation 1107/2009
- Accepted by BBPA for use on hops

Efficacy guidance
- Pymetrozine moves systemically in the plant and acts by preventing feeding leading to death by starvation in 1-4 d. There is no immediate knockdown
- Aphids controlled include those resistant to organophosphorus and carbamate insecticides
- To prevent development of resistance do not use continuously or as the sole method of control
- Best results achieved by starting spraying as soon as aphids seen in crop and repeating as necessary.
- To limit spread of persistent viruses such as potato leaf roll virus, apply from 90% crop emergence

Restrictions
- Maximum total dose equivalent to two full dose treatments on ware potatoes; three full dose treatments on seed potatoes, leaf spinach [2]; four full dose treatments on cucumbers, ornamentals
- Do not apply to potatoes when crop is in flower
- Consult processors before use on potatoes for processing

FOR FULL CONDITIONS OF USE ALWAYS READ THE PRODUCT LABEL

pyraclostrobin

- Check tolerance of ornamental species before large scale use. See label for list of species known to have been treated without damage. Visible spray deposits may be seen on leaves of some species

Crop-specific information
- HI cucumbers 3 d; potatoes, leaf spinach 7 d

Environmental safety
- High risk to bees (outdoor use only). Do not apply to crops in flower or to those in which bees are actively foraging. Do not apply when flowering weeds are present
- Avoid spraying within 6 m of field boundaries to reduce effects on non-target insects or arthropods. Risk to certain non-target insects and arthropods.

Hazard classification and safety precautions
Hazard Harmful
UN Number N/C
Risk phrases H351 [1]; R40 [2]
Operator protection A, H; U05a, U20c
Environmental protection E12a, E12e, E15a [1, 2]; H410 [1]
Storage and disposal D01, D02, D09a, D11a

380 pyraclostrobin

A protectant and curative strobilurin fungicide for cereals
FRAC mode of action code: 11

See also boscalid + epoxiconazole + pyraclostrobin
  boscalid + pyraclostrobin
  dimethomorph + pyraclostrobin
  dithianon + pyraclostrobin
  epoxiconazole + fenpropimorph + pyraclostrobin
  epoxiconazole + fluxapyroxad + pyraclostrobin
  epoxiconazole + kresoxim-methyl + pyraclostrobin
  epoxiconazole + pyraclostrobin
  fenpropimorph + pyraclostrobin
  fluxapyroxad + pyraclostrobin

Products

1	Comet 200	BASF	200 g/l	EC	12639
2	Flyer	BASF	250 g/l	EC	12654
3	Flyer 200	BASF	200 g/l	EC	17293
4	Halley	AgChem Access	200 g/l	EC	15916
5	Mascot Eland	Rigby Taylor	20% w/w	WG	14549
6	Tucana	BASF	250 g/l	EC	10899
7	Vanguard	Sherriff Amenity	20% w/w	WG	13838
8	Vivid	BASF	250 g/l	EC	10898

Uses
- Brown rust in **spring barley**, **spring wheat**, **winter barley**, **winter wheat** [1-4, 6, 8]
- Crown rust in **spring oats**, **winter oats** [1-4, 6, 8]
- Disease control in **forage maize**, **grain maize** [2, 3]
- Dollar spot in **managed amenity turf** *(useful reduction)* [5, 7]
- Eyespot in **forage maize**, **grain maize** [1-3]; **forage maize** *(off-label)* [8]
- Foliar disease control in **durum wheat** *(off-label)*, **grass seed crops** *(off-label)*, **spring rye** *(off-label)*, **triticale** *(off-label)*, **winter rye** *(off-label)* [6, 8]; **ornamental plant production** *(off-label)* [8]
- Fusarium patch in **managed amenity turf** *(moderate control)* [5]; **managed amenity turf** *(moderate control only)* [7]
- Glume blotch in **spring wheat**, **winter wheat** [1-4, 6, 8]
- Net blotch in **spring barley**, **winter barley** [1-4, 6, 8]

SEE SECTION 3 FOR PRODUCTS ALSO REGISTERED

Pesticide Profiles

- Northern leaf blight in **forage maize**, **grain maize** [1]; **forage maize** *(moderate control)*, **grain maize** *(moderate control)* [3]
- Red thread in **managed amenity turf** [5, 7]
- Rhynchosporium in **spring barley** *(moderate control)*, **winter barley** *(moderate control)* [1-4, 6, 8]
- Septoria leaf blotch in **spring wheat**, **winter wheat** [1-4, 6, 8]
- Yellow rust in **spring barley**, **spring wheat**, **winter barley**, **winter wheat** [1-4, 6, 8]

Extension of Authorisation for Minor Use (EAMUs)
- **durum wheat** *20062652* [6], *20062649* [8]
- **forage maize** *20131684* [8]
- **grass seed crops** *20062652* [6], *20062649* [8]
- **ornamental plant production** *20082884* [8]
- **spring rye** *20062652* [6], *20062649* [8]
- **triticale** *20062652* [6], *20062649* [8]
- **winter rye** *20062652* [6], *20062649* [8]

Approval information
- Pyraclostrobin included in Annex I under EC Regulation 1107/2009
- Accepted by BBPA for use on malting barley (before ear emergence only) and on hops

Efficacy guidance
- For best results apply at the start of disease attack on cereals [1, 2, 6, 8]
- For Fusarium Patch treat early as severe damage to turf can occur once the disease is established [5, 7]
- Regular turf aeration, appropriate scarification and judicious use of nitrogenous fertiliser will assist the control of Fusarium Patch [5, 7]
- Best results on Septoria glume blotch achieved when used as a protective treatment and against Septoria leaf blotch when treated in the latent phase [1, 2, 6, 8]
- Yield response may be obtained in the absence of visual disease symptoms [1, 2, 6, 8]
- Pyraclostrobin is a member of the QoI cross resistance group. Product should be used preventatively and not relied on for its curative potential
- Use product as part of an Integrated Crop Management strategy incorporating other methods of control, including where appropriate other fungicides with a different mode of action. Do not apply more than two foliar applications of QoI containing products to any cereal crop or to grass
- There is a significant risk of widespread resistance occurring in *Septoria tritici* populations in UK. Failure to follow resistance management action may result in reduced levels of disease control [1, 2, 6, 8]
- On cereal crops product must always be used in mixture with another product, recommended for control of the same target disease, that contains a fungicide from a different cross resistance group and is applied at a dose that will give robust control [1, 2, 6, 8]
- Late application to maize crops can give a worthwhile increase in biomass

Restrictions
- Maximum number of treatments 2 per crop on cereals [1, 2, 6, 8]
- Maximum total dose on turf equivalent to two full dose treatments [5, 7]
- Do not apply during drought conditions or to frozen turf [5, 7]

Crop-specific information
- Latest use: before grain watery ripe (GS 71) for wheat; up to and including emergence of ear just complete (GS 59) for barley and oats [1, 2, 6, 8]
- Avoid applying to turf immediately after cutting or 48 h before mowing [5, 7]

Environmental safety
- Dangerous for the environment
- Very toxic to aquatic organisms
- LERAP Category B

Hazard classification and safety precautions
Hazard Harmful, Dangerous for the environment [1-8]; Harmful if swallowed, Harmful if inhaled, Very toxic to aquatic organisms [1-3]
Transport code 6.1 [1-4, 6, 8]; 9 [5, 7]

FOR FULL CONDITIONS OF USE ALWAYS READ THE PRODUCT LABEL

pyraflufen-ethyl

Packaging group III
UN Number 2902 [1-4, 6, 8]; 3077 [5, 7]
Risk phrases H304, H315, H317, H319 [1-3]; R20, R50 [4-8]; R22a, R38, R53a [4, 6, 8]
Operator protection A [1-8]; D [5, 7]; U05a, U14, U20b [1-4, 6, 8]
Environmental protection E15a [7]; E15b, E16a, E38 [1-8]; E16b [1-6, 8]; E34 [1-4, 6, 8]; H410 [1-3]
Storage and disposal D01, D02 [1-4, 6, 8]; D08, D09a, D10c, D12a [1-8]
Medical advice M03, M05a [1-4, 6, 8]

381 pyraflufen-ethyl

A phenylpyrazole herbicide for potatoes
HRAC mode of action code: E

See also glyphosate + pyraflufen-ethyl

Products
Gozai   Belchim   26.5 g/l   EC   17381

Uses
- Annual dicotyledons in **potatoes**
- Desiccation in **potatoes**

Approval information
- Pyraflufen-ethyl included in Annex I under EC Regulation 1107/2009

Efficacy guidance
- Use with a methylated vegetable oil adjuvant

Following crops guidance
- After cultivation to at least 20 cms winter cereals can be sown as a following crop but the safety to following broad-leaved crops has not yet been established.

Environmental safety
- LERAP Category B

Hazard classification and safety precautions
Hazard Harmful, Dangerous for the environment, Harmful if inhaled, Very toxic to aquatic organisms
Transport code 9
Packaging group III
UN Number 3082
Risk phrases H304, H315, H317, H318
Operator protection A, C, H; U05a, U09a, U11, U14, U15
Environmental protection E15b, E16a, E34, H410
Storage and disposal D01, D02, D05, D09a, D10a, D12a, D12b
Medical advice M03, M05b

382 pyrethrins

A non-persistent, contact acting insecticide extracted from Pyrethrum
IRAC mode of action code: 3

Products
1	Dairy Fly Spray	B H & B	0.75 g/l	AL	H5579
2	Pyblast	Agropharm	30 g/l	UL	H7485
3	Pyrethrum 5 EC	Agropharm	50 g/l	EC	12685
4	Spruzit	Certis	4.59 g/l	EC	13438

SEE SECTION 3 FOR PRODUCTS ALSO REGISTERED

Pesticide Profiles

Uses
- Aphids in **all edible crops (outdoor), all non-edible crops (outdoor), protected crops** [4]; **broccoli, brussels sprouts, bush fruit, cabbages, calabrese, cane fruit, cauliflowers, lettuce, ornamental plant production, protected broccoli, protected brussels sprouts, protected bush fruit, protected cabbages, protected calabrese, protected cane fruit, protected cauliflowers, protected cayenne peppers** (off-label), **protected chilli peppers** (off-label), **protected lettuce, protected ornamentals, protected peppers** (off-label), **protected tomatoes, tomatoes (outdoor)** [3]
- Caterpillars in **all edible crops (outdoor), all non-edible crops (outdoor), protected crops** [4]; **broccoli, brussels sprouts, bush fruit, cabbages, calabrese, cane fruit, cauliflowers, lettuce, ornamental plant production, protected broccoli, protected brussels sprouts, protected bush fruit, protected cabbages, protected calabrese, protected cane fruit, protected cauliflowers, protected lettuce, protected ornamentals, protected tomatoes, tomatoes (outdoor)** [3]
- Flea beetle in **broccoli, brussels sprouts, cabbages, calabrese, cauliflowers, protected broccoli, protected brussels sprouts, protected cabbages, protected calabrese, protected cauliflowers, protected tomatoes, tomatoes (outdoor)** [3]; **protected crops** [4]
- Insect pests in **dairies, farm buildings, livestock houses, poultry houses** [1, 2]
- Macrolophus caliginosus in **protected tomatoes, protected tomatoes** (off-label) [3]
- Mealybugs in **protected crops** [4]; **protected tomatoes, protected tomatoes** (off-label) [3]
- Scale insects in **protected crops** [4]
- Spider mites in **all edible crops (outdoor), all non-edible crops (outdoor), protected crops** [4]
- Thrips in **all edible crops (outdoor), all non-edible crops (outdoor), protected crops** [4]
- Whitefly in **broccoli, brussels sprouts, cabbages, calabrese, cauliflowers, protected broccoli, protected brussels sprouts, protected cabbages, protected calabrese, protected cauliflowers, protected tomatoes, tomatoes (outdoor)** [3]; **protected crops** [4]

Extension of Authorisation for Minor Use (EAMUs)
- *protected cayenne peppers* 20091005 [3]
- *protected chilli peppers* 20091005 [3]
- *protected peppers* 20091005 [3]
- *protected tomatoes* 20063026 [3]

Approval information
- Pyrethrins have been included in Annex 1 under EC Regulation 1107/2009
- Some products formulated for ULV application [1, 2]. May be applied through fogging machine or sprayer [1, 2]. See label for details
- Accepted by BBPA for use in empty grain stores, malting barley and hops

Efficacy guidance
- For indoor fly control close doors and windows and spray or apply fog as appropriate [1, 2]
- For best fly control outdoors spray during early morning or late afternoon and evening when conditions are still [1, 2]
- For all uses ensure good spray coverage of the target area or plants by increasing spray volume where necessary
- Best results on outdoor or protected crops achieved from treatment at first signs of pest attack in early morning or evening [4]
- To avoid possibility of development of resistance do not spray more frequently than once per week

Restrictions
- Maximum number of treatments on edible crops 3 per crop or season [3]
- For use only by professional operators [1, 2]
- Do not allow spray to contact open food products or food preparing equipment or utensils [1, 2]
- Remove exposed milk and collect eggs before application [1, 2]
- Do not treat plants [1, 2]
- Avoid direct application to open flowers [4]
- Do not mix with, or apply closely before or after, products containing dithianon or tolylfluanid [4]

FOR FULL CONDITIONS OF USE ALWAYS READ THE PRODUCT LABEL

pyridate

- Do not use space sprays containing pyrethrins or pyrethroid more than once per week in intensive or controlled environment animal houses in order to avoid development of resistance. If necessary, use a different control method or product [1, 2]
- Store away from strong sunlight [3]

Crop-specific information
- HI: 24 h for edible crops [3]
- Some plant species including *Ageratum*, ferns, *Ficus, Lantana, Poinsettia, Petroselium crispum*, and some strawberries may be sensitive especially where more than one application is made. Test before large scale treatment [4]

Environmental safety
- Dangerous for the environment [2-4]
- Very toxic to aquatic organisms [1, 3]; toxic to aquatic organisms [2, 4]
- High risk to bees. Do not apply to crops in flower or to those in which bees are actively foraging. Do not apply when flowering weeds are present [3]
- Risk to non-target insects or other arthropods [3]
- Do not apply directly to livestock and exclude all persons and animals during treatment [1, 2]
- Wash spray equipment thoroughly after use to avoid traces of pyrethrum causing damage to susceptible crops sprayed later [3]
- LERAP Category A [3]

Hazard classification and safety precautions
Hazard Harmful [1]; Irritant [1-3]; Dangerous for the environment [2-4]
Transport code 9
Packaging group III
UN Number 3082
Risk phrases R22a, R36, R38 [1]; R41 [2, 3]; R50 [3]; R51 [2, 4]; R53a [2-4]
Operator protection A [1-4]; B [1]; C, D [2]; E [1, 2]; H [2-4]; U05a [1, 3]; U09a [1]; U11 [2, 3]; U15 [2]; U19a [1-3]; U20b [1, 2]; U20c [3]; U20d [4]
Environmental protection E02c [2]; E05a [1, 2]; E12a, E12e, E16c, E16d, E22c [3]; E12c, E15b [4]; E13c [1]; E15a [2, 3]; E19b [2, 4]; E38 [2-4]
Consumer protection C04, C06 [1]; C07, C08, C09, C10, C11 [1, 2]; C12 [2]
Storage and disposal D01, D02, D09a [1-4]; D05 [3]; D11a [1, 3, 4]; D12a [2-4]
Medical advice M05a [2]

383 pyridate

A contact phenylpyridazine herbicide for bulb onions and brassicas
HRAC mode of action code: C3

Products
Lentagran WP   Belchim   45% w/w   WP   14162

Uses
- Annual dicotyledons in **asparagus** (off-label), **broccoli** (off-label), **calabrese** (off-label), **cauliflowers** (off-label), **chives** (off-label), **collards** (off-label), **fodder rape** (off-label), **game cover** (off-label), **garlic** (off-label), **kale** (off-label), **leeks** (off-label), **lupins** (off-label), **oilseed rape** (off-label), **salad onions** (off-label), **shallots** (off-label), **spring greens** (off-label)
- Black nightshade in **asparagus** (off-label), **broccoli** (off-label), **brussels sprouts**, **bulb onions**, **cabbages**, **calabrese** (off-label), **cauliflowers** (off-label), **chives** (off-label), **collards** (off-label), **fodder rape** (off-label), **game cover** (off-label), **garlic** (off-label), **kale** (off-label), **leeks** (off-label), **lupins** (off-label), **oilseed rape** (off-label), **salad onions** (off-label), **shallots** (off-label), **spring greens** (off-label)
- Cleavers in **asparagus** (off-label), **broccoli** (off-label), **brussels sprouts**, **bulb onions**, **cabbages**, **calabrese** (off-label), **cauliflowers** (off-label), **chives** (off-label), **collards** (off-label), **fodder rape** (off-label), **game cover** (off-label), **garlic** (off-label), **kale** (off-label), **leeks** (off-label), **lupins** (off-label), **oilseed rape** (off-label), **salad onions** (off-label), **shallots** (off-label), **spring greens** (off-label)

SEE SECTION 3 FOR PRODUCTS ALSO REGISTERED

Pesticide Profiles

- Fat hen in **asparagus** *(off-label)*, **broccoli** *(off-label)*, **brussels sprouts**, **bulb onions**, **cabbages**, **calabrese** *(off-label)*, **cauliflowers** *(off-label)*, **chives** *(off-label)*, **collards** *(off-label)*, **fodder rape** *(off-label)*, **game cover** *(off-label)*, **garlic** *(off-label)*, **kale** *(off-label)*, **leeks** *(off-label)*, **lupins** *(off-label)*, **oilseed rape** *(off-label)*, **salad onions** *(off-label)*, **shallots** *(off-label)*, **spring greens** *(off-label)*
- Fumitory in **asparagus** *(off-label)*, **broccoli** *(off-label)*, **calabrese** *(off-label)*, **cauliflowers** *(off-label)*, **chives** *(off-label)*, **collards** *(off-label)*, **fodder rape** *(off-label)*, **game cover** *(off-label)*, **garlic** *(off-label)*, **kale** *(off-label)*, **leeks** *(off-label)*, **lupins** *(off-label)*, **oilseed rape** *(off-label)*, **salad onions** *(off-label)*, **shallots** *(off-label)*, **spring greens** *(off-label)*
- Groundsel in **asparagus** *(off-label)*, **broccoli** *(off-label)*, **calabrese** *(off-label)*, **cauliflowers** *(off-label)*, **chives** *(off-label)*, **collards** *(off-label)*, **fodder rape** *(off-label)*, **game cover** *(off-label)*, **garlic** *(off-label)*, **kale** *(off-label)*, **leeks** *(off-label)*, **lupins** *(off-label)*, **oilseed rape** *(off-label)*, **salad onions** *(off-label)*, **shallots** *(off-label)*, **spring greens** *(off-label)*

Extension of Authorisation for Minor Use (EAMUs)
- *asparagus* 20091039
- *broccoli* 20090786
- *calabrese* 20090786
- *cauliflowers* 20090786
- *chives* 20120267
- *collards* 20090785
- *fodder rape* 20093230
- *game cover* 20090788
- *garlic* 20092862
- *kale* 20090785
- *leeks* 20090784
- *lupins* 20090787
- *oilseed rape* 20093230
- *salad onions* 20120267
- *shallots* 20092862
- *spring greens* 20090785

Approval information
- Pyridate included in Annex I under EC Regulation 1107/2009

Efficacy guidance
- Best results achieved by application to actively growing weeds at 6-8 leaf stage when temperatures are above 8°C before crop foliage forms canopy

Restrictions
- Do not apply in mixture with or within 14 d of any other product which may result in dewaxing of crop foliage
- Do not use on crops suffering stress from frost, drought, disease or pest attack

Crop-specific information
- Latest use: before 6 true leaf stage of Brussels sprouts and cabbage; before 7 true leaves for onions; before flower buds visible for oilseed rape
- Apply to cabbages and Brussels sprouts after 4 fully expanded leaf stage. Allow 2 wk after transplanting before treating

Environmental safety
- Dangerous for the environment
- Toxic to aquatic organisms

Hazard classification and safety precautions
Hazard Irritant, Dangerous for the environment
Transport code 9
Packaging group III
UN Number 3077
Risk phrases R43, R51, R53a
Operator protection A, C; U05a, U14, U22a
Environmental protection E15a, E38
Storage and disposal D01, D02, D09a, D10c, D12a

FOR FULL CONDITIONS OF USE ALWAYS READ THE PRODUCT LABEL

384 pyrimethanil

An anilinopyrimidine fungicide for apples and strawberries
FRAC mode of action code: 9

See also chlorothalonil + pyrimethanil

Products

Scala	BASF	400 g/l	SC	15222

Uses
- Botrytis in **bilberries** *(off-label)*, **blackberries** *(off-label)*, **blackcurrants** *(off-label)*, **blueberries** *(off-label)*, **cranberries** *(off-label)*, **forest nurseries** *(off-label)*, **gooseberries** *(off-label)*, **grapevines** *(off-label)*, **ornamental plant production** *(off-label)*, **pears** *(off-label)*, **protected aubergines** *(off-label)*, **protected baby leaf crops** *(off-label)*, **protected blackberries** *(off-label)*, **protected herbs (see appendix 6)** *(off-label)*, **protected lettuce** *(off-label)*, **protected raspberries** *(off-label)*, **protected strawberries**, **protected tomatoes** *(off-label)*, **quinces** *(off-label)*, **raspberries** *(off-label)*, **redcurrants** *(off-label)*, **strawberries**, **top fruit** *(off-label)*, **vaccinium spp.** *(off-label)*, **whitecurrants** *(off-label)*
- Disease control in **forest nurseries** *(off-label)*, **top fruit** *(off-label)*
- Grey mould in **protected strawberries** *(moderate control)*, **strawberries**
- Scab in **apples**, **pears** *(off-label)*, **quinces** *(off-label)*

Extension of Authorisation for Minor Use (EAMUs)
- *bilberries* 20110291
- *blackberries* 20110293
- *blackcurrants* 20110291
- *blueberries* 20110291
- *cranberries* 20110291
- *forest nurseries* 20122054
- *gooseberries* 20110291
- *grapevines* 20110283
- *ornamental plant production* 20111315
- *pears* 20110295
- *protected aubergines* 20110282
- *protected baby leaf crops* 20110287
- *protected blackberries* 20110292
- *protected herbs (see appendix 6)* 20110287
- *protected lettuce* 20110287
- *protected raspberries* 20110292
- *protected tomatoes* 20110282
- *quinces* 20110295
- *raspberries* 20110293
- *redcurrants* 20110291
- *top fruit* 20122054
- *vaccinium spp.* 20110291
- *whitecurrants* 20110291

Approval information
- Pyrimethanil included in Annex I under EC Regulation 1107/2009.

Efficacy guidance
- On apples a programme of sprays will give early season control of scab. Season long control can be achieved by continuing programme with other approved fungicides
- In strawberries product should be used as part of a programme of disease control treatments which should alternate with other materials to prevent or limit development of less sensitive strains of grey mould

Restrictions
- Maximum number of treatments 4 per yr for apples and pears; 2 per yr for cane fruit, bush fruit, strawberries, tomatoes
- Product does not taint apples. Processors should be consulted before use on strawberries

SEE SECTION 3 FOR PRODUCTS ALSO REGISTERED

Crop-specific information
- Latest use: before end of flowering for apples
- HI protected cane fruit, strawberries 1 d; aubergines, tomatoes 3 d; protected lettuce 14 d; bush fruit, grapevines 21 d
- All varieties of apples and strawberries may be treated
- Treat apples from bud burst at 10-14 d intervals
- In strawberries start treatments at white bud to give maximum protection of flowers against grey mould and treat every 7-10 d. Product should not be used more than once in a 3 or 4 spray programme

Environmental safety
- Harmful to aquatic organisms
- Product has negligible effect on hoverflies and lacewings. Limited evidence indicates some margin of safety to *Typhlodromus pyri*
- Broadcast air-assisted LERAP (20 m); LERAP Category B

Hazard classification and safety precautions
UN Number N/C
Risk phrases R52, R53a
Operator protection A, H, M; U08, U20b
Environmental protection E15a, E16a; E17f (20 m)
Storage and disposal D01, D02, D05, D08, D09a, D10b, D12a

385 pyriofenone

A protectant fungicide with some curative activity against mildew in the latent phase
FRAC mode of action code: U8

Products
Property 180 SC   Belchim   180 g/l   SC   15661

Uses
- Powdery mildew in **spring wheat**, **winter wheat**

Approval information
- Pyrifenone is included in Annex 1 under EC Regulation 1107/2009

Efficacy guidance
- Best results obtained from application at the start of disease attack. Acts preventatively against powdery mildew with some curative activity during the latent phase. Apply before disease spreads on to new growth with a follow-up after 2 - 4 weeks according to climatic conditions.
- Should be used as part of a mildew control programme that incorporates other methods of control including fungicides active against mildew via a different mode of action.

Restrictions
- Only two applications of benzoylpyridine fungicides are permitted per crop per season. Where a second application is required it should be applied in mixture with fungicides using a different mode of action such as fenpropimorph.

Crop-specific information
- Latest use: mid-flowering (GS65) in spring and winter wheat

Following crops guidance
- Oilseed rape, maize, potato, sugar beet, kidney bean, soybean, pea, onion, turnip, lettuce, flax, oat, barley and wheat may be sown as a following crop.

Hazard classification and safety precautions
Hazard Harmful, Dangerous for the environment
Transport code 9
Packaging group III
UN Number 3082
Risk phrases H351, R53a

FOR FULL CONDITIONS OF USE ALWAYS READ THE PRODUCT LABEL

quinmerac

Operator protection A, H; U05a, U20b
Environmental protection E15b, E34, E38, H411
Storage and disposal D01, D02, D05, D09a, D12a, D20
Medical advice M05a

386   quinmerac

A residual herbicide available only in mixtures
HRAC mode of action code: O

See also chloridazon + quinmerac
dimethenamid-p + metazachlor + quinmerac
dimethenamid-p + quinmerac
imazamox + metazachlor + quinmerac
imazamox + quinmerac
metamitron + quinmerac
metazachlor + quinmerac

387   quinoclamine

A selective moss killer for moderate control of moss in managed amenity turf

Products
Mogeton	Certis	25% w/w	WP	15837

Uses
- Moss in **managed amenity turf** *(moderate control)*, **protected ornamentals** *(off-label)*

Extension of Authorisation for Minor Use (EAMUs)
- *protected ornamentals* 20152273

Approval information
- Quinoclamine included in Annex 1 under EC Regulation 1107/2009

Environmental safety
- When treating managed amenity turf, do not apply within 20 m of a static or flowing water body
- LERAP Category A

Hazard classification and safety precautions
Hazard Harmful, Dangerous for the environment
Transport code 9
Packaging group III
UN Number 3077
Risk phrases R22a, R36, R43, R48, R50, R53a, R63
Operator protection A, C, G, H; U05a, U11, U19a, U19e
Environmental protection E15a, E16c
Storage and disposal D01, D02, D12a
Medical advice M03, M05a

388   quinoxyfen

A systemic protectant quinoline fungicide for cereals
FRAC mode of action code: 13

See also cyproconazole + quinoxyfen
fenpropimorph + quinoxyfen

Products
Fortress	Dow	500 g/l	SC	08279

SEE SECTION 3 FOR PRODUCTS ALSO REGISTERED

Uses
- Powdery mildew in **blackcurrants** *(off-label)*, **blueberries** *(off-label)*, **cranberries** *(off-label)*, **durum wheat**, **forest nurseries** *(off-label)*, **gooseberries** *(off-label)*, **hops** *(off-label)*, **ornamental plant production** *(off-label)*, **protected forest nurseries** *(off-label)*, **protected ornamentals** *(off-label)*, **protected strawberries** *(off-label)*, **redcurrants** *(off-label)*, **spring barley**, **spring oats**, **spring rye**, **spring wheat**, **strawberries** *(off-label)*, **sugar beet**, **triticale**, **whitecurrants** *(off-label)*, **winter barley**, **winter oats**, **winter rye**, **winter wheat**

Extension of Authorisation for Minor Use (EAMUs)
- *blackcurrants* 20111459
- *blueberries* 20111459
- *cranberries* 20111459
- *forest nurseries* 20082852
- *gooseberries* 20111459
- *hops* 20061579
- *ornamental plant production* 20082852
- *protected forest nurseries* 20082852
- *protected ornamentals* 20082852
- *protected strawberries* 20041923
- *redcurrants* 20111459
- *strawberries* 20041923
- *whitecurrants* 20111459

Approval information
- Quinoxyfen included in Annex I under Regulation 1107/2009
- Accepted by BBPA for use on malting barley (before ear emergence only) and on hops

Efficacy guidance
- For best results treat at early stage of disease development before infection spreads to new crop growth. Further treatment may be necessary if disease pressure remains high
- Product not curative and will not control latent or established disease infections
- For broad spectrum control in cereals use in tank mixtures - see label
- Product rainfast after 1 h
- Systemic activity may be reduced in severe drought
- Use as part of an integrated crop management (ICM) strategy incorporating other methods of control or fungicides with different modes of action

Restrictions
- Maximum total dose equivalent to two full dose treatments in cereals; see labels for split dose recommendation in sugar beet
- On cereals apply only in the spring from mid-tillering stage (GS 25)

Crop-specific information
- Latest use: first awns visible stage (GS 49) for cereals
- HI sugar beet 28 d

Environmental safety
- Dangerous for the environment
- Very toxic to aquatic organisms
- LERAP Category B

Hazard classification and safety precautions
Hazard Irritant, Dangerous for the environment
Transport code 9
Packaging group III
UN Number 3082
Risk phrases R43, R50, R53a
Operator protection A, C, H; U05a, U14
Environmental protection E15a, E16a, E16b, E34, E38
Storage and disposal D01, D02, D12a

FOR FULL CONDITIONS OF USE ALWAYS READ THE PRODUCT LABEL

389 quizalofop-P-ethyl

An aryl phenoxypropionic acid post-emergence herbicide for grass weed control
HRAC mode of action code: A

Products

1	Pilot Ultra	Nissan	50 g/l	SC	17136
2	Targa Max	Nissan	100 g/l	EC	17135
3	Targa Super	Nissan	50 g/l	EC	17134

Uses
- Annual grasses in **combining peas, linseed, spring field beans, vining peas, winter field beans** [1]; **fodder beet, mangels, potatoes, red beet, spring oilseed rape, sugar beet, winter oilseed rape** [1-3]
- Couch in **combining peas, linseed, spring field beans, vining peas, winter field beans** [1]; **fodder beet, mangels, potatoes, red beet, spring oilseed rape, sugar beet, winter oilseed rape** [1-3]
- Perennial grasses in **combining peas, linseed, spring field beans, vining peas, winter field beans** [1]; **fodder beet, mangels, potatoes, red beet, spring oilseed rape, sugar beet, winter oilseed rape** [1-3]
- Volunteer cereals in **combining peas, linseed, spring field beans, vining peas, winter field beans** [1]; **fodder beet, mangels, potatoes, red beet, spring oilseed rape, sugar beet, winter oilseed rape** [1-3]

Approval information
- Quizalofop-P-ethyl has been included in Annex 1 under EC Regulation 1107/2009

Efficacy guidance
- Best results achieved by application to emerged weeds growing actively in warm conditions with adequate soil moisture
- Weed control may be reduced under conditions such as drought that limit uptake and translocation
- Annual meadow-grass is not controlled
- For effective couch control do not hoe beet crops within 21 d after spraying
- At least 2 h without rain should follow application otherwise results may be reduced
- Quizalofop-P-ethyl is an ACCase inhibitor herbicide. To avoid the build up of resistance do not apply products containing an ACCase inhibitor herbicide more than twice to any crop. In addition do not use any product containing quizalofop-P-ethyl in mixture or sequence with any other product containing the same ingredient
- Use these products as part of a resistance management strategy that includes cultural methods of control and does not use ACCase inhibitors as the sole chemical method of grass weed control
- Applying a second product containing an ACCase inhibitor to a crop will increase the risk of resistance development; only use a second ACCase inhibitor to control different weeds at a different timing
- Always follow WRAG guidelines for preventing and managing herbicide resistant weeds. See Section 5 for more information

Restrictions
- Maximum number of treatments 1 on all recommended crops
- Do not spray crops under stress from any cause or in frosty weather
- Consult processor before use on crops recommended for processing
- An interval of at least 3 d must elapse between treatment and use of another herbicide on beet crops, 14 d on oilseed rape, 21 d on linseed and other recommended crops
- Avoid drift onto neighbouring crops

Crop-specific information
- HI 16 wk for beet crops; 11 wk for oilseed rape, linseed; 8 wk for field beans; 5 wk for peas
- In some situations treatment can cause yellow patches on foliage of peas, especially vining varieties. Symptoms usually rapidly and completely outgrown
- May cause taint in peas

SEE SECTION 3 FOR PRODUCTS ALSO REGISTERED

Pesticide Profiles

Following crops guidance
- In the event of failure of a treated crop broad-leaved crops may be resown after a minimum interval of 2 wk, and cereals after 2-6 wk depending on dose applied
- Onions, leeks and maize are not recommended to follow a failed treated crop

Environmental safety
- Dangerous for the environment
- Toxic to aquatic organisms
- Flammable

Hazard classification and safety precautions
Hazard Dangerous for the environment [1-3]; Harmful if swallowed, Very toxic to aquatic organisms [3]
Transport code 9
Packaging group III
UN Number 3082
Risk phrases H304, H317 [3]; H318 [2, 3]
Operator protection A, C, H; U09a, U20b
Environmental protection E13b, E34, E38 [1-3]; H410 [3]
Storage and disposal D09a, D10c, D12a

390　quizalofop-P-tefuryl

An aryloxyphenoxypropionate herbicide for grass weed control
HRAC mode of action code: A

Products

1	Panarex	Certis	40 g/l	EC	16945
2	Rango	Certis	40 g/l	EC	17006

Uses
- Blackgrass in *combining peas, fodder beet, potatoes, spring field beans, spring linseed, spring oilseed rape, sugar beet, winter field beans, winter linseed, winter oilseed rape*
- Couch in *combining peas, fodder beet, potatoes, spring field beans, spring linseed, spring oilseed rape, sugar beet, winter field beans, winter linseed, winter oilseed rape*
- Italian ryegrass in *combining peas, fodder beet, potatoes, spring field beans, spring linseed, spring oilseed rape, sugar beet, winter field beans, winter linseed, winter oilseed rape*
- Perennial ryegrass in *combining peas, fodder beet, potatoes, spring field beans, spring linseed, spring oilseed rape, sugar beet, winter field beans, winter linseed, winter oilseed rape*
- Volunteer cereals in *combining peas, fodder beet, potatoes, spring field beans, spring linseed, spring oilseed rape, sugar beet, winter field beans, winter linseed, winter oilseed rape*
- Wild oats in *combining peas, fodder beet, potatoes, spring field beans, spring linseed, spring oilseed rape, sugar beet, winter field beans, winter linseed, winter oilseed rape*

Approval information
- Quizalofop-P-tefuryl has been included in Annex 1 under EC Regulation 1107/2009

Efficacy guidance
- Best results on annual grass weeds when growing actively and treated from 2 leaves to the start of tillering
- Treat cover crops when they have served their purpose and the threat of wind blow has passed
- Best results on couch achieved when the weed is growing actively and commencing new rhizome growth
- Grass weeds germinating after treatment will not be controlled
- Treatment quickly stops growth and visible colour changes to the leaf tips appear after about 7 d. Complete kill takes 3-4 wk under good growing conditions
- Quizalofop-P-tefuryl is an ACCase inhibitor herbicide. To avoid the build up of resistance do not apply products containing an ACCase inhibitor herbicide more than twice to any crop. In addition

FOR FULL CONDITIONS OF USE ALWAYS READ THE PRODUCT LABEL

do not use any product containing quizalofop-P-tefuryl in mixture or sequence with any other product containing the same ingredient
- Use these products as part of a resistance management strategy that includes cultural methods of control and does not use ACCase inhibitors as the sole chemical method of grass weed control
- Applying a second product containing an ACCase inhibitor to a crop will increase the risk of resistance development; only use a second ACCase inhibitor to control different weeds at a different timing
- Always follow WRAG guidelines for preventing and managing herbicide resistant weeds. See Section 5 for more information

Restrictions
- Maximum number of treatments 1 per crop
- Consult processors before use on peas or potatoes for processing
- Do not treat crops and weeds growing under stress from any cause

Crop-specific information
- HI 60 d for all crops
- Treat oilseed rape from the fully expanded cotyledon stage
- Treat linseed, peas and field beans from 2-3 unfolded leaves
- Treat sugar beet from 2 unfolded leaves

Following crops guidance
- In the event of failure of a treated crop any broad-leaved crop may be planted at any time. Cereals may be drilled from 4 wk after treatment

Environmental safety
- Dangerous for the environment
- Very toxic to aquatic organisms
- Risk to certain non-target insects or other arthropods. For advice on risk management and use in Integrated Pest Management (IPM) see directions for use
- Avoid spraying within 6 m of the field boundary to reduce effects on non-target insects and other arthropods

Hazard classification and safety precautions
Hazard Toxic [2]; Irritant, Very toxic to aquatic organisms [1]; Dangerous for the environment [1, 2]
Transport code 9
Packaging group III
UN Number 3082
Risk phrases H304, H315, H317, H319, H336, H341, H360 [1]; R41, R43, R51, R53a, R61, R65, R68 [2]
Operator protection A, C, H; U02a, U04a, U05a, U10, U11, U14, U15, U19a, U20b
Environmental protection E15a, E22b, E38 [1, 2]; H410, H411 [1]
Storage and disposal D01, D02, D09a, D10b, D12a
Medical advice M04a [2]; M05b [1, 2]

391 rimsulfuron

A selective systemic sulfonylurea herbicide
HRAC mode of action code: B

Products

1	Caesar	AgChem Access	25% w/w	SG	15272
2	Clayton Bramble	Clayton	25% w/w	SG	15281
3	Clayton Nero	Clayton	25% w/w	SG	15511
4	Clayton Rasp	Clayton	25% w/w	SG	15952
5	Titus	Adama	25% w/w	SG	15050

SEE SECTION 3 FOR PRODUCTS ALSO REGISTERED

Pesticide Profiles

Uses
- Annual dicotyledons in **forage maize, potatoes** [1-5]; **forest nurseries** *(off-label)*, **ornamental plant production** *(off-label)* [5]
- Charlock in **forage maize, potatoes** [4, 5]
- Chickweed in **forage maize, potatoes** [4, 5]
- Cleavers in **forage maize, potatoes** [4, 5]
- Hemp-nettle in **forage maize, potatoes** [4, 5]
- Red dead-nettle in **forage maize, potatoes** [4, 5]
- Redshank in **forage maize, potatoes** [4, 5]
- Scentless mayweed in **forage maize, potatoes** [4, 5]
- Small nettle in **forage maize, potatoes** [4, 5]
- Volunteer oilseed rape in **forage maize, potatoes** [1-5]

Extension of Authorisation for Minor Use (EAMUs)
- *forest nurseries* *20141911* [5], *20141912* [5]
- *ornamental plant production* *20141911* [5], *20141912* [5]

Approval information
- Rimsulfuron included in Annex I under EC Regulation 1107/2009

Efficacy guidance
- Product should be used with a suitable adjuvant or a suitable herbicide tank-mix partner. See label for details
- Product acts by foliar action. Best results obtained from good spray cover of small actively growing weeds. Effectiveness is reduced in very dry conditions
- Weed spectrum can be broadened by tank mixture with other herbicides. See label for details
- Susceptible weeds cease growth immediately and symptoms can be seen 10 d later
- Rimsulfuron is a member of the ALS-inhibitor group of herbicides

Restrictions
- Maximum number of treatments 1 per crop
- Do not treat maize previously treated with organophosphorus insecticides
- Do not apply to potatoes grown for certified seed
- Consult processor before use on crops grown for processing
- Avoid high light intensity (full sunlight) and high temperatures on the day of spraying
- Do not treat during periods of substantial diurnal temperature fluctuation or when frost anticipated
- Do not apply to any crop stressed by drought, water-logging, low temperatures, pest or disease attack, nutrient or lime deficiency
- Do not apply to forage maize treated with organophosphate insecticides
- Do not apply to forage maize undersown with grass or clover

Crop-specific information
- Latest use: before most advanced potato plants are 25 cm high; before 4-collar stage of fodder maize
- All varieties of ware potatoes may be treated, but variety restrictions of any tank-mix partner must be observed
- Only certain named varieties of forage maize may be treated. See label

Following crops guidance
- Only winter wheat should follow a treated crop in the same calendar yr
- Only barley, wheat or maize should be sown in the spring of the yr following treatment
- In the second autumn after treatment any crop except brassicas or oilseed rape may be drilled

Environmental safety
- Dangerous for the environment
- Toxic to aquatic organisms
- Extremely dangerous to fish or other aquatic life. Do not contaminate surface waters or ditches with chemical or used container
- Herbicide is very active. Take particular care to avoid drift onto plants outside the target area

FOR FULL CONDITIONS OF USE ALWAYS READ THE PRODUCT LABEL

- Spraying equipment should not be drained or flushed onto land planted, or to be planted, with trees or crops other than potatoes or forage maize and should be thoroughly cleansed after use - see label for instructions

Hazard classification and safety precautions
Hazard Dangerous for the environment
Transport code 9
Packaging group III
UN Number 3077
Risk phrases R51, R53a
Operator protection A [3-5]; U08, U19a, U20b
Environmental protection E13a, E34, E38
Storage and disposal D01, D09a, D11a, D12a

392 silthiofam

A thiophene carboxamide fungicide seed dressing for cereals
FRAC mode of action code: 38

Products

1	Latitude XL	Monsanto	125 g/l	FS 16467
2	Meridian	AgChem Access	125 g/l	FS 14676

Uses
- Take-all in **spring wheat** (seed treatment), **winter barley** (seed treatment), **winter wheat** (seed treatment)

Approval information
- Silthiofam included in Annex I under EC Regulation 1107/2009
- Accepted by BBPA for use on malting barley

Efficacy guidance
- Apply using approved seed treatment equipment which has been accurately calibrated
- Apply simultaneously (i.e. not in mixture) with a standard seed treatment
- Drill treated seed in the season of purchase. The viability of treated seed and fungicide activity may be reduced by physical storage
- Drill treated seed at 2.5-4 cm into a well prepared firm seedbed
- As precaution against possible development of disease resistance do not treat more than three consecutive susceptible cereal crops in any one rotation

Restrictions
- Maximum number of treatments one per batch of seed
- Do not use on seed with more than 16% moisture content, or on sprouted, cracked or skinned seed
- Test germination of all seed batches before treatment

Crop-specific information
- Latest use: immediately prior to drilling

Hazard classification and safety precautions
UN Number N/C
Operator protection A, H; U20b
Environmental protection E03, E15a, E34
Storage and disposal D01, D02 [1]; D05, D09a, D10a [1, 2]
Treated seed S01, S02, S04b, S05, S06a, S07

SEE SECTION 3 FOR PRODUCTS ALSO REGISTERED

Pesticide Profiles

393 S-metolachlor

A chloroacetamide residual herbicide for use in forage maize and grain maize
HRAC mode of action code: K3

See also mesotrione + s-metolachlor

Products

1	Clayton Smelter	Clayton	960 g/l	EC	14937
2	Dual Gold	Syngenta	960 g/l	EC	14649

Uses

- Annual dicotyledons in **baby leaf crops** (off-label), **forest nurseries** (off-label), **garlic** (off-label), **herbs (see appendix 6)** (off-label), **lettuce** (off-label), **onions** (off-label), **ornamental plant production** (off-label), **red beet** (off-label), **shallots** (off-label), **swedes** (off-label), **turnips** (off-label) [2]
- Annual grasses in **begonias** (off-label), **broccoli** (off-label), **brussels sprouts** (off-label), **cabbages** (off-label), **calabrese** (off-label), **cauliflowers** (off-label), **chinese cabbage** (off-label), **collards** (off-label), **kale** (off-label) [2]; **chicory** (off-label), **dwarf beans** (off-label), **endives** (off-label), **french beans** (off-label), **runner beans** (off-label), **strawberries** (off-label) [1, 2]; **chicory root** (off-label) [1]
- Annual meadow grass in **baby leaf crops** (off-label), **edible podded peas** (off-label), **forest nurseries** (off-label), **garlic** (off-label), **herbs (see appendix 6)** (off-label), **lettuce** (off-label), **onions** (off-label), **ornamental plant production** (off-label), **red beet** (off-label), **shallots** (off-label), **spring field beans** (off-label), **swedes** (off-label), **turnips** (off-label), **vining peas** (off-label) [2]; **forage maize**, **grain maize** [1, 2]
- Black nightshade in **chicory** (off-label), **endives** (off-label) [1, 2]; **chicory root** (off-label) [1]
- Chickweed in **forage maize** (moderately susceptible), **grain maize** (moderately susceptible) [1, 2]; **spring field beans** (off-label) [2]
- Fat hen in **forage maize** (moderately susceptible), **grain maize** (moderately susceptible) [1, 2]; **spring field beans** (off-label) [2]
- Field speedwell in **forage maize**, **grain maize** [1, 2]; **spring field beans** (off-label) [2]
- Mayweeds in **forage maize**, **grain maize** [1, 2]; **spring field beans** (off-label) [2]
- Red dead-nettle in **forage maize**, **grain maize** [1, 2]
- Sowthistle in **forage maize**, **grain maize** [1, 2]

Extension of Authorisation for Minor Use (EAMUs)

- **baby leaf crops** *20120594* [2]
- **begonias** *20132574* [2]
- **broccoli** *20111258* [2]
- **brussels sprouts** *20111259* [2]
- **cabbages** *20111259* [2]
- **calabrese** *20111258* [2]
- **cauliflowers** *20111258* [2]
- **chicory** *20111262* [1], *20111257* [2]
- **chicory root** *20103107* [1]
- **chinese cabbage** *20111260* [2]
- **collards** *20111260* [2]
- **dwarf beans** *20103103* [1], *20132572* [2]
- **edible podded peas** *20140164* [2]
- **endives** *20103107* [1], *20111262* [1], *20111257* [2]
- **forest nurseries** *20120501* [2]
- **french beans** *20103103* [1], *20132572* [2]
- **garlic** *20110840* [2]
- **herbs (see appendix 6)** *20120594* [2]
- **kale** *20111260* [2]
- **lettuce** *20120594* [2]
- **onions** *20110840* [2]
- **ornamental plant production** *20120501* [2]
- **red beet** *20111006* [2]

FOR FULL CONDITIONS OF USE ALWAYS READ THE PRODUCT LABEL

sodium hypochlorite (commodity substance)

- *runner beans* 20103103 [1], 20132572 [2]
- *shallots* 20110840 [2]
- *spring field beans* 20140163 [2]
- *strawberries* 20103104 [1], 20132573 [2]
- *swedes* 20111006 [2]
- *turnips* 20111006 [2]
- *vining peas* 20140164 [2]

Approval information
- S-metolachlor included in Annex I under EC Regulation 1107/2009

Environmental safety
- LERAP Category B

Hazard classification and safety precautions
    Hazard Irritant, Dangerous for the environment
    Transport code 9
    Packaging group III
    UN Number 3082
    Risk phrases R43, R50, R53a
    Operator protection A [1, 2]; C [2]; H [1]; U05a, U09a, U20b
    Environmental protection E15b, E16a, E16b, E38
    Storage and disposal D01, D02, D09a, D10c, D11a, D12a

394   sodium hypochlorite (commodity substance)

An inorganic horticultural bactericide for use in mushrooms

Products
    sodium hypochlorite   various   100% w/v   ZZ   -

Uses
- Bacterial blotch in **mushrooms**

Approval information
- Sodium hypochlorite included in Annex 1 under EC Regulation 1107/2009
- Approval as a commodity substance valid until 31 Aug 2019

Restrictions
- Maximum concentration 315 mg/litre of water
- Mixing and loading must only take place in a ventilated area
- Must only be used by suitably trained and competent operators

Crop-specific information
- HI 1 d

Environmental safety
- Harmful to fish or other aquatic life. Do not contaminate surface waters or ditches with chemical or used container

Hazard classification and safety precautions
    Operator protection A, C, H
    Environmental protection E13c

395   spearmint oil

A hot fogging treatment for post-harvest sprout suppression in potatoes

Products
    Biox-M   Juno   100% w/w   HN   16021

SEE SECTION 3 FOR PRODUCTS ALSO REGISTERED

Uses
- Sprout suppression in **potatoes**

Approval information
- Not listed in Annex 1 under EC Regulation 1107/2009 since it is a natural product.

Hazard classification and safety precautions
Hazard Harmful, Dangerous for the environment
UN Number N/C
Risk phrases R22b, R43, R50, R53a
Operator protection A, C, D, H, J, M
Storage and disposal D12a

396 spinosad

A selective insecticide derived from naturally occurring soil fungi (naturalyte)
IRAC mode of action code: 5

Products
1	Conserve	Fargro	120 g/l	SC	12058
2	Tracer	Landseer	480 g/l	SC	12438

Uses
- Asparagus beetle in **asparagus** (off-label) [2]
- Cabbage moth in **brussels sprouts**, **cabbages**, **cauliflowers**, **collards** (off-label), **kale** (off-label), **kohlrabi** (off-label), **oriental cabbage** (off-label) [2]
- Cabbage root fly in **collards** (off-label), **kale** (off-label), **kohlrabi** (off-label), **oriental cabbage** (off-label) [2]
- Cabbage white butterfly in **brussels sprouts**, **cabbages**, **cauliflowers**, **collards** (off-label), **kale** (off-label), **kohlrabi** (off-label), **oriental cabbage** (off-label) [2]
- Caterpillars in **celery (outdoor)** (off-label), **collards** (off-label), **fennel** (off-label), **kale** (off-label), **kohlrabi** (off-label), **oriental cabbage** (off-label) [2]
- Celery fly in **celery (outdoor)** (off-label), **fennel** (off-label) [2]
- Celery leaf miner in **celery (outdoor)** (off-label), **fennel** (off-label) [2]
- Codling moth in **apples**, **pears** [2]
- Diamond-back moth in **brussels sprouts**, **cabbages**, **cauliflowers**, **collards** (off-label), **kale** (off-label), **kohlrabi** (off-label), **oriental cabbage** (off-label) [2]
- Insect pests in **blackberries** (off-label), **blackcurrants** (off-label), **celery leaves** (off-label), **chard** (off-label), **cress** (off-label), **endives** (off-label), **frise** (off-label), **grapevines** (off-label), **herbs (see appendix 6)**, **hops** (off-label), **lamb's lettuce** (off-label), **leaf brassicas** (off-label), **lettuce** (off-label), **ornamental specimens** (off-label), **protected blackberries** (off-label), **protected herbs (see appendix 6)** (off-label), **protected raspberries** (off-label), **radicchio** (off-label), **raspberries** (off-label), **red mustard** (off-label), **redcurrants** (off-label), **rocket** (off-label), **scarole** (off-label), **soft fruit** (off-label), **spinach** (off-label), **spinach beet** (off-label), **top fruit** (off-label), **whitecurrants** (off-label) [2]; **protected tomatoes** (off-label) [1]
- Large white butterfly in **collards** (off-label), **kale** (off-label), **kohlrabi** (off-label), **oriental cabbage** (off-label) [2]
- Light brown apple moth in **blueberries** (off-label), **cranberries** (off-label), **gooseberries** (off-label), **haskap** (off-label), **protected blackcurrants** (off-label), **protected blueberry** (off-label), **protected gooseberries** (off-label), **protected haskap** (off-label), **protected redcurrants** (off-label) [2]
- Liriomyza bryoniae in **protected tomatoes** (off-label) [1]
- Moths in **hops** (off-label), **ornamental specimens** (off-label), **soft fruit** (off-label), **top fruit** (off-label) [2]
- Sawflies in **blackcurrants** (off-label), **redcurrants** (off-label), **whitecurrants** (off-label) [2]
- Silver Y moth in **celery (outdoor)** (off-label), **dwarf beans** (off-label), **fennel** (off-label), **french beans** (off-label) [2]
- Small white butterfly in **brussels sprouts**, **cabbages**, **cauliflowers**, **collards** (off-label), **kale** (off-label), **kohlrabi** (off-label), **oriental cabbage** (off-label) [2]

FOR FULL CONDITIONS OF USE ALWAYS READ THE PRODUCT LABEL

spinosad

- Spotted wing drosophila in **blueberries** *(off-label)*, **cranberries** *(off-label)*, **damsons** *(off-label)*, **gooseberries** *(off-label)*, **haskap** *(off-label)*, **plums** *(off-label)*, **protected blackcurrants** *(off-label)*, **protected blueberry** *(off-label)*, **protected gooseberries** *(off-label)*, **protected haskap** *(off-label)*, **protected redcurrants** *(off-label)*, **rubus hybrids** *(off-label)*, **strawberries** *(off-label)* [2]
- Thrips in **asparagus** *(off-label)*, **blackberries** *(off-label)*, **bulb onions**, **celery (outdoor)** *(off-label)*, **fennel** *(off-label)*, **leeks, protected blackberries** *(off-label)*, **protected raspberries** *(off-label)*, **raspberries** *(off-label)*, **salad onions** [2]; **protected tomatoes** *(off-label)* [1]
- Tortrix moths in **apples**, **pears** [2]
- Western flower thrips in **protected aubergines**, **protected cucumbers**, **protected ornamentals, protected peppers, protected tomatoes** [1]
- Winter moth in **blackcurrants** *(off-label)*, **redcurrants** *(off-label)*, **whitecurrants** *(off-label)* [2]

Extension of Authorisation for Minor Use (EAMUs)
- **asparagus** *20140301* [2]
- **blackberries** *20092118* [2]
- **blackcurrants** *20113223* [2]
- **blueberries** *20150775* [2]
- **celery (outdoor)** *20131946* [2]
- **celery leaves** *20081290* [2]
- **chard** *20081290* [2]
- **collards** *20150102* [2]
- **cranberries** *20150775* [2]
- **cress** *20081290* [2]
- **damsons** *20161562* [2]
- **dwarf beans** *20150103* [2]
- **endives** *20081290* [2]
- **fennel** *20131946* [2]
- **french beans** *20150103* [2]
- **frise** *20081290* [2]
- **gooseberries** *20150775* [2]
- **grapevines** *20122222* [2]
- **haskap** *20150775* [2]
- **hops** *20082908* [2]
- **kale** *20150102* [2]
- **kohlrabi** *20150102* [2]
- **lamb's lettuce** *20081290* [2]
- **leaf brassicas** *20081290* [2]
- **lettuce** *20081290* [2]
- **oriental cabbage** *20150102* [2]
- **ornamental specimens** *20082908* [2]
- **plums** *20161562* [2]
- **protected blackberries** *20130245* [2]
- **protected blackcurrants** *20150775* [2]
- **protected blueberry** *20150775* [2]
- **protected gooseberries** *20150775* [2]
- **protected haskap** *20150775* [2]
- **protected herbs (see appendix 6)** *20081290* [2]
- **protected raspberries** *20130245* [2]
- **protected redcurrants** *20150775* [2]
- **protected tomatoes** *20130325* [1]
- **radicchio** *20081290* [2]
- **raspberries** *20092118* [2]
- **red mustard** *20081290* [2]
- **redcurrants** *20113223* [2]
- **rocket** *20081290* [2]
- **rubus hybrids** *20142018* [2]
- **scarole** *20081290* [2]
- **soft fruit** *20082908* [2]

SEE SECTION 3 FOR PRODUCTS ALSO REGISTERED

Pesticide Profiles

- ***spinach*** *20081290* [2]
- ***spinach beet*** *20081290* [2]
- ***strawberries*** *20161540* [2]
- ***top fruit*** *20082908* [2]
- ***whitecurrants*** *20113223* [2]

Approval information
- Spinosad included in Annex I under EC Regulation 1107/2009

Efficacy guidance
- Product enters insects by contact from a treated surface or ingestion of treated plant material therefore good spray coverage is essential
- Some plants, for example Fuchsia flowers, can provide effective refuges from spray deposits and control of western flower thrips may be reduced [1]
- Apply to protected crops when western flower thrip nymphs or adults are first seen [1]
- Monitor western flower thrip development carefully to see whether further applications are necessary. A 2 spray programme at 5-7 d intervals may be needed when conditions favour rapid pest development [1]
- Treat top fruit and field crops when pests are first seen or at very first signs of crop damage [2]
- Ensure a rain-free period of 12 h after treatment before applying irrigation [2]
- To reduce possibility of development of resistance, adopt resistance management measures. See label and Section 5 for more information

Restrictions
- Maximum number of treatments 4 per crop for brassicas, onions, leeks [2]; 1 pre-blossom and 3 post blossom for apples, pears [2]
- Apply no more than 2 consecutive sprays. Rotate with another insecticide with a different mode of action or use no further treatment after applying the maximum number of treatments
- Establish whether any incoming plants have been treated and apply no more than 2 consecutive sprays. Maximum number of treatments 6 per structure per yr
- Avoid application in bright sunlight or into open flowers [1]

Crop-specific information
- HI 3 d for protected cucumbers, brassicas; 7 d for apples, pears, onions, leeks
- Test for tolerance on a small number of ornamentals or cucumbers before large scale treatment
- Some spotting of african violet flowers may occur

Environmental safety
- Dangerous for the environment
- Very toxic to aquatic organisms
- Whenever possible use an Integrated Pest Management system. Spinosad presents low risk to beneficial arthropods
- Product has low impact on many insect and mite predators but is harmful to adults of most parasitic wasps. Most beneficials may be introduced to treated plants when spray deposits are dry but an interval of 2 wk should elapse before introduction of parasitic wasps. See label for details
- Treatment may cause temporary reduction in abundance of insect and mite predators if present at application
- Exposure to direct spray is harmful to bees but dry spray deposits are harmless. Treatment of field crops should not be made in the heat of the day when bees are actively foraging [2]
- Broadcast air-assisted LERAP [2] (40 m); LERAP Category B [2]

Hazard classification and safety precautions
 Hazard Dangerous for the environment
 Transport code 9
 Packaging group III
 UN Number 3082
 Risk phrases R50, R53a
 Operator protection A, H [2]; U05a [1]; U08, U20b [2]
 Environmental protection E15a, E16a [2]; E15b [1]; E17b [2] (40 m); E34, E38 [1, 2]
 Storage and disposal D01, D05 [1]; D02, D10b, D12a [1, 2]

FOR FULL CONDITIONS OF USE ALWAYS READ THE PRODUCT LABEL

397 spirodiclofen

A tetronic acid which inhibits lipid biosynthesis
IRAC mode of action code: 23

Products

1	Envidor	Bayer CropScience	240 g/l	SC	13947
2	Envidor	Bayer CropScience	240 g/l	SC	17518

Uses
- Mussel scale in *apples*, *pears* [1, 2]
- Pear sucker in *pears* [1, 2]
- Red spider mites in *apples*, *pears* [1, 2]; *hops* (off-label) [1]
- Rust mite in *apples*, *pears* [1, 2]; *plums* (off-label) [2]
- Spider mites in *apricots* (off-label), *blackcurrants* (off-label), *gooseberries* (off-label), *nectarines* (off-label), *peaches* (off-label), *plums* (off-label), *protected chilli peppers* (off-label), *protected cucumbers* (off-label), *protected peppers* (off-label), *protected tomatoes* (off-label), *redcurrants* (off-label), *whitecurrants* (off-label) [2]; *cherries* (off-label), *ornamental plant production* (off-label) [1, 2]; *chillies* (off-label), *cucumbers* (off-label), *hops* (off-label), *peppers* (off-label), *protected ornamentals* (off-label), *protected strawberries* (off-label), *soft fruit* (off-label), *strawberries* (off-label), *tomatoes* (off-label) [1]
- Two-spotted spider mite in *apples*, *pears* [1, 2]; *protected strawberries* (off-label), *strawberries* (off-label) [1]

Extension of Authorisation for Minor Use (EAMUs)
- *apricots* 20160999 [2]
- *blackcurrants* 20161000 [2]
- *cherries* 20151932 expires 31 Dec 2017 [1], 20160998 [2]
- *chillies* 20162025 [1]
- *cucumbers* 20162025 [1]
- *gooseberries* 20161000 [2]
- *hops* 20120253 expires 31 Dec 2017 [1], 20131654 expires 31 Dec 2017 [1]
- *nectarines* 20160999 [2]
- *ornamental plant production* 20120254 expires 31 Dec 2017 [1], 20160997 [2]
- *peaches* 20160999 [2]
- *peppers* 20162025 [1]
- *plums* 20160995 [2]
- *protected chilli peppers* 20160996 [2]
- *protected cucumbers* 20160996 [2]
- *protected ornamentals* 20120254 expires 31 Dec 2017 [1]
- *protected peppers* 20160996 [2]
- *protected strawberries* 20120255 expires 31 Dec 2017 [1]
- *protected tomatoes* 20160996 [2]
- *redcurrants* 20161000 [2]
- *soft fruit* 20120253 expires 31 Dec 2017 [1]
- *strawberries* 20120255 expires 31 Dec 2017 [1]
- *tomatoes* 20162025 [1]
- *whitecurrants* 20161000 [2]

Approval information
- Spirodiclofen included in Annex 1 under EC Regulation 1107/2009

Restrictions
- Consult processor on crops grown for processing or for cider production

Environmental safety
- Broadcast air-assisted LERAP (18 m)

Hazard classification and safety precautions
Hazard Harmful
UN Number N/C

SEE SECTION 3 FOR PRODUCTS ALSO REGISTERED

Pesticide Profiles

Risk phrases H317, H351 [2]; R40, R43, R52, R53a [1]
Operator protection A, H; U05a, U20c
Environmental protection E12c, E12f, E15b, E22c, E34, E38 [1, 2]; E17b [1, 2] (18 m); H410 [2]
Storage and disposal D01, D02, D09a, D10c, D12a
Medical advice M05a

398 spirotetramat

Tetramic acid derivative that acts on lipid synthesis
IRAC mode of action code: 23

Products
Movento   Bayer CropScience   150 g/l   OD   14446

Uses
- Aphids in **blackcurrants** *(off-label)*, **blueberries** *(off-label)*, **broccoli**, **brussels sprouts**, **cabbages**, **calabrese**, **cauliflowers**, **chard** *(off-label)*, **chinese cabbage** *(off-label)*, **choi sum** *(off-label)*, **cress** *(off-label)*, **endives** *(off-label)*, **frise** *(off-label)*, **gooseberries** *(off-label)*, **herbs (see appendix 6)** *(off-label)*, **hops** *(off-label)*, **lamb's lettuce** *(off-label)*, **leaf brassicas** *(off-label)*, **lettuce**, **mustard** *(off-label)*, **pak choi** *(off-label)*, **protected chard** *(off-label)*, **protected cress** *(off-label)*, **protected endives** *(off-label)*, **protected frise** *(off-label)*, **protected lamb's lettuce** *(off-label)*, **protected leaf brassicas** *(off-label)*, **protected rocket** *(off-label)*, **protected scarole** *(off-label)*, **protected spinach** *(off-label)*, **red mustard** *(off-label)*, **redcurrants** *(off-label)*, **rocket** *(off-label)*, **salad greens** *(off-label)*, **scarole** *(off-label)*, **sorrel** *(off-label)*, **spinach** *(off-label)*, **spinach beet** *(off-label)*, **tatsoi** *(off-label)*, **whitecurrants** *(off-label)*, **wine grapes** *(off-label)*
- Carrot willow aphid in **carrots** *(off-label)*, **parsnips** *(off-label)*, **swedes** *(off-label)*, **turnips** *(off-label)*
- Damson-hop aphid in **hops** *(off-label)*
- Insect pests in **forest nurseries** *(off-label)*, **ornamental plant production** *(off-label)*
- Parsnip aphid in **carrots** *(off-label)*, **parsnips** *(off-label)*, **swedes** *(off-label)*, **turnips** *(off-label)*
- Peach-potato aphid in **carrots** *(off-label)*, **parsnips** *(off-label)*, **swedes** *(off-label)*, **turnips** *(off-label)*
- White tip in **chard** *(off-label)*
- Whitefly in **broccoli**, **brussels sprouts**, **cabbages**, **calabrese**, **cauliflowers**, **chinese cabbage** *(off-label)*, **choi sum** *(off-label)*, **cress** *(off-label)*, **endives** *(off-label)*, **frise** *(off-label)*, **herbs (see appendix 6)** *(off-label)*, **lamb's lettuce** *(off-label)*, **leaf brassicas** *(off-label)*, **mustard** *(off-label)*, **pak choi** *(off-label)*, **protected chard** *(off-label)*, **protected cress** *(off-label)*, **protected endives** *(off-label)*, **protected frise** *(off-label)*, **protected lamb's lettuce** *(off-label)*, **protected leaf brassicas** *(off-label)*, **protected rocket** *(off-label)*, **protected scarole** *(off-label)*, **protected spinach** *(off-label)*, **red mustard** *(off-label)*, **rocket** *(off-label)*, **salad greens** *(off-label)*, **scarole** *(off-label)*, **sorrel** *(off-label)*, **spinach** *(off-label)*, **spinach beet** *(off-label)*, **tatsoi** *(off-label)*
- Willow parsnip aphid in **carrots** *(off-label)*, **parsnips** *(off-label)*, **swedes** *(off-label)*, **turnips** *(off-label)*

Extension of Authorisation for Minor Use (EAMUs)
- *blackcurrants 20121401*
- *blueberries 20121401*
- *carrots 20161305*
- *chard 20102410*
- *chinese cabbage 20101095*
- *choi sum 20101095*
- *cress 20102410, 20140394*
- *endives 20102410, 20140394*
- *forest nurseries 20111987*
- *frise 20102410*
- *gooseberries 20121401*
- *herbs (see appendix 6) 20140394*
- *hops 20102117*
- *lamb's lettuce 20102410*

FOR FULL CONDITIONS OF USE ALWAYS READ THE PRODUCT LABEL

- *leaf brassicas* 20102410
- *mustard* 20102410
- *ornamental plant production* 20111987
- *pak choi* 20101095
- *parsnips* 20161305
- *protected chard* 20102410
- *protected cress* 20102410
- *protected endives* 20102410
- *protected frise* 20102410
- *protected lamb's lettuce* 20102410
- *protected leaf brassicas* 20102410
- *protected rocket* 20102410
- *protected scarole* 20102410
- *protected spinach* 20102410
- *red mustard* 20140394
- *redcurrants* 20121401
- *rocket* 20102410
- *salad greens* 20140394
- *scarole* 20102410
- *sorrel* 20140394
- *spinach* 20102410
- *spinach beet* 20140394
- *swedes* 20161305
- *tatsoi* 20101095
- *turnips* 20161305
- *whitecurrants* 20121401
- *wine grapes* 20113085

Approval information
- Spirotetramat is included in Annex 1 under EC Regulation 1107/2009
- Accepted by BBPA for use on hops

Hazard classification and safety precautions
Hazard Irritant, Dangerous for the environment
Transport code 9
Packaging group III
UN Number 3082
Risk phrases H317, H319, H361
Operator protection A; U05a, U14, U20b
Environmental protection E12c, E15b, E22c, E34, E38, H411
Storage and disposal D02, D09a, D10c, D12a
Medical advice M03

399 spiroxamine

A spiroketal amine fungicide for cereals
FRAC mode of action code: 5

See also bixafen + prothioconazole + spiroxamine
prothioconazole + spiroxamine
prothioconazole + spiroxamine + tebuconazole

Products
Torch   Bayer CropScience   500 g/l   EW   11258

Uses
- Brown rust in **spring barley**, **spring rye**, **spring wheat**, **winter barley**, **winter rye**, **winter wheat**
- Powdery mildew in **forest nurseries** (off-label), **ornamental plant production** (off-label), **spring barley**, **spring rye**, **spring wheat**, **winter barley**, **winter rye**, **winter wheat**

SEE SECTION 3 FOR PRODUCTS ALSO REGISTERED

Pesticide Profiles

- Rhynchosporium in **spring barley** *(reduction)*, **spring rye** *(reduction)*, **winter barley** *(reduction)*, **winter rye** *(reduction)*
- Yellow rust in **spring barley**, **spring rye**, **spring wheat**, **winter barley**, **winter rye**, **winter wheat**

Extension of Authorisation for Minor Use (EAMUs)
- *forest nurseries* 20151897 expires 30 Jun 2017
- *ornamental plant production* 20151897 expires 30 Jun 2017

Approval information
- Spiroxamine included in Annex I under EC Regulation 1107/2009
- Accepted by BBPA for use on malting barley
- Approval expiry 30 Jun 2017 [1]

Efficacy guidance
- For best results treat at an early stage of disease development before infection spreads to new growth
- To reduce the risk of development of resistance avoid repeat treatments on diseases such as powdery mildew. If necessary tank mix or alternate with other non-morpholine fungicides

Restrictions
- Maximum total dose equivalent to two full dose treatments

Crop-specific information
- Latest use: before caryopsis watery ripe (GS 71) for wheat, rye; ear emergence complete (GS 59) for barley

Environmental safety
- Dangerous for the environment
- Very toxic to aquatic organisms
- LERAP Category B

Hazard classification and safety precautions
    Hazard Harmful, Dangerous for the environment
    Risk phrases R20, R22a, R38, R41, R43, R50, R53a
    Operator protection A, C, H; U05a, U11, U13, U14, U15, U20a
    Environmental protection E13b, E16a, E34, E38
    Storage and disposal D01, D02, D05, D09a, D10b, D12a
    Medical advice M03

400 starch, protein, oil and water

natural plant extracts for control of moss, liverworts and algae

Products
    MossKade   Fargro   n/a g/l   SL   00000

Uses
- Algae in **hard surfaces**
- Liverworts in **hard surfaces**
- Moss in **hard surfaces**

Approval information
- No listing required on Annex 1

Efficacy guidance
- Treat surfaces at first sign of moss, lichen, liverworts or algae since young growth is more susceptible.
- Must dry on the treated surface before it begins to work
- Summer growth is less susceptible than growth developing in autumn, winter or spring
- Repeat applications may be necessary for full efficacy

FOR FULL CONDITIONS OF USE ALWAYS READ THE PRODUCT LABEL

sulfosulfuron

Hazard classification and safety precautions
UN Number N/C
Operator protection U08, U09a, U19a
Storage and disposal D06c, D07

401 sulfosulfuron

A sulfonylurea herbicide for grass and broad-leaved weed control in winter wheat
HRAC mode of action code: B

See also glyphosate + sulfosulfuron

Products

Monitor	Interfarm	80% w/w	WG	17695

Uses
- Brome grasses in **winter wheat** *(moderate control of barren brome)*
- Chickweed in **winter wheat**
- Cleavers in **winter wheat**
- Couch in **winter wheat** *(moderate control only)*
- Loose silky bent in **winter wheat**
- Mayweeds in **winter wheat**
- Onion couch in **winter wheat** *(moderate control only)*

Approval information
- Sulfosulfuron included in Annex I under EC Regulation 1107/2009

Efficacy guidance
- For best results treat in early spring when annual weeds are small and growing actively. Avoid treatment when weeds are dormant for any reason
- Best control of onion couch is achieved when the weed has more than two leaves. Effects on bulbils or on growth in the following yr have not been examined
- An extended period of dry weather before or after treatment may result in reduced control
- The addition of a recommended surfactant is essential for full activity
- Specific follow-up treatments may be needed for complete control of some weeds
- Use only where competitively damaging weed populations have emerged otherwise yield may be reduced
- Sulfosulfuron is a member of the ALS-inhibitor group of herbicides. To avoid the build up of resistance do not use any product containing an ALS-inhibitor herbicide with claims for control of grass weeds more than once on any crop
- Use these products as part of a resistance management strategy that includes cultural methods of control and does not use ALS inhibitors as the sole chemical method of grass weed control

Restrictions
- Maximum total dose equivalent to one treatment at full dose
- Do not treat crops under stress
- Apply only after 1 Feb and from 3 expanded leaf stage
- Do not treat durum wheat or any undersown wheat crop
- Do not use in mixture or in sequence with any other sulfonyl urea herbicide on the same crop

Crop-specific information
- Latest use: flag leaf ligule just visible for winter wheat (GS 39)

Following crops guidance
- In the autumn following treatment winter wheat, winter rye, winter oats, triticale, winter oilseed rape, winter peas or winter field beans may be sown on any soil, and winter barley on soils with less than 60% sand
- In the next spring following application crops of wheat, barley, oats, maize, peas, beans, linseed, oilseed rape, potatoes or grass may be sown
- In the second autumn following application winter linseed may be sown, and winter barley on soils with more than 60% sand

SEE SECTION 3 FOR PRODUCTS ALSO REGISTERED

Pesticide Profiles

- Sugar beet or any other crop not mentioned above must not be drilled until the second spring following application
- Where winter oilseed rape is to be sown in the autumn following treatment soil cultivation to a minimum of 10 cm is recommended

Environmental safety
- Dangerous for the environment
- Very toxic to aquatic organisms
- Extremely dangerous to fish or other aquatic life. Do not contaminate surface waters or ditches with chemical or used container
- Take extreme care to avoid drift onto broad leaved plants or other crops, or onto ponds, waterways or ditches.
- Follow detailed label instructions for cleaning the sprayer to avoid damage to sensitive crops during subsequent use
- LERAP Category B

Hazard classification and safety precautions
Hazard Dangerous for the environment, Very toxic to aquatic organisms
UN Number N/C
Operator protection U20b
Environmental protection E15a, E16a, E16b, E34, E38, H410
Storage and disposal D09a, D11a, D12a

402 sulphur

A broad-spectrum inorganic protectant fungicide, foliar feed and acaricide
FRAC mode of action code: M2

Products

1	Kumulus DF	BASF	80% w/w	WG	04707
2	Microthiol Special	UPL Europe	80% w/w	MG	16989

Uses
- Disease control/foliar feed in **bilberries** *(off-label)*, **blueberries** *(off-label)*, **cranberries** *(off-label)*, **redcurrants** *(off-label)*, **whitecurrants** *(off-label)* [1]
- Foliar feed in **spring barley**, **spring wheat**, **winter barley**, **winter wheat** [1, 2]; **spring oats**, **sugar beet**, **swedes**, **turnips**, **winter oats** [2]; **winter oilseed rape** [1]
- Gall mite in **blackcurrants** [1]
- Powdery mildew in **apples**, **gooseberries**, **hops**, **strawberries**, **sugar beet** [1, 2]; **blackcurrants**, **borage for oilseed production** *(off-label)*, **combining peas** *(off-label)*, **fodder beet** *(off-label)*, **grapevines**, **grass seed crops** *(off-label)*, **parsnips** *(off-label)*, **protected aubergines** *(off-label)*, **protected cayenne peppers** *(off-label)*, **protected chilli peppers** *(off-label)*, **protected cucumbers** *(off-label)*, **protected herbs (see appendix 6)** *(off-label)*, **protected peppers** *(off-label)*, **protected tomatoes** *(off-label)*, **redcurrants**, **spring barley**, **spring oats**, **spring wheat**, **swedes**, **turnips**, **vining peas** *(off-label)*, **winter barley**, **winter oats**, **winter wheat** [2]; **wine grapes** [1]
- Scab in **apples** [1]

Extension of Authorisation for Minor Use (EAMUs)
- *bilberries 20061042* [1]
- *blueberries 20061042* [1]
- *borage for oilseed production 20151521* [2]
- *combining peas 20151524* [2]
- *cranberries 20061042* [1]
- *fodder beet 20151526* [2]
- *grass seed crops 20151523* [2]
- *parsnips 20151527* [2]
- *protected aubergines 20151520* [2]
- *protected cayenne peppers 20151520* [2]
- *protected chilli peppers 20151520* [2]

FOR FULL CONDITIONS OF USE ALWAYS READ THE PRODUCT LABEL

tau-fluvalinate

- *protected cucumbers* 20151520 [2]
- *protected herbs (see appendix 6)* 20151522 [2]
- *protected peppers* 20151520 [2]
- *protected tomatoes* 20151520 [2]
- *redcurrants* 20061042 [1]
- *vining peas* 20151524 [2]
- *whitecurrants* 20061042 [1]

Approval information
- Sulphur is included in Annex 1 under EC Regulation 1107/2009
- Accepted by BBPA for use on malting barley and hops (before burr)

Efficacy guidance
- Apply when disease first appears and repeat 2-3 wk later. Details of application rates and timing vary with crop, disease and product. See label for information
- Sulphur acts as foliar feed as well as fungicide and with some crops product labels vary in whether treatment recommended for disease control or growth promotion
- In grassland best results obtained at least 2 wk before cutting for hay or silage, 3 wk before grazing
- Treatment unlikely to be effective if disease already established in crop

Restrictions
- Maximum number of treatments normally 2 per crop for grassland, sugar beet, parsnips, swedes, hops, protected herbs; 3 per yr for blackcurrants, gooseberries; 4 per crop on apples, pears but labels vary
- Do not use on sulphur-shy apples (Beauty of Bath, Belle de Boskoop, Cox's Orange Pippin, Lanes Prince Albert, Lord Derby, Newton Wonder, Rival, Stirling Castle) or pears (Doyenne du Comice)
- Do not use on gooseberry cultivars Careless, Early Sulphur, Golden Drop, Leveller, Lord Derby, Roaring Lion, or Yellow Rough
- Do not use on apples or gooseberries when young, under stress or if frost imminent
- Do not use on fruit for processing, on grapevines during flowering or near harvest on grapes for wine-making
- Do not use on hops at or after burr stage
- Do not spray top or soft fruit with oil or within 30 d of an oil-containing spray

Crop-specific information
- Latest use: before burr stage for hops; before end Sep for parsnips, swedes, sugar beet; fruit swell for gooseberries; milky ripe stage for cereals
- HI cutting grass for hay or silage 2 wk; grazing grassland 3 wk

Environmental safety
- Sulphur products are attractive to livestock and must be kept out of their reach
- Do not empty into drains

Hazard classification and safety precautions
  UN Number N/C
  Operator protection U20b [2]; U20c [1]
  Environmental protection E15a
  Storage and disposal D01 [2]; D09a, D11a [1, 2]

403 tau-fluvalinate

A contact pyrethroid insecticide for cereals and oilseed rape
IRAC mode of action code: 3

Products

1	Greencrop Malin	Greencrop	240 g/l	EW	11787
2	Mavrik	Adama	240 g/l	EW	10612

SEE SECTION 3 FOR PRODUCTS ALSO REGISTERED

Uses
- Aphids in **borage for oilseed production** (off-label), **durum wheat** (off-label), **evening primrose** (off-label), **grass seed crops** (off-label), **honesty** (off-label), **linseed** (off-label), **mustard** (off-label), **rye** (off-label), **spring linseed** (off-label), **spring rye** (off-label), **triticale** (off-label), **winter rye** (off-label) [2]; **spring barley**, **spring oilseed rape**, **spring wheat**, **winter barley**, **winter oilseed rape**, **winter wheat** [1, 2]
- Barley yellow dwarf virus vectors in **winter barley**, **winter wheat** [1, 2]
- Cabbage stem flea beetle in **winter oilseed rape** [2]
- Insect pests in **grass seed crops** (off-label), **mustard** (off-label) [1]
- Pollen beetle in **spring oilseed rape**, **winter oilseed rape** [1, 2]

Extension of Authorisation for Minor Use (EAMUs)
- *borage for oilseed production* 20061366 [2], 20141329 [2]
- *durum wheat* 20061365 [2], 20141328 [2]
- *evening primrose* 20061366 [2], 20141329 [2]
- *grass seed crops* 20061368 [1], 20061365 [2], 20141328 [2]
- *honesty* 20061366 [2], 20141329 [2]
- *linseed* 20141329 [2]
- *mustard* 20061367 [1], 20061366 [2], 20141329 [2]
- *rye* 20141328 [2]
- *spring linseed* 20061366 [2]
- *spring rye* 20061365 [2]
- *triticale* 20061365 [2], 20141328 [2]
- *winter rye* 20061365 [2]

Approval information
- Tau-fluvalinate included in Annex 1 under EC Regulation 1107/2009
- Accepted by BBPA for use on malting barley

Efficacy guidance
- For BYDV control on winter cereals follow local warnings or spray high risk crops in mid-Oct and make repeat application in late autumn/early winter if aphid activity persists
- For summer aphid control on cereals spray once when aphids present on two thirds of ears and increasing
- On oilseed rape treat peach potato aphids in autumn in response to local warning and repeat if necessary
- Best control of pollen beetle in oilseed rape obtained from treatment at green to yellow bud stage and repeat if necessary
- Good spray cover of target essential for best results
- Do not mix with boron or products containing boron [2]

Restrictions
- Maximum total dose equivalent to two full dose treatments on oilseed rape. See label for dose rates on cereals
- A minimum of 14 d must elapse between applications to cereals

Crop-specific information
- Latest use: before caryopsis watery ripe (GS 71) for barley; before flowering for oilseed rape; before kernel medium milk (GS 75) for wheat

Environmental safety
- Dangerous for the environment
- Very toxic to aquatic organisms
- High risk to non-target insects or other arthropods. Do not spray within 6 m of the field boundary [1, 2]
- Avoid spraying oilseed rape within 6 m of field boundary to reduce effects on certain non-target species or other arthropods
- Must not be applied to cereals if any product containing a pyrethroid insecticide or dimethoate has been sprayed after the start of ear emergence (GS 51)
- LERAP Category A

FOR FULL CONDITIONS OF USE ALWAYS READ THE PRODUCT LABEL

tebuconazole

Hazard classification and safety precautions
Hazard Irritant [1]; Dangerous for the environment [1, 2]
Transport code 9
Packaging group III
UN Number 3082
Risk phrases R36, R38, R51 [1]; R50 [2]; R53a [1, 2]
Operator protection A, C, H; U05a, U10, U11, U19a, U20a
Environmental protection E15a, E16c, E16d, E22a
Storage and disposal D01, D02, D05, D10c [1, 2]; D09a [2]; D09b [1]

404   tebuconazole

A systemic triazole fungicide for cereals and other field crops
FRAC mode of action code: 3

See also azoxystrobin + tebuconazole
          bixafen + prothioconazole + tebuconazole
          bromuconazole + tebuconazole
          carbendazim + tebuconazole
          chlorothalonil + tebuconazole
          clothianidin + prothioconazole + tebuconazole + triazoxide
          difenoconazole + fludioxonil + tebuconazole
          fenpropidin + prochloraz + tebuconazole
          fenpropidin + propiconazole + tebuconazole
          fenpropidin + tebuconazole
          fludioxonil + tebuconazole
          fluopyram + prothioconazole + tebuconazole
          fluoxastrobin + prothioconazole + tebuconazole
          imidacloprid + tebuconazole + triazoxide
          prochloraz + proquinazid + tebuconazole
          prochloraz + tebuconazole
          propiconazole + tebuconazole
          prothioconazole + spiroxamine + tebuconazole
          prothioconazole + tebuconazole
          prothioconazole + tebuconazole + triazoxide
          spiroxamine + tebuconazole

Products

1	Clayton Tebucon EW	Clayton	250 g/l	EW	14824
2	Deacon	Adama	200 g/l	EW	17604
3	Erase	Headland	200 g/l	EC	17507
4	Fathom	Headland	200 g/l	EC	17452
5	Fezan	Sipcam	250 g/l	EW	17082
6	Folicur	Bayer CropScience	250 g/l	EW	16731
7	Hecates	Ventura	430 g/l	SC	17403
8	Odin	Rotam	250 g/l	EW	13468
9	Orius	Adama	200 g/l	EW	17414
10	Savannah	Rotam	250 g/l	EW	14821
11	Sokol	Becesane	430 g/l	SC	15782
12	Spekfree	Rotam	430 g/l	SC	16004
13	Starpro	Rotam	250 g/l	EW	14823
14	TEB250	Goldengrass	250 g/l	EW	15755
15	Tebucur 250	Globachem	255.1 g/l	EW	13975
16	Tebuzol 250	UPL Europe	250 g/l	EW	17417
17	Teson	Belchim	250 g/l	EW	17128
18	Tesoro	Belchim	250 g/l	EW	17222
19	Tharsis	Generica	250 g/l	EW	17201
20	Toledo	Rotam	430 g/l	SC	14036

SEE SECTION 3 FOR PRODUCTS ALSO REGISTERED

Pesticide Profiles

Products – continued

21	Tubosan	Belchim	250 g/l	EW	17127
22	Ulysses	Rotam	430 g/l	SC	16052

Uses
- Alternaria in **cabbages**, **carrots**, **horseradish** [1, 3, 4, 7, 8, 10-14, 18, 20, 22]; **parsnips** [8, 14]; **spring oilseed rape**, **winter oilseed rape** [1-4, 6-18, 20-22]; **spring wheat**, **winter wheat** [7, 11, 12, 20, 22]
- Blight in **blackberries** (off-label), **raspberries** (off-label), **rubus hybrids** (off-label) [8, 10, 13]
- Botrytis in **daffodils grown for galanthamine production** (off-label) [8, 10, 13, 20]; **linseed** [7, 11, 12, 20, 22]; **linseed** (reduction) [1, 3, 4, 10, 13, 18]
- Botrytis fruit rot in **blackberries** (off-label), **loganberries** (off-label), **raspberries** (off-label), **rubus hybrids** (off-label) [6]
- Brown rust in **rye** [7, 11, 12, 20, 22]; **spring barley**, **spring wheat**, **winter barley**, **winter wheat** [1-22]; **spring rye** [1-5, 8-10, 13-15, 17-19, 21]; **triticale** [3-5, 7, 11, 12, 20, 22]; **triticale** (off-label) [8]; **winter rye** [1-6, 8-10, 13-19, 21]
- Cane blight in **blackberries** (off-label), **raspberries** (off-label), **rubus hybrids** (off-label) [6, 8, 10, 13, 20]; **loganberries** (off-label) [6]
- Canker in **apples** (off-label), **chestnuts** (off-label), **pears** (off-label), **quinces** (off-label), **walnuts** (off-label) [6, 8, 10, 13, 20]; **cob nuts** (off-label), **crab apples** (off-label) [8, 10, 13, 20]; **hazel nuts** (off-label) [6]
- Chocolate spot in **spring field beans**, **winter field beans** [1, 3, 4, 6-8, 10-16, 18, 20, 22]
- Cladosporium in **spring wheat**, **winter wheat** [7, 11, 12, 20, 22]
- Crown rust in **spring oats** [2-5, 7, 8, 11, 12, 14, 15, 19, 20, 22]; **spring oats** (reduction), **winter oats** (reduction) [1, 6, 10, 13, 16-18, 21]; **triticale** [19]; **winter oats** [2-5, 7-9, 11, 12, 14, 15, 19, 20, 22]
- Dark leaf spot in **spring oilseed rape**, **winter oilseed rape** [7, 11, 12, 20, 22]
- Disease control in **borage for oilseed production** (off-label), **canary flower (echium spp.)** (off-label), **evening primrose** (off-label), **grass seed crops** (off-label), **honesty** (off-label), **mallow (althaea spp.)** (off-label), **mustard** (off-label), **parsley root** (off-label), **triticale** (off-label) [15]; **chives** (off-label) [8]; **linseed** [6]
- Ear diseases in **spring wheat**, **winter wheat** [7, 11, 12, 20, 22]
- Fusarium in **spring wheat**, **winter wheat** [7, 11, 12, 20, 22]
- Fusarium ear blight in **spring wheat**, **winter wheat** [1-4, 6, 8-10, 13-18, 21]; **triticale** [3, 4]
- Glume blotch in **spring wheat**, **winter wheat** [1-22]; **triticale** [3-5, 7, 11, 12, 20, 22]; **triticale** (off-label) [8]
- Light leaf spot in **cabbages** [1, 3, 4, 7, 8, 10-14, 18, 20, 22]; **spring oilseed rape**, **winter oilseed rape** [1-22]
- Lodging control in **spring oilseed rape**, **winter oilseed rape** [1, 3, 4, 6, 9, 10, 13, 16-18, 21]
- Net blotch in **spring barley**, **winter barley** [1, 3-22]
- Phoma in **spring oilseed rape**, **winter oilseed rape** [1, 3-6, 8-10, 13-18, 21]; **spring oilseed rape** (good reduction), **winter oilseed rape** (good reduction) [19]
- Phoma leaf spot in **spring oilseed rape**, **winter oilseed rape** [7, 11, 12, 20, 22]
- Powdery mildew in **apples** (off-label), **blackberries** (off-label), **chestnuts** (off-label), **pears** (off-label), **quinces** (off-label), **raspberries** (off-label), **rubus hybrids** (off-label), **walnuts** (off-label) [6, 8, 10, 13, 20]; **borage** (off-label) [8]; **borage for oilseed production** (off-label) [10, 13, 20]; **broad beans** (off-label), **bulb onions** (off-label), **canary flower (echium spp.)** (off-label), **chives** (off-label), **choi sum** (off-label), **cob nuts** (off-label), **crab apples** (off-label), **daffodils grown for galanthamine production** (off-label), **dwarf beans** (off-label), **evening primrose** (off-label), **french beans** (off-label), **garlic** (off-label), **grass seed crops** (off-label), **honesty** (off-label), **kohlrabi** (off-label), **mallow (althaea spp.)** (off-label), **mustard** (off-label), **pak choi** (off-label), **parsley root** (off-label), **runner beans** (off-label), **salad onions** (off-label), **shallots** (off-label) [8, 10, 13, 20]; **bulb onion sets** (off-label) [20]; **cabbages**, **carrots**, **parsnips**, **swedes**, **turnips** [1, 3, 4, 7, 8, 10-14, 18, 20, 22]; **hazel nuts** (off-label), **loganberries** (off-label) [6]; **linseed** [1, 3, 4, 7, 8, 10-15, 18, 20, 22]; **rye** [7, 11, 12, 20, 22]; **spring barley**, **spring wheat**, **winter barley**, **winter wheat** [1, 3-22]; **spring barley** (moderate control), **spring wheat** (moderate control), **winter barley** (moderate control), **winter wheat** (moderate control) [2]; **spring oats**, **winter oats** [1-22]; **spring rye** [1-5, 8-10, 13-15, 17-19, 21]; **triticale** [3-5, 19]; **triticale** (off-label) [8, 10, 13]; **winter rye** [1-6, 8-10, 13-19, 21]

FOR FULL CONDITIONS OF USE ALWAYS READ THE PRODUCT LABEL

tebuconazole

- Rhynchosporium in **rye** [7, 11, 12, 20, 22]; **spring barley, winter barley** [1, 3-22]; **spring barley** *(moderate control)*, **spring rye** *(moderate control)*, **winter barley** *(moderate control)*, **winter rye** *(moderate control)* [2]; **spring rye** [1, 3-5, 8-10, 13-15, 17-19, 21]; **winter rye** [1, 3-6, 8-10, 13-19, 21]
- Ring spot in **broccoli** *(off-label)*, **calabrese** *(off-label)*, **cauliflowers** *(off-label)*, **choi sum** *(off-label)*, **pak choi** *(off-label)* [8, 10, 13, 20]; **cabbages** [1, 3, 4, 7, 8, 10-14, 18, 20, 22]; **spring oilseed rape, winter oilseed rape** [7, 11, 12, 20, 22]; **spring oilseed rape** *(reduction)*, **winter oilseed rape** *(reduction)* [1, 3, 4, 6, 9, 10, 13, 16-18, 21]
- Rust in **borage for oilseed production** *(off-label)*, **canary flower (echium spp.)** *(off-label)*, **evening primrose** *(off-label)*, **grass seed crops** *(off-label)*, **honesty** *(off-label)*, **mallow (althaea spp.)** *(off-label)*, **mustard** *(off-label)*, **parsley root** *(off-label)*, **triticale** *(off-label)* [15]; **broad beans** *(off-label)*, **dwarf beans** *(off-label)*, **french beans** *(off-label)*, **runner beans** *(off-label)* [6, 8, 10, 13, 20]; **chives** *(off-label)* [10, 13, 20]; **leeks** [1, 7, 8, 10-14, 18, 20, 22]; **linseed** [6]; **spring field beans, winter field beans** [1, 3, 4, 6-8, 10-16, 18, 20, 22]
- Sclerotinia in **carrots** [1, 3, 4, 7, 8, 10-14, 18, 20, 22]; **parsnips** [8, 14]; **spring oilseed rape, winter oilseed rape** [2, 5, 7, 11, 12, 20, 22]; **spring oilseed rape** *(reduction)*, **winter oilseed rape** *(reduction)* [19]
- Sclerotinia stem rot in **spring oilseed rape, winter oilseed rape** [1, 3, 4, 6, 8-10, 13-18, 21]
- Septoria leaf blotch in **spring wheat, winter wheat** [1, 3-22]; **triticale** [3-5, 7, 11, 12, 20, 22]; **triticale** *(off-label)* [8]
- Sooty moulds in **spring wheat, winter wheat** [1-4, 6, 8-10, 13-18, 21]; **triticale** [3, 4]
- Stem canker in **spring oilseed rape, winter oilseed rape** [1, 3, 4, 6-18, 20-22]
- White rot in **bulb onion sets** *(off-label)*, **garlic** *(off-label)*, **salad onions** *(off-label)*, **shallots** *(off-label)* [8, 10, 13, 20]; **bulb onions** *(off-label)* [6, 8, 10, 13, 20]
- Yellow rust in **rye** [7, 11, 12, 20, 22]; **spring barley, spring wheat, winter barley, winter wheat** [1-22]; **spring rye** [1-5, 8-10, 13-15, 17-19, 21]; **triticale** [3-5]; **triticale** *(off-label)* [8]; **winter rye** [1-6, 8-10, 13-19, 21]

Extension of Authorisation for Minor Use (EAMUs)

- **apples** *20162647* [6], *20090329* [8], *20101183* [10], *20101221* [13], *20090572* [20]
- **blackberries** *20162648* [6], *20090304* [8], *20101179* [10], *20101225* [13], *20090568* [20]
- **borage** *20090331* [8]
- **borage for oilseed production** *20101170* [10], *20101229* [13], *20122270* [15], *20090563* [20]
- **broad beans** *20162203* [6], *20090321* [8], *20101175* [10], *20101228* [13], *20090567* [20]
- **broccoli** *20141885* [8], *20141890* [10], *20141889* [13], *20141888* [20]
- **bulb onion sets** *20090323* [8], *20101207* [10], *20101223* [13], *20090569* [20]
- **bulb onions** *20162045* [6], *20090320* [8], *20101186* [10], *20101213* [13], *20090562* [20]
- **calabrese** *20141885* [8], *20141890* [10], *20141889* [13], *20141888* [20]
- **canary flower (echium spp.)** *20090331* [8], *20101170* [10], *20101229* [13], *20122270* [15], *20090563* [20]
- **cauliflowers** *20141885* [8], *20141890* [10], *20141889* [13], *20141888* [20]
- **chestnuts** *20162647* [6], *20090324* [8], *20101177* [10], *20101219* [13], *20090573* [20]
- **chives** *20090303* [8], *20101209* [10], *20101224* [13], *20090574* [20]
- **choi sum** *20090325* [8], *20101178* [10], *20101220* [13], *20090576* [20]
- **cob nuts** *20090324* [8], *20101177* [10], *20101219* [13], *20090573* [20]
- **crab apples** *20090329* [8], *20101183* [10], *20101221* [13], *20090572* [20]
- **daffodils grown for galanthamine production** *20090322* [8], *20101176* [10], *20101217* [13], *20090575* [20]
- **dwarf beans** *20162203* [6], *20090332* [8], *20101206* [10], *20101222* [13], *20090565* [20]
- **evening primrose** *20090331* [8], *20101170* [10], *20101229* [13], *20122270* [15], *20090563* [20]
- **french beans** *20162203* [6], *20090332* [8], *20101206* [10], *20101222* [13], *20090565* [20]
- **garlic** *20090320* [8], *20101186* [10], *20101213* [13], *20090562* [20]
- **grass seed crops** *20090318* [8], *20101180* [10], *20101226* [13], *20122273* [15], *20090570* [20]
- **hazel nuts** *20162647* [6]
- **honesty** *20090331* [8], *20101170* [10], *20101229* [13], *20122270* [15], *20090563* [20]
- **kohlrabi** *20090326* [8], *20101185* [10], *20101210* [13], *20090571* [20]
- **loganberries** *20162648* [6]
- **mallow (althaea spp.)** *20090330* [8], *20101184* [10], *20101211* [13], *20122272* [15], *20090566* [20]

SEE SECTION 3 FOR PRODUCTS ALSO REGISTERED

Pesticide Profiles

- **mustard** *20090331* [8], *20101170* [10], *20101229* [13], *20122270* [15], *20090563* [20]
- **pak choi** *20090325* [8], *20101178* [10], *20101220* [13], *20090576* [20]
- **parsley root** *20090330* [8], *20101184* [10], *20101211* [13], *20122272* [15], *20090566* [20]
- **pears** *20162647* [6], *20090329* [8], *20101183* [10], *20101221* [13], *20090572* [20]
- **quinces** *20162647* [6], *20090329* [8], *20101183* [10], *20101221* [13], *20090572* [20]
- **raspberries** *20162648* [6], *20090304* [8], *20101179* [10], *20101225* [13], *20090568* [20]
- **rubus hybrids** *20162648* [6], *20090304* [8], *20101179* [10], *20101225* [13], *20090568* [20]
- **runner beans** *20162203* [6], *20090332* [8], *20101206* [10], *20101222* [13], *20090565* [20]
- **salad onions** *20090319* [8], *20101208* [10], *20101212* [13], *20090577* [20]
- **shallots** *20090320* [8], *20101186* [10], *20101213* [13], *20090562* [20]
- **triticale** *20090305* [8], *20101173* [10], *20101218* [13], *20122271* [15]
- **walnuts** *20162647* [6], *20090324* [8], *20101177* [10], *20101219* [13], *20090573* [20]

Approval information
- Tebuconazole is included in Annex 1 under EC Regulation 1107/2009
- Accepted by BBPA for use on malting barley

Efficacy guidance
- For best results apply at an early stage of disease development before infection spreads to new crop growth
- To protect flag leaf and ear from Septoria diseases apply from flag leaf emergence to ear fully emerged (GS 37-59). Earlier application may be necessary where there is a high risk of infection
- Improved control of established cereal mildew can be obtained by tank mixture with fenpropimorph
- For light leaf spot control in oilseed rape apply in autumn/winter with a follow-up spray in spring/summer if required
- For control of most other diseases spray at first signs of infection with a follow-up spray 2-4 wk later if necessary. See label for details
- For disease control in cabbages a 3-spray programme at 21-28 d intervals will give good control
- Tebuconazole is a DMI fungicide. Resistance to some DMI fungicides has been identified in Septoria leaf blotch which may seriously affect performance of some products. For further advice contact a specialist advisor and visit the Fungicide Resistance Action Group (FRAG)-UK website

Restrictions
- Maximum total dose equivalent to 1 full dose treatment on linseed; 2 full dose treatments on wheat, barley, rye, oats, field beans, swedes, turnips, onions; 2.25 full dose treatments on cabbages; 2.5 full dose treatments on oilseed rape; 3 full dose treatments on leeks, parsnips, carrots
- Do not treat durum wheat
- Apply only to listed oat varieties (see label) and do not apply to oats in tank mixture
- Do not apply before swedes and turnips have a root diameter of 2.5 cm, or before heart formation in cabbages, or before button formation in Brussels sprouts
- Consult processor before use on crops for processing
- Newer authorisations for tebuconazole products require application to cereals only after GS 30 and applications to oilseed rape and linseed after GS 20 - check label

Crop-specific information
- Latest use: before grain milky-ripe for cereals (GS 71); when most seed green-brown mottled for oilseed rape (GS 6,3); before brown capsule for linseed
- HI field beans, swedes, turnips, linseed, winter oats 35 d; market brassicas, carrots, horseradish, parsnips 21 d; leeks 14 d
- Some transient leaf speckling on wheat or leaf reddening/scorch on oats may occur but this has not been shown to reduce yield response to disease control

Environmental safety
- Dangerous for the environment
- Toxic to aquatic organisms
- LERAP Category B [3-6, 9, 15-19, 21]

FOR FULL CONDITIONS OF USE ALWAYS READ THE PRODUCT LABEL

Hazard classification and safety precautions

Hazard Harmful, Dangerous for the environment [1, 6-8, 10-22]; Irritant [2-5, 9]; Harmful if swallowed, Harmful if inhaled [6, 16]; Very toxic to aquatic organisms [6]
Transport code 9
Packaging group III
UN Number 3082
Risk phrases H317 [3, 4]; H318 [3-6, 15-18, 21]; H319 [2, 9]; H361 [2-7, 9, 12, 15-18, 21, 22]; R20 [1, 10, 13, 19]; R22a, R41 [1, 8, 10, 13, 14, 19]; R38 [8, 14]; R43, R63 [11, 20]; R51, R53a [1, 8, 10, 11, 13, 14, 19, 20]
Operator protection A, C, H; U05a, U20a [1-22]; U09a, U12 [7, 11, 12, 20, 22]; U11 [1-6, 8-10, 13-19, 21]; U14, U15 [2-5, 7, 9, 11, 12, 20, 22]; U19a [8, 14, 15]
Environmental protection E15a [1-17, 20, 22]; E16a [3-6, 9, 15-19, 21]; E34 [1-22]; E38 [1, 6, 7, 10-13, 16-22]; H410 [2-4, 6, 9, 16]; H411 [5, 7, 12, 15, 17, 18, 21, 22]
Storage and disposal D01, D02, D09a [1-22]; D05, D10b [1, 6, 8, 10, 13-19, 21]; D10c [7, 11, 12, 20, 22]; D12a [1, 6, 7, 10-13, 16-22]
Medical advice M03 [1, 6-8, 10-22]; M05a [6, 7, 11, 12, 16, 20, 22]

405 tebuconazole + trifloxystrobin

A conazole and stobilurin fungicide mixture for wheat and vegetable crops
FRAC mode of action code: 3 + 11

See also trifloxystrobin

Products

1	Clayton Bestow	Clayton	200:100 g/l	SC	17185
2	Dedicate	Bayer Environ.	200:100 g/l	SC	17003
3	Fortify	Rigby Taylor	200:100 g/l	SC	17718
4	Fusion	Rigby Taylor	200:100 g/l	SC	17225
5	Nativo 75 WG	Bayer CropScience	50:25% w/w	WG	16867

Uses

- Alternaria in *broccoli, brussels sprouts, cabbages, calabrese, carrots, cauliflowers* [5]
- Anthracnose in *amenity grassland, managed amenity turf* [4]
- Dollar spot in *amenity grassland, managed amenity turf* [1-4]
- Fusarium in *amenity grassland, managed amenity turf* [1-3]
- Fusarium patch in *amenity grassland, managed amenity turf* [4]
- Leaf spot in *horseradish* (off-label), *parsley root* (off-label), *parsnips* (off-label), *salsify* (off-label) [5]
- Light leaf spot in *broccoli, brussels sprouts, cabbages, calabrese, cauliflowers* [5]
- Melting out in *amenity grassland, managed amenity turf* [1-3]
- Phoma leaf spot in *broccoli, brussels sprouts, cabbages, calabrese, cauliflowers* [5]
- Powdery mildew in *broccoli, brussels sprouts, cabbages, calabrese, carrots, cauliflowers, horseradish* (off-label), *parsley root* (off-label), *parsnips* (off-label), *salsify* (off-label) [5]
- Red thread in *amenity grassland, managed amenity turf* [1-4]
- Ring spot in *broccoli, brussels sprouts, cabbages, calabrese, cauliflowers* [5]
- Rust in *amenity grassland, managed amenity turf* [1-3]
- Sclerotinia in *horseradish* (off-label), *parsley root* (off-label), *parsnips* (off-label), *salsify* (off-label) [5]
- Sclerotinia rot in *carrots* [5]
- White blister in *broccoli, brussels sprouts, cabbages, calabrese, cauliflowers* [5]

Extension of Authorisation for Minor Use (EAMUs)

- *horseradish* 20151959 [5]
- *parsley root* 20151959 [5]
- *parsnips* 20151959 [5]
- *salsify* 20151959 [5]

Approval information

- Tebuconazole and trifloxystrobin included in Annex I under EC Regulation 1107/2009

SEE SECTION 3 FOR PRODUCTS ALSO REGISTERED

Pesticide Profiles

Efficacy guidance
- Best results obtained from treatment at early stages of disease development. Further treatment may be needed if disease attack is prolonged
- Applications to established infections of any disease are likely to be less effective
- Treatment may give some control of *Stemphylium botryosum* and leaf blotch on leeks
- Tebuconazole is a DMI fungicide. Resistance to some DMI fungicides has been identified in Septoria leaf blotch which may seriously affect performance of some products. For further advice contact a specialist advisor and visit the Fungicide Resistance Action Group (FRAG)-UK website
- Trifloxystrobin is a member of the QoI cross resistance group. Product should be used preventatively and not relied on for its curative potential
- Use product as part of an Integrated Crop Management strategy incorporating other methods of control, including where appropriate other fungicides with a different mode of action. Do not apply more than two foliar applications of QoI containing products to any cereal crop, broccoli, calabrese or cauliflower. Do not apply more than three applications to Brussels sprouts, cabbage, carrots or leeks
- There is a significant risk of widespread resistance occurring in *Septoria tritici* populations in UK. Failure to follow resistance management action may result in reduced levels of disease control
- Strains of wheat powdery mildew resistant to QoIs are common in the UK. Control of wheat mildew can only be relied on from the triazole component
- Where specific control of wheat mildew is required this should be achieved through a programme of measures including products recommended for the control of mildew that contain a fungicide from a different cross-resistance group and applied at a dose that will give robust control
- Do not apply to grass during drought conditions nor to frozen turf

Restrictions
- Maximum number of treatments 2 per crop for broccoli, calabrese, cauliflowers; 3 per crop for Brussels sprouts, cabbages, carrots, leeks
- In addition to the maximum number of treatments per crop a maximum of 3 applications may be applied to the same ground in one calendar year
- Consult processor before use on vegetable crops for processing
- Newer authorisations for tebuconazole products require application to cereals only after GS 30 and applications to oilseed rape and linseed after GS 20 - check label

Crop-specific information
- Latest use: before milky ripe stage on winter wheat
- HI 35 d for wheat; 21 d for all other crops
- Performance against leaf spot diseases of brassicas, *Alternaria* leaf blight of carrots, and rust in leeks may be improved by mixing with an approved sticker/wetter

Environmental safety
- Dangerous for the environment
- Very toxic to aquatic organisms
- Buffer zone requirement 12 m [5]
- LERAP Category B [4]

Hazard classification and safety precautions
    Hazard Harmful, Dangerous for the environment [1-5]; Very toxic to aquatic organisms [1-3]
    Transport code 9
    Packaging group III
    UN Number 3077 [5]; 3082 [1-4]
    Risk phrases H317, H319 [5]; H361 [1-3, 5]; R37 [1-4]; R50, R53a, R63 [4]; R70 [1-3]
    Operator protection A [1-5]; H [4, 5]; P [1-3]; U05a [1-3, 5]; U09a [1-3]; U09b [4, 5]; U19a [1-4]; U20b [1-5]
    Environmental protection E15a, E34, E38 [1-5]; E16a [4]; H410 [1-3, 5]
    Storage and disposal D01, D02, D09a, D10b, D12a [1-5]; D05 [4]
    Medical advice M03

FOR FULL CONDITIONS OF USE ALWAYS READ THE PRODUCT LABEL

406 tebufenpyrad

A pyrazole mitochondrial electron transport inhibitor (METI) aphicide and acaricide
IRAC mode of action code: 21

Products

Masai	BASF	20% w/w	WB	13082

Uses
- Bud mite in **almonds** *(off-label)*, **chestnuts** *(off-label)*, **hazel nuts** *(off-label)*, **walnuts** *(off-label)*
- Damson-hop aphid in **hops**, **plums** *(off-label)*
- Gall mite in **bilberries** *(off-label)*, **blackberries** *(off-label)*, **blackcurrants** *(off-label)*, **blueberries** *(off-label)*, **cranberries** *(off-label)*, **gooseberries** *(off-label)*, **protected rubus hybrids** *(off-label)*, **raspberries** *(off-label)*, **redcurrants** *(off-label)*, **vaccinium spp.** *(off-label)*, **whitecurrants** *(off-label)*
- Red spider mites in **apples**, **blackberries** *(off-label)*, **pears**, **protected rubus hybrids** *(off-label)*, **raspberries** *(off-label)*
- Two-spotted spider mite in **hops, protected roses, strawberries**

Extension of Authorisation for Minor Use (EAMUs)
- **almonds** 20080133
- **bilberries** 20080131
- **blackberries** 20100204
- **blackcurrants** 20080131
- **blueberries** 20080131
- **chestnuts** 20080133
- **cranberries** 20080131
- **gooseberries** 20080131
- **hazel nuts** 20080133
- **plums** 20080132
- **protected rubus hybrids** 20100204
- **raspberries** 20100204
- **redcurrants** 20080131
- **vaccinium spp.** 20080131
- **walnuts** 20080133
- **whitecurrants** 20080131

Approval information
- Tebufenpyrad included in Annex 1 under EC Regulation 1107/2009
- Accepted by BBPA for use on hops

Efficacy guidance
- Acts on eggs (except winter eggs) and all motile stages of spider mites up to adults
- Treat spider mites from 80% egg hatch but before mites become established
- For effective control total spray cover of the crop is required
- Product can be used in a programme to give season-long control of damson-hop aphids coupled with mite control
- Where aphids resistant to tebufenpyrad occur in hops control is unlikely to be satisfactory and repeat treatments may result in lower levels of control. Where possible use different active ingredients in a programme

Restrictions
- Maximum total dose equivalent to one full dose treatment on apples, pears, strawberries; 3 full dose treatments on hops
- Other mitochondrial electron transport inhibitor (METI) acaricides should not be applied to the same crop in the same calendar yr either separately or in mixture
- Do not treat apples before 90% petal fall
- Small-scale testing of rose varieties to establish tolerance recommended before use
- Inner liner of container must not be removed

SEE SECTION 3 FOR PRODUCTS ALSO REGISTERED

Pesticide Profiles

Crop-specific information
- Latest use: end of burr stage for hops
- HI strawberries 3 d; apples, bilberries, blackcurrants, blueberries, cranberries, gooseberries, pears, redcurrants, whitecurrants 7d; blackberries, plums, raspberries 21 d
- Product has no effect on fruit quality or finish

Environmental safety
- Dangerous for the environment
- Very toxic to aquatic organisms
- High risk to bees. Do not apply to crops in flower or to those in which bees are actively foraging. Do not apply when flowering weeds are present
- Broadcast air-assisted LERAP (18 m); LERAP Category B

Hazard classification and safety precautions
Hazard Harmful, Dangerous for the environment
Transport code 9
Packaging group III
UN Number 3077
Risk phrases R20, R22a, R50, R53a
Operator protection A; U02a, U05a, U09a, U13, U14, U20b
Environmental protection E12a, E12e, E15a, E16a, E16b, E38; E17b (18 m)
Storage and disposal D01, D02, D09a, D11a, D12a
Medical advice M05a

407 tefluthrin

A soil acting pyrethroid insecticide seed treatment
IRAC mode of action code: 3

See also fludioxonil + tefluthrin

Products

Force ST	Syngenta	200 g/l	CF	11752

Uses
- Bean seed fly in **baby leaf crops** *(off-label - seed treatment)*, **bulb onions** *(off-label)*, **chard** *(off-label - seed treatment)*, **chives** *(off-label - seed treatment)*, **green mustard** *(off-label - seed treatment)*, **herbs (see appendix 6)** *(off-label - seed treatment)*, **leaf spinach** *(off-label - seed treatment)*, **leeks** *(off-label)*, **miduna** *(off-label - seed treatment)*, **mizuna** *(off-label - seed treatment)*, **pak choi** *(off-label - seed treatment)*, **parsley** *(off-label - seed treatment)*, **red mustard** *(off-label - seed treatment)*, **salad onions** *(off-label)*, **shallots** *(off-label)*, **spinach beet** *(off-label - seed treatment)*, **tatsoi** *(off-label - seed treatment)*
- Carrot fly in **carrots** *(off-label - seed treatment)*, **parsnips** *(off-label - seed treatment)*
- Millipedes in **fodder beet** *(seed treatment)*, **sugar beet** *(seed treatment)*
- Onion fly in **bulb onions** *(off-label)*, **leeks** *(off-label)*, **salad onions** *(off-label)*, **shallots** *(off-label)*
- Pygmy beetle in **fodder beet** *(seed treatment)*, **sugar beet** *(seed treatment)*
- Springtails in **fodder beet** *(seed treatment)*, **sugar beet** *(seed treatment)*
- Symphylids in **fodder beet** *(seed treatment)*, **sugar beet** *(seed treatment)*
- Wireworm in **chicory** *(off-label)*, **chicory root** *(off-label)*, **witloof** *(off-label)*

Extension of Authorisation for Minor Use (EAMUs)
- *baby leaf crops* (seed treatment) 20050545
- *bulb onions* 20120604
- *carrots* (seed treatment) 20050547
- *chard* (seed treatment) 20050545
- *chicory* 20142248
- *chicory root* 20142248
- *chives* (seed treatment) 20050545
- *green mustard* (seed treatment) 20050545
- *herbs (see appendix 6)* (seed treatment) 20050545
- *leaf spinach* (seed treatment) 20050545

FOR FULL CONDITIONS OF USE ALWAYS READ THE PRODUCT LABEL

tefluthrin

- **leeks** 20120604
- **miduna** (seed treatment) 20050545
- **mizuna** (seed treatment) 20050545
- **pak choi** (seed treatment) 20050545
- **parsley** (seed treatment) 20050545
- **parsnips** (seed treatment) 20050547
- **red mustard** (seed treatment) 20050545
- **salad onions** 20120604
- **shallots** 20120604
- **spinach beet** (seed treatment) 20050545
- **tatsoi** (seed treatment) 20050545
- **witloof** 20142248

Approval information
- Tefluthrin included in Annex 1 under EC Regulation 1107/2009
- Accepted by BBPA for use on malting barley

Efficacy guidance
- Apply during process of pelleting beet seed. Consult manufacturer for details of specialist equipment required
- Micro-capsule formulation allows slow release to provide a protection zone around treated seed during establishment

Restrictions
- Maximum number of treatments 1 per batch of seed
- Sow treated seed as soon as possible. Do not store treated seed from one drilling season to next
- If used in areas where soil erosion by wind or water likely, measures must be taken to prevent this happening
- Can cause a transient tingling or numbing sensation to exposed skin. Avoid skin contact with product, treated seed and dust throughout all operations in the seed treatment plant and at drilling

Crop-specific information
- Latest use: before drilling seed
- Treated seed must be drilled within the season of treatment

Environmental safety
- Dangerous for the environment
- Very toxic to aquatic organisms
- Keep treated seed secure from people, domestic stock/pets and wildlife at all times during storage and use
- Treated seed harmful to game and wild life. Bury spillages
- In the event of seed spillage clean up as much as possible into the related seed sack and bury the remainder completely
- Do not apply treated seed from the air
- Keep livestock out of areas drilled with treated seed for at least 80 d

Hazard classification and safety precautions
Hazard Harmful, Dangerous for the environment
Transport code 9
Packaging group III
UN Number 3082
Risk phrases R20, R22b, R43, R50, R53a
Operator protection A, D, E, H; U02a, U04a, U05a, U08, U20b
Environmental protection E06a (80 d); E15a, E34, E38
Storage and disposal D01, D02, D05, D09a, D11a, D12a
Treated seed S02, S03, S04b, S05, S07
Medical advice M05b

SEE SECTION 3 FOR PRODUCTS ALSO REGISTERED

408 terbuthylazine

A triazine herbicide available only in mixtures
HRAC mode of action code: C1

See also bromoxynil + terbuthylazine
isoxaben + terbuthylazine
mesotrione + terbuthylazine
pendimethalin + terbuthylazine

409 thiabendazole

A systemic, curative and protectant benzimidazole (MBC) fungicide
FRAC mode of action code: 1

See also imazalil + thiabendazole
metalaxyl + thiabendazole

Products
Tezate 220 SL   UPL Europe   220 g/l   SL   16979

Uses
- Basal stem rot in **narcissi** *(off-label)*
- Dry rot in **potatoes**, **seed potatoes**
- Fusarium basal neck rot in **narcissi**
- Gangrene in **potatoes**, **seed potatoes**
- Neck rot in **narcissi** *(off-label)*
- Silver scurf in **potatoes**, **seed potatoes**
- Skin spot in **potatoes**, **seed potatoes**

Extension of Authorisation for Minor Use (EAMUs)
- *narcissi* 20151528

Approval information
- Thiabendazole included in Annex I under EC Regulation 1107/2009
- Use of thiabendazole products on ware potatoes requires that the discharge of thiabendazole to receiving water from washing plants is kept within emission limits set by the UK monitoring authority

Efficacy guidance
- For best results tuber treatments should be applied as soon as possible after lifting and always within 24 hr
- Dust treatments should be applied evenly over the whole tuber surface
- Thiabendazole should only be used on ware potatoes where there is a likely risk of disease during the storage period and in combination with good storage hygiene and maintenance
- Benzimidazole tolerant strains of silver scurf and skin spot are common in UK and tolerant strains of dry rot have been reported. To reduce the chance of such strains increasing benzimidazole based products should not be used more than once in the cropping cycle
- On narcissi treat as the crop goes into store and only if fungicidal treatment is essential. Adopt a resistance management strategy

Restrictions
- Maximum number of treatments 1 per batch for ware or seed potato tuber treatments; 1 per yr for narcissi bulbs
- Treated seed potatoes must not be used for food or feed
- Do not remove treated potatoes from store for sale, processing or consumption for at least 21 d after application
- Do not mix with any other product
- Off-label use as post-lifting cold water dip or pre-planting hot water dip not to be used on narcissus bulbs grown in the Isles of Scilly

thiacloprid

Crop-specific information
- Latest use: 21 d before removal from store for sale, processing or consumption for ware potatoes
- Apply to potatoes as soon as possible after harvest using suitable equipment and always within 2 wk of lifting provided the skins are set. See label for details
- Potatoes should only be treated by systems that provide an accurate dose to tubers not carrying excessive quantities of soil
- Use as a post-lifting treatment to reduce basal and neck rot in narcissus bulbs. Ensure bulbs are clean

Environmental safety
- Dangerous for the environment
- Toxic to aquatic organisms

Hazard classification and safety precautions
Hazard Dangerous for the environment
Transport code 9
Packaging group III
UN Number 3082
Risk phrases R51, R53a
Operator protection A, C, D, H; U05a, U19a, U20b
Environmental protection E15b, E38
Storage and disposal D01, D02, D09a, D10c, D12a
Treated seed S03, S04a, S05

410 thiacloprid

A chloronicotinyl insecticide for use in agriculture and horticulture
IRAC mode of action code: 4A

Products

1	Agrovista Reggae	Agrovista	480 g/l	SC	13706
2	Biscaya	Bayer CropScience	240 g/l	OD	15014
3	Calypso	Bayer CropScience	480 g/l	SC	11257
4	Exemptor	Bayer Environ.	10% w/w	GR	15615
5	Sonido	Bayer CropScience	400 g/l	SC	16368
6	Standon Zero Tolerance	Standon	240 g/l	OD	13546
7	Zubarone	AgChem Access	240 g/l	OD	15798

Uses
- Aphids in **all edible seed crops grown outdoors** (off-label), **all non-edible seed crops grown outdoors** (off-label), **celeriac** (off-label), **chinese cabbage** (off-label), **choi sum** (off-label), **pak choi** (off-label), **swedes** (off-label), **tatsoi** (off-label), **turnips** (off-label), **winter wheat** [2]; **bedding plants, hardy ornamental nursery stock, ornamental plant production, pot plants, protected ornamentals** [4]; **carrots, parsnips, peas, potatoes** [2, 7]; **cherries** (off-label), **courgettes** (off-label), **gherkins** (off-label), **hops** (off-label), **leaf brassicas** (off-label), **marrows** (off-label), **mirabelles** (off-label), **protected forest nurseries** (off-label), **soft fruit** (off-label), **top fruit** (off-label) [3]; **leaf brassicas** (off-label - baby leaf production) [1]; **ornamental plant production** (off-label) [2, 3]; **protected herbs (see appendix 6)** (off-label), **protected lamb's lettuce** (off-label), **protected lettuce** (off-label) [1, 3]; **seed potatoes, ware potatoes** [2, 6, 7]
- Asparagus beetle in **asparagus** (off-label) [3]
- Beetles in **bedding plants, hardy ornamental nursery stock, ornamental plant production, pot plants, protected ornamentals** [4]
- Blueberry gall midge in **protected blueberry** (off-label) [3]
- Bud mite in **almonds** (off-label), **chestnuts** (off-label), **cob nuts** (off-label), **hazel nuts** (off-label), **walnuts** (off-label) [1, 3]; **filberts** (off-label) [3]
- Cabbage aphid in **broccoli, brussels sprouts, cabbages, calabrese, cauliflowers** [2, 7]; **chinese cabbage** (off-label), **choi sum** (off-label), **collards** (off-label), **kale** (off-label), **pak choi** (off-label), **spring greens** (off-label), **tatsoi** (off-label) [2]
- Capsids in **strawberries** (off-label) [1, 3]

SEE SECTION 3 FOR PRODUCTS ALSO REGISTERED

Pesticide Profiles

- Common green capsid in **blackberries** *(off-label)*, **protected blackberries** *(off-label)*, **protected raspberries** *(off-label)*, **raspberries** *(off-label)*, **rubus hybrids** *(off-label)* [3]
- Damson-hop aphid in **plums** *(off-label)* [1, 3]
- Gall midge in **protected blueberry** *(off-label)* [1]
- Insect pests in **bilberries** *(off-label)*, **blackberries** *(off-label)*, **blackcurrants** *(off-label)*, **blueberries** *(off-label)*, **cherries** *(off-label - under temporary protective rain covers)*, **courgettes** *(off-label)*, **cranberries** *(off-label)*, **gherkins** *(off-label)*, **gooseberries** *(off-label)*, **marrows** *(off-label)*, **mirabelles** *(off-label - under temporary protective rain covers)*, **pears** *(off-label)*, **protected blackberries** *(off-label)*, **protected raspberries** *(off-label)*, **protected strawberries** *(off-label)*, **raspberries** *(off-label)*, **redcurrants** *(off-label)*, **rubus hybrids** *(off-label)*, **whitecurrants** *(off-label)* [1]; **hops** *(off-label)*, **ornamental plant production** *(off-label)*, **protected forest nurseries** *(off-label)*, **soft fruit** *(off-label)*, **top fruit** *(off-label)* [1, 3]
- Leaf miner in **protected aubergines** *(off-label)*, **protected courgettes** *(off-label)*, **protected cucumbers** *(off-label)*, **protected ornamentals** *(off-label)*, **protected peppers** *(off-label)*, **protected tomatoes** *(off-label)* [1]
- Light brown apple moth in **protected blueberry** *(off-label)* [3]
- Mangold fly in **sugar beet** *(off-label)* [2]
- Mealy aphid in **broccoli**, **brussels sprouts**, **cabbages**, **calabrese**, **cauliflowers** [2, 7]; **chinese cabbage** *(off-label)*, **choi sum** *(off-label)*, **collards** *(off-label)*, **kale** *(off-label)*, **pak choi** *(off-label)*, **spring greens** *(off-label)*, **swedes** *(off-label)*, **tatsoi** *(off-label)*, **turnips** *(off-label)* [2]
- Moths in **protected blueberry** *(off-label)* [1]
- Palm thrips in **protected aubergines** *(off-label)*, **protected courgettes** *(off-label)*, **protected cucumbers** *(off-label)*, **protected ornamentals** *(off-label)*, **protected peppers** *(off-label)*, **protected tomatoes** *(off-label)* [3]
- Pea midge in **peas** [2, 7]
- Peach-potato aphid in **leaf brassicas** *(off-label)* [3]; **leaf brassicas** *(off-label - baby leaf production)* [1]; **protected herbs (see appendix 6)** *(off-label)*, **protected lamb's lettuce** *(off-label)*, **protected lettuce** *(off-label)* [1, 3]
- Pear leaf midge in **pears** *(off-label)* [3]
- Pear midge in **pears** *(off-label)* [1, 3]
- Pollen beetle in **mustard** [2, 7]; **spring oilseed rape**, **winter oilseed rape** [2, 6, 7]
- Rosy apple aphid in **apples** [1, 3]
- Sciarid flies in **protected ornamentals** *(off-label)* [3]
- South American leaf miner in **protected aubergines** *(off-label)*, **protected courgettes** *(off-label)*, **protected cucumbers** *(off-label)*, **protected ornamentals** *(off-label)*, **protected peppers** *(off-label)*, **protected tomatoes** *(off-label)* [3]
- Tarnished plant bug in **blackberries** *(off-label)*, **protected blackberries** *(off-label)*, **protected raspberries** *(off-label)*, **protected strawberries** *(off-label)*, **raspberries** *(off-label)*, **rubus hybrids** *(off-label)* [3]
- Thrips in **chives** *(off-label)*, **leeks** *(off-label)*, **salad onions** *(off-label)* [3]; **protected aubergines** *(off-label)*, **protected courgettes** *(off-label)*, **protected cucumbers** *(off-label)*, **protected ornamentals** *(off-label)*, **protected peppers** *(off-label)*, **protected tomatoes** *(off-label)* [1, 3]
- Tobacco whitefly in **protected aubergines** *(off-label)*, **protected courgettes** *(off-label)*, **protected cucumbers** *(off-label)*, **protected ornamentals** *(off-label)*, **protected peppers** *(off-label)*, **protected tomatoes** *(off-label)* [3]
- Vine weevil in **bedding plants**, **hardy ornamental nursery stock**, **ornamental plant production**, **pot plants**, **protected ornamentals** [4]; **protected ornamentals** *(off-label)* [3]
- Western flower thrips in **protected aubergines** *(off-label)*, **protected courgettes** *(off-label)*, **protected cucumbers** *(off-label)*, **protected ornamentals** *(off-label)*, **protected peppers** *(off-label)*, **protected tomatoes** *(off-label)* [1, 3]
- Wheat-blossom midge in **spring wheat**, **winter wheat** [2, 7]
- Whitefly in **bedding plants**, **hardy ornamental nursery stock**, **ornamental plant production**, **pot plants**, **protected ornamentals** [4]; **protected aubergines** *(off-label)*, **protected courgettes** *(off-label)*, **protected cucumbers** *(off-label)*, **protected ornamentals** *(off-label)*, **protected peppers** *(off-label)*, **protected tomatoes** *(off-label)* [1]
- Willow aphid in **horseradish** *(off-label)*, **parsley root** *(off-label)*, **red beet** *(off-label)*, **salsify** *(off-label)* [2]
- Wireworm in **fodder maize**, **grain maize**, **sweetcorn** [5]

FOR FULL CONDITIONS OF USE ALWAYS READ THE PRODUCT LABEL

thiacloprid

- Woolly currant scale in **bilberries** *(off-label)*, **blackcurrants** *(off-label)*, **blueberries** *(off-label)*, **cranberries** *(off-label)*, **gooseberries** *(off-label)*, **redcurrants** *(off-label)*, **whitecurrants** *(off-label)* [3]

Extension of Authorisation for Minor Use (EAMUs)
- **all edible seed crops grown outdoors** *20142041* [2]
- **all non-edible seed crops grown outdoors** *20142041* [2]
- **almonds** *20080471* [1], *20142137* [3]
- **asparagus** *20142142* [3]
- **bilberries** *20080466* [1], *20142133* [3]
- **blackberries** *20080475* [1], *20142138* [3]
- **blackcurrants** *20080466* [1], *20142133* [3]
- **blueberries** *20080466* [1], *20142133* [3]
- **celeriac** *20152075* [2]
- **cherries** *(under temporary protective rain covers)* *20080469* [1], *20142135* [3]
- **chestnuts** *20080471* [1], *20142137* [3]
- **chinese cabbage** *20142063* [2]
- **chives** *20142150* [3]
- **choi sum** *20142063* [2]
- **cob nuts** *20080471* [1], *20142137* [3]
- **collards** *20142249* [2]
- **courgettes** *20102032* [1], *20142140* [3]
- **cranberries** *20080466* [1], *20142133* [3]
- **filberts** *20142137* [3]
- **gherkins** *20102032* [1], *20142140* [3]
- **gooseberries** *20080466* [1], *20142133* [3]
- **hazel nuts** *20080471* [1], *20142137* [3]
- **hops** *20102034* [1], *20142149* [3]
- **horseradish** *20142062* [2]
- **kale** *20142249* [2]
- **leaf brassicas** *(baby leaf production)* *20080470* [1], *20142136* [3]
- **leeks** *20142150* [3]
- **marrows** *20102032* [1], *20142140* [3]
- **mirabelles** *(under temporary protective rain covers)* *20080469* [1], *20142135* [3]
- **ornamental plant production** *20102034* [1], *20142041* [2], *20142149* [3]
- **pak choi** *20142063* [2]
- **parsley root** *20142062* [2]
- **pears** *20080464* [1], *20142130* [3]
- **plums** *20080468* [1], *20142134* [3]
- **protected aubergines** *20080474* [1], *20142151* [3]
- **protected blackberries** *20080467* [1], *20142139* [3]
- **protected blueberry** *20102035* [1], *20142141* [3]
- **protected courgettes** *20080474* [1], *20142151* [3]
- **protected cucumbers** *20080474* [1], *20142151* [3]
- **protected forest nurseries** *20102034* [1], *20142149* [3]
- **protected herbs (see appendix 6)** *20080472* [1], *20142152* [3]
- **protected lamb's lettuce** *20080472* [1], *20142152* [3]
- **protected lettuce** *20080472* [1], *20142152* [3]
- **protected ornamentals** *20080474* [1], *20142151* [3], *20142153* [3]
- **protected peppers** *20080474* [1], *20142151* [3]
- **protected raspberries** *20080467* [1], *20142139* [3]
- **protected strawberries** *20102033* [1], *20142132* [3]
- **protected tomatoes** *20080474* [1], *20142151* [3]
- **raspberries** *20080475* [1], *20142138* [3]
- **red beet** *20142062* [2]
- **redcurrants** *20080466* [1], *20142133* [3]
- **rubus hybrids** *20080475* [1], *20142138* [3]
- **salad onions** *20142150* [3]
- **salsify** *20142062* [2]

SEE SECTION 3 FOR PRODUCTS ALSO REGISTERED

Pesticide Profiles

- **soft fruit** 20102034 [1], 20142149 [3]
- **spring greens** 20142249 [2]
- **strawberries** 20080465 [1], 20142131 [3]
- **sugar beet** 20161556 [2]
- **swedes** 20142061 [2]
- **tatsoi** 20142063 [2]
- **top fruit** 20102034 [1], 20142149 [3]
- **turnips** 20142061 [2]
- **walnuts** 20080471 [1], 20142137 [3]
- **whitecurrants** 20080466 [1], 20142133 [3]

Approval information
- Thiacloprid included in Annex I under EC Regulation 1107/2009

Efficacy guidance
- Best results in apples obtained by a programme of sprays commencing pre-blossom at the first sign of aphids [3]
- For best pear midge control treat under warm conditions (mid-day) when the adults are flying (temperatures higher than 12°C). Applications on cool damp days are unlikely to be successful [3]
- Best results in field crops obtained from treatments when target pests reach threshold levels. For wheat blossom midge use pheromone or sticky yellow traps to monitor adult activity in crops at risk [2, 6]
- Treatment of ornamentals and protected ornamentals is by incorporation into peat-based growing media prior to sowing or planting using suitable automated equipment [4]
- Ensure thorough mixing in the compost to achieve maximum control. Top dressing is ineffective [4]
- Unless transplants have been treated with a suitable pesticide prior to planting in treated compost, full protection against target pests is not guaranteed [4]
- Minimise the possibility of the development of resistance by alternating insecticides with different modes of action in the programme
- In dense canopies and on larger trees increase water volume to ensure full coverage
- It can give effective contact activity against adult CSFB "present and active" in oilseed rape at the time of an application to control Myzus persicae [2]

Restrictions
- Maximum number of treatments 2 per yr on apples, seed potatoes; 1 per yr on other field crops and for compost incorporation
- Consult processor before use on crops for processing [2, 6]
- Do not mix treated compost with any other bulky materials such as perlite or bark [4]
- Use treated compost as soon as possible, and within 4 wk of mixing [4]

Crop-specific information
- Latest use: up to and including flowering just complete (GS 69) for wheat; before sowing or planting for treated compost
- HI apples and potatoes 14 d; oilseed rape and mustard 30 d
- Test tolerance of ornamental species before large scale use [4]

Environmental safety
- Dangerous for the environment
- Very toxic or toxic to aquatic organisms
- Risk to certain non-target insects or other arthropods. See directions for use [3]
- Broadcast air-assisted LERAP [1] (18 m); Broadcast air-assisted LERAP [3] (30 m)

Hazard classification and safety precautions
Hazard Harmful [1-4, 6, 7]; Dangerous for the environment [1-7]
Transport code 6.1 [1, 3]; 9 [4, 5]
Packaging group III [1, 3-5]
UN Number 2902 [1, 3]; 3077 [4]; 3082 [5]; N/C [2, 6, 7]
Risk phrases R20, R52 [1, 3]; R22a [1-3, 5-7]; R36, R38, R51 [2, 6, 7]; R40 [1-4, 6, 7]; R40b [5]; R43 [1, 3, 4]; R50 [4, 5]; R53a [1-7]

FOR FULL CONDITIONS OF USE ALWAYS READ THE PRODUCT LABEL

Operator protection A [1-7]; C [2, 6, 7]; D [4, 5]; H [1, 3, 5]; U05a [1-7]; U11, U19a [2, 6, 7]; U14 [1, 3, 5]; U16a [2, 7]; U20b [1-3, 5-7]; U20c [4]
Environmental protection E10b [5]; E13c, E22b [3]; E15a [4]; E15b [1, 2, 5-7]; E17b [1] (18 m); E17b [3] (30 m); E22c [1]; E34 [1-4, 6, 7]; E38 [4, 5]
Storage and disposal D01, D02 [1-7]; D09a [1-4, 6, 7]; D09b, D10d, D12b, D14 [5]; D10b [1, 3]; D10c [2, 6, 7]; D11a [4]; D12a [1, 2, 6, 7]
Treated seed S01, S02, S03, S04d, S05, S07, S08 [5]; S04a [4]
Medical advice M03 [1-7]; M05a [1, 5]; M05b [3]

411 thiamethoxam

A neonicotinoid insecticide for apples, pears, potatoes and beet crops with a 2 year ban from Dec 2013 on use in many crops due to concerns over effects on bees.
IRAC mode of action code: 4A

See also fludioxonil + metalaxyl-M + thiamethoxam

Products

1	Actara	Syngenta	25% w/w	WG	13728
2	Cruiser SB	Syngenta	600 g/l	FS	15012

Uses

- Aphids in **fodder beet**, **sugar beet** [2]; **potatoes, potatoes grown for seed** [1]
- Beet leaf miner in **fodder beet** *(seed treatment)*, **sugar beet** *(seed treatment)* [2]
- Cabbage root fly in **kale** *(off-label)*, **swedes** *(off-label)* [2]
- Cabbage stem flea beetle in **kale** *(off-label)*, **swedes** *(off-label)* [2]
- Flea beetle in **fodder beet** *(seed treatment)*, **sugar beet** *(seed treatment)*, **swedes** *(off-label)* [2]
- Millipedes in **fodder beet** *(seed treatment)*, **sugar beet** *(seed treatment)* [2]
- Peach-potato aphid in **kale** *(off-label)*, **swedes** *(off-label)* [2]
- Pygmy beetle in **fodder beet** *(seed treatment)*, **sugar beet** *(seed treatment)* [2]
- Springtails in **fodder beet** *(seed treatment)*, **sugar beet** *(seed treatment)* [2]
- Symphylids in **fodder beet** *(seed treatment)*, **sugar beet** *(seed treatment)* [2]
- Virus yellows vectors in **kale** *(off-label)*, **swedes** *(off-label)* [2]
- Western flower thrips in **protected ornamentals** *(off-label)* [1]
- Wireworm in **fodder beet** *(seed treatment - reduction)*, **sugar beet** *(seed treatment - reduction)* [2]

Extension of Authorisation for Minor Use (EAMUs)
- **kale** *20131877* [2]
- **protected ornamentals** *20140186* [1]
- **swedes** *20111292* [2], *20131877* [2]

Approval information
- Thiamethoxam included in Annex I under EC Regulation 1107/2009

Efficacy guidance
- Apply to sugar beet or fodder beet seed as part of the normal commercial pelleting process using special treatment machinery [2]
- Seed drills must be suitable for use with polymer-coated seeds. Standard drill settings should not need to be changed [2]
- Drill treated seed into a firm even seedbed. Poor seedbed quality or seedbed conditions may results in delayed emergence and poor establishment [2]
- Where very high populations of soil pests are present protection may be inadequate to achieve an optimum plant stand [2]
- Use in line with latest IRAG guidelines

Restrictions
- Maximum number of treatments 1 per seed batch [2]
- Avoid deep or shallow drilling which may adversely affect establishment and reduce the level of pest control [2]
- Do not use herbicides containing lenacil pre-emergence on treated crops [2]

SEE SECTION 3 FOR PRODUCTS ALSO REGISTERED

Pesticide Profiles

Crop-specific information
- Latest use: before drilling for fodder beet, sugar beet [2]

Environmental safety
- Dangerous for the environment
- Very toxic to aquatic organisms

Hazard classification and safety precautions
  Hazard Dangerous for the environment
  UN Number N/C
  Risk phrases R50, R53a
  Operator protection A, D, H [2]; U05a, U07 [2]; U05b, U20c [1]
  Environmental protection E03, E34, E36a [2]; E12c, E12e [1]; E15b, E38 [1, 2]
  Storage and disposal D01, D02, D09a, D12a [1, 2]; D05, D11a, D14 [2]; D10c [1]
  Treated seed S02, S04b, S05, S06a, S06b, S08 [2]

412   thifensulfuron-methyl

A translocated sulfonylurea herbicide
HRAC mode of action code: B

See also carfentrazone-ethyl + thifensulfuron-methyl
  flupyrsulfuron-methyl + thifensulfuron-methyl
  fluroxypyr + metsulfuron-methyl + thifensulfuron-methyl
  fluroxypyr + thifensulfuron-methyl + tribenuron-methyl
  metsulfuron-methyl + thifensulfuron-methyl
  nicosulfuron + thifensulfuron-methyl

Products
  Pinnacle   DuPont   50% w/w   SG   12285

Uses
- Annual dicotyledons in **soya beans** *(off-label)*
- Docks in **grassland**
- Green cover in **land temporarily removed from production**

Extension of Authorisation for Minor Use (EAMUs)
- *soya beans* 20112063

Approval information
- Thifensulfuron-methyl included in Annex I under EC Regulation 1107/2009
- Accepted by BBPA for use on malting barley

Efficacy guidance
- Best results achieved from application to small emerged weeds when growing actively. Broad-leaved docks are susceptible during the rosette stage up to onset of stem extension
- Ensure good spray coverage and apply to dry foliage
- Susceptible weeds stop growing almost immediately but symptoms may not be visible for about 2 wk
- Product sold in twin pack with mecoprop-P to provide option for improved control of cleavers
- Only broad-leaved docks (*Rumex obtusifolius*) are controlled; curled docks (*Rumex crispus*) are resistant
- Docks with developing or mature seed heads should be topped and the regrowth treated later
- Established docks with large tap roots may require follow-up treatment
- High populations of docks in grassland will require further treatment in following yr
- Thifensulfuron-methyl is a member of the ALS-inhibitor group of herbicides and products should be used in a planned Resistance Management strategy. See Section 5 for more information

Restrictions
- Maximum number of treatments 1 per crop for cereals or one per yr for grassland and cereals. Must only be applied from 1 Feb in yr of harvest
- Do not treat new leys in year of sowing

FOR FULL CONDITIONS OF USE ALWAYS READ THE PRODUCT LABEL

thifensulfuron-methyl + tribenuron-methyl

- Do not treat where nutrient imbalances, drought, waterlogging, low temperatures, lime deficiency, pest or disease attack have reduced crop or sward vigour
- Do not roll or harrow within 7 d of spraying
- Do not graze grass crops within 7 d of spraying
- Specific restrictions apply to use in sequence or tank mixture with other sulfonylurea or ALS-inhibiting herbicides. See label for details
- Only one application of a sulfonylurea product may be applied per calendar yr to grassland and green cover on land temporarily removed from production

Crop-specific information
- Latest use: before 1 Aug on grass; flag leaf fully emerged stage (GS 39) on cereals
- On grass apply 7-10 d before grazing and do not graze for 7 d afterwards
- Product may cause a check to both sward and clover which is usually outgrown

Following crops guidance
- Only grass or cereals may be sown within 4 wk of application to grassland or setaside, or in the event of failure of any treated crop
- No restrictions apply after normal harvest of a treated cereal crop

Environmental safety
- Dangerous for the environment
- Very toxic to aquatic organisms
- Keep livestock out of treated areas for at least 7 d following treatment
- Take extreme care to avoid drift onto broad-leaved plants outside the target area or onto surface waters or ditches, or land intended for cropping
- Spraying equipment should not be drained or flushed onto land planted, or to be planted, with trees or crops other than cereals and should be thoroughly cleansed after use - see label for instructions

Hazard classification and safety precautions
Hazard Dangerous for the environment, Very toxic to aquatic organisms
Transport code 9
Packaging group III
UN Number 3077
Operator protection U19a, U20b
Environmental protection E06a (7 d); E15b, E38, H410
Storage and disposal D09a, D11a, D12a

413 thifensulfuron-methyl + tribenuron-methyl

A mixture of two sulfonylurea herbicides for cereals
HRAC mode of action code: B + B

See also tribenuron-methyl

Products

1	Hiatus	Rotam	40:15% w/w	WG	16059
2	Inka SX	DuPont	25:25% w/w	SG	13601
3	Parana	Certis	33.3:16.7% w/w	SG	16258
4	Ratio SX	DuPont	40:10% w/w	SG	12601
5	Seduce	Certis	33.3:16.7% w/w	SG	16261

Uses
- Annual dicotyledons in *game cover* (off-label), *spring oats*, *spring rye* [2]; *spring barley*, *spring wheat*, *winter barley*, *winter wheat* [1-5]; *triticale*, *winter rye* [1, 2]; *winter oats* [2, 4]
- Black bindweed in *spring barley*, *spring wheat*, *triticale*, *winter barley*, *winter rye*, *winter wheat* [1]
- Charlock in *spring barley*, *spring wheat*, *winter barley*, *winter wheat* [2-5]; *spring oats*, *spring rye*, *triticale*, *winter rye* [2]; *winter oats* [2, 4]
- Chickweed in *spring barley*, *spring wheat*, *winter barley*, *winter wheat* [1-5]; *spring oats*, *spring rye* [2]; *triticale*, *winter rye* [1, 2]; *winter oats* [2, 4]

SEE SECTION 3 FOR PRODUCTS ALSO REGISTERED

Pesticide Profiles

- Docks in **spring barley**, **spring oats**, **spring rye**, **spring wheat**, **triticale**, **winter barley**, **winter oats**, **winter rye**, **winter wheat** [2]
- Fat hen in **spring barley**, **spring wheat**, **triticale**, **winter barley**, **winter rye**, **winter wheat** [1]
- Mayweeds in **spring barley**, **spring wheat**, **winter barley**, **winter wheat** [1-5]; **spring oats**, **spring rye** [2]; **triticale**, **winter rye** [1, 2]; **winter oats** [2, 4]
- Poppies in **spring barley**, **spring wheat**, **triticale**, **winter barley**, **winter rye**, **winter wheat** [1]

Extension of Authorisation for Minor Use (EAMUs)
- *game cover* 20101408 [2]

Approval information
- Thifensulfuron-methyl and tribenuron-methyl included in Annex I under EC Regulation 1107/2009
- Accepted by BBPA for use on malting barley

Efficacy guidance
- Apply when weeds are small and actively growing
- Apply after 1 Feb [4]
- Apply after end of Feb in year of harvest [2, 3, 5]
- Ensure good spray cover of the weeds
- Susceptible weeds cease growth almost immediately after application and symptoms become evident 2 wk later
- Effectiveness reduced by rain within 4 h of treatment and in very dry conditions
- Various tank mixtures recommended to broaden weed control spectrum
- Thifensulfuron-methyl and tribenuron-methyl are members of the ALS-inhibitor group of herbicides and products should be used in a planned Resistance Management strategy. See Section 5 for more information

Restrictions
- Maximum number of treatments 1 per crop
- Do not apply to cereals undersown with grass, clover or other legumes, or any other broad-leaved crop
- Do not apply within 7 d of rolling
- Specific restrictions apply to use in sequence or tank mixture with other sulfonylurea or ALS-inhibiting herbicides. See label for details
- Do not apply to any crop suffering from stress
- Consult contract agents before use on crops grown for seed

Crop-specific information
- Latest use: before flag leaf ligule first visible (GS 39) for all crops

Following crops guidance
- Only cereals, field beans, grass or oilseed rape may be sown in the same calendar year as harvest of a treated crop.
- In the event of failure of a treated crop sow only a cereal crop within 3 mth of product application and after ploughing and cultivating to at least 15 cm. After 3 mth field beans or oilseed rape may also be sown [2]
- Only cereals, oilseed rape or field beans may be sown in the same calendar year after harvest. In the following spring only cereals, oilseed rape or sugar beet may be sown. In case of crop failure for any reason, sow only spring small grain cereal. Before sowing, soil should be ploughed and cultivated to a depth of at least 15 cm[1]

Environmental safety
- Dangerous for the environment
- Very toxic to aquatic organisms
- Buffer zone requirement 6m [1]
- Spraying equipment should not be drained or flushed onto land planted, or to be planted, with trees or crops other than cereals and should be thoroughly cleansed after use - see label for instructions
- Take particular care to avoid damage by drift onto broad-leaved plants outside the target area or onto surface waters or ditches
- LERAP Category B

FOR FULL CONDITIONS OF USE ALWAYS READ THE PRODUCT LABEL

thiophanate-methyl

Hazard classification and safety precautions
   Hazard Irritant [3-5]; Dangerous for the environment [1-5]; Very toxic to aquatic organisms [3, 5]
   Transport code 9
   Packaging group III
   UN Number 3077
   Risk phrases R43 [4]
   Operator protection A, H [3-5]; U05a, U08, U20a [3-5]; U14 [4]; U19a [1]; U20c [2]
   Environmental protection E15a [3-5]; E15b [1, 2]; E16a, E38, H410 [1-5]; E16b [3, 5]
   Storage and disposal D01, D02 [1, 3-5]; D09a, D11a, D12a [1-5]

414   thiophanate-methyl

A thiophanate fungicide with protectant and curative activity
FRAC mode of action code: 1

See also iprodione + thiophanate-methyl

Products

1	Cercobin WG	Certis	70% w/w	WG	13854
2	Taurus	Certis	70% w/w	WG	15508
3	Topsin WG	Certis	70% w/w	WG	13988

Uses
- Disease control in **hops** (off-label - do not harvest for human or animal consumption (including idling) within 12 months of treatment.), **soft fruit** (off-label) [1]
- Fusarium in **durum wheat**, **spring oilseed rape**, **winter oilseed rape** [3]; **durum wheat** (reduction), **spring wheat** (reduction), **triticale** (reduction), **winter wheat** (reduction) [2]; **spring wheat**, **triticale**, **winter wheat** [1, 3]
- Fusarium diseases in **protected ornamentals** (off-label) [1]
- Mycotoxins in **durum wheat** (reduction) [2, 3]; **spring oilseed rape** (reduction), **winter oilseed rape** (reduction) [3]; **spring wheat** (reduction), **triticale** (reduction), **winter wheat** (reduction) [1-3]
- Root diseases in **container-grown ornamentals** (off-label), **protected ornamentals** (off-label) [1]
- Sclerotinia stem rot in **spring oilseed rape**, **winter oilseed rape** [2]
- Verticillium wilt in **protected tomatoes** (off-label) [1]

Extension of Authorisation for Minor Use (EAMUs)
- *container-grown ornamentals* 20111655 [1]
- *hops* (do not harvest for human or animal consumption (including idling) within 12 months of treatment.) 20082833 [1]
- *protected ornamentals* 20111887 [1]
- *protected tomatoes* 20091969 [1]
- *soft fruit* 20082833 [1]

Approval information
- Thiophanate-methyl included in Annex I under EC Regulation 1107/2009

Efficacy guidance
- To avoid the development of resistance, a maximum of 2 applications of any MBC product (thiophanate-methyl or carbendazim) are allowed in any one crop. Avoid using MBC fungicides alone.

Environmental safety
- Dangerous for the environment
- Very toxic to aquatic organisms
- LERAP Category B

Hazard classification and safety precautions
   Hazard Harmful, Dangerous for the environment [2]
   Transport code 9

SEE SECTION 3 FOR PRODUCTS ALSO REGISTERED

Pesticide Profiles

Packaging group III
UN Number 3077
Risk phrases R20, R22a, R43, R51, R53a, R68 [2]
Operator protection A; U19a [2, 3]
Environmental protection E15b, E16a, E16b [1-3]; E38 [2, 3]
Storage and disposal D01, D02, D12a [2, 3]; D10b [1-3]
Medical advice M05a [2, 3]

415 thiram

A protectant dithiocarbamate fungicide
FRAC mode of action code: M3

See also carboxin + thiram
prochloraz + thiram

Products

1	Agrichem Flowable Thiram	Agrichem	600 g/l	FS	10784
2	Thyram Plus	Agrichem	600 g/l	FS	10785

Uses

- Damping off in **broad beans** *(seed treatment)*, **bulb onions** *(seed treatment)*, **cabbages** *(seed treatment)*, **cauliflowers** *(seed treatment)*, **combining peas** *(seed treatment)*, **dwarf beans** *(seed treatment)*, **edible podded peas** *(seed treatment)*, **forage maize** *(seed treatment)*, **grass seed** *(seed treatment)*, **kale** *(off-label)*, **leeks** *(seed treatment)*, **millet** *(off-label)*, **radishes** *(seed treatment)*, **runner beans** *(seed treatment)*, **salad onions** *(seed treatment)*, **sorghum** *(off-label)*, **soya beans** *(seed treatment - qualified minor use)*, **spring field beans** *(seed treatment)*, **spring oilseed rape** *(seed treatment)*, **turnip seed crops** *(off-label)*, **turnips** *(seed treatment)*, **vining peas** *(seed treatment)*, **winter field beans** *(seed treatment)*, **winter oilseed rape** *(seed treatment)* [1, 2]; **corn gromwell** *(off-label)*, **cress** *(off-label)*, **endives** *(off-label)*, **frise** *(off-label)*, **garlic** *(off-label - seed treatment)*, **lamb's lettuce** *(off-label)*, **lettuce** *(off-label)*, **lupins** *(off-label - seed treatment)*, **poppies for morphine production** *(off-label - seed treatment)*, **quinoa** *(off-label)*, **radicchio** *(off-label)*, **rhubarb** *(off-label - seed treatment)*, **scarole** *(off-label)*, **shallots** *(off-label - seed treatment)*, **swedes** *(off-label - seed treatment)* [1]; **kale seed crops** *(off-label)*, **turnips** *(off-label)* [2]
- Pythium in **poppies for morphine production** *(off-label - seed treatment)* [1]
- Root rot in **corn gromwell** *(off-label)* [1]
- Seed-borne diseases in **carrots** *(seed soak)*, **celery (outdoor)** *(seed soak)*, **fodder beet** *(seed soak)*, **mangels** *(seed soak)*, **parsley** *(seed soak)*, **red beet** *(seed soak)*, **sugar beet** *(seed soak)* [1]; **garlic** *(off-label - seed treatment)*, **lupins** *(off-label - seed treatment)*, **shallots** *(off-label - seed treatment)*, **swedes** *(off-label - seed treatment)* [2]

Extension of Authorisation for Minor Use (EAMUs)

- **corn gromwell** *20141813* [1]
- **cress** *20070792* [1]
- **endives** *20070792* [1]
- **frise** *20070792* [1]
- **garlic** *(seed treatment) 20052395* [1], *(seed treatment) 20052391* [2]
- **kale** *20120996* [1], *20120724* [2], *20121006* [2]
- **kale seed crops** *20121006* [2]
- **lamb's lettuce** *20070792* [1]
- **lettuce** *20070792* [1]
- **lupins** *(seed treatment) 20052394* [1], *(seed treatment) 20052389* [2]
- **millet** *20120996* [1], *20120724* [2], *20121006* [2]
- **poppies for morphine production** *(seed treatment) 20040681* [1]
- **quinoa** *20140482* [1]
- **radicchio** *20070792* [1]
- **rhubarb** *(seed treatment) 20052396* [1]

FOR FULL CONDITIONS OF USE ALWAYS READ THE PRODUCT LABEL

thiram

- **scarole** *20070792* [1]
- **shallots** *(seed treatment) 20052395* [1], *(seed treatment) 20052391* [2]
- **sorghum** *20120996* [1], *20120724* [2], *20121006* [2]
- **swedes** *(seed treatment) 20052397* [1], *(seed treatment) 20052390* [2]
- **turnip seed crops** *20120996* [1], *20121006* [2]
- **turnips** *20120724* [2]

Approval information
- Thiram included in Annex I under EC Regulation 1107/2009

Efficacy guidance
- Spray before onset of disease and repeat every 7-14 d. Spray interval varies with crop and disease. See label for details
- Seed treatments may be applied through most types of seed treatment machinery, from automated continuous flow machines to smaller batch treating apparatus [1, 2]
- Co-application of 175 ml water per 100 kg of seed likely to improve evenness of seed coverage [1, 2]

Restrictions
- Maximum number of treatments 3 per crop for protected winter lettuce (thiram based products only; 2 per crop if sequence of thiram and other EBDC fungicides use); 2 per crop for protected summer lettuce; 1 per batch of seed for seed treatments
- Do not apply to hydrangeas
- Notify processor before dusting or spraying crops for processing
- Do not dip roots of forestry transplants
- Do not treat seed of tomatoes, peppers or aubergines
- Soya bean seed treatments restricted to crops that are to be harvested as a mature pulse crop only
- Treated seed should not be stored from one season to the next

Crop-specific information
- Latest use: pre-drilling for seed treatments and seed soaks [1, 2]; 21 d after planting out or 21 d before harvest, whichever is earlier, for protected winter lettuce; 14 d after planting out or 21 d before harvest, whichever is earlier, for spraying protected summer lettuce
- Seed to be treated should be of satisfactory quality and moisture content [1, 2]
- Follow label instructions for treating small quantities of seed [1, 2]
- For use on tulips, chrysanthemums and carnations add non-ionic wetter

Environmental safety
- Dangerous for the environment [1, 2]
- Very toxic to aquatic organisms [1, 2]
- Do not use treated seed as food or feed [1, 2]
- Treated seed harmful to game and wildlife [1, 2]
- A red dye is available from manufacturer to colour treated seed, but not recommended as part of the seed steep. See label for details [1]

Hazard classification and safety precautions
Hazard Harmful, Dangerous for the environment
Transport code 9
Packaging group III
UN Number 3082
Risk phrases R20, R22a, R36, R38, R43, R48, R50, R53a
Operator protection A; U02a, U05a, U08, U11, U20b
Environmental protection E15a, E34
Storage and disposal D01, D02, D05, D09a, D10c, D12b
Treated seed S01, S02, S03, S04a, S04b, S05, S06a, S07
Medical advice M04a

SEE SECTION 3 FOR PRODUCTS ALSO REGISTERED

416   tolclofos-methyl

A protectant nitroaniline fungicide for soil-borne diseases
FRAC mode of action code: 14

Products

Basilex	Everris Ltd	50% w/w	WP	16243

Uses
- Bottom rot in **protected lettuce**
- Damping off in **broccoli** *(off-label)*, **leaf brassicas** *(off-label)*, **radishes** *(off-label)*
- Damping off and wirestem in **broccoli**, **brussels sprouts**, **cabbages**, **calabrese**, **cauliflowers**, **chinese cabbage**
- Foot rot in **ornamental plant production**
- Rhizoctonia in **protected celery** *(off-label)*, **protected radishes** *(off-label)*, **radishes** *(off-label)*, **swedes** *(off-label)*, **turnips** *(off-label)*
- Root rot in **ornamental plant production**, **radishes** *(off-label)*
- Wirestem in **broccoli** *(off-label)*, **leaf brassicas** *(off-label)*

Extension of Authorisation for Minor Use (EAMUs)
- *broccoli* 20131781
- *leaf brassicas* 20131782
- *protected celery* 20131778
- *protected radishes* 20131779
- *radishes* 20131780
- *swedes* 20131777
- *turnips* 20131777

Approval information
- Tolclofos-methyl included in Annex I under EC Regulation 1107/2009
- In 2006 CRD required that all products containing this active ingredient should carry the following warning in the main area of the container label: "Tolclofos-methyl is an anticholinesterase organophosphate. Handle with care"
- May be applied by misting equipment mounted over roller table. See label for details

Efficacy guidance
- Apply dust to seed potatoes during hopper loading
- To control Rhizoctonia in vegetables and ornamentals apply as drench before sowing, pricking out or planting [1]
- On established seedlings and pot plants apply as drench and rinse off foliage [1]

Restrictions
- Tolclofos-methyl is an atypical organophosphorus compound which has weak anticholinesterase activity. Do not use if under medical advice not to work with such compounds
- Maximum number of treatments 1 per batch for seed potatoes; 1 per crop for lettuce, brassicas, swedes, turnips; 1 at each stage of growth (ie sowing, pricking out, potting) to a maximum of 3, for ornamentals
- Only to be used with automatic planters
- Not recommended for use on seed potatoes where hot water treatment used or to be used
- Do not apply as overhead drench to vegetables or ornamentals when hot and sunny [1]
- Do not use on heathers [1]
- Must not be used via hand-held equipment
- Application to protected crops must only be made where the operator is outside the structure at the time of treatment

Crop-specific information
- Latest use: at planting of seed potatoes; before transplanting for lettuce, brassicas, ornamental specimens [1]; before 2 true lvs for swedes, turnips [1]

Environmental safety
- Dangerous for the environment
- Very toxic to aquatic organisms

FOR FULL CONDITIONS OF USE ALWAYS READ THE PRODUCT LABEL

tri-allate

- Treated tubers to be used as seed only, not for food or feed
- LERAP Category B

Hazard classification and safety precautions
Hazard Irritant, Dangerous for the environment
Transport code 9
Packaging group III
UN Number 3077
Risk phrases R36, R37, R38, R51, R53a
Operator protection A, D, H; U19a, U20b
Environmental protection E16a
Storage and disposal D09a, D11a
Medical advice M01

417 tri-allate

A soil-acting thiocarbamate herbicide for grass weed control
HRAC mode of action code: N

Products
Avadex Excel 15G   Gowan   15% w/w   GR   16998

Uses
- Annual grasses in *canary seed* (off-label), *linseed* (off-label), *miscanthus* (off-label), *spring barley*, *winter barley*, *winter wheat*
- Annual meadow grass in *canary seed* (off-label), *linseed* (off-label), *miscanthus* (off-label)
- Blackgrass in *canary seed* (off-label), *linseed* (off-label), *miscanthus* (off-label), *spring barley*, *winter barley*, *winter wheat*
- Meadow grasses in *spring barley*, *winter barley*, *winter wheat*
- Wild oats in *canary seed* (off-label), *linseed* (off-label), *miscanthus* (off-label), *spring barley*, *winter barley*, *winter wheat*

Extension of Authorisation for Minor Use (EAMUs)
- *canary seed* 20152324
- *linseed* 20152471
- *miscanthus* 20152325

Approval information
- Tri-allate included in Annex 1 under EC Regulation 1107/2009
- Accepted by BBPA for use on malting barley

Efficacy guidance
- Apply to soil surface pre-emergence
- For maximum activity apply to well-prepared moist seedbeds
- Do not use on soils with more than 10% organic matter
- Wild oats controlled up to 2-leaf stage
- Do not apply with spinning disc granule applicator; see label for suitable types
- Do not apply to cloddy seedbeds
- Use sequential treatments to improve control of barren brome and annual dicotyledons (see label for details)

Restrictions
- Maximum number of treatments 1 per crop
- Drill wheat well below treated layer of soil (see label for safe drilling depths)
- Do not apply to shallow-drilled wheat crops
- Do not undersow grasses into treated crops
- Do not sow oats or grasses within 1 yr of treatment

SEE SECTION 3 FOR PRODUCTS ALSO REGISTERED

Pesticide Profiles

Crop-specific information
- Latest use: pre-drilling for beet crops; before crop emergence for field beans, spring barley, peas, forage legumes; before first node detectable stage (GS 31) for winter wheat, winter barley, durum wheat, triticale, winter rye

Following crops guidance
- Pre-emergence use in winter wheat and winter barley
- Pre-drilling or pre-emergence use in spring barley

Environmental safety
- Irritating to eyes and skin
- May cause sensitization by skin contact
- Harmful to fish or other aquatic life. Do not contaminate surface waters or ditches with chemical or used container
- Buffer zone requirement 10 m [1]
- Avoid drift on to non-target plants outside the treated area
- LERAP Category B

Hazard classification and safety precautions
Hazard Harmful, Dangerous for the environment, Harmful if swallowed
Transport code 9
Packaging group III
UN Number 3077
Risk phrases H304, H373
Operator protection A, C, H; U02a, U05a, U19a, U20a
Environmental protection E16a, H410, H411
Storage and disposal D01, D02, D09a, D11a, D19
Medical advice M05a

418 tribenuron-methyl

A foliar acting sulfonylurea herbicide with some root activity for use in cereals
HRAC mode of action code: B

See also florasulam + tribenuron-methyl
   fluroxypyr + thifensulfuron-methyl + tribenuron-methyl
   metsulfuron-methyl + tribenuron-methyl
   thifensulfuron-methyl + tribenuron-methyl

Products
1	Thor	Nufarm UK	50% w/w	WG	15239
2	Triad	Headland	50% w/w	TB	12751

Uses
- Annual dicotyledons in **durum wheat**, **farm forestry** *(off-label)*, **forest nurseries** *(off-label)*, **green cover on land temporarily removed from production**, **miscanthus** *(off-label)*, **spring oats**, **spring wheat**, **triticale**, **winter barley**, **winter oats**, **winter rye**, **winter wheat** [1]; **grassland**, **grassland** *(off-label)*, **undersown spring barley** [2]; **spring barley** [1, 2]
- Chickweed in **grassland**, **grassland** *(off-label)*, **spring barley**, **undersown spring barley** [2]

Extension of Authorisation for Minor Use (EAMUs)
- *farm forestry* 20122064 [1]
- *forest nurseries* 20122064 [1]
- *grassland* 20122493 [2]
- *miscanthus* 20122064 [1]

Approval information
- Tribenuron-methyl included in Annex I under EC Regulation 1107/2009
- Accepted by BBPA for use on malting barley

Efficacy guidance
- Best control achieved when weeds small and actively growing

FOR FULL CONDITIONS OF USE ALWAYS READ THE PRODUCT LABEL

Trichoderma asperellum (Strain T34)

- Good spray cover must be achieved since larger weeds often become less susceptible
- Susceptible weeds cease growth almost immediately after treatment and symptoms can be seen in about 2 wk
- Weed control may be reduced when conditions very dry
- Tribenuron-methyl is a member of the ALS-inhibitor group of herbicides and products should be used in a planned Resistance Management strategy. See Section 5 for more information

Restrictions
- Maximum number of treatments 1 per crop
- Specific restrictions apply to use in sequence or tank mixture with other sulfonylurea or ALS-inhibiting herbicides. See label for details
- Do not apply to crops undersown with grass, clover or other broad-leaved crops
- Do not apply to any crop suffering stress from any cause or not actively growing
- Do not apply within 7 d of rolling

Crop-specific information
- Latest use: up to and including flag leaf ligule/collar just visible (GS 39)
- Apply in autumn or in spring from 3 leaf stage of crop

Following crops guidance
- Only cereals, field beans or oilseed rape may be sown in the same calendar yr as harvest of a treated crop
- In the event of crop failure sow only a cereal within 3 mth of application. After 3 mth field beans or oilseed rape may also be sown

Environmental safety
- Dangerous for the environment
- Very toxic to aquatic organisms
- Take extreme care to avoid drift onto broad-leaved plants outside the target area or onto surface waters or ditches, or land intended for cropping
- Spraying equipment should not be drained or flushed onto land planted, or to be planted, with trees or crops other than cereals and should be thoroughly cleansed after use - see label for instructions

Hazard classification and safety precautions
    Hazard Irritant, Dangerous for the environment
    Transport code 9
    Packaging group III
    UN Number 3077
    Risk phrases H317 [1]; R43, R50, R53a [2]
    Operator protection A [1, 2]; H [1]; U05a, U08, U20b
    Environmental protection E07b [2] (3 weeks); E07e [2]; E15a [1, 2]; E38, H410 [1]
    Storage and disposal D01, D02, D09a, D11a [1, 2]; D12a [1]

419 Trichoderma asperellum (Strain T34)

A biological control agent

Products
    T34 Biocontrol   Fargro   10.832% w/w   WP   17290

Uses
- Fusarium in **protected baby leaf crops** (off-label), **protected bilberries** (off-label), **protected blackberries** (off-label), **protected blackcurrants** (off-label), **protected blueberry** (off-label), **protected broccoli** (off-label), **protected brussels sprouts** (off-label), **protected cabbages** (off-label), **protected calabrese** (off-label), **protected cauliflowers** (off-label), **protected choi sum** (off-label), **protected collards** (off-label), **protected courgettes** (off-label), **protected cranberries** (off-label), **protected cucumbers** (off-label), **protected elderberries** (off-label), **protected forest nurseries** (off-label), **protected gherkins** (off-label), **protected gooseberries** (off-label), **protected herbs (see appendix 6)** (off-label), **protected kale** (off-label), **protected kohlrabi** (off-label), **protected loganberries** (off-label), **protected melons** (off-label), **protected**

SEE SECTION 3 FOR PRODUCTS ALSO REGISTERED

Pesticide Profiles

  mulberry (off-label), **protected oriental cabbage** (off-label), **protected ornamentals** (off-label), **protected pumpkins** (off-label), **protected raspberries** (off-label), **protected redcurrants** (off-label), **protected rose hips** (off-label), **protected rubus hybrids** (off-label), **protected strawberries** (off-label), **protected summer squash** (off-label), **protected winter squash** (off-label)
- Fusarium foot rot and seedling blight in **aubergines**, **chillies**, **ornamental plant production**, **peppers**, **tomatoes**
- Pythium in **protected baby leaf crops** (off-label), **protected bilberries** (off-label), **protected blackberries** (off-label), **protected blackcurrants** (off-label), **protected blueberry** (off-label), **protected broccoli** (off-label), **protected brussels sprouts** (off-label), **protected cabbages** (off-label), **protected calabrese** (off-label), **protected cauliflowers** (off-label), **protected choi sum** (off-label), **protected collards** (off-label), **protected courgettes** (off-label), **protected cranberries** (off-label), **protected cucumbers** (off-label), **protected elderberries** (off-label), **protected forest nurseries** (off-label), **protected gherkins** (off-label), **protected gooseberries** (off-label), **protected herbs (see appendix 6)** (off-label), **protected kale** (off-label), **protected kohlrabi** (off-label), **protected loganberries** (off-label), **protected melons** (off-label), **protected mulberry** (off-label), **protected oriental cabbage** (off-label), **protected ornamentals** (off-label), **protected pumpkins** (off-label), **protected raspberries** (off-label), **protected redcurrants** (off-label), **protected rose hips** (off-label), **protected rubus hybrids** (off-label), **protected strawberries** (off-label), **protected summer squash** (off-label), **protected winter squash** (off-label)

Extension of Authorisation for Minor Use (EAMUs)
- *protected baby leaf crops* 20161809
- *protected bilberries* 20161808
- *protected blackberries* 20161808
- *protected blackcurrants* 20161808
- *protected blueberry* 20161808
- *protected broccoli* 20161809
- *protected brussels sprouts* 20161809
- *protected cabbages* 20161809
- *protected calabrese* 20161809
- *protected cauliflowers* 20161809
- *protected choi sum* 20161809
- *protected collards* 20161809
- *protected courgettes* 20161805
- *protected cranberries* 20161808
- *protected cucumbers* 20161805
- *protected elderberries* 20161808
- *protected forest nurseries* 20161810
- *protected gherkins* 20161805
- *protected gooseberries* 20161808
- *protected herbs (see appendix 6)* 20161809
- *protected kale* 20161809
- *protected kohlrabi* 20161809
- *protected loganberries* 20161808
- *protected melons* 20161805
- *protected mulberry* 20161808
- *protected oriental cabbage* 20161809
- *protected ornamentals* 20161810
- *protected pumpkins* 20161805
- *protected raspberries* 20161808
- *protected redcurrants* 20161808
- *protected rose hips* 20161808
- *protected rubus hybrids* 20161808
- *protected strawberries* 20161808
- *protected summer squash* 20161805
- *protected winter squash* 20161805

FOR FULL CONDITIONS OF USE ALWAYS READ THE PRODUCT LABEL

Approval information
- Trichoderma asperellum (Strain T34) included in Annex 1 under EC Regulation 1107/2009

Efficacy guidance
- May be applied by spraying, through an irrigation system or by root dipping.

Restrictions
- Efficacy has been demonstrated on peat and coir composts but should be checked on the more unusual composts before large scale use.

Crop-specific information
- Crop safety has been demonstrated on a range of carnation species. It is advisable to check the safety to other species on a sample of the population before large-scale use.

Hazard classification and safety precautions
Hazard Harmful
UN Number N/C
Risk phrases H317
Operator protection A, D, H; U05a, U09a, U19a, U20b
Environmental protection E15a
Storage and disposal D01, D02, D05, D09a, D09b, D10c
Medical advice M03, M04a

420 trifloxystrobin

A protectant strobilurin fungicide for cereals and managed amenity turf
FRAC mode of action code: 11

See also cyproconazole + trifloxystrobin
fluopyram + trifloxystrobin
fluoxastrobin + prothioconazole + trifloxystrobin
iprodione + trifloxystrobin
propiconazole + trifloxystrobin
prothioconazole + trifloxystrobin
tebuconazole + trifloxystrobin

Products
1	Martinet	AgChem Access	500 g/l	SC	14614
2	Mascot Defender	Rigby Taylor	500 g/l	SG	14065
3	Pan Aquarius	Pan Agriculture	50% w/w	WG	13018
4	PureFloxy	Pure Amenity	50% w/w	WG	15473
5	Scorpio	Bayer Environ.	50% w/w	WG	12293
6	Swift SC	Bayer CropScience	500 g/l	SC	11227

Uses
- Brown rust in **spring barley, winter barley, winter wheat** [1, 6]
- Foliar disease control in **durum wheat** *(off-label)*, **grass seed crops** *(off-label)*, **ornamental specimens** *(off-label)*, **spring rye** *(off-label)*, **triticale** *(off-label)*, **winter rye** *(off-label)* [6]
- Fusarium patch in **amenity grassland, managed amenity turf** [2-5]
- Glume blotch in **winter wheat** [1, 6]
- Net blotch in **spring barley, winter barley** [1, 6]
- Red thread in **amenity grassland, managed amenity turf** [2-5]
- Rhynchosporium in **spring barley, winter barley** [1, 6]
- Septoria leaf blotch in **winter wheat** [1, 6]

Extension of Authorisation for Minor Use (EAMUs)
- *durum wheat* 20061287 [6]
- *grass seed crops* 20061287 [6]
- *ornamental specimens* 20082882 [6]
- *spring rye* 20061287 [6]
- *triticale* 20061287 [6]
- *winter rye* 20061287 [6]

SEE SECTION 3 FOR PRODUCTS ALSO REGISTERED

Pesticide Profiles

Approval information
- Trifloxystrobin included in Annex I under EC Regulation 1107/2009
- Accepted by BBPA for use on malting barley

Efficacy guidance
- Should be used protectively before disease is established in crop. Further treatment may be necessary if disease attack prolonged
- Treat grass after cutting and do not mow for at least 48 h afterwards to allow adequate systemic movement [3, 5]
- Trifloxystrobin is a member of the QoI cross resistance group. Product should be used preventatively and not relied on for its curative potential
- Use product as part of an Integrated Crop Management strategy incorporating other methods of control, including where appropriate other fungicides with a different mode of action. Do not apply more than two foliar applications of QoI containing products to any cereal crop
- There is a significant risk of widespread resistance occurring in *Septoria tritici* populations in UK. Failure to follow resistance management action may result in reduced levels of disease control
- On cereal crops product must always be used in mixture with another product, recommended for control of the same target disease, that contains a fungicide from a different cross resistance group and is applied at a dose that will give robust control
- Strains of barley powdery mildew resistant to QoIs are common in the UK

Restrictions
- Maximum number of treatments 2 per crop per yr
- Do not apply to turf during dought conditions or to frozen turf [3, 5]

Crop-specific information
- HI barley, wheat 35 d [6]

Environmental safety
- Dangerous for the environment
- Very toxic to aquatic organisms
- LERAP Category B [2-5]

Hazard classification and safety precautions
    Hazard Irritant [2-5]; Dangerous for the environment [1-6]; Very toxic to aquatic organisms [5, 6]
    Transport code 9
    Packaging group III
    UN Number 3077 [3-5]; 3082 [1, 2, 6]
    Risk phrases H317 [5]; R43 [2-4]; R50, R53a [1-4]
    Operator protection A, H; U02a, U09a, U19a [1, 6]; U05a, U20b [1-6]; U11, U13, U14, U15 [2-5]
    Environmental protection E13b [1, 6]; E15b, E16a, E16b, E34 [2-5]; E38 [1-6]; H410 [5, 6]
    Consumer protection C02a [1, 6] (35 d)
    Storage and disposal D01, D02, D09a, D10c, D12a
    Medical advice M03 [2-5]

421 triflusulfuron-methyl

A sulfonyl urea herbicide for beet crops
HRAC mode of action code: B

See also lenacil + triflusulfuron-methyl

Products

1	Debut	DuPont	50% w/w	WG	07804
2	Shiro	UPL Europe	50% w/w	WG	17439
3	Tricle	UPL Europe	50% w/w	WG	17448
4	Upbeet	DuPont	50% w/w	WG	17544

Uses
- Annual dicotyledons in **chicory** *(off-label)* [1]; **fodder beet**, **sugar beet** [1-4]
- Charlock in **red beet** *(off-label)* [1]

FOR FULL CONDITIONS OF USE ALWAYS READ THE PRODUCT LABEL

triflusulfuron-methyl

- Cleavers in **red beet** *(off-label)* [1]
- Flixweed in **red beet** *(off-label)* [1]
- Fool's parsley in **red beet** *(off-label)* [1]
- Nipplewort in **red beet** *(off-label)* [1]
- Wild chrysanthemum in **red beet** *(off-label)* [1]

Extension of Authorisation for Minor Use (EAMUs)
- *chicory* 20142535 [1]
- *red beet* 20142536 [1]

Approval information
- Triflusulfuron-methyl included in Annex I under EC Regulation 1107/2009

Efficacy guidance
- Product should be used with a recommended adjuvant or a suitable herbicide tank-mix partner - see label for details
- Product acts by foliar action. Best results obtained from good spray cover of small actively growing weeds
- Susceptible weeds cease growth immediately and symptoms can be seen 5-10 d later
- Best results achieved from a programme of up to 4 treatments starting when first weeds have emerged with subsequent applications every 5-14 d when new weed flushes at cotyledon stage
- Weed spectrum can be broadened by tank mixture with other herbicides. See label for details
- Product may be applied overall or via band sprayer
- Triflusulfuron-methyl is a member of the ALS-inhibitor group of herbicides

Restrictions
- Maximum number of treatments 4 per crop
- Do not apply to any crop stressed by drought, water-logging, low temperatures, pest or disease attack, nutrient or lime deficiency

Crop-specific information
- Latest use: before crop leaves meet between rows
- HI 4 wk for red beet
- All varieties of sugar beet and fodder beet may be treated from early cotyledon stage until the leaves begin to meet between the rows

Following crops guidance
- Only winter cereals should follow a treated crop in the same calendar yr. Any crop may be sown in the next calendar yr
- After failure of a treated crop, sow only spring barley, linseed or sugar beet within 4 mth of spraying unless prohibited by tank-mix partner

Environmental safety
- Dangerous for the environment
- Very toxic to aquatic organisms
- Extremely dangerous to fish or other aquatic life. Do not contaminate surface waters or ditches with chemical or used container
- Take extreme care to avoid drift onto broad-leaved plants outside the target area or onto surface waters or ditches, or land intended for cropping
- Spraying equipment should not be drained or flushed onto land planted, or to be planted, with trees or crops other than sugar beet and should be thoroughly cleansed after use - see label for instructions
- LERAP Category B

Hazard classification and safety precautions
Hazard Irritant, Dangerous for the environment [1-4]; Very toxic to aquatic organisms [2-4]
Transport code 9
Packaging group III
UN Number 3077
Risk phrases H351 [2, 3]; R43, R50, R53a [1]
Operator protection A; U05a, U08, U19a, U20a

SEE SECTION 3 FOR PRODUCTS ALSO REGISTERED

Pesticide Profiles

Environmental protection E13a, E15a, E16a, E16b, E38 [1-4]; H410 [2-4]
Storage and disposal D01, D02, D05, D09a, D11a, D12a

422 trinexapac-ethyl

A novel cyclohexanecarboxylate plant growth regulator for cereals, turf and amenity grassland

See also prohexadione-calcium + trinexapac-ethyl

Products

1	Clipless	Headland Amenity	120 g/l	ME	15435
2	Clipless NT	Headland	120 g/l	ME	17558
3	Confine	Headland	250 g/l	EC	16109
4	Cutaway	Syngenta	121 g/l	SL	14445
5	Freeze NT	Headland	250 g/l	EC	17478
6	Gyro	UPL Europe	250 g/l	EC	17426
7	Iceni	UPL Europe	250 g/l	EC	16835
8	Life Scientific Trinexapac 250	Life Scientific	250 g/l	EC	16047
9	Limitar	Belcrop	250 g/l	EC	16301
10	Modan 250 EC	Belchim	250 g/l	EC	17060
11	Moddus	Syngenta	250 g/l	EC	15151
12	Mowless	ProKlass	121 g/l	SL	17086
13	Moxa	Globachem	250 g/l	EC	16105
14	Moxa 250 EC	Belchim	250 g/l	EC	16176
15	Optimus	Adama	175 g/l	EC	15249
16	Pan Tepee	Pan Agriculture	250 g/l	EC	15867
17	Plaza	Goldengrass	250 g/l	EC	16855
18	Primo Maax II	Syngenta	121 g/l	SL	17509
19	Primo Maxx	Syngenta	121 g/l	SL	14780
20	Pure Max	Pure Amenity	121 g/l	SL	14952
21	Shorten	Becesane	250 g/l	EC	15377
22	Shrink	AgChem Access	250 g/l	EC	15660
23	Tacet	Goldengrass	250 g/l	EC	15258
24	Tempo	Syngenta	250 g/l	EC	15170
25	Tridus	Globachem	250 g/l	EC	16938

Uses
- Growth regulation in **canary seed** *(off-label)*, **spring wheat** *(off-label)* [15]; **durum wheat, spring barley, spring oats, triticale, winter barley, winter oats, winter wheat** [6, 7, 9, 21-23]; **forest nurseries** *(off-label)*, **ornamental plant production** *(off-label)* [4, 11]; **grass seed crops** [6, 9]; **lettuce** *(off-label)* [4]; **rye, spring wheat** [9]; **ryegrass seed crops, spring rye, winter rye** [6, 7, 21-23]; **spring red wheat** *(off-label)* [3]
- Growth retardation in **amenity grassland** [1, 2, 4, 12, 18-20]; **managed amenity turf** [1, 2, 12, 18-20]; **spring red wheat** *(off-label)* [24]
- Lodging control in **durum wheat** [3, 5, 8, 11, 13-16, 24, 25]; **grass seed crops** [3, 5, 8, 13, 14, 25]; **rye** [5, 13, 14, 25]; **ryegrass seed crops** [11, 15, 16, 24]; **spring barley, spring oats, triticale, winter barley, winter oats, winter wheat** [3, 5, 8, 10, 11, 13-17, 24, 25]; **spring red wheat** *(off-label)* [3, 11, 24]; **spring rye, winter rye** [3, 8, 10, 11, 15-17, 24]; **spring wheat** [5, 11, 13-15, 25]

Extension of Authorisation for Minor Use (EAMUs)
- **canary seed** *20111263* [15], *20141226* [15]
- **forest nurseries** *20162140* [4], *20122055* [11]
- **lettuce** *20162140* [4]
- **ornamental plant production** *20162140* [4], *20103062* [11]
- **spring red wheat** *20130322* [3], *20121385* [11], *20113201* [24]
- **spring wheat** *20111264* [15], *20141227* [15]

FOR FULL CONDITIONS OF USE ALWAYS READ THE PRODUCT LABEL

trinexapac-ethyl

Approval information
- Trinexapac-ethyl included in Annex I under EC Regulation 1107/2009
- Accepted by BBPA for use on malting barley

Efficacy guidance
- Best results on cereals and ryegrass seed crops obtained from treatment from the leaf sheath erect stage [11, 16]
- Best results on turf achieved from application to actively growing weed free turf grass that is adequately fertilized and watered and is not under stress. Adequate soil moisture is essential [20]
- Turf should be dry and weed free before application [20]
- Environmental conditions, management and cultural practices that affect turf growth and vigour will influence effectiveness of treatment [20]
- Repeat treatments on turf up to the maximum approved dose may be made as soon as growth resumes [20]

Restrictions
- Maximum total dose equivalent to one full dose on cereals, ryegrass seed crops [11, 16]
- Maximum number of treatments on turf equivalent to five full dose treatments [20]
- Do not apply if rain or frost expected or if crop wet. Products are rainfast after 12 h
- Only use on crops at risk of lodging [11, 16]
- Do not apply within 12 h of mowing turf [20]
- Do not treat newly sown turf [20]
- Not to be used on food crops [20]
- Do not compost or mulch grass clippings [20]
- Some products stipulate that application should not be made before GS30

Crop-specific information
- Latest use: before 2nd node detectable (GS 32) for oats, ryegrass seed crops; before 3rd node detectable (GS 33) for durum wheat, spring barley, triticale, rye; before flag leaf sheath extending (GS 41) for winter barley, winter wheat
- On wheat apply as single treatment between leaf sheath erect stage (GS 30) and flag leaf fully emerged (GS 39) [11, 16]
- On barley, rye, triticale and durum wheat apply as single treatment between leaf sheath erect stage (GS 30) and second node detectable (GS 32), or on winter barley at higher dose between flag leaf just visible (GS 37) and flag leaf fully emerged (GS 39) [11, 16]
- On oats and ryegrass seed crops apply between leaf sheath erect stage (GS 30) and first node detectable stage (GS 32) [11, 16]
- Treatment may cause ears of cereals to remain erect through to harvest [11, 16]
- Turf under stress when treated may show signs of damage [20]
- Any weed control in turf must be carried out before application of the growth regulator [20]

Environmental safety
- Dangerous for the environment [11, 16]
- Toxic to aquatic organisms [11, 16]

Hazard classification and safety precautions
Hazard Irritant [3, 5-8, 10, 11, 13-17, 21-25]; Dangerous for the environment [3, 5-8, 10, 11, 13-16, 21-25]; Flammable liquid and vapour [13, 25]; Harmful if inhaled [13, 18, 25]; Very toxic to aquatic organisms [6]
Transport code 9 [1-3, 5-8, 10, 11, 13-17, 21-25]
Packaging group III [1-3, 5-8, 10, 11, 13-17, 21-25]
UN Number 3082 [1-3, 5-8, 10, 11, 13-17, 21-25]; N/C [4, 9, 12, 18-20]
Risk phrases H315 [6, 15]; H317 [5-7, 10, 14, 15, 18]; H318 [6, 10]; H319 [5, 9, 13, 15, 25]; H335 [9, 10, 13, 25]; R36 [3]; R37, R41 [17]; R43, R51, R53a [3, 8, 11, 16, 17, 21-24]
Operator protection A [1-25]; C [1-5, 8-25]; H [1, 2, 4-7, 10, 12, 17-20]; K [1, 2, 4, 12, 18-20]; U05a [1-8, 10-16, 18-25]; U08, U09c, U20a [9]; U10, U14 [10, 17]; U11, U19a [9, 10, 17]; U15, U20c [3, 5-8, 10, 11, 13-16, 21-25]; U20b [1, 2, 4, 12, 18-20]
Environmental protection E15a [3, 5-8, 10, 11, 13-16, 21-25]; E15b, E34 [9]; E38 [3, 5-11, 13-17, 21-25]; H411 [7, 9, 10, 13, 14, 25]; H412 [2, 5, 15, 18]
Consumer protection C01 [1, 2, 4, 12, 18-20]

SEE SECTION 3 FOR PRODUCTS ALSO REGISTERED

Pesticide Profiles

Storage and disposal D01 [1-25]; D02, D05 [1-8, 10-25]; D09a [3, 5-11, 13-16, 21-25]; D10c [1-8, 10-16, 18-25]; D11a, D12b [9]; D12a [3, 5-11, 13-17, 21-25]
Medical advice M03 [9]; M05a [17]

423 urea

Commodity substance for fungicide treatment of cut tree stumps

424 zeta-cypermethrin

A contact and stomach acting pyrethroid insecticide
IRAC mode of action code: 3

Products

1	Angri	AgChem Access	100 g/l	EW	13730
2	Chimpanzee	Pan Agriculture	100 g/l	EW	16043
3	Fury 10 EW	Headland	100 g/l	EW	17255
4	Minuet EW	Headland	100 g/l	EW	17250

Uses
- Aphids in **broad beans** *(off-label)*, **durum wheat** *(off-label)*, **fodder beet** *(off-label)*, **grass seed crops** *(off-label)*, **lupins** *(off-label)*, **rye** *(off-label)*, **triticale** *(off-label)* [3]; **spring barley, spring oats, spring wheat, winter barley, winter oats, winter wheat** [1-4]
- Barley yellow dwarf virus vectors in **spring barley, spring wheat, winter barley, winter wheat** [1-4]
- Cabbage seed weevil in **spring oilseed rape, winter oilseed rape** [1-4]
- Cabbage stem flea beetle in **spring oilseed rape, winter oilseed rape** [1-4]
- Cutworms in **potatoes, sugar beet** [1-4]
- Flax flea beetle in **linseed** [1-4]
- Flea beetle in **spring oilseed rape, winter oilseed rape** [1-4]
- Insect pests in **broad beans** *(off-label)*, **durum wheat** *(off-label)*, **fodder beet** *(off-label)*, **grass seed crops** *(off-label)*, **lupins** *(off-label)*, **rye** *(off-label)*, **triticale** *(off-label)* [3]
- Large flax flea beetle in **linseed** [1-4]
- Pea and bean weevil in **combining peas, spring field beans, vining peas, winter field beans** [1-4]
- Pea aphid in **combining peas, vining peas** [1-4]
- Pea moth in **combining peas, vining peas** [1-4]
- Pod midge in **spring oilseed rape, winter oilseed rape** [1-4]
- Pollen beetle in **spring oilseed rape, winter oilseed rape** [1-4]
- Rape winter stem weevil in **spring oilseed rape, winter oilseed rape** [1-4]

Extension of Authorisation for Minor Use (EAMUs)
- *broad beans* 20160125 [3]
- *durum wheat* 20160124 [3]
- *fodder beet* 20160126 [3]
- *grass seed crops* 20160124 [3]
- *lupins* 20160127 [3]
- *rye* 20160124 [3]
- *triticale* 20160124 [3]

Approval information
- Zeta-cypermethrin included in Annex I under EC Regulation 1107/2009
- Accepted by BBPA for use on malting barley
- Approval expiry 31 Mar 2017 [1]

Efficacy guidance
- On winter cereals spray when aphids first found in the autumn for BYDV control. A second spray may be required on late drilled crops or in mild conditions
- For summer aphids on cereals spray when treatment threshold reached

FOR FULL CONDITIONS OF USE ALWAYS READ THE PRODUCT LABEL

- For listed pests in other crops spray when feeding damage first seen or when treatment threshold reached. Under high infestation pressure a second treatment may be necessary
- Best results for pod midge and seed weevil control in oilseed rape obtained from treatment after pod set but before 80% petal fall
- Pea moth treatments should be applied according to ADAS/PGRO warnings or when economic thresholds reached as indicated by pheromone traps
- Treatments for cutworms should be made at egg hatch and repeated no sooner than 10 d later

Restrictions
- Maximum total dose equivalent to two full dose treatments on all crops
- Consult processors before use on crops for processing

Crop-specific information
- Latest use: before 4 true leaves for linseed, before end of flowering for oilseed rape, cereals
- HI potatoes, field beans, peas 14 d; sugar beet 60 d

Environmental safety
- Dangerous for the environment
- Very toxic to aquatic organisms
- High risk to non-target insects or other arthropods. Do not spray within 6 m of the field boundary
- LERAP Category A

Hazard classification and safety precautions
Hazard Harmful, Dangerous for the environment [1-4]; Harmful if swallowed, Harmful if inhaled [3, 4]; Very toxic to aquatic organisms [3]
Transport code 6.1 [1, 2]; 9 [3, 4]
Packaging group III
UN Number 3082 [3, 4]; 3352 [1, 2]
Risk phrases H317 [3, 4]; R20, R22a, R43, R50, R53a [1, 2]
Operator protection A, C, H; U05a, U08, U14, U15, U19a, U20b
Environmental protection E15b, E16c, E16d, E22a, E34, E38 [1-4]; H410 [3, 4]
Storage and disposal D01, D02, D09a, D10b, D12a
Medical advice M05a

425 zoxamide

A substituted benzamide fungicide available only in mixtures
FRAC mode of action code: 22

See also cymoxanil + zoxamide
  dimethomorph + zoxamide
  mancozeb + zoxamide

SEE SECTION 3 FOR PRODUCTS ALSO REGISTERED

SECTION 3
PRODUCTS ALSO REGISTERED

Products also Registered

Products listed in the table below have not been notified for inclusion in Section 2 of this edition of the *Guide*. These products may legally be stored and used in accordance with their label until their approval expires, but they may not still be available for purchase.

Product	Approval holder	MAPP No.	Expiry Date
1 abamectin			
Abam	CMI	15535	31 Jul 2017
Abamex 18 EC	MAC	15806	31 Dec 2021
Abamite Pro	CMI	16138	31 Dec 2021
Agrimec	Syngenta	14491	31 Jul 2017
Biotrine	Russell	16367	31 Dec 2021
Dynamec	Syngenta	13331	31 Dec 2021
Hi-Mectin	Hockley	15420	31 Jul 2017
2 acetamiprid			
Aceta 20 SG	Euro	16919	31 Oct 2019
Acetamex 20 SP	MAC	15888	31 Oct 2019
Gazelle	Certis	12909	31 Oct 2019
Persist	RAAT	17633	31 Oct 2019
3 acetic acid			
Natural Weed Spray NO. 1	Punya	12328	28 Feb 2017
New-Way Weed Spray	Punya	15319	31 Aug 2019
OWK	UK Organic	15363	28 Feb 2022
4 acibenzolar-S-methyl			
Bion	Syngenta	09803	31 Dec 2018
Inssimo	Syngenta	16870	15 Jan 2019
5 adoxophyes orana gv			
Capex	Andermatt	16351	06 Jan 2018
Capex	Sentomol	16493	06 Jan 2018
6 alpha-cypermethrin			
Al-cyper Ec	Harvest	17801	31 Jan 2020
Alert	BASF	16785	31 Jan 2020
Alpha C 6 ED	Techneat	13611	31 Jan 2020
Contest	BASF	16764	31 Jan 2020
Fastac	BASF	16761	31 Jan 2020
7 aluminium ammonium sulphate			
Curb S	Sphere	13564	28 Feb 2017
Guardsman	Chiltern	05494	31 Dec 2021
Rezist Liquid	Barrett	14643	31 Dec 2021
8 aluminium phosphide			
Detia Gas Ex-P	Igrox	17036	28 Feb 2022
Detia Gas-Ex-T	Igrox	17034	28 Feb 2022
Quickphos Pellets 56% GE	UPL Europe	16987	28 Feb 2022
11 ametoctradin + dimethomorph			
Resplend	BASF	14975	31 Jan 2021
12 ametoctradin + mancozeb			
Diablo	BASF	16084	13 Jan 2019

Products also Registered

Product	Approval holder	MAPP No.	Expiry Date
16 aminopyralid			
Pro-Banish	Dow	14730	31 Dec 2018
17 aminopyralid + fluroxypyr			
Halcyon	Dow	14709	31 Dec 2018
19 aminopyralid + propyzamide			
Dymid	BASF	17585	31 Dec 2018
Galactic Pro	Chem-Wise	17180	31 Dec 2018
Milepost	Terrechem	17537	31 Dec 2018
Milestone	PSI	17494	31 Dec 2018
Pyzamid Universe	Euro	16956	31 Dec 2018
20 aminopyralid + triclopyr			
Speedline Pro	Bayer Environ.	16208	31 Dec 2018
21 amisulbrom			
Leimay	Nissan	17412	31 Dec 2017
Sanblight	Nissan	17411	31 Dec 2017
Shinkon	Nissan	17498	31 Dec 2017
Shinkon	Procam	13722	30 Sep 2017
24 Ampelomyces quisqualis (Strain AQ10)			
AQ 10	Fargro	15518	31 May 2019
26 azoxystrobin			
5504	Syngenta	12351	30 Jun 2024
Astrobin 250	Euro	15336	30 Jun 2024
Astrobin 250	Euro	15354	30 Jun 2024
Azaka	Headland	16422	30 Jun 2024
Chamane	UPL Europe	15922	31 Dec 2017
Cleancrop Celeb	Agrii	16519	30 Jun 2024
Euclid	Clayton	15926	31 Jan 2018
Harness	ChemSource	16083	30 Jun 2024
Harness	ChemSource	14803	30 Jun 2024
Hi Strobin 25	Hockley	15468	30 Jun 2024
Kingdom Turf	Mitsui	13833	30 Jun 2024
Life Scientific Azoxystrobin	Life Scientific	15341	30 Jun 2017
Life Scientific Azoxystrobin	Life Scientific	16782	31 Jan 2020
Mirador	Adama	15557	30 Jun 2017
MS Sansa	Micromix	17147	30 Jun 2024
Ortiva	Syngenta	10542	30 Jun 2024
Panama	Pan Amenity	13950	30 Jun 2024
Priori	Syngenta	10543	30 Jun 2024
Puma	Ascot Pro-G	17678	30 Jun 2024
PureAzoxy	Pure Amenity	15136	30 Jun 2024
Reconcile	Becesane	15379	30 Jun 2024
RouteOne Roxybin 25	Albaugh UK	15289	30 Jun 2024
Sinstar	Sinon EU	16852	16 Feb 2019
Xylem	EuroChem	17677	30 Jun 2024
Zoxis	Agriphar	15483	31 Dec 2018
27 azoxystrobin + chlorothalonil			
Mount	Euro	16878	30 Apr 2020
Perseo	Sipcam	17230	17 Feb 2020
Quadris Opti	Syngenta	17752	30 Apr 2020
30 azoxystrobin + fenpropimorph			
Amistar Pro	Syngenta	10513	30 Jun 2017
Aspect	Syngenta	10516	30 Jun 2017

Products also Registered

Product	Approval holder	MAPP No.	Expiry Date
RouteOne Roxypro FP	Albaugh UK	15291	30 Jun 2017

34 azoxystrobin + tebuconazole

Seraphin	Adama	16248	28 Feb 2022

35 Bacillus firmus I - 1582

Flocter	Bayer CropScience	16480	31 Mar 2026

36 Bacillus subtilis

Serenade ASO	BASF	15625	31 Mar 2017
Solani	Russell	16585	31 Oct 2020

37 Bacillus thuringiensis

Biocure	Russell	16432	30 Sep 2017
Dipel DF	Interfarm	14119	30 Sep 2017

38 Bacillus thuringiensis aizawai GC-91

Agree 50 WG	Mitsui	17502	31 Oct 2021

42 benalaxyl + mancozeb

Galben M	Headland	17247	26 Jul 2020
Intro Plus	Headland	17253	26 Jul 2020
Tairel	Headland	17254	26 Jul 2020

44 bentazone

Agriguard Bentazone 480	AgriGuard	11569	31 Dec 2019
Agrotech Bentazone	Me2	12143	31 Dec 2019
Barclay Dingo	Barclay	07642	31 Dec 2019
Basagran	BASF	00188	31 Dec 2019
Benta 48	RouteOne	13469	31 Dec 2019
Benta 480 SL	Nufarm UK	17355	31 Dec 2019
Bentazone 480	Goldengrass	14423	31 Dec 2019
Bently	Chem-Wise	14210	31 Dec 2019
Buddy	EZCrop	15727	31 Dec 2019
Euro Benta 480	Euro	14707	31 Dec 2019
Hockley Bentazone 48	Hockley	15543	31 Dec 2019
I T Bentazone 480	I T Agro	07458	31 Dec 2019
IT Bentazone	I T Agro	13132	31 Dec 2019
Landgold Bentazone SL	Landgold	10841	31 Dec 2019
Landgold Bentazone SL	Teliton	13418	31 Dec 2019
Landgold Bentazone SL	Teliton	12111	31 Dec 2019
Mac-Bentazone 480 SL	AgChem Access	13598	31 Dec 2019
Master Zone	Generica	16441	31 Dec 2019
Milentus Bentazone	Milentus	12257	31 Dec 2019
Osiris	PSI	07953	31 Dec 2019
Pulse	AgriGuard	12764	31 Dec 2019
RouteOne Benta 48	Albaugh UK	13662	31 Dec 2019
RouteOne Benta 48	Albaugh UK	15266	31 Dec 2019
RouteOne Bentazone 48	Albaugh UK	15269	31 Dec 2019
Standon Bentazone	Standon	06895	31 Dec 2019
Standon Bentazone	Standon	09204	31 Dec 2019
Standon Bentazone S	Standon	10124	31 Dec 2019
Top Farm Bentazone	Top Farm	08026	31 Dec 2019
Troy 480	UPL Europe	12341	30 Jun 2018
UPL B Zone	UPL Europe	14808	31 Dec 2019
Zone 48	Mitsui	13165	31 Dec 2019

48 bentazone + pendimethalin

Impuls	BASF	13372	31 Dec 2019

Products also Registered

Product	Approval holder	MAPP No.	Expiry Date
50 benthiavalicarb-isopropyl + mancozeb			
En-Garde	Certis	14901	31 Jul 2020
51 benzoic acid			
MENNO Florades	Brinkman	15091	31 Jul 2019
56 bifenazate			
Inter Bifenazate 240 SC	Iticon	14543	31 Jan 2020
57 bifenox			
Wolf	RAAT	17378	31 Dec 2021
62 bixafen			
Bixafen EC125	Bayer CropScience	15951	31 Mar 2026
65 bixafen + prothioconazole			
Saltri	Euro	16999	31 Jan 2021
69 boscalid			
Boscalid 50	RouteOne	13450	31 Jan 2021
Bosco WG	Goldengrass	14455	31 Jan 2021
Coli	Gemini	17616	31 Jan 2021
Fulmar	AgChem Access	14073	31 Jan 2021
RouteOne Boscalid 50	Albaugh UK	13660	31 Jan 2021
71 boscalid + epoxiconazole			
Whistle	BASF	16093	31 Jan 2021
74 boscalid + pyraclostrobin			
Daisy	RAAT	17492	31 Jul 2019
Insignis	Euro	17116	31 Jul 2019
76 bromadiolone			
Romax B Rat & Mouse Killer	Zapi	UK14-0814	
Romax Bromablock	Zapi	UK13-0777	
Romax Bromadiolone Whole Wheat	PelGar	UK12-0640	
Romax Bromapasta Bait	Zapi	UK13-0747	
77 bromoxynil			
Akocynil 225 EC	Aako	15688	31 Jan 2020
Alpha Bromolin 225 EC	Adama	14864	31 Jan 2020
Flagon 400 EC	Adama	14921	31 Jan 2020
Xinca	Nufarm UK	16833	31 Jan 2020
86 bromoxynil + prosulfuron			
Jester	Syngenta	13500	31 Dec 2019
87 bromoxynil + terbuthylazine			
Alpha Bromotril PT	Adama	09435	30 Jun 2017
Cleancrop Amaize	Agrii	11990	30 Jun 2017
Templar	Adama	10254	30 Jun 2017
91 buprofezin			
Applaud 25 SC	Certis	17196	30 Jun 2023
Applaud 25 WP	Certis	17197	30 Jun 2023
92 Candida oleophila Strain O			
Nexy 1	BioNext	16961	31 Mar 2026

Products also Registered

Product	Approval holder	MAPP No.	Expiry Date
93 captan			
Akotan 80 WG	Aako	16307	31 Jan 2021
Malvin WG	Arysta	16308	31 Jan 2021
Orthocide WG	Arysta	16309	31 Jan 2021
Ratan 80 WG	RAAT	17252	31 Jan 2021
95 carbendazim			
Caste Off	Nufarm UK	12832	31 Aug 2017
Clayton Am-Carb	Clayton	11906	31 Aug 2017
Cleancrop Curve	United Agri	11774	31 Aug 2017
Mascot Systemic	Everris Ltd	09132	31 Aug 2017
Turf Systemic Fungicide	Barclay	13291	31 Aug 2017
Turfclear	Everris Ltd	07506	31 Aug 2017
103 carbetamide			
Kartouch 60 WG	Aako	16897	30 Nov 2023
106 carboxin + thiram			
Anchor	Chemtura	08684	31 Dec 2021
Anchor	Certis	16488	31 Dec 2021
107 carfentrazone-ethyl			
Aurora	Headland	11613	31 Jan 2020
Aurora 40 WG	Headland	11614	31 Jan 2019
Carfen	Mitsui	13005	31 Jan 2020
Carzone 60 ME	Euro	16469	31 Jan 2020
Platform	Belchim	11615	31 Jan 2020
Shark	Belchim	12762	31 Jan 2019
Spotlight Plus	Belchim	12436	31 Jan 2019
108 carfentrazone-ethyl + flupyrsulfuron-methyl			
Lexus Class	DuPont	10809	31 Dec 2019
109 carfentrazone-ethyl + mecoprop-P			
Jewel	Everris Ltd	14327	31 Jul 2017
Platform S	Belchim	14380	31 Jan 2019
110 carfentrazone-ethyl + metsulfuron-methyl			
Ally Express	DuPont	08640	31 Dec 2018
Asset Express	Mitsui	14378	31 Dec 2018
Standon Carfentrazone MM	Standon	09159	31 Dec 2018
113 chloridazon			
Better Flowable	Sipcam	15740	30 Jun 2021
Goiana	Agroquimicos	15778	30 Sep 2017
Parador	UPL Europe	16892	15 Dec 2018
117 chloridazon + quinmerac			
Fiesta T	BASF	11734	31 Mar 2017
118 chlormequat			
AgriGuard Chlormequat 750	Mitsui	14144	31 May 2020
Barleyquat B	Taminco	16066	30 Jun 2017
Bettaquat B	Taminco	16078	30 Jun 2017
CCC 750	Becesane	16948	31 May 2020
CCC 750	Harvest	17753	31 May 2020
CS Chlormequat	ChemSource	14946	31 May 2020
Jadex Plus	SFP Europe	17524	31 May 2022
K2	Taminco	16069	30 Jun 2017
Manipulator	Taminco	16075	30 Jun 2017
Mequat 750	ChemSource	16143	31 May 2020

Products also Registered

Product	Approval holder	MAPP No.	Expiry Date
Selon	Taminco	16074	30 Jun 2017
Silk 750 SL	Terrechem	17496	31 May 2020
Stefes CCC 720	Stefes	17731	31 May 2022

119 chlormequat + 2-chloroethylphosphonic acid

BOGOTA UK	SFP Europe	16318	31 Jan 2021
Socom	SFP Europe	17155	31 Jan 2021

124 2-chloroethylphosphonic acid

Coryx	SFP Europe	17620	31 Jan 2021
Hi-Phone 48	Hockley	15506	31 Jan 2021
Padawan	SFP Europe	15577	31 Jan 2021
Rogan	Novastar	17186	31 Jan 2021
Telsee	SFP Europe	16325	31 Jan 2021

128 chlorothalonil

Alternil 500	Agriphar	16813	30 Apr 2020
Alternil Excel 720	Arysta	17434	30 Apr 2020
Asterix	Barclay	17383	30 Apr 2020
Balear 720 SC	Arysta	15545	30 Apr 2020
Banko 720	Arysta	17728	30 Apr 2020
Barclay Chloroflash	Barclay	16651	30 Apr 2020
Chlorthalis	Agriphar	16513	30 Apr 2020
Claw 500	Gemini	17332	30 Apr 2020
Cleancrop Wanderer 2	Agrii	16302	19 Sep 2017
Daconil	Syngenta	17778	30 Apr 2020
Doolin	Barclay	17676	30 Apr 2020
Duster	JT Agro	17517	30 Apr 2020
FSA Chlorothalonil	Synergy	17182	30 Apr 2020
GF Chlorothalonil	Genfarm	17011	30 Apr 2020
IT Chlorothalonil	I T Agro	15845	30 Apr 2020
Life Scientific Chlorothalonil 500	Life Scientific	15813	30 Apr 2020
LS Chlorothalonil XL	Agrovista	15443	30 Apr 2020
MS Catelyn	Micromix	17148	30 Apr 2020
Multi-Star 500	Albaugh UK	16784	30 Apr 2020
Optio 500	Novastar	17216	30 Apr 2020
Pronto	EZCrop	16371	31 Jan 2018
Renew Chlorothalonil 500SC	Renew	16706	30 Apr 2020
Sinconil	Agrii	16268	19 Sep 2017
Spirodor	Globachem	17710	30 Apr 2020
Standonil	Standon	16882	19 Sep 2017
Supreme	Certis	16505	30 Apr 2020
Synergy Chlorothalonil	Synergy	17215	30 Apr 2020
Thalonil 500	Euro	14926	30 Apr 2020
Tymoon 500	Novastar	17336	30 Apr 2020
X-SEPT	Albaugh UK	15431	31 Oct 2018

130 chlorothalonil + cyproconazole

Bravo Xtra	Syngenta	11824	31 Dec 2021
Cleancrop Cyprothal	United Agri	09580	31 Dec 2021

131 chlorothalonil + cyproconazole + propiconazole

Apache	Syngenta	14255	31 Dec 2021

142 chlorothalonil + propiconazole

Barclay Avoca Premium	Barclay	15619	31 Jul 2019
Oxana	Arysta	15177	31 Jul 2019
SIP 313	Sipcam	14755	31 Jul 2019

Products also Registered

Product	Approval holder	MAPP No.	Expiry Date
145 chlorothalonil + tebuconazole			
Cigal Plus	Rotam	17229	30 Apr 2020
Confucius	Rotam	17257	30 Apr 2020
Nectar	Nufarm UK	16872	30 Apr 2020
146 chlorothalonil + tetraconazole			
Eminent Star	Isagro	10447	31 Dec 2021
151 chlorpropham			
Aliacine 400 EC	Arysta	16283	31 Jan 2020
BL 500	UPL Europe	14387	31 Aug 2017
CIPC Gold	Dormfresh	15674	31 Aug 2017
Cleancrop Amigo	United Agri	15292	31 Jan 2020
Gro-Stop 100	Certis	14182	31 Aug 2017
Gro-Stop Electro	Certis	17445	26 Jul 2020
Gro-Stop Fog	Certis	14183	31 Aug 2017
Gro-Stop HN	Certis	14146	31 Aug 2017
Gro-Stop Innovator	Certis	14147	31 Aug 2017
MSS CIPC 50 M	UPL Europe	14388	31 Aug 2017
MSS Sprout Nip	UPL Europe	15676	31 Aug 2017
Neo-Stop 120 RTU	Agriphar	17420	31 Jan 2020
Neo-Stop 300 HN	Agriphar	15261	31 Aug 2017
Neo-Stop Starter	Agriphar	15651	31 Jan 2020
Pro-Long	UPL Europe	14389	31 Aug 2017
Tuberprop Easy	UPL Europe	15701	31 Jan 2020
154 chlorpyrifos			
Akofos 480 EC	Aako	14211	31 Dec 2021
Ballad	Headland	11659	31 Dec 2021
Chlorpyrat 480 EC	RAAT	17333	31 Dec 2021
Clayton Pontoon	Clayton	13219	31 Dec 2021
EA Pyrifos	European Ag	15005	31 Dec 2021
Pirisect	Mitsui	13371	31 Dec 2021
Pyrinex 48 EC	Adama	13534	31 Dec 2021
155 chlorpyrifos-methyl			
Garrison	Pan Agriculture	15657	30 Sep 2017
158 citronella oil			
Barrier H	Barrier	10136	31 Mar 2017
159 clethodim			
Centurion Max	Arysta	16189	21 May 2017
Clayton Gatso	Clayton	16244	21 May 2017
Clayton Gatso	Clayton	17746	09 Nov 2020
Magpie	Goldengrass	16279	21 May 2017
Select Prime	Arysta	16304	09 Nov 2020
Spartan	Unique Marketing	16312	21 May 2017
Standon Clethodim	Standon	16276	21 May 2017
160 clodinafop-propargyl			
Epee	Harvest	17601	31 Oct 2020
Life Scientific Clodinafop 240	Life Scientific	17181	31 Jan 2020
Ravena	Headland	17313	31 Oct 2020
Viscount	Syngenta	17495	31 Oct 2020
Wildcat	Matrix	17317	31 Oct 2020
161 clodinafop-propargyl + cloquintocet-mexyl			
Ciclope	Industrias Afrasa	17226	29 Jul 2019

SECTION 3

Products also Registered

Product	Approval holder	MAPP No.	Expiry Date
163 clodinafop-propargyl + pinoxaden			
Traxos Pro	Syngenta	16713	30 Jun 2019
164 clodinafop-propargyl + prosulfocarb			
Auxiliary	BASF	14576	30 Apr 2021
166 clofentezine			
Acaristop 500 SC	Aako	14405	28 Feb 2017
Acaristop 500 SC	Aako	17232	30 Jun 2021
Ariane	RAAT	17296	30 Jun 2021
167 clomazone			
Blanco	Adama	16704	02 Nov 2018
Centium 360 CS	Belchim	16237	12 May 2018
Cirrus CS	Belchim	16579	12 May 2018
Cleancrop Chicane	Belchim	17015	12 May 2018
Cleancrop Covert	Belchim	17014	12 May 2018
Clomate	Albaugh UK	15565	30 Apr 2021
Clozone	Euro	16826	12 May 2018
Czar	Terrechem	17423	30 Apr 2021
Gambit	Harvest	17747	12 May 2018
Gamit 36 CS	Belchim	16563	12 May 2018
Hobby	Ascot Pro-G	17667	30 Apr 2021
Mazone 360	Euro	16827	12 May 2018
Noble	Rovogate	17162	30 Apr 2021
Notion	AgChem Access	16773	30 Apr 2021
Regal	Matrix	16617	30 Apr 2021
Regal Gold	Matrix	17065	12 May 2018
Retribute	Agform	16739	12 May 2018
Zulkon	Unisem	17610	30 Apr 2021
169 clomazone + metazachlor			
Circuit Synctec	Headland	17118	26 Jul 2019
170 clomazone + metazachlor + napropamide			
Colzor SyncTec	Syngenta	17113	28 Jul 2019
Tribeca Sync Tec	Headland	17312	28 Jul 2019
172 clomazone + napropamide			
Altiplano DAMtec	Headland	16957	31 Jul 2019
173 clomazone + pendimethalin			
Stallion Sync TEC	Belchim	16200	30 Jun 2017
174 clopyralid			
Bariloche	Proplan-Plant	17577	31 Oct 2020
Cliophar 400	Arysta	15008	31 Oct 2020
Cliophar 600 SL	Agriphar	16735	31 Oct 2020
GF-2895	Dow	16821	31 Oct 2020
Glopyr 400	Globachem	15009	31 Oct 2020
Lontrel 72SG	Dow	15235	31 Oct 2020
Vivendi 200	UPL Europe	15017	31 Mar 2019
177 clopyralid + florasulam			
GF-2463	Dow	16364	17 Dec 2017
178 clopyralid + florasulam + fluroxypyr			
Mogul	Aremie	17289	30 Jun 2018
179 clopyralid + fluroxypyr + MCPA			
Interfix	Iticon	15762	30 Apr 2020

Products also Registered

Product	Approval holder	MAPP No.	Expiry Date
180 clopyralid + fluroxypyr + triclopyr			
Pastor	Dow	11168	30 Jun 2017
181 clopyralid + picloram			
Curlew	Goldengrass	16791	10 Mar 2018
Pyralid Extra	Euro	16606	10 Mar 2018
183 clopyralid + triclopyr			
Headland Flail 2	Headland	16007	31 Oct 2020
Tor	Dow	17777	31 Oct 2020
185 clothianidin			
Hinder	ChemSource	16117	31 Jul 2020
NipsIT Inside	Interfarm	14744	31 Jul 2020
195 coumatetralyl			
Romax Rat CP	Bayer Environ.	UK16-1003	
197 cyazofamid			
Rithfir	DuPont	14816	31 Jan 2020
RouteOne Roazafod	Albaugh UK	15271	31 Jan 2020
200 cycloxydim			
Laser	BASF	12930	31 May 2017
201 Cydia pomonella GV			
Cyd-X	Certis	17019	31 Oct 2021
Cyd-X Duo	Certis	16779	31 Oct 2021
202 cyflufenamid			
Cyflu EW	Euro	16472	30 Sep 2022
Diego	Star	16980	30 Sep 2022
NF-149 SC	Certis	15835	30 Sep 2022
203 cymoxanil			
Cymostraight 45	Belchim	14615	31 Jul 2017
Cymostraight 45	Belchim	17437	28 Feb 2022
Dauphin 45	SFP Europe	16434	28 Feb 2022
205 cymoxanil + fluazinam			
Kunshi	Syngenta	15631	31 Mar 2019
Shirlan Forte	Syngenta	15542	31 Aug 2021
Tezuma	Belchim	17396	31 Aug 2021
206 cymoxanil + fludioxonil + metalaxyl-M			
Wakil XL	Syngenta	10562	31 Mar 2017
207 cymoxanil + mancozeb			
Globe	Sipcam	10339	31 Mar 2017
Globe	Sipcam	17175	17 Sep 2019
Manconil 725	Euro	15997	31 Jul 2020
Master Cyman	Generica	16686	31 Jul 2020
Nautile DG	UPL Europe	14692	31 Aug 2017
Nautile WP	UPL Europe	16468	31 Jul 2020
Palmas WP	SFP Europe	16632	31 Jul 2020
Solace Max	Nufarm UK	16697	31 Jul 2020
Video	UPL Europe	14850	31 Aug 2017
Zetanil	Sipcam	11993	30 Apr 2017
Zetanil	Sipcam	17285	17 Sep 2019

SECTION 3

Products also Registered

Product	Approval holder	MAPP No.	Expiry Date
209 cymoxanil + propamocarb			
Axidor	Agriphar	16830	31 Jan 2021
Fracarb	Frarimpex	16445	30 Sep 2017
Proxanil	Agriphar	16664	31 Jan 2021
210 cymoxanil + zoxamide			
Reboot	Gowan	16909	31 Jul 2019
211 cypermethrin			
Afrisect 10	Agriphar	13159	30 Apr 2017
Afrisect 10	Agriphar	17069	17 Mar 2019
Afrisect 500 EC	Agriphar	17137	16 Mar 2019
Cetan 100 EC	CMI	16708	30 Apr 2017
Cyper 100	AgChem Access	15936	30 Apr 2017
Cyperkill 10	Agriphar	13157	30 Apr 2017
Cyperkill 10	Agriphar	17068	17 Mar 2019
Cythrin Max	Agriphar	17138	16 Mar 2019
Landgold Cyper 100	Goldengrass	14105	30 Apr 2017
MAC-Cypermethrin 100 EC	MAC	14934	30 Apr 2017
Permasect 500 EC	Agriphar	17132	16 Mar 2019
Permasect C	Nufarm UK	13158	31 Mar 2017
Permasect C	Agriphar	16994	17 Mar 2019
Signal 300 ES	Certis	16481	30 Apr 2020
Signal 300 ES	Chemtura	15949	30 Apr 2020
Supasect	Agriphar	15536	30 Apr 2017
Supasect	Agriphar	17070	17 Mar 2019
Supasect 500 EC	Agriphar	17131	16 Mar 2019
Syper 100	Goldengrass	14194	30 Apr 2017
Talisma EC	Agriphar	16541	30 Apr 2020
Talisma UL	Agriphar	16542	30 Apr 2020
Toppel 100	UPL Europe	15696	30 Apr 2017
Toppel 100	UPL Europe	17071	17 Mar 2019
215 cyproconazole + penthiopyrad			
Cielex	DuPont	17758	30 Nov 2023
216 cyproconazole + picoxystrobin			
Furlong	DuPont	13818	31 Mar 2017
217 cyproconazole + propiconazole			
Alto Xtra	Syngenta	14437	30 Nov 2017
219 cyproconazole + trifloxystrobin			
Escolta	Bayer CropScience	13923	30 Jun 2017
220 cyprodinil			
Unix	Syngenta	14846	31 Oct 2020
221 cyprodinil + fludioxonil			
Button	Euro	16842	31 Oct 2020
Reversal	RAAT	17359	31 Oct 2020
222 cyprodinil + isopyrazam			
Bontima	Syngenta	14899	30 Sep 2017
Cebara	Syngenta	16035	30 Sep 2017
Concorde	Syngenta	16225	30 Sep 2017
223 cyprodinil + picoxystrobin			
Acanto Prima	DuPont	14971	30 Apr 2020

Products also Registered

Product	Approval holder	MAPP No.	Expiry Date
224 2,4-D			
2,4-D Amine 500	Nufarm UK	14360	31 Dec 2018
Damine	Agriphar	13366	31 Dec 2018
Depitox 500	Nufarm UK	17597	31 Dec 2018
Growell 2,4-D Amine	GroWell	13146	31 Dec 2018
Herboxone 60	Headland	14080	31 Dec 2018
Maton	Headland	13234	31 Dec 2018
Zip	UPL Europe	14806	30 Jun 2018
Zip	UPL Europe	16968	31 Dec 2018
226 2,4-D + dicamba			
Compo Floranid + Herbicide	Compo	14503	31 Dec 2018
Landscaper Pro Weed Control + Fertilizer	Everris Ltd	16609	30 Jun 2018
229 2,4-D + dicamba + MCPA + mecoprop-P			
Dicophar	Agriphar	16460	18 Mar 2018
230 2,4-D + dicamba + triclopyr			
Broadshot	Agriphar	16655	15 Sep 2018
Cleancrop Broadshot	Agriphar	16756	15 Sep 2018
Kaskara	Arysta	16757	15 Sep 2018
233 2,4-D + glyphosate			
Rasta	Nufarm UK	16795	30 Jun 2018
Rasto	Nufarm UK	16795	30 Jun 2018
234 2,4-D + MCPA			
Agroxone Combi	Nufarm UK	14907	31 Dec 2018
237 2,4-D + triclopyr			
Genoxone ZX EC	Agriphar	17016	22 Mar 2019
238 daminozide			
B-Nine SG	Chemtura	14434	30 Apr 2020
Dazide Enhance	Fargro	14433	28 Feb 2017
239 dazomet			
Basamid	Kanesho	12895	31 Dec 2021
240 2,4-DB			
Butoxone DB	Nufarm UK	14100	30 Apr 2020
Butoxone DB	Nufarm UK	14840	30 Apr 2020
242 2,4-DB + MCPA			
Butoxone DB Extra	Nufarm UK	15741	30 Apr 2020
243 deltamethrin			
CMI Delta 2.5 EC	CMI	16695	30 Apr 2020
Decis Forte	Bayer CropScience	16110	30 Apr 2020
Deltason-D	DAPT	17735	30 Apr 2020
GAT Decline 2.5 EC	Headland	17703	30 Apr 2020
GAT Decline 2.5 EC	GAT Micro	15769	30 Apr 2019
Grain-Tect ULV	Limagrain	15253	30 Apr 2020
MiteStop	ChemSource	15775	30 Apr 2020
247 desmedipham + ethofumesate + phenmedipham			
Betanal Elite	Bayer CropScience	17721	31 Jan 2020
Betanal Expert	Bayer CropScience	14034	31 Jan 2020
Beta-Team	UPL Europe	15423	31 Jan 2019
Bison 400 SE	Aako	16213	31 Jan 2020

SECTION 3

Products also Registered

Product	Approval holder	MAPP No.	Expiry Date
Conqueror	UPL Europe	16972	31 Jan 2020
Conqueror	AgriChem BV	16157	31 Jan 2019

248 desmedipham + phenmedipham

Betanal Maxxim	Bayer CropScience	14186	31 Jan 2020

252 dicamba + MCPA + mecoprop-P

Hyprone-P	UPL Europe	09125	31 Jul 2017
Hysward-P	Agrichem	09052	30 Nov 2017
Nomix Tribute Turf	Nomix Enviro	16313	31 Jul 2019
Relay Turf Elite	Headland	17192	31 Jul 2019

253 dicamba + mecoprop-P

Dimeco	UPL Europe	17190	31 Mar 2018
Headland Swift	Headland	11945	31 Mar 2018
Mircam	Nufarm UK	11707	31 Mar 2018
Optica Forte	Nufarm UK	14845	31 Mar 2018

254 dicamba + prosulfuron

Clayton Spook	Clayton	15846	31 Dec 2019

260 dichlorprop-P + MCPA + mecoprop-P

Hymec Triple	Agrichem	15753	31 Jul 2019
Optica Trio	Nufarm UK	15717	31 Mar 2017

263 difenacoum

Romax D Block	PelGar	UK11-0077	
Romax Difenacoum Rat & Mouse Block	Zapi	UK14-0813	
Romax Difenacoum Whole Wheat Bait	PelGar	UK11-0109	
Romax DP	Rentokil	UK13-0804	
Romax DP Pasta Sachets	Zapi	UK14-0843	
Romax Mouse DP	PelGar	UK11-0089	
Romaz D Rat & Mouse Killer	Rentokil	UK14-0808	

265 difenoconazole

Alternet	Belcrop	17689	30 Jun 2021
Bogard	Syngenta	17310	30 Jun 2021
Bogard	Syngenta	16260	31 May 2017
Change	RAAT	17400	30 Jun 2021
Diconazol 250	Euro	16387	30 Jun 2021
Difcon 250	Becesane	15985	30 Jun 2021
Difenostar	Life Scientific	17598	28 Jun 2020
EA Difcon	European Ag	14965	30 Jun 2021
MAC-Difenoconazole 250 EC	MAC	14935	31 Dec 2017
Plover	Syngenta	11763	31 Dec 2017
Sandpiper	ChemSource	15004	30 Jun 2021
Septuna	Novastar	16379	31 Dec 2017
Septuna	Novastar	17390	30 Jun 2021
Slick	Syngenta	17331	30 Jun 2021
Slick	Syngenta	16253	31 Dec 2017
Turnstone	AgChem Access	14019	31 Dec 2017

267 difenoconazole + fludioxonil

Difend Extra	Globachem	17739	30 Apr 2021

268 difenoconazole + fludioxonil + tebuconazole

Celest Trio	Syngenta	15510	30 Apr 2021

Products also Registered

Product	Approval holder	MAPP No.	Expiry Date
272 diflubenzuron			
Diflox Flow	Hockley	15640	31 Dec 2021
Dimilin Flo	Chemtura	08769	31 Dec 2021
273 diflufenican			
Beeper 500 SC	Aako	15358	30 Jun 2021
Dican	Albaugh UK	16922	30 Jun 2021
Diecot	Gemini	17191	30 Jun 2021
Diflufen 500	Euro	16687	30 Jun 2021
Diflufenican GL 500	Globachem	17084	30 Jun 2021
Dina 50	Globachem	17083	30 Jun 2021
Flyflo	Gemini	17694	30 Jun 2021
Inter Diflufenican	UPL Europe	17342	29 Feb 2020
Overlord	Adama	13521	30 Jun 2021
Prefect	Novastar	17713	30 Jun 2021
Sempra	AgriChem BV	16824	31 Mar 2019
Solo D500	Sipcam	16098	30 Jun 2021
Terrier	Terrechem	17761	30 Jun 2021
Tornado	Matrix	17425	30 Jun 2021
Twister	Unique Marketing	16374	30 Jun 2021
Twister 500	Harvest	17627	30 Jun 2021
274 diflufenican + florasulam			
Bow	Headland	16845	30 Jun 2018
275 diflufenican + flufenacet			
Adept	Agform	17628	30 Apr 2020
Amaranth	PSI	16949	30 Apr 2020
Clayton Vista	Clayton	16705	30 Apr 2020
Cleancrop Marauder	United Agri	15933	30 Apr 2020
Deliverer	Euro	16692	30 Apr 2020
Diflufenastar	Life Scientific	17611	27 Jun 2020
Diflufenastar 100	Life Scientific	17475	30 Apr 2020
Duke	Chem-Wise	16738	30 Apr 2020
Fenican	Euro	16790	30 Apr 2020
Firebird	Bayer CropScience	14826	30 Apr 2020
Flufenican	Euro	15386	30 Apr 2020
Life Scientific Flufenacet + DFF	Life Scientific	16805	30 Apr 2019
Navigate	Headland	17506	09 Mar 2020
Pincer Plus	Agform	17306	30 Apr 2020
Plumage	Goldengrass	15085	30 Apr 2020
Saviour	Chem-Wise	16702	30 Apr 2020
276 diflufenican + flufenacet + flurtamone			
Flufenamone	Euro	17260	30 Apr 2020
Voyage	Chem-Wise	17238	30 Apr 2020
279 diflufenican + glyphosate			
Pistol	Bayer Environ.	12173	28 Feb 2018
Pistol Rail	Bayer Environ.	17450	14 Feb 2020
Proshield	Everris Ltd	16118	31 Mar 2018
280 diflufenican + iodosulfuron-methyl-sodium + mesosulfuron-methyl			
Kalenkoa	Bayer CropScience	17733	16 Aug 2020
281 diflufenican + isoproturon			
Blutron	Agform	16712	30 Sep 2017
Blutron Plus	Agform	16711	30 Sep 2017

SECTION 3

Products also Registered

Product	Approval holder	MAPP No.	Expiry Date
285 dimethachlor			
Clayton Saxon	Clayton	16044	31 May 2017
Teridox	Syngenta	15876	30 Jun 2022
287 dimethenamid-p + metazachlor			
Impala	Euro	16650	30 Sep 2017
288 dimethenamid-p + metazachlor + quinmerac			
Luna	Goldengrass	17053	30 Apr 2020
Medirac	Euro	17214	30 Apr 2020
Meta Turbo	Euro	16688	30 Sep 2017
Titanic	Agroquimicos	16191	30 Sep 2017
291 dimethoate			
Danadim	Headland	15960	03 Sep 2017
Rogor L40	Headland	15918	20 Sep 2017
Sector	Headland	15959	03 Sep 2017
292 dimethomorph			
Morph	Adama	15121	31 Jan 2021
Murphy 500 SC	Aako	15203	31 Jan 2021
Rigel WP	Servem	17347	31 Mar 2020
294 dimethomorph + mancozeb			
Saracen	BASF	15250	31 Jul 2020
295 dimethomorph + pyraclostrobin			
Optimo Tech	BASF	16455	31 Jul 2019
300 diquat			
A1412A2	Syngenta	13440	31 Dec 2019
Agriguard Diquat	AgriGuard	10836	31 Dec 2019
Balista	Novastar	15186	31 Dec 2019
Barclay DQ-200	Barclay	15260	31 Dec 2019
Barclay D-Quat	Barclay	14833	31 Dec 2019
BD-200	Barclay	14918	31 Dec 2019
Brogue	Certis	16421	31 Dec 2019
Brogue	Mitsui	15638	28 Feb 2018
Brogue	Mitsui	13328	31 Dec 2019
Buckler	AgriGuard	13080	31 Dec 2019
ChemSource Diquat	ChemSource	14533	31 Dec 2019
Clayton Diquat	Clayton	10730	31 Dec 2019
Clayton Diquat	Clayton	12739	31 Dec 2019
Clayton Diquat 200	Clayton	13942	31 Dec 2019
Clayton IQ	Clayton	14519	31 Dec 2019
CleanCrop Flail	United Agri	14870	31 Dec 2019
CropStar Diquat	CropStar	17679	31 Dec 2019
Diquanet	Belcrop	16868	31 Dec 2019
Diquash	Agrichem	14849	31 Dec 2019
Di-Quattro	Agriphar	15712	31 Dec 2019
D-kwat 200	ChemSource	16144	31 Dec 2019
Hockley Diquat 20	Hockley	15585	31 Dec 2019
I.T. Diquat	I T Agro	13557	30 Jun 2018
I.T. Diquat	UPL Europe	16970	31 Dec 2019
Inter Diquat	UPL Europe	16986	31 Dec 2019
Inter Diquat	I T Agro	14196	30 Jun 2018
Kalahari	Barclay	14917	31 Dec 2019
Knoxdoon	AgChem Access	13987	31 Dec 2019
Landgold Diquat	Teliton	12116	31 Dec 2019
Landgold Diquat	Landgold	10649	31 Dec 2019

Products also Registered

Product	Approval holder	MAPP No.	Expiry Date
Life Scientific Diquat 200	Life Scientific	15811	31 Dec 2019
Mission 200 SL	UPL Europe	16475	30 Jun 2018
Quad S	Q-Chem	13578	31 Dec 2019
Quat 200	Euro	15042	31 Dec 2019
Roquat 20	Q-Chem	15703	31 Dec 2019
Standon Diquat	Standon	10674	31 Dec 2019
Standon Diquat SL	Standon	10732	31 Dec 2019
Standon Googly	Standon	13281	30 Jun 2017
Standon Yorker	Standon	14751	30 Jun 2017
Synchem Diquat 200	Synchem	16817	31 Dec 2019
Tuber Mission	UPL Europe	16965	31 Dec 2019
Tuber Mission	UPL Europe	14542	30 Jun 2018
UPL Diquat	UPL Europe	14944	30 Jun 2018
Waterloo	Interfarm	11196	31 Dec 2019
Waterloo	Alfa	10839	31 Dec 2019

301 dithianon

Rathianon 70 WG	RAAT	17471	30 Nov 2023

308 dodine

Radspor 400	Agriphar	16001	31 Dec 2021
Syllit 400 SC	Agriphar	13363	31 Dec 2021

309 epoxiconazole

Amber	Adama	16285	31 Oct 2021
Aska	Terrechem	17461	31 Oct 2021
Barret 125 SL	Agrifarm	16709	31 Oct 2021
Bassoon	BASF	14402	31 Oct 2021
Bowman	Adama	16286	31 Oct 2021
Corral	Headland	16942	31 Oct 2021
Cube	Becesane	16876	31 Oct 2021
Epzole 125	Unique Marketing	16412	31 Dec 2018
Intercord	Iticon	16155	31 Oct 2021
Life Scientific Epoxiconazole	Life Scientific	16907	03 Mar 2019
Ollyx	Novastar	16380	31 Oct 2021
Opus	BASF	12057	31 Oct 2021
Paladin	Novastar	16397	31 Oct 2021
Regent	Matrix	17087	31 Oct 2021
Voodoo	Sipcam	16072	31 Oct 2021
Warlock	Matrix	17026	31 Oct 2021

310 epoxiconazole + fenpropimorph

Eclipse	BASF	11731	31 Dec 2021

312 epoxiconazole + fenpropimorph + metrafenone

Polaca	Euro	16436	31 Dec 2021
Stiletto	BASF	14729	31 Dec 2021

313 epoxiconazole + fenpropimorph + pyraclostrobin

Diamant	BASF	14149	31 Dec 2021
Sapphire	Euro	16648	31 Dec 2021

314 epoxiconazole + fluxapyroxad

Adexar	BASF	15474	31 May 2017
Adexar	PSI	17441	31 Oct 2021
Apex Pro	Chem-Wise	17497	31 Oct 2021
Apex Pro	Chem-Wise	17632	31 Oct 2021
Clayton Index	Clayton	16145	31 May 2017
Cougar	Matrix	17442	31 Oct 2021
Epoflux	Pan Agriculture	15889	31 May 2017

Products also Registered

Product	Approval holder	MAPP No.	Expiry Date
Morex	BASF	17307	31 Oct 2021
Pexan	BASF	15584	31 May 2017
Redax Extra	Euro	16575	31 May 2017
Standon Coolie	Standon	16091	31 May 2017
ZORO	Terrechem	17462	31 Oct 2021
Zoro	Terrechem	17621	31 Oct 2021

315 epoxiconazole + fluxapyroxad + pyraclostrobin

Ceriax	BASF	15639	31 May 2017
Ceriax	PSI	17527	31 Jul 2019
Chusan	Terrechem	17528	31 Jul 2019
Smaragdin	PSI	17532	31 Jul 2019
Vortex	BASF	15663	31 May 2017

317 epoxiconazole + isopyrazam

Clayton Kudos	Clayton	16580	28 Feb 2017

320 epoxiconazole + metconazole

Ettu	Euro	16435	31 Oct 2020

322 epoxiconazole + prochloraz

Ennobe	BASF	14340	30 Jun 2017

323 epoxiconazole + pyraclostrobin

Ambassador	Euro	16410	31 Jul 2019
Bugle	BASF	17007	31 Jul 2019

324 esfenvalerate

Barclay Alphasect	Barclay	16053	31 Aug 2019
Clayton Cajole	Clayton	14995	31 Aug 2019
Clayton Slalom	Clayton	15054	31 Aug 2019
Clayton Vindicate	Clayton	15055	31 Aug 2019

326 ethanol

Ethy-Gen II	Ripe Rite	15839	31 Dec 2021
Restrain Fuel	Restrain	14520	31 Dec 2021

327 ethephon + mepiquat chloride

Gunbar	Terrechem	17460	31 Jan 2021
Mepicame	Unisem	17541	31 Jan 2021
Riggid	Gemini	17522	31 Jan 2021
Terpal	PSI	17440	31 Jan 2021

328 ethofumesate

Alpha Ethofumesate 50SC	Adama	13055	31 Jan 2020
Barclay Keeper 500 SC	Barclay	14016	31 Jan 2020
Barclay Keeper 500 SC	Barclay	13430	31 Jan 2020
Ethosat 500	Adama	13050	31 Jan 2020
Kubist Flo	Nufarm UK	12987	31 Jan 2020
Master Ethos	Generica	16442	31 Jan 2020
Nortron Flo	Bayer CropScience	12986	31 Jan 2020
Oblix 500	UPL Europe	12349	30 Sep 2018
Stelga 500 SC	Novastar	14737	31 Jan 2020
Trojan 500 SC	Nuvaros	13946	31 Jan 2020

329 ethofumesate + metamitron

Goltix Plus	Aako	16538	31 Jan 2020
Goltix Super	Adama	16537	31 Jan 2020
Metafol Super	UPL Europe	16881	31 Jan 2020

Products also Registered

Product	Approval holder	MAPP No.	Expiry Date
330 ethofumesate + metamitron + phenmedipham			
Phemo	UPL Europe	16352	31 Jan 2019
331 ethofumesate + phenmedipham			
Betosip Combi FL	Sipcam	14403	31 Jan 2020
Duo 400 SC	Mitsui	14323	31 Jan 2020
Fenlander 2	Adama	14030	31 Jan 2020
Gemini	EZCrop	15670	31 Jan 2020
Magic Tandem	Bayer CropScience	17358	31 Jan 2020
Powertwin	Adama	14004	31 Jan 2020
Teamforce	UPL Europe	14923	31 Jan 2019
Teamforce SE	UPL Europe	15403	31 Jan 2019
Teamforce SE	UPL Europe	16983	31 Jan 2020
Thunder	Adama	14031	31 Jan 2020
332 ethoprophos			
Mocap 15G	Certis	16562	31 Dec 2017
338 fatty acids			
Finalsan	Certis	13102	31 Dec 2021
Fytoclean	Russell	16788	28 Feb 2017
Hydro Coco Houseplant Pest & Spider Mite Killer	151 Products Ltd	16898	31 Dec 2021
NEU 1170 H	Sinclair	15754	31 Dec 2021
Organic Weed Killer RTU	Growing Success	13142	31 Dec 2021
Safers Insecticidal Soap	Woodstream	07197	31 Dec 2021
Savona	Koppert	06057	28 Feb 2017
344 fenazaquin			
Matador 200 SC	Margarita	11058	31 Dec 2018
Matador 200 SC	Gowan	16875	31 Dec 2021
345 fenbuconazole			
Indar 5 EW	Whelehan	16014	31 Aug 2017
348 fenhexamid			
RouteOne Fenhex 50	Albaugh UK	15282	30 Jun 2018
358 fenpropimorph			
Cleancrop Fenpro	United Agri	09885	31 Dec 2021
Cleancrop Fenpropimorph	United Agri	09445	31 Dec 2021
Marnoch Phorm	Me2	11087	31 Dec 2021
366 ferric phosphate			
Aristo IP	De Sangosse	17166	22 Jun 2019
Ferramol Max	Certis	14463	30 Jun 2018
Ferramol Slug Killer	Growing Success	12274	30 Jun 2018
Ferrox	Neudorff	14356	30 Jun 2018
NEU 1181 M	Neudorff	14355	30 Jun 2018
NEU 1185	Neudorff	14736	30 Jun 2018
Slucks	AgriGuard	15355	31 Jul 2018
Sluggo	Omex	14788	30 Jun 2018
Sluxx	Certis	14462	31 Jul 2018
367 ferrous sulphate			
Aitken Lawn Sand Mosskiller	Aitken	17044	08 Mar 2019
Aitken's Lawn Sand	Aitken	05253	28 Feb 2017
Elliott's Lawn Sand	Elliott	04860	28 Feb 2017
Elliott's Mosskiller	Elliott	04909	28 Feb 2017
Ferromex Mosskiller Concentrate	Omex	16896	28 Feb 2022

SECTION 3

Products also Registered

Product	Approval holder	MAPP No.	Expiry Date
Ferromex Mosskiller Concentrate	Omex	13180	28 Feb 2017
Greenmaster Autumn	Everris Ltd	15808	28 Feb 2017
Greentec Mosskiller	Headland Amenity	12518	28 Feb 2017
Greentec Mosskiller Pro	Headland Amenity	14645	28 Feb 2017
Landscaper Pro Moss Control + Fertilizer	Everris Ltd	15795	28 Feb 2017
LSTF Pro	LSTF	16992	08 Mar 2019
Maxicrop Moss Killer & Conditioner	Maxicrop	04635	30 Nov 2017
Maxicrop No. 2 Moss Killer and Conditioner	Maxicrop	17300	02 Nov 2019
Moss Control Plus Lawn Fertilizer	Miracle	07912	28 Feb 2017
Moss-Kill Pro	Bioservices	17029	08 Mar 2019
No More Moss Lawn Feed	Wolf	10754	28 Feb 2017
Rigby Taylor Turf Moss Killer	Rigby Taylor	17219	28 Feb 2022
Scotts Moss Control	Everris Ltd	12928	28 Feb 2017
SHL Lawn Sand	Sinclair	05254	28 Feb 2017
Sinclair Lawn Sand	Sinclair	17028	08 Mar 2019
Sinclair Lawn Sand	Westland Horticulture	17562	08 Mar 2019
Taylors Lawn Sand	Rigby Taylor	04451	28 Feb 2017

370 flazasulfuron

Product	Approval holder	MAPP No.	Expiry Date
Chikara	Nomix Enviro	13775	31 Jul 2019
Clayton Apt	Clayton	15157	31 Jul 2019
Flazasulf 25	Euro	15723	31 Jul 2019
Paradise	Pan Agriculture	14504	31 Jul 2019
PureFlazasul	Pure Amenity	15135	31 Jul 2019
Railtrax	PSI	15139	31 Jul 2019

372 flonicamid

Product	Approval holder	MAPP No.	Expiry Date
Primeman	RAAT	17656	28 Feb 2023

373 florasulam

Product	Approval holder	MAPP No.	Expiry Date
Barton WG	Dow	13284	30 Jun 2018
Boxer	Dow	09819	30 Jun 2018
Flora 50	Euro	15700	30 Jun 2018
Life Scientific Florasulam	Life Scientific	16581	30 Jun 2019
RouteOne Florasul 50	Albaugh UK	15285	30 Jun 2018
Solstice	Headland	17164	30 Jun 2018
Troller	Nufarm UK	16627	14 Aug 2018

374 florasulam + fluroxypyr

Product	Approval holder	MAPP No.	Expiry Date
Flatline	Aremie	17422	30 Jun 2018
Flurosulam XL	Euro	15236	30 Jun 2018
Hiker	Dow	11451	30 Jun 2017
Nevada	Dow	17349	30 Jun 2018
Starane Gold	Dow	10879	30 Jun 2017
Starane Vantage	Dow	10922	30 Jun 2017

375 florasulam + pinoxaden

Product	Approval holder	MAPP No.	Expiry Date
Axial One	Syngenta	15213	30 Nov 2017
Axial One	Syngenta	17262	30 Jun 2019

377 florasulam + tribenuron-methyl

Product	Approval holder	MAPP No.	Expiry Date
Bolt	Headland	17363	27 Sep 2019
Paramount Max	Headland	17271	27 Sep 2019

379 fluazifop-P-butyl

Product	Approval holder	MAPP No.	Expiry Date
A12791B	Syngenta	16178	31 Dec 2021
Clayton Crowe	Clayton	12572	31 Dec 2021
Cleancrop Clifford	United Agri	13597	30 Jun 2017

Products also Registered

Product	Approval holder	MAPP No.	Expiry Date
Fusilade 250 EW	Syngenta	10525	30 Jun 2017
RouteOne Fluazifop +	Albaugh UK	13641	31 Dec 2021
RouteOne Fluazifop +	Generica	15679	30 Jun 2017

380 fluazinam

Product	Approval holder	MAPP No.	Expiry Date
Boyano	Belchim	16621	31 Aug 2021
Float	Headland	16718	31 Aug 2021
Fluazinova	Barclay	17625	31 Aug 2021
FOLY 500 SC	Aako	16486	31 Aug 2021
Frowncide	ISK Biosciences	16619	31 Aug 2021
Ibiza 500	Belchim	16620	31 Aug 2021
Legacy	Syngenta	16622	31 Aug 2021
Ohayo	Syngenta	16588	31 Aug 2021
Shirlan Programme	Syngenta	16623	31 Aug 2021
Smash	RAAT	17488	31 Aug 2021
Winby	ISK Biosciences	16618	31 Aug 2021
Zinam 500	Euro	16831	31 Aug 2021
Zinam II	Euro	16886	31 Aug 2021

386 fludioxonil + tebuconazole

Product	Approval holder	MAPP No.	Expiry Date
Fountain	Adama	17708	30 Apr 2021

387 fludioxonil + tefluthrin

Product	Approval holder	MAPP No.	Expiry Date
Austral Plus	Bayer CropScience	13376	30 Jun 2017

388 flufenacet

Product	Approval holder	MAPP No.	Expiry Date
Ranger	Headland	17123	07 Jul 2019
Realm	Headland	17198	07 Jul 2019

390 flufenacet + isoxaflutole

Product	Approval holder	MAPP No.	Expiry Date
Cadou Star	Bayer CropScience	13242	31 Jan 2020
RouteOne Oxanet 481	Albaugh UK	13828	31 Jan 2020
RouteOne Oxanet 481	Albaugh UK	15309	31 Jan 2020

392 flufenacet + pendimethalin

Product	Approval holder	MAPP No.	Expiry Date
Clayton Glacier	Clayton	15920	31 Jan 2020
Cleancrop Hector	Agrii	15932	31 Jan 2020
Frozen	Harvest	17691	31 Jan 2020
Ice	BASF	13930	31 Jan 2020
Kruos	Chem-Wise	15880	31 Jan 2020
Latice	Euro	16693	31 Jan 2020
Standon Sparkle	Standon	17114	31 Jan 2020

393 flufenacet + picolinafen

Product	Approval holder	MAPP No.	Expiry Date
Quirinus	BASF	17711	03 Aug 2020

394 flumioxazin

Product	Approval holder	MAPP No.	Expiry Date
Guillotine	Interfarm	13562	31 Dec 2019

396 fluopicolide + propamocarb hydrochloride

Product	Approval holder	MAPP No.	Expiry Date
Itofin	Euro	16843	31 Jan 2021

398 fluopyram + prothioconazole

Product	Approval holder	MAPP No.	Expiry Date
Clayton Repel	Clayton	17463	31 Dec 2017
Pronapse	Euro	16838	31 Dec 2017

400 fluopyram + trifloxystrobin

Product	Approval holder	MAPP No.	Expiry Date
Luna Sensation	Bayer Environ.	15793	31 Jan 2020

401 fluoxastrobin

Product	Approval holder	MAPP No.	Expiry Date
Bayer UK 831	Bayer CropScience	12091	31 Jul 2018

Products also Registered

Product	Approval holder	MAPP No.	Expiry Date
402 fluoxastrobin + prothioconazole			
Clayton Edge	Clayton	12911	31 Dec 2019
Curuni	Euro	16860	31 Jan 2020
Exploit	Euro	16851	31 Jan 2021
Firefly	Bayer CropScience	13692	31 Jan 2021
Prostrob 20	Chem-Wise	15698	31 Jan 2021
405 flupyrsulfuron-methyl			
Ductis SX	DuPont	15426	31 Dec 2019
Oklar SX	Nufarm UK	15037	31 Dec 2019
Oriel 50SX	Nufarm UK	14640	31 Dec 2019
Spelio SX	DuPont	15424	31 Dec 2019
408 flupyrsulfuron-methyl + pyroxsulam			
GF-2070	Dow	15414	31 Aug 2017
GF-2070	Dow	17472	31 Dec 2019
Unite	Dow	15758	31 Jul 2017
409 flupyrsulfuron-methyl + thifensulfuron-methyl			
Lancer	Headland	13031	31 Dec 2019
410 fluquinconazole			
Flamenco	BASF	11699	30 Jun 2017
Galmano	Bayer CropScience	11650	30 Jun 2017
Jockey Solo	BASF	14281	30 Jun 2017
Triplex	BASF	12443	30 Jun 2017
411 fluquinconazole + prochloraz			
Baron	BASF	11686	30 Jun 2017
Epona	BASF	12444	30 Jun 2017
Foil	BASF	11700	30 Jun 2017
Jockey	BASF	11689	30 Jun 2017
412 fluroxypyr			
Agrostar	Agro Trade	14925	30 Apr 2017
Agrotech Fluroxypyr	Agrotech-Trading	12294	30 Jun 2017
Awac	Globachem	13296	30 Sep 2017
Awac	Globachem	17501	02 Feb 2020
Barclay Hudson	Barclay	13606	30 Jun 2017
Cascade	Chem-Wise	13939	30 Jun 2017
Casino	Mitsui	14352	30 Jun 2017
Casino	Certis	16570	31 Oct 2017
Casino	Certis	17592	30 Jun 2024
Cerastar	Agro Trade	17066	30 Jun 2017
Cerastar	Cera Chem	13183	30 Jun 2017
Clean Crop Gallifrey 200	Globachem	17503	02 Feb 2020
Clean Crop Gallifrey 200	Globachem	13813	30 Sep 2017
Crescent	Certis	16568	31 Oct 2017
Crescent	Mitsui	13330	30 Jun 2017
Flixfit 20 EC	GAT Micro	15684	30 Jun 2017
Flurostar 200	Globachem	12903	31 Aug 2017
Flurox 180	Stockton	13959	09 May 2020
Fluroxy 200	Goldengrass	14421	30 Jun 2017
Fluxcon 20 EC	GAT Micro	15941	30 Jun 2017
Fluxyr 200 EC	UPL Europe	14086	30 Jun 2017
Fluxyr 200 EC	UPL Europe	16958	30 Jun 2017
Gala	Dow	12019	30 Jun 2017
GAL-GONE	Globachem	13821	30 Sep 2017
GAT Stake 20 EC	GAT Micro	15399	30 Jun 2017
Gf-1784	Dow	16850	11 May 2018

Products also Registered

Product	Approval holder	MAPP No.	Expiry Date
Greencrop Reaper	Greencrop	12261	30 Jun 2017
Hatchet 200 EC	Certis	16525	30 Jun 2017
Hatchet 200 EC	Mitsui	15097	30 Jun 2017
Hatchet Xtra	Mitsui	13213	30 Jun 2017
Hatchet Xtra	Certis	16526	30 Sep 2017
Hockley Fluroxypyr 20	Hockley	15448	30 Sep 2017
Hudson 200	Barclay	16391	31 Aug 2018
Hudson 200	Barclay	17749	04 Aug 2020
Hurler	Barclay	13458	31 Aug 2018
Klever	Globachem	17504	02 Feb 2020
Klever	Globachem	13295	30 Sep 2017
Skora	AgChem Access	13422	30 Jun 2017
Standon Homerun	Standon	13177	30 Jun 2017
Starane 2	Dow	12018	30 Jun 2017
Taipan	Agro Trade	17063	30 Jun 2024
Taipan	Cera Chem	14155	30 Jun 2017
Tandus	Nufarm UK	13432	30 Jun 2017
Tomahawk	Adama	09249	30 Jun 2017
Tomahawk 2	Adama	17468	30 Jun 2024
Tomahawk 2	Adama	15231	31 Aug 2017
Toska EC	Novastar	17709	30 Jun 2024
Toska EC	Novastar	16311	09 May 2020
Triton	EZCrop	15673	09 May 2020

416 fluroxypyr + metsulfuron-methyl

Croupier	Certis	16543	30 Jun 2017
Croupier OD	Certis	17630	28 Jun 2020

417 fluroxypyr + metsulfuron-methyl + thifensulfuron-methyl

Omnera LQM	DuPont	17769	28 Sep 2020

419 fluroxypyr + triclopyr

Pas	Dow	17772	31 Oct 2020

422 flutolanil

NNF-136	Nichino	14302	31 Aug 2021

423 flutriafol

Impact	Headland	12776	31 Dec 2021

424 fluxapyroxad

BAS 700	BASF	15552	31 May 2017
BAS 700	BASF	17106	30 Jun 2025
Flux	Chem-Wise	17602	30 Jun 2025
Flux	Chem-Wise	16470	31 May 2017
Imtrex	BASF	15509	31 May 2017

425 fluxapyroxad + metconazole

Clayton Tardis	Clayton	16577	31 May 2017
Librax	BASF	15887	31 May 2017
Wolverine	Dow	17729	31 Oct 2020

429 foramsulfuron + iodosulfuron-methyl-sodium

Logo	Bayer CropScience	17624	31 Jan 2020

430 fosetyl-aluminium

Plant Trust	Everris Ltd	15779	31 Oct 2020

436 garlic extract

Eagle Green Care	ECOspray	14989	26 Nov 2018
EGCÂGranules	Rigby Taylor	17233	12 Nov 2018

Products also Registered

Product	Approval holder	MAPP No.	Expiry Date
Grenadier	Certis	17617	26 Nov 2018
NEMguard A PCN Granules	ECOspray	17377	12 Nov 2018
NEMguard granules	ECOspray	15254	28 Feb 2022

438 gibberellins

Berelex	Interfarm	14123	28 Feb 2017
GIBB 3	Globachem	12290	28 Feb 2017
GIBB Plus	Globachem	11875	31 Mar 2017
Gibb Plus 10 SL	Q-Chem	14968	30 Sep 2017
Regulex 10 SG	Interfarm	14116	28 Feb 2017

439 Gliocladium catenulatum

Prestop	Fargro	15103	30 Sep 2019
Prestop Mix	Fargro	15104	31 Jan 2020

440 glufosinate-ammonium

Clayton Reap	Clayton	17055	28 Feb 2017
Diminish	Matrix	17205	28 Feb 2017
Finale 150	Bayer CropScience	16733	31 Jan 2021
Halta	Belcrop	17315	31 Jan 2021
Harvest	Bayer CropScience	07321	28 Feb 2017
Kaspar	Certis	11214	28 Feb 2017
Kur Tail	Pure Amenity	15329	28 Feb 2017
Kurtail Gold	Progreen	17292	31 Mar 2020
Kurtail Gold	Progreen	17489	31 Jan 2021
Marengo	Pan Agriculture	16296	28 Feb 2017
Pearl	Agrigem	17531	31 Jan 2021
Weedex	Novastar	17764	31 Jan 2021
Weedex	Novastar	15115	28 Feb 2017
Whippet	AgChem Access	16556	28 Feb 2017
Whippet	AgChem Access	17669	31 Jan 2021

441 glyphosate

Accelerate	Headland	13390	30 Jun 2018
Acrion	Bayer Environ.	12677	30 Jun 2018
Alekto Plus TF	Belchim	17143	29 Oct 2017
Alekto Plus TF	Helm	16478	29 Oct 2017
Ardee	Barclay	17167	30 Jun 2018
Ardee XL	Barclay	17515	31 Dec 2018
Azural	Monsanto	14361	30 Sep 2017
Barbarian XL	Barclay	17665	31 Dec 2018
Barbarian XL	Barclay	16911	31 Dec 2018
Barclay Barbarian	Barclay	12714	30 Jun 2018
Barclay Gallup Biograde 360	Barclay	17612	31 Dec 2018
Barclay Gallup Biograde Amenity	Barclay	12716	30 Jun 2020
Barclay Gallup Hi-Aktiv	Barclay	17646	31 Dec 2018
Barclay Garryowen	Barclay	12715	30 Jun 2018
Boom Efekt	Albaugh UK	17588	31 Dec 2018
Boom efekt	Pinus	15606	31 Dec 2018
Cleancrop Bronco	Belchim	17056	29 Oct 2017
Cleancrop Corral	United Agri	15220	31 Dec 2018
Cleancrop Corral	Agrii	17554	31 Dec 2018
CleanCrop Hoedown	United Agri	15404	30 Jun 2018
Cleancrop Hoedown	United Agri	17563	31 Dec 2018
Cleancrop Hoedown	United Agri	12913	31 Dec 2018
Cleancrop Tungsten	United Agri	13049	31 Dec 2018
Cleancrop Tungsten	United Agri	17553	31 Dec 2018
Clinic	Nufarm UK	12678	30 Jun 2018
Clipper	Adama	14820	30 Jun 2018
Cosmic NG	Arysta	15646	30 Jun 2018

Products also Registered

Product	Approval holder	MAPP No.	Expiry Date
Credit DST	Nufarm UK	13822	30 Jun 2018
Credit DST	Nufarm UK	14066	30 Jun 2018
Dow Agrosciences Glyphosate 360	Dow	12720	30 Jun 2018
Ecoplug Max	Monsanto	14741	31 Dec 2018
Egret	Monsanto	17510	31 Dec 2018
Etna	UPL Europe	16962	30 Jun 2018
Etna	UPL Europe	14674	30 Jun 2018
Euro Glyfo 360	Euro	14691	30 Jun 2018
Euro Glyfo 450	Euro	14936	30 Jun 2018
Excel DF Gold	Excel	17413	28 Feb 2020
Figaro	Belchim	16947	30 Jun 2018
Fozat 360 SL	Agro-Chemie	16171	30 Jun 2018
Gallup Hi- Aktiv Amenity	Barclay	12898	30 Jun 2020
Gallup Hi-Aktiv Amenity	Barclay	17681	30 Jun 2020
Gladiator	Barclay	16641	30 Jun 2018
Glister	Sinon EU	12990	30 Jun 2018
Glister Ultra	Sinon EU	15209	30 Jun 2018
Glycel	Excel	15068	30 Jun 2018
Glydate	Nufarm UK	12679	30 Jun 2018
Glyder	Agform	15442	30 Jun 2018
Glyder 450 TF	Agform	16520	29 Oct 2017
Glyfo Star	Q-Chem	17457	31 Dec 2018
Glyfo_TDI	Q-Chem	14743	30 Jun 2018
Glyfos Gold	Nomix Enviro	10570	31 Dec 2018
Glyfos Gold ECO	Headland Amenity	16489	30 Jun 2018
Glyfos Monte	Headland	15069	30 Jun 2018
Glyfo-TDI	Q-Chem	13940	30 Jun 2018
Glymark	Nomix Enviro	13870	30 Jun 2018
Glyper	SBM	14383	30 Jun 2018
Glyphogan	Adama	12668	30 Jun 2018
Glypho-Rapid 450	Barclay	17647	31 Dec 2018
Glypho-Rapid 450	Barclay	13882	31 Dec 2018
Glyphosate 360	Monsanto	12669	30 Jun 2018
Glystar 360	Albaugh UK	16179	30 Jun 2018
Glyweed	Sabero	14444	30 Jun 2018
Habitat	Barclay	17662	31 Dec 2018
Habitat	Barclay	15199	31 Dec 2018
Hauberk	Synergy	16163	30 Jun 2018
Helosate 450 TF	Helm	16326	29 Oct 2017
Hi-Fosate	Hockley	15062	30 Jun 2018
HY-GLO 360	Agrichem	15194	30 Jun 2018
HY-Gly 360	Agrichem	14919	31 May 2017
Jali 450	DAPT	15930	30 Jun 2018
Klean G	Ventura	17645	31 Dec 2018
KN 540	Monsanto	12009	30 Jun 2018
Kraken	Bayer CropScience	15175	30 Jun 2018
Landmaster	Albaugh UK	15759	30 Jun 2018
Landmaster 360 TF	Albaugh UK	17683	30 Jun 2020
Manifest	Headland	11041	30 Jun 2018
Mascot Hi-Aktiv Amenity	Rigby Taylor	17696	30 Jun 2020
Mentor	Monsanto	16508	30 Jun 2018
MON 76473	Monsanto	15348	30 Jun 2018
MON 79351	Monsanto	12860	30 Jun 2018
MON 79376	Monsanto	12664	30 Jun 2018
MON 79545	Monsanto	12663	30 Jun 2018
MON 79632	Monsanto	14841	30 Jun 2018
MON 79991	Monsanto	16300	25 Nov 2017
Monosate G	Monsanto	17449	31 Dec 2018
Monsanto Amenity Glyphosate	Monsanto	15227	30 Jun 2018

SECTION 3

Products also Registered

Product	Approval holder	MAPP No.	Expiry Date
Montana	Sapec	14843	30 Jun 2018
MS Theon	Micromix	17150	30 Jun 2018
NASA	Agria SA	17237	30 Jun 2018
Nomix Conqueror	Nomix Enviro	12370	28 Feb 2017
Nomix Frontclear	Nomix Enviro	15180	30 Jun 2018
Nomix G	Nomix Enviro	13872	30 Jun 2018
Nomix Nova	Nomix Enviro	13873	30 Jun 2018
Nomix Prolite	Frontier	16132	30 Jun 2018
Nomix Revenge	Nomix Enviro	13874	30 Jun 2018
Nufosate	Nufarm UK	12699	30 Jun 2018
Nufosate Ace	Nufarm UK	14959	30 Jun 2018
Nufosate Ace	Nufarm UK	13794	30 Jun 2018
Onslaught	Procam	14564	31 Dec 2018
Onslaught	Agrii	17607	31 Dec 2018
Ormond	Barclay	17712	30 Jun 2020
Ovation	Monsanto	17476	31 Dec 2018
Oxalis NG	Arysta	15612	30 Jun 2018
Pitch	Nufarm UK	15485	30 Jun 2018
Pontil 360	Barclay	14322	30 Jun 2018
Preline	Linemark	12756	30 Jun 2018
Rodeo	Monsanto	16242	30 Jun 2018
Rodeo	Monsanto	15242	30 Sep 2017
Romany	Greencrop	12681	30 Jun 2018
Rosate 36	Albaugh UK	14459	30 Jun 2018
Rosate 36 SL	Albaugh UK	15263	30 Jun 2018
Rosate 360 TF	Albaugh UK	17682	30 Jun 2020
Rosate Green	Albaugh UK	15122	30 Jun 2018
Roundup	Monsanto	12645	30 Jun 2018
Roundup Ace	Monsanto	12772	30 Jun 2018
Roundup Advance	Monsanto	15540	30 Jun 2018
Roundup Amenity	Monsanto	12672	30 Jun 2018
Roundup Assure	Monsanto	15537	30 Jun 2018
Roundup Biactive	Monsanto	10320	30 Jun 2018
Roundup Biactive 3G	Monsanto	13409	30 Jun 2018
Roundup Biactive Dry	Monsanto	12646	30 Jun 2018
Roundup Bio	Monsanto	15538	30 Jun 2018
Roundup Express	Monsanto	12526	30 Jun 2018
Roundup Gold	Monsanto	10975	30 Jun 2018
Roundup Klik	Monsanto	12866	30 Jun 2018
Roundup Max	Monsanto	12952	30 Jun 2018
Roundup Metro	Monsanto	14842	30 Jun 2018
Roundup Pro Biactive	Monsanto	10330	30 Jun 2018
Roundup Proactive	Monsanto	17380	31 Dec 2018
Roundup ProBiactive 450	Monsanto	12778	30 Jun 2018
Roundup ProBio	Monsanto	15539	31 Dec 2018
Roundup Pro-Green	Monsanto	11907	30 Jun 2018
Roundup Provide	Monsanto	12953	30 Jun 2018
Roundup Rail	Monsanto	12671	30 Jun 2018
Roundup Sonic	Monsanto	17152	30 Jun 2018
Roundup Star	Monsanto	15470	30 Jun 2018
Roundup Ultimate	Monsanto	12774	30 Jun 2018
Roundup Vista	Monsanto	17156	30 Jun 2018
RouteOne Rosate 36	Albaugh UK	13889	30 Jun 2018
RouteOne Rosate 36	Albaugh UK	15280	30 Jun 2018
RouteOne Rosate 360	Albaugh UK	15283	30 Jun 2018
RouteOne Rosate 360	Albaugh UK	15284	30 Jun 2018
Rustler Pro-Green	ChemSource	15454	31 Dec 2018
RVG Glyphosate 360	Rovogate	17072	30 Jun 2018
Samurai	Monsanto	12674	30 Sep 2017

Products also Registered

Product	Approval holder	MAPP No.	Expiry Date
Scorpion	Monsanto	17516	31 Dec 2018
Scorpion	Cardel	12673	31 Aug 2017
Silvio	Rotam	14626	30 Jun 2018
Snapper	Nufarm UK	15489	30 Jun 2018
Spear & Jackson Professional Super Strong Concentrated Weed Killer	Assured Products	17305	30 Jun 2018
Stacato	Sipcam	12675	30 Jun 2018
Stamen	Mitsui	14895	30 Jun 2020
Surrender T/F	Agrovista	17575	31 Dec 2018
Symbol	UPL Europe	14769	30 Jun 2018
Tamba Bio	Maxwell	17653	31 Dec 2018
Tambora	Generica	16861	31 Dec 2018
Tangent	Headland Amenity	11872	31 Dec 2018
Task 360	Albaugh UK	14570	30 Jun 2018
Toro 360	Sipcam	15652	30 Jun 2018
Total 360 WDC	Extreme Green	14958	31 Dec 2018
Total 360 WDC	Contrapak	16020	31 Dec 2018
Total 360 WDC	Dynamite	16267	31 Dec 2018
Trustee Amenity	Barclay	17697	30 Jun 2020
Typhoon 360	Adama	14817	31 Dec 2018
Ultramax	PSI	15931	30 Jun 2018
Vesuvius Green	Ventura	17650	31 Dec 2018
Vival	Belchim	14550	30 Jun 2018
Vival	Belchim	17048	30 Jun 2018
Vival	Belchim	17224	30 Jun 2018
Wither 36	Albaugh UK	15768	30 Jun 2018
Wopro Glyphosate	B.V. Industrie	15462	30 Jun 2018

442 glyphosate + pyraflufen-ethyl

Product	Approval holder	MAPP No.	Expiry Date
Hammer	Everris Ltd	15060	31 Mar 2018
Thunderbolt	Nichino	14102	30 Jun 2018

443 glyphosate + sulfosulfuron

Product	Approval holder	MAPP No.	Expiry Date
Nomix Blade	Nomix Enviro	13907	30 Jun 2018
Nomix Duplex	Nomix Enviro	14953	30 Jun 2018

447 halauxifen-methyl

Product	Approval holder	MAPP No.	Expiry Date
GF-2573	Dow	17540	05 Feb 2028

449 imazalil

Product	Approval holder	MAPP No.	Expiry Date
Fungazil 50LS	Certis	14069	30 Jun 2024
Sphinx	BASF	11764	30 Jun 2017

450 imazalil + ipconazole

Product	Approval holder	MAPP No.	Expiry Date
Rancona i-MIX	Chemtura	15574	31 Aug 2018
Rancona i-MIX	Certis	16492	31 Aug 2018

452 imazalil + thiabendazole

Product	Approval holder	MAPP No.	Expiry Date
Storite Super	Syngenta	13135	30 Jun 2017

456 imazamox + metazachlor + quinmerac

Product	Approval holder	MAPP No.	Expiry Date
Clesima	BASF	16275	31 Jan 2020

457 imazamox + pendimethalin

Product	Approval holder	MAPP No.	Expiry Date
Nirvana	BASF	13220	31 Jan 2020

461 imidacloprid

Product	Approval holder	MAPP No.	Expiry Date
Couraze	Solufeed	13862	28 Feb 2017
Couraze	Solufeed	17183	31 Jan 2022

Products also Registered

Product	Approval holder	MAPP No.	Expiry Date
Imidasect 5GR	Fargro	14574	28 Feb 2017
Merit Forest	Bayer CropScience	15583	31 May 2017
Mido 70% WDG	Sharda	14174	28 Feb 2017
Mido 70% WDG	Sharda	16812	23 Aug 2019

464 4-indol-3-ylbutyric acid

Chryzoplus Grey 0.8%	Fargro	09094	30 Nov 2017
Chryzoplus Grey 0.8%	Rhizopon	17569	30 Nov 2023
Chryzopon Rose 0.1%	Rhizopon	17566	30 Nov 2023
Chryzopon Rose 0.1%	Fargro	09092	30 Nov 2017
Chryzosan White 0.6%	Fargro	09093	31 Dec 2017
Chryzotek Beige	Fargro	09081	30 Nov 2017
Chryzotek Beige 0.4%	Rhizopon	17568	30 Nov 2023
Chryzotop Green	Fargro	09085	30 Nov 2017
Chryzotop Green 0.25%	Rhizopon	17567	30 Nov 2023
Rhizopon AA Powder (0.5%)	Fargro	09082	30 Nov 2017
Rhizopon AA Powder (1%)	Fargro	09084	30 Nov 2017
Rhizopon AA Powder (2%)	Fargro	09083	30 Nov 2017
Rhizopon AA Tablets	Fargro	09086	31 Dec 2021

466 indoxacarb

Picard 300 WG	Agrifarm	16714	30 Apr 2020

467 iodosulfuron-methyl-sodium

Hussar	Bayer CropScience	12364	30 Apr 2020
Klaxon	Bayer CropScience	15174	31 Aug 2017
Klaxon OD	Bayer CropScience	16173	31 Aug 2017

468 iodosulfuron-methyl-sodium + mesosulfuron-methyl

Diplomat	Euro	16691	31 Jul 2019
Hatra	Bayer CropScience	14524	31 May 2017
Horus	Bayer CropScience	14541	30 Jun 2017
Idosem 36	Chem-Wise	15913	31 Jul 2019
Iodomeso OD	Euro	16501	31 Jul 2019
Jerico	Terrechem	17477	31 Jul 2019
Mesiodo 35	Euro	15305	31 Jul 2019
Mesoiodostar WG	Life Scientific	17760	31 Jul 2019
Ocelot	Matrix	17261	31 Jul 2019
Qdos Rockall 2	Me2	12962	31 Jul 2019
RouteOne Seafarer	Albaugh UK	15317	31 Jul 2019
RouteOne Seafarer	Albaugh UK	13654	31 Jul 2019
Seafarer	RouteOne	13303	31 Jul 2019
Teliton Ocean	Teliton	13163	31 Jul 2019

469 iodosulfuron-methyl-sodium + propoxycarbazone-sodium

Caliban Duo	Headland	14283	31 Jul 2019

471 ipconazole

Rancona 15 ME	Chemtura	16136	28 Feb 2027

472 iprodione

Acamar WG	Servem	17329	30 Apr 2020
Advance Green	Matrix	16675	30 Apr 2020
Advance Green	Matrix	17088	30 Apr 2020
Cavron	Belcrop	16168	30 Apr 2020
Emerald Turf	Agrigem	17590	30 Apr 2020
Hi-Prodione 50	BASF	15498	30 Apr 2020
Iprodione 500 SC	Sinon EU	16159	31 Aug 2017
MAC-Iprodione 255 SC	MAC	14781	30 Apr 2020
Mascot Rayzor	Rigby Taylor	14329	30 Apr 2020

Products also Registered

Product	Approval holder	MAPP No.	Expiry Date
Prime Turf	ChemSource	16134	30 Apr 2020
Prime-Turf	ChemSource	14617	30 Apr 2020
Progress Green	Matrix	16930	30 Apr 2020
Sinpro	Sinon EU	16236	30 Apr 2020
Superturf	Mitsui	14495	30 Apr 2020

477 isopyrazam

A15149W	Syngenta	15046	30 Sep 2025
Zulu	Syngenta	16028	30 Sep 2017

481 kresoxim-methyl

Beem WG	AgChem Access	13807	30 Nov 2017
Fermor 500 WG	Agrifarm	16724	30 Nov 2017
Stroby WG	BASF	08653	30 Nov 2017
Strong WG	RAAT	17470	30 Jun 2024

482 lambda-cyhalothrin

Astec Lambda CY	Agrii	16319	31 Dec 2018
Block	Agrii	16319	31 Dec 2018
Clayton Lambada	Clayton	13231	31 Dec 2018
Clayton Sparta	Clayton	13457	31 Dec 2018
CleanCrop Corsair	Clayton	14124	31 Dec 2018
CM Lambaz 50 EC	CMI	16241	31 Dec 2018
Colt 10 CS	Headland	17430	31 Dec 2018
EA Lambda 50	European Ag	15041	31 Dec 2018
EA Lambda C	European Ag	15093	31 Dec 2018
Eminentos 10 CS	GAT Micro	15854	31 Dec 2018
Eminentos 10 CS	Headland	17690	31 Dec 2018
Euro Lambda 100 CS	Euro	14794	31 Dec 2018
Hockley Lambda 5EC	Hockley	15516	31 Dec 2018
Ingot Z	AgChem Access	13929	31 Dec 2018
IT Lambda	Inter-Trade	15119	31 Dec 2018
Jackpot 2	Mitsui	12888	31 Dec 2018
Karis 10 CS	Headland	15312	30 Jun 2018
Kendo	Syngenta	15562	31 Dec 2018
Kusti	Syngenta	16656	31 Dec 2018
Laidir 10 CS	GAT Micro	15461	31 Dec 2018
Laidir 10 CS	Headland	17693	31 Dec 2018
Lambda C	RouteOne	13471	31 Dec 2018
Lambd'Africa	Astec	15697	31 Dec 2018
Landgold Lambda-Z	Teliton	13533	31 Dec 2018
Life Scientific Lambda-Cyhalothrin	Life Scientific	15342	31 Dec 2018
Life Scientific Lambda-Cyhalothrin 50	Life Scientific	16889	24 Feb 2019
MAC-Lambda-Cyhalothrin 50 EC	MAC	14941	31 Dec 2018
Major	Adama	16023	31 May 2017
Markate 50	Agrovista	13529	31 Dec 2018
Martlet	Goldengrass	17052	31 Dec 2018
Me2 Lambda2	Me2	12883	31 Dec 2018
Ninja 5CS	Syngenta	16417	31 Dec 2018
Osmoze 50 CS	Aako	16416	31 Dec 2018
Reparto	Sipcam	15452	31 Dec 2018
RouteOne Lambda C	Albaugh UK	13663	31 Dec 2018
RVG Lambda-cyhalothrin	Rovogate	16806	31 Dec 2018
Sergeant Major	Adama	16175	31 Dec 2018
Stealth	Syngenta	14551	31 Dec 2018
Triumph	Matrix	16929	31 Dec 2018
Warrior	Syngenta	13857	31 Dec 2018

SECTION 3

Products also Registered

Product	Approval holder	MAPP No.	Expiry Date
484 laminarin			
Vacciplant	Goemar	13260	31 Jan 2020
486 lenacil			
Venzar 80 WP	DuPont	09981	31 Dec 2021
488 linuron			
Aredios 45 SC	Novastar	14586	31 Jan 2020
Datura	UPL Europe	14915	31 Jan 2019
Kaedyn 500	DAPT	14739	31 Jan 2020
Linurex 50 SC	Adama	14652	31 Jan 2020
Messidor 45 SC	Novastar	14710	31 Jan 2020
489 magnesium phosphide			
Magtoxin	Detia Degesch	16996	31 Jan 2020
Magtoxin Pellets	Detia Degesch	17376	28 Feb 2022
490 maleic hydrazide			
Cleancrop Malahide	United Agri	13629	30 Apr 2020
Fazor	Chemtura	13617	30 Apr 2020
Gro-Slo	Agriphar	15851	30 Apr 2020
Himalaya	Agriphar	15214	30 Apr 2020
Himalaya XL	Arysta	17511	30 Apr 2020
Shaver	ChemSource	15335	30 Apr 2020
Source II	Chiltern	13618	30 Apr 2020
492 maltodextrin			
Eradicoat	Certis	13724	28 Feb 2017
Majestik	Certis	14831	28 Feb 2017
Terminus	Certis	17319	03 Aug 2019
493 mancozeb			
Agria Mancozeb 75WDG	Agria SA	16807	16 Nov 2018
Agria Mancozeb 80WP	Agria SA	16636	16 Nov 2018
Cleancrop Mandrake	United Agri	14792	31 Jul 2020
Dithane 945	Indofil	17269	31 Jul 2020
Dithane 945	Interfarm	15715	30 Sep 2019
Dithane Dry Flowable Neotec	Indofil	15720	30 Sep 2019
Dithane Dry Flowable Neotec	Indofil	17268	31 Jul 2020
Emzeb 80 WP	Sabero	17714	02 Aug 2020
Laminator 75 WG	Interfarm	15667	31 Jul 2020
Manfil 75 WG	Interfarm	15958	31 Jul 2020
Manfil 80 WP	Indofil	15956	31 Jul 2020
Penncozeb 80 WP	UPL Europe	14718	28 Feb 2019
Penncozeb WDG	UPL Europe	14719	31 Jan 2019
Trimanzone	UPL Europe	14726	31 Oct 2018
Trimanzone	UPL Europe	16759	31 Jul 2020
Zebra WDG	Headland	14875	31 Jul 2020
496 mancozeb + zoxamide			
Unikat 75WG	Gowan	14200	31 Jul 2019
497 mandipropamid			
Mandimid	Euro	16917	31 Oct 2017
Mandimid	Euro	17651	31 Jan 2026
Pergardo Uni	Fargro	17513	31 Jan 2026
Pergardo Uni	Fargro	15734	31 Aug 2017
Revus	Syngenta	13484	31 Oct 2017
Standon Mandor	Standon	16095	31 Oct 2017
Standon Mandor	Standon	17536	31 Jan 2026

Products also Registered

Product	Approval holder	MAPP No.	Expiry Date
499 MCPA			
Agrichem MCPA 500	UPL Europe	16964	30 Apr 2020
Agritox 50	Nufarm UK	14814	30 Apr 2020
Dow MCPA Amine 50	Dow	14911	30 Apr 2020
Go-Low Power	DuPont	13812	30 Apr 2020
MCPA 50	UPL Europe	14908	30 Apr 2020
Nufarm MCPA 750	Nufarm UK	14892	30 Apr 2020
POL-MCPA 500 SL	Zaklady	14912	30 Apr 2020
Tasker 75	Headland	14913	30 Apr 2020
501 MCPA + mecoprop-P			
Cleanrun Pro	Everris Ltd	15073	31 Jul 2019
Greenmaster Extra	Everris Ltd	15817	31 Jul 2019
503 mecoprop-P			
Clenecorn Super	Nufarm UK	14628	31 Jul 2019
Isomec	Nufarm UK	14385	31 Jul 2019
508 mepiquat chloride + metconazole			
Mepicon	Euro	16844	31 Oct 2020
509 mepiquat chloride + prohexadione-calcium			
Medax Top	BASF	16574	31 Aug 2021
510 meptyldinocap			
Convivial	ChemSource	15865	31 Mar 2019
512 mesosulfuron-methyl + propoxycarbazone-sodium			
Monolith	Bayer CropScience	17687	28 Sep 2020
513 mesotrione			
A12739A	Syngenta	16158	31 Jan 2020
Barracuda	Albaugh UK	17433	21 Mar 2020
Callisto	PSI	17512	31 Jan 2020
Cuter	Industrias Afrasa	17668	21 Mar 2020
Evolya	Syngenta	17490	31 Jan 2020
Hockley Mesotrione 10	Hockley	15519	31 Jan 2020
Kalypstowe	AgChem Access	13844	31 Jan 2020
Labyrinth	EZCrop	15702	31 Jan 2020
Life Scientific Mesotrione	Life Scientific	17092	31 Jan 2019
Marvel	ChemSource	14508	31 Jan 2020
Meristo	Syngenta	16865	31 Jan 2020
Mesotrion 100SC	Euro	16362	31 Jan 2020
Minerva	Terrechem	17514	31 Jan 2020
Osorno	Globachem	17321	31 Jan 2020
Raikiri	Sumi Agro	17670	31 Jan 2019
RouteOne Trione 10	Albaugh UK	14765	31 Jan 2020
RouteOne Trione 10	Albaugh UK	13765	31 Jan 2020
514 mesotrione + nicosulfuron			
Choriste	Syngenta	16899	31 Jan 2020
Clayton Aspen	Clayton	16114	31 Jan 2020
515 mesotrione + s-metolachlor			
Camix	Syngenta	17722	10 Aug 2020
516 mesotrione + terbuthylazine			
Callistar	Syngenta	16864	31 Dec 2021
Orion	EZCrop	16049	31 Dec 2021
Riscala	Euro	16839	31 Dec 2021
RouteOne Mesot	Albaugh UK	13576	31 Dec 2021

SECTION 3

Products also Registered

Product	Approval holder	MAPP No.	Expiry Date
RouteOne Mesot	Albaugh UK	15302	31 Dec 2021

518 metalaxyl-M

Divot	ChemSource	15413	31 Dec 2019
Fongarid Gold	Syngenta	12547	31 Dec 2019
Shams 465	DAPT	16107	31 Dec 2019

519 metaldehyde

Appeal	Chiltern	12022	31 Dec 2021
Attract	Chiltern	12023	31 Dec 2021
Cargo 3 GB	Aako	15784	31 Dec 2021
Doff Horticultural Slug Killer Blue Mini Pellets	Certis	11463	31 Dec 2021
TDS Metarex Amba	De Sangosse	13461	31 Dec 2021
Tremolo	Chiltern	16582	31 Dec 2021

520 metamitron

Atlas Too-Da Loo	Nufarm UK	17394	28 Feb 2022
Betra SC	Novastar	17244	28 Feb 2022
Bettix WG	UPL Europe	16496	31 Mar 2018
Celmitron 70% WDG	UPL Europe	16990	31 Dec 2021
Celmitron 70% WDG	UPL Europe	14408	31 Mar 2019
Defiant WG	UPL Europe	16544	31 Mar 2018
Goltix 90	Adama	16497	28 Feb 2022
Goltix Compact	Aako	16545	28 Feb 2022
Goltix WG	Adama	16853	28 Feb 2022
Inter Metron 700 SC	UPL Europe	17389	28 Feb 2022
Inter Metron 700 SC	Inter-Phyto	17101	31 Oct 2019
Meta WDG	ChemSource	15010	31 Mar 2019
Target SC	UPL Europe	13307	31 Mar 2019

521 metamitron + quinmerac

Goltix Titan	Adama	17301	28 Feb 2022

522 metam-sodium

Discovery	UPL Europe	10416	31 Dec 2017
Fumetham	Chemical Nutrition	10047	31 Dec 2017
Metam 510	Taminco	09796	31 Dec 2017
Metam 510	Taminco	16079	30 Jun 2018
Sistan 51	Unicrop	10046	31 Dec 2017

523 Metarhizium anisopliae

Met52 OD	Fargro	17367	31 Oct 2021

524 metazachlor

Clayton Buzz	Clayton	16928	31 Jan 2022
Fuego	Adama	11177	31 Jul 2017
Fuego	Adama	16679	31 Jan 2022
Inter Metazachlor	UPL Europe	17323	28 Jan 2019
Makila 500 SC	Novastar	16752	31 Jan 2022
Metaza	Euro	17168	31 Jan 2022
Mikado	Matrix	17004	31 Jan 2022
Standon New Metazachlor	Standon	17473	31 Jan 2022
Taza 500	Becesane	17284	31 Jan 2022

525 metazachlor + quinmerac

Metamerac	Euro	17139	31 Jan 2022
Metamerac-II	Euro	16810	31 Jan 2022
Metazerac	Becesane	17744	31 Jan 2022
Vesta	Goldengrass	17051	31 Jan 2022

Products also Registered

Product	Approval holder	MAPP No.	Expiry Date
526 metconazole			
Caramba	BASF	15337	31 Oct 2020
Life Scientific Metconazole	Life Scientific	16022	31 Oct 2020
Life Scientific Metconazole 60	Life Scientific	17037	29 Feb 2020
Metal	Arysta	17265	31 Oct 2020
Metal 60	Agriphar	17074	31 Oct 2020
Rambaca 60	Euro	16841	31 Oct 2020
532 1-methylcyclopropene			
EthylBloc Tabs	Smither	16252	29 Feb 2020
EthylBloc Tabs	Oasis	17487	30 Apr 2020
Ethylene Buster	Chrysal	16222	30 Apr 2020
RipeLock Tabs	Rohm & Haas	16339	29 Feb 2020
Ripelock Tabs	AgroFresh	17480	30 Apr 2020
Ripelock Tabs 2.0	AgroFresh	17486	30 Apr 2020
RipeLock Tabs 2.0	Rohm & Haas	16547	07 Jul 2018
Ripelock VP	AgroFresh	17484	30 Apr 2020
RipeLock VP	Rohm & Haas	16338	29 Feb 2020
SmartFresh ProTabs	Landseer	16546	30 Apr 2020
SmartFresh SmartTabs	Landseer	12684	30 Apr 2020
Stay Fresh	ChemSource	15760	30 Apr 2020
537 metribuzin			
Discus	RAAT	17552	31 Jan 2021
Python	Adama	16010	31 Jan 2021
538 metsulfuron-methyl			
Accurate	Headland	13224	31 Dec 2018
Agriguard Metsulfuron	AgriGuard	13324	31 Dec 2018
Alias	DuPont	13397	31 Dec 2018
Ally SX	DuPont	12059	31 Dec 2018
Aztec	EZCrop	16073	31 Dec 2018
Clayton Peak	Clayton	13547	31 Dec 2018
Cleancrop Mondial	United Agri	13353	31 Dec 2018
Finy	UPL Europe	12855	30 Jun 2018
Firecrest	European Ag	14486	31 Dec 2018
Forge	Interfarm	12848	31 Dec 2018
Goldron-M	Goldengrass	13466	31 Dec 2018
Landgold Metsulfuron	Teliton	13421	31 Dec 2018
Life Scientific Metsulfuron-Methyl	Life Scientific	16654	29 Sep 2021
Metro 20	Certis	16503	31 Dec 2018
Metro 20	Mitsui	15633	30 Apr 2018
Metro Clear	Mitsui	15624	31 Dec 2018
Metro Clear	Certis	16234	31 Dec 2018
Minx PX	Nufarm UK	13337	31 Dec 2018
Pike	Nufarm UK	12746	31 Dec 2018
Pike PX	Nufarm UK	13249	31 Dec 2018
Quest	EZCrop	15659	31 Dec 2018
Revenge 20	Agform	14449	31 Dec 2018
Revenge 20 SG	Agform	13633	31 Dec 2018
Ricorso Premium	Rotam	16389	31 Dec 2018
RouteOne Romet 20	Albaugh UK	15277	31 Dec 2018
Standon Gusto Xtra	Standon	13886	31 Dec 2018
Tuko	Terrechem	15611	31 Dec 2018
Victor	EZCrop	15669	31 Dec 2018
539 metsulfuron-methyl + thifensulfuron-methyl			
Accurate Extra	Headland	14167	31 Dec 2018
Aria	Mitsui	14771	31 Dec 2018
Choir	Nufarm UK	14081	31 Dec 2018

SECTION 3

Products also Registered

Product	Approval holder	MAPP No.	Expiry Date
Mozaic SX	DuPont	16108	31 Dec 2018

540 metsulfuron-methyl + tribenuron-methyl

Met-Tribe 286	Euro	16572	31 Dec 2018

541 myclobutanil

Aristocrat	Mitsui	12343	30 Sep 2017
Clayton Lithium	Clayton	15803	30 Sep 2017
Crassus	Landseer	13899	30 Sep 2017
CS Myclobutanil	ChemSource	15434	30 Sep 2017
Hithane	ChemSource	14852	30 Sep 2017
PureMyclobutanil	Pure Amenity	15370	30 Sep 2017
PureThane	Pure Amenity	15134	30 Sep 2017
Robut 20	RouteOne	13437	30 Sep 2017
RouteOne Butanil 20	Albaugh UK	14773	30 Sep 2017
RouteOne Robut 20	Albaugh UK	13646	30 Sep 2017
RouteOne Robut 20	Albaugh UK	15273	30 Sep 2017
RouteOne Robut 20	Generica	15685	30 Sep 2017
Systhane 20EW	Landseer	09396	30 Sep 2017
Systhane 20EW	PSI	16837	30 Sep 2017
Tanil 200	ChemSource	16148	30 Sep 2017

542 1-naphthylacetic acid

Fixor	Belcrop	17428	30 Jun 2024
Rhizopon B Powder (0.1%)	Fargro	09089	30 Jun 2017
Rhizopon B Powder (0.2%)	Fargro	09090	30 Jun 2017
Rhizopon B Tablets	Fargro	09091	30 Jun 2017
Tipoff	Globachem	15728	30 Jun 2017

544 napropamide

AC651	UPL Europe	16227	31 Dec 2017
Associate	Headland	17279	31 Dec 2017
Associate	Headland	13933	31 Dec 2017
Banweed	UPL Europe	09376	31 Dec 2017
Devrinol	UPL Europe	09375	31 Dec 2017
MAC-Napropamide 450 SC	AgChem Access	13972	31 Dec 2021
Nappa	ChemSource	14881	31 Dec 2021
Naprop	Globachem	13402	31 Dec 2017

545 nicosulfuron

Bandera	Rotam	16576	11 May 2018
Fornet 4 SC	Belchim	16082	30 Jun 2021
Hill-Agro 40	Hillfield	17654	18 Jul 2020
Nico Pro 4SC	Syngenta	15645	28 Feb 2018
Samson	Syngenta	12141	28 Feb 2018
Samson	Belchim	16425	30 Jun 2021

546 nicosulfuron + thifensulfuron-methyl

Collage	DuPont	16820	31 Dec 2019

550 paclobutrazol

A10784A	Syngenta	17172	30 Nov 2023
Bonzi	Syngenta Bioline	17095	30 Apr 2020

553 penconazole

Rapas	RAAT	17740	30 Jun 2022

555 pendimethalin

Akolin 330 EC	Aako	14282	31 Jan 2020
Alpha Pendimethalin 330 EC	Adama	13815	31 Jan 2020

Products also Registered

Product	Approval holder	MAPP No.	Expiry Date
Aquarius	Adama	14712	31 Jan 2020
Atlas Pendimethalin 400	Nufarm UK	16815	31 Jan 2020
Blazer M	Headland	15084	31 Jan 2020
Bunker	Adama	13816	31 Jan 2020
Campus 400 CS	Aako	15366	31 Jan 2020
Cinder	Adama	14526	31 Jan 2020
CleanCrop Stomp	United Agri	13443	31 Jan 2020
Clinker	ChemSource	15453	31 Jan 2020
Convey 330EC	Mitsui	14535	31 Jan 2020
Eximus	Gemini	17174	31 Jan 2020
Fastnet	Sipcam	14068	31 Jan 2020
Hockley Pendi 330	Hockley	15513	31 Jan 2020
PDM 330 EC	BASF	13406	31 Jan 2020
Pendi 330	Euro	15807	31 Jan 2020
Pendimet 400	Goldengrass	14715	31 Jan 2020
Pendragon	Mitsui	15512	31 Aug 2017
Pendragon	Certis	16262	31 Jan 2020
Penta	Sharda	17680	31 Jan 2020
Pre-Empt	Euro	16649	31 Jan 2020
Quarry	Gemini	17789	31 Jan 2020
RouteOne Penthal 330	Albaugh UK	15314	31 Jan 2020
RouteOne Penthal 330	Albaugh UK	13768	31 Jan 2020
Sherman	Adama	13859	31 Jan 2020
Shuttle	AgriGuard	14041	31 Jan 2020
Sovereign	BASF	13442	31 Jan 2020
Stamp 330 EC	AgChem Access	13995	31 Jan 2020
Stomp 400 SC	BASF	13405	31 Jan 2020
Yellow Hammer	Goldengrass	16533	31 Jan 2020

556 pendimethalin + picolinafen

Product	Approval holder	MAPP No.	Expiry Date
Flight	BASF	12534	31 Dec 2019
PicoStomp	BASF	13455	31 Dec 2019
Sienna	BASF	14991	31 Dec 2019
Sienna Pro	BASF	17047	31 Dec 2019

557 pendimethalin + pyroxsulam

Product	Approval holder	MAPP No.	Expiry Date
Broadway Sunrise	Dow	14960	31 Jan 2020

558 pendimethalin + terbuthylazine

Product	Approval holder	MAPP No.	Expiry Date
Bullet XL	Adama	14469	30 Jun 2017

559 penflufen

Product	Approval holder	MAPP No.	Expiry Date
Emesto Prime FS	Bayer CropScience	15446	31 Jul 2026

560 penthiopyrad

Product	Approval holder	MAPP No.	Expiry Date
DP 747	DuPont	17530	31 Oct 2026
Intellis	DuPont	16186	26 Mar 2017
Sentenza	Terrechem	17529	31 Oct 2026
Vertisan	DuPont	16104	31 Aug 2017

562 pepino mosaic virus strain CH2 isolate 1906

Product	Approval holder	MAPP No.	Expiry Date
PMV-01	De Ceuster	17587	28 Feb 2033

565 pethoxamid

Product	Approval holder	MAPP No.	Expiry Date
Successor	Headland	17287	18 Oct 2019

567 phenmedipham

Product	Approval holder	MAPP No.	Expiry Date
Agrichem Phenmedipham EC	UPL Europe	16224	31 Jan 2020
Agrichem PMP EC	UPL Europe	16963	31 Jan 2020
Agrichem PMP EC	AgriChem BV	15230	31 Mar 2019

SECTION 3

Products also Registered

Product	Approval holder	MAPP No.	Expiry Date
Agrichem PMP SE	AgriChem BV	16226	31 Mar 2019
Agrichem PMP SE	UPL Europe	16971	31 Jan 2020
Alpha Phenmedipham 320 SC	Adama	14070	31 Jan 2020
Betamax	Mitsui	14371	31 Jan 2020
Betanal Flow	Bayer CropScience	13893	31 Jan 2020
Corzal	UPL Europe	14192	28 Feb 2019
Corzal SC	UPL Europe	17751	31 Jan 2020
Dancer Flow	Sipcam	14395	31 Jan 2020
Galcon SC	Novastar	14607	31 Jan 2020
Herbasan Flow	Nufarm UK	13894	31 Jan 2020
Mandolin Flow	Nufarm UK	13895	31 Jan 2020
Parish	Adama	17464	31 Jan 2020
Pump	Adama	14087	31 Jan 2020
Rubie	EZCrop	15655	31 Jan 2020
Shrapnel	Adama	14088	31 Jan 2020

573 picolinafen

Product	Approval holder	MAPP No.	Expiry Date
AC 900001	BASF	10714	31 Dec 2019
Vixen	BASF	13621	31 Dec 2019

574 picoxystrobin

Product	Approval holder	MAPP No.	Expiry Date
Acanto	DuPont	13043	30 Apr 2020

575 pinoxaden

Product	Approval holder	MAPP No.	Expiry Date
A13814D	Syngenta	17386	30 Jun 2019
Aoraki	AgChem Access	13528	30 Jun 2019
Clayton Tonto	Clayton	14436	30 Jun 2019
Liberate	ChemSource	14872	30 Jun 2019
Quail	ChemSource	14506	30 Jun 2019
Roaxe 100	Chem-Wise	16111	30 Jun 2019
Roaxe 100	RouteOne	13304	30 Jun 2019
RouteOne Roaxe 100	Albaugh UK	13655	30 Jun 2019
RouteOne Roaxe 100	Albaugh UK	15267	30 Jun 2019
RouteOne Roaxe 100	Chem-Wise	15630	30 Jun 2019
Symmetry	Mitsui	13577	30 Jun 2019
Wren	Matrix	16700	30 Jun 2019

576 pirimicarb

Product	Approval holder	MAPP No.	Expiry Date
Aphox	Syngenta	10515	31 Jul 2017
Clayton Pirimicarb	Clayton	17759	31 Oct 2020
Clayton Pirimicarb 50	Clayton	12910	31 Jul 2017
Cleancrop Miricide	United Agri	11776	31 Jul 2017
Hockley Pirimicarb WG	Hockley	15529	31 Jul 2017
Milentus Pirimicarb	Milentus	12268	31 Jul 2017
Phantom	Syngenta	11954	31 Jul 2017
Piri 50	Euro	14956	31 Jul 2017
Pirimate 500	Becesane	15995	31 Jul 2017
Pirimate 500	Becesane	17657	31 Oct 2020
Pirimex 50 WG	MAC	15964	31 Jul 2017
Pirimicarb 50	Goldengrass	14295	31 Jul 2017
RouteOne Primro 50 WG	Albaugh UK	15315	31 Jul 2017
Standon Pirimicarb 50	Standon	13290	30 Jun 2017

577 pirimiphos-methyl

Product	Approval holder	MAPP No.	Expiry Date
PHOBI SMOKE PRO90	Lodi UK	17117	31 Jan 2021

580 Potassium salts of fatty acids

Product	Approval holder	MAPP No.	Expiry Date
Jaboland	Invest A. B.	17324	28 Feb 2022
Jabolim	Quimicas	17482	28 Feb 2022
Nakar	Seipasa	17483	28 Feb 2022

Products also Registered

Product	Approval holder	MAPP No.	Expiry Date
Tec-bom	Iberfol	17455	28 Feb 2022

581 prochloraz

Product	Approval holder	MAPP No.	Expiry Date
Cloraz 400	Goldengrass	14657	30 Jun 2017
Mirage 40 EC	Adama	06770	30 Jun 2017
Mirage 45 SC	Adama	10894	30 Jun 2017
Octave	BASF	11741	30 Jun 2017
Poraz	BASF	11701	30 Jun 2017
Prelude 20 LF	BASF	11904	30 Jun 2017
Prelude 20LF	Agrichem	04371	30 Jun 2017
Scotts Octave	Everris Ltd	15756	30 Jun 2017
Sporgon 50 WP	Sylvan	17700	26 Jul 2020
Sporgon 50 WP	Sylvan	08802	31 Jan 2018
Sportak 45 EW	BASF	11696	31 Dec 2017
Sportak 45 HF	BASF	11697	30 Jun 2017

582 prochloraz + propiconazole

Product	Approval holder	MAPP No.	Expiry Date
Bumper P	Adama	08548	31 Dec 2021

584 prochloraz + tebuconazole

Product	Approval holder	MAPP No.	Expiry Date
Agate	Adama	14482	30 Jun 2017
Astute	Sherriff Amenity	14897	30 Jun 2017
Cyclone 400 EW	Aako	16045	30 Jun 2017
Foremost	Rigby Taylor	16440	30 Jun 2017
Pan Proteb	Pan Amenity	14896	30 Jun 2017
Phaser	Indigrow	16473	30 Jun 2017
PureProteb	Pure Amenity	15469	30 Jun 2017
Throttle	Headland Amenity	13900	30 Jun 2017

586 prochloraz + triticonazole

Product	Approval holder	MAPP No.	Expiry Date
Kinto	BASF	12038	30 Jun 2017

587 prohexadione-calcium

Product	Approval holder	MAPP No.	Expiry Date
Regalis	BASF	12414	30 Jun 2017

589 propamocarb hydrochloride

Product	Approval holder	MAPP No.	Expiry Date
Edipro	Agriphar	15564	31 Jan 2021
Promess	Agriphar	16008	31 Jan 2021
Propamex-I 604 SL	MAC	15901	31 Jan 2021
Proplant	Agriphar	15422	31 Jan 2021

590 propaquizafop

Product	Approval holder	MAPP No.	Expiry Date
Flanoc	Euro	16816	31 May 2022

591 propiconazole

Product	Approval holder	MAPP No.	Expiry Date
Anode	Adama	14447	31 Jul 2019
Apache 250 EC	Aako	17454	31 Jul 2019
Atlas Propiconazole 250	Gemini	16377	31 Jul 2019
Atlas Propiconazole 250	Whyte Agrochemicals	17416	31 Jul 2019
Barclay Propizole	Barclay	15113	09 Feb 2020
Hockley Propicon 25	Hockley	15517	31 Jul 2019
Matsuri	Sumi Agro	15544	31 Jul 2019
Profiol 250	PSI	15393	31 Jul 2019
Zolex	ITACA	15713	31 Jul 2019

596 propyzamide

Product	Approval holder	MAPP No.	Expiry Date
Careca	UPL Europe	14948	31 Mar 2019
Cleancrop Rumble	Agrii	17600	31 Jul 2019
Dennis	Interfarm	15081	31 Jul 2019
Edge 400	Albaugh UK	14683	31 Jul 2019

Products also Registered

Product	Approval holder	MAPP No.	Expiry Date
Engage	Interfarm	14233	31 Jul 2019
Flomide	Interfarm	14223	31 Jul 2019
Hockley Propyzamide 40	Hockley	15449	31 Jul 2019
KeMiChem - Propyzamide 400 SC	KeMiChem	14706	31 Jul 2019
Kerb 50 W	Dow	13715	31 Jul 2019
MAC-Propyzamide 400 SC	MAC	14786	31 Jul 2019
Master Lean 500 SC	Generica	16231	31 Jul 2019
Master Prop 40	Generica	15829	31 Jul 2019
Menace 80 EDF	Dow	13714	31 Jul 2019
MS Eddard	Micromix	17149	31 Jul 2019
Pizamide	ChemSource	14684	31 Jul 2019
Ponder	ChemSource	14693	31 Jul 2019
Propyzamide 400	European Ag	14685	31 Jul 2019
Propyzamide Flo	Albaugh UK	14732	30 Nov 2017
Prova	Frontier	15641	31 Jul 2019
PureFlo	Pure Amenity	15138	31 Jul 2019
Pyzamid 400 SC	Euro	15257	31 Jul 2019
Relva Granules	Belcrop	16744	31 Jul 2019
RouteOne Zamide Flo	Albaugh UK	14559	31 Jul 2019
Setanta 50 WP	Certis	16518	31 Jul 2019
Setanta 50 WP	Mitsui	15642	31 May 2017
Setanta Flo	Mitsui	14493	31 Jul 2019
Shamal	Rotam	15686	31 Jul 2019
Solitaire 50 WP	Certis	16521	31 Jul 2019
Solitaire 50 WP	Mitsui	15636	31 May 2017
Solitaire 50 WP	Mitsui	16177	30 Apr 2018
Verdah 400	DAPT	14680	31 Jul 2019
Verge 400	Albaugh UK	14682	31 Jul 2019
Zamide 80 WG	Albaugh UK	14723	31 Jul 2019
Zamide Flo	Albaugh UK	14679	31 Jul 2019
Zammo	Headland	15313	31 Jul 2019

597 proquinazid

Product	Approval holder	MAPP No.	Expiry Date
Augment	DuPont	17200	31 Jan 2023
Proquin 200	Euro	16019	31 Jan 2023
Talius	PSI	16912	31 Jan 2023
Zorkem	Unisem	16906	31 Jan 2023

598 prosulfocarb

Product	Approval holder	MAPP No.	Expiry Date
A8545G	Syngenta	16204	30 Apr 2021
Atlas Pro	Whyte Agrochemicals	17734	30 Apr 2021
Clayton Comply	Clayton	17702	30 Apr 2021
Clayton Heed	Clayton	17701	30 Apr 2021
Elude	Chem-Wise	17195	30 Apr 2021
Fidox	Syngenta	16209	30 Jun 2021
IC1574	Syngenta	16205	30 Apr 2021
Jade	Syngenta	16203	30 Apr 2021
Nuron	Gemini	17757	30 Apr 2021
NYX	Terrechem	17737	30 Apr 2021
Prosulfocarbstar	Life Scientific	17431	31 Mar 2020
Prosulfostar	Life Scientific	17542	30 Apr 2021
Wicket	Syngenta	17555	30 Apr 2021

600 prothioconazole

Product	Approval holder	MAPP No.	Expiry Date
Banguy	PSI	15140	31 Jul 2018
Proline	Bayer CropScience	12084	31 Jul 2018
Redigo	Bayer CropScience	12085	31 Jan 2021

601 prothioconazole + spiroxamine

Product	Approval holder	MAPP No.	Expiry Date
Pikazole 46	Chem-Wise	15691	31 Jan 2021

Products also Registered

Product	Approval holder	MAPP No.	Expiry Date
Pro-Spirox	Euro	16437	31 Jan 2021

603 prothioconazole + tebuconazole
Harvest Proteb	Harvest	17692	31 Jan 2021
Pro Fit 25	Chem-Wise	17040	31 Jan 2021
Pro-Tebu 250	Euro	17142	31 Jan 2021
Proteus	Matrix	17210	31 Jan 2021

605 prothioconazole + trifloxystrobin
Biomus	Euro	16408	31 Jan 2020

606 Pseudomonas chlororaphis MA 342
Cerall	Chemtura	14546	31 Oct 2019

607 pymetrozine
Chess WG	Syngenta Bioline	13310	31 Dec 2019
MyZus	ChemSource	15903	31 Dec 2019
Quorum	Chem-Wise	16707	31 Dec 2019
Rozine	Direct	16946	31 Dec 2019
Shogi	ChemSource	15736	31 Dec 2019
Standon LBW	Standon	17213	31 Dec 2019

608 pyraclostrobin
BAS 500 06F	BASF	12338	31 Jul 2019
BASF Insignia	BASF	11900	31 Jul 2019
Comet	BASF	10875	31 Jul 2019
Insignia	Vitax	11865	31 Jul 2019
LEY	BASF	14774	31 Jul 2019
Platoon	BASF	12325	31 Jul 2019
Platoon 250	BASF	12640	31 Jul 2019
Tucana 200	BASF	17294	31 Jul 2019
Vivid 200	BASF	17295	31 Jul 2019

609 pyraflufen-ethyl
OS159	Ceres	12723	31 Dec 2018
Quickdown	Certis	15396	31 Dec 2018

611 pyridate
Diva	Belchim	17774	30 Jun 2033

612 pyrimethanil
Penbotec 400 SC	Janssen	17384	31 Oct 2020
Pyrimala	Servem	17325	31 Oct 2020
Pyrus 400 SC	Agriphar	16254	31 Oct 2020
Spectrum	RAAT	17469	31 Oct 2020

613 pyriofenone
Property 180SC	Belchim	17666	12 Sep 2020

614 pyroxsulam
Avocet	Dow	14829	31 Oct 2026

615 Pythium oligandrum M1
Polyversum	De Sangosse	17456	31 Oct 2021

618 quinoxyfen
Apres	Dow	08881	31 Oct 2019

619 quizalofop-P-ethyl
Leopard 5EC	Adama	16915	31 May 2022

SECTION 3

Products also Registered

Product	Approval holder	MAPP No.	Expiry Date
621 rimsulfuron			
Rimsulf 250	Euro	15802	31 Oct 2020
623 silthiofam			
Latitude	Monsanto	10695	30 Apr 2019
627 spinosad			
Exocet	Matrix	17146	31 Oct 2020
628 spirodiclofen			
Spirodor	Q-Chem	15401	31 Mar 2018
629 spiromesifen			
Oberon	Certis	11819	30 Apr 2017
Spry	Pan Agriculture	17474	30 Apr 2017
630 spirotetramat			
Spiro OD	Euro	16840	30 Apr 2019
632 spiroxamine + tebuconazole			
Thyme	Interfarm	14745	30 Jun 2017
634 streptomyces griseoviridis strain K61			
Mycostop	Verdera	16637	31 Oct 2021
635 sulfosulfuron			
Fosulfuron	Euro	15164	30 Jun 2018
Monitor	Monsanto	12236	30 Jun 2018
636 sulfuryl fluoride			
ProFume	Douglas	12035	31 Jan 2018
ProFume	Douglas	16396	30 Nov 2019
ProFume	Douglas	17309	30 Apr 2023
637 sulphur			
Microthiol Special	UPL Europe	06268	31 Mar 2019
POL-Sulphur 80 WG	Ciech	17613	30 Jun 2022
POL-Sulphur 800 SC	Ciech	17614	30 Jun 2022
Solfa WG	Nufarm UK	11602	31 Dec 2021
638 tau-fluvalinate			
Klartan	Adama	11074	31 Dec 2021
Revolt	Adama	13383	31 Dec 2021
639 tebuconazole			
Chem-Wise Tebu 25	Chem-Wise	16142	31 Oct 2017
Deacon	Adama	14270	30 Apr 2018
Erase	Headland	16135	30 Sep 2017
Euro Tebu 250	Euro	14754	30 Sep 2017
Fathom	Headland	15328	31 Aug 2017
Fezan	Sipcam	13681	31 Oct 2017
Folmex 250 EW	MAC	15962	30 Sep 2017
Gizmo	Nufarm UK	16502	22 Apr 2018
Greencrop Tabloid	Greencrop	11969	30 Sep 2017
Harvest Teb 250	Harvest	17706	28 Feb 2022
Hockley Tebucon 25	Hockley	15500	31 Dec 2021
Inter-Tebu	UPL Europe	17322	31 Jan 2020
Manicure	Chem-Wise	13906	30 Sep 2017
Master Teb 25	Generica	16141	31 Oct 2017
Mitre	Adama	17606	28 Feb 2022
Mitre	Adama	12901	31 Oct 2017

Products also Registered

Product	Approval holder	MAPP No.	Expiry Date
Orian	ChemSource	14798	31 Mar 2018
Orius 20 EW	Adama	12311	31 Mar 2018
Rosh	Harvest	17673	28 Feb 2022
RouteOne Tebbro 25	Albaugh UK	15301	30 Sep 2017
Santal	Novastar	16378	30 Sep 2017
Standon Beamer	Standon	17350	28 Feb 2022
Tamok	Matrix	16997	28 Feb 2022
Tamok	Matrix	17264	28 Feb 2022
Teb 250	Unique Marketing	16530	28 Feb 2022
Tebcon	Goldengrass	14300	30 Sep 2017
Tesoro	Belchim	17222	28 Feb 2022
Tharsis	Albaugh UK	17201	28 Feb 2022
Tharsis	Generica	16466	31 Oct 2017

642 tebuconazole + trifloxystrobin

Product	Approval holder	MAPP No.	Expiry Date
Inter Tebloxy	Iticon	17311	31 Jan 2020

643 tebufenpyrad

Product	Approval holder	MAPP No.	Expiry Date
Clan	RAAT	17491	31 Dec 2021

645 tefluthrin

Product	Approval holder	MAPP No.	Expiry Date
A13219F	Syngenta	16161	31 Dec 2021
Force ST	Bayer CropScience	11671	30 Jun 2017

646 tembotrione

Product	Approval holder	MAPP No.	Expiry Date
Laudis	Bayer CropScience	17302	31 Oct 2026

651 thiabendazole

Product	Approval holder	MAPP No.	Expiry Date
Hykeep	Agrichem	12704	31 Dec 2019
Storite Clear Liquid	Frontier	12706	31 Dec 2019
Storite Excel	Frontier	12705	31 Dec 2019
Tezate 220 SL	UPL Europe	14007	30 Jun 2018

652 thiacloprid

Product	Approval holder	MAPP No.	Expiry Date
Pintail	Matrix	16729	17 Aug 2018
Rana SC	Servem	17211	17 Aug 2018
Scabiya	Euro	16787	17 Aug 2018
Standon Zero Tolerance	Standon	17551	17 Aug 2018
Thiaclomex 480 SC	MAC	15900	31 Oct 2019

653 thiamethoxam

Product	Approval holder	MAPP No.	Expiry Date
Centric	Syngenta	13954	31 Oct 2020
Cruiser 70 WS	Syngenta	17338	31 Oct 2020
Leptom 250 WDG	Agrifarm	16717	31 Oct 2020

654 thifensulfuron-methyl

Product	Approval holder	MAPP No.	Expiry Date
Harmony SX	DuPont	12181	31 Dec 2019
Prospect SX	DuPont	12212	31 Dec 2019

655 thifensulfuron-methyl + tribenuron-methyl

Product	Approval holder	MAPP No.	Expiry Date
Calibre SX	Certis	15032	31 Dec 2019

656 thiophanate-methyl

Product	Approval holder	MAPP No.	Expiry Date
Thiofin WG	Q-Chem	15384	30 Apr 2020

657 thiram

Product	Approval holder	MAPP No.	Expiry Date
Thiraflo	Chemtura	13338	31 Dec 2021
Thiraflo	Certis	16499	31 Dec 2021

658 tolclofos-methyl

Product	Approval holder	MAPP No.	Expiry Date
Basilex	Everris Ltd	07494	31 Jul 2017

SECTION 3

Products also Registered

Product	Approval holder	MAPP No.	Expiry Date
Rizolex 10D	Interfarm	14204	31 Dec 2021
Rizolex 50 WP	Interfarm	14217	31 Dec 2021
Rizolex Flowable	Interfarm	14207	31 Dec 2021

662 tri-allate
Avadex Excel 15G	Gowan	16998	30 Jun 2022

665 tribenuron-methyl
Helmstar A 75 WG	Rotam	15070	30 Apr 2019
Helmstar A 75 WG	Belchim	17077	30 Apr 2020
Nuance	Headland	14813	30 Apr 2020
Quantum	DuPont	15190	30 Apr 2020
Quantum SX	DuPont	15189	30 Apr 2020
Taxi	Nufarm UK	15733	30 Apr 2020
TBM 75 WG	Sharda	17326	10 Nov 2019
Toscana	Proplan-Plant	17726	30 Apr 2020
Tribe 500	Euro	16888	30 Apr 2020
Tribun 75 WG	Helm	16995	31 May 2019
Tribun 75 WG	Belchim	17090	30 Apr 2020
Trimeo 75 WG	Belchim	17076	30 Apr 2020
Trimeo 75 WG	Helm	16719	30 Apr 2019

666 Trichoderma asperellum (Strain T34)
T34 Biocontrol	Fargro	15603	30 Sep 2017

667 Trichoderma harzianum (Strain T22)
Trianum G	Koppert	16740	31 Oct 2021
Trianum P	Koppert	16741	31 Oct 2021

668 triclopyr
Topper	Agriphar	15719	31 Oct 2020

669 trifloxystrobin
Flint	Bayer CropScience	11259	31 Jan 2020
Micro-Turf	Mitsui	12988	31 Jan 2020
Pan Tees	Pan Agriculture	12894	31 Jan 2020

671 triflusulfuron-methyl
Safari	DuPont	17547	30 Jun 2022

672 trinexapac-ethyl
A17600C	Syngenta	17548	31 Oct 2020
Cleancrop Alatrin	Agrii	15196	31 Oct 2020
Cleancrop Alatrin Evo	Agrii	17763	31 Oct 2020
Cleancrop Cutlass	Agrii	16046	31 Oct 2020
Confine NT	Zantra	17770	09 May 2020
Freeze	Headland	15871	31 May 2018
Inter Trinex	UPL Europe	17334	31 Jan 2020
Life Scientific Trinexapac	Life Scientific	17125	31 Oct 2020
Maintain	Headland	15626	31 Oct 2020
Modan	Helm	16628	15 Jul 2018
Moddus ME	Syngenta	17179	31 Oct 2020
Paket 250 EC	Arysta	17436	31 Oct 2020
Scitec	Syngenta	15588	31 Oct 2020
Scythe	Adama	16359	28 Jan 2018
Seize	Headland	16780	31 Oct 2020
Sonis	Syngenta	16891	31 Oct 2020
Standon Pygmy	Standon	15169	30 Jun 2017
Staylow	Unique Marketing	16295	31 Oct 2020
TP 100	Certis	17605	23 May 2020

Products also Registered

Product	Approval holder	MAPP No.	Expiry Date
Trexstar	Belchim	17062	15 Jul 2018
Trexstar	Helm	16677	15 Jul 2018
Trexxus	Belchim	17059	15 Jul 2018
Trexxus	Helm	16781	15 Jul 2018
Trinex 222	Euro	15322	31 Oct 2020
Trinexis	Agriphar	16429	31 Oct 2020
Upkeep	Adama	15225	31 Oct 2020
UPL Trinexapac	UPL Europe	16645	31 Oct 2020
Zira	Terrechem	17467	31 Oct 2020

676 zeta-cypermethrin

Product	Approval holder	MAPP No.	Expiry Date
Fury 10 EW	Belchim	12248	30 Sep 2019
Minuet EW	Belchim	12304	31 Mar 2017
RouteOne Zeta 10	Albaugh UK	15308	31 Dec 2021

SECTION 4
ADJUVANTS

Adjuvants

Adjuvants are not themselves classed as pesticides and there is considerable misunderstanding over the extent to which they are legally controlled under the Food and Environment Protection Act. An adjuvant is a substance other than water which enhances the effectiveness of a pesticide with which it is mixed. Consent C(i)5 under the Control of Pesticides Regulations allows that an adjuvant can be used with a pesticide only if that adjuvant is authorised and on a list published on the HSE adjuvant database: https//secure.pesticides.gov.uk/adjuvants/search.asp. An authorised adjuvant has an *adjuvant number* and may have specific requirements about the circumstances in which it may be used.

Adjuvant product labels must be consulted for full details of authorised use, but the table below provides a summary of the label information to indicate the area of use of the adjuvant. Label precautions refer to the keys given in Appendix 4, and may include warnings about products harmful or dangerous to fish. The table includes all adjuvants notified by suppliers as available in 2017.

Product	Supplier	Adj. No.	Type
Abacus	De Sangosse	A0543	vegetable oil
Contains	20% w/w alkoxylated alcohols -EAC 1, 9% w/w oil (tall oil fatty acids - EAC 2), 53.43% w/w oil (rapeseed fatty acid esters - EAC 3)		
Use with	All approved pesticides on all edible crops when used at half their recommended dose or less, and on all non-edible crops up to their full recommended dose. Also at a maximum concentration of 0.1% with approved pesticides on listed crops up to specified growth stages		
Protective clothing	A, C, H		
Precautions	R36, U05a, U11, U14, U19a, U20b, E15a, E19b, E34, D01, D02, D05, D10a, D12a, H04		
Activator 90	De Sangosse	A0547	non-ionic surfactant/wetter
Contains	375 g/kg alkoxylated alcohols (EAC 1), 375 g/kg alkoxylated alcohols (EAC 2), 150 g/kg oil (tall oil fatty acids - EAC 3)		
Use with	All approved pesticides on all edible crops when used at half their recommended dose or less, and on all non-edible crops up their full recommended dose. Also at a maximum concentration of 0.1% with approved pesticides on listed crops up to specified growth stages		
Protective clothing	A, C, H		
Precautions	R36, R38, R53a, U02a, U04a, U05a, U10, U11, U20b, E15a, E19b, D01, D02, D05, D10a, H04		
Addit	Koppert	A0693	spreader/sticker/wetter
Contains	780.2 g/l oil (rapeseed triglycerides)		
Use with	Mycotal at 0.25% spray solution and all approved pesticides at half or less than half the approved pesticide rate		
Protective clothing	A, F, H		
Precautions	R20, R21, R36, R51, R53a		
Adigor	Syngenta	A0522	wetter
Contains	47% w/w methylated rapeseed oil		
Use with	Topik, Axial, Trazos and Amazon on cereals in accordance with recommendations on the respective herbicide labels		
Protective clothing	A, C, H		
Precautions	R43, R51, R53a, U02a, U05a, U09a, U20b, E15b, E38, D01, D02, D05, D09a, D10c, D12a, H04, H11		

SECTION 4

Adjuvants

Product	Supplier	Adj. No.	Type
AdjiFe	Amega	A0797	wetter
Contains	\multicolumn{3}{l	}{345 g/l ammonium iron (III) citrate (EAC 1), 300 g/l alkyl polyglycosides (EAC 2) and 104.5 g/l ethylene oxide-propylene oxide copolymers (EAC 3)}
Use with	\multicolumn{3}{l	}{All authorised plant protection herbicides and fungicides}
Protective clothing	\multicolumn{3}{l	}{A}
Precautions	\multicolumn{3}{l	}{U02a, U04a, U05a, U14, U15, U19a, E13b, E34, D01, D02, D09a, D10a, M03}
AdjiMin	Amega	A0812	adjuvant
Contains	\multicolumn{3}{l	}{920 g/kg oil (petroleum oils) (EAC 1)}
Use with	\multicolumn{3}{l	}{All approved pesticides at half the approved maximum dose on edible crops and at the full approved dose on non-edible crops and non-crop production at 1% spray volume.}
Protective clothing	\multicolumn{3}{l	}{A}
Precautions	\multicolumn{3}{l	}{U02a, U04a, U05a, U14, U15, U19a, E13b, E34, D01, D02, D09a, D10a, M03}
AdjiSil	Amega	A0763	spreader/wetter
Contains	\multicolumn{3}{l	}{83% w/w trisiloxane organosilicone copolymers (EAC 1)}
Use with	\multicolumn{3}{l	}{All approved pesticides at half the approved maximum dose on edible crops and at the full approved dose on non-edible crops and non-crop production at 0.15% spray volume.}
Protective clothing	\multicolumn{3}{l	}{A, C, H}
Precautions	\multicolumn{3}{l	}{U02a, U04a, U05a, U14, U15, U19a, E13b, E34, D01, D02, D09a, D10a, M03, H11}
AdjiVeg	Amega	A0756	adjuvant
Contains	\multicolumn{3}{l	}{90% w/w fatty acid esters (EAC 1)}
Use with	\multicolumn{3}{l	}{All approved pesticides at half the approved maximum dose on edible crops and at the full approved dose on non-edible crops and non-crop production at 1% spray volume.}
Protective clothing	\multicolumn{3}{l	}{A, C, H}
Precautions	\multicolumn{3}{l	}{U02a, U04a, U05a, U11, U14, U15, E15a, E34, D01, D02, D09a, D10a, M03, H04}
Admix-P	De Sangosse	A0301	wetter
Contains	\multicolumn{3}{l	}{80% w/w trisiloxane organosilicone copolymers (EAC 1)}
Use with	\multicolumn{3}{l	}{A wide range of pesticides applied as corm, tuber, onion and other bulb treatments in seed production and in seed potato treatment}
Protective clothing	\multicolumn{3}{l	}{A, C, H}
Precautions	\multicolumn{3}{l	}{R20, R21, R22a, R36, R43, R48, R51, R58, U11, U15, U19a, E15a, E19b, D01, D02, D05, D10a, D12a, H03, H11}
Amber	Interagro	A0367	vegetable oil
Contains	\multicolumn{3}{l	}{95% w/w methylated rapeseed oil}
Use with	\multicolumn{3}{l	}{Sugar beet herbicides, oilseed rape herbicides, cereal graminicides and a wide range of other pesticides that have a label recommendation for use with authorised adjuvant oils on specified crops. See label for details}
Protective clothing	\multicolumn{3}{l	}{A, C}
Precautions	\multicolumn{3}{l	}{U05a, U20b, E15a, D01, D02, D05, D09a, D10a}

Adjuvants

Product	Supplier	Adj. No.	Type
Arma	Interagro	A0306	penetrant
Contains	500 g/l alkoxylated fatty amine + 500 g/l polyoxyethylene monolaurate		
Use with	Cereal growth regulators, cereal herbicides, cereal fungicides, oilseed rape fungicides and a wide range of other pesticides on specified crops		
Protective clothing	A, C		
Precautions	R51, R58, U05a, E15a, E34, E37, D01, D02, D05, D09a, D10a, H11		
Asu-Flex	Greenaway	A0677	spreader/sticker/wetter
Contains	10.0% w/w rapeseed oil		
Use with	Asulox, Greencrop Found, I T Asulam, Inter Asulam and Spitfire		
Protective clothing	A, C		
Precautions	U11, U12, U15, U20b, E15a, E34, D02		
BackRow	Interagro	A0472	mineral oil
Contains	60% w/w refined paraffinic petroleum oil		
Use with	Pre-emergence herbicides. Refer to label or contact supplier for further details		
Protective clothing	A		
Precautions	R22b, R38, U02a, U05a, U08, U20b, E15a, D01, D02, D05, D09a, D10a, M05b, H03		
Ballista	Interagro	A0524	spreader/wetter
Contains	10% w/w alkoxylated triglycerides		
Use with	All approved pesticides and plant growth regulators for use in winter and spring cereals applied at full rate up to and including GS 52, and at half the approved rate thereafter. May be applied with approved pesticides for other specified crops at full rate up to specified growth stages and at half rate thereafter		
Protective clothing	A		
Precautions	U05a, U20b, E15a, E34, D01, D02, D09a		
Banka	Interagro	A0245	spreader/wetter
Contains	14.6 % w/w alkyl pyrrolidone copolymers and 14.6 % w/w alkyl pyrrolidones		
Use with	Potato fungicides and a wide range of other pesticides on specified crops		
Protective clothing	A, C		
Precautions	R38, R41, R52, R58, U02a, U05a, U11, U19a, U20b, E15a, E34, E37, D01, D02, D05, D09a, H04		
Barramundi	Interagro	A0376	mineral oil
Contains	95% w/w mineral oil		
Use with	Sugar beet herbicides, oilseed rape herbicides, cereal graminicides and a wide range of other pesticides that have a label recommendation for use with authorised adjuvant oils on specified crops. Refer to label or contact supplier for further details		
Protective clothing	A		
Precautions	R22b, R38, U02a, U05a, U08, U20b, E15a, D01, D02, D05, D09a, D10a, M05b, H03		
Binder	Intracrop	A0598	spreader/wetter
Contains	30.0 % w/w alkoxylated alcohols (EAC 1)		
Use with	All approved pesticides at half or less than half the approved pesticide rate and all approved formulations of glyphosate		
Protective clothing	A		
Precautions	U02a, U04a, U05a, U14, U15, U19a, E13b, E34, D01, D02, D09a, D10a, M03		

SECTION 4

Adjuvants

Product	Supplier	Adj. No.	Type
Bio Syl	Intracrop	A0773	spreader/sticker/wetter
Contains	32.67% w/w alkoxylated alcohols (EAC 1) and 1.0% w/w trisiloxane organosilicone copolymers (EAC 2)		
Use with	Recommended rates of approved pesticides on non-crop and non-edible crops; when used on edible crops the dose of the approved pesticide must be half or less than half the approved dose rate		
Protective clothing	A, C		
Precautions	R36, R38, U05a, U08, U20c, E15a, D01, D02, D10a, H04		
Bioduo	Intracrop	A0606	wetter
Contains	700 g/l alkoxylated alcohols (EAC 1) and 150 g/l rapeseed fatty acids (EAC 2)		
Use with	A wide range of pesticides used in grassland, agriculture and horticulture and with pesticides used in non-crop situations		
Protective clothing	A, C		
Precautions	R22a, R36, R38, U05a, U08, U20c, E13c, E34, D01, D02, D09a, D10a, M03, H03, H04, H08		
Biofilm	Intracrop	A0634	anti-drift agent/anti-transpirant/sticker/UV screen/wetter
Contains	96.0% w/w pinene oligomers (EAC 1)		
Use with	All approved fungicides and insecticides on edible crops and all approved formulations of glyphosate used pre-harvest on wheat, barley, oilseed rape, stubble, and in non-crop situations and grassland destruction		
Protective clothing	A, C		
Precautions	U19a, U20c, E13c, D09a, D11a		
BioPower	Bayer CropScience	A0617	wetter
Contains	6.7% w/w 3,6-dioxaeicosylsulphate sodium salt and 20.2% w/w 3,6-dioxaoctadecylsulphate sodium salt		
Use with	Atlantis and all other approved cereal herbicides		
Protective clothing	A, C		
Precautions	R36, R38, U02a, U05a, U08, U13, U19a, U20b, E13c, E34, D01, D02, D05, D09a, D10a, H04		
Biothene	Intracrop	A0633	anti-drift agent/anti-transpirant/sticker/UV screen/wetter
Contains	96.0% w/w pinene oligomers (EAC 1);		
Use with	All approved fungicides and insecticides on edible crops up to 30 d before harvest, and all approved formulations of glyphosate used pre-harvest on wheat, barley, oilseed rape, stubble, and in non-crop situations and grassland destruction. Must not be used in mixture with adjuvant oils or surfactants		
Precautions	U19a, U20c, E13c, D09a, D11a		

Adjuvants

Product	Supplier	Adj. No.	Type
Bond	De Sangosse	A0556	extender/sticker/wetter
Contains	10% w/w alkoxylated alcohols (EAC 1), 45% w/w styrene-butadiene copolymers (EAC 2)		
Use with	All approved potato blight fungicides. Also with all approved pesticides on all edible crops when used at half their recommended dose or less, and on all non-edible crops up their full recommended dose. Also at a maximum concentration of 0.14% with approved pesticides on listed crops up to specified growth stages		
Protective clothing	A, C, H		
Precautions	R36, R38, U11, U14, U16b, U19a, E15a, E19b, D01, D02, D05, D10a, D12a, H04		
Broad-Flex	Greenaway	A0680	spreader/sticker/wetter
Contains	10.0% w/w rapeseed oil		
Use with	Broad Sword or Green Guard		
Protective clothing	A, C		
Precautions	U11, U12, U15, U20b, E15a, E34, D02		
Byo-Flex	Greenaway	A0545	sticker/wetter
Contains	10.0% w/w rapeseed oil		
Use with	GLY 490 (MAPP 12718)		
Protective clothing	A, C		
Precautions	U11, U12, U15, U20b, E34		
C-Cure	Interagro	A0467	mineral oil
Contains	60% w/w refined mineral oil		
Use with	Pre-emergence herbicides		
Protective clothing	A		
Precautions	R22b, R38, U02a, U05a, U08, U20b, E15a, D01, D02, D05, D09a, D10a, M05b, H03		
Ceres Platinum	Interagro	A0445	penetrant/spreader
Contains	500 g/l alkoxylated fatty amine + 500 g/l polyoxyethylene monolaurate		
Use with	Cereal growth regulators, cereal herbicides, cereal fungicides, oilseed rape fungicides and a wide range of other pesticides on specified crops		
Protective clothing	A		
Precautions	R51, R58, U05a, E15a, E34, D01, D02, D05, D09a, D10a, H11		
Clayton Astra	Clayton	A0724	penetrant/spreader
Contains	500 g/l alkoxylated sorbitan esters and 500 g/l alkoxylated tallow amines		
Use with	All approved pesticides up to a maximum conc of 0.15% of the total spray volume		
Protective clothing	A, H		
Precautions	R51, R53a, U05a, E15a, E34, D01, D02, D05, D09a, D10b, D12a, H11		
Clayton Dolmen	Clayton	A0725	adjuvant/spreader
Contains	64% w/w trisiloxane organosilicone copolymers		
Use with	All approved pesticides up to a maximum conc of 0.2% of the total spray volume		
Protective clothing	A, C, H		
Precautions	R20, R21, R36, R38, R51, R53a, U02a, U05a, U09a, U11, U16b, U19a, U20b, D01, D02, D05, D09a, D10a, H03, H11		

SECTION 4

Adjuvants

Product	Supplier	Adj. No.	Type
Clayton Union	Clayton	A0686	adjuvant
Contains	47% w/w methylated rapeseed oil		
Use with	Clayton Tonto as 0.5% of total spray volume		
Protective clothing	A, H		
Precautions	R43, R50, R53a, U02a, U05a, U09a, U20b, E15b, D01, D02, D05, D10c, D12a, H04, H11		
Codacide Oil	Microcide	A0629	vegetable oil
Contains	95% w/w oil (rapeseed triglycerides) (EAC 1)		
Use with	All approved pesticides and tank mixes. See label for details		
Protective clothing	A, C		
Precautions	U20b, D09a, D10b		
Companion Gold	Agrovista	A0723	acidifier/buffering agent/drift retardant/extender
Contains	16% w/w ammonium sulphate + 0.95% w/w polyacrylamide		
Use with	Diquat or glyphosate on oilseed rape at 0.5% solution; glyphosate on wheat, rye or triticale at 0.5% solution and with all approved pesticides on non-edible crops at 1% solution or all approved pesticides at 50% dose rate or less on edible crops		
Protective clothing	A, C, H		
Precautions	U05a, U14, U15, U19a, U19c, U20b, E15a, D02, D05, D09a, D10a		
Compliment	United Agri	A0705	spreader/vegetable oil/wetter
Contains	75% w/w mixed fatty acid esters of rapeseed oil		
Use with	All approved pesticides in non-edible crops, all approved pesticides on edible crops when used at half or less their recommended rate, morpholine or triazine fungicides on cereals, and with all pesticides on listed crops up to specified growth stages		
Protective clothing	A, C		
Precautions	R43, R52, R53a, U14, E38, D01, D05, D07, D08, H04		
Contact Plus	Interagro	A0418	mineral oil
Contains	95% w/w mineral oil		
Use with	Sugar beet herbicides, oilseed rape herbicides, cereal graminicides and a wide range of other pesticides that have a label recommendation for use with authorised adjuvant oils on specified crops. Refer to label or contact supplier for further details		
Protective clothing	A, C		
Precautions	R22b, R38, U02a, U05a, U08, U20b, E15a, D01, D02, D09a, D10a, M05b, H03		
County Mark	Greenaway	A0689	sticker/wetter
Contains	10% w/w rapeseed oil		
Use with	'Greenaway Gly-490' (MAPP 12718) and all approved 490 g/l glyphosate products		
Protective clothing	A, C		
Precautions	U11, U12, U15, U20b, E34		

Adjuvants

Product	Supplier	Adj. No.	Type
Cropspray 11-E	Petro-Lube	A0537	adjuvant/mineral oil
Contains	99% w/w oil (petroleum oils) (EAC 1)		
Use with	All approved pesticides on all edible crops when used at half their recommended dose or less, and on all non-edible crops up to their full recommended dose. Also with listed herbicides on a range of specified crops and with all pesticides on a range of specified crops up to specified growth stages.		
Protective clothing	A, C		
Precautions	R22a, U10, U16b, U19a, E15a, E19b, E34, E37, D01, D02, D05, D10a, D12a, M05b, H03		
Dash HC	BASF	A0729	non-ionic surfactant/wetter
Contains	348.75 g/l oil (fatty acid esters) and 209.25 g/l alkoxylated alcohols-phosphate esters		
Use with	For use with Cleranda at 1.0 l/ha in a water volume of 100 - 400 l/ha on Clearfield oilseed rape		
Protective clothing	A, C, H		
Precautions	U05a, U12, U14, D01, D02, D10c, M05b, H03		
Designer	De Sangosse	A0660	drift retardant/extender/sticker/wetter
Contains	25% w/w styrene-butadiene copolymers (EAC 1), 7.1% w/w trisiloxane organosilicone copolymers (EAC 2)		
Use with	A wide range of fungicides, insecticides and trace elements for cereals and specified agricultural and horticultural crops		
Protective clothing	A, C, H		
Precautions	R36, R38, R52, U02a, U11, U14, U15, U19a, E15a, E37, D01, D02, D05, D09a, D10a, H04		
Desikote Max	Taminco	A0666	anti-transpirant/extender/sticker/wetter
Contains	400 g/l di-1-p-menthene		
Use with	Approved pesticides on all edible crops (except herbicides on peas) and as an anti-transpirant on vegetables before transplanting, evergreens, deciduous trees, shrubs, bushes and on turf		
Protective clothing	A		
Precautions	R38, R50, R53a, U05a, U14, U20b, E15a, E34, E37, D01, D09a, D10b, D12b, M03, H04, H11		
Diagor	AgChem Access	A0671	wetter
Contains	47% w/w methylated rapeseed oil		
Use with	Topik, Axial, Trazos, Amazon and Viscount on cereals in accordance with recommendations on the respective herbicide labels		
Protective clothing	A, C, H		
Precautions	R43, R51, R53a, U02a, U05a, U09a, U20b, E15b, E38, D01, D02, D05, D09a, D10c, D12a, H04, H11		

SECTION 4

Adjuvants

Product	Supplier	Adj. No.	Type
Drill	De Sangosse	A0544	adjuvant
Contains	\multicolumn{3}{l	}{15% w/w alkoxylated alcohols (EAC 1), 7.5% w/w oil (tall oil fatty acids - EAC 2), 63.34% w/w oil (rapeseed fatty acid esters) (EAC 3)}	
Use with	\multicolumn{3}{l	}{All approved pesticides on all edible crops when used at half their recommended dose or less, and on all non-edible crops up to their full recommended dose. Also at a maximum concentration of 0.1% with approved pesticides on listed crops up to specified growth stages}	
Protective clothing	A, C, H		
Precautions	\multicolumn{3}{l	}{R36, U05a, U11, U14, U19a, U20b, E15a, E19b, E34, D01, D02, D05, D10a, D12a, H04}	
Eco-flex	Greenaway	A0696	sticker/wetter
Contains	\multicolumn{3}{l	}{10% w/w refined rapeseed oil}	
Use with	\multicolumn{3}{l	}{Approved formulations of glyphosate, 2,4-D}	
Protective clothing	A, C		
Precautions	\multicolumn{3}{l	}{U11, U12, U15, U20b, E34}	
Elan Xtra	Intracrop	A0735	spreader/wetter
Contains	\multicolumn{3}{l	}{530 g/l ethylene oxide-propylene oxide copolymers (EAC 1) and 415 g/l trisiloxane organosilicone copolymers (EAC 2)}	
Use with	\multicolumn{3}{l	}{All approved pesticides on non-edible crops and non-crop production and with half dose of all approved pesticides on edible crops}	
Protective clothing	A		
Emerald	Intracrop	A0636	anti-drift agent/anti-transpirant/extender/UV screen
Contains	\multicolumn{3}{l	}{96.0% w/w pinene oligomers (EAC 1)}	
Use with	\multicolumn{3}{l	}{Recommended rates of approved pesticides up to the growth stage indicated for specified crops, and with half or less than the recommended rate on these crops after the stated growth stages. Also for use alone on transplants, turf, fruit crops, glasshouse crops and Christmas trees. Must not be used in mixture with adjuvant oils or surfactants}	
Protective clothing	A, H		
Precautions	\multicolumn{3}{l	}{U19a, U20c, E13c, E37, D09a, D11a}	
Euroagkem Pen-e-trate	Intracrop	A0564	spreader/wetter
Contains	\multicolumn{3}{l	}{350 g/l propionic acid (EAC 1)}	
Use with	\multicolumn{3}{l	}{All approved pesticides which have a recommendation for use with a wetting agent; pesticides must be used at half approved dose rate or less on edible crops, up to full dose rate on non-edible crops}	
Protective clothing	A, C		
Precautions	\multicolumn{3}{l	}{U05a, U08, U11, U14, U15, U19a, U20c, E13e, E15a, D01, D02, D09a, D10b, M03, H05}	
Felix	Intracrop	A0178	spreader/wetter
Contains	\multicolumn{3}{l	}{600 g/l alkoxylated alcohols (EAC 1)}	
Use with	\multicolumn{3}{l	}{Mecoprop, 2,4-D in cereals and amenity turf, and a range of grass weedkillers in agriculture. See label for details}	
Protective clothing	A, C		
Precautions	\multicolumn{3}{l	}{R36, R38, U05a, U08, U20a, E15a, D01, D02, D09a, D10a, H04, H08}	

Adjuvants

Product	Supplier	Adj. No.	Type
Firebrand	Barclay	-	fertiliser/water conditioner
Contains	500 g/l ammonium sulphate		
Use with	Use at 0.5% v/v in the spray solution with glyphosate		
Precautions	U08, U20a, E15a, D09a, D10b		
Gateway	United Agri	A0651	extender/sticker/wetter
Contains	73% w/v synthetic latex solution and 8.5% w/v polyether modified trisiloxane		
Use with	All approved pesticides in crops not destined for human or animal consumption, all approved pesticides on edible crops when used at half or less their recommended rate, and with all pesticides on listed crops up to specified growth stages		
Protective clothing	A, C, H		
Precautions	R41, R52, R53a, U11, U15, E37, E38, D01, D05, D07, D08, H04		
Gly-Flex	Greenaway	A0588	spreader/sticker/wetter
Contains	95% w/w refined rapeseed oil		
Use with	GLY-490 (MAPP 12718) or any approved formulations of 490 g/l glyphosate		
Protective clothing	A, C		
Precautions	R36, U11, U12, U15, U20b, E13c, E34, E37, H04		
Gly-Plus A	Greenaway	A0736	spreader/sticker/wetter
Contains	10.0% w/w oil (rapeseed triglycerides) (EAC 1)		
Use with	Any approved 360 g/l glyphosate		
Protective clothing	A, C		
Precautions	R36, U05a, U08, U11, U12, U15, U20b, E13c, E34, E37, E38, H04		
Green Gold	Intracrop	A0250	spreader/wetter
Contains	950 g/l oil (rapeseed triglycerides) (EAC 1)		
Use with	All pesticides which have a recommendation for the addition of a wetter/spreader		
Precautions	U08, U20b, E13c, E34, D09a, D10b		
Grenadier	Intracrop	A0608	spreader/sticker/wetter
Contains	700 g/l alkoxylated alcohols (EAC 1) and 150 g/l rapeseed fatty acids (EAC 2);		
Use with	All approved pesticides on edible and non-edible crops when used at half their approved dose or less.		
Protective clothing	A, C		
Grounded	Helena	A0456	mineral oil
Contains	732 g/l petroleum oils		
Use with	All approved pesticides on edible and non-edible crops when used at half their approved dose or less. Also with approved pesticides on specified crops, up to specified growth stages, at up to their full approved dose		
Protective clothing	A, C		
Precautions	R53a, U02a, U05a, U08, U20b, E15a, E34, E37, D01, D02, D05, D09a, D10c, H11		
Headland Fortune	Headland	A0703	penetrant/spreader/vegetable oil/wetter
Contains	75% w/w mixed methylated fatty acid esters of seed oil and N-butanol		
Use with	Herbicides and fungicides in a wide range of crops. See label for details		
Protective clothing	A, C		
Precautions	R43, U02a, U05a, U14, U20a, E15a, E34, E38, D01, D02, D05, D09a, D10b, H04		

Adjuvants

Product	Supplier	Adj. No.	Type
Headland Guard 2000	Headland	A0652	extender/sticker
Contains	10% w/w styrene/butadiene co-polymers		
Use with	All approved pesticides on all edible crops when used at half their recommended dose or less, and on all non-edible crops up to their full recommended dose. Also at a maximum concentration of 0.1% with approved pesticides on listed crops up to specified growth stages		
Protective clothing	A, C		
Precautions	E15a, D01, D02		
Headland Guard Pro	Headland Amenity	A0653	extender/sticker
Contains	52% w/w synthetic latex solution		
Use with	All approved pesticides, micronutrients and sea-weed based plant growth stimulants in amenity grass, managed amenity turf and amenity vegetation		
Protective clothing	A		
Precautions	U05a, U08, U19a, U20b, E13b, E34, E37, D01, D02, D05, D10b		
Headland Intake	Headland	A0074	penetrant
Contains	450 g/l propionic acid		
Use with	All approved pesticides on any crop not intended for human or animal consumption, and with all approved pesticides on beans, peas, edible podded peas, oilseed rape, linseed, sugar beet, cereals (except triazole fungicides), maize, Brussels sprouts, potatoes, cauliflowers. See label for detailed advice on timing on these crops		
Protective clothing	A, C		
Precautions	R34, U02a, U05a, U10, U11, U14, U15, U19a, E34, D01, D02, D05, D09b, D10b, M04a, H05		
Headland Rhino	Headland	A0328	wetter
Contains	85% w/w polyalkylene oxide modified heptamethyl siloxane		
Use with	Any herbicide, systemic fungicide, systemic insecticide or plant growth regulator (except any product applied in or near water) where the use of a wetting, spreading and penetrating surfactant is recommended to improve foliar coverage		
Protective clothing	A, C		
Precautions	R21, R22a, R38, R41, R43, R51, R58, U02a, U05a, U08, U11, U14, U19a, U20b, E13c, E34, D01, D02, D05, D09a, D10b, M03, H03, H11		
Herbi-AKtiv	Global Adjuvants	A0821	spreader/wetter
Contains	57.0 % w/w ethylene oxide-propylene oxide copolymers (EAC 1)		
Use with	All authorised plant protection products at half or less than half the authorised plant protection product rate		
Protective clothing	A, C, H, M		
Precautions	U05a, U20c, E34, D01, D02, D10a, M04a		
Impala	Intracrop	A0607	spreader/sticker/wetter
Contains	700 g/l alkoxylated alcohols (EAC 1) and 150 g/l rapeseed fatty acids (EAC 2)		
Use with	All approved pesticides at half or less than half the approved pesticide rate on edible crops and all approved pesticides on non-edible crops		
Protective clothing	A, H		

Adjuvants

Product	Supplier	Adj. No.	Type
Intracrop Agwet GTX	Intracrop	A0646	spreader/wetter

Contains	500 g/l alkoxylated alcohols
Use with	All approved pesticides on non-edible crops, all approved pesticides on edible crops at half or less than half the pesticide rate and all approved pesticides with a recommendation for use with a non-ionic or wetting agent.
Protective clothing	A, C, H
Precautions	H318, U05a, U11, U14, U15, U19a, U20a, M04c, H226, H302

Product	Supplier	Adj. No.	Type
Intracrop BLA	Intracrop	A0655	anti-drift agent/anti-transpirant/extender/sticker/UV screen

Contains	22.0% w/w styrene-butadiene copolymers (EAC 1)
Use with	All potato blight fungicides. Also with recommended rates of approved pesticides in certain non-crop situations and up to the growth stage indicated for specified crops, and with half or less than the recommended rate on these crops after the stated growth stages. Also with pesticides in grassland at half or less their recommended rate
Protective clothing	A, C
Precautions	U08, U20b, E13c, E34, E37, D01, D05, D09a, D10a

Product	Supplier	Adj. No.	Type
Intracrop Bla-Tex	Intracrop	A0656	extender/sticker

Contains	22.0% w/w styrene-butadiene copolymers (EAC 1)
Use with	All approved pesticides on non-edible crops and non crop production, with half dose of all approved pesticides on edible crops and with blight fungicides in potatoes
Protective clothing	A, C
Precautions	U08, U20b, E13c, E34, E37, D01, D05, D09a, D10a

Product	Supplier	Adj. No.	Type
Intracrop Boost	Intracrop	A0774	spreader/sticker/wetter

Contains	32.67% w/w alkoxylated alcohols (EAC 1) and 1.0% w/w trisiloxane organosilicone copolymers (EAC 2)
Use with	All approved pesticides at half or less than half the approved pesticide rate on edible crops and all approved pesticides on non-edible crops
Protective clothing	A, C

Product	Supplier	Adj. No.	Type
Intracrop Brigand	Intracrop	A0718	spreader/wetter

Contains	30.0% w/w alkoxylated alcohols (EAC 1);
Use with	All approved pesticides at half or less than half the approved pesticide rate and all approved formulations of glyphosate
Protective clothing	A, C

Product	Supplier	Adj. No.	Type
Intracrop Cogent	Intracrop	A0775	spreader/sticker/wetter

Contains	32.67% w/w alkoxylated alcohols (EAC 1) and 1.0% w/w trisiloxane organosilicone copolymers (EAC 2)
Use with	All approved pesticides at half or less than half the approved pesticide rate on edible crops and all approved pesticides on non-edible crops
Protective clothing	A, C

Product	Supplier	Adj. No.	Type
Intracrop Dictate	Intracrop	A0673	adjuvant

Contains	91.0% w/w oil (rapeseed fatty acid esters) (EAC 1)
Use with	All approved pesticides on non-edible crops; all approved pesticides applied at half or less than half dose on edible crops
Protective clothing	A, C
Precautions	U19a, U20b, D05, D09a, D10b

SECTION 4

Adjuvants

Product	Supplier	Adj. No.	Type
Intracrop Evoque	Intracrop	A0754	spreader/wetter

Contains: 652 g/l oil (rapeseed fatty acid esters) (EAC 1), 112 g/l trisiloxane organosilicone copolymers (EAC 2), 22.5 g/l alkoxylated alcohols (EAC 3) and 19.1 g/l alkoxylated alcohols (EAC 4)

Use with: All authorised plant protection products on non-edible crops and all authorised plant protection products at half or less than half the authorised plant protection product rate on edible crops. Max conc is 0.2% of spray volume.

Protective clothing: A, C

Intracrop F16	Intracrop	A0752	spreader/wetter

Contains: 652 g/l oil (rapeseed fatty acid esters) (EAC 1), 112 g/l trisiloxane organosilicone copolymers (EAC 2), 22.5 g/l alkoxylated alcohols (EAC 3) and 19.1 g/l alkoxylated alcohols (EAC 4)

Use with: All authorised plant protection products on non-edible crops and all authorised plant protection products at half or less than half the authorised plant protection product rate on edible crops. Max conc is 0.2% of spray volume.

Protective clothing: A, C

Intracrop Green Oil	Intracrop	A0262	wetter

Contains: 950 g/l oil (rapeseed triglycerides) (EAC 1)

Use with: Recommended rates of approved pesticides up to the growth stage indicated for specified crops, and with half or less than the recommended rate on these crops after the stated growth stages. Also for use with pesticides on grassland at half their recommended rate

Protective clothing: A, C

Precautions: U08, U20b, E13c, E37, D09a, D10b

Intracrop Impetus	Intracrop	A0647	spreader/wetter

Contains: 50.0 % w/w alkoxylated alcohols

Use with: All approved pesticides on non-edible crops, all approved pesticides on edible crops at half or less than half the pesticide rate and all approved pesticides with a recommendation for use with a non-ionic or wetting agent.

Protective clothing: A, C, H

Precautions: H318, U05a, U11, U14, U15, U19a, U20a, M04c, H226, H302

Intracrop Inca	Intracrop	A0784	adjuvant

Contains: 840 g/l oil (rapeseed fatty acid esters) (EAC 1)

Use with: All approved pesticides at half or less than half the approved pesticide rate on edible crops and all approved pesticides on non-edible crops

Protective clothing: A, C

Intracrop Incite	Intracrop	A0785	adjuvant

Contains: 840 g/l oil (rapeseed fatty acid esters) (EAC 1)

Use with: All approved pesticides at half or less than half the approved pesticide rate on edible crops and all approved pesticides on non-edible crops

Protective clothing: A, C

Intracrop Mica AF	Intracrop	A0808	spreader/sticker/wetter

Contains: 250 g/l ethylene oxide-propylene oxide copolymers (EAC 1), 70.2 g/l styrene-butadiene copolymers (EAC 2) and 41.5 g/l trisiloxane organosilicone copolymers (EAC 3)

Use with: All approved pesticides at half or less than half the approved rate

Protective clothing: A, C

Precautions: U05a, U08, U20c, E13e, D01, D02, D09a, D10a

Adjuvants

Product	Supplier	Adj. No.	Type
Intracrop Neotex	Intracrop	A0657	anti-drift agent/anti-transpirant/extender/sticker/UV screen

Contains	22.0 % w/w styrene-butadiene copolymers
Use with	Recommended rates of approved pesticides up to specified growth stages of a wide range of agricultural arable and horticultural crops, and for non-crop uses. Use on edible crops beyond specified growth stages, and in grass, should only be with half recommended rates of the pesticide or less. See label for details of growth stage restrictions. In addition may be used with all potato blight fungicides at their recommended rates of use up to the latest recommended timing of the fungicide
Protective clothing	A, C
Precautions	U08, U20b, E13c, E34, E37, D01, D05, D09a, D10a

Intracrop Novatex	Intracrop	A0658	anti-drift agent/anti-transpirant/extender/sticker/UV screen

Contains	22.0 % w/w styrene-butadiene copolymers
Use with	Recommended rates of approved pesticides up to the growth stage indicated for specified crops, and with half or less than the recommended rate on these crops after the stated growth stages. Also for use with recommended rates of pesticides on grassland and specified non-crop situations
Protective clothing	A, C
Precautions	U08, U20b, E13c, E34, E37, D01, D09a, D10a

Intracrop Perm-E8	Intracrop	A0565	spreader/wetter

Contains	42.0% w/w propionic acid (EAC 1)
Use with	All approved formulations of chlormequat, all approved formulations of glyphosate, diquat, fenoxaprop-P-ethyl, tralkoxydim, clodinafop-propargyl, fluazifop-P-butyl, cycloxydim and propaquizafop at half or less than half the approved pesticide rate in edible crops, at full rate in non-edible crops.
Protective clothing	A, C
Precautions	U02a, U04a, U05a, U08, U10, U11, U13, U14, U15, U19a, U20b, D01, D02, D09a, D10a, H05

Intracrop Predict	Intracrop	A0503	adjuvant/vegetable oil

Contains	91.0% w/w oil (rapeseed fatty acid esters) (EAC 1)
Use with	All approved pesticides on non-edible crops, and all pesticides approved for use on growing edible crops when used at half recommended dose or less. On specified crops product may be used at a maximum spray concentration of 1% with approved pesticides at their full approved rate up to the growth stages shown in the label
Protective clothing	A, C
Precautions	U19a, U20b, D05, D09a, D10b

Intracrop Quad	Intracrop	A0753	spreader/wetter

Contains	652 g/l oil (rapeseed fatty acid esters) (EAC 1), 112 g/l trisiloxane organosilicone copolymers (EAC 2), 22.5 g/l alkoxylated alcohols (EAC 3) and 19.1 g/l alkoxylated alcohols (EAC 4)
Use with	All approved pesticides on non-edible crops, and all pesticides approved for use on growing edible crops when used at half recommended dose or less.
Protective clothing	A, C

SECTION 4

Adjuvants

Product	Supplier	Adj. No.	Type
Intracrop Quartz	Intracrop	A0776	spreader/sticker/wetter
Contains	32.67% w/w alkoxylated alcohols (EAC 1) and 1.0% w/w trisiloxane organosilicone copolymers (EAC 2)		
Use with	All approved pesticides on non-edible crops, and all pesticides approved for use on growing edible crops when used at half recommended dose or less.		
Protective clothing	A, C		
Intracrop Questor	Intracrop	A0495	activator/non-ionic surfactant/spreader
Contains	750 g/l ethylene oxide-propylene oxide copolymers (EAC1)		
Use with	All approved pesticides on non-edible crops and pesticides used in non-crop production, and all pesticides approved for use on growing edible crops when used at half recommended dose or less. On specified crops product may be used at a maximum spray concentration of 0.3% with approved pesticides at their full approved rate up to the growth stages shown in the label		
Protective clothing	A, C		
Precautions	R36, R38, U04a, U05a, U08, U19a, E13b, D01, D02, D09a, D10b, M03, M05a, H04		
Intracrop Rapide Beta	Intracrop	A0672	wetter
Contains	350 g/l propionic acid (EAC 1) and 100 g/l alkoxylated alcohols (EAC 2)		
Use with	All approved pesticides which have a recommendation for use with a wetting agent; pesticides must be used at half approved dose rate or less on edible crops, up to full dose rate on non-edible crops		
Protective clothing	A, C		
Precautions	U02a, U04a, U05a, U08, U10, U11, U13, U14, U15, U19a, U20b, E15a, D01, D02, D09a, D10a, M05a, H05		
Intracrop Retainer NF	Intracrop	A0711	adjuvant
Contains	91.0 % w/w oil (rapeseed fatty acid esters)		
Use with	All approved pesticides at half or less than half the approved pesticide rate on edible crops and all approved pesticides on non-edible crops		
Protective clothing	A, C		
Intracrop Rigger	Intracrop	A0783	adjuvant/vegetable oil
Contains	840 g/l oil (rapeseed fatty acid esters)		
Use with	All approved pesticides on non-edible crops, and all pesticides approved for use on growing edible crops when used at half recommended dose or less. On specified crops product may be used at a maximum spray concentration of 1.78% with approved pesticides at their full approved rate up to the growth stages shown in the label		
Protective clothing	A, C		
Precautions	U19a, U20b, D05, D09a, D10b		
Intracrop Rustler	Intracrop	A0777	spreader/sticker/wetter
Contains	32.67% w/w alkoxylated alcohols (EAC 1) and 1.0% w/w trisiloxane organosilicone copolymers (EAC 2)		
Use with	All approved pesticides on non-edible crops, and all pesticides approved for use on growing edible crops when used at half recommended dose or less.		
Protective clothing	A, C		

Adjuvants

Product	Supplier	Adj. No.	Type
Intracrop Salute AF	Intracrop	A0809	spreader/sticker/wetter
Contains	\multicolumn{3}{l	}{250 g/l ethylene oxide-propylene oxide copolymers (EAC 1), 70.2 g/l styrene-butadiene copolymers (EAC 2) and 41.5 g/l trisiloxane organosilicone copolymers (EAC 3)}
Use with	\multicolumn{3}{l	}{All approved pesticides on non-edible crops, and all pesticides approved for use on growing edible crops when used at half recommended dose or less.}
Protective clothing	\multicolumn{3}{l	}{A, C}
Precautions	\multicolumn{3}{l	}{U05a, U08, U20c, E13e, E40c, D01, D02, D09a, D10a}
Intracrop Sapper AF	Intracrop	A0810	spreader/sticker/wetter
Contains	\multicolumn{3}{l	}{250 g/l ethylene oxide-propylene oxide copolymers (EAC 1), 70.2 g/l styrene-butadiene copolymers (EAC 2) and 41.5 g/l trisiloxane organosilicone copolymers (EAC 3)}
Use with	\multicolumn{3}{l	}{All approved pesticides on non-edible crops, and all pesticides approved for use on growing edible crops when used at half recommended dose or less.}
Protective clothing	\multicolumn{3}{l	}{A, C}
Precautions	\multicolumn{3}{l	}{U05a, U08, U20c, E13e, E40c, D01, D02, D09a, D10a}
Intracrop Saturn	Intracrop	A0494	activator/non-ionic surfactant/spreader
Contains	\multicolumn{3}{l	}{750 g/l ethylene oxide-propylene oxide copolymers}
Use with	\multicolumn{3}{l	}{All approved pesticides on non-edible crops and pesticides used in non-crop production, and all pesticides approved for use on growing edible crops when used at half recommended dose or less. On specified crops product may be used at a maximum spray concentration of 0.3% with approved pesticides at their full approved rate up to the growth stages shown in the label}
Protective clothing	\multicolumn{3}{l	}{A, C}
Precautions	\multicolumn{3}{l	}{R36, R38, U04a, U05a, U08, U19a, E13b, E40c, D01, D02, D09a, D10b, M03, M05a, H04}
Intracrop Signal XL	Intracrop	A0659	anti-drift agent/anti-transpirant/extender/sticker/UV screen
Contains	\multicolumn{3}{l	}{22.0% w/w styrene-butadiene copolymers (EAC 1)}
Use with	\multicolumn{3}{l	}{All approved pesticides on non-edible crops and non crop production uses; All approved pesticides at half or less than half the approved pesticide rate on edible crops}
Protective clothing	\multicolumn{3}{l	}{A, C}
Precautions	\multicolumn{3}{l	}{U05a, U19a, U20b, E13c, E34, D01, D02, D09a, D10b, M04a, M05a}
Intracrop Sprinter	Intracrop	A0513	spreader/wetter
Contains	\multicolumn{3}{l	}{19.0 % w/v alkoxylated alcohols}
Use with	\multicolumn{3}{l	}{Recommended rates of approved pesticides up to the growth stage indicated for specified crops, and with half or less than the recommended rate on these crops after the stated growth stages. Also with herbicides on managed amenity turf at recommended rates, and on grassland at half or less than recommended rates}
Protective clothing	\multicolumn{3}{l	}{A, C}
Precautions	\multicolumn{3}{l	}{R36, R38, R41, U05a, U08, U11, U14, U15, U20c, E15a, D01, D02, D09a, D10b, M03, H04}

SECTION 4

Adjuvants

Product	Supplier	Adj. No.	Type
Intracrop Status	Intracrop	A0506	adjuvant/vegetable oil

Contains	91.0 % w/w oil (rapeseed fatty acid esters)
Use with	All approved pesticides on non-edible crops, and all pesticides approved for use on growing edible crops when used at half recommended dose or less. On specified crops product may be used at a maximum spray concentration of 1.0% with approved pesticides at their full approved rate up to the growth stages shown in the label
Protective clothing	A, C
Precautions	U19a, U20b, D05, D09a, D10b

Intracrop Stay-Put	Intracrop	A0507	vegetable oil

Contains	91.0 % w/w oil (rapeseed fatty acid esters)
Use with	Recommended rates of approved pesticides for non-crop uses and with recommended rates of approved pesticides up to the growth stage indicated for specified crops, and with half or less than the recommended rate on these crops after the stated growth stages
Protective clothing	A, C
Precautions	U19a, U20b, D05, D09a, D10b

Intracrop Super Rapeze MSO	Intracrop	A0782	adjuvant/vegetable oil

Contains	840 g/l oil (rapeseed fatty acid esters) (EAC 1)
Use with	All approved pesticides on non-edible crops, and all pesticides approved for use on growing edible crops when used at half recommended dose or less. On specified crops product may be used at a maximum spray concentration of 1.78% with approved pesticides at their full approved rate up to the growth stages shown in the label
Protective clothing	A, C
Precautions	U19a, U20b, D05, D09a, D10b

Intracrop Tonto	Intracrop	A0778	spreader/sticker/wetter

Contains	32.67% w/w alkoxylated alcohols (EAC 1) and 1.0% w/w trisiloxane organosilicone copolymers (EAC 2)
Use with	All approved pesticides at half or less than half the approved pesticide rate on edible crops and all approved pesticides on non-edible crops
Protective clothing	A, C
Precautions	U05a, U08, U20c, E13e, E40c, D01, D02, D09a, D10a

Intracrop Vantage	Intracrop	A0719	spreader/wetter

Contains	30.0% w/w alkoxylated alcohols (EAC 1)
Use with	All approved pesticides at half or less than half the approved pesticide rate and all approved formulations of glyphosate
Protective clothing	A, C
Precautions	U05a, U08, U20c, E13e, E40c, D01, D02, D09a, D10a

Adjuvants

Product	Supplier	Adj. No.	Type
Intracrop Warrior	Intracrop	A0514	spreader/wetter
Contains	19.0 % w/v alkoxylated alcohols		
Use with	Recommended rates of approved pesticides up to the growth stage indicated for specified crops, and with half or less than the recommended rate on these crops after the stated growth stages. Also with herbicides on managed amenity turf at recommended rates, and on grassland at half or less than recommended rates		
Protective clothing	A, C		
Precautions	R36, R38, R41, U05a, U08, U11, U14, U15, U20c, E15a, D01, D02, D09a, D10b, M03, H04		
Intracrop Zenith AF	Intracrop	A0811	spreader/sticker/wetter
Contains	250 g/l ethylene oxide-propylene oxide copolymers, 70.2 g/l styrene-butadiene copolymers and 41.5 g/l trisiloxane organosilicone copolymers.		
Use with	All approved pesticides on non-edible crops, all approved pesticides on edible crops at half or less than half the pesticide rate and all approved pesticides with a recommendation for use on non crop production.		
Protective clothing	A, C, H		
Precautions	H318, U05a, U11, U14, U15, U19a, U20a, M04c, H226, H302		
Kantor	Interagro	A0623	spreader/wetter
Contains	790 g/l alkoxylated triglycerides		
Use with	All approved pesticides		
Protective clothing	A, C		
Precautions	U05a, U20b, E15b, E19b, E34, D01, D02, D09a, D10a, D12a		
Katalyst	Interagro	A0450	penetrant/water conditioner
Contains	90% w/w alkoxylated fatty amine		
Use with	Glyphosate and a wide range of other pesticides on specified crops. Refer to label or contact supplier for further details		
Protective clothing	A, C		
Precautions	R22a, R36, R38, R50, R58, U02a, U05a, U08, U19a, U20b, E15a, E34, E37, D01, D02, D05, D09a, D10a, M03, H03, H11		
Kinetic	Helena	A0252	spreader/wetter
Contains	80% w/w ethylene oxide and propylene oxide copolymers, 16% w/w trisiloxane copolymers		
Use with	Approved pesticides on non-edible crops and cereals and stubbles of all edible crops when used at full recommended dose, and with pesticides approved for use on growing edible crops when used at half recommended dose or less. On specified crops product may be used at a maximum spray concentration of 0.2% with approved pesticides at their full approved rate up to the growth stages shown in the label		
Protective clothing	A, H		
Precautions	R38, R41, U02a, U05a, U08, U19a, U20b, E13c, E34, E37, D01, D02, D05, D09a, D10c, M03, H04		

SECTION 4

Adjuvants

Product	Supplier	Adj. No.	Type
Klipper	Amega	A0260	spreader/wetter
Contains	600 g/l alkoxylated alcohols (EAC 1) and isobutanol		
Use with	Listed herbicides, fungicides and chlorpyrifos on managed amenity turf or amenity grassland (see label for details).		
Protective clothing	A, C, H		
Precautions	R22a, R41, R67, U02a, U04a, U05a, U11, U14, U15, E15a, E34, D01, D02, D09a, D10a, M03, H03, H08		
Kotek	Agrovista	A0746	spreader/wetter
Contains	83.0% w/w trisiloxane organosilicone copolymers (EAC 1)		
Use with	All approved pesticides at half or less than half the approved rate on edible crops; all approved pesticides on non-edible crops or non-production targets. Max conc 0.15% of spray volume		
Protective clothing	A, H		
Precautions	R20, R41, R51, R53a, U11		
Leaf-Koat	Helena	A0511	spreader/wetter
Contains	80% w/w ethylene oxide and propylene oxide copolymers, 16% w/w trisiloxane copolymers		
Use with	Approved pesticides on non-edible crops and cereals and stubbles of all edible crops when used at full recommended dose, and with pesticides approved for use on growing edible crops when used at half recommended dose or less. On specified crops product may be used at a maximum spray concentration of 0.2% with approved pesticides at their full approved rate up to the growth stages shown in the label		
Protective clothing	A, H		
Precautions	R38, R41, U02a, U05a, U08, U19a, U20b, E13c, E34, E37, D01, D02, D05, D09a, D10c, M03, H04		
Level	United Agri	A0654	extender/sticker
Contains	10% w/w styrene-butadiene copolymers (EAC 1)		
Use with	All approved pesticides in non-edible crops, all approved pesticides on edible crops when used at half or less of their recommended rate, potato blight fungicides and with all pesticides on listed crops up to specified growth stages		
Protective clothing	A, C		
Precautions	E37, D01, D02, D05		
Li-700	De Sangosse	A0529	acidifier/drift retardant/penetrant
Contains	9.39% w/w alkoxylated alcohols (EAC 1), 35.0% w/w propionic acid (EAC 2), 35.0% w/w soybean phospholipids (EAC 3)		
Use with	All approved pesticides on all edible crops when used at half their recommended dose or less, and on all non-edible crops up to their full recommended dose. Also with morpholine fungicides on cereals and with pirimicarb on legumes when used at full dose; with iprodione on brassicas when used at 75% dose, and at a maximum concentration of 0.5% with all pesticides on listed crops at full dose up to specified growth stages		
Protective clothing	A, C, H		
Precautions	R36, R38, U11, U14, U15, U19a, E15a, E19b, D01, D02, D05, D10a, D12a, H04		

Adjuvants

Product	Supplier	Adj. No.	Type
Logic	Microcide	A0288	vegetable oil
Contains	95% w/w oil (rapeseed triglycerides) (EAC 1)		
Use with	All approved pesticides on edible and non-edible crops for ground or aerial application		
Protective clothing	A, C		
Precautions	U19a, U20b, D09a, D10b		
Logic Oil	Microcide	A0630	vegetable oil
Contains	95% w/w oil (rapeseed triglycerides) (EAC 1)		
Use with	All approved pesticides on edible and non-edible crops for ground or aerial application		
Protective clothing	A, C		
Precautions	U19a, U20b, D09a, D10b		
Low Down	Helena	A0459	mineral oil
Contains	732 g/l petroleum oils		
Use with	All approved pesticides on edible and non-edible crops when used at half their approved dose or less. Also with approved pesticides on specified crops, up to specified growth stages, at up to their full approved dose		
Protective clothing	A, C		
Precautions	R53a, U02a, U05a, U08, U20b, E15a, E34, E37, D01, D02, D05, D09a, D10c, H11		
Master Sil	Global Adjuvants	A0822	spreader/wetter
Contains	83% trisiloxane organosilicone copolymers		
Use with	All approved pesticides on all edible crops when used at half their recommended dose or less, and on all non-edible crops up to their full recommended dose.		
Protective clothing	A, C, H		
Precautions	H319, E34, H411, D01, D02, D09a, D10a		
Meco-Flex	Greenaway	A0678	spreader/sticker/wetter
Contains	10.0% w/w oil (rapeseed triglycerides) (EAC 1)		
Use with	Re-Act, Headland Relay Depitox		
Protective clothing	A, C		
Precautions	U11, U12, U15, U20b, E15a, E34, D02		
Mero	Bayer CropScience	A0818	wetter
Contains	81.4% w/w rapeseed fatty acid esters		
Use with	All approved cereal and maize herbicides and all approved brassica insecticides		
Protective clothing	A, C, H		
Precautions	R38, U05a, U09a, U13, U20b, E34, D01, D02, D09a, D10a, M05a		

SECTION 4

Adjuvants

Product	Supplier	Adj. No.	Type
Mixture B NF	Amega	A0570	non-ionic surfactant/spreader/wetter

Contains	37% w/w alkoxylated alcohols (EAC 1) and 41% w/w alkoxylated alcohols (EAC 2) and isopropanol
Use with	All approved pesticides on non-edible crops at their full recommended rate. Also with Timbrel (MAPP 05815) and all approved formulations of glyphosate in non-crop situations. Also with a range of specified herbicides when used at less than half their recommended rate on forest and grassland, and with all approved pesticides at full rate on a wide range of specified edible crops up to specified growth stages
Protective clothing	A, C, H
Precautions	U02a, U04a, U05a, U11, U14, U15, E13a, E34, D01, D02, D09a, D10a, M03, H03, H11

Product	Supplier	Adj. No.	Type
Nelson	Agrovista	A0796	spreader/wetter

Contains	300 g/l alkoxylated alcohols (EAC 1) and 300 g/l alkoxylated alcohols (EAC 2)
Use with	All approved pesticides at half the approved maximum dose on edible crops and at the full approved dose on non-edible crops and non-crop production at 0.5% spray volume.
Protective clothing	A, C, H
Precautions	H318, U02a, U05a, U08, U09c, U11, U13, U14, U15, U20d, M03a, M05b, H302

Product	Supplier	Adj. No.	Type
Newman Cropspray 11E	De Sangosse	A0530	adjuvant

Contains	99% w/w oil (petroleum oils) (EAC 1)
Use with	All approved pesticides on all edible crops when used at half their recommended dose or less, and on all non-edible crops up to their full recommended dose. Also with listed herbicides on a range of specified crops and with all pesticides on a range of specified crops up to specified growth stages.
Protective clothing	A, C
Precautions	R22a, U10, U16b, U19a, E15a, E19b, E34, E37, D01, D02, D05, D10a, D12a, M05b, H03

Product	Supplier	Adj. No.	Type
Newman's T-80	De Sangosse	A0192	spreader/wetter

Contains	800 g/l alkoxylated tallow amines (EAC 1)
Use with	Glyphosate
Protective clothing	A, C, H
Precautions	R22a, R37, R38, R41, R50, R58, R67, U05a, U11, U14, U19a, U20a, E15a, E19b, E34, E37, D01, D02, D05, D10a, D12a, M03, M04a, H03, H08, H11

Product	Supplier	Adj. No.	Type
Nion	Amega	A0760	spreader/wetter

Contains	90% w/w alkoxylated alcohols
Use with	All approved pesticides at half the approved maximum dose on edible crops and at the full approved dose on non-edible crops and non-crop production at 1% spray volume.
Protective clothing	A, C
Precautions	U02a, U04a, U05a, U11, U14, U15, E15a, E34, D01, D02, D09a, D10a, M03, H03

Adjuvants

Product	Supplier	Adj. No.	Type
Nu Film P	Intracrop	A0635	anti-drift agent/anti-transpirant/sticker/UV screen/wetter
Contains	96.0 % w/w pinene oligomers		
Use with	Glyphosate and many other pesticides and growth regulators for which a protectant is recommended. Do not use in mixture with adjuvant oils or surfactants		
Protective clothing	A, C		
Precautions	U19a, U20b, E13c, D09a, D10a		
Pan Oasis	Pan Agriculture	A0411	vegetable oil
Contains	95% w/w methylated rapeseed oil		
Use with	A wide range of pesticides that have a label recommendation for use with authorised adjuvant oils. Contact distributor for further details		
Protective clothing	A		
Precautions	R36, U02a, U05a, U08, U20b, E15a, D01, D02, D05, D09a, D10a, H04		
Pan Panorama	Pan Agriculture	A0412	mineral oil
Contains	95% w/w mineral oil		
Use with	A wide range of pesticides that have a label recommendation for use with adjuvant oils. Contact distributor for details		
Protective clothing	A		
Precautions	R22b, R38, U02a, U05a, U08, U20b, E13c, D01, D02, D05, D09a, D10a, M05b, H03		
Phase II	De Sangosse	A0622	vegetable oil
Contains	95.2% w/w oil (rapeseed fatty acid esters) (EAC 1)		
Use with	Pesticides approved for use in sugar beet, oilseed rape, cereals (for grass weed control), and other specified agricultural and horticultural crops		
Protective clothing	A, C, H		
Precautions	U19a, U20b, E15a, E19b, E34, D01, D02, D05, D10a, D12a		
Pin-o-Film	Intracrop	A0637	anti-drift agent/anti-transpirant/sticker/UV screen/wetter
Contains	96.0 % w/w pinene oligomers		
Use with	Recommended rates of approved pesticides up to the growth stage indicated for specified crops, and with half or less than the recommended rate on these crops after the stated growth stages. Also for use with pesticides on grassland at half or less than their recommended rates. Must not be used in mixture with adjuvant oils or surfactants		
Protective clothing	A, C		
Precautions	U19a, U20c, E13c, C02a, D09a, D11a		
Planet	Intracrop	A0605	non-ionic surfactant/spreader/wetter
Contains	700 g/l alkoxylated alcohols and 150 g/l rapeseed fatty acids		
Use with	Any spray for which additional wetter is recommended		
Protective clothing	A, C		
Precautions	R22a, R36, U05a, U08, U19a, U20b, E13c, D01, D09a, D10a, D11a, H03, H04, H08		

SECTION 4

Adjuvants

Product	Supplier	Adj. No.	Type
Prima	De Sangosse	A0531	mineral oil
Contains	99% w/w oil (petroleum oils) (EAC 1)		
Use with	All pesticides on all crops when used up to 50% of maximum approved dose for that use (mixtures with Roundup must only be used for treatment of stubbles). Also with all approved pesticides at full dose on listed crops (see label). Also with listed herbicides on specified crops		
Protective clothing	A, C		
Precautions	R22a, U10, U16b, U19a, E15a, E19b, E34, E37, D01, D02, D05, D10a, D12a, M05b, H03		
Profit Oil	Microcide	A0631	extender/sticker/wetter
Contains	95% w/w oil (rapeseed triglycerides) (EAC 1)		
Use with	All approved pesticides		
Protective clothing	A, C		
Precautions	U19a, U20b, D09a, D10b		
Pryz-Flex	Greenaway	A0679	spreader/sticker/wetter
Contains	10.0% w/w oil (rapeseed triglycerides) (EAC 1)		
Use with	'Greenaway Gly-490' (MAPP 12718) and all approved 490 g/l glyphosate products on permeable surfaces overlying soil or with 'Kerb Flo' (MAPP 13716) in forest situations.		
Protective clothing	A, C		
Precautions	U11, U12, U15, U20b, E34, E37		
Remix	Agrovista	A0765	wetter
Contains	732 g/l petroleum (paraffin) oil		
Use with	All approved pesticides at half or less than half the approved rate on edible crops; all approved pesticides on non-edible crops or non-production targets. Can also be used with all approved pesticides pre-emergence on bulb and stem vegetables, legumes, carrots, parsnips, swedes and turnips. Max conc for all uses is 1% of spray solution.		
Protective clothing	A, C, H		
Precautions	U02a, U05a, U19a, E15a, D01, D02, D05, D09a, D10a, D12a		
Respond	Amega	A0486	spreader/wetter
Contains	57.0 % w/w ethylene oxide-propylene oxide copolymers (EAC 1)		
Use with	All approved pesticides on all edible crops when used at half their recommended dose or less, and on all non-edible crops up to their full recommended dose. Also at a maximum concentration of 0.5%		
Protective clothing	A		
Precautions	U02a, U04a, U05a, U14, U15, U19a, E13b, E34, D01, D02, D09a, D10a, M03		
Reward Oil	Microcide	A0632	extender/sticker/wetter
Contains	95% w/w oil (rapeseed triglycerides) (EAC 1)		
Use with	All approved pesticides		
Protective clothing	A, C		
Precautions	U19a, U20b, D09a, D10b		

Adjuvants

Product	Supplier	Adj. No.	Type
Roller	Agrovista	A0748	spreader/wetter
Contains	\multicolumn{3}{l	}{832 g/l ethylene oxide-propylene oxide copolymers (EAC 1) and 169.3 g/l trisiloxane organosilicone copolymers (EAC 2)}
Use with	\multicolumn{3}{l	}{All approved pesticides up to their maximum dose at a max conc of 0.2% v/v.}
Protective clothing	\multicolumn{3}{l	}{A, C, H}
Precautions	\multicolumn{3}{l	}{R36, R38, R41, R52, R53a, U02a, U05a, U08, U11, U20b, E15a, E34, E40c, D01, D02, D05, D09a, D10a, D12b, H04}

Product	Supplier	Adj. No.	Type
Saracen	Interagro	A0368	vegetable oil
Contains	\multicolumn{3}{l	}{95% w/w methylated vegetable oil}
Use with	\multicolumn{3}{l	}{Cereal graminicides and a wide range of other pesticides that have a recommendation for use with authorised adjuvant oils on specified crops. Refer to label or contact supplier for further details}
Protective clothing	\multicolumn{3}{l	}{A, C}
Precautions	\multicolumn{3}{l	}{U05a, U20b, E15a, D01, D02, D09a, D10a}

Product	Supplier	Adj. No.	Type
SAS 90	Intracrop	A0740	spreader/wetter
Contains	\multicolumn{3}{l	}{83% w/w trisiloxane organosilicone copolymers}
Use with	\multicolumn{3}{l	}{All approved herbicides at a maximum concentration of 0.025% spray volume and all approved fungicides and insecticides at a maximum concentration of 0.05% spray volume}
Protective clothing	\multicolumn{3}{l	}{A, C}
Precautions	\multicolumn{3}{l	}{U02a, U08, U19a, U20b, E13c, E34, D09a, D10a}

Product	Supplier	Adj. No.	Type
Saturn Plus	Intracrop	A0813	spreader/wetter
Contains	\multicolumn{3}{l	}{70.0% w/w ethylene oxide-propylene oxide copolymers (EAC 1)}
Use with	\multicolumn{3}{l	}{All approved pesticides on non-edible crops, and all pesticides approved for use on growing edible crops when used at half recommended dose or less.}
Protective clothing	\multicolumn{3}{l	}{A, C}
Precautions	\multicolumn{3}{l	}{U19a, U20c, E13c, C02a, D09a, D11a}

Product	Supplier	Adj. No.	Type
Siltex AF	Intracrop	A0803	spreader/sticker/wetter
Contains	\multicolumn{3}{l	}{250 g/l ethylene oxide-propylene oxide copolymers (EAC 1), 70.2 g/l styrene-butadiene copolymers (EAC 2) and 41.5 g/l trisiloxane organosilicone copolymers (EAC 3)}
Use with	\multicolumn{3}{l	}{All approved pesticides on non-edible crops, and all pesticides approved for use on growing edible crops when used at half recommended dose or less.}
Protective clothing	\multicolumn{3}{l	}{A, C}
Precautions	\multicolumn{3}{l	}{U05a, U08, U20c, E13c, E37, D01, D02, D09a, D10a}

Product	Supplier	Adj. No.	Type
Silwet L-77	De Sangosse	A0640	drift retardant/spreader/wetter
Contains	\multicolumn{3}{l	}{80% w/w trisiloxane organosilicone copolymers (EAC 1).}
Use with	\multicolumn{3}{l	}{All approved fungicides on winter and spring sown cereals; all approved pesticides applied at 50% or less of their full approved dose. A wide range of other uses. See label or contact supplier for details}
Protective clothing	\multicolumn{3}{l	}{A, C, H}
Precautions	\multicolumn{3}{l	}{R20, R21, R22a, R36, R43, R48, R51, R58, U11, U15, U19a, E15a, E19b, D01, D02, D05, D10a, D12a, H03, H11}

SECTION 4

Adjuvants

Product	Supplier	Adj. No.	Type
Slippa	Interagro	A0206	spreader/wetter
Contains	64.0 % w/w trisiloxane organosilicone copolymers		
Use with	Cereal fungicides and a wide range of other pesticides and trace elements on specified crops		
Protective clothing	A, C, H		
Precautions	R20, R38, R41, R43, R48, R51, R58, U02a, U05a, U08, U11, U19a, U20a, E15a, E34, E37, D01, D02, D05, D09a, D10a, M03, H03, H11		
Solar	Intracrop	A0225	activator/non-ionic surfactant/spreader
Contains	750 g/l ethylene oxide-propylene oxide copolymers		
Use with	Foliar applied plant growth regulators		
Protective clothing	A, C		
Precautions	R36, R38, U04a, U05a, U08, U20a, E13c, E34, D01, D02, D09a, D10b, D11a, M03, H04		
Solar Plus	Intracrop	A0802	spreader/wetter
Contains	70.0% w/w ethylene oxide-propylene oxide copolymers (EAC 1)		
Use with	All approved pesticides on non-edible crops, and all pesticides approved for use on growing edible crops when used at half recommended dose or less.		
Protective clothing	A, C		
Precautions	R36, R38, U04a, U05a, U08, U20a, E13c, E34, D01, D02, D09a, D10b, D11a, M03, H04		
Spartan	Interagro	A0375	penetrant/water conditioner
Contains	500 g/l alkoxylated tallow amines and 400 g/l alkoxylated sorbitan esters		
Use with	Cereal growth regulators, cereal herbicides, oilseed rape fungicides and a wide range of other pesticides on specified crops. Refer to label or contact supplier for further details		
Protective clothing	A, C		
Precautions	R22a, R36, R38, R51, R58, U02a, U05a, U19a, U20b, E15a, E34, E37, D01, D02, D05, D09a, D10a, M03, H03, H11		
Speed	Headland Amenity	A0710	spreader/surfactant
Contains	18.5% w/w trisiloxane organosilicone copolymers (EAC1).		
Use with	All approved pesticides on amenity grassland and managed amenity turf at 0.25% spray solution		
Protective clothing	A, C, H		
Speedway Total	Everris Ltd	A0820	extender/spreader/wetter
Contains	41.0 % w/w alkoxylated alcohols (EAC 1) and 37.0 % w/w alkoxylated alcohols (EAC 2)		
Use with	All authorised plant protection products on non-edible crops and non-crop production uses; all authorised plant protection products at half or less than half the authorised plant protection product rate on edible crops.		
Protective clothing	A		
Precautions	U02a, U04a, U05a, U14, U15, U19a, E13b, E34, D01, D02, D09a, D10a, M03		

Adjuvants

Product	Supplier	Adj. No.	Type
Sprayfast	Taminco	A0668	extender/sticker/wetter
Contains	334 g/l di-1-p-menthene		
Use with	Glyphosate and other pesticides, growth regulators or nutrients for which a coating agent is approved and recommended		
Protective clothing	A, C		
Precautions	U20b, E13c, D09a, D10a		
Spray-fix	De Sangosse	A0559	extender/sticker/wetter
Contains	10% w/w alkoxylated alcohols (EAC 1), 45% w/w styrene-butadiene copolymers (EAC 2)		
Use with	All approved potato blight fungicides. Also with all approved pesticides on all edible crops when used at half their recommended dose or less, and on all non-edible crops up their full recommended dose. Also at a maximum concentration of 0.14% with approved pesticides on listed crops up to specified growth stages		
Protective clothing	A, C, H		
Precautions	R36, R38, U11, U14, U16b, U19a, E15a, E19b, D01, D02, D05, D10a, D12a, H04		
Spraygard	Taminco	A0732	extender/sticker/wetter
Contains	400 g/l pinene oligomers (EAC 1)		
Use with	Approved pesticides on all edible crops (except herbicides on peas) and as an anti-transpirant on vegetables before transplanting, evergreens, deciduous trees, shrubs, bushes and on turf		
Protective clothing	A		
Precautions	R38, R50, R53a, U05a, U14, U20b, E15a, E34, E37, D01, D09a, D10b, D12b, M03, H04, H11		
Spraymac	De Sangosse	A0549	acidifier/non-ionic surfactant
Contains	100g/l alkoxylated alcohols (EAC 1), 350g/l propionic acid (EAC 2)		
Use with	All approved pesticides on all edible crops when used at half their recommended dose or less, and on all non-edible crops up to their full recommended dose. Also at a maximum concentration of 0.5% with approved pesticides on listed crops up to specified growth stages		
Protective clothing	A, C, H		
Precautions	R34, U04a, U10, U11, U14, U19a, U20b, E15a, E19b, E34, D01, D02, D05, D10a, D12a, M04a, H05		
Stamina	Interagro	A0202	penetrant
Contains	100% w/w alkoxylated fatty amine		
Use with	Glyphosate and a wide range of other pesticides on specified crops		
Protective clothing	A, C		
Precautions	R22a, R38, R50, R58, U02a, U05a, U20a, E15a, E34, D01, D02, D05, D09a, D10a, M03, H03, H11		
Standon Shiva	Standon	A0816	wetter
Contains	6.7% w/w 3,6-dioxaeicosylsulphate sodium salt and 20.2% w/w 3,6-dioxaoctadecylsulphate sodium salt		
Use with	Standon Mimas WG and all other approved cereal herbicides		
Protective clothing	A, C		
Precautions	R36, R38, U02a, U05a, U08, U13, U19a, U20b, E13c, E34, D01, D02, D05, D09a, D10a, H04		

SECTION 4

Adjuvants

Product	Supplier	Adj. No.	Type
Standon Wicket	Standon	A0781	wetter
Contains	47% w/w rapeseed fatty acid esters		
Use with	Clodinafop-propargyl, pinoxaden and mixtures of the two actives in cereal crops		
Protective clothing	A, H		
Precautions	R43, R50, R53a, U02a, U05a, U09b, U20b, E15b, E38, D01, D02, D05, D09a, D10c, D12a, H04, H11		
Stika	De Sangosse	A0557	extender/sticker/wetter
Contains	10% w/w alkoxylated alcohols (EAC 1), 22.5% w/w styrene-butadione copolymers (EAC 2)		
Use with	All approved potato blight fungicides. Also with all approved pesticides on all edible crops when used at half their recommended dose or less, and on all non-edible crops up to their full recommended dose. Also at a maximum concentration of 0.14% with approved pesticides on listed crops up to specified growth stages		
Protective clothing	A, C, H		
Precautions	R36, R38, U11, U14, U16b, U19a, E15a, E19b, D01, D02, D05, D10a, D12a, H04		
SU Wett	Global Adjuvants	A0807	spreader/wetter
Contains	57.0 % w/w ethylene oxide-propylene oxide copolymers (EAC 1)		
Use with	All authorised plant protection products at half or less than half the authorised plant protection product rate		
Protective clothing	A, C, H, M		
Precautions	U05a, U20c, E34, D01, D02, D10a, M04a		
Super Nova	Interagro	A0364	penetrant
Contains	500 g/l alkoxylated fatty amine + 500 g/l polyoxyethylene monolaurate		
Use with	Cereal growth regulators, cereal herbicides, cereal fungicides, oilseed rape fungicides and a wide range of other pesticides on specified crops		
Protective clothing	A, C		
Precautions	R51, R58, U05a, E15a, E34, D01, D02, D05, D09a, D10a, H11		
Surfer	Dow	A0800	wetter
Contains	72.0 % w/w alkoxylated alcohols (EAC 1)		
Use with	All approved pesticides on all edible crops when used at half their recommended dose or less, and on all non-edible crops up to their full recommended dose.		
Protective clothing	A, C, H		
Precautions	H318, U05a, U11, U14, U15, U20a, E15a, E34, D01, D02, D09a, D10a		
Sward	Amega	A0747	spreader/wetter
Contains	400 g/l ethylene oxide-propylene oxide copolymers (EAC 1) and 58.1 g/l trisiloxane organosilicone copolymers (EAC 2)		
Use with	For use with all approved pesticides on managed amenity turf and amenity grassland at 0.2% v/v		
Protective clothing	A, C		
Precautions	U02a, U04a, U05a, U11, U14, U15, E15a, E34, D01, D02, D09a, D10a, D12b, M03, H04		

Adjuvants

Product	Supplier	Adj. No.	Type
TM 1088	Intracrop	A0675	wetter
Contains	750 g/l ethylene oxide-propylene oxide copolymers		
Use with	All approved pesticides on non-edible crops and non crop production uses; All approved pesticides at half or less than half the approved pesticide rate on edible crops		
Protective clothing	A, C		
Precautions	R36, R38, U04a, U05a, U08, U19a, U20b, E13b, E34, D01, D02, D09a, D10a, M03, M05a, H04		
Toil	Interagro	A0248	vegetable oil
Contains	95% w/w methylated rapeseed oil		
Use with	Sugar beet herbicides, oilseed rape herbicides, cereal graminicides and a wide range of other pesticides that have a label recommendation for use with authorised adjuvant oils on specified crops. Refer to label or contact supplier for further details		
Protective clothing	A		
Precautions	U05a, U20b, E15a, D01, D02, D05, D09a, D10a		
Torpedo-II	De Sangosse	A0541	penetrant/wetter
Contains	190g/kg alkoxylated alcohols (EAC 1), 190g/kg alkoxylated alcohols (EAC 2), 75g/kg oil (tall oil fatty acids) (EAC 3), 210 g/kg alkoxylated tallow amines (EAC 4)		
Use with	All approved pesticides on all edible crops when used at half their recommended dose or less, and on all non-edible crops up their full recommended dose. Also at a maximum concentration of 0.1% with approved pesticides on listed crops up to specified growth stages		
Protective clothing	A, C, H		
Precautions	R36, R38, R41, R52, R58, U05a, U10, U11, U19a, E15a, E19b, E34, E37, D01, D02, D05, D10a, D12a, M03, H04		
Transact	United Agri	A0584	penetrant
Contains	40% w/w propionic acid		
Use with	Plant growth regulators in cereals, all approved pesticides in non-edible crops, all approved pesticides on edible crops when used at half or less their recommended rate, and with all pesticides on listed crops up to specified growth stages		
Protective clothing	A, C		
Precautions	R34, U10, U11, U14, U15, U19a, D01, D05, D09b, M04a, H05		
Transcend	Helena	A0333	adjuvant/vegetable oil
Contains	80% w/w oil (soybean fatty acid esters), 12% w/w trisiloxane organosilicone copolymers.		
Use with	Approved pesticides on non-edible crops, cereals and stubbles of all edible crops, grassland (destruction), and pesticides approved for use on growing edible crops when used at half recommended dose or less. On specified crops product may be used at a maximum spray concentration of 0.5% with approved pesticides at their full approved rate up to the growth stages shown in the label		
Protective clothing	A, H		
Precautions	R38, R41, U02a, U05a, U08, U11, U19a, U20b, E13c, E34, E37, D01, D02, D05, D09a, M03, H04		

SECTION 4

Adjuvants

Product	Supplier	Adj. No.	Type
Validate	De Sangosse	A0500	adjuvant/wetter

Contains	25% w/w alkoxylated alcohols (EAC 1), 25% w/w oil (soybean fatty acid esters) (EAC 2), 50% w/w soybean phospholipids (EAC 3)
Use with	Approved pesticides on non-edible crops where the addition of a wetter/spreader or adjuvant oil is recommended on the pesticide label, and pesticides approved for use on growing edible crops when used at half recommended dose or less. On specified crops product may be used at a maximum spray concentration of 0.5% with approved pesticides at their full approved rate up to the growth stages shown in the label
Protective clothing	A, C, H
Precautions	R51, R53a, U05a, U11, U14, U19a, U20b, E15a, E19b, E34, D01, D02, D10a, D12a, H11

Velocity	Agrovista	A0697	adjuvant

Contains	745g/l oil (rapeseed fatty acid esters), 103g/l trisiloxane organosilicone copolymers.
Use with	All approved pesticides up to their full approved rate at a maximum conc of 0.5% vol/vol.
Protective clothing	A, C
Precautions	R41, R52, R53a, U02a, U05a, U08, U11, U14, U15, U20b, E13c, E15a, E34, E40c, D01, D02, D09a, D10a, H04

Velvet	Global Adjuvants	A0814	spreader/sticker/wetter

Contains	83% trisiloxane organosilicone copolymers
Use with	All approved pesticides on all edible crops when used at half their recommended dose or less, and on all non-edible crops up to their full recommended dose.
Protective clothing	A, C, H
Precautions	H319, E34, H411, D01, D02, D09a, D10a

Verdant	Agrovista	A0714	wetter

Contains	8% w/w alkoxylated alcohol, 63% w/w alkoxylated alkyl polyglucoside and 10% w/w alkyl polyglucoside
Use with	All approved pesticides at 50% or less on edible crops and with all approved pesticides on non-edible crops, non-crop production and stubbles of edible crops at 0.25% spray sloution.
Protective clothing	A, C, H
Precautions	R38, R41, U02a, U05a, U11, U14, U15, U20b, E15a, E34, D02, D05, D09a, D10a, H04

Wetcit	Plant Solutions	A0586	penetrant/wetter

Contains	8.15% w/w alcohol ethoxylate
Use with	All approved pesticides on all edible crops when used at half their recommended dose or less, and on all non-edible crops up to their full recommended dose. Also at a maximum concentration of 0.25% with approved pesticides on listed crops up to specified growth stages at full rate, and at half rate thereafter
Protective clothing	A, C, H
Precautions	R22a, R38, R41, U08, U11, U19a, U20b, E13c, E15a, E34, E37, D05, D08, D09a, D10b, H03

Adjuvants

Product	Supplier	Adj. No.	Type
X-Wet	De Sangosse	A0548	non-ionic surfactant/wetter

Contains: 375g/kg alkoxylated alcohols (EAC 1), 375g/kg alkoxylated alcohols (EAC 2), 150g/kg oil (tall oil fatty acids) (EAC 3)

Use with: All approved pesticides on all edible crops when used at half their recommended dose or less, and on all non-edible crops up their full recommended dose. Also at a maximum concentration of 0.1% with approved pesticides on listed crops up to specified growth stages

Protective clothing: A, C, H

Precautions: R36, R38, R53a, U02a, U04a, U05a, U10, U11, U20b, E15a, E19b, D01, D02, D05, D10a, H04

Zarado	De Sangosse	A0516	vegetable oil

Contains: 70% w/w oil (rapeseed oil fatty acid esters) as emulsifiable concentrate (EAC 1).

Use with: All approved pesticides on edible and non-edible crops when used at half their approved dose or less. Also with approved pesticides on specified crops, up to specified growth stages, at up to their full approved dose

Protective clothing: A, C, H

Precautions: R43, R52, U05a, U13, U19a, U20b, E15a, E19b, E34, D01, D02, D05, D10a, M05a, H04

Zeal	Interagro	A0685	water conditioner/wetter

Contains: 60.0 % w/w alkoxylated tallow amines and 37.0 % w/w alkoxylated sorbitan esters

Use with: Trace elements such as calcium, copper, manganese and sulphur and with macronutrients such as phosphates

Protective clothing: A, C, H

Precautions: R22a, R36, R38, R51, R53a, U05a, U09a, U11, U14, U15, U19a, E15b, E34, D01, D02, D09a, D10a, M03, H03, H11

Zigzag	Headland	A0692	extender/sticker/wetter

Contains: 36.9% w/v styrene/butadiene co-polymer and 6.375% w/v polyether-modified trisiloxane

Use with: Approved glyphosate formulations on cereal crops; All approved pesticides applied at half-rate or less; All approved pesticides for use on crops not destined for human or animal consumption; With all approved pesticides on edible crops up to the latest growth stage specified on the label.

Protective clothing: C, H

Precautions: U11, U15, E38, D01

Zinzan	Certis	A0600	extender/spreader/wetter

Contains: 70% w/w 1,2 bis (2-ethylhexyloxycarbonyl) ethanesulphonate

Use with: All approved pesticides on non-edible crops, and all pesticides approved for use on growing edible crops when used at half recommended dose or less. On specified edible crops product may be used at a maximum spray concentration of 0.075% with approved pesticides at their full approved rate up to the growth stages shown in the label

Protective clothing: A, C, P

Precautions: R38, R41, U11, U14, U20c, D01, D09a, D10a, H04

SECTION 4

SECTION 5
USEFUL INFORMATION

Pesticide Legislation

Anyone who advertises, sells, supplies, stores or uses a pesticide is bound by legislation, including those who use pesticides in their own homes, gardens or allotments. There are numerous UK statutory controls, but the major legal instruments are outlined below.

Regulation 1107/2009 — The Replacement for EU 91/414

Regulation 1107/2009 entered into force within the EU on 14 Dec 2009 and was applied to new approval applications from 14th June 2011. Since that date all pesticides that held a current approval under 91/414 have been deemed to be approved under 1107/2009 and the new criteria for approval will only be applied when the active substance comes up for review. The new regulation largely mirrors the previous regulation but requires additional approval criteria to be met to weed out the more hazardous chemicals. These include assessing the effect on vulnerable groups, taking into account known synergistic and cumulative effects, considering the effect on coastal and estuarine waters, effects on the behaviour of non-target organisms, biodiversity and the ecosystem. However, these effects will only be evaluated when scientific methods approved by EFSA (European Food Safety Authority) have been developed.

The additional criteria will require that no CMR's (substances which are Carcinogens, Mutagens or toxic to Reproduction), no endocrine disruptors and no POP (Persistent Organic Pollutants), PBT (chemicals that are Persistent, Bioaccumulative or Toxic) or vPVB (very Persistent and very Bioaccumulative) will gain approval unless the exposure is negligible. However, note that for EFSA the term 'negligible' has yet to receive a precise definition. A list of 66 chemicals that pose an endocrine disruptor risk has been produced. It contained maneb and vinclozolin and both actives have now been withdrawn.

If an active substance passes these hurdles, the first approval for basic substances will be granted for 10 years (15 years for 'low risk' substances, 7 years for candidates for substitution – see below) and then at each subsequent renewal a further 15 years will be granted. Actives will be divided between 5 categories – low risk substances (e.g. pheromones, semiochemicals, micro-organisms and natural plant extracts), basic substances, candidates for substitution and then safeners/synergists and co-formulants. Candidates for substitution will be those chemicals which have just scraped past the approval thresholds but are deemed to still carry some degree of hazard. They will be withdrawn when significantly safer alternatives are available. This includes physical methods of control but any alternative must not have significant economic or practical disadvantages, must minimise the risk of the development of resistance and the consequences for any 'minor uses' of the original product must also be considered.

The Water Framework Directive

This became law in the UK in 2003 but is being introduced over a number of years to allow systems to acclimatise to the new rules and requires member states of the EU to achieve 'good status' in ground and surface water bodies by 22 December 2015 with reviews every six years thereafter. To achieve 'good status' as far as pesticides are concerned the concentration of any individual pesticide must not exceed 0.1 micrograms per litre and the total quantity of all pesticides must not exceed 0.5 micrograms per litre. These levels are not based on the toxicity of the pesticides but on the level of detection that can be reliably achieved. Failure to meet these limits could jeopardise the approval of the problem active ingredients and so every effort must be made to avoid contamination of surface and ground water bodies with pesticides. The pesticides that are currently being found in water samples above the 0.1 ppb limit are metaldehyde, carbetamide, propyzamide, MCPA, mecoprop, glyphosate, bentazone, terbutryne, trietazine, clopyralid, metazachlor, chlorotoluron, chloridazon and ethofumesate. These actives are identified as a risk to water in the pesticide profiles in section 2 and great care should be taken when applying them, particularly metaldehyde which is almost impossible to remove from water at treatment plants.

Useful Information

Classification, Labelling and Packaging (CLP) Regulation

The CLP Regulation (EC No 1272/2008) governs the classification of substances and mixtures and is intended to introduce the United Nations Globally Harmonised System (GHS) of classification to Europe. This new system and the hazard icons that appear on pesticide labels is very similar to the old system with the most obvious difference being that the hazard icons have a white background rather than an orange one. The new system of classification became mandatory for new supplies from 1st June 2015 in the UK but for products already in the supply chain the compliance date is deferred to 1st June 2017. The biggest change will be the replacement of the old HARMFUL icon **'X'** with a new one **'!'** to notify that caution is required when handling or using a product. This brings Europe and the UK in line with the rest of the world. Details of the new phrases are given in Appendix 4.

The Food and Environment Protection Act 1986 (FEPA)

FEPA introduced statutory powers to control pesticides with the aims of protecting human beings, creatures and plants, safeguarding the environment, ensuring safe, effective and humane methods of controlling pests and making pesticide information available to the public. This was supplemented by Control of Pesticides Regulations 1986 (COPR) which has now been replaced by EC Regulation 1107/2009 – see above. Details are given on the websites of the Chemicals Regulation Directorate (CRD) and the Health and Safety Executive (HSE). Together these regulations mean that:

- Only approved products may be sold, supplied, stored, advertised or used.
- No advertisement may contain any claim for safety beyond that which is permitted in the approved label text.
- Only products specifically approved for the purpose may be applied from the air.
- A recognised Storeman's Certificate of Competence is required by anyone who stores for sale or supply pesticides approved for agricultural use.
- A recognised Certificate of Competence is required by anyone who gives advice when selling or supplying pesticides approved for agricultural use.
- Users of pesticides must comply with the Conditions of Approval relating to use.
- A recognised Certificate of Competence is required for all contractors and persons applying pesticides approved for agricultural use (unless working under direct supervision of a certificate holder).
- Only those adjuvants authorised by CRD may be used.
- For tank-mixes of convenience, when the purpose is to reduce the number of passes in the crop, no efficacy data is required but physical and chemical compatibility data are required in line with EU guidance. For mixtures that claim positive benefits efficacy data remains necessary in addition to the physical and chemical compatibility data. Note however that ALS herbicides (HRAC mode of action code 'B') may only be mixed or applied in sequence with other ALS herbicides when a label statement permits this. Mixtures of anticholinesterase products are not permitted unless CRD has fully assessed the toxicological effects.

Dangerous Preparations Directive (1999/45/EC)

The Dangerous Preparations Directive (DPD) came into force in the UK for pesticide and biocidal products on 30 July 2004. Its aim is to achieve a uniform approach across all Member States to the classification, packaging and labelling of most dangerous preparations, including crop protection products. The Directive is implemented in the UK under the Chemicals (Hazard Information and Packaging for Supply) Regulations 2002, often referred to under the acronym CHIP3. In most cases the Regulations have led to additional hazard symbols, and associated risk and safety phrases relating to environmental and health hazards, appearing on the label. All products affected by the DPD entering the supply chain from the implementation date above must be so labelled. The new environmental hazard classifications and risk phrases are included.

Under the DPD, the labels of all plant protection products now state 'To avoid risks to man and the environment, comply with the instructions for use'. As the instruction applies to all products, it has not been repeated in each fact sheet.

The Review Programme

The review programme of existing active substances will continue under EC Directive 1107/2009. The programme is designed to ensure that all available plant protection products are supported by up-to-date information on safety and efficacy. As this will now be under 1107/2009, the new standards for approval will be applied at the due date and any that are deemed to be hazardous chemicals will be withdrawn. Any that are just within the safety standards yet to be defined will be called 'candidates for substitution' and granted only a seven-year renewal to allow time for significantly safer alternatives to be developed. If these are available at the time of renewal, the approval will be withdrawn. Chemicals deemed to be low risk, such as natural plant extracts, microorganisms, semiochemicals and pheromones, will be renewed for 15 years, while most active substances will be renewed for a further 10 years.

Control of Substances Hazardous to Health Regulations 1988 (COSHH)

The COSHH regulations, which came into force on 1 October 1989, were made under the Health and Safety at Work Act 1974, and are also important as a means of regulating the use of pesticides. The regulations cover virtually all substances hazardous to health, including those pesticides classed as Very Toxic, Toxic, Harmful, Irritant or Corrosive, other chemicals used in farming or industry, and substances with occupational exposure limits. They also cover harmful micro-organisms, dusts and any other material, mixture, or compound used at work which can harm people's health.

The original Regulations, together with all subsequent amendments, have been consolidated into a single set of regulations: The Control of Substances Hazardous to Health Regulations 1994 (COSHH 1994).

The basic principle underlying the COSHH regulations is that the risks associated with the use of any substance hazardous to health must be assessed before it is used, and the appropriate measures taken to control the risk. The emphasis is changed from that pertaining under the Poisonous Substances in Agriculture Regulations 1984 (now repealed) – whereby the principal method of ensuring safety was the use of protective clothing – to the prevention or control of exposure to hazardous substances by a combination of measures. In order of preference the measures should be:

(a) substitution with a less hazardous chemical or product
(b) technical or engineering controls (e.g. the use of closed handling systems, etc.)
(c) operational controls (e.g. operators located in cabs fitted with air-filtration systems, etc.)
(d) use of personal protective equipment (PPE), which includes protective clothing.

Consideration must be given as to whether it is necessary to use a pesticide at all in a given situation and, if so, the product posing the least risk to humans, animals and the environment must be selected. Where other measures do not provide adequate control of exposure and the use of PPE is necessary, the items stipulated on the product label must be used as a minimum. It is essential that equipment is properly maintained and the correct procedures adopted. Where necessary, the exposure of workers must be monitored, health checks carried out, and employees instructed and trained in precautionary techniques. Adequate records of all operations involving pesticide application must be made and retained for at least 3 years.

Biocidal Products Regulations

These regulations concern disinfectants, preservatives and pest control products but only the latter are listed in this book. The regulation aims to harmonise the European market for biocidal products, to provide a high level of safety for humans, animals and the environment and to ensure that the products are effective against the target organisms. Products that have been reviewed and included under these regulations are issued with a new number of the format UK-2012-xxxx and these are summarised in this book as UK12-xxxx.

Certificates of Competence – the roles of BASIS and NPTC

COPR, COSHH and other legislation places certain obligations on those who handle and use pesticides. Minimum standards are laid down for the transport, storage and use of pesticides, and the law requires those who act as storekeepers, sellers and advisors to hold recognised Certificates of Competence.

BASIS

BASIS is an independent Registration Scheme for the pesticide industry. It is responsible for organising training courses and examinations to enable such staff to obtain a Certificate of Competence.

In addition, BASIS undertakes an annual assessment of pesticide supply stores, enabling distributors, contractors and seedsmen to meet their obligations under the Code of Practice for Suppliers of Pesticides. Further information can be obtained from BASIS. Since 26 November 2015 all businesses must have sufficient number of staff with BASIS certificates to advise customers at **the time of sale** where previously it was just at the point of sale

Certificates of Competence

Storage

- BASIS Certificate of Competence in the Storage and Handling of Crop Protection Products

Sale and supply

- BASIS Certificate in Crop Protection (Agriculture)
- BASIS Certificate in Crop Protection (Commercial Horticulture)
- BASIS Certificate in Crop Protection (Amenity Horticulture)
- BASIS Certificate in Crop Protection (Forestry)
- BASIS Certificate in Crop Protection (Seed Treatment)
- BASIS Certificate in Crop Protection (Seed Sellers)
- BASIS Certificate in Crop Protection (Field Vegetables)
- BASIS Certificate in Crop Protection (Potatoes)
- BASIS Certificate in Crop Protection (Aquatic)
- BASIS Certificate in Crop Protection (Indoor Landscaping)
- BASIS Certificate in Crop Protection (Grassland and Forage Crops)
- BASIS/LEAF ICM Certificate
- BASIS Advanced Certificate
- FACTS (Fertiliser Advisers Certification and Training Scheme) Certificate

NPTC

Certain spray operators also require certificates of competence under the Control of Pesticides Regulations. NPTC's Pesticides Award is a recognised Certificate of Competence under COPR and it is aimed principally at those people who use pesticide products approved for use in agriculture, horticulture (including amenity horticulture) and forestry. All contractors must possess a certificate if they spray such products unless they are working under the direct supervision of a certificate holder.

Because the required spraying skills vary widely among the uses listed above, candidates are assessed under one (or more) modules that are most appropriate for their professional work. Assessments are carried out by an approved NPTC or Scottish Skills Testing Service Assessor. All candidates must first complete a foundation module (PA1), for which a certificate is not issued, before taking one of the specialist modules. Certificate holders who change their work so that a different specialist module becomes more appropriate may need to obtain a new certificate under the new module.

Holders are required to produce on demand their Certificate of Competence for inspection to any authorised person. Further information can be obtained from NPTC.

Certificates of Competence for spray operators

- PA1 Foundation Module
- PA2 Ground Crop Sprayers – mounted or trailed
- PA3 Broadcast Air Blast Sprayer. Variable Geometry Boom Air Assisted Sprayer

- PA4 Granule Applicator – Mounted or Trailed
- PA5 Boat Mounted Applicators
- PA6 Hand Held Applicators
- PA7 Aerial Application
- PA8 Mixer/Loader
- PA9 Fogging, Misting and Smokes
- PA10 Dipping Bulbs, Corms, Plant Material or Containers
- PA11 Seed Treating Equipment
- PA12 Application of Pesticides to Material as a Continuous or Batch Process

Maximum Residue Levels

A small number of pesticides are liable to leave residues in foodstuffs, even when used correctly. Where residues can occur, statutory limits, known as maximum residue levels (MRLs), have been established. MRLs provide a check that products have been used as directed; **they are not safety limits.** However, they do take account of consumer safety because they are set at levels that ensure normal dietary intake of residues presents no risk to health. Wide safety margins are built in, and eating food containing residues above the MRL does not automatically imply a risk to health. Nevertheless, it is an offence to put into circulation any produce where the MRL is exceeded.

The surrounding legislation is complex. MRLs may be specified by several different bodies. The UK has set statutory MRLs since 1988. The European Union intends eventually to introduce MRLs for all pesticide/commodity combinations. These are being introduced initially by a series of priority lists, but will subsequently be covered by the review programme under EC Regulation 1107/2009. However, in cases where no information is available, EU Regulation 2000/42/EC requires many MRLs to be set at the limit of determination (LOD). As a result, certain approvals are being withdrawn where such use would leave residues above the MRL set in the Directive.

MRLs apply to imported as well as home-produced foodstuffs. Details of those that have been set have been published in *The Pesticides (Maximum Residue Levels in Crops, Food and Feeding Stuffs) Regulations 1994*, and successive amendments to these Regulations. These Statutory Instruments are available from The Stationery Office (www.tso.co.uk).

MRLs are set for many chemicals not currently marketed in Britain. Because of this and the ever-changing information on MRLs, direct access is provided to the comprehensive MRL databases on the Chemicals Regulation Directorate website (www.pesticides.gov.uk). This online database sets out in table form the levels specified by UK Regulations, EC Directives, and the Codex Alimentarius for each commodity.

Approval (On-label and Off-label)

Only officially approved pesticides can be marketed and used in the UK. Approvals are granted by UK Government Ministers in response to applications that are supported by satisfactory data on safety, efficacy and, where relevant, humaneness. The Chemicals Regulation Directorate (CRD, www.pesticides.gov.uk) comes under the Health and Safety Executive (HSE, www.hse.gov.uk), and is the UK Government Agency for regulating agricultural pesticides and plant protection products. The HSE currently fulfils the same role for other pesticides, with the two organisations now merged. The main focus of the regulatory process in both bodies is the protection of human health and the environment.

Statutory Conditions of Use

Approvals are normally granted only in relation to individual products and for specified uses. It is an offence to use non-approved products or to use approved products in a manner that does not comply with the statutory conditions of use, except where the crop or situation is the subject of an off-label extension of use (see below).

Statutory conditions have been laid down for the use of individual products and may include:

- field of use (e.g. agriculture, horticulture etc.)
- crop or situations for which treatment is permitted
- maximum individual dose
- maximum number of treatments or maximum total dose
- maximum area or quantity which may be treated
- latest time of application or harvest interval
- operator protection or training requirements
- environmental protection
- any other specific restrictions relating to particular pesticides.

Products must display these statutory conditions in a boxed area on the label entitled 'Important Information', or words to that effect. At the bottom of the boxed area must be shown a bold text statement: **'Read the label before use. Using this product in a manner that is inconsistent with the label may be an offence. Follow the Code of Practice for Using Plant Protection Products'.** This requirement came into effect in October 2006, and replaced the previous 'Statutory Box'.

Types of Approval

Where there were once three levels of approval (full; provisional; experimental permit), there will now be just two levels (authorisation for use; limited approval for research and development). Provisional approval used to be granted while further data were generated to justify label claims, but under the new regulations, decisions on authorisation will be reached much more quickly, and this 'half-way stage' is deemed to be unnecessary. The official list of approved products, excluding those approved for research and development only, are shown on the websites of CRD and the Health and Safety Executive (HSE).

Withdrawal of Approval

Product approvals may be reviewed, amended, suspended or revoked at any time. Revocation may occur for various reasons, such as commercial withdrawal, or failure by the approval holder to meet data requirements.

From September 2007, where an approval is being revoked for purely administrative, 'housekeeping' reasons, the existing approval will be revoked and in its place will be issued:

- an approval for advertisement, sale and supply by any person for 24 months, and
- an approval for storage and use by any person for 48 months.

Where an approval is replaced by a newer approval, such that there are no safety concerns with the previous approval, but the newer updated approval is more appropriate in terms of

Approval (On-label and Off-label)

reflecting the latest regulatory standard, the existing approval will be revoked and in its place will be issued:

- an approval for advertisement, sale and supply by any person for 12 months, and
- an approval for storage and use by any person for 24 months.

Where there is a need for tighter control of the withdrawal of the product from the supply chain, for example failure to meet data submission deadlines, the current timelines will be retained:

- immediate revocation for advertisement, sale or supply for the approval holder
- approval for 6 months for advertisement, sale and supply by 'others', and
- approval for storage and use by any person for 18 months.

Immediate revocation and product withdrawal remain an option where serious concerns are identified, with immediate revocation of all approvals and approval for storage only by anyone for 3 months, to allow for disposal of product.

The expiry date shown in the product fact sheet is the final date of legal use of the product.

Off-label Extension of Use

Products may legally be used in a manner not covered by the printed label in several ways:

- In accordance with an Extension of Authorisation for Minor Use (EAMU formerly known as SOLA). EAMUs are uses for which individuals or organisations other than the manufacturers have sought approval. The Notices of Approval are published by CRD and are widely available from ADAS or NFU offices. Users of EAMUs must first obtain a copy of the relevant Notice of Approval and comply strictly with the conditions laid down therein. Users of this Guide will find details of extant EAMUs and direct links to the CRD website, where EAMU notices can also be accessed.
- In tank mixture with other approved pesticides in accordance with Consent C(i) made under FEPA. Full details of Consent C(i) are given in Annex A of Guide to Pesticides on the CRD website, but there are two essential requirements for tank mixes. First, all the conditions of approval of all the components of a mixture must be complied with. Second, no person may mix or combine pesticides that are cholinesterase compounds unless allowed by the label of at least one of the pesticides in the mixture.
- In conjunction with authorised adjuvants.
- In reduced spray volume under certain conditions.
- In the use of certain herbicides on specified set-aside areas subject to restrictions, which differ between Scotland and the rest of the UK.
- By mutual recognition of a use fully approved in another Member State of the European Union and authorised by CRD.

Although approved, off-label uses are not endorsed by manufacturers and such treatments are made entirely at the risk of the user.

Using Crop Protection Chemicals

Use of Herbicides In or Near Water

Products in this Guide approved for use in or near water are listed in Table 5.1. Before use of any product in or near water, the appropriate water regulatory body (Environment Agency/Local Rivers Purification Authority; or in Scotland the Scottish Environmental Protection Agency) must be consulted. Guidance and definitions of the situation covered by approved labels are given in the Defra publication Guidelines for the Use of Herbicides on Weeds in or near Watercourses and Lakes. Always read the label before use.

Table 5.1 Products approved for use in or near water

Chemical	Product
glyphosate	Asteroid, Asteroid Pro, Asteroid Pro 450, Barclay Gallup 360, Barclay Gallup Amenity, Barclay Gallup Biograde 360, Barclay Gallup Biograde Amenity, Barclay Gallup Hi-Aktiv, Barclay Glyde 144, Buggy XTG, Discman Biograde, Envision, Glyfos Dakar, Glyfos Dakar Pro, Glyfos Proactive, Kernel, Mascot Hi-Aktiv Amenity, Palomo Green, Pure Glyphosate 360, Reaper Green, Roundup Biactive GL, Roundup POWERMAX, Sherlock, Surrender, Trustee Amenity

Use of Pesticides in Forestry

Table 5.2 Products in this *Guide* approved for use in forestry

Chemical	Product	Use
aluminium ammonium sulphate	Curb Crop Spray Powder, Liquid Curb Crop Spray, Sphere ASBO	Animal deterrent/repellent
chlorpyrifos	Dursban WG, Equity, Pyrifos 480	Acaricide, Insecticide
cycloxydim	Laser	Herbicide
cypermethrin	Forester	Insecticide
diflubenzuron	Dimilin Flo	Insecticide
ferric phosphate	Derrex, Iroxx, Slugger, Sluxx HP	Molluscicide
fluazifop-P-butyl	Clayton Maximus, Fusilade Max, Greencrop Bantry, Salvo	Herbicide
glyphosate	Amega Duo, Azural, Barclay Gallup 360, Barclay Gallup Amenity, Barclay Gallup Biograde 360, Barclay Gallup Biograde Amenity, Barclay Gallup Hi-Aktiv, Barclay Glyde 144, Buggy XTG, Charger C, Clinic Ace, Clinic TF, Credit, Crestler, Discman Biograde, Ecoplug Max, Envision, Glyfos, Glyfos Dakar Pro, Glyfos Proactive, Glyfos Supreme, Glyfosat 36, Hilite, Kernel, Liaison, Mascot Hi-Aktiv Amenity, Mentor, Monsanto Amenity Glyphosate, Motif, Nomix Conqueror Amenity, Palomo Green, Pitch, Pure Glyphosate 360, Rattler, Reaper, Reaper Green, Rodeo, Samurai, Sherlock, Shyfo, Snapper, Stirrup, Surrender, Tanker, Trustee Amenity	Herbicide
imidacloprid	Merit Forest	Insecticide, Seed treatment
metazachlor	Rapsan 500 SC, Stalwart, Sultan 50 SC	Herbicide
Phlebiopsis gigantea	PG Suspension	Biological agent
propaquizafop	Clayton Enigma, Clayton Satchmo, Falcon, Longhorn, Shogun, Snipe	Herbicide
propyzamide	Barclay Propyz, Careca, Cohort, Kerb Granules, Master Prop 40, Pizza 400 SC, Proper Flo, Setanta 50 WP, Standon Santa Fe 50 WP	Herbicide
pyrethrins	Spruzit	Insecticide

Useful Information

Pesticides Used as Seed Treatments

Information on the target for these products can be found in the relevant pesticide profile in Section 2.

Table 5.3 Products used as seed treatments (including treatments on seed potatoes)

Chemical	Product	Formulation	Crop(s)
beta-cyfluthrin + clothianidin	Poncho Beta	FS	Fodder beet, Sugar beet
clothianidin	Deter	FS	Durum wheat, Triticale, Winter barley, Winter oats, Winter rye, Winter wheat
clothianidin + prothioconazole	Redigo Deter	FS	Durum wheat, Triticale, Winter barley, Winter oats, Winter rye, Winter wheat
cymoxanil + fludioxonil + metalaxyl-M	Wakil XL	WS	Combining peas, Vining peas
difenoconazole	Difend	FS	Winter wheat
difenoconazole + fludioxonil	Celest Extra	FS	Winter oats, Winter rye, Winter wheat
fludioxonil	Beret Gold	FS	Rye, Spring barley, Spring oats, Spring wheat, Triticale, Winter barley, Winter oats, Winter wheat
	Maxim 100FS	FS	Potatoes, Seed potatoes
	Maxim 480FS	FS	Baby leaf crops, Broccoli, Brussels sprouts, Bulb onions, Cabbages, Carrots, Cauliflowers, Celeriac, Collards, Fennel, Kale, Kohlrabi, Oriental cabbage, Parsley root, Parsnips, Protected broccoli, Protected Brussels sprouts, Protected cauliflowers, Protected celeriac, Protected collards, Protected fennel, Protected kale, Protected kohlrabi, Protected oriental cabbage, Protected parsley root, Protected parsnips, Protected radishes, Radishes, Red beet, Salad onions, Shallots, Spinach
fludioxonil + metalaxyl-M	Maxim XL	FS	Forage maize
fludioxonil + tefluthrin	Austral Plus	FS	Spring barley, Spring oats, Spring wheat, Triticale seed crop, Winter barley, Winter oats, Winter wheat

Using Crop Protection Chemicals

Chemical	Product	Formulation	Crop(s)
fluopyram + prothioconazole + tebuconazole	Raxil Star	FS	Spring barley, Winter barley
flutolanil	Rhino DS	DS	Potatoes
metalaxyl-M	Apron XL	ES	Beetroot, Brassica leaves and sprouts, Broccoli, Brussels sprouts, Bulb onions, Cabbages, Calabrese, Cauliflowers, Chard, Chinese cabbage, Herbs (see appendix 6), Kohlrabi, Ornamental plant production, Radishes, Shallots, Spinach
methiocarb	Mesurol	FS	Fodder maize, Grain maize, Sweetcorn
pencycuron	Monceren DS	DS	Potatoes
	Solaren	DS	Potatoes
physical pest control	Silico-Sec	DS	Stored grain
prochloraz + thiram	Agrichem Hy-Pro Duet	FS	Winter oilseed rape
prothioconazole + tebuconazole	Redigo Pro	FS	Durum wheat, Rye, Spring barley, Spring oats, Spring wheat, Triticale, Winter barley, Winter oats, Winter wheat
silthiofam	Latitude XL	FS	Spring wheat, Winter barley, Winter wheat
	Meridian	FS	Spring wheat, Winter barley, Winter wheat
thiamethoxam	Cruiser SB	FS	Fodder beet, Kale, Sugar beet, Swedes

SECTION 5

Useful Information

Chemical	Product	Formulation	Crop(s)
thiram	Agrichem Flowable Thiram	FS	Broad beans, Bulb onions, Cabbages, Carrots, Cauliflowers, Celery (outdoor), Combining peas, Corn Gromwell, Cress, Dwarf beans, Edible podded peas, Endives, Fodder beet, Forage maize, Frise, Garlic, Grass seed, Kale, Lamb's lettuce, Leeks, Lettuce, Lupins, Mangels, Millet, Parsley, Poppies for morphine production, Quinoa, Radicchio, Radishes, Red beet, Rhubarb, Runner beans, Salad onions, Scarole, Shallots, Sorghum, Soya beans, Spring field beans, Spring oilseed rape, Sugar beet, Swedes, Turnip seed crops, Turnips, Vining peas, Winter field beans, Winter oilseed rape
	Thyram Plus	FS	Broad beans, Bulb onions, Cabbages, Cauliflowers, Combining peas, Dwarf beans, Edible podded peas, Forage maize, Garlic, Grass seed, Kale, Kale seed crops, Leeks, Lupins, Millet, Radishes, Runner beans, Salad onions, Shallots, Sorghum, Soya beans, Spring field beans, Spring oilseed rape, Swedes, Turnip seed crops, Turnips, Vining peas, Winter field beans, Winter oilseed rape

Using Crop Protection Chemicals

Aerial Application of Pesticides

New aerial spraying permit arrangements came into force in June 2012 and those undertaking aerial applications must ensure that the spraying is done in line with an Approved Application Plan that is subject to approval by CRD. Plans will only be approved where there is no viable alternative method of application or where aerial application results in reduced impact on human health and/or the environment compared to land-based application. Thus applications to bracken, forestry and possibly blight sprays to potatoes may meet these requirements but few others are likely to do so. Template application plans are available from the CRD web site and once all the necessary details have been received by CRD they undertake to approve or reject the plan within 10 working days.

Table 5.4 Products approved for aerial application

Chemical	Product	Crop(s)
diflubenzuron	Dimilin Flo	Forest
mancozeb	Laminator Flo	Potatoes
thifensulfuron-methyl + tribenuron-methyl	Ratio SX	Spring barley, Spring wheat, Winter barley, Winter oats, Winter wheat

Useful Information

Resistance Management

Pest species are, by definition, adaptable organisms. The development of resistance to some crop protection chemicals is just one example of this adaptability. Repeated use of products with the same mode of action will clearly favour those individuals in the pest population able to tolerate the treatment. This leads to a situation where the tolerant (or resistant) individuals can dominate the population and the product becomes ineffective. In general, the more rapidly the pest species reproduces and the more mobile it is, the faster is the emergence of resistant populations, although some weeds seem able to evolve resistance more quickly than would be expected. In the UK, key independent research organisations, chemical manufacturers and other organisations have collaborated to share knowledge and expertise on resistance issues through three action groups. Participants include **ADAS,** the **Chemicals Regulation Directorate,** universities, colleges and the **Home Grown Cereals Authority**. The groups have a common aim of monitoring resistance in the UK and devising and publishing management strategies designed to combat it where it occurs. The groups are:

- **The Weed Resistance Action Group (WRAG)**, formed in 1989 (Secretary: Dr Stephen Moss, Rothamsted Research, Harpenden, Herts AL5 2JQ *Tel: 01582 7631330 extension 2521*).
- **The Fungicide Resistance Action Group (FRAG),** formed in 1995 (Secretary: Mr Paul Ashby, HSE, Email: paul.ashby@hse.gsi.gov.uk).
- **The Insecticide Resistance Action Group (IRAG)**, formed in 1997 (Secretary: Dr Sacha White ADAS Boxworth, Battlegate Road, Cambridge, Cambs, CB3 8NN. *Tel: 01954 267666*, Email: sacha.white@adas.co.uk).

The above groups publish detailed advice on resistance management relevant for each sector and, in some cases, specific to a pest problem. This information, together with further details about the function of each group, can be obtained from the **Chemicals Regulation Directorate** website.

The speed of appearance of resistance depends on the mode of action of the crop protection chemicals, as well as the manner in which they are used. Resistance among insects and fungal diseases has been evident for much longer than weed resistance to herbicides, but examples in all three categories are now widespread and increasing. This has created a need for agreement on the advice given for the use of crop protection chemicals in order to reduce the likelihood of the development of resistance and to avoid the loss of potentially valuable products in the chemical armoury. Mixing or alternating modes of action is one of the guiding principles of resistance management. To assist appropriate product choices, mode of action codes, published by the international Resistance Action Committees (see below), are shown in the respective active ingredient fact sheets. The product label and/or a professional advisor should always be consulted before making decisions. The general guidelines for resistance management are similar for all three problem areas:

Preparation in advance

- Be aware of the factors that favour the development of resistance, such as repeated annual use of the same product, and assess the risk.
- Plan ahead and aim to integrate all possible means of control.
- Use cultural measures such as rotations, stubble hygiene, variety selection and, for fungicides, removal of primary inoculum sources, to reduce reliance on chemical control.
- Monitor crops regularly.
- Keep aware of local resistance problems.
- Monitor effectiveness of actions taken and take professional advice, especially in cases of unexplained poor control.

Using crop protection products

- Optimise product efficacy by using it as directed, at the right time, in good conditions.
- Treat pest problems early.
- Mix or alternate chemicals with different modes of action.
- Avoid repeated applications of very low doses.
- Keep accurate field records.

Label guidance depends on the appropriate strategy for the product. Most frequently it consists of a warning of the possibility of poor performance due to resistance, and a restriction on the number of treatments that should be applied in order to minimise the development of resistance. This information is summarised in the profiles in the active ingredient fact sheets, but detailed guidance must always be obtained by reading the label itself before use.

International Action Committees

Resistance to crop protection products is an international problem. Agrochemical industry collaboration on a global scale is via three action committees whose aims are to support a coordinated industry approach to the management of resistance worldwide. In particular they produce lists of crop protection chemicals classified according to their mode of action. These lists and other information can be obtained from the respective websites:

Herbicide Resistance Action Committee (HRAC) – www.hracglobal.com

Fungicide Resistance Action Committee (FRAC) – http://www.frac.info/

Insecticide Resistance Action Committee (IRAC) – http://www.irac-onine.org/

Poisons and Poisoning

Chemicals Subject to the Poison Law

Certain products are subject to the provisions of the Poisons Act 1972. The Poisons Rules 1982 have been revoked under the Deregulation Act 2015 (Poisons and Explosive Precursors) and replaced by The Control of Poisons and Explosive Precursors Regulations 2015 with appropriate amendments to The Poisons Act 1972. These Rules include provisions for the storage and sale and supply of listed non-medicine poisons. Details can be accessed on the HSE website. The nature of the formulation and the concentration of the active ingredient allow some products to be exempted from the Rules, while others with the same active ingredient are included (see below). The chemicals approved for use in the UK are specified under Parts I and II of the Poisons List as follows.

Part I Poisons (sale restricted to registered retail pharmacists and to registered non-pharmacy businesses provided sales do not take place on retail premises):

- aluminium phosphide
- chloropicrin
- magnesium phosphide.

Part II Poisons (sale restricted to registered retail pharmacists and listed sellers registered with a local authority):

- formaldehyde
- oxamyl (a).

Notes

(a) Granular formulations that do not contain more than 12% w/w of this or a combination of similarly flagged poisons are exempt.

Occupational Exposure Limits

A fundamental requirement of the COSHH Regulations is that exposure of employees to substances hazardous to health should be prevented or adequately controlled. Exposure by inhalation is usually the main hazard, and in order to measure the adequacy of control of exposure by this route various substances have been assigned occupational exposure limits.

There are two types of occupational exposure limits defined under COSHH: Occupational Exposure Standards (OES) and Maximum Exposure Limits (MEL). The key difference is that an OES is set at a level at which there is no indication of risk to health; for a MEL a residual risk may exist and the level takes socio-economic factors into account. In practice, MELs have been most often allocated to carcinogens and to other substances for which no threshold of effect can be identified and for which there is no doubt about the seriousness of the effects of exposure.

OESs and MELs are set on the recommendation of the Advisory Committee on Toxic Substances (ACTS). Full details are published by HSE in EH 40/2005 available at: http://www.hse.gov.uk/pubns/books/eh40.htm

As far as pesticides are concerned, OESs and MELs have been set for relatively few active ingredients. This is because pesticide products usually contain other substances in their formulation, including solvents, which may have their own OES/MEL. In practice inhalation of solvent may be at least, or more, important than that of the active ingredient. These factors are taken into account by the regulators when approving a pesticide product under the Control of Pesticides Regulations. This indicates one of the reasons why a change of pesticide formulation usually necessitates a new approval assessment under COPR.

First Aid Measures

If pesticides are handled in accordance with the required safety precautions, as given on the container label, poisoning should not occur. It is difficult, however, to guard completely against the occasional accidental exposure. Thus, if a person handling, or exposed to, pesticides becomes ill, it is a wise precaution to apply first aid measures appropriate to pesticide poisoning even though the cause of illness may eventually prove to have been quite different. An employer has a legal duty to make adequate first aid provision for employees. Regular pesticide users should consider appointing a trained first aider even if numbers of employees are not large, since there is a specific hazard.

The first essential in a case of suspected poisoning is for the person involved to stop work, to be moved away from any area of possible contamination and for a doctor to be called at once. If no doctor is available the patient should be taken to hospital as quickly as possible. In either event it is most important that the name of the chemical being used should be recorded and preferably the whole product label or leaflet should be shown to the doctor or hospital concerned.

Some pesticides, which are unlikely to cause poisoning in normal use, are extremely toxic if swallowed accidentally or deliberately. In such cases get the patient to hospital as quickly as possible, with all the information you have. Some labels now include Material Safety Data Sheets and these contain valuable information for both the first aider and medical staff. If not included on the label, MSDS are available from company websites.

General Measures

Measures appropriate in all cases of suspected poisoning include the following:

- Remove any protective or other contaminated clothing (taking care to avoid personal contamination).
- Wash any contaminated areas carefully with water or with soap and water if available.
- In cases of eye contamination, flush with plenty of clean water for at least 15 minutes.
- Lay the patient down, keep at rest and under shelter. Cover with one clean blanket or coat, etc. Avoid overheating.
- Monitor level of consciousness, breathing and pulse rate.
- If consciousness is lost, place the casualty in the recovery position (on his/her side with head down and tongue forward to prevent inhalation of vomit).

Reporting of Pesticide Poisoning

Any cases of poisoning by pesticides must be reported without delay to an HM Agricultural Inspector of the Health and Safety Executive. In addition any cases of poisoning by substances named in schedule 2 of The Reporting of Injuries, Diseases and Dangerous Occurrences Regulations 1985, must also be reported to HM Agricultural Inspectorate (this includes organophosphorus chemicals, mercury and some fumigants).

Cases of pesticide poisoning should also be reported to the manufacturer concerned.

Additional Information

General advice on the safe use of pesticides is given in a range of Health and Safety Executive leaflets available from HSE Books. The major agrochemical companies are able to provide authoritative medical advice about their own pesticide products. Useful information is now available from the Material Safety Data Sheet available from the manufacturer or the manufacturer's website.

A useful booklet, *Guidelines for Emergency Measures in Cases of Pesticide Poisoning*, is available from CropLife International.

New arrangements for the provision of information to doctors about poisons and the management of poisonings have been introduced as part of the modernisation of the National Poisons Information Service (NPIS). The Service provides a year-round, 24-hour-a-day service for healthcare staff on the diagnosis, treatment and management of patients who may have been poisoned. The new arrangements are aimed at moving away from the telephone as the

Useful Information

first point of contact for poisons information, to the use by doctors of an online database, supported by a second-tier, consultant-led information service for more complex clinical advice. NPIS no longer provides direct information on poisoning to members of the public. **Anyone suspecting poisoning by pesticides should seek professional medical help immediately via their GP or NHS Direct (www.nhsdirect.nhs.uk) or NHS 24 (www.nhs24.com).**

Environmental Protection

Environmental Land Management: 8 simple steps for arable farmers

Choosing the right measures, putting them in the right place, and managing them in the right way will make all the difference to your farm environment. The general principles given here should be considered in conjunction with local priorities for soil and water protection and wildlife conservation. This approach complements best practice in soil, crop, fertiliser and pesticide management.

An improved farm environment can be achieved by good management of around 4% of the arable area where high quality habitats are maintained or created. However, the actual area required on your farm will depend on factors such as the area of vulnerable soils and length of watercourses.

If you need further advice then consult a competent environmental adviser.

What you can do

It is important to have a balance of environmental measures that contribute to each of the relevant points below to achieve improved environmental benefits.

1. **Look after established wildlife habitats**
   Start by assessing what you already have on the farm! Maintaining, or where necessary, restoring any existing wildlife habitats, such as woodland, ponds, flower-rich grassland or field margins, is critical to the survival of much of the wildlife on the farm, and may count towards some of the following measures without the need to create new habitats. Unproductive land can be used to create new habitats to complement what you already have.
2. **Maximise the environmental value of field boundaries**
   Hedgerow management and ditch management on a 2–3 year rotation boosts flowers, fruit and refuges for wildlife. This is most suited to hedges dominated by hawthorn and blackthorn, and ditches where rotational management will not compromise the drainage function. Establish new hedgerow trees to maintain or restore former numbers within the landscape.
3. **Create a network of grass margins**
   The highest priority is to buffer watercourses, ideally with a minimum of 5 m buffer strips. Grass margins can also be used to boost beneficial insects and small mammals, and buffer hedges, ponds and other environmental features. Beetle banks can be used to reduce soil erosion and run-off on slopes greater than 1:20 and boost beneficial insects in fields greater than 20 ha.
4. **Establish flower rich habitats**
   Evidence suggests that a network of flower-rich margins on 1% of arable land will support beneficial insects and a wealth of wildlife that feeds on insects. Assess whether this is best done by allowing arable plants in the seed bank to germinate, establishing perennial margins with a grass and wildflower mix, or using nectar flower mixtures. Improving the linkages between these features on the farm will also help wildlife move across the landscape.
5. **Provide winter food for birds with weedy over-wintered stubbles or wild bird cover**
   Provision of seed for wildlife is best achieved by leaving over-wintered stubbles unsprayed and uncultivated until at least mid-February on at least 5% of arable land, or growing seed-rich crops as wild bird cover on 2% of arable land.
6. **Use of spring cropping or in-field measures to help ground-nesting birds**
   Spring crops provide better habitat for a range of plants and insects, and birds such as lapwings and skylarks. Use rotational fallows, skylark plots in winter cereals or (if breeding lapwings occur) fallow plots to support ground-nesting birds where spring cropping forms less than 25% of the arable area. Fallow plots should not be created on land liable to runoff or erosion. Evidence suggests that at least 20 skylark plots or a 1 ha fallow plot per 100 ha would support ground-nesting birds.

Useful Information

7. **Use winter cover crops to protect water**
   You should consider whether a winter cover crop (e.g. mustard) is necessary to capture residual nitrogen on cultivated land left fallow through the winter. This is not necessary if the stubble is retained until at least mid-February and forms a green cover.
8. **Establish in-field grass areas to reduce soil erosion and run-off**
   Land liable to act as channels for soil erosion or run-off (e.g. steep slopes or field corners) should be converted to in-field grass areas.

Remember

- Right measures
- Right place
- Right management

This guidance has been produced by The Voluntary Initiative and the Campaign for the Farmed Environment in conjunction with RSPB, Natural England, GWCT, Plantlife, Butterfly Conservation, BCPC and Buglife

Protection of Bees

Honey bees

Honey bees are a source of income for their owners and important to farmers and growers as pollinators of their crops. It is irresponsible and unnecessary to use pesticides in such a way that may endanger them. Pesticides vary in their toxicity to bees, but those that present a special hazard carry a specific warning in the precautions section of the label. They are indicated in this *Guide* in the hazard classification and safety precautions section of the pesticide profile.

Product labels indicate the necessary environmental precautions to take, but where use of an insecticide on a flowering crop is contemplated the British Beekeepers Association have produced the following guidelines for growers:

- Target insect pests with the most appropriate product.
- Choose a product that will cause minimal harm to beneficial species.
- Follow the manufacturer's instructions carefully.
- Inspect and monitor crops regularly.
- Avoid spraying crops in flower or where bees are actively foraging.
- Keep down flowering weeds.
- Spray late in the day, in still conditions.
- Avoid excessive spray volume and run-off.
- Adjust sprayer pressure to reduce production of fine droplets and drift.
- Give local beekeepers as much warning of your intention as possible.

Wild bees

Wild bees also play an important role. Bumblebees are useful pollinators of spring flowering crops and fruit trees because they forage in cool, dull weather when honey bees are inactive. They play a particularly important part in pollinating field beans, red and white clover, lucerne and borage. Bumblebees nest and overwinter in field margins and woodland edges. Avoidance of direct or indirect spray contamination of these areas, in addition to the creation of hedgerows and field margins, and late cutting or grazing of meadows, all help the survival of these valuable insects.

BeeConnected

It is best practice to notify local beekeepers before using certain crop protection products where there is a risk to bees. The online BeeConnected tool (www.beeconnected.org.uk) can help you do this.

The Campaign Against Accidental or Illegal Poisoning of Wildlife

The Campaign Against Accidental or Illegal Poisoning (CAIP) of Wildlife, aimed at protecting some of Britain's rarest birds of prey and wildlife whilst also safeguarding domestic animals, was launched in March 1991 by the (then) Ministry of Agriculture, Fisheries and Food and the Department of the Environment, Transport and the Regions. The main objective is to deter those who may be considering using pesticides illegally.

The Campaign is supported by a range of organisations associated with animal welfare, nature preservation, field sports and gamekeeping including the RSPB, Natural England, the Countryside Alliance and the Game and Wildlife Conservation Trust.

The three objectives are:

- To advise farmers, gamekeepers and other land managers on legal ways of controlling pests;
- To advise the public on how to report illegal poisoning incidents and to respect the need for legal alternatives;
- To investigate incidents and prosecute offenders.

A freephone number (0800 321 600) is available to make it easier for the public to report incidents, and numerous leaflets, posters, postcards and stickers have been created to publicise the existence of the Campaign. A video has also been produced to illustrate the many talks, demonstrations and exhibitions.

The Campaign arose from the results of the Wildlife Incident Investigation Scheme for the investigation of possible cases of illegal poisoning. Under this scheme, all reported incidents are considered and thoroughly investigated where appropriate. Enforcement action is taken wherever sufficient evidence of an offence can be obtained and numerous prosecutions have been made since the start of the Campaign.

Further information about the Campaign is available from: Chemicals Regulation Directorate, Research Co-ordination and Environmental Policy Branch, Room 317, Mallard House, 3 Peasholme Green, York YO1 7PX, or from the website: www.caip-uk.info WIIS hotline: 0800 321600 E-mail: wiis@hse.gsi.gov.uk

Water Quality

Even when diluted, some pesticides are potentially dangerous to fish and other aquatic life. Not only this, but many watercourses and groundwaters are sources of drinking water, and it requires only a tiny amount of contamination to breach the stringent European Union water quality standards. The EEC Drinking Water Directive sets a maximum admissible level in drinking water for any pesticide, regardless of its toxicity, at 1 part in 10,000 million. As little as 250 grams could be enough to cause the daily supply to a city the size of London to exceed the permitted levels (although it would be very unlikely to present a health hazard to consumers).

The protection of groundwater quality is therefore vital. The Food and Environment Protection Act 1985 (FEPA) places a special obligation on users of pesticides to "safeguard the environment and in particular avoid the pollution of water". Under the Water Resources Act 1991 it is an offence to pollute any controlled waters (watercourses or groundwater), either deliberately or accidentally. Protection of controlled waters from pollution is the responsibility of the Environment Agency (in England and Wales) and the Scottish Environmental Protection Agency. Addresses for both organisations can be found under Useful Contacts.

Users of pesticides therefore have a duty to adopt responsible working practices and, unless they are applying herbicides in or near water, to prevent them getting into water. Guidance on how to achieve this is given in the Defra Code of Good Agricultural Practice for the Protection of Water.

The duty of care covers not only the way in which a pesticide is sprayed, but also its storage, preparation and disposal of surplus, sprayer washings and the container. Products that are a major hazard to fish, other aquatic life or aquatic higher plants carry one of several specific label precautions in their fact sheet, depending on the assessed hazard level.

Where to get information

For general enquiries telephone 03708 506 506 (Environment Agency) or 0131 449 7296 (SEPA). Both Agencies operate a 24-hour emergency hotline for reporting all environmental incidents relating to air, land and water:

Useful Information

0800 80 70 60 (for incidents in England and Wales) 0345 73 72 71 (for incidents in Scotland)

In addition, printed literature concerning the protection of water is available from the Environment Agency and the Crop Protection Association (see Useful Contacts).

Protecting surface waters

Surface waters are particularly vulnerable to contamination. One of the best ways of preventing those pesticides that carry the greatest risk to aquatic wildlife from reaching surface waters is to prohibit their application within a boundary adjacent to the water. Such areas are known as no-spray, or buffer, zones. Certain products are restricted in this way and have a legally binding label precaution to make sure the potential exposure of aquatic organisms to pesticides that might harm them is minimised.

Before 1999 the protected zones were measured from the edge of the water. The distances were 2 metres for hand-held or knapsack sprayers, 6 metres for ground crop sprayers, and a variable distance (but often 18 metres) for broadcast air-assisted applications, such as in orchards. The introduction of LERAPs (see below) has changed the method of measuring buffer zones.

'Surface water' includes lakes, ponds, reservoirs, streams, rivers and watercourses (natural or artificial). It also includes temporarily or seasonally dry ditches, which have the potential to carry water at different times of the year. Buffer zone restrictions do not necessarily apply to all products containing the same active ingredient. Those in formulations that are not likely to contaminate surface water through spray drift do not pose the same risk to aquatic life and are not subject to the restrictions.

Local Environmental Risk Assessment for Pesticides (LERAPs)

Local Environment Risk Assessments for Pesticides (LERAPs) were introduced in March 1999, and revised guidelines were issued in January 2002. They give users of most products currently subject to a buffer zone restriction the option of continuing to comply with the existing buffer zone restriction (using the new method of measurement), or carrying out a LERAP and possibly reducing the size of the buffer zone as a result. In either case, there is a new legal obligation for the user to record his decision, including the results of the LERAP.

The scheme has changed the method of measuring the buffer zone. Previously the zone was measured from the edge of the water, but it is now the distance from the top of the bank of the watercourse to the edge of the spray area. In 2013 new rules were introduced which allowed LERAP buffer zones greater than the normal 5 metres for LERAP B pesticides to avoid the loss of some important actives. The products concerned are identified in the Environmental safety section of the affected pesticide profiles.

The LERAP provides a mechanism for taking into account other factors that may reduce the risk, such as dose reduction, the use of low drift nozzles, and whether the watercourse is dry or flowing. The previous arrangements applied the same restriction regardless of whether there was actually water present. Now there is a standard zone of 1 m from the top of a dry ditch bank.

Other factors to include in a LERAP that may allow a reduction in the buffer zone are:

- The size of the watercourse, because the wider it is, the greater the dilution factor, and the lower the risk of serious pollution.
- The dose applied. The lower the dose, the less is the risk.
- The application equipment. Sprayers and nozzles are star-rated according to their ability to reduce spray drift fallout. Equipment offering the greatest reductions achieves the highest rating of three stars. The scheme was originally restricted to ground crop sprayers; new, more flexible rules introduced in February 2002 included broadcast air-assisted orchard and hop sprayers.
- Other changes introduced in 2002 allow the reduction of a buffer zone if there is an appropriate living windbreak between the sprayed area and a watercourse.

Not all products that had a label buffer zone restriction are included in the LERAP scheme. The option to reduce the buffer zone does not apply to organophosphorus or synthetic pyrethroid insecticides. This group are classified as Category A products. All other products that had a label buffer zone restriction are classified as Category B. In addition some products have a 'Broadcast air-assisted LERAP' classification. The wording of the buffer zone label precautions has been amended for all products to take account of the new method of measurement, and whether or not the particular product qualifies for inclusion in the LERAP scheme.

Products in this *Guide* that are in Category A or B, or have a broadcast air-assisted LERAP, are identified with an appropriate icon on the product fact sheet. Updates to the list are published regularly by CRD and details can be obtained from its website at www.pesticides.gov.uk.

The introduction of LERAPs was an important step forward because it demonstrated a willingness to reduce the impact of regulation on users of pesticides by allowing flexibility where local conditions make it safe to do so. This places a legal responsibility on the user to ensure the risk assessment is done either by himself or by the spray operator or by a professional consultant or advisor. It is compulsory to record the LERAP and make it available for inspection by enforcement authorities. In 2013 new rules were introduced which allowed LERAP buffer zones greater than 5 m to avoid the loss of some important actives. These products are identified in the section of Environmental safety for the pesticide profiles affected.

More details of the LERAP arrangements and guidance on how to carry out assessments are published in the Ministry booklets PB5621 *Local Environmental Risk Assessment for Pesticides – Horizontal Boom Sprayers,* and booklet PB6533 *Local Environment Risk Assessment for Pesticides – Broadcast Air-Assisted Sprayers.* Additional booklets, PB2088 *Keeping Pesticides Out of Water,* and PB3160 *Is Your Sprayer Fit for Work* give general practical guidance.

Groundwater Regulations

Groundwater Regulations were introduced in 1999 to complete the implementation in UK of the EU Groundwater Directive (Protection of Groundwater Against Pollution Caused by Certain Dangerous Substances – 80/68/EEC). These Regulations help prevent the pollution of groundwater by controlling discharges or disposal of certain substances, including all pesticides.

Groundwater is defined under the Regulations as any water contained in the ground below the water table. Pesticides must not enter groundwater unless it is deemed by the appropriate Agency to be permanently unsuitable for other uses. The Agricultural Waste Regulations were introduced in May 2006 and apply to the disposal of all farm wastes, including pesticides. With certain exemptions farmers are required to obtain a waste management licence for most waste disposal activities.

As far as pesticides are concerned the new Regulations made little difference to existing controls, except for the disposal of empty containers and the disposal of sprayer washings and rinsings. It remains an offence to dispose of pesticides onto land without official authorisation. Normal use of a pesticide in accordance with product approval does not require authorisation. This includes spraying the washings and rinsings back on the crop provided that, in so doing, the maximum approved dose for that product on that crop is not exceeded. However, those wishing to use a lined biobed for this purpose need to obtain an exemption from the Agency.

In practice the best advice to farmers and growers is to plan to use all diluted spray within the crop and to dispose of all washings via the same route making sure that they stay within the conditions of approval of the product. The enforcing agencies for these Regulations are the Environment Agency (in England and Wales) and the Scottish Environment Protection Agency. The Agencies can serve notice at any time to modify the conditions of an authorisation where necessary to prevent pollution of groundwater.

Integrated Farm Management (IFM)

Integrated farm management is a method of farming that balances the requirements of running a profitable farming business with the adoption of responsible and sensitive environmental management. It is a whole-farm, long-term strategy that combines the best of modern technology with some basic principles of good farming practice. It is a realistic way forward that addresses the justifiable concerns of the environmental impact of modern farming practices, at the same time as ensuring that the industry remains viable and continues to provide

Useful Information

wholesome, affordable food. IFM embraces arable and livestock management. The phrase 'integrated crop management' is sometimes used where a farm is wholly arable.

Pest control is essential in any management system. Much can be done to minimise the incidence and impact of pests, but their presence is almost inevitable. IFM ensures that, where a pest problem needs to be contained, the action taken is the best combination of all available options. Pesticides are one of these options, and form an essential, but by no means exclusive, part of pest control strategy.

Where chemicals are to be used, the choice of product should be made not only with the pest problem in mind, but with an awareness of the environmental and social risks that might accompany its use. The aim should be to use as much as necessary, but as little as possible. A major part of the approval process aims to safeguard the environment, so that any approved product, when used as directed, will not cause long-term harm to wildlife or the environment. This is achieved by specifying on the label detailed rules for the way in which a product may be used.

The skill in implementing an IFM pest control strategy is the decision on how easily these rules may be complied with in the particular situation where use is contemplated. Although many chemical options may be available, some are likely to be more suitable than others. This *Guide* lists, under the heading **Special precautions/Environmental safety**, the key label precautions that need to be considered in this context, although the actual product label must always be read before a product is used.

Campaign for the Farmed Environment

With the demise of set-aside, it was realised that the environmental benefits achieved by this scheme could be lost. The Campaign for the Farmed Environment (CFE) scheme seeks to retain or even exceed the environmental benefits achieved by set-aside by voluntary measures.

It has three main themes:

- farming for cleaner water and a healthier soil
- helping farmland birds thrive
- improving the environment for farm wildlife.

The scheme encourages farmers to sign up for Entry Level Stewardship (ELS) schemes and to target their already extensive knowledge of habitat management to greater effect. If successful, it will help to fend off the proposal that farmers of cultivated land in England will have to adopt environmental management options on up to 6% of their land.

SECTION 6
APPENDICES

Appendix 1
Suppliers of Pesticides and Adjuvants

Aceto:
Aceto Agricultural Chemicals
Corporation (UK) Ltd
2nd Floor, Refuge House
33-37 Watergate Row
Chester
Cheshire
CH1 2LE
Tel: (516) 627-6000 ext.
Email: ftrudwig@aceto.com
Web: www.aceto.com

Adama:
Adama Agricultural Solutions UK Ltd
Unit 15
Thatcham Business Village
Colthrop Way
Thatcham
Berks.
RG19 4LW
Tel: (01635) 860555
Fax: (01635) 861555
Email: ukenquiries@adama.com
Web: www.adama.com

AgChem Access: AgChemAccess Limited
Cedar House,
41 Thorpe Road
Norwich
Norfolk
NR1 1ES
Tel: 0845 459 9413
Fax: 0207 149 9815
Email: thomas@agchemaccess.com
Web: www.agchemaccess.com

Agform: Agform Ltd
Maidstone Heath
Blundell Lane
Bursledon
Southampton
SO31 1AA
Tel: 023 8040 7831
Fax: 023 8040 7198
Email: info@agform.com

Agrichem: Agrichem (International) Ltd
Industrial Estate
Station Road
Whittlesey
Cambs.
PE7 2EY
Tel: (01733) 204019
Fax: (01733) 204162
Email: admin@agrichem.co.uk
Web: www.agrichem.co.uk

Agrii: United Agri Products Ltd
Throws Farm
Stebbing
Great Dunmow
Essex
DM6 3AQ
Web: www.agrii.co.uk

Agropharm: Agropharm Limited
Buckingham Place
Church Road
Penn
High Wycombe
Bucks.
HP10 8LN
Tel: (01494) 816575
Fax: (01494) 816578
Email: sales@agropharm.co.uk
Web: www.agropharm.co.uk

Agroquimicos : Agroquimicos Genericos
Calle San Vicente Martir
no 85, 8, 46007
Valencia,
Spain
Email: thomas@agchemaccess.com

Agrovista: Agrovista UK Ltd
Rutherford House
Nottingham Science Park
University Boulevard
Nottingham
NG7 2PZ
Tel: (0115) 939 0202
Fax: (0115) 921 8498
Email: enquiries@agrovista.co.uk
Web: www.agrovista.co.uk

Albaugh UK: Albaugh UK Ltd
 Manor Farm
 Eddlethorpe
 Malton
 North Yorkshire
 YO17 9QT
 UK

Amega: Amega Sciences
 Lanchester Way
 Royal Oak Industrial Estate
 Daventry
 Northants.
 NN11 5PH
 Tel: (01327) 704444
 Fax: (01327) 71154
 Email: admin@amega-sciences.com
 Web: www.amega-sciences.com

Arysta: Arysta Life Sciences
 Route d'Artix, BP80
 64150
 Nogueres
 France
 Email: Don.pendergrast@arysta.com

B H & B: Battle Hayward & Bower Ltd
 Victoria Chemical Works
 Crofton Drive
 Allenby Road Industrial Estate
 Lincoln
 LN3 4NP
 Tel: (01522) 529206
 Fax: (01522) 538960
 Email: orders@battles.co.uk
 Web: www.battles.co.uk

Barclay:
 Barclay Chemicals Manufacturing Ltd
 Damastown Way
 Damastown Industrial Park
 Mulhuddart
 Dublin 15
 Ireland
 Tel: (+353) 1 811 2900
 Fax: (+353) 1 822 4678
 Email: info@barclay.ie
 Web: www.barclay.ie

Barrier: Barrier BioTech Ltd
 36/37 Haverscroft Industrial Estate
 New Road
 Attleborough
 Norfolk
 NR17 1YE
 Tel: (01953) 456363
 Fax: (01953) 455594
 Email: sales@barrier-biotech.com
 Web: www.barrier-biotech.com

BASF: BASF plc
 Agricultural Divison
 PO Box 4, Earl Road
 Cheadle Hulme
 Cheshire
 SK8 6QG
 Tel: (0845) 602 2553
 Fax: (0161) 485 2229
 Web: www.agricentre.co.uk

Bayer CropScience:
 Bayer CropScience Limited
 230 Cambridge Science Park
 Milton Road
 Cambridge
 CB4 0WB
 Tel: (01223) 226500
 Fax: (01223) 426240
 Web: www.bayercropscience.co.uk

Bayer Environ.:
 Bayer Environmental Science
 230 Cambridge Science Park
 Milton Road
 Cambridge
 CB4 0WB
 Tel: (01223) 226680
 Fax: (01223) 226635
 Email: colin.munford@bayer.com
 Web: www.bayer-escience.co.uk

Becesane : Becesane s.r.o.
 Rohacova, 188/377
 130 00
 PRAHA 3
 Czech Republic
 Email: thomas@agchemaccess.com

Belchim: Belchim Crop Protection Ltd
 Unit 1b, Fenice Court
 Phoenix Park
 Eaton Socon
 St Neots
 Cambs.
 PE19 8EP
 Tel: (01480) 403333
 Fax: (01480) 403444
 Email: info@belchim.com
 Web: www.belchim.co.uk

Belcrop: Belcrop (UK)
 21 Victoria Rd
 Wargrave
 Berkshire
 RG10 8AD
 Tel: (0118) 940 4264
 Fax: (0118) 940 4264
 Email: JOHNHUDSON23@aol.com
 Web: www.BELCROP.com

Suppliers of Pesticides and Adjuvants

Biofresh: Biofresh Ltd
 INEX Business Centre
 Herschell Building
 Newcastle University Campus
 Newcastle on Tyne
 Tyne & Wear
 NE1 7RU
 Tel: 0191 243 0879
 Fax: 0191 243 0244
 Email: info@bio-fresh.co.uk

Certis: Certis
 Suite 5, 1 Riverside
 Granta Park
 Great Abington
 Cambs.
 CB21 6AD
 Tel: 0845 373 0305
 Fax: 01223 891210
 Email: certis@certiseurope.co.uk
 Web: www.certiseurope.co.uk

Chemtura: Chemtura Europe Ltd
 Kennet House
 4 Langley Quay
 Slough
 Berks.
 SL3 6EH
 Tel: (01753) 603000
 Fax: (01753) 603077
 Web: www.chemtura.com

Chem-Wise: Chem-Wise Ltd
 Westwood
 Rattar Mains
 Thurso
 KW14 8XW

Chiltern: Chiltern Farm Chemicals Ltd
 East Mellwaters
 Stainmore
 Bowes
 Barnard Castle
 Co. Durham
 DL12 9RH
 Tel: (01833) 628282
 Email: chilternfarm@aol.com
 Web: www.chilternfarm.com

Ciech: CIECH Sarzyna S.A
 Chemikow 1 str
 37-310 Nowa Sarzyna
 Poland

Clayton: Clayton Plant Protection Ltd
 Bracetown Business Park
 Clonee
 Dublin 15
 Ireland
 Tel: (+353) 1 821 0127
 Fax: (+353) 81 841 1084
 Email: info@cpp.ag
 Web: www.cpp.ag

Custodian:
 Custodian Computer Services Ltd
 Plash farm
 Bedingfield
 Eye
 Suffolk
 IP23 7QD

DAPT: DAPT Agrochemicals Ltd
 14 Monks Walk
 Southfleet
 Gravesend
 Kent
 DA13 9NZ
 Tel: (01474) 834448
 Fax: (01474) 834449
 Email: rkjltd@supanet.com

De Sangosse: De Sangosse Ltd
 Hillside Mill
 Quarry Lane
 Swaffham Bulbeck
 Cambridge
 CB25 0LU
 Tel: (01223) 811215
 Fax: (01223) 810020
 Email: info@desangosse.co.uk
 Web: www.desangosse.co.uk

Douglas : Douglas BLG
 BVBA, 97
 4th Floor, Rue Royale
 1000 Brussels
 Brussels
 Belgium

Dow: Dow AgroSciences Ltd
 CPC2,
 Capital Park
 Fulbourn
 Cambridge
 Cambridgeshire
 CB21 5XE
 Tel: (01462) 457272
 Fax: (01462) 426605
 Email: ukhotline@dow.com
 Web: www.uk.dowagro.com

Appendices

DuPont: DuPont (UK) Ltd
Crop Protection Products Department
4th Floor, Kings Court
London Road
Stevenage
Herts.
SG1 2NG
Tel: (01438) 734450
Fax: (01438) 734452
Email: enquiry.agproducts@dupont.com
Web: www.cropprotection.dupont.co.uk

Elliott: Thomas Elliott Fertilisers
Selby Place
Stanley Industrial Estate
Skelmersdale
Lancashire
WN8 8EF
Tel: 01522 811981
Fax: 01522 810238
Email: christinehall@thomaselliot.demon.co.uk

Everris Ltd: Everris Limited
Epsilon House
West Road
Ipswich
Suffolk
IP3 9FJ
Tel: 01473 237100
Fax: 01473 830386
Email: prof.sales@everris.com

Exosect: Exosect Ltd
Leylands Business Park
Colden Common
Winchester
SO21 1TH
Tel: 02380 60 395
Email: sue.harris@exosect.com
Web: www.exosect.com

Fargro: Fargro Ltd
Vinery Fields
Arundel Road
Poling
Arundel
West Sussex
BN18 9PY
Tel: (01903) 721591
Fax: (01903) 883303
Email: info@fargro.co.uk
Web: www.fargro.co.uk

Fine: Fine Agrochemicals Ltd
Hill End House
Whittington
Worcester
WR5 2RQ
Tel: (01905) 361800
Fax: (01905) 361810
Email: enquire@fine.eu
Web: www.fine.eu

Forest Research: Forest Research
Alice Holt Lodge
Farnham
Surrey
GU10 4LH
Tel: 01420 22255
Fax: 01420 23653
Email: Katherine.tubby@forestry.gsi.gov.uk
Web: www.forestry.gov.uk

Generica : Generica Europa Ltd
The Granary
27C Silver Street
Buckden
St Neots
Cambridgeshire
PE19 5TS
Email: office@genericaeuropa.com

Globachem: Globachem NV
Brustem Industriepark
Lichtenberglaan 2019
3800 Sint-Truiden
Belgium
Tel: (+32) 11 69 01 73
Fax: (+32) 11 68 15 65
Email: globachem@globachem.com
Web: www.globachem.com

Global Adjuvants :
Global Adjuvants Company Ltd
20-22 Wenlock Road
London
N1 7GU

Goldengrass: Goldengrass Ltd
PO Box 280
Ware
SG12 8XY
Tel: 01920 486253
Fax: 01920 485123
Email: contact@goldengrass.uk.com

Gowan:
Gowan Comercio e Servicos Limitada
2 Leazes Cresent
Hexham
Northumberland
NE46 3JX
Tel: 07584 052323
Email: DLamb@GOWANCO.com

Greenaway: Greenaway Amenity Ltd
7 Browntoft Lane
Donington
Spalding
Lincs.
PE11 4TQ
Tel: (01775) 821031
Fax: (01775) 821034
Email: greenawayamenity@aol.com
Web: www.greenawaycda.com

Greencrop: Greencrop Technology Ltd
c/o Clayton Plant Protection Ltd
Bracetown Business Park
Clonee
Dublin 15
Ireland
Tel: (+353) 1 821 0127
Fax: (+353) 81 841 1084
Email: info@cpp.ag
Web: www.cpp.ag

Headland:
Headland Agrochemicals trading as Cheminova (UK) Ltd
Rectors Lane
Pentre
Deeside
Flintshire
CH5 2DH
Tel: (01244) 537370
Fax: (01244) 532097
Email: enquiry@headlandgroup.com
Web: www.headland-ag.co.uk

Headland Amenity:
Headland Amenity Limited
1 Burr Elm Court
Main Street
Caldecote
Cambs.
CB23 7NU
Tel: (01223) 481090
Fax: (01223) 491091
Email: info@headlandamenity.com
Web: www.headlandamenity.com

Helena: Helena Chemical Company
Cambridge House
Nottingham Road
Stapleford
Nottingham
NG9 8AB
Tel: (0115) 939 0202
Fax: (0115) 939 8031

Hermoo: Hermoo (UK)
21 Victoria Road
Wargrave
Berks.
RG10 8AD
Tel: (0118) 940 4264
Fax: (0118) 940 4264
Email: johnhudson23@aol.com
Web: www.hermoo.com

Industrias Afrasa: Industrias Afrasa, S.A
Ciudad de Sevilla, 53
Pol. Industrial Fuenta del Jarro
46988 - Paterna
Valencia
Spain

Interagro: Interagro (UK) Ltd
Thorley Wash Barn
London Road
Thorley
Bishop's Stortford
Hertfordshire
CM23 4AT
Tel: (01279) 714970
Fax: (01279) 758227
Email: info@interagro.co.uk
Web: www.interagro.co.uk

Interfarm: Interfarm (UK) Ltd
Kinghams's Place
36 Newgate Street
Doddington
Cambs.
PE15 0SR
Tel: (01354) 741414
Fax: (01354) 741004
Email: technical@interfarm.co.uk
Web: www.interfarm.co.uk

Intracrop: Intracrop
Park Chambers,
10 Hereford Road,
Abergavenny,
NP7 5PR
Tel: (01926) 634801
Fax: (01926) 634798
Email: admin@intracrop.co.uk
Web: www.intracrop.co.uk

Juno : Juno (Plant Protection) Ltd
Little Mill Farm
Underlyn Lane
Marden
Kent
TN12 9AT
Tel: 01580 765079
Email: nick@junopp.com

Appendices

Killgerm: Killgerm Chemicals Ltd
115 Wakefield Road
Flushdyke
Ossett
W. Yorks.
WF5 9AR
Tel: (01924) 268400
Fax: (01924) 264757
Email: info@killgerm.com
Web: www.killgerm.com

Koppert: Koppert (UK) Ltd
Unit 8, Tudor Rose Court
53 Hollands Road
Haverhill
Suffolk
CB9 8PJ
Tel: (01440) 704488
Fax: (01440) 704487
Email: info@koppert.co.uk
Web: www.koppert.com

Landseer: Landseer Ltd
Lodge Farm
Goat Hall Lane
Galleywood
Chelmsford
Essex
CM2 8PH
Tel: (01245) 357109
Fax: (01245) 494165
Web: www.lanfruit.co.uk

Life Scientific: Life Scientific Limited
Unit 12, NovaUCD
Belfield Innovation Park
University College Dublin
Dublin 4
Ireland
Tel: 00353 1 283 2024
Web: www.lifescientific.com

Maxwell: Maxwell Amenity Limited
T/A Amenity Land Solutions
Units 1-3
Allscot
Telford
TF6 5DY

Microcide: Microcide Ltd
Shepherds Grove
Stanton
Bury St. Edmunds
Suffolk
IP31 2AR
Tel: (01359) 251077
Fax: (01359) 251545
Email: microcide@microcide co.uk
Web: www.microcide.co.uk

Mitsui:
Mitsui AgriScience International S.A
Unit 26, Block B
Northwood House,
Northwood Business Campus,
Santry
Dublin 9
Ireland
Tel: 81-3-3285-7540
Fax: 81-3-3285-9819

Monsanto: Monsanto (UK) Ltd
PO Box 663
Cambourne
Cambridge
CB1 0LD
Tel: (01954) 717575
Fax: (01954) 717579
Email: technical.helpline.uk@monsanto.com
Web: www.monsanto-ag.co.uk

Nissan: Nissan Chemical Europe Sarl
Parc d'Affaires de Crecy
2 Rue Claude Chappe
69371 St Didier au Mont d'Or
France
Tel: (+33) 4 376 44020
Fax: (+33) 4 376 46874

Nomix Enviro:
Nomix Enviro, A Division of Frontier
Agriculture Ltd
The Grain Silos
Weyhill Road
Andover
Hampshire
SP10 3NT
Tel: 01264 388050
Fax: 01522 866176
Email: nomixenviro@frontierag.co.uk
Web: www.nomix.co.uk

Nufarm UK: Nufarm UK Ltd
Wyke Lane
Wyke
West Yorkshire
BD12 9EJ
Tel: 01274 69 1234
Fax: 01274 69 1176
Email: infouk@uk.nufarm.com
Web: www.nufarm.co.uk

Omex: Omex Agriculture Ltd
Bardney Airfield
Tupholme
Lincoln
LN3 5TP
Tel: (01526) 396000
Fax: (01526) 396001
Email: enquire@omex.com
Web: www.omex.co.uk

Suppliers of Pesticides and Adjuvants

Pan Agriculture: Pan Agriculture Ltd
8 Cromwell Mews
Station Road
St Ives
Huntingdon
Cambs.
PE27 5HJ
Tel: (01480) 467790
Fax: (01480) 467041
Email: info@panagriculture.co.uk

Pan Amenity: Pan Amenity Ltd
8 Cromwell Mews
Station Road
St Ives
Cambs.
PE27 5HJ
Tel: 01480 467790
Fax: 01480 467041

Petro-Lube: Petro-Lube (UK) Limited
Lonsto House, Suit No: 10
276 Chase Road
Southgate
London
N14 6HA

Plant Solutions: Plant Solutions
Pyports
Downside Bridge Road
Cobham
Surrey
KT11 3EH
Tel: (01932) 576699
Fax: (01932) 868973
Email: sales@plantsolutionsltd.com
Web: www.plantsolutionsltd.com

Procam: Procam Group Ltd
Industrial Estate
Saxon Way
Melbourne
Royston
Herts
S98 6DN
Web: www.procam.co.uk

Progreen:
Progreen Weed Control Solutions Limited
Kellington House
South Fen Business Park
South Fen Road
Bourne
Lincolnshire
PE10 0DN
Email: info@progreen.co.uk
Web: www.progreen.co.uk

ProKlass: ProKlass products Limited
145-157 St John Street
London
EC1V 4PW
UK
Tel: 01480 810137
Email: office@proklass-products.com

Pure Amenity : Pure Amenity Limited
Pure House
64-66 Westwick Str.
Norwich
Norfolk
NR2 4SZ
Tel: 0845 257 4710
Email: martin@pureamenity.co.uk
Web: www.pureamenity.co.uk

Q-Chem: Q-CHEM nv
Leeuwerweg 138
BE-3803
Belgium
Tel: (+32) 11 785 717
Fax: (+32) 11 681 565
Email: q-chem@skynet.be

Rentokil: Rentokil Initial plc
7-8 Foundry Court,
Foundry Lane
Horsham
W. Sussex
RH13 5PY
Tel: (01403) 214122
Fax: (01403) 214101
Email: dawn.kirby@rentokil-initial.com
Web: www.rentokil.com

Resource Chemicals:
Resource Chemicals Ltd
High Street
Brackley
NN13 7DS
Tel: 01280 843800

Rigby Taylor: Rigby Taylor Ltd
Rigby Taylor House
Crown Lane
Horwich
Bolton
Lancs.
BL6 5HP
Tel: (01204) 677777
Fax: (01204) 677715
Email: info@rigbytaylor.com
Web: www.rigbytaylor.com

Appendices

Rotam : Rotam Europe Ltd
  Hamilton House
  Mabledon Place
  London
  WC1H 9BB
  Tel: 0207 953 0447
  Web: www.rotam.com/uk

Servem: Servem Limited
  Maxboley
  Duncormick
  County Wexford
  Ireland

SFP Europe: SFP Europe SA
  Espace des Remparts 10
  1950 Sion
  Switzerland

Sharda:
  Sharda Worldwide Exports PVT Ltd
  Domnic Holm
  29th Road
  Bandra (West)
  Mumbai
  400050
  India
  Tel: 0032 02 466 4444

Sherriff Amenity:
  Sherriff Amenity Services
  The Pines
  Fordham Road
  Newmarket
  Cambs
  CB8 7LG
  Tel: (01638) 721888
  Fax: (01638) 721815
  Web: www.sherriffamenity.com

Sinclair: William Sinclair Horticulture Ltd
  Firth Road
  Lincoln
  LN6 7AH
  Tel: (01522) 537561
  Fax: (01522) 513609
  Web: www.william-sinclair.co.uk

Sipcam: Sipcam UK Ltd
  3 The Barn
  27 Kneesworth Street
  Royston
  Herts.
  SG8 5AB
  Tel: (01763) 212100
  Fax: (01763) 212101
  Email: paul@sipcamuk.co.uk
  Web: sipcamuk.co.uk

Sphere: Sphere Laboratories (London) Ltd
  c/o Mainswood
  Putley Common
  Ledbury
  Herefordshire
  HR8 2RF
  Tel: 01684 899306
  Fax: 01684 893322
  Email: Fiona.Homes@Sphere-London.co.uk

Standon: Standon Chemicals Ltd
  48 Grosvenor Square
  London
  W1K 2HT
  Tel: (020) 7493 8648
  Fax: (020) 7493 4219

Stefes: Stefes GMBH
  Wendenstrasse 21b
  20097 Hamburg
  Germany
  Tel: +49 (0)40 533083-0
  Fax: +49 (0)40 5330833-29
  Email: info@stefes.eu
  Web: www.stefes.eu

Sumi Agro: Sumi Agro Europe Limited
  Vintners' Place
  68 Upper Thames Street
  London
  EC4V 3BJ
  Tel: (020) 7246 3697
  Fax: (020) 7246 3799
  Email: summit@summit-agro.com

Synergy : Synergy Generics Limited
  64-66 Pure House
  Norwich
  Norfolk
  NR2 4SZ
  Tel: 07884 435 282
  Email: thomas@agchemaccess.com

Syngenta: Syngenta UK Limited
  CPC4
  Capital Park
  Fulbourn
  Cambridge
  CB21 5XE
  Tel: 01223 883400
  Fax: (01223) 493700
  Web: www.syngenta-crop.co.uk

Taminco: Taminco UK Ltd
  15 Rampton Drift
  Longstanton
  Cambridge
  CB24 3EH
  Tel: 01954 789941
  Email: philip.forster@taminco.com
  Web: www.taminco.com

Suppliers of Pesticides and Adjuvants

Unique Marketing : Unique Marketing SL
Calle Marques Del Duero 67 Planta 1
Puerta A
San Pedro Alcantara,
29670 Marbella,
Malaga
Spain
Tel: +34 674 514 601

United Agri: United Agri Products Ltd
The Crossways
Alconbury Hill
Huntingdon
PE28 4JH
Tel: (01480) 418000
Fax: (01480) 418010
Web: www.agrii.co.uk

UPL Europe: UPL Europe Ltd
The Centre
Birchwood Park
Warrington
Cheshire
WA3 6YN
Tel: (01925) 819999
Fax: (01925) 856075
Email: rosie.wilson@uniphos.com
Web: www.upleurope.com

Ventura: Ventura Agroscience Ltd
20-22 Wenlock Road
London
N1 7GU
Tel: 01480 810137
Email: office@ventura-agroscience.com

Vitax: Vitax Ltd
Owen Street
Coalville
Leicester
LE67 3DE
Tel: (01530) 510060
Fax: (01530) 510299
Email: info@vitax.co.uk
Web: www.vitax.co.uk

Zantra: Zantra Limited
Westwood
Rattar mains
Scarfskerry
Thurso
Caithness
KW14 8XW
Tel: 01480 861066
Fax: 01480 861099

Appendix 2
Useful Contacts

Agriculture Industries Confederation Ltd
Confederation House
East of England Showground
Peterborough PE2 6XE
Tel: (01733) 385230
Web: www.agindustries.org.uk

AHDB Horticulture
Stoneleigh Park
Kenilworth
Warwickshire CV8 2TL
Tel: (024) 7669 2051
Web: www.horticulture.ahdb.org.uk

BASIS Ltd
St Monica's House Business Centre
39 Windmill Lane
Ashbourne
Derbyshire DE6 1EY
Tel: (01335) 343945
Fax: (01335) 301205
Web: www.basis-reg.co.uk

British Beekeepers' Association
National Beekeeping Centre
Stoneleigh
Kenilworth
Warwickshire CV8 2LG
Tel: (0871) 811 2282
Web: www.bbka.org.uk

British Beer & Pub Association
Ground Floor
Brewers' Hall
Aldermanbury Square
EC2V 7HR
Tel: (0207) 627 9191
Web: www.beerandpub.com

British Crop Production Council (BCPC)
7 Omni Business Centre
Omega Park
Alton
Hampshire GU34 2QD
Tel: (01420) 593200
Fax: (01420) 593209
Web: www.bcpc.org

British Pest Control Association (BPCA)
4A Mallard Way
Pride Park
Derby DE24 8GX
Tel: (01332) 294288
Fax: (01332) 225101
Web: www.bpca.org.uk

Chemicals Regulation Directorate
Room 1A, Mallard House
King's Pool
3 Peasholme Green
York YO1 7PX
Tel: (01904) 640500 or (03459) 335577
Fax: (01904) 455733
Web: www.pesticides.gov.uk

Crop Protection Association Ltd
2 Swan Court
Cygnet Park
Hampton
Peterborough PE7 8GX
Tel: (01733) 355370
Web: www.cropprotection.org.uk

CropLife International
326 Avenue Louise Box 35
B-1050 Brussels
Belgium
Tel: (+32) 2 542 0410
Fax: (+32) 2 542 0419
Web: www.croplife.org

Department of Agriculture and Rural Development (Northern Ireland)
Pesticides Section
Dundonald House
Upper Newtownards Road
Belfast BT4 3SB
Tel: (028) 9052 4704 or (0300) 200 7850
Fax: (028) 9052 4059
Web: www.daera-ni.gov.uk

Department of Environment, Food and Rural Affairs (Defra)
Nobel House
17 Smith Square
London SW1P 3JR
Tel: (020) 7238 6000 / 03459 335577
Web: www.defra.gov.uk

Environment Agency
National Customer Contact Centre
PO Box 544
Rotherham S60 1BY
Tel: (03708) 506 506
Email: enquiries@environment-agency.gov.uk
Web: www.environment-agency.gov.uk

Useful Contacts

European Crop Protection Association (ECPA)
Avenue E van Nieuwenhuyse 6
B-1160 Brussels
Belgium
Tel: (+32) 2 663 1550
Fax: (+32) 2 663 1560
Web: www.ecpa.eu

Farmers' Union of Wales
Llys Amaeth
Plas Gogerddan
Aberystwyth
Ceredigion SY23 3BT
Tel: (01970) 820820
Fax:(01970) 820821
Web: www.fuw.org.uk

Forestry Commission
620 Bristol Business Pk
Coldharbour Lane
Bristol BS16 1EJ
Tel: (0300) 067 4000
Web: www.forestry.gov.uk

Game and Wildlife Conservation Trust (GWCT)
Head Office
Burgate Manor
Fordingbridge
Hampshire
SP6 1EF
Tel: 01425 652381
Email: info@gwct.org.uk
Web: www.gwct.org.uk/

Health and Safety Executive
Product Approvals
Chemical Regulation
Redgrave Court
Merton Road
Bootle
Merseyside L20 7HS
Tel: (0151) 951 4000
Web: www.hse.gov.uk

Health and Safety Executive – Books
PO Box 5050
Sherwood Park
Annesley,
Nottinghamshire NG15 0JD

Lantra
National Agricultural Centre
Lantra House
Stoneleigh
Kenilworth
Warwickshire CV8 2LG
Tel: (024) 7669 6996
Fax: (024) 7669 6732
Web: www.lantra.co.uk

National Association of Agricultural Contractors (NAAC)
The Old Cart Shed
Easton Lodge Farm
Old Oundle Road, Wansford
Peterborough PE8 6NP
Tel: (01780) 784631
Fax: (01780) 784933
Web: www.naac.co.uk

National Chemicals Emergency Centre
The Gemini Building
Fermi Avenue
Harwell
Didcot
Oxfordshire OX11 0RG
Tel: (01235) 753654
Fax: (01235) 753656
Email: ncec@ricardo-aea.com
Web: www.the-ncec.com

National Farmers' Union
Agriculture House Stoneleigh Park
Stoneleigh
Warwickshire CV8 2TZ
Tel: (024) 7685 8500
Web: www.nfu.org.uk

National Poisons Information Service
(Birmingham Unit)
City Hospital, Dudley Road
Birmingham B18 7QH
Tel: (0121) 507 4123
Fax: (0121) 507 5580
Web: www.npis.org

Natural England (enquiries)
County Hall
Spetchley Road
Worcester
WR5 2NP
Tel: 0300 060 3900
Suspected wildlife/pesticide poisoning
Tel: 0800 321 600
Email: enquiries@naturalengland.org.uk
Web: www.gov.uk/government/organisations/natural-england

Processors and Growers Research Organisation
The Research Station
Great North Road, Thornhaugh
Peterborough, Cambs. PE8 6HJ
Tel: (01780) 782585
Web: www.pgro.org

Appendices

Scottish Beekeepers' Association
Cowiemuir
Fochabers
Moray IV32 7PS
Email: secretary@scottishbeekeepers.org.uk
Web: www.scottishbeekeepers.org.uk

Scottish Environment Protection Agency (SEPA)
Clearwater House
Heriot Watt Research Park
Avenue North
Riccarton
Edinburgh EH14 4AP
Tel: (0131) 449 7296
Fax: (0131) 449 7277
Web: www.sepa.org.uk

TSO (The Stationery Office)
85 Buckingham Gate
London SW1E 6PD
Tel: (0333) 202 5070
Email: customer.services@tso.co.uk
Web: www.tsoshop.co.uk

Appendix 3
Keys to Crop and Weed Growth Stages

Decimal Code for the Growth Stages of Cereals
Illustrations of these growth stages can be found in the reference indicated below and in some company product manuals.

0 Germination
- 00 Dryseed
- 03 Imbibition complete
- 05 Radicle emerged from caryopsis
- 07 Coleoptile emerged from caryopsis
- 09 Leaf at coleoptile tip

1 Seedling growth
- 10 First leaf through coleoptile
- 11 First leaf unfolded
- 12 2 leaves unfolded
- 13 3 leaves unfolded
- 14 4 leaves unfolded
- 15 5 leaves unfolded
- 16 6 leaves unfolded
- 17 7 leaves unfolded
- 18 8 leaves unfolded
- 19 9 or more leaves unfolded

2 Tillering
- 20 Main shoot only
- 21 Main shoot and 1 tiller
- 22 Main shoot and 2 tillers
- 23 Main shoot and 3 tillers
- 24 Main shoot and 4 tillers
- 25 Main shoot and 5 tillers
- 26 Main shoot and 6 tillers
- 27 Main shoot and 7 tillers
- 28 Main shoot and 8 tillers
- 29 Main shoot and 9 or more tillers

3 Stem elongation
- 30 Ear at 1 cm
- 31 1st node detectable
- 32 2nd node detectable
- 33 3rd node detectable
- 34 4th node detectable
- 35 5th node detectable
- 36 6th node detectable
- 37 Flag leaf just visible
- 39 Flag leaf ligule/collar just visible

4 Booting
- 41 Flag leaf sheath extending
- 43 Boots just visibly swollen
- 45 Boots swollen
- 47 Flag leaf sheath opening
- 49 First awns visible

5 Inflorescence
- 51 First spikelet of inflorescence just visible
- 52 1/4 of inflorescence emerged
- 55 1/2 of inflorescence emerged
- 57 3/4 of inflorescence emerged
- 59 Emergence of inflorescence completed

6 Anthesis
- 60
- 61 Beginning of anthesis
- 64
- 65 Anthesis half way
- 68
- 69 Anthesis complete

7 Milk development
- 71 Caryopsis watery ripe
- 73 Early milk
- 75 Medium milk
- 77 Late milk

8 Dough development
- 83 Early dough
- 85 Soft dough
- 87 Hard dough

9 Ripening
- 91 Caryopsis hard (difficult to divide by thumb-nail)
- 92 Caryopsis hard (can no longer be dented by thumb-nail)
- 93 Caryopsis loosening in daytime

(From Tottman, 1987. *Annals of Applied Biology*, **110**, 441–454)

Stages in Development of Oilseed Rape

Illustrations of these growth stages can be found in the reference indicated below and in some company product manuals.

0 Germination and emergence

1 Leaf production
- 1,0   Both cotyledons unfolded and green
- 1,1   First true leaf
- 1,2   Second true leaf
- 1,3   Third true leaf
- 1,4   Fourth true leaf
- 1,5   Fifth true leaf
- 1,10 About tenth true leaf
- 1,15 About fifteenth true leaf

2 Stem extension
- 2,0   No internodes ('rosette')
- 2,5   About five internodes

3 Flower bud development
- 3,0   Only leaf buds present
- 3,1   Flower buds present but enclosed by leaves
- 3,3   Flower buds visible from above ('green bud')
- 3,5   Flower buds raised above leaves
- 3,6   First flower stalks extending
- 3,7   First flower buds yellow ('yellow bud')

4 Flowering
- 4,0   First flower opened
- 4,1   10% all buds opened
- 4,3   30% all buds opened
- 4,5   50% all buds opened

5 Pod development
- 5,3   30% potential pods
- 5,5   50% potential pods
- 5,7   70% potential pods
- 5,9   All potential pods

6 Seed development
- 6,1   Seeds expanding
- 6,2   Most seeds translucent but full size
- 6,3   Most seeds green
- 6,4   Most seeds green-brown mottled
- 6,5   Most seeds brown
- 6,6   Most seeds dark brown
- 6,7   Most seeds black but soft
- 6,8   Most seeds black and hard
- 6,9   All seeds black and hard

7 Leaf senescence

8 Stem senescence
- 8,1   Most stem green
- 8,5   Half stem green
- 8,9   Little stem green

9 Pod senescence
- 9,1   Most pods green
- 9,5   Half pods green
- 9.9   Few pods green

(From Sylvester-Bradley, 1985. *Aspects of Applied Biology*, **10**, 395–400)

Keys to Crop and Weed Growth Stages

Stages in Development of Peas

Illustrations of these growth stages can be found in the reference indicated below and in some company product manuals.

0   Germination and emergence
   000   Dry seed
   001   Imbibed seed
   002   Radicle apparent
   003   Plumule and radicle apparent
   004   Emergence

1   Vegetative stage
   101   First node (leaf with one pair leaflets, no tendril)
   102   Second node (leaf with one pair leaflets, simple tendril)
   103   Third node (leaf with one pair leaflets, complex tendril)
   •
   l0x   X nodes (leaf with more than one pair leaflets, complex tendril)
   •
   •
   10n   Last recorded node

2   Reproductive stage (main stem)
   201   Enclosed buds
   202   Visible buds
   203   First open flower
   204   Pod set (small immature pod)
   205   Flat pod
   206   Pod swell (seeds small, immature)
   207   Podfill
   208   Pod green, wrinkled
   209   Pod yellow, wrinkled (seeds rubbery)
   210   Dry seed

3   Senescence stage
   301   Desiccant application stage. Lower pods dry and brown, middle yellow, upper green. Overall moisture content of seed less than 45%
   302   Pre-harvest stage. Lower and middle pods dry and brown, upper yellow. Overall moisture content of seed less than 30%
   303   Dry harvest stage. All pods dry and brown, seed dry

(From Knott, 1987. *Annals of Applied Biology*, **111**, 233–244)

Appendices

Stages in Development of Faba Beans
Illustrations of these growth stages can be found in the reference indicated below and in some company product manuals.

0 Germination and emergence

    000 Dry seed
    001 Imbibed seed
    002 Radicle apparent
    003 Plumule and radicle apparent
    004 Emergence
    005 First leaf unfolding
    006 First leaf unfolded

1 Vegetative stage

    101 First node
    102 Second node
    103 Third node
    •
    •
    l0x X nodes
    •
    •
    10n N, last recorded node

2 Reproductive stage (main stem)

    201 Flower buds visible
    203 First open flowers
    204 First pod set
    205 Pods fully formed, green
    207 Pod fill, pods green
    209 Seed rubbery, pods pliable, turning black
    210 Seed dry and hard, pods dry and black

3 Pod senescence

    301 10% pods dry and black
    •
    •
    305 50% pods dry and black
    •
    •
    308 80% pods dry and black, some upper pods green
    309 90% pods dry and black, most seed dry. Desiccation stage.
    310 All pods dry and black, seed hard. Pre-harvest (glyphosate application stage)

4 Stem senescence

    401 10% stem brown/black
    •
    •
    405 50% stem brown/black
    •
    •
    409 90% stem brown/black
    410 All stems brown/black. All pods dry and black, seed hard.

(From Knott, 1990. *Annals of Applied Biology*, **116**, 391–404)

Keys to Crop and Weed Growth Stages

Stages in Development of Potato
Illustrations of these growth stages can be found in the reference indicated below and in some company product manuals.

0 Seed germination and seedling emergence
- 000 Dry seed
- 001 Imbibed seed
- 002 Radicle apparent
- 003 Elongation of hypocotyl
- 004 Seedling emergence
- 005 Cotyledons unfolded

1 Tuber dormancy
- 100 Innate dormancy (no sprout development under favourable conditions)
- 150 Enforced dormancy (sprout development inhibited by environmental conditions)

2 Tuber sprouting
- 200 Dormancy break, sprout development visible
- 21x Sprout with 1 node
- 22x Sprout with 2 nodes
- •
- •
- 29x Sprout with 9 nodes
- 21x(2) Second generation sprout with 1 node
- 22x(2) Second generation sprout with 2 nodes
- •
- •
- 29x(2) Second generation sprout with 9 nodes

Where x = 1, sprout >2 mm;
2, 2-5 mm; 3, 5-20 mm;
4, 20-30 mm; 5, 50-100 mm;
6, 100-150 mm long

3 Emergence and shoot expansion
- 300 Main stem emergence
- 301 Node 1
- 302 Node 2
- •
- •
- 319 Node 19
- Second order branch
- 321 Node 1
- •
- •
- Nth order branch
- 3N1 Node 1
- •
- •
- 3N9 Node 9

4 Flowering
Primary flower
- 400 No flowers
- 410 Appearance of flower bud
- 420 Flower unopen
- 430 Flower open
- 440 Flower closed
- 450 Berry swelling
- 460 Mature berry

Second order flowers
- 410(2) Appearance of flower bud
- 420(2) Flower unopen
- 430(2) Flower open
- 440(2) Flower closed
- 450(2) Berry swelling
- 460(2) Mature berry

5 Tuber development
- 500 No stolons
- 510 Stolon initials
- 520 Stolon elongation
- 530 Tuber initiation
- 540 Tuber bulking (>10 mm diam)
- 550 Skin set
- 560 Stolon development

6 Senescence
- 600 Onset of yellowing
- 650 Half leaves yellow
- 670 Yellowing of stems
- 690 Completely dead

(From Jefferies & Lawson, 1991. *Annals of Applied Biology*, **119**, 387–389)

Stages in Development of Linseed

Illustrations of these growth stages can be found in the reference indicated below and in some company product manuals.

0 Germination and emergence
- 00 Dry seed
- 01 Imbibed seed
- 02 Radicle apparent
- 04 Hypocotyl extending
- 05 Emergence
- 07 Cotyledon unfolding from seed case
- 09 Cotyledons unfolded and fully expanded

1 Vegetative stage (of main stem)
- 10 True leaves visible
- 12 First pair of true leaves fully expanded
- 13 Third pair of true leaves fully expanded
- 1*n* *n* leaf fully expanded

2 Basal branching
- 21 One branch
- 22 Two branches
- 23 Three branches
- 2*n* *n* branches

3 Flower bud development (on main stem)
- 31 Enclosed bud visible in leaf axils
- 33 Bud extending from axil
- 35 Corymb formed
- 37 Buds enclosed but petals visible
- 39 First flower open

4 Flowering (whole plant)
- 41 10% of flowers open
- 43 30% of flowers open
- 45 50% of flowers open
- 49 End of flowering

5 Capsule formation (whole plant)
- 51 10% of capsules formed
- 53 30% 0f capsules formed
- 55 50% of capsules formed
- 59 End of capsule formation

6 Capsule senescence (on most advanced plant)
- 61 Capsules expanding
- 63 Capsules green and full size
- 65 Capsules turning yellow
- 67 Capsules all yellow brown but soft
- 69 Capsules brown, dry and senesced

7 Stem senescence (whole plant)
- 71 Stems mostly green below panicle
- 73 Most stems 30% brown
- 75 Most stems 50% brown
- 77 Stems 75% brown
- 79 Stems completely brown

8 Stems rotting (retting)
- 81 Outer tissue rotting
- 85 Vascular tissue easily removed
- 89 Stems completely collapsed

9 Seed development (whole plant)
- 91 Seeds expanding
- 92 Seeds white but full size
- 93 Most seeds turning ivory yellow
- 94 Most seeds turning brown
- 95 All seeds brown and hard
- 98 Some seeds shed from capsule
- 99 Most seeds shed from capsule

(From Jefferies & Lawson, 1991. *(From Freer, 1991. Aspects of Applied Biology,* **28**, *33–40)*

Stages in Development of Annual Grass Weeds

Illustrations of these growth stages can be found in the reference indicated below and in some company product manuals.

0 Germination and emergence

    00 Dry seed
    01 Start of imbibition
    03 Imbibition complete
    05 Radicle emerged from caryopsis
    07 Coleoptile emerged from caryopsis
    09 Leaf just at coleoptile tip

1 Seedling growth

    10 First leaf through coleoptile
    11 First leaf unfolded
    12 2 leaves unfolded
    13 3 leaves unfolded
    14 4 leaves unfolded
    15 5 leaves unfolded
    16 6 leaves unfolded
    17 7 leaves unfolded
    18 8 leaves unfolded
    19 9 or more leaves unfolded

2 Tillering

    20 Main shoot only
    21 Main shoot and 1 tiller
    22 Main shoot and 2 tillers
    23 Main shoot and 3 tillers
    24 Main shoot and 4 tillers
    25 Main shoot and 5 tillers
    26 Main shoot and 6 tillers
    27 Main shoot and 7 tillers
    28 Main shoot and 8 tillers
    29 Main shoot and 9 or more tillers

3 Stem elongation

    31 First node detectable
    32 2nd node detectable
    33 3rd node detectable
    34 4th node detectable
    35 5th node detectable
    36 6th node detectable
    37 Flag leaf just visible
    39 Flag leaf ligule just visible

4 Booting

    41 Flag leaf sheath extending
    43 Boots just visibly swollen
    45 Boots swollen
    47 Flag leaf sheath opening
    49 First awns visible

5 Inflorescence emergence

    51 First spikelet of inflorescence just visible
    53 1/4 of inflorescence emerged
    55 1/2 of inflorescence emerged
    57 3/4 of inflorescence emerged
    59 Emergence of inflorescence completed

6 Anthesis

    61 Beginning of anthesis
    65 Anthesis half-way
    69 Anthesis complete

(From Lawson & Read, 1992. *Annals of Applied Biology*, **12**, 211–214)

Growth Stages of Annual Broad-leaved Weeds

Preferred Descriptive Phrases
Illustrations of these growth stages can be found in the reference indicated below and in some company product manuals.

Pre-emergence
Early cotyledons
Expanded cotyledons
One expanded true leaf
Two expanded true leaves
Four expanded true leaves
Six expanded true leaves
Plants up to 25 mm across/high
Plants up to 50 mm across/high
Plants up to 100 mm across/high
Plants up to 150 mm across/high
Plants up to 250 mm across/high
Flower buds visible
Plant flowering
Plant senescent

(From Lutman & Tucker, 1987. *Annals of Applied Biology*, **110**, 683–687)

Appendix 4
Key to Hazard Classifications and Safety Precautions

Every product label contains information to warn users of the risks from using the product, together with precautions that must be followed in order to minimise the risks. A hazard classification (if any) and the symbol are shown with associated risk phrases, followed by a series of safety numbers under the heading **Hazard classification and safety precautions**.

The codes are defined below, under the same sub-headings as they appear in the pesticide profiles.

Where a product label specifies the use of personal protective equipment (PPE), the requirements are listed under the sub-heading **Operator protection**, using letter codes to denote the protective items, e.g. handling the concentrate, cleaning equipment etc., but it is not possible to list them separately. The lists of PPE are therefore an indication of what the user may need to have available to use the product in different ways. **When making a COSHH assessment it is therefore essential that the product label is consulted for information on the particular use that is being assessed.**

Where the generalised wording includes a phrase such as '... for xx days', the specific requirement for each pesticide is shown in brackets after the code.

Hazard

H01	Very toxic
H02	Toxic
H03	Harmful
H04	Irritant
H05	Corrosive
H06	Extremely flammable
H07	Highly flammable
H08	Flammable
H09	Oxidising agent
H10	Explosive
H11	Dangerous for the environment
H225	Highly flammable liquid and vapour
H226	Flammable liquid and vapour
H228	Flammable solid
H260	In contact with water releases flammable gases which may ignite spontaneously
H300	Fatal if swallowed
H301	Toxic if swallowed
H302	Harmful if swallowed
H311	Toxic in contact with skin
H330	Fatal if inhaled
H331	Toxic if inhaled
H332	Harmful if inhaled
H400	Very toxic to aquatic organisms
H401	Toxic to aquatic organisms

Risk phrases

H280	Contains gas under pressure; may explode if heated
H290	May be corrosive to metals
H304	May be fatal if swallowed and enters airways
H310	Fatal in contact with skin
H312	Harmful in contact with skin
H314	Causes severe skin burns and eye damage

H315	Causes skin irritation
H317	May cause an allergic skin reaction
H318	Causes serious eye damage
H319	Causes serious eye irritation
H320	Causes eye irritation
H334	May cause allergy or asthma symptoms or breathing difficulties if inhaled
H335	May cause respiratory irritation
H336	May cause drowsiness or dizziness
H340	May cause genetic defects
H341	Suspected of causing genetic defects
H351	Suspected of causing cancer
H360	May damage fertility or the unborn child
H361	Suspected of damaging fertility or the unborn child
H370	Causes damage to organs
H371	May cause damage to organs
H372	Causes damage to organs through prolonged or repeated exposure
H373	May cause damage to organs through prolonged or repeated exposure
R08	Contact with combustible material may cause fire
R09	Explosive when mixed with combustible material
R15	Contact with water liberates extremely flammable gases
R16	Explosive when mixed with oxidising substances
R19	Fatal if swallowed
R20	Harmful by inhalation
R21	Harmful in contact with skin
R22a	Harmful if swallowed
R22b	May cause lung damage if swallowed
R22c	May be fatal if swallowed and enters airways
R23a	Toxic by inhalation
R23b	Fatal if inhaled
R24	Toxic in contact with skin
R25	Toxic if swallowed
R25b	Toxic; Danger of serious damage to health by prolonged exposure if swallowed
R26	Very toxic by inhalation
R27	Very toxic in contact with skin
R28	Very toxic if swallowed
R31	Contact with acids liberates toxic gas
R34	Causes burns
R35	Causes severe burns
R36	Irritating to eyes
R37	Irritating to respiratory system
R38	Irritating to skin
R39	Danger of very serious irreversible effects
R40	Limited evidence of a carcinogenic effect
R40b	Suspected of causing cancer
R41	Risk of serious damage to eyes
R42	May cause sensitization by inhalation
R43	May cause sensitization by skin contact
R45	May cause cancer
R46	May cause heritable genetic damage
R48	Danger of serious damage to health by prolonged exposure
R50	Very toxic to aquatic organisms
R51	Toxic to aquatic organisms
R52	Harmful to aquatic organisms
R53a	May cause long-term adverse effects in the aquatic environment
R53b	Dangerous to aquatic organisms
R54	Toxic to flora
R55	Toxic to fauna
R56	Toxic to soil organisms
R57	Toxic to bees
R58	May cause long-term adverse effects in the environment
R60	May impair fertility

Key to Hazard Classifications and Safety Precautions

R60b   May cause heritable genetic damage
R61   May cause harm to the unborn child
R62   Possible risk of impaired fertility
R63   Possible risk of harm to the unborn child
R64   May cause harm to breast-fed babies
R65   Harmful: May cause lung damage if swallowed
R66   Repeated exposure may cause skin dryness or cracking
R66b   May cause damage to organs through prolonged or repeated exposure
R67   Vapours may cause drowsiness and dizziness
R68   Possible risk of irreversible effects
R69   Danger of serious damage to health by prolonged oral exposure.
R70   May produce an allergic reaction

Operator protection

A   Suitable protective gloves (the product label should be consulted for any specific requirements about the material of which the gloves should be made)
B   Rubber gauntlet gloves
C   Face-shield
D   Approved respiratory protective equipment
E   Goggles
F   Dust mask
G   Full face-piece respirator
H   Coverall
J   Hood
K   Apron/Rubber apron
L   Waterproof coat
M   Rubber boots
N   Waterproof jacket and trousers
P   Suitable protective clothing
U01   To be used only by operators instructed or trained in the use of chemical/product/type of produce and familiar with the precautionary measures to be observed
U02a   Wash all protective clothing thoroughly after use, especially the inside of gloves
U02b   Avoid excessive contamination of coveralls and launder regularly
U02c   Remove and wash contaminated gloves immediately
U03   Wash splashes off gloves immediately
U04a   Take off immediately all contaminated clothing
U04b   Take off immediately all contaminated clothing and wash underlying skin. Wash clothes before re-use
U04c   Wash clothes before re-use
U05a   When using do not eat, drink or smoke
U05b   When using do not eat, drink, smoke or use naked lights
U06   Handle with care and mix only in a closed container
U07   Open the container only as directed (returnable containers only)
U08   Wash concentrate/dust from skin or eyes immediately
U09a   Wash any contamination/splashes/dust/powder/concentrate from skin or eyes immediately
U09b   Wash any contamination/splashes/dust/powder/concentrate from eyes immediately
U09c   If on skin, wash with plenty of soap and water
U10   After contact with skin or eyes wash immediately with plenty of water
U11   In case of contact with eyes rinse immediately with plenty of water and seek medical advice
U12   In case of contact with skin rinse immediately with plenty of water and seek medical advice
U12b   After contact with skin, take off immediately all contaminated clothing and wash immediately with plenty of water
U13   Avoid contact by mouth
U14   Avoid contact with skin
U15   Avoid contact with eyes
U16a   Ensure adequate ventilation in confined spaces
U16b   Use in a well ventilated area

Appendices

U18	Extinguish all naked flames, including pilot lights, when applying the fumigant/dust/liquid/product
U19a	Do not breathe dust/fog/fumes/gas/smoke/spray mist/vapour. Avoid working in spray mist
U19b	Do not work in confined spaces or enter spaces in which high concentrations of vapour are present. Where this precaution cannot be observed distance breathing or self-contained breathing apparatus must be worn, and the work should be done by trained operators
U19c	In case of insufficient ventilation, wear suitable respiratory equipment
U19d	Wear suitable respiratory equipment during bagging and stacking of treated seed
U19e	During fumigation/spraying wear suitable respiratory equipment
U19f	In case of accident by inhalation: remove casualty to fresh air and keep at rest
U20a	Wash hands and exposed skin before eating, drinking or smoking and after work
U20b	Wash hands and exposed skin before eating and drinking and after work
U20c	Wash hands before eating and drinking and after work
U20d	Wash hands after use
U20e	Wash hands and exposed skin after cleaning and re-calibrating equipment
U21	Before entering treated crops, cover exposed skin areas, particularly arms and legs
U22a	Do not touch sachet with wet hands or gloves/Do not touch water soluble bag directly
U22b	Protect sachets from rain or water
U23a	Do not apply by knapsack sprayer/hand-held equipment
U23b	Do not apply through hand held rotary atomisers
U23c	Do not apply via tractor mounted horizontal boom sprayers
U24	Do not handle grain unnecessarily
U25	Open the container only as directed
U26	Keep unprotected workers out of treated areas for at least 5 days after treatment.

Environmental protection

E02a	Keep unprotected persons/animals out of treated/fumigation areas for at least xx hours/days
E02b	Prevent access by livestock, pets and other non-target mammals and birds to buildings under fumigation and ventilation
E02c	Vacate treatment areas before application
E02d	Exclude all persons and animals during treatment
E03	Label treated seed with the appropriate precautions, using the printed sacks, labels or bag tags supplied
E05a	Do not apply directly to livestock/poultry
E05b	Keep poultry out of treated areas for at least xx days/weeks
E05c	Do not apply directly to animals
E06a	Keep livestock out of treated areas for at least xx days/weeks after treatment
E06b	Dangerous to livestock. Keep all livestock out of treated areas/away from treated water for at least xx days/weeks. Bury or remove spillages
E06c	Harmful to livestock. Keep all livestock out of treated areas/away from treated water for at least xx days/weeks. Bury or remove spillages
E06d	Exclude livestock from treated fields. Livestock may not graze or be fed treated forage nor may it be used for hay silage or bedding
E07a	Keep livestock out of treated areas for up to two weeks following treatment and until poisonous weeds, such as ragwort, have died down and become unpalatable
E07b	Dangerous to livestock. Keep livestock out of treated areas/away from treated water for at least xx weeks and until foliage of any poisonous weeds, such as ragwort, has died and become unpalatable
E07c	Harmful to livestock. Keep livestock out of treated areas/away from treated water for at least xx days/weeks and until foliage of any poisonous weeds such as ragwort has died and become unpalatable
E07d	Keep livestock out of treated areas for up to 4-6 weeks following treatment and until poisonous weeds, such as ragwort, have died down and become unpalatable
E07e	Do not take grass crops for hay or silage for at least 21 days after application
E07f	Keep livestock out of treated areas for up to 3 days following treatment and until poisonous weeds, such as ragwort, have died
E07g	Keep livestock out of treated areas for at least 50 days following treatment
E08a	Do not feed treated straw or haulm to livestock within xx days/weeks of spraying

Key to Hazard Classifications and Safety Precautions

E08b	Do not use on grassland if the crop is to be used as animal feed or bedding
E09	Do not use straw or haulm from treated crops as animal feed or bedding for at least xx days after last application
E10a	Dangerous to game, wild birds and animals
E10b	Harmful to game, wild birds and animals
E10c	Dangerous to game, wild birds and animals. All spillages must be buried or removed
E11	Paraquat can be harmful to hares; spray stubbles early in the day
E12a	High risk to bees
E12b	Extremely dangerous to bees
E12c	Dangerous to bees
E12d	Harmful to bees
E12e	Do not apply to crops in flower or to those in which bees are actively foraging. Do not apply when flowering weeds are present
E12f	Do not apply to crops in flower, or to those in which bees are actively foraging, except as directed on [crop]. Do not apply when flowering weeds are present
E12g	Apply away from bees
E13a	Extremely dangerous to fish or other aquatic life. Do not contaminate surface waters or ditches with chemical or used container
E13b	Dangerous to fish or other aquatic life. Do not contaminate surface waters or ditches with chemical or used container
E13c	Harmful to fish or other aquatic life. Do not contaminate surface waters or ditches with chemical or used container
E13d	Apply away from fish
E13e	Harmful to fish or other aquatic life. The maximum concentration of active ingredient in treated water must not exceed XX ppm or such lower concentration as the appropriate water regulatory body may require.
E14a	Extremely dangerous to aquatic higher plants. Do not contaminate surface waters or ditches with chemical or used container
E14b	Dangerous to aquatic higher plants. Do not contaminate surface waters or ditches with chemical or used container
E15a	Do not contaminate surface waters or ditches with chemical or used container
E15b	Do not contaminate water with product or its container. Do not clean application equipment near surface water. Avoid contamination via drains from farmyards or roads
E15c	To protect groundwater, do not apply to grass leys less than 1 year old
E16a	Do not allow direct spray from horizontal boom sprayers to fall within 5 m of the top of the bank of a static or flowing waterbody, unless a Local Environment Risk Assessment for Pesticides (LERAP) permits a narrower buffer zone, or within 1 m of the top of a ditch which is dry at the time of application. Aim spray away from water
E16b	Do not allow direct spray from hand-held sprayers to fall within 1 m of the top of the bank of a static or flowing waterbody. Aim spray away from water
E16c	Do not allow direct spray from horizontal boom sprayers to fall within 5 m of the top of the bank of a static or flowing waterbody, or within 1m of the top of a ditch which is dry at the time of application. Aim spray away from water. This product is not eligible for buffer zone reduction under the LERAP horizontal boom sprayers scheme.
E16d	Do not allow direct spray from hand-held sprayers to fall within 1 m of the top of the bank of a static or flowing waterbody. Aim spray away from water. This product is not eligible for buffer zone reduction under the LERAP horizontal boom sprayers scheme scheme.
E16e	Do not allow direct spray from horizontal boom sprayers to fall within 5 m of the top of the bank of a static or flowing water body or within 1 m from the top of any ditch which is dry at the time of application. Spray from hand held sprayers must not in any case be allowed to fall within 1 m of the top of the bank of a static or flowing water body. Always direct spray away from water. The LERAP scheme does not extend to adjuvants. This product is therefore not eligible for a reduced buffer zone under the LERAP scheme.
E16f	Do not allow direct spray/granule applications from vehicle mounted/drawn hydraulic sprayers/applicators to fall within 6 m of surface waters or ditches/Do not allow direct spray/granule applications from hand-held sprayers/applicators to fall within 2 m of surface waters or ditches. Direct spray/applications away from water

Appendices

E16g	Do not allow direct spray from train sprayers to fall within 5 m of the top of the bank of any static or flowing waterbody.
E16h	Do not spray cereals after 31st March in the year of harvest within 6 metres of the edge of the growing crop
E16i	Do not allow direct spray from horizontal boom sprayers to fall within 12 metres of the top of the bank of a static or flowing water body.
E16j	Do not allow direct spray from horizontal boom sprayers to fall within the specified distance of the top of the bank of a static or flowing waterbody,
E17a	Do not allow direct spray from broadcast air-assisted sprayers to fall within xx m of surface waters or ditches. Direct spray away from water
E17b	Do not allow direct spray from broadcast air-assisted sprayers to fall within xx m of the top of the bank of a static or flowing waterbody, unless a Local Environmental Risk Assessment for Pesticides (LERAP) permits a narrower buffer zone, or within 5 m of the top of a ditch which is dry at the time of application. Aim spray away from water
E18	Do not spray from the air within 250 m horizontal distance of surface waters or ditches
E19a	Do not dump surplus herbicide in water or ditch bottoms
E19b	Do not empty into drains
E20	Prevent any surface run-off from entering storm drains
E21	Do not use treated water for irrigation purposes within xx days/weeks of treatment
E22a	High risk to non-target insects or other arthropods. Do not spray within 6 m of the field boundary
E22b	Risk to certain non-target insects or other arthropods. For advice on risk management and use in Integrated Pest Management (IPM) see directions for use
E22c	Risk to non-target insects or other arthropods
E23	Avoid damage by drift onto susceptible crops or water courses
E34	Do not re-use container for any purpose/Do not re-use container for any other purpose
E35	Do not burn this container
E36a	Do not rinse out container (returnable containers only)
E36b	Do not open or rinse out container (returnable containers only)
E37	Do not use with any pesticide which is to be applied in or near water
E38	Use appropriate containment to avoid environmental contamination
E39	Extreme care must be taken to avoid spray drift onto non-crop plants outside the target area
E40	To protect groundwater/soil organisms the maximum total dose of this or other products containing ethofumesate MUST NOT exceed 1.0 kg ethofumesate per hectare in any three year period
E40b	To protect aquatic organisms respect an unsprayed buffer zone to surface waters in line with LERAP requirements
E40c	This product must not be used with any pesticide to be applied in or near water
E40d	To protect non-target arthropods respect an untreated buffer zone of 5m to non crop land
E40e	To protect aquatic organisms respect an unsprayed buffer zone of 20 metres to surface water bodies
E41	Hay for silage must not be cut from treated crops for at least 21 days after treatment
H410	Very toxic to aquatic life with long-lasting effects
H411	Toxic to aquatic life with long lasting effects
H412	Harmful to aquatic life with long lasting effects
H413	May cause long lasting harmful effects to aquatic life

Consumer protection

C01	Do not use on food crops
C02a	Do not harvest for human or animal consumption for at least xx days/weeks after last application
C02b	Do not remove from store for sale or processing for at least 21 days after application
C02c	Do not remove from store for sale or processing for at least 2 days after application
C04	Do not apply to surfaces on which food/feed is stored, prepared or eaten
C05	Remove/cover all foodstuffs before application
C06	Remove exposed milk before application

C07	Collect eggs before application
C08	Protect food preparing equipment and eating utensils from contamination during application
C09	Cover water storage tanks before application
C10	Protect exposed water/feed/milk machinery/milk containers from contamination
C11	Remove all pets/livestock/fish tanks before treatment/spraying
C12	Ventilate treated areas thoroughly when smoke has cleared/Ventilate treated rooms thoroughly before occupying

Storage and disposal

D01	Keep out of reach of children
D02	Keep away from food, drink and animal feeding-stuffs
D03	Store away from seeds, fertilizers, fungicides and insecticides
D04	Store well away from corms, bulbs, tubers and seeds
D05	Protect from frost
D06a	Store away from heat
D06b	Do not store near heat or open flame
D06c	Do not store in direct sunlight
D06d	Do not store above 30/35 °C
D06e	Store in a well-ventilated place. Keep container tightly closed
D06f	Keep away from sources of ignition - No smoking
D06g	Take precautionary measures against static discharges
D06h	Store at 10 - 25 °C
D07	Store under cool, dry conditions
D08	Store in a safe, dry, frost-free place designated as an agrochemical store
D09a	Keep in original container, tightly closed, in a safe place
D09b	Keep in original container, tightly closed, in a safe place, under lock and key
D09c	Store unused sachets in a safe place. Do not store half-used sachets
D10a	Wash out container thoroughly and dispose of safely
D10b	Wash out container thoroughly, empty washings into spray tank and dispose of safely
D10c	Rinse container thoroughly by using an integrated pressure rinsing device or manually rinsing three times. Add washings to sprayer at time of filling and dispose of container safely
D10d	Do not rinse out container
D11a	Empty container completely and dispose of safely/Dispose of used generator safely
D11b	Empty container completely and dispose of it in the specified manner
D11c	Ventilate empty containers until no phosphine is detected and then dispose of as hazardous waste via an authorised waste-disposal contractor
D12a	This material (and its container) must be disposed of in a safe way
D12b	This material and its container must be disposed of as hazardous waste
D13	Treat used container as if it contained pesticide
D14	Return empty container as instructed by supplier (returnable containers only)
D15	Store container in purpose built chemical store until returned to supplier for refilling (returnable containers only)
D16	Do not store below 4 degrees Centigrade
D17	Use immediately on removal of foil
D18	Place the tablets whole into the spray tank - do not break or crumble the tablets
D19	Do not empty into drains
D20	Clean all equipment after use
D21	Open the container only as directed
D22	Collect spillage

Treated Seed

S01	Do not handle treated seed unnecessarily
S02	Do not use treated seed as food or feed
S03	Keep treated seed secure from people, domestic stock/pets and wildlife at all times during storage and use
S04a	Bury or remove spillages
S04b	Harmful to birds/game and wildlife. Treated seed should not be left on the soil surface. Bury or remove spillages

S04c	Dangerous to birds/game and wildlife. Treated seed should not be left on the soil surface. Bury or remove spillages
S04d	To protect birds/wild animals, treated seed should not be left on the soil surface. Bury or remove spillages
S05	Do not reuse sacks or containers that have been used for treated seed for food or feed
S06a	Wash hands and exposed skin before meals and after work
S06b	Wash hands and exposed skin after cleaning and re-calibrating equipment
S07	Do not apply treated seed from the air
S08	Treated seed should not be broadcast
S09	Label treated seed with the appropriate precautions using printed sacks, labels or bag tags supplied
S10	Only use automated equipment for planting treated seed potatoes

Vertebrate/Rodent control products

V01a	Prevent access to baits/powder by children, birds and other animals, particularly cats, dogs, pigs and poultry
V01b	Prevent access to bait/gel/dust by children, birds and non-target animals, particularly dogs, cats, pigs, poultry
V02	Do not prepare/use/lay baits/dust/spray where food/feed/water could become contaminated
V03a	Remove all remains of bait, tracking powder or bait containers after use and burn or bury
V03b	Remove all remains of bait and bait containers/exposed dust/after treatment (except where used in sewers) and dispose of safely (e.g. burn/bury). Do not dispose of in refuse sacks or on open rubbish tips.
V04a	Search for and burn or bury all rodent bodies. Do not place in refuse bins or on rubbish tips
V04b	Search for rodent bodies (except where used in sewers) and dispose of safely (e.g. burn/bury). Do not dispose of in refuse sacks or on open rubbish tips
V04c	Dispose of safely any rodent bodies and remains of bait and bait containers that are recovered after treatment (e.g. burn/bury). Do not dispose of in refuse sacks or on open rubbish tips
V05	Use bait containers clearly marked POISON at all surface baiting points

Medical advice

M01	This product contains an anticholinesterase organophosphorus compound. DO NOT USE if under medical advice NOT to work with such compounds
M02	This product contains an anticholinesterase carbamate compound. DO NOT USE if under medical advice NOT to work with such compounds
M03	If you feel unwell, seek medical advice immediately (show the label where possible)
M03a	If exposed or concerned, get medical advice or attention
M04a	In case of accident or if you feel unwell, seek medical advice immediately (show the label where possible)
M04b	In case of accident by inhalation, remove casualty to fresh air and keep at rest
M04c	If inhaled, remove victim to fresh air and keep at rest in a position comfortable for breathing. IMMEDIATELY call Poison Centre or doctor/physician
M05a	If swallowed, seek medical advice immediately and show this container or label
M05b	If swallowed, do not induce vomiting: seek medical advice immediately and show this container or label
M05c	If swallowed induce vomiting if not already occurring and take patient to hospital immediately
M05d	If swallowed, rinse the mouth with water but only if the person is conscious
M06	This product contains an anticholinesterase carbamoyl triazole compound. DO NOT USE if under medical advice NOT to work with such compounds

Appendix 5
Key to Abbreviations and Acronyms

The abbreviations of formulation types in the following list are used in Section 2 (Pesticide Profiles) and are derived from the *Catalogue of Pesticide Formulation Types and International Coding System* (CropLife International Technical Monograph 2, 5th edn, March 2002).

1 Formulation Types

AB	Grain bait
AE	Aerosol generator
AL	Other liquids to be applied undiluted
AP	Any other powder
BB	Block bait
BR	Briquette
CB	Bait concentrate
CC	Capsule suspension in a suspension concentrate
CF	Capsule suspension for seed treatment
CG	Encapsulated granule (controlled release)
CL	Contact liquid or gel (for direct application)
CP	Contact powder (for direct application)
CR	Crystals
CS	Capsule suspension
DC	Dispersible concentrate
DP	Dustable powder
DS	Powder for dry seed treatment
EC	Emulsifiable concentrate
EG	Emulsifiable granule
EO	Water in oil emulsion
ES	Emulsion for seed treatment
EW	Oil in water emulsion
FG	Fine granules
FP	Smoke cartridge
FS	Flowable concentrate for seed treatment
FT	Smoke tablet
FU	Smoke generator
FW	Smoke pellets
GA	Gas
GB	Granular bait
GE	Gas-generating product
GG	Macrogranules
GL	Emulsifiable gel
GP	Flo-dust (for pneumatic application)
GR	Granules
GS	Grease
GW	Water soluble gel
HN	Hot fogging concentrate
KK	Combi-pack (solid/liquid)
KL	Combi-pack (liquid/liquid)
KN	Cold-fogging concentrate
KP	Combi-pack (solid/solid)
LA	Lacquer
LI	Liquid, unspecified
LS	Solution for seed treatment
ME	Microemulsion
MG	Microgranules
MS	Microemulsion for seed treatment
OD	Oil dispersion

Appendices

OL	Oil miscible liquid
PA	Paste
PC	Gel or paste concentrate
PO	Powder
PS	Seed coated with a pesticide
PT	Pellet
RB	Ready-to-use bait
RC	Ready-to-use low volume CDA sprayer
RH	Ready-to-use spray in hand-operated sprayer
SA	Sand
SC	Suspension concentrate (= flowable)
SE	Suspo-emulsion
SG	Water soluble granules
SL	Soluble concentrate
SP	Water soluble powder
SS	Water soluble powder for seed treatment
ST	Water soluble tablet
SU	Ultra low-volume suspension
TB	Tablets
TC	Technical material
TP	Tracking powder
UL	Ultra low-volume liquid
VP	Vapour releasing product
WB	Water soluble bags
WG	Water dispersible granules
WP	Wettable powder
WS	Water dispersible powder for slurry treatment of seed
WT	Water dispersible tablet
XX	Other formulations
ZZ	Not Applicable

2 Other Abbreviations and Acronyms

ACP	Advisory Committee on Pesticides
ACTS	Advisory Committee on Toxic Substances
ADAS	Agricultural Development and Advisory Service
a.i.	active ingredient
AIC	Agriculture Industries Confederation
BBPA	British Beer and Pub Association
CDA	Controlled droplet application
CPA	Crop Protection Association
cm	centimetre(s)
COPR	Control of Pesticides Regulations 1986
COSHH	Control of Substances Hazardous to Health Regulations
CRD	Chemicals Regulation Directorate
d	day(s)
Defra	Department for Environment, Food and Rural Affairs
EA	Environment Agency
EBDC	ethylene-bis-dithiocarbamate fungicide
EAMU	Extension of Authorisation for Minor Use
FEPA	Food and Environment Protection Act 1985
g	gram(s)
GS	growth stage (unless in formulation column)
h	hour(s)
ha	hectare(s)
HBN	hydroxybenzonitrile herbicide
HI	harvest interval
HSE	Health and Safety Executive
ICM	integrated crop management
IPM	integrated pest management
kg	kilogram(s)
l	litre(s)

Key to Abbreviations and Acronyms

LERAP	Local Environmental Risk Assessments for Pesticides
m	metre(s)
MBC	methyl benzimidazole carbamate fungicide
MEL	maximum exposure limit
min	minute(s)
mm	millimetre(s)
MRL	maximum residue level
mth	month(s)
NA	Notice of Approval
NFU	National Farmers' Union
OES	Occupational Exposure Standard
OLA	off-label approval
PPE	personal protective equipment
PPPR	Plant Protection Products Regulations
SOLA	specific off-label approval
ULV	ultra-low volume
VI	Voluntary Initiative
w/v	weight/volume
w/w	weight/weight
wk	week(s)
yr	year(s)

3 UN Number details

UN No.	Substance	EAC	APP	Hazards Class	Sub risks	HIN
1692	Strychnine or Strychnine salts	2X		6.1		66
1760	Corrosive Liquid, N.O.S., Packing groups II & III	2X	B	8		88
1830	Sulphuric acid, with more than 51% acid	2P		8		80
1993	Flammable Liquid, N.O.S., Packing group III	•3Y		3		30
2011	Magnesium Phosphide	4W[1]		4.3	6.1	
2588	Pesticide, Solid, TOXIC, N.O.S.	2X		6.1		66/60
2757	Carbamate Pesticide, Solid, Toxic	2X		6.1		66/60
2783	Organophosphorus Pesticide, Solid, TOXIC	2X		6.1		66/60
2902	Pesticide, Liquid, TOXIC, N.O.S., Packing Groups I, II & III	2X	B	6.1		66/60
3077	Environmentally hazardous substance, Solid, N.O.S.	2Z		9		90
3082	Environmentally hazardous substance, Solid, N.O.S.	•3Z		9		90
3265	Corrosive Liquid, Acidic, Organic, N.O.S., Packing group I, II or III	2X	B	8		80/88
3351	Pyrethroid Pesticide, Liquid, Toxic, Flammable, Flash Point 23 °C or more	•3W	A(fl)	6.1	3	663/63
3352	Pyrethroid Pesticide, Liquid, TOXIC, Packing groups I, II & III	2X	B	6.1		66/60

Ref: Dangerous Goods Emergency Action Code List 2009 (TSO).
[1] Not applicable to the carriage of dangerous goods under RID or ADR.

Appendix 6
Definitions

The descriptions used in this *Guide* for the crops or situations in which products are approved for use are those used on the approved product labels. These are now standardised in a Crop Hierarchy published by the Chemicals Regulation Directorate in which definitions are given. To assist users of this *Guide* the definitions of some of the terminology where misunderstandings can occur are reproduced below.

Rotational grass: Short-term grass crops grown on land that is likely to be growing different crops in future years (*e.g. short-term intensively managed leys for one to three years that may include clover*)

Permanent grassland: Grazed areas that are intended to be permanent in nature (e.g. permanent pasture and moorland that can be grazed).

Ornamental Plant Production: All ornamental plants that are grown for sale or are produced for replanting into their final growing position (*e.g. flowers, house plants, nursery stock, bulbs grown in containers or in the ground*).

Managed Amenity Turf: Areas of frequently mown, intensively managed, turf that is not intended to flower and set seed. It includes areas that may be for intensive public use (*e.g. all types of sports turf*).

Amenity Grassland: Areas of semi-natural or planted grassland subject to minimal management. It includes areas that may be accessed by the public (*e.g. railway and motorway embankments, airfields, and grassland nature reserves*). These areas may be managed for their botanical interest, and the relevant authority should be contacted before using pesticides in such locations.

Amenity Vegetation: Areas of semi-natural or ornamental vegetation, including trees, or bare soil around ornamental plants, or soil intended for ornamental planting. It includes areas to which the public have access. It does NOT include hedgerows around arable fields.

Natural surfaces not intended to bear vegetation: Areas of soil or natural outcroppings of rock that are not intended to bear vegetation, including areas such as sterile strips around fields. It may include areas to which the public have access. It does not include the land between rows of crops.

Hard surfaces: Man-made impermeable surfaces that are not intended to bear vegetation (*e.g. pavements, tennis courts, industrial areas, railway ballast*).

Permeable surfaces overlying soil: Any man-made permeable surface (excluding railway ballast) such as gravel that overlies soil and is not intended to bear vegetation

Green Cover on Land Temporarily Removed from Production: Includes fields covered by natural regeneration or by a planted green cover crop that will not be harvested (*e.g. green cover on setaside*). It does NOT include industrial crops.

Forest Nursery: Areas where young trees are raised outside for subsequent forest planting.

Forest: Groups of trees being grown in their final positions. Covers all woodland grown for whatever objective, including commercial timber production, amenity and recreation, conservation and landscaping, ancient traditional coppice and farm forestry, and trees from natural regeneration, colonisation or coppicing. Also includes restocking of established woodlands and new planting on both improved and unimproved land.

Farm forestry: Groups of trees established on arable land or improved grassland including those planted for short rotation coppicing. It includes mature hedgerows around arable fields.

Indoors (for rodenticide use): Situations where the bait is placed within a building or other enclosed structure, and where the target is living or feeding predominantly within that building or structure.

Definitions

Herbs: Reference to Herbs or Protected Herbs when used in Section 2 may include any or all of the following. The particular label or EAMU Notice will indicate which species are included in the approval.

Agastache spp.
Angelica
Applemint
Balm
Basil
Bay
Borage (except when grown for oilseed)
Camomile
Caraway
Catnip
Chervil
Clary
Clary sage
Coriander
Curry plant
Dill
Dragonhead
English chamomile
Fennel
Fenugreek
Feverfew
French lavender
Gingermint
Hyssop
Korean mint
Land cress
Lavandin
Lavender
Lemon balm
Lemon peppermint
Lemon thyme
Lemon verbena
Lovage
Marigold
Marjoram
Mint
Mother of thyme
Nasturtium
Nettle
Oregano
Origanum heracleoticum
Parsley root
Peppermint
Pineapplemint
Rocket
Rosemary
Rue
Sage
Salad burnet
Savory
Sorrel
Spearmint
Spike lavender
Tarragon
Thyme
Thymus camphoratus
Violet
Winter savory
Woodruff

Herbs for Medicinal Uses: Reference to Herbs for Medicinal Uses when used in Section 2 may include any or all of the following. The particular label or EAMU Notice will indicate which species are included in the approval

Black cohosh
Burdock
Dandelion
Echinacea
Ginseng
Goldenseal
Liquorice
Nettle
Valerian

Appendix 7
References

The information given in *The UK Pesticide Guide* provides some of the answers needed to assess health risks, including the hazard classification and the level of operator protection required. However, the Guide cannot provide all the details needed for a complete hazard assessment, which must be based on the product label itself and, where necessary, the Health and Safety Data Sheet and other official literature.

Detailed guidance on how to comply with the Regulations is available from several sources.

Pesticides: Code of Practice

The sale, supply, storage and use of pesticides are strictly controlled by EU law. The key acts, Regulations and Codes of Practice setting out the duties of employers, supervisors and operators are:

- The *Plant Protection Products Regulations 2011*;
- The *Plant Protection Products (Sustainable Use) Regulations 2012*;
- The *Control of Substances Hazardous to Health Regulations 2002* (COSHH) (as amended);
- Pesticides: *Code of Practice for Using Plant Protection Products 2006*, (incorporating the former 'Green and Orange Codes') with specific Codes of Practice for Scotland and Northern Ireland;
- Classification, Labelling and Packaging (CLP) Regulations.

The Code is currently being updated (as of May 2016) and may be replaced by a series of guidance notes as opposed to a Code in future. The current Code should be read in conjunction with guidance published on the HSE website on how new pesticide legislation has affected the responsibilities of the regulated community.

Further information and details of legislation relating to water, environment protection, waste management and transportation can be found in the UK National action Plan for the Sustainable Use of Pesticides (www.gov.uk/government/publications/pesticides-uk-national-action-plan).

Other Codes of Practice

The Food and Environment Protection Act and its regulations aim to protect the health of human beings, creatures and plants, safeguard the environment and secure safe, efficient and humane methods of controlling pests.

Code of Practice for Suppliers of Pesticides to Agriculture, Horticulture and Forestry (the 'Yellow Code') (Defra Booklet PB 0091)

Code of Good Agricultural Practice for the Protection of Soil (Defra Booklet PB 0617)

Code of Good Agricultural Practice for the Protection of Water (Defra Booklet PB 0587)

Code of Good Agricultural Practice for the Protection of Air (Defra Booklet PB 0618)

Approved Code of Practice for the Control of Substances Hazardous to Health in Fumigation Operations. Health and Safety Commission (ISBN 0-717611-95-7)

Code of Best Practice: Safe use of Sulphuric Acid as an Agricultural Desiccant. National Association of Agricultural Contractors, 2002. (Also available at www.naac.co.uk/?Codes/acidcode.asp)

Safe Use of Pesticides for Non-agricultural Purposes. HSE Approved Code of Practice. HSE L21 (ISBN 0-717624-88-9)

References

Other Guidance and Practical Advice

HSE (by mail order from HSE Books – See Appendix 2)

COSHH – A brief guide to the Regulations 2003 (INDG136)

A Step by Step Guide to COSHH Assessment, 2004 (HSG97) (ISBN 0-717627-85-3)

Defra (from The Stationery Office – see Appendix 2)

Local Environment Risk Assessments for Pesticides (LERAP): Horizontal Boom Sprayers.

Local Environment Risk Assessments for Pesticides (LERAP): Broadcast Air-assisted Sprayers.

Crop Protection Association and the Voluntary Initiative (see Appendix 2)

Every Drop Counts: Keeping Water Clean

Best Practice Guides. A range of leaflets giving guidance on best practice when dealing with pesticides before, during and after application.

H2OK? Best Practice Advice and Decision Trees - July 2009/10

BCPC (British Crop Production Council – see Appendix 2)

ukpesticideguide.co.uk (*The UK Pesticide Guide* online) – subscription resource

The Pesticide Manual (16th edition) (ISBN 978-1-901396-86-7)

The Pesticide Manual online (www.bcpcdata.com) – subscription resource

The GM Crop Manual (ISBN 978-1-901396-20-1)

GM Crop Manual online (www.bcpcdata.com/gm) - subscription resource

The Manual of Biocontrol Agents (5th edition) (ISBN 978-1-901396-87-4)

The Manual of Biocontrol Agents online (www.bcpcdata.com/mba) – subscription resource.

IdentiPest www.identipest.co.uk

Small Scale Spraying (ISBN 1-901396-07-X)

Field Scale Spraying – 2016 (ISBN 978-1-901396-89-8)

Using Pesticides (ISBN 978-1-901396-10-2)

Safety Equipment Handbook (ISBN 1-901396-06-1)

Spreading Fertilisers and Applying Slug Pellets (ISBN 978-1-901396-15-7)

The Environment Agency (see Appendix 2)

Best Farming Practices: Profiting from a Good Environment

Use of Herbicides in or Near Water

SECTION 7
INDEX

Index of Proprietary Names of Products

The references are to entry numbers, not to pages. Adjuvant names are referred to as 'Adj' and are listed separately in Section 4. Products also registered, i.e. not actively marketed, are entered as PAR and are listed by their active ingredient in Section 3.

2,4-D Amine 500 . . . *PAR* 224	Afalon. 301	Alpha Bromotril PT . . . *PAR* 87
3C Chlormequat 750 67	Afrisect 10 . . . . *PAR* 211	Alpha C 6 ED . . . . *PAR* 6
5504 . . . . *PAR* 26	Afrisect 500 EC . . . . *PAR* 211	Alpha Ethofumesate
A10784A. . . . . *PAR* 550	Agate . . . . *PAR* 584	50SC. . . . . *PAR* 328
A12739A. . . . . *PAR* 513	Agate EW 362	Alpha Pendimethalin
A12791B. . . . . *PAR* 379	Agree 50 WG . . . . . . *PAR* 38	330 EC . . . . *PAR* 555
A13219F . . . . *PAR* 645	Agria Mancozeb	Alpha Phenmedipham
A13814D . . . . *PAR* 575	75WDG. . . . . *PAR* 493	320 SC . . . . *PAR* 567
A1412A2. . . . . *PAR* 300	Agria Mancozeb 80WP *PAR* 493	Alternet . . . . *PAR* 265
A15149W . . . . *PAR* 477	Agrichem Flowable	Alternil 500 . . . . *PAR* 128
A17600C. . . . . *PAR* 672	Thiram. 415	Alternil Excel 720 . . . *PAR* 128
A8545G . . . . *PAR* 598	Agrichem Hy-Pro Duet . . . 363	Altiplano DAMtec. . . *PAR* 172
Abacus . . . . *Adj*	Agrichem MCPA 500 *PAR* 499	Altiplano DAMtec. 100
Abam . . . . *PAR* 1	Agrichem	Alto Elite. 74
Abamex 18 EC . . . . . . *PAR* 1	Phenmedipham EC *PAR* 567	Alto Xtra . . . . *PAR* 217
Abamite Pro . . . . *PAR* 1	Agrichem PMP EC . . *PAR* 567	Amaranth . . . . *PAR* 275
Abringo. 72	Agrichem PMP SE . . *PAR* 567	Ambassador . . . . . . *PAR* 323
Absolute 168	Agriguard Bentazone	Amber . . . . *Adj*
AC 650 339	480. . . . . *PAR* 44	Amber . . . . *PAR* 309
AC 900001 . . . . *PAR* 573	AgriGuard Chlormequat	Amega Duo 273
AC651 . . . . *PAR* 544	750. . . . . *PAR* 118	Amethyst 241
Acamar WG . . . . . . *PAR* 472	Agriguard Diquat . . . *PAR* 300	Amistar. 18
Acanto . . . . *PAR* 574	Agriguard Metsulfuron *PAR* 538	Amistar Opti 19
Acanto Prima. . . . . . *PAR* 223	Agrimec . . . . *PAR* 1	Amistar Pro. . . . . . . *PAR* 30
Acaristop 500 SC . . . *PAR* 166	Agritox 310	Amistar Top. 21
Accelerate. . . . . *PAR* 441	Agritox 50 . . . . *PAR* 499	Amphore Plus 160
Accent 340	Agrostar . . . . *PAR* 412	Anchor . . . . *PAR* 106
Accurate . . . . *PAR* 538	Agrotech Bentazone . . *PAR* 44	Angri 424
Accurate Extra . . . . . *PAR* 539	Agrotech Fluroxypyr . *PAR* 412	Anode . . . . *PAR* 591
Aceta 20 SG . . . . *PAR* 2	Agrovista 3 See 750 67	Answer SX 336
Acetamex 20 SP . . . . . *PAR* 2	Agrovista Fenamid 218	Anthem 346
Aceto Chlorpropham	Agrovista Radni 217	Antuco 151
50M 86	Agrovista Reggae 410	Aoraki . . . . *PAR* 575
Aceto Sprout Nip 86	Agrovista Trotter 331	Apache . . . . *PAR* 131
Aceto Sprout Nip Pellet . . 86	Agroxone 310	Apache 250 EC . . . . *PAR* 591
Aceto Sprout Nip Ultra . . . 86	Agroxone Combi . . . *PAR* 234	Apex Pro. . . . . *PAR* 314
Acrion. . . . . *PAR* 441	Aitken Lawn Sand	Aphox . . . . *PAR* 576
Actara. 411	Mosskiller . . . . . . *PAR* 367	Aphox. 357
Actellic 50 EC 358	Aitken's Lawn Sand . *PAR* 367	Apollo 50 SC 95
Actellic Smoke	Akocynil 225 EC . . . . *PAR* 77	Appeal . . . . *PAR* 519
Generator No. 20 358	Akofos 480 EC . . . . *PAR* 154	Appeal 324
Activator 90. . . . . *Adj*	Akolin 330 EC . . . . . *PAR* 555	Applaud 25 SC. . . . . *PAR* 91
Addit . . . . *Adj*	Akotan 80 WG. . . . . . *PAR* 93	Applaud 25 WP . . . . *PAR* 91
Adept . . . . *PAR* 275	Al-cyper Ec . . . . *PAR* 6	Apres . . . . *PAR* 618
Adexar . . . . *PAR* 314	Alekto Plus TF . . . . . *PAR* 441	Apron XL 323
Adexar 197	Alert . . . . *PAR* 6	AQ 10 . . . . *PAR* 24
Adigor . . . . *Adj*	Aliacine 400 EC . . . . *PAR* 151	AQ 10 17
AdjiFe. . . . . *Adj*	Alias . . . . *PAR* 538	Aquarius . . . . *PAR* 555
AdjiMin . . . . *Adj*	Alias SX 336	Aquarius 420
AdjiSil . . . . *Adj*	Allure 324	Ardee . . . . *PAR* 441
AdjiVeg . . . . *Adj*	Ally Express . . . . . . *PAR* 110	Ardee XL . . . . *PAR* 441
Adjust 67	Ally Max SX 338	Aredios 45 SC . . . . . *PAR* 488
Admix-P . . . . *Adj*	Ally SX . . . . *PAR* 538	Aria . . . . *PAR* 539
Advance Green . . . . *PAR* 472	Alpha Bromolin 225 EC *PAR* 77	Ariane. . . . . *PAR* 166

REFERENCES ARE TO ENTRY NUMBERS NOT PAGES

Index

Aristo IP . . . . *PAR* 366	Azural . . . . *PAR* 441	Basamid 143
Aristocrat . . . . *PAR* 541	Azural 273	BASF Insignia . . . . . *PAR* 608
Arizona 265	Azzox 18	Basilex . . . . *PAR* 658
Arma . . . . *Adj*	Bacara 169	Basilex 416
Armure 162	BackRow . . . . *Adj*	Bassoon . . . . *PAR* 309
Artax Flo 370	Balear 720 SC . . . . . *PAR* 128	Bassoon EC 193
Artemis 222	Balista . . . . *PAR* 300	Bayer UK 831 . . . . . *PAR* 401
Artist 242	Ballad . . . . *PAR* 154	BD-200 . . . . *PAR* 300
Asana 195	Ballista . . . . *Adj*	Beem WG . . . . *PAR* 481
Ascra Xpro 40	Banastar 177	Beeper 500 SC . . . . *PAR* 273
Aska . . . . *PAR* 309	Bandera . . . . *PAR* 545	Beetron 700 325
Aspect . . . . *PAR* 30	Bandu 146	Beetup Compact SC 150
Asset Express . . . . . *PAR* 110	Banguy . . . . *PAR* 600	Beetup Flo 350
Associate . . . . *PAR* 544	Banka . . . . *Adj*	Belcocel 67
Assuage 337	Banko 720 . . . . *PAR* 128	Belfry 18
Astec Lambda CY . . *PAR* 482	Banner Maxx 368	Bellis 50
Asterix . . . . *PAR* 128	Bantry 235	Bellmac Straight 312
Asteroid 273	Banweed . . . . *PAR* 544	Benta 48 . . . . *PAR* 44
Asteroid Pro 273	Barbarian XL . . . . . . *PAR* 441	Benta 480 SL . . . . . . *PAR* 44
Asteroid Pro 450 273	Barclay Alphasect . . . *PAR* 324	Benta 480 SL 30
Astrobin 250 . . . . . . *PAR* 26	Barclay Avoca 72	Bentazone 480 . . . . . *PAR* 44
AstroKerb 15	Barclay Avoca Premium . . *PAR*	Bently . . . . *PAR* 44
Astute . . . . *PAR* 584	142	Berelex . . . . *PAR* 438
Asu-Flex . . . . *Adj*	Barclay Barbarian . . . *PAR* 441	Beret Gold 237
Atilla 359	Barclay Chloroflash . . *PAR* 128	Betamax . . . . *PAR* 567
Atlantis WG 288	Barclay Dingo . . . . . . *PAR* 44	Betanal Elite . . . . . . *PAR* 247
Atlas Pendimethalin	Barclay DQ-200 . . . . *PAR* 300	Betanal Expert . . . . . *PAR* 247
400 . . . . *PAR* 555	Barclay D-Quat . . . . *PAR* 300	Betanal Flow . . . . . . *PAR* 567
Atlas Pro . . . . *PAR* 598	Barclay Gallup 360 273	Betanal Maxxim . . . . *PAR* 248
Atlas Propiconazole 250 . . *PAR*	Barclay Gallup Amenity . . 273	Betanal MaxxPro 148
591	Barclay Gallup Biograde	Betanal Turbo 150
Atlas Too-Da Loo . . . *PAR* 520	360 . . . . *PAR* 441	Betasana SC 350
Attenzo 334	Barclay Gallup Biograde	Betasana Trio SC 149
Attract . . . . *PAR* 519	360 273	Beta-Team . . . . *PAR* 247
Attract 324	Barclay Gallup Biograde	Beta-Team 149
Attribut 369	Amenity . . . . *PAR* 441	Betosip Combi FL . . *PAR* 331
Aubrac 18	Barclay Gallup Biograde	Betra SC . . . . *PAR* 520
Aubrac Optimum 19	Amenity 273	Bettaquat B . . . . *PAR* 118
Augment . . . . *PAR* 597	Barclay Gallup Hi-Aktiv *PAR* 441	Bettequat B 67
AUK 75 WG 306	Barclay Gallup Hi-Aktiv . . . 273	Better DF 64
Aurora . . . . *PAR* 107	Barclay Garryowen . . *PAR* 441	Better Flowable . . . . *PAR* 113
Aurora 40 WG . . . . . *PAR* 107	Barclay Glyde 144 273	Bettix Flo 325
Aurora 40 WG 61	Barclay Hudson . . . . *PAR* 412	Bettix WG . . . . *PAR* 520
Austral Plus . . . . *PAR* 387	Barclay Keeper 500 SC*PAR* 328	Binder . . . . *Adj*
Austral Plus 239	Barclay Propizole . . . *PAR* 591	Bio Syl . . . . *Adj*
Auxiliary . . . . *PAR* 164	Barclay Propyz 370	Biocure . . . . *PAR* 37
Avadex Excel 15G . . *PAR* 662	Bariloche . . . . *PAR* 174	Bioduo . . . . *Adj*
Avadex Excel 15G 417	Baritone 74	Biofilm . . . . *Adj*
Avatar 267	Barleyquat B . . . . . . *PAR* 118	Biofresh Safestore 210
Aviator 235 Xpro 42	Barleyquat B 67	Biomus . . . . *PAR* 605
Avocet . . . . *PAR* 614	Baron . . . . *PAR* 411	Bion . . . . *PAR* 4
Avro SX 337	Barracuda . . . . *PAR* 513	BioPower . . . . *Adj*
Awac . . . . *PAR* 412	Barramundi . . . . *Adj*	Biothene . . . . *Adj*
Axial 356	Barret 125 SL . . . . . *PAR* 309	Biotrine . . . . *PAR* 1
Axial One . . . . *PAR* 375	Barrier H . . . . *PAR* 158	Biox-M 395
Axidor . . . . *PAR* 209	Barrier H 89	BiPlay SX 338
Aylora 79	Barton WG . . . . *PAR* 373	Biretta 340
Azaka . . . . *PAR* 26	BAS 500 06F . . . . . . *PAR* 608	Biscaya 410
Azaka 18	BAS 700 . . . . *PAR* 424	Bison 400 SE . . . . . *PAR* 247
Azoxy 250 18	Basagran . . . . *PAR* 44	Bixafen EC125 . . . . . *PAR* 62
Azoxystar 18	Basagran SG 30	BL 500 . . . . *PAR* 151
Aztec . . . . *PAR* 538	Basamid . . . . *PAR* 239	Blanco . . . . *PAR* 167

REFERENCES ARE TO ENTRY NUMBERS NOT PAGES

Index

Blaster Pro 109	Byo-Flex . . . . *Adj*	Cerone 71
Blazer M . . . . *PAR* 555	Cabadex 233	Certis Metaldehyde 3 324
Block . . . . *PAR* 482	Cadou Star . . . . *PAR* 390	Cetan 100 EC . . . . . *PAR* 211
Block 297	Caesar 391	Chamane . . . . *PAR* 26
Blutron . . . . *PAR* 281	Cajole Ultra 204	Change . . . . *PAR* 265
Blutron Plus . . . . . . *PAR* 281	Calaris 322	Charger C 273
B-Nine SG . . . . *PAR* 238	Caliban Duo . . . . . . *PAR* 469	Chekker 10
B-Nine SG 142	Calibre SX . . . . *PAR* 655	Chellist 90
Bogard . . . . *PAR* 265	Callistar . . . . *PAR* 516	ChemSource Diquat . *PAR* 300
BOGOTA UK . . . . . . *PAR* 119	Callisto . . . . *PAR* 513	Chem-Wise Tebu 25 . *PAR* 639
Bokken 372	Callisto 320	Cherokee 75
Bolt . . . . *PAR* 377	Calypso 410	Chess WG . . . . *PAR* 607
Bond . . . . *Adj*	Camix . . . . *PAR* 515	Chess WG 379
Bontima . . . . *PAR* 222	Campus 400 CS . . . *PAR* 555	Chikara . . . . *PAR* 370
Bontima 134	Canopy 317	Chikara Weed Control 229
Bonzi . . . . *PAR* 550	Capalo 196	Chimera SX 337
Bonzi 342	Capex . . . . *PAR* 5	Chimpanzee 424
Boogie Xpro 43	Captan 80 WDG 57	Chipco Green 290
Boom efekt . . . . *PAR* 441	Carakol 3 324	Chlobber 87
Borneo 212	Caramba . . . . *PAR* 526	Chlormephon UK 68
Boscalid 50 . . . . *PAR* 69	Caramba 90 329	Chlorpyrat 480 EC . . *PAR* 154
Bosco WG . . . . *PAR* 69	carbon dioxide 60	Chlorthalis . . . . *PAR* 128
Botanigard WP 29	Careca . . . . *PAR* 596	Choir . . . . *PAR* 539
Boudha 338	Careca 370	Chord 47
Bounty 368	Carfen . . . . *PAR* 107	Choriste . . . . *PAR* 514
Bow . . . . *PAR* 274	Cargo 3 GB . . . . *PAR* 519	Chronicle 347
Bowman . . . . *PAR* 309	Carial Flex 124	Chryzoplus Grey 0.8% *PAR* 464
Boxer . . . . *PAR* 373	Carial Star 160	Chryzopon Rose 0.1% *PAR* 464
Boyano . . . . *PAR* 380	Carpovirusine 117	Chryzosan White 0.6% *PAR* 464
Bravo 500 72	Carpovirusine EVO 2 117	Chryzotek Beige . . . *PAR* 464
Bravo Xtra . . . . *PAR* 130	Caryx 316	Chryzotek Beige 0.4% *PAR* 464
Brazzos 284	Carzone 60 ME . . . . *PAR* 107	Chryzotop Green . . . *PAR* 464
Brevis 325	Cascade . . . . *PAR* 412	Chryzotop Green
Brigida 209	Casino . . . . *PAR* 412	0.25% . . . . *PAR* 464
Broad-Flex . . . . *Adj*	Casper 154	Chusan . . . . *PAR* 315
Broadsheet 78	Cassiopeia 184	Ciclope . . . . *PAR* 161
Broadshot . . . . *PAR* 230	Caste Off . . . . *PAR* 95	Cielex . . . . *PAR* 215
Broadway Star 234	Caveo 323	Cielex 128
Broadway Sunrise . . *PAR* 557	Cavron . . . . *PAR* 472	Cigal Plus . . . . *PAR* 145
Brogue . . . . *PAR* 300	Cavron Green 290	Cimarron 336
Bromag 51	CCC 750 . . . . *PAR* 118	Cinder . . . . *PAR* 555
Bromag Fresh Bait 51	C-Cure . . . . *Adj*	CIPC Fog 300 86
Bromag Wax Blocks 51	CDA Vanquish Biactive . . . 273	CIPC Gold . . . . *PAR* 151
Brutus 201	Ceando 202	Circuit Synctec . . . . *PAR* 169
Buckler . . . . *PAR* 300	Cebara . . . . *PAR* 222	Cirkon 360
Buctril 52	Cebara 134	Cirran 360 141
Buddy . . . . *PAR* 44	Celest Extra 159	Cirrus CS . . . . *PAR* 167
Buggy XTG 273	Celest Trio . . . . *PAR* 268	Cirrus CS 96
Bugle . . . . *PAR* 323	Cello 376	Clan . . . . *PAR* 643
Buguis 92	Celmitron 70% WDG *PAR* 520	Claw 500 . . . . *PAR* 128
Bulldog 173	Centium 360 CS . . . *PAR* 167	Claymore 346
Bullet XL . . . . *PAR* 558	Centium 360 CS 96	Clayton Abba 1
Bullion 253	Centium Plus 98	Clayton Am-Carb . . . *PAR* 95
Bumper 250 EC 368	Centric . . . . *PAR* 653	Clayton Apt . . . . *PAR* 370
Bumper P . . . . *PAR* 582	Centurion Max . . . . *PAR* 159	Clayton Aspen . . . . . *PAR* 514
Bunker . . . . *PAR* 555	Centurion Max 90	Clayton Astra . . . . *Adj*
Butisan S 327	Cerall . . . . *PAR* 606	Clayton Bastille 18
Butoxone 312	Cerastar . . . . *PAR* 412	Clayton Belfry 18
Butoxone DB . . . . . . *PAR* 240	Cercobin WG 414	Clayton Belstone 38
Butoxone DB Extra . *PAR* 242	Ceres Platinum . . . . *Adj*	Clayton Bestow 405
Butryflow 52	Ceriax . . . . *PAR* 315	Clayton Bramble 391
Button . . . . *PAR* 221	Ceriax 198	Clayton Buzz . . . . . *PAR* 524

REFERENCES ARE TO ENTRY NUMBERS NOT PAGES

Index

Clayton Buzz 327	Clayton Tine 323	Cliophar 600 SL . . . . *PAR* 174
Clayton Cajole . . . . . *PAR* 324	Clayton Tonto . . . . . . *PAR* 575	Clipless 422
Clayton Cast 131	Clayton Turret 72	Clipless NT 422
Clayton Chrome CS 96	Clayton Union . . . . *Adj*	Clipper . . . . *PAR* 441
Clayton Comply . . . . *PAR* 598	Clayton Vindicate . . . *PAR* 324	Clodinastar 91
Clayton Core 57	Clayton Vista . . . . . . *PAR* 275	Cloister 202
Clayton Crowe . . . . . *PAR* 379	Clayton Vista 166	Clomate . . . . *PAR* 167
Clayton Dent 480 30	Clayton Vitis 225	Clopic 108
Clayton Dimethachlor . . . 174	Clayton Zorro Pro 377	Cloraz 400 . . . . *PAR* 581
Clayton Diquat . . . . . *PAR* 300	Clean Crop Gallifrey 200 . . *PAR* 412	Clovermax 145
Clayton Diquat 200 . . *PAR* 300		Clozone . . . . *PAR* 167
Clayton Dolmen . . . . *Adj*	Cleancrop Affable 301	CM Lambaz 50 EC . . *PAR* 482
Clayton Edge . . . . . . *PAR* 402	Cleancrop Alatrin . . . *PAR* 672	CMI Delta 2.5 EC . . . *PAR* 243
Clayton El Nino 164	Cleancrop Alatrin Evo *PAR* 672	Codacide Oil . . . . *Adj*
Clayton Enigma 367	Cleancrop Amaize . . . *PAR* 87	Cohort 370
Clayton Faize 322	Cleancrop Amigo . . . *PAR* 151	Coli . . . . *PAR* 69
Clayton Galic 358	Cleancrop Amigo 2 86	Collage . . . . *PAR* 546
Clayton Gatso . . . . . *PAR* 159	Cleancrop Broadshot *PAR* 230	Colt 10 CS . . . . *PAR* 482
Clayton Gatso 90	Cleancrop Bronco . . . *PAR* 441	Colzamid 339
Clayton Gear 133	Cleancrop Celeb . . . . . *PAR* 26	Colzor SyncTec . . . . *PAR* 170
Clayton Glacier . . . . *PAR* 392	Cleancrop Chicane . . *PAR* 167	Comet . . . . *PAR* 608
Clayton Glacier 243	Cleancrop Chicane 96	Comet 200 380
Clayton Goldcob 320	Cleancrop Clifford . . . *PAR* 379	Companion Gold . . . . *Adj*
Clayton Heed . . . . . . *PAR* 598	Cleancrop Corral . . . *PAR* 441	Compass 291
Clayton Index . . . . . *PAR* 314	CleanCrop Corsair . . *PAR* 482	Compitox Plus 313
Clayton Index 197	Cleancrop Covert . . . *PAR* 167	Compliment . . . . *Adj*
Clayton IQ . . . . *PAR* 300	Cleancrop Covert 96	Compo Floranid +
Clayton Java 212	Cleancrop Curve . . . . *PAR* 95	Herbicide . . . . . . *PAR* 226
Clayton Kibo 373	Cleancrop Cutlass . . . *PAR* 672	Concert SX 337
Clayton Kudos . . . . . *PAR* 317	Cleancrop Cyprothal . *PAR* 130	Conclude 18
Clayton Lambada . . . *PAR* 482	Cleancrop Diode 38	Concorde . . . . *PAR* 222
Clayton Lanark 297	Cleancrop Fenpro . . . *PAR* 358	Concorde 134
Clayton Launch 178	Cleancrop	Condor 3 324
Clayton Lithium . . . . *PAR* 541	Fenpropimorph . . . *PAR* 358	Configure 34
Clayton Maritime 56	CleanCrop Flail . . . . . *PAR* 300	Confine 422
Clayton Maximus 235	Cleancrop Frizbee 335	Confine NT . . . . *PAR* 672
Clayton Nero 391	Cleancrop Gallifrey 3 256	Confluence 243
Clayton Obey 372	Cleancrop Hector . . . *PAR* 392	Conform 370
Clayton Peak . . . . . . *PAR* 538	CleanCrop Hoedown *PAR* 441	Confucius . . . . *PAR* 145
Clayton Pirimicarb . . *PAR* 576	Cleancrop Malahide . *PAR* 490	Conqueror . . . . *PAR* 247
Clayton Pirimicarb 357	Cleancrop Mandrake . *PAR* 493	Consento 216
Clayton Pirimicarb 50 *PAR* 576	Cleancrop Marauder . *PAR* 275	Conserve 396
Clayton Pontoon . . . . *PAR* 154	Cleancrop Miricide . . *PAR* 576	Consul 261
Clayton Propel Plus 15	Cleancrop Mondial . . *PAR* 538	Contact Plus . . . . *Adj*
Clayton Proud 205	Cleancrop National 346	Contans WG 112
Clayton Putter 18	Cleancrop Rumble . . *PAR* 596	Contest . . . . *PAR* 6
Clayton Rasp 391	CleanCrop Stomp . . . *PAR* 555	Convey 330EC . . . . . *PAR* 555
Clayton Reap . . . . . . *PAR* 440	CleanCrop Transit 67	Convivial . . . . *PAR* 510
Clayton Repel . . . . . *PAR* 398	Cleancrop Tungsten . *PAR* 441	Coracle 132
Clayton Rugby 356	Cleancrop Wanderer 2 *PAR* 128	Coragen 63
Clayton Satchmo 367	Cleanrun Pro . . . . . . *PAR* 501	Corbel 223
Clayton Saxon . . . . . *PAR* 285	Cleanrun Pro 311	Corinth 377
Clayton Slalom . . . . . *PAR* 324	Cleave 233	Corral . . . . *PAR* 309
Clayton Smelter 393	Clenecorn Super . . . *PAR* 503	Cortez 193
Clayton Sparta . . . . . *PAR* 482	Cleranda 280	Coryx . . . . *PAR* 124
Clayton Spigot 223	Cleravo 282	Corzal . . . . *PAR* 567
Clayton Spook . . . . . *PAR* 254	Clesima . . . . *PAR* 456	Corzal 350
Clayton Sputnik 95	Clinic . . . . *PAR* 441	Corzal SC . . . . *PAR* 567
Clayton Stun 303	Clinic Ace 273	Cosine 118
Clayton Tardis . . . . . *PAR* 425	Clinic TF 273	Cosmic NG . . . . *PAR* 441
Clayton Tardis 263	Clinker . . . . *PAR* 555	Cougar . . . . *PAR* 314
Clayton Tebucon EW 404	Cliophar 400 . . . . . . *PAR* 174	County Mark . . . . *Adj*

REFERENCES ARE TO ENTRY NUMBERS NOT PAGES

Index

Couraze . . . . *PAR* 461	Dancer Flow . . . . . . *PAR* 567	Diflufen 500 . . . . . . *PAR* 273
Crafter 83	Dash HC . . . . *Adj*	Diflufenastar . . . . . . *PAR* 275
Crassus . . . . *PAR* 541	Datura . . . . *PAR* 488	Diflufenastar 100 . . . *PAR* 275
Crawler. 59	Datura 301	Diflufenican GL 500 . *PAR* 273
Crebol 223	Dauphin 45 . . . . . . *PAR* 203	Difure Pro 162
Credit 273	Dazide Enhance . . . . *PAR* 238	Digital. 244
Credit DST . . . . *PAR* 441	Dazide Enhance 142	Dime 178
Credo 80	DB Straight 144	Dimeco. . . . . *PAR* 253
Crescent. . . . . *PAR* 412	Deacon . . . . *PAR* 639	Dimilin Flo . . . . *PAR* 272
Crescent. 256	Deacon 404	Dimilin Flo 163
Crestler 273	Debut. 421	Diminish . . . . *PAR* 440
Crew 340	Decis 146	Dina 50. . . . . *PAR* 273
Cropspray 11-E . . . . *Adj*	Decis Forte . . . . . . *PAR* 243	Dioweed 50 135
CropStar Diquat . . . . *PAR* 300	Decis Protech 146	Dipel DF . . . . *PAR* 37
Croupier . . . . *PAR* 416	Dedicate 405	Dipel DF 27
Croupier OD . . . . . . *PAR* 416	Defender 420	Diplomat . . . . *PAR* 468
Cruiser 70 WS . . . . . *PAR* 653	Defiant SC 325	Diqua 188
Cruiser SB 411	Defiant WG . . . . . . *PAR* 520	Diquanet . . . . *PAR* 300
Crystal 243	Deft Premium 336	Diquash . . . . *PAR* 300
CS Chlormequat . . . *PAR* 118	Defy 372	Di-Quattro . . . . *PAR* 300
CS Myclobutanil. . . . *PAR* 541	Degesch Fumigation	Discman Biograde 273
CTL 500 72	Tablets 6	Discovery . . . . *PAR* 522
Cube . . . . *PAR* 309	Degesch Plates 302	Discus . . . . *PAR* 537
Cuprokylt 113	Delan Pro 190	Dithane 945 . . . . . . *PAR* 493
Curator 19	Deliverer . . . . *PAR* 275	Dithane 945 306
Curb Crop Spray	Deltason-D . . . . . . *PAR* 243	Dithane Dry Flowable
Powder 5	Dennis . . . . *PAR* 596	Neotec . . . . *PAR* 493
Curb S . . . . *PAR* 7	Depitone Ultra 135	Dithane NT Dry
Curlew . . . . *PAR* 181	Depitox. 135	Flowable 306
Curuni . . . . *PAR* 402	Depitox 500 . . . . . *PAR* 224	Dithianon WG 189
Curzate M WG 123	Derrex 226	Diva . . . . *PAR* 611
Custodia 25	Designer . . . . *Adj*	Divot . . . . *PAR* 518
Cutaway 422	Desire 324	D-kwat 200 . . . . . . *PAR* 300
Cuter . . . . *PAR* 513	Dessicash 200 188	Doff Horticultural Slug
Cyclade 69	Destiny 322	Killer Blue Mini
Cyclone 400 EW . . . *PAR* 584	Deter 110	Pellets . . . . *PAR* 519
Cyclops 53	Détia Gas Ex-P . . . . . . *PAR* 8	Doolin. . . . . *PAR* 128
Cyd-X. . . . . *PAR* 201	Detia Gas-Ex-B 6	Doonbeg 2 288
Cyd-X Duo . . . . *PAR* 201	Detia Gas-Ex-T . . . . . . *PAR* 8	Dow Agrosciences
Cyd-X Xtra. 117	Devrinol . . . . *PAR* 544	Glyphosate 360 . . *PAR* 441
Cyflamid 118	Devrinol 339	Dow MCPA Amine 50 *PAR* 499
Cyflu EW . . . . *PAR* 202	Di Novia 188	Dow Shield 400 102
Cymbal 45 119	Diablo . . . . *PAR* 12	Doxstar Pro 258
Cymostraight 45 . . . *PAR* 203	Diagor . . . . *Adj*	DP 747 . . . . *PAR* 560
Cyper 100 . . . . *PAR* 211	Diamant . . . . *PAR* 313	DP911 SX 338
Cyperkill 10 . . . . *PAR* 211	Dican . . . . *PAR* 273	Dragoon 188
Cypher 349	Diconazol 250 . . . . *PAR* 265	Dragoon Gold 188
Cyren 87	Dicophar . . . . *PAR* 229	Drill . . . . *Adj*
Cythrin 500 EC 126	Diecot . . . . *PAR* 273	Drum 119
Cythrin Max . . . . . . *PAR* 211	Diego . . . . *PAR* 202	Dual Gold 393
Czar . . . . *PAR* 167	Difcon 250 . . . . . . *PAR* 265	Ductis SX . . . . *PAR* 405
D.E.P. 205 149	Difcor 250 EC 158	Ductis SX 253
D.E.P. 251 149	Difenag 157	Duke . . . . *PAR* 275
Daconil . . . . *PAR* 128	Difenag Fresh Bait 157	Duo 400 SC . . . . . . *PAR* 331
Dairy Fly Spray 382	Difenag Wax Bait 157	Duplosan KV 313
Daisy . . . . *PAR* 74	Difenag Wax Blocks 157	Dursban WG 87
Dakota 105	Difend 158	Duster . . . . *PAR* 128
Dalda 5 297	Difend Extra . . . . . . *PAR* 267	Dymid . . . . *PAR* 19
Damine. . . . . *PAR* 224	Difenostar . . . . *PAR* 265	Dynamec . . . . *PAR* 1
Damocles 72	Difference. 158	EA Difcon . . . . *PAR* 265
Danadim . . . . *PAR* 291	Diflanil 500 SC 164	EA Lambda 50 . . . . . *PAR* 482
Danadim Progress 180	Diflox Flow . . . . *PAR* 272	EA Lambda C . . . . . *PAR* 482

REFERENCES ARE TO ENTRY NUMBERS NOT PAGES

Index

EA Pyrifos	PAR 154	Ethy-Gen II	PAR 326	Firefly 155	251
Eagle	9	EthylBloc Tabs	PAR 532	Firestorm	166
Eagle Green Care	PAR 436	Ethylene Buster	PAR 532	Fixor	PAR 542
Easel	310	Etna	PAR 441	Flagon 400 EC	PAR 77
Eclipse	PAR 310	Ettu	PAR 320	Flamenco	PAR 410
Eco-flex	Adj	Euclid	PAR 26	Flanker	355
Ecoplug Max	PAR 441	Euro Benta 480	PAR 44	Flanoc	PAR 590
Ecoplug Max	273	Euro Glyfo 360	PAR 441	Flatline	PAR 374
Edge 400	PAR 596	Euro Glyfo 450	PAR 441	Flazasulf 25	PAR 370
Edipro	PAR 589	Euro Lambda 100 CS	PAR 482	Flexidor	294
Efeckt	206	Euro Tebu 250	PAR 639	Flexity	334
EGCÂGranules	PAR 436	Euroagkem Pen-e-trate	Adj	Flight	PAR 556
Egret	PAR 441	Evolya	PAR 513	Flint	PAR 669
Elan Xtra	Adj	Excalibur	168	Flixfit 20 EC	PAR 412
Electis 75WG	308	Exceed SX	253	Float	PAR 380
Elk	177	Excel DF Gold	PAR 441	Flocter	PAR 35
Elliott's Lawn Sand	PAR 367	Exemptor	410	Flomide	PAR 596
Elliott's Mosskiller	PAR 367	Exilis	34	Flora 50	PAR 373
Elude	PAR 598	Eximus	PAR 555	Floramite 240 SC	37
Elumis	321	Exocet	PAR 627	Floreo	323
Embutone 24DB	144	Exosex CM	192	Fluazinova	PAR 380
Emerald	Adj	Explicit	287	Flufenamone	PAR 276
Emerald Turf	PAR 472	Exploit	PAR 402	Flufenican	PAR 275
Emesto Prime FS	PAR 559	Falcon	367	Flurostar 200	PAR 412
Eminent Star	PAR 146	Fandango	251	Flurostar 200	256
Eminentos 10 CS	PAR 482	Fastac	PAR 6	Flurosulam XL	PAR 374
Empire	225	Fastnet	PAR 555	Flurox 180	PAR 412
Emzeb 80 WP	PAR 493	Fathom	PAR 639	Flurox 200	256
Enforcer	138	Fathom	404	Fluroxy 200	PAR 412
Engage	PAR 596	Fazor	PAR 490	Flux	PAR 424
En-Garde	PAR 50	Fazor	303	Fluxcon 20 EC	PAR 412
Ennobe	PAR 322	Felix	Adj	Fluxyr 200 EC	PAR 412
Ensis	307	Fence	240	Flycatcher	358
Entail	340	Fenican	PAR 275	Flyer	380
Enterprise	47	Fenlander 2	PAR 331	Flyer 200	380
Envidor	397	Fenomenal	215	Flyflo	PAR 273
Envision	273	Fenprop 750	223	Foil	PAR 411
Envoy	203	Fermor 500 WG	PAR 481	Folicur	404
Enzo	324	Ferramol Max	PAR 366	Folio Gold	78
Epee	PAR 160	Ferramol Slug Killer	PAR 366	Folmex 250 EW	PAR 639
Epic	193	Ferromex Mosskiller		FOLY 500 SC	PAR 380
Epoflux	PAR 314	Concentrate	PAR 367	Fongarid Gold	PAR 518
Epona	PAR 411	Ferrox	PAR 366	Fontelis	348
Epoxi 125 SC	193	Fezan	PAR 639	Force ST	PAR 645
Epzole 125	PAR 309	Fezan	404	Force ST	407
Equity	87	Fezan Plus	83	Forefront T	16
Eradicoat	PAR 492	Fidox	PAR 598	Foremost	PAR 584
Eradicoat	305	Fielder SE	82	Forester	126
Erase	PAR 639	Fiesta T	PAR 117	Forge	PAR 538
Erase	404	Fiesta T	66	Formula B Rat & Mouse	
Ergon	337	Figaro	PAR 441	Killer	51
Eribea	4	Filan	45	Fornet 4 SC	PAR 545
Escolta	PAR 219	Finale 150	PAR 440	Fornet 6 OD	340
Escolta	131	Finalsan	PAR 338	Fortify	405
ESP3	324	Finalsan Plus	304	Fortress	388
Esteem	103	Finish SX	337	Fosulfuron	PAR 635
Esteron T	139	Finy	PAR 538	Foundation	153
Ethe 480	71	Finy	336	Fountain	PAR 386
Ethefon 480	71	Firebird	PAR 275	Fox	38
Etho 500 SC	206	Firebrand	Adj	Foxtrot EW	219
Ethofol 500 SC	206	Firecrest	PAR 538	Fozat 360 SL	PAR 441
Ethosat 500	PAR 328	Firefly	PAR 402	Fracarb	PAR 209

REFERENCES ARE TO ENTRY NUMBERS NOT PAGES

Index

Freeze . . . . *PAR* 672	Gf-1784 . . . . *PAR* 412	Gozai 381
Freeze NT 422	GF-2070 . . . . *PAR* 408	Grainstore 146
Frelizon 349	GF-2463 . . . . *PAR* 177	Grain-Tect ULV . . . . . *PAR* 243
Frontsweep 273	GF-2573 . . . . *PAR* 447	Grapple 94
Frowncide . . . . *PAR* 380	GF-2895 . . . . *PAR* 174	Grazon Pro 109
Frozen . . . . *PAR* 392	GIBB 3 . . . . *PAR* 438	Grecale 121
Frupica SC 314	Gibb 3 270	Green Gold . . . . *Adj*
FSA Chlorothalonil . . *PAR* 128	GIBB Plus . . . . *PAR* 438	Green Turf King 290
Fubol Gold WG 307	Gibb Plus 270	Greencrop Bantry 235
Fuego . . . . *PAR* 524	Gibb Plus 10 SL . . . . *PAR* 438	Greencrop Cajole Ultra . . . 204
Fulmar . . . . *PAR* 69	Gizmo . . . . *PAR* 639	Greencrop Doonbeg 2 . . . 288
Fulmar 45	Gladiator . . . . *PAR* 441	Greencrop Goldcob 320
Fumesate 500 SC 206	Glam 272	Greencrop Helvick 356
Fumetham . . . . *PAR* 522	Glister . . . . *PAR* 441	Greencrop Malin 403
Fungazil 50LS . . . . . *PAR* 449	Glister Ultra . . . . *PAR* 441	Greencrop Reaper . . *PAR* 412
Furlong . . . . *PAR* 216	Globaryll 100 34	Greencrop Tabloid . . *PAR* 639
Furlong 129	Globaztar SC 18	Greencrop Twinstar 360
Fury 10 EW . . . . *PAR* 676	Globe . . . . *PAR* 207	Greenmaster Autumn *PAR* 367
Fury 10 EW 424	Glopyr 400 . . . . *PAR* 174	Greenmaster Autumn 227
Fusilade 250 EW . . . *PAR* 379	Glopyr 400 102	Greenmaster Extra . . *PAR* 501
Fusilade Max 235	Glotron 700 SC 325	Greenmaster
Fusion 405	Glycel . . . . *PAR* 441	Mosskiller 227
Fytoclean . . . . *PAR* 338	Glydate . . . . *PAR* 441	Greenor 106
Gala . . . . *PAR* 412	Glyder . . . . *PAR* 441	Greentec Mosskiller . *PAR* 367
Galactic Pro . . . . *PAR* 19	Glyder 450 TF . . . . . *PAR* 441	Greentec Mosskiller
Galaxy 105	Gly-Flex . . . . *Adj*	Pro . . . . *PAR* 367
Galben M . . . . *PAR* 42	Glyfo Star . . . . *PAR* 441	Greentec Mosskiller
Galcon SC . . . . *PAR* 567	Glyfo_TDI . . . . *PAR* 441	Pro 227
Galera 108	Glyfos 273	Grenadier . . . . *Adj*
GAL-GONE . . . . *PAR* 412	Glyfos Dakar 273	Grenadier . . . . *PAR* 436
GAL-GONE 256	Glyfos Dakar Pro 273	Gringo 329
Galileo 355	Glyfos Gold . . . . *PAR* 441	Grisu 290
Galivor 347	Glyfos Gold ECO . . . *PAR* 441	Gropper SX 336
Gallup 360 273	Glyfos Monte . . . . . . *PAR* 441	Gro-Slo . . . . *PAR* 490
Gallup Amenity 273	Glyfos Proactive 273	Gro-Stop 100 . . . . . . *PAR* 151
Gallup Hi- Aktiv	Glyfos Supreme 273	Gro-Stop Electro . . . *PAR* 151
Amenity . . . . *PAR* 441	Glyfosat 36 273	Gro-Stop Fog . . . . . . *PAR* 151
Gallup Hi-Aktiv 273	Glyfo-TDI . . . . *PAR* 441	Gro-Stop HN . . . . . . *PAR* 151
Gallup Hi-Aktiv Amenity*PAR* 441	Glymark . . . . *PAR* 441	Gro-Stop Innovator . . *PAR* 151
Gallup XL 273	Glyper . . . . *PAR* 441	Gro-Stop Ready 86
Galmano . . . . *PAR* 410	Glyphogan . . . . *PAR* 441	Gro-Stop Solid 86
Gambit . . . . *PAR* 167	Glypho-Rapid 450 . . . *PAR* 441	Grounded . . . . *Adj*
Gamit 36 CS . . . . . . *PAR* 167	Glyphosate 360 . . . . *PAR* 441	Growell 2,4-D Amine *PAR* 224
Gamit 36 CS 96	Gly-Plus A . . . . *Adj*	Guardsman . . . . *PAR* 7
Gandalf 35	Glystar 360 . . . . *PAR* 441	Guillotine . . . . *PAR* 394
Garlon Ultra 16	Glyweed . . . . *PAR* 441	Gunbar . . . . *PAR* 327
Garrison . . . . *PAR* 155	Gocha 204	Guru 77
Garryowen XL 273	Goiana . . . . *PAR* 113	Gusto 3 324
Gartrel 104	Goldcob 320	Gyro 422
GAT Decline 2.5 EC . *PAR* 243	Goldron-M . . . . *PAR* 538	Habitat . . . . *PAR* 441
GAT Stake 20 EC . . . *PAR* 412	Go-Low Power . . . . *PAR* 499	Halcyon . . . . *PAR* 17
Gateway . . . . *Adj*	Goltix 70 SC 325	Hale 375
Gavel 278	Goltix 90 . . . . *PAR* 520	Halley 380
Gazelle . . . . *PAR* 2	Goltix Compact . . . . *PAR* 520	Hallmark with Zeon
Gazelle SG 2	Goltix Plus . . . . *PAR* 329	Technology 297
Gemini . . . . *PAR* 331	Goltix Super . . . . *PAR* 329	Halta . . . . *PAR* 440
Gemstone 203	Goltix Titan . . . . *PAR* 521	Hamlet 171
Genoxone ZX EC . . . *PAR* 237	Goltix WG . . . . *PAR* 520	Hammer . . . . *PAR* 442
Geoxe 237	Googly 188	Hammer 274
GF 184 233	Gorgon 240	Harmony M SX 337
GF Chlorothalonil . . . *PAR* 128	Goshawk 164	Harmony SX . . . . . . *PAR* 654
GF Propiconazole 250 . . . 368	Governer 107	Harness . . . . *PAR* 26

REFERENCES ARE TO ENTRY NUMBERS NOT PAGES

Index

Harvest.......... *PAR* 440	Hockley Propyzamide	Intellis Plus........... 79
Harvest.............. 272	40............. *PAR* 596	Inter Bifenazate 240 SC *PAR* 56
Harvest Proteb.... *PAR* 603	Hockley Tebucon 25. *PAR* 639	Inter Diflufenican... *PAR* 273
Harvest Teb 250 ... *PAR* 639	Holster XL........... 137	Inter Diquat....... *PAR* 300
Hatchet 200 EC.... *PAR* 412	Horus........... *PAR* 468	Inter Metazachlor... *PAR* 524
Hatchet Xtra...... *PAR* 412	Horus.............. 288	Inter Metron 700 SC *PAR* 520
Hatchet Xtra......... 256	Hounddog........... 204	Inter Tebloxy...... *PAR* 642
Hatra........... *PAR* 468	Hubble............. 182	Inter Trinex...... *PAR* 672
Hatra............. 288	Hudson 200...... *PAR* 412	Intercord......... *PAR* 309
Hauberk......... *PAR* 441	Hunter............. 233	Interface........... 292
Headland Charge....... 313	Hurler.......... *PAR* 412	Interfix.......... *PAR* 179
Headland Flail 2.... *PAR* 183	Hurler............. 256	Inter-Tebu........ *PAR* 639
Headland Fortune........ *Adj*	Hurricane SC........ 164	Intracrop Agwet GTX..... *Adj*
Headland Guard 2000.... *Adj*	Hussar........ *PAR* 467	Intracrop BLA......... *Adj*
Headland Guard Pro..... *Adj*	Hyban P............ 153	Intracrop Bla-Tex..... *Adj*
Headland Intake......... *Adj*	HY-D Super........ 135	Intracrop Boost........ *Adj*
Headland Polo......... 141	Hydro Coco Houseplant	Intracrop Brigand...... *Adj*
Headland Rhino......... *Adj*	Pest & Spider Mite	Intracrop Cogent....... *Adj*
Headland Saxon........ 153	Killer........... *PAR* 338	Intracrop Dictate...... *Adj*
Headland Spear....... 310	HY-GLO 360...... *PAR* 441	Intracrop Evoque...... *Adj*
Headland Spruce...... 144	HY-Gly 360...... *PAR* 441	Intracrop F16......... *Adj*
Headland Staff 500..... 135	Hygrass P........... 153	Intracrop Green Oil..... *Adj*
Headland Swift.... *PAR* 253	Hykeep.......... *PAR* 651	Intracrop Impetus...... *Adj*
Headlite............. 61	HY-MCPA........... 310	Intracrop Inca........ *Adj*
Headway............ 24	Hymec Triple...... *PAR* 260	Intracrop Incite....... *Adj*
Hecates............ 404	Hyprone P.......... 152	Intracrop Mica AF...... *Adj*
Helix............. 375	Hyprone-P...... *PAR* 252	Intracrop Neotex...... *Adj*
Helmstar A 75 WG.. *PAR* 665	Hysward-P....... *PAR* 252	Intracrop Novatex..... *Adj*
Helosate 450 TF.... *PAR* 441	Hysward-P.......... 152	Intracrop Perm-E8..... *Adj*
Helvick............ 356	I T Bentazone 480... *PAR* 44	Intracrop Predict...... *Adj*
Herbasan Flow.... *PAR* 567	I.T. Diquat....... *PAR* 300	Intracrop Quad....... *Adj*
Herbi-AKtiv.......... *Adj*	Ibex............... 203	Intracrop Quartz...... *Adj*
Herboxone.......... 135	Ibiza 500....... *PAR* 380	Intracrop Questor...... *Adj*
Herboxone 60..... *PAR* 224	IC1574......... *PAR* 598	Intracrop Rapide Beta... *Adj*
Heritage............ 18	Icade.............. 16	Intracrop Retainer NF... *Adj*
Heritage Maxx......... 18	Icarus.............. 201	Intracrop Rigger....... *Adj*
Herold............. 166	Ice............ *PAR* 392	Intracrop Rustler...... *Adj*
Hi Strobin 25....... *PAR* 26	Iceni.............. 422	Intracrop Salute AF..... *Adj*
Hiatus.............. 413	Idosem 36....... *PAR* 468	Intracrop Sapper AF..... *Adj*
Hi-Fosate........ *PAR* 441	Ignite.............. 193	Intracrop Saturn....... *Adj*
High Load Mircam...... 153	Imidasect 5GR.... *PAR* 461	Intracrop Signal XL..... *Adj*
Highgate............ 49	Imidasect 5GR...... 285	Intracrop Sprinter..... *Adj*
Hiker........... *PAR* 374	Impact.......... *PAR* 423	Intracrop Status....... *Adj*
Hilite.............. 273	Impala............. *Adj*	Intracrop Stay-Put...... *Adj*
Hill-Agro 40....... *PAR* 545	Impala.......... *PAR* 287	Intracrop Super Rapeze
Himalaya......... *PAR* 490	Impuls........... *PAR* 48	MSO.............. *Adj*
Himalaya XL...... *PAR* 490	Imtrex......... *PAR* 424	Intracrop Tonto........ *Adj*
Hi-Mectin.......... *PAR* 1	Imtrex............. 262	Intracrop Vantage...... *Adj*
Hinder........... *PAR* 185	Indar 5 EW....... *PAR* 345	Intracrop Warrior...... *Adj*
Hi-Phone 48....... *PAR* 124	Indar 5 EW......... 217	Intracrop Zenith AF..... *Adj*
Hi-Prodione 50..... *PAR* 472	Infinito............. 246	Intro Plus........ *PAR* 42
Hithane.......... *PAR* 541	Ingot Z......... *PAR* 482	Intruder............. 86
Hobby........... *PAR* 167	Inigo............. 333	Invader............. 183
Hobson............. 323	Inka SX............ 413	Iodomeso OD..... *PAR* 468
Hockley Bentazone 48 *PAR* 44	Insegar WG......... 220	Ipanema............ 71
Hockley Diquat 20.. *PAR* 300	Insignia........ *PAR* 608	Iprodione 500 SC... *PAR* 472
Hockley Fluroxypyr 20 *PAR* 412	Insignis......... *PAR* 74	Ironmax Pro......... 226
Hockley Lambda 5EC *PAR* 482	Inssimo.......... *PAR* 4	Iroxx.............. 226
Hockley Mesotrione 10*PAR* 513	Instrata............ 76	Isomec.......... *PAR* 503
Hockley Pendi 330.. *PAR* 555	Insyst.............. 2	Isomec Ultra......... 156
Hockley Pirimicarb WG*PAR* 576	Intellis......... *PAR* 560	IT Bentazone....... *PAR* 44
Hockley Propicon 25. *PAR* 591	Intellis............ 348	IT Chlorothalonil.... *PAR* 128

REFERENCES ARE TO ENTRY NUMBERS NOT PAGES

Index

IT Lambda . . . . *PAR* 482	Kix 158	Lector Delta 165
Itofin . . . . *PAR* 396	Klartan . . . . *PAR* 638	Legacy . . . . *PAR* 380
Jaboland . . . . *PAR* 580	Klaxon . . . . *PAR* 467	Legara 108
Jabolim . . . . *PAR* 580	Klaxon OD . . . . *PAR* 467	Legion 328
Jackpot 2 . . . . *PAR* 482	Klean G . . . . *PAR* 441	Leimay . . . . *PAR* 21
Jade . . . . *PAR* 598	Klever . . . . *PAR* 412	Lenazar Flo 299
Jadex Plus . . . . *PAR* 118	Klipper . . . . *Adj*	Lentagran WP 383
Jadex-o-720 67	KN 540 . . . . *PAR* 441	Leopard 5EC . . . . . . *PAR* 619
Jali 450 . . . . *PAR* 441	Knoxdoon . . . . *PAR* 300	Lepinox Plus 27
Jatapu 56	K-Obiol EC 25 146	Leptom 250 WDG . . *PAR* 653
Jaunt 252	K-Obiol ULV 6 146	Levada 370
Jenton 224	Kotek . . . . *Adj*	Level . . . . *Adj*
Jerico . . . . *PAR* 468	Kraken . . . . *PAR* 441	Lexi 334
Jester . . . . *PAR* 86	Kruos . . . . *PAR* 392	Lexus Class . . . . *PAR* 108
Jewel . . . . *PAR* 109	Kubist Flo . . . . *PAR* 328	Lexus Millenium 255
Jewel 62	Kudos 364	Lexus SX 253
Jockey . . . . *PAR* 411	Kumulus DF 402	LEY . . . . *PAR* 608
Jockey Solo . . . . . . *PAR* 410	Kunshi . . . . *PAR* 205	Li-700 . . . . *Adj*
Joules 72	Kunshi 121	Liaison 273
Jubilee SX 336	Kur Tail . . . . *PAR* 440	Liberate . . . . *PAR* 575
Judo 370	Kurdi 251	Liberator 166
Junction 139	Kurtail Gold . . . . *PAR* 440	Librax . . . . *PAR* 425
Justice 371	Kusti . . . . *PAR* 482	Librax 263
Juventus 329	Kyleo 140	Lieto 125
K2 . . . . *PAR* 118	Labyrinth . . . . *PAR* 513	Life Scientific
K2 67	Laidir 10 CS . . . . *PAR* 482	Azoxystrobin . . . . . *PAR* 26
Kaedyn 500 . . . . *PAR* 488	Lambda 100 CS 297	Life Scientific
Kalahari . . . . *PAR* 300	Lambda C . . . . *PAR* 482	Chlorothalonil 72
Kalenkoa . . . . *PAR* 280	Lambda-C 100 297	Life Scientific
Kalypstowe . . . . *PAR* 513	Lambda-C 50 297	Chlorothalonil 500 . *PAR* 128
Kantor . . . . *Adj*	Lambd'Africa . . . . . . *PAR* 482	Life Scientific
Karamate Dry Flo	Lambdastar 297	Clodinafop 240 . . . *PAR* 160
Neotec 306	Laminator 75 WG . . . *PAR* 493	Life Scientific Clopyralid . . 102
Karate 2.5WG 297	Laminator Flo 306	Life Scientific Diquat 188
Karis 10 CS . . . . *PAR* 482	Lanark 297	Life Scientific Diquat
Karis 10 CS 297	Lancer . . . . *PAR* 409	200 . . . . *PAR* 300
Karma 359	Landgold Bentazone SL *PAR* 44	Life Scientific
Kartouch 60 WG . . . *PAR* 103	Landgold Cyper 100 . *PAR* 211	Epoxiconazole . . . *PAR* 309
Kaskara . . . . *PAR* 230	Landgold Diquat . . . . *PAR* 300	Life Scientific
Kaspar . . . . *PAR* 440	Landgold Lambda-Z . *PAR* 482	Florasulam . . . . . . *PAR* 373
Katalyst . . . . *Adj*	Landgold Metsulfuron *PAR* 538	Life Scientific
Katamaran 328	Landmaster . . . . *PAR* 441	Flufenacet + DFF . *PAR* 275
Katamaran Turbo 177	Landmaster 360 TF . *PAR* 441	Life Scientific Lambda-
Katana 229	Landscaper Pro Feed	Cyhalothrin . . . . . *PAR* 482
Kayak 132	Weed Mosskiller 228	Life Scientific Lambda-
KeMiChem -	Landscaper Pro Moss	Cyhalothrin 50 . . . *PAR* 482
Propyzamide 400 SC*PAR* 596	Control + Fertiliser 227	Life Scientific
Kendo . . . . *PAR* 482	Landscaper Pro Moss	Mesotrione . . . . . *PAR* 513
Kerb 50 W . . . . *PAR* 596	Control + Fertilizer *PAR* 367	Life Scientific
Kerb Flo 370	Landscaper Pro Weed	Metconazole . . . . *PAR* 526
Kerb Flo 500 370	Control + Fertilizer *PAR* 226	Life Scientific
Kerb Granules 370	Landscaper Pro Weed	Metconazole 60 . . *PAR* 526
Kernel 273	Control + Fertilizer 136	Life Scientific
Kestrel 377	Larke 310	Metsulfuron-Methyl *PAR* 538
Keystone 200	Laser . . . . *PAR* 200	Life Scientific
Killermite 1	Laser 116	Trinexapac . . . . . . *PAR* 672
Killgerm ULV 500 352	Latice . . . . *PAR* 392	Life Scientific
Kindred 318	Latitude . . . . *PAR* 623	Trinexapac 250 422
Kinetic . . . . *Adj*	Latitude XL 392	Limitar 422
Kingdom 47	Laudis . . . . *PAR* 646	Linford 115
Kingdom Turf . . . . . . *PAR* 26	Leaf-Koat . . . . *Adj*	Lingo 97
Kinto . . . . *PAR* 586	Lector 232	Linurex 50 SC . . . . . *PAR* 488

REFERENCES ARE TO ENTRY NUMBERS NOT PAGES

Index

Linzone	97	Martlet	PAR 482
Liquid Curb Crop Spray	5	Marvel	PAR 513
Logic	Adj	Masai	406
Logic Oil	Adj	Mascot Crossbar	137
Logo	PAR 429	Mascot Defender	420
Longbow	138	Mascot Eland	380
Longhorn	367	Mascot Exocet	368
Lontrel 72SG	PAR 174	Mascot Hi-Aktiv Amenity	PAR 441
Lorate	336	Mascot Hi-Aktiv Amenity	273
Low Down	Adj	Mascot Rayzor	PAR 472
LS Chlorothalonil XL	PAR 128	Mascot Rayzor	290
LSTF Pro	PAR 367	Mascot Systemic	PAR 95
Luna	PAR 288	Mascot Systemic	58
Luna Sensation	PAR 400	Master Cyman	PAR 207
Lupo	141	Master Ethos	PAR 328
Lynx H	324	Master Gly	273
Mac-Bentazone 480 SL	PAR 44	Master Gly 36T	273
Maccani	191	Master Lean 500 SC	PAR 596
MAC-Cypermethrin 100 EC	PAR 211	Master Prop 40	PAR 596
MAC-Difenoconazole 250 EC	PAR 265	Master Prop 40	370
MAC-Iprodione 255 SC	PAR 472	Master Sil	Adj
MAC-Lambda-Cyhalothrin 50 EC	PAR 482	Master Teb 25	PAR 639
MAC-Napropamide 450 SC	PAR 544	Master Zone	PAR 44
MAC-Propyzamide 400 SC	PAR 596	Matador 200 SC	PAR 344
Magic Tandem	PAR 331	Maton	PAR 224
Magneto	136	Matsuri	PAR 591
Magpie	PAR 159	Mavrik	403
Magtoxin	PAR 489	MaxCel	34
Magtoxin Pellets	PAR 489	Maxicrop Moss Killer & Conditioner	PAR 367
Mainman	231	Maxicrop No. 2 Moss Killer and Conditioner	PAR 367
Maintain	PAR 672	Maxim 100FS	237
Maister WG	266	Maxim 480FS	237
Majestik	PAR 492	Maxim XL	238
Majestik	305	Maximus	235
Major	PAR 482	Maya	52
Makila 500 SC	PAR 524	Mazone 360	PAR 167
Malin	403	MCPA 25%	310
Malvi	306	MCPA 50	PAR 499
Malvin WG	PAR 93	Me2 Lambda2	PAR 482
Mamba	51	Meco-Flex	Adj
Manconil 725	PAR 207	Medallion TL	237
Mandimid	PAR 497	Medax Max	365
Mandolin Flow	PAR 567	Medax Top	PAR 509
Manfil 75 WG	PAR 493	Medirac	PAR 288
Manfil 80 WP	PAR 493	Menace 80 EDF	PAR 596
Manicure	PAR 639	Menara	130
Manifest	PAR 441	MENNO Florades	PAR 51
Manipulator	PAR 118	Menno Florades	33
Manipulator	67	Mentor	PAR 441
Manitoba	199	Mentor	273
Mantra	195	Mepicame	PAR 327
Manzate 75 WG	306	Mepicon	PAR 508
Marengo	PAR 440	Mequat 750	PAR 118
Markate 50	PAR 482	Meridian	392
Markate 50	297	Meristo	PAR 513
Marnoch Phorm	PAR 358	Merit Forest	PAR 461
Martinet	420	Merit Forest	285
		Mero	Adj
Mesiodo 35	PAR 468		
Mesoiodostar WG	PAR 468		
Mesostar	320		
Mesotrion 100SC	PAR 513		
Messidor 45 SC	PAR 488		
Mesurol	330		
Met52 granular bioinsecticide	326		
Met52 OD	PAR 523		
Meta Turbo	PAR 288		
Meta WDG	PAR 520		
Metafol Super	PAR 329		
Metal	PAR 526		
Metal 60	PAR 526		
Metam 510	PAR 522		
Metamerac	PAR 525		
Metamerac-II	PAR 525		
Metaza	PAR 524		
Metazerac	PAR 525		
Metcostar	329		
Meteor	70		
Metmeth 200	336		
Metric	99		
Metro 20	PAR 538		
Metro Clear	PAR 538		
Metso	336		
Metso XT	336		
Met-Tribe 286	PAR 540		
Mewati	306		
Mexxon	336		
Mexxon xtra	336		
Micaraz	200		
Microthiol Special	PAR 637		
Microthiol Special	402		
Micro-Turf	PAR 669		
Mido 70% WDG	PAR 461		
Mikado	PAR 524		
Milagro 240 OD	340		
Milentus Bentazone	PAR 44		
Milentus Pirimicarb	PAR 576		
Milepost	PAR 19		
Milestone	PAR 19		
Minerva	PAR 513		
Minstrel	256		
Minuet EW	PAR 676		
Minuet EW	424		
Minx PX	PAR 538		
Mirador	PAR 26		
Mirage 40 EC	PAR 581		
Mirage 45 SC	PAR 581		
Mircam	PAR 253		
Mircam Plus	152		
Mission 200 SL	PAR 300		
Mission 200 SL	188		
MiteStop	PAR 243		
Mitre	PAR 639		
Mitron 700 SC	325		
Mixanil	73		
Mixture B NF	Adj		
Mobius	378		
Mocap 15G	PAR 332		
Modan	PAR 672		

REFERENCES ARE TO ENTRY NUMBERS NOT PAGES

Index

Modan 250 EC	422	Naturalis-L	29	Nortron Flo	*PAR* 328
Moddus	422	Nautile DG	*PAR* 207	Notion	*PAR* 167
Moddus ME	*PAR* 672	Nautile DG	123	Novagib	270
Mogeton	387	Nautile WP	*PAR* 207	Nu Film P	*Adj*
Mogul	*PAR* 178	Navigate	*PAR* 275	Nuance	*PAR* 665
MON 76473	*PAR* 441	Nebula XL	48	Nufarm MCPA 750	*PAR* 499
MON 79351	*PAR* 441	Nectar	*PAR* 145	Nufosate	*PAR* 441
MON 79376	*PAR* 441	Nelson	*Adj*	Nufosate Ace	*PAR* 441
MON 79545	*PAR* 441	Nemathorin 10G	268	Nuron	*PAR* 598
MON 79632	*PAR* 441	NEMguard A PCN		NYX	*PAR* 598
MON 79991	*PAR* 441	Granules	*PAR* 436	Oberon	*PAR* 629
Monceren DS	345	NEMguard DE	269	Oblix 500	*PAR* 328
Monitor	*PAR* 635	NEMguard granules	*PAR* 436	Oblix 500	206
Monitor	401	Nemo	288	Oblix MT	207
Monkey	362	Neokil	157	Oceal	151
Monolith	*PAR* 512	Neosorexa Bait Blocks	157	Ocean	288
Monosate G	*PAR* 441	Neosorexa Gold	157	Ocelot	*PAR* 468
Monsanto Amenity		Neosorexa Gold		Octave	*PAR* 581
Glyphosate	*PAR* 441	Ratpacks	157	Octolan	74
Monsanto Amenity		Neosorexa Pasta Bait	157	Odin	404
Glyphosate	273	Neo-Stop 120 RTU	*PAR* 151	Ohayo	*PAR* 380
Montana	*PAR* 441	Neo-Stop 300 HN	*PAR* 151	Oklar SX	*PAR* 405
Morex	*PAR* 314	Neo-Stop Starter	*PAR* 151	Ollyx	*PAR* 309
Morph	*PAR* 292	NEU 1170 H	*PAR* 338	Olympus	19
Moss Control Plus		NEU 1181 M	*PAR* 366	Omaha 2	173
Lawn Fertilizer	*PAR* 367	NEU 1185	*PAR* 366	Omnera LQM	*PAR* 417
MossKade	400	Nevada	*PAR* 374	Onslaught	*PAR* 441
Moss-Kill Pro	*PAR* 367	Nevada	233	Opera	203
Most Micro	346	New Way Weed Spray	3	Optica	313
Motif	273	Newman Cropspray		Optica Forte	*PAR* 253
Mount	*PAR* 27	11E	*Adj*	Optica Trio	*PAR* 260
Movento	398	Newman's T-80	*Adj*	Optica Trio	156
Movon	167	New-Way Weed Spray	*PAR* 3	Optimo Tech	*PAR* 295
Mowless	422	Nexy 1	*PAR* 92	Optimus	422
Moxa	422	NF-149 SC	*PAR* 202	Optio 500	*PAR* 128
Moxa 250 EC	422	Nico Pro 4SC	*PAR* 545	Option	119
Mozaic SX	*PAR* 539	Nico Pro 4SC	340	Opus	*PAR* 309
Mozaic SX	337	Nighthawk 330 EC	346	Opus Team	194
MS Catelyn	*PAR* 128	Nightjar	301	Oracle Duet	194
MS Eddard	*PAR* 596	Nimbus CS	98	Oranis	355
MS Sansa	*PAR* 26	Nimrod	56	Organic Weed Killer	
MS Theon	*PAR* 441	Ninja 5CS	*PAR* 482	RTU	*PAR* 338
MSS CIPC 50 M	*PAR* 151	Nion	*Adj*	Orian	*PAR* 639
MSS Sprout Nip	*PAR* 151	NipsIT Inside	*PAR* 185	Oriel 50SX	*PAR* 405
Multicap	57	Nirvana	*PAR* 457	Orient	347
Multi-Star 500	*PAR* 128	Nirvana	281	Orion	*PAR* 516
Mundua	149	NNF-136	*PAR* 422	Orius	404
Muntjac	176	No More Moss Lawn		Orius 20 EW	*PAR* 639
Murphy 500 SC	*PAR* 292	Feed	*PAR* 367	Orius P	362
Mycostop	*PAR* 634	Noble	*PAR* 167	Ormond	*PAR* 441
Mycotal	298	Nomix Blade	*PAR* 443	Orthocide WG	*PAR* 93
MyZus	*PAR* 607	Nomix Conqueror	*PAR* 441	Ortiva	*PAR* 26
Nakar	*PAR* 580	Nomix Conqueror		OS159	*PAR* 609
Nando 500SC	236	Amenity	273	Osarex W	324
Nappa	*PAR* 544	Nomix Dual	275	Osiris	*PAR* 44
Naprop	*PAR* 544	Nomix Duplex	*PAR* 443	Osiris P	201
Narita	158	Nomix Frontclear	*PAR* 441	Oskar	219
NASA	*PAR* 441	Nomix G	*PAR* 441	Osmoze 50 CS	*PAR* 482
Naspar Extra	328	Nomix Nova	*PAR* 441	Osorno	*PAR* 513
Nativo 75 WG	405	Nomix Prolite	*PAR* 441	Ossetia	164
Natural Weed Spray		Nomix Revenge	*PAR* 441	Othello	171
NO. 1	*PAR* 3	Nomix Tribute Turf	*PAR* 252	Ovation	*PAR* 441

REFERENCES ARE TO ENTRY NUMBERS NOT PAGES

Index

Overlord . . . . *PAR* 273	Pexan . . . . *PAR* 314	PMT 358
OWK . . . . *PAR* 3	Pexan 197	PMV-01 . . . . *PAR* 562
Oxalis NG . . . . *PAR* 441	PG Suspension 353	Pointer 261
Oxana . . . . *PAR* 142	Phantom . . . . *PAR* 576	Polaca . . . . *PAR* 312
Pacaya 229	Phase II . . . . *Adj*	Polax 105
Pacifica 288	Phaser . . . . *PAR* 584	Polecat 219
Pacifica Plus 11	Phemo . . . . *PAR* 330	POL-MCPA 500 SL . . *PAR* 499
Padawan . . . . *PAR* 124	Phemo 208	Polo 141
Paket 250 EC . . . . . *PAR* 672	PHOBI SMOKE PRO90 *PAR* 577	POL-Sulphur 80 WG . *PAR* 637
Paladin . . . . *PAR* 309	Phoenix 265	POL-Sulphur 800 SC . *PAR* 637
Palamares 206	Phostoxin 6	Polyversum . . . . *PAR* 615
Palio 234	Picard 300 WG . . . . *PAR* 466	Poncho Beta 36
Palmas WP . . . . *PAR* 207	Piccant 108	Ponder . . . . *PAR* 596
Palomo Green 273	PicoMax 347	Pontil 360 . . . . *PAR* 441
Pampa 4 SC 340	Picona 347	Poraz . . . . *PAR* 581
Pan Aquarius 420	PicoPro 347	potassium bicarbonate . . . 359
Pan Cradle 267	PicoStomp . . . . *PAR* 556	Powertwin . . . . *PAR* 331
Pan Glory 62	Pictor 46	PP Captan 80 WG 57
Pan Oasis . . . . *Adj*	Pikazole 46 . . . . *PAR* 601	Prairie 81
Pan Panorama . . . . *Adj*	Pike . . . . *PAR* 538	Praxim 333
Pan PMT 358	Pike PX . . . . *PAR* 538	Praxys 105
Pan Proteb . . . . *PAR* 584	Pilot Ultra 389	Pre-Empt . . . . *PAR* 555
Pan Tees . . . . *PAR* 669	Pinball 356	Prefect . . . . *PAR* 273
Pan Tepee 422	Pincer 166	Preline . . . . *PAR* 441
Pan Theta 322	Pincer Plus . . . . *PAR* 275	Prelude 20 LF . . . . . *PAR* 581
Panama . . . . *PAR* 26	Pinnacle 412	Prelude 20LF . . . . . . *PAR* 581
Panarex 390	Pin-o-Film . . . . *Adj*	Premium 500 72
Panto 290	Pintail . . . . *PAR* 652	Presidium 185
Paraat 181	Piper 72	Presite SX 337
Parade 347	Pirapama 299	Prestop . . . . *PAR* 439
Paradise . . . . *PAR* 370	Piri 50 . . . . *PAR* 576	Prestop 271
Paradise 229	Pirimate 500 . . . . *PAR* 576	Prestop Mix . . . . *PAR* 439
Parador . . . . *PAR* 113	Pirimate 500 357	Prevail 109
Paramount 232	Pirimex 50 WG . . . . . *PAR* 576	Previcur Energy 267
Paramount Max . . . . *PAR* 377	Pirimicarb 50 . . . . . . *PAR* 576	Priaxor EC 264
Parana 413	Pirisect . . . . *PAR* 154	Prima . . . . *Adj*
Parish . . . . *PAR* 567	Pirouette 342	Prime Turf . . . . *PAR* 472
Parsan Extra 328	Pistol . . . . *PAR* 279	Primeman . . . . *PAR* 372
Pas . . . . *PAR* 419	Pistol 170	Primero 340
Pastor . . . . *PAR* 180	Pistol Rail . . . . *PAR* 279	Prime-Turf . . . . *PAR* 472
Pastor Pro 107	Pitch . . . . *PAR* 441	Primo Maax II 422
PDM 330 EC . . . . . . *PAR* 555	Pitch . . . . *PAR* 273	Primo Maxx 422
Peak 373	Pivot 356	Priori . . . . *PAR* 26
Pearl . . . . *PAR* 440	Pivotal 258	Priori Xtra 20
Pelican Delta 172	Pixxaro EC 257	Pro Fit 25 . . . . *PAR* 603
Penbotec 400 SC . . . *PAR* 612	Pizamide . . . . *PAR* 596	Pro-Banish . . . . *PAR* 16
Pendi 330 . . . . *PAR* 555	Pizza 400 SC 370	Profilux WG 123
Pendimet 400 . . . . . *PAR* 555	Pizza Flo 370	Profiol 250 . . . . *PAR* 591
Pendragon . . . . *PAR* 555	Planet . . . . *Adj*	Profit Oil . . . . *Adj*
Pennant 337	Plant Invigorator 354	ProFume . . . . *PAR* 636
Penncozeb 80 WP . . *PAR* 493	Plant Trust . . . . *PAR* 430	Progress Green . . . . *PAR* 472
Penncozeb 80 WP 306	Platform . . . . *PAR* 107	Prolectus 225
Penncozeb WDG . . . *PAR* 493	Platform S . . . . *PAR* 109	Proliance Quattro 273
Penncozeb WDG 306	Platform S 62	Proline . . . . *PAR* 600
Penta . . . . *PAR* 555	Platoon . . . . *PAR* 608	Proline 275 374
Pentangle 83	Platoon 250 . . . . *PAR* 608	Pro-Long . . . . *PAR* 151
Percos 8	Plaza 422	Prolonger 86
Pergardo Uni . . . . . . *PAR* 497	Plenum WG 379	Promess . . . . *PAR* 589
Permasect 500 EC . . *PAR* 211	Plexus 121	Prompt 153
Permasect C . . . . *PAR* 211	Plover . . . . *PAR* 265	Prompto 216
Perseo . . . . *PAR* 27	Plover 158	Pronapse . . . . *PAR* 398
Persist . . . . *PAR* 2	Plumage . . . . *PAR* 275	Pronto . . . . *PAR* 128

REFERENCES ARE TO ENTRY NUMBERS NOT PAGES

Index

Pronto 72	Quat 200 . . . . PAR 300	Relay Turf Elite . . . . PAR 252
Propamex-l 604 SL. . PAR 589	Quazar 188	Reldan 22 88
Proper Flo 370	Quell Flo 306	Relva 370
Property 180 SC 385	Quest . . . . PAR 538	Relva Granules . . . PAR 596
Property 180SC . . . . PAR 613	Quickdown . . . . PAR 609	Remix . . . . Adj
Propi 25 EC 368	Quickfire 153	Renew Chlorothalonil
Propicon 250 368	Quickphos Pellets 56%	500SC . . . . PAR 128
Propimorf 750 223	GE . . . . PAR 8	Renovator Pro 228
Proplant . . . . PAR 589	Quidam 372	Reparto . . . . PAR 482
Proplant 366	Quilt Xcel 24	Rescue 356
Propulse 248	Quirinus . . . . PAR 393	Resplend . . . . PAR 11
Propyzamide 400 . . . PAR 596	Quit 188	Respond . . . . Adj
Propyzamide Flo . . . PAR 596	Quorum . . . . PAR 607	Restrain Fuel . . . . . . PAR 326
Proquin 200 . . . . PAR 597	Radspor 400 . . . . . . PAR 308	Retribute . . . . PAR 167
Prosaro 377	Raikiri . . . . PAR 513	Retro 188
Proshield . . . . PAR 279	Railtrax . . . . PAR 370	Revenge 20 . . . . PAR 538
Proshield 170	Ralos 14	Revenge 20 SG . . . . PAR 538
Prospect SX . . . . . . PAR 654	Rambaca 60 . . . . . . PAR 526	Reversal . . . . PAR 221
Pro-Spirox . . . . PAR 601	Rana SC . . . . PAR 652	Revolt . . . . PAR 638
Prostrob 20 . . . . PAR 402	Rancona 15 ME . . . . PAR 471	Revus . . . . PAR 497
Prosulfocarbstar . . . . PAR 598	Rancona 15 ME 289	Revus 309
Prosulfostar . . . . PAR 598	Rancona i-MIX . . . . PAR 450	Reward Oil . . . . Adj
Pro-Tebu 250 . . . . . . PAR 603	Ranger . . . . PAR 388	Rezist 5
Proteus . . . . PAR 603	Rango 390	Rezist Liquid . . . . PAR 7
Prova . . . . PAR 596	Ranman Top 115	Rezist Liquid 5
Prowler 324	Rapas . . . . PAR 553	Rhapsody 123
Proxanil . . . . PAR 209	Rapsan 500 SC 327	Rhino 260
Pryz-Flex . . . . Adj	Rasta . . . . PAR 233	Rhino DS 260
Pulse . . . . PAR 44	Rasto . . . . PAR 233	Rhizopon AA Powder
Puma . . . . PAR 26	Ratak Cut Wheat 157	(0.5%) . . . . PAR 464
Pump . . . . PAR 567	Ratan 80 WG . . . . . . PAR 93	Rhizopon AA Powder
Pure Glyphosate 360 273	Rathianon 70 WG . . . PAR 301	(0.5%) 286
Pure Max 422	Ratio SX 413	Rhizopon AA Powder
Pure Turf 290	Rattler 273	(1%) . . . . PAR 464
PureAzoxy . . . . PAR 26	Ravena . . . . PAR 160	Rhizopon AA Powder
PureFlazasul . . . . . . PAR 370	Raxil Star 249	(1%) 286
PureFlo . . . . PAR 596	Realm . . . . PAR 388	Rhizopon AA Powder
PureFloxy 420	Reaper 273	(2%) . . . . PAR 464
PureMyclobutanil . . . PAR 541	Reaper Green 273	Rhizopon AA Powder
PureProgress 24	Reboot . . . . PAR 210	(2%) 286
PureProteb . . . . PAR 584	Recital 248	Rhizopon AA Tablets . PAR 464
PureRecovery 356	Reconcile . . . . PAR 26	Rhizopon AA Tablets 286
PureThane . . . . PAR 541	Reconcile 18	Rhizopon B Powder
Pyblast 382	Redax Extra . . . . PAR 314	(0.1%) . . . . PAR 542
Pyralid Extra . . . . PAR 181	Redeem 103	Rhizopon B Powder
Pyramin DF 64	Redigo . . . . PAR 600	(0.2%) . . . . PAR 542
Pyrethrum 5 EC 382	Redigo Deter 111	Rhizopon B Tablets . . PAR 542
Pyrifos 480 87	Redigo Pro 377	Ricorso Premium . . . PAR 538
Pyrimala . . . . PAR 612	Redlegor 145	Rifle 150
Pyrinex 48 EC . . . . . PAR 154	Redstar 329	Rigby Taylor Turf Moss
Pyrus 400 SC . . . . . . PAR 612	Refine Max SX 337	Killer . . . . PAR 367
Python . . . . PAR 537	Refinzar 349	Rigel WP . . . . PAR 292
Pyzamid 400 SC . . . PAR 596	Reflect 293	Riggid . . . . PAR 327
Pyzamid Universe . . . PAR 19	Regal . . . . PAR 167	Rimsulf 250 . . . . PAR 621
Qdos Rockall 2 . . . . PAR 468	Regal Gold . . . . PAR 167	Ringer 58
Quad S . . . . PAR 300	Regalis . . . . PAR 587	RipeLock Tabs . . . . . PAR 532
Quad-Glob 200 SL 188	Regalis Plus 364	Ripelock Tabs 2.0 . . . PAR 532
Quadris Opti . . . . PAR 27	Regatta 166	Ripelock VP . . . . PAR 532
Quail . . . . PAR 575	Regent . . . . PAR 309	Riscala . . . . PAR 516
Quantum . . . . PAR 665	Reglone 188	Rithfir . . . . PAR 197
Quantum SX . . . . . . PAR 665	Regulex 10 SG . . . . PAR 438	Rizolex 10D . . . . PAR 658
Quarry . . . . PAR 555	Regulex 10 SG 270	Rizolex 50 WP . . . . . PAR 658

REFERENCES ARE TO ENTRY NUMBERS NOT PAGES

Index

Rizolex Flowable . . . *PAR* 658	Roundup ProVantage 273	Samson . . . . *PAR* 545
Roaxe 100 . . . . *PAR* 575	Roundup Provide . . . *PAR* 441	Samson Extra 6% 340
Robut 20. . . . . *PAR* 541	Roundup Rail. . . . . . *PAR* 441	Samurai . . . . *PAR* 441
Rock 243	Roundup Sonic . . . . *PAR* 441	Samurai 273
Rocket 328	Roundup Star . . . . . *PAR* 441	SAN 703 74
Rodeo . . . . *PAR* 441	Roundup Ultimate . . *PAR* 441	Sanblight. . . . . *PAR* 21
Rodeo 273	Roundup Ultimate 273	Sandpiper . . . . *PAR* 265
Rogan. . . . . *PAR* 124	Roundup Vista . . . . *PAR* 441	Sangue 2
Rogor L40. . . . . *PAR* 291	RouteOne Benta 48 . . *PAR* 44	Santal . . . . *PAR* 639
Roller . . . . *Adj*	RouteOne Bentazone	Sapphire . . . . *PAR* 313
Romany . . . . *PAR* 441	48. . . . . *PAR* 44	Saracen . . . . *Adj*
Romax B rat & Mouse	RouteOne Boscalid 50 *PAR* 69	Saracen . . . . *PAR* 294
Killer . . . . *PAR* 76	RouteOne Butanil 20 *PAR* 541	SAS 90 . . . . *Adj*
Romax Bromablock . . *PAR* 76	RouteOne Fenhex 50 *PAR* 348	Saturn Plus . . . . *Adj*
Romax Bromadiolone	RouteOne Florasul 50 *PAR* 373	Savannah 404
Whole Wheat . . . . . *PAR* 76	RouteOne Fluazifop + *PAR* 379	Saviour . . . . *PAR* 275
Romax Bromapasta	RouteOne Lambda C *PAR* 482	Savona . . . . *PAR* 338
Bait. . . . . *PAR* 76	RouteOne Mesot . . . *PAR* 516	Savvy 336
Romax D Block . . . . *PAR* 263	RouteOne Oxanet 481 *PAR* 390	Savvy Premium 336
Romax Difenacoum Rat	RouteOne Penthal 330 *PAR* 555	Saxon 153
& Mouse Block . . *PAR* 263	RouteOne Primro 50	SB Plant Invigorator 354
Romax Difenacoum	WG. . . . . *PAR* 576	Scabiya . . . . *PAR* 652
Whole Wheat Bait *PAR* 263	RouteOne Roaxe 100 *PAR* 575	Scala 384
Romax DP . . . . . . *PAR* 263	RouteOne Roazafod . *PAR* 197	Scholar 193
Romax DP Pasta	RouteOne Robut 20 . *PAR* 541	Scitec. . . . . *PAR* 672
Sachets. . . . . *PAR* 263	RouteOne Romet 20 *PAR* 538	Scorpio 420
Romax Mouse DP . . *PAR* 263	RouteOne Rosate 36 *PAR* 441	Scorpion . . . . *PAR* 441
Romax Rat CP . . . . . *PAR* 195	RouteOne Rosate 360 *PAR* 441	Scotia 199
Romaz D Rat & Mouse	RouteOne Roxybin 25 . *PAR* 26	Scotts Moss Control . *PAR* 367
Killer . . . . *PAR* 263	RouteOne Roxypro FP *PAR* 30	Scotts Octave . . . . . *PAR* 581
Roquat 20 . . . . *PAR* 300	RouteOne Seafarer . *PAR* 468	Scythe . . . . *PAR* 672
Rosan. 154	RouteOne Tebbro 25 *PAR* 639	Seafarer . . . . *PAR* 468
Rosate 36 . . . . *PAR* 441	RouteOne Trione 10 . *PAR* 513	Seal Z. 297
Rosate 36 SL . . . . . *PAR* 441	RouteOne Zamide Flo *PAR* 596	Sebrant. 349
Rosate 360 TF . . . . . *PAR* 441	RouteOne Zeta 10 . . *PAR* 676	Sector . . . . *PAR* 291
Rosate Green . . . . . *PAR* 441	Rover 500 72	Seduce 413
Rosh . . . . *PAR* 639	Rovral AquaFlo 290	Seguris 200
Roundup. . . . . *PAR* 441	Rovral WG 290	Seize . . . . *PAR* 672
Roundup Ace. . . . . *PAR* 441	Roxam 75WG 308	Sekator OD 10
Roundup Advance . . *PAR* 441	Rozine . . . . *PAR* 607	Select Prime . . . . . . *PAR* 159
Roundup Amenity . . *PAR* 441	Rubie . . . . *PAR* 567	Selon . . . . *PAR* 118
Roundup Assure . . . *PAR* 441	Rubric. 193	Selon 67
Roundup Biactive . . . *PAR* 441	Ruby 193	Sempra . . . . *PAR* 273
Roundup Biactive 3G *PAR* 441	Rudis 374	Sempra 164
Roundup Biactive Dry *PAR* 441	Rumo 287	Sempra XL 164
Roundup Biactive GL 273	Runner 331	Sencorex Flow. 335
Roundup Bio . . . . . *PAR* 441	Rustler Pro-Green . . *PAR* 441	Sentenza. . . . . *PAR* 560
Roundup Energy 273	RVG Glyphosate 360 *PAR* 441	Septuna . . . . *PAR* 265
Roundup Express . . . *PAR* 441	RVG Lambda-	Seraphin . . . . *PAR* 34
Roundup Flex 273	cyhalothrin. . . . . . *PAR* 482	Serenade ASO. . . . . *PAR* 36
Roundup Gold . . . . *PAR* 441	Sacron WG 119	Serenade ASO 26
Roundup Klik. . . . . *PAR* 441	Safari . . . . *PAR* 671	Sergeant Major . . . . *PAR* 482
Roundup Max . . . . *PAR* 441	Safari Lite WSB 300	Serpent 264
Roundup Metro . . . *PAR* 441	Safers Insecticidal Soap . . *PAR*	Setanta 50 WP . . . . *PAR* 596
Roundup Metro XL. 273	338	Setanta 50 WP 370
Roundup POWERMAX . . . 273	Sakarat Bromabait 51	Setanta Flo . . . . . . *PAR* 596
Roundup Pro Biactive *PAR* 441	Sakarat Bromakill 51	Setanta Flo 370
Roundup Proactive . . *PAR* 441	Sakarat D Pasta Bait. . . . 157	Shadow 177
Roundup ProBiactive	Sakarat D Wax Bait 157	Shamal . . . . *PAR* 596
450 . . . . *PAR* 441	Sakarat D Whole Wheat . . 157	Shams 465 . . . . . . *PAR* 518
Roundup ProBio . . . *PAR* 441	Saltri . . . . *PAR* 65	Shark . . . . *PAR* 107
Roundup Pro-Green . *PAR* 441	Salvo 235	Shark 61

REFERENCES ARE TO ENTRY NUMBERS NOT PAGES

Index

Shaver . . . . *PAR* 490	Snicket 336	Spruzit 382
Sherlock 273	Snipe 367	Spry . . . . *PAR* 629
Sherman . . . . *PAR* 555	Sniper 149	Squire Ultra 9
Shinkon . . . . *PAR* 21	Socom . . . . *PAR* 119	Stabilan 750 67
Shirlan 236	sodium hypochlorite 394	Stacato . . . . *PAR* 441
Shirlan Forte . . . . . . *PAR* 205	Sokol 404	Staff 500 135
Shirlan Programme . . *PAR* 380	Solace Max . . . . *PAR* 207	Staka SX 253
Shiro 421	Solani . . . . *PAR* 36	Stallion Sync TEC . . . *PAR* 173
SHL Granular Feed,	Solar . . . . *Adj*	Stallion Sync TEC 101
Weed & Mosskiller 155	Solar Plus . . . . *Adj*	Stalwart 327
SHL Lawn Sand . . . . *PAR* 367	Solaren 345	Stamen . . . . *PAR* 441
Shogi . . . . *PAR* 607	Soleil 55	Stamina . . . . *Adj*
Shogun 367	Soleto 333	Stamp 330 EC . . . . . *PAR* 555
Shooter 243	Solfa WG . . . . *PAR* 637	Standon Beamer . . . *PAR* 639
Shorten 422	Solitaire 370	Standon Bentazone . . *PAR* 44
Shotput 335	Solitaire 50 WP . . . . *PAR* 596	Standon Bentazone S . *PAR* 44
Shrapnel . . . . *PAR* 567	Solo 500 SC 164	Standon Carfentrazone
Shrike 200 188	Solo D500 . . . . *PAR* 273	MM . . . . *PAR* 110
Shrink 422	Solstice . . . . *PAR* 373	Standon Clethodim . . *PAR* 159
Shuttle . . . . *PAR* 555	Sonido 410	Standon Cobmajor 320
Shyfo 273	Sonis . . . . *PAR* 672	Standon Cobmaster 322
Shylock 61	Sorexa D 157	Standon Coolie . . . . *PAR* 314
Sienna . . . . *PAR* 556	Sorexa Gel 157	Standon Coolie 197
Sienna Pro . . . . *PAR* 556	Source II . . . . *PAR* 490	Standon Diquat . . . . *PAR* 300
Signal 300 ES . . . . . *PAR* 211	Sovereign . . . . *PAR* 555	Standon Diquat SL . . *PAR* 300
Signum 50	Span 368	Standon Diwana 243
Silico-Sec 354	Spaniel 368	Standon
Silk 750 SL . . . . *PAR* 118	Spartan . . . . *Adj*	Fenpropimorph 750 223
Siltex AF . . . . *Adj*	Spartan . . . . *PAR* 159	Standon Frontrunner
Siltra Xpro 42	Sparticus Xpro 44	Super 6 340
Silvio . . . . *PAR* 441	Sparviero 297	Standon Googly . . . *PAR* 300
Silwet L-77 . . . . *Adj*	Spatial Plus 68	Standon Googly 188
Simba SX 336	Spear 310	Standon Gusto Xtra . *PAR* 538
Sinclair Lawn Sand . . *PAR* 367	Spear & Jackson	Standon Homerun . . *PAR* 412
Sinconil . . . . *PAR* 128	Professional Super	Standon Hounddog 204
Sinpro . . . . *PAR* 472	Strong Concentrated	Standon LBW . . . . . *PAR* 607
Sinstar . . . . *PAR* 26	Weed Killer . . . . . *PAR* 441	Standon Mandor . . . *PAR* 497
SIP 313 . . . . *PAR* 142	Spectrum . . . . *PAR* 612	Standon Mandor 309
Sipcam C50 WG 119	Speed . . . . *Adj*	Standon Mastana 377
Sirena 329	Speedline Pro . . . . . . *PAR* 20	Standon Metso XT 336
Siskin 325	Speedster 137	Standon Mexxon Xtra 336
Sistan 51 . . . . *PAR* 522	Speedway Total . . . . *Adj*	Standon Mimas WG 288
Skora . . . . *PAR* 412	Spekfree 404	Standon New
Skyway 285 Xpro 44	Spelio SX . . . . *PAR* 405	Metazachlor . . . . . *PAR* 524
SL 567A 323	Sphere ASBO 5	Standon Pinoxy 356
Slalom 233	Sphinx . . . . *PAR* 449	Standon Pirimicarb 50 *PAR* 576
Slick . . . . *PAR* 265	Spigot 223	Standon Pygmy . . . . *PAR* 672
Slippa . . . . *Adj*	Spiral 375	Standon Santa Fe 50
Slucks . . . . *PAR* 366	Spiro OD . . . . *PAR* 630	WP 370
Slugger 226	Spirodor . . . . *PAR* 128	Standon Shiva . . . . *Adj*
Sluggo . . . . *PAR* 366	Spirodor . . . . *PAR* 628	Standon Soulmate 96
Sluxx . . . . *PAR* 366	Spitfire 233	Standon Sparkle . . . . *PAR* 392
Sluxx HP 226	Sporgon 50 WP . . . . *PAR* 581	Standon Wicket . . . . *Adj*
Smaragdin . . . . *PAR* 315	Sportak 45 EW . . . . *PAR* 581	Standon Yorker . . . . *PAR* 300
SmartFresh 332	Sportak 45 HF . . . . . *PAR* 581	Standon Zero Tolerance*PAR* 652
SmartFresh ProTabs . *PAR* 532	Spotlight Plus . . . . . . *PAR* 107	Standon Zero Tolerance . . 410
SmartFresh SmartTabs*PAR* 532	Spotlight Plus 61	Standonil . . . . *PAR* 128
Smartgrass 270	Sprayfast . . . . *Adj*	Starane 2 . . . . *PAR* 412
Smash . . . . *PAR* 380	Spray-fix . . . . *Adj*	Starane Gold . . . . . . *PAR* 374
Smite 1	Spraygard . . . . *Adj*	Starane Hi-Load HL 256
Snapper . . . . *PAR* 441	Spraymac . . . . *Adj*	Starane Vantage . . . . *PAR* 374
Snapper 273	Springbok 176	Starane XL 233

REFERENCES ARE TO ENTRY NUMBERS NOT PAGES

Index

Starpro 404	T34 Biocontrol . . . . . *PAR* 666	Terpal 205
Stay Fresh . . . . *PAR* 532	T34 Biocontrol 419	Terrane 166
Staylow . . . . *PAR* 672	Tacet 422	Terrier . . . . *PAR* 273
Stealth . . . . *PAR* 482	Tachigaren 70 WP 277	Teson 404
Stefes CCC 720 . . . . *PAR* 118	Taipan . . . . *PAR* 412	Tesoro . . . . *PAR* 639
Stefonil 72	Tairel . . . . *PAR* 42	Tesoro 404
Stelga 500 SC . . . . . *PAR* 328	Takron 64	Tezate 220 SL . . . . . *PAR* 651
Steward 287	Takumi SC 118	Tezate 220 SL 409
Stika . . . . *Adj*	Talisma EC . . . . *PAR* 211	Tezuma . . . . *PAR* 205
Stiletto . . . . *PAR* 312	Talisma UL . . . . *PAR* 211	Thalonil 500 . . . . *PAR* 128
Stirrup 273	Talius . . . . *PAR* 597	Tharsis . . . . *PAR* 639
Stomp 400 SC . . . . . *PAR* 555	Talius 371	Tharsis 404
Stomp Aqua 346	Talunex 6	Thiaclomex 480 SC . *PAR* 652
Storite Clear Liquid . . *PAR* 651	Tamba Bio . . . . *PAR* 441	Thiofin WG . . . . *PAR* 656
Storite Excel . . . . . . *PAR* 651	Tambora . . . . *PAR* 441	Thiraflo . . . . *PAR* 657
Storite Super . . . . . . *PAR* 452	Tambora 273	Thistlex 109
Storm Mini Bits 230	Tamok . . . . *PAR* 639	Thor 418
Storm Pasta 230	Tanaris 179	Throttle . . . . *PAR* 584
Storm Secure 230	Tanaru 30	Thrust 136
Strand 193	Tandus . . . . *PAR* 412	Thunder . . . . *PAR* 331
Stroby WG . . . . *PAR* 481	Tangent . . . . *PAR* 441	Thunderbolt . . . . *PAR* 442
Stroby WG 296	Tanil 200 . . . . *PAR* 541	Thyme . . . . *PAR* 632
Stroller 370	Tanker 273	Thyram Plus 415
Strong WG . . . . *PAR* 481	Tanos 120	Timpani 83
Stymie 370	Targa Max 389	Tine 323
SU Wett . . . . *Adj*	Targa Super 389	Tipoff . . . . *PAR* 542
Subdue 323	Target SC . . . . *PAR* 520	Titan 96
Successor . . . . *PAR* 565	Target SC 325	Titanic . . . . *PAR* 288
Sultan 50 SC 327	Task 360 . . . . *PAR* 441	Titus 391
Sumi-Alpha 204	Tasker 75 . . . . *PAR* 499	Tizca 236
Sumimax 244	Tattler 336	Toil . . . . *Adj*
Sunfire 240	Taurus 414	Toledo 404
Sunorg Pro 329	Taxi . . . . *PAR* 665	Tomahawk . . . . *PAR* 412
Supasect . . . . *PAR* 211	Taylors Lawn Sand . . *PAR* 367	Tomahawk 2 . . . . . . *PAR* 412
Supasect 500 EC . . . *PAR* 211	Taza 500 . . . . *PAR* 524	Top Farm Bentazone . . *PAR* 44
Super 3 324	Tazer 18	Topas 344
Super Nova . . . . *Adj*	TBM 75 WG . . . . *PAR* 665	Topenco 344
Superturf . . . . *PAR* 472	TDS Major 324	Topenco 100 EC 344
Supreme . . . . *PAR* 128	TDS Metarex Amba . *PAR* 519	Topik 91
Supreme 72	Teamforce . . . . *PAR* 331	Topkat 179
Surfer . . . . *Adj*	Teamforce 209	Toppel 100 . . . . *PAR* 211
Surpass Pro 290	Teamforce SE . . . . . . *PAR* 331	Topper . . . . *PAR* 668
Surrender 273	Teb 250 . . . . *PAR* 639	Toprex 161
Surrender T/F . . . . . *PAR* 441	TEB250 404	Topsin WG 414
Sven 204	Tebcon . . . . *PAR* 639	Tor . . . . *PAR* 183
Swallow 115	Tebucur 250 404	Torch 399
Sward . . . . *Adj*	Tebuzol 250 404	Torero 207
Swift SC 420	Tec-bom . . . . *PAR* 580	Tornado . . . . *PAR* 273
Swing Gold 187	Tectura 49	Toro 360 . . . . *PAR* 441
Switch 133	Teldor 218	Torpedo-II . . . . *Adj*
Sword 91	Teliton Ocean . . . . . . *PAR* 468	Toscana . . . . *PAR* 665
Syllit 400 SC . . . . . . *PAR* 308	Telsee . . . . *PAR* 124	Toska EC . . . . *PAR* 412
Symbol . . . . *PAR* 441	Templar . . . . *PAR* 87	Total 360 WDC . . . . *PAR* 441
Symetra 23	Templier 340	Touchdown Quattro 273
Symmetry . . . . *PAR* 575	Tempo 422	Tower 85
Synchem Diquat 200 *PAR* 300	Tempt 324	TP 100 . . . . *PAR* 672
Synergy Chlorothalonil *PAR* 128	Temsa SC 320	Tracer 396
Synero 13	Teppeki 231	Tracker 47
Syper 100 . . . . *PAR* 211	Teridox . . . . *PAR* 285	Trafalgar 233
System 50 240	Teridox 174	Trample 346
Systhane 20EW . . . . *PAR* 541	Terminus . . . . *PAR* 492	Transact . . . . *Adj*
T2 Green Pro 152	Terpal . . . . *PAR* 327	Transcend . . . . *Adj*

REFERENCES ARE TO ENTRY NUMBERS NOT PAGES

Index

Traton SX	338	Unite	PAR 408	Vydate 10G	341
Traxos	93	Unite	254	Wakil XL	PAR 206
Traxos Pro	PAR 163	Unix	PAR 220	Wakil XL	122
Trebon 30 EC	211	Upbeet	421	Warlock	PAR 309
Tremolo	PAR 519	Upkeep	PAR 672	Warrior	PAR 482
Treoris	79	UPL B Zone	PAR 44	Waterloo	PAR 300
Trexstar	PAR 672	UPL Chlorothalonil	72	Weedex	PAR 440
Trexxus	PAR 672	UPL Diquat	PAR 300	Wetcit	Adj
Triad	418	UPL Trinexapac	PAR 672	Whippet	PAR 440
Trianum G	PAR 667	Upright	70	Whistle	PAR 71
Trianum P	PAR 667	Vacciplant	PAR 484	Wicket	PAR 598
Tribal	85	Valbon	32	Wildcat	PAR 160
Tribe 500	PAR 665	Valiant	18	Winby	PAR 380
Tribeca Sync Tec	PAR 170	Validate	Adj	Wing - P	178
Tribun 75 WG	PAR 665	Vanguard	380	Wither 36	PAR 441
Tricle	421	Vareon	361	Wolf	PAR 57
Tridus	422	Variano Xpro	41	Wolverine	PAR 425
Trigger 3	324	Vazor Cypermax Plus	127	Woomera	188
Trilogy	149	Vazor Cypermethrin 10	126	Wopro Glyphosate	PAR 441
Trimanzone	PAR 493	Vazor DE	354	Wren	PAR 575
Trimeo 75 WG	PAR 665	Vazor Liquid Mosquito		Xerton	206
Trinex 222	PAR 672	Film	354	Xinca	PAR 77
Trinexis	PAR 672	Vectobac 12 AS	28	X-SEPT	PAR 128
Triplex	PAR 410	Vegas	118	X-Wet	Adj
Triton	PAR 412	Velocity	Adj	Xylem	PAR 26
Triumph	PAR 482	Velvet	Adj	Yellow Hammer	PAR 555
Trojan 500 SC	PAR 328	Vendetta	22	Yomp	346
Troller	PAR 373	Venzar 80 WP	PAR 486	Zamide 400	370
Trooper	243	Venzar Flowable	299	Zamide 80 WG	PAR 596
Tropotox	312	Verdah 400	PAR 596	Zamide Flo	PAR 596
Trounce	324	Verdant	Adj	Zammo	PAR 596
Troy 480	PAR 44	Verge 400	PAR 596	Zampro DM	8
Troy 480	30	Verimark 20 SC	114	Zarado	Adj
Trust	79	Vertisan	PAR 560	Zeal	Adj
Trustee Amenity	PAR 441	Vertisan	348	Zebra WDG	PAR 493
Trustee Amenity	273	Vesta	PAR 525	Zenith	193
Tuber Mission	PAR 300	Vesuvius	273	Zephyr	378
Tuberprop Easy	PAR 151	Vesuvius Green	PAR 441	Zero Tolerance	410
Tubosan	404	Victor	PAR 538	Zetanil	PAR 207
Tucana	380	Video	PAR 207	Zetanil WG	123
Tucana 200	PAR 608	Video DG	123	Zigzag	Adj
Tuko	PAR 538	Vigon	167	Zimbrail	80
Tuli	91	Viscount	PAR 160	Zinam 500	PAR 380
Turf Systemic Fungicide	PAR 95	Vival	PAR 441	Zinam II	PAR 380
Turfclear	PAR 95	Vivax	68	Zinzan	Adj
Turfmaster	152	Vivendi 200	PAR 174	Zip	PAR 224
Turnstone	PAR 265	Vivendi 200	102	Zira	PAR 672
Twin	209	Vivid	380	Zolex	PAR 591
Twinstar	360	Vivid 200	PAR 608	Zone 48	PAR 44
Twister	PAR 273	Vixen	PAR 573	Zorkem	PAR 597
Twister 500	PAR 273	Volcan Combi FL	65	ZORO	PAR 314
Tymoon 500	PAR 128	Volcano	207	Zoxis	PAR 26
Typhoon 360	PAR 441	Volley	236	Zubarone	410
Ultramax	PAR 441	Voodoo	PAR 309	Zulkon	PAR 167
Ulysses	404	Vortex	PAR 315	Zulu	PAR 477
Unicur	251	Vortex	198	Zulu	293
Unikat 75WG	PAR 496	Voyage	PAR 276		
Unikat 75WG	308	Vulcan	2		

REFERENCES ARE TO ENTRY NUMBERS NOT PAGES

NOTES

NOTES

THE UK PESTICIDE GUIDE 2017

RE-ORDERS

☐ Please send me ____ more copies of *The UK Pesticide Guide 2017* at £52.50 each

Postage: single copy £6.95; orders up to £200 £9.95… orders over £200 £12.50

Outside the UK please add £10 per order for delivery.

Name _____ Position_____

Institution _____ Department_____

Address _____

City _____ Region _____ Postcode_____ Country_____

Tel_____ Fax_____ E-mail_____

EU countries except UK – VAT No:_____

Payment (pre-payment is required):

☐ I enclose a cheque/draft for £_____ payable to BCPC. Please send me a receipt.

☐ I wish to pay by credit card: ☐ Visa   ☐ Mastercard   ☐ Amex   ☐ Switch

Please charge to my card £_____ and send me a receipt. Name of issuing bank_____

Card no. ☐☐☐☐ ☐☐☐☐ ☐☐☐☐ ☐☐☐☐

Expiry date ☐☐/☐☐ Security code ☐☐☐   Switch cards only: Start date ☐☐/☐☐ Issue no. ☐

Signature_____ Date_____

Name and address of cardholder if different from above:_____

_____

Please photocopy and return to:

BCPC BCPC Publications Sales, 7 Omni Business Centre, Omega Park, Alton, Hants GU34 2QD, UK
Tel: 01420 593 200, Fax: 01420 593 209, Email: publications@bcpc.org, Web: www.bcpc.org

FUTURE EDITIONS

☐ I wish to take out an annual order for ____ copies of each new edition of *The UK Pesticide Guide*

☐ Please send me advance price details for the 2018 edition of *The UK Pesticide Guide* when available

Order by phone: 01420 593 200 or online: www.bcpc.org/shop

Bookshop orders to:	Bulk discount:	
Marston Book Services Ltd 160 Milton Park Abingdon, Oxfordshire, OX14 4SD, UK Tel +44(0) 1235 465577 Fax: +44(0) 1235 465555 Email: direct.orders@marston.co.uk	100+ copies 50–99 10–49 List price £52.50	30% 20% 10%

www.cabi.org

Thank you for your order